Errata and Corrigenda No. 6 (vols. 1-13), Jan. 1978, have been entered in this volume.

SW

Errata & Corrigenda No 9 (vols 1-13) 4-16-81 have been entered in this volume.

THERMOPHYSICAL PROPERTIES OF MATTER

The TPRC Data Series

A Comprehensive Compilation of Data by the
Thermophysical Properties Research Center (TPRC), Purdue University

Y. S. Touloukian, Series Editor
C. Y. Ho, Series Technical Editor

Volume	1.	Thermal Conductivity–Metallic Elements and Alloys
Volume	2.	Thermal Conductivity–Nonmetallic Solids
Volume	3.	Thermal Conductivity–Nonmetallic Liquids and Gases
Volume	4.	Specific Heat–Metallic Elements and Alloys
Volume	5.	Specific Heat–Nonmetallic Solids
Volume	6.	Specific Heat–Nonmetallic Liquids and Gases
Volume	7.	Thermal Radiative Properties–Metallic Elements and Alloys
Volume	8.	Thermal Radiative Properties–Nonmetallic Solids
Volume	9.	Thermal Radiative Properties–Coatings
Volume	10.	Thermal Diffusivity
Volume	11.	Viscosity
Volume	12.	Thermal Expansion–Metallic Elements and Alloys
Volume	13.	Thermal Expansion–Nonmetallic Solids

New data on thermophysical properties are being constantly accumulated at TPRC. Contact TPRC and use its interim updating services for the most current information.

THERMOPHYSICAL PROPERTIES OF MATTER

VOLUME 10

THERMAL DIFFUSIVITY

Y. S. Touloukian
Director
Thermophysical Properties Research Center
and
Distinguished Atkins Professor of Engineering
School of Mechanical Engineering
Purdue University
and
Visiting Professor of Mechanical Engineering
Auburn University

R. W. Powell
Senior Researcher
Thermophysical Properties Research Center
Purdue University
Formerly
Senior Principal Scientific Officer
Basic Physics Division
National Physical Laboratory
England

C. Y. Ho
Head of Data Tables Division
and
Associate Senior Researcher
Thermophysical Properties Research Center
Purdue University

M. C. Nicolaou
Post Doctoral Research Associate
Thermophysical Properties Research Center
Purdue University

IFI/PLENUM • NEW YORK-WASHINGTON • 1973

Library of Congress Catalog Card Number 73-129616

ISBN (13-Volume Set) 0-306-67020-8
ISBN (Volume 10) 0-306-67030-5

IFI/Plenum Data Corporation is a subsidiary of
Plenum Publishing Corporation
227 West 17th Street, New York, N.Y. 10011

Distributed in Europe by Heyden & Son, Ltd.
Spectrum House, Alderton Crescent
London NW4 3XX, England

Printed in the United States of America

"In this work, when it shall be found that much is omitted, let it not be forgotten that much likewise is performed..."

SAMUEL JOHNSON, A.M.

From last paragraph of Preface to his two-volume *Dictionary of the English Language*, Vol. I, page 5, 1755, London, Printed by Strahan.

Foreword

In 1957, the Thermophysical Properties Research Center (TPRC) of Purdue University, under the leadership of its founder, Professor Y. S. Touloukian, began to develop a coordinated experimental, theoretical, and literature review program covering a set of properties of great importance to science and technology. Over the years, this program has grown steadily, producing bibliographies, data compilations and recommendations, experimental measurements, and other output. The series of volumes for which these remarks constitute a foreword is one of these many important products. These volumes are a monumental accomplishment in themselves, requiring for their production the combined knowledge and skills of dozens of dedicated specialists. The Thermophysical Properties Research Center deserves the gratitude of every scientist and engineer who uses these compiled data.

The individual nontechnical citizen of the United States has a stake in this work also, for much of the science and technology that contributes to his well-being relies on the use of these data. Indeed, recognition of this importance is indicated by a mere reading of the list of the financial sponsors of the Thermophysical Properties Research Center; leaders of the technical industry of the United States and agencies of the Federal Government are well represented.

Experimental measurements made in a laboratory have many potential applications. They might be used, for example, to check a theory, or to help design a chemical manufacturing plant, or to compute the characteristics of a heat exchanger in a nuclear power plant. The progress of science and technology demands that results be published in the open literature so that others may use them. Fortunately for progress, the useful data in any single field are not scattered throughout the tens of thousands of technical journals published throughout the world. In most fields, fifty percent of the useful work appears in no more than thirty or forty journals. However, in the case of TPRC, its field is so broad that about 100 journals are required to yield fifty percent. But that other fifty percent! It is scattered through more than 3500 journals and other documents, often items not readily identifiable or obtainable. Nearly 65,000 references are now in the files.

Thus, the man who wants to use existing data, rather than make new measurements himself, faces a long and costly task if he wants to assure himself that he has found all the relevant results. More often than not, a search for data stops after one or two results are found—or after the searcher decides he has spent enough time looking. Now with the appearance of these volumes, the scientist or engineer who needs these kinds of data can consider himself very fortunate. He has a single source to turn to; thousands of hours of search time will be saved, innumerable repetitions of measurements will be avoided, and several billions of dollars of investment in research work will have been preserved.

However, the task is not ended with the generation of these volumes. A critical evaluation of much of the data is still needed. Why are discrepant results obtained by different experimentalists? What undetected sources of systematic error may affect some or even all measurements? What value can be derived as a "recommended" figure from the various conflicting values that may be reported? These questions are difficult to answer, requiring the most sophisticated judgment of a specialist in the field. While a number of the volumes in this Series do contain critically evaluated and recommended data, these are still in the minority. The data are now being more intensively evaluated by the staff of TPRC as an integral part of the effort of the National Standard Reference Data System (NSRDS). The task of the National Standard Reference Data System is to organize and operate a comprehensive program to prepare compilations of critically evaluated data on the properties of substances. The NSRDS is administered by the National Bureau of Standards under a directive from the Federal Council for Science and Technology, augmented by special legislation of the Congress of the United States. TPRC is one of the national resources participating in the National Standard Reference Data System in a united effort to satisfy the needs of the technical community for readily accessible, critically evaluated data.

As a representative of the NBS Office of Standard Reference Data, I want to congratulate Professor Touloukian and his colleagues on the accomplishments represented by this Series of reference data books. Scientists and engineers the world over are indebted to them. The task ahead is still an awesome one and I urge the nation's private industries and all concerned Federal agencies to participate in fulfilling this national need of assuring the availability of standard numerical reference data for science and technology.

EDWARD L. BRADY
Associate Director for Information Programs
National Bureau of Standards

Preface

Thermophysical Properties of Matter, the TPRC Data Series, is the culmination of fifteen years of pioneering effort in the generation of tables of numerical data for science and technology. It constitutes the restructuring, accompanied by extensive revision and expansion of coverage, of the original *TPRC Data Book*, first released in 1960 in loose-leaf format, 11″ × 17″ in size, and issued in June and December annually in the form of supplements. The original loose-leaf *Data Book* was organized in three volumes: (1) metallic elements and alloys; (2) nonmetallic elements, compounds, and mixtures which are solid at N.T.P., and (3) nonmetallic elements, compounds, and mixtures which are liquid or gaseous at N.T.P. Within each volume, each property constituted a chapter.

Because of the vast proportions the *Data Book* began to assume over the years of its growth and the greatly increased effort necessary in its maintenance by the user, it was decided in 1967 to change from the loose-leaf format to a conventional publication. Thus, the December 1966 supplement of the original *Data Book* was the last supplement disseminated by TPRC.

While the manifold physical, logistic, and economic advantages of the bound volume over the loose-leaf oversize format are obvious and welcome to all who have used the unwieldy original volumes, the assumption that this work will no longer be kept on a current basis because of its bound format would not be correct. Fully recognizing the need of many important research and development programs which require the latest available information, TPRC has instituted a *Data Update Plan* enabling the subscriber to inquire, by telephone if necessary, for specific information and receive, in many instances, same-day response on any new data processed or revision of published data since the latest edition. In this context, the TPRC Data Series departs drastically from the conventional handbook and giant multivolume classical works, which are no longer adequate media for the dissemination of numerical data of science and technology without a continuing activity on contemporary coverage. The loose-leaf arrangements of many works fully recognize this fact and attempt to develop a combination of bound volumes and loose-leaf supplement arrangements as the work becomes increasingly large. TPRC's *Data Update Plan* is indeed unique in this sense since it maintains the contents of the TPRC Data Series current and live on a day-to-day basis between editions. In this spirit, I strongly urge all purchasers of these volume to complete in detail and return the *Volume Registration Certificate* which accompanies each volume in order to assure themselves of the continuous receipt of annual listing of corrigenda during the life of the edition.

The TPRC Data Series consists initially of 13 independent volumes. The initial seven volumes were published in 1970. Volumes 8, 9, and 10 are planned for 1972, Volume 11 for 1973, and Volumes 12 and 13 for 1974. It is also contemplated that subsequent to the first edition, each volume will be revised, updated, and reissued in a new edition approximately every fifth year. The organization of the TPRC Data Series makes each volume a self-contained entity available individually without the need to purchase the entire Series.

The coverage of the specific thermophysical properties represented by this Series constitutes the most comprehensive and authoritative collection of numerical data of its kind for science and technology.

Whenever possible, a uniform format has been used in all volumes, except when variations in presentation were necessitated by the nature of the property or the physical state concerned. In spite of the wealth of data reported in these volumes, it should be recognized that all volumes are not of the same degree of completeness. However, as additional data are processed at TPRC on a continuing basis, subsequent editions will become increasingly more complete and up to date. Each volume in the Series basically comprises three sections, consisting of a text, the body of numerical data with source references, and a material index.

The aim of the textual material is to provide a complementary or supporting role to the body of numerical data rather than to present a treatise on the subject of the property. The user will find a basic theoretical treatment, a comprehensive presentation of selected works which constitute reviews, or compendia of empirical relations useful in estimation of the property when there exists a paucity of data or when data are completely lacking. Established major experimental techniques are also briefly reviewed.

The body of data is the core of each volume and is presented in both graphical and tabular formats for convenience of the user. Every single point of numerical data is fully referenced as to its original source and no secondary sources of information are used in data extraction. In general, it has not been possible to critically scrutinize all the original data presented in these volumes, except to eliminate perpetuation of gross errors. However, in a significant number of cases, such as for the properties of liquids and gases and the thermal conductivity of all the elements, the task of full evaluation, synthesis, and correlation has been completed. It is hoped that in subsequent editions of this continuing work, not only new information will be reported but the critical evaluation will be extended to increasingly broader classes of materials and properties.

The third and final major section of each volume is the material index. This is the key to the volume, enabling the user to exercise full freedom of access to its contents by any choice of substance name or detailed alloy and mixture composition, trade name, synonym, etc. Of particular interest here is the fact that in the case of those properties which are reported in separate companion volumes, the material index in each of the volumes also reports the contents of the other companion volumes.* The sets of companion volumes are as follows:

Thermal conductivity:	Volumes 1, 2, 3
Specific heat:	Volumes 4, 5, 6
Radiative properties:	Volumes 7, 8, 9
Thermal expansion:	Volumes 12, 13

The ultimate aims and functions of TPRC's Data Tables Division are to extract, evaluate, reconcile, correlate, and synthesize all available data for the thermophysical properties of materials with the result of obtaining internally consistent sets of property values, termed the "recommended reference values." In such work, gaps in the data often occur, for ranges of temperature, composition, etc. Whenever feasible, various techniques are used to fill in such missing information, ranging from empirical procedures to detailed theoretical calculations. Such studies are resulting in valuable new estimation methods being developed which have made it possible to estimate values for substances and/or physical conditions presently unmeasured or not amenable to laboratory investigation. Depending on the available information for a particular property and substance, the end product may vary from simple tabulations of isolated values to detailed tabulations with generating equations, plots showing the concordance of the different values, and, in some cases, over a range of parameters presently unexplored in the laboratory.

The TPRC Data Series constitutes a permanent and valuable contribution to science and technology. These constantly growing volumes are invaluable sources of data to engineers and scientists, sources in which a wealth of information heretofore unknown or not readily available has been made accessible. We look forward to continued improvement of both format and contents so that TPRC may serve the scientific and technological community with ever-increasing excellence in the years to come. In this connection, the staff of TPRC is most anxious to receive comments, suggestions, and criticisms from all users of these volumes. An increasing number of colleagues are making available at the earliest possible moment reprints of their papers and reports as well as pertinent information on the more obscure publications. I wish to renew my earnest request that this procedure become a universal practice since it will prove to be most helpful in making TPRC's continuing effort more complete and up to date.

It is indeed a pleasure to acknowledge with gratitude the multisource financial assistance received from over fifty of TPRC's sponsors which has made the continued generation of these tables possible. In particular, I wish to single out the sustained major support being received from the Air Force Materials Laboratory–Air Force Systems Command, the Office of Standard Reference Data–National Bureau of Standards, and the Office of Advanced Research and Technology–National Aeronautics and Space Administration. TPRC is indeed proud to have been designated as a National Information Analysis Center

*For the first edition of the Series, this arrangement was not feasible for Volumes 7 and 8 due to the sequence and the schedule of their publication. This situation will be resolved in subsequent editions.

for the Department of Defense as well as a component of the National Standard Reference Data System under the cognizance of the National Bureau of Standards.

While the preparation and continued maintenance of this work is the responsibility of TPRC's Data Tables Division, it would not have been possible without the direct input of TPRC's Scientific Documentation Division and, to a lesser degree, the Theoretical and Experimental Research Divisions. The authors of the various volumes are the senior staff members in responsible charge of the work. It should be clearly understood, however, that many have contributed over the years and their contributions are specifically acknowledged in each volume. I wish to take this opportunity to personally thank those members of the staff, research assistants, graduate research assistants, and supporting graphics and technical typing personnel without whose diligent and painstaking efforts this work could not have materialized.

Y. S. Touloukian

Director
Thermophysical Properties Research Center
Distinguished Atkins Professor of Engineering

Purdue University
Lafayette, Indiana
June 1972

Introduction to Volume 10

This volume of *Thermophysical Properties of Matter*, the TPRC Data Series, follows the general format of the volumes on thermal conductivity of solids and is one of the most comprehensive in covering the available data from the literature.

The volume comprises three major sections: the front text on *theory, estimation, and measurement* together with its bibliography, the main body of *numerical data* and its references, and the *material index.*

The text material is intended to assume a role complementary to the main body of numerical data, the presentation of which is the primary purpose of this volume. It is felt that a moderately detailed discussion of the theoretical nature of the property under consideration together with an overview of predictive procedures and recognized experimental methods and techniques will be appropriate in a major reference work of this kind. The extensive reference citations given in the text should lead the interested reader to sufficient literature for a more comprehensive study. It is hoped, however, that enough detail is presented for this volume to be self-contained for the practical user.

The main body of the volume consists of the presentation of numerical data compiled over the years in a most comprehensive and meticulous manner. The scope of coverage includes the metallic and nonmetallic elements, ferrous and nonferrous alloys, intermetallic, inorganic, and organic compounds, glasses, minerals, composites, mixtures, polymers, and foods and biological materials. The extraction of all data directly from their original sources ensures freedom from errors of transcription. Furthermore, some gross errors appearing in the original source documents have been corrected. The organization and presentation of the data together with other pertinent information for the use of the tables and figures are discussed in detail in the introductory material to the section entitled *Numerical Data*.

The part of the data tables covering the elements deserves special mention. We wish to point out that the original literature data have been critically evaluated and analyzed, and "recommended reference values" or "provisional values" are presented. Experimental thermal diffusivity data are available in the world literature for only 39 elements, but recommended or provisional values are given in this volume for 75 elements. Most of the values were first derived from the recommended values of thermal conductivity and selected values of specific heat and density, and then compared with the evaluated experimental thermal diffusivity data whenever available.

Such recommended values are those that were considered to be the most probable when assessments were made of the information available in early 1972. Their inclusion adds a unique feature that is designed to provide the user with acceptable values. It should be realized, however, that these recommended values are not necessarily the final true values and that changes directed toward this end will often become necessary as more data become available. Future editions will contain these changes and will provide similar recommendations made for an increasing number of materials.

As stated earlier, all data have been obtained from their original sources and each data set is so referenced. TPRC has in its files all documents cited in this volume. Those that cannot readily be obtained elsewhere are available from TPRC in microfiche form.

This volume has grown out of activities made possible through the principal support of the Air Force Materials Laboratory–Air Force Systems Command, under the monitorship of Mr. John H. Charlesworth. Over the years, several assistant researchers have contributed to the preparation of this volume for varying periods under the authors' supervision. In chronological order of their association with TPRC, we wish to acknowledge the contributions of Messrs. C. Y. Wang, K. Y. Wu, and T. Y. R. Lee.

Inherent in the character of this work is the fact that in the preparation of this volume, we have drawn most heavily upon the scientific literature and feel a debt of gratitude to the authors of the referenced articles. While their often discordant results have caused us much difficulty in reconciling their findings, we consider this to be our challenge and our contribution to negative entropy of information, as an effort is made to create from the randomly distributed data a condensed, more orderly state.

While this volume is primarily intended as a reference work for the designer, researcher, experimentalist, and theoretician, the teacher at the graduate level may also use it as a teaching tool to point out to his students the topography of the state of knowledge on the thermal diffusivity of materials. We believe there is also much food for reflection by the specialist and the academician concerning the meaning of "original" investigation and its "information content."

The authors and their contributing associates are keenly aware of the possibility of many weaknesses in a work of this scope. We hope that we will not be judged too harshly and that we will receive the benefit of suggestions regarding references omitted, additional material groups needing more detailed treatment, improvements in presentation or in recommended values, and, most important, any inadvertent errors. If the *Volume Registration Certificate* accompanying this volume is returned, the reader will assure himself of receiving annually a list of corrigenda as possible errors come to our attention.

Lafayette, Indiana
June 1972

Y. S. Touloukian
R. W. Powell
C. Y. Ho
M. C. Nicolaou

Contents

Numerical Data

Material Index

GROUPING OF MATERIALS AND LIST OF FIGURES AND TABLES

1. ELEMENTS

* Number marked with an asterisk indicates that recommended or provisional values are also reported for this material on a separate figure and table of the same number followed by the letter R.

1. ELEMENTS (continued)

* Number marked with an asterisk indicates that recommended or provisional values are also reported for this material on a separate figure and table of the same number followed by the letter R.

2. NONFERROUS BINARY ALLOYS

3. NONFERROUS MULTIPLE ALLOYS

† No figure given.

†No figure given.

† No figure given.

9. NONOXIDE COMPOUNDS

10. MIXTURES OF NONOXIDES

†No figure given.

10. MIXTURES OF NONOXIDES (continued)

11. MINERAL AND ROCKS

12. SYSTEMS

13. REFRACTORY MATERIALS, PROCESSED COMPOSITES, AND GLASSES

14. ORGANIC COMPOUNDS

† No figure given.

15. POLYMERS

16. FOODS AND BIOLOGICAL MATERIALS

†No figure given.

Theory, Estimation, and Measurement

Notation

a — Heat loss parameter [equations (77) and (80)]
A — Area; Numerical coefficient [equations (82) and (84)]
A_1, A_2 — Areas of container and fluid, respectively [equation (96)]
A_n — Coefficient of series expansion [equation (54)]
b — Numerical coefficient [equations (77) and (80)], approximately equal to $a^2/4$, with a given by equation (80)
B — Constant; Coefficient [equations (75) and (76)]
C_p — Specific heat at constant pressure
C_1, C_2 — Specific heat of container and fluid, respectively [equation (96)]
C — Capacitance; Constant [equation (9)]
d — Density
d_1, d_2 — Density of container and fluid, respectively [equation (96)]
d_0, d_r — Density at T_0 and T, respectively
E — Electromotive force
f — Frequency
F — Numerical coefficient [equation 82)]
g — Small thickness
h — Heat transfer coefficient
H — Flux of absorbed energy
H_m — Mean flux
H_0 — Magnetic field intensity
ΔH — Latent heat of vaporization
I — Electric current
I_0 — Bessel function of first kind and zeroth order; Amplitude of sinusoidally varying electric current
J_0 — Bessel function of zeroth order
k — Thermal conductivity
k_0 — Thermal conductivity at $T = 0$
L — Length or thickness; Lorenz function
m — Modulation coefficient
n — Integer; Number of time intervals
p — Perimeter
P — Power
Pr — Prandtl number
q — Logarithmic attenuation [equation (68)]; Amplitude ratio
q_1, q_2 — Amplitude ratios
q' — Phase shift [equation (69)]
Q — Heat quantity
Q_0, Q_1, Q_2, Q_3 — Coefficients [equation (79)]
r — Radius; Recovery factor
r_1, r_2 — Radii
r_h — Distance for half maximum temperature difference
R — Resistance; Radius of sample
R_k — Radius of coil
s — Mean error [equation (14)]
S — Slope
t — Time
t_c — Characteristic temperature rise time $= L^2/\pi^2\alpha$
t_i — Time axis intercept
t_1, t_2 — Time at two positions
T — Temperature, K
T_0 — Initial temperature; Ambient temperature
T_e — Effective temperature
T_f — Temperature maximum of front face
T_m — Maximum temperature
T_1, T_2 — Temperatures at two positions or times
ΔT — Temperature difference
ΔT_0 — Initial temperature difference
ΔT_f — Final temperature difference
v — Velocity
v_1, v_2 — Two different velocities
V — Voltage
V_1, V_2 — Voltage readings of two thermocouples
x — Distance; Position; Length

x_h, y_h	Coordinates of position for half maximum temperature difference
x_1, x_2	x coordinates of two positions
X_a, X_p	Horizontal distances such that $X_p = X_a \sin \phi$
Y_a, Y_p	Vertical distances such that $Y_p = Y_a \sin \phi$
α	Thermal diffusivity
α^*	Measured thermal diffusivity
α_1	Thermal diffusivity of container
α_2	Thermal diffusivity of liquid
β	Coefficient [equation (17)]
β_1, β_2	Values of β for same temperature interval and velocities v_1 and v_2
γ	$(\mu\sigma\omega/2)^{-1/2}$
δ	Effective thickness of skin layer
Δ	Correction factor [equation (95)]
ϵ	Emissivity; Deformation coefficient
ϵ_i	Temperature error at time t_i
η	Viscosity; also $\delta/2R$
θ	Debye characteristic temperature; Amplitude
θ_n	Expression [equations (54) and (56)]
Θ	Variable $= (1/k_0)\int_0^T k\, dT$ [equation (6)]
μ	Magnetic permeability
ν	Kinematic viscosity
π	Equal to $3.14159 \cdots$
ρ	Electrical resistivity
ρ_0	Residual electrical resistivity
σ	Electrical conductivity; Stefan–Boltzmann constant
τ	Wave period; Pulse time
τ_1, τ_2	Wave periods at two positions
ϕ	Phase shift
ϕ_1, ϕ_2	Phase differences between modulated beam and front and rear surfaces
$\varkappa$	Equal to $\omega r^2/\alpha$ [equation (84)]; Equal to $R(\Omega/\alpha)^{0.5}$ [equation (88)]
Ψ, Ψ_1	Thomson functions [equation (90)]
ω	Angular frequency
Ω	Modulation frequency

Thermal Diffusivity Data and Relationship to Other Properties

1. INTRODUCTION

Thermal diffusivity, the quantity dealt with in this volume, is the quantity which enters into certain equations relating to the heat flow under nonsteady-state conditions. The name *temperature conductivity* is also used for this quantity, particularly in some European literature, but in the present volume only the term *thermal diffusivity* will be used.

For steady-state conditions of unidirectional heat flow in an isotropic medium the well-known Fourier–Biot equation holds, namely,

$$\frac{Q}{A} = k\left(\frac{\partial T}{\partial x}\right) \tag{1}$$

where Q is the quantity of heat flowing in unit time through an area A under the influence of a temperature gradient $\partial T/\partial x$, and k is the thermal conductivity. The thermal conductivity may therefore be defined as the quantity of heat transmitted, due to unit temperature gradient, in unit time under steady-state conditions in a direction normal to a surface of unit area. When the thermal properties are independent of temperature but the temperature varies with time, t, the equation becomes

$$dC_p\frac{\partial T}{\partial t} = k\frac{d^2T}{dx^2} \tag{2}$$

where d is the density and C_p the specific heat at constant pressure. Equation (2) can be written in the form

$$\frac{\partial T}{\partial t} = \alpha\frac{d^2T}{dx^2} \tag{3}$$

where α, representing k/dC_p, is the quantity which Maxwell [1] called the thermometric conductivity and Thomson (Lord Kelvin) [2] the thermal diffusivity. This last name is appropriate since all diffusion processes can be represented by an equation similar to equation (3), and since α has the dimensions $(\text{length})^2\ (\text{time})^{-1}$, which are the dimensions of a diffusion coefficient.

It will be noted that in the one-dimensional case represented by equations (1) and (2), the thermal properties k, d, and C_p are treated as constants, independent of both position and temperature. This approximation is often acceptable. However, when the thermal conductivity is not a constant, equation (2) becomes

$$dC_p\frac{\partial T}{\partial t} = \nabla\cdot k\nabla T \tag{4}$$

The temperature dependency of d and C_p must be taken into account when equation (4) is solved. For Cartesian coordinates and isotropic materials, equation (4) may be rewritten as

$$dC_p\frac{\partial T}{\partial t} = \frac{\partial}{\partial x}\left(k\frac{\partial T}{\partial x}\right) + \frac{\partial}{\partial y}\left(k\frac{\partial T}{\partial y}\right) + \frac{\partial}{\partial z}\left(k\frac{\partial T}{\partial z}\right) \tag{5}$$

Carslaw and Jaeger [3] show how equation (5) can be reduced to

$$\frac{\partial\Theta}{\partial t} = \alpha\left(\frac{\partial^2\Theta}{\partial x^2} + \frac{\partial^2\Theta}{dy^2} + \frac{\partial^2\Theta}{\partial z^2}\right) \tag{6}$$

where

$$\Theta = \frac{1}{k_0}\int_0^T k\,dT \tag{7}$$

in which k_0 is the value of k when $T = 0$, and where in equation (6) $\alpha = k/dC_p$ is a function of the new variable Θ.

For more complex cases, such as anisotropic materials where the conductivity is treated as a second-order tensor, or when large temperature gradients are present or very short times are considered, the thermal diffusivity (and, for the latter two cases, the thermal conductivity) becomes ill-defined.

The allowance for k to be a function of temperature therefore gives rise to greater complications in equations relating to the transient state than for the steady state. When making experimental determinations of thermal diffusivity, however, the normal practice has involved relatively small ranges of temperature, and the assumption, within this range, of a constant thermal conductivity has invariably been made. This assumption has usually been well justified and is consistent with the few percent uncertainty that is claimed for most thermal diffusivity determinations.

Those working in this field, and particularly those engaged in problems in which large temperature differences can be established, and for which there may be considerable temperature variation of the thermal and physical properties, must however be watchful for conditions for which the simpler equation could become quite inapplicable. This could well be the case, for instance, in problems relating to the re-entry of space vehicles into the earth's atmosphere, and possibly also to the flash method (see Section 1.H of the next chapter), if operated on poorly conducting materials with intense heating.

Since these are mid-twentieth century developments, the present text and collected data will be found to relate almost entirely to methods in which constancy of the thermal properties has been assumed.

The fact that the thermal diffusivity is defined as k/dC_p means that this property differs fundamentally from that of the other volumes of this series in involving properties belonging to eight of the thirteen volumes. It is most closely related to thermal conductivity, the subject of Volumes 1 to 3, but also involves specific heat, which is the subject of Volumes 4 to 6, and thermal expansion, the subject of Volumes 12 and 13, since this property determines the temperature variation of density.

While no case is known in which knowledge of thermal diffusivity has led to the evaluation of a thermal expansion coefficient, and only one instance will be described in which use has been made of a thermal diffusivity method for specific heat determinations, considerable use has been made of thermal diffusivity methods for the determination of thermal conductivity. Indeed the primary requirement of the investigation has frequently been thermal conductivity rather than thermal diffusivity. Many investigations of this type have been included in Volumes 1 and 2 of this series, and some may tend to regard a treatise on thermal diffusivity as superfluous since any required thermal diffusivity value should be obtainable from the information given in these other volumes. Such derived thermal diffusivity values are indeed included in this volume for many elements and serve to extend considerably the temperature ranges for which direct determinations of this quantity have been made.

Despite its being a derived quantity, interrelating three other properties, thermal diffusivity is regarded as an important physical quantity in that it determines the rate of heat propagation in transient-state processes. Furthermore, the development of improved measurement techniques and their application to transient experimental methods is leading to an increasing use of these methods as a means of deriving thermal conductivity values. This is particularly true at high temperatures, where physicochemical changes in samples often necessitate rapid experimental methods. Kaspar and Zehms [4] have gone further by recognizing the measurement of thermal diffusivity in solids to be more than an indirect method for the determination of thermal conductivity. They point out that under certain conditions thermal diffusivity measurements are capable of revealing thermal conduction phenomena which would not be apparent by steady-state measurement methods. Transient or steady periodic methods should, for instance, be capable of detecting and determining time-dependent property values for which only a final value would be obtained by steady-state methods.

In view of these facts, in addition to the recognized position which thermal diffusivity has long occupied, a separate volume devoted to this subject is considered highly desirable. It is well justified on historical grounds and should be welcomed by those who are concerned with modern heat transfer problems.

Thermal diffusivity can be determined by any of the many experimental methods involving a temperature–time dependence for which equation (3) is applicable. Only some of these methods will be described in detail, and the following general considerations should be mentioned.

All materials conduct heat to a greater or lesser extent, and, just as with the determination of thermal conductivity, the methods employed for the determination of α are troubled by errors due to unwanted heat transfers. These are naturally reduced when the experimental time is decreased, and much shorter times can be used in transient, variable-state experiments than when a periodic temperature oscillation or a steady-state temperature distribution has to be established. This has long been realized, but the satisfactory application of transient methods has awaited technological developments which have allowed temperature changes to be more accurately determined over short measured time intervals. Improved technology has thus led to a revival of interest in transient methods for determining thermal properties, particularly at high temperatures. In the following sections these newer methods will be described in rather more detail than those originally used.

Other accounts, often giving more details of the older experimental methods and related mathematical treatments, will be found in the works of Carslaw and Jaeger [3], Callendar [5], Preston [6], Ingersoll *et al.* [7], Jakob [8], Gröber and Erk [9], and, for fluids only, Tsederberg [10]. More recent accounts are those of Sherwood [11] and of Danielson and Sidles [12].

2. AVAILABLE THERMAL DIFFUSIVITY DATA

A. The Chemical Elements

The purpose of the present volume is to make available under one cover the available experimental data for the thermal diffusivity of materials, at least insofar as such data have been extracted from the original publications and collected at the Thermophysical Properties Research Center (TPRC).

It will be found that this volume contains fewer data than do the volumes relating to thermal conductivity. This is because the property of thermal diffusivity has been investigated to a much more limited extent. For instance, experimental thermal diffusivity data are available for only some 39 of the 105 elements whereas experimental thermal conductivity data are available for at least 82. Even for most of these 39 elements the measurements often cover only a small temperature region near room temperature or to a few hundred degrees above. Experimental determinations below room temperature and for the molten phase are very scarce.

In order to remedy this situation, provisional thermal diffusivity values have been published by Ho *et al.* [13] for 75 of the elements for which the recommended (or provisional) thermal conductivity values had already been derived at TPRC [14–16]. These recommended thermal conductivity values (as of 1968) were used for k; C_p was obtained from TPRC and other compilations, d was taken from the literature or computed from thermal expansion data, and α was then calculated from k/dC_p.

Subsequently the recommended thermal conductivity values were completely updated and revised for publication in the *Journal of Physical and Chemical Reference Data.* In the course of preparing this volume, the thermal diffusivity values were recalculated using the revised thermal conductivity values. On the other hand, the experimental thermal diffusivity data for the elements were critically evaluated, analyzed, and synthesized. The derived thermal diffusivity values were then compared with the evaluated experimental data, whenever available, to generate the recommended (or provisional) thermal diffusivity values for the elements presented in this volume. These values serve also to increase substantially the range of temperatures for which information on thermal diffusivity is available.

Figure 1 shows available values of the thermal diffusivity of the elements at 300 K plotted against the atomic number of the particular element. A fair amount of systematic variation is evident, and from this figure very tentative values can be derived for those elements for which more direct information is lacking.

B. Titanium at High Temperatures

The important low-density high-melting-point metal titanium happens to be a metal for which, until recently, thermal diffusivity had provided the only experimental information for the thermal conductivity at high temperatures, indeed for the β-phase of this element.

A linear plot of all available information [17–31, 253] on the thermal conductivity of titanium is shown in Fig. 2. The curves which relate to direct thermal conductivity measurements are shown as continuous lines. Our present interest centers on the high-temperature data, and it will be seen that although above room temperature some twelve sets of direct determinations have been made, the upper limit of these determinations except one is about 975 K. Ho *et al.* [15] deduced the probable course of the heavy line of Fig. 2 from consideration of values of the electrical resistivity [32] and assuming the Lorenz function to decrease from 3.06×10^{-8} to 2.44×10^{-8} V^2 K^{-2} over the range 773 to 1673 K with most of the decrease occurring in the region of the alpha-to-beta transformation. The derived curve is shown to increase steadily with increase in temperature, although the possibility of a small discontinuity in the thermal conductivity in the region of 1155 K was fully realized.

Rudkin *et al.* [31] have made determinations of the thermal diffusivity of titanium for the range 746 to 1598 K, and Zinov'ev *et al.* [29] have since made similar determinations from 1006 to 1498 K. It is of

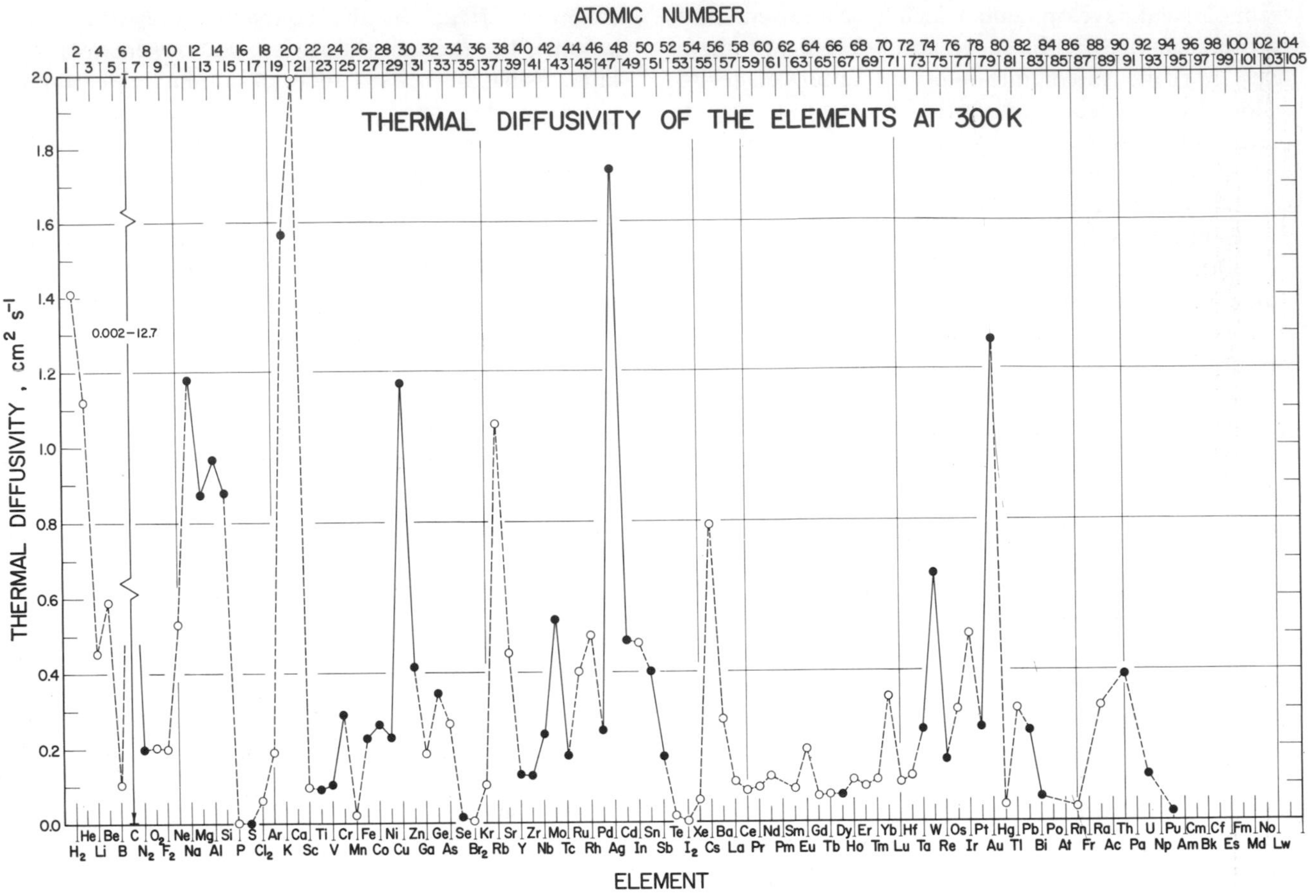

Fig. 1. Thermal diffusivity of the elements at 300 K. Open circles indicate estimated values for those elements for which no experimental data are available.

interest that these latest measurements indicated increases of 12 and 17 percent, respectively, to occur at the transformation temperature. From these thermal diffusivity values the curves shown in Fig. 2 as dashed lines have been obtained. In the case of the measurements of Rudkin *et al.* [31] the thermal conductivity values were derived at TPRC by using the specific heat data of Hultgren *et al.* [33]. Zinov'ev *et al.* themselves made this derivation, and the five sets of thermal conductivity data attributed to them in Fig. 2 were obtained by using five different sets of specific heat data. For curves 27 and 25 the same specific heat values have been used, namely, those of Hultgren *et al.* [33], so that these two curves indicate the real difference between the thermal diffusivity values of Rudkin *et al.* and of Zinov'ev *et al.* Those of the former workers are greater, the differences increasing from 18 percent at 1010 K to 40 percent at 1500 K. Doubt as to the true specific heat clearly adds to the uncertainty of any thermal conductivity value derived from a thermal diffusivity measurement.

This example shows that despite the manner in which modern thermal diffusivity measurements are helping to increase knowledge of thermal properties at high temperatures, large uncertainties still exist and every effort should still be directed toward the achievement of greater accuracy in all these measurements.

C. Experimental Data as Exemplified by Copper and Tungsten

The compilation of experimental data for the thermal diffusivity of various materials, which forms the major part of the present volume, aims to include all published data. One consequence of this is that the data are not selective, irrespective of the uncertainty of the data. Any reported reliability or error limits included are the estimations given in the original publications. Only for the elements, critical evaluation of the validity and accuracy of the experimental data have been made, and recommended or provisional values have been generated.

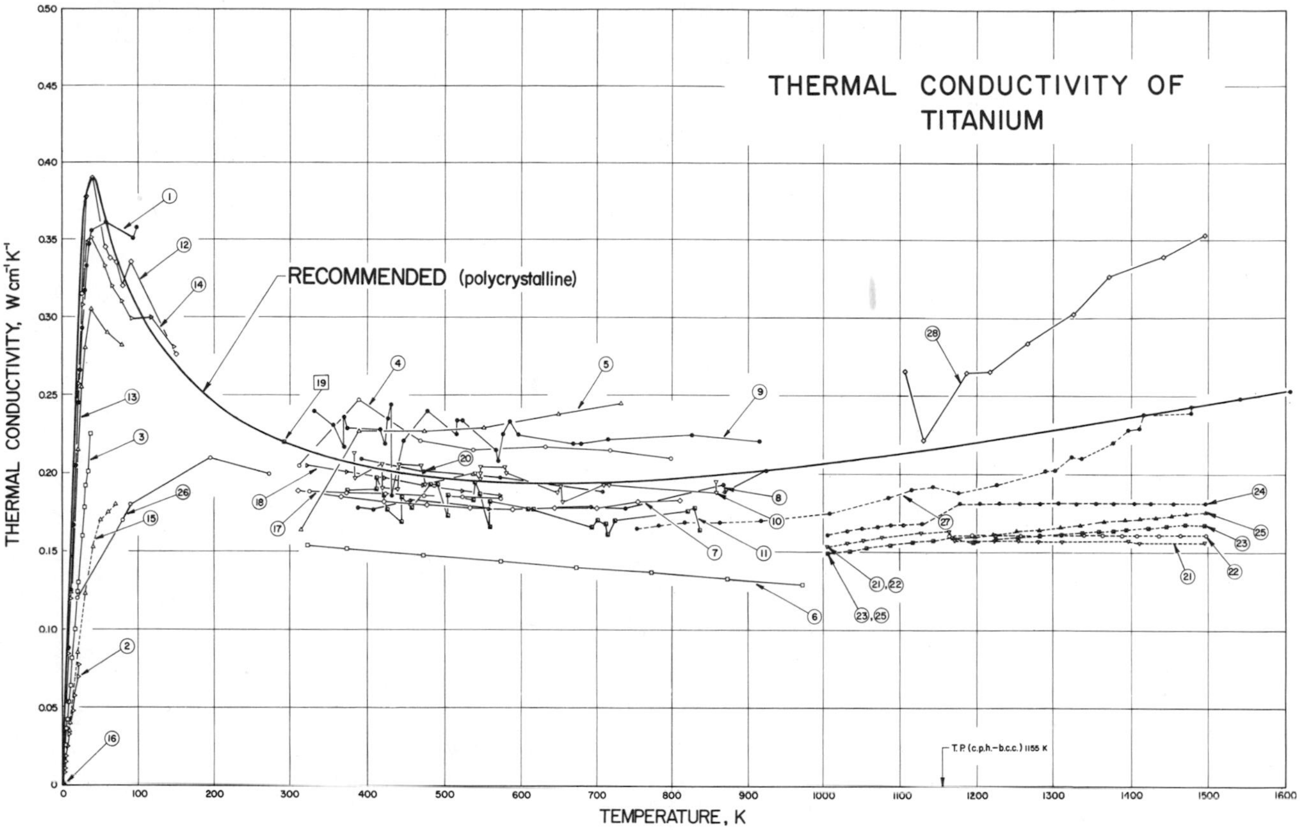

Fig. 2. Thermal conductivity of titanium. 1, Rosenberg [17]; 2 and 3, Mendelssohn and Rosenberg [18]; 4 and 5, Mikryukov [19]; 6, Silverman [20]; 7, Deem, Wood, and Lucks [21]; 8, Krzhizhanovskii [22]; 9–11, Loewen [23]; 12–14, White and Woods [24]; 15, Gladun and Holzhäuser [25]; 16, Davey and Mendelssohn [26]; 17–19, Powell and Tye [27]; 20, Kuprovskii and Gel'd [28]; 21–25, Zinov'ev, Krentsis, and Gel'd [29]; 26, Rigney and Bockstahler [30]; 27, Rudkin, Parker, and Jenkins [31]; 28, Unvala and Goel [253]. (15, 21–25, and 27 have been derived from thermal diffusivity measurements.)

Two metals, copper and tungsten, will now be treated in some detail by way of illustration, not only of the type of agreement obtained between the experimental and the derived values of α, but of the useful purpose served by these derived values.

a. Copper

Copper is one of the metals on which many thermal diffusivity measurements have been made. At the time that this section was prepared the TPRC data files contained 78 sets of experimental thermal diffusivity data for copper, and these cover the range 200 to 1730 K. No measurements had been recorded below 200 K. To avoid confusion only 12 representative sets of values [34–40] have been plotted in Fig. 3 where logarithmic scales are used. The spread is considerable and, on the basis of these measurements, it would be very difficult to predict acceptable values to higher or lower temperatures.

Figure 4 presents 111 of the 218 sets of thermal conductivity data available for copper of better than 99.5 percent purity. The heavy line drawn in this figure is the curve recommended as most probable for a high-purity sample of this metal on the basis of the information then available. The low-temperature portion of this curve, $T < 100$ K approximately, is strongly dependent on sample purity, and the recommended curve is for a well-annealed 99.999^+ percent Cu sample having a residual electrical resistivity, ρ_0, of 0.579×10^{-9} ohm cm.

Figure 5 is Fig. 14 of Volume 4, which shows 12 of the 55 sets of data available for the specific heat of copper. From this figure and consideration of data given in other sources [33, 41, 42], a specific heat-temperature curve has been selected.

Experimental density values [43] for solid and liquid copper are considered slightly low. Consequently, the density d_0 of solid copper at temperature

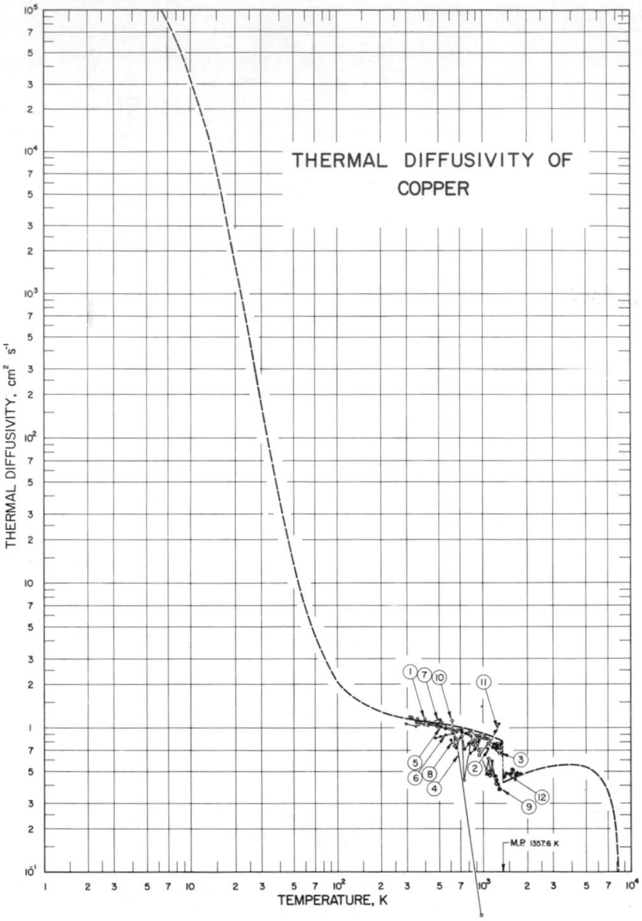

Fig. 3. Thermal diffusivity of copper. 1–3, Butler and Inn [34]; 4, Sheer *et al.* [35]; 5 and 6, El-Hifni and Chao [36]; 7, Sonnenschein and Winn [37]; 8, Dennis *et al.* [38]; 9–11, Sheer *et al.* [39]; 12, Mardykin and Filippov [40]. The dashed curve was derived from $\alpha = k/dC_p$.

T_0 is taken from [32] and d_T is derived from expansion data [32]. The density of molten copper at various temperatures was selected from the literature [44–46].

The thermal diffusivity curve derived from the recommended thermal conductivity, selected specific heat and liquid density, and the calculated solid density, is shown as the heavy broken line in Fig. 3. It is interesting to note that for copper of the particular purity chosen, the thermal diffusivity increases one hundred twenty-seven thousand times as the temperature decreases from room temperature to 5 K. This compares with about a forty-nine-fold increase in thermal conductivity at 5 K, and about a sixty-two-fold increase at 9 K where the thermal conductivity attains a maximum value. In the experimentally determined range from room temperature to the melting point, the derived curve lies close to the higher experimental curves and decreases by only about 30 percent.

Considerable uncertainty still exists as to the true thermal conductivity curve for molten copper. While this uncertainty must also be reflected in the thermal diffusivity and will render uncertain the value near the melting point, the initial temperature coefficient, and the position of the maximum, the general form of the curve must be much as shown, since it is conditioned by the sharp decrease in thermal conductivity which occurs as the critical temperature is approached. At the critical temperature, the thermal conductivity of the liquid is assumed to merge toward that of the vapor, an assumption which ignores the sharp increase that is believed to occur in a very narrow temperature region close to the critical point.

b. Tungsten

Figure 6 represents the main sets of experimental thermal diffusivity data available for tungsten. While

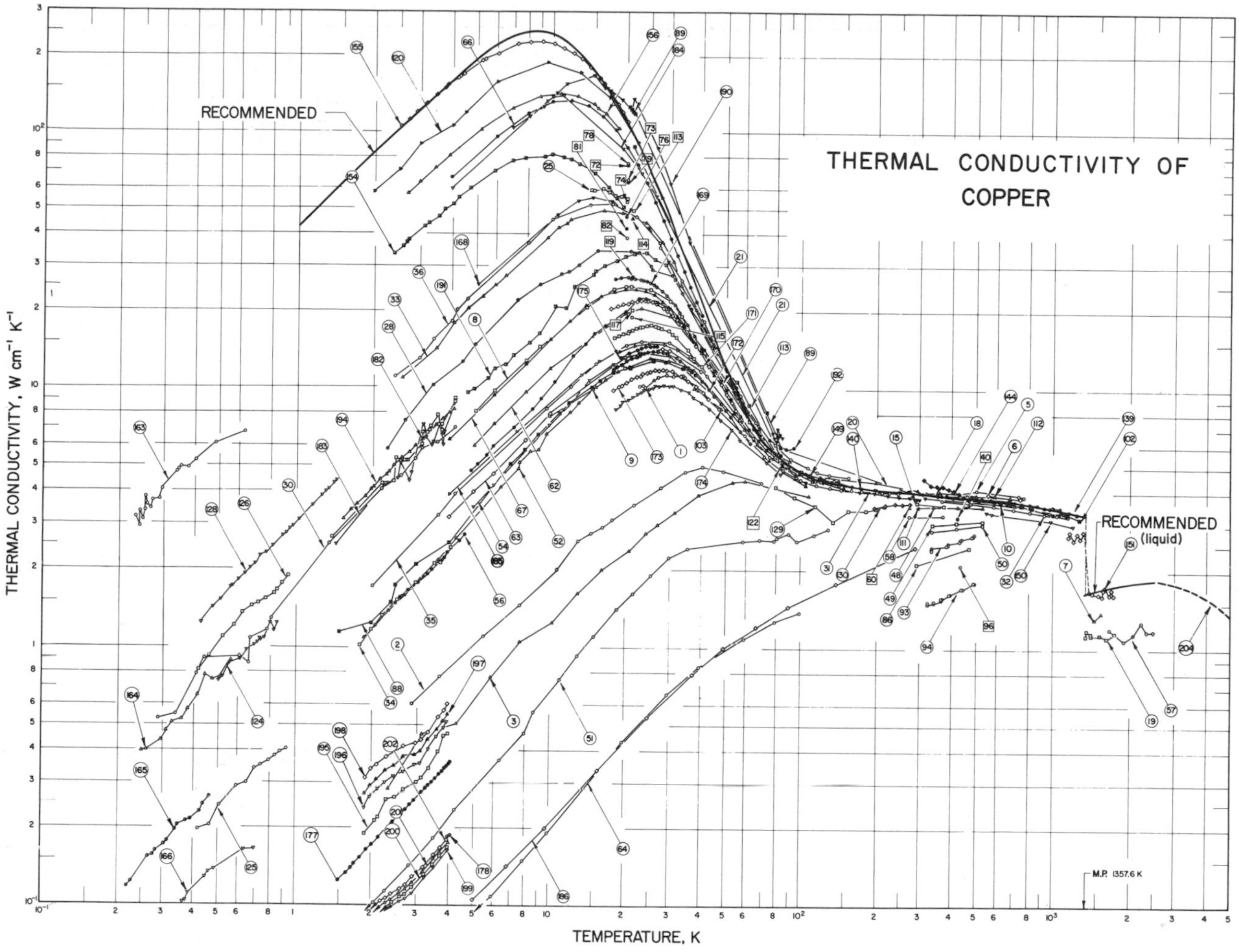

Fig. 4. Thermal conductivity of copper.

these extend up to a temperature of 3234 K, it seems that until quite recently no determination had been made on tungsten below 407 K. Furthermore, this determination is one of several due to Sheer *et al.* [39] which lie very much below both the other determinations and the heavy broken curve derived [13] from the curve of most probable thermal conductivity [14]. Tungsten was chosen as representing an important metal for which the derived value of the thermal diffusivity at normal temperatures was much to be preferred to the nearest experimental value available in 1968. At that time an engineer in need of the thermal diffusivity of tungsten at room temperature and deciding to use the value of Sheer *et al.* [39] for 407 K, would be using a value that is only about one-fifth of the true value. This is indicative of the caution necessary before accepting literature values as furnishing equally reliable and useful data. Unfortunately, some reported values have proved to be much removed from the truth.

Another reason for discussing the results for tungsten is that, because of its high melting point, tungsten is a strong candidate for adoption as a thermal conductivity reference material for use at high temperatures. It is therefore a good material for tests by methods regarded as suitable for determinations to very high temperatures. Consideration of the available results serves to indicate the order of agreement that is being obtained.

Figure 7 is intended to show this more clearly. In this figure linear scales are used and individual points have been included both to indicate the considerable number of observations furnished by these methods and their degree of scatter. The obviously low data of Sheer *et al.* [39] have been omitted from this figure but the values derived by Ho *et al.* [13]

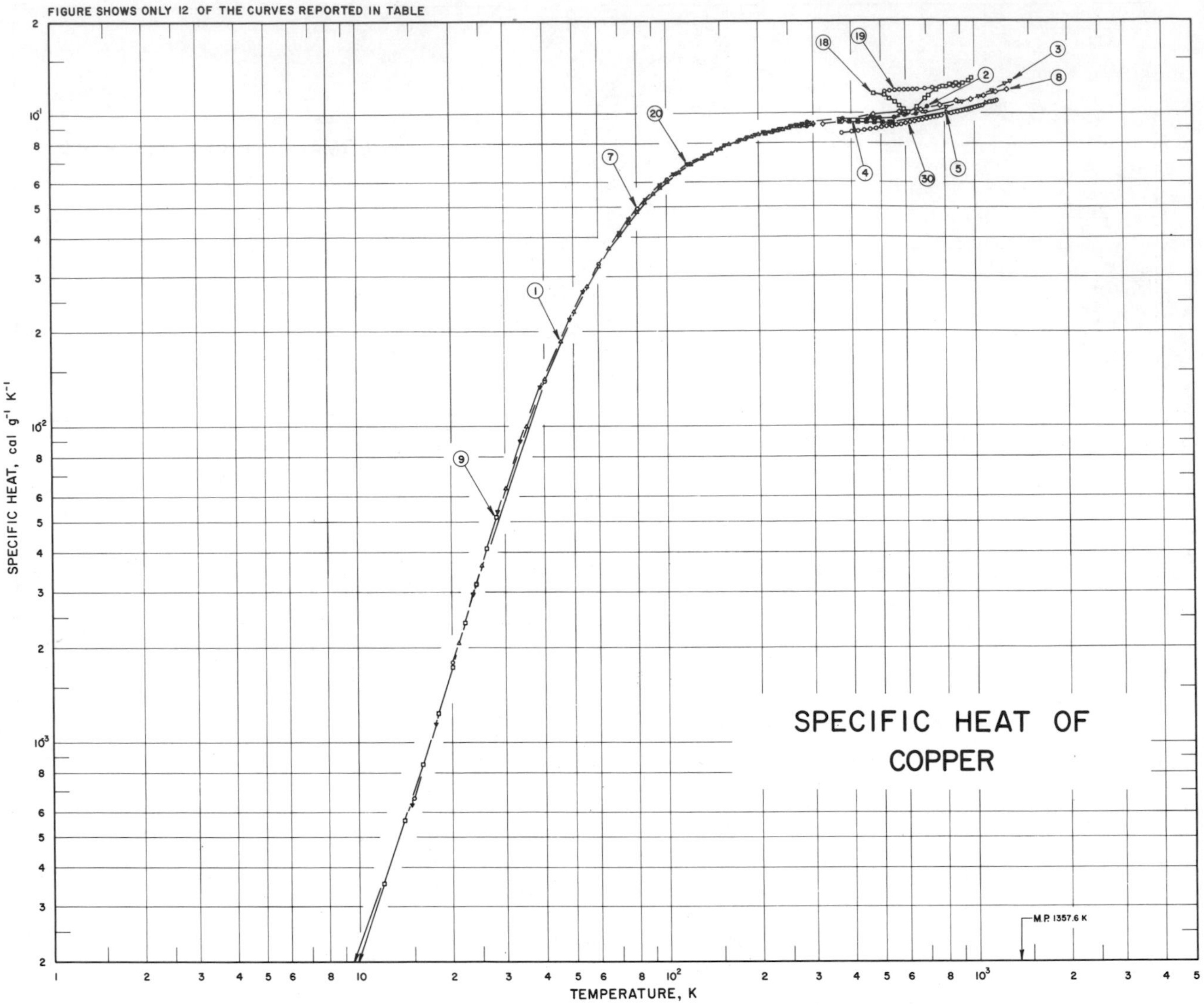

Fig. 5. Specific heat of copper (Fig. 14 of Volume 4 of TPRC Data Series).

from the relation $\alpha = k/C_p d$ are again shown by the heavy broken line.

The main interest centers in the six sets of data which are grouped on either side of the derived curve. The two preliminary values derived by Pigal'skaya and Filippov [50] at 1673 K, from amplitude and phase measurements at several frequencies, are, respectively, about 9 and 11.5 percent above this curve and most of the subsequent data of Pigal'skaya *et al.* [51] are within 12 percent above. The other Russian measurements of Kraev and Stel'makh [47, 52] lie about a corresponding amount below the curve up to about 2500 K but continue to diverge and are about 24 percent below at 3200 K. The values of Wheeler [48] agree more closely with the derived curve particularly at their upper range of 2600 to 2900 K. This order of agreement is encouraging, particularly as Pigal'skaya *et al.* [50, 51] used peripheral heating of a cylindrical specimen, while the other measurements were made on thin disks which were heated on one face. Rawuka and Gaz [53], using laser pulse-heating of a sample 1.8 mm thick, obtained a value at 1593 K which is about 4.5 percent above the broken curve; however, nearer 1000 K their values are some 10 percent below this curve.

The Battelle Memorial Institute [54] and Atomics International [54] also used the laser-flash method. The former's values ranged from 10 percent below the derived curve at 1270 K to 3 percent below at 2150 K. The three values due to Atomics International

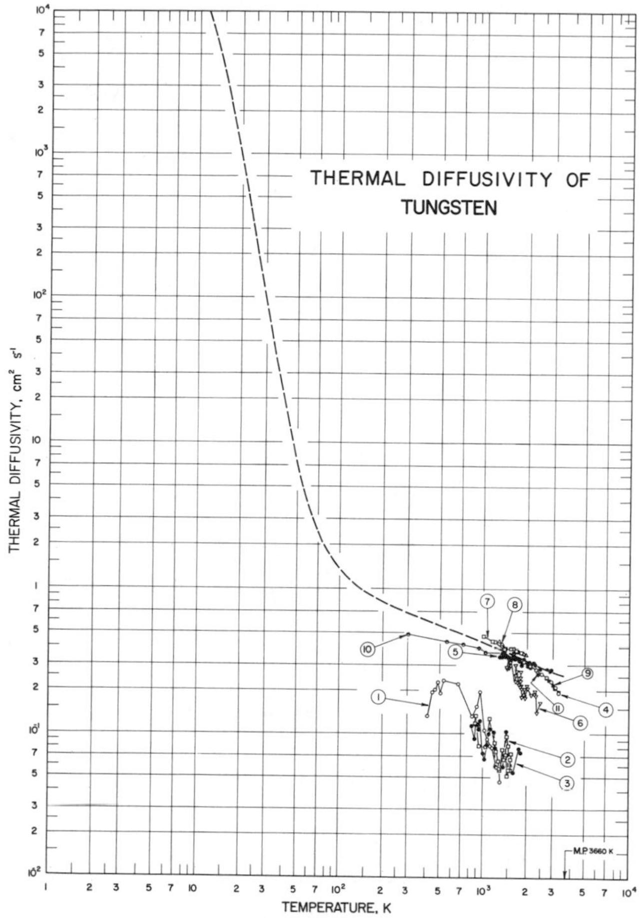

Fig. 6. Thermal diffusivity of tungsten. 1–3, Sheer *et al.* [39]; 4, Kraev and Stel'makh [47]; 5, Wheeler [48]; 6, Taylor and Nakata [49]; 7 and 8, Pigal'skaya, Filippov, and Borisov [51]; 9, Kraev and Stel'makh [52]; 10, Rawuka and Gaz [53]; 11, Battelle Memorial Institute [54]. The dashed curve was derived from $\alpha = k/dC_p$, Ho *et al.* [13].

were for the narrow range of 1757 to 1825 K and lie from 8.5 to 25 percent below the derived curve. These values tend to support those of Taylor and Nakata [49], reported in a progress report and considered doubtful by the authors. In a later paper Cape *et al.* [55] reproduce from Reference 49 the results for tantalum, but omit those for tungsten. The tantalum values agreed with calculated values from $\alpha = k/C_p d$ and were accepted as confirming the adopted radial heat flow method to be satisfactory for materials with thermal diffusivity values of well under 0.2 cm^2/sec. This would rule out tungsten, and the method was subsequently applied for carbides of zirconium and tantalum.

These considerations serve to emphasize two facts which should be borne in mind. Although a method may be checked and found to be satisfactory for one material, it need not necessarily be satisfactory for all materials. Nor should all reported data be accepted as reliable. The disregarding of the results for tungsten of Taylor and Nakata [49] as well as those of Sheer *et al.* [39] are examples of the selective assessment which comes from a more detailed study of available data, and such assessment will not yet have been applied to most of the experimental data contained in the Data Section of this volume. At the present stage, the aim has been to approach completion of the compilation and to generate recommended values for the elements, the user being left to make his own assessment of the available data for the other materials.

D. Thermal Diffusivity of Foodstuffs and Biological Materials

Food substances and biological materials constitute two important groups of materials for which readers could expect to find thermal diffusivity data

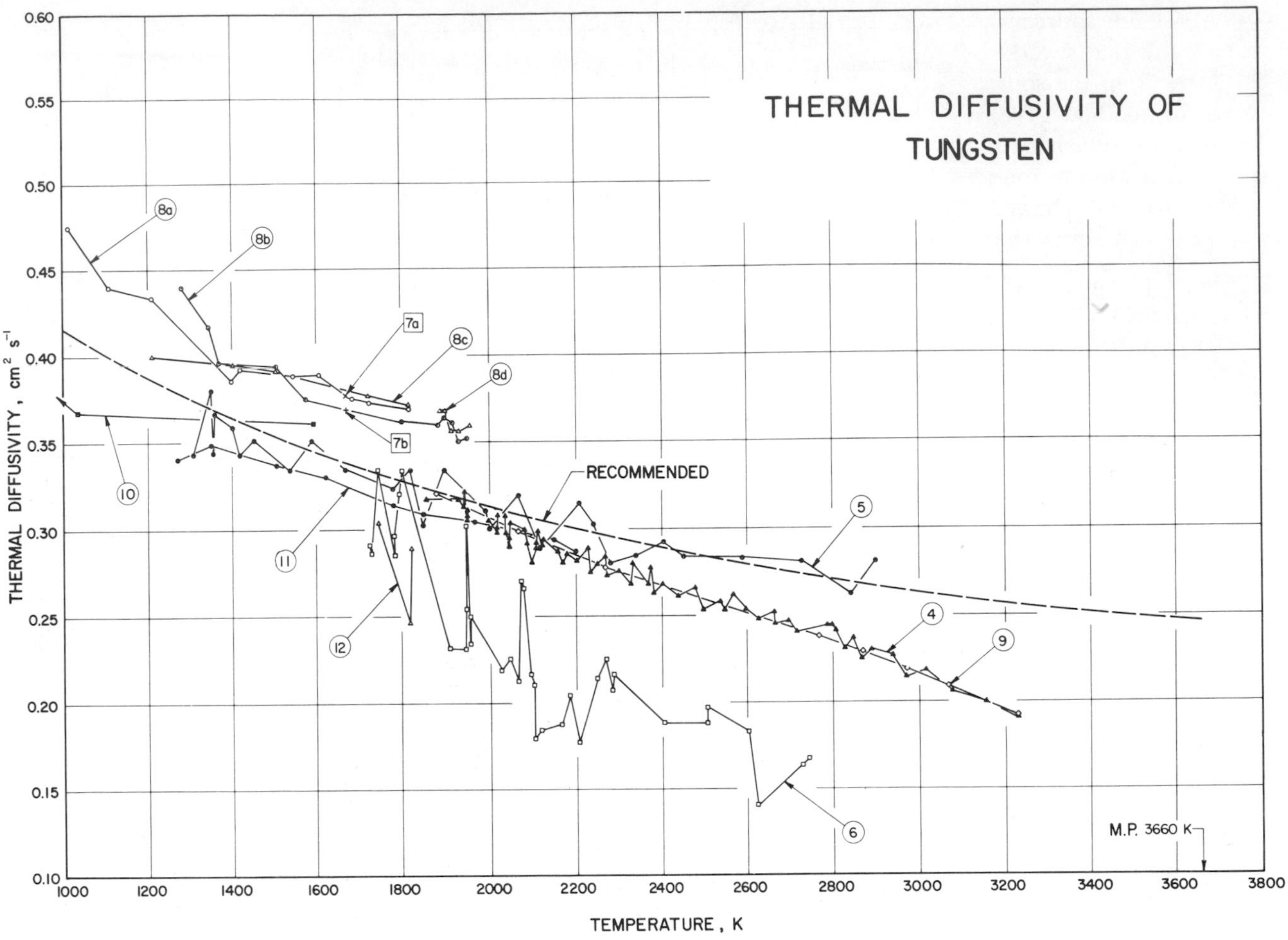

Fig. 7. Thermal diffusivity of tungsten at high temperatures. 4, Kraev and Stel'makh [47]; 5, Wheeler [48]; 6, Taylor and Nakata [49]; 7, Pigal'skaya and Filippov [50] (7a phase method, 7b amplitude method), 8, Pigal'skaya, Filippov, and Borisov [51]; (8a phase method, equal periods, 8b phase method, unequal periods, 8c amplitude method, equal periods, 8d amplitude method, unequal periods), 9, Kraev and Stel'makh [52]; 10, Rawuka and Gaz [53]; 11, Battelle Memorial Institute [54]; 12, Atomics International [54]. The dashed curve was derived from $\alpha = k/dC_p$, Ho *et al.* [13].

in this volume. Unfortunately, these materials are dealt with only partially. It is only recently that the TPRC has been able to commence any detailed work on these substances, and much effort will need to be expended before a comprehensive coverage of this information can be made. Published values often differ, and complete information regarding such factors as the content of moisture, fat, protein, blood flow rate, and so on of the sample is often lacking. Then again, many of these materials have anisotropic thermal properties, and it is important that the direction of the heat flow with regard to muscle and other fiber directions be stated.

Chato [56], in a survey paper of 1966, collected together the available data for the thermal conductivity and thermal diffusivity of biological materials and emphasized the need for more attention to be directed to this field. Two Masters theses, dealing mainly with the thermal conductivity of foodstuffs, have since appeared. These are by Reidy [57] and by Qashou [58]. A separate report by Reidy [59] is also available. Direct determinations of thermal diffusivity are dealt with in some of the referenced papers, notably those of Hurwicz and Tischer [60], Dickerson [61], and of Wadsworth and Spadaro [62], the latter for sweet potatoes. Qashou [58] also mentions a paper by Nix *et al.* [63] which describes a useful extension to the normal heated line-source thermal conductivity method that allows simultaneous determinations to be made of both thermal conductivity and thermal diffusivity. See Section 1.F of the next chapter for details. Trezek *et al.* [64] used a rather similar radial heat flow method when determining the thermal diffusivity of cat brain tissue. Their technique differed in that the line source was suddenly cooled instead of heated and that several thermocouples at increasing distances from the source were used instead of only two. The method of data evaluation also differed. The thermal diffusivity of cat brain tissue immediately following death was found to be close to that of water but subsequent changes occurred and $2\frac{1}{2}$ hr after death the value was about 50 percent lower.

3. RELATIONSHIP OF THERMAL DIFFUSIVITY TO OTHER PROPERTIES

As had already been explained, $\alpha = k/dC_p$ and much of the motivation for the determination of α has been that the relation $k = \alpha dC_p$ provides a simpler, and hopefully a more accurate, means of deriving k. However, before k can be derived from this expression, it is necessary to know C_p, and at times obtaining the true specific heat for a particular material can present difficulties. Data for the specific heat are contained in Volumes 4, 5, and 6 of the present series, and some methods leading to approximate values in cases where no data are found should also be noted (see also the text of Volumes 4, 5, and 6).

According to the law of Dulong and Petit, at about normal temperature, the atomic heat, which is the product of the specific heat and the atomic weight, is the same for all substances and is equal to $3R =$ 5.96 cal g-atom^{-1} C^{-1}, or close to 25 J g-atom^{-1} C^{-1}. An analogous law, that of Kopp and Neumann, holds approximately for chemical compounds. According to this law the molecular heat of a solid chemical compound is equal to the sum of the atomic heats of the constituent atoms. Many solid solutions and alloys behave similarly. Below the Debye temperatures, the specific heat decreases and tends toward zero at absolute zero, whereas at higher temperatures increases of up to about 30 percent can occur.

A reduced plotting of the ratio of the specific heat at a temperature T to the specific heat at the Debye temperature θ, against T/θ, obtained by using available data, can serve as a means for estimating the specific heats of similar substances for which no data are available. A plot of this type has for example been used by Steigmeier and Kudman [65] to derive specific heat data at high temperatures for certain Group III–V compounds.

At all temperatures and for substances for which the coefficient of thermal expansion is known, use can be made of Grüneisen's law, according to which for any one substance the coefficient of expansion is proportional to the atomic or specific heat. Caution is of course needed in applying such a method, particularly for materials in which the true specific heat is augmented by an anomalous reaction which needs to be included.

Awbery [66] examined the Grüneisen law in connection with experimental data obtained for a series of 21 steels. By plotting the derived Grüneisen constants against the atomic percentage of elements other than iron in each steel, he showed clearly that iron when in the gamma phase possessed about a 50 percent greater Grüneisen constant than alpha iron. Another conclusion from this work related to the values of the atomic heats which should be used for the main constituents, iron and carbon, when attempting to estimate the atomic heat of a steel from its chemical composition. At 100 C the derived atomic heats of alpha iron, gamma iron, and carbon were found to be, respectively, 6.52, 6.93, and 3.68 cal g-atom^{-1} C^{-1}.

So far in this section, only the derivation of k from the product αdC_p has been considered, but, since the ratio k/dC_p is also used to evaluate α, means for the estimation of k might therefore be helpful. In this connection reference should be made to the close connection between thermal and electrical conductivity and the use of the expressions

$$k = LT/\rho \tag{8}$$

or

$$k = L'T/\rho + C \tag{9}$$

where L is the theoretical Lorenz function with a value of 2.443×10^{-8} V^2 K^{-2} and ρ is the electrical resistivity. In equation (9) L' and C are the constants obtained for the straight lines fitting plots of the thermal and electrical conductivity data for various metals and their alloys. A paper by Powell [67] tabulates values of L' and C for alloys of Al, Cu, Fe (both ferrous and austenitic), Mg, Ni, Ti, and Zr. Another paper by Williams and Fulkerson [68] should be consulted for further details regarding the subdivision of the thermal conductivity of metals into electronic and lattice components (see also the text of Volume 1 of this Series).

In the case of fluids, thermal diffusivity is related to the Prandtl number, Pr, and the kinematic viscosity, ν, by the equation

$$\alpha = \nu/\mathrm{Pr} \tag{10}$$

and this receives further consideration in Section 3 of the third chapter.

Measurement of Thermal Diffusivity of Solids

According to the nature of the temperature–time variation that is imparted to the specimen under investigation, the methods used for the measurement of thermal diffusivity fall into two main classes:

1. Transient heat-flow methods
2. Periodic heat-flow methods

These methods can be further classified according to the originator or to the nature of the technique used, yielding the following:

1. Transient heat-flow methods
   - A. Long bar, heated at center or at one end
     - a. Forbes' bar method
     - b. Other bar methods
   - B. Moving heat-source method
   - C. Small-area-contact method
   - D. Thermoelectric effect method
   - E. Semi-infinite plate method
   - F. Radial heat-flow methods, including line source
   - G. High-intensity arc method
   - H. Flash methods
     - a. Flash heating by xenon lamp
     - b. Flash heating by laser beam
     - c. Flash heating by electron beam
     - d. Flash method applied to composite samples
     - e. Flash method applied for direct determination of thermal conductivity
   - I. Electrically-heated rod methods
2. Periodic heat-flow methods
   - A. Ångström's method
   - B. Temperature wave velocity method
   - C. Temperature wave amplitude-decrement method
   - D. Modified Ångström's methods
     - a. Method of Sidles and Danielson
     - b. Method of Abeles *et al.*
   - E. Phase-lag methods
   - F. Thermoelectric methods
   - G. Radial-wave method
   - H. Cryogenic method of Howling, Mendoza, and Zimmerman

Descriptions of these methods follow, but those which are of little more than historic interest are only briefly treated.

More attention is given to the flash methods, modified Ångström's methods, and some other methods introduced relatively recently which possess the merits of high speed and considerable freedom from the influence of heat losses, and so become valuable methods for use under extreme conditions, for instance at very high temperatures where a short test-time is often desirable to avoid heat losses and complications due to any structural and chemical changes occurring in the specimen. Thus at high temperatures thermal diffusivity determinations are tending to supplement and even to displace many of the steady-state thermal conductivity methods. For the direct determination of the thermal conductivity of an electrically conducting solid to high temperatures, one of the most promising methods is that of Taylor *et al.* [69] in which a wire or rod is directly heated by the passage of an electric current (see also Section 1.I). Mention of the investigation into this method made at the TPRC, but mainly in relation to thermal conductivity, is most appropriate since the apparatus used and the test procedure developed have been shown by Powell and Taylor [70] to be capable of producing values for a dozen or so properties for the one temperature-equilibriated specimen, including those from which the thermal diffusivity can be derived.

The equation for a long thin rod of density d and cross-sectional area A, heated in an enclosure at temperature T_0 by a current I, to have temperature T at location Z and time t is

$$k\left(\frac{d^2T}{dZ^2}\right) + \frac{dk}{dT}\left(\frac{dT}{dZ}\right)^2 + \frac{I^2\rho}{A^2} - \frac{p\epsilon_H\sigma(T^4 - T_0^4)}{A} - \frac{\mu I}{A}\left(\frac{dT}{dZ}\right) = C_p d\left(\frac{dT}{dt}\right) \qquad (11)$$

where p is the perimeter, σ the Stefan–Boltzmann constant, and k, ρ, ϵ_H, μ, and C_p, the thermal conductivity, electrical resistivity, total hemispherical emittance, Thomson coefficient, and specific heat at constant pressure, are functions of temperature.

In the work so far described [69, 70] temperature profiles have been measured for the equilibrium condition in which the term on the right-hand side of equation (11) is zero. For this condition, and using different sample lengths, these workers show that in the case of tungsten evaluations can be made to 2700 K for electrical resistivity to well within 1 percent, total hemispherical and spectral emissivity to within 2 percent, and the thermal conductivity to well within 5 percent. The Thomson coefficient is also evaluated.

In later work provision will be made to include measurements of the thermal expansion during this steady-state experiment for a long rod by direct observation of the distance between two fiducial markers attached to the rod. This will allow d to be determined as a function of T. Also the specific heat will be determined for a variable state experiment in which dT/dt is observed for the original temperature obtained with current I_1 when the long rod is heated by a different current I_2 and passes through this temperature. Then

$$C_p = \frac{\rho(I_1^2 - I_2^2)}{dA^2(dT/dt)} \qquad (12)$$

The thermal diffusivity is then obtainable from the relation $\alpha = kdC_p$.

In other instances, and especially for very thin coatings and for electrical nonconductors, resort to a thermal diffusivity determination by one of the methods described in Sections 1.H or 2.D is probably best, with the thermal conductivity then derived from $k = \alpha dC_p$.

1. TRANSIENT HEAT-FLOW METHODS

A. Long Bar Heated at Center or at End

a. Forbes' Bar Method

The method introduced by Forbes [71, 72] about a century ago for determining the thermal conductivity of a wrought iron bar is a two-part experiment which actually yields the thermal diffusivity. The first is a steady-state experiment in which the temperature profile is determined for a horizontal bar when heated at one end by being brought into contact with a molten metal bath; the rest of the bar is unheated and wrapped with thermal insulation. In the second experiment a cooling curve is obtained for the same bar after being uniformly heated to the bath temperature and then exposed to the same ambient conditions as in the previous experiment.

From the heat-flow equations for the two experiments, and, making the assumption that the rate of heat loss from an element of the rod at a particular temperature is the same when the temperature is uniform as when a gradient is imposed, it is deduced that

$$\alpha = \frac{k}{dC_p} = \int_{x=L}^{x=x} \frac{\partial T}{\partial t}\, dx \Big/ \left(\frac{dT}{dx}\right)_{x=x} \qquad (13)$$

Hence the original work of Forbes actually gave k/dC_p, and was multiplied by the product dC_p to give the thermal conductivity, k.

A comment appears to be appropriate. This is to mention that workers who subsequently used this method, notably Bidwell [73, 74] and his associates [75, 76], clearly regarded the Forbes' bar method as being used for the determination of k and not α. In their later modifications [75, 76] the quantities density and specific heat do not even appear, since, by heating the sample rod electrically to an approximately uniform excess temperature, the thermal conductivity of the surrounding insulation was derived. This modification allowed evaluation of the loss of heat between the two locations at which the temperature gradients were observed and enabled k to be directly calculated.

b. Other Bar Methods

Before moving on to consider other transient methods, it seems appropriate to mention the method used by Ingersoll and Koepp [77] for thermal diffusivity determinations on soils and by Frazier [78–80] for determinations on nickel and zinc. In this method a sudden temperature change was made at one end of a rod of uniform cross section and initially in thermal equilibrium with its surroundings. The thermal diffusivity was obtained from measurement of the rate of change of the temperature difference between two properly chosen points on the rod. The treatment assumes no lateral heat exchange from the rod so a guard tube is required. In this respect it resembles the set-up and method used more recently by Kennedy [81] as shown in Fig. 8. Kennedy's technique also requires the temperature profiles to be observed in a rod following the application of heat to one end. In this method use is made of a high-speed computer to

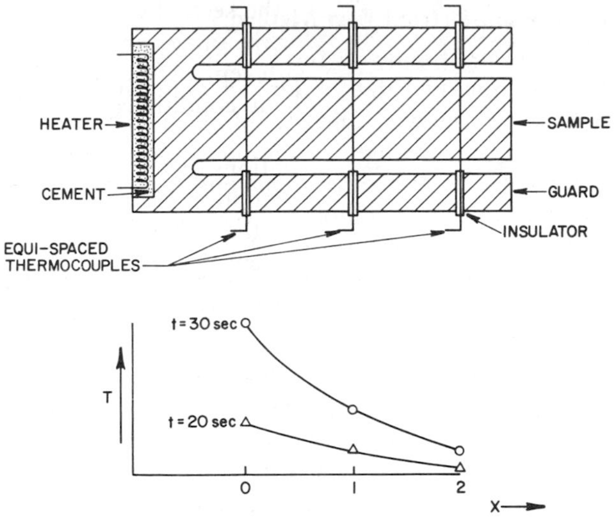

Fig. 8. Kennedy's method using guarded sample [81].

solve the one-dimensional equation (3) with boundary conditions applying to a finite rod. The surrounding guard tube is fabricated from the same block of material as the sample and the heater is common to both. Three thermocouples are attached to the sample at equally spaced intervals. After the heat is applied, the reading of each thermocouple is recorded as a function of time. The temperatures at the two outer thermocouples determine the boundary conditions for the heat-flow equation, which is then solved for the midpoint temperature by assuming various values of the thermal diffusivity. The calculated temperatures are then compared with the measured value in order to determine which value of α best describes the thermal behavior of the sample. The squared mean error

$$s^2 = \sum_{i=1}^{n} \epsilon_i^2/n \tag{14}$$

is computed over the time range $0 < t < t_{\max}$ for each trial value of thermal diffusivity, the parameter ϵ_i being the difference between the computed and observed temperatures for the middle thermocouple for a particular time t_i. The value of α which makes the squared mean error a minimum is assumed to be the best estimate of the thermal diffusivity. The method has been used by Kennedy [81] to measure the diffusivity of Armco iron. It has been also applied by Kennedy *et al.* [82], and by Shanks *et al.* [83, 84] to measure the diffusivity of Armco iron and silicon. The measurements on silicon extended to 1400 K. As was shown by Kennedy, methods of this type can readily be programmed for computer data reduction.

Jones and Chisholm [85] have described an end-cooling method, specifically a modification of the Jominy end-quench test normally employed for hardenability studies on steels and therefore a method which can be applied in most steel laboratories. For the diffusivity measurement the test samples, which are 2.5 cm in diameter and 10 cm in length, are brought to a uniform temperature of about 100 C and the lower end is quenched in water. A fine-wire thermocouple is welded a known distance L, of about 4 cm, above the quench level, and, following the quenching, the time is observed for it to attain the mean temperature between the initial bar and water temperatures. The authors show how the thermal diffusivity can be derived to a relative accuracy of about ± 2 percent from L and observations of this half-temperature time. Corrections can be made for side losses, but these have only about a 1 percent effect on the thermal diffusivity for the cases studied.

B. Moving Heat-Source Method

The theory of this method, which stems from temperature observations made by Bornefeld [86] during fusion welding, has been developed by Rosenthal [87] and applied by Rosenthal and others [88, 89] for determinations on copper and some aluminum alloys. On the assumption that the heated surface loses heat, hT, according to Newton's law of cooling, except at the location of the small heat source, then, if v is the velocity of the source and T the temperature rise of any point in the bar above its surroundings, the relevant differential equation is

$$\frac{1}{v^2}\frac{d^2T}{dt^2} + \frac{1}{v^2}\frac{d(\ln k)}{dT}\left(\frac{dT}{dt}\right)^2 = \frac{1}{\alpha}\frac{dT}{dt} + hT \tag{15}$$

So long as k varies slowly with T the second term can be omitted giving

$$\frac{1}{v^2}\frac{d^2T}{dt^2} = \frac{1}{\alpha}\frac{dT}{dt} + hT \tag{16}$$

which is satisfied by $T = B\exp(\beta T)$, and yields

$$\frac{\beta^2}{v^2} - \frac{\beta}{\alpha} - h = 0 \tag{17}$$

β is a function of both T and v. By obtaining two different values β_1 and β_2 for the same interval of temperature, but for two different velocities, v_1 and v_2,

$$\alpha = (\beta_1 - \beta_2)\left[\left(\frac{\beta_1}{v_1}\right)^2 - \left(\frac{\beta_2}{v_2}\right)^2\right]^{-1} \tag{18}$$

A fundamental difficulty of this method would appear to involve obtaining a satisfactory point source.

C. Small-Area-Contact Method

A method has been used by Cutler [90] for thermal diffusivity measurements on germanium which possesses similarities to the necked-down-sample method used for determinations of thermal conductivity (see Volume 1). This is the so-called small-area-contact method, and in this instance the heat flow into the sample occurs at the small area where a wire is joined and serves as a current-carrying electrode. In the necked-down-sample method, only one metal is involved and the length and diameter of the constricted region is made sufficiently small to render negligible any radiation or other lateral heat losses. Under these conditions, the maximum temperature rise of the constriction T_m due to the passage of an electric current is a function only of V, the voltage drop across it, and of the electrical and thermal conductivities of the material. Hence the ratio k/σ, and therefore the Lorenz function, $k/\sigma T$, can be evaluated in terms of T_m and V, and no knowledge of sample dimensions is necessary. This was the method of Hopkins [91] and Hopkins and Griffith [92]. By treating the sample as a resistance thermometer, T_m can be derived from the temperature coefficient of resistance. This additional knowledge enabled Holm and Störmer [93], Cutler *et al.* [94], and Flynn and O'Hagan [95] to derive values of the thermal conductivity.

Cutler [90] has considered the additional thermoelectric heating introduced when the contact is made with dissimilar metals, and the two-part method which he employed to determine the thermal diffusivity of germanium involved first the measurement of the power P required to maintain steady conditions for the determination of k, and then switching off this power and observing the time variation of the temperature of the contact as indicated by the recording of the residual thermoelectric voltage. The temperature–time decay curve for a hemispherical contact of radius, r, is then given by

$$\partial T(t) = \frac{P}{2\pi k r}\exp\left(\frac{\alpha t}{r^2}\right)\operatorname{erfc}\left(\frac{\alpha t}{r^2}\right)^{1/2} \tag{19}$$

As $\alpha t/r^2$ is made small and much less than 1, equation (19) reduces to

$$\partial T(t) = \frac{P}{2\pi k(\pi\alpha t)^{1/2}} \tag{20}$$

The method maintains the merit of the necked-down-sample method in that the radiation losses from the small heated area are relatively small.

D. Thermoelectric Effect Method

A rather similar method to the above, but for a short sample of length x and of uniform diameter, was used by Hérinckx and Monfils [96] and by Pinnow *et al.* [97] to measure the thermal diffusivity of bismuth telluride and other semiconducting materials. In order to ensure uniform current distribution, the ends of the sample are nickel plated and wires to serve as either current or potential leads are spot welded to the plated ends. The passage of an electric current through the sample, when this is suspended in an evacuated enclosure, heats the specimen to a small excess temperature. At time $t = 0$ a switch is thrown which breaks the current circuit and connects the leads from the specimen to a potential recorder. The potential difference between the leads is proportional to the temperature difference and this decays exponentially with time. It is shown that for a small temperature interval, in which α can be assumed to be constant, a plot of log $\Delta T(t)$ against t yields a straight line of slope s, and α is calculated from the equation

$$\alpha = -s\left(\frac{x}{\pi}\right)^2 \tag{21}$$

E. Semi-Infinite Plate and Related Methods

The term "semi-infinite" has been used by mathematicians to define a sample in which one-directional flow of heat normal to the surface can be ensured. In practice, by suitable thermal guarding to achieve the required boundary conditions, quite small areas and thicknesses can often approximate to this requirement. The familiar guarded-plate method and many other steady- and variable-state methods relate to a sample of this kind. Consider such a sample of thickness L to be adiabatically insulated over the face $x = 0$ and heated uniformly by radiation or some other means over the other face $x = L$, from time $t = 0$. Then, if H is the absorbed energy, the temperature distribution T at position x and time t is given by

$$T = T_0 + \frac{HL}{k}\left[\frac{\alpha t}{L^2} + \frac{3x^2 - L^2}{6L^2} - \frac{2}{\pi^2}\sum_{n=1}^{\infty}\frac{(-1)^n}{n^2} \times \cos\frac{n\pi x}{L}\exp-\left(\frac{n\pi}{L}\right)^2\alpha t\right] \tag{22}$$

where T_0 is the initial temperature of the slab.

After a time greater than $0.54\,L^2/\alpha$, which is sufficient for the series term of equation (22) to be

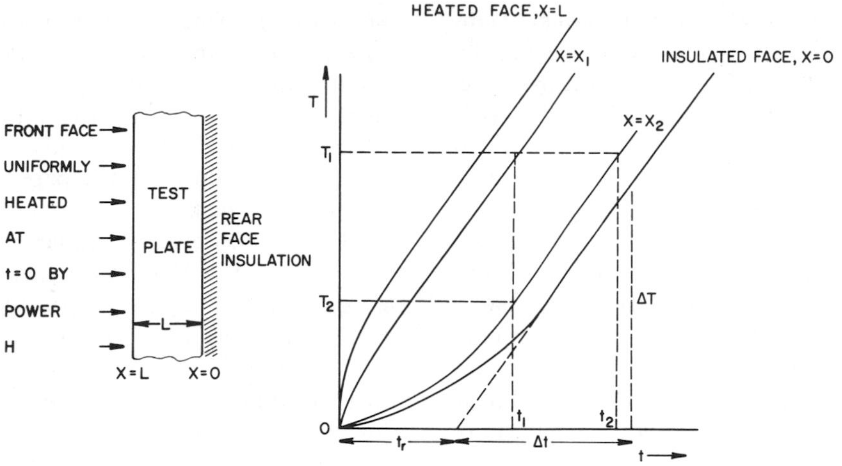

Fig. 9. Temperature–time curves for a plate, after uniform heating is applied to one face (methods of Sonnenschein and Winn [37] and Butler and Inn [34]).

neglected,

$$T = T_0 + \frac{HL}{k}\left[\frac{\alpha t}{L^2} + \frac{3x^2 - L^2}{6L^2}\right] \qquad (23)$$

and the temperature–time curve as shown in Fig. 9 becomes linear for all values of x.

This equation gives the so-called long-time solution from which the following three experimental treatments have been suggested for the derivation of α, none of which requires any determination of the absorbed heat flux, H.

a. Method Based on Relaxation Time

If t_r is the intercept on the axis when $T = T_0$, equation (23) gives

$$t_r = (L^2 - 3x^2)/6\alpha \qquad (24)$$

The quantity t_r is the so-called relaxation time defined as the time interval from the time of application of heat source H to the intercept with the time axis of the negatively extrapolated linear portion of the temperature–time curve for the particular location, x.

Of special interest is the case when $x = 0$, that is, when the temperature change is measured at a point in the remote insulated face. In this case,

$$\alpha = L^2/6t_r \qquad (25)$$

This method is the single-point relaxation-time method that has been used by Sonnenschein and Winn [98] to determine α for several materials, including Armco iron, to temperatures of about 1000 C.

As they point out, this measurement has been reduced to quite a simple nondestructive type test for samples that are thin but for which L is accurately known. Providing that arrangements are made to heat the exposed face uniformly, all that is necessary is a temperature sensor of low heat capacity such as a fine-wire thermocouple held in good thermal contact with the adiabatically insulated face.

b. Method Based on Time Interval

Differentiation of equation (23) with respect to t, yields the same slope

$$\frac{dT}{dt} = \frac{H\alpha}{Lk} = \frac{H}{dC_pL} \qquad (26)$$

for all values of x. Thus, if H is known, the specific heat can be evaluated, while, by making temperature observations at two positions x_1 and x_2, and noting the time interval, $t_1 - t_2$, for equal temperatures, then α is given independently of H by the equation

$$\alpha = (x_1^2 - x_2^2)/2(t_1 - t_2) \qquad (27)$$

c. Method Based on Ratio of Slope to Temperature Difference

Again by using the foregoing observations, but noting the temperature difference $T_1 - T_2$ at any time for which there is a linear temperature rise for two locations x_1 and x_2, then

$$\alpha = \frac{(dT/dt)(x_1^2 - x_2^2)}{2(T_1 - T_2)} \qquad (28)$$

This last is the method used by Butler and Inn [99] for determinations on copper, and some alloys of iron, aluminum, titanium, and magnesium. Temperatures of the order of 1000 C were reached by heating with radiant energy from a high intensity arc.

The same principle has been used by Smith [100] to obtain thermal diffusivity data to 1600 F for much poorer conducting samples of spacemetal. This is a thin cellular panel formed in 301 stainless steel. Smith mounted back to back two similar panels each of thickness L and arranged for the exposed faces to be heated at equal rates by two power-regulated batteries of quartz lamps. Attached differentially connected thermocouples measured the constant temperature difference ΔT established for the linear temperature rise condition. The thermal diffusivity was then given by

$$\alpha = \frac{(dT/dt)L^2}{2\Delta T} \qquad (29)$$

Smith referred to the above as a single sandwich method as he also used a double sandwich modification in which the two test panels were place between two outer layers of another material having known thermal properties. By adopting this procedure it became possible to evaluate the product dC_p for the unknown test material and hence to determine its thermal conductivity from the observed thermal diffusivity. Determinations were made by Smith on laminated phenolic-asbestos materials under consideration for use as missile nose cones, with transite as the known material.

d. Comparative Method

One of the earlier of the more modern attempts to develop a transient method that would enable a thermal diffusivity value to be obtained from simple observations lasting only about half a minute was that of Hsu [101]. This method requires two identical sets of standard and test plates. The chosen standard was a nickel cylinder, 2.5 in. in diameter and 3 in. in length of known density, specific heat, and thermal conductivity. The test plates were of the same diameter and 2 to 10 mm in thickness depending on the order of thermal diffusivity to be measured. By using a graphite paste insert, a test and standard sample is brought into good thermal contact and a thermocouple is mounted centrally at the interface. One of these combined units is kept at room temperature and the other is heated to attain a uniform temperature. The two units are then forced together with the two test samples contacting each other. A high pressure and graphite paste are used to improve the thermal contact, and the alignment is such as to bring the full faces of the test samples into contact. The ensuing transient temperatures are measured at certain times, say, 5 and 25 sec after contact, by the thermocouples located between the test and standard specimens. For the theory and methods of evaluating the thermal diffusivity, the papers of Hsu [101, 102] should be consulted. Hsu made determinations on steel and considered the values to be good to about ± 3 percent. The assumption is made that the contact is perfect and that the temperature of the contacted surfaces changes instantly to a common mean value. This method can be regarded as a precursor of the method due to Parker *et al.* [103] in which the instantaneous change in the surface temperature of a sample is promoted by a flash of high-intensity radiation (see Section 1.H).

e. Methods for Poor Thermal Conductors

Methods of the semi-infinite plate type have also been used for determinations on samples of low thermal conductivity intended for use as thermal insulation. One advantage of such variable-state methods over the more conventional steady-state method has been that the appreciable reduction in the test time would help in reducing errors associated with moisture migration in any test materials having significant moisture content.

A method proposed by Vernotte [104] was used by Clarke and Kingston [105, 106] for determinations on woods. The required boundary conditions were met by placing eight similar test slabs in a pile and weaving a heating strip of copper foil having a width equal to that of the slabs between each successive pair of slabs. A thermocouple with the smallest possible heat capacity was located at the central point between the two middle slabs, the location of the required nonconducting plane, and was used to record the change in temperature with time. Methods of this type have since been used by Krischer and Esdorn [107], Harmathy [108], and Pratt and Ball [109] whose paper included good accounts of their method and of the relevant equations. Pratt and Ball were interested in building constructional materials and, by intercomparing these results with others obtained for the same samples by the steady-state method, concluded that the variable-state method, taking only about 5 minutes, yielded results of only a little lower accuracy than the normal steady-state methods which require a few hours.

Levine [110] and later Paladino *et al.* [111] considered the radiant heating of bodies of other geometric forms and developed a method for determining the thermal diffusivity of a cylindrical sample of alumina in the temperature range 1500 to 1800 C.

f. Methods for Glasses and Ceramics

A modified version of the semi-infinite plate method was introduced by Plummer *et al.* [112] to measure to high temperatures the thermal diffusivity of glasses, glass ceramics, and magnesia and alumina ceramics. Their temperature measurements were restricted to observations for $x = 0$ and $x = L$ and were actually made in the flat faces of the bounding heater and sink. Hence, good thermal contact is essential between these surfaces and the plane surfaces of the test sample. When heated by a constant energy flux, H, the temperature rise, ΔT, for such a system is given by

$$\Delta T = \frac{2H}{k}\sqrt{\alpha t}\; i\,\mathrm{erfc}\left(\frac{x}{2\sqrt{\alpha t}}\right) \tag{30}$$

Hence the ratio of the temperature rise for the two observed locations is

$$R = \frac{\Delta T(L)}{\Delta T(o)} = \sqrt{\pi}\; i\,\mathrm{erfc}\left(\frac{L}{2\sqrt{\alpha t}}\right) \tag{31}$$

By evaluating this ratio in terms of $\sqrt{\pi}$ times the integrated error function for a series of values of $L/2\sqrt{\alpha t}$, a curve can be drawn from which αt can be evaluated from the experimentally determined ratios for particular values of t.

F. Radial-Heat Flow Method

If the outer surface of a long hollow cylindrical specimen of internal and external radii, r_1 and r_2, is heated at a constant rate, then the temperature difference $T_2 - T_1$ established between r_2 and r_1 is

$$T_2 - T_1 = \frac{1}{2\alpha}\frac{\partial T}{\partial t}\left[\frac{1}{2}(r_2^2 - r_1^2) - r_1^2 \ln\frac{r_2}{r_1}\right] \tag{32}$$

This equation assumes that the specimen is isotropic and homogeneous, with α independent of T. It also assumes $\partial T/\partial t$ to be constant and that there is no internal loss of heat, so that $\partial T/\partial t$ is zero when $r = r_1$.

Fitzsimmons [113] described a method of measuring thermal diffusivities in which long solid cylinders are heated or quenched by rapid immersion in a well-stirred fluid, and the subsequent temperature changes at the center of the rod were observed.

Equation (32) was the basis of a rather better controlled method described by Flieger and Ginnings [114] and by Flieger *et al.* [115] with the difference that the two radii r_2 and r_1 at which temperatures are observed were located within the body of a hollow cylinder of inner radius r_0. Then the appropriate equation becomes

$$\alpha = \frac{1}{2(T_2 - T_1)}\frac{\partial T}{\partial t}\left[\frac{1}{2}(r_2^2 - r_1^2) - r_0^2 \ln\frac{r_2}{r_1}\right] \tag{33}$$

Small changes in the radial-flow method were made by Lehman [116] and later by Cape and Taylor [117], Taylor and Nakata [49], and Cape, Lehman, and Nakata [55].

In Lehman's method the cylindrical sample was placed within a heated enclosure and fitted with end guards to ensure that all heat flowed radially inwards. The sample was heated rapidly by a constant source of power and temperatures were measured, either by thermocouples or by an optical pyrometer, at the bottom of two holes drilled axially to the longitudinal center of the sample with radii of r_1 and r_2. For times sufficiently long for a linear rate of temperature increase to be established

$$\alpha = \frac{r_2^2 - r_1^2}{4(t_2 - t_1)} \tag{34}$$

where $t_2 - t_1$ is the time interval between the attainment of a specific temperature at r_2 and r_1.

After Lehman had successfully used the method for Armco iron, the method was further modified by Cape and Taylor [117]. These later investigators [49, 54, 117] located the temperature observation holes to make $r_1 = 0$ and $r_2 = r/\sqrt{2}$, where r is the radius of the sample. Then,

$$\alpha = \frac{r^2}{8(t_2 - t_1)} \tag{35}$$

By using a rapid-response automatic optical pyrometer, α was determined to high temperatures for several materials, including tantalum, tungsten, and the carbides of tantalum, titanium, zirconium, and hafnium. As mentioned previously in Section 2.C.b of the last chapter, the method did not prove suitable for tungsten, owing to the high value of α.

G. High-Intensity Arc Method

This is the method developed by Sheer *et al.* [35, 39] and used to measure the thermal diffusivity of copper, graphite, molybdenum, and tungsten. Cylindrical specimens were thermally insulated except for one flat face which was heated directly and by radiation from the plasma of the tail-flame of an electric arc. While heated in this way, temperatures were recorded of a thermocouple located in each of two holes drilled normal to the axis at specific locations along the length of the specimen. For a perfectly insulated specimen, the thermal diffusivity is derived from

$$\alpha = \frac{x(\partial T/\partial t)}{(\partial T/\partial x)} \tag{36}$$

where x is the mean distance of the two thermocouples from the unheated back face, and $\partial T/\partial t$ and $\partial T/\partial x$ are obtained from the recorded temperature–time curves and the distance separating the two thermocouples. Further temperature–time curves taken during cooling and with the arc removed, allow the cooling rate $(\partial T/\partial t)_c$ to be evaluated. The thermal diffusivity is then determined to a closer approximation from the equation

$$\alpha = \frac{x\left[\dfrac{\partial T}{\partial t} + \left(\dfrac{\partial T}{\partial t}\right)_c\right]}{\dfrac{\partial T}{\partial x}} \tag{37}$$

An previously indicated, the results obtained for tungsten by this method are much too low.

H. Flash Methods

A very simple method for the determination of thermal diffusivity is one in which a sample in the form of a small disk is brought to a steady uniform temperature in a suitable furnace or cryostat. A flash of thermal energy is then supplied to the front surface of the sample within a time interval which is short compared with the resulting thermal transient for which temperatures are observed on the back surface. This method has become very popular since it was first described in 1960 by Parker *et al.* [103]. Quite small samples can be used and heating of the front face, $x = 0$, has been by xenon lamps, solid-state laser, and electron beams. The energy absorption is maximized by blackening the exposed surface. Sensitive recording apparatus is used to give the temperature of the rear face, $x = L$, as a function of time, and heat-losses are minimized by making the measurements in a time so short that little cooling can occur.

According to Carslaw and Jaeger [3], if such a sample has a steady initial temperature distribution, $T(x, 0)$, then the temperature $T(x, t)$ at any time, t, after receiving the flash at $t = 0$, is given by

$$T(x, t) = \frac{1}{L}\int_0^L T(x, 0)\,dx + \frac{2}{L}\sum_{n=1}^{\infty} \exp\left(\frac{-n^2\pi^2\alpha t}{L^2}\right) \times \cos\frac{n\pi x}{L}\int_0^L T(x, 0)\cos\frac{n\pi x}{L}\,dx \tag{38}$$

Assuming the pulse of energy, Q, to be instantaneously and uniformly absorbed in a small depth g at the surface $x = 0$, then, at that instant, the temperature distribution is

$$T(x, 0) = Q/dcg \quad \text{for} \quad 0 < x < g$$

$$T(x, 0) = 0 \quad \text{for} \quad g < x < L$$

For these initial conditions equation (38) becomes

$$T(x, t) = \frac{Q}{dcL}\left[1 + 2\sum_{n=1}^{\infty} \cos\frac{n\pi x}{L}\,\frac{\sin(n\pi g/L)}{(n\pi g/L)} \times \exp\left(\frac{-n^2\pi^2}{L^2}\alpha t\right)\right] \tag{39}$$

For opaque materials, g is sufficiently small for $\sin(n\pi g/L) \approx n\pi g/L$, and with this approximation the temperature at the rear face becomes

$$T(L, t) = \frac{Q}{dcL}\left[1 + 2\sum_{n=1}^{\infty} (-1)^n \exp\left(\frac{-n\pi^2}{L^2}\alpha t\right)\right] \tag{40}$$

Equation (40) indicates the maximum temperature of the rear face to be

$$T_{L\,\max} = \frac{Q}{dcL} \tag{41}$$

and hence, at any time, t, the rear face will rise to a fraction of its maximum rise which is given by

$$\frac{T_{(L,t)}}{T_{(L\,\max)}} = 1 + 2\sum_{n=1}^{\infty} (-1)^n \exp\left(\frac{-n^2\pi^2}{L^2}\alpha t\right) \tag{42}$$

Two convenient means have been proposed for the determination of α from equation (42).

When $T_{(L,t)}/T_{(L\,\max)} = 1/2$, the dimensionless quantity $\pi^2\alpha t/L^2$ has a value of approximately 1.37, and then

$$\alpha = \frac{1.37\,L^2}{\pi^2 t_{1/2}} = \frac{0.139\,L^2}{t_{1/2}} \tag{43}$$

where $t_{1/2}$ is the time required for the rear surface to attain half its maximum increase in temperature. Of course, the assumption has been made that no loss of heat occurs, and the effect of any neglected heat loss will increase with time and will tend to reduce $T_{L\,max}$.

The second method allows the observations to be restricted to shorter times. It requires the temperature–time curve to be determined at the rear face for a time sufficient for the linear rate of rise to become established. As illustrated in Fig. 10, it can be shown that this straight line portion extrapolates back to give the initial temperature at a time, t_i, where $t_i = 0.48\,L^2/\pi^2\alpha$ and hence

$$\alpha = \frac{0.48\,L^2}{\pi^2 t_i} \tag{44}$$

where t_i is the time-axis intercept of the linear extrapolation.

It must be appreciated that this method, in which a strong energy pulse is concentrated in a very thin surface layer, first causes an appreciable, almost instantaneous, temperature increase in this thin layer. Then, as heat is conducted toward the rear face, the temperature of the front face must decrease until it reaches $T_{L\,max}$, when the sample will be at a uniform temperature. These temperature variations promote questions regarding the maximum front surface temperature, T_f, and the effective temperature, T_e, to which the measurement of α should be assigned. Both questions were dealt with by Parker *et al.* [~~102~~], [103] who concluded that to a fair approximation

$$T_f = 38\,L T_{L\,max}\alpha^{-0.5} \tag{45}$$

$$T_e = 1.6\,T_{L\,max} \tag{46}$$

For a copper sample tested by Parker *et al.* [103] L was 0.312 cm, for which equation (45) would indicate T_f to be about 20 K. In the case of titanium, despite

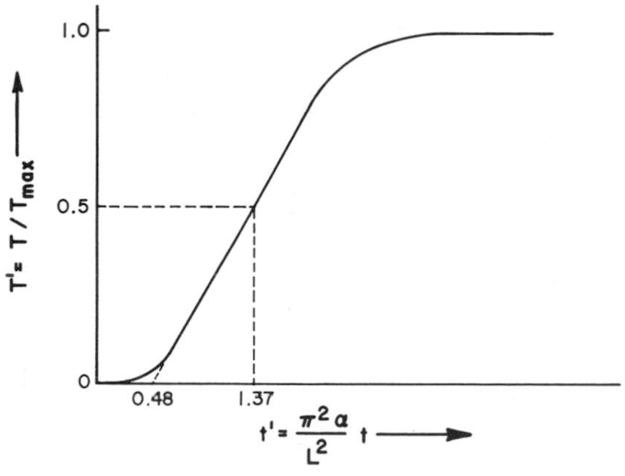

Fig. 10. Flash method of Parker *et al.* [103]; dimensionless plot of rear-face temperature *vs* time.

L having been reduced to 0.1 cm, the indicated value of T_f for this metal of lower thermal conductivity is about 160 K. It is important to ensure that no phase changes are included; also, to avoid errors due to radiation losses, T_f should not rise unduly. Radiation losses reduce the numerical coefficient of equation (43). On the other hand, the requirement for the flash time to be small, compared with the time for the heat pulse to travel through the material, imposes a limit on the amount by which L can be reduced before necessitating an increase in the numerical coefficient of equation (43). Theoretical treatments regarding radiation losses and finite pulse time corrections have been given by Cowan [118, 119], Mendelsohn [120], Cape and Lehman [121], Taylor and Cape [122], Watt [123], and Larson and Koyama [124].

The results of the analysis by Cape and Lehman [121] confirm that, for all but the poorer conductors, radiation losses can be neglected provided the sample thickness is of the order of 0.1 cm. The required coefficient reduction in equation (43) can exceed 10 percent when $4\sigma\epsilon T_0^3 L/k$ is greater than 0.1; the reduction however remains less than 5 percent for values of $k/\epsilon L$ of not less than 0.1 and T_0 of up to 1000 K. In these expressions, σ is the Stefan–Boltzmann constant, T_0 the ambient temperature, and ϵ the surface emissivity.

In attempting to reduce this correction by reducing L, care needs to be exercised, or the finite-pulse time correction will need to be applied. If t_c is the characteristic rise time, $L^2/\pi^2\alpha$, for the rear face, and τ is the pulse time, then, provided that all heat losses are negligible, Taylor and Cape [122] show that, so long as $t_c \geq 50\tau$, then equation (43) would be expected to hold good to about 1 percent. Since for the normally operated optical sources of power then available, τ was of the order of 1 or 2 msec, it followed that for such a source t_c should not be less than 50 to 100 msec. As a means for dealing with those cases where it is not possible for τ and t_c to obey these restrictions, Taylor and Cape [122] make use of an equation given by Cape and Lehman [121] to construct curves of τ/t_c against $t_{1/2}/t_c$ for a finite saw-tooth pulse of base-width τ and for a square wave of width τ. These curves are reproduced in Fig. 11. When τ and $t_{1/2}$ are determined experimentally, the appropriate graph can be used to derive t_c and hence $t_{1/2}/t_c$, which is then used as the numerical coefficient of equation (43) for the determination of α.

Taylor and Cape [122], who indicate that computer methods can be used to derive appropriate

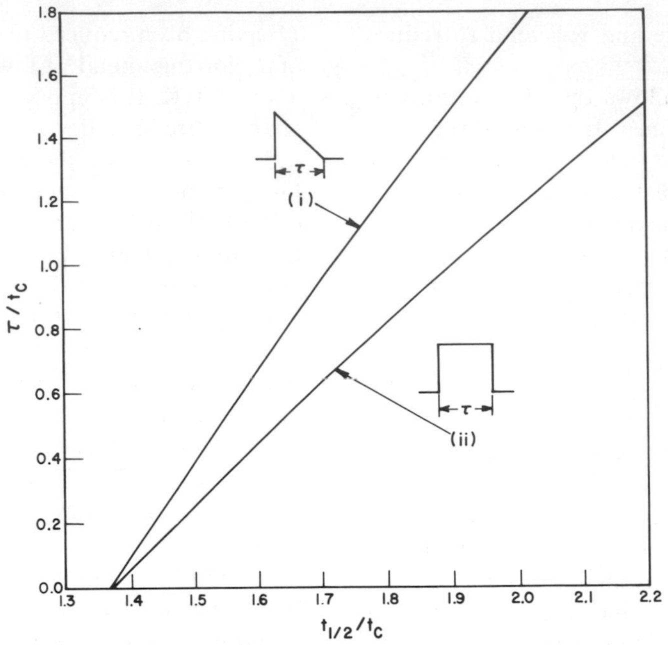

Fig. 11. τ/t_c against $t_{1/2}/t_c$ after Taylor and Cape [122] for (i) sawtooth and (ii) square-wave pulse.

curves for any arbitrary pulse shape, tested out this method from observations made on Armco iron samples of different thicknesses and from measurements of the thermal diffusivity of the same sample at different temperatures. This last procedure was satisfactory since the values of α for iron fall rapidly with increase in T, and hence t_c is strongly temperature dependent. The results of these tests are illustrated in Fig. 12 where the open symbols represent uncorrected data as derived by means of equation (43); the solid symbols represent values obtained by use of the more appropriate numerical coefficient in this equation, as derived from curve (ii) of Fig. 11 which applied to the laser beam used. The continuous curve of Fig. 12 has been derived using the thermal conductivity values considered by Powell [125] to be the most probable for Armco iron. The uncorrected values were clearly much too low, but are seen to conform well after being corrected by the proposed method.

The paper by Larson and Koyama [124] is a rather more detailed treatment yielding very similar correction methods which were found to give satisfactory results when tested for very thin samples of copper as well as for Armco iron.

White and Koyama [126] used the flash method to determine the thermal diffusivity of HWLC-Mo graphite. This is a graphite produced by hot-working with a dispersed liquid-carbide phase containing additions of molybdenum. The thermal diffusivity perpendicular to the hot-working direction was notable in being about three times that of copper; in the parallel direction it was about half the value for

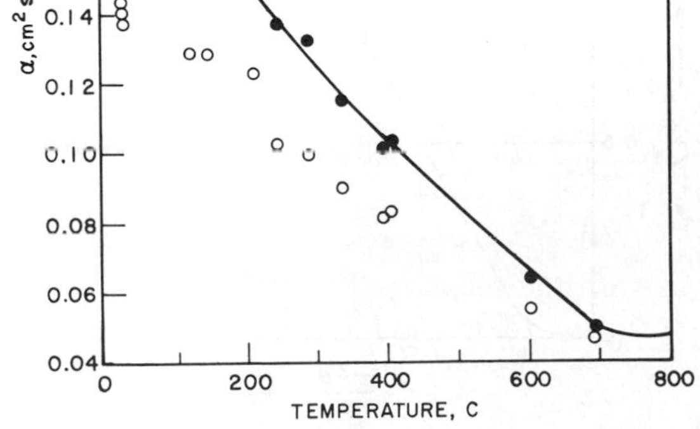

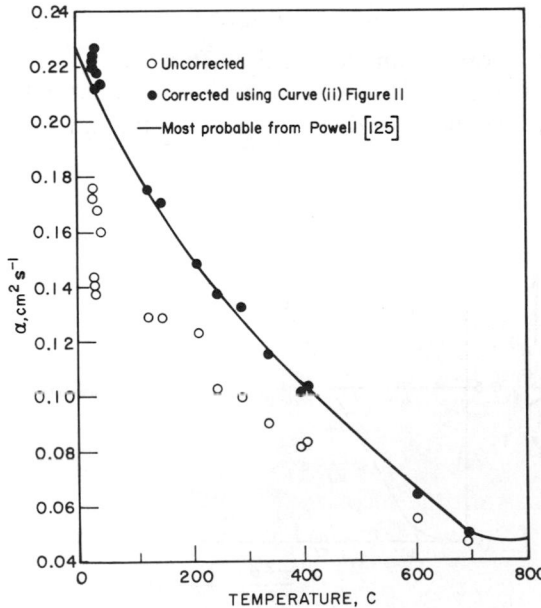

Fig. 12. Armco iron: corrected and uncorrected values of α, after Taylor and Cape [122].

copper. White and Koyama used the pulse-time correction method of Larson and Koyama [124] and also included tests on Armco iron and OFHC copper. The fact that their corrected values for both were within 2 percent of the accepted values for these two standards gives a fair degree of confidence in the values obtained for the graphite samples.

An investigation by Beedham and Dalrymple [127] into the errors introduced by departures from the assumed boundary conditions has included a rather overdue examination of those associated with spatial nonuniformity of the heating pulse. A photographic method was developed to determine such intensity variations for the energy pulse from a ruby laser. The transient heat flows through disk samples of 1, 0.5, and 0.25 mm thickness have been evaluated for a 2-D heat flow code involving inversion of a matrix of implicit finite difference equations [128]. Computer methods have been used to evaluate data for samples of graphite having an assumed thermal diffusivity of 0.026 $cm^2 s^{-1}$. The conclusions reached were that temperature measurements made in regions of high or low intensity led to positive or negative errors in the derived thermal diffusivity, but that the errors for the cases examined did not exceed 8 percent, and would be reduced for methods in which average rather than point rear face temperatures were observed. These workers also considered the effect of finite pulse times and found that when $\tau \leq t_{1/2}$ the errors became less than 2 percent when zero time was assumed as the mid point of a symmetrical heat-pulse or for an asymmetrical pulse, by use of the time at which half the energy has entered the sample.

Similar investigations of the influence of laser beam uniformity on the derived thermal diffusivity are currently being undertaken by K. W. R. Johnson and J. F. Kerrisk at the Los Alamos Scientific Laboratory, Los Alamos, New Mexico.

A brief survey follows on some further contributions relating to the various flash methods, and includes two important new applications by which determinations can be made on thin composite materials, including oxide films *in situ* and by which the thermal conductivity can be directly determined.

a. Flash Heating by Xenon Lamp

As already mentioned, the xenon flash method was the first flash method used by Parker *et al.* [103]. The method was later used by Parker and Jenkins [129] to measure the thermal diffusivity of bismuth telluride, by Jenkins and Westover [130], and by Jenkins and Parker [131] for metals, stainless steels, and other alloys. Others to use the method have been Rudkin *et al.* [31], Rudkin [132], Baker [133], Moser and Kruger [134], Taylor [135, 136], Wagner and Dauelsberg [137, 138, 139], and Morrison [140]. While the foregoing have mostly been interested in high-temperature determinations, Taylor [135] covered the range 100 to 900 K, and Wagner and Dauelsberg [138] the range 173 to 573 K, with a good link-on with the results of a steady-state method at the higher temperature. Makarounis and Jenkins [141] have described the use of the method for measurements over the range −93 to 473 K. This last work included determinations on a single-crystal sample of magnesium dioxide. As this sample was a nonconductor of electricity it was found to be desirable to coat the rear surface with a very thin film of liquid platinum which served to complete the circuit across the tips of the two applied thermocouple wires. Furthermore, in addition to coating the front surface with the customary thin layer of Parson's black, experience showed that direct radiation through the crystal was reduced by also applying a coating of liquid platinum to the front surface before applying the Parson's black. Iacobelli and Moretti [142] used several different flash sources in the course of their thermal diffusivity measurements on uranium oxide to a temperature of 1400 C. It is interesting to note that their derived values for thermal conductivity agree well with the radial heat flow determinations of Godfrey *et al.* [143], whose values up to 1300 C had been chosen as standard reference data for this material [144]. Larson and Koyama [145] measured composite samples of explosively bonded metals (see Section d below).

b. Flash Heating by Laser Beam

A laser beam can deliver more energy to a small-area specimen than would be possible with a xenon flash lamp. Such a source of energy has the added advantage that it can operate from much greater distances. The beam can, for instance, pass through a window to be absorbed by a sample maintained at a high temperature within an evacuated enclosure. Furthermore, by using a Q-switching mode it becomes possible to reduce the dissipation time to less than 1 μsec.

Use of the laser-beam method for a thermal diffusivity determination appears to have been reported almost simultaneously by Carpenter [146] and by Deem and Wood [147]. Subsequently it was used by Taylor and Nakata [148], and by Taylor and Morreale [149] to measure to high temperatures the thermal diffusivity of titanium carbide and titanium nitride, and by Taylor [150] for a zirconium alloy. Measurements using this method have been reported also by Méndez Peñalosa [151] who, like Carpenter, had worked originally with Dr. R. E. Taylor, for uranium carbide and for nitrides and carbides of titanium and zirconium, by Namba *et al.* [152] for nickel, by Nasu and Kikuchi [153] for uranium nitride, by Rawuka and Gaz [53] for synthetic graphites to 2500 C, by Lagedrost *et al.* [154] for determinations on plutonium oxide in the temperature range 250 to 1200 C, by Gilchrist [155] for determinations to 800 C on the oxide films formed on stainless steel (see also Section d below), and by Ferro, Morretti, and Patimo [156] for mixed uranium and thorium oxides in the range 800 to 1350 C. These last used a lead sulfide cell as the transient temperature detector.

c. Flash Heating by Electron Beam

In this method the front face of a thin wafer-like sample is heated by a square-wave pulse obtained by focusing on it the beam of electrons emitted from a modulated electron gun. Mustacchi and Giuliana [157] used this form of the flash diffusivity method for measurements on uranium carbide in the 1000 to 2000 C range.

Walter, Dell, and Burgess [158] have also described the use of square pulses of electrons of duration 0.1 to 4 msec obtained from a tungsten filament electron gun and focused onto the sample for thermal diffusivity determinations for the range 40 to 1400 C on iron and uranium dioxide. The electron pulse is regarded as more flexible than the laser beam and therefore adaptable to a wider range of materials, since adjustments can readily be made to the pulse duration, the accelerating voltage, and the current density.

d. Flash Method Applied to Composite Samples

Larson and Koyama [145] have extended the theoretical treatment to cover the case of two-layer composite samples and have made what appears to be satisfactory tests for layers of stainless steel explosively bonded to either copper or brass. In view of the wide range of composite materials, e.g., refractory-clad materials, oxidized metals, etc., for which thermal properties are required, this offers a most useful test method, as has been shown by an application by Gilchrist [155].

Gilchrist [155] gives details of the use of the flash diffusivity method to determine the thermal diffusivity for the range 30 to 800 C in the first instance of very thin films of stainless steel and later for the oxides formed on stainless steel at temperatures of 750, 850, and 900 C. A Q-switched ruby laser was employed to give a pulse dissipation time of about 20×10^{-9} sec which was suitable for specimens for which $t_{1/2}$ was of the order of 1 msec, that is, for sample thicknesses of less than 0.5 mm. To allow measurements to commence at room temperature the thermal radiation from the central area of the rear face of the sample was focused onto an indium antimonide, infrared detector cell cooled with liquid nitrogen. In order to apply the method to the two-layer composite samples of oxides on steel use was made of the analysis by Larson and Koyama [145] which assumes each layer to be homogeneous but neglects any interfacial contact resistance.

Murfin [159] has in hand developments of the flash diffusivity method which will allow the contact resistance h_c^{-1} to be measured, and which at the same time has led to an interesting variant of the test method for homogeneous materials.

Murfin shows that for two materials of properties $k_1\alpha_1C_{p_1}d_1$ and $k_2\alpha_2C_{p_2}d_2$, of thickness, x, equal to L_1 and L_2 and having a contact conductance k_c at their interface coinciding with $x = 0$, that if zero heat losses are assumed, the temperature at any point in the specimen in material 1 is

$$T_1(x, t) = T_\infty\left\{1 + 2\sum_{n=1}^{\infty}(A_n/\sin\gamma_n)\cos\gamma_n \times\left(1 + \frac{x}{L_1}\right)\exp(-\alpha_1\gamma_n^2 t/L_1^2)\right\} \tag{47}$$

and in material 2, is

$$T_2(x, t_1) = T_\infty\left\{1 + 2\sum_{n=1}^{\infty}(aA_n/\sin b\gamma_n)\cos b\gamma_n \times(1 - x/L_2)\exp(-\alpha_1\gamma_n^2 t/L_1^2)\right\} \tag{48}$$

where the γ_n are the roots of

$$k_1/L_1h_c = \cot\gamma + a\cot b\gamma \tag{49}$$

$$A_n = (C_{p_1}d_1L_1 + C_{p_2}d_2L_2)/C_{p_1}d_1\sin\gamma_n \times\left(\frac{1}{h_c} - \frac{L_1}{k_1\sin^2\gamma_n} - \frac{L_2}{k_2\sin^2 b\gamma_n}\right) \tag{50}$$

$$a = (C_{p_1}^2 d_1^2\alpha_1/C_{p_2}^2 d_2^2\alpha_2)^{1/2}$$

and

$$b = (L_2\alpha_1^2/L_1\alpha_2^2) \tag{51}$$

Once the γ's are known, equation (49) provides a method for the determination of h_c.

An experimental setup is proposed for the determination of h_c comprising a lead-sulfide cell and an arrangement of mirrors including a toothed rotating mirror acting as a chopper, which system serves to measure the temperature differential between the opposite faces of a compound cylindrical sample.

For homogeneous disks Murfin has described a new method which is based on measurements of the time constant of the transient rather than on the usual $t_{1/2}$ temperature observation. It is shown that the flash produces a temperature difference between any two points on the specimen which decays exponentially to zero with a time constant of the form $l^2/\alpha\beta^2$, where β is a constant and l is the distance between the points. Hence an observation of the time constant for any two points enables α to be derived. In Murfin's work the temperature difference was observed for points near the center and edge of the rear face. For samples of copper and of beryllia the method was stated to have yielded results that were reproducible to ± 5 percent.

e. Flash Method Applied for Direct Determination of Thermal Conductivity

As can be seen from equation (41), the original treatment of Parker *et al.* [103] offers a means for the evaluation of dC_p provided the energy absorbed by the front surface, Q, is known. di Novi [160] substituted a disk of Armco iron of known thermal properties as a means of deriving Q for a standard thermal diffusivity setup that was to be used for other materials. The front face of all materials was coated with colloidal graphite to ensure that Q remained constant for a fixed power emission from the flash lamp. The thermal conductivity was then derived from

$$k = \frac{\alpha Q}{L T_m} \tag{52}$$

Measurements were reported for samples of 347 stainless steel to 800 C, α being considered accurate to ± 5 percent and k to ± 10 percent. Peggs and Mills [161] arranged to directly measure the amount of energy received at the sample position from a laser beam and subtracted from this the amount reflected from the colloidal graphite coating on the front face of the sample. Preliminary results for Armco iron and Pyroceram 9606 samples at 298 K agreed to within about 10 percent with expected values for the thermal conductivities of these materials.

The motivation for both these direct determinations of k is directed toward tests on irradiated ceramic samples for which property changes resulting from the irradiation would render uncertain any derived values.

I. Electrically-Heated Rod Methods

The high temperature use of the electrically heated rod method for the determination of several properties including k, C_p, d, and hence α, has been dealt with in the introduction to this chapter. The particular development now to be described relates to one of the few methods that have been used for thermal diffusivity determinations at cryogenic temperatures. Another is a periodic method which will be treated in Section 2.H.

Erdmann and Jahoda [162] were interested in studying the deformation of metal specimens under load at the temperatures of liquid helium, and in simultaneously measuring their thermal and electrical conductivities, thermal diffusivities and heats of deformation. For wire samples 1 to 3 mm in diameter and some 10 cm long they chose to use an arrangement and method of the Kohlrausch–Diesselhorst type [163, 164] in which the ends of the rod and its surroundings are kept at a constant temperature and passage of an electric current through the wire raises its temperature. The ratio of the thermal conductivity, k, to the electrical conductivity, σ, is then given by

$$\frac{k}{\sigma} = \frac{V^2}{2(\Delta T_f - \Delta T_0)} \tag{53}$$

where, for two positions on the rod, one at the center and the other near one end, having a small initial temperature difference ΔT_0, ΔT_f is the final steady-state temperature difference and V is the corresponding voltage difference. This is the normal expression of Kohlrausch [163] and the authors also give a modified expression which allows for the dimensional changes induced by the applied strain.

The thermal diffusivity was deduced by them for a sample of constantan, an alloy containing 60 percent of copper with 40 percent of nickel, from observations of the transition from one state of temperature equilibrium to another. For instance, after measuring the thermal conductivity according to equation (53), if the current is switched off at time $t = 0$, the temperature difference for $t > 0$ decays according to

$$\Delta T - \Delta T_0 = (\Delta T_f - \Delta T_0) \sum_{n=0}^{\infty} A_n \exp(-\theta_n t) \tag{54}$$

with

$$A_n = 8\pi^{-3}(L_0/d_0)^2(-1)^n \times \{1 - \cos[(2n+1)\pi d_0/L_0]\} \tag{55}$$

and

$$\theta_n = [(2n + 1)\pi/L]^2\alpha \tag{56}$$

L_0 being the initial length and $L = L_0(1 + \epsilon)$ the deformed length. Equation (54) contains only odd-order terms and, since those belonging to $n > 0$ decay rapidly, a plot of $\log(\Delta T_f - \Delta T_0)$ against t yields the straight line

$$\log(\Delta T_f - \Delta T_0) = -\pi^2 L^{-2}\alpha t \tag{57}$$

from which the thermal diffusivity α is derived.

The ratio of thermal conductivity to thermal diffusivity is found to be constant to within 3 percent for all conditions of strain. This presumably is an indication of the order of reliability of the determinations of these quantities, since both the density and the specific heat should be unaffected by the deformation. One advantage of operating the method at cryogenic temperatures is that radiation losses are negligible. Of course this method would be inapplicable to wires which become superconducting. It need not, however, be limited to cryogenic temperatures, and could be more generally applied. Indeed, a note on the use of the method appears in the report by Taylor and Nakata [148], who applied it to a tantalum wire. Their value is 0.181 cm^2/sec for the range 0 to 656 C. This low value can probably be attributed to neglected radiation losses.

2. PERIODIC HEAT-FLOW METHODS

The many experimental methods which involve the periodic or cyclic heating and cooling of a test sample all stem from the pioneer work of Ångström [165–167].

The following is a brief survey of these methods, which are presented for the most part in chronological order and have been given distinctive names relating either to the variant of the method or to its introducer.

A. Ångström's Method

Ångström [165] in his first experiments subjected the middle of a long metal bar to periodic heating and cooling by encasing this section with a vessel through which steam or cold water could be passed alternatively for equal time periods. A long rod could equally well be heated and cooled periodically at one end as has been done in most later applications of the method. Thermometers were mounted by Ångström in wells along the length of the bar to measure temperatures at two points on the same side of its center, and here again, thermocouples have since been used. After the cyclic temperature variations had been continued for a sufficient time for the temperature wave to become stationary and reproducible, then, provided the bar is of such a length that the temperature of its ends has remained unchanged, Ångström showed that

$$\frac{k}{dC_p} = \frac{\pi L^2}{\tau\phi \log q} \tag{58}$$

where L is the distance between the two observation points, τ is the period of the wave, ϕ is the phase lag, or difference in phase between the temperature fluctuations at the two observation points, and q is the amplitude ratio of the temperatures at these points. Ångström, who determined the thermal conductivities of copper and iron by multiplying the results obtained from equation (58) by the known specific heats and densities, had established this method in advance of the naming of the quantity first derived as the thermal diffusivity.

Many versions of the periodic method have since been reported. An early modification designed to accommodate samples of finite length was one used by Weber [168], who, following a suggestion by F. Neumann, impressed periodic temperature oscillations of the same period and amplitude but of opposite phase on the two ends of a short rod. For iron and German silver rods of length L, the thermal diffusivity was evaluated by this method from observations of the variation with time of the temperature difference between points on the rod at conveniently selected distances of $L/6$ and $4L/6$ from one end.

Two much later developments will be described in the next two sections and then some important methods will be considered which relate more closely to Ångström's original conception, but which involve advances made possible by subsequent technological developments.

B. Temperature Wave Velocity Method

King [169] made two modifications to Ångström's method. He designed a special heater in which the current could be varied according to the relation $I = I_0 \sin(\omega t/2)$, and thus was able to ensure that the temperature oscillations throughout his specimen were truly sinusoidal. For such sinusoidal oscillations, $\omega = 2\pi/\tau = 2\pi f$, where ω is the angular frequency, f is the oscillation frequency, and τ is the wave period. King made experiments for two different wave periods, τ_1 and τ_2, and corresponding propagation velocities, v_1 and v_2, from which he evaluated α from the expression

$$\alpha = \frac{\tau_1 \tau_2 v_1 v_2}{4\pi} \left[\frac{v_1^2 - v_2^2}{\tau_2^2 v_2^2 - \tau_1^2 v_1^2} \right]^{0.5} \tag{59}$$

The use of two periods was adopted as a means of eliminating the need to know the coefficient of surface heat loss, which was assumed to remain the same throughout both stages of the experiment.

King used this method for determinations at room temperature on copper and tin and obtained what appeared to be satisfactory values. Later the method was used for determinations by Ellis *et al.* [170], again at room temperature, on copper, nickel, and several nickel alloys, and then by Sager [171] to about 700 C for copper and nickel and several binary alloys of these metals. At higher temperatures the values tended to be too high. No explanation is given, but it seems likely that the temperature measurements were at fault, owing to heat being conducted from the 2- to 3-mm-diameter samples to the No. 26 (B and S) gauge chromel and alumel wires of 0.405 mm diameter, that were attached to serve as thermocouples.

C. Temperature Wave Amplitude-Decrement Method

An alternative method, originally suggested by Ångström [165–167], has been used by Starr [172] for measurements on nickel. Starr again used two different wave periods as a means of eliminating the lateral heat-loss term, but observed the amplitude ratios q_1 and q_2 for two chosen points on the rod. The equation for the thermal diffusivity is

$$\alpha = \frac{\pi L^2}{\tau_1 \ln q_1 \ln q_2} \left[\frac{(\tau_1/\tau_2)^2 - (\ln q_2/\ln q_1)^2}{(\ln q_2/\ln q_1) - 1} \right]^{0.5} \tag{60}$$

For negligible lateral heat losses and high ratios of ω/α, the logarithm of the amplitude ratio is approximately equal to the phase lag, and then,

$$\alpha = \frac{\omega L^2}{2(\ln q)^2} = \frac{\omega L^2}{2\phi^2} \tag{61}$$

Thus, under these conditions, measurement of either the amplitude decrement, q, or the phase difference, ϕ, enables the thermal diffusivity to be determined. Nii [173] and Kanai and Nii [174] applied both methods for determinations on bismuth telluride, lead telluride, and indium antimonide. Nii [173], however, considered that more reliable results were obtained from the amplitude ratio.

Kevane [175] made use of a solar furnace as a heat source and established a periodic thermal wave by moving his samples into and out of the focal spot, while Mustacchi and Giuliani [156] generated a sinusoidal thermal wave by means of a modulated electron gun. The latter employed a wide range of frequencies, from 0.01 to 100 cps, and obtained a straight line when $\ln q$, with q the amplitude ratio for the front to rear surface, was plotted against $f^{0.5}$. If S is the slope of this line, then

$$\alpha = 0.59\, L^2/S^2 \tag{62}$$

Filippov and Nurumbetov [176] and Filippov [177] used a cylindrical metal specimen a few centimeters in length with the upper end heated by electron bombardment. The modulation was achieved by periodically switching on and off the several hundred volts potential between the specimen as anode and the electron emitting cathode. The periods used were within the range 5–20 sec. Several thermocouples welded to the sample at regularly increasing distances of some 3 mm from the top enabled temperature recordings to be obtained.

Keeping the cyclic frequency, ω, constant, the amplitude of the temperature oscillations, θ, for the various locations is plotted against distance, x. The thermal diffusivity is then obtained from the slope of this line and ω according to

$$\alpha = \frac{\omega}{2(d \ln \theta/dx)^2} \tag{63}$$

To increase the precision, data were obtained for several frequencies. This method was used for Armco iron at 300 and 370 C and gave good agreement with existing values. By arranging to include measurements of the variable component of the actual heat flux these workers were able to extend the capabilities of the method to include determinations of the specific heat and thermal conductivity. Khusainova and Filippov [178] subsequently used the method to determine all three properties for a molybdenum single crystal cylindrical rod in the course of a single experiment covering the temperature range 450 to 1200 K. They estimated the errors in the thermal diffusivity and thermal conductivity measurements to be 4 and 7 percent, respectively.

Anger *et al.* [179] used yet another modification of the amplitude-decrement method when determining the thermal diffusivity of bismuth telluride. Their sample was in the form of a cylindrical rod, one end of which was sinusoidally heated and the other end held at a constant temperature by water cooling.

Temperatures were measured by means of thermocouples attached at two points on the rod, separated by a distance, L. Assuming both thermocouples to have the same Seeback coefficient and to give readings V_1 and V_2 at a particular time, then, if q is the coefficient of attenuation, or the amplitude ratio, they show that

$$q = (\omega/2\alpha)^{0.5}[1 + (\alpha h/\omega r)] \qquad (64)$$

where r is the radius and h represents the lateral heat loss coefficient; also,

$$q = \frac{1}{L} \ln \frac{V_1}{V_2} \qquad (65)$$

By means of equation (65) q is determined for several frequencies of the applied thermal wave. A plot of $q(2/\omega)^{0.5}$ as ordinate against $1/\omega$ as abscissa then yields a straight line according to equation (64), for which the intercept on the ordinate is $\alpha^{-0.5}$.

D. Modified Ångström Methods

The two main methods to be described under this heading are those of Sidles and Danielson [180, 181] and of Abeles *et al.* [182, 183].

a. Method of Sidles and Danielson

Sidles and Danielson [180] followed the original Ångström method by measuring the velocity, v, and amplitude decrement, q, for one suitably chosen wave period. By restricting the observations to the one equilibrium condition the determination could be made more quickly than for either of the foregoing versions of King [169] or Starr [172]; also this procedure avoided troubles which might arise in their two-stage experiments from changes in thermocouple calibration or surface heat loss.

According to the original theory of Ångström

$$\alpha = \frac{Lv}{2 \ln q} \qquad (66)$$

and this equation is fundamental to the work now to be described.

In the earlier experiments of Sidles and Danielson [180] a sinusoidal heat input having a period of about 120 sec is impressed on one end of a cylindrical sample, 3 to 9 mm in diameter and 10 to 30 cm long, the shorter lengths and larger diameters being used for the poorer conducting metals such as iron and thorium. The temperatures are measured at two points, L apart, the first point being much closer to the hot end than the cold. Special techniques are described for recording the readings of these two thermocouples, which include bucking off a dc component about equal to the ambient temperature readings and amplification of the remaining signal. From this recording and a knowledge of the chart speed, both v and q can be derived: α is then calculated according to equation (64).

Sidles and Danielson [180] developed a very neat technique for use with this method. By connecting the suitably bucked-off voltages of the two thermocouples to the X and Y directions of an X–Y plotter an ellipse is recorded. Figure 13 indicates the use of such an ellipse to obtain the relevant parameters X_a, Y_a, and ϕ that are required for the evaluation of α from the equation

$$\alpha = \frac{\pi L^2}{\tau\phi \ln(X_a/Y_a)} \qquad (67)$$

which has been derived from equation (66) by substitution of $v = 2\pi L/\tau\phi$ and $q = X_a/Y_a$.

The ratio X_a/Y_a of the length of the sides of the circumscribed rectangle were shown to be equal to q_1 the amplitude decrement, while ϕ, the phase angle between the X and Y displacements, is equal to the angle whose sine is X_p/X_a or Y_p/Y_a. The agreement of these two derivations is a check on the experimental setup.

This method has been used by Sidles and Danielson [180] to determine α for several metals over the approximate range 320 to 1230 K. Subsequent measurements by Shanks *et al.* [184] on Armco iron not only extended the range to 1373 K, but showed the method to be of particular value in the region of the alpha–gamma phase transition. This is the region where property discontinuities occur and where direct thermal conductivity determinations by methods involving a temperature gradient tend to meet difficulties. Martin *et al.* [185] used the method

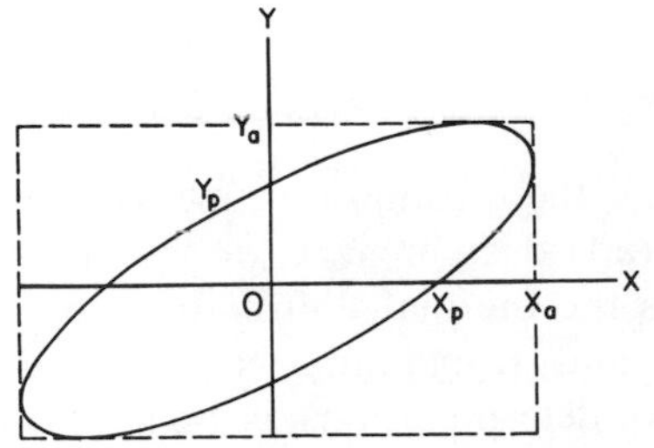

Fig. 13. Ellipse given by X–Y recorder used by Sidles and Danielson [180] to derive α from $\alpha = \pi L^2/\tau\phi \ln(X_a/Y_a)$.

for measurements on several samples of platinum to 1200 K, and Ciszek [186], working in the same laboratory made preliminary determinations on platinum to 1600 K (see Fig. 14). The foregoing applications serve to demonstrate the value of the method as a means for making determinations of a thermal property to temperatures well above 1000 C. Advantages are that absolute values are not essential for either the heat flow or the observed temperatures and that the temperature observations are all made close to a mean temperature.

Ginnings [187] has described measurements made at the National Bureau of Standards in the course of a study of the Sidles and Danielson method, which sought to check the influence, if any, of the increased radiation which occurs when the method is applied at high temperatures. If such an effect is present, it will be large when the period of the sine wave is large and small when it is small, hence, the values obtained for the thermal diffusivity would become dependent on the frequency. The check measurements, however, were made on a sample of Armco iron but only for a temperature of 186 C, and for sine-wave periods which could be varied from 28 to 113 sec. To achieve this range of variation, one end of the iron rod was welded to a short length of Inconel rod which could be heated electrically or water-cooled in accordance with a prearranged program. Values of α, so determined, agreed to within ± 1 percent. Furthermore, tests made for the 28-sec period with a saw-tooth instead of a sine-wave input yielded a value of α that was believed to be in error by only about 2 percent. The conclusion from these experiments was that use of a sine wave having a period of 15 sec or less should allow the method to give values of thermal diffusivity at 1000 C or higher that would be of comparable accuracy to those obtained at lower temperature.

It might be mentioned that Ciszek [186], for his determination on platinum to 1600 K, used a period of 30 sec; Martin *et al.* [185] for measurements to 1200 K appear to have used mainly a period of 60 sec, with some checks (temperature unspecified) using one of 120 sec. For the common range of temperature good agreement had been obtained.

b. Method of Abeles et al.

The method of Abeles *et al.* [182] is a very similar modification of the original Ångström method, but one which is regarded by Danielson and Sidles [12] as possessing some significant theoretical and experimental improvements.

From the separate expressions derived for q, the logarithmic attenuation

$$q = \left[\frac{h + (h^2 + n^2\omega^2)^{0.5}}{2\alpha}\right]^{0.5} \tag{68}$$

and for q', the phase shift

$$q' = \left[\frac{-h + (h^2 + n^2\omega^2)^{0.5}}{2\alpha}\right]^{0.5} \tag{69}$$

where h is the lateral heat transfer per unit length, it follows that

$$qq' = \frac{\omega}{2\alpha} \tag{70}$$

and

$$\alpha = \frac{\omega}{2qq'} \tag{71}$$

which is independent of any lateral heat loss.

Furthermore, when $h = 0$,

$$q = q' = \left(\frac{\omega}{2\alpha}\right)^{0.5} \tag{72}$$

For high-temperature measurements, when the main lateral heat transfer is by radiation they consider that the method will yield satisfactory values provided that

$$\omega r \gg 8\sigma\epsilon T^3/dC_p \tag{73}$$

and

$$\exp(-2qL) \ll 1 \tag{74}$$

The modulation frequencies of the cam-driven sinusoidal heater used by Abeles *et al.* [182] ranged from 0.02 to 0.5 Hz. The Y-axis of an X–Y recorder was fed by a similar sinusoidal voltage and served as a reference voltage. The phase of this voltage was adjusted to coincide with that of each thermocouple output when applied in succession to the X-axis of the recorder. From the X amplitudes of the resulting figure-of-eight Lissajous' figures, the ratio q was obtained, while their cross-over points yielded the phase shifts with respect to the heater signal. This information gave values for q and q', and hence α could be calculated. The three equally spaced thermocouples attached to the sample were used in pairs, and therefore provided three independent values.

By using appropriate frequencies, samples only about 5 cm in length could be used. These were short enough to avoid trouble due to radiation and determinations to high temperatures have been reported for materials having thermal conductivities of 0.005 to 4 W cm^{-1} K^{-1}. Such determinations include those of Cody *et al.* [188] for Armco iron to about 1300 K, Beers *et al.* [189] for germanium and silicon to 1100 and 1200 K, respectively, Abeles *et al.* [183] for germanium–silicon alloys to 1200 K, Steigmeier and Kudman [65] for indium arsenide to 900 K, the same authors [190] for gallium antimonide, aluminum antimonide, and gallium phosphide to 910, 930, and 550 K, respectively, and of Habachi *et al.* [191] for copper, cobalt, and alloys of copper with arsenic, bismuth, and silver to about 850 K.

It will be noted that the thermal diffusivity measurements on iron and cobalt have been made for temperature ranges which include the Curie temperature of iron and the phase-transition temperatures of both metals. Sidles and Danielson [180] had made determinations on iron and nickel* and had pointed out that the Curie temperature is located more definitely by the strong dip observed in a plot of thermal diffusivity against temperature than by a plot of electrical resistance against temperature. However, Powell and Hickman [192] had noted earlier that it is the rate of change of resistance with temperature for which a peak can often be observed. Thermal diffusivity measurements by the present method can be accepted as having an advantage over most thermal conductivity determinations, in that the sample can be maintained at a more uniform temperature and hence observations can be made much nearer to a transformation temperature before changes at this temperature can influence the measurement. Care is required, however, and clearly, the smaller the amplitude of the temperature oscillation, the more closely the transformation can be approached before its influence is detected. At many such transformations, as well as a property change, there is an associated latent heat, and the sign of this latent heat will depend on whether the transformation occurs as the temperature rises or falls. Hysteresis also frequently occurs and on cooling, the transformation may be delayed to an appreciably lower temperature. In the case of cobalt, the electrical resistivity determination by Powell [193] indicated the phase transition to commence at about 440 C on heating and at about 400 C on cooling. It would seem therefore that experimental determinations of thermal diffusivity made in the region of a transformation are likely to yield values which will depend not only on both the amplitude and frequency of the temperature wave but also on the thermal history of the sample investigated. The measurements by Habachi *et al.* [191] on a 99.999 percent sample of cobalt represent the first determinations of either the thermal diffusivity or thermal conductivity to be made on this metal to a temperature which includes the alpha-to-beta transformation. These workers used a satisfactorily small temperature oscillation of only ± 1 K yet they regarded their measurements as indicating a transformation temperature of 410 ± 10 C. As this temperature is much lower than that obtained from other property studies, a possible explanation could be that this result was obtained for the sample when in the supercooled beta phase. This could happen, for instance, if, following a reading at about 400 C, the sample had been heated inadvertently to 450 C or above, before the thermal diffusivity at about 415 C had been measured. These considerations clearly show the need for the exact thermal treatment to be specified.

*For thermal diffusivity determinations on cobalt near the Curie temperature, see also the later determinations by Zinov'ev *et al.* [203].

E. Phase-Lag Methods

The use of phase-lag measurements as an independent means for determining thermal diffusivity has already been referred to in Section C for the conditions under which the logarithm of the amplitude decrement and the phase-lag could be regarded as approximately equal. A method of the phase-lag type has also been used by McIntosh [194] and by McIntosh *et al.* [195]. The uncertainty which they claim is only about 6 percent, but their reported values for Armco iron are clearly in error by much more than this. Their determinations only cover the range 0 to 100 C, but use of literature data for the density and specific heat yield thermal conductivity values for this increasing temperature range which increase by as much as 37 percent, whereas the accepted values for Armco iron decrease by about 9 percent.

Subsequent developments of the phase-lag method for use with thin slabs or disks of material have been fairly widely adopted for determinations to high temperatures. Hirschman *et al.* [196, 38] investigated Armco iron, copper, and brass specimens in the form of small flat disks about 6.4 mm in diameter and from 0.77 to 1.48 mm thick for the temperature range 400 to 1000 K. These were irradiated with a chopped beam from a carbon-arc image furnace. The phase-lag between the square-wave irradiance impinging on the front surface, as reflected to a photovoltaic cell, and the temperature oscillations of the back surface, as measured by a thermoelectric probe, is given by

$$\phi = \tan^{-1}\left[\frac{\sinh B \cos B + \cosh B \sin B}{\sinh B \cos B - \cosh B \sin B}\right] \tag{75}$$

where,

$$B = L(\omega/2\alpha)^{0.5} \tag{76}$$

and heat loss from the rear face is neglected.

They also derived expressions in which allowance is made for surface heat losses, but regarded equation (75) as adequate to temperatures of the order of 2000 K, when using frequencies to 10 Hz, and to still higher temperatures by the use of higher frequencies.

Cowan [118] is responsible for the full theoretical treatment used, for instance, by Wheeler [48] to evaluate observations made on several high-temperature materials when heated to incandescence by a modulated electron beam. Cowan [118] shows that the phase difference, ϕ_L, between the temperature fluctuations of the face at $x = L$, and the electron beam when modulated to produce a sine-wave is

$$\phi_L = \tan^{-1}\left[\frac{2B^3Q_1 + 2B^2aQ_2(1 + r)^{-1} + bBQ_3r^{-1}}{2aB^2Q_0 + b(1 + 2r)BQ_1r^{-1} + abQ_2(1 + r)^{-1} + 2B^3Q_3}\right] \tag{77}$$

and that the phase difference, ϕ_0, between the temperature fluctuations of the other face at $x = 0$ and the sine-wave modulated beam is

$$\phi_0 = \tan^{-1}\left[\frac{b(\tan B - \tanh B) + 2aB \tan B \tanh B + 2B^2(\tan B + \tanh B)}{b(\tan B + \tanh B) + 2aB - 2B^2(\tan B - \tanh B)}\right] \tag{78}$$

where B is given by equation (76) and

$$\begin{aligned} Q_0 &= \cosh^2 B + \sinh^2 B \sin^2 B \\ Q_1 &= \cosh B \sinh B + \cos B \sin B \\ Q_2 &= \cosh^2 B \sin^2 B + \sinh^2 B \cos^2 B \\ Q_3 &= \cosh B \sinh B - \cos B \sin B \end{aligned} \tag{79}$$

For radiation losses proportional to T^4, the two heat loss parameters a and r are given in Wheeler's experiments by approximately

$$\begin{aligned} r &\approx (T_{x=L}/T_{x=0})^3 \approx 1 \\ a &\approx 4\sigma L(1 + r)\epsilon T^4 k^{-1} \end{aligned} \tag{80}$$

and

$$b \approx a^2/4$$

With $\phi = \phi_0 - \phi_L$ it is further shown that to a close approximation,

$$\phi = 1.45\ L^2\omega(\pi\alpha)^{-1} \tag{81}$$

The more exact solution requires $\log(\phi_0 - \phi_L)\pi^{-1}$ as calculated from equations (77) and (78) to be plotted against $\log B\pi^{-1}$ for different values of a. When $(\phi_0 - \phi_L)\pi^{-1}$ is less than about 0.1, each curve may be represented by a straight line from which

$$\begin{aligned} (\phi_0 - \phi_L)\pi^{-1} = \phi\pi^{-1} &= A[B\pi^{-1}]^F \\ &= A\left[\frac{L}{\pi}\left(\frac{\omega}{2\alpha}\right)^{0.5}\right]^F \end{aligned} \tag{82}$$

The numerical coefficients A and F are functions of a, but for small values of a, F can be taken as equal to 2 and A as 2.9, so yielding equation (81).

Wheeler [48] measured ϕ experimentally by means of a circuit containing two photocells so disposed that

$$\phi = 2\tan^{-1}(\omega CR) \tag{83}$$

where C and R are a capacitance and a phase-shift resistance in this circuit.

Knowing ϕ, an approximate value for the thermal diffusivity, α, was obtained by means of equation (81). With d and C_p known, this enabled an approximate value of the thermal conductivity, k, to be evaluated and a could then be derived from equation (80). This value of a gave the constants A and F, and, knowing these constants, equation (82) was used to give a more exact value for α, and hence for k, the procedure being repeated until stable values were derived for both quantities. Wheeler found the first approximate value of k to be invariably within 8 percent of the final value and that two or three iterations always proved sufficient.

Kraev and Stel'makh [47, 52] have used a very similar method for thermal diffusivity measurements on tungsten between 1600 and 2960 K [47] and on tantalum, molybdenum, and niobium at temperatures above 1800 K [52]. As stated in Section 2.C.b, the values of Wheeler and of Kraev and Stel'makh agree well for tungsten near 1900 K but those of the latter become about 20 percent lower at 2900 K.

Much the same procedure was also used by Cerceo and Childers [197] when making thermal diffusivity determinations on thin plates of alumina in the range 1289 to 1409 K and of carbon at 1189 K, while Ainscough and Wheeler [198] used it for one of the most comprehensive studies made on uranium oxide. On behalf of the UKAEA, some 30 determinations were made in the range 970 to 2020 K on each of 25 samples of 2.7 percent enriched uranium dioxide, and from these data thermal conductivity values were derived.

In a subsequent paper by Wheeler [199] several modifications are described, the chief of which involves the use of infrared-sensitive devices and of calcium fluoride windows to allow temperature fluctuations of the test-sample surface to be recorded for temperatures as low as 250 C. The lower limit attainable in this way is a function of the specimen emissivity. Thus for the reported measurements on brightly polished iron of 99.85 percent purity this limit was 550 C, but, after oxidizing the surface by heating in air, temperature measurements down to 280 C became possible. Good agreement was obtained between the derived thermal conductivity values and the direct measurements of this property reported by the National Physical Laboratory [200] and by Fulkerson *et al.* [201]. This agreement related to Wheeler's determinations of thermal diffusivity on a sample only 0.128-cm thick. The other sample tested of 0.201 cm thickness gave about 10 percent higher values just below the alpha-to-gamma phase transformation and 14 percent higher values above this transformation. This finding, in one of the most recent papers, is again indicative of the care and need for cross checking which is still necessary with these methods. These determinations and those which follow again confirmed the merit of thermal diffusivity as providing means for investigating the changes associated with phase transitions.

The phase-shift method of Kraev and Stel'makh [47] was subsequently employed by Zinov'ev *et al.* [202] to determine the thermal diffusivity of nickel in the range 920 to 1670 K and by Zinov'ev *et al.* [203] for similar measurements on cobalt for the range 950 to 1710 K. The cobalt measurements included the Curie point, 1390 K, and the curve again has a deep trough in this region, similar to that noticed previously by Sidles and Danielson [180] for nickel and iron.

The subsequent use of this method by Zinov'ev *et al.* [204] to determine the thermal diffusivity of platinum over the temperature range 1040 to 2000 K calls for special comment. Platinum has been regarded as a strong candidate for consideration as a thermal conductivity standard, particularly for use at temperatures above about 1000 K. Much of the uncertainty up to 1200 K appears to have been resolved by the careful measurements of Flynn and O'Hagan [95], whose values are regarded as acceptable to that temperature, and possibly to 1373 K (O'Hagan [205]).

Figure 14 is an attempt to portray all the data above 1000 K relating to the thermal conductivity of platinum. In this figure the directly determined thermal conductivity values are shown as broken lines [93–95] [205–210], the dotted line represents values derived from the Lorenz function determinations of Hopkins and Griffith [92] together with the electrical resistivity data of Vines [211], and the full lines represent values derived from thermal diffusivity determinations [48, 53, 185, 186, 204, 212]. The data derived in the last mentioned manner are seen to lie in about the center of this spread of values and to conform reasonably well with the Flynn and O'Hagan [95, 205] values. This is particularly true of the preliminary data of Ciszek [186] which, together with the Hopkins and Griffith curve tend to suggest that the thermal conductivity of platinum does continue to increase with increase in temperature until the melting point is reached. On the other hand, the values derived from two of the most recent sets of thermal diffusivity determinations, those of Zinov'ev *et al.* [204] and of Rawuka and Gaz [53], both indicate the thermal conductivity of platinum to attain a broad maximum in the region of 1600 to 1700 K and to decrease to their respective upper temperature limits of 1994 and 1844 K. Thus the true curve for platinum still remains uncertain and its determination should present a challenge to future experimentalists. It does, however, seem desirable to point out that for other metals including titanium (see Fig. 2), the high-temperature thermal diffusivity determinations of Zinov'ev *et al.* [29] led to comparatively low thermal conductivity values and even to Lorenz functions which were thought to be unacceptably low.

F. Thermoelectric Methods

These methods comprise several techniques in which the temperature changes are produced by thermoelectric means. Since thermoelectric effects tend to be large for semiconductors it follows that methods of this class have been applied to such materials.

In one such technique use is made of the Peltier heating or cooling generated at the junction between an electrically conducting specimen and a current-carrying electrode of a different material. Periodic reversal of the current changes the sign of the Peltier heat and symmetrical temperature variations are

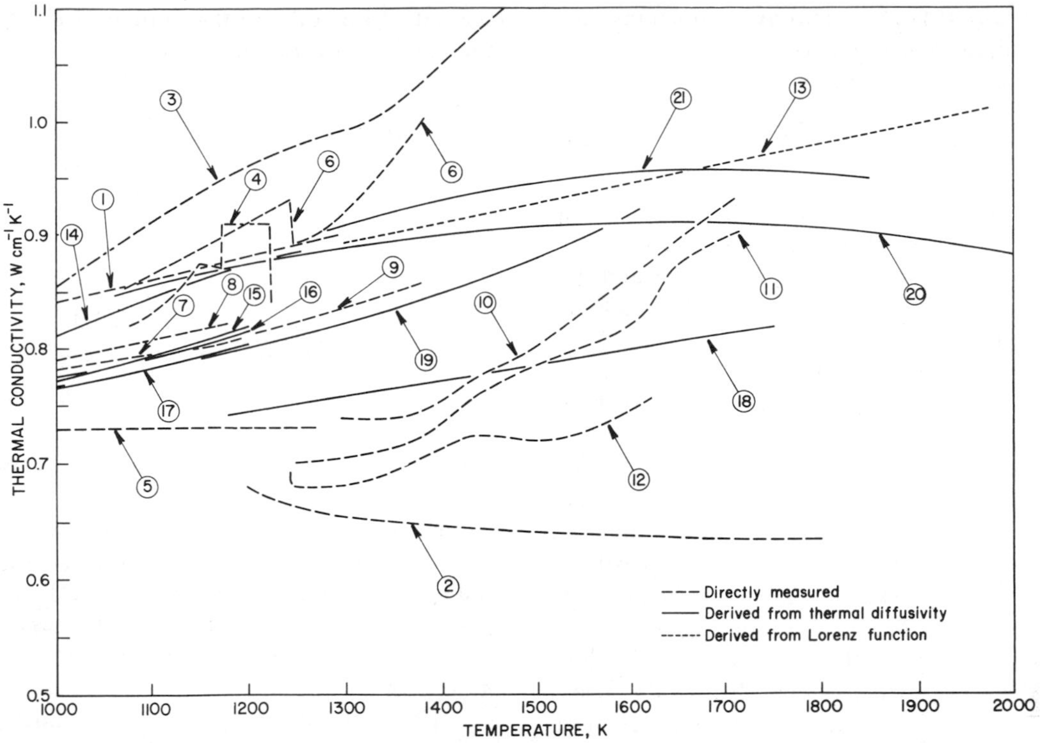

Fig. 14. Thermal conductivity of platinum at high temperatures. 1, Holm and Störmer [93]; 2, Krishnan and Jain [206]; 3, Cutler *et al.* [94]; 4, Bode [207]; 5, Powell and Tye [208]; 6, Kobushko *et al.* [209]; 7 and 8, Flynn and O'Hagan [95]; 9, O'Hagan [205]; 10, 11, and 12, Jain, Goel, and Narayan [210]; 13, Hopkins and Griffith [92] (derived from Lorenz function); 14, Martin and Sidles [212]; 15–17, Martin, Sidles, and Danielson [185]; 18, Wheeler [48]; 19, Ciszek [186]; 20, Zinov'ev, Krentsis, and Gel'd [204]; 21, Rawuka and Gaz [53] (14–21 have been derived from thermal diffusivity measurements).

thereby established. If the temperature changes are recorded for two locations sufficiently removed from the end $x = 0$, where the periodic heating takes place, the upper harmonics of the temperature wave will be sufficiently damped to give the sensibly sinusoidal temperature variations required for Ångström's original method. Hence, from the observed amplitude ratio and phase-lag, the thermal diffusivity can be derived by means of equation (57). Methods of this type were used by Green and Cowles [213] to measure the thermal diffusivity of single-crystal bismuth telluride and by McNeill [214] to measure the thermal diffusivity of lead telluride doped with iodine.

Other techniques which are applicable to semiconducting materials and which involve Seebeck, Nernst, and bolometric effects are those proposed by Becker [215] and used by Sochard and Becker [216], Timberlake *et al.* [217], Davis *et al.* [218], and by Leroux-Hugon and Weill [219].

G. Radial-Wave Method

This method has been developed by Yurchak and Filippov [220–222] and applied for thermal diffusivity determinations on metals in both solid and liquid phases. It is also described by Filippov [177], Filippov and Pigal'skaya [223], and Mardykin and Filippov [40]. A cylindrical sample in equilibrium with an enclosure at constant temperature has a sinusoidal temperature fluction established within it by periodic heating of the external curved surface. The solution of the appropriate heat-flow equation is

$$T = AJ_0(\varkappa\sqrt{i}) \tag{84}$$

where A is a constant of integration, and $J_0(\varkappa\sqrt{i})$ is the zeroth-order Bessel function of the complex argument $\varkappa^2 = \omega\alpha^{-1}r^2$, ω being the angular frequency of the temperature oscillations and r the radius at which the temperature is being measured. From equation (82) the ratio of the temperature amplitudes at two radii, or at one radius and the axis, can be obtained giving

$$\frac{q_r}{q_0} = f(\varkappa) \tag{85}$$

or

$$\frac{q_r}{q_0} = (B_1^2(\varkappa) + B_2^2\varkappa)^{1/2} \tag{86}$$

where $B_1(\varkappa)$ and $B_2(\varkappa)$ are Thomson functions, i.e., real and imaginary parts of $J_0(\varkappa\sqrt{i})$.

The phase shift is given by

$$\phi = \arc\tan \frac{B_2(\varkappa)}{B_1(\varkappa)} \tag{87}$$

Hence, once again, the thermal diffusivity can be obtained from observations of either the amplitude ratio or the phase shift. Both have been used, and the use of internal heating of a thick-walled cylinder is another variant of the method employed by these Russian workers. See also Filippov [177].

A further development by this group of workers has been to employ induction heating and by repeatedly switching the power on and off, to vary periodically the surface temperature of the cylinder. This was the method of Pigal'skaya *et al.* [51] and the detailed theory of the method is given in a paper by Filippov and Pigal'skaya [223]. Subsequently, Filippov and Makarenko [224] announced an important further development. They showed that by also measuring the power which is absorbed by the specimen from the induction heating it then becomes possible to evaluate the thermal conductivity and specific heat as well as the thermal diffusivity from the same set of experimental data. The surface emissivity can also be determined. This is a major advance as all four quantities will apply to the same sample and sample condition.

The relevant equations are

$$\theta = \frac{Hm}{2\pi k}\Psi(\varkappa) = \frac{Hm}{\pi R^2 \Omega d C_p}\Psi_1(\varkappa) \tag{88}$$

$$\phi = \phi(\varkappa) \tag{89}$$

where θ is the amplitude of the temperature fluctuations, ϕ the phase difference between the power fluctuations and the temperature, H the power absorbed by unit length of the sample, m the modulation coefficient, Ω the modulation frequency, R the radius of the sample, k, α, C_p, and d have the usual meanings, $\varkappa = R(\Omega/\alpha)^{0.5}$, and Thompson functions $\Psi(\varkappa)$ and $\Psi_1(\varkappa)$ are connected by the relationship

$$\Psi_1(\varkappa) = \Psi(\varkappa)\varkappa^2/2 \tag{90}$$

The power H is determined by measuring the magnetic field intensity, H_0, close to the sample in terms of the electromotive force setup in a loop of thin wire encircling the sample. This method assumes the electromagnetic field to be uniform over the length of the sample and over the radius in the gap between the induction coil and the sample. If E is the electromotive force of the loop,

$$E = i\omega\pi\mu H_0 \left\{(R_k^2 - R^2) + \frac{2\int_0^R I_0[(r\sqrt{2}/\gamma)\sqrt{i}]r\,dr}{I_0(R\sqrt{2}/\gamma)\sqrt{i}}\right\} \tag{91}$$

and

$$H = \frac{2E^2\sigma\eta^3}{\pi} \frac{1 - \eta - \eta^2/4}{[(R_k/R)^2 - 1 + 2\eta(1 + \eta^2/4)]^2 + 4\eta^2(1 - \eta - \eta^2/4)^2} \tag{92}$$

where R_k is the radius of the coil, I_0 is a Bessel function of the first kind and zero order expressed as a Thompson function, μ is the magnetic permeability, ω is the angular frequency of the generator of the induction furnace, σ is the electrical conductivity, $\gamma = (\mu\sigma\omega/2)^{-1/2}$ is the effective thickness of the skin layer, r is the distance from the axis of the sample, and $\eta = \gamma/2R$.

From the sample temperature fluctuations, as recorded by a contactless photoelectric pyrometer and equations (88) to (92), α, k, and C_p can be determined. Also, at a known mean temperature, T, the emissivity, ϵ, can be derived from the equation

$$\epsilon = H_m/2\pi R\sigma T^4 \tag{93}$$

where H_m is the mean power over the heating period and σ is the constant of the Stefan–Boltzmann law.

Use of different values of Ω allows either a combination of the values of α and C_p to be obtained, so yielding k, or a combination of α and the thermal activity, $(kdC_p)^{1/2}$, so yielding k and C_p. The error in the determination of α is in the range 4 to 8 percent and over a wide variation of Ω, it is stated that k and C_p can be determined to within 1 to 3 percent. Determinations made on niobium at 1660 K are given in the paper and the indications are that the method is being used for similar determinations on other high-melting-point metals. The quoted results for niobium show that for four different values of Ω which increase by ratios of 1.3, 1.8, and 3.0, the values of α are 0.243, 0.246, 0.250, and 0.248 cm^2/sec and those of k are 0.683, 0.687, 0.691, and 0.699 W cm^{-1} K^{-1}. These values of α can be compared with an earlier value of 0.246 cm^2/sec obtained for the same sample with alternating electronic heating, the value of k then obtained being 0.750 W cm^{-1} K^{-1}. The authors do not comment on this difference of 7 to 9 percent.

H. Cryogenic Method of Howling, Mendoza, and Zimmerman

A method designed for measurements at very low temperatures of both the thermal diffusivity and thermal conductivity was developed by Howling, Mendoza, and Zimmerman [225] in 1955 and applied to aluminum for the range 1.7 to 4.1 K. It is interesting that the true purpose of this work related to the subsequent evaluation of the specific heat from the expression $C_p = k/\alpha d$, since at very low temperatures, below say 1 K, the methods that were then normally employed for specific heat determinations tended to become less suitable. For instance, a normal requirement is for the sample to be thermally isolated, and this is a condition which could not well be met when the cooling was by adiabatic demagnetization and the sample was in contact with a paramagnetic salt. The development (Hall *et al.* [226]) and fairly general use of helium dilution refrigerators has since removed the thermal isolation problem and specific heats can readily be measured down to temperatures of about 0.04 K that can readily be obtained by this means.

The method used by Howling *et al.* [225] was of the Ångström type, but, since short rods were used, allowance was necessary for the reflection and transmission of thermal waves where contact was made with the cooling medium. At intervals along the rod were a heater and two thermometers having very small response times. The heater could be operated to provide either a sinusoidal or steady source of power. In the first mode frequency ranges of 16 to 1800 cps were employed and the thermal diffusivity was determined from observation of the velocity of propagation of the thermal waves as deduced from the phase difference or amplitude attenuation for the two thermometer locations. The steady heating of the second experiment allowed the thermal conductivity to be derived from the power input and the resulting temperature gradient. Not only could the specific heat per unit volume be derived from the ratio of k to α, but knowledge of k also proved helpful in assessing the net purity of the sample.

The above work is mentioned as one instance in which C_p has been determined from observations of α and k, rather than the more usual procedure of deriving k from observations of α and C_p, or of α and an assumed value for C_p. The original paper should be consulted for full details of the experimental method and the relevant equations.

Measurement of Thermal Diffusivity of Fluids

1. THERMAL DIFFUSIVITY OF LIQUIDS

Whereas some sixty or more works are available which describe methods used for determining the thermal conductivity of water, during the twentieth century relatively few, those of Soonawala [227], Hurt *et al.* [228], Filippov and Pashenkova [229], and Bryngdahl [230] deal with the experimental determination of the thermal diffusivity of water. This is fairly typical of the small amount of direct experimental information contained in the literature relating to the thermal diffusivity of fluids. For only a few fluids have experimental determinations of thermal diffusivity been undertaken, and hence correspondingly few entries will be found in the data section which follows. Should information on the thermal diffusivity of liquids be required it will often be necessary to calculate values of α from the equation $\alpha = k/dC_p$. In fact, if other experimental results are to be judged by those of the first two determinations referenced above, then such a check might with advantage be applied to all reported values for the thermal diffusivity of fluids.

Both Soonawala [227] and Hurt *et al.* [228] employed a periodic method of the Ångström type, the periodic heating being applied to the surface of a vertical column of water and temperature–time curves obtained by means of thermocouples located at two positions on the axis and below the surface. Soonawala's derived thermal conductivities, obtained near room temperature, were of the correct order of magnitude, but the six reported values showed large scatter, the extremes being $+8$ and -12 percent from the mean value. The value which Hurt *et al.* [228] obtained was much too small, being only about one-third of that derived from k/dC_p. Convection would seem to be a likely source of trouble, but in neither case does any serious effort appear to have been made to determine the cause of the uncertainties, nor to improve the method. The work of Hurt *et al.* [228] was really directed toward determinations on a few oils, and as water and some other liquids of low viscosity gave low values, whereas for glycerol the value agreed closely with that derived independently, the method was assumed to be suitable for these more viscous oils.

Filippov and Pashenkova [229] describe a method for liquids to which equation (28) applies. The liquid sample is held in a flat disk-shaped cell and heated at a constant rate. For this condition, the temperature difference is observed for two locations at distances of x_1 and x_2 from the middle of the liquid layer. Results are given at 34 C for water and seven organic liquids (CH_3OH, CCl_4, C_6H_6, C_6H_{14}, C_7H_{16}, C_8H_{16}, and C_8H_{18}) and these are stated to be within about ± 5 percent of literature data for the thermal diffusivities of these liquids. Their reported measurement precision was within 4 to 5 percent.

The possibility of using the initial transient of the hot wire line source method as a means for the determination of thermal diffusivity had been considered by Weishaupt [231]. This was in 1940, and, with the measurement techniques then available, he regarded the method as not suitable for this determination because of overlarge experimental uncertainties. Some two decades later, Bryngdahl [230], by the introduction of a highly sensitive optical interferometric method for studying the time variation of the temperature gradient set up in a liquid column by a vertical heated wire, was able to make thermal diffusivity determinations to an estimated accuracy of about ± 0.5 percent. His determinations were made on water, glycerol, octyl alcohol, and castor oil. From the same experiment, but using the customary plot of temperature rise *vs* the logarithm of time to give a straight line of slope $Q/4\pi k$, the thermal conductivity is also obtainable from the knowledge of this slope and the applied power per unit length of the line source. Bryngdahl's values for both k and α were in good agreement with other accepted values.

It is interesting to note that for water there is good agreement between the values of Bryngdahl [230] and of Filippov and Pashenkova [229], that of the latter at 34 C being 5.18 m^2/hr or 1.44_4 m^2/sec, while the mean of three values by Bryngdahl at 25.5 C is 1.46_4 m^2/sec.

A paper by Shashkov and San'ko [232] also considers the use of the heated probe variable-state method for the determination of thermal diffusivities as well as of thermal conductivities. The probe used consisted of a glass tube of 0.95 mm OD, 0.79 mm ID, and 91 mm length, containing liquid polysiloxane, a heater, and a resistance thermometer. The heater and thermometer were each composed of 20 lengths of 0.07-mm-diameter insulated wires, twisted together to pack tightly into the closed-ended capillary tube. The heater was of manganin and the thermometer of copper; the respective resistances were of the order of 340 and 8 ohms. The paper contains details of the instrumentation and of the relevant equations for evaluation of k and α for any medium into which the probe is immersed. The equations involve three constants but these can be reduced to two by assuming there is to be no thermal resistance between the probe heater and the test medium. Test measurements are briefly reported for four fluids, petroleum jelly, toluene, glycerol, and water. By using glycerol and water as reference fluids of assumed thermal properties to evaluate the constants, measurements on the other two fluids are stated to give thermal diffusivity coefficients for which the likely error was thought to be 12 to 15 percent.

Theoretical analyses relating to variable-state methods available for the determination of liquid (and gaseous) thermal diffusivities are also found in the treatise by Carslaw and Jaeger [3], and in a paper by Fischer [233]. The latter deals both with the application of a sinusoidal temperature wave to the original hot-wire method of Schleiermacher [234] and to the normal hot-plate method. One application of the radial heat-flow variable-state method has been by van Zee and Babcock [235] who determined the thermal diffusivities of two samples of molten glass for the range 700 to 1400 C from observations of the sinusoidal temperature and the time lag for heat transfer from the curved surface to the axis of a 6.5-in.-diameter cylindrical sample. In this work no attempt was made to separate the radiation component so this was included in the evaluated data.

Many of the experimental determinations of α will be seen to relate to liquid metals. Ångström [167] included mercury at 323 K in the substances to which his original method was applied, and obtained a thermal conductivity value that was only some 16 percent below the currently accepted value [14, 16]. The method was applied later by Weber [236] whose results were similar. Also for mercury Istrati [237] used a transient method in which a previously heated plate of iron was introduced to rest on and cover the top surface of a column of the liquid metal. A centrally disposed iron *vs* mercury thermocouple located a distance x below the plate was used to determine the time, t, that elapsed after the plate had been introduced for the temperature to reach a maximum value. The thermal diffusivity, α, was then calculated from the equation

$$\alpha = \frac{x^2}{2t} \tag{94}$$

The thermal conductivity value derived by Istrati from the result of these measurements agrees well with the curve of Powell *et al.* [14].

Methods of the Ångström type have been used more recently for determinations to quite high temperatures on sodium, potassium, and a 78 percent potassium–22 percent sodium alloy by Novikov *et al.* [238] and on lithium and sodium by Rudnev *et al.* [239].

Since in these experiments the liquid metal needs to be held in a thin-walled tube, usually of stainless steel, the thermal diffusivity is calculated from the equation

$$\alpha = \alpha^*(1 + \Delta) \tag{95}$$

where α^* is the measured thermal diffusivity for the liquid metal and its container and Δ is a correction factor given by

$$\Delta = \frac{d_1 C_{p_1} A_1}{d_2 C_{p_2} A_2}\left(1 + \frac{\alpha_1}{\alpha^*}\right) \tag{96}$$

where d_1, C_{p_1}, A_1, and α_1 are the density, specific heat, cross-sectional area, and thermal diffusivity of the container and d_2, C_{p_2}, and A_2 are the corresponding parameters for the liquid metal under investigation. In this technique the amplitude ratio and phase shift of the temperature waves are usually measured by thermocouples welded at two points on the tube. As with all such measurements on fluids, the avoidance of convection is essential, and for this purpose special baffles are often installed. In the radial-wave method of Yurchak and Filippov [220] and Filippov [177] the liquid metal was contained in a vertical cylindrical vessel of tantalum of 8 mm internal diameter and 23.6 mm external diameter, fitted with horizontal baffles of 0.1-mm tantalum sheet, spaced

10 to 15 mm apart. In the course of their determinations on liquid lead to 1355 K, tin to 1600 K, bismuth to about 1023 K, and cadmium, a range of frequencies was used for periodic heating applied either to the internal or to the external curved surfaces and consistent results were obtained both from the observed amplitude of the temperature oscillations at two radial distances and from the phase difference between them. A similar method has been used by Mardykin and Filippov [40] for determinations on liquid copper and antimony. Filippov [177] has discussed the possible sources of error, and an accuracy of 4 percent has been claimed. He does not, however, appear to mention any correction to allow for the baffles that were present.

These Russian measurements have been commented on by Powell [240, 241] since they yield results which are in marked disagreement with prevailing ideas that the thermal conductivities of most molten metals can be estimated with a fair amount of certainty from the electrical resistivity and an assumed value for the Lorenz function close to the theoretical value. Indeed, on this basis, Grosse [242] had predicted the thermal conductivities of several liquid metals right up to their critical points. At 1000 K the Lorenz functions as determined by Filippov [177] for tin and lead are, respectively, about 25 and 11 percent below the theoretical value and still appear to be decreasing with further increase in temperature. Similar behavior is also noticed for thermal conductivity determinations made on gallium by Yurchak and Smirnov [243] using quite a different method, a modification of the necked-down-sample method, and in the results of measurements for lead and for indium by Duggin [244, 245] at the Australian National Standards Laboratory.

There is a strong need at the present time for both theoretical and careful experimental investigations in this field, which will help to determine the true temperature behavior of the Lorenz functions on liquid metals. It would seem that thermal diffusivity determinations could play an important role in these investigations since they can be operated with only small temperature oscillations and do not require exact knowledge of the amount of energy supplied.

2. THERMAL DIFFUSIVITY OF GASES

Still fewer determinations have been reported for the thermal diffusivity of gases, for which many of the same methods are clearly applicable.

Bomelburg [246] described a set of measurements to determine the thermal diffusivity of atmospheric air. These are based on measuring the phase angle of periodic waves generated by a wire when heated by an alternating electric current. The sinusoidal temperature wave spreading out into space is picked up by another fine diameter heated wire which acts as a rapid-response resistance thermometer and allows the amplitude and phase of the received waves to be determined. The thermal diffusivity can be derived from the fact that a plot of the phase-lag ϕ *vs* the distance between the axes of the two parallel wires yields a straight line. This phase difference, $\phi_1 - \phi_2$ for two spacings, r_1 and r_2, is expressed by the equation

$$\phi_1 - \phi_2 = (r_1 - r_2)\left(\frac{\omega}{2\alpha}\right)^{1/2} \tag{97}$$

In Bomelburg's experiment, the hot transmitting wire was 10 μ in diameter and the receiving wire 1 μ. Both were of platinum and had lengths of 1 mm.

A rather different technique was used by Harrison [247] for determinations on water vapor and by Harrison *et al.* [248] for nitrogen. This method involved imposing a sinusoidal temperature wave in the wall of a cylindrical tube and measuring the phase shift between the response curves obtained at positions near the tube wall and at the axis. The electrical resistances of platinum wires were again used to record the temperature fluctuations at these locations. Accuracy appeared to be of the anticipated order near room temperature, but the need for further refinements became apparent at higher temperatures since the value at 500 C proved to be high by a factor of about two. Troubles due to radiation and convection seemed likely.

A more promising method, reported by Westenberg and de Haas [249], has been applied to nitrogen for the range 300 to 1100 K and is claimed to have a precision of ± 2 percent and to give derived thermal conductivity data in good agreement with the main literature values. This is a line-source method, in which the line source is mounted to lie at right angles to a laminar stream of uniformly heated gas that is flowing over the wire with a uniform velocity, v. Three methods were studied for the derivation of the thermal diffusivity from observations of temperature measurements made in the wake downstream of the heated line source. The method finally adopted was one which did not require a knowledge of the power dissipated, and hence could be applied to high temperatures without complications arising from radiation and end losses. All that is required is for differential thermocouple observations to be obtained in a plane at a series of distances x downstream of a heated wire located along the z axis. This enables a plot of the increase in gas temperature, ΔT, due to the power generated in the line source to be plotted for each value of x against the corresponding y coordinate. From this plot the distances r_h, equal to $(x_h^2 + y_h^2)^{1/2}$ and corresponding to the position x_h, y_h for which $\Delta T = \Delta T_m/2$, are obtained, where ΔT_m is the maximum temperature in each transverse plane at a height x above the wire. The thermal diffusivity is then calculated from

$$\alpha = \frac{v}{2}\left[\frac{r_h - x}{\ln(2\sqrt{x/r_h})}\right] \tag{98}$$

The term in brackets should be a constant for each series of measurements, a fact which is of assistance in determining the effective origin of the symmetrical temperature profile. The only other quantity that is required is the gas velocity and Westenberg and de Haas devised a novel method for the measurement of low-speed gas velocities of the order of 100 cm/sec which required the heated wire that provided the line source to be pulsed with a low-frequency signal of about 100 cps and the phase of this signal to be observed for two downstream positions a known distance apart. The sensing probe consisted of 1-mil Pt–Ag Wollaston wire with the silver dissolved off over a short central length to expose the 0.1-mil core. This was mounted on the same support as the traversing thermocouple and could be moved into position downstream of the source wire. Small amounts of heating of the 0.5-mil platinum wire line source were used to keep negligible any viscosity and convection effects. Appropriate power supplies were 40 mA of 100 cps AC superimposed on 25 mA of DC. The wire was mounted about 0.1 in. above a series of precision screens which served to give a uniform laminar velocity to the stream of gas (nitrogen) issuing from a 1.4-in.-diameter furnace tube.

3. USE OF PRANDTL NUMBER TO ESTIMATE THERMAL DIFFUSIVITY OF FLUIDS

In connection with the thermal diffusivity of fluids it might be useful to recollect that the dimensionless group known as the Prandtl number, Pr, is given by

$$\mathrm{Pr} = \nu/\alpha = \eta C_p/k \tag{99}$$

where ν is the kinematic viscosity.

Hence it is possible to calculate α from a knowledge of the ratio of these two parameters, ν/Pr. Furthermore, since, according to simple kinetic theory, the value of Pr for simple low-pressure gases is 2/3 and is independent of temperature, it follows that for these monatomic gases $\alpha = 1.5\,\nu$. For linear nonpolar gases, nonlinear nonpolar gases, highly polar gases, and for steam or ammonia the numerical coefficient decreases to about 1.37, 1.27, 1.16, and 1.0, respectively. Useful guidance on the estimation of viscosity η and of Pr can be obtained from a series of papers by Gambill [250].

Eckert and Irvine [251] describe a method for determining the Prandtl number of a gas which makes use of the fact that a unique relation $\mathrm{Pr} = r^2$ exists between this quantity and the recovery factor for laminar high-velocity boundary layer flow. From temperature measurements made in the high-velocity gas flow through a rotationally symmetrical nozzle and in a region of low gas flow they are able to determine Pr. For air at atmospheric pressure and for the temperature range 60 F to 350 F the values so obtained are considered accurate to ± 0.5 percent. These workers then derived the thermal conductivity from the relation $k = \eta C_p/\mathrm{Pr}$ using literature values for η and C_p. They regarded these values as of greater accuracy than that of some direct determinations. From the same data the thermal diffusivity, α, could equally well have been derived by using $\alpha = \eta/d\mathrm{Pr}$.

In the case of liquids, values of Pr vary considerably and η is a strong function of temperature. Denbigh [252] has suggested the following relation:

$$\log_{10} \mathrm{Pr} = a\frac{\Delta H}{RT} - b \tag{100}$$

where the logarithm is to base 10, R is the gas constant, T is the temperature in degrees K, ΔH is the molal latent heat of vaporization at the boiling point, and a and b are constants. This expression may be of value in certain cases, but should be used with caution, since some quite large departures were apparent. With the thermal quantities expressed in calories, Denbigh found that for water $a = 0.2$, and $b = -1.8$. Somewhat different constants appear necessary for organic liquids and it seems likely that some advantage could result from the use of different values for different groups of liquids. It should also be noted that Denbigh, when deriving values of Pr, used thermal conductivity data which are believed to have been of low accuracy.

References to Text

1. Maxwell, J. C., *Theory of Heat*, Longmans, Green and Co., London, 4th Edition, p. 255, 1875.
2. Thomson, W. (Lord Kelvin), "Heat," in *The Encyclopaedia Britannica*, 9th Edition, Volume 11, Section 82, 1880; See also, *Mathematical and Physical Papers*, University Press, Cambridge, Volume III, p. 205, 1884.
3. Carslaw, H. S. and Jaeger, J. C., *Conduction of Heat in Solids*, The Clarendon Press, Oxford, 2nd Edition, p. 510, 1959.
4. Kaspar, J. and Zehms, E. H., "A Diffusivity Measurement Technique for very High Temperatures," Air Force Rept. No. SAMSO-TR-70, Aerospace Rept. No. TR-0059(9250-02)-1, 31 pp., 30 August 1970.
5. Callendar, H. L., in *The Encyclopaedia Britannica*, The Encyclopaedia Britannica Co., New York, 11th Edition, Volume 6, 890–6, 1910.
6. Preston, T., *The Theory of Heat* (Editor, J. R. Cotter), Macmillan, London, 4th Edition, 1929.
7. Ingersoll, L. R., Zobel, O. J., and Ingersoll, A. C., *Heat Conduction*, McGraw-Hill, New York, 1948.
8. Jakob, M., *Heat Transfer*, Wiley, New York, 1949.
9. Gröber, H. and Erk, S., *Fundamentals of Heat Transfer*, 3rd Edition (in German), revised by U. Grigull (1955); English translation by J. R. Moszynski, McGraw-Hill, New York, 1961.
10. Tsederberg, N. V., *Thermal Conductivity of Gases and Liquids*, an English translation (of the Russian book) by Scripta Technica (Editor, R. D. Cess), MIT Press, Cambridge, Mass., U.S.A., 246 pp., 1965.
11. Sherwood, E. M., "Determination of Thermophysical Properties: Thermal Diffusivity," in *High Temperature Materials and Technology* (Editors, I. D. Campbell and E. M. Sherwood), Wiley and Sons, New York, 945–61, 1967.
12. Danielson, G. C. and Sidles, P. H., "Thermal Diffusivity and Other Non-Steady-State Methods," Chapter 3 in *Thermal Conductivity*, Volume 2 (Editor, R. P. Tye), Academic Press, London and New York, 149–201, 1969.
13. Ho, C. Y., Powell, R. W., and Wu, K. Y., "Thermal Diffusivity of the Elements," in *Thermal Conductivity: Proceedings of the Eighth Conference* (Editors, C. Y. Ho and R. E. Taylor), Plenum Press, New York, 971–98, 1969.
14. Powell, R. W., Ho, C. Y., and Liley, P. E., "Thermal Conductivity of Selected Materials," National Standard Reference Data Series—National Bureau of Standards, NSRDS-NBS-8, U.S. Govt. Printing Office, Washington, D.C., November, 1966.
15. Ho, C. Y., Powell, R. W., and Liley, P. E., "Thermal Conductivity of Selected Materials, Part 2," National Standard Reference Data Series—National Bureau of Standards, NSRDS-NBS-16, U.S. Govt. Printing Office, Washington, D.C., February, 1968.
16. Ho, C. Y., Powell, R. W., and Liley, P. E., "Standard Reference Data on the Thermal Conductivity of Selected Materials (Part 3)," Thermophysical Properties Research Center, Final Report on NBS-NSRDS Contract CST-1346, 1–435, 1968.
17. Rosenberg, H. M., "The Thermal Conductivity of Metals at Low Temperatures," *Phil. Trans. Roy. Soc., London*, **247A**, 441–97, 1955.
18. Mendelssohn, K. and Rosenberg, H. M., "The Thermal Conductivity of Metals at Low Temperatures, I. The Elements of Groups 1, 2 and 3. II. The Transition Elements," *Proc. Phys. Soc., London*, **65A**, 385–94, 1952.
19. Mikryukov, V. E., "Temperature Dependence of the Heat Conductivity and Electrical Resistance of Ti, Zr and Zr Alloys," *Vestnik Moskov. Univ., Ser. Mat. Mekh. Astron. Fiz. Khim.*, **12**(5), 73–80, 1957. (In Russian.)
20. Silverman, L., "Therman Conductivity Data Presented for Various Metals and Alloys up to 900 Degrees," *J. Metals*, **5**, 631–2, 1953.
21. Deem, H. W., Wood, W. D., and Lucks, C. F., "The Relation between Electrical and Thermal Conductivities of Titanium Alloys," *Trans. Met. Soc., AIME*, **212**, 520–3, 1958.
22. Krzhizhanovskii, R. E., "The Thermophysical Properties of Titanium and the Thermal Conductivity of Its Alloys with Tin and Aluminum," *Teplo. Vys. Temp.*, **2**, 392–6, 1964 (In Russian); English translation: *High Temperature*, **2**, 359–62, 1964.
23. Loewen, E. G., "Thermal Properties of Titanium Alloys and Selected Tool Materials," *Trans. Am. Soc. Mech. Engrs.*, **78**, 667–70, 1956.
24. White, G. K. and Woods, S. B., "Electrical and Thermal Resistivity of the Transition Elements at Low Temperatures," *Phil. Trans. Roy. Soc.., London*, **251A**, 273–302, 1959.
25. Gladun, C. and Holzhäuser, W., "Studies in Heat Conductivity at Low Temperatures," *Monatsber. Deut. Akad. Wiss., Berlin*, **6**(4), 310–3, 1964. (In German.)
26. Davey, G. and Mendelssohn, K., "Heat Conductivity of Pure Metals below 1 K," *Phys. Letters (Netherlands)*, **7**, 183–4, 1963.
27. Powell, R. W. and Tye, R. P., "The Thermal Conductivity of Titanium and its Alloys," *J. Less-Common Metals*, **3**, 226–33, 1961.
28. Kuprovskii, B. B. and Gel'd, P. V., "Thermal Conductivity of α-Titanium," *Tr. Ural'sk. Politekhn. Inst.*, Sb. No. 114, 153–4, 1961. (In Russian.)
29. Zinov'ev, V. E., Krentsis, R. P., and Gel'd, P. V., "Thermal Diffusivity of Titanium at High Temperatures," *Teplo. Vys. Temp.*, **6**, 927–8, 1968 (In Russian); English translation, *High Temperature*, **6**, 888–90, 1968.

30. Rigney, C. J. and Bockstahler, L. I., "The Thermal Conductivity of Titanium between 20 and 273 degrees Kelvin," *Phys. Rev.*, **83**, 220, 1951.
31. Rudkin, R. L., Parker, W. J., and Jenkins, R. J., "Thermal Diffusivity Measurements on Metals and Ceramics at High Temperatures," *Rev. Sci. Instrum.*, **33**, 21–4, 1963.
32. Touloukian, Y. S. (Editor), *Thermophysical Properties of High Temperature Solid Materials*, Vol. 1: *Elements*, The Macmillan Co., New York, 1270 pp., 1967.
33. Hultgren, R., Orr, R. L., Anderson, P. D., and Kelley, K. K., *Selected Values of Thermodynamic Properties of Metals and Alloys*, John Wiley and Sons, New York, 963 pp., 1963.
34. Butler, C. P. and Inn, E. C. Y., "Thermal Diffusivity of Metals at Elevated Temperatures," USNRDL-TR-177, 1–27, 1957. [AD 143 863]
35. Sheer, C., Mead, L. H., Rothacker, D. L., and Johnson, L. H., "Measurement of Thermal Diffusivity of Various Materials by Means of the High Intensity Electric Arc Technique," WADC-TR-57-226, 1–53, 1957. [AD 142 093]
36. El-Hifni, M. A. and Chao, B. T., "Measuring the Thermal Diffusivity of Metals at Elevated Temperatures," *Trans. ASME*, **78**, 813–21, 1956.
37. Sonnenschein, G. and Winn, R. A., "A Relaxation Time Technique for Measurement of Thermal Diffusivity," WADC-TR-59-273, 1–23, 1960. [AD 236 600]
38. Dennis, J. E., Hirshman, A., Derksen, W. L., and Monahan, T. I., "A Method to Determine the Thermal Diffusivity of Metals at High Temperatures," Naval Material Laboratory, DASA-1187, 1–10, 1960. [AD 242 669]. See also Reference 196.
39. Sheer, C., Fitz., C. D., Mead, L. H., Holmgren, J. D., Rothacker, D. L., and Allmand, D., "Investigation of the High Intensity Arc Technique for Materials Testing," WADC-TR-58-142, 75–96, 1958. [AD 205 364]
40. Mardykin, I. P. and Filippov, L. P., "Thermal Properties of Liquid Metals. I. Copper and Antimony," *Fiz. Khim. Obrab. Mater.*, **1**, 110–12, 1968. (In Russian.)
41. Furukawa, G. T. and Douglas, T. B., "Heat Capacities," Section 4e in *American Institute of Physics Handbook*, McGraw-Hill, New York, 2nd Edition, pp. (4-47)–(4-63), 1963.
42. Furukawa, G. T., Saba, W. G., and Reilly, M. L., "Critical Analysis of the Heat-Capacity Data of the Literature and Evaluation of Thermodynamic Properties of Copper, Silver, and Gold from 0 to 300 K," National Standard Reference Data Series—National Bureau of Standards, NSRDS-NBS-18, 1–49, 1968.
43. Bornemann, K. and Sauerwald, F., "Density Measurements of Metals and Alloys at High Temperatures, with Special Consideration of the Liquid State—Measurements with the Buoyancy Method. The Systems Cu–Sn and Cu–Al," *Z. Metallk.*, **14**, 145–59, 1922. (In German.)
44. Cahill, J. A. and Kirshenbaum, A. D., "The Density of Liquid Copper from its Melting Point (1356 K) to 2500 K and an Estimate of its Critical Constants," *J. Phys. Chem.*, **66**, 1080–2, 1962.
45. Lucas, L. D., "Density of Silver, Copper, Palladium and Platinum in Liquid State," *Compt. Rend.*, **253**, 2526–8, 1961.
46. Gebhardt, E., Becker, M., and Schäfer, S., "On the Properties of Metallic Melts. V. The Viscosity of Liquid Copper–Tin Alloys," *Z. Metallk.*, **43**, 292–6, 1952. (In German.)
47. Kraev, O. A. and Stel'makh, A. A., "Thermal Diffusivity in Tungsten at Temperatures Between 1600 and 2960 C," *High Temperature*, **1**, 5–8, 1963.
48. Wheeler, M. J., "Thermal Diffusivity at Incandescent Temperatures by a Modulated Electron Beam Technique," *Brit. J. Appl. Phys.*, **16**, 365–76, 1965.
49. Taylor, R. E. and Nakata, M. M., "Thermal Properties of Refractory Materials," WADD-TR-60-581, Part 4, 1–109, 1963. [AD 428 669, AD 441 079]
50. Pigal'skaya, L. A. and Filippov, L. P., "Measurement of the Thermal Diffusivity of Metals at High Temperatures: II. Experimental Method of Periodic Heating in a High Frequency Furnace," *High Temperature*, **2**, 501–4, 1964.
51. Pigal'skaya, L. A., Filippov, L. P., and Borisov, V. D., "Thermal Diffusivity of Tungsten at High Temperatures," *High Temperature*, **4**, 290–2, 1966.
52. Kraev, O. A. and Stel'makh, A. A., "Thermal Diffusivity and Thermal Conductivity of Metals at High Temperatures," in *High Temperature Research*, Academy of Sciences of the Siberian Section, USSR, 55–74, 1966.
53. Rawuka, A. C. and Gaz, R. A., "A Pulse Technique for Thermal Diffusivity Determination with Particular Reference to Instrumentation," in *Temperature Measurements Society: Sixth Conference and Exhibit*, Western Periodicals Co., North Hollywood, California, 55–67, 1969.
54. Arthur D. Little, Inc., "Development of High Temperature Thermal Conductivity Standards," Technical Report AFML-TR-69-2, 162–3, June 1969.
55. Cape, J. A., Lehman, G. W., and Nakata, M. M., "Transient Thermal Diffusivity Technique for Refractory Solids," *J. Appl Phys.*, **34**, 3550–5, 1963.
56. Chato, J. C., "A Survey of Thermal Conductivity and Diffusivity Data on Biological Materials," ASME Paper 66-WA/HT-37, Contribution to Heat Transfer Winter Annual Meeting and Energy Systems Exposition, Nov. 27–Dec. 1, 9 pp., 1966.
57. Reidy, G. A., "Thermal Properties of Foods and Methods of Their Determinations," Michigan State University, East Lansing, Mich., U.S.A., M.S. Thesis, 1968.
58. Qashou, M. S., "Compilation of Thermal Conductivity of Foods," Auburn University, Alabama, M.S. Thesis, December 1970.
59. Reidy, G. A., "I. Methods for Determining Thermal Conductivity and Thermal Diffusivity of Foods; II. Values for Thermal Properties of Foods Gathered from the Literature," Department of Food Science, College of Agriculture and Natural Resources, Michigan State University, East Lansing, Mich., U.S.A., June 1968.
60. Hurwicz, H. and Tischer, R. G., "Heat Processing of Beef. II. Development of Isothermal and Isochronal Distributions during Heat Processing of Beef," *Food Research*, **17**, 518, 1952; "VI. Thermal Diffusivity and 'Slopes' of Heating and Cooling Curves for the High Temperature Process," *Food Research*, **21**, 147, 1956.
61. Dickerson, R. W., "An Apparatus for the Measurement of Thermal Diffusivity of Foods," *Food Technology*, **19**, 880, 1965.
62. Wadsworth, J. I. and Spadaro, J. J., "Transient Temperature Distribution in Whole Sweetpotato Roots during Immersion Heating. I. Thermal Diffusivity of Sweetpotatoes," *Food Technology*, **23**, 85–9, 1969.

63. Nix, G. H., Lowery, G. W., Vachon, R. I., and Tanger, G. E., "Direct Determination of Thermal Diffusivity and Conductivity with Refined Line-Source Techniques," in *Thermophysics of Spacecraft and Planetary Bodies* (Editor, G. B. Heller), *Progress in Astronautics and Aeronautics*, Vol. 20, Academic Press, New York, 865–78, 1967.
64. Trezek, G. J., Jewett, D. L., and Cooper, T. E., "Measurements of In-Vivo Thermal Diffusivity of Cat Brain," in *Thermal Conductivity: Proceedings of the Seventh Conference* (Editors, D. R. Flynn and B. A. Peavy, Jr.), National Bureau of Standards Special Publication 302, 749–54, 1968.
65. Steigmeier, E. F. and Kudman, I., "Thermal Conductivity of III–V Compounds at High Temperatures," *Phys. Rev.*, **132**, 508–12, 1963.
66. Awbery, J. H., "The Physical Properties of a Series of Steels, Part II Section 1A, Specific Heat up to about 900 C," *J. Iron and Steel Inst.*, **154**, 84–90, 1946.
67. Powell, R. W., "Correlation of Metallic Thermal and Electrical Conductivities for both Solid and Liquid Phases," *Intnl. J. Heat and Mass Transfer*, **8**, 1033–45, 1965.
68. Williams, R. K. and Fulkerson, W., "Separation of the Electronic and Lattice Contributions to the Thermal Conductivity of Metals and Alloys," in *Thermal Conductivity, Proceedings of the Eighth Conference* (Editors, C. Y. Ho and R. E. Taylor), Plenum Press, New York, 389–456, 1969.
69. Taylor, R. E., Davis, F. E., and Powell, R. W., "Direct Heating Methods for Measuring Thermal Conductivity of Solids at High Temperatures," *High Temperatures—High Pressures*, **1**, 663–73, 1969.
70. Powell, R. W. and Taylor, R. E., "Multi-Property Apparatus and Procedure for High Temperature Determinations," *Rev. Hautes Tempér. et Réfract.*, **7**, 298–304, 1970. (In English.)
71. Forbes, J. D., "Experimental Inquiry into the Laws of the Conduction of Heat in Bars, and into the Conducting Power of Wrought Iron," *Trans. Roy. Soc.* (*Edinburgh*), **23**, 133–46, 1864.
72. Forbes, J. D., "Experimental Inquiry into the Laws of the Conduction of Heat in Bars. Part II. On the Conductivity of Wrought Iron Deduced from the Experiments in 1851," *Trans. Roy. Soc.* (*Edinburgh*), **24**, 73–110, 1865.
73. Bidwell, C. C., "The Thermal Conductivity of Li and Na by a Modification of the Forbes Bar Method," *Phys. Rev.*, **28**, 584–97, 1926.
74. Bidwell, C. C., "A Precise Method of Measuring Heat Conductivity Applicable to Either Molten or Solid Metals. Thermal Conductivity of Zinc," *Phys. Rev.*, **56**, 594–8, 1939.
75. Hogan, C. L. and Sawyer, R. B., "The Thermal Conductivity of Metals at High Temperatures," *J. Appl. Phys.*, **23**, 177–80, 1952.
76. Raezer, S. D., "Thermal and Electrical Conductivities of AISI C-1010 Steel in the Range 25 to 800 C," Lehigh University, Inst. of Research, Tech. Rept. No. 1, Office of Ordinance Research, Contract No. DA-36-034-ORD-1475, Project No. TB-2,0001-(151), 36 pp., 31 December 1954.
77. Ingersoll, L. R. and Koepp, O. A., "Thermal Diffusivity and Conductivity of Some Soil Materials," *Phys. Rev.*, **24**, 92, 1924.
78. Frazier, R. H., "A Precision Method for Determining the Thermal Diffusivity of Solids," *Phys. Rev.*, **39**, 515–24, 1932.
79. Frazier, R. H., "Further Data on the Thermal Diffusivity of Nickel," *Phys. Rev.*, **40**, 592–5, 1932.
80. Frazier, R. H., "A Precise Determination of the Thermal Diffusivity of Zinc," *Phys. Rev.*, **43**, 135–6, 1933.
81. Kennedy, W. L., "An IBM Computer Program for Determining the Thermal Diffusivity of Finite-Length Samples," USAEC IS-137, 1–59, 1960.
82. Kennedy, W. L., Sidles, P. H., and Danielson, G. C., "Thermal Diffusivity Measurements on Finite Samples," *Advanced Energy Conversion*, **2**, 53–8, 1962.
83. Shanks, H. R., Klein, A. H., and Danielson, G. C., "Thermal Properties of Armco Iron," *J. Appl. Phys.*, **38**, 2885–92, 1967.
84. Shanks, H. R., Maycock, P. D., Sidles, P. H., and Danielson, G. C., "Thermal Conductivity of Silicon from 300 to 1400 K," *Phys. Rev.*, **130**, 1743–8, 1963.
85. Jones, F. W. and Chisholm, P. J., "Thermal Conductivity and Diffusivity of Steel," *J. Iron and Steel Inst.*, **209**, 210–14, 1971.
86. Bornefeld, H., "Temperature Measurements in Fusion Welding," *Tech. Zentrablatt für praktische Metallbearbeitung*, **43**, 14–8, 1933. (In German.)
87. Rosenthal, D., "The Theory of Moving Sources and its Application to Metal Treatment," *Trans. ASME*, **68**, 849–66, 1946.
88. Rosenthal, D. and Ambrosio, A., "A New Method of Determining Thermal Diffusivity of Solids at Various Temperatures," *Trans. ASME*, **73**, 971–4, 1951.
89. Rosenthal, D. and Friedmann, N. E., "The Determination of Thermal Diffusivity of Aluminum Alloys at Various Temperatures by Means of a Moving Heat Source," *Trans. ASME*, **78**, 1175–80, 1956.
90. Cutler, M., "Thermoelectric Measurements at Small Area Contacts," *J. Appl. Phys.*, **32**, 1075–82, 1961.
91. Hopkins, M. R., "The Thermal and Electrical Conductivities of Metals at High Temperatures," *Z. Phys.*, **147**, 148–60, 1957.
92. Hopkins, M. R. and Griffith, R. Ll., "The Determination of the Lorenz Number at High Temperatures," *Z. Phys.*, **150**, 325–31, 1958.
93. Holm, R. and Störmer, R., "Measurement of the Thermal Conductivity of a Platinum Sample in the Temperature Range 19–1020 C," *Wiss. Veröffentl. Siemens-Konzern*, **9**, 312–22, 1930. (In German.)
94. Cutler, M., Snodgrass, H. R., Cheney, G. T., Appel, J., Mallon, C. E., and Meyer, C. H., Jr., "Thermal Conductivity of Reactor Materials," Final Rept., USAEC, GS-1939, General Atomics Division, General Dynamics Corp., San Diego, 30 January 1961.
95. Flynn, D. R. and O'Hagan, M. E., "Measurements of the Thermal Conductivity and Electrical Resistivity of Platinum from 100 to 900 C," *J. Res. National Bureau of Standards*, **71C**, 255–84, 1967.
96. Hérinckx, C. and Monfils, A., "Electrical Determination of the Thermal Parameters of Semiconducting Thermoelements," *Brit. J. Appl. Phys.*, **10**, 235–6, 1959.
97. Pinnow, D. A., Li, C. V., and Spencer, C. W., "Determination of Thermal Diffusivity by Utilization of the Thermoelectric Effect," *Rev. Sci. Instr.*, **32**, 1417–8, 1961.
98. Sonnenschein, G. and Winn, R. A., "A Relaxation Time Technique for Measurement of Thermal Diffusivity," WADC Tech. Rept., 59–273, 1–23, February 1960.
99. Butler, C. P. and Inn, E. C. Y., "Thermal Diffusivity of Metals at Elevated Temperatures," in *Thermodynamic and Transport*

Properties of Gases, Liquids and Solids, *Trans. ASME*, New York, 377–90, 1959.

100. Smith, W. K., "Measurement of Thermal Properties at High Temperatures," U.S. Navy Rept., NOTS-T-P-2624, August 1966.
101. Hsu, S. T., "Theory of a New Apparatus for Determining the Thermal Conductivity of Metals," *Rev. Sci. Instr.*, **28**, 333–6, 1957.
102. Hsu, S. T., "Determination of the Thermal Conductivity of Metals by Measuring Transient Temperatures in Semi-Infinite Solids," *Trans. ASME*, **79**, 1197–1203, 1957.
103. Parker, W. J., Jenkins, R. J., Butler, C. P., and Abbott, G. L., "A Flash Method of Determining Thermal Diffusivity, Heat Capacity and Thermal Conductivity," U.S. Naval Radiological Defense Lab. Tech. Rept., USNRDL-TR-424, 1960; *J. Appl. Phys.*, **32**, 1679–84, 1961.
104. Vernotte, P., "Simultaneous Determination of Specific Heat and Thermal Conductivity of Insulators," *Compt. Rend.*, **204**, 563–5, 1937. (In French.)
105. Clarke, L. N. and Kingston, R. S. T., "Equipment for the Simultaneous Determination of Thermal Conductivity and Diffusivity of Insulating Materials using a Variable-State Method," *Austral. J. Appl. Sci.*, **1**, 172–87, 1950.
106. Clarke, L. N. and Kingston, R. S. T., "Further Investigation of Some Errors in a Dynamic Method for the Determination of Thermal Conductivity and Diffusivity of Insulating Materials," *Austral. J. Appl. Sci.*, **2**, 235–42, 1951.
107. Krischer, O. and Esdorn, H., "Heat Transfer in Damp Porous Materials of Various Structures," *Forsch. Geb. Ingenieurw.*, **22**, 1–8, 1956. (In German.)
108. Harmathy, T. Z., "Variable State Methods of Measuring the Thermal Properties of Solids," *J. Appl. Phys.*, **35**, 1190–1200, 1964.
109. Pratt, A. W. and Ball, J. M. E., "Thermal Conductivity of Building Materials, Methods of Determination and Results," *J. Inst. Heating Ventilating Engrs.*, **24**, 201–26, 1956.
110. Levine, H. S., "An Unsteady-State Method for Measuring Thermal Diffusivity at Elevated Temperatures," ATI-78715, 1–32, 1950.
111. Paladino, A. E., Swarts, E. L., and Crandall, W. B., "Unsteady-State Method of Measuring Thermal Diffusivity and Biot's Modulus for Alumina between 1500 and 1800 C," *J. Amer. Ceramic Soc.*, **40**, 340–5, 1957.
112. Plummer, W. A., Campbell, D. E., and Comstock, A. A., "Method of Measurement of Thermal Diffusivity to 1000 C," *J. Amer. Ceramic Soc.*, **45**, 310–6, 1962.
113. Fitzsimmons, E. S., "Thermal Diffusivity of Refractory Oxides," *J. Amer. Ceramic Soc.*, **33**, 327–32, 1950.
114. Flieger, H. W., Jr. and Ginnings, D. C., "Physical Properties of High Temperature Materials. Part IV. Thermal Diffusivity Apparatus for 100 to 1500 C," WADC-TR-57-374 (Pt. IV), 1–9, 1957. [AD 205 797]
115. Flieger, H. W., Jr., Knudsen, F. P., and Ginnings, D. C., "Physical Properties of High Temperature Materials. Part V. Thermal Diffusivity of Magnesia Stabilized Zirconium Oxide at High Temperatures," WADC-TR-57-374 (Pt. V), 1–14, 1960. [AD 249 385]
116. Lehman, G. W., "Thermal Properties of Refractory Materials," WADD-TR-60-581, 1–19, 1960. [AD 247 411]
117. Cape, J. A. and Taylor, R. E., "Thermal Properties of Refractory Materials," WADD-TR-60-581 (Pt. 2), 1–22, 1961. [AD 264 288, AD 284 464]
118. Cowan, R. D., "Proposed Method of Measuring Thermal Diffusivity at High Temperatures," *J. Appl. Phys.*, **32**, 1363–70, 1961.
119. Cowan, R. D., "Pulse Method of Measuring Thermal Diffusivity at High Temperatures," *J. Appl. Phys.*, **34**, 926–7, 1963.
120. Mendelsohn, A. R., "The Effect of Heat Loss on the Flash Method of Determining Thermal Diffusivity," *Appl. Phys. Letters*, **2**, 19–21, 1963.
121. Cape, J. A. and Lehman, G. W., "Temperature and Finite Pulse-Time Effects in the Flash Method for Measuring Thermal Diffusivity," *J. Appl. Phys.*, **34**, 1909–13, 1963.
122. Taylor, R. E. and Cape, J. A., "Finite Pulse Time Effect in the Flash Diffusivity Method," *Appl. Phys. Letters*, **5**, 212–3, 1964.
123. Watt, D. A., "Theory of Thermal Diffusivity by Pulse Technique," *Brit. J. Appl. Phys.*, **17**, 231–40, 1966.
124. Larson, K. B. and Koyama, K., "Correction for Finite Pulse-Time Effects in very Thin Samples using the Flash Method of Measuring Thermal Diffusivity," *J. Appl. Phys.*, **38**, 465–74, 1967.
125. Powell, R. W., "Armco Iron as a Thermal Conductivity Standard: Review of Published Data," in *Progress in International Research on Thermodynamic and Transport Properties* (Editors, J. F. Masi and D. H. Tsai), ASME, Academic Press, New York, 454–65, 1962.
126. White, J. L. and Koyama, K., "Graphite Materials Hot-Worked with a Dispersed Liquid Carbide: Thermal and Electrical Conductivity," *J. Amer. Ceramic Soc.*, **51**, 394–8, 1968.
127. Beedham, K. and Dalrymple, I. P., "The Measurement of Thermal Diffusivity by the Flash Method. An Investigation into Errors Arising from the Boundary Conditions," *Rev. Int. Hautes Tempér et Réfract.*, **7**, 278–83, 1970.
128. National Physical Laboratory, "Modern Computer Methods," Notes on Applied Science No. 16, H.M.S.O., London, 2nd Edition, 1961.
129. Parker, W. J. and Jenkins, R. J., "Thermal Conductivity Measurements on Bismuth Telluride in the Presence of a 2 MEV Electron Beam," U.S. Naval Radiological Defense Lab. Tech. Rept., USNRDL-TR-462, 1960; *Advanced Energy Conversion*, **2**, 87–103, 1962.
130. Jenkins, R. J. and Westover, R. W., "The Thermal Diffusivity of Stainless Steel over the Temperature Range 20 C–1000 C," USNRDL-TR-484, 1–13, 1960. [AD 249 578]
131. Jenkins, R. J. and Parker, W. J., "A Flash Method for Determining Thermal Diffusivity over a Wide Temperature Range," WADD-TR-61-95, 1–29, 1961. [AD 268 752]
132. Rudkin, R. L., "Thermal Diffusivity Measurements on Metals and Ceramics at High Temperatures," ASD-TDR-62-24 (Pt. II), 1–16, 1963. [AD 415 005]
133. Baker, D. E., "Thermal Conductivity of Irradiated Graphite by a Rapid Thermal-Pulse Method," *J. Nucl. Mater.*, **12**, 120–4, 1964.
134. Moser, J. B. and Kruger, O. L., "Heat Pulse Measurements on Uranium Compounds," *J. Nucl. Mater.*, **17**, 153–8, 1965.
135. Taylor, R., "An Investigation of the Heat Pulse Method for Measuring Thermal Diffusivity," *Brit. J. Appl. Phys.*, **16**, 509–15, 1965.
136. Taylor, R., "Thermal Conductivity of Pyrolytic Graphite," *Phil. Mag.*, **13**, 157–66, 1966.

137. Wagner, P. and Dauelsberg, L. B., "The Thermal Conductivity of ZTA Graphite," *Carbon*, **5**, 271–9, 1967.
138. Wagner, P. and Dauelsberg, L. B., "The Thermal Conductivity of SX-5 Graphite," *Carbon*, **6**, 373–80, 1968.
139. Wagner, P. and Dauelsberg, L. B., "Some Thermal Properties of a Polyfurfuryl Alcohol Bonded Graphite," *Carbon*, **7**, 273–8, 1969.
140. Morrison, B. H., "Thermal Diffusivity of SX-5 Graphite from 800 to 2800 C," in *Thermal Conductivity, Proceedings of the Eighth Conference* (Editors, C. Y. Ho and R. E. Taylor), Plenum Press, New York, 1031–49, 1969.
141. Makarounis, O. and Jenkins, R. J., "Thermal Diffusivity and Heat Capacity Measurements at Low Temperatures by the Flash Method," USNRDL-TR-599, 1–24, 1962. [AD 295 887].
142. Iacobelli, R. and Moretti, S., "Thermal Diffusivity and Conductivity Measurements on Metals and Oxides at High Temperatures," *Rev. Hautes Tempér et Réfract.*, **3**, 215–28, 1966. (In English.)
143. Godfrey, T. G., Fulkerson, W., Kollie, T. G., Moore, J. P., and McElroy, D. L., "Thermal Conductivity of Uranium Dioxide and Armco Iron by an Improved Radial Heat Flow Technique," ORNL-3556, June 1964.
144. Report of the Panel on Thermal Conductivity of Uranium Dioxide, IAEA Technical Report Series, No. 59 (IAEA, Vienna, 1966).
145. Larson, K. B. and Koyama, K., "Measurement by the Flash Method of Thermal Diffusivity, Heat Capacity, and Thermal Conductivity in Two-Layer Composite Samples," *J. Appl. Phys.*, **39**, 4408–16, 1968.
146. Carpenter, R. S., "Flash Diffusivity Apparatus," NAA-SR-TDR-7643, 1–20, 1962.
147. Deem, H. W. and Wood, W. D., "Flash Thermal-Diffusivity Measurements Using a Laser," *Ref. Sci. Instr.*, **33**, 1107–9, 1962.
148. Taylor, R. E. and Nakata, M. M., "Thermal Properties of Refractory Materials," WADD-TR-60-581 (Pt. III), 1962 and (Pt. IV), 1963. [AD 428, 669, AD 441 079]
149. Taylor, R. E. and Morreale, J., "Thermal Conductivity of Titanium Carbide, Zirconium Carbide and Titanium Nitride at High Temperatures," *J. Amer. Ceramic Soc.*, **47**, 69–73, 1964.
150. Taylor, R. E., "Thermal Conductivity of 3Z1," NAA-SR-TDR-9334, 1–14, December 1963.
151. Méndez Peñalosa, R., "Contribution to the Study of the Thermal Properties of the Carbides and Nitrides of Uranium and the Transition Elements to Elevated Temperatures," Faculty of Science, University of Madrid, Doctoral Thesis, 1967. (In Spanish.)
152. Namba, S., Kim, P. H., and Arai, T., "Measurement of Thermal Diffusivity by Laser Pulse," *Japan J. Appl. Phys.*, **6**, 1019, 1967.
153. Nasu, S. and Kikuchi, T., "Thermal Diffusivity of UN from 20 to 1000 C by Laser Pulse Method," *J. Nucl. Sci. Technol.* (Tokyo), **5**, 318–9, 1968.
154. Lagedrost, J. F., Askey, D. F., Storhok, V. W., and Gates, J. E., "Thermal Conductivity of PuO_2 as Determined from Thermal Diffusivity Measurements," *Nucl. Appl.*, **4**, 54–61, 1968.
155. Gilchrist, K. E., "Measurement of the Thermal Conductivity of Ultra Thin Single or Double Layer Samples," Bundesministerium für Bildung und Wisenschaft BMBW-FBK70-01, European (Baden-Baden, Nov., 1968), *Conference on Thermophysical Properties of Solids at High Temperatures*, Published Zentralstelle für Atomkernenergie-Dokumentation, Karlsruhe, W. Germany, February 1970, 368–92. (In English.)
156. Ferro, C., Moretti, S., and Patimo, C., "Thermal Diffusivity and Conductivity of Sintered Uranium–Thorium Mixed Oxides," in *Thermal Conductivity, Proceedings of the Eighth Conference* (Editors, C. Y. Ho and R. E. Taylor), Plenum Press, New York, 815–22, 1969.
157. Mustacchi, C. and Giuliani, S., "Development of Methods for the Determination of the High Temperature Thermal Diffusivity of UC," European Atomic Energy Community EURATOM, EUR-337e, 1–27, 1963.
158. Walter, A. J., Dell, R. M., and Burgess, P. C., "The Measurement of Thermal Diffusivity using a Pulsed Electron Beam," *Rev. Int. Hautes Tempér et Réfract.*, **7**, 271–7, 1970.
159. Murfin, D., "Developments in the Flash Method for the Measurement of Thermal Diffusivity," *Rev. Int. Hautes Tempér et Réfract.*, **7**, 284–9, 1970.
160. di Novi, R. A., "Application of the Pulse Method to a Specific Heat and Density-Independent Measurement of Thermal Conductivity," *J. Sci. Instrum.* (*J. of Physics, E*), Ser. 2, **1**, 379–83, 1968.
161. Peggs, I. D. and Mills, R. W., "The Direct Determination of Thermal Conductivity by the Flash Technique," *Rev. Int. Hautes Tempér et Réfract.*, **7**, 264–7, 1970.
162. Erdmann, J. C. and Jahoda, J. A., "Apparatus for Low-Temperature Deformation and Simultaneous Measurements of Thermal Properties of Metals," *Rev. Sci. Instrum.*, **34**, 172–9, 1963.
163. Kohlrausch, F., "On the Stationary Temperature State of a Conductor Heated by an Electric Current," *Sitz. Berlin Akad.*, **38**, 711–8, 1899. (In German.)
164. Diesselhorst, H., "The Problem of an Electrically Heated Conductor," *Ann der Phys.*, **1**(4), 312–25, 1900. (In German.)
165. Ångström, A. J., "A New Method of Determining the Thermal Conductivity of Bodies," *Ann. der Phys. u. Chem.* (*Pogg. Ann.*), **114**, 513–30, 1861; *Phil. Mag.*, **25**, 130–42, 1863 (English translation).
166. Ångström, A. J., "On the Conducting-Power of Copper and Iron for Heat at Different Temperatures," *Ann. der Phys. u. Chem.* (*Pogg. Ann.*), **118**, 423–31, 1863; *Phil. Mag.*, **26**, 161–7, 1863 (English translation).
167. Ångström, A. J., "Supplement to the Paper: New Method for Determining the Thermal Conductivity of Solid Substances," *Ann. der Phys. u. Chem.* (*Pogg. Ann.*), **123**, 628–40, 1864. (In German.)
168. Weber, H., "On the Heat Conducting Power of Iron and German Silver," *Ann. der Phys. u. Chem.* (*Pogg. Ann.*), **146**, 257–83, 1872; *Phil. Mag.*, **44**, 481–500, 1872 (English translation).
169. King, R. W., "A Method of Measuring Heat Conductivities," *Phys. Rev.*, **6**, 437–45, 1915.
170. Ellis, W. C., Morgan, F. L., and Sager, G. F., "Thermal Conductivities of Copper, Nickel, and Some Alloys of Nickel," Rensselaer Polytech. Inst. Bull., Eng. and Sci. Ser., No. 21, 1–23, 1928.
171. Sager, G. F., "Investigation of the Thermal Conductivity of the System Copper–Nickel," Rensselaer Polytech. Inst. Bull., Eng. and Sci. Ser., No. 27, 3–48, 1930.
172. Starr, C., "An Improved Method for the Determination of Thermal Diffusivities," *Rev. Sci. Instr.*, **8**, 61–4, 1937.

173. Nii, R., "Measurement of the Thermal Conductivity of Semiconductors," *J. Phys. Soc., Japan*, **13**, 769–70, 1958.
174. Kanai, Y. and Nii, R., "Experimental Studies of the Thermal Conductivity of Semiconductors," *J. Phys. Chem. Solids*, **8**, 338–9, 361–2, 1959.
175. Kevane, C. J., "Report on the Measurement of Thermal Diffusivity Using a Solar Furnace," 1–29, 1958. [AD 207 634]
176. Filippov, L. P. and Nurumbetov, A. N., "Apparatus for the Measurement of the Thermal Diffusivity of Metals," Izd. Gos. Nauch-Issled. In-ta Nauch. Tekh. Inform. No. 18-65-406/36, 1–9, 1965. (In Russian.)
177. Filippov, L. P., "Methods of Simultaneous Measurement of Heat Conductivity, Heat Capacity and Thermal Diffusivity of Solid and Liquid Metals at High Temperatures," *Int. J. Heat and Mass Transfer*, **9**, 681–91, 1966.
178. Khusainova, B. N. and Filippov, L. P., "Thermal Properties of Single Crystal Molybdenum at High Temperatures," *High Temp.*, **6**, 891–2, 1968 (English translation of *Teplo. Vys. Temp.*, **6**, 929–30, 1968).
179. Anger, H., Baumberger, C., and Guennec, H., "Method of Measuring the Thermal Diffusivity of Solids: Application to Some Semiconducting Compounds," *Annales de Radioélectricité*, **17**, 13–23, 1962. (In French.)
180. Sidles, P. H. and Danielson, G. C., "Thermal Diffusivity of Metals at High Temperatures," *J. Appl. Phys.*, **25**, 58–66, 1954.
181. Sidles, P. H. and Danielson, G. C., "Thermal Diffusivity Measurements at High Temperatures," in *Thermoelectricity* (Editor, P. H. Egli), Wiley, New York, Chapter 16, 270–87, 1960.
182. Abeles, B., Cody, G. D., and Beers, D. S., "Apparatus for the Measurement of the Thermal Diffusivity of Solids at High Temperatures," *J. Appl. Phys.*, **31**, 1585–92, 1960.
183. Abeles, B., Beers, D. S., Cody, G. D., and Dismukes, J. P., "Thermal Conductivity of Ge–Si Alloys at High Temperatures," *Phys. Rev.*, **125**, 44–6, 1962.
184. Shanks, H. R., Klein, A. H., and Danielson, G. C., "Thermal Properties of Armco Iron," *J. Appl. Phys.*, **38**, 2885–92, 1967.
185. Martin, J. J., Sidles, P. H., and Danielson, G. C., "Thermal Diffusivity of Platinum from 300 to 1200 K," *J. Appl. Phys.*, **38**, 3075–8, 1967.
186. Ciszek, T. F., "The Thermal Diffusivity and Electrical Resistivity of Platinum at Temperatures above 1000 K," Iowa State University, Ames, Iowa, U.S.A., Masters Thesis, 1966.
187. Ginnings, D. C., "Standards of Heat Capacity and Thermal Conductivity," in *Thermoelectricity* (Editor, P. H. Egli), Wiley, New York, Chapter 20, 320–41, 1960.
188. Cody, G. D., Abeles, B., and Beers, D. S., "Thermal Diffusivity of Armco Iron," *Trans. Am. Inst. Metals*, **221**, 25–7, 1961.
189. Beers, D. S., Cody, G. D., and Abeles, B., "Thermal Conductivity of Germanium, Silicon and III–V Compounds at High Temperatures," in *Proceedings of the International Conference on the Physics of Semiconductors, Exeter, England*, Inst. Phys. and the Phys. Soc., London, 41–8, 1962.
190. Steigmeier, E. F. and Kudman, I., "Acoustical-Optical Phonon Scattering in Ge, Si and III–V Compounds," *Phys. Rev.*, **141**, 767–74, 1966.
191. Habachi, M., Azou, P., and Bastien, P., "Contributions to the Study of the Thermal Diffusivity of Metals and Metallic Alloys," *Compt. Rend.*, **261**(15), Group 7, 2899–2902, 1965. (In French.)
192. Powell, R. W. and Hickman, M. J., "Physical Properties of a Series of Steels: Part II, Section IIIA, Electrical Resistivities up to 1300 C," *J. Iron and Steel Inst.*, **154**, 99–104, 1946.
193. Powell, R. W., "Some Preliminary Measurements of the Thermal Conductivity and Electrical Resistivity of Cobalt," *Cobalt*, No. 24, 145–50, 1964.
194. McIntosh, G. E., "Thermal Diffusivity of Metals," Purdue University, Lafayette, Indiana, U.S.A., Ph.D. Thesis, 1–45, 1952.
195. McIntosh, G. E., Hamilton, D. C., and Sibbitt, W. L., "Rapid Measurements of Thermal Diffusivity," *Trans. ASME*, **76**, 407–10, 1954.
196. Hirschman, A., Dennis, J., Derksen, W., and Monahan, T., "An Optical Method for Measuring the Thermal Diffusivity of Solids," in *International Developments in Heat Transfer, Part IV*, ASME, New York, 863–9, 1961.
197. Cerceo, M. and Childers, H. M., "Thermal Diffusivity by Electron Bombardment Heating," *J. Appl. Phys.*, **34**, 1445–9, 1963.
198. Ainscough, J. B. and Wheeler, M. J., "The High-Temperature Thermal Conductivity of Sintered Uranium Dioxide," *Brit. J. Appl. Phys.*, *2*, **1**, 859–68, 1968.
199. Wheeler, M. J., "Thermal Diffusivity Measurements by the Modulated Electron-Beam Method: Thermal Diffusivity of Iron between 280 and 1100 C," *High Temperature and High Pressures*, **1**, 13–20, 1969.
200. National Physical Laboratory Report, "The Thermal Conductivity of Iron," 128–30, 1964, Her Majesty's Stationery Office, London, 1965.
201. Fulkerson, W., Moore, J. P., and McElroy, D. L., "Comparison of the Thermal Conductivity, Electrical Resistivity and Seebeck Coefficient of a High-Purity Iron and an Armco Iron to 1000 C," *J. Appl. Phys.*, **37**, 2639–53, 1966.
202. Zinov'ev, V. E., Krentsis, R. P., and Gel'd, P. V., "Thermal Diffusivity and Conductivity of Nickel at High Temperatures," *Phys. Metal and Metallog.*, **25**(6), 188–90, 1968 (English translation of *Fiz. Metal Metalloved.*, **25**(6), 1137–9 1968).
203. Zinov'ev, V. E., Krentsis, R. P., Petrova, L. N., and Gel'd, P. V., "High-Temperature Thermal Diffusivity and Conductivity of Cobalt," *Phys. Metal and Metallog.*, **26**(1), 57–63, 1968 (English translation of *Fiz. Metal. Metalloved.*, **26**(1), 60–5, 1968).
204. Zinov'ev, V. E., Krentsis, R. P., and Gel'd, P. V., "Thermal Conductivity and Thermal Diffusivity of Platinum at High Temperatures," *Soviet Physics–Solid State*, **10**, 2228–30, 1969 (English translation of *Fiz. Tvedogo Tela*, **10**, 2826–8, 1968).
205. O'Hagan, M. E., "Measurements of the Thermal Conductivity and Electrical Resistivity of Platinum from 373 to 1373 K," The George Washington Univ., Washington, D.C., Doctoral Dissertation, August 1966.
206. Krishnan, K. S. and Jain, S. C., "Determination of Thermal Conductivities at High Temperatures," *Brit. J. Appl. Phys.*, **5**, 426–30, 1954.
207. Bode, K. H., "A New Method to Measure the Thermal Conductivity of Metals at High Temperatures," *Allgemeine Wärmetechnik*, **10**, 110–20, and 125–42, 1961. (In German.)
208. Powell, R. W. and Tye, R. P., "The Promise of Platinum as a High-Temperature Thermal Conductivity Reference Material," *Brit. J. Appl. Phys.*, **14**, 662–6, 1963.

209. Kobushko, V. S., Merisov, B. A., and Khomkevich, V. I., "A Method for Determining the Thermal Conductivity of Metals at High Temperatures," *International Chemical Engineering*, **5**, 485–8, 1965.
210. Jain, S. C., Goel, T. C., and Narayan, V., "Thermal Conductivity of Metals at High Temperatures by the Jain and Krishnan Method. III. Platinum," *Brit. J. Appl. Phys.* (*J. Phys. D*), **2**, 109–13, 1969.
211. Vines, R. F., *The Platinum Metals and Their Alloys*, International Nickel Co., New York, 19–20, 1941.
212. Martin, J. J. and Sidles, P. H., "Thermal Diffusivity of Platinum from 300 to 1300 K," USAEC, Ames Laboratory, IS-1018, 1964.
213. Green, A. and Cowles, L. E. J., "Measurement of Thermal Diffusivity of Semiconductors by Angström's Method," *J. Sci. Instrum.*, **37**, 349–51, 1960.
214. McNeill, D. J., "Measurement of the Thermal Diffusivity of Thermoelectric Materials," *J. Appl. Phys.*, **33**, 597–600, 1962.
215. Becker, J. H., "Several New Methods to Measure the Thermal Diffusivity of Semiconductors," *J. Appl. Phys.*, **31**, 612–3, 1960.
216. Sochard, I. I. and Becker, J. H., "Measurements of Thermal Diffusivity in Semiconductors: InSb and Mg_2Sn" (Abstract only of paper presented to meeting of the American Physical Society, held March 1959), *Bull. Am. Phys. Soc., Ser. II*, **4**, 134, 1959.
217. Timberlake, A. B., Davis, P. W., and Shilliday, T. S., "Thermal Diffusivity Measurements on Small Samples," *Advanced Energy Conversion*, **2**, 45–51, 1962.
218. Davis, P. W., Timberlake, A. B., and Shilliday, T. S., "A Method for Measuring Thermal Diffusivity in Small Semiconducting Samples," *J. Appl. Phys.*, **33**, 765–6, 1962.
219. Leroux-Hugon, P. and Weill, G., "Measurement of Thermal Diffusivities at Acoustic Frequencies," *J. Phys. Radium*, **23**, 215–16A, 1962.
220. Yurchak, R. P. and Filippov, L. P., "Measurement of the Thermal Diffusivity of Metals by the Method of Radial Temperature Waves," *Inzh. Fiz. Zh.*, **7**(4), 84–9, 1964. (In Russian.)
221. Yurchak, R. P. and Filippov, L. P., "Measurement of the Thermal Diffusivity of Liquid Metals," *Teplofiz. Vys. Temp.*, **2**, 696–704, 1964; English translation: *High Temp.*, **2**, 628–30, 1964.
222. Yurchak, R. P. and Filippov, L. P., "Apparatus for Measuring the Thermal Diffusivity of Solid and Liquid Metals," *Zavodsk. Lab.*, **31**, 1142–4, 1965. (In Russian.)
223. Filippov, L. P. and Pigal'skaya, L. A., "Measurement of the Thermal Diffusivity of Metals at High Temperatures: I. Theory of the Method of Periodic Heating in a High-Frequency Furnace," *High Temp.*, **2**, 351–8, 1964.
224. Filippov, L. P. and Makarenko, I. N., "Method of Measuring the Complex Thermal Characteristics of Metals at High Temperatures," *Teplo. Vys. Temp.*, **6**, 149–55, 1968; English translation: *High Temp.*, **6**, 143–9, 1968.
225. Howling, D. H., Mendoza, E., and Zimmerman, J. E., "Preliminary Experiments on the Temperature-Wave Method of Measuring Specific Heats of Metals at Low Temperatures," *Proc. Roy. Soc., London*, **A229**, 86–109, 1955.
226. Hall, H. E., Ford, P. J., and Thompson, K., "A Helium-3 Dilution Refrigerator," *Cryogenics*, **6**, 80–8, 1966.
227. Soonawala, M. F., "Thermal Conductivity of Water," *Indian J. Phys.*, **18**, 71–3, 1944.
228. Hurt, J. E., Kohnke, E. E., and Schmidt, A. R., "Unsteady-State Method for the Determination of Thermal Conductivities of Oils," *Proc. Oklahoma Acad. Sci.*, **38**, 94–103, 1957.
229. Filippov, L. P. and Pashenkova, I. G., "Measurement of the Coefficient of Thermal Diffusivity of Liquids," *Inzh. Fiz. Zh.*, **1**, 84–8, 1958. (In Russian.)
230. Bryngdahl, O., "Accurate Determination of the Thermal Conducting Properties of Liquids by a Shear-Interferometric Method," *Ark. f. Fys.* (Sweden), **21**, 289–369, 1962. (In German.)
231. Weishaupt, J., "Method of Determining Thermal Conductivity of Liquids by the Nonstationary Method," *Forsch. Geb. Ing.*, **11**, 20–35, 1940. (In German.)
232. Shashkov, A. G. and San'ko, Yu. P., "Determining the Thermal Conductivities and Thermal Diffusivities of Liquids by the 'Probe' Method," *Heat Transfer—Soviet Research*, **1**(5), 119–25, 1969. (English translation of paper in *Vesti Ak. Nauk Bel. SSR*, No. 3, 1968.)
233. Fischer, J., "To Determine Thermal Conductivity and Thermal Diffusivity from the Equilibrium Procedures of the Schleiermacher-Tube and the Plate Methods," *Ann. der Phys.*, **34**, 669–88, 1939. (In German.)
234. Schleiermacher, A., "On the Thermal Conductivity of Gases," *Wied. Ann.*, **34**, 623–46, 1888. (In German.)
235. van Zee, A. F. and Babcock, C. L., "A Method for the Measurement of the Thermal Diffusivity of Molten Glass," *J. Amer. Ceramic Soc.*, **34**, 244–50, 1951.
236. Weber, H. F., "Investigations on Heat Conduction in Liquids," *Ann. Phys.*, **10**, 472–500, 1880. (In German.)
237. Istrati, M. I., "On the Determination of the Coefficient of the Thermal Conductivity of Mercury," *Ann. Sci. Univ. Jassy*, **14**, 23–7, 1926. (In French.)
238. Novikov, I. I., Soloviev, A. N., Khabakhnasheva, E. M., Gruzdev, V. A., Pridantzev, A. I., and Vasenina, M. Ya., "The Heat-Transfer and High-Temperature Properties of Liquid Alkali Metals," *Soviet J. Nucl. Energy*, **4**(3), 387–408, 1957 (English translation from *Atomnaya Energiya*, **1**(4), 92–106, 1956).
239. Rudnev, I. I., Lyashenko, V. S., and Abramovich, M. D., "Thermal Diffusivity of Sodium and Lithium," *Soviet J. Atomic Energy*, **11**, 877–80, 1962 (English translation from *Atomnaya Energiya*, **11**(3), 230–2, 1961).
240. Powell, R. W., "The Thermal and Electrical Conductivities of Molten Metals," in *Thermal Conductivity, Proceedings of the Eighth Conference* (Editors, C. Y. Ho and R. E. Taylor), Plenum Press, New York, 357–365, 1969.
241. Powell, R. W., "Thermal Conductivity: A Review of Some Important Developments," *Contemporary Physics*, **10**, 579–600, 1969.
242. Grosse, A. V., "The Thermal Conductivity of Liquid Metals over their Entire Liquid Range, i.e., from Melting Point to Critical Point, and the Containment of Metallic Substances up to 5000 K for Substantial Periods of Time," *Rev. Hautes Tempér. et Réfract.*, **3**, 115–46, 1966.
243. Yurchak, R. P. and Smirnov, B. P., "Thermal Conductivity and Lorenz Number of Solid and Liquid Gallium," *Soviet Physics–Solid State*, **10**, 1065–6, 1968 (English translation of *Fiz. Tverd. Tela*, **10**, 1340–2, 1968).
244. Duggin, M. J., "The Thermal Conductivity of Liquid Lead," *J. Physics D*, in press.
245. Duggin, M. J., private communication.

246. Bomelburg, H. J., "A Direct Method to Measure the Thermal Diffusivity of Gases," Ballistic Research Labs., Aberdeen Proving Ground, Maryland, BRL Rept. 1058, 1–28, 1958. [AD 209 438]
247. Harrison, W. B., "A Discussion of the Cyclic Heat Transfer Method for Determination of the Thermal Diffusivity of Water Vapor," Fourth International Conf. on the Properties of Steam, Philadelphia, Pa., 1954.
248. Harrison, W. B., Boteler, W. C., and Spurlock, J. M., "Thermal Diffusivity of Nitrogen as Determined by the Cyclic Heat Transfer Method," in *Thermodynamic and Transport Properties of Gases, Liquids and Solids*, ASME, New York, 304–12, 1959.
249. Westenberg, A. A. and de Haas, N., "High Temperature Gas Thermal Diffusivity Measurement with the Line Source Technique," in *Progress in International Research on Thermodynamic and Transport Properties* (Editors, J. F. Masi and D. H. Tsai), ASME, Academic Press, New York and London, 412–7, 1962.
250. Gambill, W. R., "Estimate Engineering Properties, Part V, Viscosity and Prandtl Number," *Chemical Engineering*, 121–4, August 25, 1958; 169–72, Sept. 22, 1958; 157–62, Oct. 20, 1958; 157–60, Nov. 17, 1958; 127–30, Jan. 12, 1959; 123–6, Feb. 9, 1959; 151–2, March 9, 1959.
251. Eckert, E. R. G. and Irvine, T. F., Jr., "A New Method to Measure Prandtl Number and Thermal Conductivity of Fluids," *J. Apppl. Mech.*, **24**, 25–8, 1957.
252. Denbigh, K. G., "Estimating the Prandtl Number of Liquids," *J. Soc. Chem. Ind.*, **65**, 61–3, 1946.
253. Unvala, B. A. and Goel, T. C., "Thermal Conductivity, Electrical Resistivity and Total Emittance of Titanium at High Temperatures," *Rev. Int. Hautes Tempér. et Réfract.*, **7**(4), 341–5, 1970.

Numerical Data

Data Presentation and Related General Information

1. SCOPE OF COVERAGE

Presented in this volume are the thermal diffusivity data for 75 elements, 63 alloy systems, 86 compounds and mixtures, and many entries of other kinds of materials including composites, glasses, minerals, polymers, and foods and biological materials. These data were obtained by processing over 800 research documents on thermal diffusivity dated from 1861 to 1970, of which 310 contain usable data. Materials within each group are arranged in alphabetical order by name, as listed in the *Grouping of Materials and List of Figures and Tables* in the front of the volume. In all, this volume reports 1733 sets of data on 445 materials, which are listed in the *Material Index* at the end of the volume.

The data for the elements have been critically evaluated, analyzed, and synthesized, and recommended reference values or provisional values for each element are presented. Experimental thermal diffusivity data are available in the world literature for only 39 elements. However, the recommended or provisional values are given in this volume for 75 elements. Since the available experimental thermal diffusivity data for most of these 39 elements often cover only a small temperature range, most of the recommended or provisional thermal diffusivity values were therefore first derived from the recommended values of thermal conductivity [311], selected values of specific heat [312–315], and selected values of density or calculated density values from thermal expansion data, and then compared with the critically evaluated experimental thermal diffusivity data, whenever available, to generate the final values given in this volume. Future editions of this volume will contain recommended values for an increasing number of materials.

2. PRESENTATION OF DATA

The thermal diffusivity data and information on test specimens for each material are generally presented in three sections arranged in the following order: Original Data Plot, Specification Table, and Data Table. For the elements, a Graph and Table of Recommended Values for each element is added to be the first section preceding the Original Data Plot. However, for each of those elements for which no experimental data are available, this graph and table of recommended (or provisional) values is the only page presented. Furthermore, for a number of materials for which there exists only a small number of data, the Original Data Plot may be omitted.

The Original Data Plot is a full-page linear scale graphical presentation of the original thermal diffusivity data as a function of temperature. When several sets of data are too close together to be distinguishable, some of the data sets may be omitted from the plot for the sake of clarity. They are, however, presented in the Specification Table and Data Table.

The Specification Table provides in a concise form the comprehensive information on the test specimens for which the data are reported. The curve numbers in the Specification Table correspond exactly to the numbers which also appear in the Original Data Plot and in the Data Table. The Specification Tables gives for each set of data the reference number which corresponds to the number in the list of References to Data Sources, the authors of the publication, the year of publication of the data, the temperature range, the reported estimate of error of the data, the particular name of the material other than that appearing in the title of the table and the specimen designation, and the specimen composition, characterization, and test conditions. The information of the last category, which is reported to the extent provided in the original source document, includes the following:

(1) purity, chemical composition
(2) type of crystal, crystal axis orientation
(3) microstructure, grain size, inhomogeneity, and additional phases
(4) specimen shape and dimensions, method and procedure of fabrication
(5) thermal history and cold work history, heat treatment, mechanical, irradiative, and other treatments
(6) manufacturer and supplier, stock number, and catalog number
(7) test environment, degree of vacuum or pressure, heat flow direction
(8) pertinent physical properties such as density, porosity, hardness, electrical resistivity (residual, ratio, and temperature variations), Lorenz function, transition temperatures, etc.
(9) form in which the extracted data are presented in the original source document other than raw data points
(10) additional information obtained directly from the author

Unfortunately, in the majority of cases the authors do not report in their research papers all the necessary pertinent information to fully characterize and identify the materials for which their data are reported. This is particularly true for the authors of earlier investigations. Consequently, the amount of information on specimen characterization reported in the Specification Tables varies greatly from specimen to specimen.

In the Data Table, tabular presentation is given for all the data described in the Specification Table and shown or not shown in the Original Data Plot. Attempts have often been made to contact the authors for tabular data whenever the original data are given in the research paper only in a figure too small to allow accurate data extraction compatible with the reported accuracy of the measurement.

The recommended or provisional values for each element are presented also in both graphical and tabular formats in a separate graph and table preceding the Original Data Plot. Special remarks on material characterization and identification and the estimated accuracy of the values are noted in the table.

In this volume, the thermal diffusivity data are presented in cm^2/sec, and the temperatures in kelvins. To convert the values given in cm^2/sec to values in other units, the conversion factors for units of thermal diffusivity given in the table may be used.

3. CLASSIFICATION OF MATERIALS

The classification scheme as shown in the table for the elements, alloys, compounds, and mixtures is based strictly upon the chemical composition of the material. This scheme is mainly for the convenience of materials grouping and data organization, and is not intended to be used as basic definitions for the various material groups.

4. SYMBOLS AND ABBREVIATIONS USED IN THE FIGURES AND TABLES

Symbols and abbreviations used in the figures and/or tables are as follows:

atm	Atmosphere
b.c.c.	Body-centered cubic
c.	Cubic
cm	Centimeter
c.p.h.	Close-packed hexagonal
C.T.	Critical temperature
d	Density
d.	Diamond (crystal structure)
Decomp.	Decomposition
f.c.c.	Face-centered cubic
f.c.t.	Face-centered tetragonal
g	Gram
h.	Hexagonal
hr	Hour(s)
I.D.	Inside diameter
in.	Inch(es)
Max.	Maximum
Min.	Minimum
M.P.	Melting point
monocl.	Monoclinic
NTP	Normal temperature and pressure
O.D.	Outside diameter
orthorh.	Orthorhombic
r.	Rhombohedral
s	Second
s.c.	Superconducting
Subl.	Sublimation
T	Temperature
t.	Tetragonal
Temp.	Temperature
T.P.	Transition point

Conversion Factors for Units of Thermal Diffusivity

MULTIPLY ↓ by appropriate factor to OBTAIN →	cm^2/sec	cm^2/hr	m^2/sec	m^2/hr	$in.^2/sec$	ft^2/sec	ft^2/hr
cm^2/sec	1	3.6×10^3	1×10^{-4}	3.6×10^{-1}	1.55000×10^{-1}	1.07639×10^{-3}	3.87501
cm^2/hr	2.77778×10^{-4}	1	2.77778×10^{-8}	1×10^{-4}	4.30556×10^{-5}	2.98998×10^{-7}	1.07639×10^{-3}
m^2/sec	1×10^4	3.6×10^7	1	3.6×10^3	1.55000×10^3	10.7639	3.87501×10^4
m^2/hr	2.77778	1×10^4	2.77778×10^{-4}	1	4.30556×10^{-1}	2.98998×10^{-3}	10.7639
$in.^2/sec$	6.45160	2.32258×10^4	6.45160×10^{-4}	2.32258	1	6.94444×10^{-3}	25
ft^2/sec	9.29030×10^2	3.34451×10^6	9.29030×10^{-2}	3.34451×10^2	1.44×10^2	1	3.6×10^3
ft^2/hr	2.58064×10^{-1}	9.29030×10^2	2.58064×10^{-5}	9.29030×10^{-2}	4×10^{-2}	2.77778×10^{-4}	1

Vit.	Vitreous
α	Thermal diffusivity
ρ	Electrical resistivity
μ	Micro
$>$	Greater than
$<$	Less than
$\sim$	Approximately
③	Curve number
[4]	Single data point number

5. CONVENTION FOR BIBLIOGRAPHIC CITATION

For the following types of documents the bibliographic information is cited in the sequences given below.

Journal Article

a. Author(s)—The names and initials of all authors are given. The last name is written first, followed by initials.
b. Title of article.
c. Journal name—The abbreviated name of the journal as used in *Chemical Abstracts* is given.
d. Series, volume, and number—If the series is designated by a letter, no comma is used between the letter for series and the numeral for volume, and they are underlined together. If the series is also designated by a numeral, a comma is used between the numeral for series and the numeral for volume, and only the numeral representing volume is underlined. No comma is used between the numerals representing volume and number. The numeral for number is enclosed in parentheses.
e. Pages—The inclusive page numbers of the article are given.
f. Year—The year of publication.

Report

a. Author(s)
b. Title of report
c. Name of the responsible organization
d. Report, or bulletin, circular, technical note, etc.
e. Number
f. Part
g. Pages
h. Year
i. ASTIA's AD number—This is given in square brackets whenever available

Book

a. Author(s)
b. Title
c. Volume
d. Edition
e. Publisher
f. City, state—Address of the publisher
g. Pages
h. Year

Classification of Materials

Classification			Limits of composition (weight percent)*			
			X_1	$X_1 + X_2$	X_2	X_3
1. Elements			>99.5	–	<0.2	<0.2
2. Nonferrous alloys ($X_1 \neq$ Fe)	A. Binary alloys		–	≥99.5	≥0.2	≤0.2
	B. Multiple alloys		–	≥99.5	>0.2	>0.2
			–	<99.5	≥0.2	≤0.2
			–	<99.5	>0.2	>0.2
			≤99.5	–	<0.2	<0.2
			X_1	X_2	X_3	Mn, P, S, or Si
3. Ferrous alloys (X_1 = Fe)	A. Carbon steels	Group I	Fe	C ≤ 2.0	≤0.2	≤0.6
		Group II	Fe	C ≤ 2.0	≤0.2	>0.6
			Fe	C ≤ 2.0	>0.2	≤0.6
			Fe	C ≤ 2.0	>0.2	>0.6
	B. Cast irons	Group I	Fe	C > 2.0	≤0.2	≤0.6
		Group II	Fe	C > 2.0	≤0.2	>0.6
			Fe	C > 2.0	>0.2	≤0.6
			Fe	C > 2.0	>0.2	>0.6
	C. Alloy steels†	Group I	Fe	≠C	≤0.2 and C ≤ 2.0	≤0.6
		Group II	Fe	≠C	≤0.2	>0.6
			Fe	≠C	>0.2	≤0.6
			Fe	≠C	>0.2	>0.6
			X_1	$X_1 + X_2$	X_2	X_3
4. Compounds			>95.0	–	<2.0	<2.0
5. Mixtures (or solutions) of compounds	A. Binary		–	≥95.0	≥2.0	≤2.0
	B. Multiple		–	≥95.0	>2.0	>2.0
			–	<95.0	≥2.0	≤2.0
			–	<95.0	>2.0	>2.0
			≤95.0	–	<2.0	<2.0

*$X_1 \geq X_2 \geq X_3 \geq X_4 \cdots$
†In case Mn, P, S, or Si represents X_2, this particular element is dropped from the last column. Alloy cast irons are also included in Group II of this category.

6. CRYSTAL STRUCTURES, TRANSITION TEMPERATURES, AND PERTINENT PHYSICAL CONSTANTS OF THE ELEMENTS

The table on the following pages contains information on the crystal structures, transition temperatures, and certain pertinent physical constants of the elements. This information is very useful in data analysis and synthesis. For example, the thermal diffusivity of a material generally changes abruptly when the material undergoes any transformation. One must therefore be extremely cautious in attempting to extrapolate the thermal diffusivity values across any phase, state, magnetic, or superconducting transition temperature, as given in the table.

No attempt has been made to critically evaluate the temperatures/constants given in the table and they should not be considered recommended values. This table has an independent series of numbered references which immediately follows the table.

CRYSTAL STRUCTURES, TRANSITION TEMPERATURES, AND PERTINENT PHYSICAL CONSTANTS OF THE ELEMENTS

Name	Atomic Number	Atomic Weight[a]	Density,[b] $kg\,m^{-3}\cdot 10^{-3}$	Crystal Structure	Phase Transition Temp., K	Superconducting Transition Temp., K	Curie Temp., K	Néel Temp., K	Debye Temperature at 0 K, K	Debye Temperature at 298 K, K	Melting Point, K	Boiling Point, K	Critical Temp., K
Actinium	89	(227)	10.07[1c]	f.c.c.[2]					124[3]	100[4] (at~50 K)	1323[5]	3200±300[6]	
Aluminum	13	26.9815	2.702[5]	f.c.c.[7]		1.196[5] 1.17[8] 1.18[9]			423±5[3]	390[3]	933.52[123]	2723[29]	8650[11] 7740[109]
Americium	95	(243)	11.7[5]	Double c.p.h.[2]							1473[29]	2880[108]	
Antimony	51	121.75	6.684[29]	r.[2] (?) ? (?) ? (?)	367.8[13] (?-?) 690[13] (?-?)	2.6[8] (Sb II, high-pressure modification)			150[3]	200[14]	903.89[123]	1907±10[3]	2989[15]
Argon	18	39.948	0.0017824[29] (at 273.2 K and 1 atm)	f.c.c.[16]						90[4] (at~45 K)	83.8[17]	87.29[13]	151[15]
Arsenic	33	74.9216	5.73[29] (gray, at 287.2 K) 4.7[29] (black) 2.0[29] (yellow)	r.[7] (gray) c.[5] (yellow)					236[3]	275[18]	1090[13] (35.8 atm) subl. 886[5]	1090[13] (35.8 atm)	
Astatine	85	(210)									573.2[19]	650[20]	
Barium	56	137.34	3.5[29]	b.c.c.[2] (α) ? (β)[13]	648[13, 21] (α-β)				110.5±1.8[22]	116[23]	998.2[5]	1910[3]	3663[15] 3920[109]
Berkelium	97	(249)											
Beryllium	4	9.0122	1.85[29]	c.p.h.[2] (α) b.c.c.[2] (β)	1533[24] (α-β)	~6[108] ~0.026[125]			1160[25]	1031[3]	1550[26]	3142±100[3]	6153[15]
Bismuth	83	208.980	9.78[29]	r.[2]		3.9[8] (Bi II, at 25 kbar) 7.2[8] (Bi III, at 27 kbar)			119±2[3]	116±5[3]	544.592[123]	1824±8[3]	4620[27]
Boron	5	10.811	2.50[42]	Simple r.[2] (α) r.[2] (β)	1473[2] (α-β)				1315[53]	1362[3]	2573[5]	4050±100[30]	
Bromine	35	79.909	3.119[29]	orthorh.[16]							266.0[17]	331.93[29]	584[15]

[a] Atomic weights are based on $^{12}C = 12$ as adopted by the International Union of Pure and Applied Chemistry in 1961; those in parentheses are the mass numbers of the isotopes of longest known half-life.

[b] Density values are given at 293.2 K unless otherwise noted.

[c] Superscript numbers designate references listed at the end of the table.

Name	Atomic Number	Atomic Weight[a]	Density,[b] kg m^{-3}·10^{-3}	Crystal Structure	Phase Transition Temp., K	Superconducting Transition Temp., K	Curie Temp., K	Néel Temp., K	Debye Temperature at 0 K, K	Debye Temperature at 298 K, K	Melting Point, K	Boiling Point, K	Critical Temp., K
Cadmium	48	112.40	8.65[29]	c.p.h.[2] b.c.c.[4] (?)		0.56[5] 0.52[9]			252 ± 48[3]	221[3] 170[4] (b.c.c., at ~85 K)	594.258[123] Subl. 594.1[13] (at 0.11 mm Hg)	1038[3]	1903[15] 3560[109]
Calcium	20	40.08	1.55[29]	f.c.c.[7] (α) b.c.c.[7] (β)	737[62] (α–β)				234 ± 5[3]	230[3]	1123[19] Subl. 1123[13] (at 0.35 mm Hg)	1765[3]	3267[15]
Californium	98	(251)											
Carbon (amorphous)	6	12.01115	1.8~2.1[29]								Subl. 3925–3970[5]	4473[5]	
Carbon (diamond)	6	12.01115	3.51[29]	d.[16]					2240 ± 5[31]	1874[3]	> 3823[5]	5100[5]	
Carbon (graphite)	6	12.01115	2.26[29] (α)	h.[2] (α) r.[2] (β)					402 ± 11[3]	1550[3]	Subl. 3925–3970[5]	4473[5]	
Cerium	58	140.12	6.90[29]	f.c.c. (α)[32] Double c.p.h.?[8] (β) f.c.c. (γ)[32] b.c.c. (δ)[32]	103 ± 5[33] (α–β) 263 ± 5[33] (β–γ) 1003[32] (γ–δ)	1.7[126] (above 50 kbar)		13[32]	146[3]	138[34]	1077[26]	3972[3]	10400[109]
Cesium	55	132.905	1.873[29]	b.c.c.[2]					40 ± 5[3]	43[23]	301.9[29] Subl. 301.9[13] (at 1.2 μHg)	939[35]	2060[113,114,115] 1900[109]
Chlorine	17	35.453	0.003214[29] (at 273.2 K)	t.[16]						115[4,36] (at ~58 K)	172.2[26]	239.10[13]	417[15]
Chromium	24	51.996	7.16[42]	c.p.h.[17,d] (α) b.c.c.[7] (β)	~299[17] (α–β)[d]			311[37]	598 ± 32[3]	424[3]	2118[38]	2918 ± 35[3]	
Cobalt	27	58.9332	8.862[42]	c.p.h.[7] (α) f.c.c.[17] (β)	690[39] (α–β)		1394[129] (α) 1130[129] (β)		452 ± 17[3]	386[3]	1767[123]	3229[3]	
Copper	29	63.54	8.933[29]	f.c.c.[2]					342 ± 2[3]	310[3]	1357.6[123]	2811 ± 20[41]	8500[11] 8280[109]
Curium	96	(247)	7[42]	Double c.p.h.[8]									
Dysprosium	66	162.50	8.556[42]	c.p.h.[2] (α) b.c.c.[2] (β)	Near m.p.[2] (α–β)		83.5[43] (ferro.–antiferromag.)	174[43]	172 ± 35[3]	158[44]	1773[12]	3011[44]	7640[109]

[d]Close-packed hexagonal crystalline modification of chromium may be formed by electrodeposition below 293 K under special conditions of deposition process. This c.p.h. form is unstable and will irreversibly transform into b.c.c. form on heating.

Name	Atomic Number	Atomic Weight[a]	Density,[b] kg m^{-3} · 10^{-3}	Crystal Structure	Phase Transition Temp., K	Superconducting Transition Temp., K	Curie Temp., K	Neel Temp., K	Debye Temperature at 0 K, K	Debye Temperature at 298 K, K	Melting Point, K	Boiling Point, K	Critical Temp., K
Einsteinium	99	(254)											
Erbium	68	167.26	9.06[42]	c.p.h.[2] (α) b.c.c.[2] (β)	1643[2] (α-β)		19[4] (ferro.-antiferromag.)	80[4]	134 ± 10[45]	163[44]	1770[26]	3000[3]	7250[109]
Europium	63	151.96	5.245[28]	b.c.c.[7]				~90[4][130] 88	127[3]		1099[5]	1971[46]	4600[109]
Fermium	100	(253)											
Fluorine	9	18.9984	0.001695[29] (at 273.2 K and 1 atm)	c.[108] (β-F_2)							53.58[5]	85.24[13]	144[15]
Francium	87	(223)							39[3]		300.2[19]	879[108]	
Gadolinium	64	157.25	7.87[42]	c.p.h.[2] (α) b.c.c.[2] (β)	1535[32] (α-β)		292[40]		170[3]	155 ± 3[3]	1579[19]	3540[3]	8670[109]
Gallium	31	69.72	5.91[29]	orthorh.[4] (α) t.[4] (β)	275.6[13] (α-β) (at 8.86 x 10^6 mm Hg)	1.091[5] 7.2[38] (Ga II, high-pressure modification)			317[3]	240[14] 125[4] (tetra at ~63 K)	302.93[5] 275.6[13] (at 8.86 x 10^6 mm Hg)	2510[3]	7620[27]
Germanium	32	72.59	5.36[29]	d.[7]		5.5[47] (at ~118 kbar) 8.4[108]			378 ± 22[3]	403[3]	1210.6[5]	3100[3]	5642[15]
Gold	79	196.967	19.3[42]	f.c.c.[7]					165 ± 1[3]	178 ± 8[3]	1337.58[123]	3240[3]	9500[11] 8060[109]
Hafnium	72	178.49	13.28[42]	c.p.h.[48] (α) b.c.c.[48] (β)	2023 ± 20[48] (α-β)	0.16[9] 0.35[108]			256 ± 5[3]	213[23]	2495[19]	4575 ± 150[49]	
Helium	2	4.0026	0.0001785[29] (at 273.2 K and 1 atm)	c.p.h.[16]						30[4] (at ~15 K)	3.45[29] 1.8 ± 0.2[17] (at 30 atm)	4.216[13] 4.22[23]	5.3[15]
Holmium	67	164.930	8.80[29]	c.p.h.[2] (α) b.c.c.[2] (β)	Near m.p.[50] (α-β)		20[4] (ferro.-antiferromag.)	132[4]	114 ± 7[45]	161[44]	1734[19]	3228[51]	
Hydrogen	1	1.00797	0.00008987[29] (at 273.2 K and 1 atm)	c.p.h.[16]						116[36] (para., at ~58 K) 105[36] (ortho., at ~53 K)	13.8 ± 0.1[17]	20.397[123]	33.3[15]
Indium	49	114.82	7.3[29]	f.c.t.[7]		3.4035[5]			108.8 ± 0.3[3]	129[14]	429.784[123]	2279 ± 6[3]	4377[15] 7050[109]
Iodine	53	126.9044	4.93[29]	orthorh.[16]						105[4] (at ~53 K)	386.8[29] subl. 298.16[13] (at 0.31 mm Hg)	457.50[29]	785[15] 819[134]
Iridium	77	192.2	22.5[42]	f.c.c.[7]		0.14[5,9]			425 ± 5[3]	228[3]	2720[123]	4820 ± 30[3]	

Name	Atomic Number	Atomic[a] Weight	Density,[b] kg m^{-3} . 10^{-3}	Crystal Structure	Phase Transition Temp., K	Superconducting Transition Temp., K	Curie Temp., K	Néel Temp., K	Debye Temperature at 0 K, K	Debye Temperature at 298 K, K	Melting Point, K	Boiling Point, K	Critical Temp., K
Iron	26	55.847	7.87[28]	b.c.c.-ferromag.[7] (α) b.c.c.-paramag.[7] (β) f.c.c.[7] (γ) b.c.c.[7] (δ)	1183[2] (β–γ) 1673[13] (γ–δ)		1043[40]		457 ± 12[3]	373[3]	1810[19]	3160[20]	6750[132] 9400[109]
Krypton	36	83.80	0.003708[29] (at 273.2 K and 1 atm)	f.c.c.[16]						60[4] (at ~30 K)	116.6[5]	119.93[13]	209.4[15]
Lanthanum	57	138.91	6.18[42]	Double c.p.h.[8] (α) f.c.c.[2] (β) b.c.c.[2] (γ)	583[32] (α–β) 1141[32] (β–γ)	4.9[8] (α) 6.3[8] (β)			142 ± 3[52]	135 ± 5[44]	1193[5]	3713 ± 70[3]	10500[109]
Lawrencium	103	(257)											
Lead	82	207.19	11.34[29]	f.c.c.[2]		7.193[5]			102 ± 5[3]	87 ± 1[3]	600.652[123]	2022 ± 10[41]	5400[27] 4760[109]
Lithium	3	6.939	0.534[29]	b.c.c.[7]	Martensitic transformation at low temp.[56]				352 ± 17[3]	448[3]	453.7[19]	1599[13]	4150[11] 3720[109]
Lutetium	71	174.97	9.85[29]	c.p.h.[2] (α) b.c.c.[2] (β)	Near m.p.[50] (α–β)				210[54]	116[3]	1923[19]	4140[3]	
Magnesium	12	24.312	1.74[29]	c.p.h.[7]					396 ± 54[3]	330[3]	923[55]	1385[3]	3530[109]
Manganese	25	54.9380	7.43(α)[28] 7.29(β)[28] 7.18(γ)[28]	~~c. (α)~~ + b.c.c. 93(α) c.[7] (β) ~~f.c.t. (γ)~~ + f.c.c. 93(γ) b.c.c.[7] (δ)	1000[13] (α–β) 1374[13] (β–γ) 1410[13] (γ–δ)			95[5] (α) ~580[124] (β) 660[129] (γ)	418 ± 32[3]	363[3]	1517 ± 3[5]	2360[13]	6050[109]
Mendelevium	101	(256)											
Mercury	80	200.59	13.546[29] 14.19[29] (at 234.25 K)	r.[7] (α) b.c.t.-pressure induced structure (β)	Martensitic transformation at low temp.[56]	4.153[5] (α) 3.949[5] (β)			~ 75[58]	92 ± 8[3]	234.288[123]	629.81[123]	1733[27] 1705[109]
Molybdenum	42	95.94	10.24[42]	b.c.c.[2]		0.92[5,9]			459 ± 11[3]	377[3]	2894 ± 10[124]	5785 ± 175[3]	17000[11] 16800[109]
Neodymium	60	144.24	7.007[29]	Double c.p.h.[8] (α) b.c.c.[32] (β)	1135[32] (α–β)			8[4] (ordinary) 19[4] (special)	159[3]	148 ± 8[3]	1292[19]	2956[60]	7900[109]
Neon	10	20.183	0.0009002[29] (at 273.2 K and 1 atm)	f.c.c.[16]						60[4] (at ~30 K)	24.48[5]	27.102[123]	44.5[15]

Name	Atomic Number	Atomic Weight[a]	Density,[b] kg m^{-3} · 10^{-3}	Crystal Structure	Phase Transition Temp., K	Superconducting Transition Temp., K	Curie Temp., K	Néel Temp., K	Debye Temperature at 0 K, K	Debye Temperature at 298 K, K	Melting Point, K	Boiling Point, K	Critical Temp., K
Neptunium	93	(237)	20.46[42]	orthorh.[2] (α) t.[2] (β) b.c.c.[2] (γ)	551[2] (α–β) 847 ± 5[133] (β–γ)				121[3]	163[3]	913.2[5]	4150[3]	
Nickel	28	58.71	8.90[42]	f.c.c.[7]			631[40]		427 ± 14[3]	345[3]	1728[123]	3055[63]	6294[15] 11750[109]
Niobium	41	92.906	8.57[42]	b.c.c.[7]		9.13[5] 9.09[8] 9.1[9]			241 ± 13[3]	260[64]	2741 ± 27[3] 2688[65]	4813[66]	19000[109] ·
Nitrogen	7	14.0067	0.0012506[29]	c.[16] (α) h.[107] (β)	35.62[13] (α–β)					70[4] (at ~ 35 K)	63.29[5]	77.348[123]	126.2[15]
Nobelium	102	(254)											
Osmium	76	190.2	22.48[29]	c.p.h.[2]		0.655[5] 0.65[8]			500[67]	400[68]	3283 ± 10[69]	5300 ± 100[70]	
Oxygen	8	15.9994	0.001429[29] (at 273.2 K and 1 atm)	b.c. orthorh.[7] (α) r.[7] (β) c.[7] (γ)	23.876 ± 0.01[112] (α–β) 43.818 ± 0.01[112] (β–γ)					250[4] (at ~ 125 K) 500[36] (at ~ 250 K)	54.8[5]	90.188[123]	154.8[15]
Palladium	46	106.4	12.02[28]	f.c.c.[2]					283 ± 16[3]	275[14]	1827[123]	3200[3]	
Phosphorus	15	30.9738	1.82[29] (β) 2.22[29] (γ) 2.69[29] (δ)	h. ?[7] (α) b.c.c.[7] (β) c.[7] (γ) f.c. orthorh.[17] (δ)	196[71] (α–β) 298.16[13] (β–γ) 298.16[13] (β–δ)	5.8[127] (P II, at 180 kbar) 5.4[127] P IV, at 230 kbar)			193[3] (white) 325[3] (red)	576[3] (white) 800[3] (red)	317.3[5] (white) 1300[72] (black)	553[13]	993.8[15]
Platinum	78	195.09	21.45[29]	f.c.c.[2]					234 ± 1[3]	225 ± 5[3]	2045[123]	4100[3]	8280[15]
Plutonium	94	(242)	19.737[29] (at 298.2 K)	Simple monocl.[2] (α) b.c. monocl.[2] (β) f.c. orthorh.[2] (γ) f.c.c.[2] (δ) b.c.t.[2] (δ') b.c.c.[2] (ϵ)	396.7[73] (α–β) 475[73] (β–γ) 591.4[73] (γ–δ) 729[73] (δ–δ') 757 ± 3[73] (δ'–ϵ)			60[8, 131]	(α) 171[74]	176[74]	912.7[5]	3727[75]	
Polonium	84	(210)	9.3[29] (α) 9.5[29] (β)	Simple c.[7] (α) r.[7] (β)	327 ± 1.5[76] (α–β)				81[3]		527.2[5]	1235[20]	2281[15]
Potassium	19	39.102	0.86[29]	b.c.c.[7]					89.4 ± 0.5[3]	100[3]	336.8[5]	1027[35]	2450[11] 2140[109]
Praseodymium	59	140.907	6.769[29]	Double c.p.h.[8] (α) b.c.c.[2] (β)	1071[32] (α–β)			25[77]	85 ± 1[45]	138[78]	1192 ± 2[79]	3616[80]	8900[109]

Name	Atomic Number	Atomic Weight [a]	Density [b], kg m⁻³ · 10⁻³	Crystal Structure	Phase Transition Temp., K	Superconducting Transition Temp., K	Curie Temp., K	Néel Temp., K	Debye Temperature at 0 K, K	Debye Temperature at 298 K, K	Melting Point, K	Boiling Point, K	Critical Temp., K
Promethium	61	(145)		h. [7](α) b.c.c. [120](β)	1185 [120] (α-β)			6 [120]			1353 ± 10 [81]	2730 [3]	
Protactinium	91	(231)	15.37 [42]	b.c.t. [2]		1.4 [9]			159 [3]	262 [3]	1503 [5]	4680 [3]	
Radium	88	(226)	5 [29]						89 [3]		973.2 [5]	1900 [3]	
Radon	86	(222)	0.00973 [29] (at 273.2 K and 1 atm)	f.c.c. [7]						400 [4] (at ~200 K)	202.2 [5]	211 [13]	377.16 [15]
Rhenium	75	186.2	21.1 [42]	c.p.h. [2]		1.698 [26]			429 ± 22 [3]	275 [23]	3453 [5]	6035 ± 135 [3]	20000 [11]
Rhodium	45	102.905	12.45 [42]	f.c.c. [7]	possible transformation at 1373–1473 K [57]				480 ± 32 [3]	350 [3]	2236 [123]	3960 ± 60 [3]	
Rubidium	37	85.47	1.53 [29]	b.c.c. [2]					54 ± 4 [3]	59 [23]	312.04 [5]	959 [35]	2100 [113,115,116] 2030 [109]
Ruthenium	44	101.07	12.2 [29]	c.p.h. [7](α) ? (β) ? (γ) ? (δ)	1308 [13, 121] (α-β) 1473 [13, 121] (β-γ) 1773 [13, 121] (γ-δ)	0.49 [5, 9]			600 [67]	415 [3]	2523 ± 10 [69]	4325 ± 25 [3]	
Samarium	62	150.35	7.54 [29]	r. [32](α) b.c.c. [32](β)	1190 [32](α-β)		14 [8] (ferro.-antiferromag.)	106 [8]	116 [45]	184 ± 4 [3]	1345.2 [83]	2140 [3]	5400 [109]
Scandium	21	44.956	3.00 [42]	c.p.h. [2](α) b.c.c. [2](β)	1607 [2](α-β)				470 ± 80 [52]	476 [3]	1812 [5]	3537 ± 30 [3]	
Selenium	34	78.96	4.50 [29](α) 4.80 [29](β)	monocl. [7](α) h. [7](β) amorphous [7]	304 [84,117] (vitrification) 398 [13](vit.-β) 423 [13](α-β)	7.3 [85] (at ~118 kbar)			151.7 ± 0.4 [86]	89 [36] (at ~45 K) 150 [4] (at ~75 K)	490.2 [5]	1009 [13](Se_6) 958.0 [13] ($Se_{4.37}$) 1027 [13](Se_2)	1757 [15]
Silicon	14	28.086	2.33 [42]	d. [7]		7.5 [47] (at 118–128 kbar)			647 ± 11 [3]	692 [3]	1685 ± 2 [87]	2753 [28]	5159 [15]
Silver	47	107.870	10.5 [29]	f.c.c. [2]					228 ± 3 [3]	221 [3]	1235.08 [123]	2468 ± 15 [41]	7460 [11]
Sodium	11	22.9898	0.9712 [29]	b.c.c. [2]	Martensitic transformation at low temp. [56]				157 ± 1 [3]	155 ± 5 [3]	371.0 [13]	1154 [35]	2800 [11] 2400 [109]
Strontium	38	87.62	2.60 [28]	f.c.c. [88](α) c.p.h. [7](β) b.c.c. [7](γ)	488 [88](α-β) 878 [88](β-γ)				147 ± 1 [22]	148 [23]	1042 [5]	1645 [3]	3059 [15] 3810 [109]
Sulfur	16	32.064	2.07 [29](α) 1.96 [29](β)	r. [7](α) monocl. [7](β)	368.6 [13](α-β)				200 [3](β)	527 [89](α) 250 [89](α, at 40 K)	386.0 [5](α) 392.2 [5](β) Subl. 368.6 [13] (at 0.0047 mm Hg)	717.824 [123]	1313 [15]
Tantalum	73	180.948	16.6 [42]	b.c.c. [2]		4.483 [5] 4.48 [9]			247 ± 13 [3]	225 [14]	3269 [5]	5760 ± 60 [3]	22000 [11]

Name	Atomic Number	Atomic Weight[a]	Density,[b] kg m^{-3} · 10^{-3}	Crystal Structure	Phase Transition Temp., K	Superconducting Transition Temp., K	Curie Temp., K	Neel Temp., K	Debye Temperature at 0 K, K	Debye Temperature at 298 K, K	Melting Point, K	Boiling Point, K	Critical Temp., K
Technetium	43	(99)	11.50[29]	c.p.h.[2]		8.22[5] 11.2[9]			351[3]	422[3]	2473 ± 50[5]	5300[3]	
Tellurium	52	127.60	6.24[29] (α) 6.00[5] (amorph.)	h.[7] (α) ? (β)[7] amorph.[5]	621[13] (α–β)	3.3[8] (Te II, at 56 kbar)			141 ± 12[3]		722.7[5]	1163 ± 1[3]	2329[15]
Terbium	65	158.924	8.25[29]	c.p.h.[2, 32] (α) b.c.c.[2] (β)	Near m.p.[2] (α–β)		219[90] (ferro.-antiferromag.)	230[90]	150[91]	158[44]	1629[19]	3810[3]	
Thallium	81	204.37	11.85[29]	c.p.h.[2] (α) b.c.c.[2] (β)	508.3[5] (α–β)	2.39[5] 2.38[8] 2.37[9]			88 ± 1[3]	96[14]	576.2[19]	1939[92]	3219[15]
Thorium	90	232.038	11.7[42]	f.c.c.[2] (α) b.c.c.[2] (β)	1673 ± 25[93] (α–β)	1.368[5] 1.37[9]			170[94]	100[14]	2023[19]	4500[20]	14550[109]
Thulium	69	168.934	9.32[29]	c.p.h.[2] (α) b.c.c.[2] (β)	Near m.p.[50] (α–β)		22[95] (ferro.-antiferromag.)	53[96]	127 ± 1[45]	167[44]	1818[5]	2266[97]	6430[109]
Tin	50	118.69	5.750[29] (α) 7.31[29] (β)	f.c.c.[7] (α) b.c.t.[7] (β) r.[29] (?)	286.2 ± 3[98] (α–β)	3.722[5] (β)			236 ± 24[3] (gray) 196 ± 9[3] (white)	254[3] (gray) 170[14] (white)	505.1181[123]	2766 ± 14[3]	8000[11] 9300[109]
Titanium	22	47.90	4.5[29]	c.p.h.[7] (α) b.c.c.[7] (β)	1155[13] (α–β)	0.39[5, 9]			426 ± 5[3]	380[14]	1953[99]	3586[100]	
Tungsten	74	183.85	19.3[29]	b.c.c.[2]		0.011[122]			388 ± 17[3]	312 ± 3[3]	3660[123]	6000 ± 200[3]	23000[11]
Uranium	92	238.03	19.07[28]	orthorh.[7] (α) t.[7] (β) b.c.c.[7] (γ)	37 ± 2[118] ($α_0$–α) 938[13] (α–β) 1049[13] (β–γ)	0.68[5] (α) 0.8[128] (β) 1.80[5] (γ)			200[94]	300[3]	1405.6 ± 0.6[101]	3950 ± 250[102]	12500[27] 12000[109]
Vanadium	23	50.942	6.1[28]	b.c.c.[2]		5.3[5] 5.03[9]			326 ± 54[3]	390[14]	2192 ± 2[61]	3582 ± 42[3]	11200[109]
Xenon	54	131.30	0.005851[29] (at 273.2 K and 1 atm)	f.c.c.[16]							161.2[26]	165.1[13]	289.75[15]
Ytterbium	70	173.04	7.02[42]	f.c.c.[32] (α) b.c.c.[32] (β)	1071[2, 5] (α–β)				118[103]		1097[12]	1970[3]	4420[109]
Yttrium	39	88.905	4.47[29]	c.p.h.[32] (α) b.c.c.[32] (β)	1753[119] (α–β)				268 ± 32[3]	214[104]	1798[119]	3670[105]	8950[109]
Zinc	30	65.37	7.140[29]	c.p.h.[2]		0.875[5] 0.85[9]			316 ± 20[3]	237 ± 3[3]	692.73[123]	1175[106]	2169[15] 2910[109]
Zirconium	40	91.22	6.57[59]	c.p.h.[7] (α) b.c.c.[7] (β)	1135[13] (α–β)	0.546[5] 0.55[9]			289 ± 24[3]	250[14]	2125[19]	4650[20]	12300[109]

REFERENCES

(Crystal Structures, Transition Temperatures, and Other Pertinent Physical Constants of the Elements)

1. Farr, J.D., Giorgi, A.L., and Bowman, M.G., USAEC Rept. LA-1545, 1-13, 1953.
2. Elliott, R.P., Constitution of Binary Alloys, 1st Suppl., McGraw-Hill, 1965.
3. Gschneider, K.A, Jr., Solid State Physics (Sietz, F. and Turnbull, D., Editors), 16, 275-426, 1964.
4. Gopal, E.S.R., Specific Heat at Low Temperatures, Plenum Press, 1966.
5. Weast, R.C. (Editor), Handbook of Chemistry and Physics, 47th Ed., The Chemical Rubber Co., 1966-67.
6. Foster, K.W. and Fauble, L.G., J. Phys. Chem., 64, 958-60, 1960.
7. The Institution of Metallurgists, Annual Yearbook, pp. 68-73, 1960-61.
8. Meaden, G.T., Electrical Resistance of Metals, Plenum Press, 1965.
9. Matthias, B.T., Geballe, T.H., and Compton, V.B., Rev. Mod. Phys., 35, 1-22, 1963.
10. Stimson, H.F., J. Res. NBS, 42, 209, 1949.
11. Grosse, A.V., Rev. Hautes Tempér. et Réfract., 3, 115-46, 1966.
12. Spedding, F.H. and Daane, A.H., J. Metals, 6 (5), 504-10, 1954.
13. Rossini, F.D., Wagman, D.D., Evans, W.H., Levine, S., and Jaffe, I., NBS Circ. 500, 537-822, 1952.
14. deLaunay, J., Solid State Physics, 2, 219-303, 1956.
15. Gates, D.S. and Thodos, G., AIChE J., 6 (1), 50-4, 1960.
16. Gray, D.E. (Coordinating Editor), American Institute of Physics Handbook, McGraw-Hill, 1957.
17. Sasaki, K. and Sekito, S., Trans. Electrochem. Soc., 59, 437-60, 1931.
18. Anderson, C.T., J. Am. Chem. Soc., 52, 2296-300, 1930.
19. Trombe, F., Bull. Soc. Chim. (France), 20, 1010-2, 1953.
20. Stull, D.R. and Sinke, G.C., Thermodynamic Properties of the Elements in Their Standard State, American Chemical Soc., 1956.
21. Rinck, E., Ann. Chim. (Paris), 18 (10), 455-531, 1932.
22. Roberts, L.M., Proc. Phys. Soc. (London), B70, 738-43, 1957.
23. Zemansky, M.W., Heat and Thermodynamics, 4th Ed., McGraw-Hill, 1957.
24. Martin, A.J. and Moore, A., J. Less-Common Metals, 1, 85, 1959.
25. Hill, R.W. and Smith, P.L., Phil. Mag., 44 (7), 636-44, 1953.
26. Moffatt, W.G., Pearsall, G.W., and Wulff, J., The Structure and Properties of Materials, Vol. I, pp. 205-7, 1964.
27. Grosse, A.V., Temple Univ. Research Institute Rept., 1-40, 1960.
28. Lyman, T. (Editor), Metals Handbook, Vol. 1, 8th Ed., American Soc. for Metals, 1961.
29. Lange, N.A. (Editor), Handbook of Chemistry, Revised 10th Edition, McGraw-Hill, 1967.
30. Paule, R.C., Dissertation Abstr., 22, 4200, 1962.
31. Burk, D.L. and Friedberg, S.A., Phys. Rev., 111 (5), 1275-82, 1958.
32. Spedding, F.H. and Daane, A.H. (Editors), The Rare Earths, John Wiley, 1961.
33. McHargue, C.J., Yakel, H.L., and Letter, C.K., ACTA Cryst., 10, 832-33, 1957.
34. Arajs, S. and Colvin, R.V., J. Less-Common Metals, 4, 159-68, 1962.
35. Bonilla, C.F., Sawhney, D.L., and Makansi, M.M., Trans. Am. Soc. Metals, 55, 877, 1962.
36. Rosenberg, H.M., Low Temperature Solid State Physics, Oxford at Clarendon Press, 1965.
37. Arajs, S., J. Less-Common Metals, 4, 46-51, 1962.
38. Edwards, A.R. and Johnstone, S.T.M., J. Inst. Metals, 84 (8), 313-7, 1956.
39. Lagneborg, R. and Kaplow, R., ACTA Metallurgica, 15 (1), 13-24, 1967.
40. Kittel, C., Introduction to Solid State Physics, 3rd Ed., John Wiley, 1967.
41. Kirshenbaum, A.D. and Cahill, J.A., J. Inorg. and Nucl. Chem., 25 (2), 232-34, 1963.
42. Touloukian, Y.S. (Ed.), Thermophysical Properties of High Temperature Solid Materials, MacMillan, Vol. 1, 1967.
43. Griffel, M., Skochdopole, R.E., and Spedding, F.H., J. Chem. Phys., 25 (1), 75-9, 1956.
44. Gschneidner, K.A., Jr., Rare Earth Alloys, Van Nostrand, 1961.
45. Dreyfus, B., Goodman, B.B., Lacaze, A., and Trolliet, G., Compt. Rend., 253, 1764-6, 1961.

46. Spedding, F.H., Hanak, J.J., and Daane, A.H., Trans. AIME, 212, 379, 1958.
47. Buckel, W. and Wittig, J., Phys. Letters (Netherlands), 17(3), 187-8, 1965.
48. Deardorff, D.K. and Kata, H., Trans. AIME, 215, 876-7, 1959.
49. Panish, M.B. and Reif, L., J. Chem. Phys., 38 (1), 253-6, 1963.
50. Miller, A.E. and Daane, A.H., Trans. AIME, 230, 568-72, 1964.
51. Spedding, F.H. and Daane, A.H., USAEC Rept. IS-350, 22-4, 1961.
52. Montgomery, H. and Pells, G.P., Proc. Phys. Soc. (London), 78, 622-5, 1961.
53. Kaufman, L. and Clougherty, E.V., ManLabs, Inc., Semi-Annual Rept. No. 2, 1963.
54. Lounasmaa, O.V., Proc. 3rd Rare Earth Conf., 1963, Gordon and Breach, New York, 1964.
55. Baker, H., WADC TR 57-194, 1-24, 1957.
56. Reed, R.P. and Breedis, J.F., ASTM STP 387, pp. 60-132, 1966.
57. Hansen, M., Constitution of Binary Alloys, 2nd Edition, McGraw-Hill, p. 1268, 1958.
58. Smith, P.L., Conf. Phys. Basses Temp., Inst. Intern. du Froid, Paris, 281, 1956.
59. Powell, R.W. and Tye, R.P., J. Less-Common Metals, 3, 202-15, 1961.
60. Yamamoto, A.S., Lundin, C.E., and Nachman, J.F., Denver Res. Inst. Rept., NP-11023, 1961.
61. Oriena, R.A. and Jones, T.S., Rev. Sci. Instr., 25, 248-51, 1954.
62. Smith, J.F., Carlson, O.N., and Vest, R.W., J. Electrochem. Soc., 103, 409-13, 1956.
63. Edwards, J.W. and Marshal, A.L., J. Am. Chem. Soc., 62, 1382, 1940.
64. Morin, F.J. and Maita, J.P., Phys. Rev., 129 (3), 1115-20, 1963.
65. Pendleton, W.N., ASD-TDR-63-164, 1963.
66. Woerner, P.F. and Wakefield, G.F., Rev. Sci. Instr., 33 (12), 1456-7, 1962.
67. Walcott, N.M., Conf. Phys. Basses Temp., Inst. Intern. du Froid, Paris, 286, 1956.
68. White, G.K. and Woods, S.B., Phil. Trans. Roy. Soc. (London), A251 (995), 273-302, 1959.
69. Douglass, R.W. and Adkins, E.F., Trans. AIME, 221, 248-9, 1961.
70. Panish, M.B. and Reif, L., J. Chem. Phys., 37 (1), 128-31, 1962.
71. Bridgman, P.W., J. Am. Chem. Soc., 36 (7), 1344-63, 1914.
72. Slack, G.A., Phys. Rev., A139 (2), 507-15, 1965.
73. Sandenaw, T.A. and Gibney, R.B., J. Phys. Chem. Solids, 6 (1), 81-8, 1958.
74. Sandenaw, T.A., Olsen, C.E., and Gibney, R.B., Plutonium 1960, Proc. 2nd Intern. Conf. (Grison, E., Lord, W.B.H., and Fowler, R.D., Editors), 66-79, 1961.
75. Mulford, R.N.R., USAEC Rept. LA-2813, 1-11, 1963.
76. Goode, J.M., J. Chem. Phys., 26 (5), 1269-71, 1957.
77. Cable, J.W., Moon, R.M., Koehler, W.C., and Wollan, E.O., Phys. Rev. Letters, 12 (20), 553-5, 1964.
78. Murao, T., Progr. Theoret. Phys. (Kyoto), 20 (3), 277-86, 1958.
79. Grigor'ev, A.T., Sokolovskaya, E.M., Budennaya, L.D., Iyutina, I.A., and Maksimona, M.V., Zhur. Neorg. Khim., 1, 1052-63, 1956.
80. Daane, A.H., USAEC AECD-3209, 1950.
81. Weigel, F., Angew. Chem., 75, 451, 1963.
82. Nassau, K. and Broyer, A.M., J. Am. Ceram. Soc., 45 (10), 474-8, 1962.
83. McKeown, J.J., State Univ. of Iowa, Ph.D. Dissertation, 1-113, 1958.
84. Abdullaev, G.B., Mekhtiyeva, S.I., Abdinov, D.Sh., and Aliev, G.M., Phys. Letters, 23 (3), 215-6, 1966.
85. Wittig, J., Phys. Rev. Letters, 15 (4), 159, 1965.
86. Fukuroi, T. and Muto, Y., Tohoku Univ. Res. Inst. Sci. Rept., A8, 213-22, 1956.
87. Olette, M., Compt. Rend., 244, 1033-6, 1957.
88. Sheldon, E.A., and King, A.J., ACTA Cryst., 6, 100, 1953.
89. Eastman, E.D. and McGavock, W.C., J. Am. Chem. Soc., 59, 145-51, 1937.
90. Arajs, S. and Colvin, R.V., Phys. Rev., A136 (2), 439-41, 1964.
91. Roach, P.R. and Lounasmaa, O.V., Bull. Am. Phys. Soc., 7, 408, 1962.
92. Shchukarev, S.A., Semenov, G.A., and Rat'kovskii, I.A., Zh. Neorgan. Khim., 7, 469, 1962.
93. Pearson, W.B., A Handbook of Lattice Spacings and Structures of Metals and Alloys, Pergamon Press, 1958.

94. Smith, P.L. and Walcott, N.M., Conf. Phys. Basses Temp., Inst. Intern. du Froid, 283, 1956.

95. Davis, D.D. and Bozorth, R.M., Phys. Rev., 118 (6), 1543-5, 1960.

96. Aliev, N.G. and Volkenstein, N.V., Soviet Physics - JETP, 22 (5), 997-8, 1966.

97. Spedding, F.H., Barton, R.J., and Daane, A.H., J. Am. Chem. Soc., 79, 5160, 1957.

98. Raynor, G.V. and Smith, R.W., Proc. Roy. Soc. (London), A244, 101-9, 1958.

99. Savitskii, E.M. and Burhkanov, G.S., Zhur. Neorg. Khim., 2, 2609-16, 1957.

100. Argent, B.B. and Milne, J.G.C., Niobium, Tantalum, Molybdenum and Tungsten, Elsevier Publ. Co. (Quarrell, A.G., Editor), pp. 160-8, 1961.

101. Argonne National Laboratory, USAEC Rept. ANL-5717, 1-67, 1957.

102. Holden, A.N., Physical Metallurgy of Uranium, Addison-Wesley, 1958.

103. Lounasmaa, O.V., Phys. Rev., 129, 2460-4, 1963.

104. Jennings, L.D., Miller, R.E., and Spedding, F.H., J. Chem. Phys., 33 (6), 1849-52, 1960.

105. Ackerman, R.J. and Rauh, E.G., J. Chem. Phys., 36 (2), 448-52, 1962.

106. Rosenblatt, G.M. and Birchenall, C.E., J. Chem. Phys., 35 (3), 788-94, 1961.

107. Streib, W.E., Jordan, T.H., and Lipscomb, W.N., J. Chem. Phys., 37 (12), 2962-5, 1962.

108. Samsonov, G.V. (Editor), Handbook of the Physicochemical Properties of the Elements, Plenum Press, 1968.

109. Kopp, I.Z., Russ. J. Phys. Chem., 41 (6), 782-3, 1967.

110. Stimson, H.F., in Temperature, Its Measurement and Control in Science and Industry (Herzfeld, C.M., Ed.), Vol. 3, Part 1, Reinhold, New York, pp. 59-66, 1962.

111. McLaren, E.H., in Temperature, Its Measurement and Control in Science and Industry (Herzfeld, C.M., Ed.), Vol. 3, Part 1, Reinhold, New York, pp. 185-98, 1962.

112. Orlova, M.P., in Temperature, Its Measurement and Control in Science and Industry (Herzfeld, C.M., Ed.), Vol. 3, Part 1, Reinhold, New York, pp. 179-83, 1962.

113. Grosse, A.V., J. Inorg. Nucl. Chem., 28, 2125-9, 1966.

114. Hochman, J.M. and Bonilla, C.F., in Advances in Thermophysical Properties at Extreme Temperatures and Pressures (Gratch, S., Ed.), ASME 3rd Symposium on Thermophysical Properties, Purdue University, March 22-25, 1965, ASME, pp. 122-30, 1965.

115. Dillon, I.G., Illinois Institute of Technology, Ph.D. Thesis, June 1965.

116. Hochman, J.M., Silver, I.L., and Bonilla, C.F., USAEC Rept. CU-2660-13, 1964.

117. Abdullaev, G.B., Mekhtieva, S.I., Abdinov, D.Sh., Aliev, G.M., and Alieva, S.G., Phys. Status Solidi, 13(2), 315-23, 1966.

118. Fisher, E.S. and Dever, D., Phys. Rev., 2, 170(3), 607-13, 1968.

119. Beaudry, B.J., J. Less-Common Metals, 14(3), 370-2, 1968.

120. Williams, R.K. and McElroy, D.L., USAEC Rept. ORNL-TM 1424, 1-32, 1966.

121. Jaeger, F.M. and Rosenbaum, E., Proc. Nederland Akademie van Wetenschappen, 44, 144-52, 1941.

122. Gibson, J.W. and Hein, R.A., Phys. Letters, 12(25), 688-90, 1964.

123. "The International Practical Temperature Scale of 1968, adopted by the Comité International des Poids et Mesures," Metrologia, 5 (2), 35-44, 1969.

124. Cezairliyan, A., Morse, M.S. and Beckett, C.W., Rev. Int. Hautes Tempér. et Réfract., 7 (4), 382-8, 1970.

125. Falge, R.L., Phys. Letters (Netherlands), 24A, 579-80, 1967.

126. Wittig, J., Phys. Rev. Letters, 21, 1250-2, 1968.

127. Berman, I.V. and Brandt, N.B., J.E.T.P. Letters, 7 (11), 323-6, 1968.

128. Matthias, B.T., Geballe, T.H., Corenzwit, E., Andres, K., and Hall, G.W., Science, 151, 985-6, 1966.

129. Meaden, G.T., Metallurgical Rev., 13 (125), 97-114, 1968.

130. Meaden, G.T. and Sze, N.H., "Fluctuations and Critical Indices Close to the Néel Temperature of Europium Metal," presented at the International Colloquium on the Rare Earth Elements, Paris-Grenoble, May 1969.

131. Meaden, G.T., Proc. Roy. Soc., 276, 553-70, 1963.

132. Grosse, A.V., Research Institute of Temple Univ., Report on USAEC Contract No. AT (30-1)-2082, 1-71, 1965.

133. Lee, J.A., "A Review of the Physical Metallurgy of Neptunium," in Progress in Nuclear Energy, Series V. Metallurgy and Fuels, Vol. 3, Pergamon Press, New York, 453-67, 1961.

134. Mathews, J.F., Chem. Rev., 72(1), 71-100, 1972.

Numerical Data on Thermal Diffusity

1. ELEMENTS

FIGURE AND TABLE 1R. RECOMMENDED THERMAL DIFFUSIVITY OF ALUMINUM

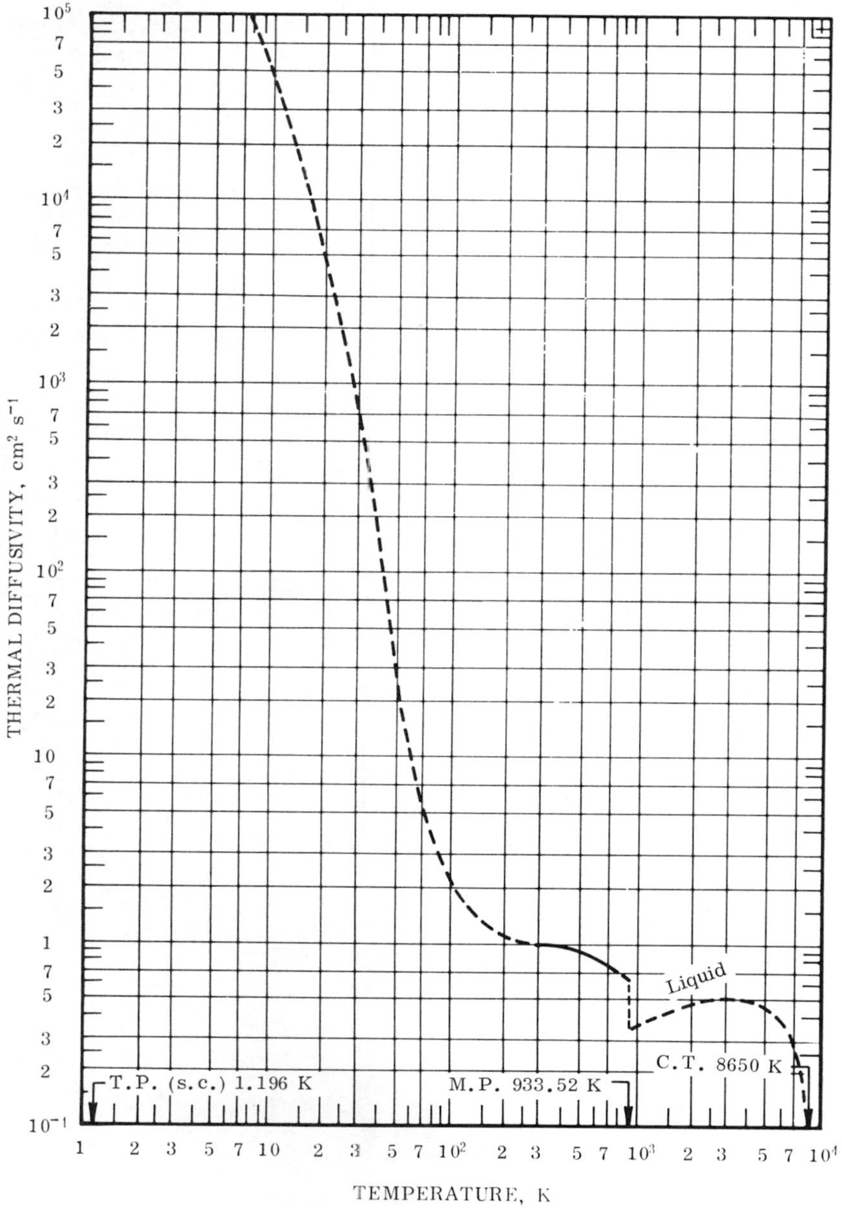

RECOMMENDED VALUES †

[Temperature, T, K; Thermal Diffusivity, α, cm² s⁻¹]

SOLID				LIQUID	
T	α	T	α	T	α
2	193000	40	150*	933.52	0.352*
3	234000	45	71.2*	1000	0.364*
4	208000	50	35.8*	1100	0.380*
5	175000*	60	12.2*	1200	0.395*
6	143000*	70	6.19*	1300	0.410*
7	115000*	80	3.94*	1400	0.424*
8	91800*	90	2.87*	1500	0.436*
9	72600*	100	2.28*	1600	0.448*
10	56500*	150	1.32*	1700	0.459*
11	44000*	200	1.09*	1800	0.468*
12	34100*	250	1.00*	1900	0.477*
13	26300*	273.2	0.982*	2000	0.485*
14	20300*	300	0.968	2200	0.498*
15	15700*	350	0.952	2400	0.508*
16	12300*	400	0.936	2600	0.516*
18	7920*	500	0.888	2800	0.522*
20	5100*	600	0.837	3000	0.527*
25	1900*	700	0.784	3500	0.527*
30	765*	800	0.736	4000	0.515*
35	333*	900	0.692*	4500	0.497*
		933.52	0.680*	5000	0.473*
				6000	0.402*
				7000	0.297*
				8000	0.140*

REMARKS

The recommended values are for well-annealed high-purity aluminum and are thought to be accurate to within ± 7% of the true values at temperatures below room temperature and ± 4% above. For molten aluminum near the melting point the values are probably good to within ± 8%. Values above 1200 K are provisional. The thermal diffusivity values below 150 K are applicable only to a specimen having residual electrical resistivity of 0.000594 μΩ cm.

†Values above 1200 K are provisional.

*In temperature range where no experimental data are available.

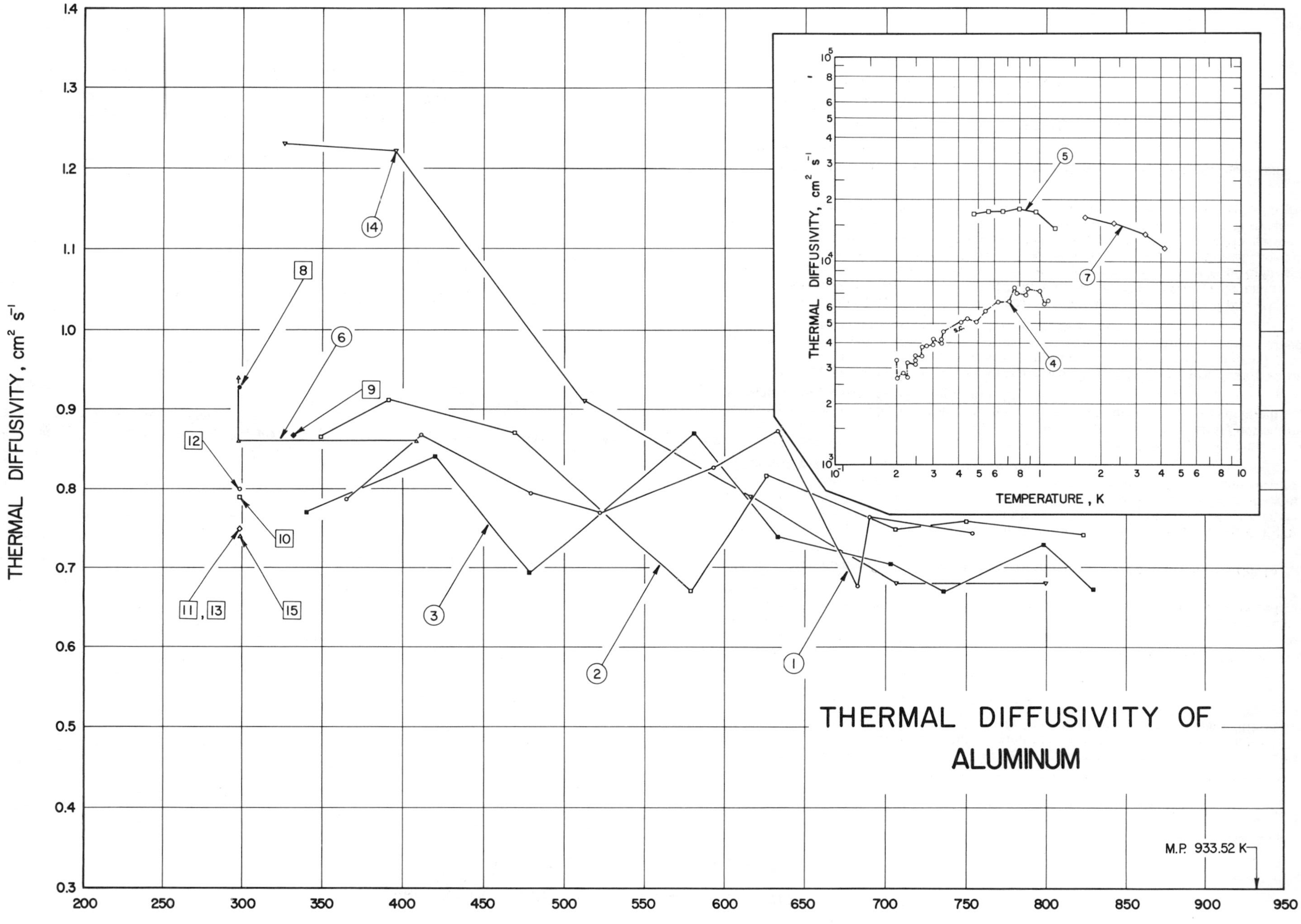
THERMAL DIFFUSIVITY OF ALUMINUM
THERMAL DIFFUSIVITY, cm² s⁻¹
TEMPERATURE, K
M.P. 933.52 K
1
2
3
4
5
6
7
8
9
10
11, 13
12
14
15
s.c.

FIGURE 1

SPECIFICATION TABLE 1. THERMAL DIFFUSIVITY OF ALUMINUM

(Impurity < 0.20% each; total impurities < 0.50%)

Cur. No.	Ref. No.	Author(s)	Year	Temp. Range, K	Reported Error, %	Name and Specimen Designation	Composition (weight percent), Specifications, and Remarks
1	1	Sonnenschein, G. and Winn, R.A.	1960	365-754			99.97 Al, 0.01 Cu, 0.01 Fe, and 0.01 Si; cylindrical specimen 0.635 cm in diameter; front surface covered with fine film of lamp black; measured under a vacuum of $\sim 10^{-4}$ mm Hg; one-dimensional heat flow; flash method used to measure diffusivity.
2	1	Sonnenschein, G. and Winn, R.A.	1960	349-823			Above specimen measured again for diffusivity.
3	1	Sonnenschein, G. and Winn, R.A.	1960	340-829			Above specimen measured again for diffusivity.
4	2	Zavaritskii, N.V.	1958	0.20-1.1		Al-2	0.0001 impurity; large crystals; cylindrical specimen ~1.5 mm in dia and 100 mm long; annealed in vacuum at ~873.2 K for 4 hrs before measurement; experiment carried out over a period of several days; critical temp 1.17 K; Debye temp. 375 K; measured in the superconducting state and in a magnetic field which as compensated to 0.2 oersted.
5	2	Zavaritskii, N.V.	1958	0.48-1.20		Al-2	Above specimen measured in the normal state in a field of 115 oersted parallel to the axis of the specimen.
6	4	Jenkins, R.J. and Parker, W.J.	1961	295-408	± 5		Square specimen 1.9 cm side and 0.352 cm thick; high intensity short duration light pulse absorbed in front surface of thermally insulated specimen coated with camphor black; 408.2 K measurements obtained by heating specimen holder and specimen with infrared lamp; both data points at 295.2 K obtained from measurements using different equations for data reduction.
7	5	Howling, D.H., Mendoza, E., and Zimmerman, J.E.	1955	1.7-4.1	~1	Al I	99.998 pure; cylindrical specimen 2.00 mm in diameter and 9.88 cm long; mass ~0.84 g; annealed at 773.2 K for 5 hrs in vacuum; Debye temp. 375 K; diffusivity determined from measured velocity of propagation of sinusoidal temp. waves in specimen; measured in vacuum of $\sim 2 \times 10^{-6}$ mm Hg.
8	6	Moser, J.B. and Kruger, O.L.	1963	298.2			Plate specimen with surface area lying in the range from 1 to 4 cm^2 and thickness in the range from 0.1 to 0.3 cm; front surface thinly coated with colloidal graphite; irradiated with a pulse of thermal energy of short duration; diffusivity determined from measured history of the back surface temperature.
9	161	Batalov, V.S. and Peletskii, V.E.	1968	331.8			Wire specimen 27 cm long; upper end of specimen embedded in bottom of plexiglass vessel serving as a channel for flow of heat-exchanging fluid while lower end is free to expand in a vacuum chamber; mirror reflecting a coherent ray of light from source of Michelson interferometer attached to specimen 10 cm from embedded end; thermal diffusivity determined from measured time dependence of the elongation rate of specimen; measured in a vacuum of $\sim 10^{-3}$ mm Hg; temperature of measurement not reported by authors but assumed to be equal to that of the heat-exchanging fluid.
10	88	Jacovelli, P.B. and Zinke, O.H.	1966	298.2			Foil specimen 0.013 cm thick; heat pulse generated by electric current; measuring temperature assumed 25 C.
11	88	Jacovelli, P.B. and Zinke, O.H.	1966	298.2			Similar to above but specimen thickness 0.020 cm.
12	88	Jacovelli, P.B. and Zinke, O.H.	1966	298.2			Similar to above but specimen thickness 0.025 cm.
13	88	Jacovelli, P.B. and Zinke, O.H.	1966	298.2			Similar to above but specimen thickness 0.038 cm.

SPECIFICATION TABLE 1. THERMAL DIFFUSIVITY OF ALUMINUM (continued)

Cur. No.	Ref. No.	Author(s)	Year	Temp. Range, K	Reported Error, %	Name and Specimen Designation	Composition (weight percent), Specifications, and Remarks
14	54	Mikryukov, V.E. and Karagezyan, A.G.	1961	326-799			No details reported.
15	56	Steinberg, S., Larsen, R.E., and Kydd, A.R.	1963	298.2			Specimen size 2.0 cm^2 x 0.406 cm; measured by a flash method; measuring temperature not reported, here assumed as 25 C.

DATA TABLE 1. THERMAL DIFFUSIVITY OF ALUMINUM

(Impurity < 0.20% each; total impurities < 0.50%)

[Temperature, T, K; Thermal Diffusivity, α, $cm^2\ s^{-1}$]

T	α
CURVE 1	
365.2	0.787
411.2	0.867
479.2	0.795
522.2	0.770
593.2	0.827
633.2	0.873
683.2	0.677
690.2	0.765
754.2	0.745
CURVE 2	
349.2	0.865
391.2	0.910
469.2	0.870
579.2	0.670
626.2	0.817
706.2	0.750
750.2	0.760
823.2	0.743
CURVE 3	
340.2	0.770
420.2	0.840
478.2	0.693
581.2	0.870
633.2	0.740
703.2	0.705
736.2	0.670
798.2	0.730
829.2	0.673
CURVE 4	
0.200	3300
0.200	2650
0.215	2850
0.225	2700
0.225	3200
0.245	3150
0.245	3300
CURVE 4 (cont.)	
0.245	3500
0.265	3450
0.265	3850
0.280	3900
0.300	3900
0.300	4200
0.330	4000
0.330	4250
0.335	4600
0.410	5100
0.440	5300
0.490	5100
0.540	5800
0.620	6400
0.710	6400
0.750	7500
0.770	7100
0.855	6900
0.875	7400
1.00	7200
1.05	6300
1.10	6500
CURVE 5	
0.475	17000
0.560	17500
0.660	17500
0.790	18000
0.960	17500
1.200	14500
CURVE 6	
295.2	0.94
295.2	0.86
408.2	0.86
CURVE 7	
1.68	16500
2.34	15500
3.31	13600
4.14	11600
CURVE 8	
298.2	0.928
CURVE 9	
331.8	0.866
CURVE 10	
298.2	0.79
CURVE 11	
298.2	0.75
CURVE 12	
298.2	0.80
CURVE 13	
298.2	0.75
CURVE 14	
326	1.23
395	1.22
512	0.91
616	0.79
706	0.68
799	0.68
CURVE 15	
298.2	0.74

FIGURE AND TABLE 2R. PROVISIONAL THERMAL DIFFUSIVITY OF ANTIMONY

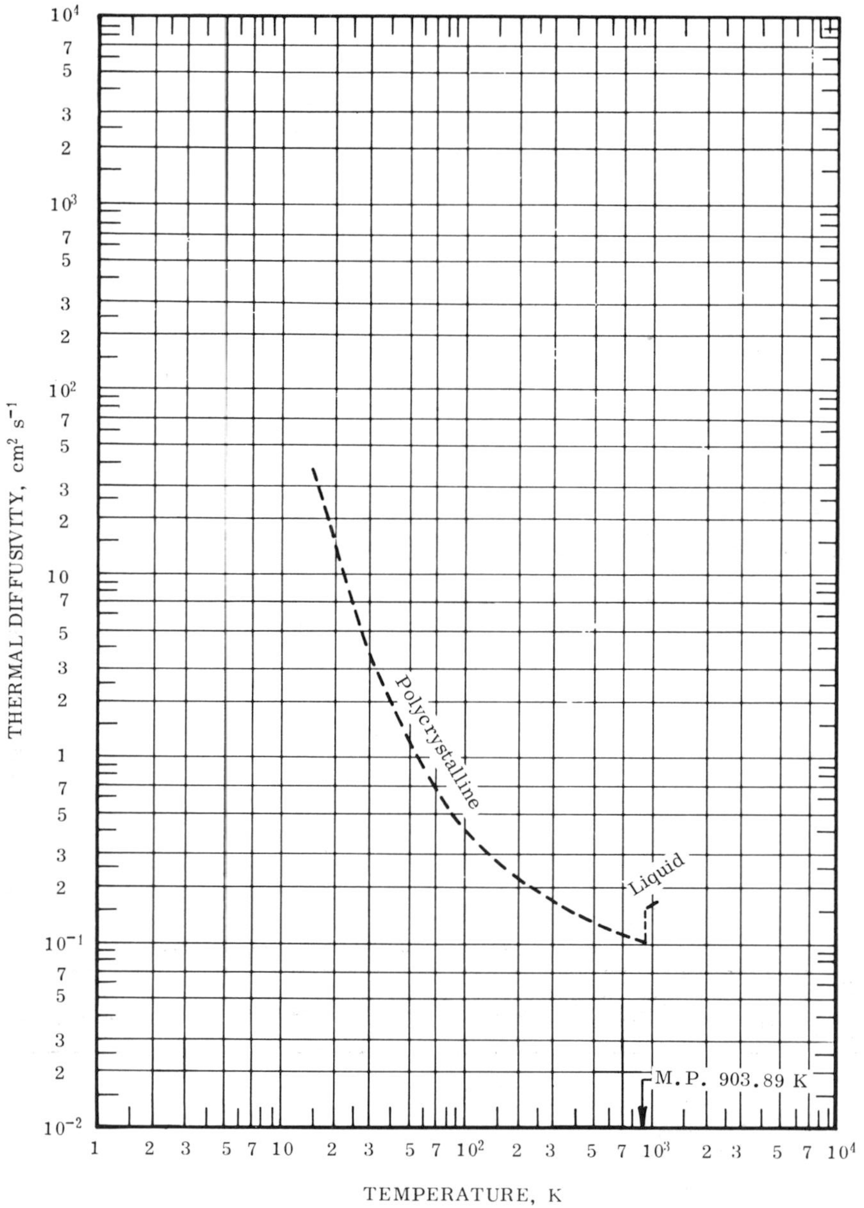

PROVISIONAL VALUES†

[Temperature, T, K; Thermal Diffusivity, α, $cm^2\ s^{-1}$]

SOLID		LIQUID	
T	α	T	α
15	37.0*	903.89	0.155*
16	29.2*	1000	0.163*
18	19.6*	1100	0.170
20	13.9*		
25	6.92*		
30	4.01*		
35	2.71*		
40	1.98*		
45	1.53*		
50	1.23*		
60	0.861*		
70	0.661*		
80	0.540*		
90	0.462*		
100	0.408*		
150	0.278*		
200	0.225*		
250	0.195*		
273.2	0.185*		
300	0.175*		
350	0.161*		
400	0.150*		
500	0.134*		
600	0.123*		
700	0.115*		
800	0.108*		
900	0.102*		
903.89	0.102*		

REMARKS

The values above 100 K are for well-annealed high-purity antimony and are considered accurate to within ±20% of the true values at moderate temperatures and ±30% near the melting point and above. The values below 100 K only represent a typical curve serving to indicate the general trend of the low-temperature behavior of the thermal diffusivity.

† Values below 100 K are merely typical values.

* In temperature range where no experimental data are available.

SPECIFICATION TABLE 2. THERMAL DIFFUSIVITY OF ANTIMONY

(Impurity < 0.20% each; total impurities < 0.50%)

Cur. No.	Ref. No.	Author(s)	Year	Temp. Range, K	Reported Error, %	Name and Specimen Designation	Composition (weight percent), Specifications, and Remarks
1*	95	Dutchak, Ya.I. and Panasyuk, P.V.	1966	1073.2			In liquid state; electrical conductivity 8.50, 8.28, and 8.00 x 10^2 Ω^{-1} cm^{-1} at 620, 700, and 800 C, respectively.

DATA TABLE 2. THERMAL DIFFUSIVITY OF ANTIMONY

(Impurity < 0.20% each; total impurities < 0.50%)

[Temperature, T, K; Thermal Diffusivity, α, cm^2 s^{-1}]

T	α
CURVE 1*	
1073.2	0.183

* No figure given.

FIGURE AND TABLE 3R. PROVISIONAL THERMAL DIFFUSIVITY OF ARSENIC

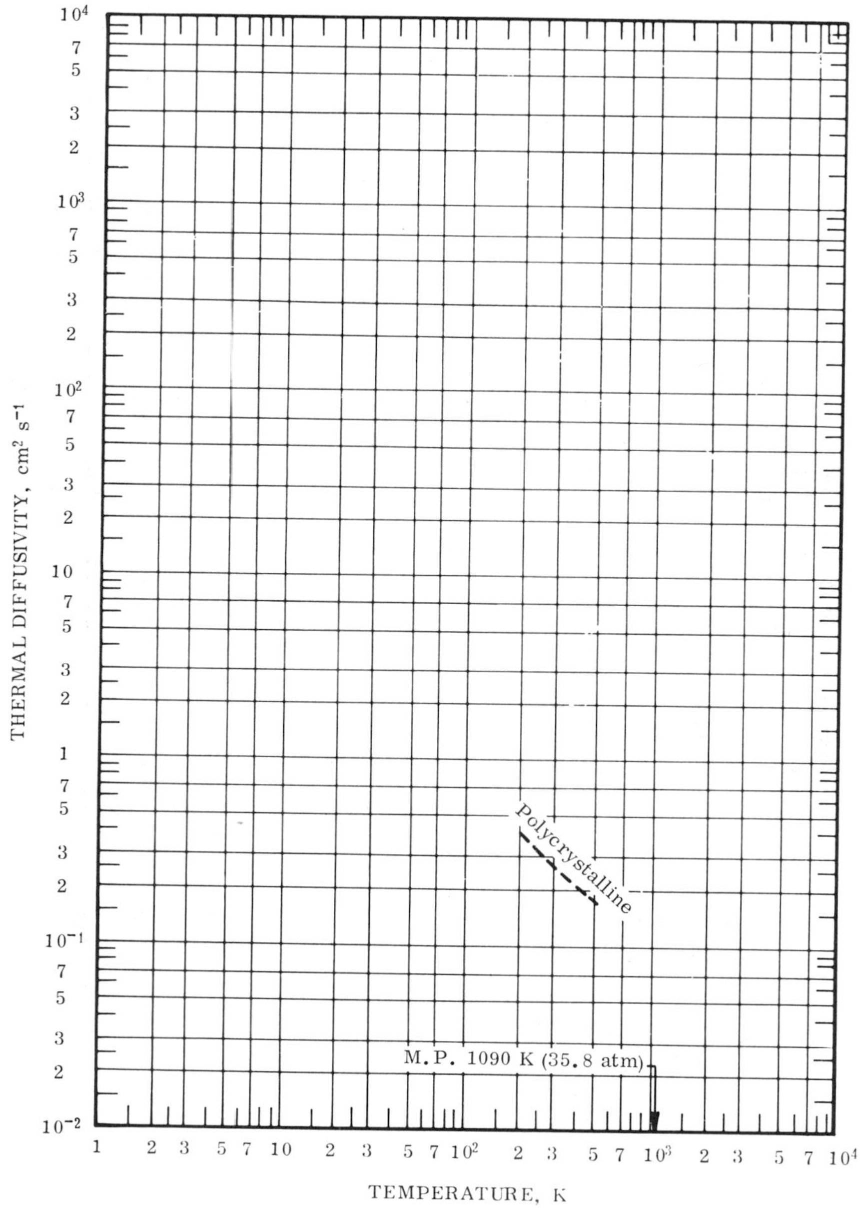

PROVISIONAL VALUES*

[Temperature, T, K; Thermal Diffusivity, α, $cm^2\ s^{-1}$]

SOLID (Gray, polycrystalline)

T	α
200	0.396
250	0.314
273.2	0.288
300	0.265
350	0.231
400	0.206
500	0.172

REMARKS

The provisional values are for well-annealed high-purity polycrystalline gray arsenic and should be good to within ±20%.

*All values are estimated.

FIGURE AND TABLE 4R. PROVISIONAL THERMAL DIFFUSIVITY OF BARIUM

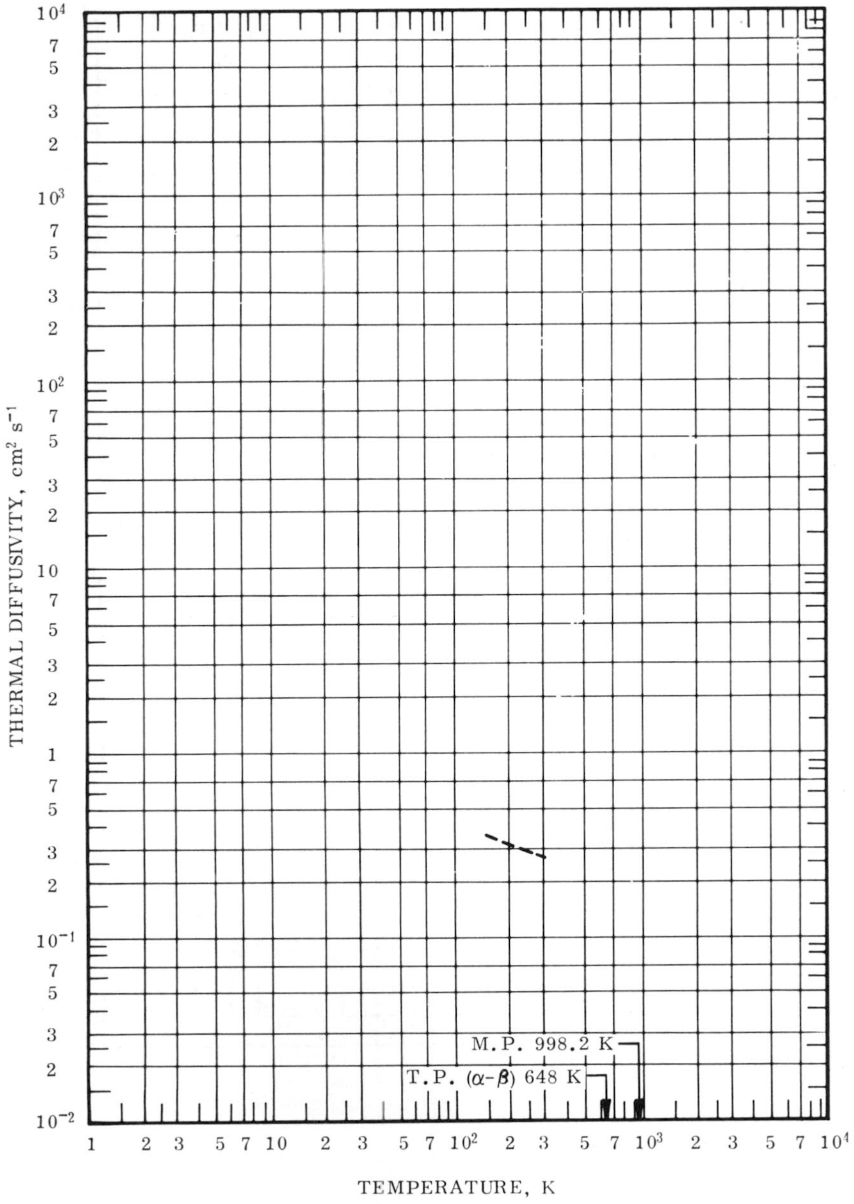

PROVISIONAL VALUES*

[Temperature, T, K; Thermal Diffusivity, α, cm^2 s^{-1}]

SOLID

T	α
150	0.352
200	0.314
250	0.287
273.2	0.279
298.2	0.274

REMARKS

The provisional values are for well-annealed high-purity barium and should be good to ± 25%.

* All values are estimated.

FIGURE AND TABLE 5R. PROVISIONAL THERMAL DIFFUSIVITY OF BERYLLIUM

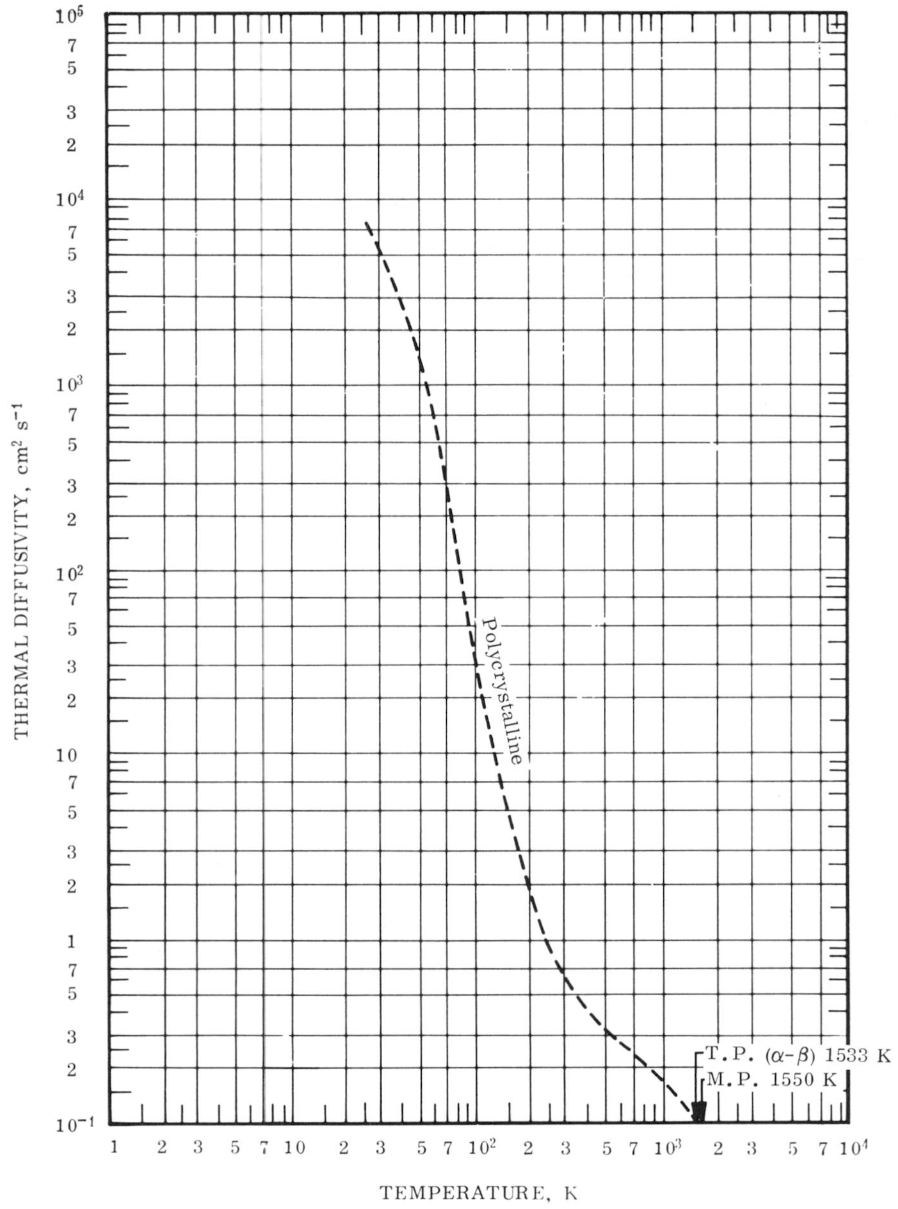

PROVISIONAL VALUES*

[Temperature, T, K; Thermal Diffusivity, α, cm^2 s^{-1}]

SOLID (Polycrystalline)

T	α	T	α
5	33900	60	416
6	32800	70	235
7	31100	80	110
8	29200	90	50.5
9	27200	100	26.2
10	25200	150	3.83
11	23500	200	1.46
12	21700	250	0.831
13	20100	273.2	0.692
14	18600	300	0.590
15	17200	350	0.471
16	15900	400	0.398
18	13700	500	0.315
20	11700	600	0.269
25	7720	700	0.235
30	5190	800	0.209
35	4580	900	0.187
40	2500	1000	0.168
45	1720	1100	0.152
50	1160	1200	0.138
		1300	0.126
		1400	0.115

REMARKS

The provisional values are for well-annealed high-purity beryllium and their uncertainty is thought to be of the order of ± 15% above room temperature and ± 25% below. The values below room temperature are only applicable to beryllium having residual electrical resistivity of 0.0135 μΩ cm.

*All values are estimated.

FIGURE AND TABLE 6R. RECOMMENDED THERMAL DIFFUSIVITY OF BISMUTH

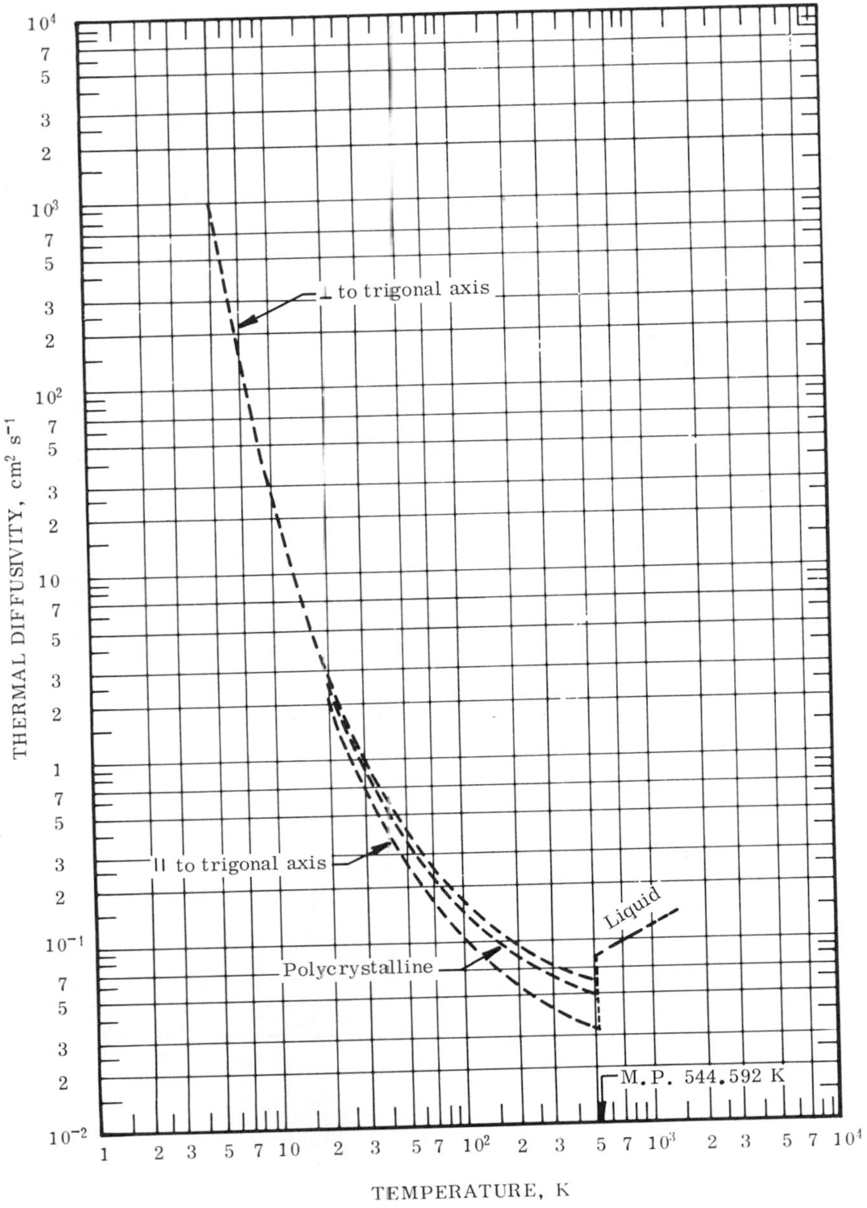

RECOMMENDED VALUES†

[Temperature, T, K; Thermal Diffusivity, α, cm² s⁻¹]

SOLID

T	∥ to trigonal axis α	⊥ to trigonal axis α	Poly-crystalline α
1		7610*	
2		8660*	
3		6930*	
4		3460*	
5		936*	
6		342*	
7		151*	
8		79.8*	
9		46.9*	
10		29.6*	
11		20.2*	
12		14.4*	
13		10.7*	
14		8.20*	
15		6.55*	
16		5.32*	
18		3.76*	
20	2.00*	2.86*	2.57*
25	1.16*	1.67*	1.50*
30	0.761*	1.12*	1.00*
35	0.550*	0.822*	0.727*
40	0.428*	0.639*	0.562*
45	0.346*	0.518*	0.459*
50	0.289*	0.436*	0.384*
60	0.215*	0.328*	0.287*
70	0.172*	0.264*	0.230*
80	0.144*	0.221*	0.194*
90	0.125*	0.193*	0.170*
100	0.110*	0.172*	0.151*
150	0.0720*	0.119*	0.103*
200	0.0566*	0.0952*	0.0823*
250	0.0485*	0.0827*	0.0714*
273.2	0.0458*	0.0788*	0.0680*
300	0.0434*	0.0748*	0.0646*
350	0.0398*	0.0695*	0.0596*
400	0.0373*	0.0654*	0.0562*
500	0.0337	0.0597	0.0510
544.592	0.0326*	0.0578*	0.0493*

LIQUID

T	α
544.592	0.0809*
600	0.0859
700	0.0940
800	0.102
900	0.109
1000	0.116*
1100	0.122*
1200	0.128*
1300	0.135*

REMARKS

The values are for well-annealed high-purity bismuth. The probable uncertainty of the recommended values (those for solid above 20 K) is of the order of ±10% at room temperature and above and ±15% below room temperature. The provisional values for the molten state are probably good to ±20%. The values below 20 K only represent a typical curve serving to indicate the general trend of the low-temperature behavior of the thermal diffusivity.

† Values for the molten state are provisional and those below 20 K are merely typical values.
* In temperature range where no experimental data are available.

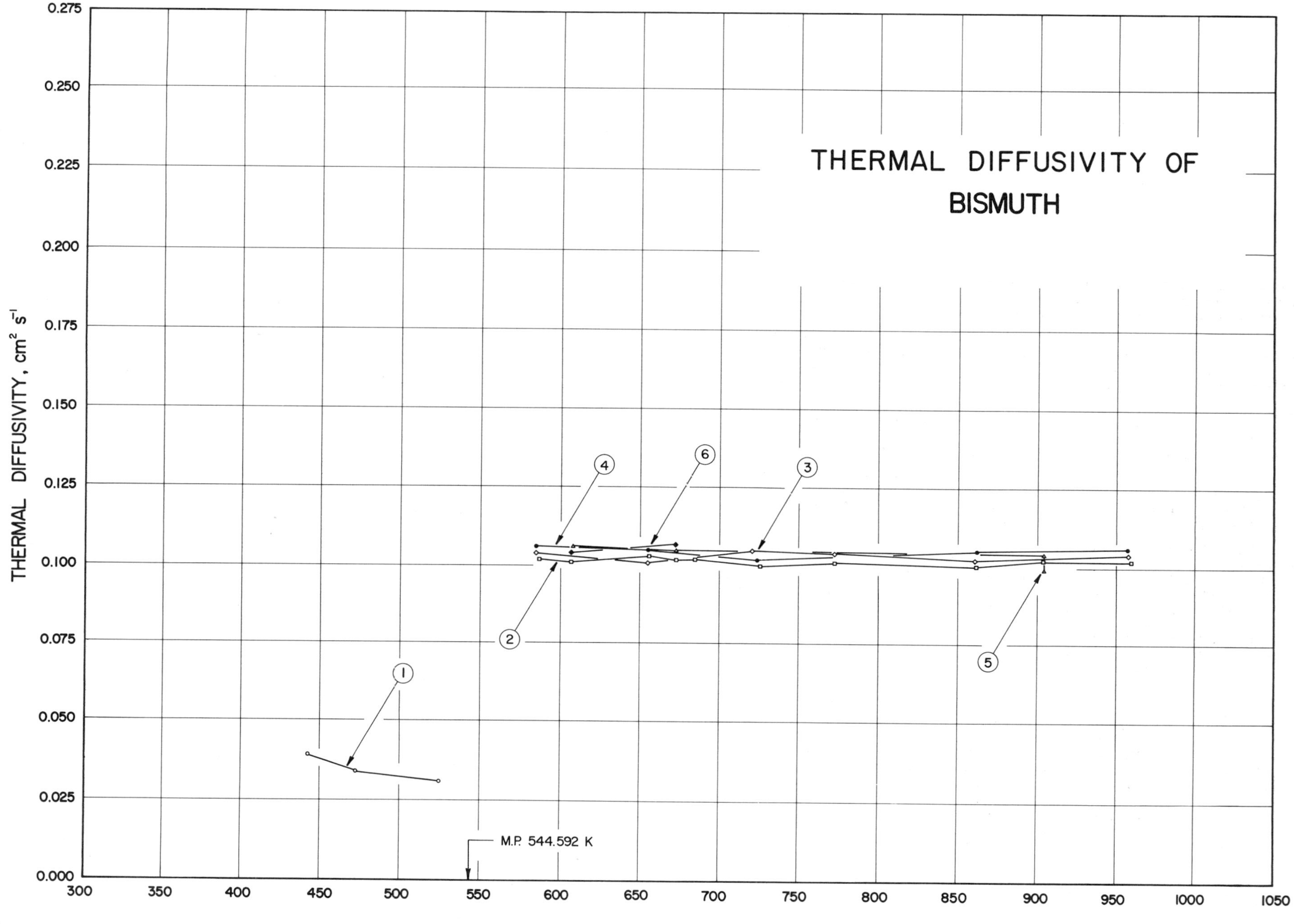
THERMAL DIFFUSIVITY OF BISMUTH
THERMAL DIFFUSIVITY, cm² s⁻¹
TEMPERATURE, K
0.000
0.025
0.050
0.075
0.100
0.125
0.150
0.175
0.200
0.225
0.250
0.275
300
350
400
450
500
550
600
650
700
750
800
850
900
950
1000
1050
M.P. 544.592 K
1
2
3
4
5
6

FIGURE 6

SPECIFICATION TABLE 6. THERMAL DIFFUSIVITY OF BISMUTH

(Impurity < 0.20% each; total impurities < 0.50%)

Cur. No.	Ref. No.	Author(s)	Year	Temp. Range, K	Reported Error, %	Name and Specimen Designation	Composition (weight percent), Specifications, and Remarks
1	109	Yurchak, R. P. and Filippov, L. P.	1964	443-525	7		Pure; cylindrical specimen; radial wave method used to measure diffusivity; diffusivity determined from measurements of temp. at two points on the specimen using a heating period of 6.6 sec; electrical resistivity measured and reported as 294.2, 276.1, 274.0, and 270.7 μohm cm at 293.2, 442.2, 468.2, and 500.2 K, respectively.
2	109	Yurchak, R. P. and Filippcv, L. P.	1964	587-960	7		Pure; in molten state; sample consists of a cylindrical tantalum crucible containing metal to be measured; horizontal partitions of tantalum plate impede convective mixing of liquid; outer surface of crucible subjected to periodic heating using a heating period of 6.6 sec; radial wave method used to measure diffusivity; diffusivity determined from measurements of temp. at two points on specimen; electrical resistivity measured and reported as 131.9, 132.4, 133.3, 138.4, 139.3, 143.7, 146.0, and 155.1 μohm cm at 565.2, 578.2, 593.2, 698.2, 728.2, 783.2, 848.2, and 998.2 K, respectively; Lorenz number reported as 3.38, 3.13, 2.99, 2.84, 2.69, 2.54, 2.42, 2.31, 2.22, and 2.20 x 10^{-8} V^2 K^{-2} at 573.2, 623.2, 673.2, 723.2, 773.2, 823.2, 873.2, 923.2, 973.2, and 1023.2 K, respectively.
3	109	Yurchak, R. P. and Filippov, L. P.	1964	585-958	7		Pure; in molten state; sample consists of a cylindrical tantalum crucible containing metal to be measured; diffusivity determined from the phase difference between the temp. waves measured at two points on specimen using a heating period of 13.2 sec; other conditions same as above.
4	109	Yurchak, R. P. and Filippov, L. P.	1964	585-957	7		Pure; in molten state; sample consists of a cylindrical tantalum crucible containing metal to be measured; diffusivity determined from the amplitude ratio of the temp. waves measured at two points on specimen using a heating period of 13.2 sec; other conditions same as above.
5	109	Yurchak, R. P. and Filippov, L. P.	1964	608-904	7		Pure; in molten state; sample consists of a cylindrical tantalum crucible containing metal to be measured; diffusivity determined from the amplitude ratio of the temp. waves measured at two points on specimen after resetting the thermocouples; other conditions same as above.
6	109	Yurchak, R. P. and Filippov, L. P.	1964	607, 673	7		Pure; in molten state; sample consists of a cylindrical tantalum crucible containing metal to be measured; diffusivity determined from the phase difference between the temp. waves measured at two points on specimen after resetting the thermocouples; other conditions same as above.

DATA TABLE 6. THERMAL DIFFUSIVITY OF BISMUTH

(Impurity < 0.20% each; total impurities < 0.50%)

[Temperature, T, K; Thermal Diffusivity, α, $cm^2\ s^{-1}$]

T	α	T	α
CURVE 1		CURVE 6	
443.2	0.039	607.2	0.104
473.2	0.034	673.2	0.107
525.2	0.031		
CURVE 2			
587.2	0.102		
607.2	0.101		
656.2	0.103		
673.2	0.102		
685.2	0.102		
726.2	0.100		
773.2	0.101		
862.2	0.100		
904.2	0.102		
960.2	0.102		
CURVE 3			
585.2	0.104		
655.2	0.101		
721.2	0.105		
773.2	0.104		
861.2	0.102		
958.2	0.104		
CURVE 4			
585.2	0.106		
655.2	0.105		
724.2	0.102		
862.2	0.105		
957.2	0.106		
CURVE 5			
608.2	0.106		
673.2	0.105		
904.2	0.104		
904.2	0.099		

FIGURE AND TABLE 7R. PROVISIONAL THERMAL DIFFUSIVITY OF BORON

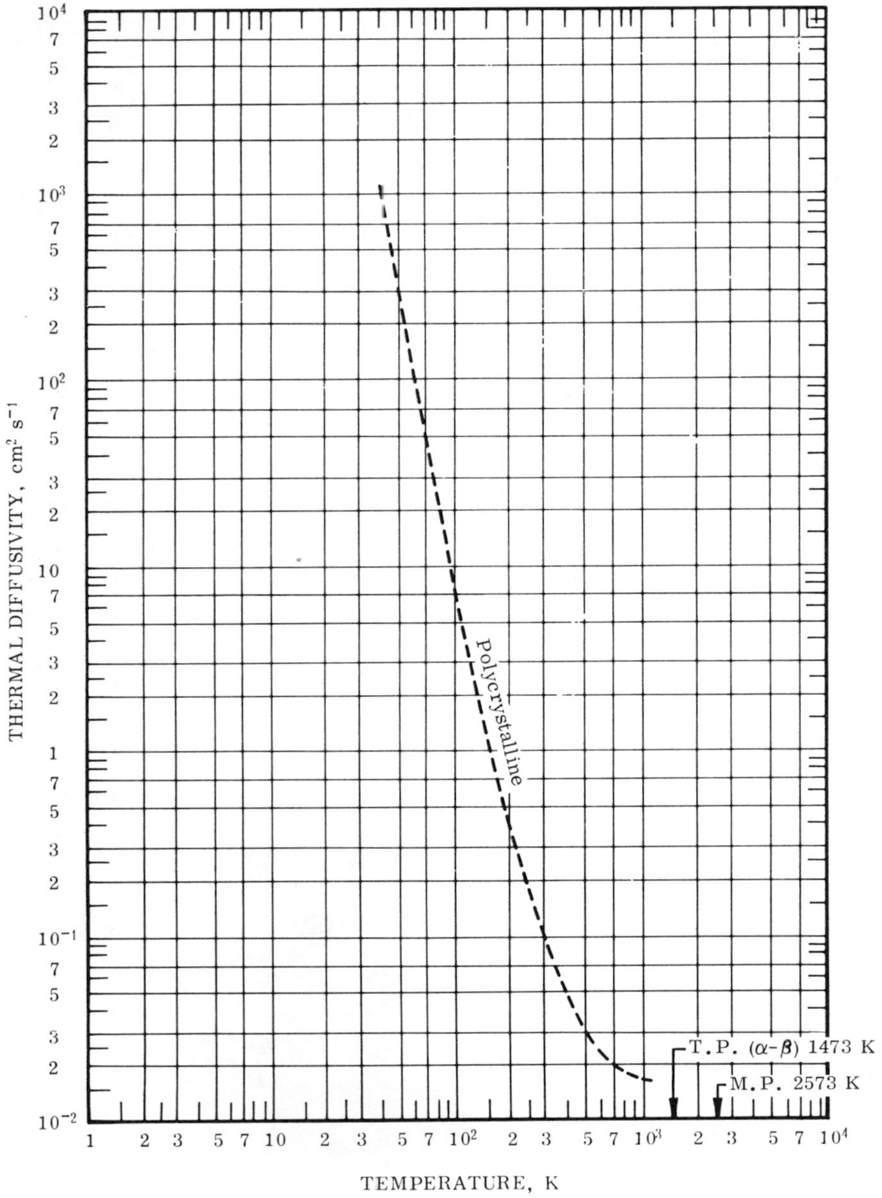

PROVISIONAL VALUES*

[Temperature, T, K; Thermal Diffusivity, α, cm^2 s^{-1}]

SOLID
(Polycrystalline)

T	α
40	1130
45	553
50	395
60	107
70	47.0
80	23.5
90	12.9
100	7.63
150	1.25
200	0.389
250	0.180
273.2	0.135
300	0.104
350	0.0693
400	0.0500
500	0.0319
600	0.0241
700	0.0205
800	0.0188
900	0.0178
1000	0.0175
1100	0.0174

REMARKS

The values are for well-annealed high-purity boron. The provisional values (those above 200 K) are probably accurate to within ± 15 to ± 20%. The values below 200 K only represent a typical curve serving to indicate the general trend of the low-temperature behavior of the thermal diffusivity.

*All values are estimated and those below 200 K are merely typical values.

FIGURE AND TABLE 8R. RECOMMENDED THERMAL DIFFUSIVITY OF CADMIUM

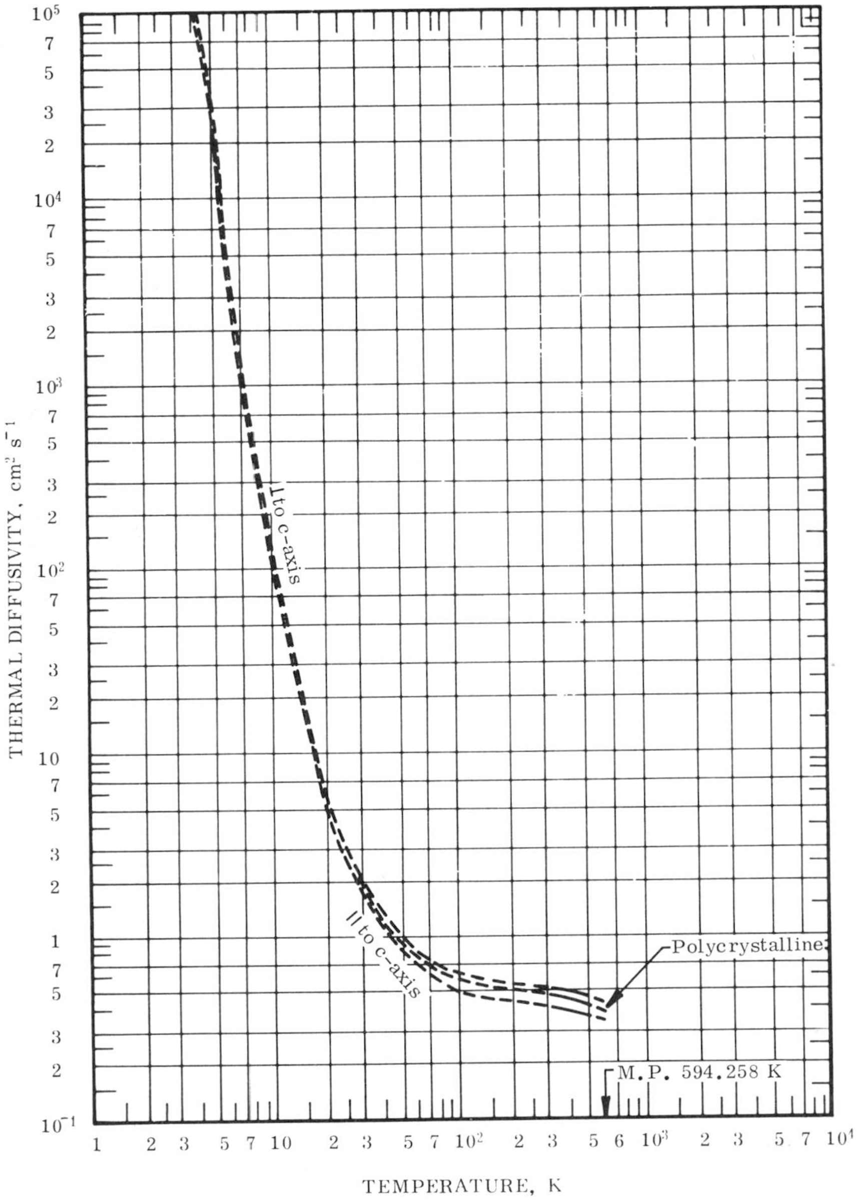

RECOMMENDED VALUES†

[Temperature, T, K; Thermal Diffusivity, α, cm^2 s^{-1}]

SOLID

T	∥ to ~~trigonal axis~~ c-axis α	⊥ to ~~trigonal axis~~ c-axis α	Poly-crystalline α	T	∥ to ~~trigonal axis~~ c-axis α	⊥ to ~~trigonal axis~~ c-axis α	Poly-crystalline α
1	2450000*	3220000*	3010000*	35	1.33*	1.65*	1.56*
2	1180000*	1520000*	1430000*	40	1.07*	1.33*	1.26*
3	448000*	548000*	522000*	45	0.908*	1.13*	1.09*
4	120000*	143000*	137000*	50	0.898*	0.994*	0.943*
5	27700*	29400*	29000*	60	0.668*	0.831*	0.787*
6	5800*	6310*	6860*	70	0.594*	0.741*	0.696*
7	1690*	1510*	1790*	80	0.550*	0.685*	0.646*
8	644*	742*	715*	90	0.523*	0.650*	0.614*
9	305*	338*	327*	100	0.503*	0.626*	0.591*
10	157*	175*	168*	150	0.456*	0.570*	0.537*
11	88.5*	99.0*	95.2*	200	0.437*	0.546*	0.515*
12	53.0*	59.0*	58.2*	250	0.424*	0.532*	0.498*
13	33.9*	38.0*	37.3*	273.2	0.418*	0.526*	0.492*
14	22.8*	25.7*	25.1*	300	0.412	0.519	0.486
15	16.2*	18.2*	17.9*	350	0.402	0.506	0.474
16	12.0*	13.6*	13.2*	400	0.392	0.492	0.460
18	7.25*	8.31*	8.06*	500	0.371	0.463	0.430
20	4.89*	5.75*	5.55*	594.258	0.346*	0.432*	0.398*
25	2.69*	3.21*	3.10*				
30	1.80*	2.18*	2.08*				

REMARKS

The values are for well-annealed high-purity cadmium and are thought to be accurate to within ± 25% at low temperatures, ± 6% at normal and moderate temperatures, and ± 10% near the melting point. The provisional values below 100 K for $\alpha_{\parallel}$, $\alpha_{\perp}$, and α_{poly} are applicable only to samples having residual electrical resistivities of 0.000134, 0.000103, and 0.000112 $\mu\Omega$ cm, respectively.

† Values below 100 K are provisional.

* In temperature range where no experimental data are available.

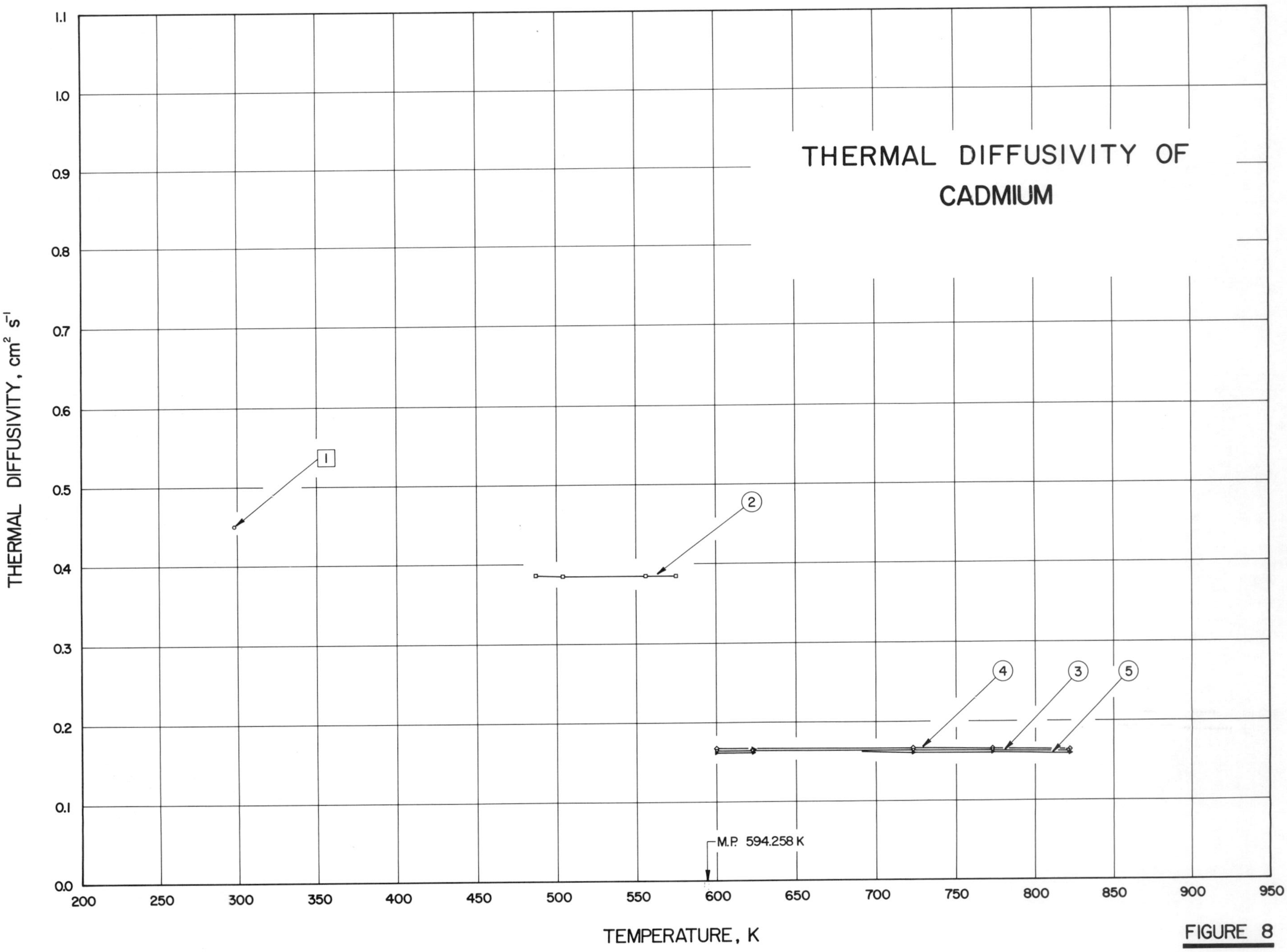
THERMAL DIFFUSIVITY OF
CADMIUM
THERMAL DIFFUSIVITY, cm^2 s^{-1}
TEMPERATURE, K
M.P. 594.258 K
0.0
0.1
0.2
0.3
0.4
0.5
0.6
0.7
0.8
0.9
1.0
1.1
200
250
300
350
400
450
500
550
600
650
700
750
800
850
900
950
1
2
3
4
5

FIGURE 8

SPECIFICATION TABLE 8. THERMAL DIFFUSIVITY OF CADMIUM

(Impurity < 0.20% each; total impurities < 0.50%)

Cur. No.	Ref. No.	Author(s)	Year	Temp. Range, K	Reported Error, %	Name and Specimen Designation	Composition (weight percent), Specifications, and Remarks
1	7	diNovi, R. A.	1963	298	10		Specimen with thickness lying in the range from 1 to 2 mm; front surface uniformly irradiated by a very short pulse of radiant energy supplied by a xenon flash tube; diffusivity determined from measured history of the back surface temp; temp. at which specimen was measured not given by author but assumed to be room temp.
2	109	Yurchak, R. P. and Filippov, L. P.	1964	487-575	7		Of analytical purity; cylindrical specimen; radial wave method used to measure diffusivity; diffusivity determined from measurements of temp. at two points on the specimen using a heating period of 6.6 sec.
3	109	Yurchak, R. P. and Filippov, L. P.	1964	600-821	7		Of analytical purity; in molten state; sample consists of a cylindrical tantalum crucible containing the metal to be measured; horizontal partitions of tantalum plate impede convective mixing of liquid; outer surface of crucible subjected to periodic heating using a heating period of 6.6 sec; radial wave method used to measure diffusivity; diffusivity determined from measurements of temp. at two points on specimen.
4	109	Yurchak, R. P. and Filippov, L. P.	1964	600-822	7		Of analytical purity; in molten state; sample consists of a cylindrical tantalum crucible containing the metal to be measured; horizontal partitions of tantalum plate impede convective mixing of liquid; outer surface of crucible subjected to periodic heating; radial wave method used to measure diffusivity; diffusivity determined from the phase difference between the temp. waves measured at two points on specimen using a heating period of 13.2 sec.
5	109	Yurchak, R. P. and Filippov, L. P.	1964	600-822	7		Of analytical purity; in molten state; sample consists of a cylindrical tantalum crucible containing the metal to be measured; horizontal partitions of tantalum plate impede convective mixing of liquid; outer surface of crucible subjected to periodic heating; radial wave method used to measure diffusivity; diffusivity determined from the amplitude ratio of the temp. waves measured at two points on specimen using a heating period of 13.2 sec.

DATA TABLE 8. THERMAL DIFFUSIVITY OF CADMIUM

(Impurity < 0.20% each; total impurities < 0.50%)

[Temperature, T, K; Thermal Diffusivity, α, $cm^2\ s^{-1}$]

T	α	T	α	T	α	T	α	T	α
CURVE 1		CURVE 2		CURVE 3		CURVE 4		CURVE 5	
298	0.450	487	0.386	600	0.164	600	0.166	600	0.162
		504	0.385	623	0.164	723	0.166	623	0.162
		556	0.385	723	0.163	773	0.165	623	0.166
		575	0.385	773	0.163	822	0.164	723	0.161
				821	0.162			773	0.161
								822	0.160

FIGURE AND TABLE 9R. PROVISIONAL THERMAL DIFFUSIVITY OF CALCIUM

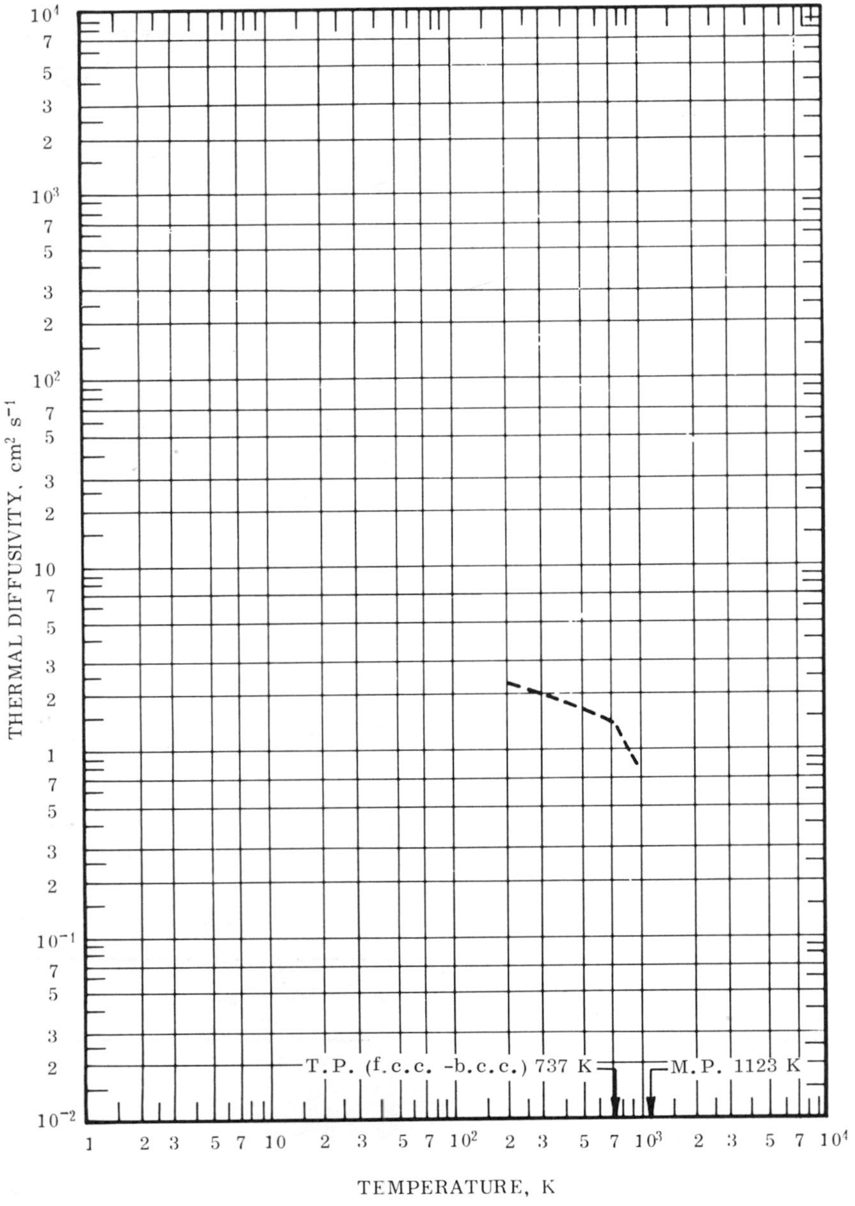

PROVISIONAL VALUES*

[Temperature, T, K; Thermal Diffusivity, α, $cm^2\ s^{-1}$]

SOLID

T	α
200	2.30
250	2.13
273.2	2.06
300	1.99
350	1.88
400	1.78
500	1.63
600	1.52
700	1.43
737.2	1.52
737.2	1.36
800	1.25
900	0.942
1000	0.810

REMARKS

The provisional values are probably good tc ±30%.

*All values are estimated.

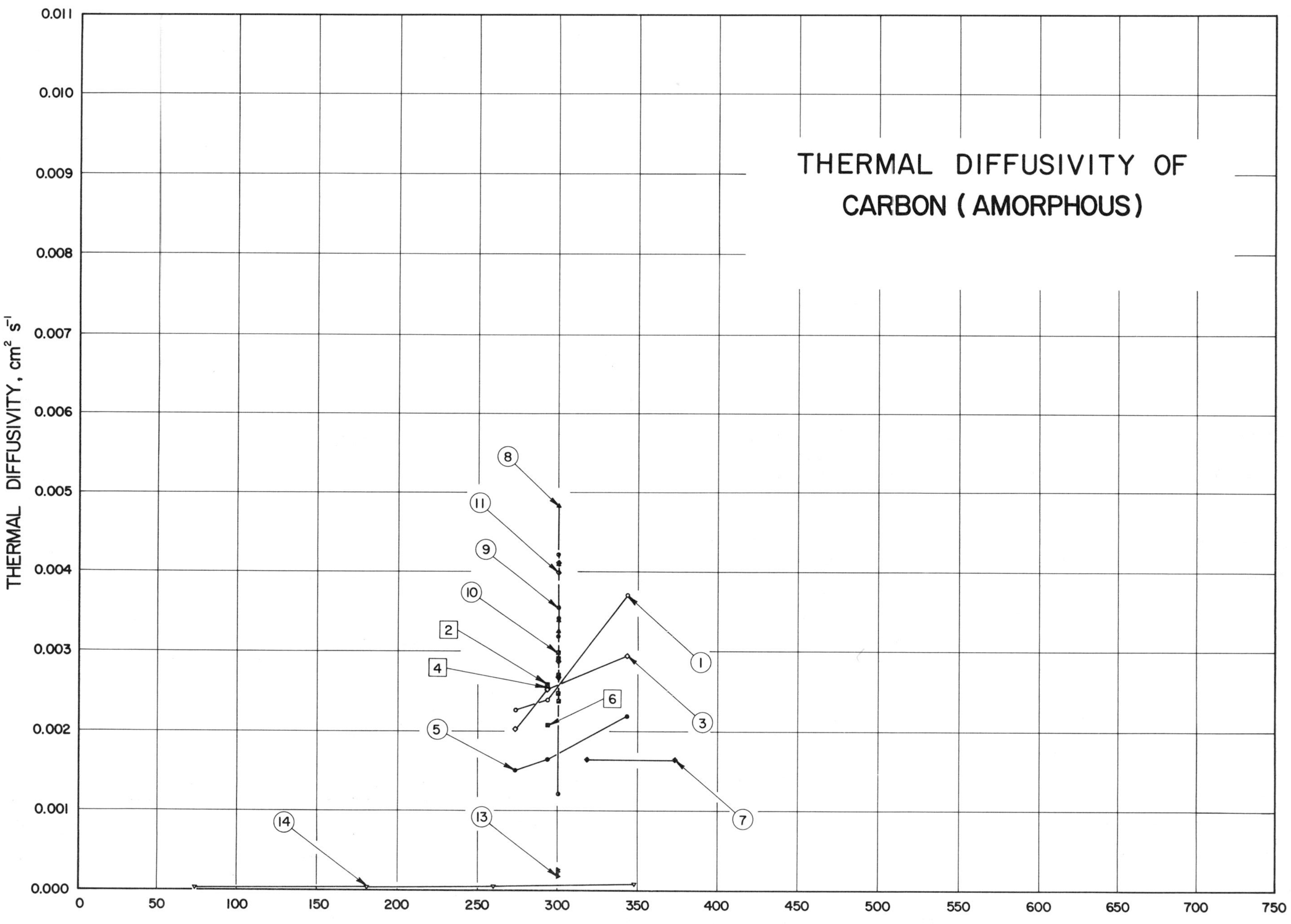
THERMAL DIFFUSIVITY OF
CARBON (AMORPHOUS)
THERMAL DIFFUSIVITY, cm² s⁻¹
TEMPERATURE, K
0.011
0.010
0.009
0.008
0.007
0.006
0.005
0.004
0.003
0.002
0.001
0.000
0
50
100
150
200
250
300
350
400
450
500
550
600
650
700
750
1
2
3
4
5
6
7
8
9
10
11
13
14

FIGURE 10

SPECIFICATION TABLE 10. THERMAL DIFFUSIVITY OF CARBON (AMORPHOUS)

(Impurity <0.20% each; total impurities <0.50%)

Cur. No.	Ref. No.	Author(s)	Year	Temp. Range, K	Reported Error, %	Name and Specimen Designation	Composition (weight percent), Specifications, and Remarks
1	164	Simonova, L.K.	1943	273-343		No. 1	Small granules; density 0.474 g cm^{-3}.
2	164	Simonova, L.K.	1943	293.2		No. 1	The above specimen powdered; density 0.686 g cm^{-3}.
3	164	Simonova, L.K.	1943	273-343		No. 2	Small granules; density 0.700 g cm^{-3}.
4	164	Simonova, L.K.	1943	293.2		No. 2	The above specimen powdered; density 0.810 g cm^{-3}.
5	164	Simonova, L.K.	1943	273-343		No. 3	Small granules; density 0.513 g cm^{-3}.
6	164	Simonova, L.K.	1943	293.2		No. 3	The above specimen powdered; density 0.630 g cm^{-3}.
7	166	Williams, I.	1923	318-373			Gas black; density 2.00 g cm^{-3}.
8	191	Zamoluev, V.K.	1960	300	< 1	Anthracite	Grain size < 0.5 mm; measured as a function of the heat-treating temperature, τ, ranging from 1273 to 2119 K.
9	210	Zamoluev, V.K., Kasatochkin V.I., Koverov, A.T., and Usenbaer, K.	1960	300	< 1	Petroleum coke	Grain size < 0.5 mm; density 1.405 g cm^{-3}; measured as a function of the heat-treating temperature, τ, ranging from room temperature to 2119 K.
10	191	Zamoluev, V.K.	1960	300		Gas coal	Grain size < 0.5 mm; measured as a function of the heat-treating temperature, τ, ranging from 1277 to 2118 K.
11	191	Zamoluev, V.K.	1960	300		Brown coal	Similar to above.
12*	191	Zamoluev, V.K.	1960	300		Channel black	Grain size < 1 μ; measured as a function of the heat-treating temperature, τ, ranging from room temperature to 1985 K.
13	210	Kasatochkin, V.I., Zamoluev, V.K., Kaverov, A.T., and Usenbaer, K.	1960	300		Thermal black	Grain size < 1 μ; measured as a function of the heat-treating temperature, τ, ranging from 1273 to 2373 K.
14	240	Kropschot, R.H., Knight, B.L., and Timmerhaus, K.D.	1968	74-298		Nerofil	Finely divided powder.

* Not shown in figure.

DATA TABLE 10. THERMAL DIFFUSIVITY OF CARBON (AMORPHOUS)

(Impurity < 0.20% each; total impurities < 0.50%)

[Temperature, T, K; Thermal Diffusivity, α, cm^2 s^{-1}]

T	α
CURVE 1	
273.2	0.002275
293.2	0.002381
343.2	0.003706
CURVE 2	
293.2	0.002582
CURVE 3	
273.2	0.002029
293.2	0.002523
343.2	0.002939
CURVE 4	
293.2	0.002575
CURVE 5	
273.2	0.001503
293.2	0.001641
343.2	0.002185
CURVE 6	
293.2	0.002067
CURVE 7	
318.2	0.00164
373.2	0.00164

τ(K)	α
CURVE 8 (T = 300 K)	
1273	0.00325
1370	0.00339
1775	0.00409
2119	0.00483
CURVE 9 (T = 300 K)	
300	0.00120
1280	0.00318
1475	0.00340
1775	0.00354
2119	0.00422
CURVE 10 (T = 300 K)	
1277	0.00237
1371	0.00246
1577	0.00269
1768	0.00290
2118	0.00297
CURVE 11 (T = 300 K)	
1273	0.00267
1368	0.00287
1773	0.00398
2119	0.00410
CURVE 12* (T = 300 K)	
300	0.00223
1273	0.00227
1768	0.00314
1985	0.00351
CURVE 13 (T = 300 K)	
1273	0.000172
1573	0.000172*
1973	0.000219
2373	0.000258

T	α
CURVE 14	
74	0.000020
181	0.000039
260	0.000049
298	0.000063

* Not shown in figure.

FIGURE AND TABLE 11R. TYPICAL THERMAL DIFFUSIVITY OF DIAMOND

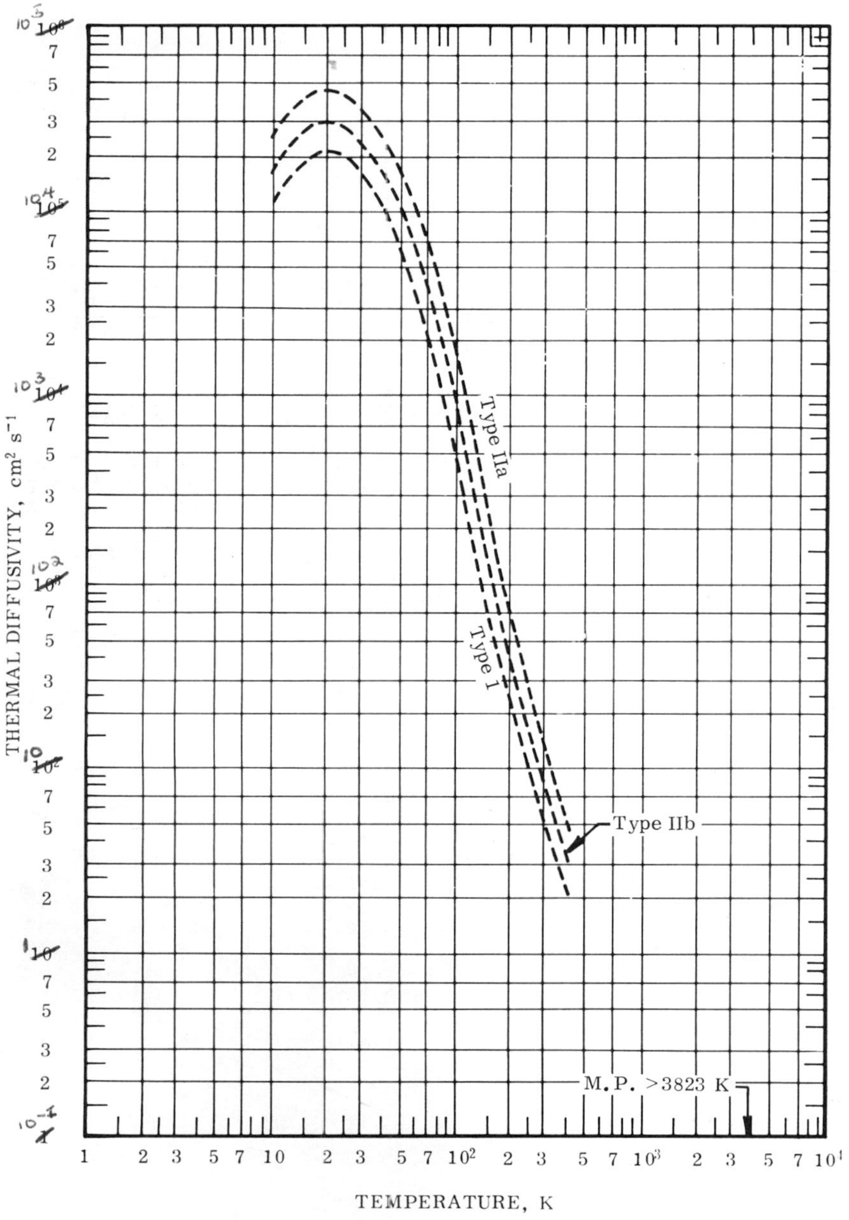

TYPICAL VALUES*

[Temperature, T, K; Thermal Diffusivity, α, cm^2 s^{-1}]

SOLID

T	Type I α	Type IIa α	Type IIb α
10	11500	25900	16600
11	13500	30000	19300
12	15300	33800	21900
13	16900	37100	24200
14	18300	39800	26100
15	19400	42100	27600
16	20200	43600	28900
18	21300	45600	30300
20	21500	45800	30500
25	19400	41500	27400
30	16000	34900	23200
35	12600	28500	19100
40	9700	23100	15400
45	7580	18800	12300
50	5780	15100	9700
60	3300	9860	5900
70	1890	6080	3530
80	1110	3600	2070
90	673	2200	1230
100	416	1390	751
150	66.2	207	116
200	20.6	59.0	33.1
250	9.09	24.5	14.0
273.2	6.70	17.9	10.2
300	4.96	12.7	7.43
350	3.10	7.66	4.60
400	2.17	5.14	3.11

REMARKS

The 3 sets of thermal diffusivity values only represent 3 typical curves serving to indicate the general trend of the thermal diffusivity of the three types of diamond.

*All values are estimated.

FIGURE AND TABLE 12R. RECOMMENDED THERMAL DIFFUSIVITY OF ATJ GRAPHITE

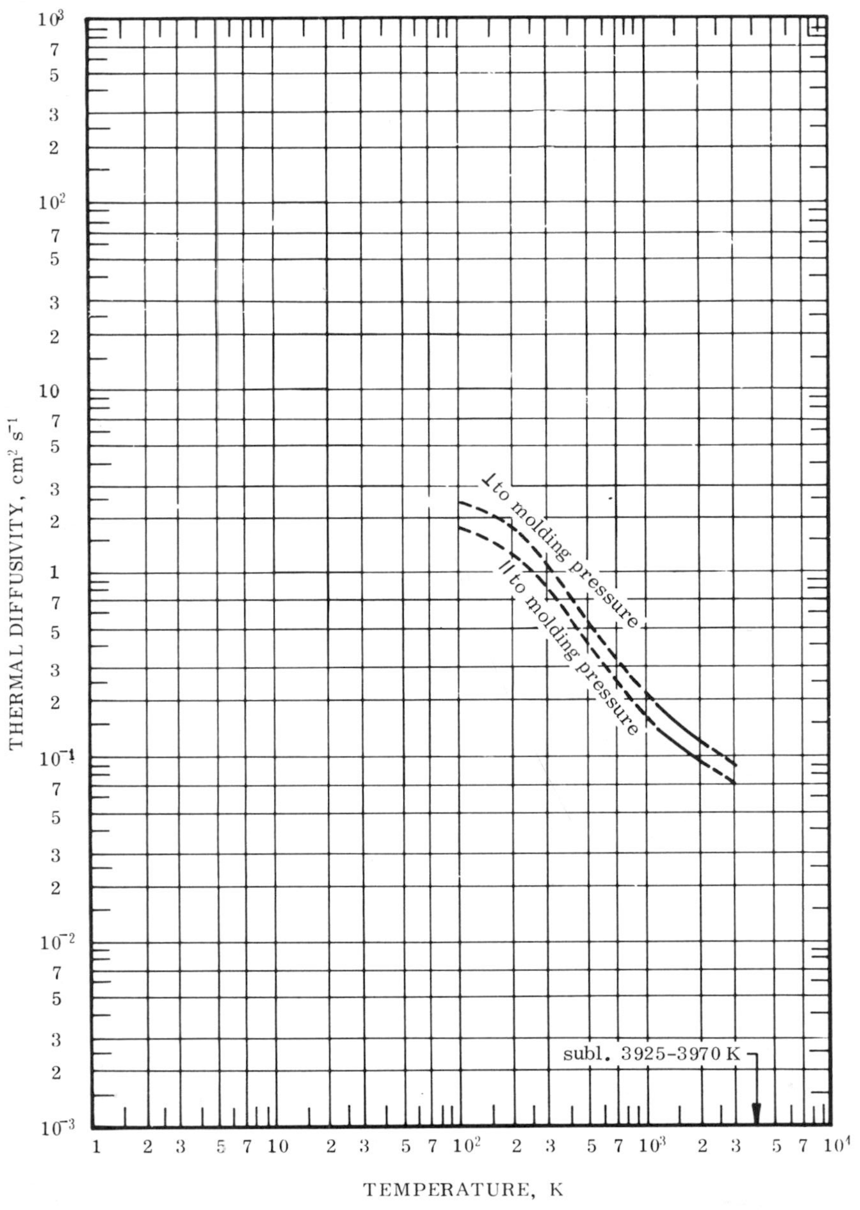

RECOMMENDED VALUES†

[Temperature, T, K; Thermal Diffusivity, α, cm^2 s^{-1}]

SOLID

T	to molding pressure α	to molding pressure α	T	to molding pressure α	to molding pressure α
100	1.70*	2.33*	1200	0.134	0.174
150	1.44*	2.02*	1300	0.125	0.160
200	1.20*	1.66*	1400	0.117	0.150
250	0.980*	1.32*	1500	0.111	0.141
273.2	0.888*	1.18*	1600	0.105	0.134
300	0.789*	1.04*	1700	0.101	0.128
350	0.638*	0.832*	1800	0.0974	0.123
400	0.526*	0.685*	1900	0.0939	0.118
500	0.383*	0.498*	2000	0.0911	0.114
600	0.298*	0.389*	2200	0.0868	0.108
700	0.244*	0.320*	2400	0.0827*	0.103*
800	0.207*	0.271*	2600	0.0785*	0.0970*
900	0.180*	0.235*	2800	0.0742*	0.0915*
1000	0.161*	0.210*	3000	0.0696*	0.0857*
1100	0.147	0.189			

REMARKS

The values at temperatures from 400 to 1500 K are recommended values for ATJ graphite and are thought to be accurate to within ±12%. Above 1500 K the values are provisional and are probably good to ±25%. The values below 400 K are merely typical values.

†Values at temperatures above 1500 K are provisional and below 400 K are merely typical values.
*In temperature range where no experimental data are available.

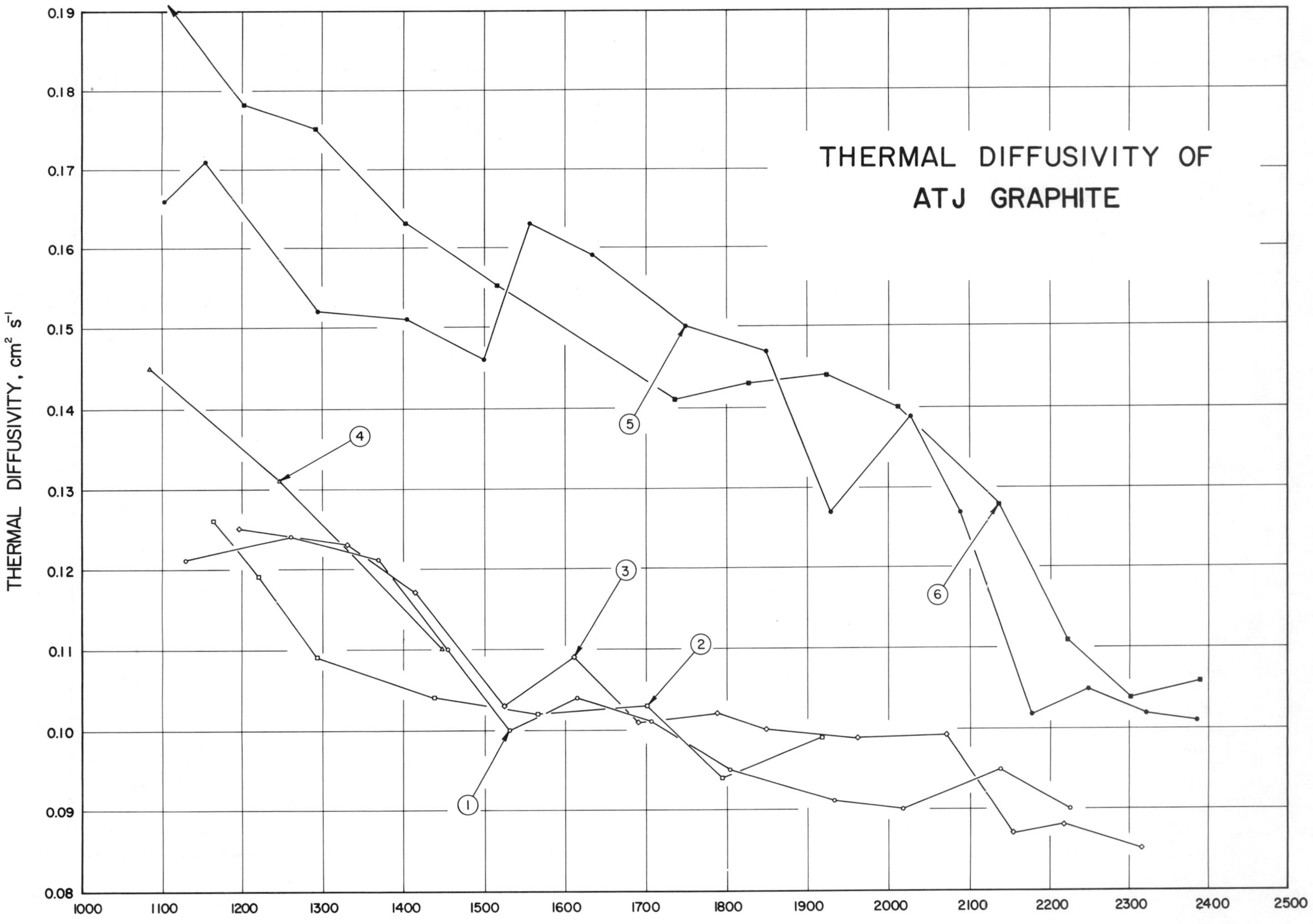
THERMAL DIFFUSIVITY OF
ATJ GRAPHITE
THERMAL DIFFUSIVITY, cm² s⁻¹
TEMPERATURE, K
0.08
0.09
0.10
0.11
0.12
0.13
0.14
0.15
0.16
0.17
0.18
0.19
1000
1100
1200
1300
1400
1500
1600
1700
1800
1900
2000
2100
2200
2300
2400
2500
1
2
3
4
5
6

FIGURE 12

SPECIFICATION TABLE 12. THERMAL DIFFUSIVITY OF ATJ GRAPHITE

(Impurity < 0.20% each; total impurities < 0.50%)

Cur. No.	Ref. No.	Author(s)	Year	Temp. Range, K	Reported Error, %	Name and Specimen Designation	Composition (weight percent), Specifications, and Remarks
1	213	Morrison, B.H., Klein, D.J., and Cowder, L.R.	1965	1129-2225		ATJ	Thin disk specimen 0.5 in. diameter and 0.030 in. thick; obtained from National Carbon Co.; bulk density 1.73 g cm^{-3}; heat flow across grain; measured by a flash method using laser pulse technique.
2	213	Morrison, B.H., et al.	1965	1164-1918		ATJ	Similar to above but specimen 0.040 in. thick.
3	213	Morrison, B.H., et al.	1965	1196-2315		ATJ	Similar to above but specimen 0.050 in. thick.
4	213	Morrison, B.H., et al.	1965	1083-1447		ATJ	Similar to above but specimen 0.080 in. thick.
5	213	Morrison, B.H., et al.	1965	1103-2385		ATJ	Thin disk specimen 0.5 in. diameter and 0.040 in. thick; obtained from National Carbon Co.; bulk density 1.73 g cm^{-3}; heat flow with grain; measured by a flash method using laser pulse technique.
6	213	Morrison, B.H., et al.	1965	1081-2389		ATJ	Similar to above but specimen 0.060 in. thick.

DATA TABLE 12. THERMAL DIFFUSIVITY OF ATJ GRAPHITE

(Impurity < 0.20% each; total impurities < 0.50%)

[Temperature, T, K; Thermal Diffusivity, α, cm^2 s^{-1}]

T	α	T	α	T	α	T	α	T	α
CURVE 1		CURVE 2		CURVE 3 (cont.)		CURVE 5		CURVE 6	
1129	0.121	1164	0.126	1787	0.102	1103	0.166	1081	0.194*
1260	0.124	1220	0.119	1849	0.100	1154	0.171	1202	0.178
1368	0.121	1292	0.109	1961	0.099	1294	0.152	1291	0.175
1454	0.110	1436	0.104	2070	0.093	1403	0.151	1402	0.163
1530	0.100	1565	0.102	2153	0.087	1499	0.146	1515	0.155
1614	0.104	1700	0.103	2218	0.088	1556	0.163	1736	0.141
1706	0.101	1794	0.094	2315	0.085	1633	0.159	1828	0.143
1804	0.095	1918	0.099			1748	0.150	1924	0.144
1933	0.091			CURVE 4		1849	0.147	2012	0.140
2017	0.090	CURVE 3				1929	0.127	2136	0.128
2138	0.095			1083	0.145	2027	0.139	2222	0.111
2225	0.090	1196	0.125	1245	0.131	2089	0.127	2301	0.104
		1330	0.123	1447	0.110	2177	0.102	2389	0.106
		1413	0.117			2248	0.105		
		1523	0.103			2321	0.102		
		1610	0.109			2385	0.101		
		1689	0.101						

* Not shown in figure.

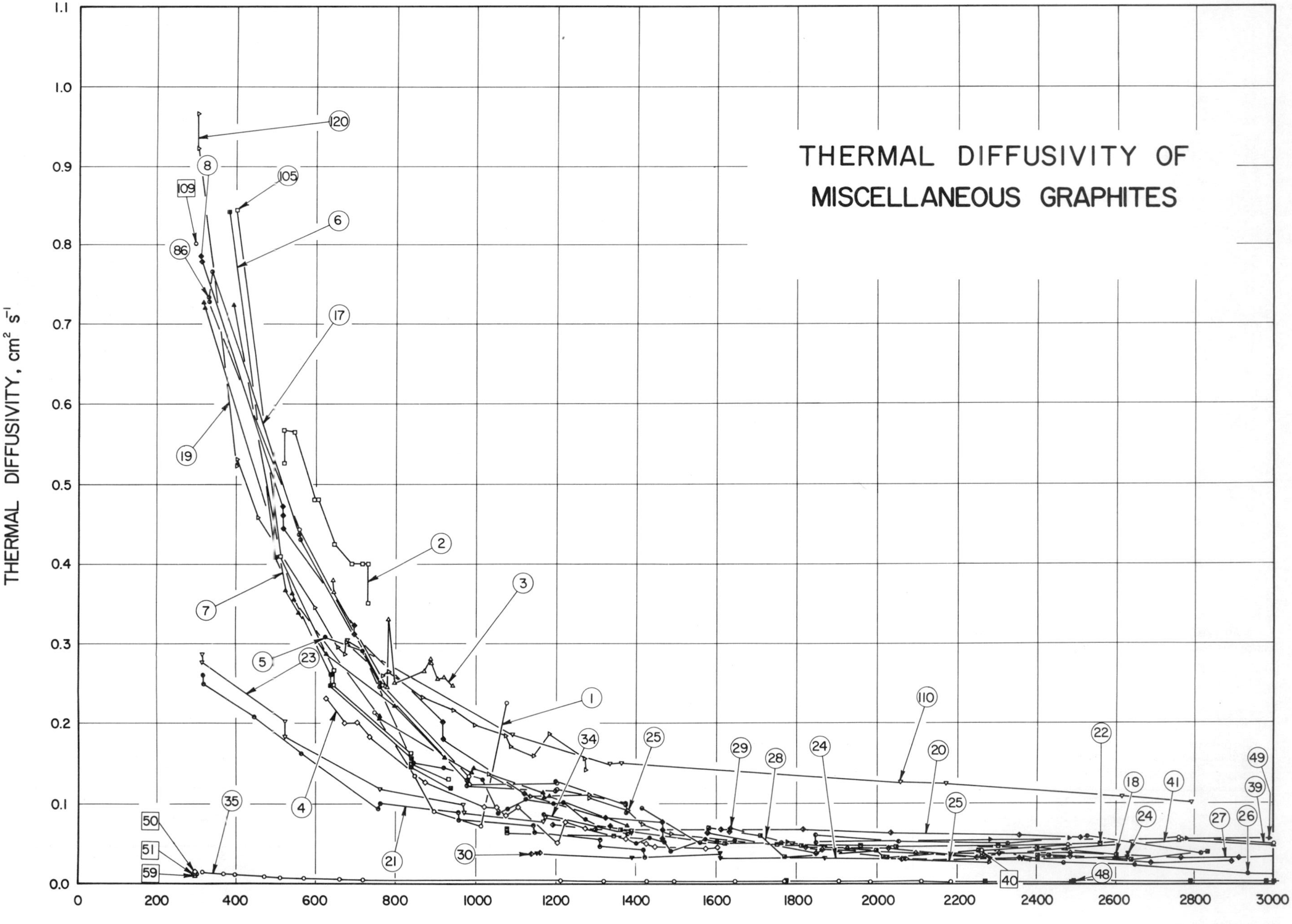
THERMAL DIFFUSIVITY OF
MISCELLANEOUS GRAPHITES
THERMAL DIFFUSIVITY, cm² s⁻¹
1.1
1.0
0.9
0.8
0.7
0.6
0.5
0.4
0.3
0.2
0.1
0.0
0
200
400
600
800
1000
1200
1400
1600
1800
2000
2200
2400
2600
2800
3000
TEMPERATURE, K
120
8
109
105
86
6
17
19
2
3
7
5
23
1
34
25
29
28
24
110
20
22
18
41
49
39
26
27
24
25
4
21
30
35
50
51
59
40
48

FIGURE 13

SPECIFICATION TABLE 13. THERMAL DIFFUSIVITY OF MISCELLANEOUS GRAPHITES

(Impurity <0. 20% each; total impurities <0. 50%)

Cur. No.	Ref. No.	Author(s)	Year	Temp. Range, K	Reported Error, %	Name and Specimen Designation	Composition (weight percent), Specifications, and Remarks
1	8	Sheer, C., Mead, L. H., Rothacker, D. L., and Johnson, L. H.	1957	748-1076	± 20		Cylindrical specimen 0. 625 in. in dia. and 1. 375 in. long; radial heat flow inhibited by wrapping specimen with asbestos tape before being mounted in its graphite housing.
2	9	Deem, H. W. and Wood, W. D.	1962	520-730		7087	Cylindrical specimen 0. 25 in. max in dia. and ≤0. 25 in. long; specimen measured in air; laser beam used to generate heat pulse.
3	9	Deem, H. W. and Wood, W. D.	1962	643-941		7087	Above specimen measured in vacuum.
4	10	Sheer, C., Fitz, C. D., Mead, L. H., Holmgren, J. D., Rothacker, D. L., and Allmand, D.	1958	627-1605			Cylindrical specimen 1. 0 cm in dia. and 4. 5 cm long; machined to dimensions given; insulated on the sides and over one end face; other end face suddenly exposed to a constant heat flow from the plasma of a high intensity arc; measured in vacuum chamber under an ambient pressure of 0. 1 atmosphere; diffusivity data computed assuming conditions of zero heat flow across the lateral and rear end surfaces.
5	10	Sheer, C., et al.	1958	623-1646			Above specimen allowed to cool in chamber after being heated during above measurements; exposed to the arc to measure diffusivity again.
6	11	Mrozowski, S., Andrew, J. F., Juul, N., Okada, J., Strauss, H. E., and Wobschall, D. C.	1960	385-936		Type CS	Rod specimen; made of National Carbon Co. Graphite; sinusoidal temp. variation impressed at one end of specimen; diffusivity calculated from data of the amplitudes and phase shifts obtained from recorded temp. waves; period of temp. wave 8. 7 min; a minimum of five complete cycles for each temp. was recorded.
7	11	Mrozowski, S., et al.	1960	394-922		Type CS	Above specimen measured again using a period of 2. 5 min for the temp. wave.
8	12	Juul, N., Sato, S., and Strauss, H. E.	1963	311-1466			Soft graphite 3273. 2 K heat treated specimen; measured parallel to the extrusion axis.
9*	13	Childers, H. M. and Cerceo, J. M.	1961	1193			Disc specimen 0. 5 cm thick; density 1. 51 g cm^{-3}; specimen heated by electron bombardment; sinusoidal temp. wave imposed on one face of specimen; measured in vacuum of $\sim 10^{-5}$ mm Hg; radiation losses from back face neglected; temp. wave phase shift front to back faces ϕ = 0. 326 radians.
10*	13	Childers, H. M. and Cerceo, J. M.	1961	1185			Above specimen measured for diffusivity again under same conditions as above except that ϕ = 0. 329 radians.
11*	13	Childers, H. M. and Cerceo, J. M.	1961	1185			Above specimen measured for diffusivity again under same conditions as above except that ϕ = 0. 518 radians.
12*	13	Childers, H. M. and Cerceo, J. M.	1961	1194			Above specimen measured for diffusivity again under same conditions as above except that ϕ = 0. 722 radians.
13*	13	Childers, H. M. and Cerceo, J. M.	1961	1193			Above specimen measured for diffusivity again assuming radiation losses from back face; phase shift angle ϕ = 0. 326 radians.
14*	13	Childers, H. M. and Cerceo, J. M.	1961	1185			Above specimen measured for diffusivity again under same conditions as above except that ϕ = 0. 329 radians.
15*	13	Childers, H. M. and Cerceo, J. M.	1961	1185			Above specimen measured for diffusivity again under same conditions as above except that ϕ = 0. 518 radians.
16*	13	Childers, H. M. and Cerceo, J. M.	1961	1194			Above specimen measured for diffusivity again under same conditions as above except that ϕ = 0. 722 radians.

* Not shown in figure.

SPECIFICATION TABLE 13. THERMAL DIFFUSIVITY OF MISCELLANEOUS GRAPHITES (continued)

Cur. No.	Ref. No.	Author(s)	Year	Temp. Range, K	Reported Error, %	Name and Specimen Designation	Composition (weight percent), Specifications, and Remarks
17	14	Mrozowski, S., Andrew, J. F., Juul, N., Sato, S., Strauss, H. E., and Tsuzuku, T.	1963	330-1375		U. B. Carbon "R" (SF-SB)	Highly graphitized; rod specimen 1 in. in dia; laboratory prepared; soft filler-soft binder; extruded; initially heat treated to 3073.2 K; density 1.49 g cm^{-3}; anisotropic; steady sinusoidal technique used to measure diffusivity; measured parallel to the extrusion axis.
18	14	Mrozowski, S., et al.	1963	1361-2603		U. B. Carbon "R" (SF-SB)	Highly graphitized; rod specimen 0.5 in. in dia; laboratory prepared; soft filler-soft binder; extruded; initially heat treated to 3073.2 K; density 1.55 g cm^{-3}; anisotropic; transient heat flow method used to measure diffusivity; measured perpendicular to extrusion axis.
19	14	Mrozowski, S., et al.	1963	316-1378		U. B. Carbon "A" (SF-SB)	Highly graphitized; rod specimen 1 in. in dia; laboratory prepared; soft filler-soft binder; extruded; initially heat treated to 3073.2 K; density 1.20 g cm^{-3}; anisotropic; steady sinusoidal technique used to measure diffusivity; measured parallel to the extrusion axis.
20	14	Mrozowski, S., et al.	1963	1192-2511		U. B. Carbon "A" (SF-SB)	Highly graphitized; rod specimen 0.5 in. in dia; laboratory prepared; soft filler-soft binder; extruded; initially heat treated to 3073.2 K; density 1.33 g cm^{-3}; anisotropic; transient heat flow method used to measure diffusivity; measured perpendicular to the extrusion axis.
21	14	Mrozowski, S., et al.	1963	318-1420		Thermax "W"	Poorly graphitized; rod specimen 1 in. in dia; hard filler-hard binder; extruded; initially heat treated to 3073.2 K; density 1.84 g cm^{-3}; anisotropic; steady sinusoidal technique used to measure diffusivity; measured parallel to the extrusion axis.
22	14	Mrozowski, S., et al.	1963	1077-2560		Thermax "W"	Poorly graphitized; rod specimen 0.5 in. in dia; hard filler-hard binder; extruded; initially heat treated to 3073.2 K; density 1.86 g cm^{-3}; anisotropic; transient heat flow method used to measure diffusivity; measured perpendicular to the extrusion axis.
23	14	Mrozowski, S., et al.	1963	316-1389		U. B. Carbon "Z" (HF-HB)	Poorly graphitized; rod specimen 1 in. in dia; laboratory prepared; hard filler-hard binder; extruded; initially heat treated to 3073.2 K; density 1.23 g cm^{-3}; anisotropic; steady sinusoidal technique used to measure diffusivity; measured parallel to the extrusion axis.
24	14	Mrozowski, S., et al.	1963	1152-2631		U. B. Carbon "Z" (HF-HB)	Poorly graphitized; rod specimen 0.5 in. in dia; laboratory prepared; hard filler-hard binder; extruded; initially heat treated to 3073.2 K; density 1.32 g cm^{-3}; anisotropic; transient heat flow method used to measure diffusivity; measured perpendicular to the extrusion axis.
25	15	Mrozowski, S., Andrew, J. F., Juul, N., Strauss, H. E., and Wobschall, D. C.	1961	1031-2280			Rod specimen 0.75 in. in dia. and 6 in. long; obtained from National Carbon Co.; density 1.63 g cm^{-3}; heated by passing electric current through rod; temp. measured using thermocouples; transient heat flow method used to measure diffusivity.
26	15	Mrozowski, S., et al.	1961	1415-3094			Rod specimen 0.5 in. in dia. and 6 in. long; obtained from National Carbon Co.; density 1.63 g cm^{-3}; heated by passing electric current through rod; temp. measured by optical and radiamatic pyrometer; transient heat flow method used to measure diffusivity.

SPECIFICATION TABLE 13. THERMAL DIFFUSIVITY OF MISCELLANEOUS GRAPHITES (continued)

Cur. No.	Ref. No.	Author(s)	Year	Temp. Range, K	Reported Error, %	Name and Specimen Designation	Composition (weight percent), Specifications, and Remarks
27	15	Mrozowski, S., Andrew, J. F., Juul, N., Strauss, H. E., and Wobschall, D. C.	1961	1577-3048			Rod specimen 0.5 in. in dia. and 6 in. long; obtained from National Carbon Co.; density 1.63 g cm^{-3}; heated by passing electric current through rod.
28	15	Mrozowski, S., et al.	1961	1580-2835			Rod specimen 0.5 in. in dia. and 6 in. long; obtained from Graphite Specialties Co.; density 1.87 g cm^{-3}; heated by passing electric current through rod.
29	15	Mrozowski, S., et al.	1961	1634-2895		G-1	Rod specimen 0.5 in. in dia. and 6 in. long; obtained from U. B. Carbon Laboratory; density 1.54 g cm^{-3}; heated by passing electric current through rod.
30	16	Mrozowski, S., Andrew, J. F., Juul, N., Strauss, H. E., Tsuzuku, T., and Wobschall, D. C.	1962	1138-2638		U. B. Graphite "Z" (HF-HB)	Composition after extrusion and before heat treatment: phenol formaldehyde used as filler and phenol benzaldehyde as binder, filler size: 50 parts 100/150 and 50 parts 270, binder content: 43 parts; rod specimen; extruded; initially heat treated to ~3273.2 K; density 1.32 g cm^{-3} after heat treatment; measured for diffusivity in direction perpendicular to the extrusion axis and in an argon pressure exceeding atmospheric by 3 cm Hg.
31*	16	Mrozowski, S., et al.	1962	1073-2568		Graphite Specialties Co. "W"	Composition after extrusion and before heat treatment: thermax used as filler; rod specimen; extruded; initially heat treated to ~3273.2 K; density 1.86 g cm^{-3} after heat treatment; measured for diffusivity in direction perpendicular to the extrusion axis and in an argon pressure exceeding atmospheric by 3 cm Hg.
32*	16	Mrozowski, S., et al.	1962	1348-2928		U. B. Graphite "R" (SF-SB)	Composition after extrusion and before heat treatment: Texas Coke used as filler and M-30 Coal Tar Pitch as binder, filler size: 50 parts 65/100 and 50 parts 200/270, binder content: 40 parts; rod specimen; extruded; initially heat treated to ~3273.2 K; density 1.55 g cm^{-3} after heat treatment; measured for diffusivity in direction perpendicular to the extrusion axis and in an argon pressure exceeding atmospheric by 3 cm Hg.
33*	16	Mrozowski, S., et al.	1962	1188-3038		U. B. Graphite "A" (SF-SB)	Composition after extrusion and before heat treatment: Texas Coke used as filler and M-30 Coal Tar Pitch as binder, filler size: 100 parts 28/35, binder content: 44 parts; rod specimen; extruded; initially heat treated to ~3273.2 K; density 1.33 g cm^{-3} after heat treatment; measured for diffusivity in direction perpendicular to the extrusion axis and in an argon pressure exceeding atmospheric by 3 cm Hg.
34	16	Mrozowski, S., et al.	1962	1168-2818		U. B. Graphite "G" (SF-SB)	Composition after extrusion and before heat treatment: Texas Coke used as filler and M-30 Coal Tar Pitch as binder, filler size: 100 parts 200/270, binder content: 50 parts; rod specimen; extruded; initially heat treated to ~3273.2 K; density 1.55 g cm^{-3} after heat treatment; measured for diffusivity in direction perpendicular to the extrusion axis and in an argon pressure exceeding atmospheric by 3 cm Hg.

* Not shown in figure.

SPECIFICATION TABLE 13. THERMAL DIFFUSIVITY OF MISCELLANEOUS GRAPHITES (continued)

Cur. No.	Ref. No.	Author(s)	Year	Temp. Range, K	Reported Error, %	Name and Specimen Designation	Composition (weight percent), Specifications, and Remarks
35*	135	Morrison, B.H., Klein, K.J., and Cowder, L.R.	1966	303-2185		Pyrolytic Graphite	Commercial pyrolytic graphite; disc specimen 0.020 in. thick; manufactured by General Electric Co.; deposition temp. 2373.2 K with no regraphitization; anisotropic; measured in the c-direction (perpendicular to the basal or deposition plane ab); measured in one atmosphere of helium in the temp. range 303-753 K; high temp. measurements conducted in vacuum; ruby laser used as pulse energy source; diffusivity determined from measured temp.-time response of back face; data points reported corrected for heat loss and finite pulse time.
36*	135	Morrison, B.H., et al.	1966	304-2240		Pyrolytic Graphite	Commercial pyrolytic graphite; disc specimen 0.020 in. thick; manufactured by General Electric Co.; deposition temp. 2373.2 K with no regraphitization; anisotropic; measured in the c-direction (perpendicular to the basal or deposition plane ab); measured in one atmosphere of helium in the temp. range 303-753 K; high temp. measurements conducted in vacuum; ruby laser used as pulse energy source; diffusivity determined from measured temp.-time response of back face; data points reported corrected for heat loss and finite pulse time.
37*	135	Morrison, B.H., et al.	1966	300-2064		Pyrolytic Graphite	Commercial pyrolytic graphite; sleeve specimen 0.358 in. I.D., 0.458 in. O.D., and 0.75 in. long; manufactured by High Temp Corp.; deposition temp. 2373.2 K with no regraphitization; anisotropic; measured in the c-direction (perpendicular to the basal or deposition plane ab); measured in one atmosphere of helium in the temp. range 300-681 K; high temp. measurements conducted in vacuum; ruby laser used as pulse energy source; diffusivity determined from temp-time response measured for an area behind that upon which laser pulse impinged; data points reported corrected for heat loss and finite pulse time.
38	138	Macqueron, J.-L., Sinicki, G., Durand, G., and Rinaldi, D.	1967	79-303		Carbone Pyrolytique (Carbone Lorraine)	Flat specimen; one face exposed to thermal pulse of short duration generated by a flash tube; diffusivity determined from resulting temp. evolution of rear face measured using a thermoelectric couple with a large figure of merit; measured in the principal crystallographic direction; measured in a vacuum of 10^{-3} mm Hg.
39	139	Kaspar, J. and Zehms, E.H.	1964	2260-3000		CEP-Type Graphite	Specimen composed of two discs ~1 in. in dia and 2 mm thick each, and spaced 1 mm apart; obtained from National Carbon Co.; well stabilized within the temp. region between 2000 and 3000 K and of quasi-isotropic character; specimen placed in tube furnace and exposed to a sinusoidally modulated heat flux resulting from chopping and re-imaging the radiation emanating from the specimen discs onto themselves using reflective optical systems; diffusivity determined from measured phase shift between recorded temp signals received from front and rear faces of one of the discs only; measured using a frequency of 0.19 cycles per sec.
40	139	Kaspar, J. and Zehms, E.H.	1964	2260		CEP-Type Graphite	Above specimen measured for diffusivity again using a frequency of 0.33 cycles per sec.
41	139	Kaspar, J. and Zehms, E.H.	1964	2260-3000		CEP-Type Graphite	Above specimen measured for diffusivity again using a frequency of 0.375 cycles per sec.
42*	139	Kaspar, J. and Zehms, E.H.	1964	2260-3000		CEP-Type Graphite	Above specimen measured for diffusivity again using a frequency of 0.47 cycles per sec.
43*	139	Kaspar, J. and Zehms, E.H.	1964	2260-3000		CEP-Type Graphite	Above specimen measured for diffusivity again using a frequency of 0.625 cycles per sec.

* Not shown in figure.

SPECIFICATION TABLE 13. THERMAL DIFFUSIVITY OF MISCELLANEOUS GRAPHITES (continued)

Cur. No.	Ref. No.	Author(s)	Year	Temp. Range, K	Reported Error, %	Name and Specimen Designation	Composition (weight percent), Specifications, and Remarks
44*	152	Kaspar, J. and Zehms, E.H.	1967	2477		Supertemp PG	Pyrolytic graphite; disc specimen 0.65 mm thick; annealed at 3300 K; strongly anisotropic; measured for diffusivity with no radiation shield at rear face; measured in the c-direction; placed in a tube furnace and exposed to a periodic heat flux on one face resulting from chopping and re-imaging the radiation emanating from front face of specimen onto itself using reflective optical system; diffusivity determined from phase lag between the measured temperature variations of front and rear faces; measured using various frequencies of f, ranging from 0.053 to 0.628 cps.
45*	152	Kaspar, J. and Zehms, E.H.	1967	2274		Supertemp PG	Above specimen measured for diffusivity again with 9 ATJ radiation shield elements 0.38 mm thick at the rear face; measured using various frequencies f, ranging from 0.026 to 0.625 cps.
46*	152	Kaspar, J. and Zehms, E.H.	1967	2477		Supertemp PG	Above specimen measured for diffusivity again using no radiation shield at rear face; measured using various frequencies, f, ranging from 0.052 to 0.627 cps.
47*	152	Kaspar, J. and Zehms, E.H.	1967	2491		Supertemp PG	Pyrolytic graphite; disc specimen 0.25 mm thick; annealed at 3300 K; strongly anisotropic; measured for diffusivity with 5 TaC radiation shield elements 0.025 mm thick at the rear face; measured using various frequencies, f, ranging from 0.024 to 0.628 cps.
48	152	Kaspar, J. and Zehms, E.H.	1967	1767-3280		Supertemp PG Graphite	Above specimen measured for diffusivity at various temperatures; same radiation shield and same measurement technique as above.
49	152	Kaspar, J. and Zehms, E.H.	1967	1776-2990		CEP type Graphite	Disc specimen 2 mm thick; quasi-isotropic carbon; measured for diffusivity with 9 ATJ radiation shield elements 0.46 mm thick at the rear face; same measurement technique as above.
50	184	Taylor, R.	1965	298.2	<±5		Heavily irradiated; cylindrical specimen 0.25 in. in diameter and 1.269 cm long; front face exposed to heat pulse from xenon flash tube; thermal diffusivity calculated from measured time necessary for the rear face to reach one-half the maximum temperature rise $t_{0.5}$ by employing the equation $\alpha = 1.37\ L^2/\pi^2 t_{0.5}$; temperature of measurement not given by author but assumed to be room temperature.
51	184	Taylor, R.	1965	298.2	<±5		Above specimen measured for diffusivity again after being shortened to a length of 1.168 cm; other conditions same as above.
52*	184	Taylor, R.	1965	298.2	<±5		Above specimen measured for diffusivity again after being shortened to a length of 1.088 cm; other conditions same as above.
53*	184	Taylor, R.	1965	298.2	<±5		Above specimen measured for diffusivity again after being shortened to a length of 0.997 cm; other conditions same as above.
54*	184	Taylor, R.	1965	298.2	<±5		Above specimen measured for diffusivity again after being shortened to a length of 0.886 cm; other conditions same as above.
55*	184	Taylor, R.	1965	298.2	<±5		Above specimen measured for diffusivity again after being shortened to a length of 0.798 cm; other conditions same as above.

* Not shown in figure.

SPECIFICATION TABLE 13. THERMAL DIFFUSIVITY OF MISCELLANEOUS GRAPHITES (continued)

Cur. No.	Ref. No.	Author(s)	Year	Temp. Range, K	Reported Error, %	Name and Specimen Designation	Composition (weight percent), Specifications, and Remarks
56*	184	Taylor, R.	1965	298.2	<± 5		Above specimen measured for diffusivity again after being shortened to a length of 0.697 cm; other conditions same as above.
57*	184	Taylor, R.	1965	298.2	<± 5		Above specimen measured for diffusivity again after being shortened to a length of 0.597 cm; other conditions same as above.
58*	184	Taylor, R.	1965	298.2	<± 5		Above specimen measured for diffusivity again after being shortened to a length of 0.497 cm; other conditions same as above.
59*	184	Taylor, R.	1965	298.2	<± 5		Above specimen measured for diffusivity again after being shortened to a length of 0.396 cm; other conditions same as above.
60*	184	Taylor, R.	1965	298.2	<± 5		Above specimen measured for diffusivity again after being shortened to a length of 0.298 cm; other conditions same as above.
61*	184	Taylor, R.	1965	298.2	<± 5		Above specimen measured for diffusivity again after being shortened to a length of 0.218 cm; other conditions same as above.
62*	184	Taylor, R.	1965	298.2	<± 5		Above specimen measured for diffusivity again after being shortened to a length of 0.149 cm; other conditions same as above.
63*	184	Taylor, R.	1965	298.2	± 5		Same specimen and same measurement pertaining to curve 50 above; diffusivity calculated from time intercept t_x of measured temperature - time curve of rear face by employing the equation $\alpha = 0.48\ L^2/\pi^2 t_x$; temperature of measurement not given by author but assumed to be room temperature.
64*	184	Taylor, R.	1965	298.2	± 5		Same specimen and same measurement pertaining to curve 51 above; other conditions same as above.
65*	184	Taylor, R.	1965	298.2	± 5		Same specimen and same measurement pertaining to curve 52 above; other conditions same as above.
66*	184	Taylor, R.	1965	298.2	± 5		Same specimen and same measurement pertaining to curve 53 above; other conditions same as above.
67*	184	Taylor, R.	1965	298.2	± 5		Same specimen and same measurement pertaining to curve 54 above; other conditions same as above.
68*	184	Taylor, R.	1965	298.2	± 5		Same specimen and same measurement pertaining to curve 55 above; other conditions same as above.
69*	184	Taylor, R.	1965	298.2	± 5		Same specimen and same measurement pertaining to curve 56 above; other conditions same as above.
70*	184	Taylor, R.	1965	298.2	± 5		Same specimen and same measurement pertaining to curve 57 above; other conditions same as above.
71*	184	Taylor, R.	1965	298.2	± 5		Same specimen and same measurement pertaining to curve 58 above; other conditions same as above.
72*	184	Taylor, R.	1965	298.2	± 5		Same specimen and same measurement pertaining to curve 59 above; other conditions same as above.

* Not shown in figure.

SPECIFICATION TABLE 13. THERMAL DIFFUSIVITY OF MISCELLANEOUS GRAPHITES (continued)

Cur. No.	Ref. No.	Author(s)	Year	Temp. Range, K	Reported Error, %	Name and Specimen Designation	Composition (weight percent), Specifications, and Remarks
73*	184	Taylor, R.	1965	298.2	± 5		Same specimen and same measurement pertaining to curve 60 above; other conditions same as above.
74*	184	Taylor, R.	1965	298.2	± 5		Same specimen and same measurement pertaining to curve 61 above; other conditions same as above.
75*	184	Taylor, R.	1965	298.2	± 5		Same specimen and same measurement pertaining to curve 62 above; other conditions same as above.
76*	191	Zamoluev, V.K.	1960	300	<1	Petroleum coke	Grain size <0.5 mm; measured as a function of heat-treating temperature, τ, ranging from 2000 to 2800 C.
77*	191	Zamoluev, V.K.	1960	300	<1	Petroleum coke	Similar to the above specimen but heat-treated at 2000 C and measured as a function of heat-treatment time, t, ranging from 1.5 to 180 min.
78*	191	Zamoluev, V.K.	1960	300	<1	Petroleum coke	Similar to above but heat-treated at 2300 C with heat-treatment time, t, ranging from 10.5 to 180 min.
79*	191	Zamoluev, V.K.	1960	300	<1	Petroleum coke	Similar to above but heat-treated at 2800 C with heat-treatment time, t, ranging from 5.5 to 180 min.
80*	191	Zamoluev, V.K.	1960	300	<1	Gas coal	Grain size <0.5 mm; heat-treated at 2300 C.
81*	191	Zamoluev, V.K.	1960	300	<1	Brown coal	Grain size <0.5 mm; heat-treated at 2350 C.
82*	191	Zamoluev, V.K.	1960	300	<1	Channel black	Grain size <0.5 mm; measured as a function of heat-treating temperature, τ, ranging from 2000 to 3000 C.
83*	191	Zamoluev, V.K.	1960	300	<1	Channel black	Grain size <0.5 mm; heat-treated at 2500 C; measured as a function of heat-treatment time, t, ranging from 2.2 to 180 min.
84*	191	Zamoluev, V.K.	1960	300	<1	Channel black	Similar to above but heat-treated at 3000 C with heat-treatment time, t, ranging from 2.0 to 180 min.
85*	210	Kabatochkin, V.I., Zamoluev, V.K., Kaverov, A.T., and Usenbaev, K.	1960	300		Thermal black	Grain size <1μ; measured as a function of the heat-treating tempeature, τ, ranging from 2300 to 3000 C.
86*	158	Juul, N.H.	1964	329-1373		U.B. carbon "R"	Thin tube specimen; extruded; density 1.49 g cm^{-3}; heat flow parallel to extrusion axis; electrical resistivity 8.34, 9.00, 9.70, 10.33, 10.86, 11.33, and 11.99 mΩ cm at 1094, 1351, 1556, 1785, 2028, 2332, and 2665 C, respectively.
87*	158	Juul, N.H.	1964	1367-2604		U.B. carbon "R"	Similar to the above specimen but density 1.55 g cm^{-3} and heat flow perpendicular to extrusion axis.
88*	158	Juul, N.H.	1964	324-1381		U.B. carbon "A"	Thin tube specimen; extruded; density 1.20 g cm^{-3}; heat flow parallel to extrusion axis; electrical resistivity 10.55, 11.04, 11.94, 12.69, 13.47, 14.34, 14.65, 15.43, and 15.43 mΩ cm at 922, 1134, 1351, 1547, 1769, 2088, 2246, 2509, and 2772 C, respectively.
89*	158	Juul, N.H.	1964	1193-2517		U.B. carbon "A"	Thin tube specimen; extruded; density 1.33 g cm^{-3}; heat flow perpendicular to extrusion axis.

* Not shown in figure.

SPECIFICATION TABLE 13. THERMAL DIFFUSIVITY OF MISCELLANEOUS GRAPHITES (continued)

Cur. No.	Ref. No.	Author(s)	Year	Temp. Range, K	Reported Error, %	Name and Specimen Designation	Composition (weight percent), Specifications, and Remarks
90*	158	Juul, N. H.	1964	320-1412		U. B. carbon "Z"	Thin tube specimen; extruded; density 1.23 g cm^{-3}; heat flow parallel to extrusion axis; electrical resistivity 27.77, 27.64, 25.95, 25.41, 25.24, 24.23, 23.24, 22.41, 21.47, and 21.19 mΩ cm at 883, 889, 1130, 1350, 1350, 1616, 1801, 2096, 2363, and 2365 C, respectively.
91*	158	Juul, N. H.	1964	1148-2631		U. B. carbon "Z"	Thin tube specimen; extruded; density 1.32 g cm^{-3}; heat flow perpendicular to extrusion axis.
92*	158	Juul, N. H.	1964	320-1390		Thermax "W"	Thin tube specimen; extruded; density 1.84 g cm^{-3}; heat flow parallel to extrusion axis; electrical resistivity 19.11, 18.64, 18.45, 18.40, 17.86, and 16.70 mΩ cm at 800, 1074, 1493, 1694, 2030, and 2298 C, respectively.
93*	158	Juul, N. H.	1964	1073-2562		Thermax "W"	Thin tube specimen; extruded; density 1.86 g cm^{-3}; heat flow perpendicular to extrusion axis.
94*	212	Van der Berg, M. and Schmidt, H. E.	1965	1476-2185			5 to 6 mm in diameter and 1 to 2 mm thick; measured in a vacuum of 5 x 10^{-6} mm Hg.
95*	212	Van der Berg, M. and Schmidt, H. E.	1965	1884-2209			Similar to above.
96*	213	Morrison, B. H., Klein, D. J., and Cowder, L. R.	1965	1102-1723		EP 192 C	0.5 in. diameter x 0.020 in. thick; obtained from Pure Oil Co.; density 1.55 g cm^{-3}.
97*	213	Morrison, B. H., et al.	1965	1085-2438		EP 192 C	Similar to the above specimen but 0.030 in. thick.
98*	213	Morrison, B. H., et al.	1965	1077-2430		EP 192 C	Similar to the above specimen but 0.050 in. thick.
99*	213	Morrison, B. H., et al.	1965	1110-2432		TS-699	0.5 in. diameter x 0.020 in. thick; obtained from National Carbon Co.; density 1.85 g cm^{-3}.
100*	213	Morrison, B. H., et al.	1965	1100-2438		TS-699	Similar to the above specimen but 0.030 in. thick.
101*	213	Morrison, B. H., et al.	1965	1079-2375		TS-699	Similar to the above specimen but 0.040 in. thick.
102*	213	Morrison, B. H., et al.	1965	1075-2433		TS-699	Similar to the above specimen but 0.050 in. thick.
103*	229	Juul, N. H.	1961	401-931		CS	Tube specimen; obtained from National Carbon Co.; heat-treated at 3000 C; density 1.63 g cm^{-3}; heat flow parallel to extrusion axis.
104*	229	Juul, N. H.	1961	411-916		CS	The above specimen.
105*	229	Juul, N. H.	1961	1176-2818		U.B. graphite "G"	Cylindrical specimen; heat-treated at 3000 C; density 1.55 g cm^{-3}; heat flow perpendicular to extrusion axis; electrical resistivity 6.95, 7.32, 8.06, 8.61, 8.61, 9.12, 9.65, 10.11, and 10.53 mΩ cm at 910, 1104, 1317, 1568, 1579, 1773, 2042, 2247, and 2547 C, respectively. (Wrongly reported data adjusted 10 times lower.)
106*	229	Juul, N. H.	1961	506-931		CS	Tube specimen; obtained from National Carbon Co.; heat-treated at 3000 C; density 1.63 g cm^{-3}. (Author's wrongly reported data adjusted 10 times lower.)
107*	229	Juul, N. H.	1961	1022-1483		CS	Similar to above but specimen dimensions 1.9 cm diameter x 15 cm long.
108*	229	Juul, N. H.	1961	1414-3085		CS	Similar to above but specimen diameter 1.25 cm.

* Not shown in figure.

SPECIFICATION TABLE 13. THERMAL DIFFUSIVITY OF MISCELLANEOUS GRAPHITES (continued)

Cur. No.	Ref. No.	Author(s)	Year	Temp. Range, K	Reported Error, %	Name and Specimen Designation	Composition (weight percent), Specifications, and Remarks
109	214, 215	Moser, J.B. and Kruger, O.L.	1964	298.2	5		1.9 cm in diameter and 0.1 to 0.3 cm thick; density 1.73 g cm^{-3}.
110	216	Cunnington, G.R., Smith, F.J., and Bradshaw, W.	1966	679-2793		CS	9.6 mm diameter x 2.38 mm thick; machined; density 1.63 g cm^{-3}.
111*	216	Cunnington, G.R., et al.	1966	651-1586		Pyrolytic graphite 1	Disc specimen 1.01 mm thick; density 2.17 g cm^{-3}; heat flow parallel to c-axis.
112*	216	Cunnington, G.R., et al.	1966	1089-1417		Pyrolytic graphite 1	Machined from the same slab as the above specimen; 2.03 mm thick; density 2.12 g cm^{-3}; heat flow parallel to c-axis.
113*	216	Cunnington, G.R., et al.	1966	1148-1826		Expanded pyrolytic graphite 1	1.98 mm thick; bulk density 0.26 g cm^{-3}.
114*	216	Cunnington, G.R., et al.	1966	1089-1522		Expanded pyrolytic graphite 1	Taken from the same piece of material as the above specimen; 2.13 mm thick; bulk density 0.31 g cm^{-3}.
115*	217	Morrison, B.H.	1968	1072-2610		SX-5	0.5 in. disc specimen; fabricated from calcined coke and coal tar pitch and graphitized at a Max. temperature of 2800 C by extrusion; electrical resistivity 9.81 mΩ cm at room temperature; density 1.66 g cm^{-3}; heat flow along x-axis (extrusion direction); measured in a vacuum of 10^{-4} torr.
116*	217	Morrison, B.H.	1968	1091-3046		SX-5	The above specimen measured in helium at 1 atm.
117*	217	Morrison, B.H.	1968	1081-3085		SX-5	Similar to the above specimen but cut from another slab.
118*	217	Morrison, B.H.	1968	1073-3073		SX-5	Similar to the above specimen but with heat flow along y-axis.
119*	217	Morrison, B.H.	1968	1073-3073		SX-5	Similar to the above specimen but with heat flow along z-axis.
120	162	Kobayasi, K. and Kumada, T.	1968	303-1273		Reactor grade	12.8 mm diameter x 14.4 mm long; heat-treated at 2800 C; density 1.646 g cm^{-3}.
121*	162	Kobayasi, K. and Kumada, T.	1968	293-1287		Reactor grade	The above specimen irradiated by a dose of 4.68 x 10^{18} nvt.
122*	219	Null, M.R. and Lozier, W.W.	1969	2500		Pyrolytic graphite	12.7 mm diameter x 0.280 mm thick; stress annealed; bulk density 2.223 g cm^{-3}; heat flow parallel to the c-axis.
123*	219	Null, M.R. and Lozier, W.W.	1969	1600		CEP	12.7 mm diameter x 0.280 mm thick; bulk density 1.614 g cm^{-3}; electrical resistivity 4.4 mΩ cm at room temperature; heat flow across grain.
124*	219	Null, M.R. and Lozier, W.W.	1969	2000		CEP	Similar to above but specimen 0.761 mm thick.
125*	219	Null, M.R. and Lozier, W.W.	1969	2200-2800		CEP	Similar to above but specimen 0.790 mm thick and bulk density 1.618 g cm^{-3}.
126*	219	Null, M.R. and Lozier, W.W.	1969	1600		CEP	12.7 mm diameter x 0.778 mm thick; bulk density 1.621 g cm^{-3}; electrical resistivity 5.2 mΩ cm at room temperature; heat flow with grain.
127*	219	Null, M.R. and Lozier, W.W.	1969	2000		CEP	Similar to above but specimen 0.764 mm thick.
128*	219	Null, M.R. and Lozier, W.W.	1969	2200, 2800		CEP	Similar to above but specimen 0.784 mm thick and bulk density 1.622 g cm^{-3}.
129*	219	Null, M.R. and Lozier, W.W.	1969	2500		CEP	Similar to above but specimen 0.791 mm thick.
130*	267	Rawuka, A.C. and Gaz, R.A.	1969	1481-2658		AXM-5Q1 Spec. No. 1	Specimen 0.12 cm in thickness; supplied by Poco Graphite, Inc.; density 1.77 g cm^{-3}; diffusivity measured using pulse technique.

* Not shown in figure.

SPECIFICATION TABLE 13. THERMAL DIFFUSIVITY OF MISCELLANEOUS GRAPHITES (continued)

Cur. No.	Ref. No.	Author(s)	Year	Temp. Range, K	Reported Error, %	Name and Specimen Designation	Composition (weight percent), Specifications, and Remarks
131*	267	Rawuka, A. C. and Gaz, R. A.	1969	1896-2689		RVD-Type Spec. No. 1	Specimen 0.10 cm in thickness; supplied by Union Carbide Corp.; density 1.86 g cm^{-3}; diffusivity measured using pulse technique.
132*	267	Rawuka, A. C. and Gaz, R. A.	1969	1313-2703		RVD-Type Spec. No. 2	Similar to the above specimen.
133*	211	Cerceo, J. M. and Childers, H. M.	1963	1189			Density 0.51 g cm^{-3}.
134*	211	Cerceo, J. M. and Childers, H. M.	1963	1189			The above specimen.

* Not shown in figure.

DATA TABLE 13. THERMAL DIFFUSIVITY OF MISCELLANEOUS GRAPHITES

(Impurity < 0.20% each; total impurities < 0.50%)

[Temperature, T, K; Thermal Diffusivity, α, $cm^2\ s^{-1}$]

T	α
CURVE 1	
748.2	0.213
848.2	0.133
896.2	0.090
1013.2	0.072
1076.2	0.225
CURVE 2	
520.2	0.525
520.2	0.567
546.2	0.565
596.2	0.480
606.2	0.480
646.2	0.425
690.2	0.400
716.2	0.400
730.2	0.400
730.2	0.350
CURVE 3	
643.2	0.380
646.2	0.365
779.2	0.245
780.2	0.330
796.2	0.250
870.2	0.265
886.2	0.280
888.2	0.275
903.2	0.255
920.2	0.257
941.2	0.247
CURVE 4	
627.2	0.231
673.2	0.200
703.2	0.200
735.2	0.182
873.2	0.126
1020.2	0.0950
1051.2	0.0950
CURVE 4 (cont.)	
1076.2	0.0850
1105.2	0.0950
1203.2	0.0510
1223.2	0.0770
1273.2	0.0700
1373.2	0.0565
1425.2	0.0500
1448.2	0.0465
1475.2	0.0430
1573.2	0.0450
1605.2	0.0445
CURVE 5	
623.2	0.308
716.2	0.291
843.2	0.149
919.2	0.143
1016.2	0.128
1054.2	0.0870
1078.2	0.0935
1123.2	0.103
1193.2	0.0990
1218.2	0.100
1273.2	0.0800
1400.2	0.0500
1433.2	0.0575
1470.2	0.0510
1486.2	0.0405
1573.2	0.0550
1601.2	0.0530
1646.2	0.0535
CURVE 6	
384.8	0.841
499.8	0.408
638.7	0.261
638.7	0.246
838.7	0.155
838.7	0.145
935.9	0.119
CURVE 7	
394.3	0.723
524.8	0.367
558.2	0.338
627.6	0.286
797.1	0.221
922.1	0.157
CURVE 8	
311.2	0.785
313.2	0.778
518.2	0.472
518.2	0.462
518.2	0.445
696.2	0.312
696.2	0.323
916.2	0.201
918.2	0.180
1120.2	0.112
1323.2	0.0825
1335.2	0.0720
1466.2	0.0775
1466.2	0.0675
1466.2	0.0575
CURVE 9*	
1193	0.0473
CURVE 10*	
1185	0.0468
CURVE 11*	
1185	0.0723
CURVE 12*	
1194	0.0992
CURVE 13*	
1193	0.0407
CURVE 14*	
1185	0.0403
CURVE 15*	
1185	0.0622
CURVE 16*	
1194	0.0854
CURVE 17	
330.2	0.727
339.2	0.765
559.2	0.437
561.2	0.430
760.2	0.250
760.2	0.245
976.2	0.128
976.2	0.121
1199.2	0.115
1199.2	0.127
1374.2	0.098
1375.2	0.089
CURVE 18	
1361.2	0.067
1621.2	0.051
1812.2	0.048
2047.2	0.047
2402.2	0.045
2603.2	0.037
CURVE 19	
316.2	0.727
319.2	0.720
CURVE 19 (cont.)	
542.2	0.364
546.2	0.356
760.2	0.208
760.2	0.206
990.2	0.138
994.2	0.144
1167.2	0.113
1170.2	0.107
1184.2	0.108
1378.2	0.083
CURVE 20	
1192.2	0.074
1297.2	0.068
1611.2	0.068
1816.2	0.068
2035.2	0.064
2356.2	0.062
2511.2	0.057
CURVE 21	
318.2	0.260
318.2	0.249
445.2	0.207
563.2	0.161
756.2	0.093
762.2	0.100
956.2	0.089
956.2	0.078
1143.2	0.073
1147.2	0.064
1309.2	0.055
1309.2	0.047
1417.2	0.042
1420.2	0.035
CURVE 22	
1077.2	0.063
1077.2	0.068
CURVE 22 (cont.)	
1343.2	0.060
1752.2	0.049
1960.2	0.048
2302.2	0.048
2560.2	0.050
CURVE 23	
316.2	0.286
316.2	0.276
524.2	0.201
524.2	0.182
761.2	0.117
969.2	0.098
971.2	0.088
1170.2	0.077
1174.2	0.082
1389.2	0.067
1389.2	0.064
CURVE 24	
1152.2	0.037
1389.2	0.032
1611.2	0.032
1611.2	0.038
1871.2	0.031
2068.2	0.031
2362.2	0.031
2630.2	0.033
2631.2	0.038
CURVE 25	
1031.2	0.135
1135.2	0.108
1282.2	0.111
1282.2	0.107
1377.2	0.091
1416.2	0.074
1488.2	0.062
2072.2	0.032
2280.2	0.027
CURVE 26	
1415.2	0.094
1559.2	0.051
1761.2	0.051*
1926.2	0.044*
2469.2	0.026
2649.2	0.024
2935.2	0.014
3094.2	0.013*
CURVE 27	
1577.2	0.0645
1849.2	0.0382
1863.2	0.0438
2096.2	0.0380
2306.2	0.0385
2487.2	0.0386
2767.2	0.0331
2907.2	0.0339
3048.2	0.0377*
CURVE 28	
1580.2	0.0696
1710.2	0.0607
1770.2	0.0635
1998.2	0.0417
2017.2	0.0355
2024.2	0.0388
2254.2	0.0420
2403.2	0.0360
2403.2	0.0316
2430.2	0.0363
2640.2	0.0316
2835.2	0.0400
CURVE 29	
1634.2	0.0680
1634.2	0.0665
1822.2	0.0468
2028.2	0.0445
CURVE 29 (cont.)	
2250.2	0.0338
2486.2	0.0345
2691.2	0.0277
2895.2	0.0287
CURVE 30	
1138.2	0.0370
1158.2	0.0382
1388.2	0.0317*
1608.2	0.0333*
1608.2	0.0343*
1873.2	0.0305*
2063.2	0.0312
2358.2	0.0325
2638.2	0.0337*
2638.2	0.0325*
CURVE 31*	
1073.2	0.0620
1073.2	0.0640
1333.2	0.0590
1753.2	0.0478
1756.2	0.0495
1958.2	0.0496
2296.2	0.0482
2568.2	0.0498
CURVE 32*	
1348.2	0.0675
1613.2	0.0515
1818.2	0.0495
2058.2	0.0473
2398.2	0.0435
2603.2	0.0373
2928.2	0.0420

* Not shown in figure.

DATA TABLE 13. THERMAL DIFFUSIVITY OF MISCELLANEOUS GRAPHITES (continued)

T	α
CURVE 33*	
1188.2	0.0717
1398.2	0.0665
1618.2	0.0683
1828.2	0.0677
2043.2	0.0610
2363.2	0.0600
2518.2	0.0547
2778.2	0.0487
3038.2	0.0477
CURVE 34	
1168.2	0.0860
1378.2	0.0613
1578.2	0.0443*
1848.2	0.0605
1848.2	0.0527
2053.2	0.0530
2328.2	0.0507
2528.2	0.0600
2818.2	0.0395
CURVE 35	
303.2	0.012
316.2	0.014
369.2	0.012
398.2	0.011
472.2	0.0081
514.2	0.0067
572.2	0.0063
660.2	0.0054
718.2	0.0045
1211.2	0.0037
1320.2	0.0037
1427.2	0.0035
1495.2	0.0035
1648.2	0.0033
1769.2	0.0031
1905.2	0.0034
1983.2	0.0029
2110.2	0.0034
2185.2	0.0036
CURVE 36*	
304.2	0.014
334.2	0.013
CURVE 36 (cont.)*	
416.2	0.010
482.2	0.0084
526.2	0.0073
555.2	0.0072
666.2	0.0059
678.2	0.0058
751.2	0.0051
1179.2	0.0039
1324.2	0.0039
1432.2	0.0035
1523.2	0.0031
1709.2	0.0030
1815.2	0.0032
1922.2	0.0033
2087.2	0.0034
2240.2	0.0035
CURVE 37*	
300.2	0.013
322.2	0.012
335.2	0.011
362.2	0.010
400.2	0.0087
429.2	0.0074
467.2	0.0076
486.2	0.0059
560.2	0.0051
664.2	0.0048
681.2	0.0045
1189.2	0.0029
1336.2	0.0025
1439.2	0.0025
1522.2	0.0025
1721.2	0.0023
1907.2	0.0023
2026.2	0.0025
2064.2	0.0026
CURVE 38	
79.2	0.0340
87.2	0.0321
97.2	0.0286
123.2	0.0231
149.2	0.0192
173.2	0.0153
197.2	0.0121
CURVE 38 (cont.)	
223.2	0.0107
249.2	0.0087
273.2	0.0082
288.2	0.0077
303.2	0.0077
CURVE 39	
2260	0.042
2480	0.049
2760	0.058
3000	0.051
CURVE 40	
2260	0.041
CURVE 41	
2260	0.036
2480	0.045
2760	0.054
3000	0.050
CURVE 42*	
2260	0.047
2480	0.041
2760	0.074
3000	0.046
CURVE 43*	
2260	0.043
2480	0.046
2760	0.062
3000	0.054
f (cps)	**α**
CURVE 44* (T = 2477K)	
0.053	0.00424
0.108	0.00576
0.190	0.00623
0.331	0.00614
0.470	0.00639
0.628	0.00649
CURVE 45* (T = 2474 K)	
0.026	0.00144
0.053	0.00238
0.108	0.00363
0.192	0.00396
0.331	0.00452
0.470	0.00475
0.625	0.00529
CURVE 46* (T = 2477 K)	
0.052	0.00418
0.111	0.00574
0.165	0.00616
0.333	0.00614
0.469	0.00638
0.627	0.00650
CURVE 47* (T = 2491 K)	
0.024	0.00166
0.052	0.00223
0.052	0.00251
0.105	0.00336
0.190	0.00311
0.333	0.00297
0.628	0.00312
T	**α**
CURVE 48	
1767	0.00292
2272	0.00268
2487	0.00297
2497	0.00306
2786	0.00316
2982	0.00310
3001	0.00310
3280	0.00167*
CURVE 49	
1776	0.0524
2281	0.0552
2483	0.0570
CURVE 49 (cont.)	
2780	0.0576
2990	0.0579
CURVE 50	
298.2	0.0142
CURVE 51	
298.2	0.0125
CURVE 52*	
298.2	0.0127
CURVE 53*	
298.2	0.0118
CURVE 54*	
298.2	0.0114
CURVE 55*	
298.2	0.0108
CURVE 56*	
298.2	0.0105
CURVE 57*	
298.2	0.0101
CURVE 58*	
298.2	0.0102
CURVE 59*	
298.2	0.0097
CURVE 60*	
298.2	0.0092
CURVE 61*	
298.2	0.0096
CURVE 62*	
298.2	0.0099
CURVE 63*	
298.2	0.096
CURVE 64*	
298.2	0.092
CURVE 65*	
298.2	0.093
CURVE 66*	
298.2	0.097
CURVE 67*	
298.2	0.097
CURVE 68*	
298.2	0.100
CURVE 69*	
298.2	0.0096
CURVE 70*	
298.2	0.0093
CURVE 71*	
298.2	0.0097
CURVE 72*	
298.2	0.0096
CURVE 73*	
298.2	0.0093
CURVE 74*	
298.2	0.0099
CURVE 75*	
298.2	0.0104
τ(C)	**α**
CURVE 76* (T = 300 K)	
2000	0.00544
2300	0.00609
2600	0.00636
2800	0.00644
t(min)	**α**
CURVE 77* (T = 300 K)	
1.5	0.00357
31.0	0.00541
92.0	0.00546
180	0.00550
CURVE 78* (T = 300 K)	
10.5	0.00541
31.0	0.00602
92.0	0.00626
180	0.00670
CURVE 79* (T = 300 K)	
5.5	0.00568
10.5	0.00616
22.0	0.00644
92.0	0.00666
180	0.00693
T	**α**
CURVE 80*	
300	0.00437

* Not shown in figure.

DATA TABLE 13. THERMAL DIFFUSIVITY OF MISCELLANEOUS GRAPHITES (continued)

T	α
CURVE 81*	
300	0.00427
τ(C)	α
CURVE 82* (T = 300 K)	
2000	0.00366
2500	0.00389
2700	0.00397
3000	0.00213
t(min)	α
CURVE 83* (T = 300 K)	
2.2	0.00285
10.0	0.00396
30.0	0.00396
180	0.00397
CURVE 84* (T = 300 K)	
2.0	0.00483
10.0	0.00286
30.0	0.00286
120	0.00286
180	0.00286
τ(K)	α
CURVE 85* (T = 300 K)	
2300	0.000270
2500	0.000278
3000	0.000285
T	α
CURVE 86	
329	0.733
336	0.768
557	0.443
557	0.436*

T	α
CURVE 86 (cont.)	
765	0.255
765	0.248
983	0.133
983	0.123
1201	0.125
1201	0.116
1373	0.101
CURVE 87*	
1367	0.067
1624	0.055
1818	0.050
2052	0.048
2402	0.046
2604	0.038
CURVE 88*	
324	0.733
324	0.714
528	0.374
528	0.356
773	0.214
773	0.209
998	0.150
998	0.139
1175	0.117
1175	0.113
1381	0.088
CURVE 89*	
1193	0.079
1307	0.069
1622	0.070
1826	0.070
2044	0.067
2361	0.067
2517	0.060
CURVE 90*	
320	0.264
320	0.250
445	0.206
558	0.164
756	0.091

T	α
CURVE 90 (cont.)*	
760	0.101
955	0.086
955	0.077
1148	0.071
1148	0.064
1312	0.061
1312	0.051
1412	0.048
1412	0.037
CURVE 91*	
1148	0.038
1385	0.033
1608	0.041
1609	0.037
1867	0.034
2062	0.035
2356	0.035
2631	0.039
2631	0.037
CURVE 92*	
320	0.291
320	0.281
526	0.203
526	0.186
763	0.121
764	0.102
970	0.101
970	0.095
1175	0.082
1390	0.071
1390	0.067
CURVE 93*	
1073	0.073
1073	0.066
1344	0.066
1752	0.052
1958	0.052
2304	0.052
2562	0.052

T	α
CURVE 94*	
1476	0.1072
1642	0.0886
1774	0.0876
2185	0.0433
CURVE 95*	
1884	0.0700
2031	0.0680
2086	0.0566
2209	0.0571
CURVE 96*	
1102	0.079
1119	0.076
1174	0.070
1254	0.069
1385	0.068
1496	0.064
1585	0.065
1615	0.065
1691	0.064
1723	0.061
CURVE 97*	
1085	0.081
1176	0.082
1248	0.077
1351	0.072
1428	0.069
1528	0.064
1636	0.064
1756	0.062
1874	0.060
1979	0.060
2050	0.059
2158	0.056
2242	0.055
2338	0.053
2438	0.052
CURVE 98*	
1077	0.086
1115	0.087
1139	0.084

T	α
CURVE 98 (cont.)*	
1173	0.079
1227	0.079
1312	0.074
1410	0.067
1478	0.065
1586	0.067
1729	0.064
1812	0.065
1915	0.064
2021	0.056
2108	0.056
2188	0.054
2303	0.053
2378	0.053
2430	0.051
CURVE 99*	
1110	0.087
1151	0.087
1261	0.092
1282	0.094
1311	0.093
1399	0.084
1485	0.076
1642	0.076
1813	0.074
1955	0.077
2122	0.077
2235	0.077
2339	0.075
2379	0.080
2432	0.087
CURVE 100*	
1100	0.109
1138	0.096
1223	0.103
1384	0.102
1519	0.095
1660	0.096
1771	0.096
1869	0.095
1976	0.034
2116	0.038
2220	0.032
2295	0.032

T	α
CURVE 100 (cont.)*	
2321	0.079
2400	0.075
2438	0.075
CURVE 101*	
1079	0.120
1153	0.105
1178	0.112
1284	0.110
1421	0.101
1619	0.094
1741	0.093
1805	0.089
1956	0.083
2006	0.081
2104	0.078
2225	0.080
2260	0.080
2338	0.072
2375	0.074
CURVE 102*	
1075	0.125
1158	0.130
1168	0.128
1269	0.131
1339	0.113
1405	0.119
1485	0.107
1560	0.108
1625	0.104
1787	0.098
1907	0.097
2022	0.088
2152	0.081
2271	0.082
2296	0.083
2366	0.080
2393	0.081
2433	0.080
CURVE 103	
401	0.843
511	0.409
645	0.265

T	α
CURVE 103 (cont.)	
645	0.247
837	0.161
837	0.148
931	0.130
CURVE 104*	
411	0.729
532	0.373
565	0.339
658	0.289
798	0.222
916	0.161
CURVE 105*	
1176	0.0869
1384	0.0616
1586	0.0453
1850	0.0617
1850	0.0540
2055	0.0545
2331	0.0524
2818	0.0415
CURVE 106*	
506	0.408
532	0.369
570	0.337
653	0.247
665	0.256
672	0.291
820	0.220
863	0.155
863	0.140
931	0.161
CURVE 107*	
1022	0.135
1128	0.110
1273	0.114
1273	0.111
1369	0.092
1404	0.078
1483	0.064

* Not shown in figure.

DATA TABLE 13. THERMAL DIFFUSIVITY OF MISCELLANEOUS GRAPHITES (continued)

T	α
CURVE 108*	
1414	0.096
1559	0.051
1758	0.053
1928	0.046
2068	0.036
2283	0.028
2473	0.028
2655	0.030
2936	0.014
3085	0.013
CURVE 109	
298.2	0.80
CURVE 110	
679	0.304
1089	0.1845
1333	0.1489
1362	0.1494
2059	0.1262
2172	0.1246
2616	0.1089
2793	0.1014
CURVE 111*	
651	0.00431
913	0.00307
1058	0.00245
1171	0.00227
1286	0.00217
1364	0.00199
1586	0.00199
CURVE 112*	
1089	0.00237
1417	0.00217
CURVE 113*	
1148	0.00877
1326	0.00849
1826	0.0118
CURVE 114*	
1089	0.00661
1089	0.00612
1126	0.00632
1522	0.00710
CURVE 115*	
1072	0.1732
1084	0.1792
1099	0.1671
1110	0.1771
1113	0.1831
1122	0.1753
1122	0.1562
1182	0.1640
1211	0.1521
1219	0.1633
1219	0.1612
1221	0.1452
1308	0.1442
1314	0.1461
1319	0.1477
1370	0.1521
1393	0.1506
1400	0.1525
1475	0.1428
1582	0.1408
1582	0.1368
1601	0.1351
1662	0.1386
1681	0.1319
1690	0.1288
1709	0.1269
1776	0.1235
1789	0.1249
1799	0.1229
2033	0.1108
2046	0.1077
2070	0.1046
2110	0.1146
2280	0.1058
2280	0.1019
2311	0.0991
2402	0.0985
2405	0.1007
2405	0.0960
2423	0.0958
2585	0.0926
2610	0.0868
CURVE 116*	
1091	0.1831
1094	0.1721
1098	0.1622
1139	0.1603
1421	0.1421
1442	0.1460
1549	0.1441
1561	0.1421
1708	0.1319
1731	0.1290
1919	0.1198
1928	0.1187
2173	0.1056
2173	0.1038
2211	0.1065
2222	0.0988
2309	0.1038
2331	0.1038
2560	0.0900
2577	0.0943
2622	0.0882
2638	0.0943
2790	0.0900
2799	0.0876
2852	0.0877
2871	0.0898
3023	0.0817
3032	0.0803
3046	0.0825
3046	0.0805
CURVE 117*	
1081	0.1722
1097	0.1659
1118	0.1682
1171	0.1500
1201	0.1601
1253	0.1470
1309	0.1499
1391	0.1517
1391	0.1389
1401	0.1269
1474	0.1510
1510	0.1379
1595	0.1393
1683	0.1292
1752	0.1232
1766	0.1139
CURVE 117 (cont.)*	
1881	0.1139
2003	0.1058
2123	0.1118
2223	0.1070
2332	0.0942
2369	0.1000
2409	0.1038
2553	0.0924
2565	0.0944
2687	0.0960
2721	0.0851
2842	0.0826
2856	0.0813
2916	0.0850
3062	0.0822
3085	0.0846
CURVE 118*	
1073	0.2154
1173	0.2041
1273	0.1946
1373	0.1865
1473	0.1795
1573	0.1733
1673	0.1679
1773	0.1631
1873	0.1589
1973	0.1550
2073	0.1515
2173	0.1484
2273	0.1455
2373	0.1429
2473	0.1405
2573	0.1382
2673	0.1362
2773	0.1342
2873	0.1325
2973	0.1308
3073	0.1292
CURVE 119*	
1073	0.1734
1173	0.1605
1273	0.1497
1373	0.1404
1473	0.1324
1573	0.1254
CURVE 119 (cont.)*	
1673	0.1192
1773	0.1138
1873	0.1089
1973	0.1045
2073	0.1006
2173	0.0970
2273	0.0937
2373	0.0907
2473	0.0879
2573	0.0854
2673	0.0830
2773	0.0808
2873	0.0788
2973	0.0769
3073	0.0751
CURVE 120	
303	0.965
307	0.922
401	0.522
410	0.531
453	0.458
597	0.345
656	0.295
673	0.286
675	0.302
767	0.259
783	0.264
867	0.232
943	0.216
996	0.197
1072	0.184
1087	0.170
1142	0.159
1182	0.186
1271	0.155
1273	0.141
CURVE 121*	
293	0.146
300	0.172
306	0.149
359	0.130
359	0.117
379	0.119
401	0.128
465	0.149
CURVE 121 (cont.)*	
475	0.149
482	0.171
543	0.169
546	0.197
579	0.176
579	0.197
664	0.227
674	0.213
774	0.217
873	0.195
975	0.184
975	0.170
1070	0.165
1080	0.155
1182	0.167
1273	0.167
1287	0.150
CURVE 122*	
2500	0.00452
CURVE 123*	
1600	0.0563
CURVE 124*	
2000	0.0505
CURVE 125*	
2200	0.0510
2500	0.0476
2800	0.0482
CURVE 126*	
1600	0.0589
CURVE 127*	
2000	0.0570
CURVE 128*	
2200	0.0556
2800	0.0523
CURVE 129*	
2500	0.0532
CURVE 130*	
1481	0.115
1881	0.102
2289	0.094
2658	0.084
CURVE 131*	
1896	0.069
2289	0.063
2689	0.060
CURVE 132*	
1313	0.090
1495	0.078
1873	0.072
2289	0.065
2703	0.062
CURVE 133*	
1189	0.0854
CURVE 134*	
1189	0.0992

* Not shown in figure.

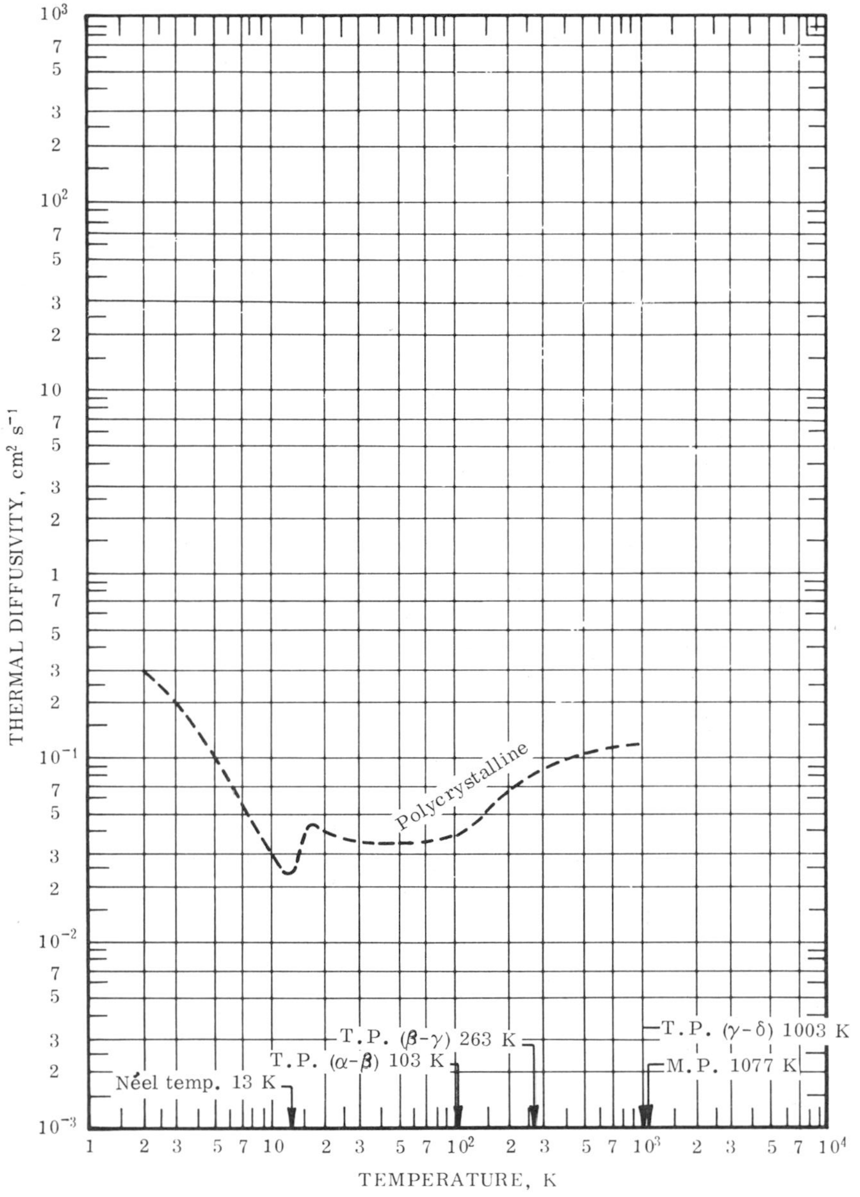

PROVISIONAL VALUES*

[Temperature, T, K; Thermal Diffusivity, α, cm² s⁻¹]

SOLID

(Polycrystalline)

T	α	T	α
2	0.297	35	0.0348
3	0.200	40	0.0346
4	0.137	45	0.0346
5	0.0968	50	0.0347
6	0.0712	60	0.0350
7	0.0545	70	0.0355
8	0.0455	80	0.0360
9	0.0356	90	0.0364
10	0.0301	100	0.0367
11	0.0264	103	0.0368
12	0.0259	150	0.0540
12.5	0.0253	200	0.0706
13	0.0243	250	0.0797
14	0.0298	263	0.0798
15	0.0369	273.2	0.0815
16	0.0430	300	0.0855
18	0.0413	350	0.0918
20	0.0391	400	0.0973
25	0.0365	500	0.0105
30	0.0353	600	0.110
		700	0.114
		800	0.116
		900	0.118
		1000	0.119

REMARKS

Near room temperature the uncertainty of the provisional values is probably of the order of ± 20% but it will be greater at lower temperatures on account of the phase changes and the magnetic transformation. The values below 270 K are applicable only to cerium having electrical resistivity ratio ρ (293K)/ρ (20K) = 1.93.

*All values are estimated.

FIGURE AND TABLE 15R. PROVISIONAL THERMAL DIFFUSIVITY OF CESIUM

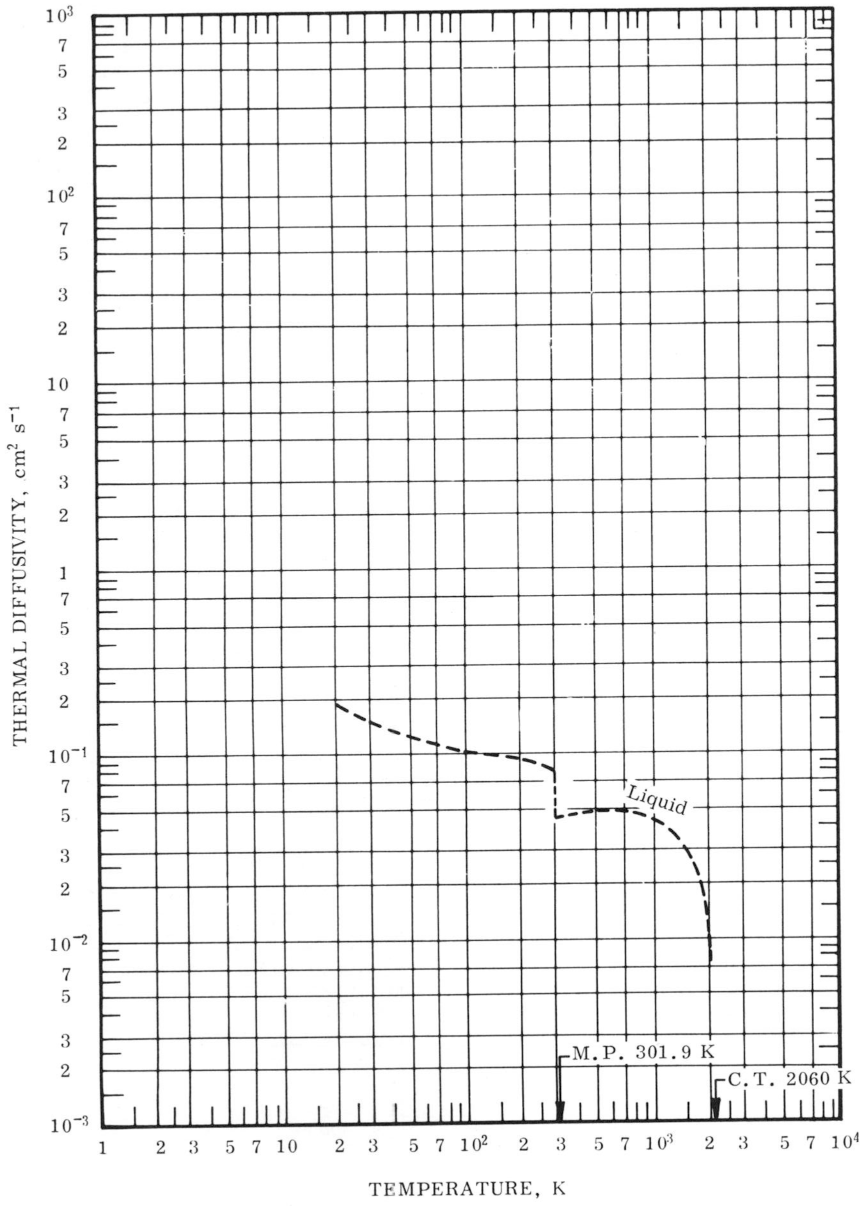

PROVISIONAL VALUES*

[Temperature, T, K; Thermal Diffusivity, α, $cm^2\ s^{-1}$]

SOLID T	SOLID α	LIQUID T	LIQUID α
20	1.85	301.9	0.446
25	1.61	350	0.458
30	1.46	400	0.468
35	1.37	500	0.481
40	1.30	600	0.486
45	1.25	700	0.484
50	1.22	800	0.474
60	1.16	900	0.460
70	1.12	1000	0.440
80	1.08	1100	0.416
90	1.06	1200	0.387
100	1.03	1300	0.354
150	0.960	1400	0.319
200	0.926	1500	0.284
250	0.878	1600	0.250
273.2	0.846	1700	0.215
300	0.794	1800	0.179
301.9	0.789	1900	0.138
		2000	0.0798

REMARKS

The provisional values are for high-purity cesium and are thought to be accurate to within ± 15% of the true values at temperatures from room temperature to about 1500 K and ± 25% at other temperatures.

* All values are estimated.

FIGURE AND TABLE 16R. RECOMMENDED THERMAL DIFFUSIVITY OF CHROMIUM

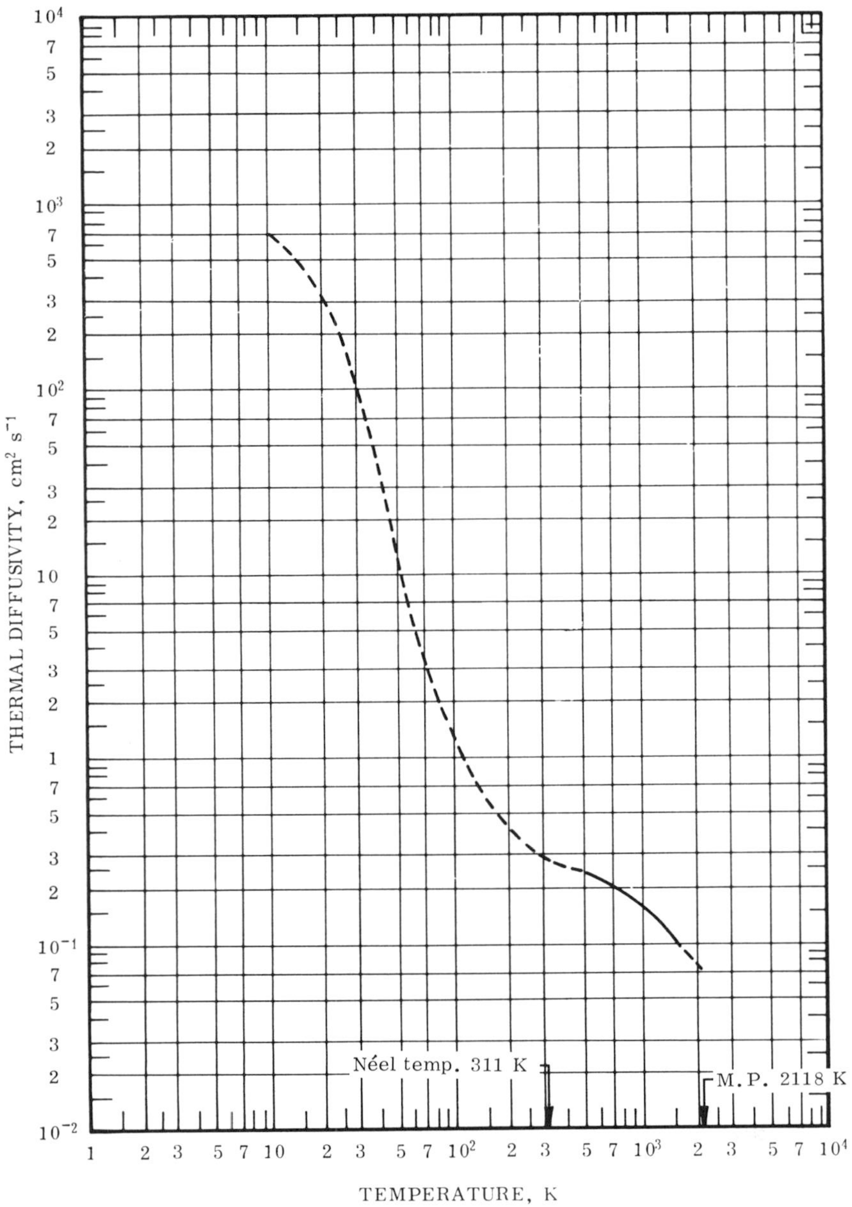

RECOMMENDED VALUES

[Temperature, T, K; Thermal Diffusivity, α, cm^2 s^{-1}]

SOLID

T	α	T	α
10	664*	150	0.564*
11	609*	200	0.400*
12	558*	250	0.328*
13	512*	273.2	0.308*
14	469*	300	0.290*
15	429*	350	0.268*
16	392*	400	0.254*
18	328*	500	0.233
20	275*	600	0.215
25	170*	700	0.198
30	99.0*	800	0.182
35	56.4*	900	0.167
40	31.5*	1000	0.153
45	18.6*	1100	0.141
50	11.7*	1200	0.130
60	5.60*	1300	0.120
70	3.20*	1400	0.112
80	2.10*	1500	0.104
90	1.50*	1600	0.0972
100	1.15*	1700	0.0909*
		1800	0.0855*
		1900	0.0808*
		2000	0.0765*
		2100	0.0725*

REMARKS

The recommended values are for well-annealed high-purity chromium and are thought to be accurate to within ±13% of the true values at temperatures below 150 K and above 700 K and ±5% from 150 to 700 K except possibly near the Néel temperature. At low temperatures the values are highly conditioned by impurity and imperfection and those below 150K are applicable only to a specimen having residual electrical resistivity of 0.0608 $\mu\Omega$ cm.

*In temperature range where no experimental data are available.

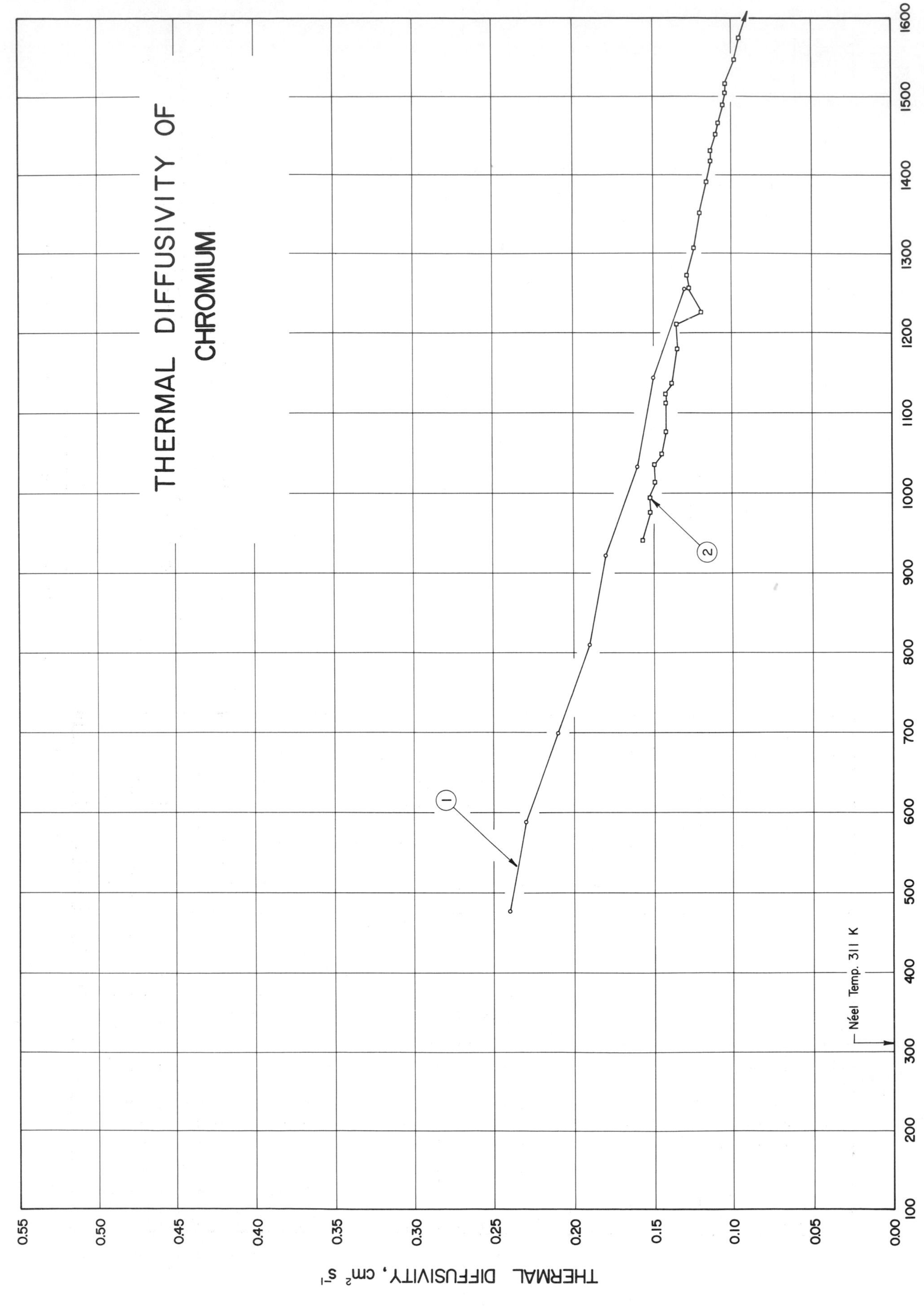
THERMAL DIFFUSIVITY OF CHROMIUM
THERMAL DIFFUSIVITY, cm² s⁻¹
TEMPERATURE, K
0.00
0.05
0.10
0.15
0.20
0.25
0.30
0.35
0.40
0.45
0.50
0.55
100
200
300
400
500
600
700
800
900
1000
1100
1200
1300
1400
1500
1600
Néel Temp. 311 K
1
2

FIGURE 16

SPECIFICATION TABLE 16. THERMAL DIFFUSIVITY OF CHROMIUM

(Impurity < 0.20%; total impurities < 0.50%)

Cur. No.	Ref. No.	Author(s)	Year	Temp. Range, K	Reported Error, %	Name and Specimen Designation	Composition (weight percent), Specifications, and Remarks
1	17	Lucks, C. F. and Deem, H. W.	1958	478-1255			Ductile, commercially pure; supplied by Bureau of Mines; thermal diffusivity calculated from measured conductivity, specific heat, and density.
2	200	Zinov'ev, V. E., Krentsis, R. P., and Gel'd, P. V.	1970	941-1670	< 5		Specimen containing less than 0.05 impurity; fabricated from plate electrolytic Cr of surface area 8 x 8 mm, ground to thickness of 0.214 mm, annealed at 900 K for 3 hrs in vacuum chamber, then heated to 1600 K for a short time; electrical resistivity ratio $\rho(293K)/\rho(4.2K) = 65$; diffusivity measured by using temperature wave method in a plane-parallel plate at frequency of 168.8 cps.

DATA TABLE 16. THERMAL DIFFUSIVITY OF CHROMIUM

(Impurity < 0.20%; total impurities < 0.50%)

[Temperature, T, K; Thermal Diffusivity, α, $cm^2\ s^{-1}$]

T	α	T	α	T	α
CURVE 1		CURVE 2 (cont.)		CURVE 2 (cont.)	
477.6	0.24	1035	0.1496	1417	0.1123
588.7	0.23	1049	0.1447	1430	0.1127
699.8	0.21	1076	0.1412	1451	0.1092
810.9	0.19	1106	0.1417	1465	0.1079
922.1	0.18	1123	0.1417	1488	0.1048
1033.2	0.16	1136	0.1378	1504	0.1030
1144.3	0.15	1179	0.1339	1516	0.1030
1255.4	0.13	1210	0.1343	1546	0.0975
		1225	0.1286	1574	0.0948
CURVE 2		1256	0.1263	1612	0.0886*
		1272	0.1276	1640	0.0811*
941	0.1564	1306	0.1233	1670	0.0781*
976	0.1513	1351	0.1199		
994	0.1520	1391	0.1151		
1013	0.1488				

* Not shown in figure.

FIGURE AND TABLE 17R. RECOMMENDED THERMAL DIFFUSIVITY OF COBALT

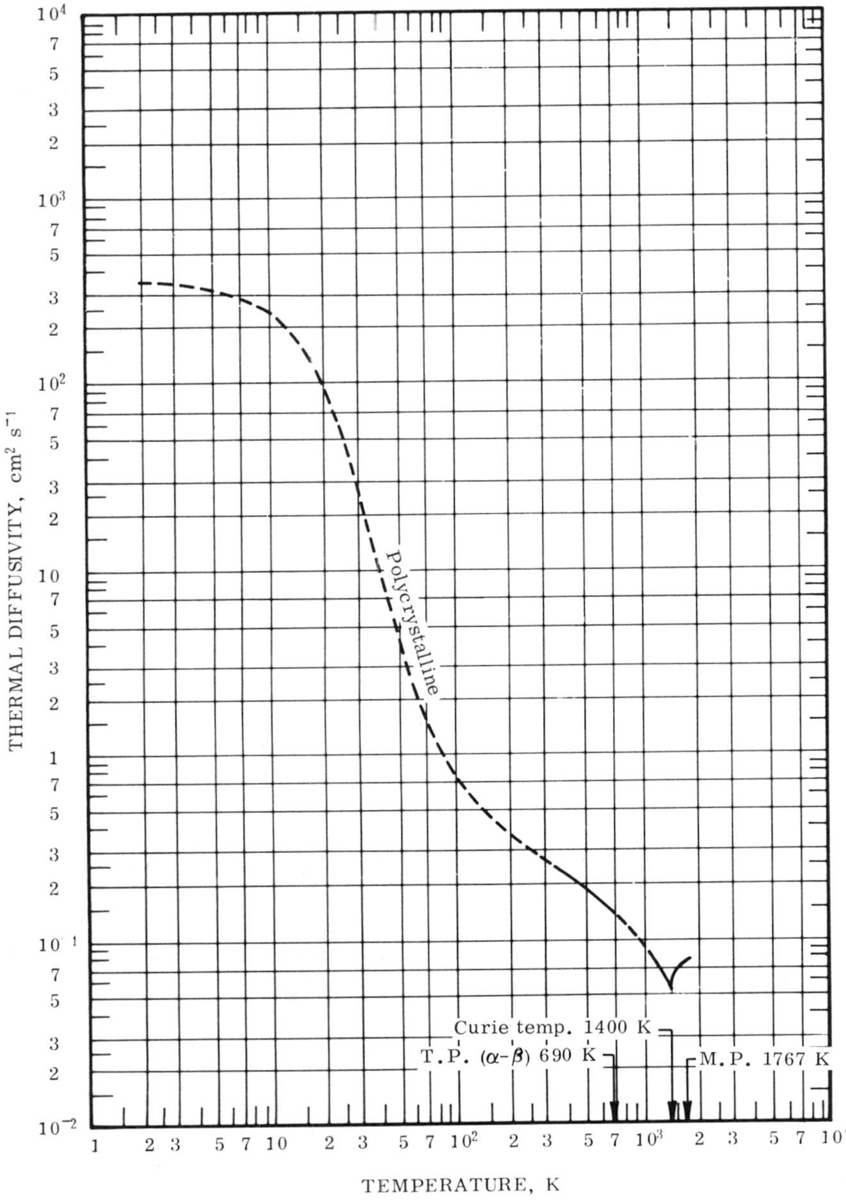

RECOMMENDED VALUES†

[Temperature, T, K; Thermal Diffusivity, α, cm² s⁻¹]

SOLID
(Polycrystalline)

T	α	T	α
2	347*	80	1.20*
3	342*	90	0.950*
4	330*	100	0.795*
5	317*	150	0.474*
6	303*	200	0.362*
7	288*	250	0.302*
8	272*	273.2	0.283*
9	257*	300	0.263*
10	241*	350	0.235
11	226*	400	0.213
12	211*	500	0.179
13	195*	600	0.155
14	180*	700	0.136
15	164*	800	0.120*
16	148*	900	0.107*
18	119*	1000	0.0952
20	94.0*	1100	0.0850
25	54.2*	1200	0.0750
30	30.8*	1300	0.0649
35	17.5*	1400	0.0528
40	10.5*	1500	0.0735
45	6.68*	1600	0.0757
50	4.55*	1700	0.0779
60	2.55*	1767	0.0792*
70	1.66*		

REMARKS

The values are for well-annealed high-purity cobalt and are considered accurate to within ± 7% of the true values near room temperature, ± 13% at low temperatures and up to 1000 K and ± 25% from 1000 K to the melting point. The values above 1000 K are provisional. At low temperatures the values are highly conditioned by impurity and imperfection and those below 200 K are applicable only to a specimen having residual electrical resistivity of 0.09075 μΩ cm.

† Values above 1000 K are provisional.
* In temperature range where no experimental data are available.

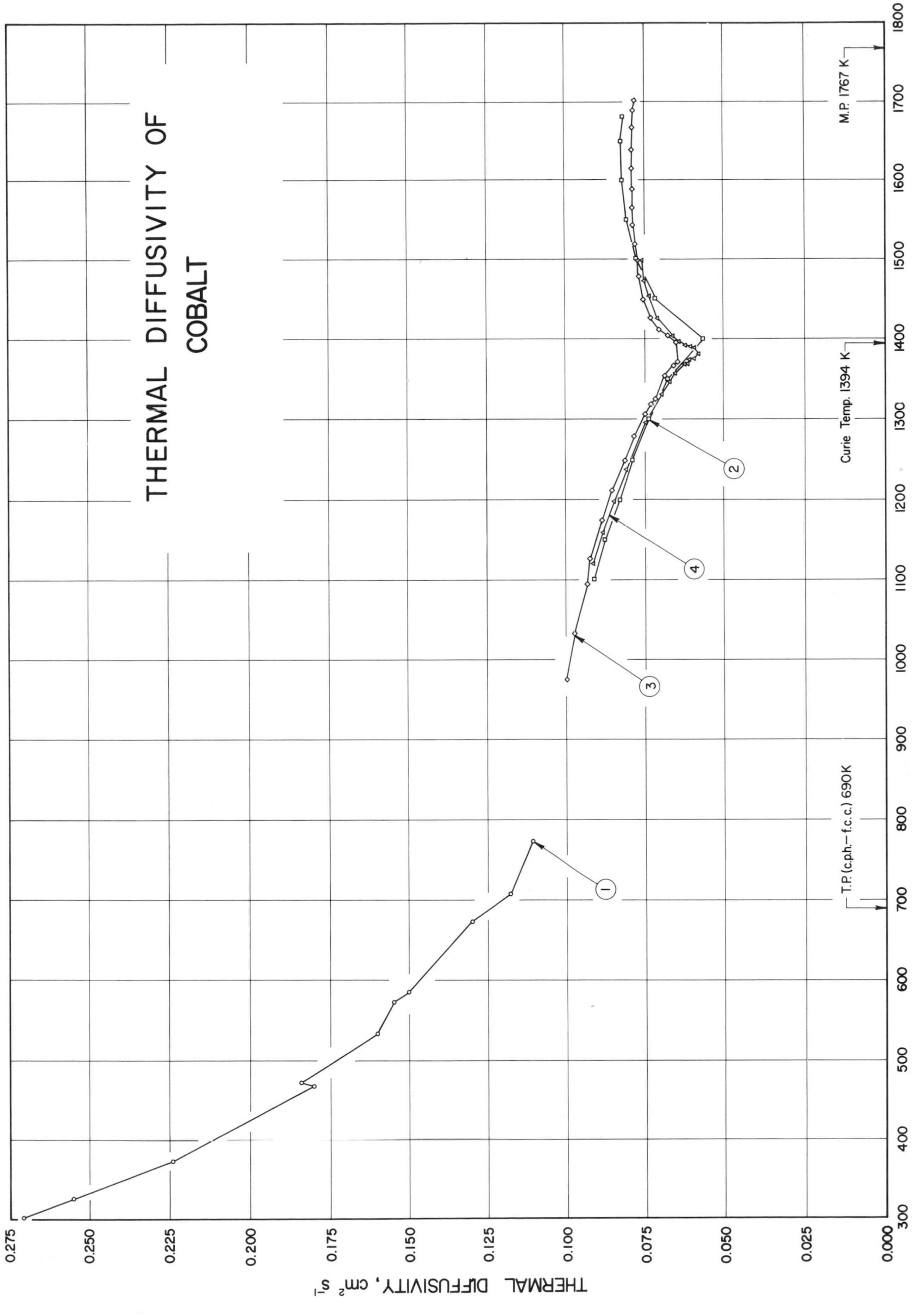
THERMAL DIFFUSIVITY OF COBALT
THERMAL DIFFUSIVITY, cm² s⁻¹
TEMPERATURE, K
0.275
0.250
0.225
0.200
0.175
0.150
0.125
0.100
0.075
0.050
0.025
0.000
300
400
500
600
700
800
900
1000
1100
1200
1300
1400
1500
1600
1700
1800
T.P. (c.p.h.—f.c.c.) 690K
Curie Temp. 1394 K
M.P. 1767 K
1
2
3
4

FIGURE 17

SPECIFICATION TABLE 17. THERMAL DIFFUSIVITY OF COBALT

(Impurity < 0.20%; total impurities < 0.50%)

Cur. No.	Ref. No.	Author(s)	Year	Temp. Range, K	Reported Error, %	Name and Specimen Designation	Composition (weight percent), Specifications, and Remarks
1	18	Habachi, M., Azou, P., and Bastien, P.	1965	301-773			99.999 pure; cylindrical specimen; method based on measuring phase shift and logarithmic attenuation between two points on specimen separated by a distance of 1 cm; Max. amplitude of temperature wave limited to 1 K; pulsation of wave lying in the range from 3 to 30 radian per min; specimen heated on one end and cooled on the other end; measured under a vacuum of 10^{-6} mm Hg.
2	201	Krentsis, R.P., Zinov'yev, V.Ye., Andreyeva, L.P., and Gel'd, P.V.	1970	1100-1681	< 5		99.99 pure; sintered in hydrogen atmosphere; fabricated from flat pieces with area 8 x 8 mm, ground to thickness of 0.18-0.27 mm, annealed at 1000 K for 3-4 hr; diffusivity measured by using temperature plane wave method at frequency of 168.8 cps and at pressure of the order of 1 x 10^{-5} mm Hg.
3	199	Zinov'yev, V.E., Krentsis, R.P., Petrova, L.N., and Gel'd, P.V.	1968	975-1702			Specimen 0.205 x 8 x 8 mm; fabricated by grinding rolled cobalt containing less than 0.05 impurities; resistivity ration $\rho(298K)/\rho(4.2K) = 86$; specimen annealed in vacuum chamber at 1200 K for 7 hrs and at pressure around 1 x 10^{-5} mm Hg; diffusivity measured at 178.3 cps.
4	199	Zinov-yev, V.E., et al.	1968	1120-1498			Specimen 0.198 x 8 x 8 mm; diffusivity measured at 168.8 cps; other conditions same as above.

DATA TABLE 17. THERMAL DIFFUSIVITY OF COBALT

(Impurity < 0.20%; total impurities < 0.50%)

[Temperature, T, K; Thermal Diffusivity, α, cm^2 s^{-1}]

T	α	T	α	T	α	T	α	T	α	T	α	T	α
CURVE 1		CURVE 2		CURVE 3		CURVE 3 (cont.)		CURVE 3 (cont.)		CURVE 4 (cont.)		CURVE 4 (cont.)	
301.2	0.271	1100	0.0910	975	0.0999	1371	0.0645	1638	0.0789	1347	0.0672	1454	0.0735
325.2	0.255	1150	0.0875	1032	0.0973	1396	0.0650	1667	0.0789	1357	0.0652	1474	0.0751
373.2	0.224	1200	0.0829	1095	0.0933	1404	0.0679	1688	0.0786	1369	0.0624	1498	0.0759
468.2	0.180	1250	0.0789	1126	0.0923	1412	0.0705	1702	0.0778	1369	0.0613		
472.2	0.184	1300	0 0736	1174	0.0884	1427	0.0730			1375	0.0608		
533.2	0.160	1350	0.0678	1212	0.0851	1449	0.0752	CURVE 4		1376	0.0593		
573.2	0.155	1400	0.0565	1248	0.0812	1478	0.0766			1382	0.0577		
585.2	0.150	1450	0.0719	1279	0.0784	1499	0.0774	1120	0.0912	1389	0.0593		
673.2	0.130	1500	0.0779	1307	0.0749	1519	0.0779	1159	0.0883	1390	0.0606		
708.2	0.118	1550	0.0805	1319	0.0730	1543	0.0786	1197	0.0847	1392	0.0621		
773.2	0.111	1600	0.0820	1325	0.0716	1565	0.0787	1237	0.0809	1397	0.0640		
		1650	0.0821	1354	0.0689	1589	0.0787	1296	0.0746	1404	0.0661		
		1681	0.0817	1367	0.0660	1615	0.0789	1330	0.0699	1426	0.0707		

FIGURE AND TABLE 18R. RECOMMENDED THERMAL DIFFUSIVITY OF COPPER

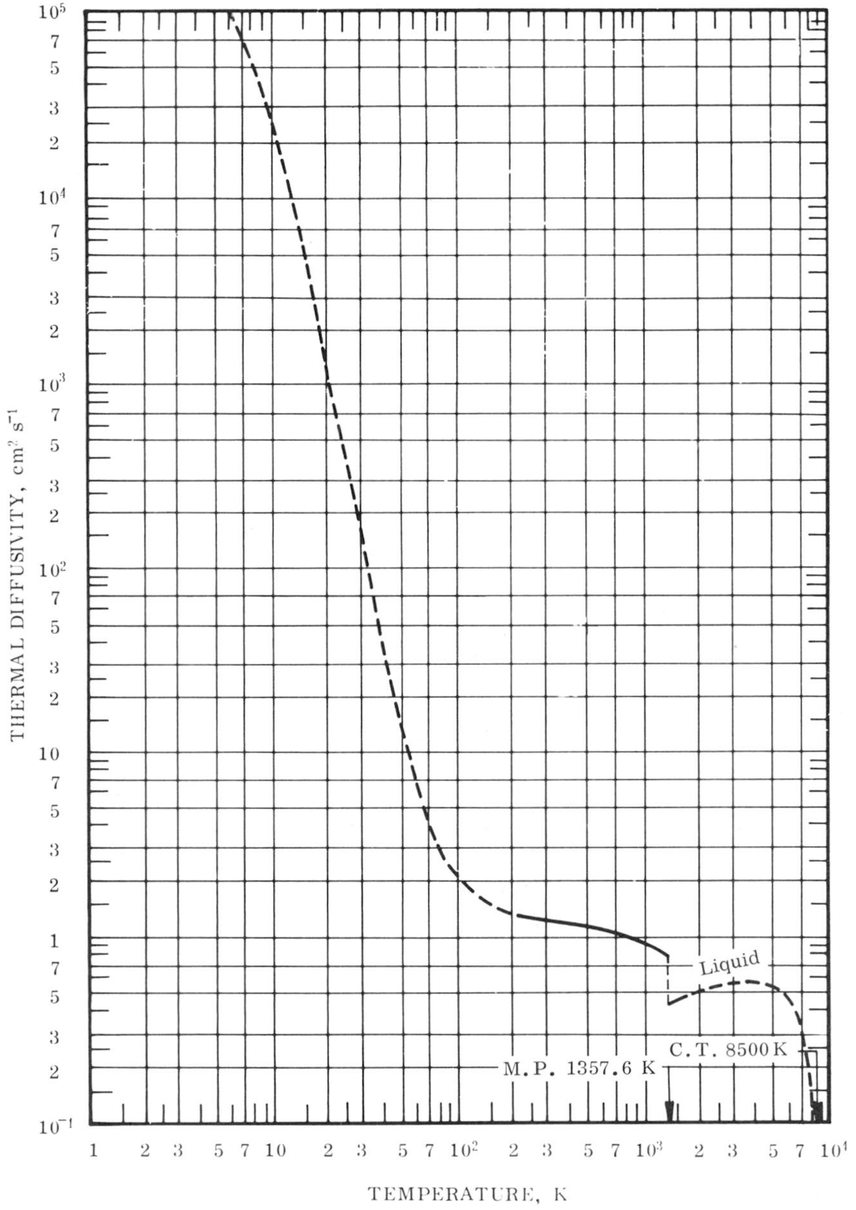

RECOMMENDED VALUES †

[Temperature, T, K; Thermal Diffusivity, α, $cm^2\ s^{-1}$]

SOLID				LIQUID	
T	α	T	α	T	α
1	406000 *	70	4.11 *	1357.6	0.421
2	337000 *	80	3.04 *	1400	0.427
3	264000 *	90	2.46 *	1500	0.440
4	198000 *	100	2.10 *	1600	0.452
5	149000 *	150	1.47 *	1700	0.463
6	100000 *	200	1.30	1800	0.474 *
7	79900 *	250	1.21	1900	0.484 *
8	58200 *	273.2	1.19	2000	0.493 *
9	42700 *	300	1.17	2200	0.509 *
10	31500 *	350	1.14	2400	0.523 *
11	23100 *	400	1.11	2600	0.533 *
12	16700 *	500	1.07	2800	0.541 *
13	12200 *	600	1.03	3000	0.547 *
14	8830 *	700	1.00	3500	0.555 *
15	6580 *	800	0.968	4000	0.554 *
16	4880 *	900	0.935	4500	0.546 *
18	2750 *	1000	0.903	5000	0.530 *
20	1620 *	1100	0.870	6000	0.468 *
25	487 *	1200	0.838	7000	0.350 *
30	180 *	1300	0.806	8000	0.160 *
35	79.4 *	1357.6	0.787	8500	0.032
40	40.4 *				
45	23.0 *				
50	14.0 *				
60	6.52 *				

REMARKS

The recommended values are for well-annealed high-purity copper and are considered accurate to within ±4% of the true values near room temperature and ±6% at low and high temperatures. The values for molten copper are provisional and those up to about 2000 K should be good to ±20%. At low temperatures the values are highly conditioned by impurity and imperfection and those below 100 K are applicable only to a specimen having residual electrical resistivity of 0.000579 $\mu\Omega$ cm.

† Values for molten copper are provisional.
* In temperature range where no experimental data are available.

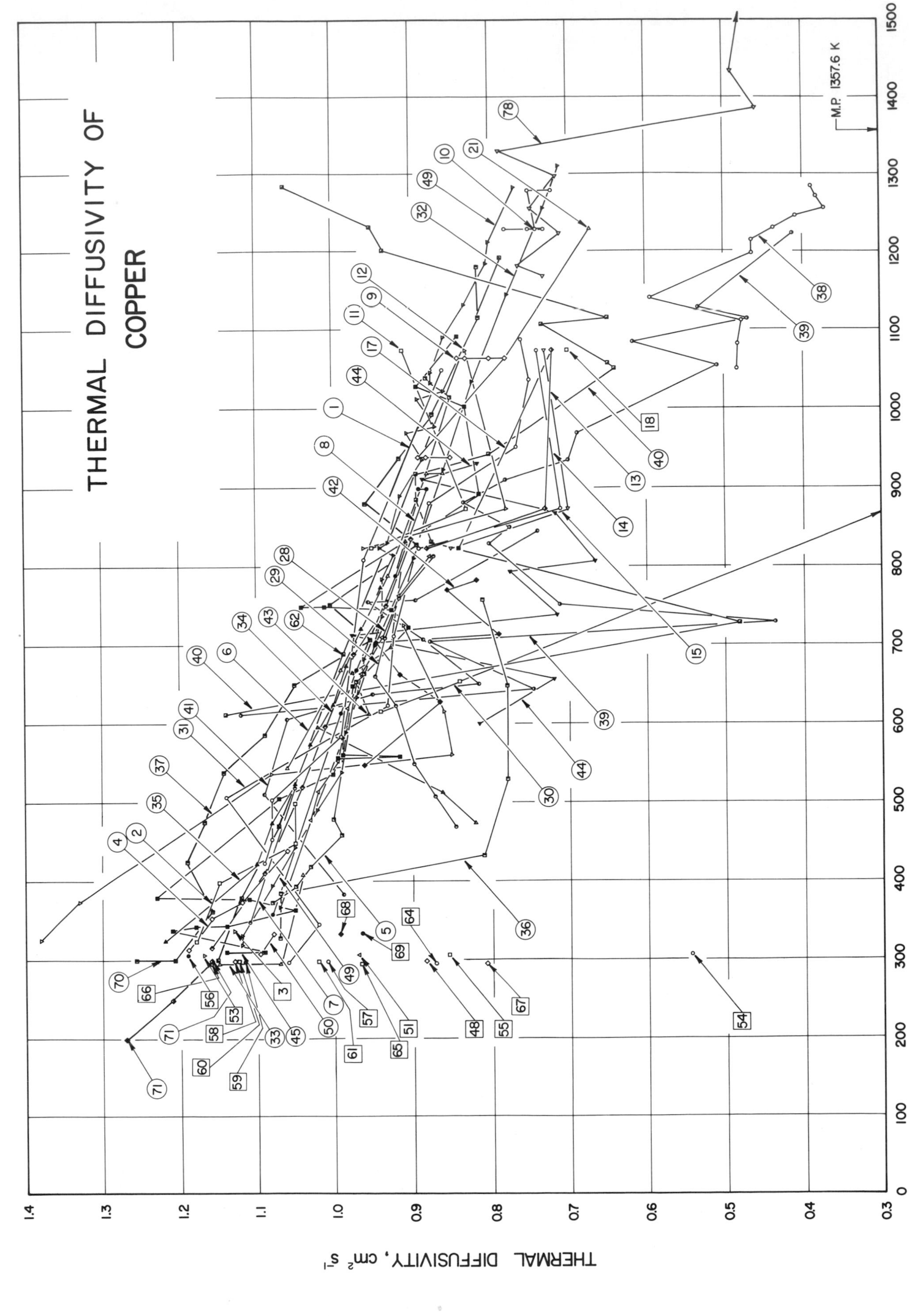
THERMAL DIFFUSIVITY OF COPPER
THERMAL DIFFUSIVITY, cm² s⁻¹
TEMPERATURE, K
M.P. 1357.6 K
1.4
1.3
1.2
1.1
1.0
0.9
0.8
0.7
0.6
0.5
0.4
0.3
0
100
200
300
400
500
600
700
800
900
1000
1100
1200
1300
1400
1500

FIGURE 18

SPECIFICATION TABLE 18. THERMAL DIFFUSIVITY OF COPPER

(Impurity < 0.20% each; total impurities < 0.50%)

Cur. No.	Ref. No.	Author(s)	Year	Temp. Range, K	Reported Error, %	Name and Specimen Designation	Composition (weight percent), Specifications, and Remarks
1	19, 20	Butler, C. P. and Inn, E. C. Y.	1957	423-1048	5-10	OFHC	Commercially pure oxygen free high conductivity copper; cylindrical specimen 0.375 in. in dia. and length lying in the range from 1 to 2.5 cm; subjected to irradiance from carbon arc lamp heat source; spectral distribution approximates that of a 5700 K black body source; specimen blackened with camphor black; measured under a vacuum of ~5 microns.
2	19, 20	Butler, C. P. and Inn, E. C. Y.	1957	323-498	5-10	OFHC	Above specimen allowed to cool and then exposed to the arc lamp to measure diffusivity again.
3*	19, 20	Butler, C. P. and Inn, E. C. Y.	1957	338.2	5-10	OFHC	Above specimen allowed to cool and then exposed to the arc lamp to measure diffusivity again.
4	19, 20	Butler, C. P. and Inn, E. C. Y.	1957	313-438	5-10	OFHC	Above specimen allowed to cool and then exposed to the arc lamp to measure diffusivity again.
5	19, 20	Butler, C. P. and Inn, E. C. Y.	1957	373-1038	5-10	OFHC	Above specimen allowed to cool and then exposed to the arc lamp to measure diffusivity again.
6	19, 20	Butler, C. P. and Inn, E. C. Y.	1957	543-918	5-10	OFHC	Above specimen allowed to cool and then exposed to the arc lamp to measure diffusivity again.
7	19, 20	Butler, C. P. and Inn, E. C. Y.	1957	348, 518	5-10	OFHC	Above specimen allowed to cool and then exposed to the arc lamp to measure diffusivity again.
8	19, 20	Butler, C. P. and Inn, E. C. Y.	1957	358-898	5-10	OFHC	Above specimen allowed to cool and then exposed to the arc lamp to measure diffusivity again.
9	19, 20	Butler, C. P. and Inn, E. C. Y.	1957	938-1063	5-10	OFHC	Above specimen exposed to the arc lamp to measure diffusivity again.
10	19, 20	Butler, C. P. and Inn, E. C. Y.	1957	1228-1278	5-10	OFHC	Above specimen exposed to the arc lamp to measure diffusivity again.
11	19, 20	Butler, C. P. and Inn, E. C. Y.	1957	823-1073	5-10	OFHC; A	Commerically pure oxygen free high conductivity copper; cylindrical specimen 0.375 in. in dia. and length lying in the range from 1 to 2.5 cm; subjected to irradiance from carbon arc lamp heat source; spectral distribution approximates that of a 5700 K black body source; specimen blackened with camphor black; measured under a vacuum of ~5 microns; subjected to a large number of measurements prior to present experiment.
12	19, 20	Butler, C. P. and Inn, E. C. Y.	1957	823-1073	5-10	OFHC; A	Above specimen allowed to cool and then exposed to the arc lamp to measure diffusivity again.
13	19, 20	Butler, C. P. and Inn, E. C. Y.	1957	823-1073	5-10	OFHC; A	Above specimen allowed to cool and then exposed to the arc lamp to measure diffusivity again.
14	19, 20	Butler, C. P. and Inn, E. C. Y.	1957	823-1073	5-10	OFHC; A	Above specimen allowed to cool and then exposed to the arc lamp to measure diffusivity again.
15	19, 20	Butler, C. P. and Inn, E. C. Y.	1957	823-1073	5-10	OFHC; A	Above specimen allowed to cool and then exposed to the arc lamp to measure diffusivity again.
16*	19, 20	Butler, C. P. and Inn, E. C. Y.	1957	823, 1073	5-10	OFHC; A	Above specimen allowed to cool and then exposed to the arc lamp to measure diffusivity again.

* Not shown in figure.

SPECIFICATION TABLE 18. THERMAL DIFFUSIVITY OF COPPER (continued)

Cur. No.	Ref. No.	Author(s)	Year	Temp. Range, K	Reported Error, %	Name and Specimen Designation	Composition (weight percent), Specifications, and Remarks
17	19, 20	Butler, C. P. and Inn, E. C. Y.	1957	823, 1073	5-10	OFHC; A	Above specimen allowed to cool and then exposed to the arc lamp to measure diffusivity again.
18	19, 20	Butler, C. P. and Inn, E. C. Y.	1957	1073. 2	5-10	OFHC; A	Above specimen allowed to cool and then exposed to the arc lamp to measure diffusivity again.
19*	19, 20	Butler, C. P. and Inn, E. C. Y.	1957	1073. 2	5-10	OFHC; A	Above specimen allowed to cool and then exposed to the arc lamp to measure diffusivity again.
20*	19, 20	Butler, C. P. and Inn, E. C. Y.	1957	1073. 2	5-10	OFHC; A	Above specimen allowed to cool and then exposed to the arc lamp to measure diffusivity again.
21	19, 20	Butler, C. P. and Inn, E. C. Y.	1957	938-1228	5-10	OFHC; B	Specimen similar to above specimen but prepared from a different stock sample and subjected to less extensive prior heat treatment; measured under the same conditions as above specimen.
22*	19, 20	Butler, C. P. and Inn, E. C. Y.	1957	938-1228	5-10	OFHC; B	Above specimen allowed to cool and then exposed to the arc lamp to measure diffusivity again.
23*	19, 20	Butler, C. P. and Inn, E. C. Y.	1957	938-1228	5-10	OFHC; B	Above specimen allowed to cool and then exposed to the arc lamp to measure diffusivity again.
24*	19, 20	Butler, C. P. and Inn, E. C. Y.	1957	938-1228	5-10	OFHC; B	Above specimen allowed to cool and then exposed to the arc lamp to measure diffusivity again.
25*	19, 20	Butler, C. P. and Inn, E. C. Y.	1957	938-1228	5-10	OFHC; B	Above specimen allowed to cool and then exposed to the arc lamp to measure diffusivity again.
26*	19, 20	Butler, C. P. and Inn, E. C. Y.	1957	1063, 1228	5-10	OFHC; B	Above specimen allowed to cool and then exposed to the arc lamp to measure diffusivity again.
27*	19, 20	Butler, C. P. and Inn, E. C. Y.	1957	1228. 2	5-10	OFHC; B	Above specimen allowed to cool and then exposed to the arc lamp to measure diffusivity again.
28	21	El-Hifni, M. A. and Chao, B. T.	1956	478-814	2-3	Electrolytic tough-pitch	99. 90 pure; tubular specimen 0. 875 in. O. D. and 5 in. long, relatively thin walled; another tube of similar material mounted concentrically with specimen to minimize heat losses; one end of specimen-and-shield assembly immersed in a liquid heating bath, while the other end supported by a transite disc for insulation; cyclic varying current generates required temp.wave at heating bath; heat supplied by electric current removed by forced draft of cooling air; one-dimensional heat flow; a minimum of three complete temp.waves recorded.
29	21	El-Hifni, M. A. and Chao, B. T.	1956	469-814	2-3	Phosphorized	99. 90 Cu and 0. 005-0. 02 P; nominal composition from Metals Handbook, Vol. 1, 8th ed., p. 961, 1961; tubular specimen 0. 875 in. O. D. and 5 in. long, relatively thin walled; another tube of similar material mounted concentrically with specimen to minimize heat losses; one end of specimen-and-shield assembly immersed in a liquid heating bath, while the other end supported by a transite disc for insulation; cyclic varying current generates required temp.wave at heating bath; heat supplied by electric current removed by forced draft of cooling air; one-dimensional heat flow; a minimum of three complete temp.waves recorded.

* Not shown in figure.

Cur. No.	Ref. No.	Author(s)	Year	Temp. Range, K	Reported Error, %	Name and Specimen Designation	Composition (weight percent), Specifications, and Remarks
30	8	Sheer, C., Mead, L.H., Rothacker, D.L., and Johnson, L.H.	1957	615-965	±20	OFHC	Commercially pure; cylindrical specimen 0.625 in. in dia. and 1.375 in. long; radial heat flow inhibited by wrapping specimen with asbestos tape before mounting in its graphite housing.
31	1	Sonnenschein, G. and Winn, R.A.	1960	297-1087			99.48 Cu, 0.1 Be, 0.1 Co, 0.1 Mg, 0.1 Sn, 0.01 Cr, and traces of As, B, Fe, P, and Sb; cylindrical specimen 0.635 cm in dia; heated several times to 873.2 K prior to actual measurement; front surface of specimen covered with fine film of lamp black; measured under a vacuum of $\sim 10^{-4}$ mm Hg; one-dimensional transient heat flow; flash method used to measure diffusivity.
32	17	Lucks, C.F. and Deem, H.W.	1958	367-1311			Electrolytic tough pitch; Federal Specification QQC-502; supplied by Williams and Co. (Revere Copper and Brass, Inc.); cold drawn; thermal diffusivity calculated from measured conductivity, specific heat and density.
33	4	Jenkins, R.J. and Parker, W.J.	1961	295-408	± 5	OFHC	Oxygen free high conductivity copper; square specimen 1.9 cm side and 0.312 cm thick; high intensity short duration light pulse absorbed in front surface of thermally insulated specimen coated with camphor black; 408.2 K measurement obtained by heating specimen holder and specimen with an infrared lamp; both data points at 295.2 K obtained from measurements using different equations for data reduction.
34	22, 23, 114, 131	Sidles, P.H. and Danielson, G.C.	1953	316-835			High purity copper; 0.05 max Ag and traces of Si, and Pb; sample $\sim$0.125 in. in dia. and 50 cm min long; specimen measured in small vacuum furnace under a vacuum of 10^{-4} mm Hg; modified Angström method used for measuring diffusivity.
35	18	Habachi, M., Azou, P., and Bastien, P.	1965	324-773			99.999 pure; cylindrical specimen; method based on measuring phase shift and logarithmic attenuation between two points on specimen separated by a distance of 1 cm; max amplitude of temp. wave limited to 1 K; pulsation of wave lying in the range from 3 to 30 radians per min; specimen heated on one end and cooled on the other end; measured under a vacuum of 10^{-6} mm Hg.
36	18	Habachi, M., et al.	1965	328-757		15	99.99 Cu (by difference), and 0.01 Bi; cylindrical specimen; tempered at temp. above 1023.2 K; Lorenz number reported as 5.819, 5.823, 5.822, 5.815, 5.829, 5.830, 5.831, 5.831, and 5.831 x 10^{-9} cal s^{-1} ohm K^{-2} at 298.2, 326.2, 373.2, 423.2, 473.2, 523.2, 573.2, 623.2, 673.2, and 723.2 K, respectively; method based on measuring phase shift and logarithmic attenuation between two points on specimen separated by a distance of 1 cm; max amplitude of temp wave limited to 1 K; pulsation of wave lying in the range from 3 to 30 radians per min; specimen heated on one end and cooled on the other end; measured under a vacuum of 10^{-6} mm Hg; specimen measured during first cycle.
37	18	Habachi, M., et al.	1965	326-725		15	Above specimen measured again during second cycle.
38	10	Sheer, C., Fitz, C.D., Mead, L.H., Holmgren, J.D., Rothacker, D.L., and Allmand, D.	1958	1050-1284			Cylindrical specimen 1.0 cm in dia. and 4.5 cm long; machined to dimensions given; insulated on the sides and over one end face; other end face suddenly exposed to a constant heat flow from the plasma of a high intensity arc; measured in vacuum chamber under an ambient pressure of 0.1 atmosphere; diffusivity data computed assuming conditions of zero heat flow across the lateral and rear end surfaces.
39	10	Sheer, C., et al.	1958	611-1223			Above specimen allowed to cool in chamber after being heated during above measurements; exposed to the arc to measure diffusivity again.

SPECIFICATION TABLE 18. THERMAL DIFFUSIVITY OF COPPER (continued)

Cur. No.	Ref. No.	Author(s)	Year	Temp. Range, K	Reported Error, %	Name and Specimen Designation	Composition (weight percent), Specifications, and Remarks
40	10	Sheer, C., Fitz, C. D., Mead, L. H., Holmgren, J. D., Rothacker, D. L., and Allmand, D.	1958	611-1283			Diffusivity data for above specimen calculated using a first order correction factor to account for heat losses through the lateral and end surfaces.
41	24	Dennis, J. E., Hirschman, A., Derksen, W. L., and Monahan, T. I.	1960	383-845			Disc specimen 0.65 cm in dia. and 0.205 cm in thickness; irradiated with a chopped beam in a carbon-arc image furnace; thermal diffusivity determined from measured phase lag between the square wave irradiance impinging upon the front face of the specimen and the resultant sinusoidal temp. of the rear face; error in calculating diffusivity (due to the use of a square instead of a sinusoidal heat input) estimated to be -3.2%.
42	24	Dennis, J. E., et al.	1960	547-782			Disc specimen 0.65 cm in dia. and 0.254 cm in thickness; irradiated with a chopped beam in a carbon-arc image furnace; thermal diffusivity determined from measured phase lag between the square wave irradiance impinging upon the front face of the specimen and the resultant sinusoidal temp. of the rear face; error in calculating diffusivity (due to the use of a square instead of a sinusoidal heat input) estimated to be -3.2%.
43	24	Dennis, J. E., et al.	1960	473-832			Disc specimen 0.65 cm in dia. and 0.305 cm in thickness; irradiated with a chopped beam in a carbon-arc image furnace; thermal diffusivity determined from measured phase lag between the square wave irradiance impinging upon the front face of the specimen and the resultant sinusoidal temp. of the rear face; error in calculating diffusivity (due to the use of a square instead of a sinusoidal heat input) estimated to be -3.2%.
44	24	Dennis, J. E., et al.	1960	600-930			Disc specimen 0.65 cm in dia. and 0.364 cm in thickness; irradiated with a chopped beam in a carbon-arc image furnace; thermal diffusivity determined from measured phase lag between the square wave irradiance impinging upon the front face of the specimen and the resultant sinusoidal temp of the rear face; error in calculating diffusivity (due to the use of a square instead of a sinusoidal heat input) estimated to be -3.2%.
45	14	Mrozowski, S., Andrew, J. F., Juul, N., Sato, S., Strauss, H. E., and Tsuzuku, T.	1963	310-1091			Diffusivity determined from data of amplitude ratio and phase shift measured at two longitudinally located points on specimen.
46*	6	Moser, J. B. and Kruger, O. L.	1963	298.2			Plate specimen with surface area lying in the range from 1 to 4 cm^2 and thickness in the range from 0.1 to 0.3 cm; front surface thinly coated with colloidal graphite; irradiated with a pulse of thermal energy of short duration; diffusivity determined from measured history of the back surface temp.

* Not shown in figure.

SPECIFICATION TABLE 18. THERMAL DIFFUSIVITY OF COPPER (continued)

Cur. No.	Ref. No.	Author(s)	Year	Temp. Range, K	Reported Error, %	Name and Specimen Designation	Composition (weight percent), Specifications, and Remarks
47	7	diNovi, R. A.	1963	298.2	10		Specimen with thickness lying in the range from 1 to 2 mm; front surface uniformly irradiated by a very short pulse of radiant energy supplied by a xenon flash tube; diffusivity determined from measured history of the back surface temp; temp. at which specimen was measured not given by author but assumed to be room temp.
48	7	diNovi, R. A.	1963	298.2	10		Uniformly bonded specimen consisting of two 0.063 cm thick pieces joined with a solder braze; front surface uniformly irradiated by a very short pulse of radiant energy supplied by a xenon flash tube; diffusivity determined from measured history of the back surface temp; temp. at which specimen was measured not given by author but assumed to be room temp; data point given is the average of the range of data given by author.
49	113	Sidles, P. H. and Danielson, G. C.	1960	319-1282		Spectrographically Standardized Copper	Spectrographically standardized copper; obtained from Johnson, Matthey and Co.; diffusivity measured using modified Angström method.
50	143	King, R. W.	1915	308, 333			Wire specimen 0.25 cm in dia. and 30 cm long; greater portion of specimen coiled up so that it occupies a space of ~7 cm long; other end inserted in small resistance heater supplied with sinusoidally varying electric current; diffusivity determined from measurement of the velocities at which the impressed heat waves travel along specimen; data points reported obtained from conductivity data reported by author divided by a density of 8.93 g cm^{-3} and specific heat values of 0.0928 and 0.0938 cal g^{-1} K^{-1} at 308.2 and 333.2 K, respectively; each data point represents average result of two independent measurements.
51	144	Larson, K. B. and Koyama, K.	1967	306.2		OFHC copper	Disk specimen 0.79 cm in dia. and 9.60 x 10^{-2} cm in thickness; machined from solid stock with the flat faces parallel to within 0.1 degrees of arc; specimen initially at room temp; effective temp of measurement calculated from the equation of Parker and others (J. Appl. Phys. 32 (1679), 1961); thermal energy pulse from a xenon flash tube used to irradiate front face; diffusivity determined from measured temp history of rear face assuming instantaneous heat pulse.
52*	144	Larson, K. B. and Koyama, K.	1967	306.2		OFHC copper	Diffusivity determined again from above measurement for above specimen assuming a sawtooth heat pulse.
53	144	Larson, K. B. and Koyama, K.	1967	306.2		OFHC copper	Diffusivity determined again from above measurement for above specimen employing an empirical function closely describing the actual waveform of the heat pulse.
54	144	Larson, K. B. and Koyama, K.	1967	306.2		OFHC copper	Disk specimen 0.79 cm in dia. and 4.06 x 10^{-2} cm in thickness; machined from solid stock with the flat faces parallel to within 0.1 degrees of arc; diffusivity determined from measured temp. history of rear face assuming instantaneous heat pulse; other conditions same as above.
55	144	Larson, K. B. and Koyama, K.	1967	306.2		OFHC copper	Diffusivity determined again from above measurement for above specimen assuming a sawtooth heat pulse.
56	144	Larson, K. B. and Koyama, K.	1967	306.2		OFHC copper	Diffusivity determined again from above measurement for above specimen employing an empirical function closely describing the actual waveform of the heat pulse.

* Not shown in figure.

SPECIFICATION TABLE 18. THERMAL DIFFUSIVITY OF COPPER (continued)

Cur. No.	Ref. No.	Author(s)	Year	Temp. Range, K	Reported Error, %	Name and Specimen Designation	Composition (weight percent), Specifications, and Remarks
57	175	Oualid, J.	1961	298.2			Slab specimen; thermal shock method used to measure diffusivity; measured in vacuum; temperature of measurement not given by author but assumed to be room temperature.
58	184	Taylor, R.	1965	298.2			Spectrographically pure; cylindrical specimen 0.25 in. in diameter and 1.271 cm long; front face exposed to heat pulse from a xenon flash tube; thermal diffusivity calculated from measured time necessary for the rear face to reach one-half the maximum temperature rise; temperature of measurement not given by author but assumed to be room temperature.
59	184	Taylor, R.	1965	298.2			Cylindrical specimen 0.25 in. in diameter and 0.698 cm long; other conditions and specifications same as above.
60	184	Taylor, R.	1965	298.2			Cylindrical specimen 0.25 in. in diameter and 0.6375 cm long; other conditions and specifications same as above.
61	108	Steinberg, S., Larson, R.E., and Kydd, A.R.	1963	298			Specimen 20 cm square, 0.3270 cm in thickness; diffusivity measuring temperature not reported but here assumed to be 25 C.
62	156, 56	Degas, P. and Bertin, J. L.	1970	300-1192			8 mm in diameter and 2 to 5 mm thick.
63*	158	Juul, N.H.	1964	309-1090			Thin tube specimen.
64	160	Smith, R.H.	1959	295		Deoxidized	Specimen size 1 x 1 in. x 0.1246 in. thick; diffusivity measured using flash heating technique; diffusivity calculated using $\alpha = 1.37\ L^2/\pi^2\ t_{0.5}$.
65	160	Smith, R.H.	1959	295		Deoxidized	The above measurement, but using $\alpha = 0.48\ L^2/\pi^2\ t_x$.
66	160	Smith, R.H.	1959	295		Deoxidized	Another measurement; diffusivity measured using $\alpha = 1.37\ L^2/\pi^2\ t_{0.5}$.
67	160	Smith, R.H.	1959	295		Deoxidized	The above measurement, but using $\alpha = 0.48\ L^2/\pi^2\ t_x$.
68	160	Smith, R.H.	1959	333		Deoxidized	Similar to the above specimen; diffusivity measured at higher temperature; diffusivity calculated using $\alpha = 1.37\ L^2/\pi^2\ t_{0.5}$.
69	160	Smith, R.H.	1959	333		Deoxidized	The above measurement, diffusivity measured using $\alpha = 0.48\ L^2/\pi^2\ t_x$.
70	198	Emery, A.F. and Smith, J.R.	1968	300-1180			Oxygen-free copper.
71	197	Adams, C.H. and Wyman, M.E.	1969	200-300			99.9 pure specimen.
72*	196	Kobayashi, K. and Kumada, T.	1967	298-830	≤5		Disk specimen <15 mm in diameter and 2-10 mm in thickness; diffusivity measured in vacuum.
73*	195, 194	Hirschman, A., Dennis, J., Derksen, W.H. and Monahan, T.I.	1961	392-842			Disk specimen 0.65 cm in diameter and 0.205 cm thick.
74*	195, 194	Hirschman, A., et al.	1961	547-781			Disk specimen 0.65 cm in diameter and 0.254 cm thick.
75*	195, 194	Hirschman, A., et al.	1961	473-829			Disk specimen 0.65 cm in diameter and 0.305 cm thick.

* Not shown in figure.

SPECIFICATION TABLE 18. THERMAL DIFFUSIVITY OF COPPER (continued)

Cur. No.	Ref. No.	Author(s)	Year	Temp. Range, K	Reported Error, %	Name and Specimen Designation	Composition (weight percent), Specifications, and Remarks
76*	195, 194	Hirschman, A., Dennis, J., Derksen, W.H. and Monahan, T.I.	1961	599-929			Disk specimen 0.65 cm in diameter and 0.364 cm thick.
77*	162	Kobayasi, K. and Kumada, T.	1968	293-1222		Electrolytic copper	Disk specimen 15 mm in diameter and 15 mm thick; density 8.93 g cm^{-3}.
78	299	Mardykin, I.P. and Fillipov, L.P.	1968	1167-1730			In solid and liquid states; measured by a radial periodic method.
79*	306, 307	Angström, A.J.	1861	306-341			Bar specimen; 23.75 mm thick.
80*	306, 307	Angström, A.J.	1861	304-317			Bar specimen; 35 mm square and 1180 mm long.
81*	308	Angström, A.J.	1863	302-345			Specimen heated by means of either gas-flame or vapor of water.

* Not shown in figure.

DATA TABLE 18. THERMAL DIFFUSIVITY OF COPPER

(Impurity < 0.20% each; total impurities < 0.50%)

[Temperature, T, K; Thermal Diffusivity, α. $cm^2\ s^{-1}$]

T	α
CURVE 1	
423.2	1.09
453.2	1.08
503.2	1.08
583.2	0.99
808.2	0.96
1048.2	0.86
CURVE 2	
323.2	1.18
398.2	1.15
448.2	1.05
498.2	1.05
CURVE 3*	
338.2	1.13
CURVE 4	
313.2	1.19
353.2	1.16
373.2	1.12
438.2	1.06
CURVE 5	
373.2	1.08
393.2	1.05
418.2	1.03
458.2	0.99
478.2	1.00
653.2	0.97
663.2	0.96
1013.2	0.85
1038.2	0.88
CURVE 6	
543.2	1.06
668.2	0.99
788.2	0.93
CURVE 6 (cont.)	
918.2	0.88
918.2	0.86
CURVE 7	
348.2	1.11
518.2	1.05
CURVE 8	
358.2	1.08
613.2	0.99
668.2	0.97
788.2	0.92
898.2	0.88
898.2	0.89
CURVE 9	
938.2	0.85
938.2	0.88
938.2	0.88*
938.2	0.89
938.2	0.89*
1063.2	0.78
1063.2	0.80
1063.2	0.83
1063.2	0.83*
1063.2	0.84
1063.2	0.80*
CURVE 10	
1228.2	0.75
1228.2	0.74
1228.2	0.78
1228.2	0.78*
1228.2	0.73
1228.2	0.74*
1278.2	0.75
1278.2	0.72
CURVE 11	
823.2	0.95
873.2	0.83
1073.2	0.91
CURVE 12	
823.2	0.96
873.2	0.78
1073.2	0.83
CURVE 13	
823.2	0.88
873.2	0.73
1073.2	0.72
CURVE 14	
823.2	0.89
873.2	0.71
1073.2	0.74
CURVE 15	
823.2	0.85
873.2	0.70
1073.2	0.73
CURVE 16*	
823.2	0.89
1073.2	0.73
CURVE 17	
823.2	0.84
1073.2	0.72
CURVE 18	
1073.2	0.70
CURVE 19*	
1073.2	0.74
CURVE 20*	
1073.2	0.74
CURVE 21	
938.2	0.85*
1063.2	0.78*
1228.2	0.67
CURVE 22*	
938.2	0.88
1063.2	0.80
1228.2	0.75
CURVE 23*	
938.2	0.88
1063.2	0.83
1228.2	0.74
CURVE 24*	
938.2	0.89
1063.2	0.83
1228.2	0.78
CURVE 25*	
938.2	0.89
1063.2	0.84
1228.2	0.78
CURVE 26*	
1063.2	0.80
1228.2	0.73
CURVE 27*	
1228.2	0.74
CURVE 28	
477.6	1.03
513.7	1.02
552.6	0.994
619.3	0.981
663.7	0.963
710.9	0.937
763.7	0.916
813.9	0.877
CURVE 29	
469.3	0.846
507.1	0.872
548.7	0.898
622.1	0.921
660.9	0.947
709.3	0.934
760.9	0.916
813.7	0.872
CURVE 30	
615.2	0.94
653.2	0.84
965.2	0.05*
CURVE 31	
297.2	1.06
345.2	1.02
506.2	1.14
623.2	0.931
711.2	0.923
880.2	0.877
951.2	0.767
1036.2	0.750
1087.2	0.760
CURVE 32	
366.5	1.07
477.6	1.03*
588.7	0.983
699.8	0.947
810.9	0.896
922.1	0.859
1033.2	0.821
1144.3	0.777
1255.4	0.730
1310.9	0.710
CURVE 33	
295.2	1.15
295.2	1.07
408.2	1.04
CURVE 34	
316.2	1.16
410.2	1.09
519.2	1.04
596.2	1.01
688.2	0.973
750.2	0.932
835.2	0.900
CURVE 35	
324.2	1.22
422.2	1.10
473.2	1.08
523.2	1.05
573.2	1.03
623.2	1.00
673.2	0.983
720.2	0.964
773.2	0.939
CURVE 36	
328.2	1.07
385.2	1.07
432.2	0.81
529.2	0.78
648.2	0.78
757.2	0.81
CURVE 37	
326.2	1.38
373.2	1.33
536.2	1.08
560.2	0.85
615.2	0.86
725.2	0.91
CURVE 38	
1050.2	0.483
1081.2	0.482
1111.2	0.477
1140.2	0.593
1197.2	0.464
1214.2	0.464
1230.2	0.436
1246.2	0.408
1256.2	0.372
1271.2	0.382
1284.2	0.388
CURVE 39	
611.2	1.12
643.2	0.746
704.2	0.877
729.2	0.436
752.2	0.712
829.2	0.802
850.2	0.776
881.2	0.833
910.2	0.781
935.2	0.700
CURVE 39 (cont.)	
969.2	0.688
1054.2	0.508
1084.2	0.616
1113.2	0.469
1128.2	0.532
1223.2	0.412
CURVE 40	
611.2	1.14
728.2	0.481
831.2	0.874
884.2	0.893
917.2	0.893
942.2	0.801
1050.2	0.639
1057.2	0.648
1106.2	0.733
1114.2	0.648
1201.2	0.934
1231.2	0.950
1283.2	1.06
CURVE 41	
383	0.988
512	1.09
605	1.06
637	0.950
650	0.815
707	0.886
755	0.955
757	0.895
845	0.740
CURVE 42	
547	0.960
627	0.865
662	0.915
703	0.945
713	0.790

*Not shown in figure.

DATA TABLE 18. THERMAL DIFFUSIVITY OF COPPER (continued)

T	α
CURVE 42 (cont.)	
770	0.855
782	0.817
CURVE 43	
473	0.821
512	0.863
595	1.02
612	0.955
664	0.975
697	0.927
832	0.907
CURVE 44	
600	0.815
657	0.720
713	0.975
738	0.715
792	0.775
806	0.665
867	0.717
910	0.885
930	0.815
CURVE 45	
310.2	1.14
310.2	1.09
337.2	1.21
342.2	1.18
343.2	1.14
363.2	1.05
377.2	1.11
379.2	1.23
504.2	1.07
535.2	1.00
556.2	0.994
558.2	0.915
560.2	0.987
647.2	0.974
722.2	0.903
745.2	0.925
748.2	1.01
748.2	1.04
891.2	0.814
CURVE 45 (cont.)	
1002.2	0.831
1027.2	0.892
1091.2	0.840
CURVE 46*	
298.2	1.12
CURVE 47*	
298.2	1.11
CURVE 48	
298.2	0.884
CURVE 49	
319.2	1.12
330.2	1.12
394.2	1.08
443.2	1.05
490.2	1.02
538.2	0.989
582.2	0.987
633.2	0.969
682.2	0.960
736.2	0.937
784.2	0.937
814.2	0.922
824.2	0.940
830.2	0.930
889.2	0.914
935.2	0.885
969.2	0.904
977.2	0.869
1012.2	0.891
1022.2	0.858
1032.2	0.874
1045.2	0.874
1090.2	0.858
1131.2	0.831
1184.2	0.803
1212.2	0.800
1282.2	0.767
CURVE 50	
308.2	1.096
333.2	1.079
CURVE 51	
306.2	0.969
CURVE 52*	
306.2	1.09
CURVE 53	
306.2	1.17
CURVE 54	
306.2	0.545
CURVE 55	
306.2	0.855
CURVE 56	
306.2	1.19
CURVE 57	
298.2	1.01
CURVE 58	
298.2	1.130
CURVE 59	
298	1.115
CURVE 60	
298.2	1.124
CURVE 61	
298	1.02
CURVE 62	
300	1.152
379	1.121
470	1.070
689	0.986
1192	0.785
CURVE 63*	
309	1.132
311	1.088
339	1.212
342	1.180
342	1.141
366	1.044
377	1.109
380	1.230
504	1.063
536	1.006
557	0.996
558	0.912
560	0.986
648	0.976
722	0.902
744	0.930
748	1.040
748	1.010
890	0.817
1002	0.835
1027	0.895
1090	0.842
CURVE 64	
295	0.873
CURVE 65	
295	0.966
CURVE 66	
295	1.161
CURVE 67	
295	0.808
CURVE 68	
333	0.994
CURVE 69	
333	0.966
CURVE 70	
300	1.257
300	1.207
362	1.159
424	1.191
475	1.168
538	1.142
586	1.089
650	1.049
707	0.953
751	1.003
827	0.892
880	0.958
937	0.915
993	0.873
1115	0.814
1180	0.814
CURVE 71	
200	1.27
250	1.21
300	1.16
CURVE 72*	
298	1.230
298	1.093
302	1.128
487	1.098
488	1.134
488	1.067
652	1.145
652	1.114
652	1.091
652	1.063
652	1.055
652	1.000
657	1.094
CURVE 72 (cont.)*	
699	1.064
699	1.031
699	1.000
704	1.058
708	1.026
770	1.050
770	1.012
773	0.970
793	1.013
830	1.066
CURVE 73*	
392	1.075
507	1.239
606	1.194
636	1.024
650	0.827
705	0.909
710	0.909
755	1.034
757	0.940
842	0.735
CURVE 74*	
547	1.059
633	0.923
663	0.983
703	1.018
712	0.800
768	0.900
781	0.855
CURVE 75*	
473	0.836
508	0.894
593	1.141
611	1.037
664	1.070
829	0.971
CURVE 76*	
599	0.831
659	0.729
CURVE 76 (cont.)*	
715	1.061
739	0.718
791	0.777
805	0.664
865	0.721
909	0.932
929	0.827
CURVE 77*	
293	1.200
293	1.165
296	1.151
318	1.198
396	1.124
472	1.067
479	1.035
576	1.035
585	1.001
587	1.017
665	0.981
669	1.002
682	0.985
793	0.964
877	0.913
902	0 902
959	0.847
1062	0.860
1079	0.845
1094	0.825
1156	0.851
1181	0.825
1222	0.793
CURVE 78	
1167	0.731
1180	0.763
1222	0.710
1255	0.747
1296	0.714
1329	0.788
1385	0.459
1433	0.490
1508	0.479*
1558	0.485*
1579	0.511*
CURVE 78 (cont.)	
1601	0.511*
1666	0.464*
1699	0.489*
1730	0.464*
CURVE 79*	
306.2	1.132
319.7	1.083
322.2	1.098
323.1	1.073
323.2	1.057
336.1	1.067
341.1	1.034
CURVE 80*	
303.7	1.113
307.1	1.089
314.2	1.096
315.0	1.113
317.2	1.085
CURVE 81*	
302.0	1.107
305.4	1.101
314.2	1.096
315.6	1.099
328.2	1.061
329.8	1.061
344.7	1.036

* Not shown in figure.

FIGURE AND TABLE 19R. PROVISIONAL THERMAL DIFFUSIVITY OF DYSPROSIUM

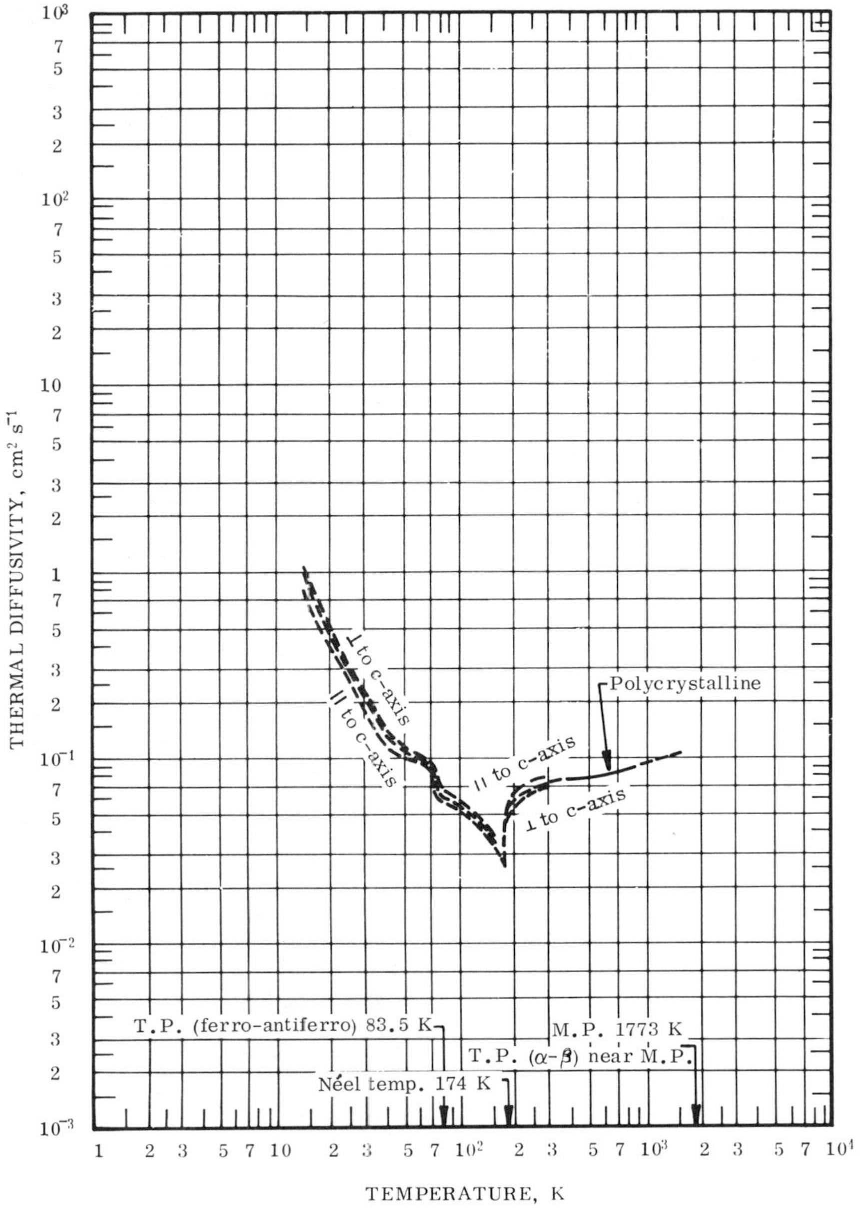

PROVISIONAL VALUES

[Temperature, T, K; Thermal Diffusivity, α, cm^2 s^{-1}]

SOLID

T	∥ to c-axis α	⊥ to c-axis α	Poly-crystalline α
15	0.809*	1.04*	0.959*
16	0.686*	0.888*	0.815*
18	0.514*	0.671*	0.614*
20	0.409*	0.527*	0.483*
25	0.258*	0.323*	0.300*
30	0.187*	0.225*	0.212*
35	0.148*	0.175*	0.165*
40	0.124*	0.144*	0.136*
45	0.110*	0.127*	0.120*
50	0.101*	0.116*	0.111*
60	0.0980*	0.109*	0.106*
70	0.0910*	0.0995*	0.0964*
80	0.0599*	0.0650*	0.0628*
83.5	0.0525*	0.0558*	0.0544*
85	0.0603*	0.0634*	0.0624*
90	0.0571*	0.0639*	0.0594*
100	0.0527*	0.0598*	0.0550*
150	0.0364*	0.0403*	0.0377*
174	0.0249*	0.0249*	0.0249*
200	0.0670*	0.0605*	0.0625*

T	∥ to c-axis α	⊥ to c-axis α	Poly-crystalline α
250	0.0748*	0.0665*	0.0694*
273.2	0.0769*	0.0681*	0.0710*
300	0.0789*	0.0696*	0.0722*
350			0.0735
400			0.0745
500			0.0764
600			0.0790
700			0.0823
800			0.0859
900			0.0890
1000			0.0919*
1100			0.0947*
1200			0.0975*
1300			0.100*
1400			0.103*
1500			0.105*

REMARKS

The provisional values are for well-annealed high-purity dysprosium and are thought to be accurate to within ± 20% of the true values at temperatures from 200 to 300 K and ± 20 to ± 30% above 300 K. At low temperatures the values are highly conditioned by impurity and imperfection, and the values below 200 K for $\alpha_{\parallel}$, $\alpha_{\perp}$, and α_{poly} are applicable only to samples having residual electrical resistivities of 5.77, 4.59, and 4.93 μΩ cm, respectively. Uncertainty limits for the values below 200 K can hardly be given.

* In temperature range where no experimental data are available.

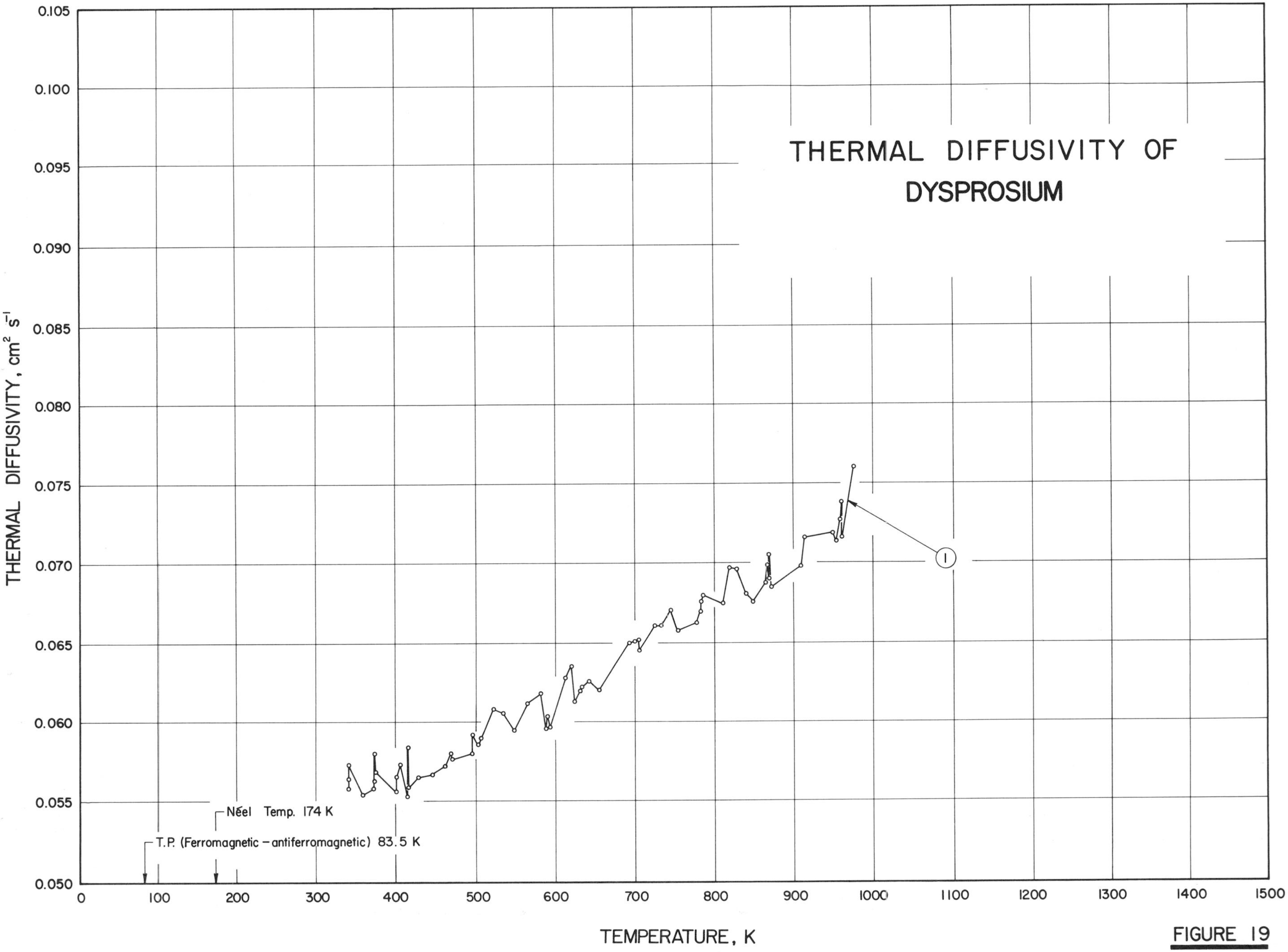
THERMAL DIFFUSIVITY OF
DYSPROSIUM
THERMAL DIFFUSIVITY, cm² s⁻¹
TEMPERATURE, K
0.105
0.100
0.095
0.090
0.085
0.080
0.075
0.070
0.065
0.060
0.055
0.050
0
100
200
300
400
500
600
700
800
900
1000
1100
1200
1300
1400
1500
1
Néel Temp. 174 K
T.P. (Ferromagnetic – antiferromagnetic) 83.5 K

FIGURE 19

SPECIFICATION TABLE 19. THERMAL DIFFUSIVITY OF DYSPROSIUM

(Impurity < 0.20% each; total impurities < 0.50%)

Cur. No.	Ref. No.	Author(s)	Year	Temp. Range, K	Reported Error, %	Name and Specimen Designation	Composition (weight percent), Specifications, and Remarks
1	193	Pollard, E.R., Jr.	1963	339-975			Cylindrical specimen about 2 in. long and 0.125 in. in diameter; diffusivity measured using modified Angström method.

DATA TABLE 19. THERMAL DIFFUSIVITY OF DYSPROSIUM

(Impurity < 0.20% each; total impurities < 0.50%)

[Temperature, T, K; Thermal Diffusivity, α, $cm^2 s^{-1}$]

T	α	T	α	T	α	T	α	T	α
CURVE 1		CURVE 1 (cont.)		CURVE 1 (cont.)		CURVE 1 (cont.)		CURVE 1 (cont.)	
339	0.0558	428	0.0565	586	0.0596	724	0.0661	866	0.0699
339	0.0564	444	0.0567	588	0.0604	732	0.0661	868	0.0690
339	0.0573	461	0.0572	592	0.0597	744	0.0671	868	0.0705
357	0.0554	467	0.0580	611	0.0628	753	0.0658	871	0.0685
372	0.0558	470	0.0576	619	0.0635	777	0.0663	908	0.0698
372	0.0563	494	0.0580	623	0.0613	782	0.0670	913	0.0716
372	0.0580	494	0.0592	629	0.0620	782	0.0676	949	0.0719
374	0.0568	502	0.0586	632	0.0622	785	0.0680	953	0.0714
400	0.0556	505	0.0590	641	0.0626	810	0.0675	957	0.0727
400	0.0565	521	0.0608	654	0.0620	818	0.0697	959	0.0738
405	0.0573	533	0.0606	692	0.0650	828	0.0696	960	0.0716
414	0.0553	547	0.0595	699	0.0651	839	0.0681	975	0.0760
414	0.0584	563	0.0612	703	0.0652	848	0.0676		
415	0.0558	580	0.0618	705	0.0646	864	0.0688		

FIGURE AND TABLE 20R. PROVISIONAL THERMAL DIFFUSIVITY OF ERBIUM

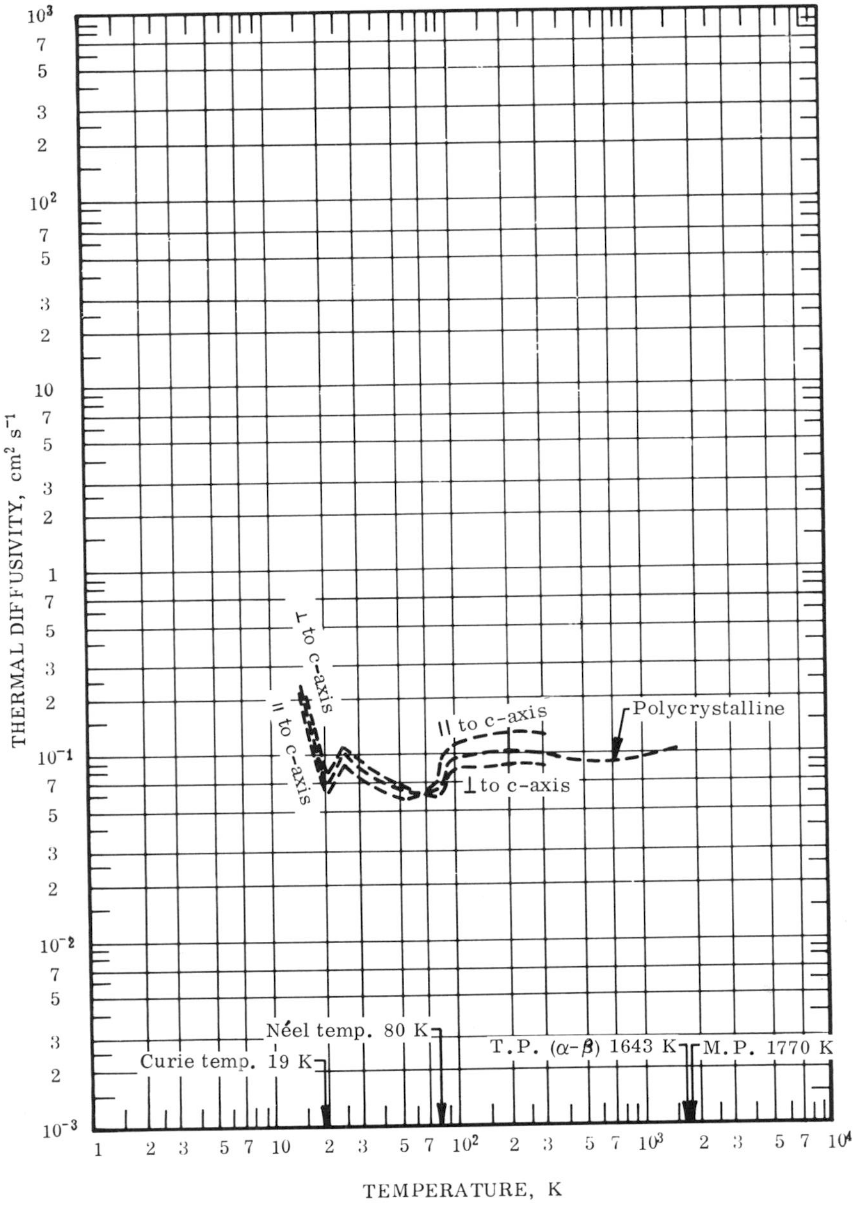

PROVISIONAL VALUES*

[Temperature, T, K; Thermal Diffusivity, α, cm² s⁻¹]

SOLID

T	∥ to c-axis α	⊥ to c-axis α	Poly-crystalline α
15	0.218	0.227	0.224
16	0.155	0.162	0.160
18	0.101	0.106	0.104
20	0.0558	0.0753	0.0682
25	0.0856	0.110	0.102
30	0.0726	0.0893	0.0833
35	0.0677	0.0792	0.0745
40	0.0655	0.0735	0.0707
45	0.0617	0.0686	0.0668
50	0.0565	0.0619	0.0601
51.5	0.0545	0.0597	0.0579
57	0.0595	0.0633	0.0618
60	0.0609	0.0631	0.0622
70	0.0638	0.0603	0.0615
80	0.0682	0.0575	0.0609
90	0.104	0.0775	0.0857
100	0.112	0.0806	0.0911
150	0.123	0.0841	0.0970
200	0.126	0.0856	0.0995
250	0.126	0.0855	0.0990

T	∥ to c-axis α	⊥ to c-axis α	Poly-crystalline α
273.2	0.124	0.0842	0.0975
300	0.121	0.0824	0.0952
350			0.0924
400			0.0902
500			0.0890
600			0.0887
700			0.0890
800			0.0897
900			0.0909
1000			0.0925
1100			0.0944
1200			0.0965
1300			0.0987
1400			0.101
1500			0.103

REMARKS

The provisional values are for well-annealed high-purity erbium and are thought to be accurate to within ±20% at temperatures from 200 to 300 K and ±20 to ±30% above 300 K. At temperatures below 200 K the values are very uncertain.

* All values are estimated.

FIGURE AND TABLE 21R. PROVISIONAL THERMAL DIFFUSIVITY OF EUROPIUM

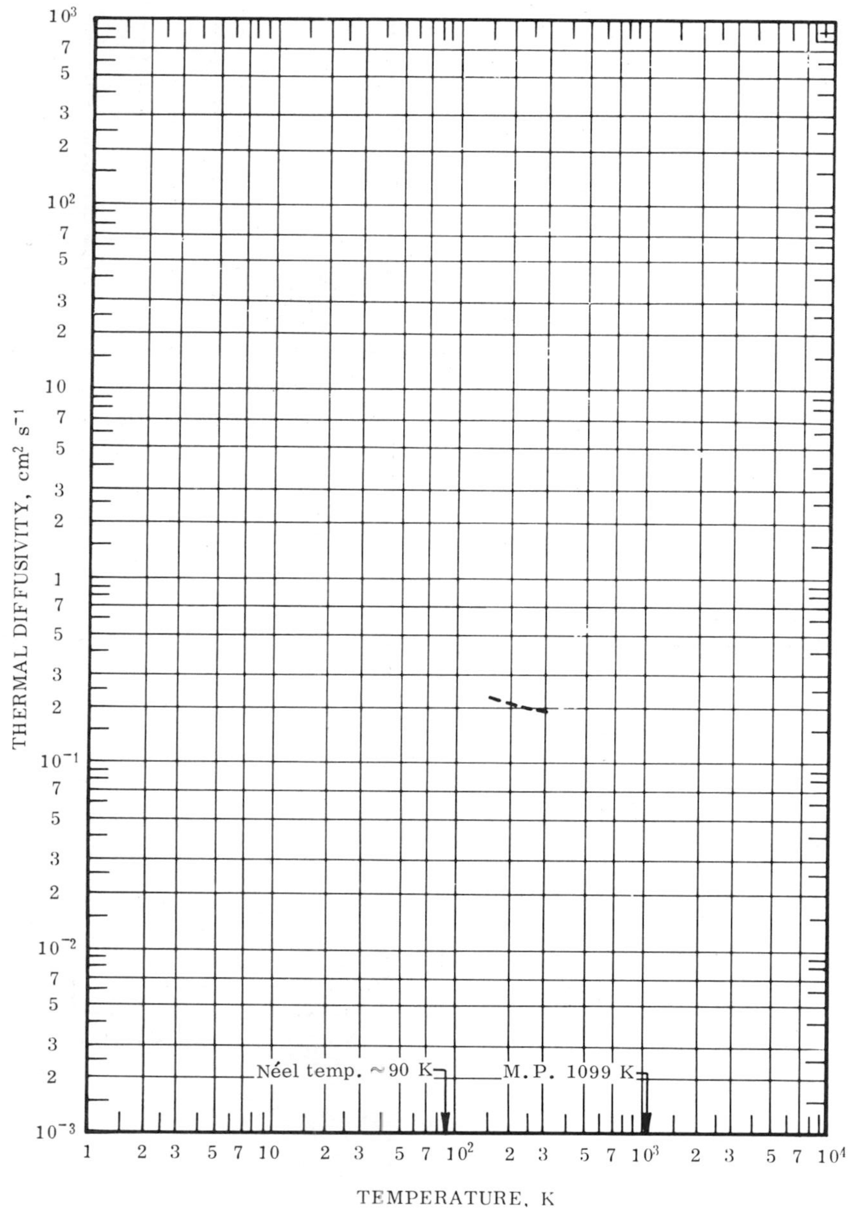

PROVISIONAL VALUES*

[Temperature, T, K; Thermal Diffusivity, α, cm^2 s^{-1}]

SOLID

T	α
150	0.230
200	0.206
250	0.196
273.2	0.195
298.2	0.194
300	0.194

REMARKS

The provisional values are for high-purity europium and are probably good to within ±25%.

* All values are estimated.

FIGURE AND TABLE 22R. PROVISIONAL THERMAL DIFFUSIVITY OF GADOLINIUM

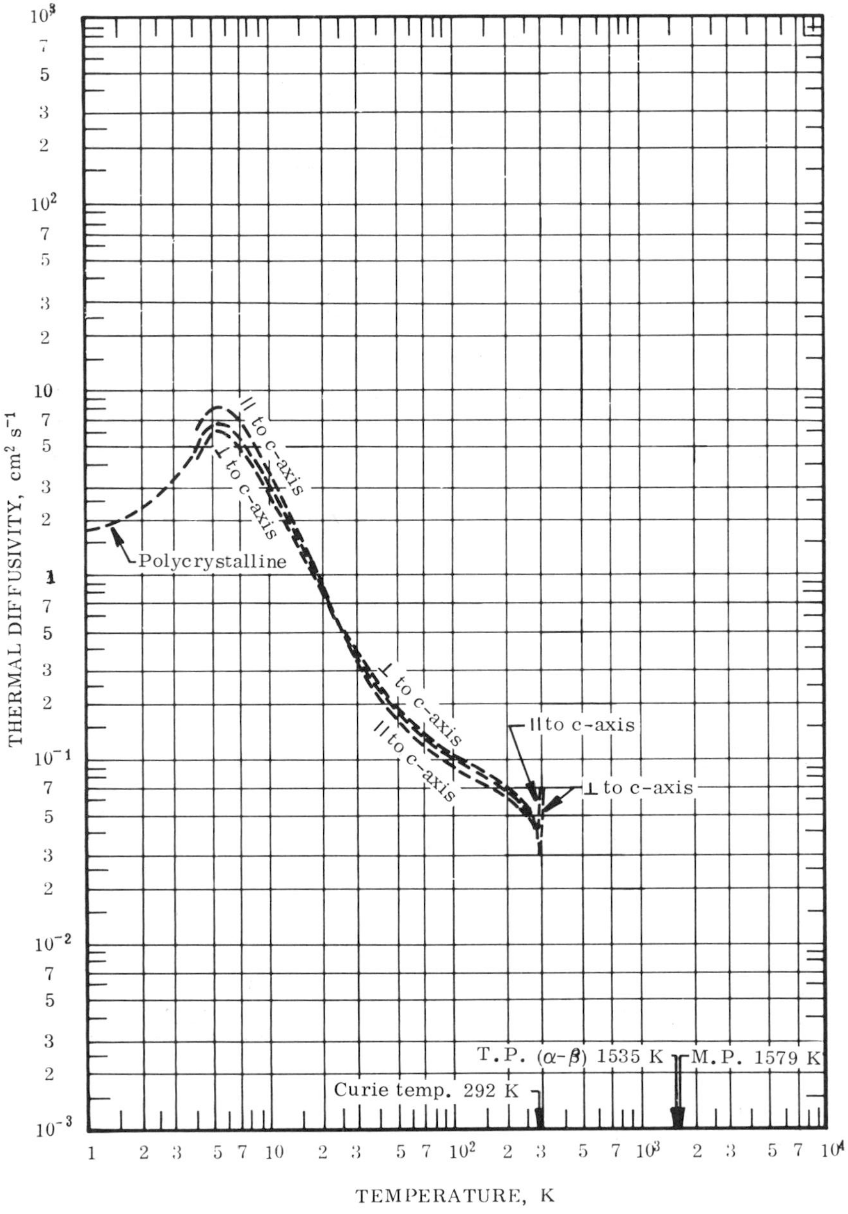

PROVISIONAL VALUES*

[Temperature, T, K; Thermal Diffusivity, α, cm^2 s^{-1}]

SOLID

T	∥ to c-axis α	⊥ to c-axis α	Poly-crystalline α	T	∥ to c-axis α	⊥ to c-axis α	Poly-crystalline α
1			1.63	35	0.259	0.300	0.283
2			2.27	40	0.209	0.241	0.227
3			3.03	45	0.178	0.203	0.193
4	5.64	4.20	4.64	50	0.156	0.178	0.170
5	8.14	6.08	6.71	60	0.130	0.148	0.143
6	7.95	5.86	6.52	70	0.116	0.131	0.126
7	6.76	5.06	5.55	80	0.106	0.119	0.115
8	5.46	4.27	4.61	90	0.0977	0.110	0.106
9	4.42	3.57	3.82	100	0.0918	0.103	0.0986
10	3.62	2.98	3.19	150	0.0738	0.0811	0.0787
11	3.03	2.52	2.69	200	0.0618	0.0657	0.0646
12	2.55	2.15	2.28	250	0.0504	0.0514	0.0510
13	2.15	1.85	1.95	273.2	0.0438	0.0406	0.0433
14	1.85	1.60	1.69	292	0.0361	0.0347	0.0351
15	1.59	1.40	1.47	295	0.0538	0.0518	0.0523
16	1.38	1.24	1.29	300	0.0597	0.0570	0.0584
18	1.07	0.985	1.02				
20	0.837	0.802	0.813				
25	0.502	0.533	0.527				
30	0.343	0.386	0.375				

REMARKS

The provisional values are for well-annealed high-purity gadolinium and are thought to be accurate to within ± 15% at temperatures above 100 K. At temperatures below 100 K the values for $\alpha_{\parallel}$, $\alpha_{\perp}$, and α_{poly} are applicable only to samples having residual electrical resistivity of 2.62, 4.43, and 3.71 μΩ cm, respectively. These values are very uncertain.

*All values are estimated.

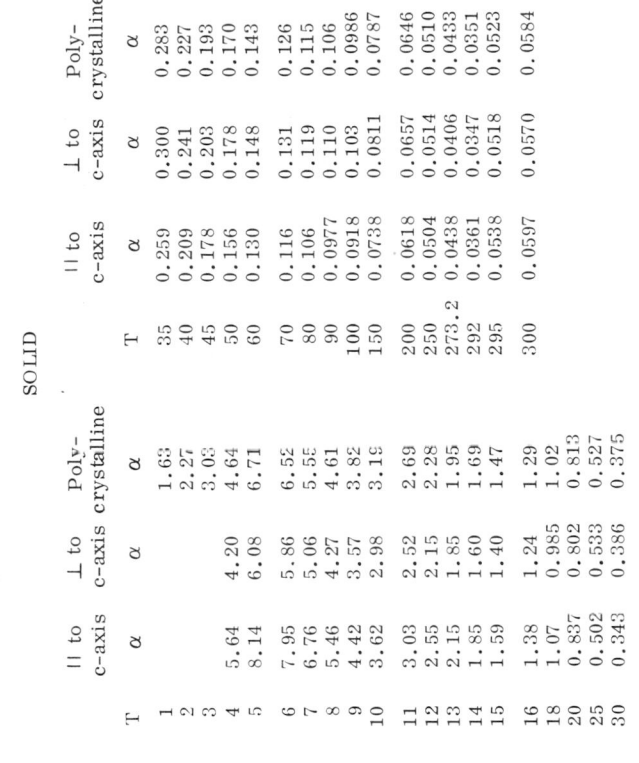

FIGURE AND TABLE 23R. RECOMMENDED THERMAL DIFFUSIVITY OF GALLIUM

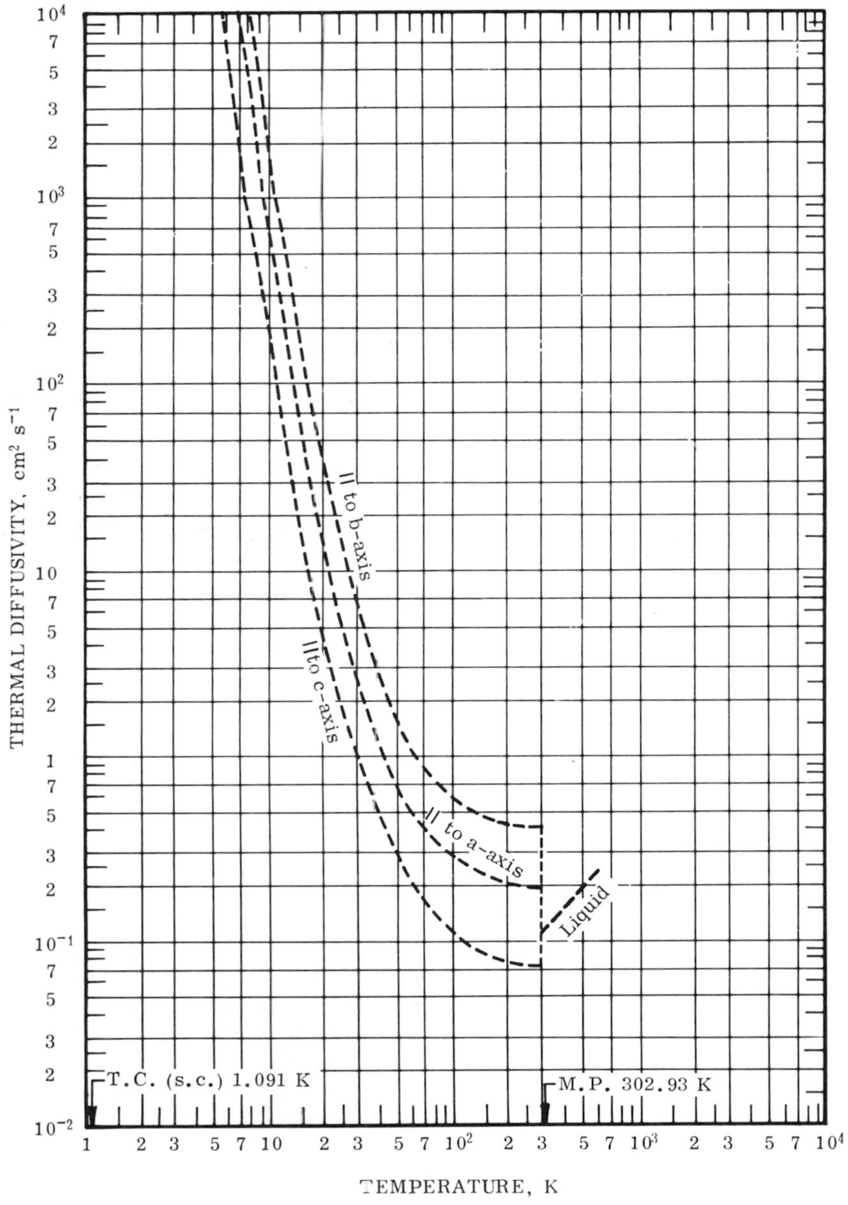

RECOMMENDED VALUES*

[Temperature, T, K; Thermal Diffusivity, α, cm^2 s^{-1}]

SOLID
(Single crystal)‡

T	|| to a-axis α	|| to b-axis α	|| to c-axis α
1	3580000	10400000	
2	2010000	5610000	513000
3	614000	1660000	184000
4	160000	438000	50100
5	50600	139000	15600
6	16400	45600	4170
7	5650	15200	1630
8	2330	6700	664
9	1120	3100	315
10	570	1560	168
11	325	890	96.8
12	192	527	58.7
13	122	329	36.8
14	79.0	216	23.9
15	55.1	149	16.3
16	39.6	107	11.6
18	22.6	59.0	6.76
20	14.1	36.9	4.33
25	5.65	15.0	1.94
30	2.92	7.64	1.10

T	|| to a-axis α	|| to b-axis α	|| to c-axis α
35	1.78	4.54	0.714
40	1.20	2.99	0.502
45	0.900	2.12	0.393
50	0.720	1.61	0.305
60	0.518	1.08	0.216
70	0.413	0.827	0.169
80	0.353	0.707	0.142
90	0.315	0.639	0.126
100	0.291	0.594	0.115
150	0.235	0.492	0.0871
200	0.209	0.449	0.0797
250	0.194	0.423	0.0754
273.2	0.190	0.414	0.0740
300	0.186	0.405	0.0724
302.93	0.186	0.404	0.0723

LIQUID

T	α
302.93	0.116
350	0.136
400	0.158
500	0.202
600	0.247

REMARKS

The values are for high-purity gallium and are thought to be accurate to within ±25% of the true values at temperatures below 10 K, ±14% from 10 K to 100 K, and ±7% from 100 K to the melting point. For liquid gallium the uncertainty of the values is probably ±15% near the melting point and increase to ±25% at the highest temperatures. The values below 10 K and those for liquid gallium are provisional. At low temperatures the values are highly conditioned by impurity and imperfection, and those below 60 K for α_a, α_b, and α_c are applicable only to specimens having residual electrical resistivities of 0.000100, 0.0000342, and 0.000425 μΩ cm, respectively.

*All values are estimated and those below 10 K and above the melting point are provisional.
‡Values for α_a are also good for polycrystalline gallium.

FIGURE AND TABLE 24R. RECOMMENDED THERMAL DIFFUSIVITY OF GERMANIUM

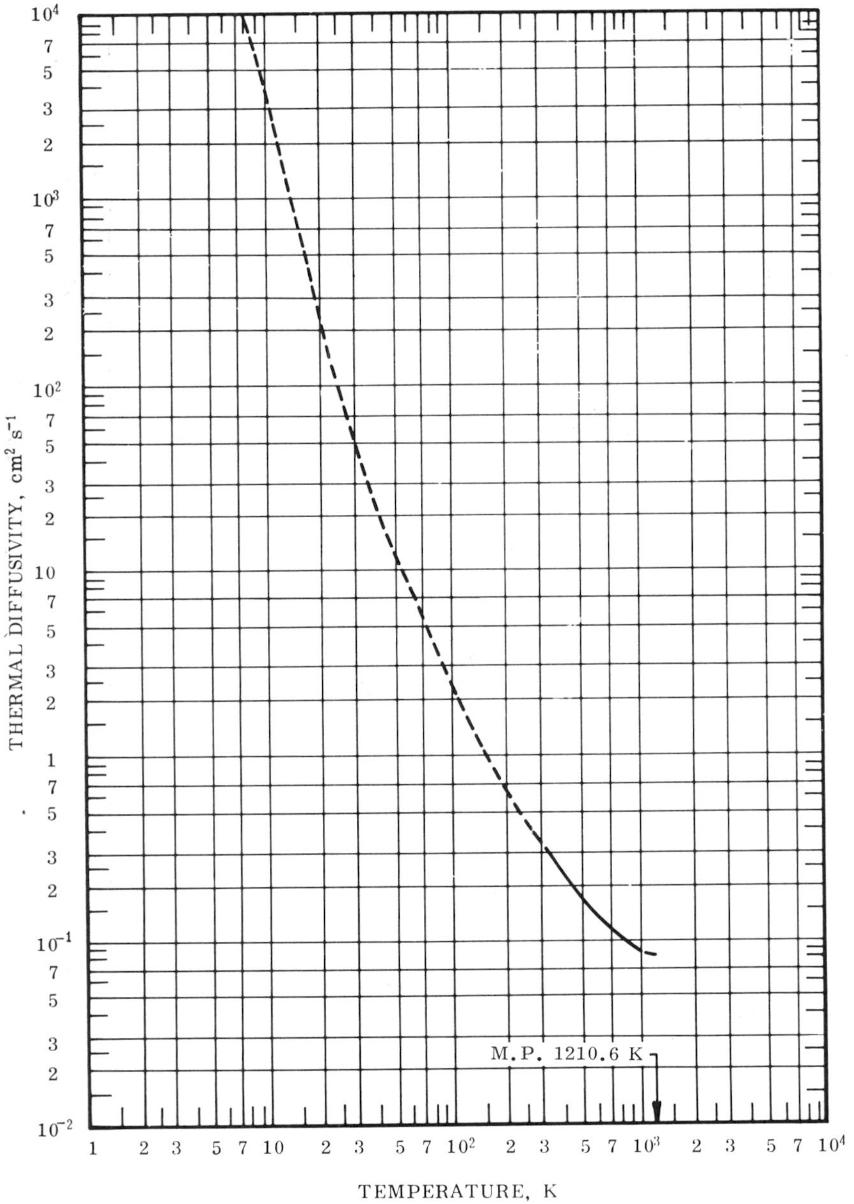

RECOMMENDED VALUES†

[Temperature, T, K; Thermal Diffusivity, α, $cm^2\ s^{-1}$]

SOLID

T	α	T	α
3	68800*	45	17.8*
4	47100*	50	13.4*
5	31400*	60	8.30*
6	21900*	70	5.57*
7	14000*	80	3.98*
8	9180*	90	2.96*
9	6100*	100	2.29*
10	4140*	150	0.972*
11	2780*	200	0.612*
12	1910*	250	0.444*
13	1350*	273.2	0.394*
14	975*	300	0.346
15	731*	350	0.282
16	555*	400	0.238
18	341*	500	0.178
20	224*	600	0.142
25	99.2*	700	0.118
30	54.9*	800	0.101
35	35.1*	900	0.0900
40	24.3*	1000	0.0851
		1100	0.0836*
		1200	0.0827*

REMARKS

The values are for well-annealed high-purity germanium. The recommended values (those at and above room temperature) are thought to be accurate to within ±13% of the true values. At low temperatures the values are highly conditioned by impurity and imperfection and those below room temperature are merely typical values representing a typical curve to indicate the general trend of the low-temperature behavior of the thermal diffusivity.

†Values below room temperature are merely typical values.
*In temperature range where no experimental data are available.

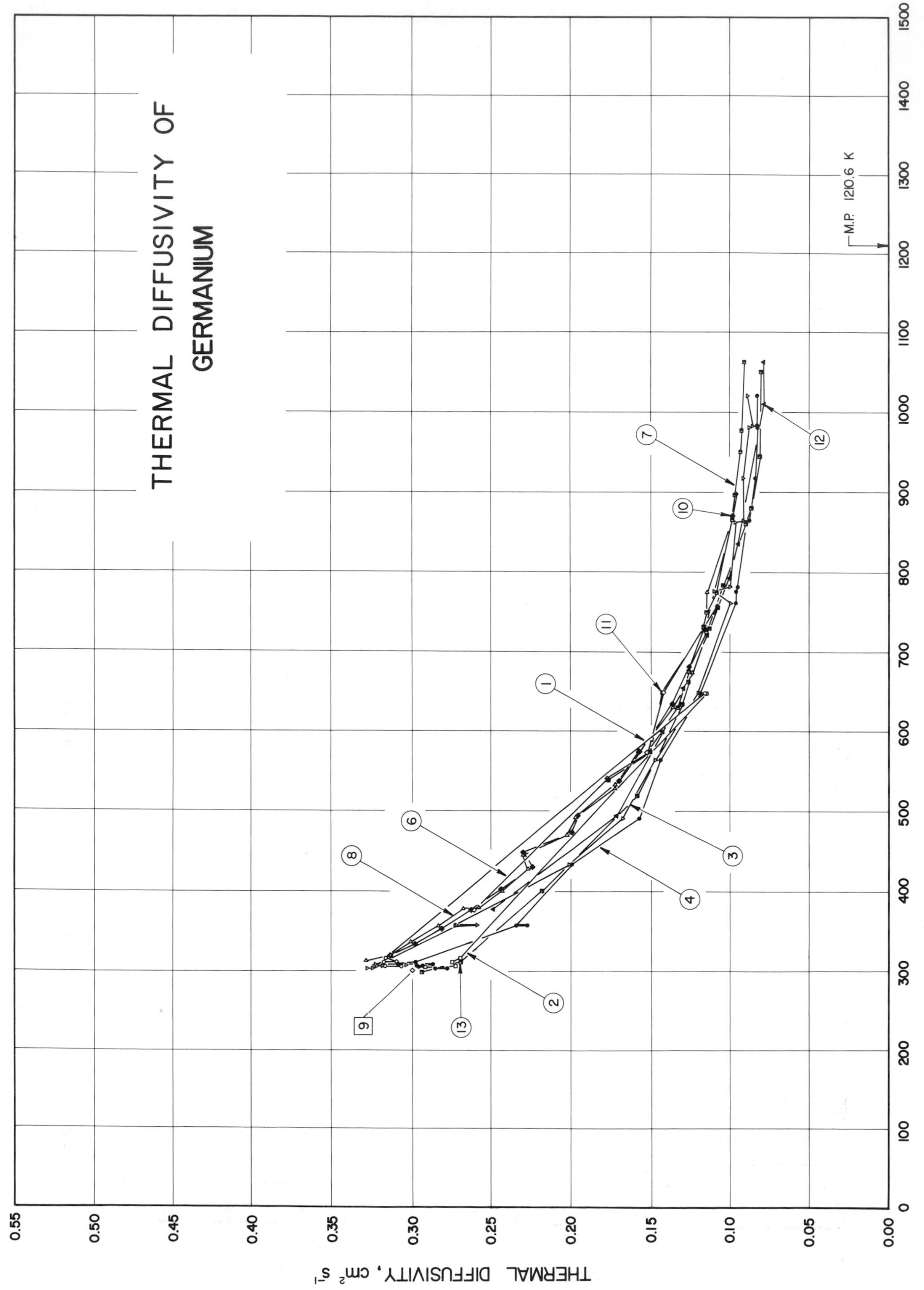
THERMAL DIFFUSIVITY OF GERMANIUM
THERMAL DIFFUSIVITY, cm² s⁻¹
TEMPERATURE, K
M.P. 1210.6 K
0.00
0.05
0.10
0.15
0.20
0.25
0.30
0.35
0.40
0.45
0.50
0.55
0
100
200
300
400
500
600
700
800
900
1000
1100
1200
1300
1400
1500
1
2
3
4
6
7
8
9
10
11
12
13

FIGURE 24

SPECIFICATION TABLE 24. THERMAL DIFFUSIVITY OF GERMANIUM

Cur. No.	Ref. No.	Author(s)	Year	Temp. Range, K	Reported Error, %	Name and Specimen Designation	Composition (weight percent), Specifications, and Remarks
1	25	Abeles, B., Bernoff, R., Cody, G. D., Hockings, E. F., and Rosi, F. D.	1959	306-648	~3	1868-B	n-type; specimen has the shape of rod of square cross section 0.75 x 0.75 cm and length lying in the range from 1 to 2 in.; electrical resistivity at room temp. 3×10^{-3} ohm cm; specimen brazed on heater with nickel and measured under a vacuum of 5×10^{-5} mm Hg; sine temp. wave supplied to specimen; specimen measured at a frequency of 0.05 cycles per sec; phase shift and attenuation assumed equal.
2	25	Abeles, B., et al.	1959	306-648	~3	1868-B	Diffusivity calculated from above measurement for above specimen using the actual measured values of phase shift and attenuation.
3	25	Abeles, B., et al.	1959	303-1021	~3	1868-B	Same as above except frequency being 0.2 cycles per sec and phase shift and attenuation being assumed equal.
4	25	Abeles, B., et al.	1959	303-1021	~3	1868-B	Diffusivity calculated from above measurement for above specimen using the actual measured values of phase shift and attenuation.
5*	26	Abeles, B., Cody, G. D., and Novak, R.	1959	302-1013		T 1868 B	n-type; electrical resistivity 0.003 ohm cm.
6	27	Cutler, M.	1961	376-758			Specimen measured using the method of small area contact.
7	28	Abeles, B., Beers, D., Cody, G., Novak, R., and Rosi, F.	1960	575-1063	± 2	T 1810	n-type As-doped single crystal with a [111] axis along specimen axis; square specimen 0.3 x 0.3 x 2 in.; electrical resistivity 0.3 ohm cm at room temp; specimen brazed to silver plated heating element at 924.2 K and a solid bond of Ag-Ge eutectic formed; measured under a vacuum of 10^{-6} mm Hg with probes at a frequency of 5 revolutions per min.
8	28	Abeles, B., et al.	1960	312-897	± 2	T 1810	Above specimen measured with thermocouples.
9	118	Perron, J. C.	1961	300	~ 8		Square specimen 3 cm long; Angström method used to measure diffusivity.
10	147, 192	Abeles, B., Cody, G.D. and Beers, D.S.	1960	319-898			Single crystal; cut to the shape of rod 0.3 in. square cross-section and 2 in. long with (111) axis along the rod axis; Angström method used to measure diffusivity.
11	205	Cutler, M.	1962	539-757			Diffusivity measured using small-area contact method.
12	206	Meddins, H.R. and Parrott, J.E.	1969	376-1063			Rectangular bar 3-5 cm in length, 0.6-0.8 cm wide, and 0.2 cm thick; diffusivity measured by Angström method using probes.
13	206	Meddins, H.R. and Parrott, J.E.	1969	299-1050			Above specimen measured with thermocouples.

* Not shown in figure.

DATA TABLE 24. THERMAL DIFFUSIVITY OF GERMANIUM

(Impurity <0.20% each; total impurities < 0.50%)

[Temperature, T, K; Thermal Diffusivity, α, $cm^2\ s^{-1}$]

T	α
CURVE 1	
305.7	0.307
306.2	0.317
311.2	0.318
311.2	0.310
316.2	0.317
648.2	0.116
CURVE 2	
305.7	0.292
306.2	0.273
311.2	0.275
311.2	0.275
316.2	0.270
648.2	0.115
CURVE 3	
303.2	0.328
303.2	0.325
305.0	0.324
306.2	0.322
306.2	0.319
307.5	0.304
308.0	0.323
311.2	0.317*
357.2	0.272
357.2	0.259
357.2	0.273
357.2	0.272
433.2	0.201
491.2	0.167
564.2	0.147
648.2	0.120
761.5	0.100
776.2	0.110
782.2	0.100
863.2	0.0966
865.2	0.0917
918.2	0.0917
981.2	0.0883
983.2	0.0858
1021.2	0.0893
CURVE 4	
303.2	0.286
303.2	0.278
305.0	0.296
306.2	0.297
306.2	0.293
307.5	0.287
308.0	0.309
311.2	0.298
357.2	0.234
357.2	0.227
357.2	0.234
357.2	0.234
433.2	0.199
491.2	0.157
564.2	0.144
648.2	0.119
761.5	0.0963
776.2	0.0963
782.2	0.0952
863.2	0.0898
865.2	0.0877
918.2	0.0844
981.2	0.0829
983.2	0.0831
1021.2	0.0833
CURVE 5*	
302.2	0.323
307.2	0.309
314.2	0.317
356.2	0.273
356.2	0.260
433.2	0.204
487.2	0.166
562.2	0.144
643.2	0.118
643.2	0.115
756.2	0.0962
770.2	0.0961
778.2	0.0950
858.2	0.0899
858.2	0.0876
CURVE 5 (cont.)*	
913.2	0.0840
975.2	0.0828
979.2	0.0832
1013.2	0.0833
CURVE 6	
376.2	0.261
541.2	0.177
573.2	0.152
649.2	0.142
758.2	0.108
CURVE 7	
575.2	0.152*
630.2	0.133
676.2	0.126
725.2	0.115
729.2	0.113
732.2	0.117
750.2	0.115
775.2	0.109
866.2	0.0986
897.2	0.0970
951.2	0.0935
977.2	0.0931
1063.2	0.0913
CURVE 8	
312.2	0.329
319.2	0.314
336.2	0.301
356.2	0.284
378.2	0.268
379.2	0.258
400.2	0.243
428.2	0.227
445.2	0.230
470.2	0.202
492.2	0.197
528.2	0.172
CURVE 8 (cont.)	
533.2	0.172
575.2	0.157
630.2	0.136
674.2	0.124
724.2	0.115*
729.2	0.113*
731.2	0.118*
749.2	0.115*
775.2	0.115
865.2	0.0980*
897.2	0.0966*
CURVE 9	
300	0.3
CURVE 10	
319	0.313
333	0.298
352	0.281
376	0.262
401	0.243
430	0.223
448	0.229
473	0.199
494	0.195
537	0.169
574	0.157
633	0.136
680	0.125
728	0.116
870	0.0982
898	0.0963
CURVE 11	
539	0.176
573	0.150
649	0.143
757	0.108
CURVE 12	
376	0.249
494	0.171
598	0.142
654	0.129
720	0.114
792	0.100
835	0.0947
1010	0.0782
1063	0.0788
CURVE 13	
299	0.293
310	0.269
400	0.217
518	0.158
633	0.130
661	0.126
783	0.104
860	0.090
879	0.0862
944	0.0809
1050	0.0804

*Not shown in figure.

FIGURE AND TABLE 25R. RECOMMENDED THERMAL DIFFUSIVITY OF GOLD

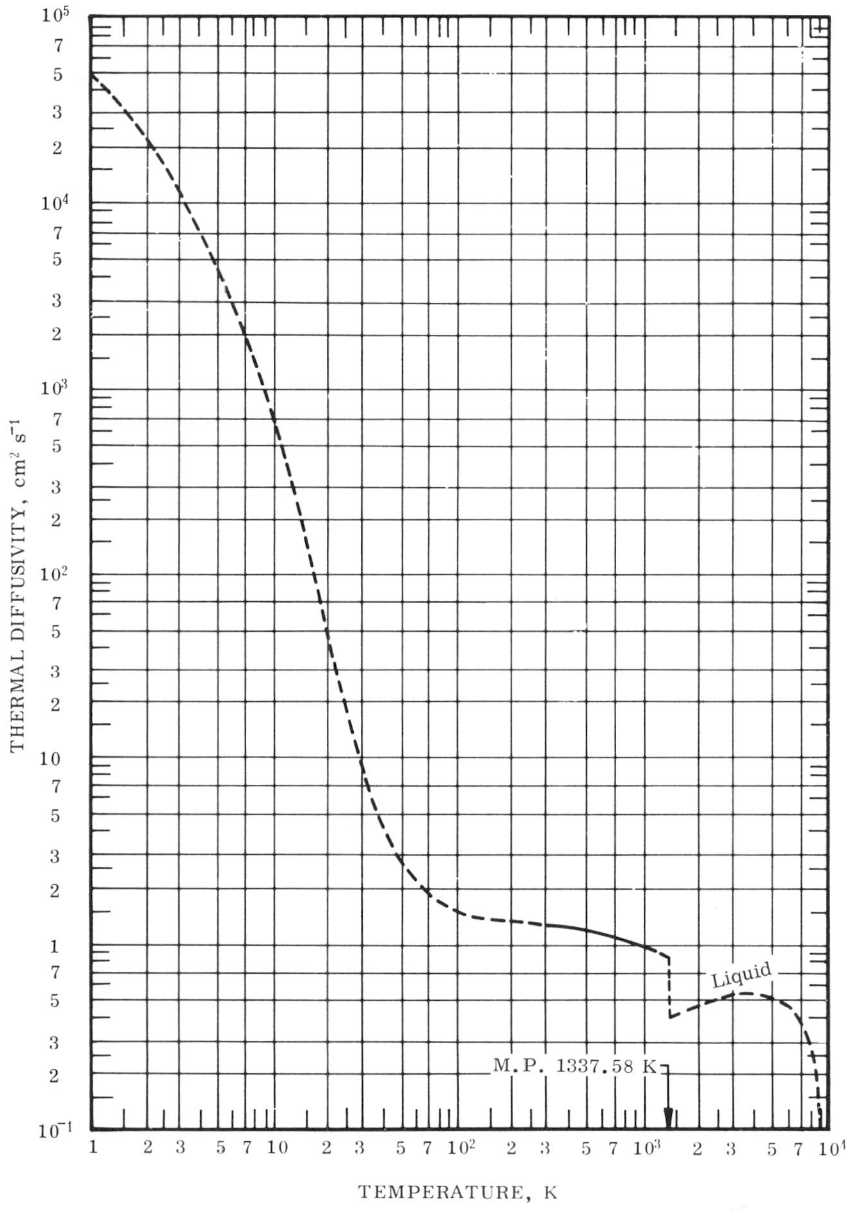

RECOMMENDED VALUES†

[Temperature, T, K; Thermal Diffusivity, α, cm² s⁻¹]

SOLID				LIQUID	
T	α	T	α	T	α
1	46300*	70	1.93*	1337.58	0.404*
2	21900*	80	1.73*	1400	0.413*
3	11600*	90	1.61*	1500	0.426*
4	6810*	100	1.55*	1600	0.438*
5	4350*	150	1.40*	1700	0.450*
6	2860*	200	1.35*	1800	0.462*
7	2000*	250	1.31*	1900	0.473*
8	1440*	273.2	1.30*	2000	0.483*
9	1040*	300	1.28*	2200	0.500*
10	760*	350	1.26	2400	0.515*
11	554*	400	1.23	2600	0.529*
12	403*	500	1.19	2800	0.540*
13	298*	600	1.15	3000	0.549*
14	221*	700	1.11	3500	0.561*
15	166*	800	1.07	4000	0.562*
16	127*	900	1.03	4500	0.555*
18	75.8*	1000	0.991	5000	0.538*
20	48.3*	1100	0.952	6000	0.481*
25	19.9*	1200	0.913	7000	0.396*
30	10.4*	1300	0.874*	8000	0.277*
35	6.50*	1337.58	0.859*	9000	0.109*
40	4.63*				
45	3.61*				
50	2.97*				
60	2.28*				

REMARKS

The recommended values are for well-annealed high-purity gold and are thought to be accurate to within ± 5% of the true values near room temperature, and ± 7% below 80 K and at 1200 K. At low temperatures the values are highly conditioned by impurity and imperfection, and those below 80 K are applicable only to a specimen having residual electrical resistivity of 0.00550 μΩ cm. The values for molten gold are provisional, and they are probably good to ± 25% from melting point to 2000 K.

†Values for molten gold are provisional.
*In temperature range where no experimental data are available.

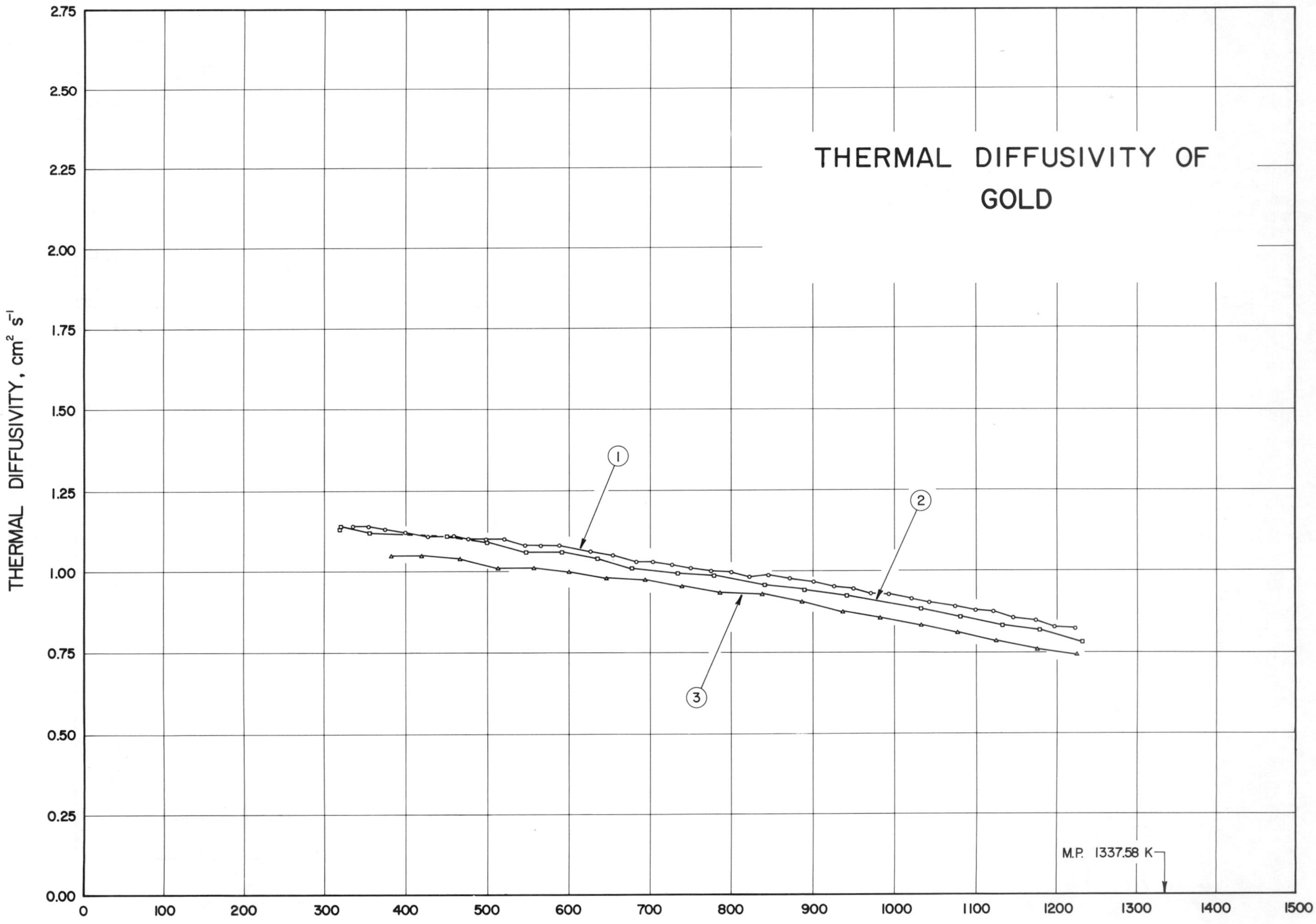

TEMPERATURE, K

FIGURE 25

SPECIFICATION TABLE 25. THERMAL DIFFUSIVITY OF GOLD

(Impurity < 0.20% each; total impurities < 0.50%)

Cur. No.	Ref. No.	Author(s)	Year	Temp. Range, K	Reported Error, %	Name and Specimen Designation	Composition (weight percent), Specifications, and Remarks
1	207, 124	Shanks, H. R., Burns, M. M., and Danielson, G. C.	~~1968~~ 1967	335-1223	~2	1	99.999 pure (as stated by manufacturer); cylindrical specimen 0.35 cm in dia. and 30 cm long; manufactured by Johnson, Matthey and Co.; electrical resistivity ratio $\rho(300\ K)/\rho(4.2\ K)$ = 500 (measured after specimen had been annealed at 1225 K for 1 hr); electrical resistivity measured and reported as 2.33, 2.49, 2.73, 3.05, 3.77, 4.13, 4.56, 5.08, 5.64, 6.27, 6.75, 7.25, 7.82, 8.32, 8.83, 9.38, 10.07, 10.64, 11.32, and 11.68 μohm cm at 302, 324, 350, 395, 477, 523, 566, 627, 686, 751, 800, 849, 907, 950, 999, 1048, 1102, 1149, 1198, and 1223 K, respectively; apparent Lorenz number reported as 2.24, 2.29, 2.32, 2.34, 2.37, 2.41, 2.43, 2.47, 2.50, and 2.52 x 10^{-8} V^2 K^{-2} at 324, 399, 500, 598, 699, 800, 900, 1000, 1099, and 1200 K, respectively; measurements made with specimen in one atmosphere of helium and with a 2.5 K amplitude and 30 sec period temp. sine wave; modified Angström method used to measure diffusivity; each data point represents average of two diffusivity measurements at each temp. obtained from two thermocouple combinations.
2	207, 124	Shanks, H. R., et al.	~~1968~~ 1967	319-1232	~2	2	99.99 pure (as stated by manufacturer); cylindrical specimen 0.35 cm in dia. and 30 cm long; manufactured by Sigmund Cohn; electrical resistivity ratio $\rho(300\ K)/\rho(4.2\ K)$ = 310 (measured after specimen had been annealed at 1225 K for 1 hr); electrical resistivity measured and reported as 2.21, 2.46, 2.71, 3.09, 3.31, 3.49, 3.87, 4.29, 4.77, 5.15, 5.57, 6.08, 6.51, 7.14, 7.68, 8.15, 9.19, 9.30, 9.79, 10.44, and 11.07 μohm cm at 300, 334, 359, 406, 428, 450, 496, 548, 596, 637, 681, 735, 787, 843, 895, 937, 1036, 1041, 1085, 1135, and 1183 K, respectively; apparent Lorenz number reported as 2.23, 2.26, 2.29, 2.31, 2.33, 2.36, 2.37, 2.40, 2.43, and 2.46 x 10^{-8} V^2 K^{-2} at 324, 399, 500, 601, 700, 801, 899, 1000, 1100, and 1200 K, respectively; measurements made with specimen in one atmosphere of helium and with a 2.5 K amplitude and 30 sec period temp. sine wave; modified Angström method used to measure diffusivity; each data point represents average of two diffusivity measurements at each temp. obtained from two thermocouple combinations.
3	207, 124	Shanks, H. R., et al.	~~1968~~ 1967	382-1225	~2	3	99.9999 pure (as stated by manufacturer); cylindrical specimen 0.35 cm in dia. and 30 cm long; manufactured by Aremco; electrical resistivity ratio $\rho(300\ K)/\rho(4.2\ K)$ = 110 (measured after specimen had been annealed at 1225 K for 1 hr); electrical resistivity measured and reported as 2.33, 2.88, 3.06, 3.51, 3.93, 4.25, 4.79, 5.26, 5.79, 6.29, 6.77, 7.21, 7.75, 8.31, 8.90, 9.39, 10.04, 10.69, 11.36, and 11.65 μohm cm at 298, 363, 385, 433, 483, 524, 580, 631, 681, 732, 784, 833, 885, 934, 984, 1034, 1084, 1133, 1182, and 1203 K, respectively; apparent Lorenz number reported as 2.15, 2.21, 2.24, 2.26, 2.29, 2.31, 2.31, 2.33, 2.33, and 2.33 x 10^{-8} V^2 K^{-2} at 322, 399, 500, 601, 698, 800, 900, 997, 1099, and 1200 K, respectively; measurements made with specimen in one atmosphere of helium and with a 2.5 K amplitude and 30 sec period temp. sine wave; modified Angström method used to measure diffusivity; each data point represents average of two diffusivity measurements at each temp. obtained from two thermocouple combinations.

DATA TABLE 25. THERMAL DIFFUSIVITY OF GOLD

(Impurity < 0.20% each; total impurities < 0.50%)

[Temperature, T, K; Thermal Diffusivity, α, $cm^2\ s^{-1}$]

T	α
CURVE 1	
335	1.14
354	1.14
374	1.13
399	1.12
427	1.11
459	1.11
476	1.10
498	1.10
521	1.10
546	1.08
565	1.08
588	1.08
626	1.06
654	1.05
683	1.03
704	1.03
727	1.02
750	1.01
775	1.00
800	0.998
822	0.982
845	0.989
872	0.977
901	0.966
926	0.952
950	0.947
971	0.931
993	0.928
1021	0.915
1043	0.902
1075	0.891
1100	0.879
1122	0.875
1146	0.856
1174	0.847
1197	0.828
1223	0.821
CURVE 2	
319	1.13
320	1.14
CURVE 2 (cont.)	
355	1.12
450	1.10
498	1.09
547	1.06
591	1.06
635	1.04
678	1.02
734	0.995
779	0.987
841	0.958
890	0.940
942	0.924
1033	0.883
1081	0.859
1133	0.831
1179	0.819
1232	0.782
CURVE 3	
382	1.05
419	1.05
466	1.04
513	1.01
557	1.01
600	0.999
646	0.978
694	0.973
739	0.954
786	0.932
838	0.927
887	0.905
937	0.873
983	0.855
1033	0.831
1078	0.810
1125	0.783
CURVE 3 (cont.)	
1176	0.758
1225	0.742

FIGURE AND TABLE 26R. PROVISIONAL THERMAL DIFFUSIVITY OF HAFNIUM

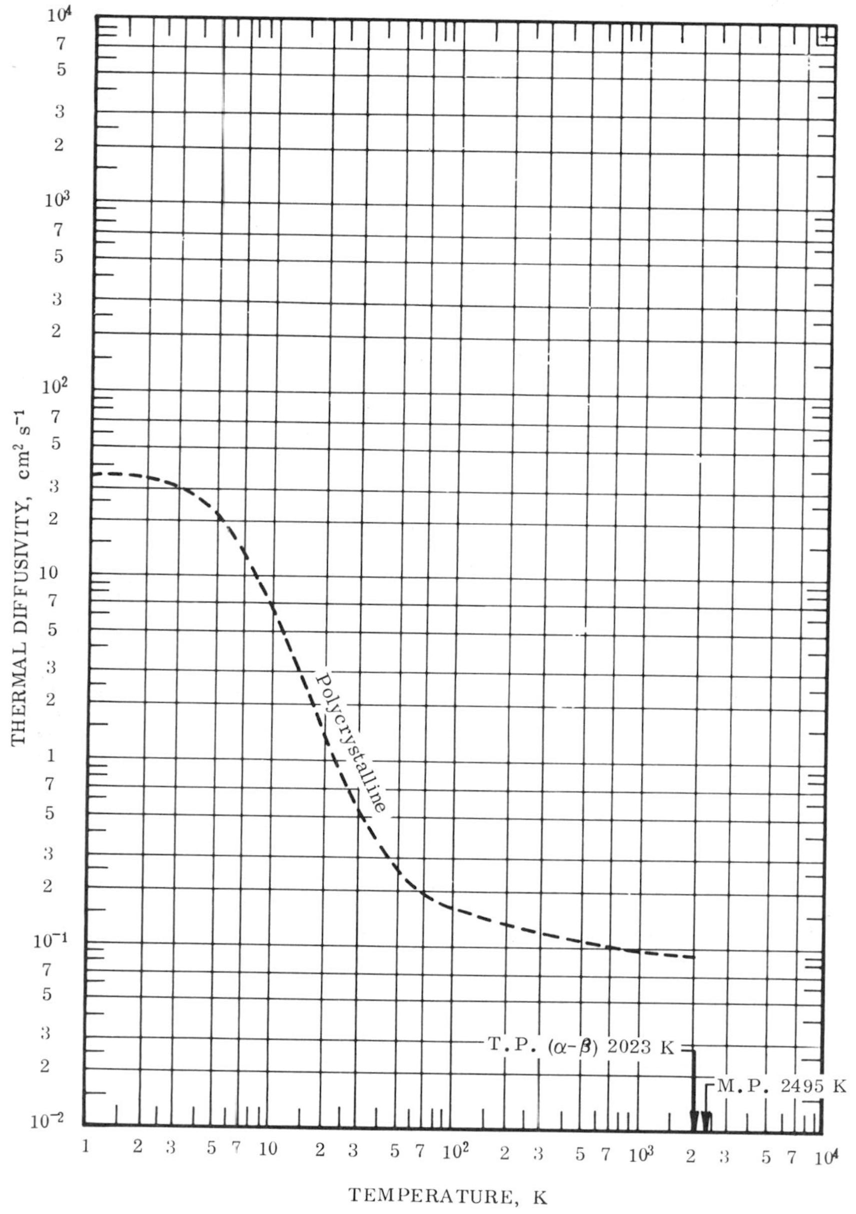

PROVISIONAL VALUES*

[Temperature, T, K; Thermal Diffusivity, α, $cm^2\ s^{-1}$]

SOLID
(Polycrystalline)

T	α	T	α
1	34.9	70	0.197
2	34.8	80	0.183
3	31.4	90	0.172
4	26.6	100	0.164
5	22.2	150	0.145
6	17.7	200	0.135
7	13.9	250	0.128
8	10.9	273.2	0.125
9	8.65	300	0.123
10	6.92	350	0.118
11	5.67	400	0.115
12	4.67	500	0.110
13	3.88	600	0.106
14	3.27	700	0.103
15	2.78	800	0.101
16	2.39	900	0.0987
18	1.80	1000	0.0971
20	1.40	1100	0.0959
25	0.840	1200	0.0949
30	0.580	1300	0.0941
35	0.442	1400	0.0934
40	0.360	1500	0.0929
45	0.309	1600	0.0925
50	0.271	1700	0.0922
60	0.225	1800	0.0921
		1900	0.0921
		2000	0.0924

REMARKS

The values are for well-annealed high-purity polycrystalline hafnium and are thought to be accurate to within ±15% of the true values at temperatures below 900 K and ±25% above 900 K. Values below 150 K are applicable only to a sample having residual electrical resistivity of 4.23 $\mu\Omega$ cm and electrical resistivity ratio $\rho(295K)/\rho_0 = 8.58$.

*All values are estimated.

FIGURE AND TABLE 27R. PROVISIONAL THERMAL DIFFUSIVITY OF HOLMIUM

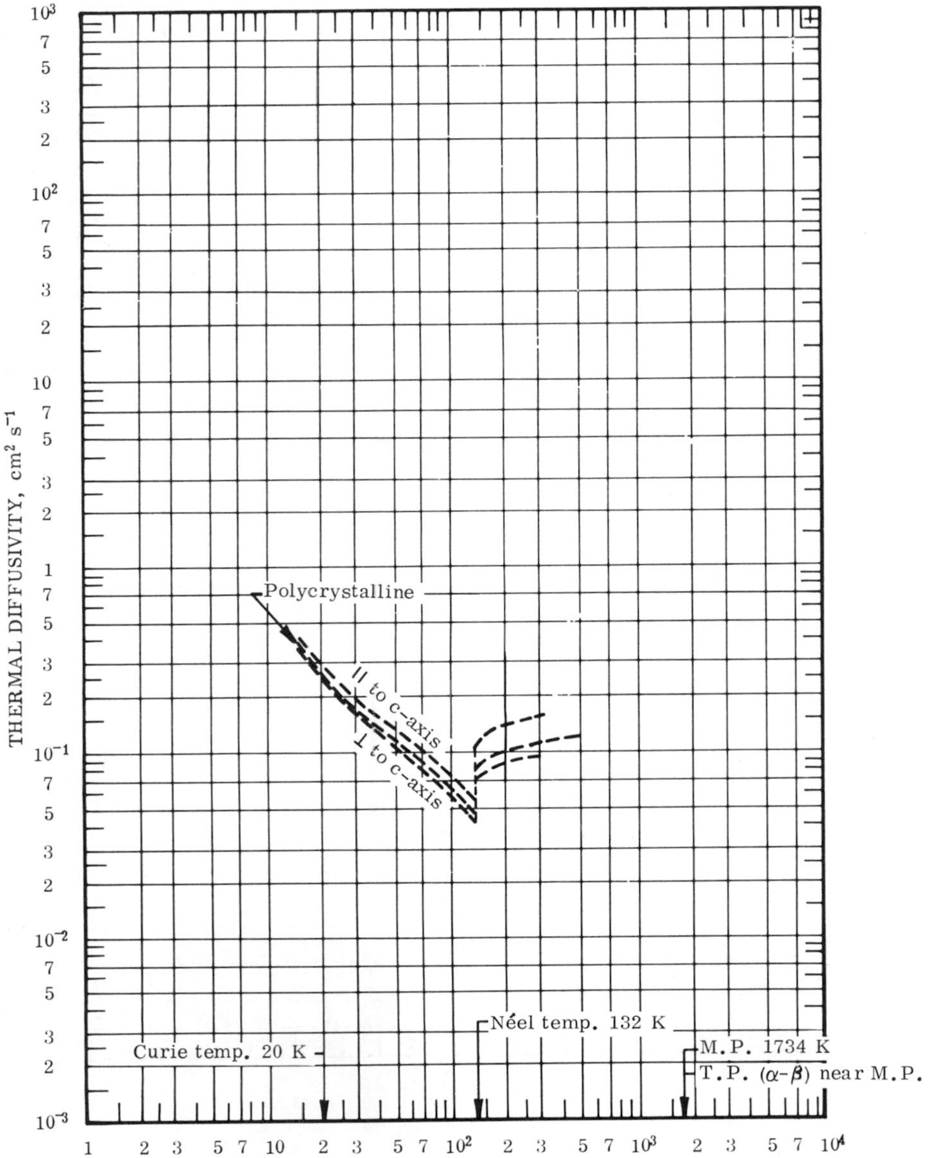

PROVISIONAL VALUES*

[Temperature, T, K; Thermal Diffusivity, α, cm^2 s^{-1}]

	SOLID (Single crystal)		
	‖ to c-axis	⊥ to c-axis	Poly-crystalline
T	α	α	α
15	0.413	0.371	0.382
16	0.376	0.336	0.345
18	0.324	0.286	0.294
20	0.288	0.252	0.260
25	0.230	0.199	0.208
30	0.196	0.168	0.176
35	0.173	0.147	0.155
40	0.156	0.132	0.140
45	0.143	0.120	0.128
50	0.132	0.111	0.118
60	0.115	0.0953	0.102
70	0.102	0.0840	0.0890
80	0.0913	0.0745	0.0787
90	0.0826	0.0662	0.0701
100	0.0750	0.0590	0.0633
132	0.0570	0.0408	0.0468
132	0.104	0.0705	0.0803
150	0.123	0.0800	0.0931
200	0.139	0.0880	0.101
250	0.147	0.0929	0.1075
273.2	0.150	0.0943	0.110
300	0.153	0.0955	0.102
350			0.105
400			0.107
500			0.111

REMARKS

The provisional values are for well-annealed high-purity holmium and are thought to to accurate to within ±20% at temperatures above 150 K. Values below 150 K for $\alpha_{\parallel}$, $\alpha_{\perp}$, and α_{poly} are applicable only to specimens having residual electrical resistivities of 3.21, 2.82, and 2.67 μΩ cm, respectively. These values are very uncertain.

---

*All values are estimated.

FIGURE AND TABLE 28R. RECOMMENDED THERMAL DIFFUSIVITY OF INDIUM

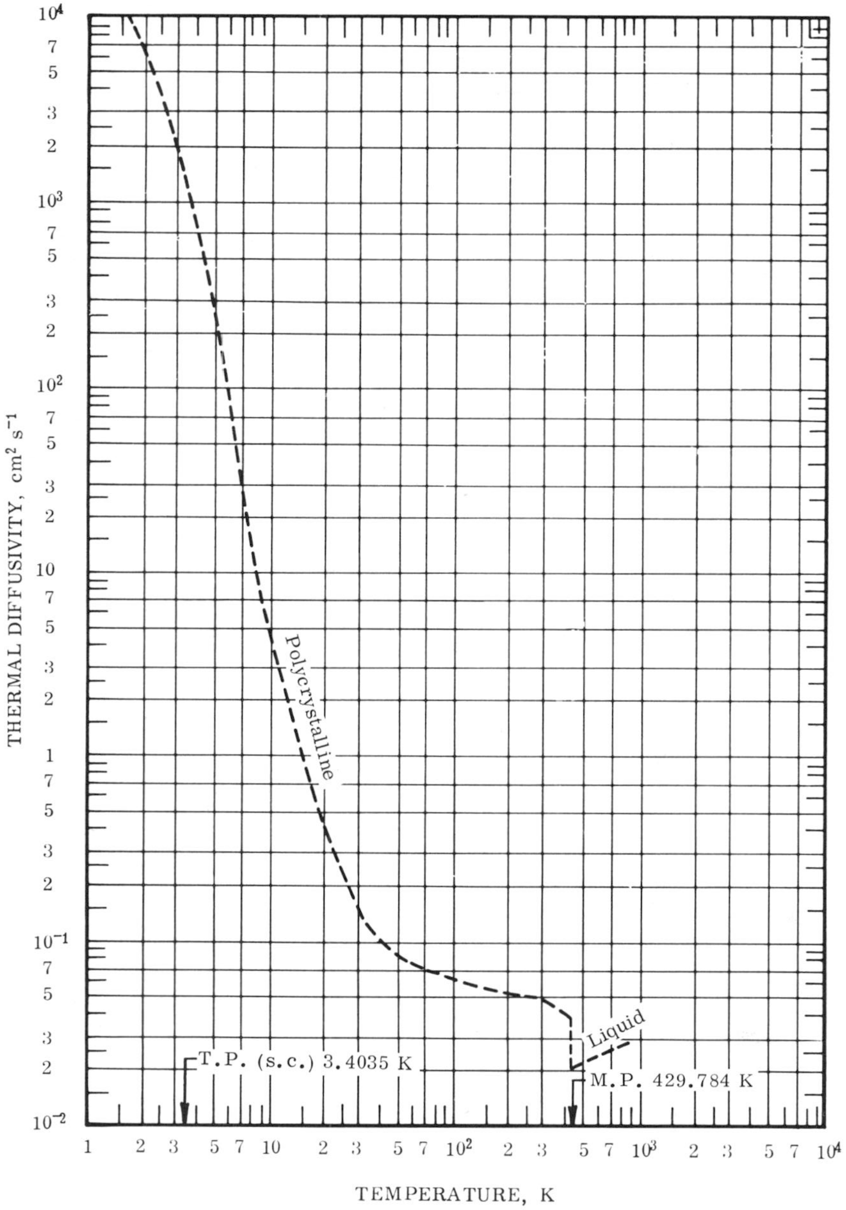

RECOMMENDED VALUES *

[Temperature, T, K; Thermal Diffusivity, α, $cm^2\ s^{-1}$]

SOLID (Polycrystalline)				LIQUID	
T	α	T	α	T	α
1	200000	35	1.25	429.784	0.211
2	63800	40	1.04	500	0.222
3	20000	45	0.928	600	0.241
4	6400	50	0.862	700	0.262
5	2160	60	0.779	800	0.282
6	778	70	0.726	900	0.295
7	325	80	0.692		
8	157	90	0.667		
9	84.8	100	0.646		
10	50.8	150	0.581		
11	34.0	200	0.540		
12	23.8	250	0.509		
13	17.5	273.2	0.495		
14	13.3	300	0.479		
15	10.4	350	0.446		
16	8.39	400	0.408		
18	5.78	429.784	0.383		
20	4.22				
25	2.36				
30	1.60				

REMARKS

The values are for well-annealed high-purity polycrystalline indium and are thought to be accurate to within ± 20% of the true values at temperatures below 100 K and ± 8% above. For liquid indium the values are probably good to ± 20%. The values below 100 K and those for liquid indium are provisional. At low temperatures the values are highly conditioned by impurity and imperfection, and those below 60 K are applicable only to a specimen having residual electrical resistivity of 0.000587 $\mu\Omega$ cm.

* All values are estimated and those below 100 K and those for liquid indium are provisional.

FIGURE AND TABLE 29R. PROVISIONAL THERMAL DIFFUSIVITY OF IODINE

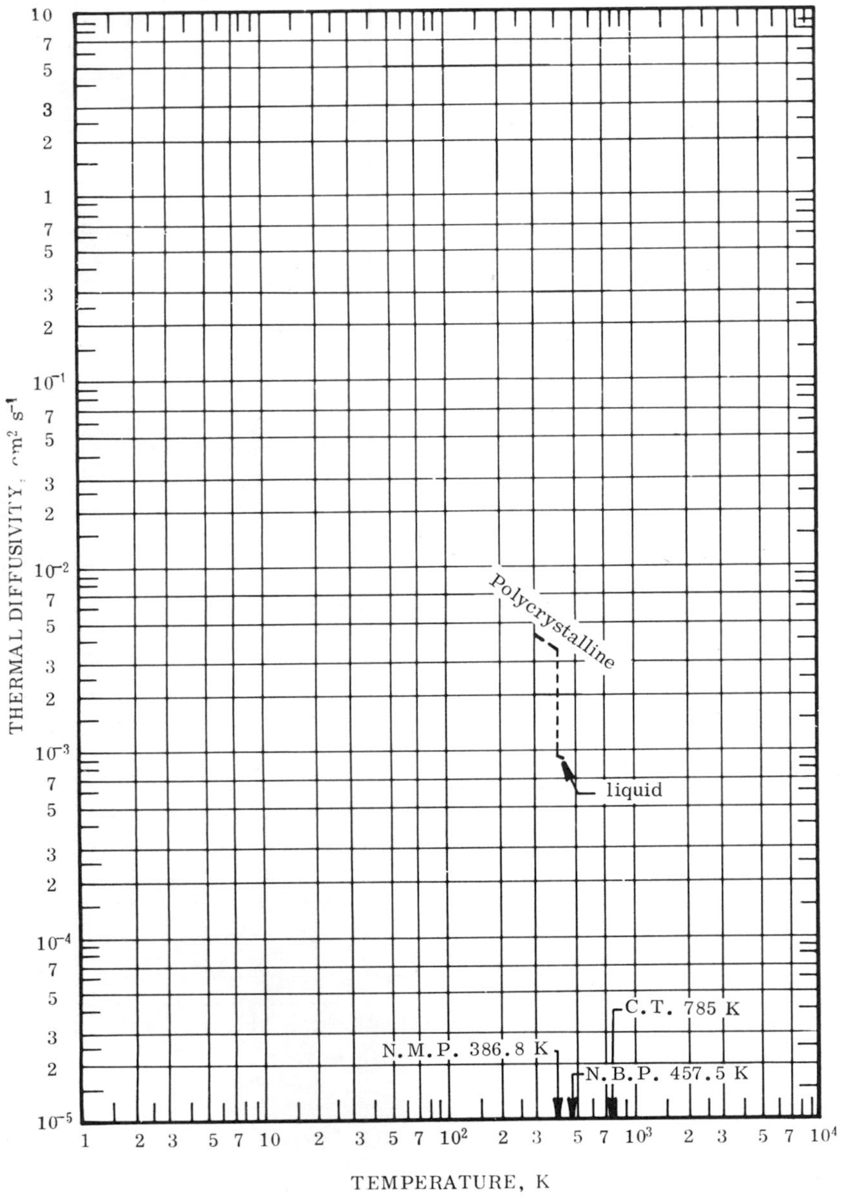

PROVISIONAL VALUES*

[Temperature, T, K; Thermal Diffusivity, α, cm^2 s^{-1}]

SOLID (Polycrystalline)		LIQUID	
T	α	T	α
298.2	0.00422	386.8	0.000910
300	0.00420	400	0.000900
350	0.00390		
386.8	0.00348		

REMARKS

The provisional values are f· high-purity iodine and are probably good to ±20%.

* All values are estimated.

FIGURE AND TABLE 30R. RECOMMENDED THERMAL DIFFUSIVITY OF IRIDIUM

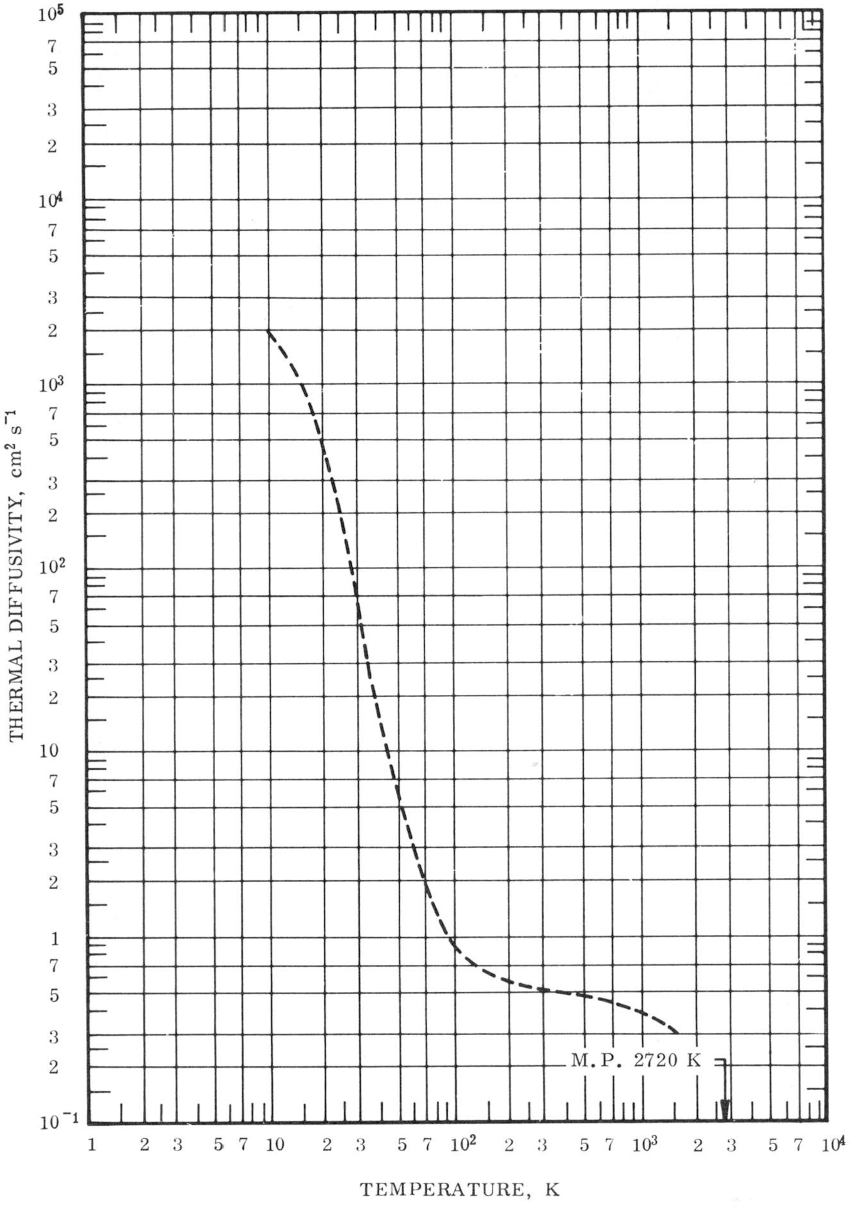

RECOMMENDED VALUES*

[Temperature, T, K; Thermal Diffusivity, α, $cm^2 s^{-1}$]

SOLID

T	α	T	α
10	1850	150	0.630
11	1650	200	0.558
12	1450	250	0.522
13	1270	273.2	0.512
14	1190	300	0.502
15	939	350	0.491
16	805	400	0.482
18	580	500	0.467
20	406	600	0.451
25	160	700	0.433
30	65.0	800	0.414
35	28.6	900	0.395
40	14.8	1000	0.377
45	8.60	1100	0.361
50	5.50	1200	0.346
60	2.78	1300	0.331
70	1.74	1400	0.318
80	1.27	1500	0.305
90	0.992		
100	0.842		

REMARKS

The values are for well-annealed high-purity iridium and are thought to be accurate to within ± 10% of the true values at temperatures below 500 K and ± 15% above. The values above 500 K are provisional. At low temperatures the values are highly conditioned by impurity and imperfection, and those below 150 K are applicable only to a specimen having residual electrical resistivity of 0.0191 $\mu\Omega$ cm.

*All values are estimated and those above 500 K are provisional.

FIGURE AND TABLE 31R. RECOMMENDED THERMAL DIFFUSIVITY OF IRON

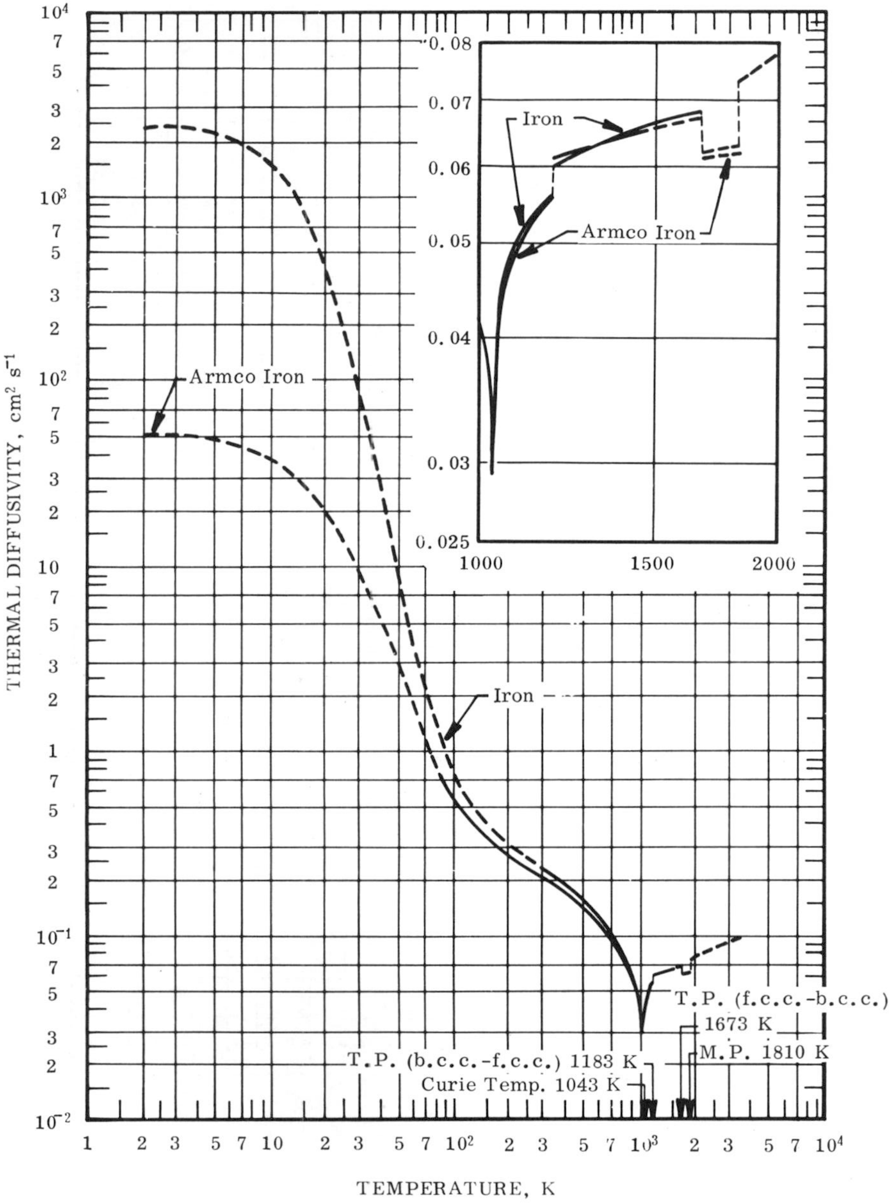

RECOMMENDED VALUES†

[Temperature, T, K; Thermal Diffusivity, α, $cm^2\ s^{-1}$]

SOLID						LIQUID	
	Iron	Armco Iron		Iron	Armco Iron		Iron
T	α	α	T	α	α	T	α
2	2360*	49.5*	250	0.260*	0.230	1810	0.0731*
3	2310*	48.8*	273.2	0.244	0.218	1900	0.0758*
4	2230*	47.4*	300	0.227	0.206	2000	0.0784*
5	2150*	46.0*	350	0.201	0.187	2200	0.0825*
6	2040*	44.3*	400	0.181	0.171	2400	0.0858*
7	1930*	42.5*	500	0.149	0.144	2600	0.0884*
8	1790*	40.6*	600	0.124	0.120	2800	0.0906*
9	1650*	38.7*	700	0.102	0.0991	3000	0.0926*
10	1500*	36.7*	800	0.0818	0.0797	3500	0.0972*
11	1370*	34.7*	900	0.0630	0.0617		
12	1230*	32.7*	1000	0.0425	0.0423		
13	1090*	30.7*	1043	0.0291	0.0289		
14	965*	28.8*	1100	0.0512	0.0505		
15	849*	26.8*	1183	0.0565	0.0557		
16	749*	25.0*	1183	0.0599	0.0610		
18	567*	21.8*	1200	0.0605	0.0612		
20	420*	18.9*	1300	0.0632	0.0624		
25	195*	13.2*	1400	0.0651	0.0638		
30	94.6*	9.21*	1500	0.0665	0.0650*		
35	48.4*	6.62*	1600	0.0677	0.0661*		
40	26.5*	4.81*	1673	0.0683*	0.0669*		
45	15.5*	3.57*	1673	0.0620*	0.0607*		
50	9.47*	2.70*	1700	0.0622*	0.0609*		
60	4.22*	1.69*	1800	0.0629*	0.0616*		
70	2.27*	1.17*	1810	0.0630*			
80	1.43*	0.858*					
90	1.01*	0.673					
100	0.782*	0.558					
150	0.404*	0.333					
200	0.309*	0.265					

REMARKS

The values are for well-annealed high-purity iron and for well-annealed Armco iron and are considered accurate to within ±10% of the true values at temperatures below 100 K, ±5% from 100 K to room temperature, ±4% from room temperature to above 1000 K, the uncertainty probably increasing to about ±15% at 1600 K and ±20% at the melting point. The values above 1600 K are provisional. At low temperatures the values are highly conditioned by impurity and imperfection and those for high-purity iron below 200 K are applicable only to a specimen having residual electrical resistivity of 0.0143 $\mu\Omega$ cm. For Armco iron the values below 200 K are applicable only to a specimen having residual electrical resistivity of 0.690 $\mu\Omega$ cm.

†Values above 1600 K are provisional.
*In temperature range where no experimental data are available.

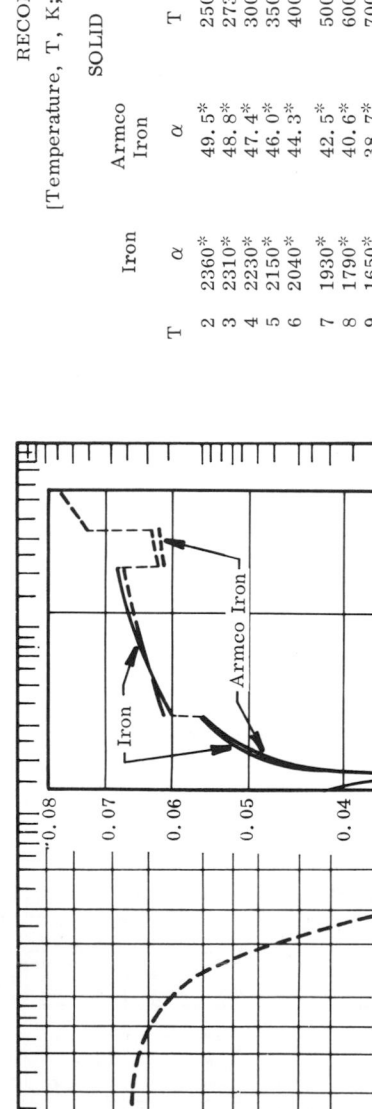

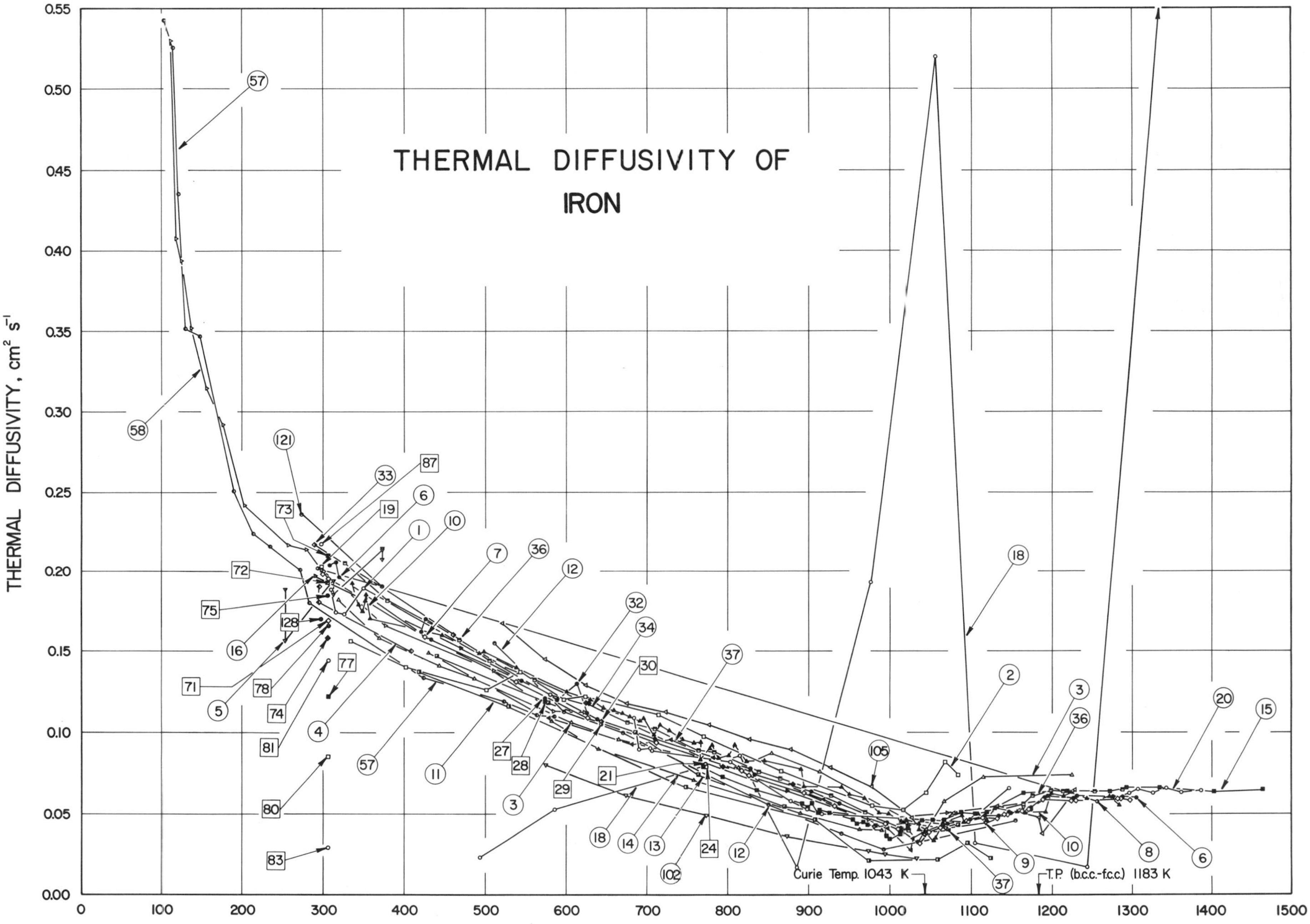
THERMAL DIFFUSIVITY OF
IRON
THERMAL DIFFUSIVITY, cm² s⁻¹
TEMPERATURE, K
Curie Temp. 1043 K
T.P. (b.c.c.-f.c.c.) 1183 K

FIGURE 31

SPECIFICATION TABLE 31. THERMAL DIFFUSIVITY OF IRON

(Impurity < 0.20% each; total impurities < 0.50%)

Cur. No.	Ref. No.	Author(s)	Year	Temp. Range, K	Reported Error, %	Name and Specimen Designation	Composition (weight percent), Specifications, and Remarks
1	1	Sonnenschein, G. and Winn, R. A.	1960	310-1146		Armco Iron; No. 1	99.94$^+$ Fe, 0.01 Cr, 0.01 Cu, 0.01 Mg, 0.01 Mn, 0.005 Ni, 0.001 Mo, and 0.001 Sn; cylindrical specimen 0.635 cm in dia; furnished by the National Bureau of Standards; front surface of specimen covered with a fine film of lamp black; measured under a vacuum of $\sim10^{-4}$ mm Hg; diffusivity measured under conditions of one-dimensional transient heat flow (flash method).
2	1	Sonnenschein, G. and Winn, R. A.	1960	334-1081		Armco Iron; No. 2	Specimen from the same batch as the above specimen but of different thickness; same composition and measuring technique as above specimen.
3	1	Sonnenschein, G. and Winn, R. A.	1960	319-1223		Armco Iron; No. 2	Above specimen measured again.
4	4	Jenkins. R. J. and Parker, W. J.	1961	295-408	±5	Armco Iron	Square specimen 1.9 cm side and 0.100 cm thick; high intensity short duration light pulse absorbed in front surface of thermally insulated specimen coated with camphor black; 408.2 K measurement obtained by heating specimen holder and specimen with an infrared lamp; both data points at 295.2 K obtained from measurements using different equations for data reduction.
5	29, 30	McIntosh, G. E.	1952	253-373	5	Armco Iron	Similar to wrought iron; rod specimen 0.125 in. in dia. and 9 in. long; surrounded by radiation shield consisting of four concentric cylinders of very thin aluminum foil, each separated by three 1/32 in. rings of balsa wood; measured after being maintained at elevated temp for several hrs; measured in vacuum; diffusivity determined from measured phase lag of the temp. wave between any two points along specimen; one-dimensional heat flow.
6	28	Abeles, B., Beers, D., Cody, G., Novak, R., and Rosi, F.	1960	308-1305	±2	Armco Iron	Cylindrical specimen 3/16 in. in dia. and 2 in. long; specimen machined from rod stock purchased from Mapes and Sprowl; sample brazed on heater with copper in a hydrogen atmosphere at 1373.2 K; Curie temp. 1043.2 K; measured under a vacuum of 10^{-6}mm Hg; sine heat input supplied to specimen at a frequency of 3 cycles per min.
7	31, 241, 147	Abeles, B., Cody, G. D., Novak, R., and Beers, D. S.	1959	300-1298		Armco Iron; V	Cylindrical specimen 3/16 in. in dia. and 2 in. long; machined from Armco stock purchased from the Mapes and Sprowl Steel Co.; electrical resistivity reported as 11.63, 11.56, and 11.66 μohmcm at 298.2, 302.2, and 303.2 K, respectively; small electrical heater brazed to one end of specimen to provide a sinusoidal heat input; brazing done with copper in a hydrogen atmosphere at 1383.2 K for ~5 min; diffusivity determined from measured attenuation and phase shift of temp. wave; measured in an evacuated oven at a pressure of $\sim5 \times 10^{-4}$ mm Hg.
8	31	Abeles, B., et al.	1959	295-1256		Armco Iron; IV	Cylindrical specimen 3/16 in. in dia. and 2 in. long; machined from Armco stock purchased from the Mapes and Sprowl Steel Co.; electrical resistivity reported as 9.78 and 9.89 μohm cm at 273.2, and 283.2 K, respectively; small electrical heater brazed to one end of specimen to provide a sinusoidal heat input; brazing done with copper in a hydrogen atmosphere at 1383.2 K for ~5 min; diffusivity determined from measured attenuation and phase shift of temp. wave; measured in an evacuated oven at a pressure of $\sim5 \times 10^{-4}$ mm Hg.

SPECIFICATION TABLE 31. THERMAL DIFFUSIVITY OF IRON (continued)

Cur. No.	Ref. No.	Author(s)	Year	Temp. Range, K	Reported Error, %	Name and Specimen Designation	Composition (weight percent), Specifications, and Remarks
9	32	Taylor, R. E. and Nakata, M. M.	1963	980-1164	7	Armco Iron	Cylindrical specimen 0.5 in. in dia. and 1 in. long; two parallel sight holes each 0.08 cm in dia. drilled to the longitudinal center of the specimen at radii $r_1 = 0$ and $r_2 = 0.449$ cm; machined to size; sight holes drilled by Elox technique; measured under a vacuum of 1×10^{-6} mm Hg; tantalum foil used as radiation shield at the ends of specimen; radial diffusivity technique used.
10	33	Chiotti, P. and Carlson, O. N.	1956	336-1283		Armco Iron	No details reported by author.
11	24	Dennis, J. E., Hirschman, A., Derksen, W. L., and Monahan, T. I.	1960	418-1123		Armco Iron	Disk specimen 0.64 cm in dia. and 0.077 cm in thickness; cut from sample of Armco Iron stock being used in the NBS for diffusivity measurements; specimen irradiated with a chopped beam in a carbon-arc image furnace; thermal diffusivity determined from measured phase lag between the square wave irradiance impinging upon the front face of the specimen and the resultant sinusoidal temp. of the rear face; error in calculating diffusivity (due to the use of a square instead of a sinusoidal heat input) estimated to be -3.2%.
12	24	Dennis, J. E., et al.	1960	511-1153		Armco Iron	Disk specimen 0.64 cm in dia. and 0.103 cm in thickness; cut from sample of Armco Iron stock being used in the NBS for diffusivity measurements; specimen irradiated with a chopped beam in a carbon-arc image furnace; thermal diffusivity determined from measured phase lag between the square wave irradiance impinging upon the front face of the specimen and the resultant sinusoidal temp of the rear face; error in calculating diffusivity (due to the use of a square instead of a sinusoidal heat input) estimated to be -3.2%.
13	24	Dennis, J. E., et al.	1960	440-1083		Armco Iron	Disk specimen 0.64 cm in dia. and 0.133 cm in thickness; cut from sample of Armco Iron stock being used in the NBS for diffusivity measurements; specimen irradiated with a chopped beam in a carbon-arc image furnace; thermal diffusivity determined from measured phase lag between the square wave irradiance impinging upon the front face of the specimen and the resultant sinusoidal temp. of the rear face; error in calculating diffusivity (due to the use of a square instead of a sinusoidal heat input) estimated to be -3.2%.
14	24	Dennis, J. E., et al.	1960	430-990		Armco Iron	Disk specimen 0.64 cm in dia. and 0.148 cm in thickness; cut from sample of Armco Iron stock being used in the NBS for diffusivity measurements; specimen irradiated with a chopped beam in a carbon-arc image furnace; thermal diffusivity determined from measured phase lag between the square wave irradiance impinging upon the front face of the specimen and the resultant sinusoidal temp. of the rear face; error in calculating diffusivity (due to the use of a square instead of a sinusoidal heat input) estimated to be -3.2%.
15	242, 141, 34	Rudkin, R. L., Parker, W. J., and Jenkins, R. J.	1961	769-1464	±5	Armco Iron	Specimen a few millimeters thick; high intensity short duration light pulse from xenon flash lamp absorbed in the front surface of thermally insulated specimen; thermal diffusivity determined from measured temp. history of the rear surface; radio frequency induction heating used for heating specimen.
16	34, 141	Rudkin, R. L., et al.	1963	290-1270	±5	Armco Iron	Above specimen measured for diffusivity in resistance furnace; measured under the same conditions as above.

SPECIFICATION TABLE 31. THERMAL DIFFUSIVITY OF IRON (continued)

Cur. No.	Ref. No.	Author(s)	Year	Temp. Range, K	Reported Error, %	Name and Specimen Designation	Composition (weight percent), Specifications, and Remarks
17*	35	Lehman, G. W.	1960	980-1164		Armco Iron	Cylindrical specimen 0.5 in. in dia. and 1 in. long; two small pyrometer holes drilled parallel to specimen axis from one end to the midpoint; specimen ends shielded against radiation by tantalum foil; diffusivity determined by measuring the time interval necessary for the temp. at a point at radius r_1 to reach that at another radius r_2.
18	36	Kevane, C. J.	1958	493-1343			Cylindrical specimen ~0.375 in. in dia. and 1 in. long; machined; holes drilled in specimen separated by a distance of 2 mm between the lines of centers along the axis; diffusivity determined from measured propagation of temp. waves set up by periodic modulation of radiation heat flux input to one surface of specimen; solar furnace used as a heat source; amplitude ratio method used to measure diffusivity; one-dimensional heat flow.
19	6	Moser, J. B. and Kruger, O. L.	1963	298.			Plate specimen with surface area lying in the range from 1 to 4 cm^2 and thickness in the range from 0.1 to 0.3 cm; front surface thinly coated with colloidal graphite; irradiated with a pulse of thermal energy of short duration; diffusivity determined from measured history of the back surface temp.
20	37, 103	Kennedy, W. L.	1960	580-1386		Armco Iron	Cylindrical rod specimen; surrounded by cylindrical guard; heater attached to one end of both specimen and guard; diffusivity determined from measured temp. variation with time at two points along specimen; one-dimensional heat flow; five different runs made.
21*	105	Yurchak, R. P. and Filippov, L. P.	1965	773	7		Cylindrical specimen 2 cm in dia. and 12 cm long, made of two halves; measured under a vacuum of $\sim 10^{-5}$ mm Hg; heat losses reduced by cylindrical shields of sheet molybdenum; radial wave method used to measure diffusivity; diffusivity determined from measured amplitude ratio using a period of 6.6 sec.
22*	105	Yurchak, R. P. and Filippov, L. P.	1965	773, 1073	7		Above specimen measured for diffusivity again using same method as above but employing a period of 13.2 sec for the temp. wave; other conditions same as above.
23*	105	Yurchak, R. P. and Filippov, L. P.	1965	1073	7		Above specimen measured for diffusivity again using same method as above but employing a period of 26.4 sec for the temp. wave; other conditions same as above.
24	105	Yurchak, R. P. and Filippov, L. P.	1965	773	7		Above specimen measured for diffusivity again; diffusivity determined from measured phase difference using a period of 6.6 sec for the radial temp. wave; other conditions same as above.
25*	105	Yurchak, R. P. and Filippov, L. P.	1965	773, 1073	7		Above specimen measured for diffusivity again using same method as above but employing a period of 13.2 sec for the temp. wave; other conditions same as above.
26*	105	Yurchak, R. P. and Filippov, L. P.	1965	1073	7		Above specimen measured for diffusivity again using same method as above but employing a period of 26.4 sec for the temp. wave; other conditions same as above.
27	107	Filippov, L. P.	1966	573	~4	Armco Iron	Cylindrical specimen having dia. lying in the range from 5 to 15 mm and a few centimeters long; specimen heated by electron bombardment using a periodically changing power of 2.52 W; period of modulation of electron beam 15.2 sec; longitudinal heat flow; measured in vacuum.

* Not shown in figure.

SPECIFICATION TABLE 31. THERMAL DIFFUSIVITY OF IRON (continued)

Cur. No.	Ref. No.	Author(s)	Year	Temp. Range, K	Reported Error, %	Name and Specimen Designation	Composition (weight percent), Specifications, and Remarks
28	107	Filippov, L. P.	1966	573	~4	Armco Iron	Cylindrical specimen having dia. lying in the range from 5 to 15 mm and a few centimeters long; specimen heated by electron bombardment using a periodically changing power of 2.52 W; period of modulation of electron beam 20.1 sec; longitudinal heat flow; measured in vacuum.
29	107	Filippov, L. P.	1966	643	~4	Armco Iron	Cylindrical specimen having dia. lying in the range from 5 to 15 mm and a few centimeters long; specimen heated by electron bombardment using a periodically changing power of 6.25 W; period of modulation of electron beam 10.1 sec; longitudinal heat flow; measured in vacuum.
30	107	Filippov, L. P.	1966	643	~4	Armco Iron	Cylindrical specimen having dia. lying in the range from 5 to 15 mm and a few centimeters long; specimen heated by electron bombardment using a periodically changing power of 6.25 W; period of modulation of electron beam 15.2 sec; longitudinal heat flow; measured in vacuum.
31*	107	Filippov, L. P.	1966	643	~4	Armco Iron	Cylindrical specimen having dia. lying in the range from 5 to 15 mm and a few centimeters long; specimen heated by electron bombardment using a periodically changing power of 6.25 W; period of modulation of electron beam 20.2 sec; longitudinal heat flow; measured in vacuum.
32	112	Carpenter, R. S., II	1962	421-625		Armco Iron	Cylindrical specimen ~0.25 in. in dia. and 0.137 in. thick; front surface irradiated with short duration heat pulse from a xenon flash tube; diffusivity determined from the measured temperature-time curve of the rear surface; measured in a vacuum of 5 x 10^{-6} mm Hg.
33	112	Carpenter, R. S., II	1962	289, 460		Armco Iron	Cylindrical specimen ~0.25 in. in dia. and 0.134 in. thick; other conditions same as above.
34	112	Carpenter, R. S., II	1962	624-880		Armco Iron	Cylindrical specimen ~0.25 in. in dia. and 0.071 in. thick; other conditions same as above.
35*	112	Carpenter, R. S., II	1962	292-479		Armco Iron	Cylindrical specimen ~0.25 in. in dia. and 0.137 in. thick; front surface irradiated with laser beam; other conditions same as above.
36	113	Sidles, P. H. and Danielson, G. C.	1960	317-1288		Armco Iron	Curie temp. 1042.2 K; α-γ transition at 1173.2 K; electrical resistivity measured during increasing and decreasing temp. and reported as 10.4, 13.2, 18.8, 25.8, 34.1, 40.2, 42.6, 48.6, 51.2, 58.5, 63.7, 68.7, 76.0, 85.6, 94.6, 106.9, 109.4, 113.4, 113.1, 116.4, and 116.8 μohm cm at 296.2, 348.2, 407.2, 502.2, 581.2, 622.2, 660.2, 702.2, 742.2, 786.2, 814.2, 856.2, 892.2, 950.2, 999.2, 1053.2, 1091.2, 1148.2, 1184.2, 1230.2, and 1273.2 K, respectively; diffusivity measured using modified Angström method.
37	119	Gibby, R. L.	1968	578-1072	±9	Armco Iron	Cylindrical specimen 0.63 cm in dia. and 0.0760 cm thick; laser beam used as the pulse energy source; diffusivity determined from measured temp. history of the rear surface; data corrected for finite-pulse-time effects and heat losses.

* Not shown in figure.

SPECIFICATION TABLE 31. THERMAL DIFFUSIVITY OF IRON (continued)

Cur. No.	Ref. No.	Author(s)	Year	Temp. Range, K	Reported Error, %	Name and Specimen Designation	Composition (weight percent), Specifications, and Remarks
38*	119	Gibby, R. L.	1968	530-1381	±9	Armco Iron	Cylindrical specimen 0.63 cm in dia. and 0.0704 cm thick; laser beam used as the pulse energy source; diffusivity determined from measured temp. history of the rear surface; data corrected for finite-pulse-time effects and heat losses.
39*	119	Gibby, R. L.	1968	525-1216	±9	Armco Iron	Above specimen measured for diffusivity during cooling; other conditions same as above.
40*	119	Gibby, R. L.	1968	621-1076	±9	Armco Iron	Cylindrical specimen 0.63 cm in dia. and 0.0711 cm thick; laser beam used as the pulse energy source; diffusivity determined from measured temp. history of the rear surface; data corrected for finite-pulse-time effects and heat losses.
41*	243, 244, 123, 245	Shanks, H. R., Klein, A. H., and Danielson, G. C.	1963	298-1174		Armco Iron; A	99.52 Fe (by difference), 0.2 Cu, 0.076 O (vacuum fusion method), 0.04 Ni, 0.035 Mn, 0.023 As, 0.02 Cr, 0.02 S, 0.0119-0.0127 C (combustion conductometric method), 0.012 Sn, 0.006 Co, 0.0051 N (vacuum fusion method), 0.005 Mo, 0.004 Ag, 0.0015 Cl, 0.0015 Ga, 0.001 Sb, 0.001 Zn, 0.0005 K, 0.0005 Na, 0.0003 Ca, 0.00007 Mg, 0.00006 Pd, 0.00006 V, 0.00001 Nb, and 0.00001 Sc (composition determined using spark-source mass spectrographic analysis); cylindrical specimen 0.25 in. in dia. and 3.5 in. long; supplied by Battelle Memorial Institute in the form of rod 1 in. in dia. previously annealed at 1143.2 K for 30 min in air; machined to final dimensions from original stock, threaded for a distance of 0.5 in. on one end, then turned into a guard cylinder and welded in place; annealed for 8 hrs at 1143.2 K at a pressure of 1 x 10^{-5} mm Hg or less; diamond pyramid hardness No. (as received from BMI): 84±4 (transverse section), 80±4 (longitudinal section); apparent electrical resistivity reported as 0.771, 0.771, 1.32, 1.59, 2.90, 4.96, 9.70, 11.00, 15.30, 22.45, 31.45, 42.20, 54.85, 69.55, 86.55, 100.85, 105.85, 112.85, 112.95, 114.20, and 115.95 μohm cm at 1.2, 4.2, 73.2, 83.2, 123.2, 173.2, 273.2, 298.2, 373.2, 473.2, 573.2, 673.2, 773.2, 873.2, 973.2, 1038.2, 1073.2, 1173.2, 1188.2, 1223.2, and 1273.2 K, respectively; electrical resistivity ratio $\rho(300\ K)/\rho(4.2\ K) = 14$; measured for diffusivity using linear finite rod method.
42*	243, 244, 123, 245	Shanks, H. R., Klein, A. H., and Danielson, G. C.	1963	301-1190		Armco Iron; A	Above specimen annealed at 1293.2 K for 2 hrs, cooled to room temp. and then measured for diffusivity again; apparent Lorenz function reported as 3.10, 3.04, 3.08, 3.13, and 3.18 x 10^{-8} V^2 K^{-2} at 1073.2, 1123.2, 1173.2, 1223.2, and 1273.2 K, respectively; annealing carried out at a pressure of 1 x 10^{-5} mm Hg or less.
43*	244, 123, 245	Shanks, H. R., Klein, A. H., and Danielson, G. C.	1964	1195-1274		Armco Iron; A	Above specimen heated to 1275.2 K and then measured for diffusivity during cooling.
44*	243, 123	Shanks, H. R., et al.	1963	966-1257		Armco Iron; A	Above specimen heated to 1373.2 K, held for 2 hr, and then measured for diffusivity again during cooling; total time specimen temperature above 1173.2 K less than 24 hr.

* Not shown in figure.

SPECIFICATION TABLE 31. THERMAL DIFFUSIVITY OF IRON (continued)

Cur. No.	Ref. No.	Author(s)	Year	Temp. Range, K	Reported Error, %	Name and Specimen Designation	Composition (weight percent), Specifications, and Remarks
45*	245, 123	Shanks, H.R., Klein, A.H., and Danielson, G.C.	1965	487-1241		Armco Iron; C	Same composition as specimen A (curve 41) above; cylindrical specimen 0.25 in. in dia. and ~12 in. long; swaged from original material supplied by Battelle Memorial Institute and annealed at 1143.2 K for 8 hrs at a pressure of 1×10^{-5} mm Hg or less; measured for diffusivity using the modified Angström method; diffusivity measurements made up to the α-γ transition; specimen then annealed at 1293.2 K and measurements made again with temp. increasing in the γ-phase; sample temp. above 1173.2 K for less than 24 hrs; data points near room temp. measured in helium gas.
46*	245, 123	Shanks, H.R., et al.	1965	399-1263		Armco Iron; B	Same composition as specimen A (curve 41) above except that carbon content decreased to 0.0075 (after annealing); cylindrical specimen 0.25 in. in dia. and ~12 in. long; swaged from original material supplied by Battelle Memorial Institute and annealed at 1623.2 K for 1 hr at a pressure of 1×10^{-5} mm Hg or less; measured for diffusivity using the modified Angström method; apparent Lorenz function reported as 2.69, 2.84, 2.94, 2.99, 2.97, 2.98, 3.05, 3.04, 3.07, 3.01, 2.79, 2.76, 2.77, and 2.82×10^{-8} V^2 K^{-2} at 273.2, 373.2, 473.2, 573.2, 673.2, 773.2, 873.2, 973.2, 1023.2, 1073.2, 1123.2, 1173.2, 1223.2, and 1273.2 K, respectively.
47*	123	Shanks, H.R., et al.	1967	1040-1272		Armco Iron; A	Data points measured during heating.
48*	123	Shanks, H.R., et al.	1967	978-1038		Armco Iron; A	Data points measured during cooling.
49*	123	Shanks, H.R., et al.	1967	978-1264		Armco Iron; B	Data points measured during cooling.
50*	123	Shanks, H.R., et al.	1967	1058-1296		Armco Iron; C	Measured for diffusivity again during heating.
51*	123	Shanks, H.R., et al.	1967	1121, 1126		Armco Iron; C	Above specimen measured for diffusivity again during cooling.
52*	123	Shanks, H.R., et al.	1967	1050-1277		Armco Iron; D	Cylindrical specimen 0.25 in. in dia. and ~12 in. long; swaged from original material supplied by Battelle Memorial Institute and annealed at 1273.2 K for 2 hrs at a pressure of 1×10^{-5} mm Hg or less; measured for diffusivity using the modified Angström method.
53*	123	Shanks, H.R., et al.	1967	1098-1277		Armco Iron; D	Above specimen measured for diffusivity during cooling.
54*	128, 246	Bates, J.L.	1968	485-970		Armco Iron	Disk specimen 0.140 cm thick; laser-pulse technique used to measure diffusivity; measured in a purified argon atmosphere under a pressure of one atmosphere; inlet argon containing <0.0001 O and <0.0005 H_2O; data corrected for heat losses.
55*	128, 246	Bates, J.L.	1968	889-1098		Armco Iron	Disk specimen 0.071 cm thick; measured under same conditions as above.
56*	128, 246	Bates, J.L.	1968	963-1096		Armco Iron	Disk specimen; measured under same conditions as above.
57	136	Morrison, B.H., Klein, D.J., and Cowder, L.R.	1966	92-585		Round Robin Armco Iron	Disk specimen 0.5 in. in diameter and having thickness in the range from 0.070 to 0.092 in.; obtained from Battelle Memorial Institute; polished with 600 grit paper and crocus cloth; front face exposed to energy pulse generated by ruby laser; diffusivity determined from measured transient time-temperature response of back face; measured in a vacuum of 1×10^{-5} mm Hg; data points reported include corrections for heat loss and finite pulse time.

* Not shown in figure.

SPECIFICATION TABLE 31. THERMAL DIFFUSIVITY OF IRON (continued)

Cur. No.	Ref. No.	Author(s)	Year	Temp. Range, K	Reported Error, %	Name and Specimen Designation	Composition (weight percent), Specifications, and Remarks
58	136	Morrison, B. H., Klein, D. J., and Cowder, L. R.	1966	93-623		LASL Armco Iron	Disk specimen 0.5 in. in dia. and having thickness in the range from 0.070 to 0.092 in.; made from bar stock Armco Iron; polished with 600 grit paper and crocus cloth; front face exposed to energy pulse generated by ruby laser; diffusivity determined from measured transient time-temp. response of back face; measured in a vacuum of 1 x 10^{-5} mm Hg; data points reported include corrections for heat loss and finite pulse time; heat flow in the direction parallel to the axis of the original stock.
59*	136	Morrison, B. H., et al.	1966	93, 298		Round Robin Armco Iron	Same specimen as that used for measurements pertaining to curve No. 57 above; each data point reported is the average of four independent measurements; other conditions same as above; measured for diffusivity again.
60*	136	Morrison, B. H., et al.	1966	93, 298		LASL Armco Iron	Same specimen as that used for measurements pertaining to curve No. 58 above; each data point reported is the average of four independent measurements; other conditions same as above; measured for diffusivity again.
61*	136	Morrison, B. H., et al.	1966	298		LASL Armco Iron	Disk-shaped; cut from same 0.75 in. dia. rod as above specimen; oriented so that heat flow measurements become perpendicular to the axis of the original stock; measured in the "as received" condition; data point reported is the average of four independent measurements; other conditions same as above.
62*	136	Morrison, B. H., et al.	1966	298		LASL Armco Iron	Above specimen measured for diffusivity again after annealing; other conditions same as above.
63*	136	Morrison, B. H., et al.	1966	303		LASL Armco Iron	Sleeve specimen 0.358 in. I. D., 0.458 in. O. D., and 2 in. long; polished to a 40 μin. finish; specimen exterior uniformly exposed to energy pulse from xenon flash tube; diffusivity determined from transient time-temp response measured by interior thermocouple probes located at center of specimen length; data point reported obtained under assumption of no heat loss; correction for finite pulse time included; data point reported is average result of five independent measurements; measured in air.
64*	136	Morrison, B. H., et al.	1966	303		LASL Armco Iron	Sleeve specimen 0.358 in. I. D., 0.558 in. O. D., and 2 in. long; polished to a 40 μin. finish; specimen exterior uniformly exposed to energy pulse from xenon flash tube; diffusivity determined from transient time-temp. response measured by interior thermocouple probes located at center of specimen length; data point reported obtained under assumption of no heat loss; correction for finite pulse time included; data point reported is the average of four independent measurements; measured in air.
65*	136	Morrison, B. H., et al.	1966	303		LASL Armco Iron	Same specimen used for measurements pertaining to curve No. 63 above measured for diffusivity again; exposed to energy pulse from laser beam; measured time-temp. response used to determine thermal diffusivity; data point reported includes corrections for heat loss and finite pulse time and represents average of four independent measurements; measured in air.

* Not shown in figure.

SPECIFICATION TABLE 31. THERMAL DIFFUSIVITY OF IRON (continued)

Cur. No.	Ref. No.	Author(s)	Year	Temp. Range, K	Reported Error, %	Name and Specimen Designation	Composition (weight percent), Specifications, and Remarks
66*	136	Morrison, B.H., Klein, D.J., and Cowder, L.R.	1966	303		LASL Armco Iron	Above specimen thermally insulated by a 0.036 in. thick boron nitride sleeve loosely fitting around it and having a 0.375 in. dia hole located to expose only the area of the iron sleeve whose interior is contacted by the thermocouples; flash tube used as the pulse energy source; thermal diffusivity determined from measured time-temp. response assuming no heat loss; measured in air.
67*	136	Morrison, B.H., et al.	1966	303		LASL Armco Iron	Thermal diffusivity determined from the above measurement assuming radial heat loss.
68*	136	Morrison, B.H., et al.	1966	303		LASL Armco Iron	Above specimen thermally insulated by a 0.072 in. thick boron nitride sleeve loosely fitting around it and having a 0.25 in. dia. hole located to expose only the area of the iron sleeve whose interior is contacted by the thermocouples; flash tube used as the pulse energy source; thermal diffusivity determined from measured time-temp response assuming no heat loss; measured in air.
69*	136	Morrison, B.H., et al.	1966	303		LASL Armco Iron	Thermal diffusivity determined from the above measurement assuming radial heat loss.
70	149	Taylor, R.E. and Cape, J.A.	1964	297-965		Armco Iron	Samples of different thicknesses used for diffusivity measurement; flash technique used to measure diffusivity; data points reported corrected for finite pulse-time effect.
71	144	Larson, K.B. and Koyama, K.	1967	306		Armco Iron	Disk specimen 0.79 cm in dia. and 4.32 x 10^{-2} cm in thickness; machined from solid stock with the flat faces parallel to within 0.1 degrees of arc; specimen initially at room temp.; effective temp. of measurement calculated from the equation of Parker and others (J. Appl. Phys. 32, 1679-84, 1961); thermal energy pulse from a xenon flash tube used to irradiate front face; diffusivity determined from measured temp. history of rear face assuming instantaneous heat pulse.
72	144	Larson, K.B. and Koyama, K.	1967	306		Armco Iron	Diffusivity determined again from above measurement for above specimen assuming a sawtooth heat pulse.
73	144	Larson, K.B. and Koyama, K.	1967	306		Armco Iron	Diffusivity determined again from above measurement for above specimen employing an empirical function closely describing the actual waveform of the heat pulse.
74	144	Larson, K.B. and Koyama, K.	1967	306		Armco Iron	Disk specimen 0.79 cm in dia. and 3.83 x 10^{-2} cm in thickness; machined from solid stock with the flat faces parallel to within 0.1 degrees of arc; diffusivity determined from measured temp. history of rear face assuming instantaneous heat pulse; other conditions same as above.
75	144	Larson, K.B. and Koyama, K.	1967	306		Armco Iron	Diffusivity determined again from above measurement for above specimen assuming a sawtooth heat pulse.
76	144	Larson, K.B. and Koyama, K.	1967	306		Armco Iron	Diffusivity determined again from above measurement for above specimen employing an empirical function closely describing the actual waveform of the heat pulse.

* Not shown in figure.

SPECIFICATION TABLE 31. THERMAL DIFFUSIVITY OF IRON (continued)

Cur. No.	Ref. No.	Author(s)	Year	Temp. Range, K	Reported Error, %	Name and Specimen Designation	Composition (weight percent), Specifications, and Remarks
77	144	Larson, K. B. and Koyama, K.	1967	306		Armco Iron	Disk specimen 0.79 cm in dia. and 2.46 x 10^{-2} cm in thickness; machined from solid stock with the flat faces parallel to within 0.1 degrees of arc; diffusivity determined from measured temp. history of rear face assuming instantaneous heat pulse; other conditions same as above.
78	144	Larson, K. B. and Koyama, K.	1967	306		Armco Iron	Diffusivity determined again from above measurement for above specimen assuming a sawtooth heat pulse.
79	144	Larson, K. B. and Koyama, K.	1967	306		Armco Iron	Diffusivity determined again from above measurement for above specimen employing an empirical function closely describing the actual waveform of the heat pulse.
80	144	Larson, K. B. and Koyama, K.	1967	306		Armco Iron	Disk specimen 0.79 cm in dia. and 1.65 x 10^{-2} cm in thickness; machined from solid stock with the flat faces parallel to within 0.1 degrees of arc; diffusivity determined from measured temp. history of rear face assuming instantaneous heat pulse; other conditions same as above.
81	144	Larson, K. B. and Koyama, K.	1967	306		Armco Iron	Diffusivity determined again from above measurement for above specimen assuming a sawtooth heat pulse.
82	144	Larson, K. B. and Koyama, K.	1967	306		Armco Iron	Diffusivity determined again from above measurement for above specimen employing an empirical function closely describing the actual waveform of the heat pulse.
83	144	Larson, K. B. and Koyama, K.	1967	306		Armco Iron	Disk specimen 0.79 cm in dia. and 7.6 x 10^{-3} cm in thickness; machined from solid stock with the flat faces parallel to within 0.1 degrees of arc; diffusivity determined from measured temp history of rear face assuming instantaneous heat pulse; other conditions same as above.
84	144	Larson, K. B. and Koyama, K.	1967	306		Armco Iron	Diffusivity determined again from above measurement for above specimen assuming a sawtooth heat pulse.
85	144	Larson, K. B. and Koyama, K.	1967	306		Armco Iron	Diffusivity determined again from above measurement for above specimen employing an empirical function closely describing the actual waveform of the heat pulse.
86	175	Oualid, J.	1961	298.2			Plate specimen 100 x 10 x 1 mm; thermal shock method used to measured diffusivity; measured in vacuum; temperature of measurement not given by author but assumed to be room temperature.
87	184	Taylor, R.	1965	298.2			Spectrographically pure; cylindrical specimen 0.25 in. in diameter and 1.270 cm long; front face exposed to heat pulse from a xenon flash tube; thermal diffusivity calculated from measured time necessary for the rear face to reach one-half the maximum temperature rise; temperature of measurement not given by author but assumed to be room temperature.
88	184	Taylor, R.	1965	298.2			Cylindrical specimen 0.25 in. in diameter and 0.634 cm long; other conditions and specifications same as above.
89	184	Taylor, R.	1965	298.2			Cylindrical specimen 0.25 in. in diameter and 0.317 cm long; other conditions and specifications same as above.

SPECIFICATION TABLE 31. THERMAL DIFFUSIVITY OF IRON (continued)

Cur. No.	Ref. No.	Author(s)	Year	Temp. Range, K	Reported Error, %	Name and Specimen Designation	Composition (weight percent), Specifications, and Remarks
90	188	Anonymous	1965	576-1278		Armco Iron	Flash method utilizing a laser used to measured diffusivity; heat pulse applied to front surface of specimen and diffusivity obtained from time dependence of temperature rise at back surface; specimen obtained from Battelle Memorial Institute.
91	204, 218, 220	Carter, R. L. and Sidles, P. H.	1964	789-1020	+1	Armco Iron Sample I	Disk specimen 4.76 cm in diameter and 1.27 cm in thickness; diffusivity measured from room temperature to just below α-γ transformation temperature using transient radial heat-flow method.
92	204, 218, 220	Carter, R. L. and Sidles, P. H.	1964	298-734	+1	Armco Iron Sample III	Disk specimen; measured under same conditions as above.
93	204, 218, 220	Carter, R. L. and Sidles, P. H.	1964	472-1169	+1	Armco Iron Sample IV	Disk specimen; measured under same conditions as above.
94	156, 56	Degas, P. and Bertin, J-L.	1970	297-1208			8 mm in diameter and 2 to 5 mm thick.
95	195, 194	Hirschman, A., Dennis, J., Derksen, W. L., and Monahen, T. I.	1961	414-1122		Armco Iron	Disk specimen, 0.64 cm in diameter and 0.077 cm in thickness; obtained from National Bureau of Standards.
96	195, 194	Hirschman, A., et al.	1961	507-1153		Armco Iron	Disk specimen, 0.64 cm in diameter and 0.103 cm in thickness.
97	195, 194	Hirschman, A., et al.	1961	436-1081		Armco Iron	Disk specimen, 0.64 cm in diameter and 0.133 cm in thickness.
98	195, 194	Hirschman, A., et al.	1961	431-992		Armco Iron	Disk specimen, 0.64 cm in diameter and 0.148 cm in thickness.
99	214, 215	Moser, J. B. and Kruger, O. L.	1964	298		Armco Iron	1.9 cm in diameter and 0.1 to 0.3 cm thick; density 7.86 g cm^{-3}.
100	216	Cunnington, G. R., Smith, F. J., and Bradshaw, W.	1966	567-1149		Armco Iron	9.6 mm in diameter and 1.98 mm thick; annealed at 1200 K for 2 hr.
101	217, 221	Morrison, B. H.	1968	1062-1350		Armco Iron	Distributed by Battelle Memorial Institute.
102	222, 223	Wittenberg, L. J. and Grove, G. R.	1964	573-1033		Armco Iron	Cylindrical specimen.
103	222, 223	Wittenberg, L. J. and Grove, G. R.	1964	473-1273		Armco Iron	The above specimen annealed at 1025 C for 1 hr.
104	224	Freeman, R. J.	1966	580-1277		Armco Iron	Disk specimen.
105	225,	Pak, M. I. and Osipova, V. A.	1967	521-1229		Armco Iron	Disk specimen 40 mm in diameter; first heating.
106	225	Pak, M. I. and Osipova, V. A.	1967	471-1231		Armco Iron	The above specimen after repeated heating and cooling in vacuo.
107	226	Gonska, H., Kierspe, W., and Kohlhaas, R.	1968	273-1273			0.064 O, 0.0027 C, 0.002 S, 0.001 each of Mn, N, and Si, and trace Cr; 0.5 cm diameter x 20 cm long; density 7.86 g cm^{-3} at 20 C.

SPECIFICATION TABLE 31. THERMAL DIFFUSIVITY OF IRON (continued)

Cur. No.	Ref. No.	Author(s)	Year	Temp. Range, K	Reported Error, %	Name and Specimen Designation	Composition (weight percent), Specifications, and Remarks
108	214	Moser, J.B. and Kruger, O. L.	1964	306-1241		Armco Iron	Cylindrical specimen.
109	227	Cooley, R.A., Janowizcki, R.J., Sonnenschein, G., Strop, H.R., and Willson, M.C.	1966	328-1151		Armco Iron	0.075 Cu, 0.040 Ni, 0.04 Mn, < 0.01 Zr, 0.008 Sn, 0.005 Co, 0.004 Mo, and 0.003 Cr; cylindrical specimen.
110	228	Wheeler, M.J.	1969	803-1276			99.85$^+$ Fe, 0.0300 N, 0.0200 C, < 0.0150 each of Cd, Ca, and Tl, < 0.0070 each of Bi, Ga, and Sn, 0.0050 Si, 0.0030 each of Pb and O, < 0.0030 each of Al, Ba, Ge, Mg, and Ti, < 0.0015 each of In, Mo, Pd, Sr, and V, 0.0010 each of Cu and Ni, < 0.0010 each of P and S, < 0.0007 each of Be and Cr, < 0.0005 each of Co and Mn, and < 0.0002 Zr; 1 cm diameter x 0.128 cm thick; density 7.886 g cm^{-3}; as received.
111	228	Wheeler, M.J.	1969	554-1234			The above specimen heated over a Bunsen flame to about 600 C in air.
112	228	Wheeler, M.J.	1969	634-1283			Similar to the above sample (oxidized) but thickness 0.201 cm and density 7.877 g cm^{-3}.
113	230	van Craeynest, J.C., Weilbacher, J.C., and Lallemont, R.	1969	366-1206	10	Armco Iron	5 mm diameter x 10 mm long; measured by the Angström method.
114	230	van Craeynest, J.C., et al.	1969	352-1184	10	Armco Iron	The above specimen; same measuring method.
115	230	van Craeynest, J.C., et al.	1969	381-1142	10	Armco Iron	The above specimen; same measuring method.
116	230	van Craeynest, J.C., et al.	1969	363-1196	10	Armco Iron	The above specimen; same measuring method.
117	231	Zinov'yev, V.Ye., Krentsis, R.P., and Gel'd, P.V.	1968	953-1445			10 x 6 x 0.180 mm; obtained from Johnson, Matthey and Co., Ltd; ground and annealed at 1300 K in vacuum; electrical resistivity ratio $\rho(298K)/\rho(4.2K) = 114$.
118	231	Zinov'yev, V.Ye., et al.	1968	991-1662			Similar to the above specimen but 0.304 mm in thickness.
119	231	Zinov'yev, V.Ye., et al.	1968	977-1313			Similar to the above specimen but 0.211 mm in thickness.
120	231	Zinov'yev, V.Ye., et al.	1968	939-1083			Similar to the above specimen but 0.230 mm in thickness.
121	232	Böhm, R. and Wachtel, E.	1969	273, 373			0.005 N, 0.004 C, and 0.003 O; cylindrical specimen; electrical resistivity 8.75 $\mu\Omega$cm at 0 C.
122	233	Walter, A.J., Dell, R.M., and Burgess, P.C.	1970	316-784		Electromagnetic grade;Hiperm	0.08 Mn, 0.06 (Cr + Ni), 0.04 Si, 0.03 C, 0.012 P, and 0.012 S; 6 mm diameter x 1 mm thick.
123	233	Walter, A.J., et al.	1970	316-1428		Electromagnetic grade;Hiperm	Similar to the above specimen but 2 mm in thickness.
124	233	Walter, A.J., et al.	1970	316-1222		Electromagnetic grade;Hiperm	Similar to the above specimen but 4 mm in thickness.
125	234	Morrison, B.H. and Sturgess, L.L.	1970	85-1265		Armco Iron	1.26 cm diameter x 0.100 cm thick.
126	235	Branscomb, T.M.	1970	473-763		Armco Iron	0.218 cm thick; machined from 1 in. rod; annealed in vacuum at 1000 C for 1 hr.
127	235	Branscomb, T.M.	1970	760-1068		Armco Iron	Similar to the above specimen but 0.143 cm in thickness.
128	108	Steinberg, S., Larson, R.E., and Kydd, A.R.	1963	298			Specimen 2.0 cm square and 0.132 cm in thickness; diffusivity measuring temperature not given but here assumed to be 25 C.

SPECIFICATION TABLE 31. THERMAL DIFFUSIVITY OF IRON (continued)

Cur. No.	Ref. No.	Author(s)	Year	Temp. Range, K	Reported Error, %	Name and Specimen Designation	Composition (weight percent), Specifications, and Remarks
129	267	Rawuka, A. C. and Gaz, R. A.	1969	300-1090		Armco Iron	Specimen 0.11 cm in thickness; hot rolled; supplied by Steel City Sales, Chicago, Ill.; density 7.74 g cm^{-3}; diffusivity measured using pulse technique.
130*	306, 307	Angström, A.J.	1861	326, 327			Bar specimen, 23.75 mm thick.
131*	308	Angström, A.J.	1863	292-317			Specimen heated by means of either gas-flame or vapor of water.

* Not shown in figure.

DATA TABLE 31. THERMAL DIFFUSIVITY OF IRON

(Impurity <0.20% each; total impurities <0.50%)

[Temperature, T, K; Thermal Diffusivity, α, $cm^2 s^{-1}$]

T	α
CURVE 1	
310.2	0.188
316.2	0.174
326.2	0.173
350.2	0.189
580.2	0.123
684.2	0.109
690.2	0.0900
704.2	0.0905
706.2	0.0890
803.2	0.0820
878.2	0.0580
898.2	0.0530
916.2	0.0500
1096.2	0.0435
1146.2	0.0660
CURVE 2	
334.2	0.156
402.2	0.140
501.2	0.126
543.2	0.137
597.2	0.120
623.2	0.122
714.2	0.111
770.2	0.0975
866.2	0.0765
978.2	0.0550
1016.2	0.0525
1044.2	0.0630
1067.2	0.0820
1081.2	0.0740
CURVE 3	
319.2	0.182
369.2	0.158
443.2	0.141
486.2	0.133
604.2	0.106
665.2	0.0960
813.2	0.0790
CURVE 3 (cont.)	
854.2	0.0875
913.2	0.0760
1044.2	0.0400
1066.2	0.0575
1114.2	0.0730
1223.2	0.0740
CURVE 4	
295.2	0.19
295.2	0.18
408.2	0.15
CURVE 5	
253.4	0.1882
253.4	0.1566
313.4	0.194
313.4	0.185
373.4	0.2136
373.4	0.2146
373.4	0.2079
CURVE 6	
308.2	0.204
316.2	0.206
320.2	0.196
433.2	0.157
545.2	0.132
638.2	0.108
710.2	0.0962
828.2	0.0781
833.2	0.0746
903.2	0.0641
938.2	0.0565
996.2	0.0448
1013.2	0.0413
1013.2	0.0405
1033.2	0.0326
1043.2	0.0397
1063.2	0.0429
CURVE 6 (cont.)	
1070.2	0.0441
1098.2	0.0465
1116.2	0.0488
1133.2	0.0488
1146.2	0.0513
1148.2	0.0505
1173.2	0.0529
1173.2	0.0541
1190.2	0.0602
1193.2	0.0613
1226.2	0.0599
1230.2	0.0610
1276.2	0.0606
1288.2	0.0606
1305.2	0.0602
CURVE 7	
300.3	0.198
300.4	0.200
307.9	0.195
425.3	0.159
538.0	0.131
626.2	0.109
712.2	0.0943
817.8	0.0772
826.2	0.0738
894.4	0.0629
932.1	0.0552
994.9	0.0442
1010.2	0.0407
1012.7	0.0395
1036.9	0.0318
1047.1	0.0388
1062.8	0.0420
1069.0	0.0431
1094.2	0.0457
1114.6	0.0476
1131.1	0.0474
1140.7	0.0498
1144.7	0.0493
1162.2	0.0515
CURVE 7 (cont.)	
1171.2	0.0529
1187.7	0.0588
1190.9	0.0602*
1223.2	0.0581
1229.8	0.0585
1281.0	0.0595
1285.2	0.0595
1297.7	0.0585
CURVE 8	
294.7	0.202
1255.9	0.0575
CURVE 9	
980.2	0.0416
1020.2	0.0444
1068.2	0.0412
1117.2	0.0447
1164.2	0.0537
CURVE 10	
336.2	0.192
344.2	0.179
348.2	0.175
352.2	0.181
353.2	0.185
388.2	0.170
446.2	0.160
492.2	0.149
498.2	0.150
538.2	0.140
600.2	0.125
629.2	0.120
646.2	0.115
659.2	0.114
669.2	0.112
678.2	0.110
691.2	0.107
695.2	0.109
CURVE 10 (cont.)	
709.2	0.099
716.2	0.105
727.2	0.101
743.2	0.0957
758.2	0.0940
766.2	0.0945
766.2	0.0837
778.2	0.0905
780.2	0.0927
788.2	0.0845
800.2	0.0863
808.2	0.0930
823.2	0.0825
845.2	0.0830
890.2	0.0777
894.2	0.0670
940.2	0.0667
950.2	0.0623
968.2	0.0597
986.2	0.0505
998.2	0.0500
1013.2	0.0455
1020.2	0.0460
1035.2	0.0435
1035.2	0.0403
1045.2	0.0393*
1054.2	0.0337
1055.2	0.0350
1061.2	0.0411*
1063.2	0.0437*
1064.2	0.0440
1088.2	0.0490
1090.2	0.0450
1108.2	0.0500
1118.2	0.0463
1128.2	0.0535
1156.2	0.0515
1166.2	0.0507
1191.2	0.0513
1198.2	0.0625
1213.2	0.0611
1223.2	0.0633
CURVE 10 (cont.)	
1232.2	0.0610*
1243.2	0.0595
1256.2	0.0587*
1271.2	0.0605
1283.2	0.0556
CURVE 11	
418	0.137
528	0.116
748	0.0665
907	0.0465
973	0.0210
1058	0.0215
1095	0.0320
1123	0.0225
CURVE 12	
511	0.155
597	0.113
670	0.100
850	0.0557
940	0.0380
992	0.0283
1153	0.0460
CURVE 13	
440	0.147
577	0.118
685	0.100
763	0.0745
870	0.0545
1083	0.0433
CURVE 14	
430	0.149
582	0.107
640	0.0900
758	0.0710
CURVE 14 (cont.)	
890	0.0505
963	0.0405
990	0.0402
CURVE 15	
769.2	0.0769
793.2	0.0730
835.2	0.0647
877.2	0.0579*
892.2	0.0565
899.2	0.0539
904.2	0.0565
911.2	0.0509
913.2	0.0526
925.2	0.0510
954.2	0.0465
959.2	0.0440
967.2	0.0453
972.2	0.0433
982.2	0.0428
993.2	0.0428*
995.2	0.0368
997.2	0.0347
1012.2	0.0376
1021.2	0.0424
1031.2	0.0432*
1039.2	0.0460
1052.2	0.0463
1066.2	0.0465
1087.2	0.0509
1104.2	0.0501
1119.2	0.0512
1163.2	0.0634
1219.2	0.0644
1271.2	0.0644
1292.2	0.0669
1334.2	0.0662
1402.2	0.0637
1464.2	0.0653
CURVE 16	
290.2	0.197
470.2	0.152
670.2	0.100*
873.2	0.0640
973.2	0.0460
1070.2	0.0505
1270.2	0.0623*
CURVE 17*	
980.2	0.0416
1020.2	0.0444
1068.2	0.0412
1117.2	0.0447
1164.2	0.0537
CURVE 18	
493.2	0.0230
585.2	0.0530
815.2	0.0860
885.2	0.0170
976.2	0.193
1056.2	0.520
1103.2	0.0320
1243.2	0.0170
1343.2	0.595*
CURVE 19	
298.2	0.203
CURVE 20	
580.2	0.126*
588.2	0.123*
598.2	0.122*
598.2	0.120*
610.2	0.117*
620.2	0.112*
628.2	0.110*
650.2	0.110*

*Not shown in figure.

DATA TABLE 31. THERMAL DIFFUSIVITY OF IRON (continued)

T	α
CURVE 20 (cont.)	
650.2	0.108*
668.2	0.105*
673.2	0.102*
693.2	0.0985*
698.2	0.100*
706.2	0.0945*
728.2	0.0917*
730.2	0.0895*
748.2	0.0895*
753.2	0.0875*
766.2	0.0835*
786.2	0.0805*
806.2	0.0780*
828.2	0.0760*
830.2	0.0720*
843.2	0.0705*
848.2	0.0705*
868.2	0.0680*
873.2	0.0665*
883.2	0.0645*
896.2	0.0655*
896.2	0.0625*
910.2	0.0595*
926.2	0.0585*
933.2	0.0565*
940.2	0.0555*
946.2	0.0540*
957.2	0.0530*
970.2	0.0505*
976.2	0.0487*
980.2	0.0483*
990.2	0.0460*
998.2	0.0450*
998.2	0.0435*
1008.2	0.0430*
1013.2	0.0410*
1018.2	0.0395*
1020.2	0.0385*
1023.2	0.0380*
1030.2	0.0370*
1033.2	0.0350*
1040.2	0.0265*
1043.2	0.0365*
1046.2	0.0390*
1050.2	0.0415*
CURVE 20 (cont.)	
1053.2	0.0420*
1056.2	0.0430*
1068.2	0.0455*
1073.2	0.0465*
1077.2	0.0483*
1088.2	0.0467*
1091.2	0.0487*
1103.2	0.0485*
1113.2	0.0505*
1126.2	0.0485*
1128.2	0.0500*
1133.2	0.0485*
1140.2	0.0520*
1150.2	0.0530*
1163.2	0.0515*
1166.2	0.0527*
1173.2	0.0540*
1178.2	0.0533*
1183.2	0.0545*
1190.2	0.059*
1190.2	0.0620*
1196.2	0.0615*
1198.2	0.0650*
1203.2	0.0590*
1210.2	0.0610*
1213.2	0.0655*
1223.2	0.0635*
1223.2	0.0615*
1231.2	0.0605*
1236.2	0.0655*
1246.2	0.0610*
1247.2	0.0645*
1260.2	0.0625*
1263.2	0.0665*
1268.2	0.0595*
1283.2	0.0620*
1296.2	0.0625
1307.2	0.0655
1326.2	0.0630
1343.2	0.0660
1361.2	0.0635
1386.2	0.0645
CURVE 21*	
773.2	0.081
CURVE 22*	
773.2	0.079
1073.2	0.044
CURVE 23*	
1073.2	0.045
CURVE 24	
773.2	0.0795
CURVE 25*	
773.2	0.0798
1073.2	0.044
CURVE 26*	
1073.2	0.044
CURVE 27	
573.2	0.121
CURVE 28	
573.2	0.119
CURVE 29	
643.2	0.105
CURVE 30	
643.2	0.107
CURVE 31*	
643.2	0.106
CURVE 32	
421.2	0.162
426.2	0.170
588.2	0.120
612.2	0.113
625.2	0.112
CURVE 33	
289.2	0.217
460.2	0.151
CURVE 34	
624.2	0.118
628.2	0.118
794.2	0.079
880.2	0.068
CURVE 35*	
292.2	0.213
333.2	0.198
479.2	0.153
CURVE 36	
317.2	0.205*
327.2	0.205
380.2	0.181
427.2	0.169*
468.2	0.157
505.2	0.144
561.2	0.132
625.2	0.118*
676.2	0.106
710.2	0.102
764.2	0.089
783.2	0.088
839.2	0.076
865.2	0.072
928.2	0.061
970.2	0.051
1012.2	0.042*
1023.2	0.040
1034.2	0.033*
CURVE 36 (cont.)	
1064.2	0.044*
1068.2	0.046*
1110.2	0.051*
1164.2	0.056
1176.2	0.061
1198.2	0.064
1213.2	0.064
1253.2	0.064
1288.2	0.065
CURVE 37	
578.2	0.109
681.2	0.0929
730.2	0.0957
782.2	0.0850
918.2	0.0587
981.2	0.0437*
997.2	0.0406
1025.2	0.0287
1027.2	0.0348
1041.2	0.0393*
1072.2	0.0397
CURVE 38*	
530.2	0.121
580.2	0.117
656.2	0.100
674.2	0.0969
719.2	0.0842
765.2	0.0797
823.2	0.0747
873.2	0.0668
928.2	0.0649
975.2	0.0497
1007.2	0.0325
1013.2	0.0380
1026.2	0.0461
1029.2	0.0421
1040.2	0.0448
1097.2	0.0538
1133.2	0.0554
1177.2	0.0776
1231.2	0.0643
CURVE 38 (cont.)*	
1274.2	0.0703
1323.2	0.0633
1381.2	0.0675
CURVE 39*	
525.2	0.129
595.2	0.108
783.2	0.0912
854.2	0.0614
929.2	0.0631
1007.2	0.0303
1064.2	0.0572
1169.2	0.0565
1216.2	0.0761
CURVE 40*	
621.2	0.0985
633.2	0.100
693.2	0.0889
830.2	0.0694
925.2	0.0554
970.2	0.0432
1024.2	0.0321
1076.2	0.0448
CURVE 41*	
298.2	0.203
308.2	0.202
320.2	0.192
346.2	0.183
363.2	0.177
373.2	0.173
392.2	0.167
439.2	0.149
465.2	0.139
502.2	0.129
557.2	0.115
585.2	0.111
632.2	0.102
636.2	0.100
655.2	0.0951
CURVE 41 (cont.)*	
687.2	0.0913
709.2	0.0875
733.2	0.0836
737.2	0.0845
762.2	0.0794
788.2	0.0780
813.2	0.0744
860.2	0.0676
911.2	0.0600
936.2	0.0546
961.2	0.0508
972.2	0.0494
978.2	0.0494
999.2	0.0447
1018.2	0.0386
1031.2	0.0348
1034.2	0.0335
1039.2	0.0373
1054.2	0.0446
1060.2	0.0459
1082.2	0.0480
1096.2	0.0510
1111.2	0.0543
1133.2	0.0527
1147.2	0.0552
1162.2	0.0552
1174.2	0.0560
CURVE 42*	
301.2	0.201
375.2	0.175
410.2	0.160
474.2	0.141
523.2	0.126
561.2	0.115
608.2	0.108
661.2	0.0972
716.2	0.0873
773.2	0.0775
822.2	0.0727
848.2	0.0687
873.2	0.0639
888.2	0.0621
893.2	0.0621
CURVE 42 (cont.)*	
920.2	0.0578
938.2	0.0549
983.2	0.0477
991.2	0.0465
1002.2	0.0431
1010.2	0.0431
1021.2	0.0392
1028.2	0.0385
1030.2	0.0356
1037.2	0.0312
1039.2	0.0403
1044.2	0.0418
1048.2	0.0436
1066.2	0.0459
1079.2	0.0490
1089.2	0.0509
1111.2	0.0509
1129.2	0.0526
1147.2	0.0537
1172.2	0.0541
1186.2	0.0658
1190.2	0.0688
CURVE 43*	
1195.2	0.0693
1195.2	0.0675
1213.2	0.0688
1226.2	0.0698
1255.2	0.0703
1274.2	0.0708
CURVE 44*	
966.2	0.0502
1034.2	0.0367
1192.2	0.0668
1201.2	0.0681
1238.2	0.0685
1257.2	0.0710
CURVE 45*	
487.2	0.139
579.2	0.113

*Not shown in figure.

DATA TABLE 31. THERMAL DIFFUSIVITY OF IRON (continued)

T	α
CURVE 45 (cont.)*	
677.2	0.0943
768.2	0.0788
868.2	0.0651
1057.2	0.0458
1082.2	0.0525
1106.2	0.0532
1130.2	0.0564
1154.2	0.0579
1196.2	0.0702
1205.2	0.0707
1215.2	0.0717
1225.2	0.0719
1241.2	0.0707
CURVE 46*	
399.2	0.163
448.2	0.147
485.2	0.140
533.2	0.126
589.2	0.112
636.2	0.103
680.2	0.0973
727.2	0.0852
776.2	0.0799
820.2	0.0747
856.2	0.0677
879.2	0.0639
913.2	0.0582
974.2	0.0477
1025.2	0.0371
1055.2	0.0394
1073.2	0.0437
1081.2	0.0440
1082.2	0.0444
1088.2	0.0455
1109.2	0.0468
1113.2	0.0482
1116.2	0.0484
1121.2	0.0484
1128.2	0.0494
1139.2	0.0505
1147.2	0.0505
1158.2	0.0505
1173.2	0.0513
1194.2	0.0604

T	α
CURVE 46 (cont.)*	
1204.2	0.0604
1208.2	0.0613
1215.2	0.0621
1217.2	0.0615
1227.2	0.0611
1238.2	0.0629
1251.2	0.0615
1263.2	0.0630
CURVE 47*	
1040.2	0.0348
1043.2	0.0408
1047.2	0.0441
1051.2	0.0443
1055.2	0.0449
1061.2	0.0456
1063.2	0.0456
1066.2	0.0463
1076.2	0.0486
1085.2	0.0478
1091.2	0.0497
1095.2	0.0504
1112.2	0.0508
1129.2	0.0527
1134.2	0.0530
1148.2	0.0536
1149.2	0.0550
1163.2	0.0545
1173.2	0.0545
1177.2	0.0555
1193.2	0.0684
1196.2	0.0682
1199.2	0.0686
1202.2	0.0685
1212.2	0.0689
1226.2	0.0694
1237.2	0.0694
1253.2	0.0704
1256.2	0.0708
1272.2	0.0705
CURVE 48*	
978.2	0.0493
983.2	0.0478

T	α
CURVE 48 (cont.)*	
993.2	0.0469
1001.2	0.0435
1002.2	0.0445
1009.2	0.0429
1020.2	0.0397
1022.2	0.0390
1025.2	0.0387
1034.2	0.0347
1034.2	0.0355
1035.2	0.0333
1038.2	0.0310
CURVE 49*	
978.2	0.0480
1027.2	0.0376
1058.2	0.0391
1075.2	0.0432
1083.2	0.0436
1085.2	0.0444
1093.2	0.0453
1112.2	0.0465
1115.2	0.0477
1125.2	0.0473
1131.2	0.0488
1143.2	0.0499
1148.2	0.0497
1166.2	0.0503
1175.2	0.0507
1195.2	0.0592
1198.2	0.0595
1206.2	0.0603
1212.2	0.0608
1216.2	0.0613
1220.2	0.0608
1230.2	0.0607
1241.2	0.0618
1252.2	0.0614
1264.2	0.0619
CURVE 50*	
1058.2	0.0459
1085.2	0.0516
1107.2	0.0531
1129.2	0.0557

T	α
CURVE 50 (cont.)*	
1154.2	0.0577
1195.2	0.0698
1206.2	0.0704
1216.2	0.0713
1224.2	0.0713
1241.2	0.0700
1247.2	0.0684
1265.2	0.0680
1296.2	0.0665
CURVE 51*	
1121.2	0.0479
1126.2	0.0486
CURVE 52*	
1050.2	0.0434
1072.2	0.0463
1097.2	0.0494
1124.2	0.0520
1153.2	0.0542
1192.2	0.0671
1200.2	0.0671
1206.2	0.0678
1216.2	0.0673
1227.2	0.0669
1234.2	0.0669
1249.2	0.0672
1257.2	0.0669
1266.2	0.0659
1275.2	0.0650
1277.2	0.0644
CURVE 53*	
1098.2	0.0458
1139.2	0.0496
1210.2	0.0620
1223.2	0.0612
1235.2	0.0621
1246.2	0.0625
1267.2	0.0620
1277.2	0.0631

T	α
CURVE 54*	
485.2	0.139
640.2	0.105
737.2	0.0887
884.2	0.0650
927.2	0.0572
951.2	0.0511
957.2	0.0544
970.2	0.0484
CURVE 55*	
889.2	0.0613
964.2	0.0509
972.2	0.0497
981.2	0.0494
988.2	0.0468
1000.2	0.0457
1002.2	0.0429
1008.2	0.0423
1010.2	0.0437
1017.2	0.0398
1025.2	0.0384
1030.2	0.0355
1036.2	0.0339
1043.2	0.0440
1052.2	0.0437
1052.2	0.0455
1058.2	0.0470
1073.2	0.0477
1078.2	0.0494
1086.2	0.0500
1089.2	0.0512
1098.2	0.0508
CURVE 56*	
963.0	0.0506
972.9	0.0500
977.7	0.0492
984.1	0.0466
992.1	0.0463
997.0	0.0450
1006.0	0.0439
1006.8	0.0424
1007.8	0.0420
1019.1	0.0394

T	α
CURVE 56 (cont.)*	
1023.9	0.0379
1035.4	0.0352
1037.5	0.0331
1045.2	0.0437
1050.2	0.0452
1053.1	0.0466
1054.1	0.0439
1058.0	0.0463
1070.8	0.0480
1072.6	0.0478
1077.6	0.0496
1080.0	0.0492
1088.4	0.0507
1092.7	0.0502
1096.3	0.0503
CURVE 57	
92.2	0.628
103.2	0.543
114.2	0.526
121.2	0.436
130.2	0.352
148.2	0.347
190.2	0.251
214.2	0.224
235.2	0.216
272.2	0.201
283.2	0.180
303.2	0.180
424.2	0.134
523.2	0.119
564.2	0.111
585.2	0.110
CURVE 58	
93.2	0.644
103.2	0.592
111.2	0.530
119.2	0.408
125.2	0.394
137.2	0.352
156.2	0.315
176.2	0.292
203.2	0.242

T	α
CURVE 58 (cont.)	
257.2	0.217
279.2	0.214
298.2	0.201
377.2	0.166
510.2	0.138
584.2	0.113
623.2	0.112
CURVE 59*	
93.2	0.628
298.2	0.180
CURVE 60*	
93.2	0.644
298.2	0.201
CURVE 61*	
298.2	0.204
CURVE 62*	
298.2	0.206
CURVE 63*	
303.2	0.190
CURVE 64*	
303.2	0.198
CURVE 65*	
303.2	0.192
CURVE 66*	
303.2	0.187
CURVE 67*	
303.2	0.188

T	α
CURVE 68*	
303.2	0.190
CURVE 69*	
303.2	0.185
CURVE 70	
297.2	0.222
297.2	0.220
298.2	0.225
301.2	0.228
301.2	0.213
304.2	0.219
310.2	0.215
393.2	0.176
415.2	0.172
482.2	0.149
518.2	0.138
559.2	0.133
608.2	0.116
667.2	0.103
679.2	0.105
874.2	0.065
965.2	0.050
CURVE 71	
306.2	0.169
CURVE 72	
306.2	0.192
CURVE 73	
306.2	0.209
CURVE 74	
306.2	0.158
CURVE 75	
306.2	0.184

* Not shown in figure.

DATA TABLE 31. THERMAL DIFFUSIVITY OF IRON (continued)

T	α
CURVE 76	
306.2	0.203
CURVE 77	
306.2	0.122
CURVE 78	
306.2	0.166
CURVE 79	
306.2	0.211
CURVE 80	
306.2	0.0855
CURVE 81	
306.2	0.144
CURVE 82	
306.2	0.200
CURVE 83	
306.2	0.0290
CURVE 84	
306.2	0.0816
CURVE 85	
306.2	0.189
CURVE 86*	
298.2	0.166
CURVE 87	
298.2	0.217
CURVE 88*	
298.2	0.219

T	α
CURVE 89*	
298.2	0.219
CURVE 90*	
576	0.1239
623	0.1136
676	0.1035
720	0.0960
772	0.0873
827	0.0778
872	0.0699
925	0.0610
977	0.0513
998	0.0468
1022	0.0410
1040	0.0341
1074	0.0358
1126	0.0507
1173	0.0532
1222	0.0627
1278	0.0643
CURVE 91*	
789	0.0778
989	0.0489
1020	0.0419
CURVE 92*	
298	0.2051
299	0.2072
336	0.1866
336	0.1836
387	0.1703
672	0.0927
674	0.0942
475	0.0962
732	0.0864
732	0.0846
734	0.0882
CURVE 93*	
472	0.1417
546	0.1199
566	0.1123
619	0.1013
691	0.0934

T	α
CURVE 93 (cont.)*	
766	0.0808
774	0.0777
775	0.0808
825	0.0728
873	0.0680
923	0.0598
935	0.0572
941	0.0572
948	0.0562
956	0.0544
969	0.0518
1028	0.0395
1029	0.0386
1032	0.0368
1035	0.0357
1036	0.0314
1039	0.0332
1061	0.0478
1099	0.0521
1101	0.0509
1169	0.0581
CURVE 94*	
297	0.1801
358	0.1615
460	0.1358
681	0.0833
685	0.0919
900	0.0504
1028	0.0377
1075	0.0450
1208	0.0545
CURVE 95*	
414	0.1364
528	0.1158
748	0.0672
907	0.0471
972	0.0220
1057	0.0220
1094	0.0331
1122	0.0234

T	α
CURVE 96*	
507	0.1555
594	0.1137
665	0.1017
852	0.0580
938	0.0388
988	0.0284
1153	0.0468
CURVE 97*	
436	0.1468
572	0.1188
678	0.1013
761	0.0757
864	0.0552
1081	0.0440
CURVE 98*	
431	0.1495
585	0.1082
633	0.0916
752	0.0722
888	0.0519
963	0.0422
992	0.0422
CURVE 99*	
298	0.20
CURVE 100*	
567	0.1223
616	0.1105
667	0.0999
762	0.0849
856	0.0681
1011	0.0395
1023	0.0369
1149	0.0501
CURVE 101*	
1062	0.0422
1073	0.0433
1083	0.0448
1092	0.0479

T	α
CURVE 101 (cont.)*	
1114	0.0481
1129	0.0501
1151	0.0527
1178	0.0559
1198	0.0609
1228	0.0670
1241	0.0689
1304	0.0584
1350	0.0589
CURVE 102	
573	0.080
673	0.062
773	0.049
873	0.036
973	0.027
993	0.025
1033	0.022
CURVE 103*	
473	0.143
573	0.130
673	0.107
773	0.086
873	0.066
973	0.048
993	0.044
1033	0.037
1038	0.035
1043	0.031
1048	0.031
1058	0.031
1073	0.031
1093	0.031
1123	0.034
1148	0.036
1173	0.039
1183	0.040
1186	0.043
1198	0.032
1223	0.039
1248	0.045
1273	0.053

T	α
CURVE 104*	
580	0.1219
626	0.1120
679	0.1022
720	0.0950
773	0.0864
827	0.0772
873	0.0685
924	0.0601
977	0.0507
998	0.0462
1021	0.0407
1038	0.0336
1073	0.0457
1124	0.0501
1173	0.0528
1221	0.0616
1277	0.0633
CURVE 105	
521	0.1678
574	0.1441
624	0.1287
675	0.1172
724	0.1121
777	0.1044
827	0.0954
878	0.0896
927	0.0780
978	0.0662
1027	0.0473
1042	0.0366
1077	0.0501
1131	0.0542
1183	0.0479
1187	0.0372
1229	0.0647
CURVE 106*	
471	0.1086
525	0.0998
577	0.0957
627	0.0905
678	0.0872
728	0.0807
777	0.0750
827	0.0675

T	α
CURVE 106 (cont.)*	
878	0.0604
927	0.0542
978	0.0466
1025	0.0357
1042	0.0288
1082	0.0398
1131	0.0443
1183	0.0414
1187	0.0328
1231	0.0508
CURVE 107*	
273	0.228
293	0.219
323	0.206
373	0.187
423	0.171
473	0.155
523	0.141
573	0.128
623	0.116
673	0.107
723	0.097
773	0.088
823	0.080
873	0.0715
923	0.0630
973	0.0525
998	0.0471
1023	0.0400
1048	0.0352
1073	0.0445
1098	0.049
1123	0.0515
1173	0.0542
1189	0.055
1189	0.0645
1223	0.0645
1273	0.065
CURVE 108*	
306	0.2000
422	0.1708
572	0.1212
651	0.0952
746	0.0866
801	0.0783

T	α
CURVE 108 (cont.)*	
902	0.0522
971	0.0400
1071	0.0433
1169	0.0665
1241	0.0665
CURVE 109*	
328	0.235
528	0.140
639	0.104
765	0.079
873	0.058
1063	0.036
1111	0.059
1151	0.069
CURVE 110*	
803	0.0786
827	0.0689
866	0.0644
885	0.0621
906	0.0566
922	0.0549
964	0.0482
988	0.0418
994	0.0396
1024	0.0394
1024	0.0366
1038	0.0374
1041	0.0391
1053	0.0437
1056	0.0443
1074	0.0468
1086	0.0488
1098	0.0507
1106	0.0507
1116	0.0512
1143	0.0529
1149	0.0529
1149	0.0523
1170	0.0566
1190	0.0633
1219	0.0642
1250	0.0656
1276	0.0667

* Not shown in figure.

DATA TABLE 31. THERMAL DIFFUSIVITY OF IRON (continued)

T	α
CURVE 111*	
554	0.1285
611	0.1210
613	0.1075
624	0.1141
632	0.1107
664	0.1045
689	0.1020
701	0.1020
719	0.0967
732	0.0944
754	0.0877
772	0.0842
772	0.0763
772	0.0750
780	0.0764
780	0.0752
844	0.0729
851	0.0702
863	0.0691
882	0.0664
891	0.0602
902	0.0583
926	0.0518
932	0.0522
953	0.0480
970	0.0453
977	0.0437
985	0.0446
1005	0.0408
1013	0.0370
1022	0.0349
1031	0.0325
1037	0.0301
1047	0.0363
1066	0.0439
1108	0.0497
1194	0.0600
1234	0.0643
CURVE 112*	
634	0.1093
653	0.1048
676	0.1045
681	0.1009
693	0.1018
701	0.0978
CURVE 112 (cont.)*	
721	0.0987
732	0.0945
737	0.0937
757	0.0899
779	0.0874
785	0.0826
795	0.0808
800	0.0790
836	0.0745
840	0.0717
864	0.0672
871	0.0656
876	0.0639
911	0.0592
918	0.0592
942	0.0592
951	0.0547
998	0.0453
1012	0.0514
1040	0.0434
1042	0.0426
1060	0.0479
1074	0.0450
1079	0.0549
1090	0.0511
1121	0.0564
1133	0.0592
1151	0.0693
1190	0.0705
1200	0.0718
1201	0.0743
1234	0.0751
1268	0.0781
1283	0.0808
CURVE 113*	
366	0.1892
650	0.1014
847	0.0687
927	0.0574
997	0.0407
1056	0.0380
1206	0.0653
CURVE 114*	
352	0.1958
526	0.1343
705	0.0898
908	0.0594
1026	0.0313
1121	0.0472
1142	0.0436
1184	0.0531
CURVE 115*	
381	0.1995
576	0.1254
774	0.0802
972	0.0429
1142	0.0489
CURVE 116*	
363	0.1995
511	0.1332
709	0.0904
888	0.0566
1017	0.0337
1063	0.0349
1196	0.0493
CURVE 117*	
953	0.0508
966	0.0499
998	0.0449
1010	0.0425
1030	0.0382
1051	0.0361
1054	0.0355
1129	0.0506
1155	0.0540
1156	0.0542
1161	0.0545
1166	0.0549
1171	0.0550
1172	0.0554
1176	0.0558
1181	0.0564
1182	0.0568
1184	0.0584
CURVE 117 (cont.)*	
1189	0.0599
1194	0.0611
1202	0.0617
1208	0.0617
1221	0.0627
1243	0.0629
1268	0.0632
1294	0.0637
1325	0.0642
1361	0.0645
1398	0.0656
1430	0.0662
1445	0.0666
CURVE 118*	
991	0.0461
1030	0.0371
1060	0.0382
1097	0.0466
1158	0.0534
1234	0.0627
1284	0.0636
1338	0.0644
1388	0.0656
1461	0.0673
1499	0.0681
1511	0.0683
1540	0.0691
1553	0.0691
1602	0.0693
1623	0.0693
1662	0.0687
CURVE 119*	
977	0.0483
1006	0.0441
1017	0.0415
1022	0.0404
1028	0.0388
1034	0.0360
1037	0.0358
1037	0.0346
1038	0.0334
1038	0.0343
1039	0.0327
CURVE 119 (cont.)*	
1044	0.0319
1046	0.0325
1048	0.0336
1049	0.0343
1055	0.0369
1058	0.0392
1061	0.0398
1081	0.0436
1099	0.0482
1121	0.0497
1137	0.0522
1257	0.0632
1313	0.0641
CURVE 120*	
939	0.0510
958	0.0493
979	0.0467
1023	0.0398
1025	0.0387
1026	0.0394
1034	0.0364
1041	0.0323
1041	0.0314
1042	0.0310
1042	0.0303
1044	0.0324
1046	0.0331
1049	0.0348
1056	0.0381
1063	0.0404
1066	0.0404
1066	0.0414
1068	0.0416
1070	0.0427
1083	0.0449
CURVE 121	
273	0.236
373	0.190
CURVE 122*	
316	0.2077
316	0.2038
CURVE 122 (cont.)*	
359	0.1940
399	0.1789
399	0.1733
517	0.1421
519	0.1379
632	0.1143
784	0.0910
784	0.0867
CURVE 123*	
316	0.1975
316	0.1881
316	0.1822
373	0.1715
452	0.1448
519	0.1355
571	0.1224
666	0.1037
754	0.0903
779	0.0827
887	0.0746
980	0.0500
1058	0.0478
1132	0.0552
1179	0.0644
1222	0.0628
1274	0.0644
1399	0.0693
1428	0.0756
CURVE 124*	
316	0.1856
363	0.1718
371	0.1640
431	0.1505
583	0.1244
649	0.1137
711	0.1003
816	0.0849
862	0.0775
936	0.0675
1003	0.0562
1092	0.0412
1098	0.0445
1158	0.0505
1222	0.0665
CURVE 125*	
84.9	0.771
88.3	0.759
100.5	0.706
111.2	0.476
114.3	0.485
135.2	0.380
139.0	0.356
177.0	0.290
184.9	0.304
197.2	0.274
200.4	0.264
214.8	0.237
244.3	0.229
269.8	0.216
276.1	0.221
300.6	0.210
376.7	0.191
452.9	0.160
520.0	0.148
520.0	0.143
575.4	0.130
636.8	0.116
669.9	0.114
822.2	0.0780
841.4	0.0750
935.4	0.0634
935.4	0.0596
968.3	0.0488
968.3	0.0458
990.8	0.0421
1014	0.0399
1014	0.0378
1026	0.0360
1045	0.0370
1045	0.0389
1062	0.0417
1062	0.0459
1069	0.0489
1112	0.0497
1169	0.0530
1227	0.0557
1265	0.0540
CURVE 126*	
473	0.140
531	0.127
CURVE 126 (cont.)*	
578	0.116
648	0.1027
703	0.0923
763	0.0826
CURVE 127*	
760	0.0819
838	0.0739
883	0.0636
938	0.0552
975	0.0482
1003	0.0404
1020	0.0366
1030	0.0330
1035	0.0325
1049	0.0429
1068	0.0460
CURVE 128	
298	0.17
CURVE 129*	
300	0.194
467	0.140
619	0.110
770	0.076
927	0.049
1090	0.044
CURVE 130*	
325.7	0.186
327.3	0.182
CURVE 131*	
292.1	0.171
316.6	0.175
316.9	0.174

* Not shown in figure.

FIGURE AND TABLE 32R. PROVISIONAL THERMAL DIFFUSIVITY OF LANTHANUM

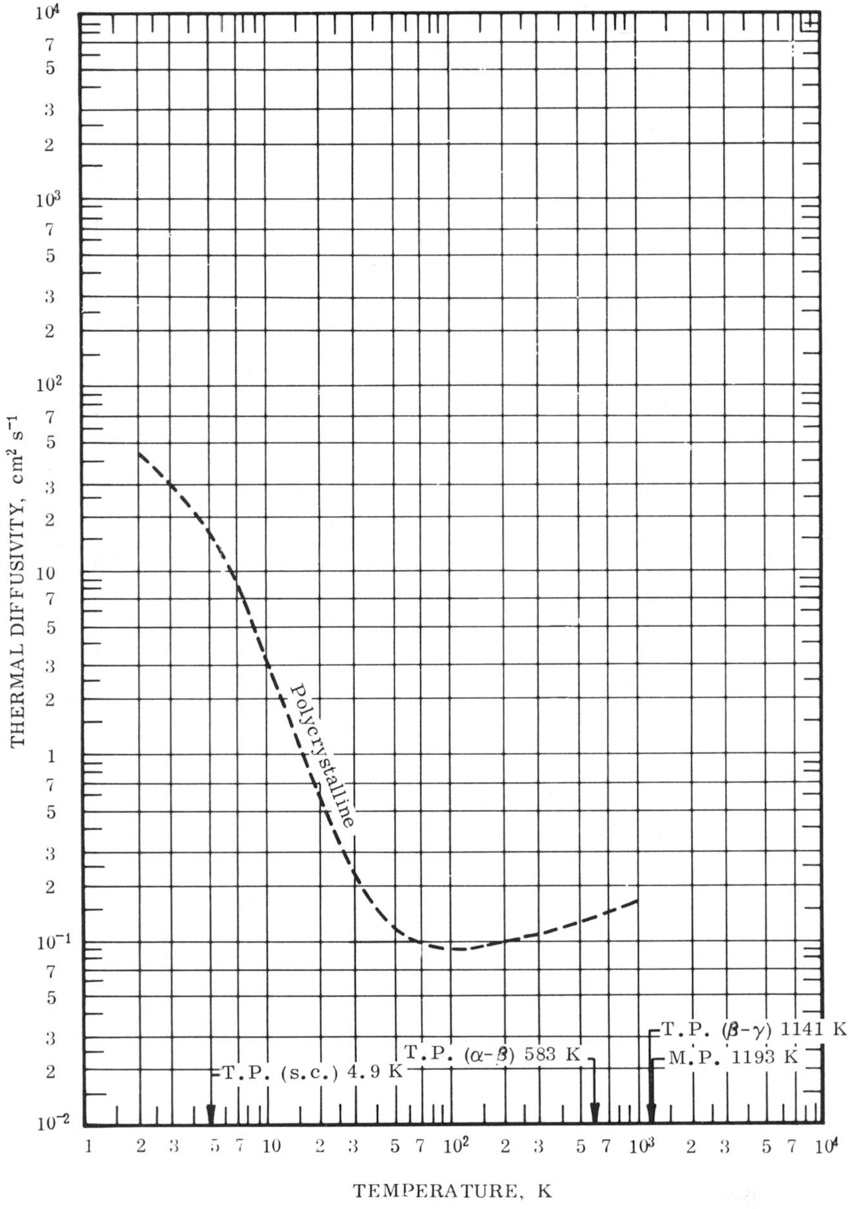

PROVISIONAL VALUES*

[Temperature, T, K; Thermal Diffusivity, α, cm^2 s^{-1}]

SOLID
(Polycrystalline)

T	α	T	α
2	43.2	80	0.0945
3	32.8	90	0.0933
4	22.9	100	0.0930
5	16.2	150	0.0960
6	11.5	200	0.100
7	8.34	250	0.104
8	6.20	273.2	0.106
9	4.68	300	0.109
10	3.63	350	0.113
11	2.89	400	0.117
12	2.30	500	0.127
13	1.875	600	0.136
14	1.55	700	0.145
15	1.29	800	0.153
16	1.09	900	0.158
18	0.790	1000	0.161
20	0.592	1100	0.160
25	0.340		
30	0.231		
35	0.176		
40	0.146		
45	0.128		
50	0.117		
60	0.103		
70	0.0978		

REMARKS

The values are for well-annealed high-purity lanthanum and are thought to be accurate to ±10% around room temperature and ±15 to ±20% at other temperatures except for those at temperatures from 10 to 80 K, which are very uncertain. At low temperatures the values are highly conditioned by impurity and imperfection, and those below 50 K are applicable only to lanthanum having residual electrical resistivity of 1.29 $\mu\Omega$ cm.

*All values are estimated and those around room temperature are recommended values.

FIGURE AND TABLE 33R. RECOMMENDED THERMAL DIFFUSIVITY OF LEAD

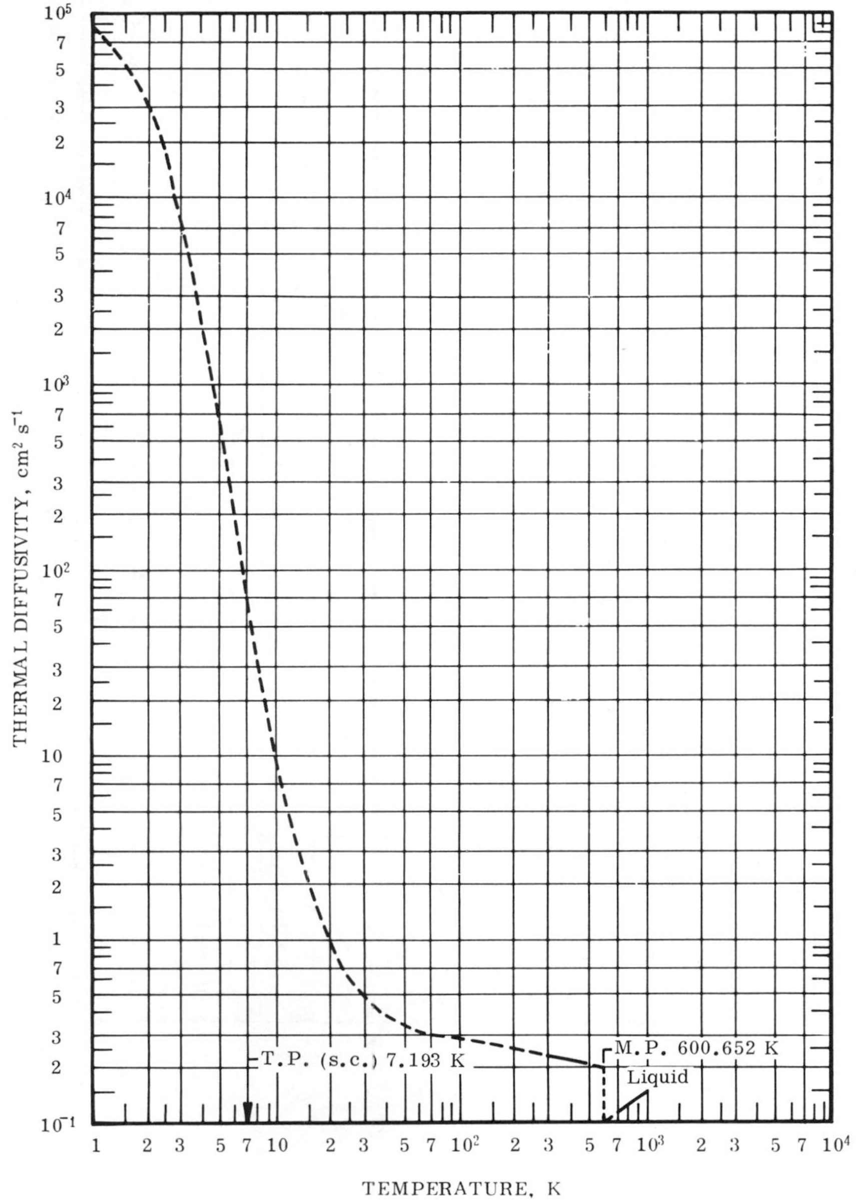

RECOMMENDED VALUES†

[Temperature, T, K; Thermal Diffusivity, α, $cm^2\ s^{-1}$]

SOLID				LIQUID	
T	α	T	α	T	α
1	85200*	35	0.437*	600.652	0.0989*
2	35100*	40	0.399*	700	0.114
3	8890*	45	0.375*	800	0.127
4	2240*	50	0.357*	900	0.139
5	731*	60	0.333*	1000	0.150
6	236*	70	0.318*		
7	86.0*	80	0.306*		
8	38.0*	90	0.298*		
9	19.6*	100	0.291*		
10	11.4*	150	0.271*		
11	7.40*	200	0.259*		
12	5.13*	250	0.250*		
13	3.73*	273.2	0.247*		
14	2.83*	300	0.243		
15	2.23*	350	0.235		
16	1.825*	400	0.228		
18	1.30*	500	0.215		
20	1.00*	600	0.201*		
25	0.645*	600.652	0.201*		
30	0.501*				

REMARKS

The recommended values are for well-annealed high-purity lead and are considered accurate to within ±5% of the true values at moderate temperatures, ±8% at high temperatures, and ±13% at low temperatures. The values for molten lead are provisional and should be good to ±15% below 800 K. At low temperatures the values are highly conditioned by impurity and imperfection, and those below 30 K are applicable only to lead in the normal state having residual electrical resistivity of 0.000862 μΩ cm.

† Values for molten lead are provisional.
* In temperature range where no experimental data are available.

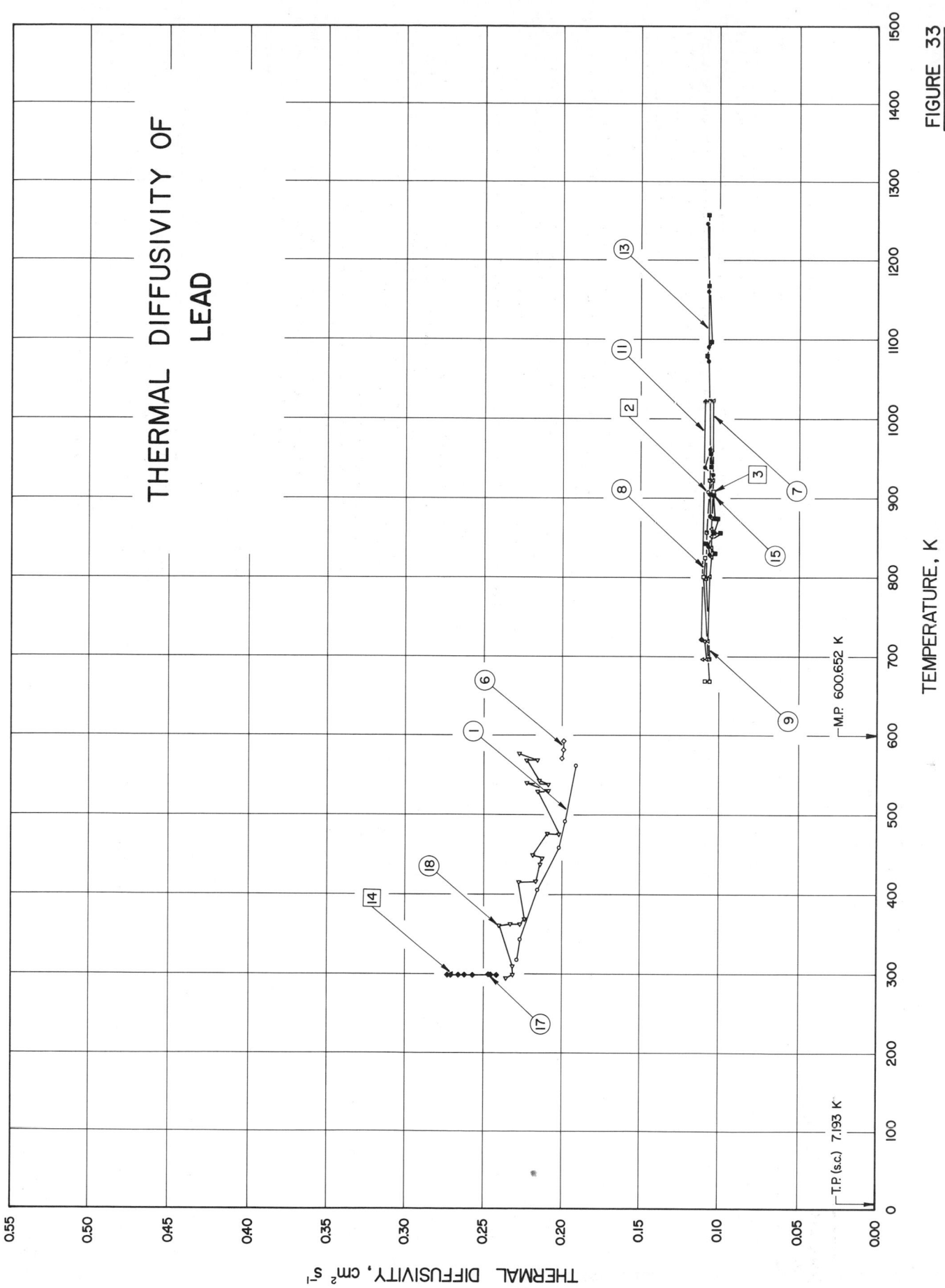

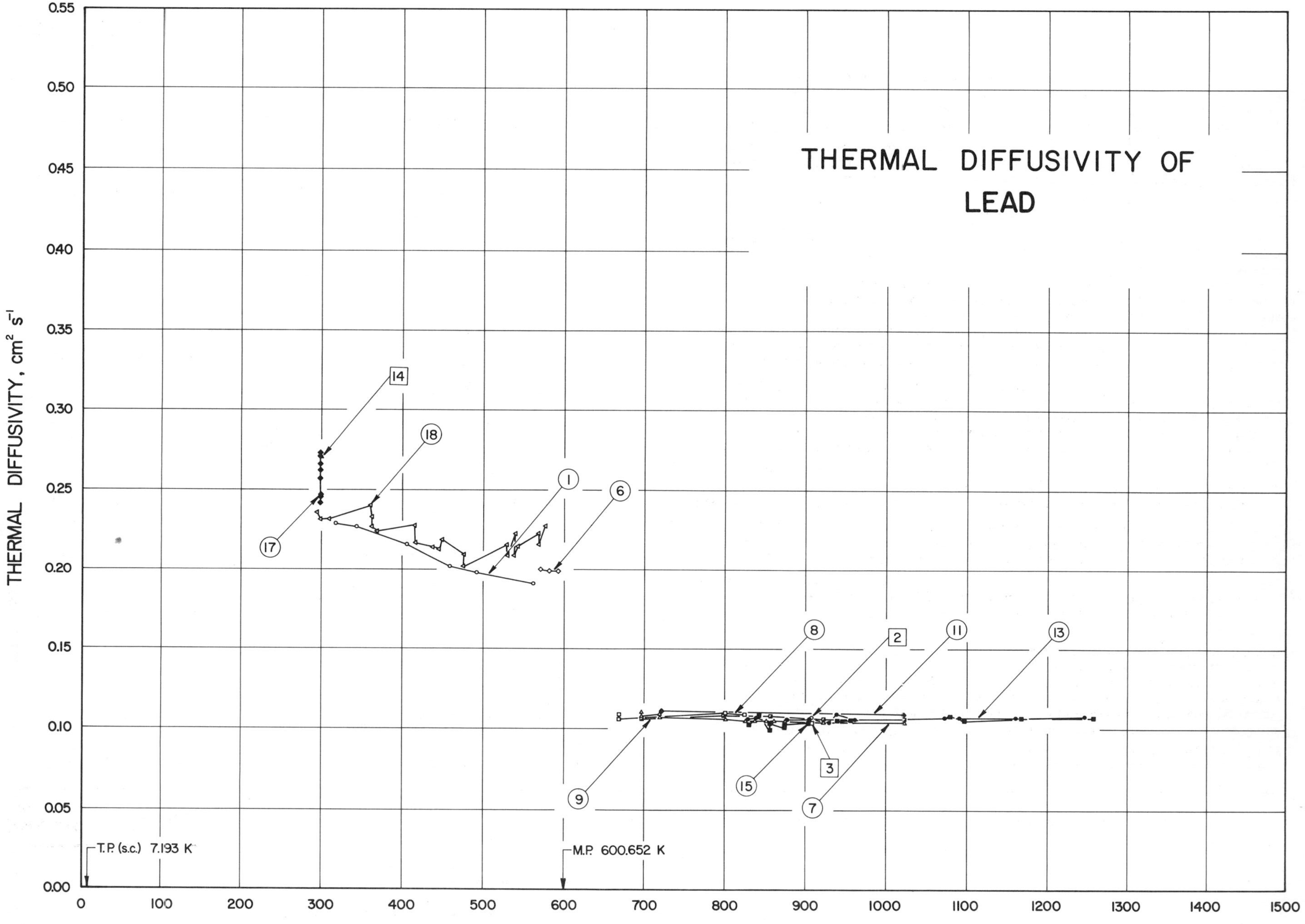
THERMAL DIFFUSIVITY OF
LEAD
THERMAL DIFFUSIVITY, cm² s⁻¹
TEMPERATURE, K
0.00
0.05
0.10
0.15
0.20
0.25
0.30
0.35
0.40
0.45
0.50
0.55
0
100
200
300
400
500
600
700
800
900
1000
1100
1200
1300
1400
1500
T.P. (s.c.) 7.193 K
M.P. 600.652 K
14
18
1
6
17
8
2
11
13
9
15
3
7

FIGURE 33

SPECIFICATION TABLE 33. THERMAL DIFFUSIVITY OF LEAD

(Impurity < 0.20% each; total impurities < 0.50%)

Cur. No.	Ref. No.	Author(s)	Year	Temp. Range, K	Reported Error, %	Name and Specimen Designation	Composition (weight percent), Specifications, and Remarks
1	38	Shvidkovskii, E. G.	1938	318-562	± 1.6/ ± 6.1		98.87 pure; cylindrical specimen 10 mm in dia. and 150 mm long; period of temp. wave 4 min; data given represent avg over a number of runs ranging from 7 to 12.
2	107	Filippov, L. P.	1966	908	4		Liquid specimen fills space between two coaxial thin-walled tubes of tantalum 24 and 8 mm in dia., respectively; horizontal plates 1 cm apart minimize convective mixing of liquid; thermocouple welded to external surface of sample; radial wave method used to measure diffusivity employing a period of 13.2 sec for the modulation of the electron beam; heated by the method of electron bombardment using a periodically changing power of 12.9 W; specific heat measured and reported as 0.0338 cal g^{-1} K^{-1}.
3	107	Filippov, L. P.	1966	908	4		Specimen measured for diffusivity under same conditions as above except that a periodic heating power of 24.0 W is employed; specific heat measured and reported as 0.0336 cal g^{-1} K^{-1}.
4*	107	Filippov, L. P.	1966	908	4		Specimen measured for diffusivity under same conditions as above except that a period of 26.4 sec for the temp. wave is used and a periodic heating power of 12.9 W is employed; specific heat measured and reported as 0.0352 cal g^{-1} K^{-1}.
5*	107	Filippov, L. P.	1966	908	4		Specimen measured for diffusivity under same conditions as above except that a periodic heating power of 24.0 W is employed; specific heat measured and reported as 0.0342 cal g^{-1} K^{-1}.
6	109, 107	Yurchak, R. P. and Filippov, L. P.	1964	571-593	7		Pure; cylindrical specimen; radial wave method used to measure diffusivity; diffusivity determined from measurements of temp. at two points on the specimen using a heating period of 6.6 sec; electrical resistivity measured and reported as 21.4, 30.8, 34.2, 42.4, 43.5, 44.3, and 47.0 μohm cm at 293.2, 411.2, 445.2, 528.2, 533.2, 553.2, and 573.2 K, respectively.
7	109, 107	Yurchak, R. P. and Filippov, L. P.	1964	697-1023	7		Pure; in molten state; sample consists of a cylindrical tantalum crucible containing the metal to be measured; horizontal partitions of tantalum plate impede convective mixing of liquid; outer surface of crucible subjected to periodic heating; radial wave method used to measure diffusivity; diffusivity determined from measurements of temp. at two points on specimen; electrical resistivity measured and reported as 97.6, 99.5, 102.6, 105.0, 107.0, 112.6, and 117.8 μohm cm at 623.2, 648.2, 698.2, 733.2, 782.2, 873.2, and 974.2 K, respectively; Lorenz number reported as 2.72, 2.6, 2.49, 2.39, 2.30, 2.21, 2.18, 2.11, and 2.09 x 10^{-8} V^2 K^{-2} at 623.2, 673.2, 723.2, 773.2, 823.2, 873.2, 923.2, 973.2, and 1023.2 K, respectively; heating period used 6.6 sec.
8	109	Yurchak, R. P. and Filippov, L. P.	1964	669-923	7		Pure; in molten state; sample consists of cylindrical tantalum crucible containing metal to be measured; diffusivity determined from the amplitude ratio of the temp. waves measured at two points on specimen using a heating period of 13.2 sec; other conditions same as above.
9	109	Yurchak, R. P. and Filippov, L. P.	1964	697-1022	7		Pure; in molten state; sample consists of cylindrical tantalum crucible containing metal to be measured; diffusivity determined from the phase difference between the temp. waves measured at two points on specimen using a heating period of 13.2 sec; other conditions same as above.

* Not shown in figure.

SPECIFICATION TABLE 33. THERMAL DIFFUSIVITY OF LEAD (continued)

Cur. No.	Ref. No.	Author(s)	Year	Temp. Range, K	Reported Error, %	Name and Specimen Designation	Composition (weight percent), Specifications, and Remarks
10*	109	Yurchak, R.P. and Filippov, L.P.	1964	840	7		Pure; in molten state; sample consists of cylindrical tantalum crucible containing metal to be measured; diffusivity determined from the amplitude ratio of the temperature waves measured at two points on specimen with second setting of the thermocouples; other conditions same as above.
11	109	Yurchak, R.P. and Filippov, L.P.	1964	722, 1022	7		Pure; in molten state; sample consists of cylindrical tantalum crucible containing metal to be measured; diffusivity determined from the phase difference between the temperature waves measured at two points on specimen; other conditions same as above.
12*	109	Yurchak, R.P. and Filippov, L.P.	1964	838	7		Pure; in molten state; sample consists of two thin walled tantalum tubes 23.7 and 8 mm in diameter containing metal to be measured; inner surface of crucible subjected to periodic heating using a heating period of 13.2 sec; diffusivity determined from the phase difference between the temperature waves measured at two points on specimen; measured in vacuum; other conditions same as above.
13	107	Filippov, L.P.	1966	828-1246			Liquid specimen; diffusivity calculated from measured thermal conductivity and heat capacity for the same specimen using steady heating for conductivity measurement and pulsating heating for measurement of the heat capacity.
14	118	Perron, J.C.	1961	300	~8		Square specimen 3 cm long; Angström method used to measure diffusivity.
15	209	Yurchak, R.P. and Filippov, L.P.	1965	830-1257	3-5		Specimen in molten state; heating cycle 13.2 sec.
16*	209	Yurchak, R.P. and Filippov, L.P.	1965	842-1255	3-5		Similar to the above specimen; heating cycle 26.4 sec.
17	196	Kobayasi, K. and Kumada, T.	1967	298.2	<± 5		99.99 pure; disk specimen 15 mm in diameter and 15 mm thick; diffusivity measured in air at pressures, p, ranging from 0.0018 to 584 mm Hg.
18	196	Kobayasi, K. and Kumada, T.	1967	294-577	<± 5		The above specimen measured in vacuum at various temperatures.

* Not shown in figure.

DATA TABLE 33. THERMAL DIFFUSIVITY OF LEAD

(Impurity <0. 20% each; total impurities <0. 50%)

[Temperature, T, K; Thermal Diffusivity, α, cm² s⁻¹]

T	α
CURVE 1	
318	0. 228
344	0. 227
406	0. 216
459	0. 204
492	0. 196
562	0. 191
CURVE 2	
908	0. 107
CURVE 3	
908	0. 104
CURVE 4*	
908	0. 105
CURVE 5*	
908	0. 109
CURVE 6	
571	0. 200
582	0. 199
593	0. 199
CURVE 7	
697	0. 110
697	0. 108
720	0. 109
720	0. 107
801	0. 106
826	0. 105
838	0. 105
852	0. 105
861	0. 105
923	0. 104
1023	0. 104

T	α
CURVE 8	
669	0. 109
669	0. 106
801	0. 110
825	0. 109
923	0. 109*
CURVE 9	
697	0. 106
799	0. 108
857	0. 108
923	0. 106
1022	0. 106
CURVE 10*	
840	0. 107
CURVE 11	
722	0. 111
1022	0. 109
CURVE 12*	
838	0. 109
CURVE 13	
828	0. 106
839	0. 107
858	0. 104
877	0. 106
905	0. 106
930	0. 104
939	0. 109
962	0. 106
1072	0. 107
1090	0. 107
1160	0. 107
1246	0. 108

T	α
CURVE 14	
300	0. 27
CURVE 15	
830	0. 1034
830	0. 1056*
842	0. 1082
856	0. 0990
856	0. 1040
874	0. 1013
874	0. 1031
904	0. 1036
917	0. 1069
929	0. 1035*
939	0. 1058
939	0. 1086
956	0. 1058
1079	0. 1080
1096	0. 1053
1167	0. 1075
1257	0. 1079
CURVE 16*	
842	0. 1055
856	0. 1023
871	0. 1068
904	0. 1078
915	0. 1043
929	0. 1068
939	0. 1043
956	0. 1025
1079	0. 1062
1096	0. 1076
1166	0. 1057
1255	0. 1055

p (mm Hg)	α
CURVE 17 (T = 298. 2K)	
0. 0018	0. 242
0. 0034	0. 245
0. 0041	0. 247
0. 0049	0. 247*
0. 471	0. 257
14. 6	0. 262
75. 3	0. 266
267	0. 273
584	0. 271
T	α
CURVE 18	
294	0. 236
298	0. 231
310	0. 232
361	0. 240
362	0. 233
362	0. 227
369	0. 224
416	0. 228
416	0. 217
438	0. 215
446	0. 213
449	0. 219
486	0. 210
476	0. 202
529	0. 216
529	0. 209
539	0. 223
539	0. 209
543	0. 215
568	0. 223
568	0. 216
577	0. 228

* Not shown in figure.

FIGURE AND TABLE 34R. RECOMMENDED THERMAL DIFFUSIVITY OF LITHIUM

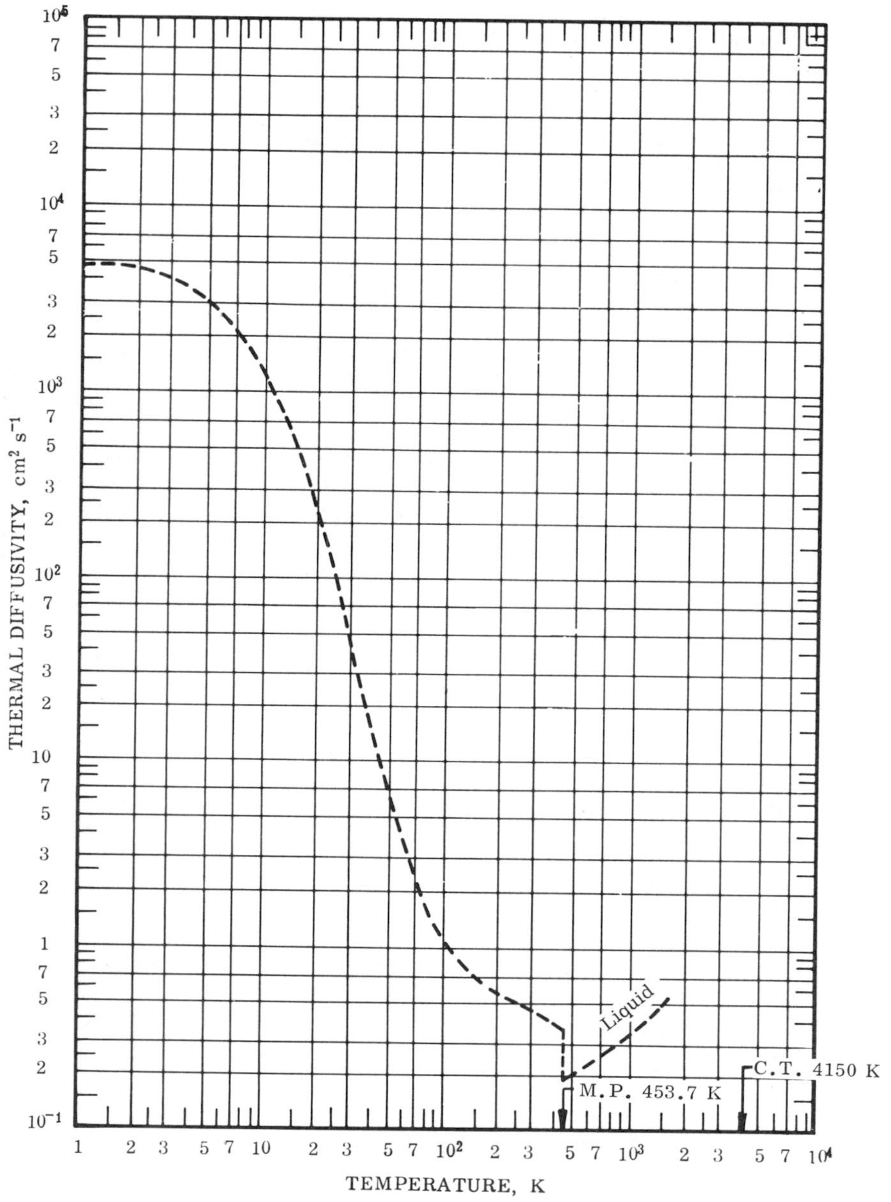

RECOMMENDED VALUES*

[Temperature, T, K; Thermal Diffusivity, α, $cm^2\ s^{-1}$]

SOLID				LIQUID	
T	α	T	α	T	α
1	4530	35	27.3	453.7	0.191
2	4400	40	16.4	500	0.203
3	3980	45	10.5	600	0.227
4	3480	50	7.13	700	0.250
5	2940	60	3.74	800	0.271
6	2450	70	2.29	900	0.292
7	2050	80	1.59	1000	0.311
8	1720	90	1.25	1100	0.329
9	1450	100	1.04	1200	0.347
10	1220	150	0.665	1300	0.364
11	1030	200	0.553	1400	0.380
12	865	250	0.498	1500	0.396
13	729	273.2	0.477	1600	0.412
14	613	300	0.454		
15	517	350	0.415		
16	437	400	0.382		
18	314	453.7	0.339		
20	226				
25	102				
30	49.7				

REMARKS

The values are for well-annealed high-purity lithium and are thought to be accurate to within about ± 8% for the solid state and for molten lithium to about 700 K. The uncertainty increases to about ± 15% by 1600 K. At low temperatures the values are highly conditioned by impurity and imperfection, and those below 150 K are applicable only to lithium having residual electrical resistivity of 0.0372 μΩ cm.

*All values are estimated.

FIGURE AND TABLE 35R. PROVISIONAL THERMAL DIFFUSIVITY OF LUTETIUM

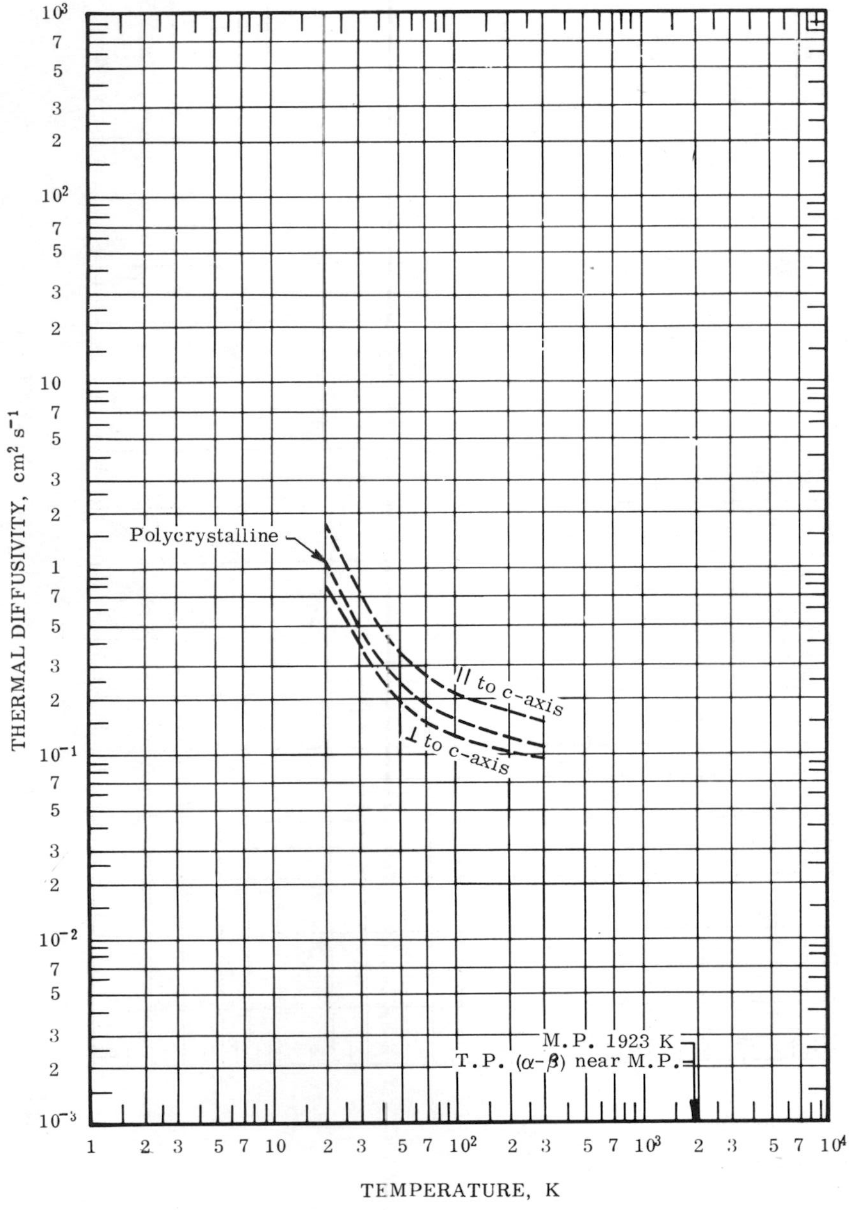

PROVISIONAL VALUES*

[Temperature, T, K; Thermal Diffusivity, α, cm^2 s^{-1}]

T	SOLID (Single crystal) ∥ to c-axis α	⊥ to c-axis α	Poly-crystalline α
20	1.75	0.795	1.04
25	1.04	0.527	0.661
30	0.711	0.382	0.473
35	0.544	0.302	0.369
40	0.448	0.250	0.305
45	0.381	0.216	0.263
50	0.338	0.193	0.233
60	0.288	0.164	0.198
70	0.260	0.149	0.179
80	0.241	0.138	0.167
90	0.228	0.131	0.158
100	0.218	0.126	0.151
150	0.188	0.110	0.131
200	0.172	0.101	0.120
250	0.162	0.0950	0.113
273.2	0.157	0.0925	0.110
300	0.153	0.0908	0.108

REMARKS

The provisional values are for well-annealed high-purity lutetium and are thought to be accurate to within ±25% of the true values at temperatures below 100 K and ±20% above. At temperatures below 100 K the values for $\alpha_{\parallel}$, $\alpha_{\perp}$, and α_{poly} are applicable only to samples having residual electrical resistivities of 0.76, 2.65, and 1.45 μΩ cm, respectively.

*All values are estimated.

FIGURE AND TABLE 36R. RECOMMENDED THERMAL DIFFUSIVITY OF MAGNESIUM

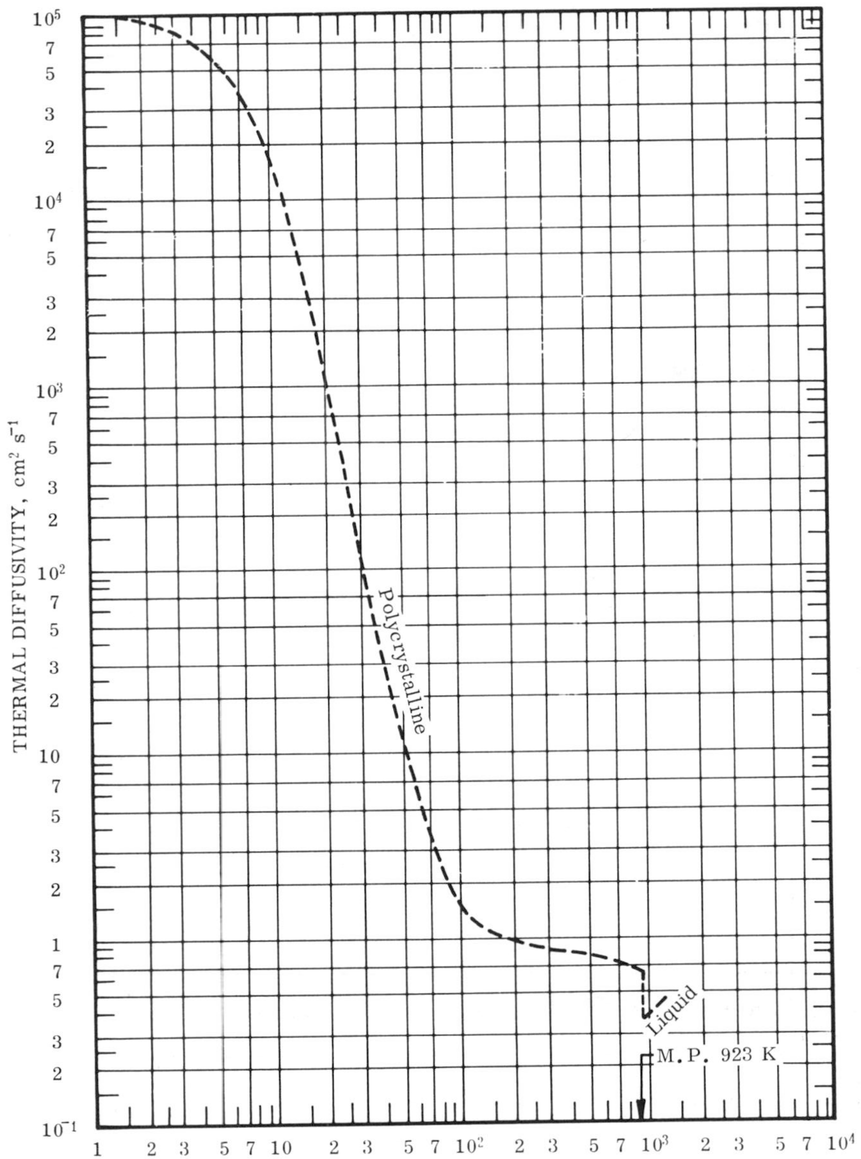

RECOMMENDED VALUES†

[Temperature, T, K; Thermal Diffusivity, α, cm^2 s^{-1}]

SOLID (Polycrystalline)

T	α	T	α
1	101000*	35	57.2*
2	93000*	40	29.4*
3	81900*	45	17.2*
4	69600*	50	11.1*
5	57000*	60	5.53*
6	45500*	70	3.28*
7	35900*	80	2.24*
8	27900*	90	1.74*
9	21600*	100	1.48*
10	16700*	150	1.09*
11	12700*	200	0.971*
12	9590*	250	0.911*
13	7170*	273.2	0.892*
14	5370*	300	0.874
15	4030*	350	0.849*
16	3040*	400	0.828*
18	1750*	500	0.792*
20	1040*	600	0.756*
25	322*	700	0.722*
30	125*	800	0.689*
		900	0.656*
		923	0.649*

LIQUID

T	α	T	α
923	0.370*	1100	0.438*
1000	0.398*	1200	0.482*

REMARKS

The values are for well-annealed high-purity magnesium and are considered accurate to within ±5% of the true values at moderate temperatures, ±13% for low temperatures and as the melting point is approached, and ±20% for the liquid state. Those for molten magnesium are provisional. At low temperatures the values are highly conditioned by impurity and imperfection, and those below 100 K are applicable only to magnesium having residual electrical resistivity of 0.00261 μΩ cm.

†Values for molten magnesium are provisional.
* In temperature range where no experimental data are available.

SPECIFICATION TABLE 36. THERMAL DIFFUSIVITY OF MAGNESIUM

(Impurity < 0.20% each; total impurities < 0.50%)

Cur. No.	Ref. No.	Author(s)	Year	Temp. Range, K	Reported Error, %	Name and Specimen Designation	Composition (weight percent), Specifications, and Remarks
1*	184	Taylor, R.	1965	298			Spectrographically pure; cylindrical specimen 0.25 in. in diameter and 1.270 cm long; front face exposed to heat pulse from a xenon flash tube; thermal diffusivity calculated from measured time necessary for the rear face to reach one-half the maximum temperature rise; temperature of measurement not given by author but assumed to be room temperature.
2*	184	Taylor, R.	1965	298			Cylindrical specimen 0.25 in. in diameter and 0.635 cm long; other conditions and specifications same as above.

DATA TABLE 36. THERMAL DIFFUSIVITY OF MAGNESIUM

(Impurity < 0.20% each; total impurities < 0.50%)

[Temperature, T, K; Thermal Diffusivity, α, $cm^2\ s^{-1}$]

T	α
CURVE 1*	
298	0.842
CURVE 2*	
298	0.839

* No figure given.

FIGURE AND TABLE 37R. PROVISIONAL THERMAL DIFFUSIVITY OF MANGANESE

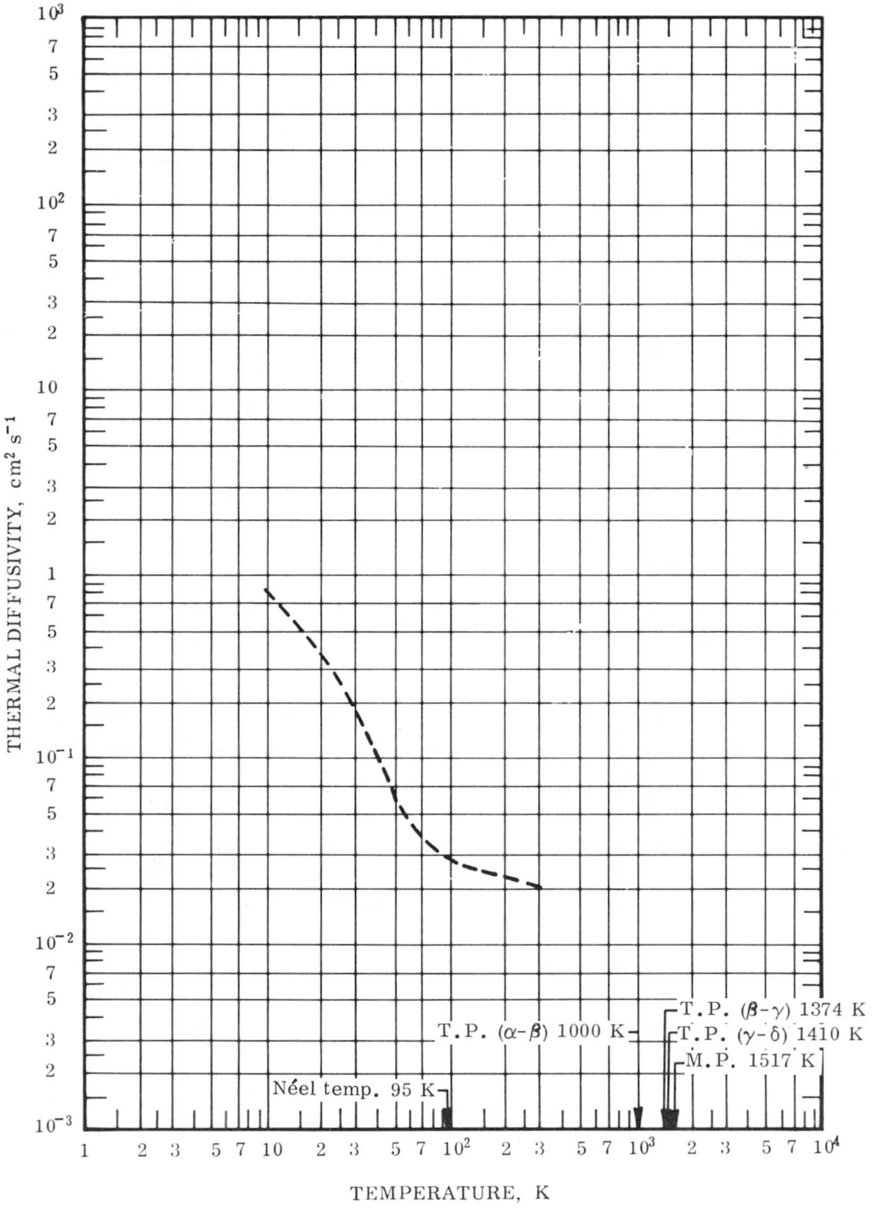

PROVISIONAL VALUES*

[Temperature, T, K; Thermal Diffusivity, α, $cm^2\ s^{-1}$]

SOLID

T	α
10	0.815
11	0.741
12	0.676
13	0.620
14	0.570
15	0.526
16	0.488
18	0.419
20	0.360
25	0.250
30	0.178
35	0.131
40	0.0990
45	0.0766
50	0.0612
60	0.0454
70	0.0378
80	0.0335
90	0.0307
100	0.0288
150	0.0247
200	0.0228
250	0.0216
273.2	0.0212
300	0.0208

REMARKS

The values are for well-annealed high-purity manganese and are thought to be accurate to within ±20%. The accuracy may be slightly better around room temperature. Values below room temperature are applicable only to manganese having residual electrical resistivity of 11.3 $\mu\Omega$ cm.

*All values are estimated.

FIGURE AND TABLE 38R. RECOMMENDED THERMAL DIFFUSIVITY OF MERCURY

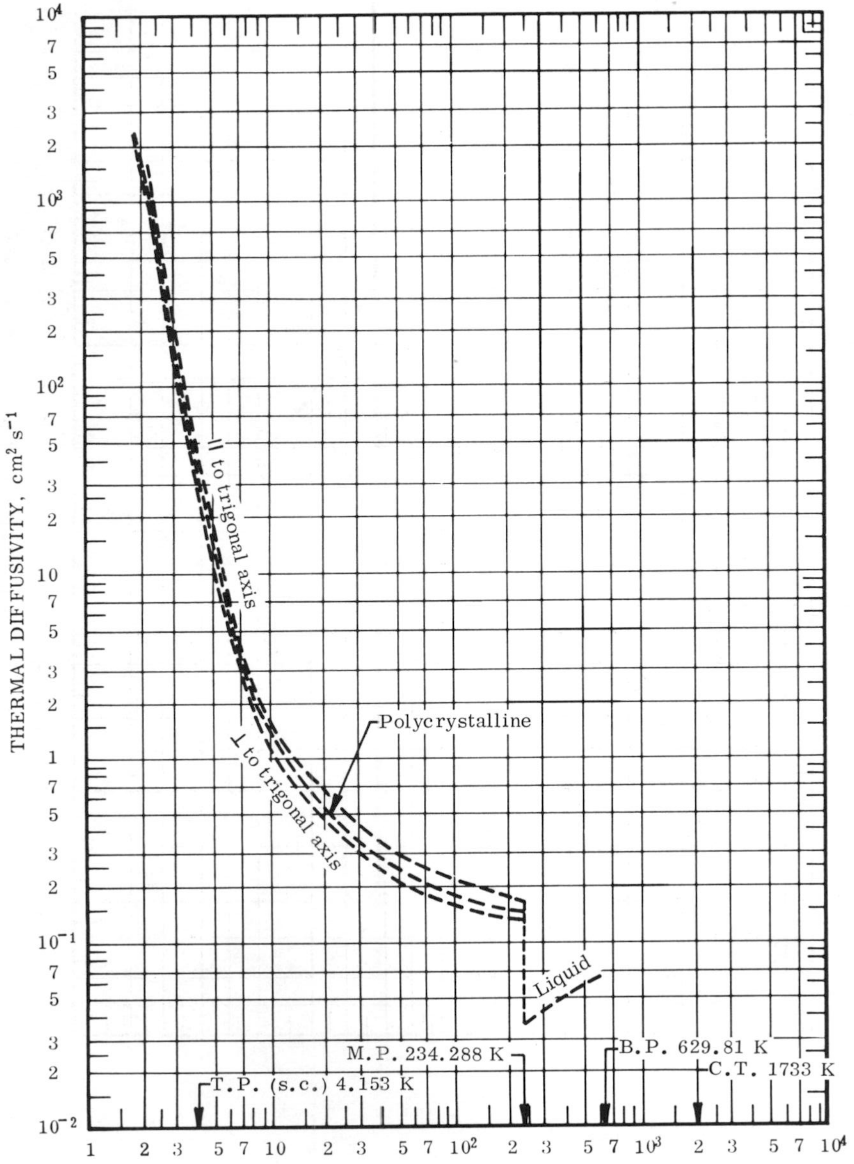

RECOMMENDED VALUES*

[Temperature, T, K; Thermal Diffusivity, α, cm^2 s^{-1}]

SOLID

T	‖ to trigonal axis α	⊥ to trigonal axis α	Polycrystalline α
2	3070	2130	2440
3	208	145	166
4	45.0	30.5	34.6
5	15.9	10.5	12.0
6	7.28	4.84	5.56
7	4.06	2.80	3.21
8	2.72	1.90	2.17
9	2.06	1.44	1.65
10	1.28	1.18	1.36
11	1.44	1.02	1.17
12	1.27	0.891	1.02
13	1.13	0.797	0.912
14	1.025	0.722	0.828
15	0.936	0.662	0.757
16	0.865	0.611	0.699
18	0.751	0.531	0.607
20	0.668	0.470	0.535
25	0.527	0.370	0.422
30	0.442	0.311	0.355
35	0.386	0.273	0.311
40	0.348	0.247	0.281
45	0.322	0.228	0.260
50	0.301	0.215	0.244
60	0.274	0.196	0.223
70	0.255	0.184	0.208
80	0.241	0.175	0.197
90	0.232	0.168	0.189
100	0.224	0.163	0.183
150	0.194	0.148	0.162
200	0.175	0.136	0.149
234.288	0.165	0.130	0.142

LIQUID

T	α
234.288	0.0362
250	0.0383
273.2	0.0410
300	0.0441
350	0.0491
400	0.0583
500	0.0600
600	0.0655

REMARKS

The values for high-purity mercury are thought to be accurate to within ± 15% of the true values at temperatures from 80 K to the melting point and ± 8% for liquid mercury. The values below 80 K are merely typical values and represent typical curves serving to indicate the general trend of the thermal diffusivity of mercury at low temperatures.

* All values are estimated and those for solid mercury above 80 K are provisional and below 80 K are merely typical values.

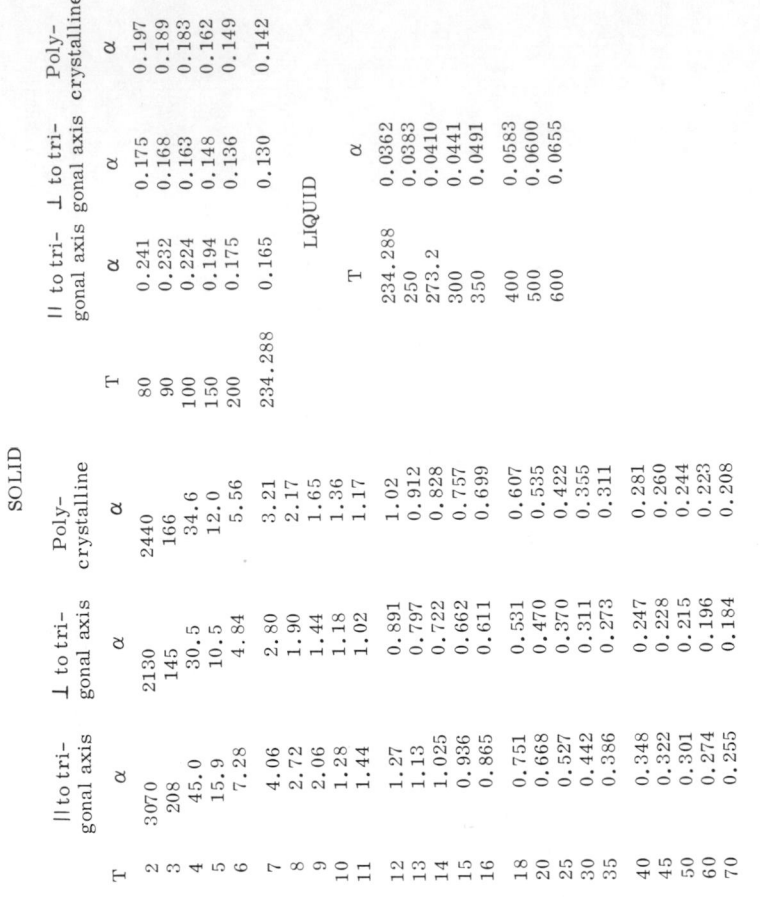

FIGURE AND TABLE 39R. RECOMMENDED THERMAL DIFFUSIVITY OF MOLYBDENUM

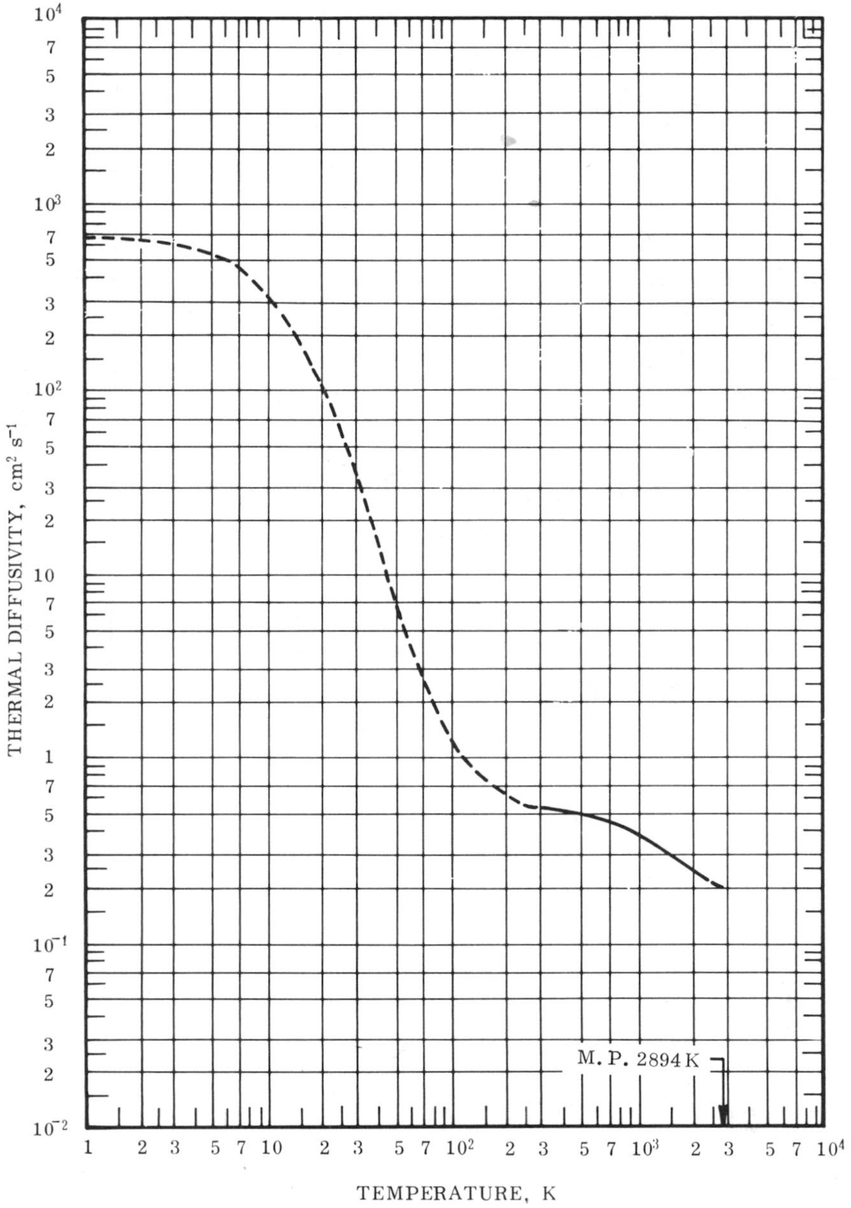

RECOMMENDED VALUES†

[Temperature, T, K; Thermal Diffusivity, α, cm² s⁻¹]

SOLID

T	α	T	α	T	α
1	652*	35	23.0*	900	0.396
2	634*	40	15.0*	1000	0.376
3	613*	45	10.25*	1100	0.357
4	585*	50	7.15*	1200	0.339
5	550*	60	4.11*	1300	0.322
6	509*	70	2.68*	1400	0.306
7	465*	80	1.93*	1500	0.293
8	415*	90	1.51*	1600	0.282
9	370*	100	1.25*	1700	0.271
10	328*	150	0.746*	1800	0.262
11	294*	200	0.630*	1900	0.253
12	262*	250	0.573*	2000	0.246
13	234*	273.2	0.558*	2200	0.232
14	209*	300	0.543	2400	0.220
15	186*	350	0.521	2600	0.210*
16	167*	400	0.503	2800	0.202*
18	134*	500	0.475	2894	0.199*
20	107*	600	0.451		
25	62.5*	700	0.432		
30	36.9*	800	0.415		

REMARKS

The values are for well-annealed high-purity molybdenum and are considered accurate to within ±13% of the true values at low temperatures, ±6% at moderate temperatures, and within ±20% as the melting point is approached. The values above 2000 K are provisional. At low temperatures the values are highly conditioned by impurity and imperfection, and those below room temperature are applicable only to molybenum having residual electrical resistivity of 0.167 μΩ cm.

† Values above 2000 K are provisional.
* In temperature range where no experimental data are available.

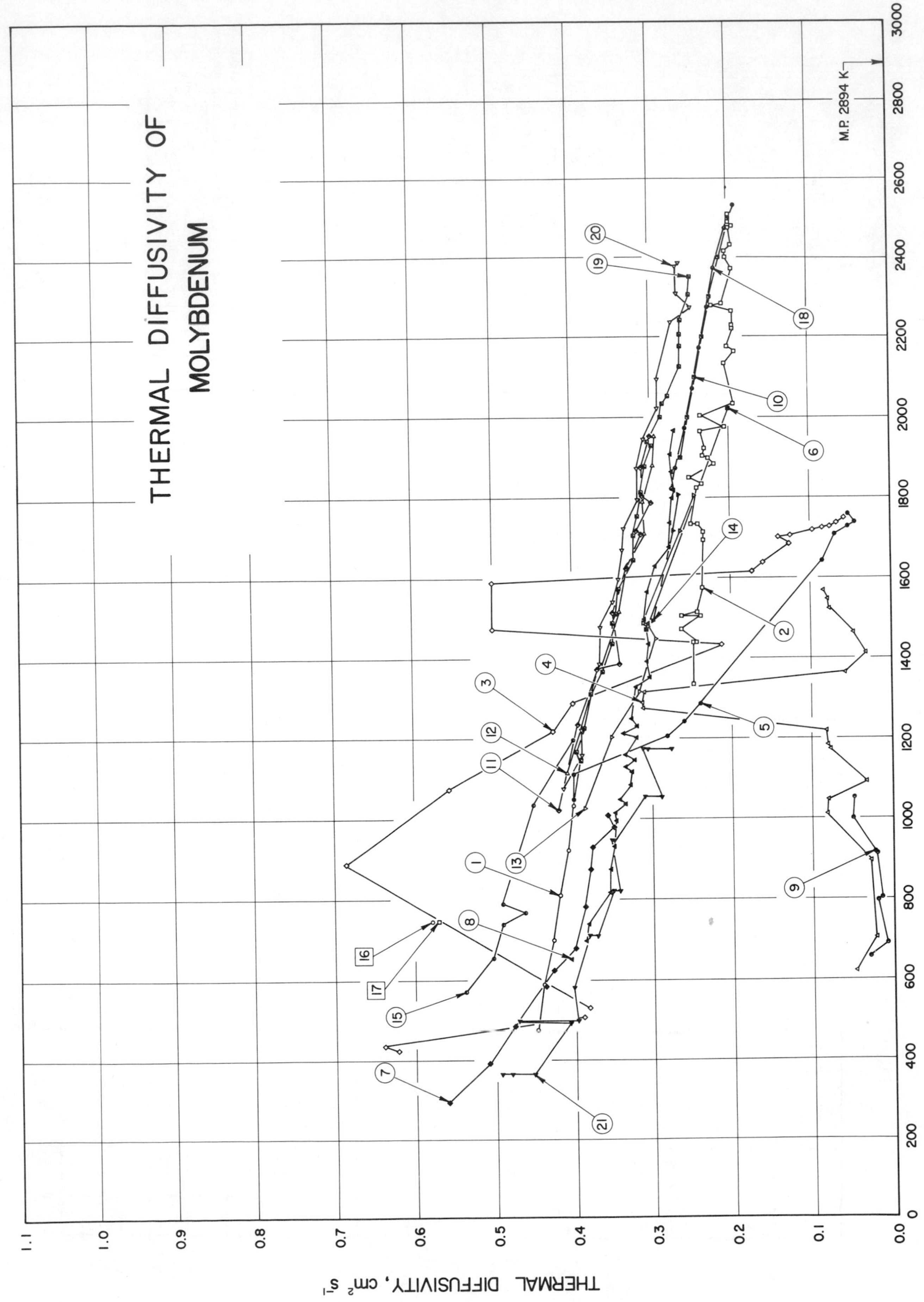
THERMAL DIFFUSIVITY OF
MOLYBDENUM
M.P. 2894 K
TEMPERATURE, K
THERMAL DIFFUSIVITY, cm² s⁻¹
0
200
400
600
800
1000
1200
1400
1600
1800
2000
2200
2400
2600
2800
3000
0.0
0.1
0.2
0.3
0.4
0.5
0.6
0.7
0.8
0.9
1.0
1.1

FIGURE 39

SPECIFICATION TABLE 39. THERMAL DIFFUSIVITY OF MOLYBDENUM

(Impurity < 0.20% each; total impurities < 0.50%)

Cur. No.	Ref. No.	Author(s)	Year	Temp. Range, K	Reported Error, %	Name and Specimen Designation	Composition (weight percent), Specifications, and Remarks
1	17	Lucks, C. F. and Deem, H. W.	1958	478-1144			Arc melted, unalloyed; supplied by Climax Molybdenum Co.; thermal diffusivity calculated from measured conductivity, specific heat, and density.
2	39	Wheeler, M. J.	1965	1335-2510			99.99$^+$ Mo (by difference), < 0.01 Fe, and traces of other elements; average grain size after testing 110 μm; rectangular specimen having top surface area lying in the range from 0.3 to 1.3 cm^2 and 0.040 in. in thickness; specimen cut from sheet of same thickness; sintered and hot rolled; supplied by Murex; density 10.3 g cm^{-3}; Lorenz function reported as 2.109, 2.117, 2.117, 2.109, 2.098, 2.089, 2.071, and 2.100 x 10^{-8} V^2 K^{-2} at 1397, 1598, 1796, 1994, 2200, 2398, 2600, and 2801 K, respectively; specimen heated to incandescence by an electron beam; intensity of beam sinusoidally moldulated at a frequency of 0.48 cycles per sec, modulation used produced temperature fluctuations in the bombarded face of the specimen of from ± 5 to ± 20 K; measured under a vacuum of ~ 5 x 10^{-6} mm Hg.
3	10	Sheer, C., Fitz, C. D., Mead, L. H., Holmgren, J. D., Rothacker, D. L., and Allmand, D.	1958	423-1751			Cylindrical specimen 1.0 cm in diameter and 4.5 cm long; machined to dimensions given; insulated on the sides and over one end face; other end suddenly exposed to a constant heat flow from the plasma of a high intensity arc; measured in vacuum chamber under an ambient pressure of 0.1 atmosphere; diffusivity values computed assuming conditions of zero heat flow across the lateral and rear end surfaces.
4	10	Sheer, C., et al.	1958	619-1568			Above specimen allowed to cool in chamber after being heated during above measurements; exposed to the arc to measure diffusivity again.
5	10	Sheer, C., et al.	1958	1048-1760			Above specimen allowed to cool in chamber after being heated during above measurements; exposed to the arc to measure diffusivity again.
6	40	Mustacchi, C. and Giuliani, S.	1963	1473-2023			Specimen measured for diffusivity using four different methods; amplitude, output face phase shift, differential phase shift, and pulse lag methods; 10 measurements made with each method.
7	242, 34, 141	Rudkin, R. L., Parker, W. J., and Jenkins, R. J.	1961	293-1008	± 5		Specimen a few millimeters thick; high intensity short duration light pulse from xenon flash lamp absorbed in the front surface of thermally insulated specimen; thermal diffusivity determined from measured temperature history of the rear surface; measured in resistance furnace.
8	242, 34, 141	Rudkin, R. L., et al.	1961	650-1968	± 5		Above specimen measured for diffusivity; radio frequency induction heating used for heating specimen; measured under same conditions as above.
9	36	Kevane, C. J.	1958	656-1051			Cylindrical specimen ~ 0.375 in. in diameter and 1 in. long; machined; holes drilled in specimen separated by a distance of 2 mm between the lines of centers along the axis; diffusivity determined from measured propagation of temperature waves set up by periodic modulation of radiation heat flux input to one surface of specimen; solar furnace used as a heat source; amplitude ratio method used to measure diffusivity; one-dimensional heat flow.
10	121, 122	Kraev, O. A. and Stel'makh, A. A.	1964	1900-2500	± 5		Disk specimen 8-9 mm in diameter and 0.3 mm thick; diffusivity measured using wave method; data points corrected for thermal expansion.

SPECIFICATION TABLE 39. THERMAL DIFFUSIVITY OF MOLYBDENUM (continued)

Cur. No.	Ref. No.	Author(s)	Year	Temp. Range, K	Reported Error, %	Name and Specimen Designation	Composition (weight percent), Specifications, and Remarks
11	150	Pigal'skaya, L.A., Yurchak, R.P., Makarenko, I.N., and Filippov, L.P.	1966	1022-1954		1	99.9 Mo, 0.01 sesquioxides, 0.001 Ni, 0.001 SiO_2, and traces of Mg and Ca oxides; cylindrical specimen 10 mm in diameter and 70 mm long; density 10.2 g cm^{-3} at 296.2 K; electrical resistivity 5.78 μohm cm at 296.2 K; diffusivity determined from measured difference in phase between the fluctuations of sample surface temperature and energy source power.
12	150	Pigal'skaya, L.A., et al.	1966	1113-1953		1	Above specimen measured for diffusivity again; diffusivity determined from measured dependence of the amplitude of temperature variation of sample surface on frequency of power source modulation.
13	150	Pigal'skaya, L.A., et al.	1966	1027-1807		2	99.54 Mo (by difference), 0.28 Ti, and 0.18 Zr; cylindrical specimen 15 mm in diameter and 70 mm long; density 10.17 g cm^{-3} at 296.2 K; electrical resistivity 6.52 μohm cm at 296.2 K; diffusivity determined from measured difference in phase between the fluctuations of sample surface temperature and energy source power.
14	150	Pigal'skaya, L.A., et al.	1966	1493-1808		2	Above specimen measured for diffusivity again; diffusivity determined from measured dependence of the amplitude of temperature variation of sample surface on frequency of power source modulation.
15	159	Khusainova, B.N. and Filippov, L.P.	1968	572-1198	4		Single crystal; cylindrical specimen 6.6 mm in diameter and 54 mm long; purified by band refining (three passages); electrical resistivity measured and reported as 5.4, 6.6, 7.7, 8.7, 10.2, 10.6, 12.7, 14.8, 18.2, and 20.6 μohm cm at 310, 352, 413, 462, 526, 534, 628, 724, 848, and 936 K, respectively (resistivity data actually reported by authors are 100 times greater than the accepted standard values, this has been considered as a printing error and the data reported here have, therefore, undergone the appropriate correction); Lorenz number reported as 3.15, 3.04, 2.99, 2.92, 3.00, and 3.00 x 10^{-8} V^2 K^{-2} at 553, 605, 698, 800, 904, and 953 K, respectively; end of specimen subjected to variable heating by electron bombardment using tantalum or tungsten cathodes; thermal diffusivity determined independently from measured amplitudes or phase difference of the temperature oscillations at two points along specimen.
16	159	Khusainova, B.N. and Filippov, L.P.	1968	547	4		Above specimen measured for diffusivity using a period of 8.05 sec for the heat wave; diffusivity determined from measured phase difference of the temperature oscillations at two points along specimen; measured using same technique as above.
17	159	Khusainova, B.N. and Filippov, L.P.	1968	547	4		Above specimen measured for diffusivity using same period and same technique as above; diffusivity determined from measured amplitudes of the temperature oscillations at two points along specimen.
18	172	Kraev, O.A. and Stel'makh, A.A.	1966	1823-2532	± 5		Disk specimen 7 to 9 mm in diameter and 0.3 mm thick; cut from plate of rolled metal; surfaces thoroughly cleaned; heated by electron bombardment; sinusoidal temperature oscillation imposed on one face of specimen; thermal diffusivity determined from phase difference between measured temperature fluctuations at front and back faces of specimen; reported data points corrected for thermal expansion and obtained from smooth curve given by authors.

SPECIFICATION TABLE 39. THERMAL DIFFUSIVITY OF MOLYBDENUM (continued)

Cur. No.	Ref. No.	Author(s)	Year	Temp. Range, K	Reported Error, %	Name and Specimen Designation	Composition (weight percent), Specifications, and Remarks
19	202	Makarenko, I. N., Trukhanova, L. N., and Filippov, L. P.	1970	1147-2352	6-8	Sample I	Single crystal; 99.99 Mo, 0.01 O_2, 0.005 C, 0.001 Mg, 0.0009 H_2, 0.0008 N_2, 0.0004 Si, 0.00003 Al, 0.00001 Fe; 10.3 mm in diameter, 75 mm in length; obtained by method of electron-beam zone melting; density at 20 C, 10.21 g cm^{-3}; specific electric resistance 5.38 $\mu\Omega$ cm; angles between axis of sample and the crystallographic directions [100] and [010] being 60° and 42° respectively; annealed in vacuum ($\sim 10^{-5}$ torr) at temperature of 2100 K for 2 hr.
20	202	Makarenko, I. N., et al.	1970	1077-2384	6-8	Sample II	Polycrystalline specimen; 99.99 Mo, 0.01 Ni, 0.01 sesquioxides, 0.001 SiO, traces of CaO and MnO; 13 mm in diameter, 75 mm in length; density at 20 C, 9.92 g cm^{-3}; specific electric resistance 5.90 $\mu\Omega$ cm; annealed in vacuum ($\sim 10^{-5}$ torr) at temperature of 2100 K for 2 hr.
21	230	van Craeynest, J. C., Weilbacher, J. C., and Lallemont, R.	1969	366-1173	10		5 mm diameter x 10 mm long; measured by Angström method.

DATA TABLE 39. THERMAL DIFFUSIVITY OF MOLYBDENUM

(Impurity < 0.20% each; total impurities < 0.50%)

[Temperature, T, K; Thermal Diffusivity, α, $cm^2 s^{-1}$]

T	α
CURVE 1	
477.6	0.449
588.7	0.439
699.8	0.426
810.9	0.418
922.1	0.408
1033.2	0.400
1144.3	0.390
CURVE 2	
1335	0.248
1440	0.247
1440	0.243
1475	0.263
1505	0.238
1505	0.263
1515	0.243
1577	0.236
1695	0.234
1715	0.234
1735	0.240
1735	0.249
1825	0.242
1837	0.235
1850	0.252
1885	0.219
1900	0.227
1907	0.234
1923	0.231
1967	0.236
1977	0.205
2005	0.237
2035	0.194
2135	0.206
2165	0.193
2175	0.202
2220	0.195
2230	0.195
2265	0.195
2277	0.221
2285	0.207
2373	0.195
CURVE 2 (cont.)	
2400	0.203
2415	0.203
2433	0.195
2475	0.199
2480	0.194
2490	0.198
2510	0.198
CURVE 3	
423	0.625
438	0.642
508	0.390
530	0.383
889	0.638
1073	0.557
1218	0.425
1288	0.400
1433	0.212
1473	0.500
1590	0.500
1618	0.173
1640	0.160
1686	0.127
1703	0.140
1708	0.125
1720	0.0975
1726	0.0850
1729	0.0750
1738	0.0675
1751	0.0575
CURVE 4	
619	0.0475
703	0.0225
890	0.0270
1010	0.0310
1046	0.0800
1090	0.0325
1173	0.0770
1193	0.0800
CURVE 4 (cont.)	
1217	0.0825
1364	0.0580
1413	0.0300
1463	0.0475
1523	0.0770
1547	0.0780
1568	0.0855
CURVE 5	
1048	0.400
1111	0.400
1205	0.282
1243	0.261
1286	0.240
1643	0.0855
1708	0.0690
1729	0.0525
1740	0.0430
1760	0.0525
CURVE 6	
1473	0.307
2023	0.200
CURVE 7	
293	0.560
390	0.507
486	0.477
596	0.437
623	0.427
678	0.400
782	0.387
873	0.379
930	0.377
978	0.350
1008	0.357
CURVE 8	
650	0.405
696	0.386
738	0.383
818	0.355
873	0.355
930	0.350
993	0.347
1013	0.348
1036	0.335
1046	0.343
1083	0.328
1116	0.327
1128	0.335
1146	0.323
1158	0.336
1203	0.320
1213	0.337
1230	0.320
1248	0.327
1276	0.312
1290	0.313
1316	0.310
1328	0.320
1353	0.303
1393	0.307
1433	0.305
1488	0.305
1498	0.309
1566	0.305
1630	0.295
1678	0.277
1738	0.275
1800	0.273
1818	0.270
1866	0.273
1908	0.275
1968	0.270
CURVE 9	
656	0.0300
688	0.0097
CURVE 9 (cont.)	
796	0.0203
801	0.0153
911	0.0203
915	0.0223
1000	0.0508
1051	0.0482
CURVE 10	
1900	0.261
2000	0.252
2100	0.243
2200	0.233
2300	0.223
2400	0.211
2500	0.198
CURVE 11	
1022	0.418*
1237	0.394
1374	0.369
1387	0.341
1514	0.349
1624	0.331
1710	0.312
1719	0.318
1789	0.299
1807	0.311
1879	0.312
1954	0.300
CURVE 12	
1113	0.408*
1516	0.341
1625	0.331*
1712	0.308
1791	0.310
1880	0.297
1953	0.295
CURVE 13	
1027	0.385*
1204	0.352
1317	0.315
1451	0.295
1483	0.305*
1488	0.311
1717	0.263
1807	0.246
CURVE 14	
1493	0.297
1717	0.271
1808	0.264
CURVE 15	
572	0.539
655	0.503
739	0.491
768	0.461
790	0.491
1034	0.450
1198	0.400
CURVE 16	
547	0.58
CURVE 17	
547	0.571
CURVE 18	
1823	0.273
1873	0.268
1973	0.256
2073	0.245
2173	0.236
2273	0.226
2373	0.217
2473	0.203
2532	0.192
CURVE 19	
1147	0.390
1168	0.395
1227	0.385
1311	0.376
1368	0.361
1439	0.349
1490	0.349
1512	0.350
1574	0.341
1647	0.322
1708	0.322
1748	0.317
1815	0.312
1879	0.307
1930	0.298
1939	0.303
2001	0.286
2035	0.284
2054	0.277
2125	0.261
2178	0.261
2207	0.261
2242	0.261
2308	0.249
2352	0.247
CURVE 20	
1077	0.413
1159	0.389
1222	0.389
1386	0.366
1477	0.365
1542	0.349
1599	0.341
1671	0.336
1723	0.335
1796	0.317
1875	0.317
1945	0.308
2021	0.291
2093	0.291
2236	0.273
CURVE 20 (cont.)	
2271	0.247
2309	0.265
2377	0.266
2384	0.262
CURVE 21	
366	0.494
366	0.481
366	0.452
494	0.405
498	0.472
498	0.397
582	0.402
710	0.383
710	0.372
823	0.352
823	0.344
947	0.353
947	0.348
1053	0.311
1053	0.290
1173	0.313
1173	0.306
1173	0.277

* Not shown in figure.

FIGURE AND TABLE 40R. PROVISIONAL THERMAL DIFFUSIVITY OF NEODYMIUM

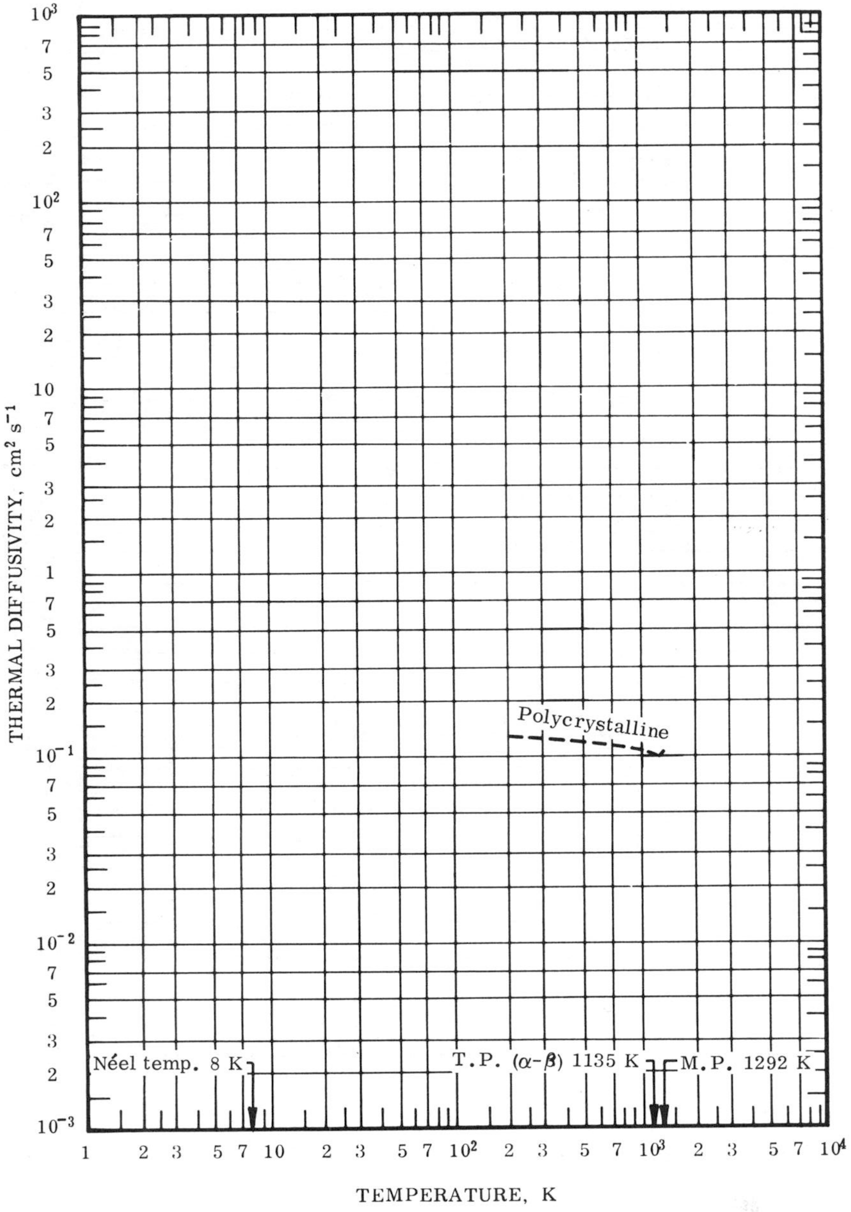

PROVISIONAL VALUES*

[Temperature, T, K; Thermal Diffusivity, α, $cm^2\ s^{-1}$]

SOLID (Polycrystalline)

T	α
200	0.128
250	0.126
273.2	0.125
300	0.124
350	0.122
400	0.120
500	0.1175
600	0.115
700	0.1135
800	0.112
900	0.109
1000	0.107
1100	0.104
1135	0.103
1135	0.101
1200	0.106

REMARKS

The provisional values are for well-annealed high-purity polycrystalline neodymium and are thought to be accurate to within ±20% of the true values near room temperature. The uncertainty increases to ±30% at the highest temperatures.

*All values are estimated.

FIGURE AND TABLE 41R. RECOMMENDED THERMAL DIFFUSIVITY OF NICKEL

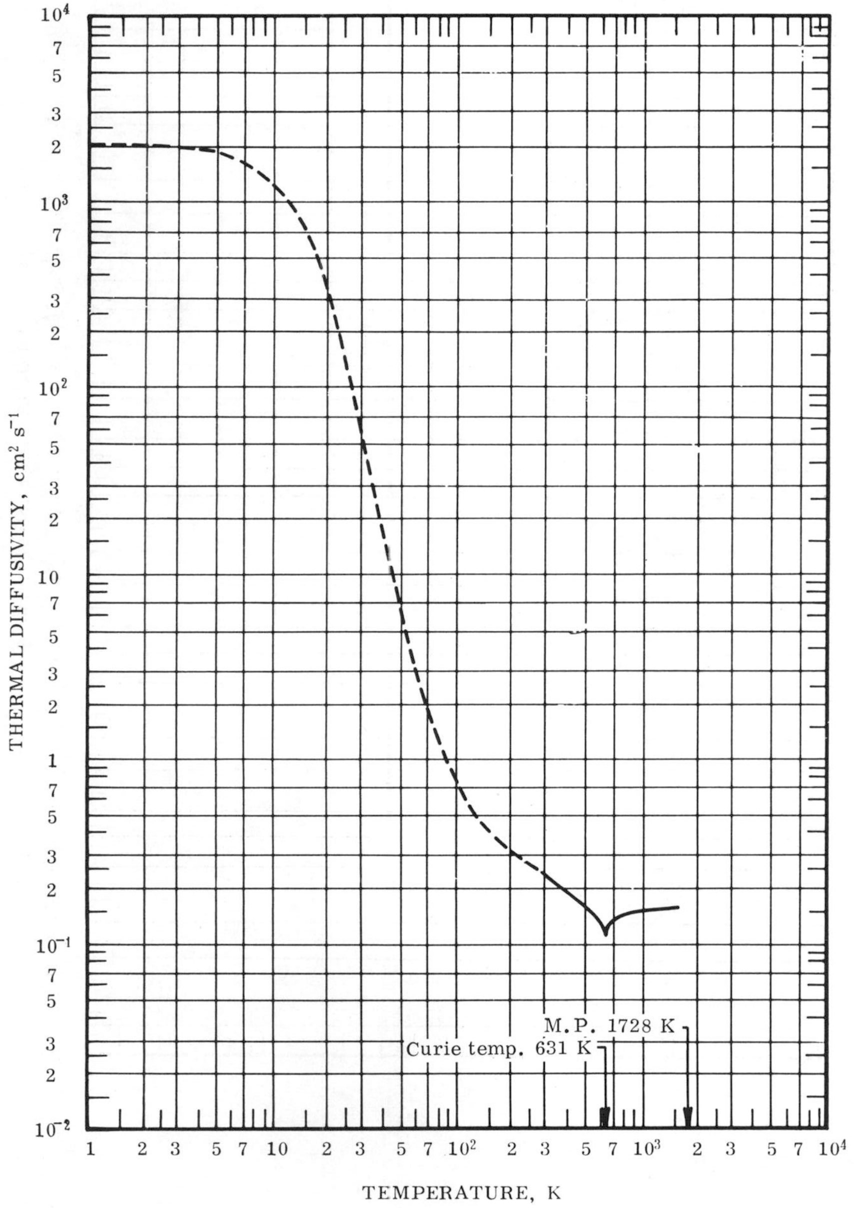

RECOMMENDED VALUES

[Temperature, T, K; Thermal Diffusivity, α, $cm^2\ s^{-1}$]

SOLID

T	α	T	α
1	2010*	70	2.03*
2	2000*	80	1.36*
3	1960*	90	0.995*
4	1900*	100	0.788*
5	1820*	150	0.415*
6	1725*	200	0.313*
7	1620*	250	0.263*
8	1490*	273.2	0.245
9	1360*	300	0.229
10	1230*	350	0.204
11	1120*	400	0.187
12	1000*	500	0.156
13	896*	600	0.126
14	798*	631	0.110
15	704*	700	0.142
16	618*	800	0.147
18	464*	900	0.148
20	335*	1000	0.150
25	141*	1100	0.152
30	63.5*	1200	0.154
35	31.6*	1300	0.156
40	17.2*	1400	0.158
45	10.2*	1500	0.160
50	6.54*		
60	3.36*		

REMARKS

The recommended values are for well-annealed high-purity nickel and are thought to be accurate to within ±10% of the true values. At low temperatures the values are highly conditioned by impurity and imperfection, and those below room temperature are applicable only to nickel having residual electrical resistivity of 0.0112 $\mu\Omega$ cm.

*In temperature range where no experimental data are available.

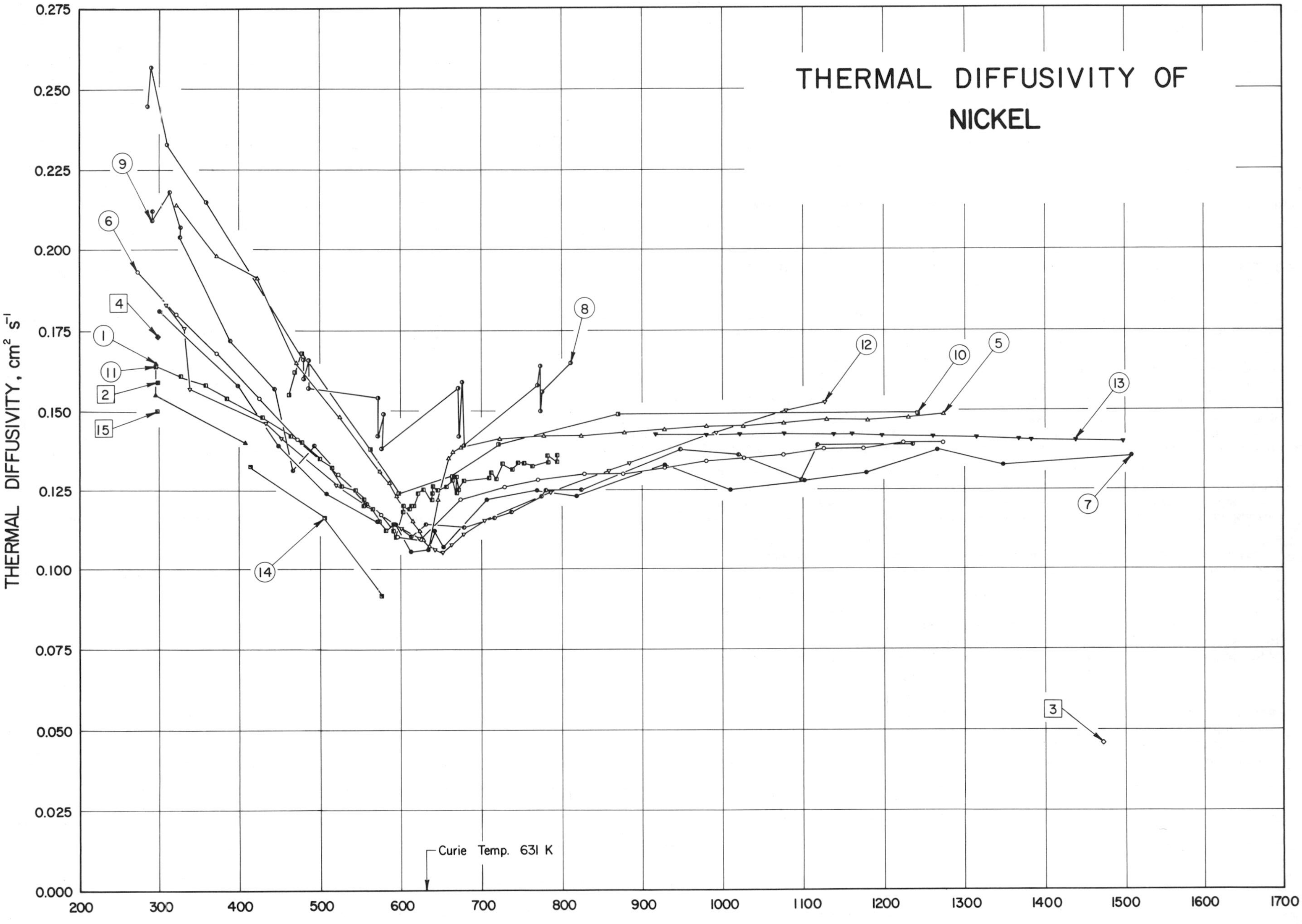
THERMAL DIFFUSIVITY OF
NICKEL
THERMAL DIFFUSIVITY, cm² s⁻¹
TEMPERATURE, K
Curie Temp. 631 K
0.275
0.250
0.225
0.200
0.175
0.150
0.125
0.100
0.075
0.050
0.025
0.000
200
300
400
500
600
700
800
900
1000
1100
1200
1300
1400
1500
1600
1700
1
2
3
4
5
6
7
8
9
10
11
12
13
14
15

FIGURE 41

SPECIFICATION TABLE 41. THERMAL DIFFUSIVITY OF NICKEL

(Impurity < 0.20% each; total impurities < 0.50%)

Cur. No.	Ref. No.	Author(s)	Year	Temp. Range, K	Reported Error, %	Name and Specimen Designation	Composition (weight percent), Specifications, and Remarks
1	4	Jenkins, R.J. and Parker, W.J.	1961	295-408	± 5		Cylindrical specimen 1.26 cm dia. and 0.100 cm thick; high intensity short duration light pulse absorbed in front surface of thermally insulated specimen coated with camphor black; 408.2 K measurement obtained by heating specimen holder and specimen with an infrared lamp; both data points at 295.2 K obtained from measurements using different equations for data reduction.
2	41	Starr, C.	1937	298			99.98$^+$ Ni, 0.009 C, and < 0.01 total Co, Cu, and Fe; wire specimen ~3 mm in dia. and 27 cm long; supplied by Research Lab. of the International Nickel Co.; annealed in hydrogen at 1143.2 K; density 8.79 g cm^{-3}, and electrical resistivity 7.21 μohm cm at 295.2 K; sinusoidal temp. impressed on one end of specimen; diffusivity determined with three different periods from measured decrement of temp. wave traveling along specimen.
3	40	Mustacchi, C. and Giuliani, S.	1963	1473			Specimen measured for diffusivity using four different methods: amplitude, output face phase shift, differential phase shift, and pulse lag methods; 10 measurements made with each method.
4	6	Moser, J.B. and Kruger, O.L.	1963	298			Plate specimen with surface area lying in the range from 1 to 4 cm^2 and thickness in the range from 0.1 to 0.3 cm; front surface thinly coated with colloidal graphite; irradiated with a pulse of thermal energy of short duration; diffusivity determined from measured history of the back surface temp.
5	113	Sidles, P.H. and Danielson, G.C.	1960	322-1273		Spectrographically Standardized Nickel	Spectrographically standardized nickel; pure; obtained from Johnson, Matthey and Co.; Curie temp. ~633.2 K; electrical resistivity reported as 7.7, 8.9, 11.9, 14.6, 17.1, 21.1, 25.4, 29.2, 32.6, 34.5, 35.8, 37.2, 38.5, 40.2, 41.5, 42.0, 43.7, 45.0, 47.5, 48.7, and 49.8 μohm cm at 294.2, 341.2, 387.2, 438.2, 480.2, 540.2, 593.2, 627.2, 662.2, 689.2, 734.2, 786.2, 831.2, 878.2, 934.2, 997.2, 1033.2, 1078.2, 1167.2, 1220.2, and 1297.2 K, respectively; diffusivity measured using modified Angström method.
6	113	Sidles, P.H. and Danielson, G.C.	1960	273-1273		High Purity Nickel	High purity nickel; less pure than above specimen; obtained from A.D. MacKay; Curie temp ~613.2 K; electrical resistivity reported as 9.3, 13.0, 15.8, 18.9, 21.2, 26.6, 29.5, 31.5, 34.3, 36.1, 37.6, 39.5, 41.2, 42.6, 44.7, 46.8, 47.3, and 49.8 μohm cm at 296.2, 381.2, 439.2, 476.2, 521.2, 591.2, 615.2, 655.2, 731.2, 766.2, 809.2, 877.2, 930.2, 993.2, 1082.2, 1152.2, 1207.2, and 1280.2 K, respectively; diffusivity measured using modified Angström method.
7	198	Emery, A.F. and Smith, J.R.	1968	301-1509		Nickel 200	Commercially pure specimen.
8	196	Kobayashi, K. and Kumada, T.	1968	286-811	≤ 5		Disk specimen, < 15 mm in diameter, 2-10 mm in thickness; diffusivity measured in vacuum.
9	162	Kobayashi, K. and Kumada, T.	1968	292-1236			99.95 pure; 15 mm diameter x 15 mm high; density 8.91 g cm^{-3}.
10	230	van Craeynest, J.C., Weilbacher, J.C., and Lallemont, R.	1969	462-1242	10		5 mm diameter x 10 mm long; measured by Angström method.
11	247	Namba, S., Kim, P.H., and Arai, T.	1967	297-794			99.4 pure; 1.5 cm x 1.5 cm square, 1.05 mm in thickness.
12	248	Kirichenko, P.I. and Mikryukov, V.E.	1964	309-1127			Pure nickel.

SPECIFICATION TABLE 41. THERMAL DIFFUSIVITY OF NICKEL (continued)

Cur. No.	Ref. No.	Author(s)	Year	Temp. Range, K	Reported Error, %	Name and Specimen Designation	Composition (weight percent), Specifications, and Remarks
13	249	Zinov'yev, V.Ye., Krentsis, R.P., and Gel'd, P.V.	1968	917-1672	5		99.95 pure; square section 8 x 8 mm machined from rolled band 0.209± 0.001 mm thick; resistivity ratio ρ (298K)/ρ (4.2K) = 80; annealed in vacuum compartment at 1200 K for 5 hr; diffusivity measured at seven frequencies in the range 150-300 cps.
14	250	Foley, E.L. and Sawyer, R.B.	1964	296-576			99.45 pure; obtained from Whitehead Metals, Inc.; diffusivity measured using heat pulse technique.
15	108	Steinberg, S., Larson, R.E., Kydd, A.R.	1963	298			Specimen 2.0 cm square, 0.1275 cm in thickness; measuring temperature not given and assumed to be 25 C.

DATA TABLE 41. THERMAL DIFFUSIVITY OF NICKEL

(Impurity < 0.20% each; total impurities < 0.50%)

[Temperature, T, K; Thermal Diffusivity, $cm^2 s^{-1}$]

T	α
CURVE 1	
295	0.16
295	0.15
408	0.14
CURVE 2	
298	0.159
CURVE 3	
1473	0.046
CURVE 4	
298	0.173
CURVE 5	
322	0.214
372	0.198
423	0.181
471	0.165
525	0.148
574	0.131
586	0.127
595	0.123
615	0.115
624	0.112
628	0.109
635	0.107
646	0.122
659	0.135
664	0.137
674	0.138
722	0.141
777	0.142
824	0.142
878	0.143
927	0.144
980	0.145
1026	0.145
1076	0.146
1129	0.147
CURVE 5 (cont.)	
1180	0.147
1230	0.148
1273	0.149
CURVE 6	
273	0.193
321	0.180
372	0.168
425	0.154
472	0.141
523	0.130
576	0.117
596	0.110
613	0.105
626	0.109
674	0.122
729	0.126
770	0.128
828	0.130
876	0.130
928	0.132
980	0.134
1027	0.135
1076	0.136
1126	0.138
1175	0.138
1224	0.140
1273	0.140
CURVE 7	
301	0.1815
398	0.1585
449	0.1392
508	0.1248
594	0.1142
613	0.1104
634	0.1112
642	0.1177
653	0.1124
707	0.1227
769	0.1253
CURVE 7 (cont.)	
818	0.1230
928	0.1330
1010	0.1256
1102	0.1283
1179	0.1308
1266	0.1388
1348	0.1334
1509	0.1367
CURVE 8	
286	0.245
291	0.257
310	0.233
359	0.215
480	0.166
480	0.160
486	0.166
486	0.157
572	0.154
572	0.142
576	0.138
578	0.149
671	0.157
672	0.142
676	0.159
678	0.139
769	0.158
773	0.164
773	0.150
775	0.156
811	0.165
CURVE 9	
292	0.2172
292	0.2098
314	0.2181
327	0.2074
327	0.2044
389	0.1721
444	0.1579
467	0.1314
CURVE 9 (cont.)	
493	0.1397
558	0.1205
561	0.1153
613	0.1108
631	0.1145
678	0.1136
716	0.1168
737	0.1185
775	0.1232
780	0.1251
824	0.1257
948	0.1382
1021	0.1365
1099	0.1281
1119	0.1397
1236	0.1397
CURVE 10	
462	0.1551
469	0.1622
478	0.1683
563	0.1389
597	0.1240
664	0.1291
721	0.1424
870	0.1491
1242	0.1494
CURVE 11	
297	0.1644
327	0.1610
358	0.1583
385	0.1545
429	0.1481
464	0.1429
478	0.1403
500	0.1358
515	0.1324
527	0.1269
544	0.1258
554	0.1224
CURVE 11 (cont.)	
554	0.1202
565	0.1196
573	0.1155
577	0.1168
582	0.1128
591	0.1141
591	0.1128
594	0.1102
603	0.1187
603	0.1209
611	0.1191
613	0.1209
617	0.1202
621	0.1245
628	0.1254
639	0.1222
639	0.1249
639	0.1261
646	0.1255
656	0.1263
664	0.1288
669	0.1249
669	0.1299
672	0.1258
678	0.1280
684	0.1278*
689	0.1288*
696	0.1284*
696	0.1304*
698	0.1292*
709	0.1288
711	0.1309
719	0.1289
726	0.1338
739	0.1315
745	0.1336
753	0.1337
763	0.1329
783	0.1338
782	0.1360
794	0.1338
794	0.1360
CURVE 12	
309	0.1833
331	0.1759
391	0.1572
433	0.1464
452	0.1413
516	0.1269
601	0.1127
623	0.1097
642	0.1060
653	0.1051
663	0.1077
678	0.1111
704	0.1151
786	0.1245
859	0.1312
884	0.1336
993	0.1431
1080	0.1500
1127	0.1526
CURVE 13	
917	0.1425
980	0.1425
1022	0.1425
1078	0.1425
1138	0.1425
1161	0.1427
1198	0.1423
1261	0.1422
1314	0.1418
1367	0.1412
1383	0.1412
1438	0.1409
1498	0.1406
1557	0.1397*
1608	0.1386*
1672	0.1365*
CURVE 14	
296	0.1653*
414	0.1327
CURVE 14 (cont.)	
506	0.1163
576	0.917
CURVE 15	
298	0.15

* Not shown in figure.

FIGURE AND TABLE 42R. RECOMMENDED THERMAL DIFFUSIVITY OF NIOBIUM

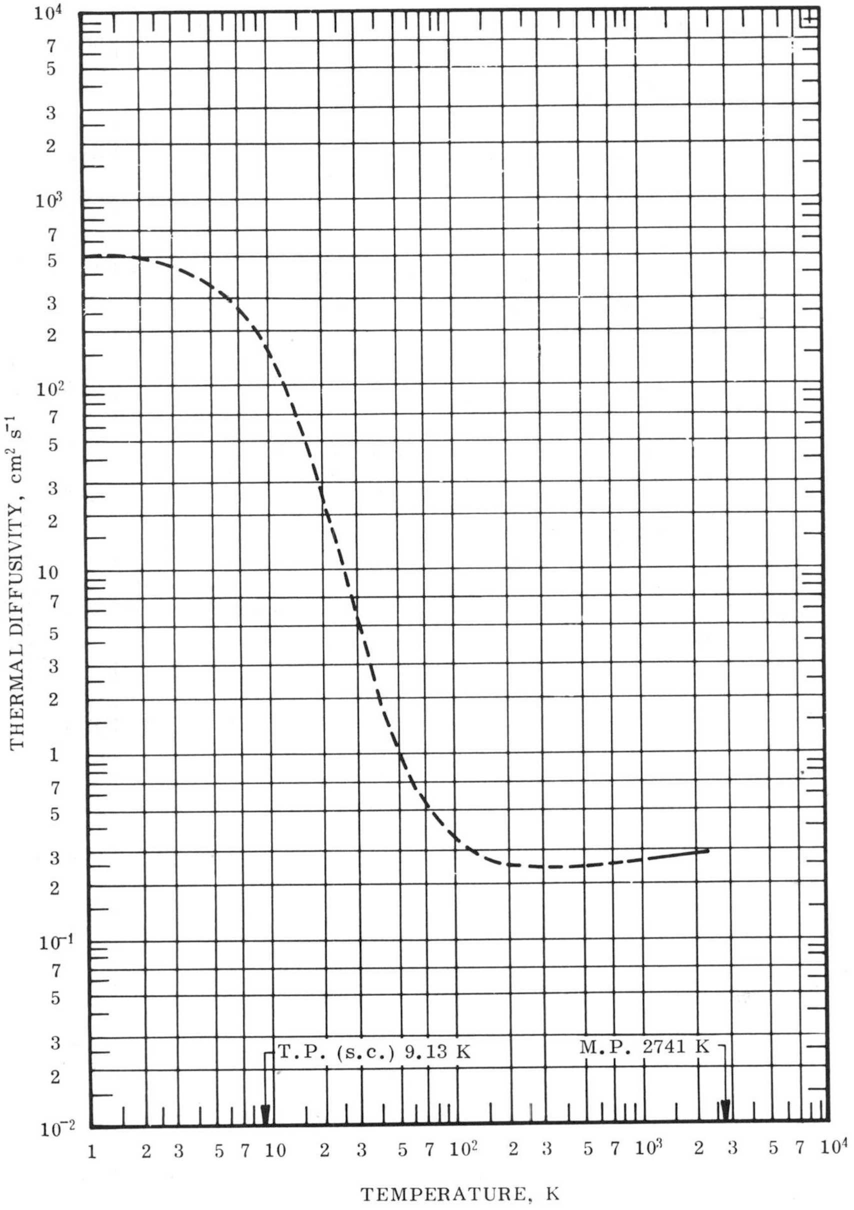

RECOMMENDED VALUES†

[Temperature, T, K; Thermal Diffusivity, α, $cm^2\ s^{-1}$]

SOLID

T	α	T	α
1	507 *	70	0.520 *
2	482 *	80	0.430 *
3	442 *	90	0.376 *
4	398 *	100	0.342 *
5	350 *	150	0.269 *
6	302 *	200	0.246 *
7	258 *	250	0.239 *
8	218 *	273.2	0.238 *
9	183 *	300	0.237 *
10	153 *	350	0.236 *
11	128 *	400	0.237 *
12	107 *	500	0.240 *
13	89.7 *	600	0.243 *
14	75.3 *	700	0.247 *
15	63.5 *	800	0.250 *
16	53.4 *	900	0.253 *
18	37.5 *	1000	0.256 *
20	26.3 *	1100	0.259
25	11.4 *	1200	0.262
30	5.54 *	1300	0.265
35	3.12 *	1400	0.268
40	1.98 *	1500	0.271
45	1.39 *	1600	0.274
50	1.04 *	1700	0.276
60	0.687 *	1800	0.278
		1900	0.280
		2000	0.282
		2200	0.285

REMARKS

The values are for well-annealed high-purity niobium and are thought to be accurate to within ±7% of the true values around room temperature and ±13% from 100 to 1500 K. The values below 100 K and above 1500 K are provisional and should be good to ±20%. At low temperatures the values are highly conditioned by impurity and imperfection, and those below 150 K are applicable only to niobium having residual electrical resistivity of 0.0679 μΩ cm.

† Values below 100 K and above 1500 K are provisional.
* In temperature range where no experimental data are available.

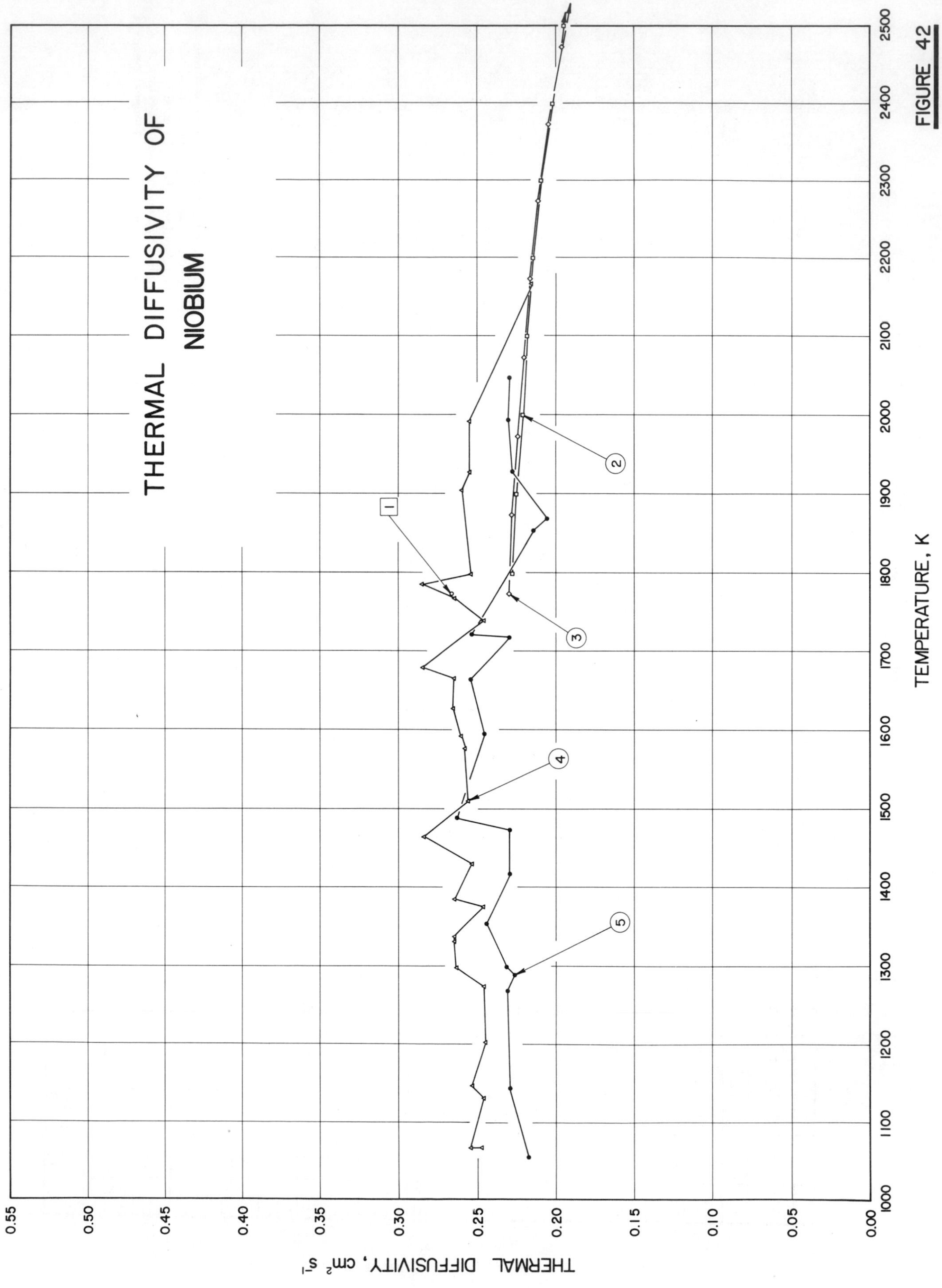
THERMAL DIFFUSIVITY OF
NIOBIUM
THERMAL DIFFUSIVITY, cm² s⁻¹
TEMPERATURE, K
0.00
0.05
0.10
0.15
0.20
0.25
0.30
0.35
0.40
0.45
0.50
0.55
1000
1100
1200
1300
1400
1500
1600
1700
1800
1900
2000
2100
2200
2300
2400
2500
1
2
3
4
5

FIGURE 42

SPECIFICATION TABLE 42. THERMAL DIFFUSIVITY OF NIOBIUM

(Impurity <0.20% each; total impurities <0.50%)

Cur. No.	Ref. No.	Author(s)	Year	Temp. Range, K	Reported Error, %	Name and Specimen Designation	Composition (weight percent), Specifications, and Remarks
1	40	Mustacchi, C. and Giuliani, S.	1963	1773			Specimen measured for diffusivity using four different methods: amplitude, output face phase shift, differential phase shift, and pulse lag methods; 10 measurements made with each method.
2	122	Kraev, O.A. and Stel'makh, A.A.	1964	1800-2600	±5		Disk specimen 8-9 mm in dia and 0.3 mm thick; diffusivity measured using wave method; data points corrected for thermal expansion.
3	172	Kraev, O.A. and Stel'makh, A.A.	1966	1773-2573	±5		Disk specimen 7 to 9 mm in diameter and 0.3 mm thick; cut from plate of rolled metal; surfaces thoroughly cleaned; heated by electron bombardment; sinusoidal temperature oscillation imposed on one face of specimen; thermal diffusivity determined from phase difference between measured temperature fluctuations at front and back faces of specimen; reported data points corrected for thermal expansion and obtained from smooth curve given by authors.
4	208	Wheeler, M.J.	1970	1067-2167		No. 1	Disk specimen 1 cm in diameter, 0.125 cm in thickness; cut from arc-melted material; diffusivity measured using modulated electron beam technique.
5	208	Wheeler, M.J.	1970	1055-2047		No. 2	Similar to the above specimen.

DATA TABLE 42. THERMAL DIFFUSIVITY OF NIOBIUM

(Impurity <0.20% each; total impurities <0.50%)

[Temperature, T, K; Thermal Diffusivity, α, $cm^2\ s^{-1}$]

T	α
CURVE 1	
1773	0.266
CURVE 2	
1800	0.228
1900	0.225
2000	0.221
2100	0.218
2200	0.214
2300	0.209
2400	0.202
2500	0.194
2600	0.183*
CURVE 3	
1773	0.230
1873	0.228
1973	0.224
2073	0.220
2173	0.216
2273	0.211
2373	0.204
2473	0.196
2573	0.186
CURVE 4	
1067	0.2474
1067	0.2540
1131	0.2[illegible]
1147	0.25[illegible]
CURVE 4 (cont.)	
1203	0.2454
1274	0.2464
1298	0.2630
1318	0.2641
1338	0.2646
1375	0.2479
1385	0.2642
1429	0.2535
1464	0.2840
1509	0.2559
1576	0.2577
1592	0.2600
1627	0.2653
1665	0.2642
1679	0.2843
1739	0.2460
CURVE 4 (cont.)	
1767	0.2643
1785	0.2848
1798	0.2530
1905	0.2593
1928	0.2547
1992	0.2547
2167	0.2153
CURVE 5	
1055	0.2179
1144	0.2292
1269	0.2306
1289	0.2261
1299	0.2315
1355	0.2444
CURVE 5 (cont.)	
1417	0.2293
1473	0.2293
1488	0.2627
1595	0.2455
1664	0.2536
1718	0.2295
1721	0.2632
1853	0.2147
1868	0.2053
1928	0.2278
1994	0.2300
2047	0.2293

* Not shown in figure.

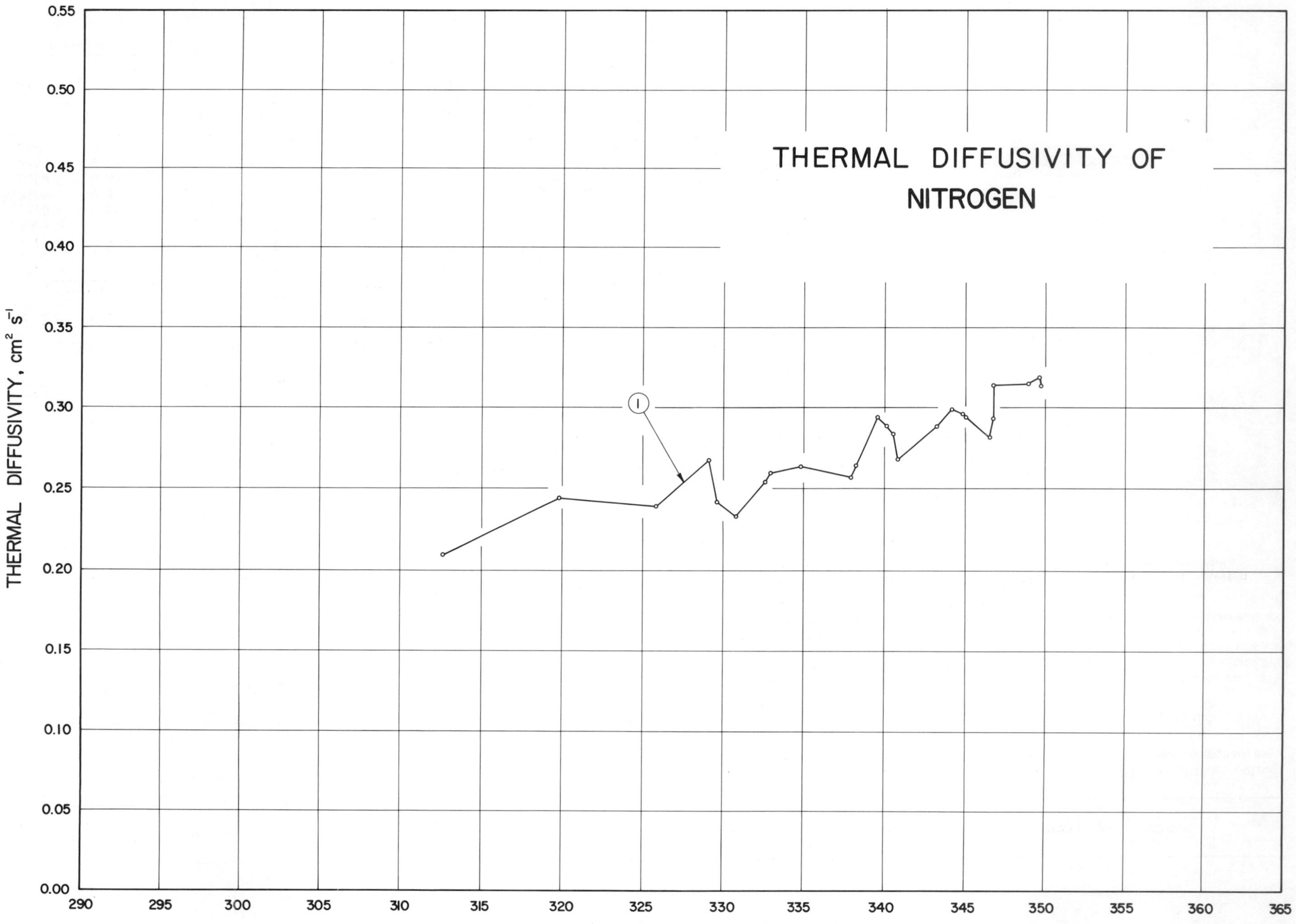
THERMAL DIFFUSIVITY OF
NITROGEN
THERMAL DIFFUSIVITY, cm² s⁻¹
TEMPERATURE, K
0.00
0.05
0.10
0.15
0.20
0.25
0.30
0.35
0.40
0.45
0.50
0.55
290
295
300
305
310
315
320
325
330
335
340
345
350
355
360
365
1

FIGURE 43

SPECIFICATION TABLE 43. THERMAL DIFFUSIVITY OF NITROGEN

(Impurity <0. 20% each; total impurities <0. 50%)

Cur. No.	Ref. No.	Author(s)	Year	Temp. Range, K	Reported Error, %	Name and Specimen Designation	Composition (weight percent), Specifications, and Remarks
1	169	Harrison, W. B., Boteler, W. C., and Spurlock, J. M.	1959	313-350			Experimental system consists of a type 310 stainless steel tube 0. 944 in. I. D., 1 in. O. D., and 33 in. long, surrounded by a concentric copper tube 3. 75 in. I. D., 4 in. O. D., and 30 in. long; assembly mounted vertically and annular space between the tubes filled with water; central tube heated by passing electrical current through tube wall; temp. fluctuations measured by means of two platinum wires each 0. 003 in. in dia, one of which located at the center of the tube and the other at a distance of 0. 425 in. from the center; current through test section controlled by a special function generator; each data point reported is the result of an independent series of data runs initiated by exhausting the test section tube with a vacuum pump after which the tube is purged several times and finally filled with nitrogen; technique consists of imposing a sinusoidal temp wave in the central tube and measuring the phase shift between the response curves of the two platinum wires; each data point corrected to adjust the experimental values to a pressure of one atmosphere.

DATA TABLE 43. THERMAL DIFFUSIVITY OF NITROGEN

(Impurity <0. 20% each; total impurities <0. 50%)

[Temperature, T, K; Thermal Diffusivity, α, $cm^2\ s^{-1}$]

T	α	T	α
CURVE 1		CURVE 1 (cont.)	
312. 6	0. 209	344. 1	0. 2991
319. 8	0. 244	344. 8	0. 2965
325. 8	0. 239	345. 0	0. 2947
329. 1	0. 2676	346. 5	0. 2821
329. 6	0. 242	346. 7	0. 2937
330. 8	0. 233	346. 7	0. 3146
332. 6	0. 254	348. 9	0. 3159
332. 9	0. 2596	349. 6	0. 3197
334. 8	0. 2632	349. 7	0. 3141
337. 9	0. 257		
338. 2	0. 2643		
339. 5	0. 2947		
340. 1	0. 2898		
340. 5	0. 2841		
340. 8	0. 2681		
343. 2	0. 2890		

FIGURE AND TABLE 44R. PROVISIONAL THERMAL DIFFUSIVITY OF OSMIUM

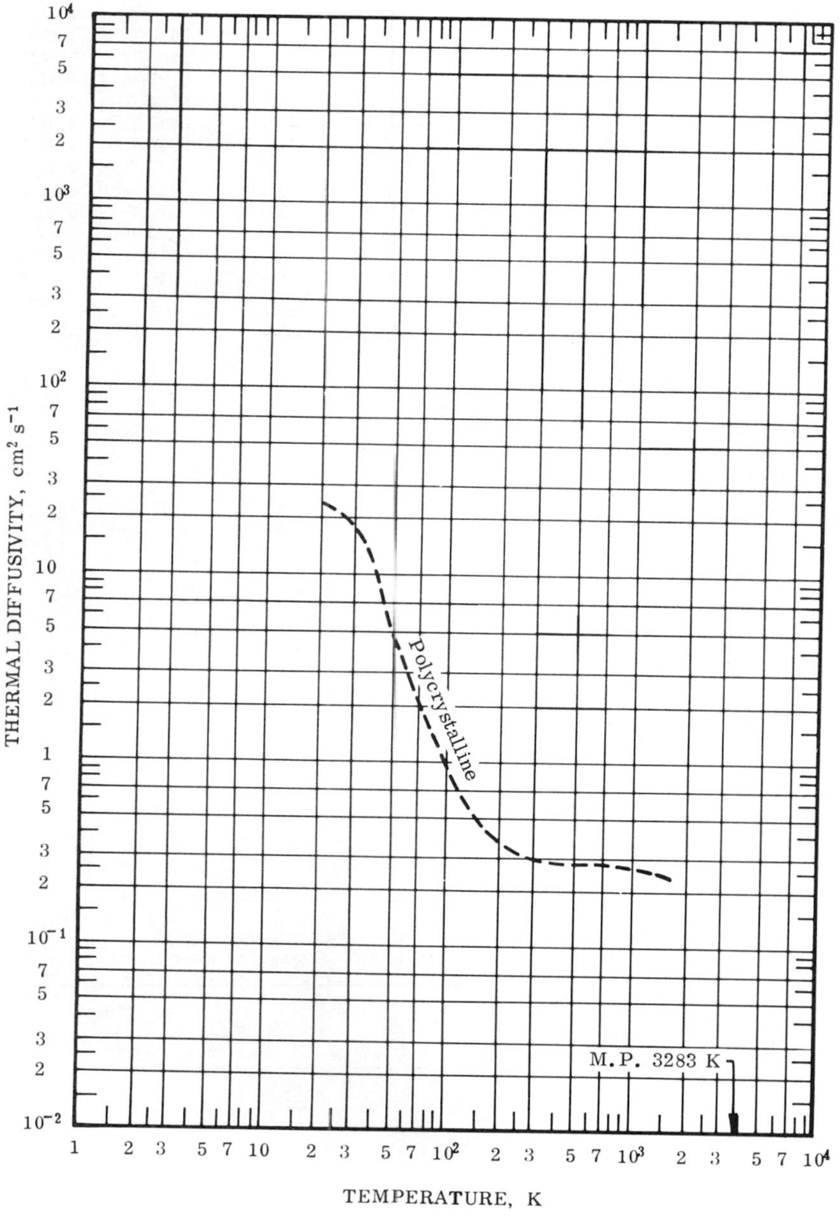

PROVISIONAL VALUES*

[Temperature, T, K; Thermal Diffusivity, α, $cm^2\ s^{-1}$]

SOLID
(Polycrystalline)

T	α	T	α
20	23.5	273.2	0.306
25	22.1	300	0.300
30	18.2	350	0.297
35	13.5	400	0.294
40	9.38	500	0.290
45	6.75	600	0.286
50	5.03	700	0.281
60	3.12	800	0.277
70	2.12	900	0.273
80	1.54	1000	0.269
90	1.17	1100	0.265
100	0.932	1200	0.261
150	0.468	1300	0.257
200	0.353	1400	0.253
250	0.315	1500	0.249
		1600	0.245

REMARKS

The values are for well-annealed high-purity polycrystalline osmium and are thought to be accurate to within ± 15% of the true values at temperatures below 250 K, ± 8% from 250 to 500 K, and ± 15% above 500 K. Those from 250 to 500 K are recommended values. At low temperatures the values are highly conditioned by impurity and imperfection, and those below 150 K are applicable only to osmium having residual electrical resistivity of 0.0234 $\mu\Omega$ cm.

*All values are estimated and those from 250 to 500 K are recommended values.

FIGURE AND TABLE 45R. RECOMMENDED THERMAL DIFFUSIVITY OF PALLADIUM

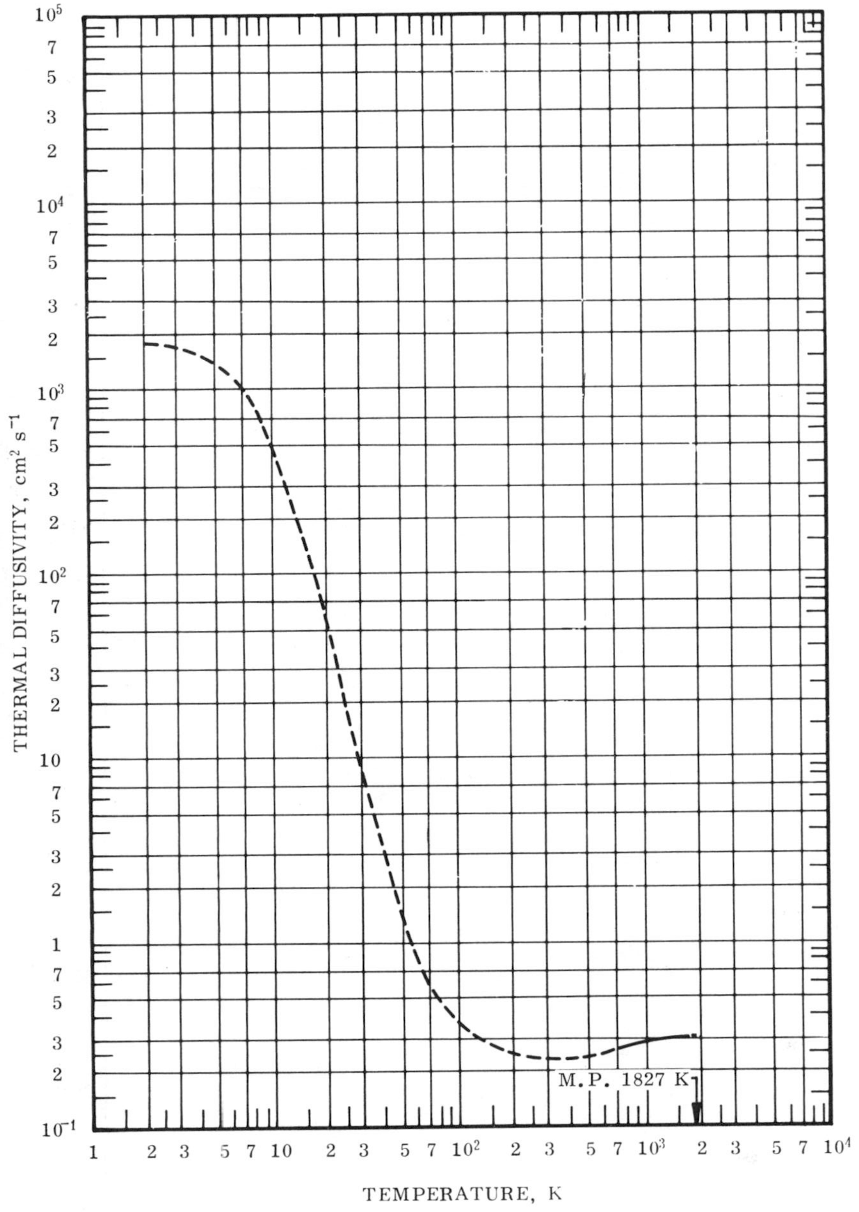

RECOMMENDED VALUES

[Temperature, T, K; Thermal Diffusivity, α, cm² s⁻¹]

SOLID

T	α	T	α
2	1750*	80	0.476*
3	1660*	90	0.416*
4	1530*	100	0.377*
5	1370*	150	0.289*
6	1170*	200	0.261*
7	968*	250	0.249*
8	785*	273.2	0.246*
9	620*	300	0.245*
10	488*	350	0.244*
11	396*	400	0.246*
12	305*	500	0.251*
13	240*	600	0.257*
14	190*	700	0.265
15	153*	800	0.271
16	123*	900	0.279
18	79.0*	1000	0.287
20	52.6*	1100	0.294
25	20.3*	1200	0.301
30	8.80*	1300	0.306
35	4.61*	1400	0.310
40	2.78*	1500	0.313
45	1.85*	1600	0.314
50	1.32*	1700	0.314
60	0.795*	1800	0.314*
70	0.583*		

REMARKS

The recommended values are for well-annealed high-purity palladium and are considered accurate to within ±7% of the true values at temperatures from room temperature to about 1000 K and ±13% below room temperature and above 1000 K. At low temperatures the values are highly conditioned by impurity and imperfection, and those below 150 K are applicable only to palladium having residual electrical resistivity of 0.0123 μΩ cm.

*In temperature range where no experimental data are available.

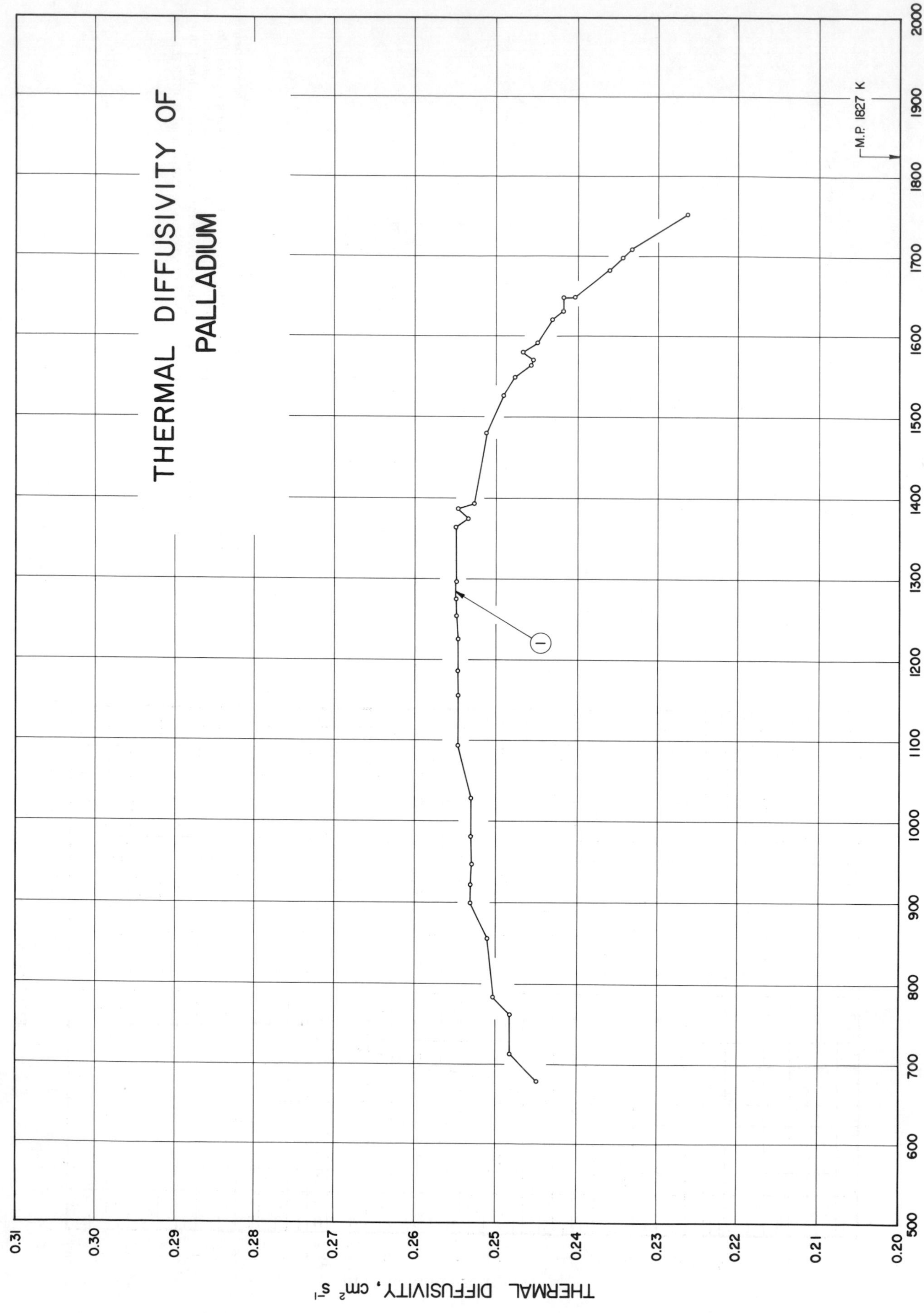
THERMAL DIFFUSIVITY OF PALLADIUM
THERMAL DIFFUSIVITY, cm² s⁻¹
TEMPERATURE, K
M.P. 1827 K
1
0.20
0.21
0.22
0.23
0.24
0.25
0.26
0.27
0.28
0.29
0.30
0.31
500
600
700
800
900
1000
1100
1200
1300
1400
1500
1600
1700
1800
1900
2000

FIGURE 45

SPECIFICATION TABLE 45. THERMAL DIFFUSIVITY OF PALLADIUM

(Impurity <0.20% each; total impurities <0.50%)

Cur. No.	Ref. No.	Author(s)	Year	Temp. Range, K	Reported Error, %	Name and Specimen Designation	Composition (weight percent), Specifications, and Remarks
1	236	Zinov'ev, V. E., Krentsis, R. P., and Gel'd, P. V.	1969	679-1752	<5		Spectroscopically pure specimen; 8 x 8 mm^2 cross-section, 0.313 mm thick; annealed for 5 hr in vacuum ($1 \cdot 10^{-5}$ mm Hg) at 1200 K and then heated to 1700 K; electrical resistivity ratio $\rho(293K)/\rho(4.2K) \approx 70$.

DATA TABLE 45. THERMAL DIFFUSIVITY OF PALLADIUM

(Impurity <0.20% each; total impurities < 0.50%)

[Temperature, T, K; Thermal Diffusivity, α, $cm^2\ s^{-1}$]

T	α	T	α	T	α
CURVE 1		CURVE 1 (cont.)		CURVE 1 (cont.)	
679	0.2449	1275	0.2549	1648	0.2417
713	0.2482	1297	0.2549	1648	0.2403
762	0.2482	1364	0.2549	1682	0.2360
784	0.2503	1375	0.2534	1697	0.2344
856	0.2510	1387	0.2545	1708	0.2333
900	0.2531	1393	0.2526	1752	0.2263
922	0.2530	1480	0.2512		
947	0.2529	1526	0.2491		
981	0.2530	1549	0.2477		
1028	0.2530	1564	0.2457		
1093	0.2547	1571	0.2454		
1155	0.2547	1580	0.2467		
1186	0.2547	1592	0.2449		
1225	0.2547	1620	0.2430		
1254	0.2549	1631	0.2417		

FIGURE AND TABLE 46R. PROVISIONAL THERMAL DIFFUSIVITY OF PHOSPHORUS (WHITE)

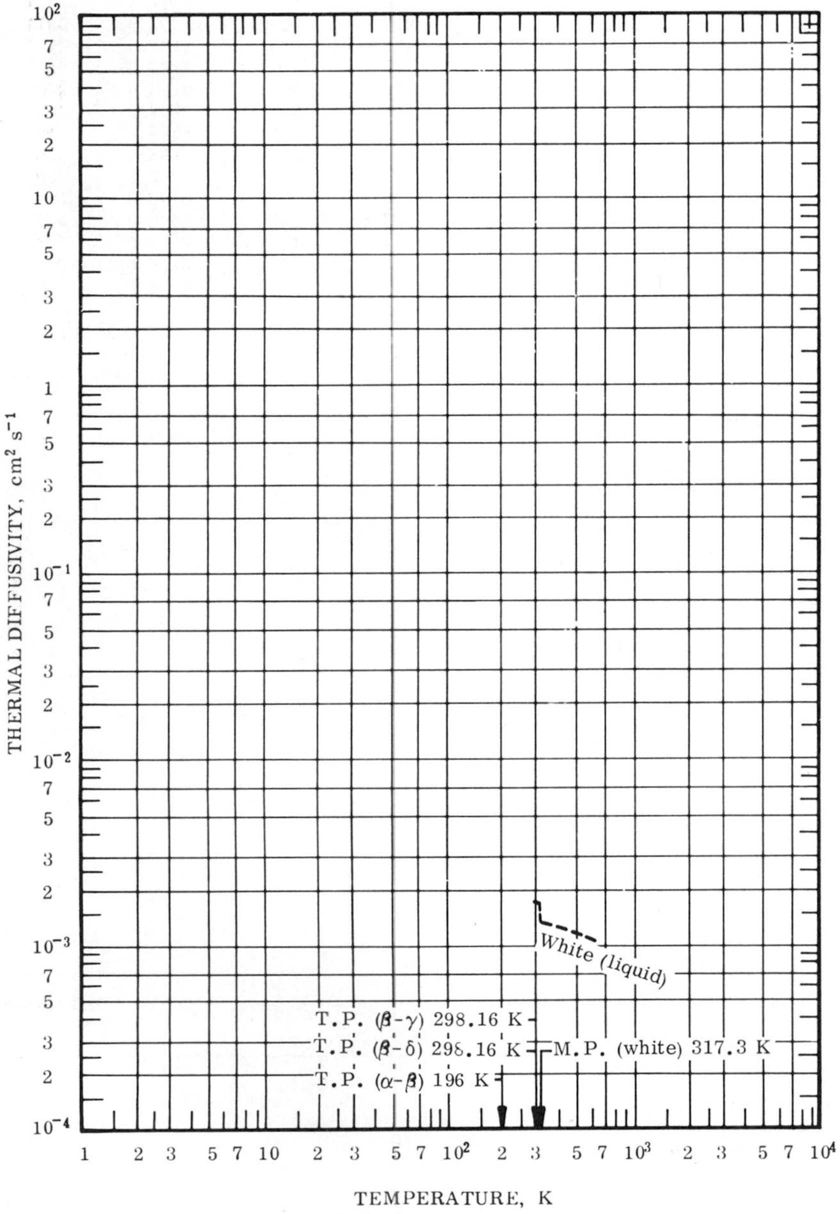

PROVISIONAL VALUES*

[Temperature, T, K; Thermal Diffusivity, α, $cm^2\ s^{-1}$]

SOLID		LIQUID	
T	α	T	α
298.2	0.00170	317.3	0.00135
		350	0.00132
		400	0.00126
		500	0.00116
		600	0.00106

REMARKS

The provisional values are for high-purity white phosphorus and are thought to be accurate to ± 18%.

*All values are estimated.

FIGURE AND TABLE 47R. RECOMMENDED THERMAL DIFFUSIVITY OF PLATINUM

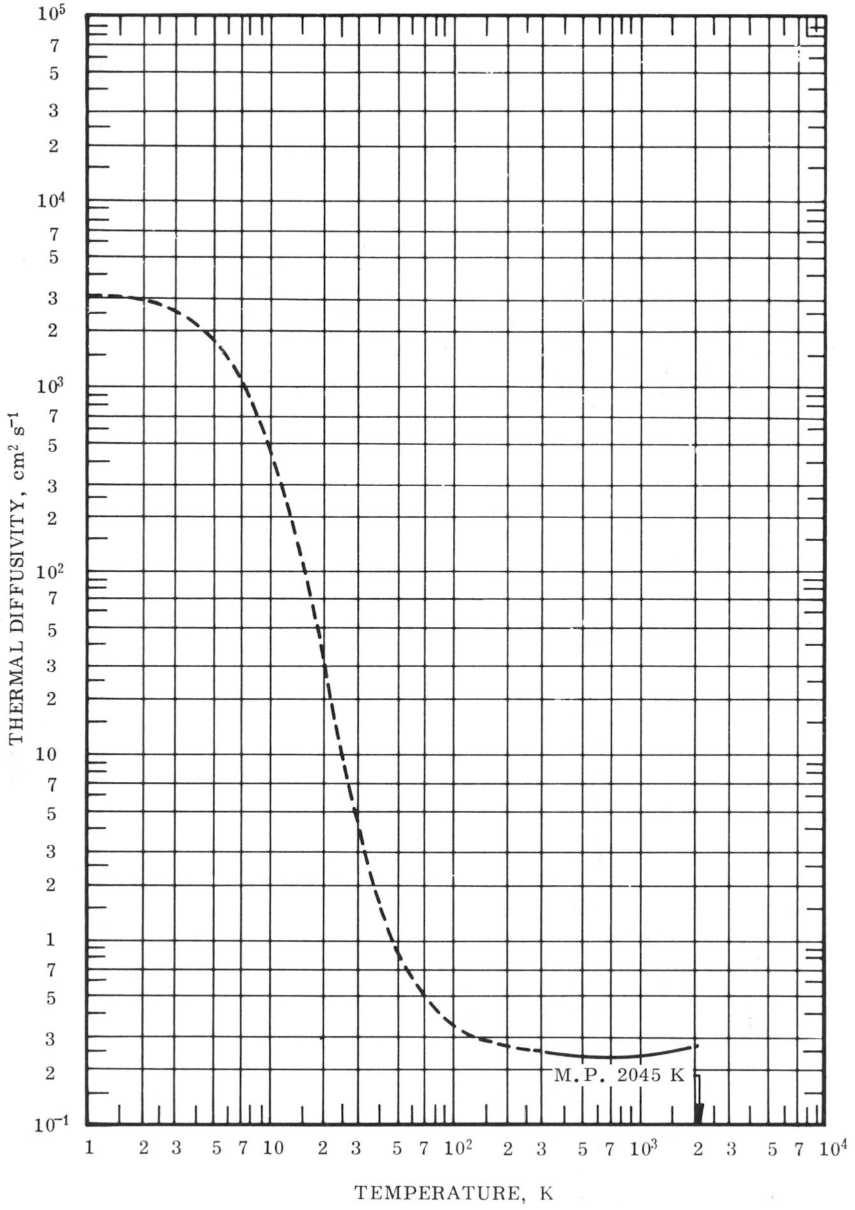

RECOMMENDED VALUES

[Temperature, T, K; Thermal Diffusivity, α, cm² s⁻¹]

SOLID

T	α	T	α
1	3050 *	70	0.513 *
2	2850 *	80	0.437 *
3	2550 *	90	0.389 *
4	2190 *	100	0.359 *
5	1800 *	150	0.291 *
6	1430 *	200	0.268 *
7	1120 *	250	0.257 *
8	866 *	273.2	0.254 *
9	654 *	300	0.252
10	484 *	350	0.249
11	358 *	400	0.247
12	263 *	500	0.245
13	195 *	600	0.244
14	145 *	700	0.244
15	110 *	800	0.244
16	82.5 *	900	0.245
18	48.7 *	1000	0.246
20	29.8 *	1100	0.248
25	10.6 *	1200	0.251
30	4.62 *	1300	0.254
35	2.55 *	1400	0.258
40	1.65 *	1500	0.262
45	1.18 *	1600	0.266
50	0.920 *	1700	0.269
60	0.648 *	1800	0.271
		1900	0.273
		2000	0.274 *
		2045	0.274 *

REMARKS

The recommended values are for well-annealed high-purity platinum and are considered accurate to within ± 5% of the true values near room temperature, ± 8% at about 100 K and 1200 K, ± 12% below 100 K and ± 15% at 2000 K. At low temperatures the values are highly conditioned by impurity and imperfection, and those below 200 K are applicable only to platinum having residual electrical resistivity of 0.0106 μΩ cm.

* In temperature range where no experimental data are available.

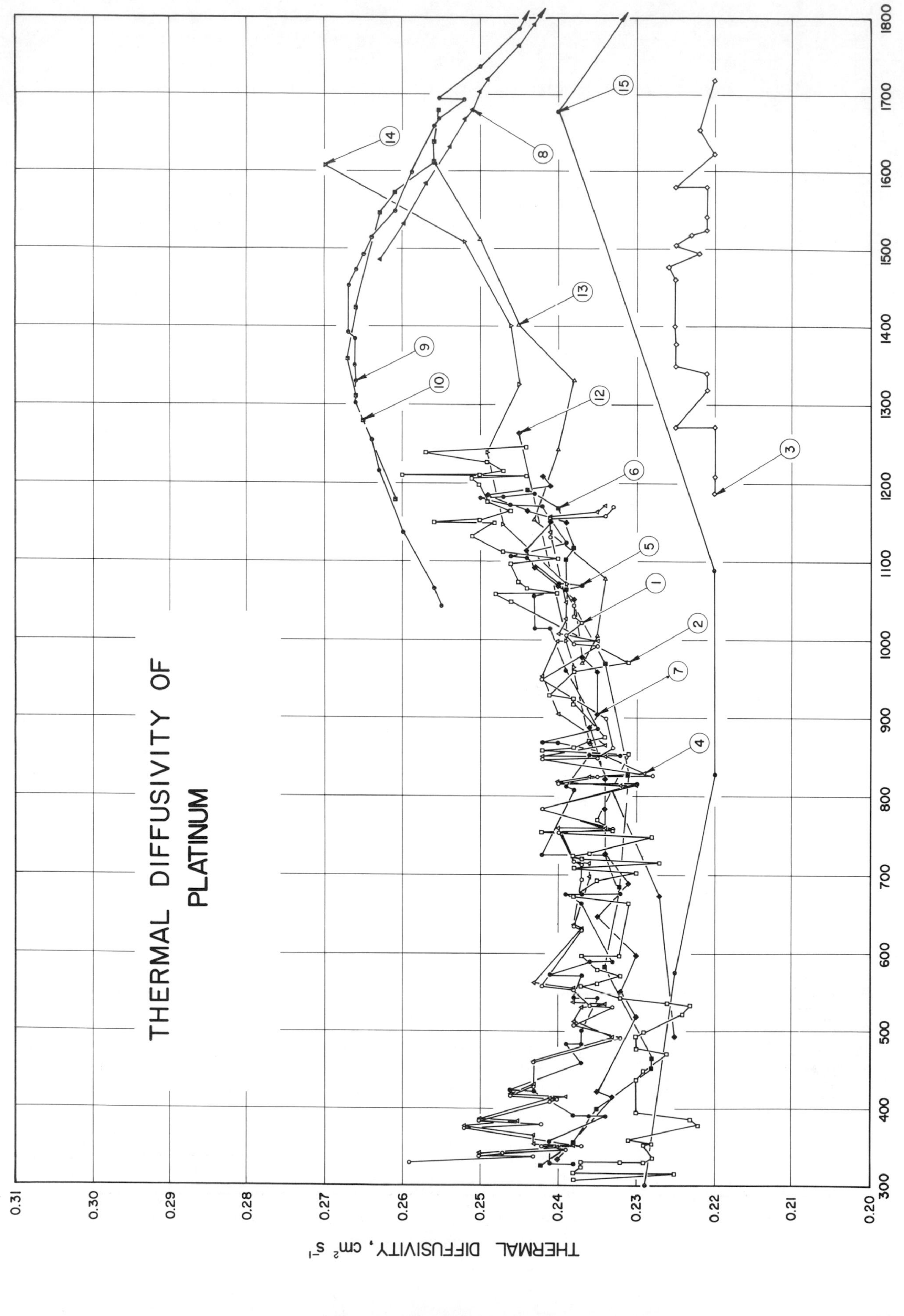
THERMAL DIFFUSIVITY OF
PLATINUM
THERMAL DIFFUSIVITY, cm² s⁻¹
TEMPERATURE, K
0.20
0.21
0.22
0.23
0.24
0.25
0.26
0.27
0.28
0.29
0.30
0.31
300
400
500
600
700
800
900
1000
1100
1200
1300
1400
1500
1600
1700
1800
1
2
3
4
5
6
7
8
9
10
12
13
14
15

FIGURE 47

SPECIFICATION TABLE 47. THERMAL DIFFUSIVITY OF PLATINUM

(Impurity < 0.20% each; total impurities < 0.50%)

Cur. No.	Ref. No.	Author(s)	Year	Temp. Range, K	Reported Error, %	Name and Specimen Designation	Composition (weight percent), Specifications, and Remarks
1	42	Martin, J.J. and Sidles, P.H.	1964	329-1170		I	99.999 pure (prior to fabrication); cylindrical specimen 3/16 in. in dia; supplied in the form of rods by J. Bishop and Co.; held at 1200 K for at least one hr; electrical resistivity ratio ρ_{273K}/ρ_{4K} = 900, measured following completion of thermal diffusivity measurements; sinusoidal temp. variation at one end of specimen with lateral heat losses; period of sinusoidal boundary condition 60 sec for most measurements and 120 sec for a few points; max amplitude of temp oscillations in specimen not exceeding 5 K; specimen measured in vacuum except at the lower temp. where a helium atmosphere was used.
2	42	Martin, J.J. and Sidles, P.H.	1964	306-1246		II	99.9 pure (prior to fabrication); cylindrical specimen 3/16 in. in dia; supplied in the form of rods by J. Bishop and Co.; held at 1200 K for at least one hr; electrical resistivity ratio ρ_{273K}/ρ_{4K} = 12, measured following completion of thermal diffusivity measurements; sinusoidal temp. variation at one end of specimen with lateral heat losses; period of sinusoidal boundary condition 60 sec for most measurements and 120 sec for a few points; max amplitude of temp. oscillations in specimen not exceeding 5 K; specimen measured in vacuum except at the lower temps. where a helium atmosphere was used.
3	39	Wheeler, M.J.	1965	1187-1719			99.95 pure; avg grain size after testing 1000 μm; rectangular specimen having top surface area lying in the range from 0.3 to 1.3 cm^2 and 1 mm in thickness; specimen cut from sheet of same thickness supplied by Johnson Matthey; density 21.5 g cm^{-3}, Lorenz function reported as 2.584, 2.602, 2.595, 2.591, 2.591, 2.601, 2.601, and 2.601 x 10^{-8} V^2 K^{-2} at 1172, 1200, 1272, 1370, 1471, 1572, 1675, and 1770 K, respectively; specimen heated to incandescence by an electron beam; intensity of beam sinusoidally modulated at a frequency of 0.48 cycles per sec, modulation used produced temp fluctuations in the bombarded face of the specimen of from ± 5 to ± 20 K; measured under a vacuum of ~5 x 10^{-6} mm Hg.
4	43	Martin, J.J., Sidles, P.H., and Danielson, G.C.	1965	341-1172		A	99.999 pure; cylindrical specimen 3/16 in. in dia. and from 10 to 12 in. long; fabricated and supplied by J. Bishop and Co.; annealed at 1200 K for at least one hr; electrical resistivity ratio $\rho_{273K}/\rho_{4.2K}$ = 900 (measured upon completion of thermal diffusivity measurements); electrical resistivity measured immediately following each measurement of diffusivity and reported as 10.90, 14.72, 18.41, 22.00, 25.50, 28.88 32.11, 35.25, and 38.25 μohm cm at 300, 400, 500, 600, 700, 800, 900, 1000, and 1100 K, respectively; Lorenz number reported as 2.54, 2.58, 2.61, 2.63, 2.64, 2.66, 2.68, 2.71, and 2.73 x 10^{-8} V^2 K^{-2} at 300, 400, 500, 600, 700, 800, 900, 1000, and 1100 K, respectively; most measurements made with specimen in a helium atmosphere and with a 60 sec sinusoidal boundary condition; a few measurements at the higher temps. made with specimen in vacuum; a few measurements also made with a 120 sec boundary condition.

SPECIFICATION TABLE 47. THERMAL DIFFUSIVITY OF PLATINUM (continued)

Cur. No.	Ref. No.	Author(s)	Year	Temp. Range, K	Reported Error, %	Name and Specimen Designation	Composition (weight percent), Specifications, and Remarks
5	43	Martin, J.J., Sidles, P.H., and Danielson, G.C.	1965	327-1186		B	99.999 pure; cylindrical specimen 3/16 in. in dia. and from 10 to 12 in. long; fabricated and supplied by Engelhard Ind.; annealed at 1200 K for at least one hr; electrical resistivity ratio $\rho_{273K}/\rho_{4.2K}$ = 100 (measured upon completion of thermal diffusivity measurements); electrical resistivity measured immediately following each measurement of diffusivity and reported as 10.95, 14.75, 18.45, 22.10, 25.64, 29.00, 32.20, 35.35, and 38.45 μohm cm at 300, 400, 500, 600, 700, 800, 900, 1000, and 1100 K, respectively; Lorenz number reported as 2.46, 2.52, 2.58, 2.62, 2.65, 2.68, 2.70, 2.74, and 2.78 x 10^{-8} V^2 K^{-2} at 300, 400, 500, 600, 700, 800, 900, 1000, and 1100 K, respectively; most measurements made with specimen in a helium atmosphere and with a 60 sec sinusoidal boundary condition; a few measurements at the higher temps. made with specimen in vacuum; a few measurements also made with a 120 sec boundary condition.
6	43	Martin, J.J., et al.	1965	326-1191		C	99.9 pure; cylindrical specimen 3/16 in. in dia. and from 10 to 12 in. long; fabricated and supplied by Engelhard Ind.; annealed at 1200 K for at least one hr; electrical resistivity ratio $\rho_{273K}/\rho_{4.2K}$ = 34 (measured upon completion of thermal diffusivity measurements); electrical resistivity measured immediately following each measurement of diffusivity and reported as 11.30, 15.13, 18.90, 22.60, 26.14, 29.51, 32.76, 35.85, and 38.89 μohm cm at 300, 400, 500, 600, 700, 800, 900, 1000, and 1100 K, respectively; Lorenz number reported as 2.55, 2.55, 2.59, 2.62, 2.65, 2.68, 2.71, 2.74, and 2.78 x 10^{-8} V^2 K^{-2} at 300, 400, 500, 600, 700, 800, 900, 1000, and 1100 K, respectively; most measurements made with specimen in a helium atmosphere and with a 60 sec sinusoidal boundary condition; a few measurements at the higher temps. made with specimen in vacuum; a few measurements also made with a 120 sec boundary condition.
7	43	Martin, J.J., et al.	1965	333-1208		D	99.999 pure; cylindrical specimen 3/16 in. in dia. and from 10 to 12 in. long; fabricated and supplied by Sigmund Cohn; annealed at 1200 K for at least one hr; electrical resistivity ratio $\rho_{273K}/\rho_{4.2K}$ = 5000 (measured upon completion of thermal diffusivity measurements); electrical resistivity measured immediately following each measurement of diffusivity and reported as 10.90, 14.68, 18.40, 21.98, 25.45, 28.82, 32.04, 35.10, and 38.13 μohm cm at 300, 400, 500, 600, 700, 800, 900, 1000, and 1100 K, respectively; Lorenz number reported as 2.47, 2.52, 2.56, 2.59, 2.61, 2.64, 2.67, 2.73, and 2.74 x 10^{-8} V^2 K^{-2} at 300, 400, 500, 600, 700, 800, 900, 1000, and 1100 K, respectively; most measurements made with specimen in a helium atmosphere and with a 60 sec sinusoidal boundary condition; a few measurements at the higher temps. made with specimen in vacuum; a few measurements also made with a 120 sec boundary condition.

SPECIFICATION TABLE 47. THERMAL DIFFUSIVITY OF PLATINUM (continued)

Cur. No.	Ref. No.	Author(s)	Year	Temp. Range, K	Reported Error, %	Name and Specimen Designation	Composition (weight percent), Specifications, and Remarks
8	171	Zinov'ev, V. E., Krentsis, R. P., and Gel'd, P. V.	1969	1485-1807	<5	1	Square specimen 8 x 8 x 0.193 mm; cut from sheet of grade PL-O pure platinum; annealed in vacuum chamber of measuring apparatus under a pressure of 10^{-5} mm Hg for 5 hr at 1973.2 K and then heated to 2123.2 K for a short time; electrical resistivity ratio $\rho(298K)/\rho(4.2K) \approx 200$, $\rho(373K)/\rho(273K) = 1.3926$; measured using radio-frequency technique; diffusivity determined from measured phase shift between temperature waves on the surfaces of specimen.
9	171	Zinov'ev, V. E., et al.	1969	1042-1998	<5	2	Square specimen 8 x 8 x 0.245 mm; other specifications and conditions same as above.
10	171	Zinov'ev, V. E., et al.	1969	1178-1679	<5	3	Square specimen 8 x 8 x 0.288 mm; other specifications and conditions same as above.
11*	203	Martin, J. J., Sidles, P. H., and Danielson, G. C.	1967	303-1234		Sample 5	99.9 pure; rod specimen 0.48 cm in diameter and 30 cm long; fabricated and supplied by J. Bishop and Co., Malvern, Pa.; electrical resistivity ratio $\rho(273K)/\rho(4.2K) = 12$; electrical resistivity measured immediately following each measurement of diffusivity by a four-wire technique; most measurements made with specimen in a helium atmosphere and with a 2.5 C amplitude 60-sec period sinusoidal wave but some made with specimen in vacuum and some with sinusoidal temperature wave having an 120-sec period; diffusivity determined by modified Angström method.
12	235	Branscomb, I. M.	1970	493-1263			0.5 in. diameter disk specimen obtained from Engelhard Industries; grain size 1 mm; annealed at 1500 C for 1 hr.
13	237	Ciszek, T. F.	1966	970-1611			<0.2 each of Au, Rh, and Ru, <0.1 each of Cu, Fe, Pd, Si, and Ag, and <0.01 each of Cr, Mn, and Ni; 0.479 cm diameter x 15 cm long; obtained from Engelhard Industries; residual electrical resistivity 10.16 $\mu\Omega$cm; electrical resistivity ratio $\rho(273K)/\rho(77K) = 4.7$, $\rho(273K)/\rho(4.2K) = 34$; electrical resistivity 15.08, 22.45, 29.37, 35.85, 41.87, 47.45, 52.58, and 57.25 $\mu\Omega$ cm at 400, 600, 800, 1000, 1200, 1400, 1600, and 1800 K, respectively; thermal diffusivity data taken with first and second thermocouples.
14	237	Ciszek, T. F.	1966	965-1607			The above specimen; data taken with first and third thermocouples.
15	267	Rawuka, A. C. and Gaz, R. A.	1969	300-1662			Commercial grade; 0.15 cm in thickness; supplied by Wilkinson Co., Westlake Village, Calif.; density 21.50 g cm^{-3}; diffusivity measured using pulse technique.

* Not shown in figure.

DATA TABLE 47. THERMAL DIFFUSIVITY OF PLATINUM

(Impurity <0.20% each; total impurities <0.50%)

[Temperature, T, K; Thermal Diffusivity, α, $cm^2\ s^{-1}$]

T	α	T	α	T	α	T	α	T	α	T	α	T	α	T	α
CURVE 1		CURVE 1 (cont.)		CURVE 2 (cont.)		CURVE 2 (cont.)		CURVE 4 (cont.)		CURVE 5 (cont.)		CURVE 6		CURVE 8	
329	0.259	1021	0.237	693	0.235	1239	0.257	431	0.243	356	0.241	326	0.242	1485	0.263
338	0.250	1030	0.238	703	0.230	1246	0.244	461	0.243	390	0.234	400	0.235	1531	0.260
338	0.243	1043	0.238	709	0.238			493	0.233	390	0.236	453	0.228	1583	0.257
340	0.247	1131	0.241	715	0.227	CURVE 3		512	0.238	391	0.238	466	0.228	1631	0.254
347	0.239	1154	0.241	720	0.237			531	0.237	423	0.243	583	0.234	1667	0.252
351	0.237	1158	0.234	724	0.238	1187	0.220	535	0.234	423	0.246	684	0.232	1679	0.251
351	0.242	1170	0.233	727	0.236	1208	0.220	537	0.238	459	0.237	826	0.231	1703	0.250
374	0.252			748	0.228	1271	0.220	556	0.238	483	0.239	969	0.234	1719	0.249
380	0.242	CURVE 2		754	0.242	1271	0.225	562	0.243	483	0.237	1063	0.239	1763	0.245
384	0.250			756	0.233	1318	0.221	631	0.237	501	0.237	1102	0.239	1791	0.243
409	0.241	306	0.238	769	0.235	1340	0.221	637	0.238	543	0.235	1117	0.238	1807	0.242*
412	0.240	315	0.225	854	0.231	1349	0.225	698	0.236	543	0.238	1167	0.240		
416	0.246	316	0.238	858	0.242	1377	0.225	714	0.236	572	0.237	1191	0.244	CURVE 9	
429	0.243	324	0.237	862	0.238	1399	0.225	717	0.238*	573	0.241				
460	0.243	330	0.237	870	0.236	1460	0.225	760	0.240	589	0.236	CURVE 7		1042	0.255
492	0.232	330	0.232	875	0.234	1475	0.226	760	0.234	589	0.233			1065	0.256
508	0.238	330	0.229	917	0.238	1493	0.222	784	0.242*	664	0.237	333	0.240	1137	0.260
532	0.233	335	0.228	924	0.238	1504	0.225	814	0.232	675	0.239	416	0.233	1215	0.263
533	0.236	352	0.229	928	0.241	1517	0.223	814	0.230	676	0.237	423	0.235	1255	0.264
554	0.238	354	0.228	959	0.238	1523	0.221	817	0.240	676	0.232	519	0.230	1302	0.266
558	0.242	359	0.231	971	0.231	1540	0.221	825	0.236	725	0.234	551	0.232	1329	0.266
630	0.237	378	0.222	1048	0.246	1579	0.221	829	0.229	725	0.242	598	0.230	1349	0.266
635	0.238	386	0.223	1058	0.248	1579	0.225	851	0.234	808	0.238	646	0.235	1383	0.266
694	0.237	395	0.230	1059	0.240	1622	0.220	851	0.242	812	0.239	688	0.231	1392	0.267
713	0.237	437	0.230	1066	0.244	1654	0.222	865	0.234	852	0.236	726	0.234	1453	0.267
717	0.238	449	0.229	1074	0.245	1719	0.220	905	0.240	852	0.232	784	0.234	1471	0.266
753	0.240	471	0.226	1097	0.246			953	0.242	868	0.240	821	0.234	1492	0.265
759	0.233	478	0.230	1104	0.240	CURVE 4		998	0.240	868	0.242	887	0.236	1514	0.264
783	0.242	494	0.230	1112	0.247			998	0.235	885	0.235	904	0.235	1548	0.261
815	0.240	500	0.229	1132	0.251	341	0.250	1007	0.240	960	0.239	959	0.235	1599	0.259
815	0.232	522	0.224	1149	0.248	342	0.247	1026	0.239	1014	0.241	977	0.237	1658	0.256
815	0.230	533	0.223	1150	0.256	350	0.240	1047	0.239	1014	0.243	1051	0.238	1668	0.254
825	0.235	537	0.226	1152	0.250	352	0.238	1137	0.241	1056	0.243	1091	0.243	1694	0.254
826	0.228	543	0.232	1165	0.246	354	0.243	1157	0.241	1069	0.237	1113	0.244	1693	0.252
847	0.242	557	0.237	1176	0.249	365	0.243	1163	0.235	1071	0.240	1150	0.241	1736	0.250
848	0.235	562	0.235	1198	0.250	376	0.252	1172	0.234	1104	0.244	1150	0.239	1786	0.245
861	0.233	572	0.232	1206	0.251	380	0.242*			1107	0.246	1163	0.244	1822	0.243*
899	0.239	579	0.235	1209	0.244	384	0.245	CURVE 5		1123	0.239	1184	0.249	1848	0.240*
949	0.242	597	0.237	1210	0.260	385	0.250			1170	0.242	1196	0.241	1871	0.239*
992	0.235	597	0.232	1211	0.250	413	0.241	327	0.238	1172	0.246	1208	0.242	1883	0.236*
994	0.238	664	0.231	1215	0.247	415	0.239	328	0.241	1181	0.250			1898	0.236*
1005	0.239	672	0.238	1227	0.249	418	0.246	355	0.238	1182	0.247			1908	0.234*
										1186	0.243				

*Not shown in figure.

DATA TABLE 47. THERMAL DIFFUSIVITY OF PLATINUM (continued)

T	α
CURVE 9 (cont.)	
1931	0.233*
1956	0.229*
1979	0.225*
1998	0.222*
CURVE 10	
1178	0.261
1279	0.265
1310	0.266
1358	0.267
1423	0.266
1545	0.263
1571	0.261
1612	0.256
1638	0.256
1679	0.254
CURVE 11*	
303	0.237
309	0.229
312	0.224
312	0.237
317	0.228
320	0.236
326	0.231
349	0.230
352	0.228
355	0.232
374	0.222
383	0.223
390	0.230
431	0.230
444	0.229
472	0.226
491	0.230
496	0.233
510	0.229
517	0.231
519	0.223
528	0.223
537	0.226
544	0.232
556	0.237
560	0.235
571	0.231
579	0.234
CURVE 11 (cont.)*	
594	0.232
596	0.237
662	0.230
668	0.234
671	0.238
691	0.235
700	0.230
706	0.238
708	0.230
716	0.228
716	0.236
720	0.230
722	0.238
726	0.236
746	0.228
755	0.231
755	0.242
758	0.234
767	0.234
854	0.231
855	0.236
855	0.241
858	0.237
868	0.236
878	0.233
879	0.238
918	0.238
926	0.241
928	0.237
939	0.236
960	0.237
974	0.232
1050	0.247
1061	0.249
1069	0.245
1098	0.246
1112	0.248
1129	0.252
1148	0.256
1148	0.248
1161	0.246
1172	0.249
1186	0.250
1202	0.252
1204	0.244
1205	0.261
1207	0.251
1211	0.249
CURVE 11 (cont.)*	
1224	0.249
1226	0.256
1234	0.257
CURVE 12	
493	0.225
673	0.227
868	0.236
1068	0.240
1263	0.245
CURVE 13	
970	0.237
1006	0.235
1078	0.234
1153	0.243
1243	0.240
1331	0.238
1402	0.245
1511	0.250
1611	0.256
CURVE 14	
965	0.238
999	0.239
1072	0.239
1147	0.247
1239	0.249
1326	0.245
1400	0.246
1508	0.252
1607	0.270
CURVE 15	
300	0.229
575	0.225
828	0.220
1089	0.220
1677	0.240
1865	0.227*
1343	0.240*
1480	0.240*
1662	0.233*

* Not shown in figure.

FIGURE AND TABLE 48R. PROVISIONAL THERMAL DIFFUSIVITY OF PLUTONIUM

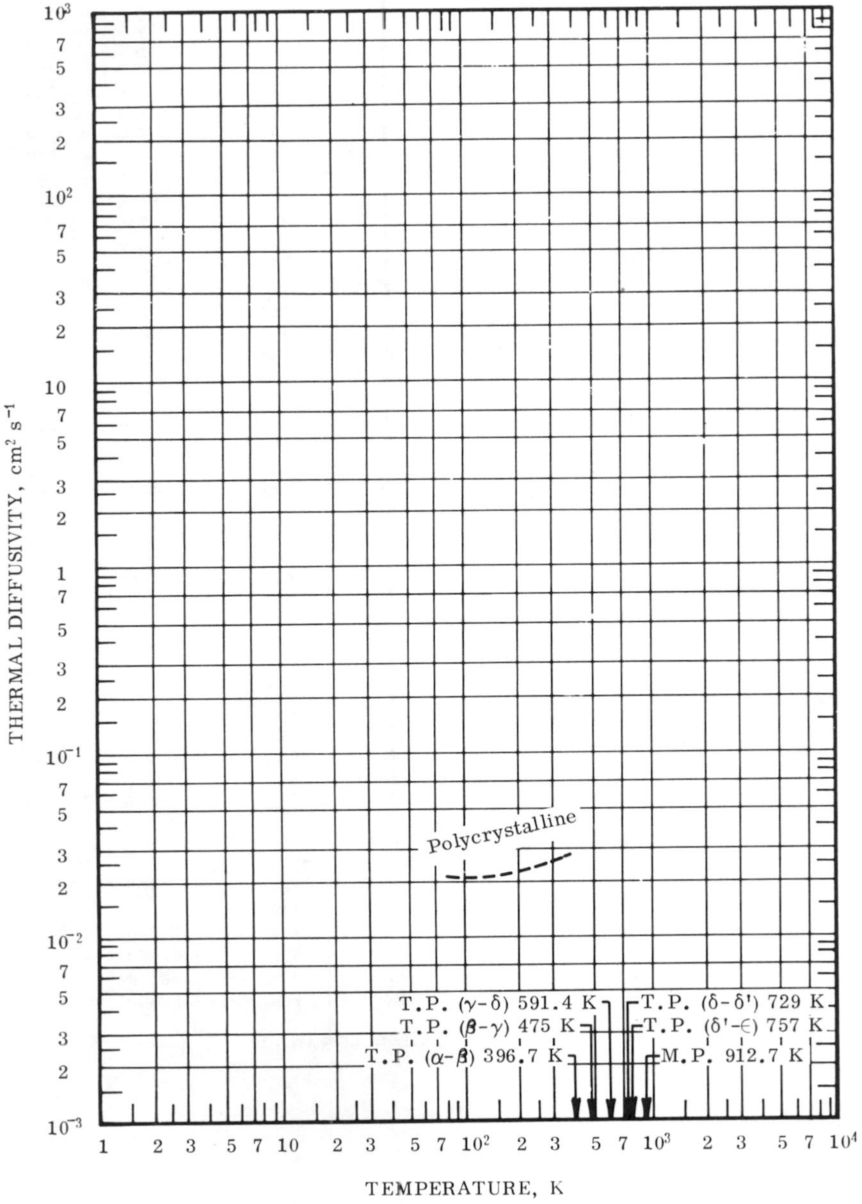

PROVISIONAL VALUES

[Temperature, T, K; Thermal Diffusivity, α, $cm^2\ s^{-1}$]

SOLID (Polycrystalline)

T	α
80	0.0210*
90	0.0210*
100	0.0211*
150	0.0214*
200	0.0223*
250	0.0238*
273.2	0.0246*
300	0.0257
350	0.0277*

REMARKS

The provisional values are for well-annealed high-purity polycrystalline plutonium. The uncertainty of the values is of the order of ±25%.

* In temperature range where no experimental data are available.

SPECIFICATION TABLE 48. THERMAL DIFFUSIVITY OF PLUTONIUM

(Impurity <0.20% each; total impurities <0.50%)

Cur. No.	Ref. No.	Author(s)	Year	Temp. Range, K	Reported Error, %	Name and Specimen Designation	Composition (weight percent), Specifications, and Remarks
1*	181	Radenac, A. and Hocheid, B.	1964	298			Unalloyed; raw-cast; cylindrical specimen 9.5 mm in diameter and 175 mm long; periodic temperature fluctuation imposed on one end of specimen; measured as a function of the period, P, ranging from 2440 to 3490 sec; measured in an atmosphere of argon.

DATA TABLE 48. THERMAL DIFFUSIVITY OF PLUTONIUM

(Impurity <0.20% each; total impurities <0.50%)

[Temperature, T, K; Thermal Diffusivity, α, $cm^2 s^{-1}$]

P(s)	α
<u>CURVE 1*</u> (T = 298K)	
2440	0.0197
2520	0.0203
3030	0.0203
3210	0.0195
3390	0.0202
3490	0.0197

* No figure given.

FIGURE AND TABLE 49R. RECOMMENDED THERMAL DIFFUSIVITY OF POTASSIUM

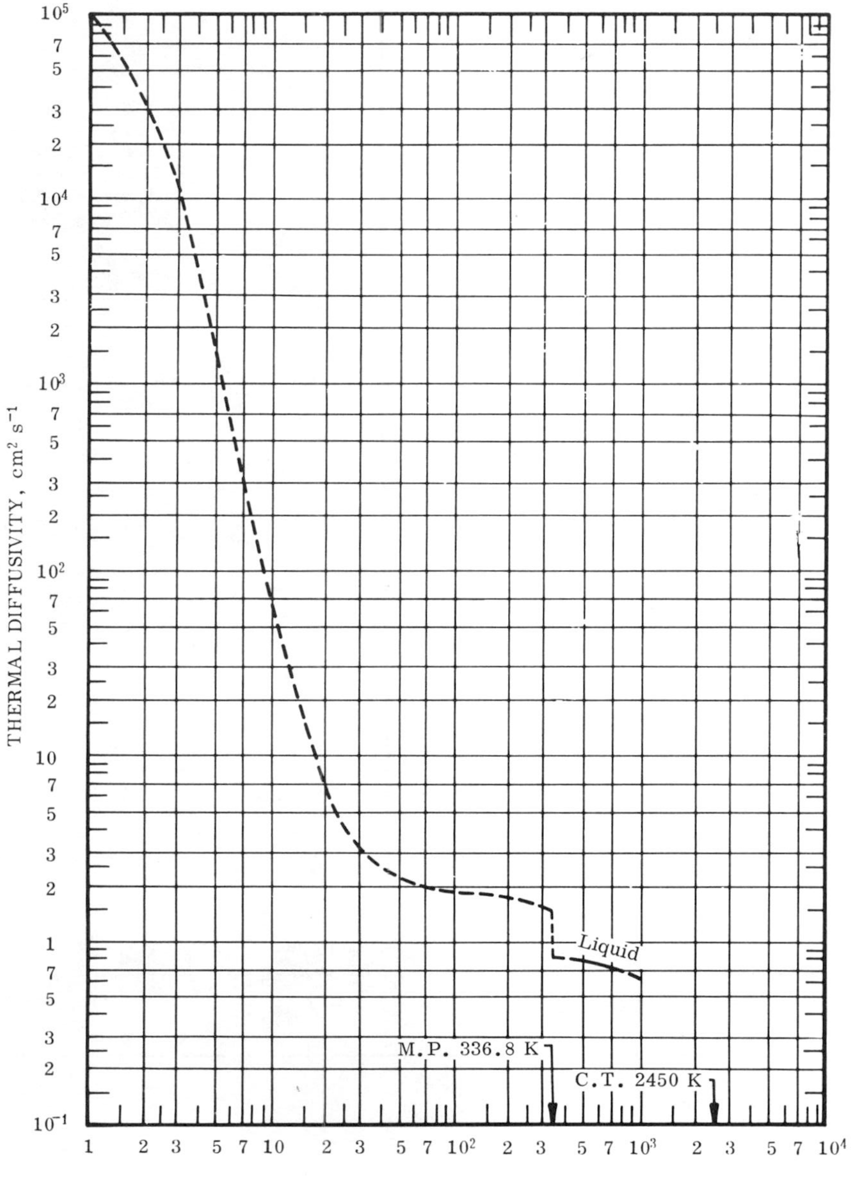

RECOMMENDED VALUES

[Temperature, T, K; Thermal Diffusivity, α, $cm^2\ s^{-1}$]

SOLID				LIQUID	
T	α	T	α	T	α
1	104000*	35	2.90*	336.8	0.814*
2	32700*	40	2.61*	350	0.811*
3	10250*	45	2.42*	400	0.800
4	3580*	50	2.29*	500	0.775
5	1410*	60	2.11*	600	0.748
6	615*	70	2.01*	700	0.715
7	310*	80	1.94*	800	0.680
8	173*	90	1.90*	900	0.642*
9	107*	100	1.88*	1000	0.603*
10	70.0*	150	1.78*		
11	48.9*	200	1.72*		
12	35.5*	250	1.66*		
13	26.7*	273.2	1.62*		
14	20.8*	300	1.57*		
15	16.7*	336.8	1.47*		
16	13.6*				
18	9.60*				
20	7.21*				
25	4.47*				
30	3.41*				

REMARKS

The recommended values are for high-purity potassium and are thought to be accurate to within ±12% of the true values for the solid state and ±8% for the liquid state. At low temperatures the values are highly conditioned by impurity and imperfection, and those below 50 K are applicable only to potassium having residual electrical resistivity of 0.00220 $\mu\Omega$ cm.

*In temperature range where no experimental data are available.

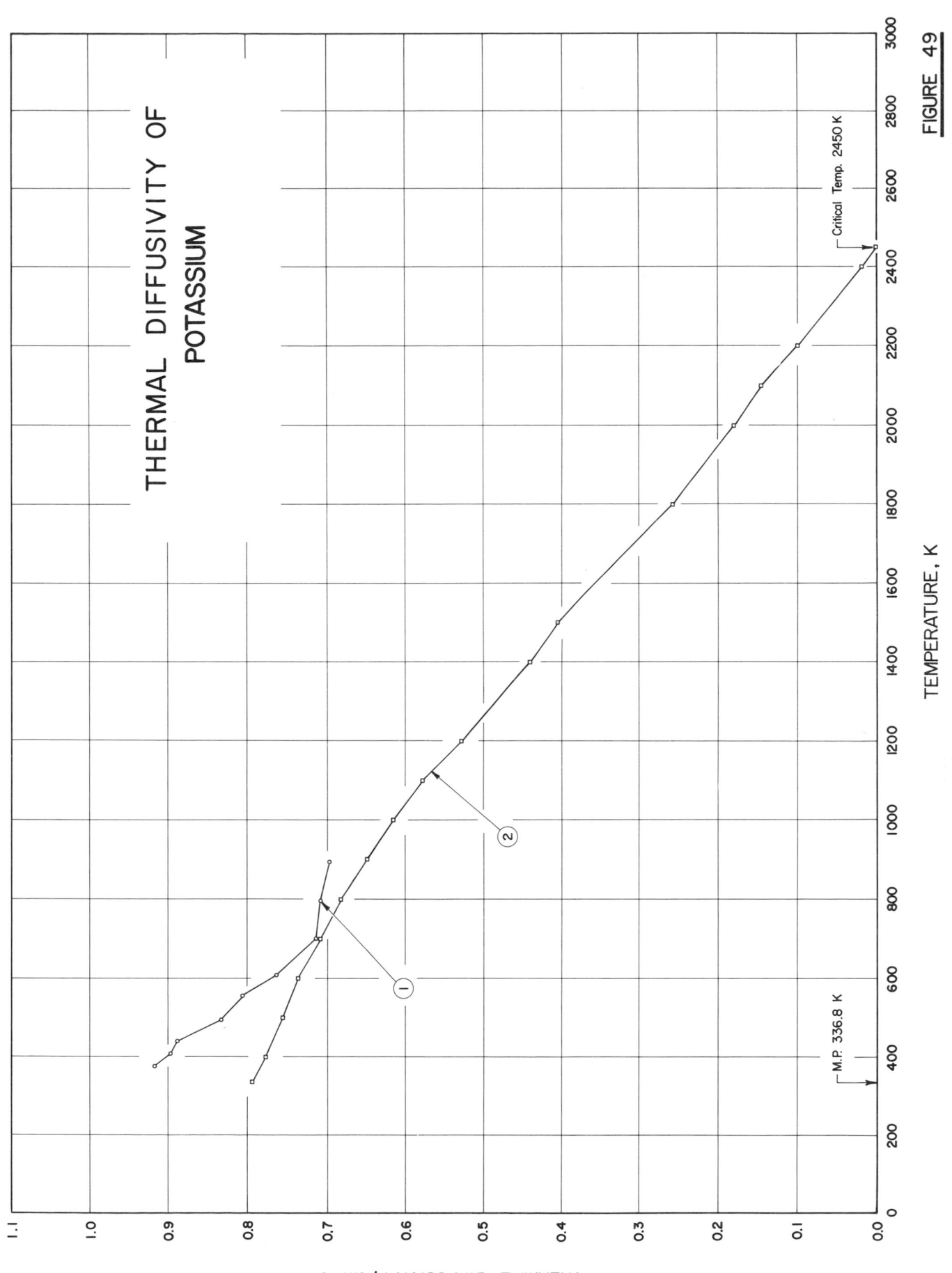

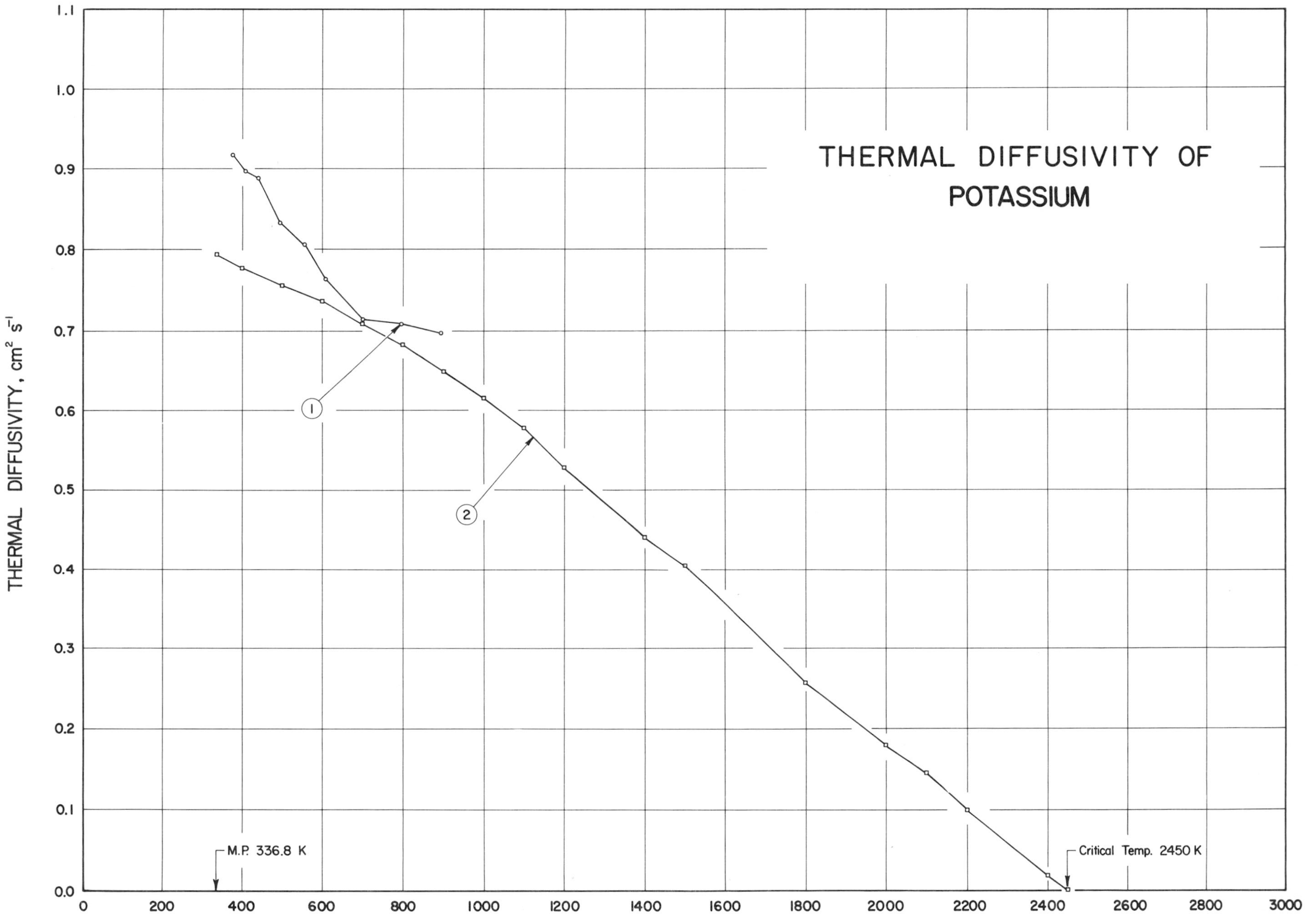
THERMAL DIFFUSIVITY OF
POTASSIUM
THERMAL DIFFUSIVITY, cm² s⁻¹
TEMPERATURE, K
M.P. 336.8 K
Critical Temp. 2450 K
1
2
0.0
0.1
0.2
0.3
0.4
0.5
0.6
0.7
0.8
0.9
1.0
1.1
0
200
400
600
800
1000
1200
1400
1600
1800
2000
2200
2400
2600
2800
3000

FIGURE 49

SPECIFICATION TABLE 49. THERMAL DIFFUSIVITY OF POTASSIUM

(Impurity < 0.20% each; total impurities < 0.50%)

Cur. No.	Ref. No.	Author(s)	Year	Temp. Range, K	Reported Error, %	Name and Specimen Designation	Composition (weight percent), Specifications, and Remarks
1	45, 140	Novikov, I. I., Solovyev, A. N., Khabakhpasheva, E. M., Gruzdev, V. A., Pridantsev, A. I., and Vasenina, M. Ya.	1956	378-893			Metal placed into a vertically positioned thin stainless steel tube; density measured and reported as 0.821, 0.818, 0.813, 0.807, 0.804, 0.798, 0.792, 0.784, 0.781, 0.770, 0.772, 0.761, 0.759, 0.753, 0.747, 0.737, 0.731, 0.728, 0.713, 0.706, 0.689, 0.684, and 0.673 g cm^{-3} at 359.2, 375.2, 391.2, 423.2, 438.2, 474.2, 493.2, 523.2, 536.2, 583.2, 583.2, 621.2, 635.2, 666.2, 684.2, 737.2, 758.2, 779.2, 844.2, 873.2, 946.2, 970.2, and 1006.2 K, respectively; in molten state; heater wound on the outside of upper part of tube; free metal surface maintained at an excess pressure of an inert gas; twenty radial copper screens distributed throughout whole length of specimen; suspended in an evacuated quartz tube; Angström's dynamic method used to measure diffusivity.
2	252	Grosse, A. V.	1968	337-2450			Thermal diffusivity values for liquid potassium calculated from chosen thermal conductivity and density using specific heat capacity values compiled by Heimel, S. (NASA Technical Note D-4165, 1967).

DATA TABLE 49. THERMAL DIFFUSIVITY OF POTASSIUM

(Impurity <0.20% each; total impurities <0.50%)

[Temperature, T, K; Thermal Diffusivity, α, cm^2 s^{-1}]

T	α	T	α	T	α
CURVE 1		CURVE 2		CURVE 2 (cont.)	
378	0.917	336	0.794	2200	0.099
408	0.897	400	0.777	2400	0.018
440	0.889	500	0.755	2450	0.000
493	0.833	600	0.736		
556	0.806	700	0.708		
608	0.764	800	0.682		
700	0.714	900	0.648		
796	0.708	1000	0.615		
893	0.694	1100	0.576		
		1200	0.527		
		1400	0.439		
		1500	0.404		
		1800	0.256		
		2000	0.179		
		2100	0.144		

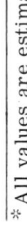

FIGURE AND TABLE 50R. PROVISIONAL THERMAL DIFFUSIVITY OF PRASEODYMIUM

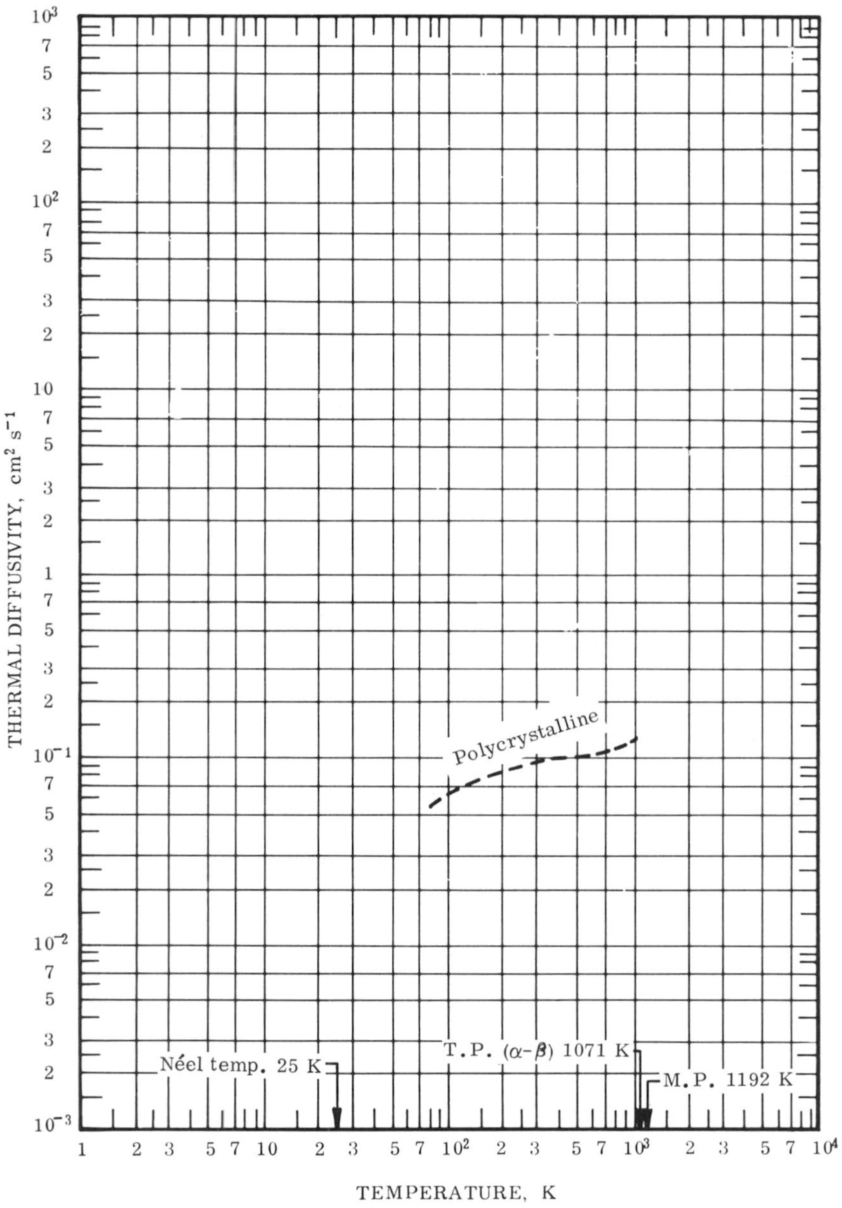

PROVISIONAL VALUES*

[Temperature, T, K; Thermal Diffusivity, α, $cm^2\ s^{-1}$]

T	α
80	0.0547
90	0.0582
100	0.0613
150	0.0739
200	0.0829
250	0.0890
273.2	0.0911
300	0.0932
350	0.0960
400	0.0980
500	0.101
600	0.104
700	0.108
800	0.113
900	0.118
1000	0.123

REMARKS

The values are for well-annealed high-purity polycrystalline praseodymium and are thought to be accurate to within ± 8% of the true values near room temperature and ± 15 to ± 20% at other temperatures.

* All values are estimated.

FIGURE AND TABLE 51R. PROVISIONAL THERMAL DIFFUSIVITY OF RADIUM

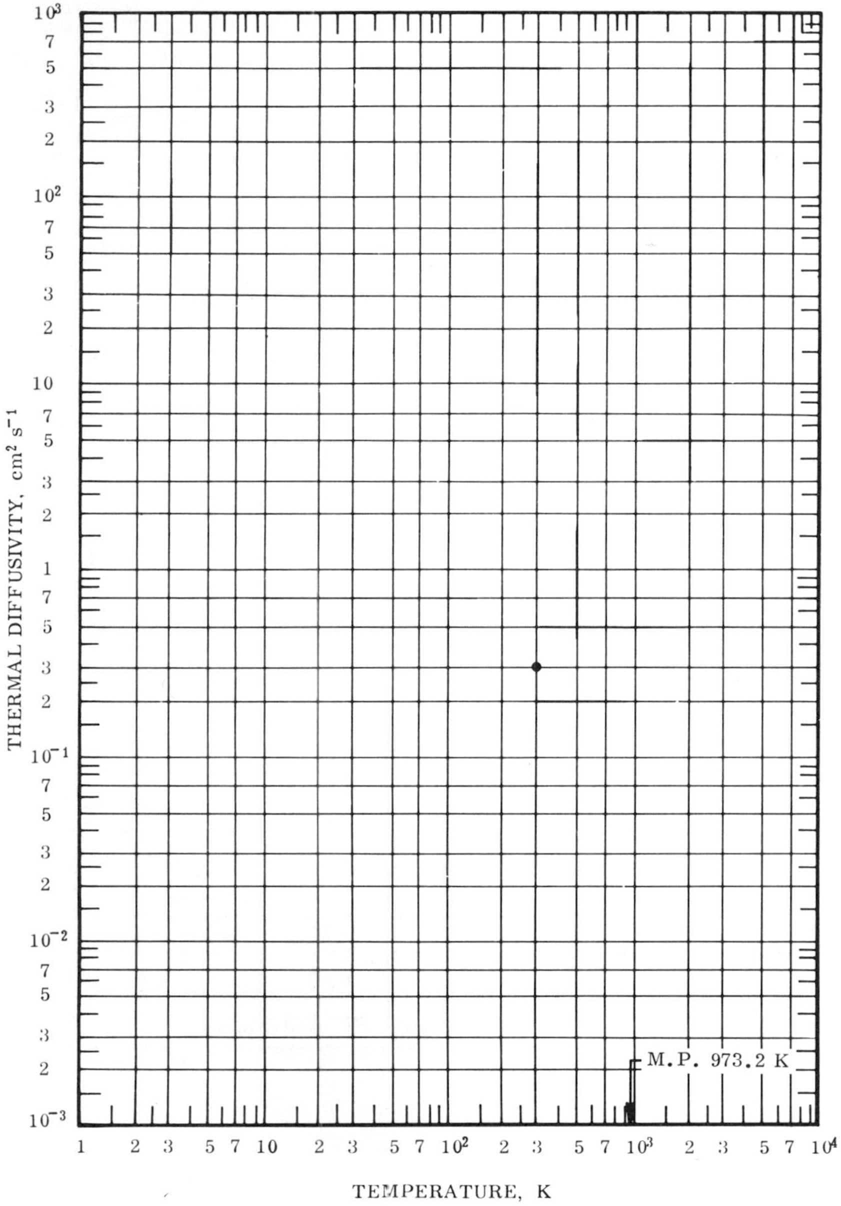

PROVISIONAL VALUES

[Temperature, T, K; Thermal Diffusivity, α, $cm^2\ s^{-1}$]

SOLID

T	α
293.2	0.31*

REMARKS

The uncertainty of this value may be as much as ± 50%.

---

* Estimated.

FIGURE AND TABLE 52R. RECOMMENDED THERMAL DIFFUSIVITY OF RHENIUM

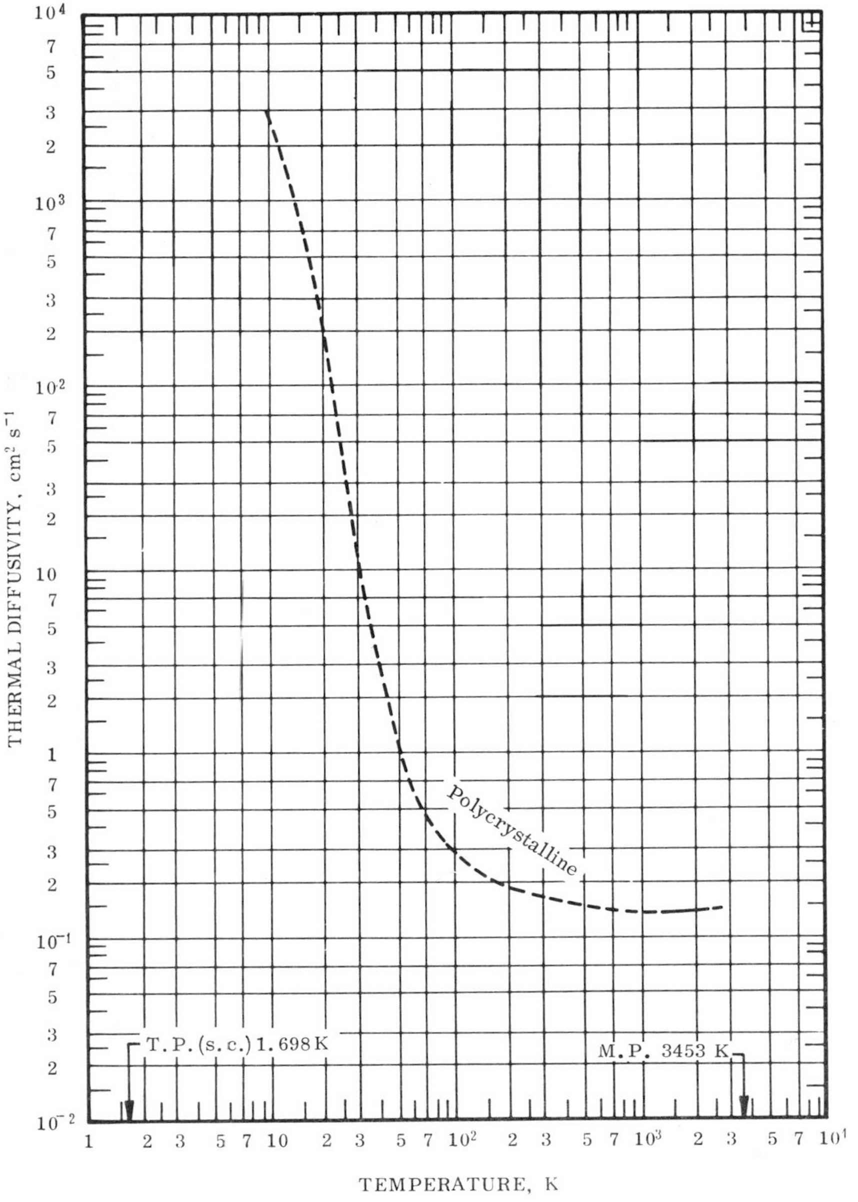

RECOMMENDED VALUES†

[Temperature, T, K; Thermal Diffusivity, α, cm^2 s^{-1}]

SOLID
(Polycrystalline)

T	α	T	α
10	2990*	350	0.159*
11	2380*	400	0.155*
12	1840*	500	0.149*
13	1390*	600	0.145*
14	1030*	700	0.142*
15	756*	800	0.140*
16	558*	900	0.139*
18	304*	1000	0.138*
20	167*	1100	0.137*
25	40.0*	1200	0.136
30	13.2*	1300	0.136
35	5.65*	1400	0.136
40	2.86*	1500	0.136
45	1.68*	1600	0.136
50	1.10*	1700	0.136
60	0.638*	1800	0.136
70	0.460*	1900	0.137
80	0.371*	2000	0.137
90	0.321*	2200	0.138
100	0.287*	2400	0.141*
150	0.212*	2600	0.146*
200	0.187*		
250	0.173*		
273.2	0.169*		
300	0.165*		

REMARKS

The values are for well-annealed high-purity polycrystalline rhenium and are thought to be accurate to within ± 13% of the true values at temperatures below 100 K, ± 7% from 100 to 500 K, and ± 20% above 500 K. Those above 500 K are provisional. At low temperatures the values are highly conditioned by impurity and imperfection, and those below 100 K are applicable only to rhenium having residual electrical resistivity of 0.00366 μΩ cm.

† Values above 500 K are provisional.
* In temperature range where no experimental data are available.

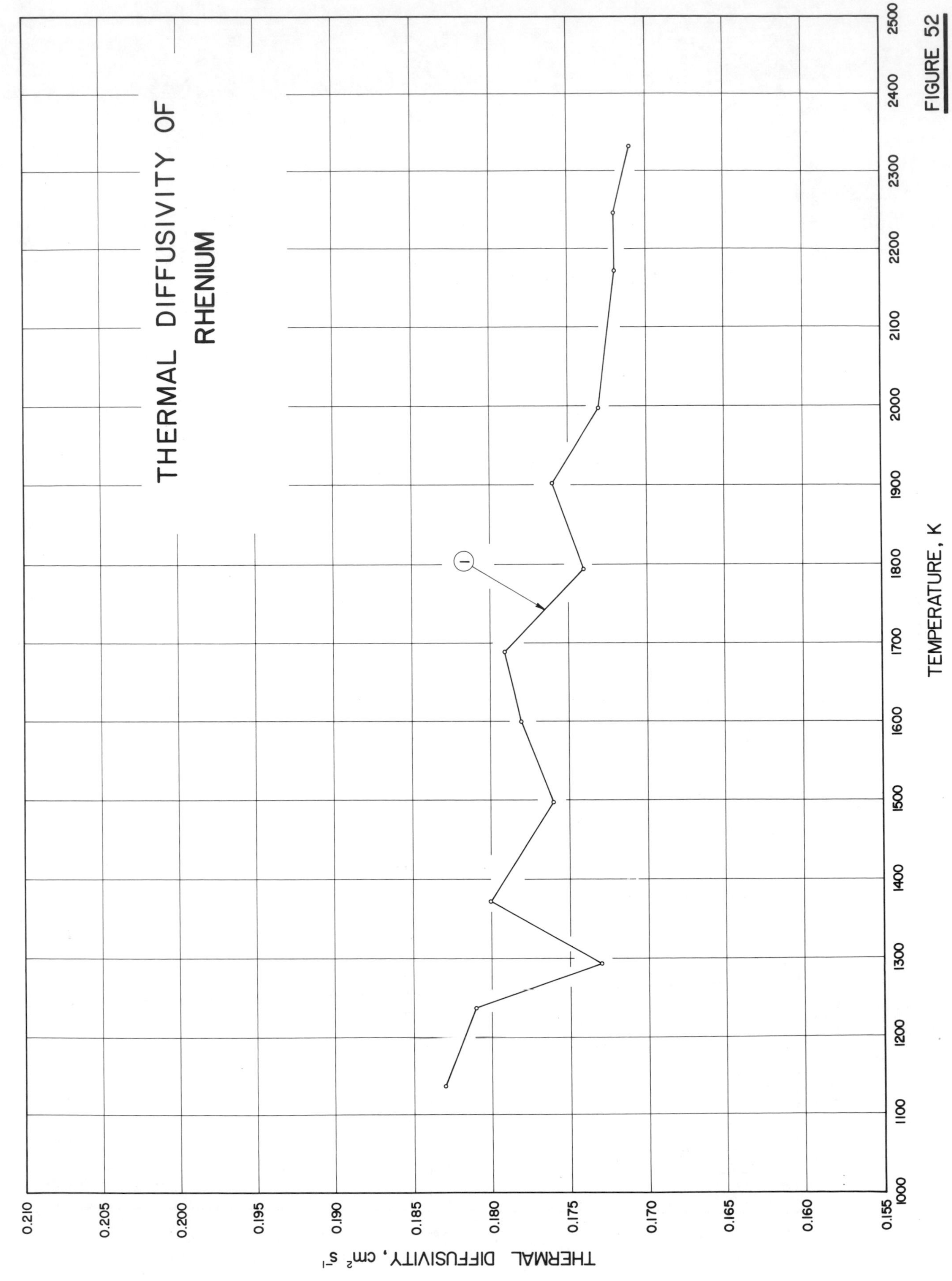
THERMAL DIFFUSIVITY OF
RHENIUM
THERMAL DIFFUSIVITY, cm² s⁻¹
0.210
0.205
0.200
0.195
0.190
0.185
0.180
0.175
0.170
0.165
0.160
0.155
TEMPERATURE, K
1000
1100
1200
1300
1400
1500
1600
1700
1800
1900
2000
2100
2200
2300
2400
2500
1

FIGURE 52

SPECIFICATION TABLE 52. THERMAL DIFFUSIVITY OF RHENIUM

(Impurity <0.20% each; total impurities <0.50%)

Cur. No.	Ref. No.	Author(s)	Year	Temp. Range, K	Reported Error, %	Name and Specimen Designation	Composition (weight percent), Specifications, and Remarks
1	238	Arutyunov, A. V. and Filippov, L. P.	1970	1137-2332			99.994 Re, 0.005 O, 0.0005 H, 0.0004 Si, 0.0002 Mo, 0.0001 Al, 0.0001 Fe, 0.00007 Mg, 0.00002 Ni, 0.00002 Cr, 0.00001 Cu, and 0.000004 Mn; single crystal; cylindrical specimen; 11 mm in diameter, 85 mm in length; produced by electron-beam zone melting; c-axis of crystal formed angle of 32° with axis of specimen; density at 20 C 21.0 g cm^{-3} and electrical resistivity 18.2 μΩ cm; before measurement, specimen heated in vacuum at 2200 K for 2 hr.

DATA TABLE 52. THERMAL DIFFUSIVITY OF RHENIUM

(Impurity <0.20% each; total impurities <0.50%)

[Temperature, T, K; Thermal Diffusivity, α, $cm^2\ s^{-1}$]

T	α
CURVE 1	
1137	0.183
1237	0.181
1293	0.173
1372	0.180
1497	0.176
1599	0.178
1688	0.179
1793	0.174
1903	0.176
1997	0.173
2172	0.172
2246	0.172
2332	0.171

FIGURE AND TABLE 53R. RECOMMENDED THERMAL DIFFUSIVITY OF RHODIUM

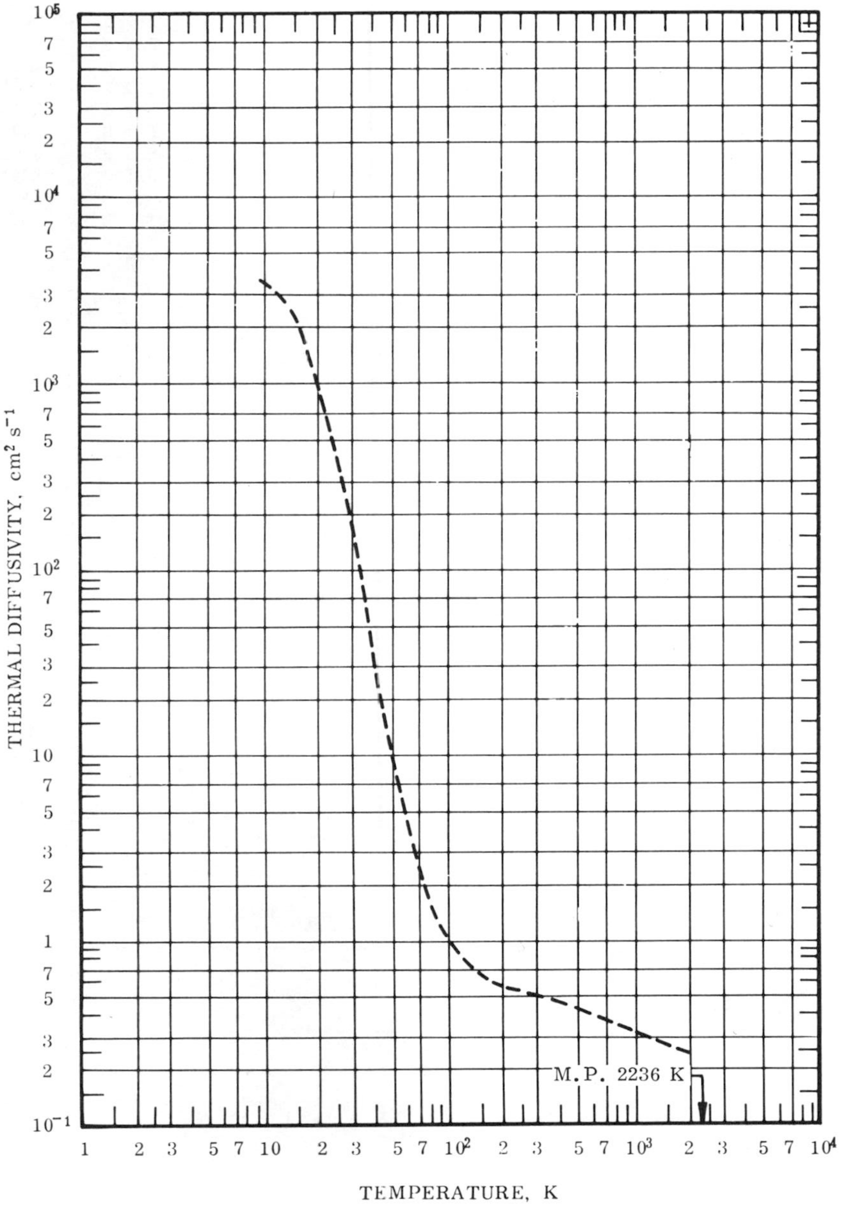

RECOMMENDED VALUES*

[Temperature, T, K; Thermal Diffusivity, α, $cm^2\ s^{-1}$]

SOLID

T	α	T	α
10	3420	150	0.646
11	3280	200	0.562
12	3060	250	0.522
13	2810	273.2	0.510
14	2560	300	0.499
15	2300	350	0.482
16	2030	400	0.465
18	1520	500	0.434
20	1080	600	0.405
25	428	700	0.379
30	173	800	0.356
35	72.6	900	0.337
40	32.7	1000	0.321
45	16.8	1100	0.307
50	9.76	1200	0.295
60	4.41	1300	0.285
70	2.54	1400	0.277
80	1.69	1500	0.269
90	1.25	1600	0.263
100	1.01	1700	0.258
		1800	0.254
		1900	0.250
		2000	0.248

REMARKS

The values are for well-annealed high-purity rhodium and are thought to be accurate to within ±12% of the true values at temperatures below 150 K, ±7% from 150 to 600 K, and ±20% above 600 K. Those above 600 K are provisional. At low temperatures the values are highly conditioned by impurity and imperfection, and those below 150 K are applicable only to rhodium having residual electrical resistivity of 0.00840 $\mu\Omega$ cm.

*All values are estimated and values above 600 K are provisional.

FIGURE AND TABLE 54R. RECOMMENDED THERMAL DIFFUSIVITY OF RUBIDIUM

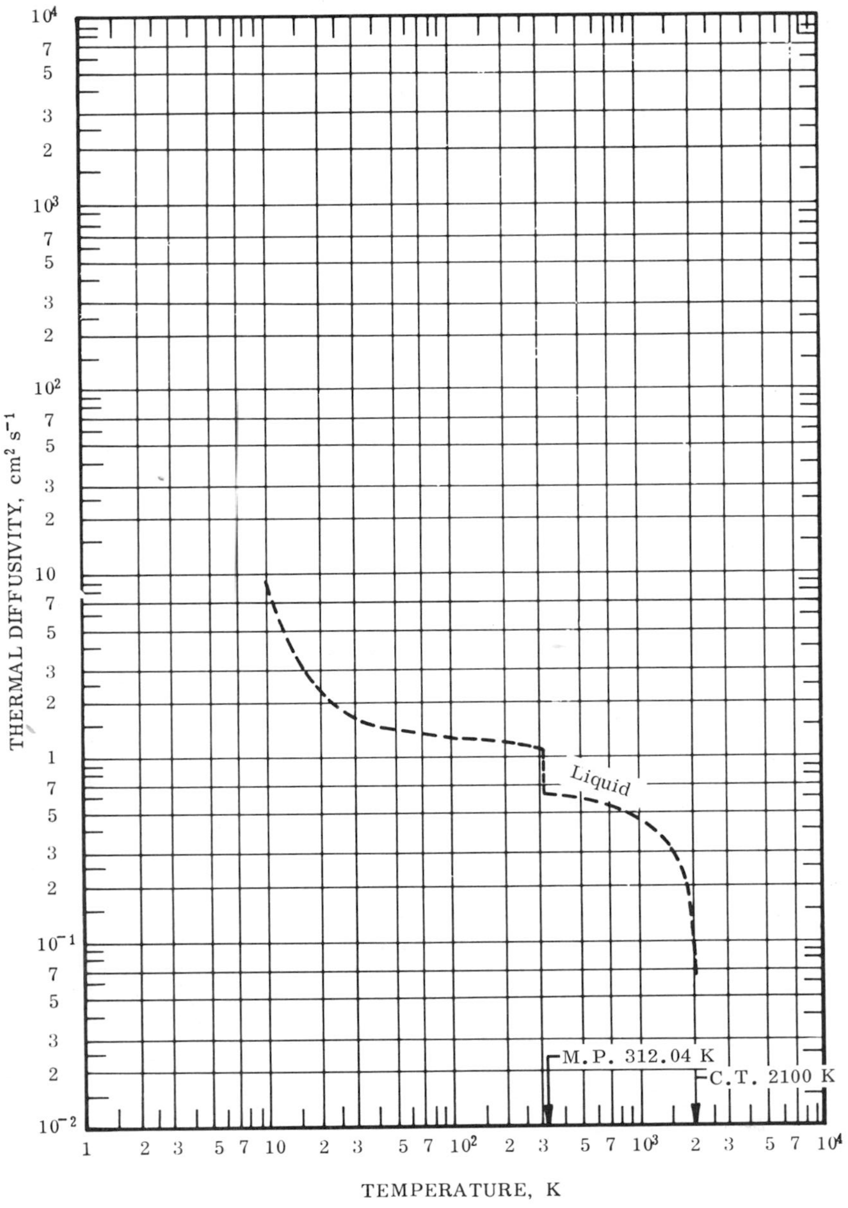

RECOMMENDED VALUES*

[Temperature, T, K; Thermal Diffusivity, α, cm^2 s^{-1}]

SOLID		LIQUID	
T	α	T	α
10	8.94	312.04	0.615
11	6.71	350	0.606
12	5.33	400	0.595
13	4.43	500	0.572
14	3.80	600	0.548
15	3.33	700	0.522
16	2.98	800	0.496
18	2.51	900	0.470
20	2.20	1000	0.442
25	1.82	1100	0.415
30	1.65	1200	0.386
35	1.55	1300	0.357
40	1.48	1400	0.325
45	1.44	1500	0.290
50	1.40	1600	0.253
60	1.36	1700	0.214
70	1.32	1800	0.169
80	1.29	1900	0.118
90	1.27	2000	0.0648
100	1.25		
200	1.21		
250	1.12		
273.2	1.09		
300	1.06		
312.04	1.03		

REMARKS

The values are for high-purity rubidium and are thought to be accurate to within ±12% of the true values at temperatures below 1000 K. The uncertainty increases at higher temperatures and those above 1000 K are provisional. At low temperatures the values are highly conditioned by impurity and imperfection, and those below 40 K are applicable only to rubidium having residual electrical resistivity of 0.0384 μΩ cm.

* All values are estimated and values above 1000 K are provisional.

FIGURE AND TABLE 55R. RECOMMENDED THERMAL DIFFUSIVITY OF RUTHENIUM

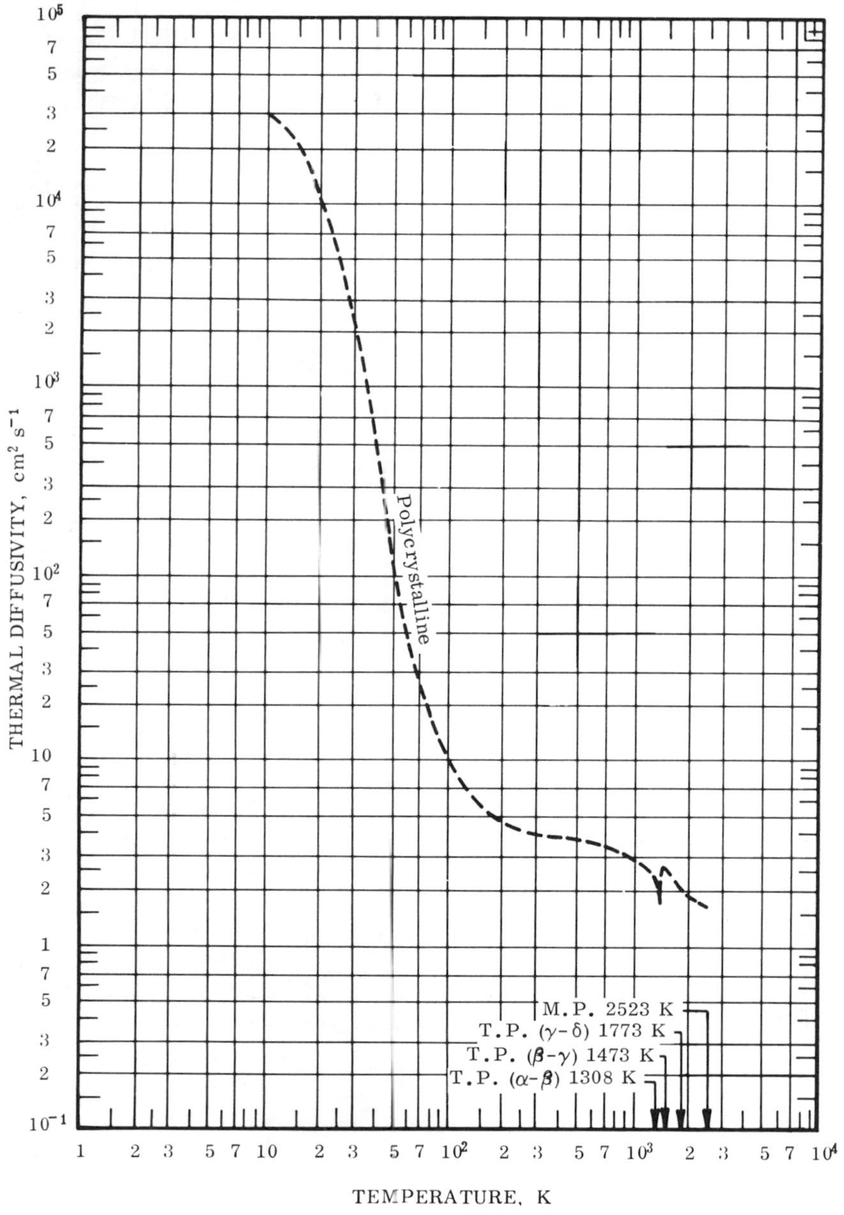

RECOMMENDED VALUES*

[Temperature, T, K; Thermal Diffusivity, α, cm^2 s^{-1}]

SOLID
(Polycrystalline)

T	α	T	α
10	2950	350	0.395
11	2780	400	0.389
12	2580	500	0.374
13	2360	600	0.356
14	2150	700	0.337
15	1970	800	0.319
16	1710	900	0.304
18	1330	1000	0.290
20	1010	1100	0.278
25	482	1200	0.266
30	215	1300	0.205
35	95.5	1308	0.177
40	44.0	1400	0.264
45	22.8	1500	0.252
50	13.0	1600	0.233
60	4.46	1700	0.211
70	2.85	1800	0.198
80	1.80	1900	0.190
90	1.28	2000	0.184
100	1.01	2200	0.176
150	0.559	2400	0.169
200	0.450		
250	0.417		
273.2	0.409		
300	0.403		

REMARKS

The values are for well-annealed high-purity polycrystalline ruthenium and are thought to be accurate to within ±12% of the true values at temperatures below 100 K, ±7% from 100 to 500 K, and ±15 to ±20% above 500 K. Those above 500 K are provisional. At low temperatures the values are highly conditioned by impurity and imperfection, and those below 200 K are applicable only to ruthenium having residual electrical resistivity of 0.0158 μΩ cm.

*All values are estimated and values above 500 K are provisional.

FIGURE AND TABLE 56R. PROVISIONAL THERMAL DIFFUSIVITY OF SAMARIUM

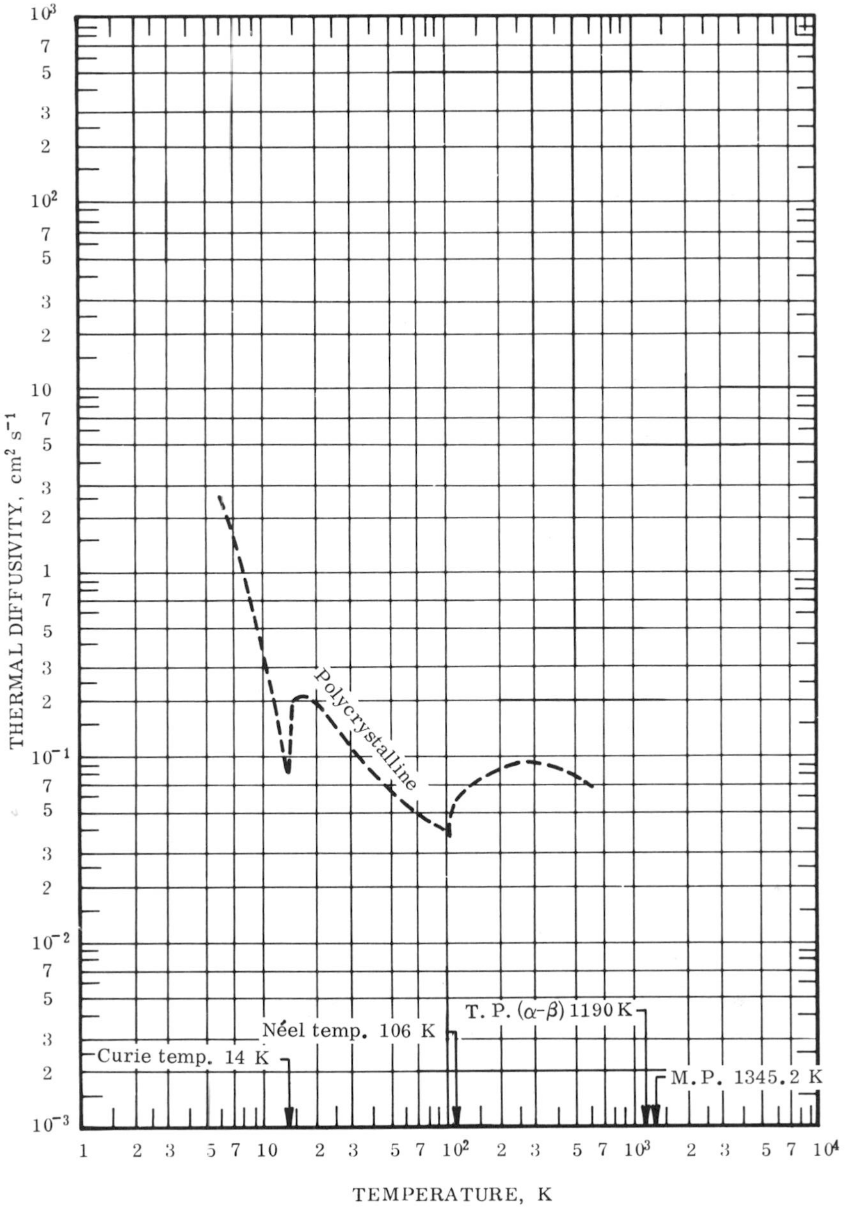

PROVISIONAL VALUES*

[Temperature, T, K; Thermal Diffusivity, α, cm^2 s^{-1}]

SOLID
(Polycrystalline)

T	α	T	α
6	2.54	70	0.0480
7	1.56	80	0.0437
8	1.00	90	0.0406
9	0.667	100	0.0386
10	0.459	104	0.0370
11	0.301	106	0.0355
12	0.182	110	0.0566
13	0.101	150	0.0701
14	0.084	200	0.0832
15	0.200	250	0.0918
16	0.211	273.2	0.0916
18	0.201	300	0.0897
20	0.186	350	0.0850
25	0.149	400	0.0804
30	0.118	500	0.0725
35	0.0964	600	0.0691
40	0.0822		
45	0.0721		
50	0.0647		
60	0.0545		

REMARKS

The values are for well-annealed high-purity polycrystalline samarium and are thought to be accurate to within ±15% of the true values near room temperature, ±20% at higher temperatures, and ±30% at lower temperatures. Values below room temperature are applicable only to samarium having electrical resistivity of 6.73 μΩ cm at 4.2 K.

* All values are estimated.

FIGURE AND TABLE 57R. PROVISIONAL THERMAL DIFFUSIVITY OF SCANDIUM

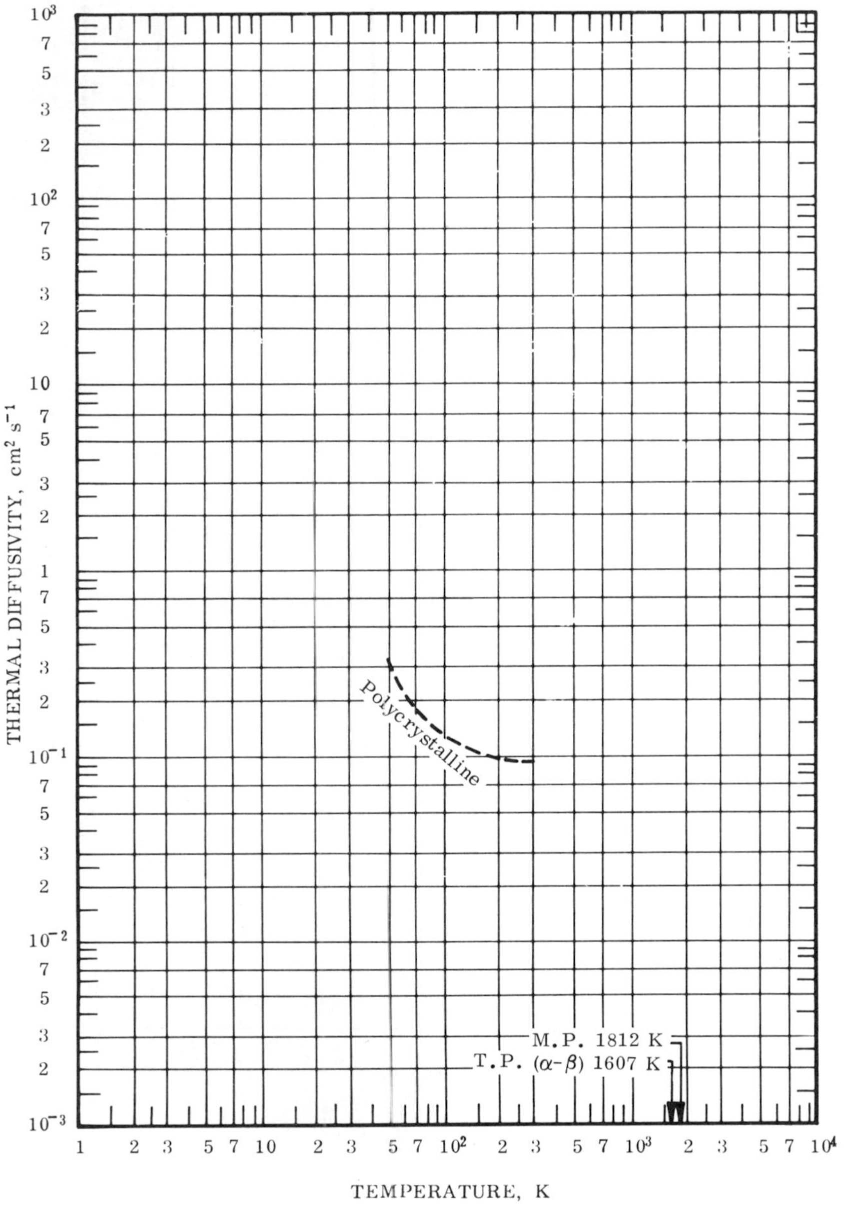

PROVISIONAL VALUES*
[Temperature, T, K; Thermal Diffusivity, α, $cm^2\ s^{-1}$]

SOLID (Polycrystalline)	
T	α
50	0.312
60	0.225
70	0.181
80	0.157
90	0.141
100	0.130
150	0.105
200	0.0977
250	0.0947
273.2	0.0936
300	0.0926

REMARKS

The values are for well-annealed high-purity polycrystalline scandium and are thought to be accurate to within ± 15% of the true values near room temperature and ± 20% below 200 K. The values below 200 K are applicable only to scandium having residual electrical resistivity of 10.6 μΩ cm.

* All values are estimated.

FIGURE AND TABLE 58R. PROVISIONAL THERMAL DIFFUSIVITY OF SELENIUM

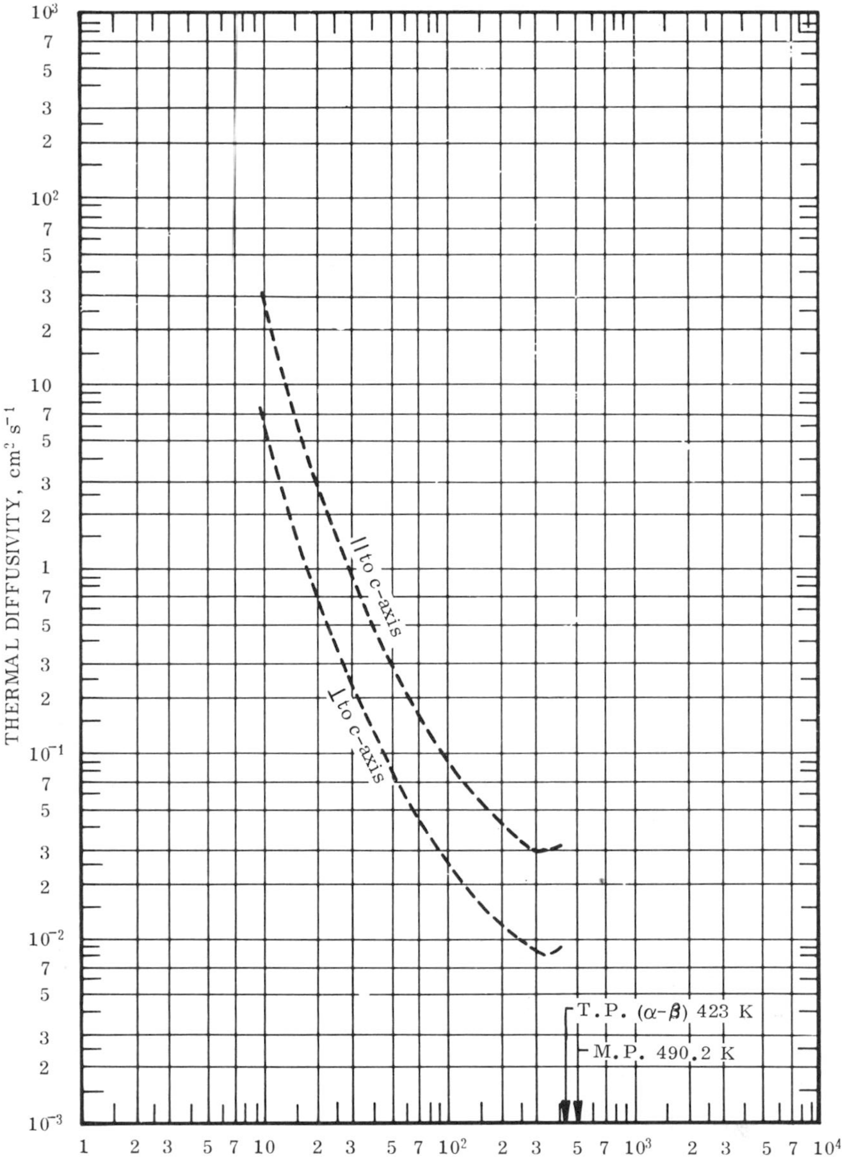

PROVISIONAL VALUES†

[Temperature, T, K; Thermal Diffusivity, α, cm² s⁻¹]

SOLID

T	∥ to c-axis α	⊥ to c-axis α	T	∥ to c-axis α	⊥ to c-axis α
10	30.3*	7.67*	60	0.216*	0.0592*
11	21.6*	5.50*	70	0.162*	0.0462*
12	15.8*	4.00*	80	0.129*	0.0374*
13	11.9*	3.03*	90	0.107*	0.0312*
14	9.16*	2.36*	100	0.0918*	0.0268*
15	7.21*	1.87*	150	0.0571*	0.0163*
16	5.80*	1.52*	200	0.0425*	0.0122*
18	3.94*	1.06*	250	0.0341*	0.00992*
20	2.95*	0.769*	273.2	0.0316	0.00913
25	1.50*	0.417*	300	0.0293	0.00838
30	0.924*	0.263*	350	0.0295*	0.00834*
35	0.649*	0.180*	400	0.0321*	0.00920*
40	0.486*	0.134*			
45	0.381*	0.103*			
50	0.307*	0.0830*			

REMARKS

The values are for high-purity selenium and those at temperatures above 80 K are thought to be accurate to within ±15 to ±25%. The values below 80 K are merely typical values and represent two typical curves serving only to indicate the general trend of the thermal diffusivity.

†Values below 80 K are merely typical values.
*In temperature range where no experimental data are available.

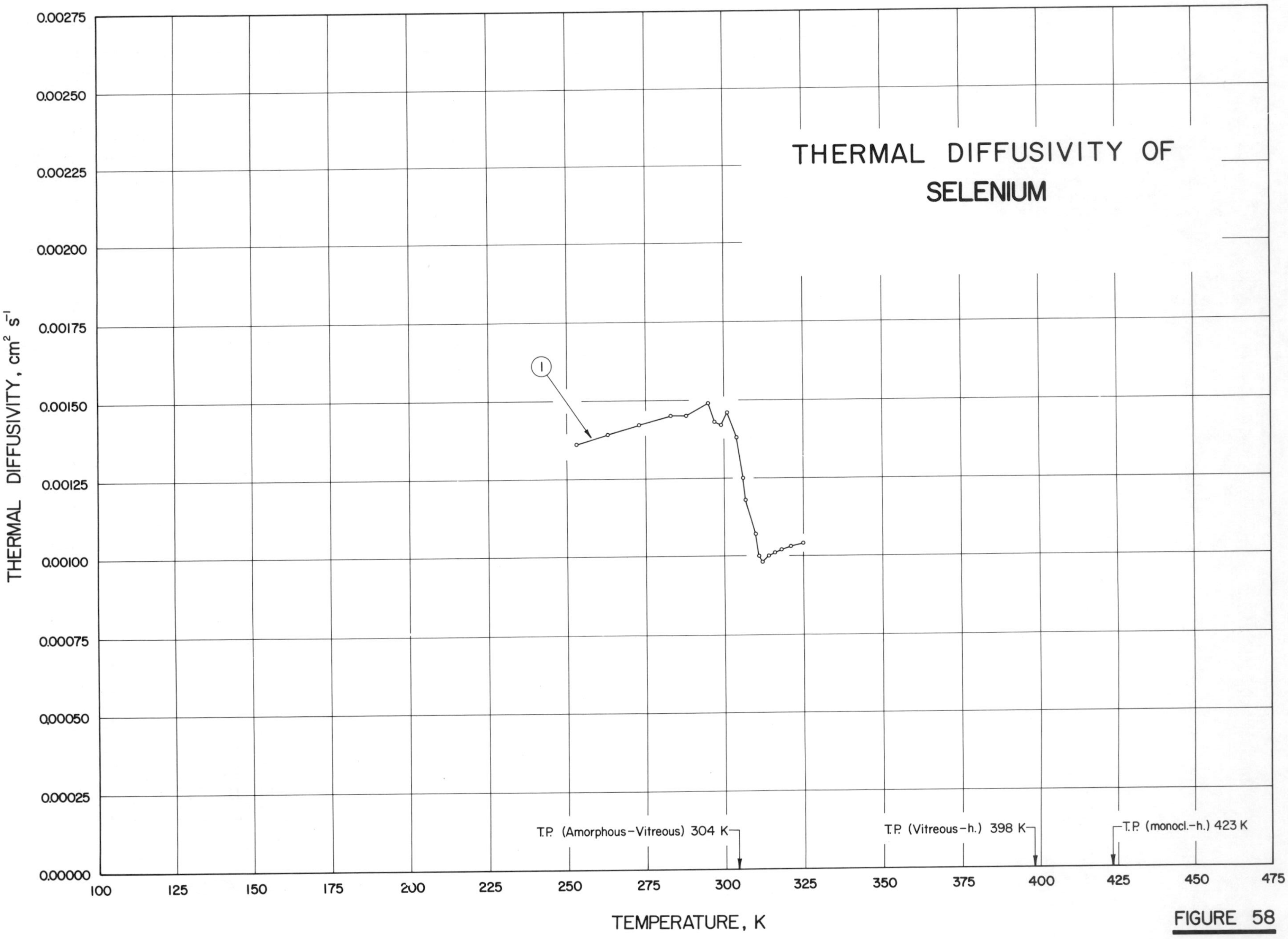

THERMAL DIFFUSIVITY OF
SELENIUM
THERMAL DIFFUSIVITY, cm² s⁻¹
TEMPERATURE, K
0.00275
0.00250
0.00225
0.00200
0.00175
0.00150
0.00125
0.00100
0.00075
0.00050
0.00025
0.00000
100
125
150
175
200
225
250
275
300
325
350
375
400
425
450
475
1
T.P. (Amorphous-Vitreous) 304 K
T.P. (Vitreous-h.) 398 K
T.P. (monocl.-h.) 423 K

FIGURE 58

SPECIFICATION TABLE 58. THERMAL DIFFUSIVITY OF SELENIUM

(Impurity < 0.20% each; total impurities < 0.50%)

Cur. No.	Ref. No.	Author(s)	Year	Temp. Range, K	Reported Error, %	Name and Specimen Designation	Composition (weight percent), Specifications, and Remarks
1	46	Orthmann, H.J. and Ueberreiter, K.	1956	253-325	± 2.2/ ± 2.9		Glassy.

DATA TABLE 58. THERMAL DIFFUSIVITY OF SELENIUM

(Impurity < 0.20% each; total impurities < 0.50%)

[Temperature, T, K; Thermal Diffusivity, α, $cm^2 s^{-1}$]

T	α	T	α
CURVE 1		CURVE 1 (cont.)	
253	0.00136	314	0.00100
263	0.00139	316	0.00101
273	0.00142	318	0.00102
283	0.00145	321	0.00103
288	0.00145	325	0.00104
295	0.00149		
297	0.00143		
299	0.00142		
301	0.00146		
304	0.00138		
306	0.00125		
307	0.00118		
310	0.00107		
311	0.00100		
312	0.00098		

FIGURE AND TABLE 59R. RECOMMENDED THERMAL DIFFUSIVITY OF SILICON

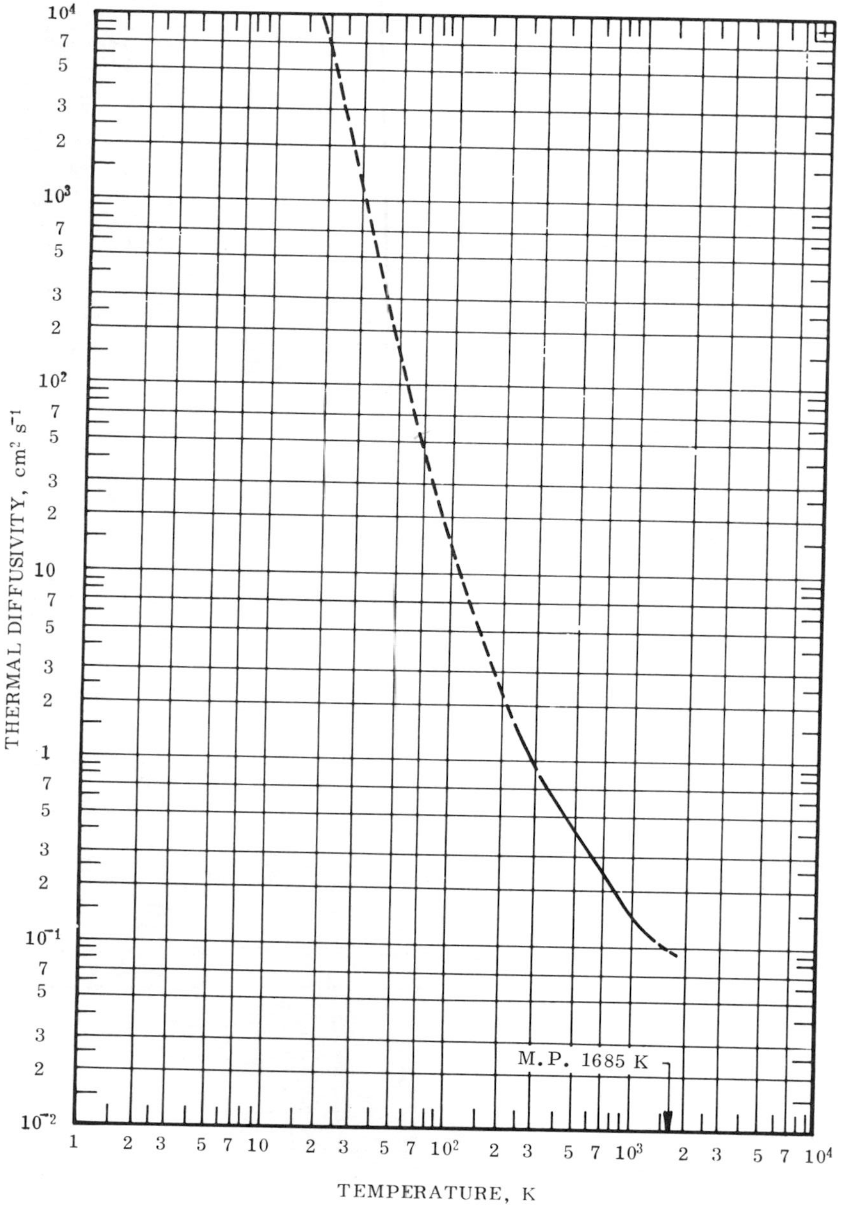

RECOMMENDED VALUES†

[Temperature, T, K; Thermal Diffusivity, α, cm^2 s^{-1}]

SOLID

T	α	T	α
10	37200*	150	4.10*
11	32700*	200	2.03*
12	28100*	250	1.26
13	23700*	273.2	1.06
14	19800*	300	0.880
15	16400*	350	0.666
16	13500*	400	0.536
18	9180*	500	0.390
20	6250*	600	0.307
25	2650*	700	0.246
30	1275*	800	0.201
35	670*	900	0.168
40	381*	1000	0.143
45	231*	1100	0.127
50	150*	1200	0.116
60	74.2*	1300	0.109
70	42.0*	1400	0.103*
80	26.2*	1500	0.0990*
90	17.6*	1600	0.0958*
100	12.7*	1685	0.0939*

REMARKS

The values are for well-annealed high-purity silicon and are thought to be accurate to within ± 7% of the true values at temperatures from room temperature to 1000 K and ± 15% at the highest temperatures. The values below room temperature are merely typical values and represent a typical curve serving to indicate the general trend of the thermal diffusivity.

†Values below **room** temperature are merely typical values.
*In temperature range where no experimental data are available.

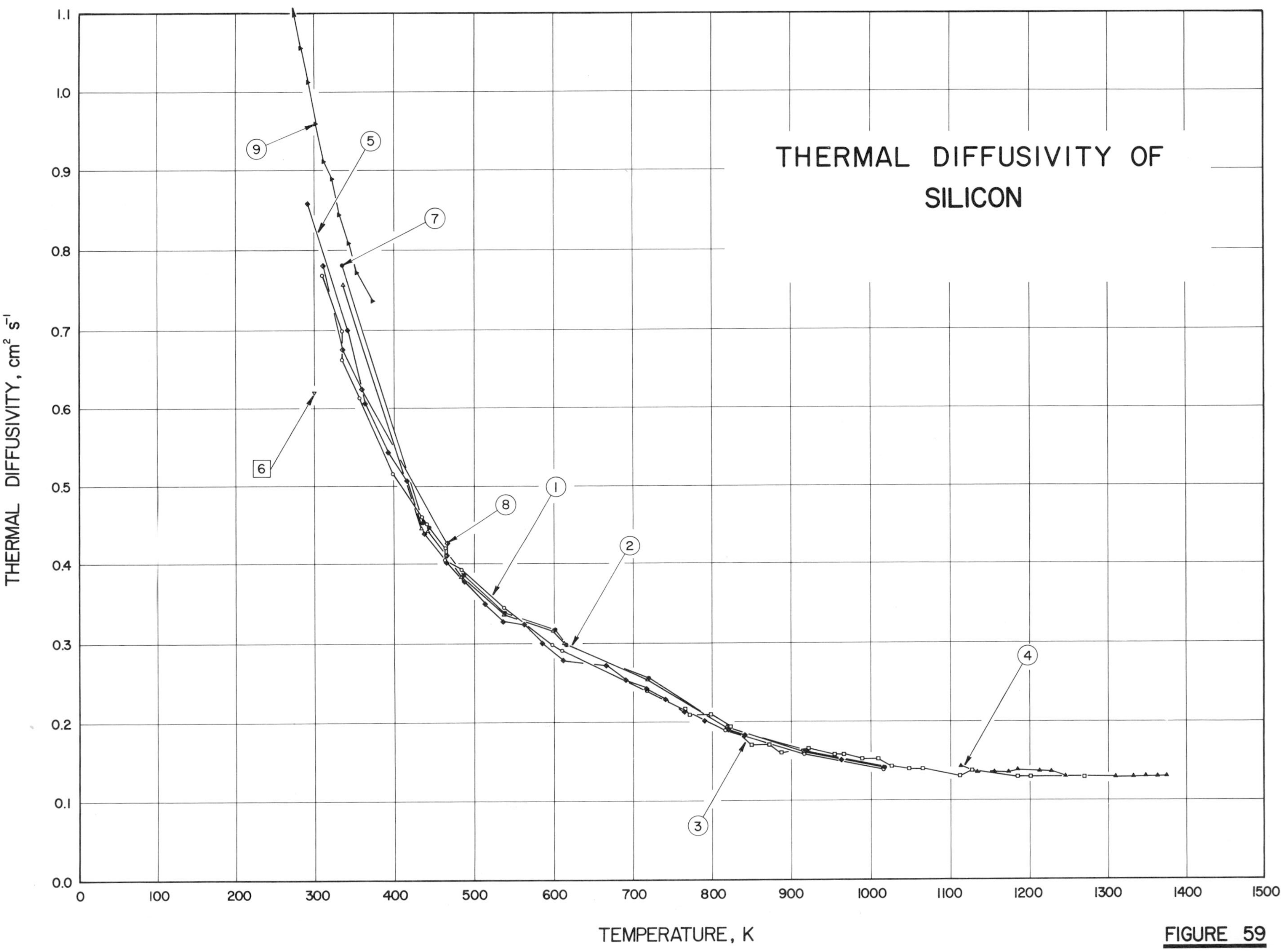
THERMAL DIFFUSIVITY OF
SILICON
THERMAL DIFFUSIVITY, cm² s⁻¹
TEMPERATURE, K
0.0
0.1
0.2
0.3
0.4
0.5
0.6
0.7
0.8
0.9
1.0
1.1
0
100
200
300
400
500
600
700
800
900
1000
1100
1200
1300
1400
1500
1
2
3
4
5
6
7
8
9

FIGURE 59

SPECIFICATION TABLE 59. THERMAL DIFFUSIVITY OF SILICON

(Impurity < 0.20% each; total impurities < 0.50%)

Cur. No.	Ref. No.	Author(s)	Year	Temp. Range, K	Reported Error, %	Name and Specimen Designation	Composition (weight percent), Specifications, and Remarks
1	47,	Abeles, B., Cheng, K. L., Cody, G. D., Baughan, B. E., Hockings, E. F., Lindenblad, N. E., and Muha, G. M.	1960	310-1016	2	Si 142	Electrical resistivity 100 ohm cm; Debye temp. 973.2 K.
2	47	Abeles, B., et al.	1960	336-1015	2	Si 142	Above specimen measured for diffusivity at different location on specimen.
3	48, 244	Shanks, H. R., Maycock, P. D., Sidles, P. H., and Danielson, G. C.	1963	766-1270		1F	Pure; n-type single crystal with axes orientation 111; cylindrical specimen 0.9 cm in dia. and 6 cm long ultrasonically cut from one end of a single crystal of silicon 2.5 cm in dia. and 7.6 cm long; outer cylinder remained joined to specimen at the heater end of rod acting as guard against radiation losses from specimen; gap between specimen and guard cylinder < 1 mm; electrical resistivity reported as 33, 29.51, 29.51, 31.99, 36.73, 44.16, 51.29, 56.89, 56.89, 58.88, 43.65, 30.20, 21.63, 17.18, 8.71, 3.72, 1.66, 0.891, and 0.132 ohm cm at 300, 302.1, 307.7, 333.3, 350.9, 378.8, 395.3, 425.5, 446.4, 469.5, 497.5, 526.3, 543.5, 578.0, 645.2, 675.7, 729.9, 781.3, and 952.4 K, respectively; measured in vacuum.
4	48, 244	Shanks, H. R., et al.	1963	1113-1375		3C	Specimen similar to above specimen but with crystal axes orientation 100; electrical resistivity reported as 1010, 1288.25, 1640.59, 1862.09, 1927.52, 1479.11, 421.70, 144.54, 31.62, 9.44, 1.17, 0.403, and 0.257 ohm cm at 300, 310.6, 336.7, 359.7, 389.1, 414.9, 450.5, 500.0, 555.6, 617.3, 769.2, 869.6, and 925.9 K, respectively; measured in vacuum.
5	48, 244	Shanks, H. R., et al.	1963	292-964		4A	Specimen similar to above specimen but p-type with crystal axes orientation 111; specific heat reported as 0.1650, 0.1840, 0.1970, 0.2025, 0.2065, 0.2105, 0.2145, 0.2180, 0.2215, 0.2250, 0.2290, and 0.2345 cal g^{-1} K^{-1} at 273, 373, 473, 573, 673, 773, 873, 973, 1073, 1173, 1273, and 1373 K, respectively; electrical resistivity reported as 125.89, and 107 ohm cm at 298.5, and 300 K, respectively; measured in vacuum.
6	118	Perron, J. C.	1961	300	~8		Square specimen 3 cm long; Angström method used to measure diffusivity.
7	192	Abeles, B., Cody, G. D., Dismukes, J. P., Hockings, E. F., Lindenblad, N. E., Richman, D., and Rosi, F. D.	1961	335-1016			Crystal specimen; electrical resistivity 100 ohm cm; diffusivity measured using modified Angström method at 5rpm and using probes.
8	192	Abeles, B., et al.	1961	311-466			Same as above except measured using thermocouples.
9	239	Ebrahimi, J.	1970	223-373			Diffusivity measured using Angström method.

DATA TABLE 59. THERMAL DIFFUSIVITY OF SILICON

(Impurity < 0.20% each; total impurities < 0.50%)

[Temperature, T, K; Thermal Diffusivity, α, cm^2 s^{-1}]

T	α
CURVE 1	
310	0.769
335	0.699
335	0.662
357	0.613
399	0.515
435	0.461
441	0.452
464	0.420
464	0.405
485	0.392
538	0.344
599	0.298
611	0.290
718	0.239
817	0.189
916	0.159
1016	0.139
CURVE 2	
336	0.758
434	0.446
442	0.442
484	0.383
537	0.336
599	0.315
613	0.300
718	0.253
818	0.192
916	0.161
1015	0.141
CURVE 3	
766	0.216
772	0.209
799	0.209
823	0.193
850	0.171
873	0.171
888	0.161
922	0.166
CURVE 3 (cont.)	
955	0.159
966	0.159
989	0.153
1009	0.153
1027	0.143
1048	0.140
1065	0.140
1112	0.131
1127	0.138
1185	0.130
1201	0.130
1270	0.130
CURVE 4	
1113	0.144
1134	0.136
1155	0.136
1173	0.136
1185	0.139
1213	0.137
1228	0.136
1246	0.131
1310	0.130
1332	0.130
1349	0.130
1363	0.130
1375	0.132
CURVE 5	
292	0.859
342	0.701
364	0.606
393	0.543
416	0.507
438	0.439
466	0.402
488	0.377
514	0.349
537	0.327
564	0.324
CURVE 5 (cont.)	
586	0.300
612	0.278
667	0.271
691	0.253
717	0.240
742	0.228
766	0.214
791	0.201
819	0.191
842	0.183
964	0.152
CURVE 6	
300	0.62
CURVE 7	
335	0.781
436	0.454
443	0.448
487	0.386
538	0.338
601	0.317
616	0.298
719	0.255
818	0.193*
919	0.162
1016	0.142
CURVE 8	
311	0.781
335	0.675
359	0.625
466	0.427
466	0.411
CURVE 9	
222.3	1.683*
233.0	1.538*
CURVE 9 (cont.)	
243.0	1.420*
252.6	1.295*
263.1	1.203*
272.8	1.106*
283.0	1.056
292.9	1.012
302.9	0.960
312.9	0.912
322.9	0.890
332.9	0.845
343.1	0.809
353.1	0.771
373.0	0.737

*- Not shown in figure.

FIGURE AND TABLE 60R. RECOMMENDED THERMAL DIFFUSIVITY OF SILVER

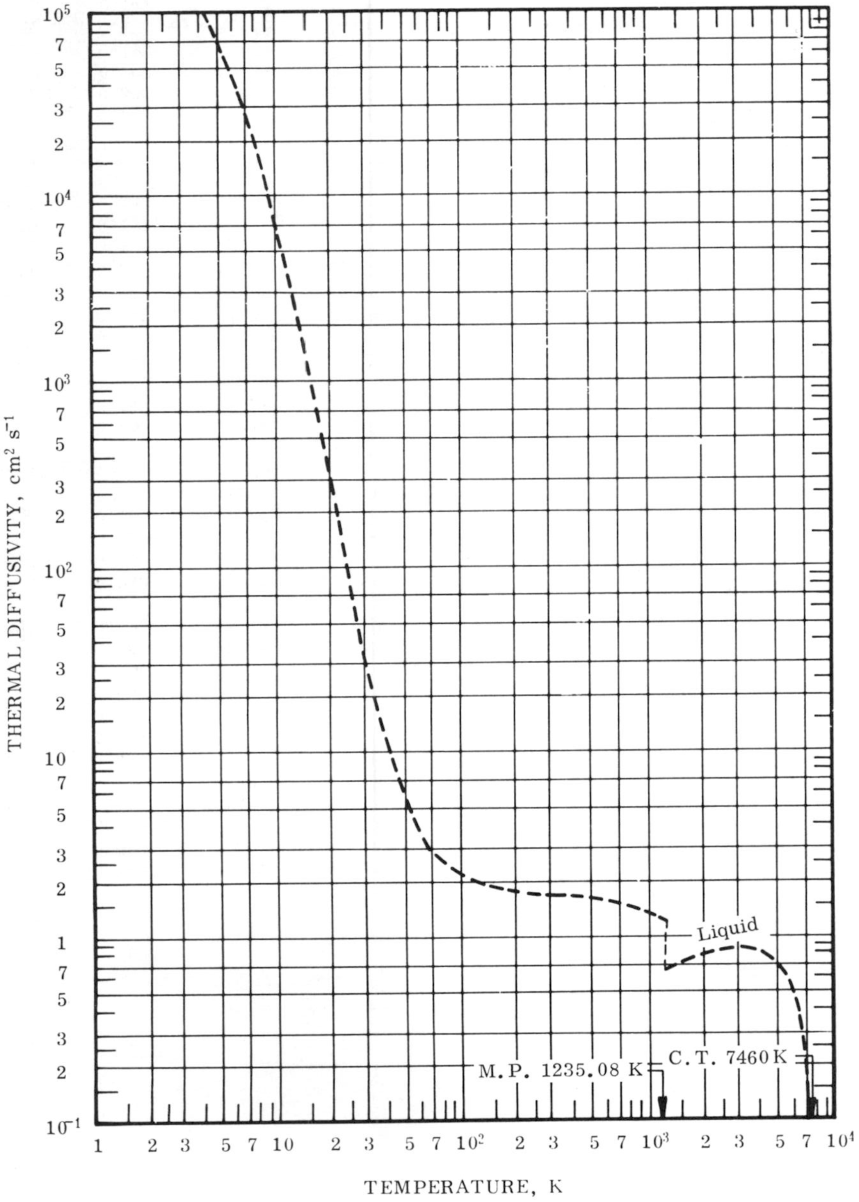

RECOMMENDED VALUES†

[Temperature, T, K; Thermal Diffusivity, α, cm² s⁻¹]

SOLID				LIQUID	
T	α	T	α	T	α
1	511000 *	70	3.12 *	1235.08	0.665 *
2	307000 *	80	2.68 *	1300	0.682 *
3	183000 *	90	2.43 *	1400	0.706 *
4	109000 *	100	2.27 *	1500	0.727 *
5	66200 *	150	1.92 *	1600	0.746 *
6	42500 *	200	1.81 *	1700	0.765 *
7	28700 *	250	1.76 *	1800	0.781 *
8	19200 *	273.2	1.75 *	1900	0.795 *
9	12800 *	300	1.74	2000	0.808 *
10	8490 *	350	1.71 *	2200	0.830 *
11	5860 *	400	1.70 *	2400	0.846 *
12	4050 *	500	1.65 *	2600	0.860 *
13	2820 *	600	1.61 *	2800	0.868 *
14	1990 *	700	1.55 *	3000	0.872 *
15	1420 *	800	1.49 *	3500	0.865 *
16	1025 *	900	1.42 *	4000	0.839 *
18	550 *	1000	1.37 *	4500	0.790 *
20	309 *	1100	1.30 *	5000	0.724 *
25	95.8 *	1200	1.24 *	6000	0.528 *
30	41.1 *	1235.08	1.23 *	7000	0.211 *
35	21.2 *				
40	12.7 *				
45	8.30 *				
50	6.11 *				
60	4.01 *				

REMARKS

The recommended values are for well-annealed high-purity silver and are considered accurate to within ± 4% of the true values near room temperature and ± 7% below 100 K and above 1000 K. At low temperatures the values are highly conditioned by impurity and imperfection, and those below 150 K are applicable only to silver having residual electrical resistivity of 0.000621 μΩ cm. The provisional values for molten silver are probably good to ± 25% from melting point to 2000 K.

†Values for molten silver are provisional.
*In temperature range where no experimental data are available.

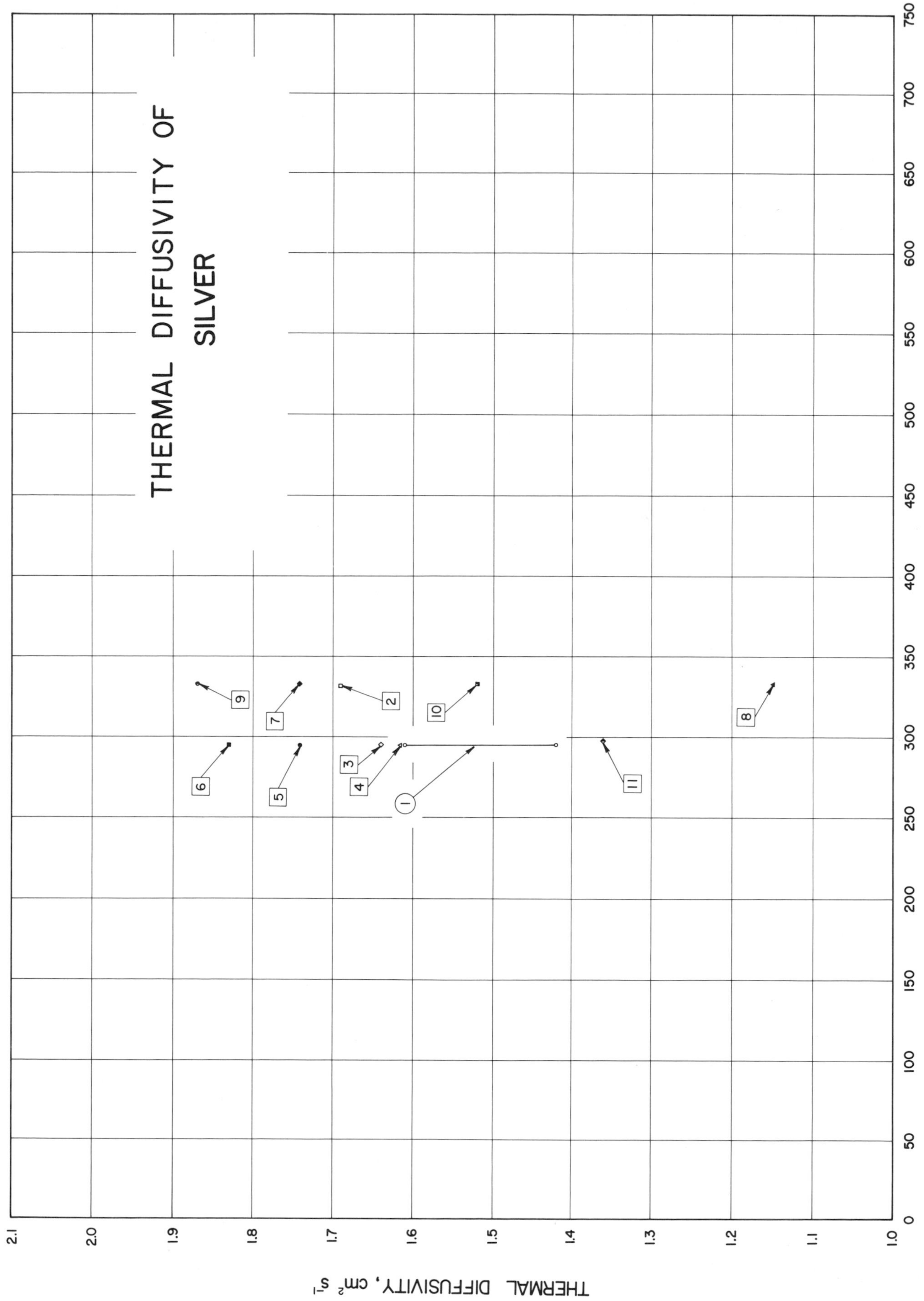
THERMAL DIFFUSIVITY OF SILVER
THERMAL DIFFUSIVITY, cm² s⁻¹
TEMPERATURE, K
0
50
100
150
200
250
300
350
400
450
500
550
600
650
700
750
1.0
1.1
1.2
1.3
1.4
1.5
1.6
1.7
1.8
1.9
2.0
2.1
1
2
3
4
5
6
7
8
9
10
11

FIGURE 60

SPECIFICATION TABLE 60. THERMAL DIFFUSIVITY OF SILVER

(Impurity <0.20% each; total impurities <0.50%)

Cur. No.	Ref. No.	Author(s)	Year	Temp. Range, K	Reported Error, %	Name and Specimen Designation	Composition (weight percent), Specifications, and Remarks
1	4, 146	Jenkins, R. J. and Parker, W. J.	1961	295	±5		Square specimen 1.9 cm side and 0.322 cm thick; high intensity short duration light pulse absorbed in front surface of thermally insulated specimen coated with camphor black; both data points obtained from measurements using different equations for data reduction.
2	161	Batalov, V. S. and Peletskii, V. E.	1968	331.8			Wire specimen 27 cm long; upper end of specimen embedded in bottom of plexiglass vessel serving as a channel for flow of heat-exchanging fluid while lower end is free to expand in a vacuum chamber; mirror reflecting a coherent ray of light from source of Michelson interferometer attached to specimen 10 cm from embedded end; thermal diffusivity determined from measured time dependence of the elongation rate of specimen; measured in a vacuum of ~10^{-3} mm Hg; temperature of measurement not reported by authors but assumed to be equal to that of the heat-exchanging fluid.
3	160	Smith, R. H.	1959	295			Specimen size 0.5 x 1 in. and 0.1215 in. thick; diffusivity measured using flash heating technique; diffusivity calculated using $\alpha = 1.37\ L^2/\pi^2\ t_{0.5}$.
4	160	Smith, R. H.	1959	295			The above measurement, but using $\alpha = 0.48\ L^2/\pi^2\ t_X$.
5	160	Smith, R. H.	1959	295			The above specimen measured again; diffusivity calculated using $\alpha = 1.37\ L^2/\pi^2\ t_{0.5}$.
6	160	Smith, R. H.	1959	295			The above measurement, but using $\alpha = 0.48\ L^2/\pi^2\ t_X$.
7	160	Smith, R. H.	1959	333			Similar to the above specimen; diffusivity measured at higher temperature using $\alpha = 1.37\ L^2/\pi^2\ t_{0.5}$.
8	160	Smith, R. H.	1959	333			The above measurement, but using $\alpha = 0.48\ L^2/\pi^2\ t_X$.
9	160	Smith, R. H.	1959	333			The above specimen measured again; diffusivity calculated using $\alpha = 1.37\ L^2/\pi^2\ t_{0.5}$.
10	160	Smith, R. H.	1959	333			The above measurement, but using $\alpha = 0.48\ L^2/\pi^2\ t_X$.
11	108	Steinberg, S., Larson, R. E., and Kydd, A. R.	1963	298			Specimen 2.0 cm square, 0.313 cm in thickness; diffusivity measuring temperature not given but assumed to be 25 C.

DATA TABLE 60. THERMAL DIFFUSIVITY OF SILVER

(Impurity <0.20% each; total impurities <0.50%)

[Temperature, T, K; Thermal Diffusivity, α, $cm^2\ s^{-1}$]

T	α	T	α	T	α	T	α	T	α	T	α
CURVE 1		CURVE 2		CURVE 4		CURVE 6		CURVE 8		CURVE 10	
295	1.61	331.8	1.69	295	1.16	295	1.83	333	1.15	333	1.52
295	1.42										
		CURVE 3		CURVE 5		CURVE 7		CURVE 9		CURVE 11	
		295	1.64	295	1.74	333	1.74	333	1.87	298	1.36

FIGURE AND TABLE 61R. RECOMMENDED THERMAL DIFFUSIVITY OF SODIUM

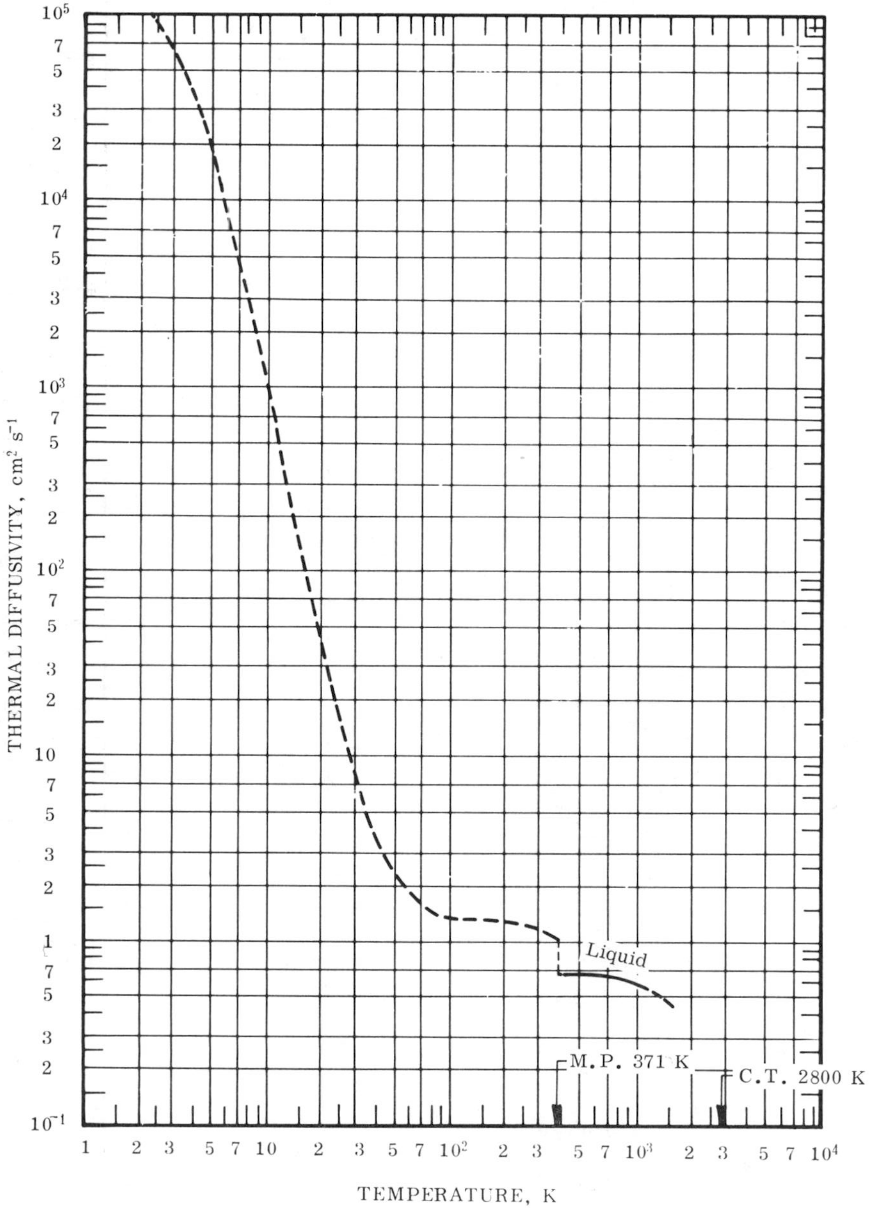

RECOMMENDED VALUES

[Temperature, T, K; Thermal Diffusivity, α, cm^2 s^{-1}]

SOLID				LIQUID	
T	α	T	α	T	α
1	208000*	35	4.79*	371	0.688*
2	109000*	40	3.40*	400	0.688
3	56000*	45	2.77*	500	0.684
4	29300*	50	2.32*	600	0.676
5	15600*	60	1.85*	700	0.663
6	8480*	70	1.61*	800	0.646
7	4690*	80	1.48*	900	0.624
8	2600*	90	1.42*	1000	0.597
9	1520*	100	1.39*	1100	0.568
10	924*	150	1.32*	1200	0.538*
11	595*	200	1.29*	1300	0.508*
12	394*	250	1.25*	1400	0.478*
13	270*	273.2	1.22*	1500	0.458*
14	191*	300	1.18*		
15	140*	350	1.07*		
16	104*	371	1.02*		
18	60.6*				
20	37.5*				
25	14.9*				
30	7.75*				

REMARKS

The values are for high-purity sodium and are thought to be accurate to within ±10% of the true values at temperatures below 60 K, ±12% from 60 K to the melting point, ±8% for the liquid state below 1000 K, and ±14% from 1000 to 1500 K. At low temperatures the values are highly conditioned by impurity and imperfections, and those below 80 K are applicable only to sodium having residual electrical resistivity of 0.00147 $\mu\Omega$ cm.

*In temperature range where no experimental data are available.

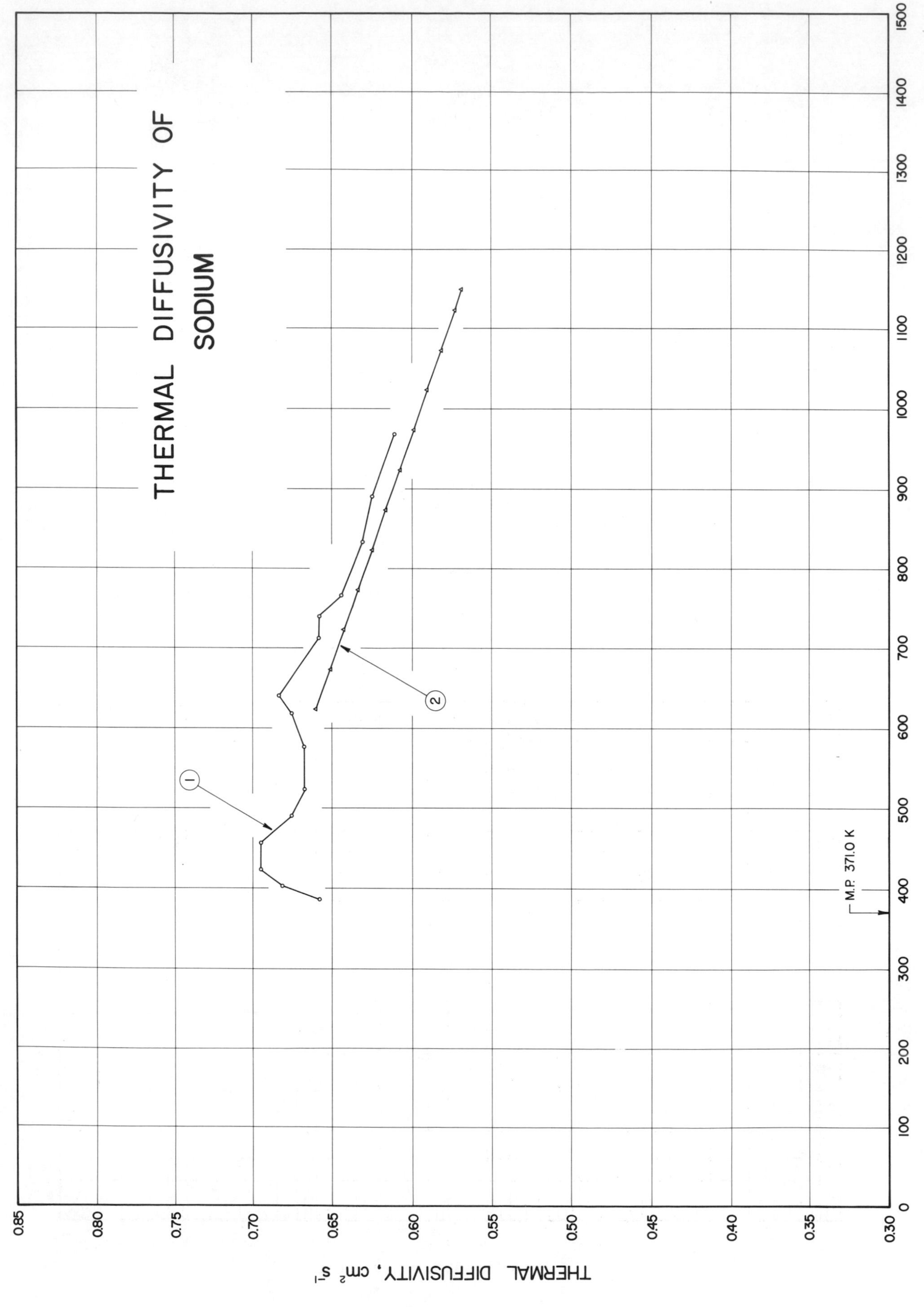

THERMAL DIFFUSIVITY OF
SODIUM
THERMAL DIFFUSIVITY, cm² s⁻¹
TEMPERATURE, K
M.P. 371.0 K
1
2
0.85
0.80
0.75
0.70
0.65
0.60
0.55
0.50
0.45
0.40
0.35
0.30
0
100
200
300
400
500
600
700
800
900
1000
1100
1200
1300
1400
1500

FIGURE 61

SPECIFICATION TABLE 61. THERMAL DIFFUSIVITY OF SODIUM

(Impurity < 0.20% each; total impurities < 0.50%)

Cur. No.	Ref. No.	Author(s)	Year	Temp. Range, K	Reported Error, %	Name and Specimen Designation	Composition (weight percent), Specifications, and Remarks
1	45, 140	Novikov, I. I., Solovyev, A. N., Khabakhpasheva, E. M., Gruzdev, V. A., Pridantsev, A. I., and Vasenina, M. Ya.	1956	386-968			Metal placed into vertically positioned thin stainless steel tube; density measured and reported as 0.920, 0.920, 0.916, 0.912, 0.909, 0.904, 0.904, 0.907, 0.904, 0.898, 0.891, 0.888, 0.880, 0.877, 0.875, 0.873, 0.871, 0.866, 0.862, 0.860, 0.857, 0.855, 0.842, 0.843, 0.834, 0.828, 0.826, 0.819, 0.815, 0.800, 0.799, and 0.790 g cm^{-3} at 393.2, 406.2, 411.2, 436.2, 456.2, 463.2, 480.2, 480.2, 490.2, 490.2, 522.2, 537.2, 601.2, 608.2, 615.2, 621.2, 634.2, 655.2, 674.2, 689.2, 692.2, 703.2, 758.2, 767.2, 798.2, 819.2, 841.2, 861.2, 882.2, 943.2, 953.2, and 992.2 K, respectively; in molten state; heater wound on the outside of upper part of tube; free metal surface maintained at an excess pressure of an inert gas; twenty radial copper screens distributed thoughout whole length of specimen; suspended in an evacuated quartz tube; Angström's dynamic method used to measure diffusivity.
2	97	Rudnev, I. I., Lyashenko, V. S., and Abramovich, M. D.	1961	623-1149			99.71^{+} Na (by difference), 0.16 Fe, 0.049 Cr, 0.04 K, 0.016 O, 0.014 Ni, <0.002 Pb, 0.0017 Mn, 0.0011 Ti, <0.001 Al, 0.00045 Cu, 0.0003 Ca, 0.00027 Mg, and 0.0002 Ag; composition obtained after completion of measurements; metal poured in a vacuum of ~1 x 10^{-2} mm Hg into a thin walled tube made of steel 1Kh 18 N9T: 8.6 mm in dia, 0.2 mm wall thickness, and 230 mm long; lower and upper steel plugs hermetically joined to the tube by argon-arc welding; necessary compensating volume provided between upper plug and metal surface; specimen heated in vertical electric furnace; measured in vacuum of ~10^{-4} mm Hg; temp versus time recorded for two points on specimen; six curves recorded for each temp. point; Angström's method used to measure diffusivity; diffusivity data calculated from equation given by author.

DATA TABLE 61. THERMAL DIFFUSIVITY OF SODIUM

(Impurity <0.20% each; total impurities < 0.50%)

[Temperature, T, K; Thermal Diffusivity, α, cm^2 s^{-1}]

T	α	T	α	T	α	T	α
CURVE 1		CURVE 1 (cont.)		CURVE 2		CURVE 2 (cont.)	
386	0.658	713	0.658	623	0.660	1023	0.591
403	0.681	740	0.658	673	0.651	1073	0.582
423	0.695	766	0.644	723	0.643	1123	0.573
456	0.695	833	0.631	773	0.634	1149	0.569
490	0.675	890	0.625	823	0.625		
523	0.667	968	0.611	873	0.617		
576	0.667			923	0.608		
618	0.675			973	0.599		
640	0.683						

FIGURE AND TABLE 62R. PROVISIONAL THERMAL DIFFUSIVITY OF STRONTIUM

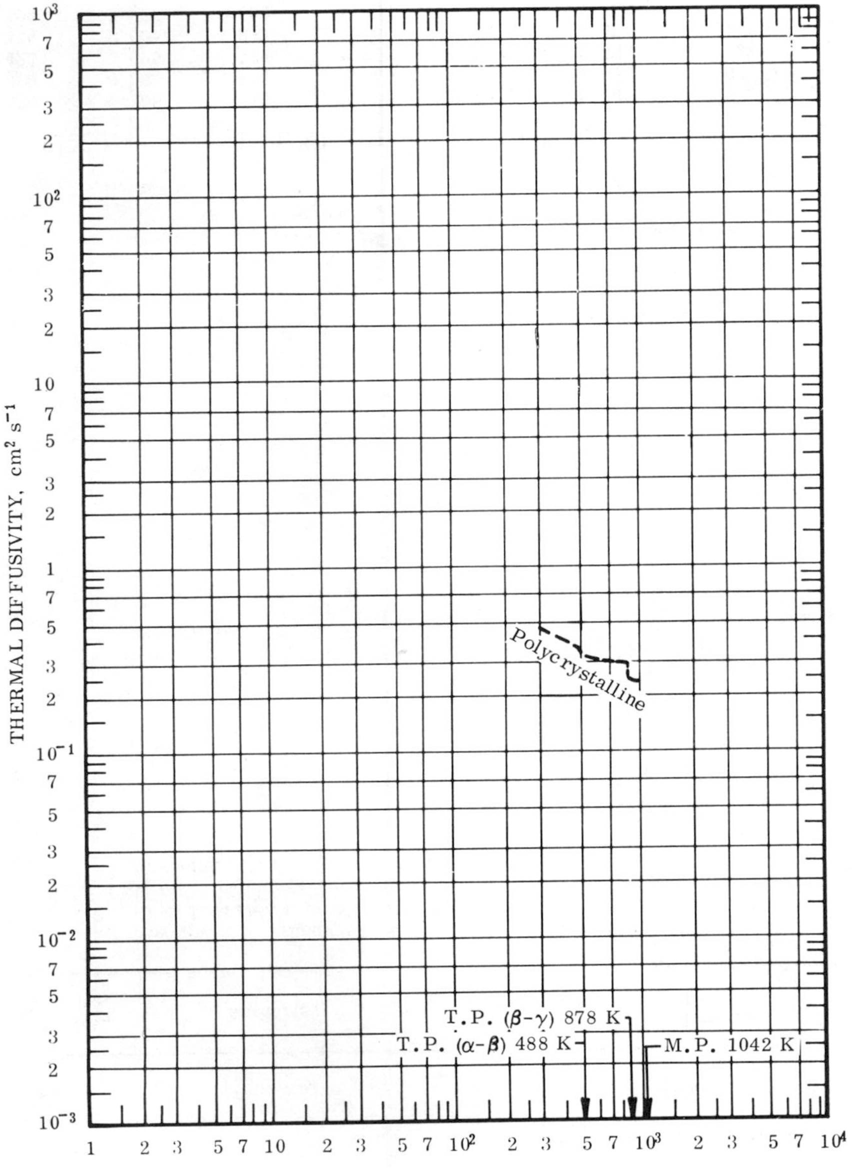

PROVISIONAL VALUES*

[Temperature, T, K; Thermal Diffusivity, α, cm^2 s^{-1}]

SOLID
(Polycrystalline)

T	α
298.2	0.453
300	0.452
350	0.418
400	0.390
488	0.351
488	0.329
500	0.326
600	0.310
700	0.300
800	0.296
878	0.295
878	0.243
900	0.237
1000	0.233

REMARKS

The provisional values are for high-purity strontium and are probably good to ±25% below 450 K. The uncertainty increases above 450 K due to the effect of the phase transformations.

*All values are estimated.

FIGURE AND TABLE 63R. RECOMMENDED THERMAL DIFFUSIVITY OF SULFUR

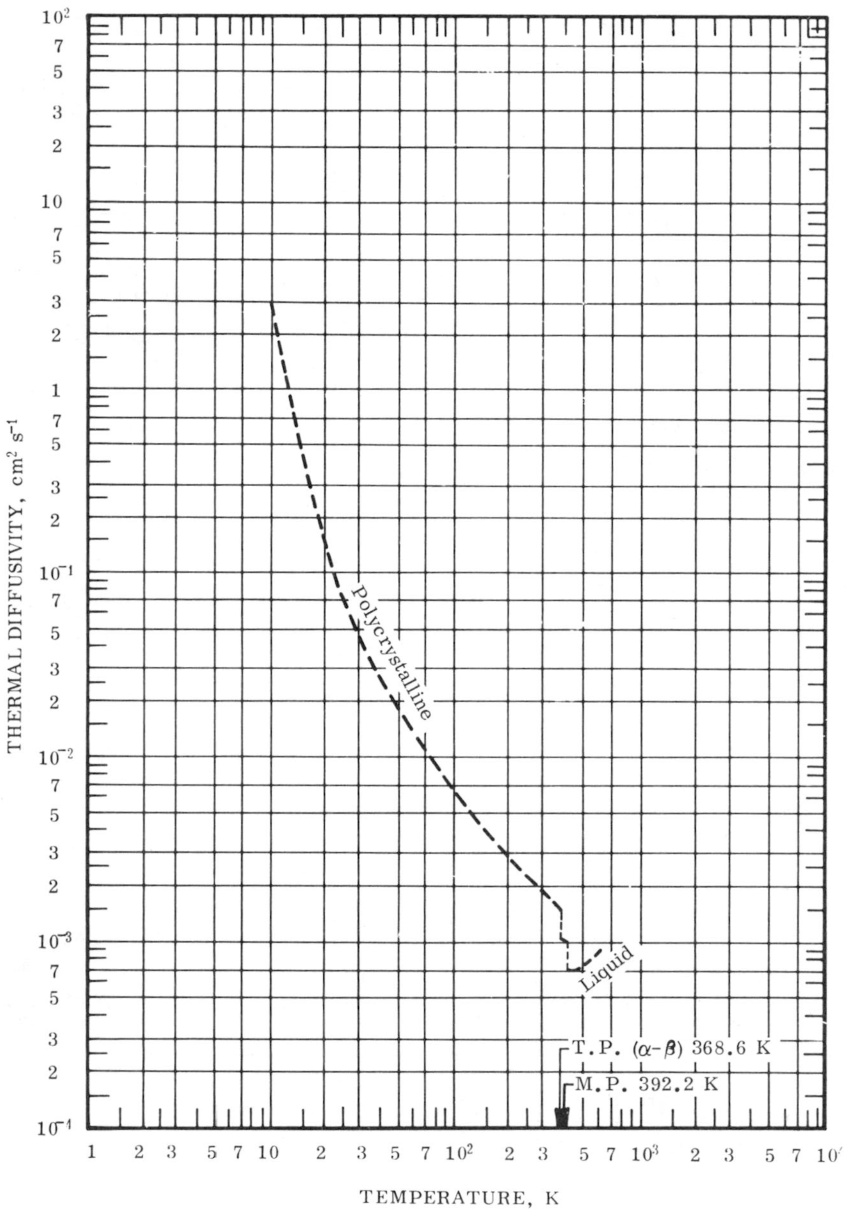

RECOMMENDED VALUES†

[Temperature, T, K; Thermal Diffusivity, α, cm^2 s^{-1}]

SOLID (Polycrystalline)				LIQUID	
T	α	T	α	T	α
10	2.90*	150	0.00390*	392.2	0.000725*
11	1.83*	200	0.00278*	400	0.000725*
12	1.20*	250	0.00219*	500	0.000765*
13	0.830*	273.2	0.00201*	600	0.000921*
14	0.598*	300	0.00184		
15	0.439*	350	0.00158		
16	0.335*	368.6	0.00150		
18	0.205*	368.6	0.00105		
20	0.137*	392.2	0.00102*		
25	0.0697*				
30	0.0460*				
35	0.0342*				
40	0.0268*				
45	0.0220*				
50	0.0185*				
60	0.0138*				
70	0.0109*				
80	0.00896*				
90	0.00760*				
100	0.00658*				

REMARKS

The values are for high-purity sulfur and are thought to be accurate to within ±13% of the true values from 70 K to room temperature and ±8% from room temperature to the melting point. The values below 70 K are merely typical values and represent a typical curve serving to indicate the general trend of the thermal diffusivity. The values for liquid sulfur are considered accurate to within ±8%.

†Values below 70 K are merely typical values.
*In temperature range where no experimental data are available.

SPECIFICATION TABLE 63. THERMAL DIFFUSIVITY OF SULFUR

(Impurity $< 0.20\%$ each; total impurities $< 0.50\%$)

Cur. No.	Ref. No.	Author(s)	Year	Temp. Range, K	Reported Error, %	Name and Specimen Designation	Composition (weight percent), Specifications, and Remarks
1*	166	Williams, J.	1923	318, 373			Density 2.00 g cm^{-3}.
2*	253	Hecht, H.	1904	298			10.8 cm cubic specimen; density 2.03 g cm^{-3}; measuring temperature not reported and here assumed to be 25 C.
3*	309	Neumann, F.	1862	323		Melted sulfur	Cubic specimen 5 to 6 in. on side, or sphere of same diameter; uniformly heated and then cooled in air; temperatures at center and surface observed by means of thermo-electric rods.

DATA TABLE 63. THERMAL DIFFUSIVITY OF SULFUR

(Impurity $< 0.20\%$ each; total impurities $< 0.50\%$)

[Temperature, T, K; Thermal Diffusivity, α, $cm^2\ s^{-1}$]

T	α
CURVE 1*	
318	0.00034
373	0.00034
CURVE 2*	
298	0.0017
CURVE 3*	
323	0.00168

* No figure given.

FIGURE AND TABLE 64R. RECOMMENDED THERMAL DIFFUSIVITY OF TANTALUM

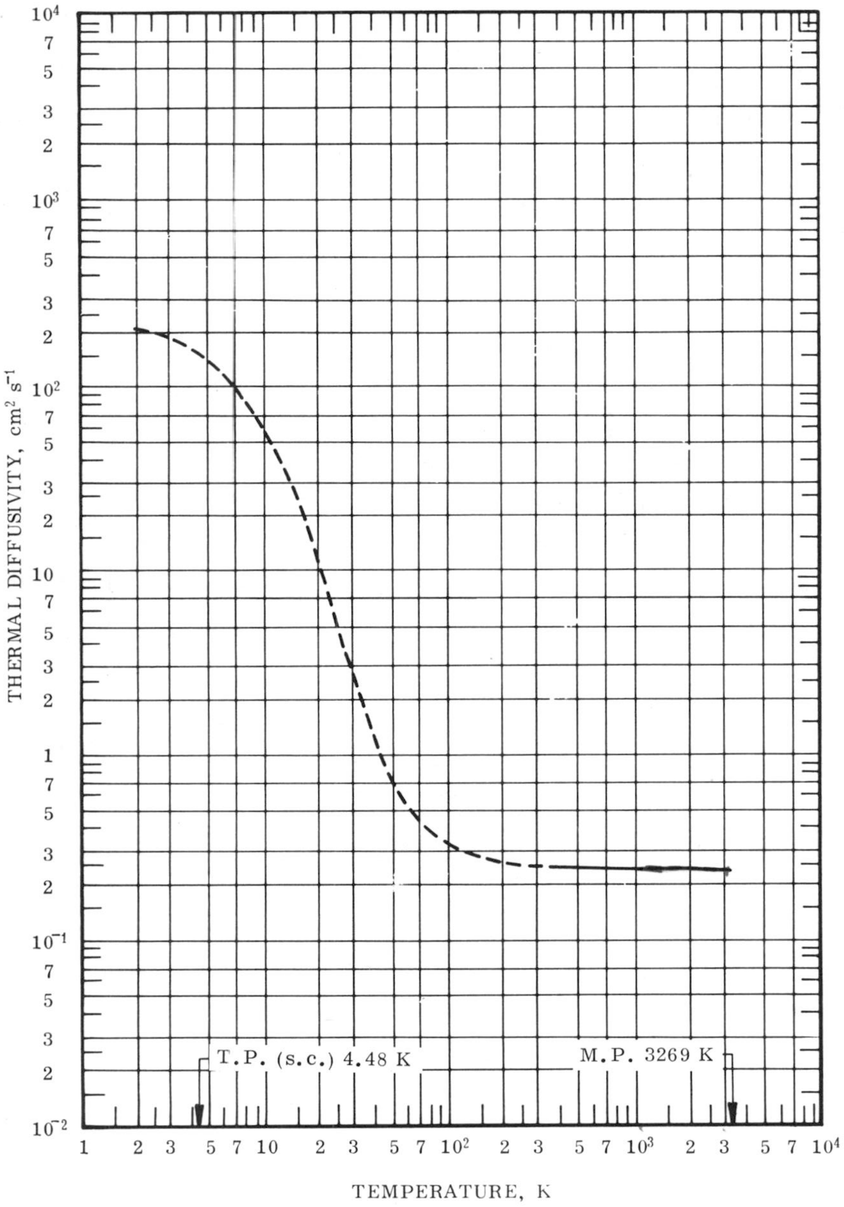

RECOMMENDED VALUES

[Temperature, T, K; Thermal Diffusivity, α, cm^2 s^{-1}]

SOLID

T	α	T	α	T	α
2	201*	40	1.21*	1000	0.242
3	180*	45	0.900*	1100	0.242
4	157*	50	0.719*	1200	0.242*
5	134*	60	0.524*	1300	0.242*
6	113*	70	0.428*	1400	0.242*
7	95.8*	80	0.374*	1500	0.242*
8	80.9*	90	0.342*	1600	0.242*
9	68.3*	100	0.323*	1700	0.242*
10	57.8*	150	0.276*	1800	0.242*
11	49.0*	200	0.260*	1900	0.242*
12	41.3*	250	0.252*	2000	0.242*
13	34.6*	273.2	0.249*	2200	0.242*
14	29.0*	300	0.247*	2400	0.242*
15	24.5*	350	0.244*	2600	0.241*
16	20.75*	400	0.243	2800	0.240*
18	15.0*	500	0.242	3000	0.238*
20	10.9*	600	0.243	3200	0.235*
25	5.36*	700	0.243		
30	2.93*	800	0.242		
35	1.80*	900	0.242		

REMARKS

The recommended values are for well-annealed high-purity tantalum and are thought to be accurate to within ± 7% of the true values at moderate temperatures and ± 13% at low and high temperatures. At low temperatures the values are highly conditioned by impurity and imperfection, and those below 100 K are applicable only to tantalum having residual electrical resistivity of 0.214 $\mu\Omega$ cm.

* In temperature range where no experimental data are available.

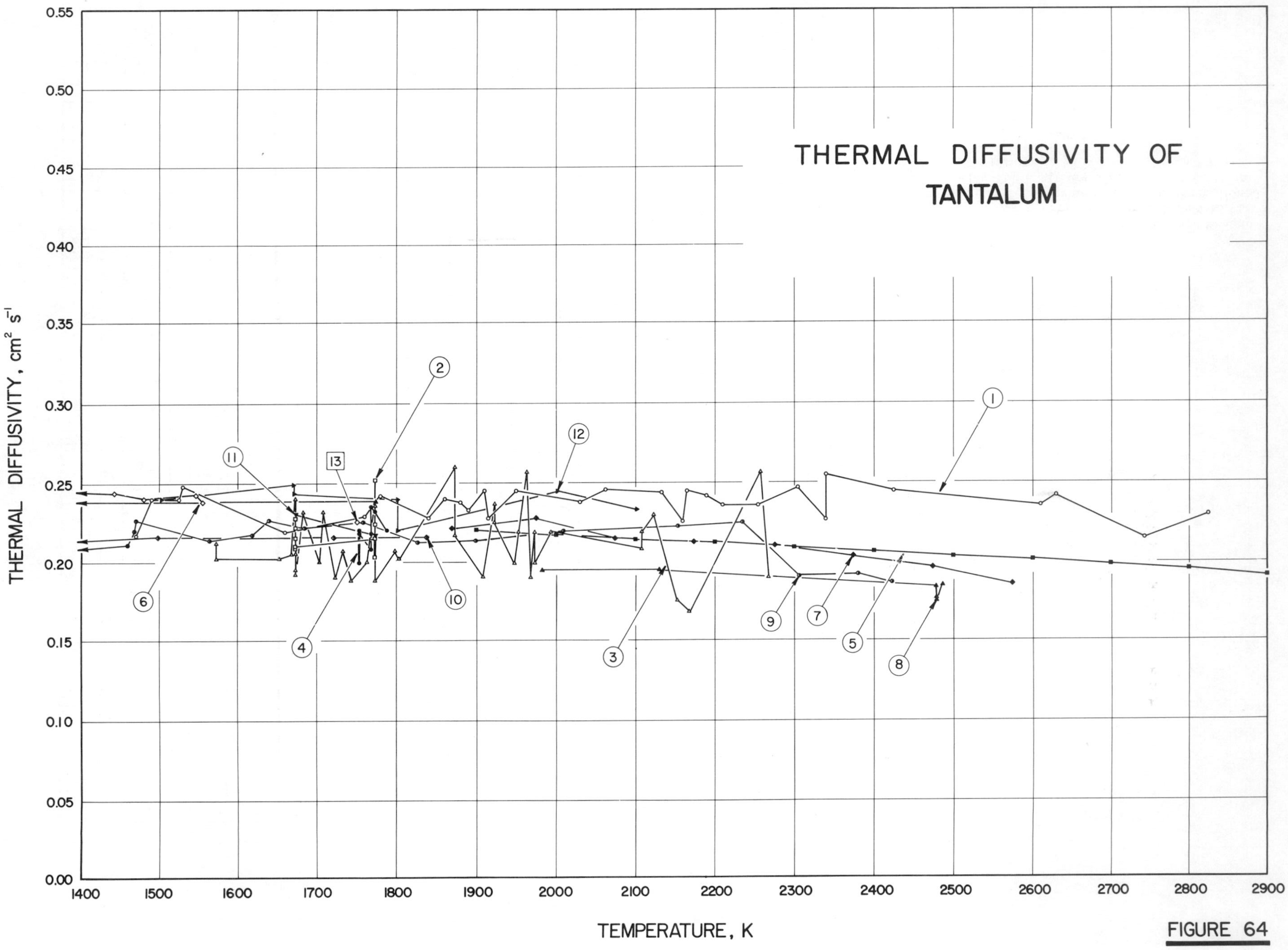
THERMAL DIFFUSIVITY OF
TANTALUM
THERMAL DIFFUSIVITY, cm² s⁻¹
TEMPERATURE, K
0.55
0.50
0.45
0.40
0.35
0.30
0.25
0.20
0.15
0.10
0.05
0.00
1400
1500
1600
1700
1800
1900
2000
2100
2200
2300
2400
2500
2600
2700
2800
2900
1
2
3
4
5
6
7
8
9
10
11
12
13

FIGURE 64

SPECIFICATION TABLE 64. THERMAL DIFFUSIVITY OF TANTALUM

(Impurity < 0.20% each; total impurities < 0.50%)

Cur. No.	Ref. No.	Author(s)	Year	Temp. Range, K	Reported Error, %	Name and Specimen Designation	Composition (weight percent), Specifications, and Remarks
1	39	Wheeler, M. J.	1965	1470-2825			99.89$^+$ Ta (by difference), <0.1 Nb, <0.01 C, and traces of other elements; avg grain size after testing 140 μm; rectangular specimen having top surface area lying in the range from 0.3 to 1.3 cm^2 and 0.040 in. in thickness; specimen cut from sheet of same thickness, vacuum beam melted and supplied by Murex; density 16.6 g cm^{-3}; Lorenz function reported as 2.643, 2.641, 2.642, 2.642, 2.649, 2.659, 2.668, and 2.670 x 10^{-8} V^2 K^{-2} at 1397, 1597, 1798, 1998, 2199, 2404, 2601, and 2804 K, respectively; specimen heated to incandescence by an electron beam; intensity of beam sinusoidally modulated; measured under a vacuum of ~5 x 10^{-6} mm Hg.
2	254, 32, 50, 182	Taylor, R. E. and Nakata, M. M.	1962	1673-1773	7	1	Metallic impurity content not determined but should be similar to that of specimens No. 2 and No. 3 below; grain size ranging from 2 mm down to ~15 μ, randomly distributed; cylindrical specimen 1.588 cm in dia. and 3.495 cm long; two parallel sight holes each 0.160 cm in dia drilled to a depth of 1.47 cm at radii r_1 = 0 and r_2 = 0.548 cm; machined out of 1 in. dia rod; sight holes drilled by Elox technique; measured after extensive heat soaking at temps. up to 2673.2 K; density 16.67 g cm^{-3}; radiation shields mounted at both ends; measured under a vacuum of 1 x 10^{-6} mm Hg; radial diffusivity technique used; specimen heated during measurement.
3	254, 32, 51, 50, 182, 255	Taylor, R. E. and Nakata, M. M.	1962	1573-2268	7	2	99.778$^+$ Ta (by difference), 0.05 Si, <0.05 Fe, <0.05 Ni, <0.05 Zr, 0.01 Ca, 0.01 Cu, and 0.002 Mg; grain size ranging from 2 mm down to ~15 μ, randomly distributed; cylindrical specimen 1.590 cm in dia. and 3.498 cm long; two parallel sight holes each 0.162 cm in dia. drilled to a depth of 1.55 cm at radii r_1 = 0 and r_2 = 0.562 cm; machined out of 1 in. dia rod; sight holes drilled by Elox technique; measured after extensive heat soaking at temps. up to 2673.2 K; density 16.66 g cm^{-3}; radiation shields mounted at both ends; measured under a vacuum of 1 x 10^{-6} mm Hg; radial diffusivity technique used; specimen heated during measurement.
4	32	Taylor, R. E. and Nakata, M. M.	1963	1753-1788	7	3	99.778$^+$ Ta (by difference), 0.05 Si, <0.05 Fe, <0.05 Ni, <0.05 Zr, 0.01 Ca, 0.01 Cu, and 0.002 Mg; grain size ranging from 2 mm down to ~15 μ, randomly distributed; cylindrical specimen 2.093 cm in dia and 3.683 cm long; two parallel sight holes each 0.149 cm in dia. drilled to a depth of 1.56 cm at radii r_1 = 0 and r_2 = 0.746 cm; machined out of 1 in. dia. rod; sight holes drilled by Elox technique; measured after extensive heat soaking at temps. up to 2673.2 K; density 16.66 g cm^{-3}; radiation shields mounted at both ends; measured under a vacuum of 1 x 10^{-6} mm Hg; radial diffusivity technique used; specimen cooled during measurement.

SPECIFICATION TABLE 64. THERMAL DIFFUSIVITY OF TANTALUM (continued)

Cur. No.	Ref. No.	Author(s)	Year	Temp. Range, K	Reported Error, %	Name and Specimen Designation	Composition (weight percent), Specifications, and Remarks
5	121, 122	Kraev, O.A. and Stel'makh, A.A.	1964	1900-3150	±5		Disk specimen 8-9 mm in dia and 0.2 mm thick; diffusivity measured using wave method; data points corrected for thermal expansion.
6	134	Denman, G. L.	1966	297-1555	<±2.3		99.94615^{+} Ta (by difference), 0.02 Nb, 0.01 Fe, <0.01 Zr, 0.0080 O, 0.0021 N, 0.002 Cr, 0.0017 C, and 0.00005 H; grain size (as received) in the range from ~0.04 to 0.06 mm; disc-shaped specimens each 0.270 in. in dia. and 0.060 in. thick; a minimum of two samples measured; supplied by Fansteel Met. Corp.; made by powder metallurgy technique; density 16.50 g cm^{-3}; ruby laser used as pulse energy source; diffusivity determined from measured history of back face temp; measured in vacuum; each data point reported represents average of four or five measurements.
7	172	Kraev, O. A. and Stel'makh, A. A.	1966	1868-2573	±5		Disk specimen 7 to 9 mm in diameter and 0.2 mm thick; cut from plate of rolled metal; surfaces thoroughly cleaned; heated by electron bombardment; sinusoidal temperature oscillation imposed on one face of specimen; thermal diffusivity determined from phase difference between measured temperature fluctuations at front and back faces of specimen; reported data points corrected for thermal expansion and obtained from smooth curve given by authors.
8	213	Morrison, B. H., Klein, D. J., and Cowder, L. R.	1965	1085-2485			<0.0040 W, 0.0015 Si, 0.0010 Fe, <0.0010 Mo, 0.0005 Ti, 0.0002 Ni, <0.0001 Cr, and <0.0001 Cu; 0.5 in. diameter x 0.030 in. thick; density 16.6 g cm^{-3}.
9	213	Morrison, B. H., et al.	1965	1160-2421			Similar to the above specimen but 0.050 in. thick.
10	213	Morrison, B. H., et al.	1965	1215-1837			Similar to the above specimen but 0.090 in. thick.
11	257	Taylor, R. E. and Nakata, M. M.	1961	1673, 1773			The same specimen and measuring technique as for curve 2.
12	258, 266	Chafik, E., Mayer, R., and Pruschek, R.	1968	1501-2100			0.04 Nb, 0.02 O, 0.01 C, 0.01 W, 0.005 each of Ca, Ni, and Mo, 0.004 N, 0.0025 Fe, 0.001 each of Cr, Cu, P, K, Si, Na, S, Ti, V, and Zr, 0.0005 each of Al, Mn, and H, and 0.0001 B; obtained from W. C. Heraeus Co.
13	219	Null, M. R. and Lozier, W. W.	1969	1750			12.7 mm diameter x 1.060 mm thick; bulk density 16.6 g cm^{-3}; electrical resistivity 13.08 $\mu\Omega$cm at room temperature.

DATA TABLE 64. THERMAL DIFFUSIVITY OF TANTALUM

(Impurity < 0.20% each; total impurities < 0.50%)

[Temperature, T, K; Thermal Diffusivity, α, $cm^2 s^{-1}$]

CURVE 1

T	α
1470	0.217
1490	0.240
1525	0.240
1530	0.248
1660	0.219
1760	0.229
1780	0.242
1840	0.228
1860	0.240
1880	0.238
1890	0.233
1910	0.245
1915	0.228
1950	0.245
2030	0.238
2063	0.246
2133	0.244
2160	0.226
2165	0.245
2190	0.242
2210	0.236
2255	0.236
2305	0.247
2340	0.227
2340	0.255
2425	0.245
2610	0.236
2630	0.242
2743	0.215
2825	0.230

CURVE 2

T	α
1673	0.215
1673	0.228
1673	0.215
1673	0.209
1673	0.222
1673	0.210
1775	0.215
1773	0.203
1773	0.203
1773	0.215
1773	0.224
1773	0.252

CURVE 3

T	α
1573	0.212
1573	0.202
1653	0.202
1668	0.205
1673	0.240
1673	0.215*
1673	0.216*
1673	0.215*
1673	0.195
1673	0.211*
1673	0.205
1673	0.192
1683	0.232
1703	0.200
1708	0.232
1723	0.190
1733	0.207
1743	0.188
1763	0.200
1773	0.236
1773	0.225*
1773	0.232*
1773	0.224*
1773	0.227*
1773	0.237*
1773	0.239*
1773	0.214*
1773	0.188
1798	0.207
1803	0.202
1873	0.260
1873	0.217
1908	0.190
1923	0.237
1923	0.225
1948	0.199
1953	0.219
1963	0.257
1968	0.190
1973	0.219
1973	0.199
1993	0.219
2108	0.208
2108	0.219
2123	0.230
2153	0.175
2168	0.168
2258	0.257
2268	0.190

CURVE 4

T	α
1753	0.220
1753	0.199
1753	0.218
1768	0.208
1768	0.218
1768	0.235
1773	0.232
1788	0.220

CURVE 5

T	α
1900	0.220
2000	0.217
2100	0.214
2200	0.212
2300	0.209
2400	0.206
2500	0.203
2600	0.201
2700	0.198
2800	0.195
2900	0.191
3000	0.186*
3100	0.178*
3150	0.175*

CURVE 6

T	α
297	0.2611*
300	0.2538*
313	0.2549*
380	0.2562*
382	0.2602*
519	0.2453*
598	0.2464*
650	0.2401*
735	0.2371*
772	0.2542*
786	0.2416*
819	0.2537*
841	0.2415*
852	0.2345*
897	0.2533*
931	0.2453*
1005	0.2511*
1092	0.2499*
1147	0.2485*
1314	0.2510*
1329	0.2447*
1442	0.2436
1480	0.2403
1547	0.2426
1555	0.2383

CURVE 7

T	α
1868	0.222
1973	0.218
2073	0.215
2173	0.213
2273	0.210
2373	0.204
2473	0.197
2573	0.186

CURVE 8

T	α
1085	0.215*
1126	0.216*
1137	0.222*
1683	0.222
1981	0.195
2129	0.195
2477	0.184
2477	0.177
2477	0.175
2485	0.185

CURVE 9

T	α
1160	0.226*
1235	0.207*
1268	0.232*
1326	0.230*
1346	0.207*
1458	0.211
1466	0.220
1469	0.227
1563	0.214
1617	0.218
1639	0.227
1757	0.226
1825	0.213
1898	0.214
2008	0.220
2152	0.223
2234	0.225
2306	0.191
2379	0.192
2421	0.187

CURVE 10

T	α
1215	0.238*
1285	0.231*
1389	0.213*
1497	0.216
1720	0.216
1837	0.216

CURVE 11

T	α
1673	0.230
1773	0.217

CURVE 12

T	α
1501	0.2400
1669	0.2490
1669	0.2439
1800	0.2400
1800	0.2202
2000	0.2451
2100	0.2336

CURVE 13

T	α
1750	0.226

*Not shown in figure.

FIGURE AND TABLE 65R. RECOMMENDED THERMAL DIFFUSIVITY OF TECHNETIUM

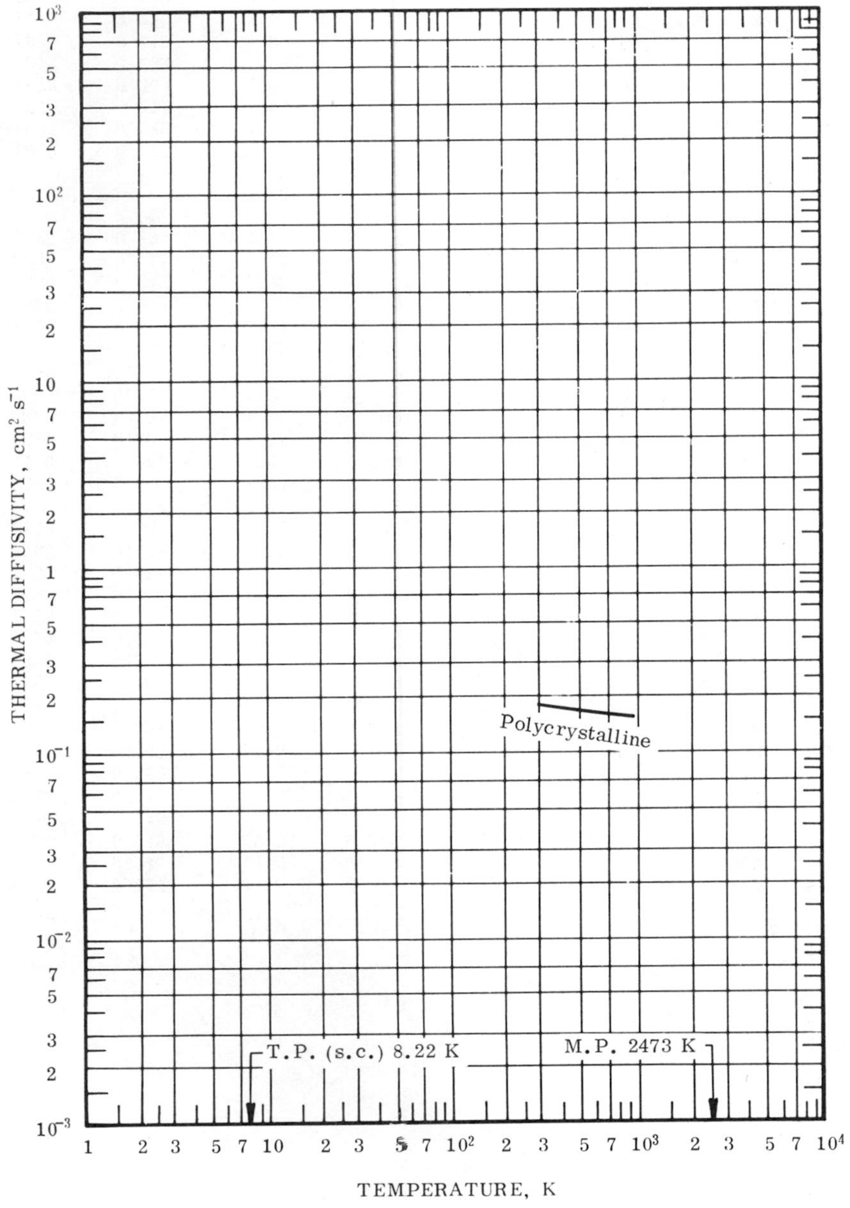

RECOMMENDED VALUES

[Temperature, T, K; Thermal Diffusivity, α, $cm^2\ s^{-1}$]

SOLID
(Polycrystalline)

T	α
298.2	0.179*
300	0.179
350	0.175
400	0.171
500	0.165
600	0.161
700	0.159
800	0.158
900	0.158*

REMARKS

The values are for well-annealed high-purity polycrystalline technetium and are thought to be accurate to within ± 10% near room temperature and ± 15% at the highest temperatures.

* Extrapolated.

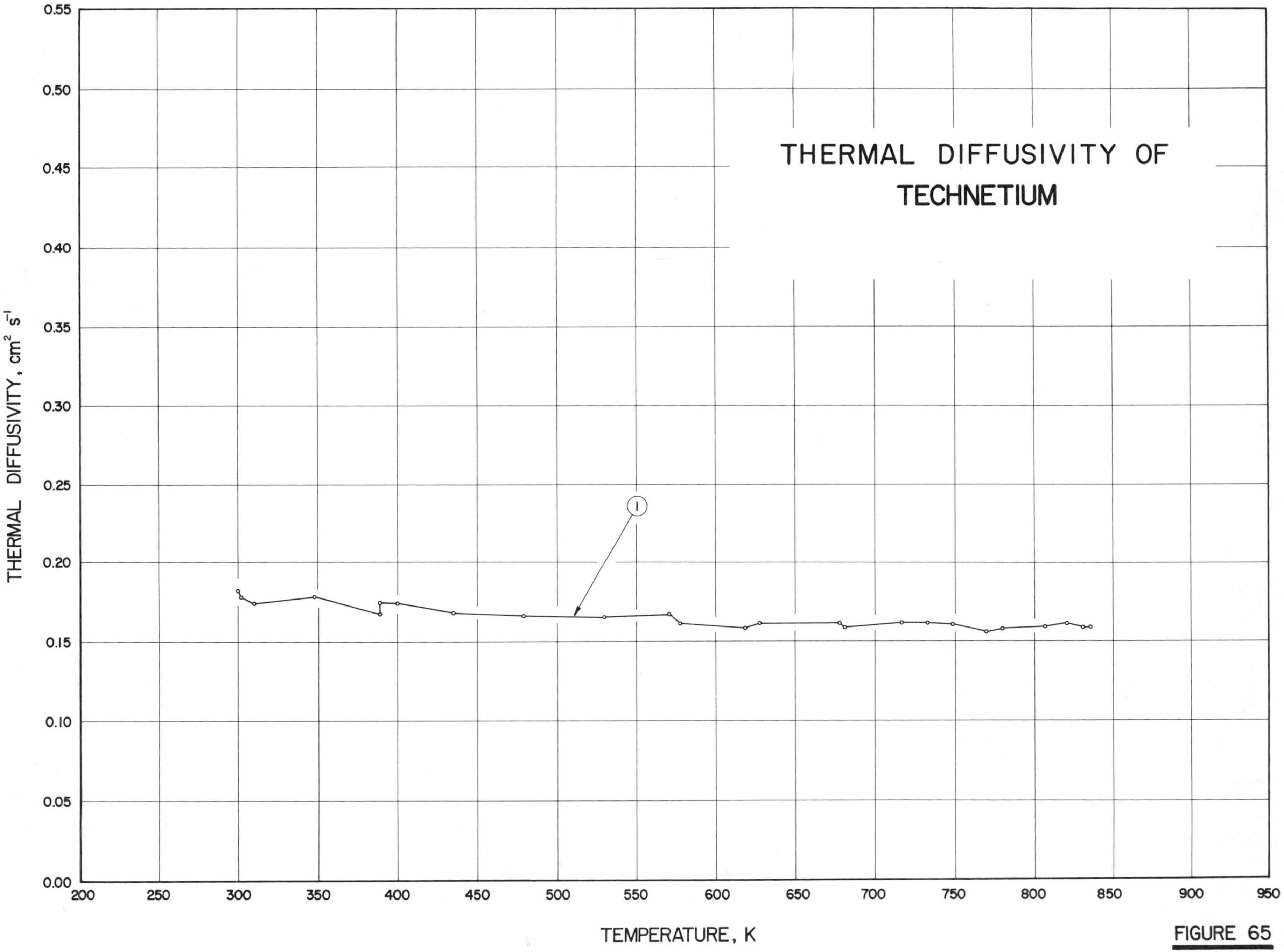
THERMAL DIFFUSIVITY OF
TECHNETIUM
THERMAL DIFFUSIVITY, cm² s⁻¹
TEMPERATURE, K
0.00
0.05
0.10
0.15
0.20
0.25
0.30
0.35
0.40
0.45
0.50
0.55
200
250
300
350
400
450
500
550
600
650
700
750
800
850
900
950
1

FIGURE 65

SPECIFICATION TABLE 65. THERMAL DIFFUSIVITY OF TECHNETIUM

(Impurity < 0.20% each; total impurities < 0.50%)

Cur. No.	Ref. No.	Author(s)	Year	Temp. Range, K	Reported Error, %	Name and Specimen Designation	Composition (weight percent), Specifications, and Remarks
1	106	Baker, D. E.	1965	300-836			0.015 total metallic impurities (principally Al and Fe) (as reduced), 0.0002 O and 0.0003 N (after melting); lattice parameters (as cast) a_0 = 2.7414 A and c_0 = 4.3997 A; wafer specimen ~2.3 cm in dia and 0.12 cm thick; isolated from fission product wastes and reduced to metal in a hydrogen furnace, eight gram button melted in an electron beam evaporator unit, button vacuum encapsulated in a heavy walled molybdenum container, heated to 1813.2 K by induction, and the assembly press forged from 2.16 to 0.95 cm height in one pass; molybdenum cladding removed chemically using HNO_3, H_2SO_4, H_2O mixture; technetium ~0.21 cm in thickness ground on both surfaces with metallographic equipment to produce flat wafer specimen with above dimensions, only partially recrystallized after this working; as-cast hardness DPH 108 to 134 (120 average); measured in a helium atmosphere; flash method used to measure diffusivity; xenon tube used to impart short pulse of radiant energy to front surface of sample.

DATA TABLE 65. THERMAL DIFFUSIVITY OF TECHNETIUM

(Impurity < 0.20% each; total impurities < 0.50%)

[Temperature, T, K; Thermal Diffusivity, α, $cm^2 s^{-1}$]

T	α	T	α
CURVE 1		CURVE 1 (cont.)	
300	0.182	681	0.158
302	0.178	717	0.161
310	0.174	733	0.161
348	0.178	749	0.160
389	0.167	770	0.155
389	0.174	780	0.157
400	0.174	807	0.158
435	0.168	821	0.160
479	0.166	831	0.158
530	0.165	836	0.158
571	0.167		
578	0.161		
619	0.158		
628	0.161		
678	0.161		

FIGURE AND TABLE 66R. PROVISIONAL THERMAL DIFFUSIVITY OF TELLURIUM

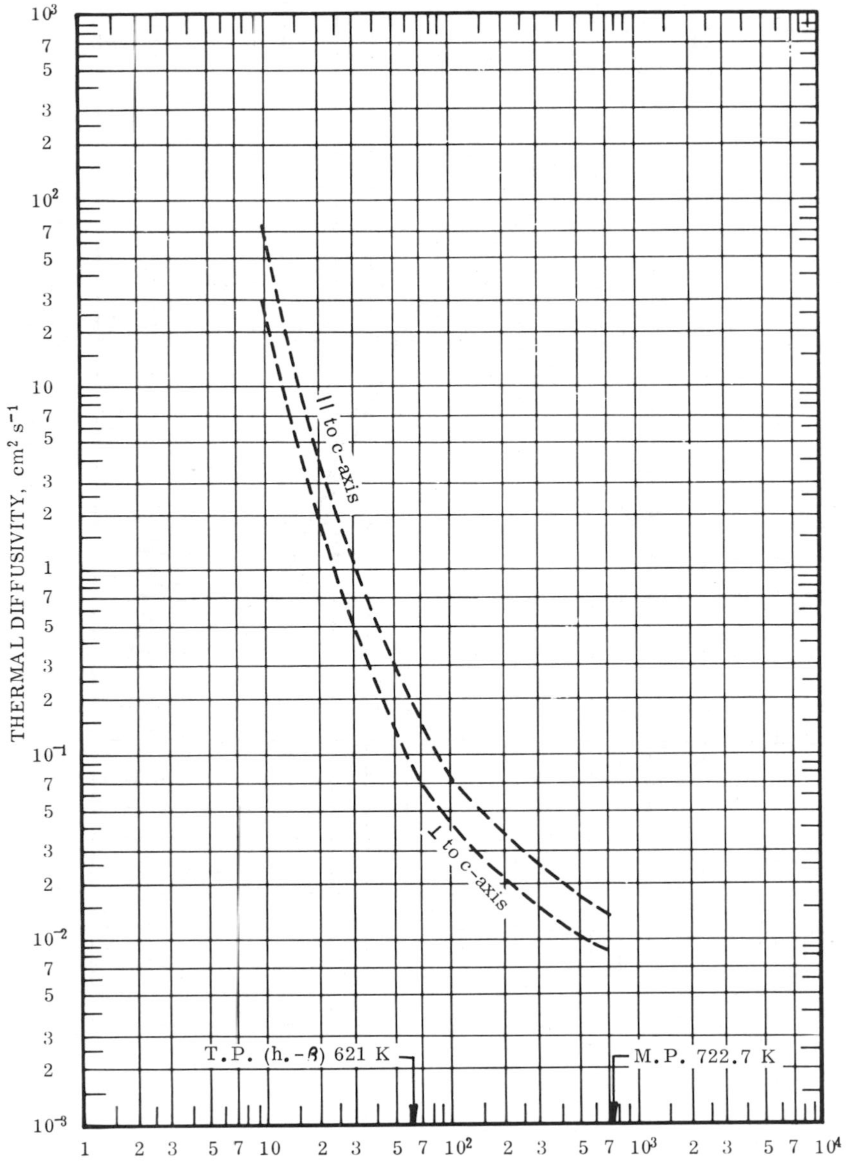

PROVISIONAL VALUES*

[Temperature, T, K; Thermal Diffusivity, α, $cm^2 s^{-1}$]

SOLID

T	∥ to c-axis α	⊥ to c-axis α	T	∥ to c-axis α	⊥ to c-axis α
10	70.9	29.8	60	0.208	0.0987
11	46.0	19.2	70	0.148	0.0732
12	31.1	12.9	80	0.114	0.0593
13	22.2	9.06	90	0.0932	0.0503
14	16.3	6.61	100	0.0799	0.0440
15	12.2	5.04	150	0.0499	0.0280
16	9.50	3.95	200	0.0377	0.0214
18	6.10	2.60	250	0.0308	0.0177
20	4.20	1.80	273.2	0.0285	0.0164
25	2.03	0.887	300	0.0262	0.0153
30	1.19	0.533	350	0.0230	0.0135
35	0.781	0.360	400	0.0206	0.0122
40	0.550	0.255	500	0.0171	0.0104
45	0.411	0.191	600	0.0150	0.00925
50	0.317	0.149	700	0.0138	0.00852
			722.7	0.0135	0.00837

REMARKS

The values are for well-annealed high-purity tellurium and are thought to be accurate to within ±15 to ±25% of the true values at temperatures above 50 K. The values below 50 K are merely typical values and represent two typical curves serving to indicate the general trend of the thermal diffusivity.

*All values are estimated and those below 50 K are merely typical values.

FIGURE AND TABLE 67R. PROVISIONAL THERMAL DIFFUSIVITY OF TERBIUM

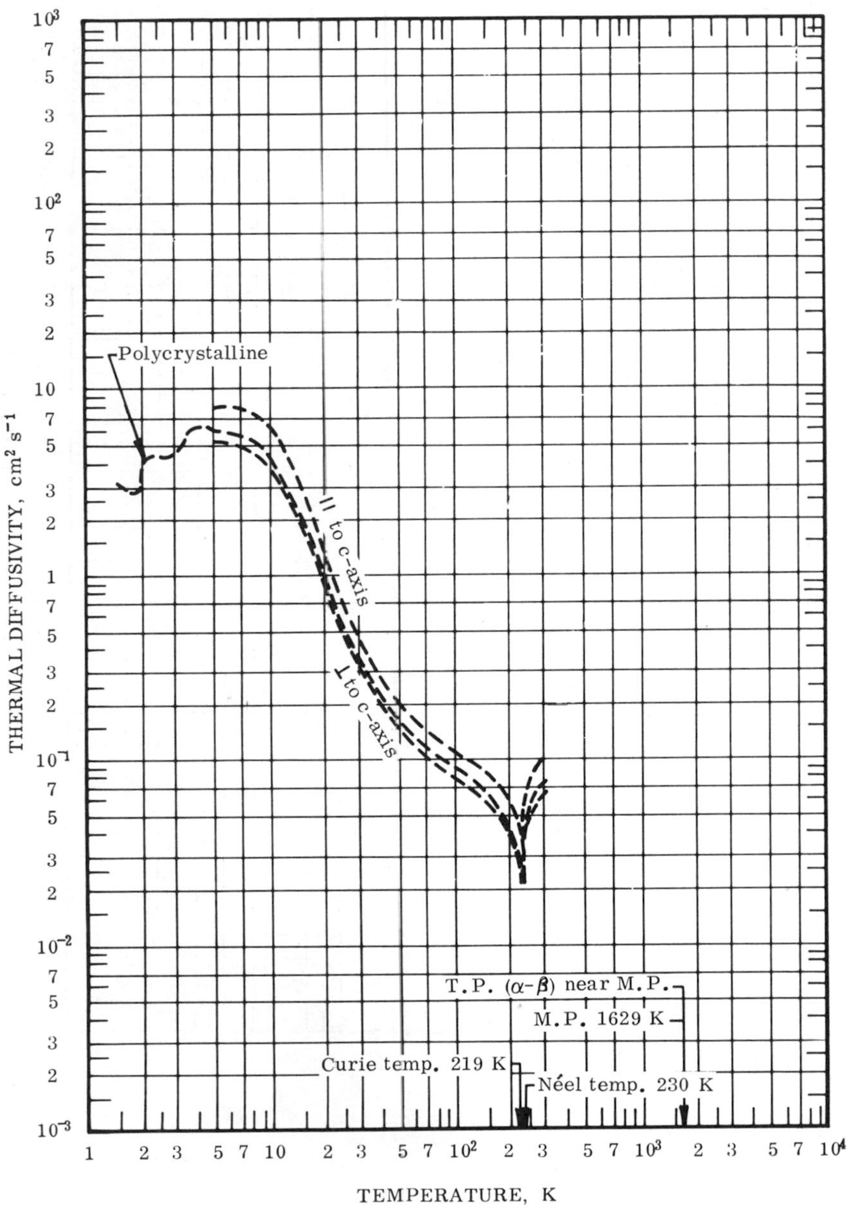

PROVISIONAL VALUES*

[Temperature, T, K; Thermal Diffusivity, α, cm² s⁻¹]

SOLID

T	// to c-axis α	⊥ to c-axis α	Poly-crystalline α	T	// to c-axis α	⊥ to c-axis α	Poly-crystalline α
1.5			3.12	18	1.79	1.16	1.28
1.97			2.78	20	1.32	0.881	0.975
2			2.98	25	0.703	0.494	0.557
2.3			4.38	30	0.456	0.334	0.373
3			4.38	35	0.332	0.247	0.275
3.2			4.52	40	0.262	0.197	0.216
3.8			6.07	45	0.219	0.165	0.181
4			6.23	50	0.192	0.144	0.158
5	7.97	5.13	5.95	60	0.161	0.119	0.132
6	7.94	5.00	5.74	70	0.141	0.1025	0.115
7	7.85	4.82	5.66	80	0.127	0.0919	0.1025
8	7.42	4.52	5.27	90	0.117	0.0838	0.0936
9	6.80	4.12	4.86	100	0.108	0.0772	0.0863
10	6.09	3.71	4.38	150	0.0784	0.0547	0.0619
11	5.42	3.32	3.94	200	0.0542	0.0370	0.0423
12	4.71	2.89	3.44	219	0.0404	0.0269	0.0310
13	4.06	2.50	2.98	230	0.0309	0.0203	0.0234
14	3.46	2.14	2.54	250	0.0798	0.0515	0.0597
15	2.89	1.82	2.12	273.2	0.0906	0.0583	0.0688
16	2.45	1.55	1.76	300	0.0982	0.0640	0.0742

REMARKS

The provisional values are for well-annealed high-purity terbium and are thought to be accurate to within ±20% of the true values near room temperature and ±25% down to 50 K. The values below 50 K are very uncertain. The values below 150 K for $\alpha_{\parallel}$, $\alpha_{\perp}$, and α_{poly} are applicable only to samples having residual electrical resistivities of 1.87, 2.37, and 2.19 $\mu\Omega$ cm, respectively.

*All values are estimated.

FIGURE AND TABLE 68R. PROVISIONAL THERMAL DIFFUSIVITY OF THALLIUM

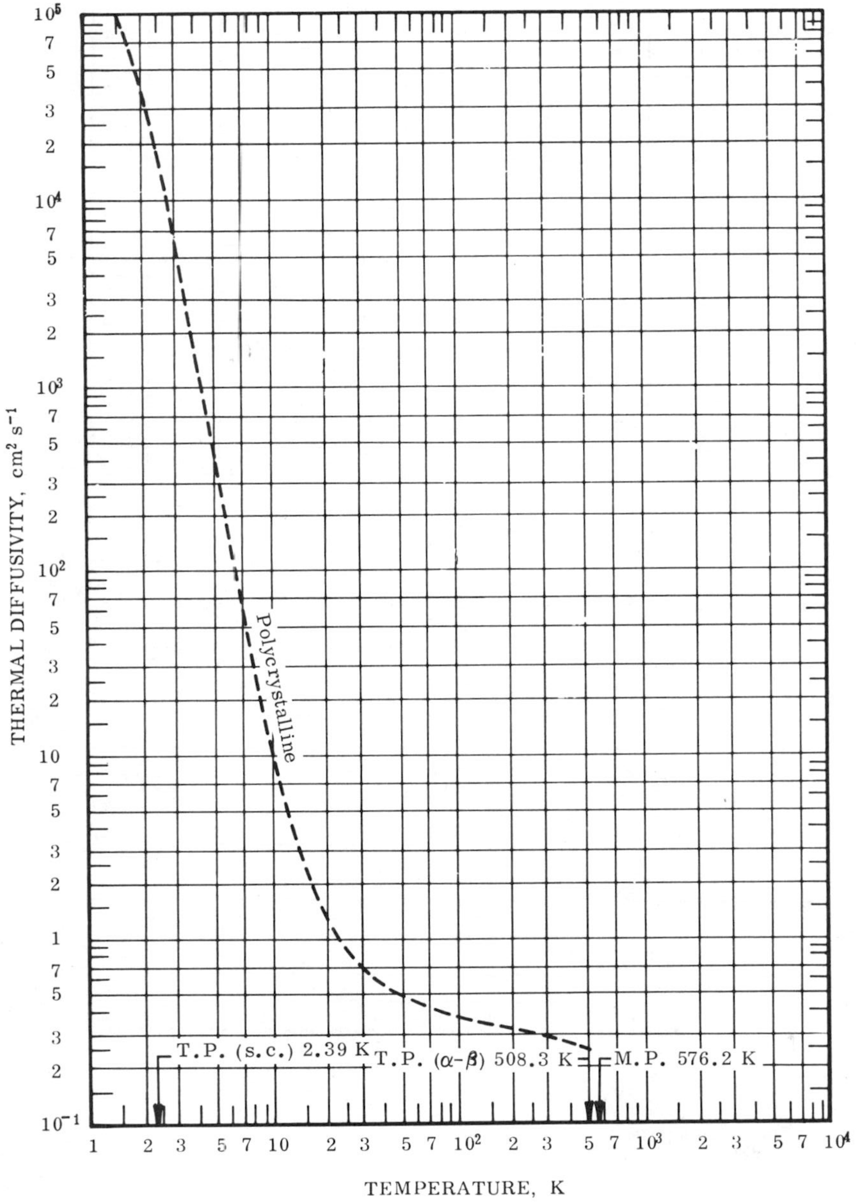

PROVISIONAL VALUES*

[Temperature, T, K; Thermal Diffusivity, α, cm^2 s^{-1}]

SOLID
(Polycrystalline)

T	α	T	α
1	251000	35	0.629
2	36900	40	0.571
3	6360	45	0.532
4	1480	50	0.501
5	448	60	0.460
6	156	70	0.433
7	66.0	80	0.415
8	31.9	90	0.398
9	17.6	100	0.387
10	10.9	150	0.349
11	7.34	200	0.328
12	5.22	250	0.314
13	3.92	273.2	0.308
14	3.14	300	0.301
15	2.56	350	0.288
16	2.15	400	0.276
18	1.62	500	0.252
20	1.31		
25	0.910		
30	0.727		

REMARKS

The values are for well-annealed high-purity polycrystalline thallium and are thought to be accurate to within ±25% of the true values at temperatures below 30 K, ±20% from 30 to 100 K and ±15% above 100 K. At low temperatures the values are highly conditioned by impurity and imperfection, and those below 30 K are applicable only to a sample having residual electrical resistivity of 0.000240 μΩ cm.

---

*All values are estimated.

FIGURE AND TABLE 69R. PROVISIONAL THERMAL DIFFUSIVITY OF THORIUM

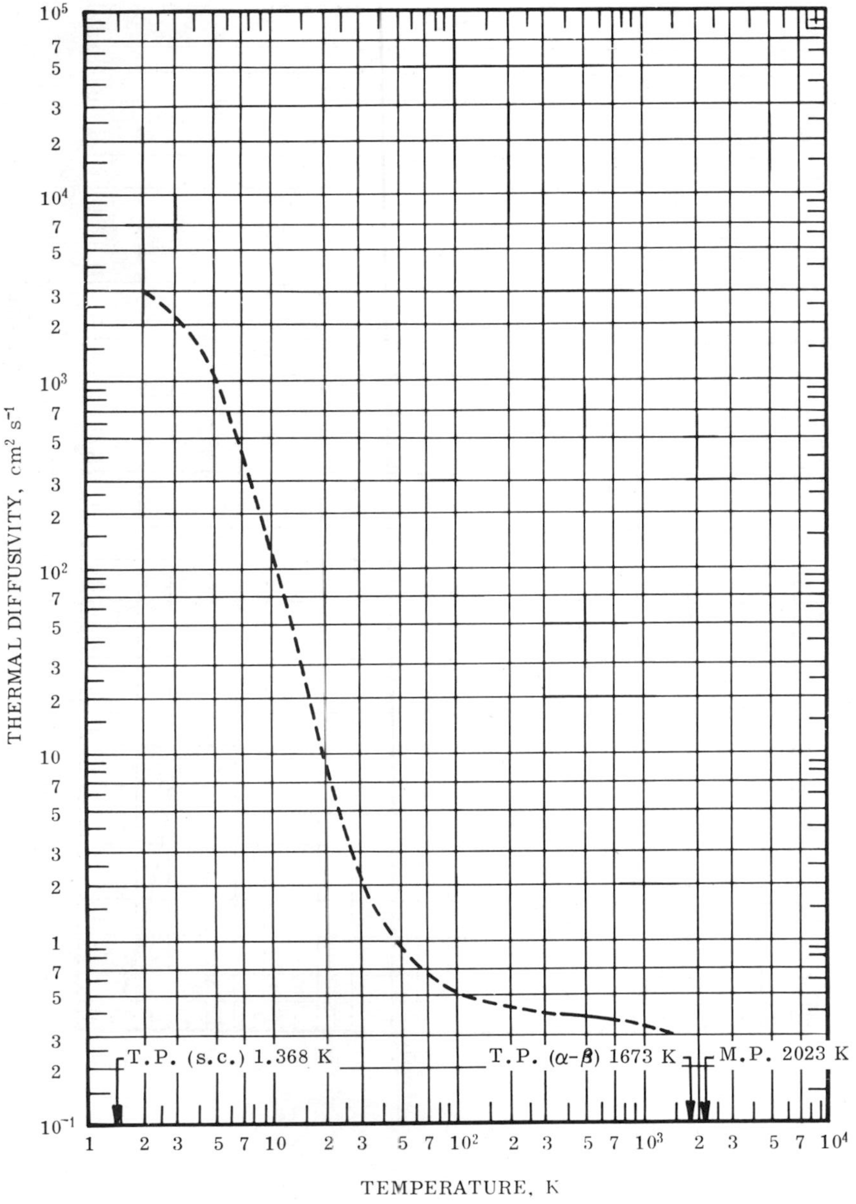

PROVISIONAL VALUES

[Temperature, T, K; Thermal Diffusivity, α, cm² s⁻¹]

SOLID

T	α	T	α
2	2930*	80	0.579*
3	2220*	90	0.541*
4	1540*	100	0.514*
5	1020*	150	0.445*
6	653*	200	0.415*
7	420*	250	0.401*
8	270*	273.2	0.396*
9	180*	300	0.392*
10	124*	350	0.384
11	87.5*	400	0.378
12	62.3*	500	0.368
13	44.8*	600	0.359
14	32.6*	700	0.350
15	24.1*	800	0.341*
16	18.1*	900	0.333*
18	11.0*	1000	0.325*
20	7.21*	1100	0.317*
25	3.29*	1200	0.310*
30	2.04*	1300	0.303*
35	1.48*		
40	1.17*		
45	0.986*		
50	0.868*		
60	0.718*		
70	0.634*		

REMARKS

The provisional values are for well-annealed high-purity thorium and their uncertainty is probably of the order of ±15% below 100 K, ±20% from 100 to 500 K, and ±30% above 500 K. At low temperatures the values are highly conditioned by impurity and imperfection, and those below 150 K are applicable only to thorium having residual electrical resistivity of 0.0268 μΩ cm.

*In temperature range where no experimental data are available.

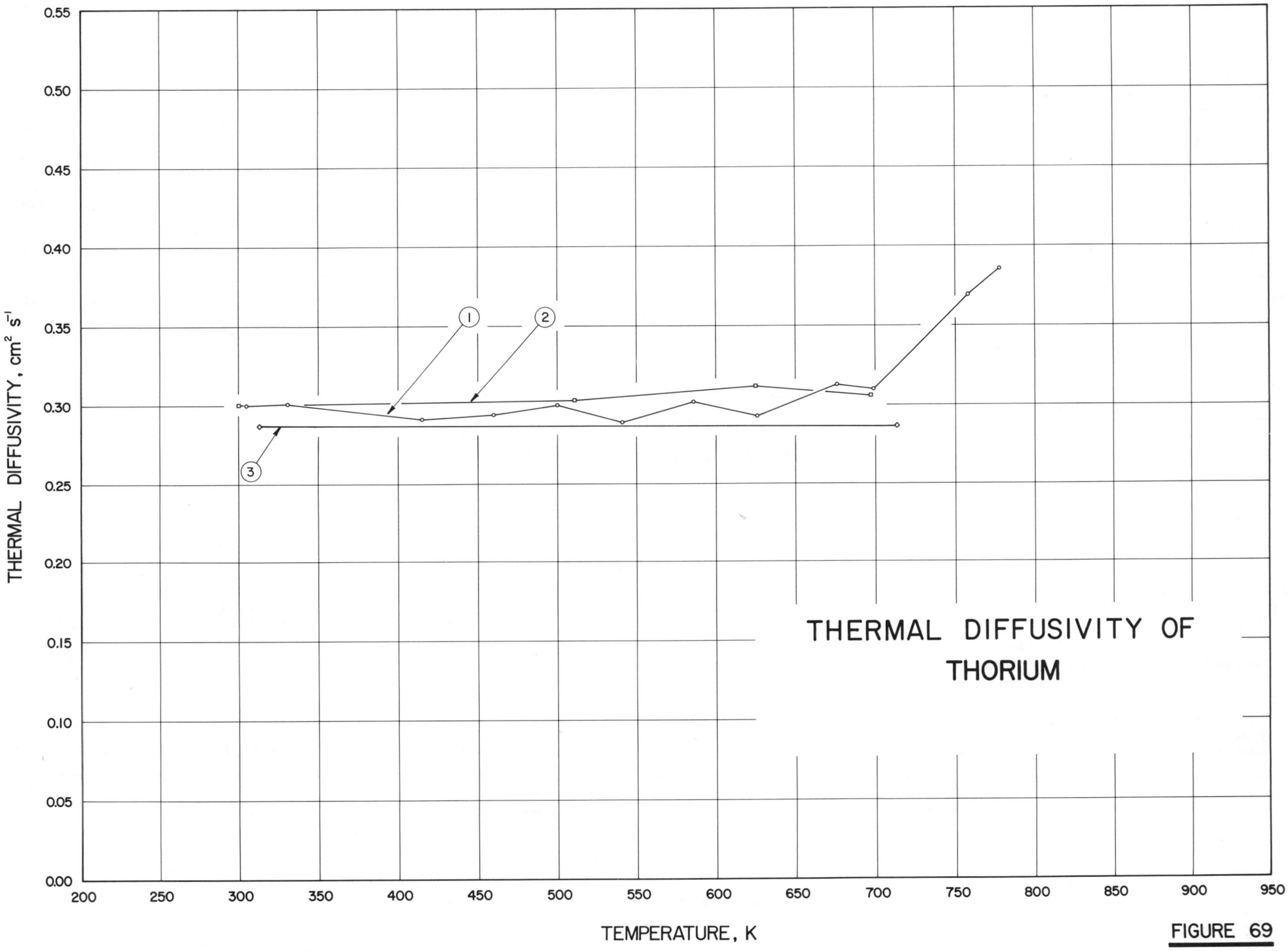
THERMAL DIFFUSIVITY OF
THORIUM
THERMAL DIFFUSIVITY, cm² s⁻¹
0.55
0.50
0.45
0.40
0.35
0.30
0.25
0.20
0.15
0.10
0.05
0.00
200
250
300
350
400
450
500
550
600
650
700
750
800
850
900
950
TEMPERATURE, K
1
2
3

FIGURE 69

SPECIFICATION TABLE 69. THERMAL DIFFUSIVITY OF THORIUM

(Impurity <0.20% each; total impurities <0.50%)

Cur. No.	Ref. No.	Author(s)	Year	Temp. Range, K	Reported Error, %	Name and Specimen Designation	Composition (weight percent), Specifications, and Remarks
1	22	Sidles, P.H. and Danielson, G.C.	1953	305-778			99.85 pure; cylindrical specimen ~0.125 in. in dia. and ≥ 50 cm in length; measured under a vacuum of 10^{-4} mm Hg; data are the result of a considerable number of measurements on six different specimens; modified Angstrom method used to measure diffusivity.
2	23, 114	Sidles, P.H. and Danielson, G.C.	1951	301-697			High purity; cylindrical specimen ~0.125 in. in dia. and ≥50 cm long; density 11.558 g cm^{-3}; heater impresses sinusoidal temp. wave at one end of specimen; measured under a vacuum of 5 x 10^{-4} mm Hg.
3	259	Spedding, F.H., Fox, G.W., Carlson, J.F., Danielson, G.C., Hudson, D.E., Jenson, E.N., Keller, J.M., Knipp, J.K., Laslett, L.J., Legvold, S., Miller, G.H., and Zaffarano, D.J.	1952	313, 713			Mean value of 15 measurements.

DATA TABLE 69. THERMAL DIFFUSIVITY OF THORIUM

(Impurity < 0.201 each; total impurities < 0.50%)

[Temperature, T, K; Thermal Diffusivity, α, $cm^2\ s^{-1}$]

T	α	T	α
CURVE 1		CURVE 2	
305	0.300	300.7	0.300
331	0.301	511.2	0.303
415	0.291	625.2	0.312
460	0.294	697.2	0.306
500	0.300		
541	0.289	CURVE 3	
586	0.302		
626	0.293	313	0.287
676	0.313	713	0.287
699	0.310		
758	0.369		
778	0.385		

FIGURE AND TABLE 70R. PROVISIONAL THERMAL DIFFUSIVITY OF THULIUM

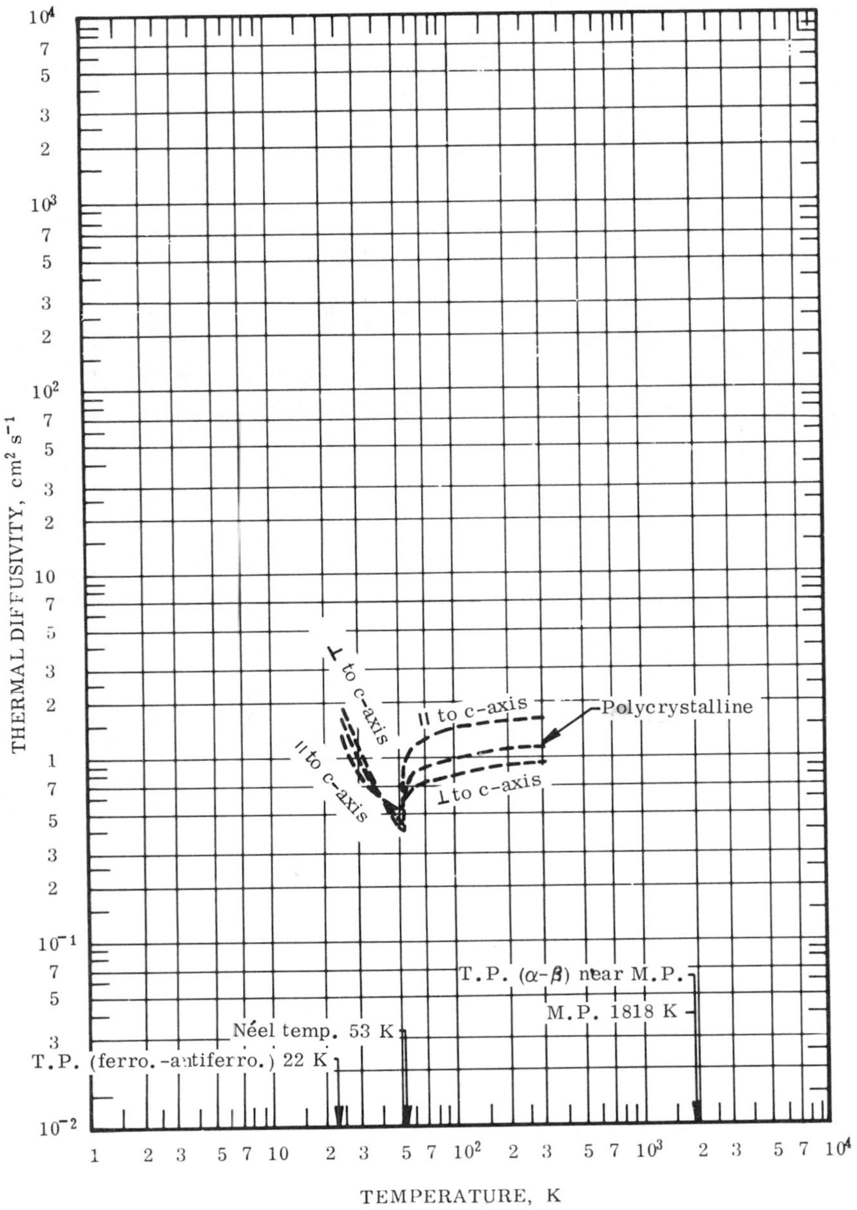

PROVISIONAL VALUES*

[Temperature, T, K; Thermal Diffusivity, α, $cm^2\ s^{-1}$]

SOLID

T	$\parallel$ to c-axis α	$\perp$ to c-axis α	Poly-crystalline α
25	0.134	0.186	0.166
30	0.0883	0.1175	0.104
35	0.0698	0.0826	0.0769
40	0.0608	0.0637	0.0623
45	0.0562	0.0521	0.0538
50	0.0535	0.0445	0.0479
53	0.0526	0.0421	0.0455
58	0.106	0.0686	0.0805
60	0.115	0.0703	0.0836
70	0.130	0.0737	0.0887
80	0.136	0.0760	0.0918
90	0.140	0.0781	0.0944
100	0.144	0.0799	0.0970
150	0.156	0.0872	0.106
200	0.161	0.0917	0.111
250	0.163	0.0940	0.113
273.2	0.163	0.0944	0.113
300	0.162	0.0947	0.113

REMARKS

The provisional values are for well-annealed high-purity thulium and are thought to be accurate to within ±20% of the true values at temperatures above 150 K. The values below 150 K are very uncertain. Values below 100 K for $\alpha_{\parallel}$, $\alpha_{\perp}$, and α_{poly} are applicable only to samples having residual electrical resistivities of 3.5, 1.7, and 0.18 μΩ cm, respectively.

* All values are estimated.

FIGURE AND TABLE 71R. RECOMMENDED THERMAL DIFFUSIVITY OF TIN

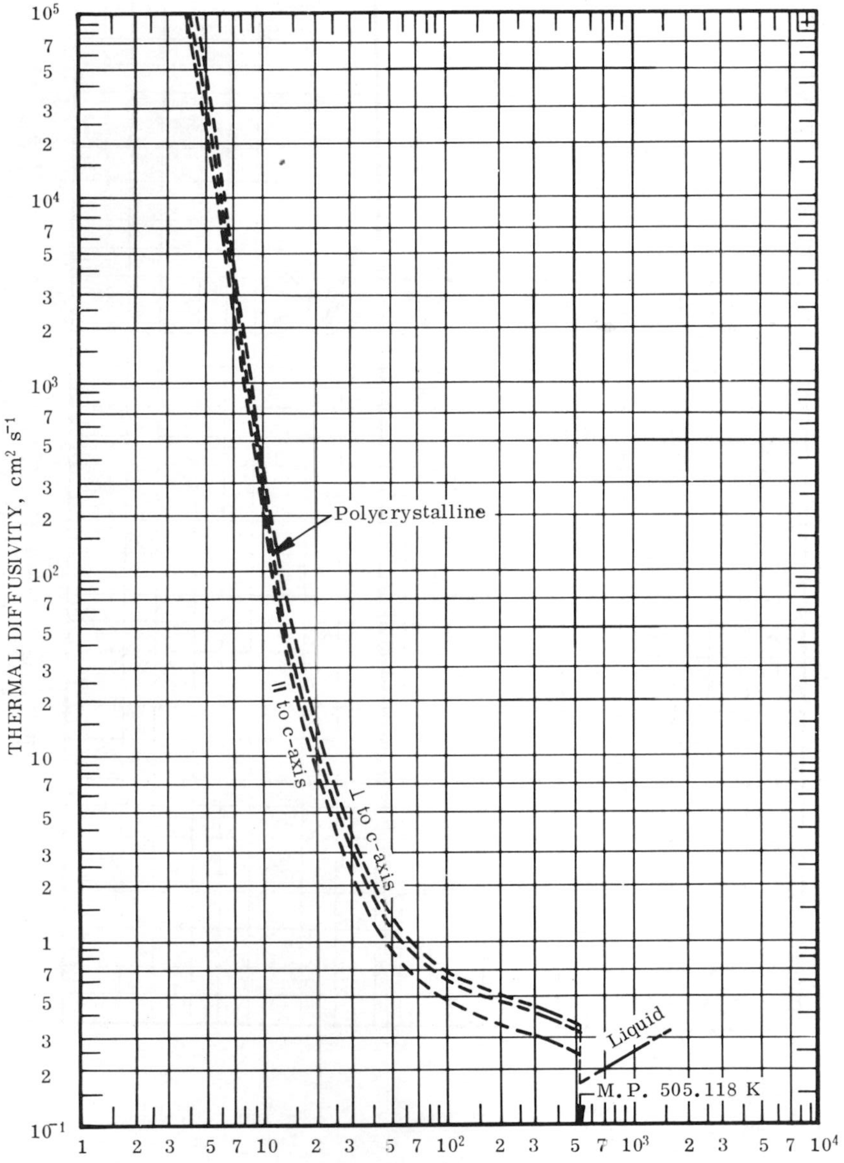

RECOMMENDED VALUES†

[Temperature, T, K; Thermal Diffusivity, α, cm^2 s^{-1}]

SOLID

T	∥ to c-axis α	⊥ to c-axis α	Poly-crystalline α
1	1140000*	1630000*	1470000*
2	723000*	1040000*	934000*
3	287000*	413000*	371000*
4	70700*	103000*	91700*
5	21000*	30300*	27300*
6	7330*	10700*	9360*
7	2810*	4100*	3600*
8	1170*	1710*	1500*
9	520*	756*	670*
10	264*	374*	330*
11	151*	215*	188*
12	93.3*	130*	113*
13	60.3*	83.5*	72.5*
14	41.1*	57.3*	49.4*
15	29.4*	41.5*	35.6*
16	21.8*	31.1*	26.8*
18	13.1*	19.1*	16.5*
20	8.67*	12.9*	11.1*
25	4.01*	4.38*	5.50*
30	2.50*	3.83*	3.29*
35	1.76*	2.62*	2.26*
40	1.35*	1.96*	1.72*
45	1.09*	1.58*	1.40*
50	0.927*	1.34*	1.20*
60	0.732*	1.06*	0.948*
70	0.624*	0.908*	0.808*
80	0.555*	0.808*	0.718*
90	0.511*	0.740*	0.660*
100	0.479*	0.689*	0.619*
150	0.397*	0.572*	0.514*
200	0.360*	0.518*	0.466*
250	0.333*	0.478*	0.431*
273.2	0.322*	0.463*	0.417*
300	0.311	0.446	0.402
350	0.292	0.419	0.376
400	0.277	0.397	0.355
500	0.250*	0.359*	0.321*
505.118	0.249*	0.357*	0.320*

LIQUID

T	α
505.118	0.173*
600	0.189
700	0.206
800	0.222
900	0.237
1000	0.253
1100	0.267
1200	0.282
1300	0.298*
1400	0.313*
1500	0.329*

REMARKS

The values are for well-annealed high-purity tin. The values for polycrystalline tin are thought to be accurate to within ±6% of the true values at temperatures above 100 K and ±15 to ±20% below. Those for $\alpha_{\parallel}$ and $\alpha_{\perp}$ of tin single crystal should be accurate to within ±10% above 100 K and ±15 to ±20% below. For molten tin the values are probably good to ±8% below 800 K, but an increasing uncertainty remains to be resolved at higher temperatures. At temperatures below 100 K the values for $\alpha_{\parallel}$, $\alpha_{\perp}$, and α_{poly} are applicable only to samples having residual electrical resistivities of 0.000170, 0.000118, and 0.000132 μΩ cm, respectively. The values below 100 K and above 800 K are provisonal.

† Values below 100 K and above 800 K are provisional.
* In temperature range where no experimental data are available.

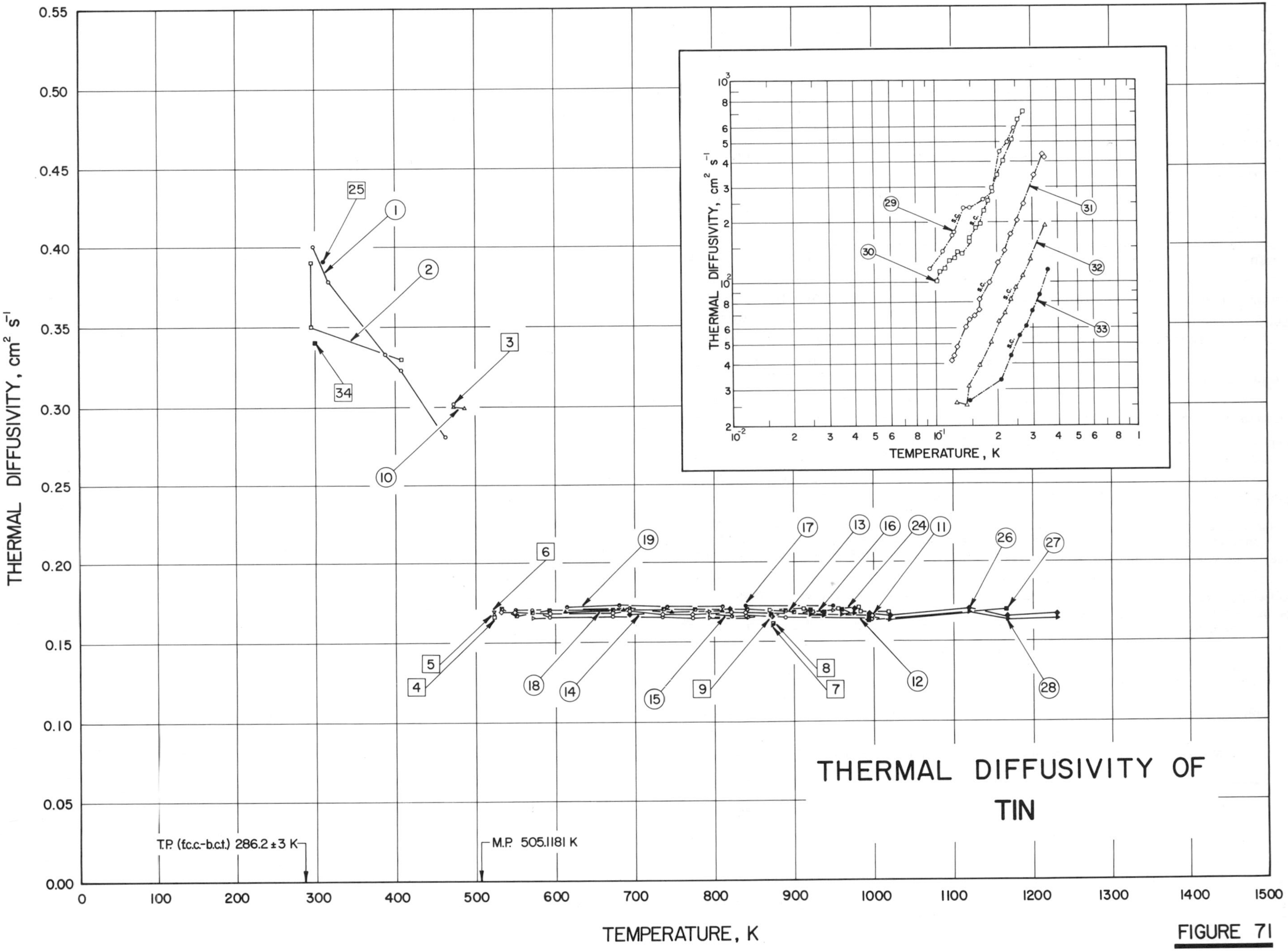
THERMAL DIFFUSIVITY OF TIN
THERMAL DIFFUSIVITY, cm² s⁻¹
TEMPERATURE, K
T.P. (f.c.c.-b.c.t.) 286.2 ± 3 K
M.P. 505.1181 K
s.c.

FIGURE 71

SPECIFICATION TABLE 71. THERMAL DIFFUSIVITY OF TIN

(Impurity < 0.20% each; total impurities < 0.50%)

Cur. No.	Ref. No.	Author(s)	Year	Temp. Range, K	Reported Error, %	Name and Specimen Designation	Composition (weight percent), Specifications, and Remarks
1	38	Shvidkovskii, E. G.	1938	298-463			Technically pure; cylindrical specimen 6 mm in dia and 150 mm long; data given represent avg over a number of runs ranging from 1 to 3.
2	4	Jenkins, R. J. and Parker, W. J.	1961	295-408	± 5		Square specimen 1.9 cm side and 0.306 cm thick; high intensity short duration light pulse absorbed in front surface of thermally insulated specimen coated with camphor black; 408.2 K measurements obtained by heating specimen holder and specimen with an infrared lamp; both data points at 295.2 K obtained from measurements using different equations for data reduction.
3	105	Yurchak, R. P. and Filippov, L. P.	1965	473	7		Cylindrical specimen 2 cm in dia and 12 cm long, made of two halves; measured under a vacuum of $\sim 10^{-5}$ mm Hg; heat losses reduced by cylindrical shields of sheet molybdenum; radial wave method used to measure diffusivity; diffusivity determined from measured amplitude ratio using a period of 6.6 sec.
4	105	Yurchak, R. P. and Filippov, L. P.	1965	523	7		Liquid specimen placed in a thin-walled tantalum crucible 2.38 cm in dia and 12 to 14 cm long; horizontal baffles prevent convective mixing of liquid; junctions of thermocouples in contact with specimen; heat losses reduced by cylindrical shields of sheet molybdenum; tantalum crucible placed in a vacuum of $\sim 10^{-5}$ mm Hg; radial wave method used to measure diffusivity; diffusivity determined from measured amplitude ratio using a period of 6.6 sec.
5	105	Yurchak, R. P. and Filippov, L. P.	1965	523	7		Specimen similar to the above; measured for diffusivity using same method as above but employing a period of 13.2 sec for the temp wave; other conditions same as above.
6	105	Yurchak, R. P. and Filippov, L. P.	1965	523	7		Specimen similar to the above; diffusivity determined from measured phase difference between temp waves; other conditions same as above.
7	105	Yurchak, R. P. and Filippov, L. P.	1965	873	7		Specimen similar to the above; diffusivity determined from measured amplitude ratio using a period of 6.6 sec; other conditions same as above.
8	105	Yurchak, R. P. and Filippov, L. P.	1965	873	7		Specimen similar to the above; diffusivity measured using same method as above but employing a period of 13.2 sec for the temp wave; other conditions same as above.
9	105	Yurchak, R. P. and Filippov, L. P.	1965	873	7		Specimen similar to the above; diffusivity determined from measured phase difference using a period of 13.2 sec for the temp wave; other conditions same as above.
10	109, 107	Yurchak, R. P. and Filippov, L. P.	1964	473, 487	7		Of analytical purity; cylindrical specimen; radial wave method used to measure diffusivity; diffusivity determined from measurements of temp at two points on the specimen using a heating period of 6.6 sec; electrical resistivity measured and reported as 11.65, 15.05, 16.9, 20.7, and 21.65 μohm cm at 293.2, 381.2, 418.2, 485.2, and 495.2 K, respectively.
11	109, 107	Yurchak, R. P. and Filippov, L. P.	1964	532-997	7		Of analytical purity; in molten state; sample consists of a cylindrical tantalum crucible containing the metal to be measured; horizontal partitions of tantalum plate impede convective mixing of liquid; outer surface of crucible subjected to periodic heating; radial wave method used to measure diffusivity; diffusivity determined from measurements of temp at two points on specimen using a heating period of 6.6 sec; electrical resistivity measured and reported as 49.1, 51.6, 53.2, 55.8, 57.6, 59.4, 61.1, and 62.5 μohm cm at 514.2, 568.2, 677.2, 758.2, 838.2, 913.2, 973.2, and 1023.2 K, respectively; Lorenz number reported as 2.82, 2.66, 2.53, 2.40, 2.28, 2.18, 2.09, 2.02, 1.94, 1.83, and 1.81 x 10^{-8} V^2 K^{-2} at 523.2, 573.2, 623.2, 673.2, 723.2, 773.2, 823.2, 873.2, 923.2, 973.2, and 1023.2 K, respectively.

SPECIFICATION TABLE 71. THERMAL DIFFUSIVITY OF TIN (continued)

Cur. No.	Ref. No.	Author(s)	Year	Temp. Range, K	Reported Error, %	Name and Specimen Designation	Composition (weight percent), Specifications, and Remarks
12	109	Yurchak, R. P. and Filippov, L. P.	1964	533-997	7		Of analytical purity; in molten state; sample consists of cylindrical tantalum crucible containing metal to be measured; diffusivity determined from the amplitude ratio of the temp waves measured at two points on specimen using a heating period of 13.2 sec; other conditions same as above.
13	109	Yurchak, R. P. and Filippov, L. P.	1964	532-999	7		Of analytical purity; in molten state; sample consists of cylindrical tantalum crucible containing metal to be measured; diffusivity determined from the phase difference between the temp waves measured at two points on specimen using a heating period of 13.2 sec; other conditions same as above.
14	109	Yurchak, R. P. and Filippov, L. P.	1964	693, 820	7		Of analytical purity; in molten state; sample consists of cylindrical tantalum crucible containing metal to be measured; diffusivity determined from the amplitude ratio of the temp waves measured at two points on specimen with second setting of the thermocouples; other conditions same as above.
15	109	Yurchak, R. P. and Filippov, L. P.	1964	550-822	7		Of analytical purity; in molten state; sample consists of cylindrical tantalum crucible containing metal to be measured; diffusivity determined from the phase difference between the temp waves measured at two points on specimen with second setting of the thermocouples; other conditions same as above.
16	109	Yurchak, R. P. and Filippov, L. P.	1964	571-948	7		Of analytical purity; in molten state; sample consists of two thin-walled tantalum tubes 23.7 and 8 mm in dia containing metal to be measured; inner surface of crucible subjected to periodic heating using a heating period of 13.2 sec; diffusivity determined from the amplitude ratio of the temp waves measured at two points on specimen; measured in vacuum; radial wave method used to measure diffusivity; thin horizontal tantalum plate partitions used to prevent convection.
17	109	Yurchak, R. P. and Filippov, L. P.	1964	839, 949	7		Of analytical purity; in molten state; sample consists of two thin-walled tantalum tubes 23.7 and 8 mm in dia containing metal to be measured; inner surface of crucible subjected to periodic heating using a heating period of 13.2 sec; diffusivity determined from the phase difference between the temp waves measured at two points on specimen; other conditions same as above.
18	109	Yurchak, R. P. and Filippov, L. P.	1964	612-809	7		Of analytical purity; in molten state; sample consists of cylindrical tantalum crucible containing metal to be measured; horizontal partitions of tantalum plate impede convective mixing of liquid; outer surface of crucible subjected to periodic heating; radial wave method used to measure diffusivity; diffusivity determined from the amplitude ratio of the temp waves measured at two points on specimen; measured with displaced outer thermocouple.
19	109	Yurchak, R. P. and Filippov, L. P.	1964	614-810	7		Of analytical purity; in molten state; sample consists of cylindrical tantalum crucible containing metal to be measured; horizontal partitions of tantalum plate impede convective mixing of liquid; outer surface of crucible subjected to periodic heating; radial wave method used to measure diffusivity; diffusivity determined from the phase difference between the temp waves measured at two points on specimen; measured with displaced outer thermocouple.

SPECIFICATION TABLE 71. THERMAL DIFFUSIVITY OF TIN (continued)

Cur. No.	Ref. No.	Author(s)	Year	Temp. Range, K	Reported Error, %	Name and Specimen Designation	Composition (weight percent), Specifications, and Remarks
20*	109	Yurchak, R. P. and Filippov, L. P.	1964	693, 843	7		Of analytical purity; in molten state; sample consists of cylindrical tantalum crucible containing metal to be measured; heated externally; diffusivity determined from amplitudes of temp waves measured at two points on specimen; radial wave method used to measure diffusivity; precautions taken to prevent convective mixing of liquid.
21*	109	Yurchak, R. P. and Filippov, L. P.	1964	693, 843	7		Of analytical purity; in molten state; sample consists of cylindrical tantalum crucible containing metal to be measured; heated externally; diffusivity determined from phases of temp waves measured at two points on specimen; other conditions same as above.
22*	109	Yurchak, R. P. and Filippov, L. P.	1964	693, 843	7		Of analytical purity; in molten state; sample consists of two thin-walled tantalum tubes 23.7 and 8 mm in dia containing metal to be measured; heated internally; diffusivity determined from amplitudes of temp waves measured at two points on specimen; other conditions same as above.
23*	109	Yurchak, R. P. and Filippov, L. P.	1964	693, 843	7		Of analytical purity; in molten state; sample consists of two thin-walled tantalum tubes 23.7 and 8 mm in dia containing metal to be measured; heated internally; diffusivity determined from phases of temp waves measured at two points on specimen; other conditions same as above.
24	107	Filippov, L. P.	1966	869-1226			Liquid specimen; diffusivity calculated from measured thermal conductivity and heat capacity for the same specimen using steady heating for conductivity measurement and pulsating heating for measurement of the heat capacity.
25	143	King, R.W.	1915	308			Wire specimen 0.25 cm in diameter and 25 cm long; greater portion of specimen coiled so that it occupies a space of ~ 7 cm long; other end inserted in small resistance heater supplied with sinusoidally varying electric current; diffusivity determined from measurement of the velocities at which the impressed heat waves travel along specimen; data point reported obtained from conductivity data reported by author divided by a density of 7.28 g cm^{-3} and a specific heat of 0.0547 cal g^{-1} K^{-1} at 308.2 K; data point reported is average result of four independent measurements.
26	209	Yurchak, R.P. and Filippov, L.P.	1965	975-1229	3-5		Specimen in molten state; heating cycle 6.2 sec.
27	209	Yurchak, R.P. and Filippov, L.P.	1965	870-1166	>>		Similar to the above specimen; heating cycle 13.2 sec.
28	209	Yurchak, R.P. and Filippov, L.P.	1965	870-1230	>>		Similar to the above specimen; heating cycle 26.4 sec.
29	260	Zavaritskii, N.V.	1965	0.093-0.25			99.9999 pure; single crystal; cylindrical specimen 2.6 mm in diameter; measured in a magnetic field; fraction of normal phase in specimen = 0.08.
30	260	Zavaritskii, N.V.	1965	0.10-0.27			The above specimen.
31	260	Zavaritskii, N.V.	1965	0.12-0.34			The above specimen; fraction of normal phase = 0.15.
32	260	Zavaritskii, N.V.	1965	0.13-0.34			The above specimen; fraction of normal phase = 0.3.
33	260	Zavaritskii, N.V.	1965	0.20-0.36			The above specimen; fraction of normal phase = 0.45.
34	108	Steinberg, S., Larson, R.E., and Kydd, A.R.	1963	298			Specimen 2.0 cm square, 0.3105 cm in thickness; diffusivity measuring temperature not reported but assumed to be 25 C.

* Not shown in figure.

DATA TABLE 71. THERMAL DIFFUSIVITY OF TIN

(Impurity < 0.20% each; total impurities < 0.50%)

[Temperature, T, K; Thermal Diffusivity, α, $cm^2\ s^{-1}$]

T	α
CURVE 1	
298	0.40
317	0.378
388	0.333
408	0.323
463	0.281
CURVE 2	
295	0.39
295	0.35
408	0.33
CURVE 3	
473	0.302
CURVE 4	
523	0.166
CURVE 5	
523	0.168
CURVE 6	
523	0.169
CURVE 7	
873	0.160
CURVE 8	
873	0.161
CURVE 9	
873	0.165
CURVE 10	
473	0.300
487	0.299
CURVE 11	
532	0.169
549	0.169
593	0.168
672	0.168
693	0.169
735	0.167
772	0.167
821	0.168
839	0.168
888	0.167
997	0.166
CURVE 12	
533	0.167
571	0.170
592	0.166
672	0.166
735	0.166
773	0.165
840	0.166
889	0.165
997	0.164
CURVE 13	
532	0.171
550	0.167
572	0.169
592	0.170
672	0.170
736	0.170
775	0.170
841	0.170
889	0.169
999	0.168
CURVE 14	
693	0.167
820	0.170
CURVE 15	
550	0.170
693	0.171
822	0.166
CURVE 16	
571	0.165
693	0.166*
793	0.165
793	0.169
842	0.165
948	0.169
CURVE 17	
839	0.172
949	0.172
CURVE 18	
612	0.169
685	0.170
746	0.169
809	0.169
CURVE 19	
614	0.172
680	0.173
743	0.172
810	0.172
CURVE 20*	
693	0.167
843	0.165
CURVE 21*	
693	0.169
843	0.169
CURVE 22*	
693	0.165
843	0.164
CURVE 23*	
693	0.170
843	0.171
CURVE 24	
869	0.169
912	0.170
923	0.168
955	0.170
981	0.171
983	0.168
1018	0.168
1114	0.170
1164	0.170
1226	0.168
CURVE 25	
308	0.3902
CURVE 26	
975	0.1702
993	0.1670
1019	0.1658
1118	0.1700
1166	0.1654
1229	0.1669
CURVE 27	
870	0.1664
920	0.1695
936	0.1685
959	0.16
CURVE 27	
870	0.1664
920	0.1695
936	0.1685
959	0.1699
993	0.1626
1166	0.1690
CURVE 28	
870	0.1694*
920	0.1669
936	0.1663
959	0.1662
975	0.1671
993	0.1650
1019	0.1631
1118	0.1678
1166	0.1638
1230	0.1640
CURVE 29	
0.0927	117
0.108	144
0.120	173
0.124	179
0.136	236
0.147	236
0.171	361
0.191	386
0.209	450
0.229	504
0.245	593
CURVE 30	
0.101	101
0.105	114
0.111	118
0.117	129
0.124	133
0.128	143
0.135	140
CURVE 30 (cont.)	
0.147	159
0.147	168
0.158	189
0.166	199
0.173	229
0.181	257
0.190	299
0.201	345
0.217	404
0.238	512
0.259	650
0.271	705
CURVE 31	
0.118	41.8
0.122	44.4
0.126	48.9
0.140	61.1
0.145	66.7
0.154	68.9
0.163	77.1
0.163	83.4
0.184	100
0.201	126
0.218	145
0.231	175
0.251	206
0.270	249
0.305	343
0.337	438
0.344	421
CURVE 32	
0.125	25.9
0.140	25.1
0.143	31.2
0.163	39.8
0.185	51.2
0.203	65.2
0.216	71.6
0.234	80.7
CURVE 32 (cont.)	
0.248	94.0
0.268	108
0.294	130
0.344	194
CURVE 33	
0.195	26.7
0.208	33.8
0.233	44.0
0.259	55.8
0.277	61.8
0.298	72.3
0.325	87.5
0.355	115
CURVE 34	
298	0.34

* Not shown in figure.

FIGURE AND TABLE 72R. RECOMMENDED THERMAL DIFFUSIVITY OF TITANIUM

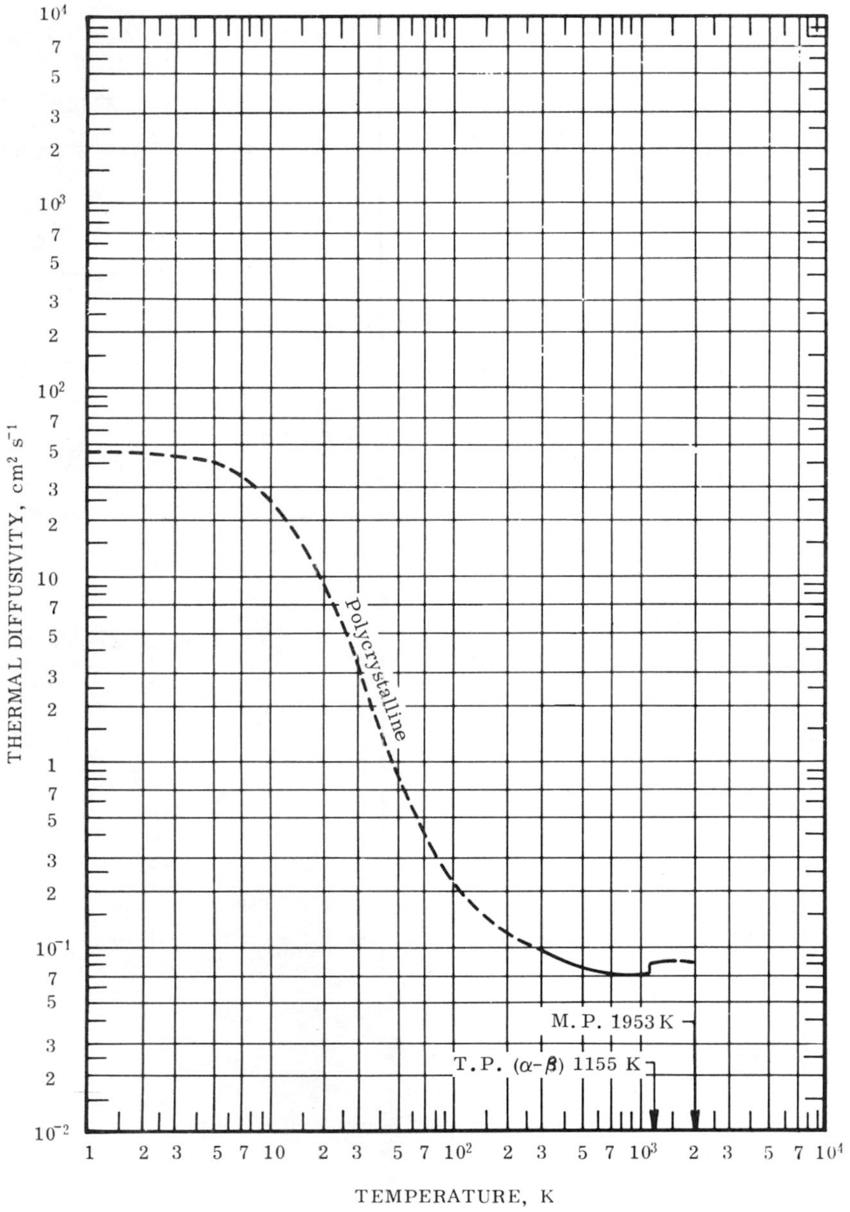

RECOMMENDED VALUES

[Temperature, T, K; Thermal Diffusivity, α, $cm^2\ s^{-1}$]

SOLID
(Polycrystalline)

T	α	T	α	T	α
1	42.5*	35	2.20*	900	0.0694
2	41.7*	40	1.54*	1000	0.0691
3	40.3*	45	1.11*	1100	0.0690
4	38.6*	50	0.842*	1155	0.0690
5	36.6*	60	0.546*	1155	0.0771
6	34.4*	70	0.398*	1200	0.0785
7	32.0*	80	0.314*	1300	0.0807
8	29.4*	90	0.270*	1400	0.0819
9	26.7*	100	0.216*	1500	0.0824
10	24.1*	150	0.1475*	1600	0.0826*
11	21.9*	200	0.117*	1700	0.0831*
12	19.8*	250	0.101*	1800	0.0830*
13	17.9*	273.2	0.0967	1900	0.0826*
14	16.2*	300	0.0925	1953	0.0823*
15	14.6*	350	0.0868		
16	13.1*	400	0.0823		
18	10.6*	500	0.0765		
20	8.56*	600	0.0729		
25	5.22*	700	0.0709		
30	3.29*	800	0.0699		

REMARKS

The values are for well-annealed high-purity polycrystalline titanium and are thought to be accurate to within ±10% of the true values at moderate temperatures and ±15% at low and high temperatures. At low temperatures the values are highly conditioned by impurity and imperfection, and those below room temperature are applicable only to titanium having residual electrical resistivity of 1.90 $\mu\Omega$ cm.

*In temperature range where no experimental data are available.

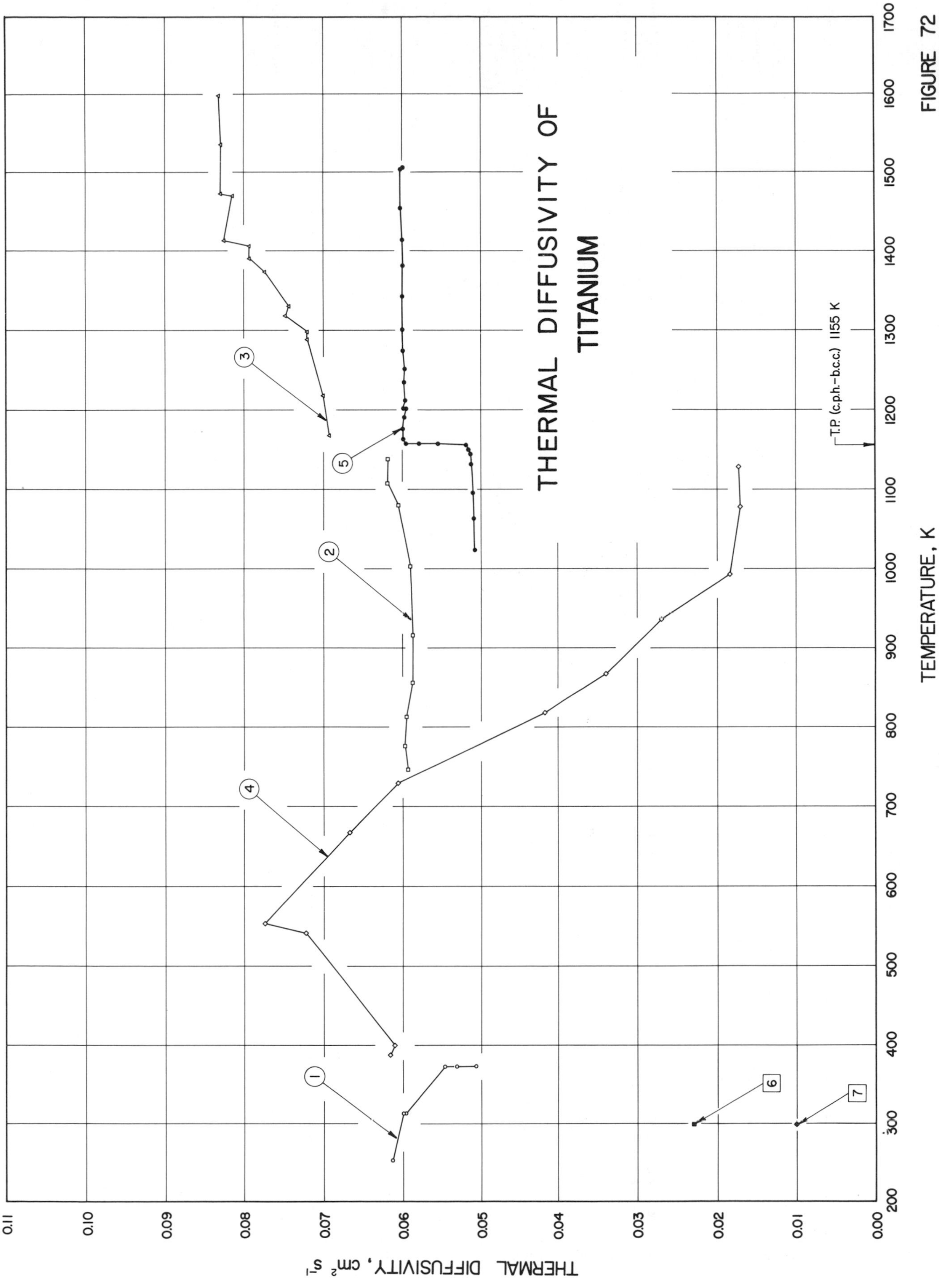
THERMAL DIFFUSIVITY OF TITANIUM
THERMAL DIFFUSIVITY, cm² s⁻¹
TEMPERATURE, K
T.P. (c.p.h.-b.c.c.) 1155 K
0.00
0.01
0.02
0.03
0.04
0.05
0.06
0.07
0.08
0.09
0.10
0.11
200
300
400
500
600
700
800
900
1000
1100
1200
1300
1400
1500
1600
1700
1
2
3
4
5
6
7

FIGURE 72

SPECIFICATION TABLE 72. THERMAL DIFFUSIVITY OF TITANIUM

(Impurity < 0.20% each; total impurities < 0.50%)

Cur. No.	Ref. No.	Author(s)	Year	Temp. Range, K	Reported Error, %	Name and Specimen Designation	Composition (weight percent), Specifications, and Remarks
1	29, 30	McIntosh, G. E.	1952	253-373	5		Commercially pure; rod specimen 0.210 in. in diameter and 7 in. long; surrounded by radiation shield consisting of our concentric cylinders of very thin aluminum foil each separated by three 1/32 in. rings of balsa wood; measured after being maintained at elevated temperature for several hours; measured in vacuum; diffusivity determined from measured phase lag of the temperature wave between any two points along specimen; one-dimensional heat flow.
2	242, 34, 141	Rudkin, R. L., Parker, W.J., and Jenkins, R.J.	1961	746-1138	±5		Specimen a few millimeters thick; high intensity short duration light pulse from xenon flash lamp absorbed in the front surface of thermally insulated specimen; thermal diffusivity determined from measured temperature history of the rear surface; radio frequency induction heating used for heating specimen.
3	242, 34, 141	Rudkin, R. L., et al.	1961	1168-1598	±5		Above specimen measured in the β-phase; measured under the same conditions as above.
4	53	Karagezyan, A. G.	1962	388-1133		α-titanium; VT-1	Pure; cylindrical rod specimen 3 mm in diameter and 300 mm long; vacuum annealed for 5 hr at 993.2 K; measured in a vacuum of $\sim 10^{-4}$ mm Hg; electrical resistivity reported as 69, 78, 79, 82, 110, 132, 139, 139, 151, 156, 164, 170, 175, and 180 μohm cm at 297.2, 356.2, 380.2, 388.2, 556.2, 645.2, 710.2, 811.2, 870.2, 938.2, 986.2, 1072.2, and 1128.2 K, respectively.
5	157	Zinov'ev, V. E., Krentsis, R. P., and Gel'd, P. V.	1968	1023-1506	7	Iodide titanium	Square specimen 8 x 8 x 0.110 mm; specimen thickness measured to within 0.002 mm; electrical resistivity ratio $\rho(298K)/\rho(4.2K) = 40$; lower surface of specimen bombarded by a current of electrons from cathode accelerated by a stabilized constant voltage generator; a sinusoidal emf from an audiofrequency generator superposed on this constant emf; thermal diffusivity determined from measured shift between the phase of the thermal current incident on the lower surface of specimen and the phase of the temperature wave at the opposite surface; measured in a vacuum of $\sim 10^{-5}$ mm Hg; measured in both α and β phases.
6	108	Steinberg, S., Larsen, R. E., and Kydd, A. R.	1963	298			2.0 cm^2 x 0.094 cm; thermal diffusivity measured by a flash method; measuring temperature not reported but assumed to be 25 C.
7	247	Namba, S., Kim, P.H., and Arai, T.	1967	298			Square specimen 1.5 x 1.5 x 0.200 cm; thermal diffusivity measured by a flash method.

DATA TABLE 72. THERMAL DIFFUSIVITY OF TITANIUM

(Impurity <0.20% each; total impurities <0.50%)

[Temperature, T, K; Thermal Diffusivity, α, $cm^2\ s^{-1}$]

T	α
CURVE 1	
253	0.06129
313	0.05997
313	0.05964
373	0.05474
373	0.05079
373	0.05321
CURVE 2	
746	0.0593
776	0.0597
813	0.0595
856	0.0587
916	0.0587
1003	0.0590
1080	0.0605
1108	0.0618
1138	0.0618
CURVE 3	
1168	0.0693
1218	0.0700
1288	0.0720
1298	0.0720
1318	0.0747
1330	0.0743
1373	0.0773
1390	0.0793
1406	0.0793
1413	0.0825
1470	0.0815
1473	0.0830
1536	0.0830
1598	0.0833
CURVE 4	
388	0.0616
400	0.0610
541	0.0722
553	0.0773
CURVE 4 (cont.)	
667	0.0667
729	0.0606
818	0.0417
867	0.0339
936	0.0269
993	0.0184
1078	0.0171
1128	0.0173
CURVE 5	
1023	0.0509
1063	0.0510
1095	0.0511
1131	0.0513
1144	0.0514
1149	0.0516
1156	0.0519
1157	0.0555
1157	0.0579
1157	0.0596
1162	0.0599
1176	0.0600
1190	0.0598
1201	0.0600
1201	0.0596
1212	0.0597
1233	0.0599
1251	0.0598
1274	0.0600
1301	0.0600
1342	0.0600
1381	0.0600
1414	0.0600
1454	0.0603
1504	0.0603
1506	0.0601
CURVE 6	
298	0.023
CURVE 7	
298	0.10

FIGURE AND TABLE 73R. RECOMMENDED THERMAL DIFFUSIVITY OF TUNGSTEN

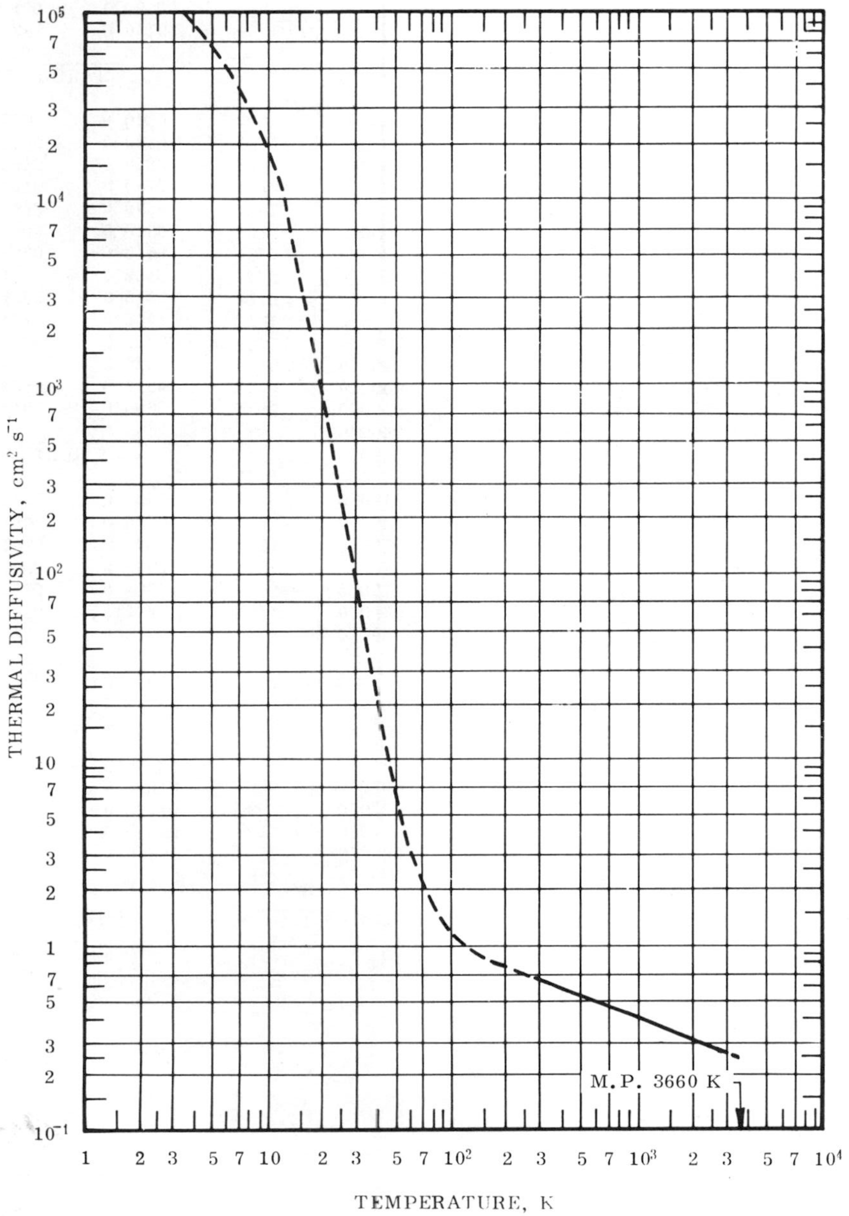

RECOMMENDED VALUES

[Temperature, T, K; Thermal Diffusivity, α, cm^2 s^{-1}]

SOLID

T	α	T	α	T	α
1	255000*	35	41.7*	900	0.434
2	163000*	40	20.5*	1000	0.416
3	116000*	45	11.4*	1100	0.401
4	86000*	50	7.03*	1200	0.388
5	66600*	60	3.53*	1300	0.375
6	52200*	70	2.26*	1400	0.364
7	41100*	80	1.68*	1500	0.354
8	32100*	90	1.39*	1600	0.345
9	24900*	100	1.21*	1700	0.336
10	19400*	150	0.873*	1800	0.328
11	15100*	200	0.764*	1900	0.320
12	11600*	250	0.703*	2000	0.313
13	8800*	273.2	0.685*	2200	0.300
14	6690*	300	0.662	2400	0.289
15	4980*	350	0.626	2600	0.279
16	3600*	400	0.595	2800	0.270
18	1960*	500	0.542	3000	0.263
20	1090*	600	0.508	3500	0.249*
25	293*	700	0.479	3660	0.246*
30	99.2*	800	0.454		

REMARKS

The recommended values are for well-annealed high-purity tungsten and are considered accurate to within ±5% of the true values at temperatures from 300 to 1500 K, ±8% from 100 to 300 K and 1500 to 3000 K, and ±14% below 100 K and above 3000 K. At low temperatures the values are highly conditioned by impurity and imperfection, and those below 200 K are applicable only to tungsten having residual electrical resistivity of 0.00170 μΩ cm.

*In temperature range where no experimental data are available.

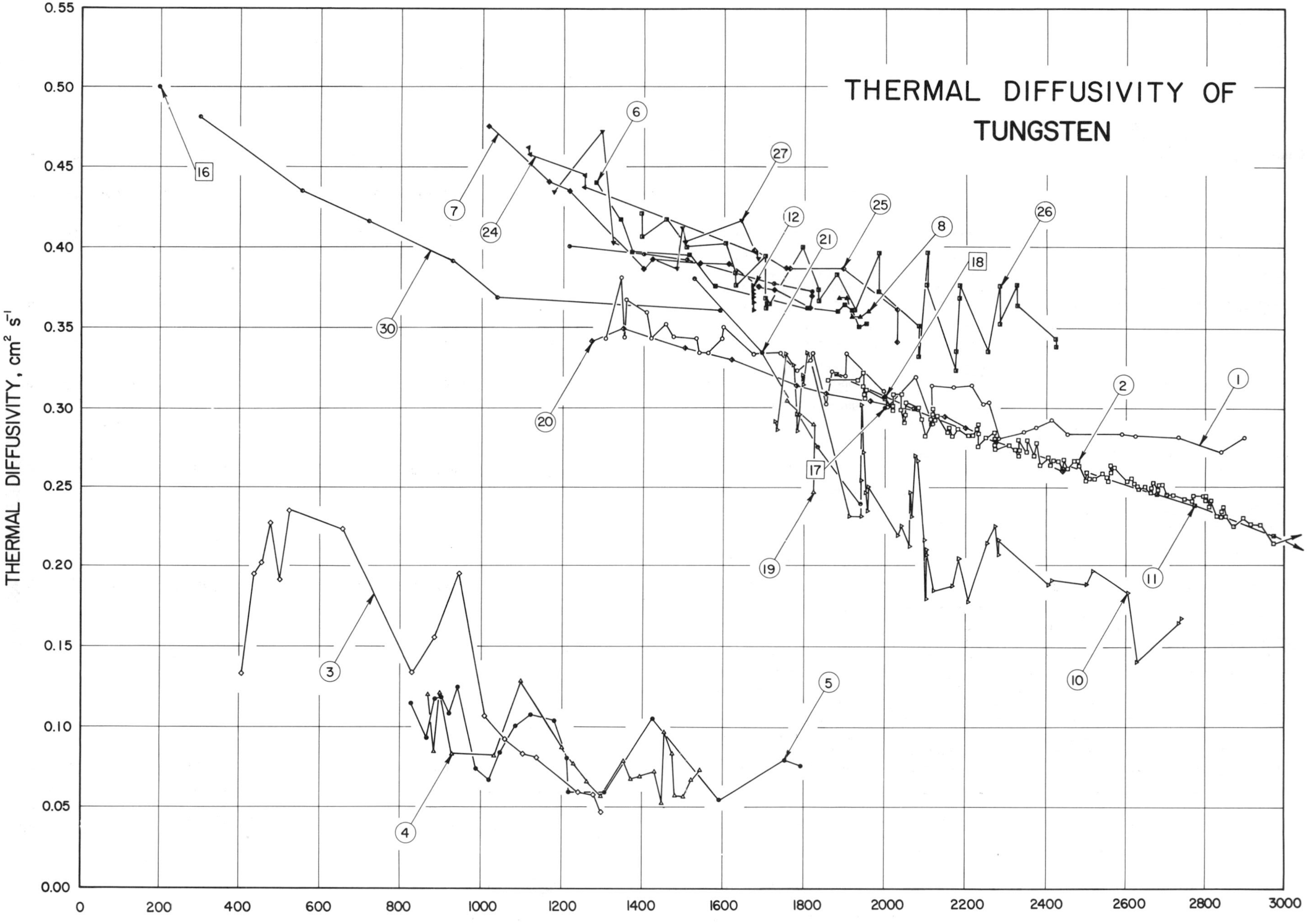
THERMAL DIFFUSIVITY OF
TUNGSTEN
THERMAL DIFFUSIVITY, cm² s⁻¹
TEMPERATURE, K
0.55
0.50
0.45
0.40
0.35
0.30
0.25
0.20
0.15
0.10
0.05
0.00
0
200
400
600
800
1000
1200
1400
1600
1800
2000
2200
2400
2600
2800
3000
1
2
3
4
5
6
7
8
10
11
12
16
17
18
19
20
21
24
25
26
27
30

FIGURE 73

SPECIFICATION TABLE 73. THERMAL DIFFUSIVITY OF TUNGSTEN

(Impurity < 0.20% each; total impurities < 0.50%)

Cur. No.	Ref. No.	Author(s)	Year	Temp. Range, K	Reported Error, %	Name and Specimen Designation	Composition (weight percent), Specifications, and Remarks
1	39	Wheeler, M. J.	1965	1308-2900			99.5 W, impurities of Fe and Mo, and traces of other elements; undoped; avg grain size after testing 46 μm; cylindrical specimen having top surface area lying in the range from 0.3 to 1.3 cm^2 and 1.5 mm in thickness; supplied by G. E. C. Osram Lamp Works; cut from a swaged rod with a diamond impregnated slitting wheel and the faces lapped parallel; density 19.3 g cm^{-3}; Lorenz function reported as 2.948, 2.956, 2.958, 2.950, 2.948, 2.948, 2.954, 2.985, 3.014, and 3.069 x 10^{-8} V^2 K^{-2} at 1200, 1401, 1601, 1802, 1999, 2200, 2401, 2601, 2805, and 3006 K, respectively; specimen heated to incandescence by an electron beam; intensity of beam sinusoidally modulated at a frequency of 0.48 cycles per sec; modulation used produced temp fluctuations in the bombarded face of the specimen of from ± 5 to ± 20 K; measured under a vacuum of ~5 x 10^{-6} mm Hg.
2	55	Kraev, O. A. and Stel'makh, A. A.	1963	1860-3232	~5		Cylindrical specimens from 7 to 8 mm in dia and 0.2 mm thick; made from plates of rolled technical tungsten and surfaces polished; specimen placed between anode and cathode and heated on one side by a flow of electrons emitted by the cathode that was heated to incandescence; periodic heat flow applied varies according to a cosine law; diffusivity determined by measuring the phase shift between first harmonic of the oscillations of the heat flow and of the temp on the other side of specimen; measured under a vacuum of the order of 10^{-5} mm Hg.
3	10	Sheer, C., Fitz, C. D., Mead, L. H., Holmgren, J. D., Rothacker, D. L., and Allmand, D.	1958	407-1298			Cylindrical specimen 1.0 cm in dia and 4.5 cm long; machined to dimensions given; insulated on the sides and over one end face; other end face suddenly exposed to a constant heat flow from the plasma of a high intensity arc; measured in vacuum chamber under an ambient pressure of 0.1 atmosphere; diffusivity data computed assuming conditions of zero heat flow across the lateral and rear end surfaces.
4	10	Sheer, C., et al.	1958	868-1543			Above specimen allowed to cool in chamber after being heated during above measurements; exposed to the arc to measure diffusivity again.
5	10	Sheer, C., et al.	1958	828-1793			Above specimen allowed to cool in chamber after being heated during above measurements; exposed to the arc to measure diffusivity again.
6	57	Pigal'skaya, L. A., Filippov, L. P., and Borisov, V. D.	1966	1282-1957			99.95 W and 0.035 Mo; cylindrical specimen 10 mm in dia and 80 mm long; forged; density 19.17 g cm^{-3} at room temp; Lorenz number reported as varying from 3.44 to 3.34 x 10^{-8} V^2 K^{-2} in the temp range from 1000 to 2000 K; data obtained by the phase method under conditions of unequal switching on and off periods.
7	57	Pigal'skaya, L. A., et al.	1966	1017-1820			Above specimen measured by the phase method under conditions of equal switching on and off periods.
8	57	Pigal'skaya, L. A., et al.	1966	1886-1959			Above specimen measured by the amplitude method under conditions of unequal switching on and off periods.
9	57	Pigal'skaya, L. A., et al.	1966	1216-1820			Above specimen measured by the amplitude method under conditions of equal switching on and off periods.

SPECIFICATION TABLE 73. THERMAL DIFFUSIVITY OF TUNGSTEN (continued)

Cur. No.	Ref. No.	Author(s)	Year	Temp. Range, K	Reported Error, %	Name and Specimen Designation	Composition (weight percent), Specifications, and Remarks
10	32	Taylor, R.E. and Nakata, M.M.	1963	1728-2743	7		99.877⁺ W (by difference), 0.05 Si, 0.03 Ca, 0.03 Ti, <0.01 Cu, 0.002 Mg, and 0.001 Ag; large grains, averaging 1 mm, with small quantities of micrograins; cylindrical specimen 1.577 cm overall dia and 3.492 cm overall length, machined in three sections: a center section 2.54 cm long and two end pieces consisting of 0.476 cm thick disks; these parts are machine fitted in such a way that a small area contact is made at the circumference only, leaving thin spaces of 0.0127 cm or less between center section and the disks that act as radiation barriers; two parallel sight holes each 0.120 cm in dia drilled through the top disk to a depth of 1.75 cm at radii $r_1 = 0$ and $r_2 = 0.563$ cm; sight holes drilled by Elox technique; bottom disk grooved to fit sample supporting pins; all parts machined out of the same material; measured after extensive heat soaking at temps up to 2673.2 K; density 18.54 g cm^{-3}; radiation shields mounted at both ends; measured under a vacuum of 1×10^{-6} mm Hg; radial diffusivity technique used.
11	172	Kraev, O.A. and Stel'makh, A.A.	1966	1881-3234	±5		Disk specimen 7 to 9 mm in diameter and 0.2 mm thick; cut from plate of rolled metal; surfaces thoroughly cleaned and polished; heated by electron bombardment; sinusoidal temperature oscillation imposed on one face of specimen; thermal diffusivity determined from phase difference bwtween measured temperature fluctuations at front and back faces of specimen; reported data points corrected for thermal expansion and obtained from smooth curve given by authors.
12	186	Pigal'skaya, L.A. and Filippov, L.P.	1964	1673.2	4-8 Max.		Cylindrical specimen 10 mm in diameter and 60 mm long; periodically modulated heat wave imposed on specimen; measured in vacuum; diffusivity determined from ratio of amplitudes of temperature fluctuation of specimen surface for two different frequencies; measured as a function of the frequency, f, ranging from 0.291 to 0.744 cps.
13*	186	Pigal'skaya, L.A. and Filippov, L.P.	1964	1673.2	4-8 Max.		Above specimen measured for diffusivity again; diffusivity determined from measured phase difference between the power modulation and the temperature fluctuation; measured in vacuum; measured as a function of the frequency, f, ranging from 0.291 to 0.744 cps.
14*	3	Dunn, S.A.	1965	298			No details reported; measuring temperature not reported but assumed to be 25 C.
15*	247	Namba, S., Kim, P.H., and Arai, T.	1967	298			Square specimen 1.5 x 1.5 x 0.109 cm; thermal diffusivity measured by a flash method.
16	261	Bettler, P.C.	1967	298			No details reported; measuring temperature not reported but assumed to be 25 C.
17	219	Null, M.R. and Lozier, W.W.	1969	2000			Disk specimen 12.7 mm in diameter and 1.060 mm thick; bulk density 19.3 g cm^{-3}; electrical resistivity 5.61 μohm cm at room temperature; thermal diffusivity measured by means of a carbon arc image furnace radiation; reported value derived from least squares extrapolation.
18	219	Null, M.R. and Lozier, W.W.	1969	2000			Similar to above but specimen 2.100 mm thick and reported value derived from nonlinear extrapolation.

*-Not shown in figure.

SPECIFICATION TABLE 73. THERMAL DIFFUSIVITY OF TUNGSTEN (continued)

Cur. No.	Ref. No.	Author(s)	Year	Temp. Range, K	Reported Error, %	Name and Specimen Designation	Composition (weight percent), Specifications, and Remarks
19	264	Nakata, M.M. (Wechsler, A.E., compiler)	1969	1757-1825	7	WA-1-1	>99.87 W, <0.1 Mo, <0.005 C, <0.0030 each of Fe, N, and O, <0.0020 each of Al, Pb, Si, and Sn, <0.0010 each of Cr, Co, Cu, Mg, Mn, Ni, Ti, and V, and <0.0005 H; disk specimen 0.259 in. in diameter and 1.025 in. thick; supplied by Climax Molybdenum Corp. and machined by Thermo Electron Engineering Co., machined to size and coated with a thin layer of aquadag by Atomic International; density 19.3 g cm^{-3}; thermal diffusivity measured by flash method in a vacuum of 10^{-6} torr.
20	274	Springer, J.R., Lagedrost, J.F., and McCann, R.A. (Wechsler, A.E., compiler)	1969	1272-2441	5-8	WB-1	From the same source as the above specimen; 0.376 in. in diameter and 0.985 in. thick; cleaned ultrasonically by BMI; density 19.3 g cm^{-3}; measured by flash method in a vacuum of 10^{-6} torr.
21	212	Van den Berg, M. and Schmidt, H.E.	1965	1527-1941			Disk specimen 5 to 6 mm in diameter and 1 to 2 mm thick; electron beam used as heat source; measured in vacuum.
22*	262	Schmidt, H.E., Van den Berg, M., and van der Hoek, L.	1969	1123-2598	7	TA-2	>99.8 W, 0.0120~0.0127 C, 0.0016 O, 0.0003~0.0004 H, and 0.0001 N; disk specimens 25 mm in diameter and 2.67 mm thick; density 18.950 g cm^{-3}; electrical resistivity 5.99 and 7.74 μohm cm at 30 and 100 C, respectively; modulated electron beam used as heat source; measured in vacuum.
23*	262	Schmidt, H.E., et al.	1969	1443-2823	7	TA-4	Similar to above but specimen thickness 2.71 mm.
24	263	Ferro, C., Patimo, C., and Piconi, C.	1970	1111-1683		Sample No.1	Specimen 10.011 ± 0.005 mm in diameter, 2.520 ± 0.005 mm in thickness; weight 3.818 g; density 19.258 g cm^{-3}; diffusivity measured under vacuum at 10^{-5} mm Hg; test No. 1.
25	263	Ferro, C., et al.	1970	1679-2032		Sample No.1	The above specimen; test No. 2.
26	263	Ferro, C., et al.	1970	1398-2425		Sample No.1	The above specimen; test No. 3.
27	263	Ferro, C., et al.	1970	1177-1686		Sample No.2	Specimen 10.000 ± 0.005 mm in diameter, 5.000 ± 0.005 mm in thickness; weight 7.559 g; density 19.260 g cm^{-3}; diffusivity measured under vacuum at 10^{-5} mm Hg; test No. 1.
28*	263	Ferro, C., et al.	1970	1136-2232		Sample No.2	The above specimen; test No. 2.
29*	263	Ferro, C., et al.	1970	1376-2106		Sample No.2	The above specimen; test No. 3.
30	267	Rawuka, A.C. and Gaz, R.A.	1969	300-1593		Spec. No.1	99.87 pure; specimen 0.18 cm in thickness; supplied by Climax Molybdenum Co., Ann Arbor, Michigan; density 19.23 g cm^{-3}; diffusivity measured by pulse technique.

* Not shown in figure.

DATA TABLE 73. THERMAL DIFFUSIVITY OF TUNGSTEN

(Impurity < 0.20% each; total impurities < 0.50%)

[Temperature, T, K; Thermal Diffusivity, α, cm² s⁻¹]

T	α
CURVE 1	
1308	0.343
1346	0.381
1354	0.344
1357	0.367
1409	0.359
1420	0.343
1456	0.352
1477	0.344
1532	0.343
1540	0.334
1562	0.334
1596	0.343
1599	0.351
1675	0.334
1741	0.334
1784	0.323
1818	0.330
1822	0.334
1853	0.302
1869	0.323
1903	0.320
1905	0.334
1996	0.310
2006	0.300
2076	0.319
2114	0.314
2119	0.289
2170	0.313
2214	0.314
2243	0.302
2258	0.303
2281	0.280
2342	0.284
2373	0.287
2413	0.292
2453	0.283
2590	0.283
2624	0.282
2732	0.281
2841	0.272
2900	0.281

T	α
CURVE 2	
1860	0.317
1933	0.317
1945	0.322
1945	0.313
1951	0.308
1951	0.305
1953	0.311
2000	0.305
2020	0.298
2020	0.301
2021	0.308
2041	0.308
2041	0.298
2049	0.290
2050	0.295
2051	0.303
2082	0.300
2091	0.292
2100	0.281
2115	0.289
2115	0.292
2117	0.295
2118	0.299
2126	0.291
2129	0.294
2157	0.284
2160	0.287
2168	0.281
2180	0.286
2207	0.282
2218	0.282
2228	0.286
2231	0.289
2231	0.283
2231	0.275
2250	0.280
2272	0.284
2272	0.280
2272	0.276
2272	0.273
2308	0.276

T	α
CURVE 2 (cont.)	
2323	0.273
2332	0.268
2335	0.273
2335	0.277
2335	0.280
2355	0.279
2355	0.271
2374	0.268
2375	0.277
2383	0.263
2404	0.268
2412	0.263
2418	0.266
2431	0.266
2441	0.261
2444	0.267
2454	0.261
2470	0.266
2480	0.266
2482	0.263
2501	0.253
2502	0.259
2510	0.255
2521	0.255
2541	0.258
2552	0.256
2552	0.253
2561	0.263
2563	0.258
2572	0.262
2604	0.253
2613	0.255
2621	0.252
2630	0.250
2632	0.248
2649	0.250
2664	0.246
2664	0.249
2669	0.252
2679	0.245
2682	0.248

T	α
CURVE 2 (cont.)	
2684	0.251
2695	0.251
2704	0.247
2720	0.240
2734	0.242
2769	0.241
2770	0.244
2793	0.244
2802	0.241
2803	0.244
2814	0.237
2816	0.241
2831	0.231
2843	0.231
2844	0.234
2846	0.237
2852	0.231
2874	0.225
2897	0.230
2916	0.226
2942	0.226
2973	0.214
3016	0.218*
3024	0.214*
3037	0.219*
3075	0.206*
3115	0.205*
3146	0.203*
3156	0.200*
3163	0.197*
3223	0.194*
3232	0.191*
CURVE 3	
407	0.132
438	0.194
458	0.202
480	0.227
503	0.191
526	0.235

T	α
CURVE 3 (cont.)	
659	0.223
830	0.133
886	0.155
943	0.195
1009	0.106
1059	0.0915
1103	0.0825
1136	0.0800
1240	0.0585
1278	0.0575
1298	0.0463
CURVE 4	
868	0.119
883	0.134
898	0.121
928	0.0825
1031	0.0813
1099	0.128
1200	0.0863
1229	0.0760
1263	0.0650
1296	0.0560
1354	0.0780
1372	0.0663
1396	0.0688
1430	0.0713
1448	0.0513
1454	0.0960
1473	0.0825
1481	0.0563
1503	0.0560
1523	0.0663
1543	0.0725
CURVE 5	
828	0.114
866	0.0925
886	0.117

T	α
CURVE 5 (cont.)	
903	0.118
921	0.108
943	0.124
988	0.0733
1020	0.0663
1048	0.0838
1085	0.100
1123	0.107
1181	0.103
1211	0.0800
1216	0.0588
1306	0.0588
1425	0.104
1591	0.0538
1753	0.0788
1793	0.0750
CURVE 6	
1282	0.440
1344	0.417
1371	0.396
1515	0.394
1579	0.375
1808	0.362
1884	0.360
1900	0.364
1919	0.361
1936	0.350
1957	0.352
CURVE 7	
1017	0.475
1166	0.440
1216	0.434
1400	0.386
1423	0.392
1543	0.389
1612	0.389
1687	0.375

T	α
CURVE 7 (cont.)	
1725	0.373
1814	0.362
1820	0.369
CURVE 8	
1886	0.368
1905	0.368
1919	0.356
1938	0.356
1959	0.360
CURVE 9	
1216	0.400
1403	0.395
1510	0.392
1725	0.377
1820	0.372
CURVE 10	
1728	0.291
1733	0.286
1753	0.334
1773	0.327
1783	0.296
1783	0.285
1788	0.296
1798	0.320
1798	0.314
1808	0.334
1913	0.231
1943	0.231
1943	0.231*
1943	0.254
1943	0.302
1948	0.271
1953	0.246
1958	0.234
1958	0.250

T	α
CURVE 10 (cont.)	
2033	0.219
2043	0.225
2063	0.212
2063	0.246
2068	0.231
2078	0.270
2083	0.266
2098	0.216
2103	0.179
2103	0.210
2103	0.207
2123	0.184
2168	0.187
2183	0.204
2208	0.177
2253	0.214
2273	0.225
2283	0.207
2283	0.216
2408	0.188
2413	0.190
2503	0.188
2518	0.197
2608	0.183
2628	0.140
2733	0.164
2743	0.167
CURVE 11	
1881	0.321
2073	0.299
2273	0.278
2673	0.245
2773	0.238
2873	0.229
2973	0.219
3073	0.209*
3234	0.192*

* Not shown in figure.

DATA TABLE 73. THERMAL DIFFUSIVITY OF TUNGSTEN (continued)

f (cps)	α
CURVE 12 (T = 1673K)	
0.291	0.371
0.344	0.368
0.417	0.366
0.477	0.370
0.572	0.376
0.678	0.361
0.744	0.373
CURVE 13* (T = 1673K)	
0.291	0.366
0.344	0.385
0.417	0.364
0.477	0.364
0.572	0.386
0.678	0.393
0.744	0.371
T	**α**
CURVE 14*	
298	0.586
CURVE 15*	
298	0.70
CURVE 16	
298	0.5
CURVE 17	
2000	0.299
CURVE 18	
2000	0.307
CURVE 19	
1757	0.304
1824	0.246
1825	0.289

T	α
CURVE 20	
1272	0.341
1352	0.349
1506	0.337
1622	0.330
1783	0.314
1855	0.309
1966	0.304
2148	0.294
2200	0.287
2441	0.260
CURVE 21	
1527	0.380
1696	0.334
1835	0.275
1941	0.239
CURVE 22*	
1123	0.383
1223	0.394
1223	0.382
1223	0.368
1333	0.371
1443	0.345
1553	0.360
1553	0.327
1663	0.327
1783	0.312
1893	0.312
1893	0.279
1951	0.263
2008	0.306
2008	0.263
2123	0.283
2123	0.277
2123	0.261
2243	0.239
2473	0.251
2598	0.245
CURVE 23*	
1443	0.344
1893	0.291
2123	0.273

T	α
CURVE 23 (cont.)	
2473	0.247
2598	0.217
2713	0.224
2823	0.211
CURVE 24	
1111	0.461
1119	0.457
1255	0.444
1255	0.437
1683	0.396
CURVE 25	
1679	0.397
1756	0.386
1764	0.386
1899	0.386
2032	0.361
2032	0.341
CURVE 26	
1398	0.421
1399	0.406
1458	0.417
1509	0.399
1605	0.402
1632	0.383
1632	0.376
1704	0.394
1704	0.368
1704	0.361
1714	0.364
1798	0.399
1834	0.373
1834	0.366
1882	0.382
1928	0.360
1986	0.396
1986	0.372
2085	0.351
2085	0.332
2105	0.396
2105	0.376
2176	0.335
2176	0.323

T	α
CURVE 26 (cont.)	
2185	0.376
2185	0.368
2256	0.335
2285	0.376
2286	0.352
2326	0.376
2328	0.363
2425	0.343
2425	0.338
CURVE 27	
1177	0.434
1299	0.471
1327	0.402
1487	0.386
1497	0.411
1506	0.402
1647	0.416
1686	0.392
CURVE 28*	
1136	0.416
1266	0.443
1344	0.431
1441	0.416
1501	0.416
1548	0.443
1648	0.386
1652	0.395
1702	0.410
1802	0.392
1809	0.382
1936	0.421
1936	0.356
2096	0.367
2111	0.360
2127	0.385
2232	0.399
CURVE 29*	
1376	0.476
1423	0.449
1436	0.371
1524	0.428

T	α
CURVE 29 (cont.)	
1554	0.438
1667	0.394
1757	0.441
1773	0.424
1955	0.377
2085	0.371
2106	0.410
CURVE 30	
300	0.481
553	0.435
720	0.416
927	0.391
1039	0.368
1593	0.361

* Not shown in figure.

FIGURE AND TABLE 74R. PROVISIONAL THERMAL DIFFUSIVITY OF URANIUM

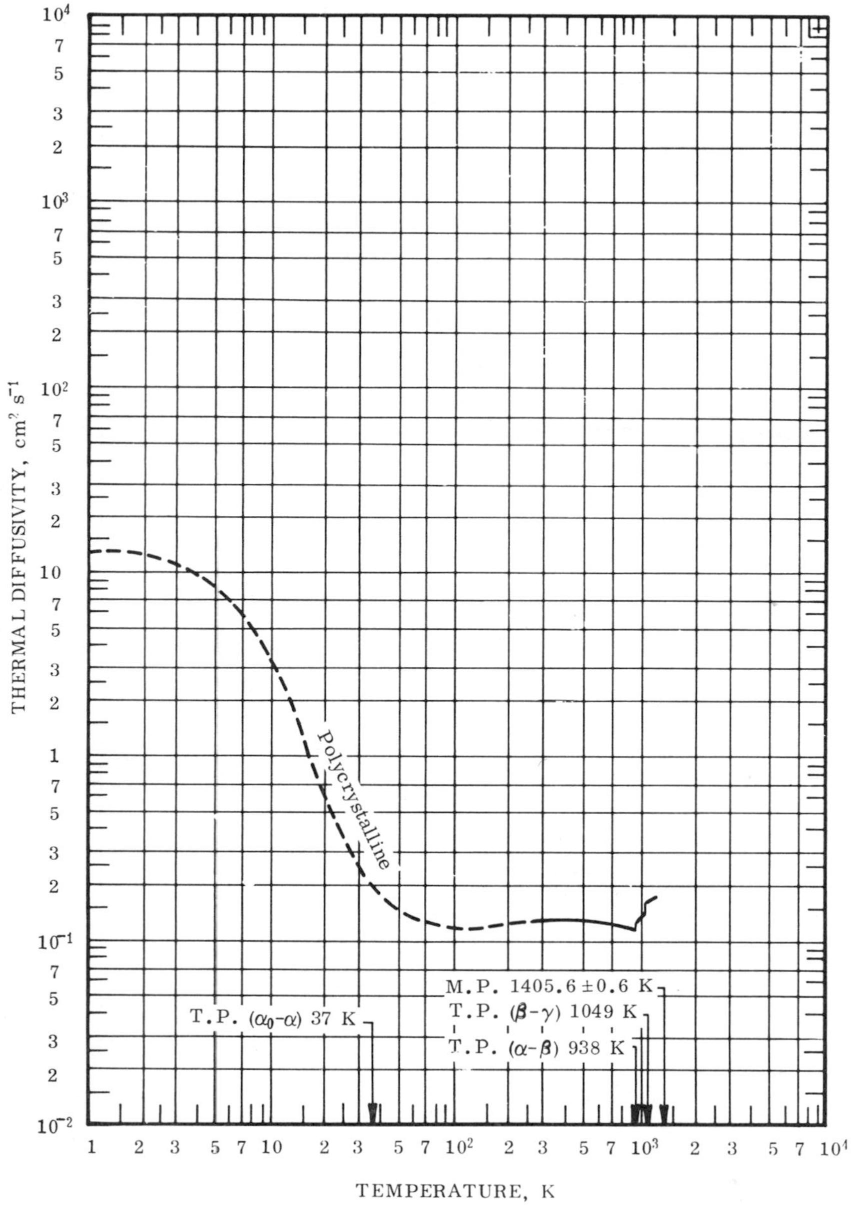

PROVISIONAL VALUES

[Temperature, T, K; Thermal Diffusivity, α, cm² s⁻¹]

SOLID
(Polycrystalline)

T	α	T	α
1	12.7*	70	0.131*
2	12.1*	80	0.126*
3	10.9*	90	0.123*
4	9.60*	100	0.121*
5	8.15*	150	0.119*
6	6.80*	200	0.121*
7	5.64*	250	0.123*
8	4.68*	273.2	0.124*
9	3.88*	300	0.125
10	3.22*	350	0.126
11	2.68*	400	0.127
12	2.22*	500	0.127
13	1.80*	600	0.125
14	1.56*	700	0.122
15	1.32*	800	0.119
16	1.13*	900	0.116
18	0.843*	938	0.114
20	0.642*	938	0.118
25	0.376*	1000	0.134
30	0.265*	1049	0.139
35	0.207*	1049	0.156
40	0.175*	1100	0.162
45	0.158*	1200	0.172
50	0.149*		
60	0.1375*		

REMARKS

The values are for well-annealed high-purity polycrystalline uranium and are thought to be accurate to within ± 13% of the true values at temperatures from room temperature to 900 K and ± 15 to ± 20% at other temperatures. The values below room temperature are applicable only to uranium having residual electrical resistivity of 2.14 μΩ cm.

*In temperature range where no experimental data are available.

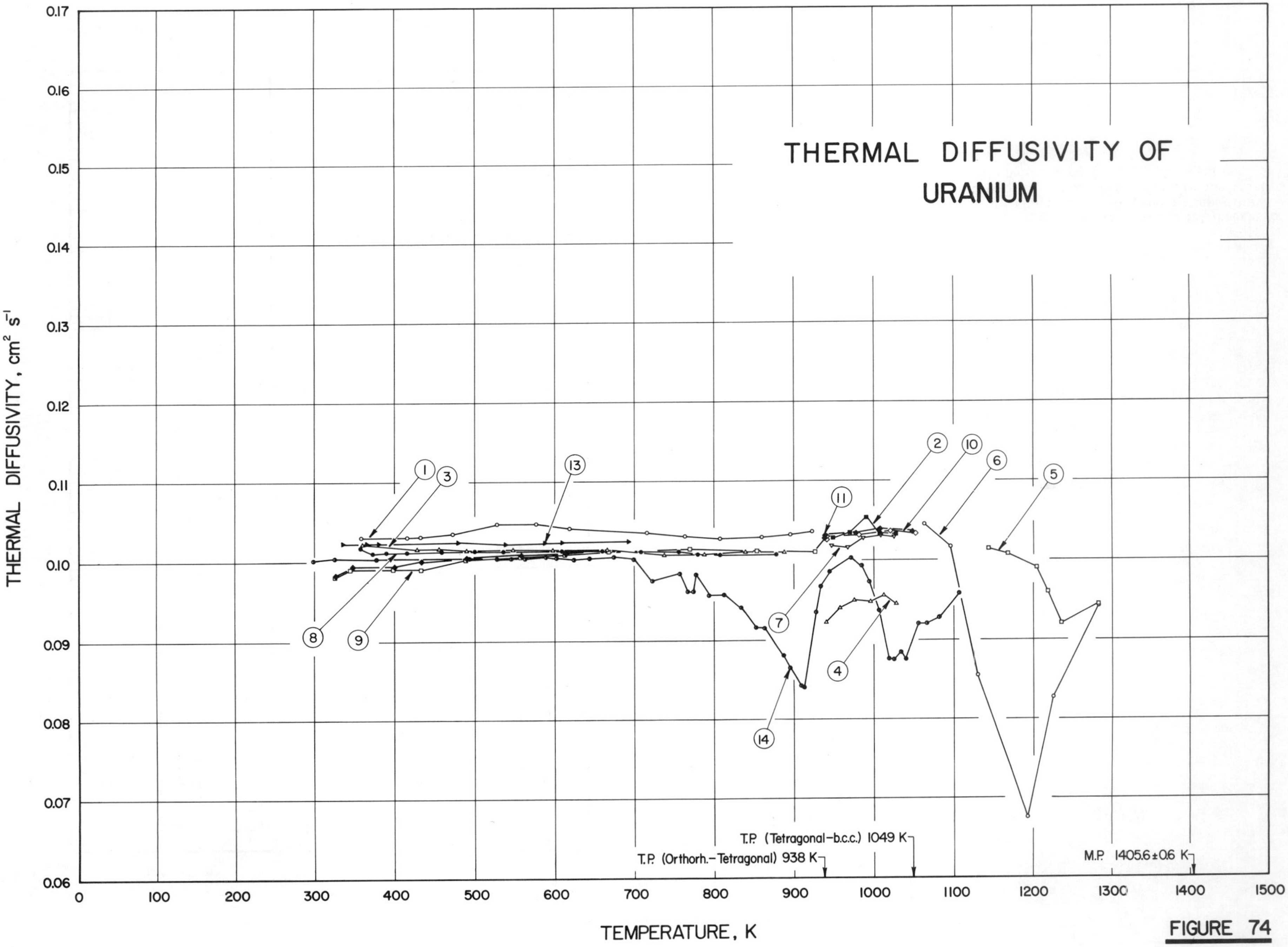
THERMAL DIFFUSIVITY OF URANIUM
THERMAL DIFFUSIVITY, cm² s⁻¹
TEMPERATURE, K
0.17
0.16
0.15
0.14
0.13
0.12
0.11
0.10
0.09
0.08
0.07
0.06
0
100
200
300
400
500
600
700
800
900
1000
1100
1200
1300
1400
1500
1
2
3
4
5
6
7
8
9
10
11
13
14
T.P. (Tetragonal–b.c.c.) 1049 K
T.P. (Orthorh.–Tetragonal) 938 K
M.P. 1405.6 ± 0.6 K

FIGURE 74

SPECIFICATION TABLE 74. THERMAL DIFFUSIVITY OF URANIUM

(Impurity < 0.20% each; total impurities < 0.50%)

Cur. No.	Ref. No.	Author(s)	Year	Temp. Range, K	Reported Error, %	Name and Specimen Designation	Composition (weight percent), Specifications, and Remarks
1	33	Chiotti, P. and Carlson, O. N.	1956	359-923			Avg grain dia.~0.03 mm (after measurement); lattice constants: a_0 = 2.8526 Å, b_0 = 5.8682 Å, and c_0 = 4.9489 Å, experimentally determined at 298.2 K; orthorhombic; annealed for 1 hr at 898.2 K; measured in the α-phase region.
2	33	Chiotti, P. and Carlson, O. N.	1956	950-1082			Above specimen cooled to 298.2 K after having been heated during above measurements; β-annealed at 973.2 K for 1 hr; measured for diffusivity again in the β-phase region; tetragonal.
3	33	Chiotti, P. and Carlson, O. N.	1956	360-888			Above specimen cooled to 298.2 K after having been heated during above measurements; α-annealed; measured for diffusivity again in the α-phase region.
4	33	Chiotti, P. and Carlson, O. N.	1956	941-1028			Above specimen measured again after having been taken into the β-phase region.
5	33	Chiotti, P. and Carlson, O. N.	1956	1144-1283			Above specimen measured again after having been taken into the γ-phase region; measured while being heated; cubic.
6	33	Chiotti, P. and Carlson, O. N.	1956	1064-1283			Above specimen measured again in the γ-phase region; measured while being cooled.
7	33	Chiotti, P. and Carlson, O. N.	1956	948-1026			Above specimen measured again after having been taken back to the β-phase region.
8	33	Chiotti, P. and Carlson, O. N.	1956	358-879			Above specimen cooled to 298.2 K; measured for diffusivity again in the α-phase region.
9	58	Sidles, P. H.	1963	326-927		Ames Uranium	Cylindrical specimen 0.125 in. in dia; two holes drilled along a diameter at different points along specimen to accomodate thermocouples; sinusoidal variation of temperature at one end and ambient temp at the other end of specimen; diffusivity determined from measured amplitudes and phase relationship between thermocouples; three independent determinations made successively at each temp; measured in the α-phase region.
10	58	Sidles, P. H.	1963	942-1053		Ames Uranium	Above specimen measured again for diffusivity in the β-phase region.
11	185	Danielson, G. C. (Chiotti, P. and Carlson, O. N.)	1954	326-1049			No details given.
12*	265	Erez, G. and Even, U.	1966	349-1068			Nuclear pure uranium with total impurity content less than 300 ppm; melted and cast in alumina-coated graphic crucible.
13	268	Danielson, G. C. (Chiotti, P. and Carlson, O. N.)	1953	337-693			Specimen 0.125 in. in diameter, 30 cm long; swaged from Hanford Uranium slug and annealed; diffusivity measured using dynamic method.
14	269	Nasu, S., Fukushima, S., Ohmichi, T., and Kikuchi, T.	1968	298-1106			99.9 pure; 1.0 cm in diameter, 0.19 cm in thickness.

* Not shown in figure.

DATA TABLE 74. THERMAL DIFFUSIVITY OF URANIUM

(Impurity < 0.20% each; total impurities < 0.50%)

[Temperature, T, K; Thermal Diffusivity, α, cm^2 s^{-1}]

T	α
CURVE 1	
359	0.130
416	0.130
433	0.131
473	0.134
528	0.146
577	0.147
619	0.141
716	0.135
764	0.130
808	0.127
860	0.129
896	0.132
923	0.136
CURVE 2	
950	0.127
971	0.134
991	0.154
1008	0.133
1028	0.132
CURVE 3	
360	0.121
428	0.115
456	0.115
490	0.113
548	0.114
598	0.113
666	0.114
738	0.107
788	0.108
840	0.111
888	0.110
CURVE 4	
941	0.0923
958	0.0940
976	0.0950
CURVE 4 (cont.)	
996	0.0948
1013	0.0956
1028	0.0945
CURVE 5	
1144	0.113
1168	0.107
1204	0.0990
1218	0.0960
1236	0.0920
1283	0.0943
CURVE 6	
1064	0.145
1096	0.116
1130	0.0855
1193	0.0675
1225	0.0827
1283	0.0943
CURVE 7	
948	0.117
968	0.115
987	0.127
1008	0.131
1026	0.128
CURVE 8	
358	0.117
373	0.110
390	0.111
416	0.111
459	0.112
500	0.112
536	0.112
609	0.113
660	0.112
709	0.111
CURVE 8 (cont.)	
756	0.111
780	0.107
808	0.107
879	0.107
CURVE 9	
326	0.098
345	0.099
399	0.099
433	0.099
489	0.102
557	0.105
613	0.108
667	0.112
770	0.114
854	0.111
927	0.110
CURVE 10	
942	0.126
942	0.129
982	0.131
1021	0.135
1053	0.133
CURVE 11	
326	0.0981
347	0.0993
400	0.0993
433	0.0999
490	0.1024
558	0.1061
614	0.1084
668	0.1124*
770	0.1140*
852	0.1112*
925	0.1104*
940	0.1263
941	0.1291
CURVE 11 (cont.)	
979	0.1318
1018	0.1360
1049	0.1341
CURVE 12*	
349	0.1175
423	0.1140
522	0.1051
584	0.1246
686	0.1452
788	0.1622
914	0.1758
951	0.2012
1033	0.1979
1050	0.1526
1068	0.2513
CURVE 13	
336.9	0.122
365.9	0.122
380.2	0.122
539.4	0.121
588.4	0.120
480.8	0.123
609.3	0.122
693.2	0.123
CURVE 14	
298	0.1033
325	0.1033
377	0.1035
492	0.1035
527	0.1034
545	0.1034
563	0.1034
589	0.1053
602	0.1039
624	0.1013
643	0.1037
CURVE 14 (cont.)	
674	0.1054
699	0.1013
723	0.0974
757	0.0984
767	0.0961
775	0.0961
777	0.0982
794	0.0956
812	0.0956
834	0.0940
852	0.0915
863	0.0915
887	0.0879
895	0.0864
909	0.0842
913	0.0839
927	0.0934
934	0.0967
945	0.0985
956	0.1011
972	0.1011
984	0.0991
994	0.0972
1006	0.0936
1018	0.0875
1025	0.0875
1033	0.0884
1040	0.0875
1056	0.0920
1066	0.0920
1081	0.0927
1096	0.0871
1106	0.0957

* Not shown in figure.

FIGURE AND TABLE 75R.   RECOMMENDED THERMAL DIFFUSIVITY OF VANADIUM

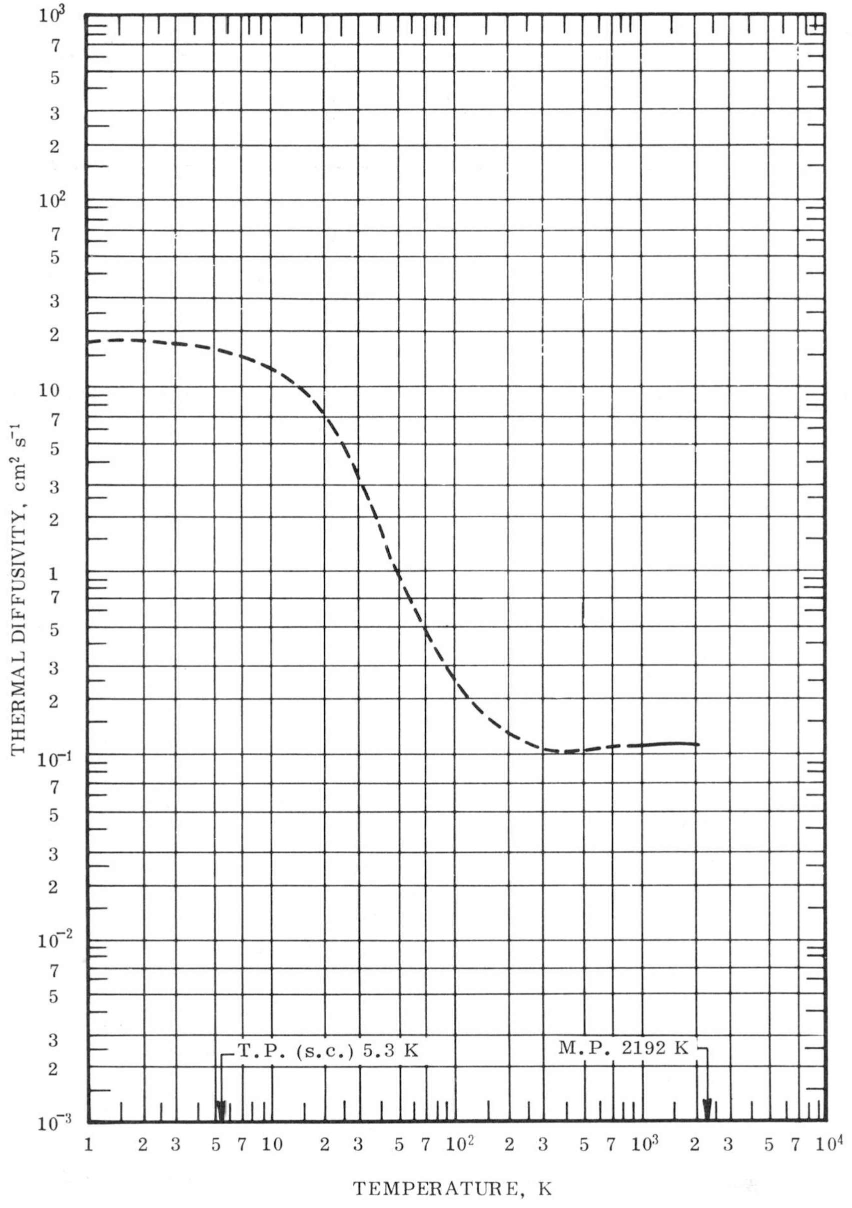

RECOMMENDED VALUES†

[Temperature, T, K; Thermal Diffusivity, α, cm² s⁻¹]

SOLID

T	α	T	α
1	16.9*	70	0.441*
2	16.8*	80	0.339*
3	16.3*	90	0.274*
4	15.8*	100	0.231*
5	15.3*	150	0.149*
6	14.7*	200	0.121*
7	14.1*	250	0.109*
8	13.6*	273.2	0.106*
9	13.0*	300	0.104*
10	12.4*	350	0.102*
11	11.8*	400	0.101*
12	11.25*	500	0.101*
13	10.6*	600	0.102*
14	9.92*	700	0.104*
15	9.30*	800	0.106*
16	8.68*	900	0.107
18	7.50*	1000	0.107
20	6.45*	1100	0.107
25	4.41*	1200	0.107
30	3.05*	1300	0.107
35	2.14*	1400	0.107
40	1.54*	1500	0.106
45	1.17*	1600	0.106
50	0.903*	1700	0.106
60	0.607*	1800	0.106
		1900	0.106
		2000	0.106

REMARKS

The values are for well-annealed high-purity vanadium and are thought to be accurate to within ±10% of the true values at room temperature and above and ±15% below 30 K. Values between 30 K and room temperature are very uncertain, and all those below room temperature are provisional. The values below 200 K are applicable only to vanadium having residual electrical resistivity of 1.72 $\mu\Omega$ cm.

†Values below room temperature are provisional.
*In temperature range where no experimental data are available.

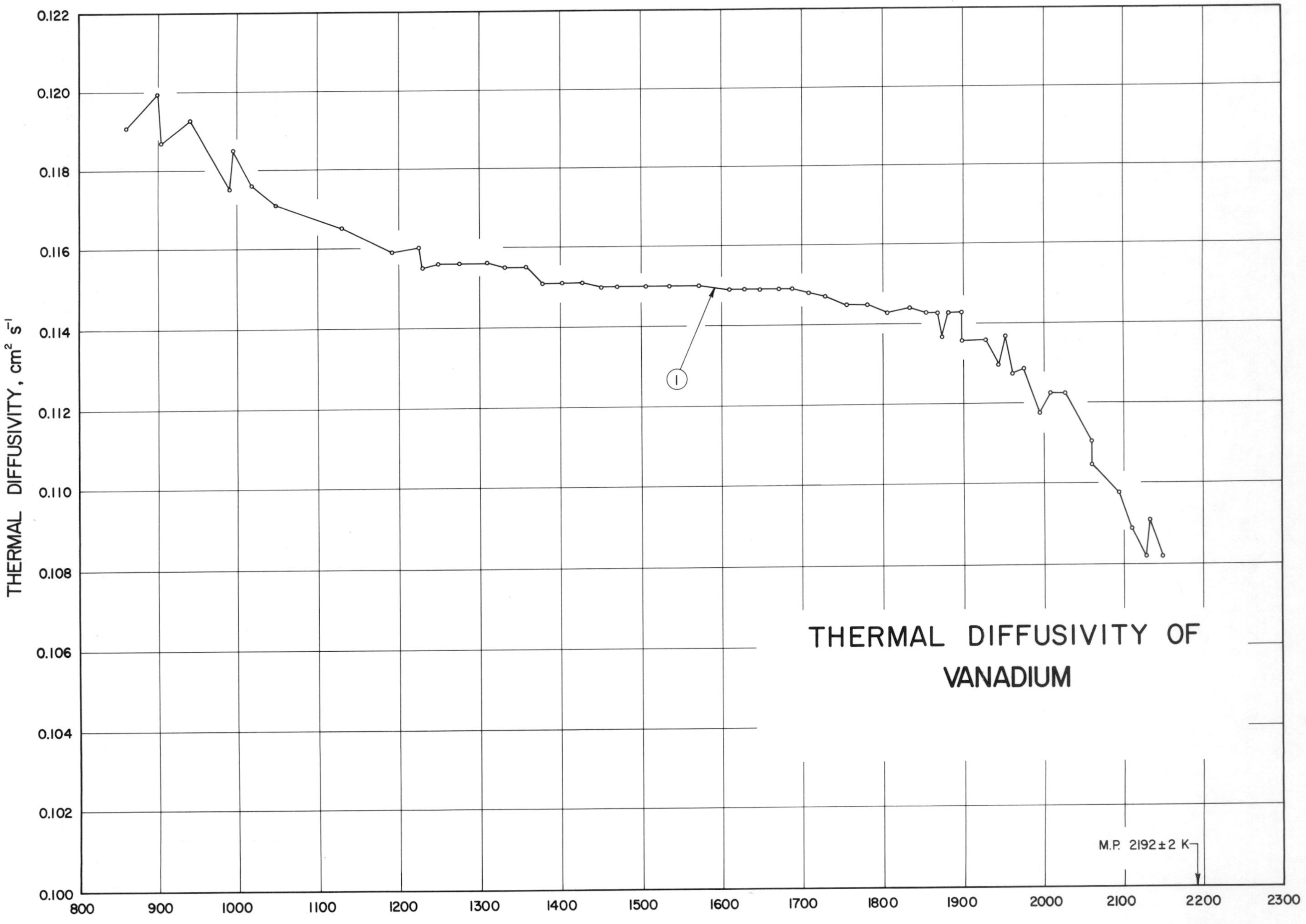
THERMAL DIFFUSIVITY OF
VANADIUM
THERMAL DIFFUSIVITY, cm^2 s^{-1}
0.122
0.120
0.118
0.116
0.114
0.112
0.110
0.108
0.106
0.104
0.102
0.100
800
900
1000
1100
1200
1300
1400
1500
1600
1700
1800
1900
2000
2100
2200
2300
1
M.P. 2192±2 K
TEMPERATURE, K

FIGURE 75

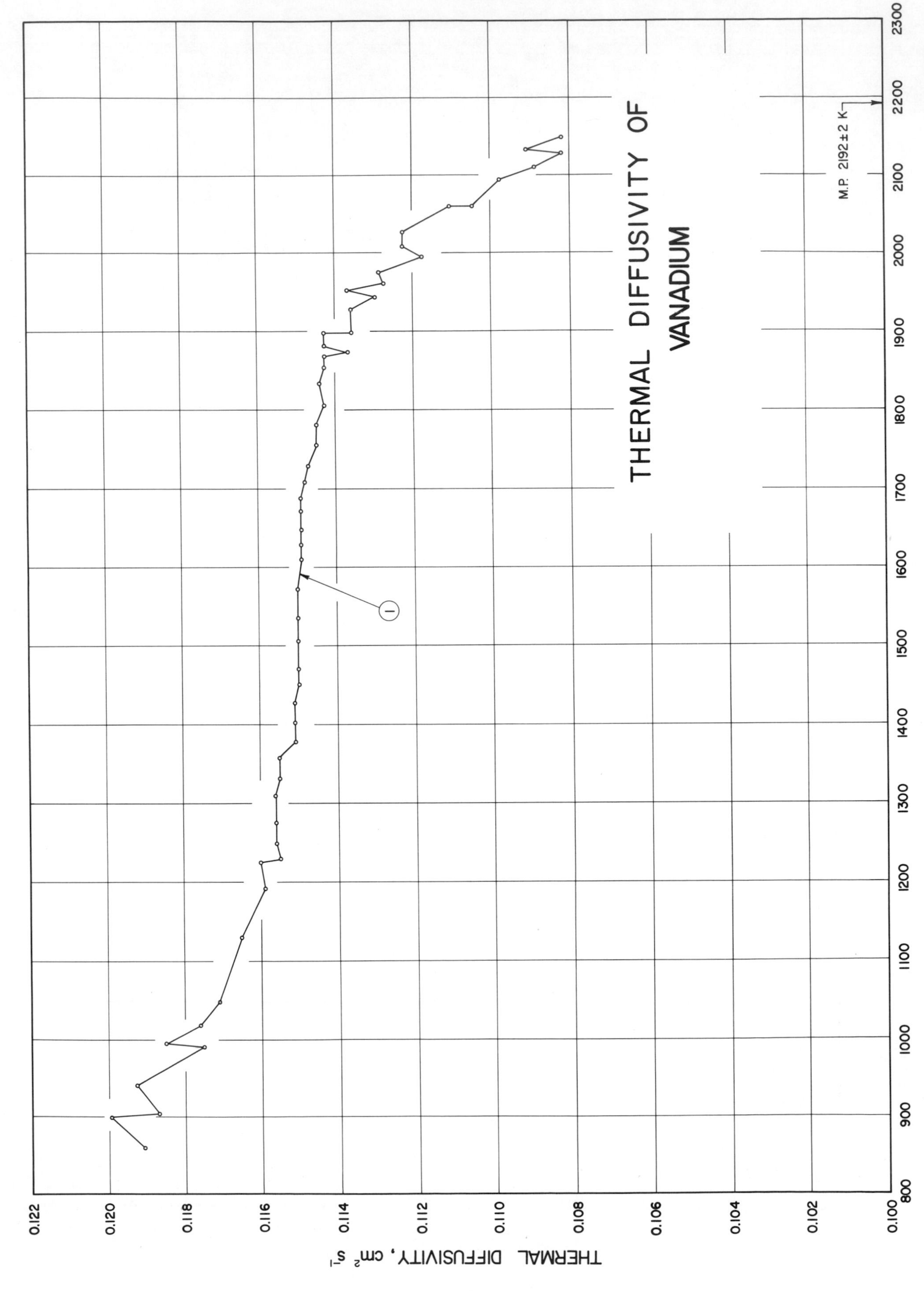

SPECIFICATION TABLE 75. THERMAL DIFFUSIVITY OF VANADIUM

(Impurity < 0.20% each; total impurities < 0.50%)

Cur. No.	Ref. No.	Author(s)	Year	Temp. Range, K	Reported Error, %	Name and Specimen Designation	Composition (weight percent), Specifications, and Remarks
1	270	Zinov'ev, V.E., Krentsis, R.P., and Gel'd, P.V.	1970	859-2149			0.05 impurities; 0.249 mm in thickness; ratio of electrical resistivities $\rho(298K)/\rho(4.2K) = 15$.

DATA TABLE 75. THERMAL DIFFUSIVITY OF VANADIUM

(Impurity < 0.20% each; total impurities < 0.50%)

[Temperature, T, K; Thermal Diffusivity, α, $cm^2\ s^{-1}$]

T	α	T	α	T	α	T	α
CURVE 1		CURVE 1 (cont.)		CURVE 1 (cont.)		CURVE 1 (cont.)	
859	0.1191	1358	0.1155	1755	0.1145	2008	0.1123
899	0.1199	1378	0.1151	1781	0.1145	2027	0.1123
904	0.1187	1402	0.1151	1805	0.1143	2060	0.1111
941	0.1193	1427	0.1151	1834	0.1144	2060	0.1105
991	0.1175	1451	0.1150	1853	0.1143	2094	0.1098
995	0.1185	1471	0.1150	1868	0.1143	2110	0.1089
1018	0.1176	1506	0.1150	1874	0.1137	2128	0.1082
1048	0.1171	1536	0.1150	1881	0.1143	2133	0.1091
1130	0.1165	1573	0.1150	1898	0.1143	2149	0.1082
1192	0.1159	1610	0.1149	1898	0.1136		
1225	0.1160	1629	0.1149	1928	0.1136		
1229	0.1155	1648	0.1149	1944	0.1130		
1249	0.1156	1671	0.1149	1952	0.1137		
1275	0.1156	1688	0.1149	1961	0.1128		
1309	0.1156	1709	0.1148	1976	0.1129		
1331	0.1155	1729	0.1147	1995	0.1118		

FIGURE AND TABLE 76R. PROVISIONAL THERMAL DIFFUSIVITY OF YTTERBIUM

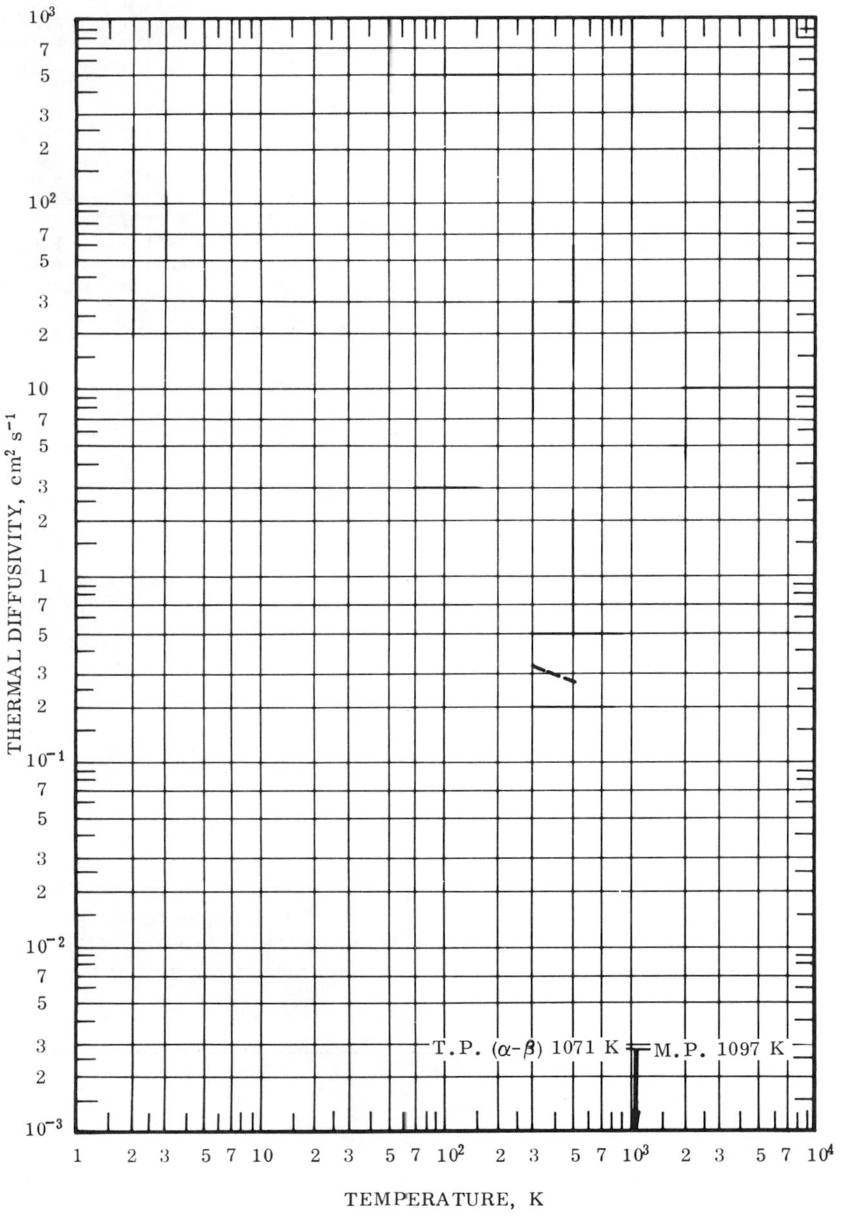

PROVISIONAL VALUES*

[Temperature, T, K; Thermal Diffusivity, α, $cm^2\ s^{-1}$]

SOLID

T	α
298.2	0.336
300	0.335
350	0.314
400	0.299
500	0.280

REMARKS

The provisional values are for well-annealed high-purity ytterbium and are probably good to ±20% near room temperature and ±30% at extreme temperatures.

*All values are estimated.

FIGURE AND TABLE 77R. PROVISIONAL THERMAL DIFFUSIVITY OF YTTRIUM

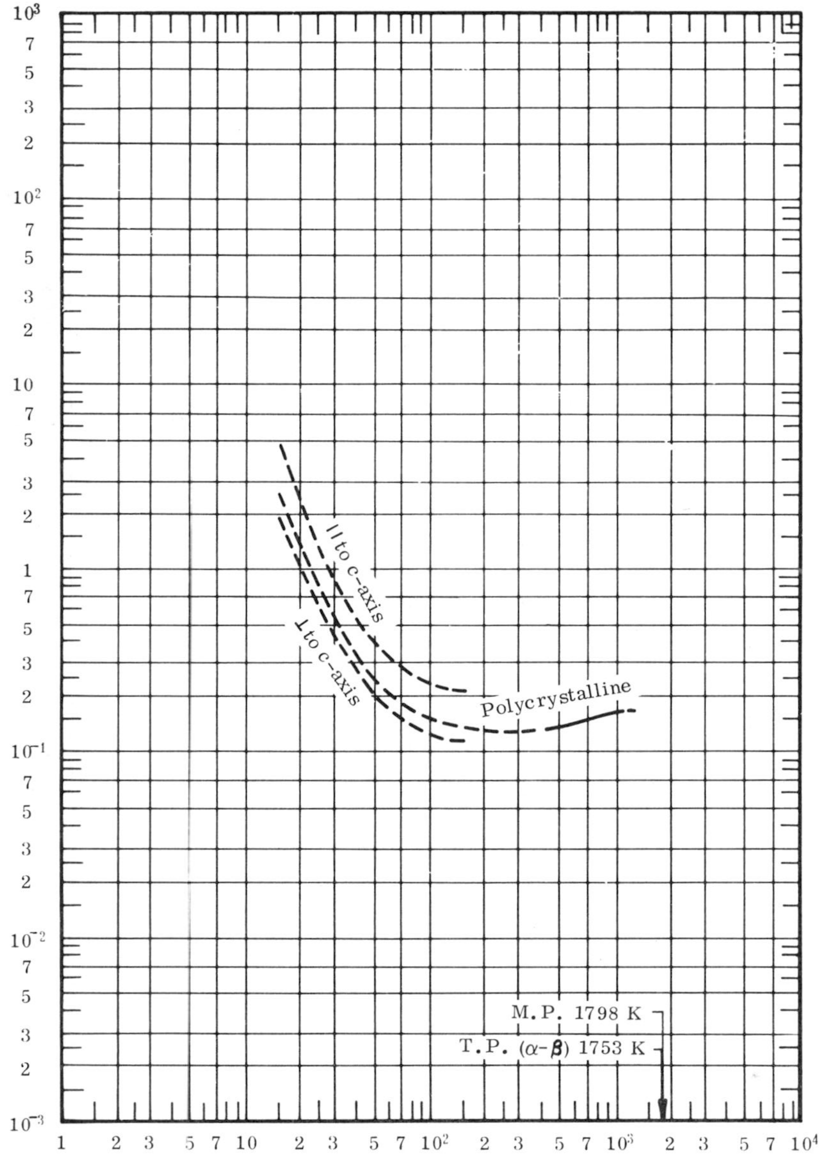

PROVISIONAL VALUES

[Temperature, T, K; Thermal Diffusivity, α, cm^2 s^{-1}]

SOLID

T	∥ to c-axis α	⊥ to c-axis α	Poly-crystalline α	T	∥ to c-axis α	⊥ to c-axis α	Poly-crystalline α
15	4.70*	1.86*	2.55*	150	0.207*	0.112*	0.137*
16	4.05*	1.61*	2.20*	200			0.132*
18	3.06*	1.24*	1.68*	250			0.130*
20	2.37*	0.986*	1.33*	273.2			0.129*
25	1.40*	0.612*	0.814*	300			0.129*
30	0.923*	0.428*	0.556*	350			0.130*
35	0.656*	0.326*	0.419*	400			0.132
40	0.525*	0.267*	0.336*	500			0.138
45	0.438*	0.228*	0.285*	600			0.144
50	0.386*	0.201*	0.249*	700			0.150
60	0.316*	0.168*	0.208*	800			0.155
70	0.279*	0.148*	0.185*	900			0.159
80	0.256*	0.135*	0.169*	1000			0.162
90	0.240*	0.1275*	0.159*	1100			0.163
100	0.230*	0.122*	0.152*	1200			0.162*

REMARKS

The provisional values are for well-annealed high-purity yttrium and are probably good to ±15% near room temperature and ±20 to ±25% at other temperatures. The values below 100 K for $\alpha_{\parallel}$, $\alpha_{\perp}$, and α_{poly} are applicable only to samples having residual electrical resistivity of 2.30, 8.70, and 5.54 μΩ cm, respectively.

*In temperature range where no experimental data are available.

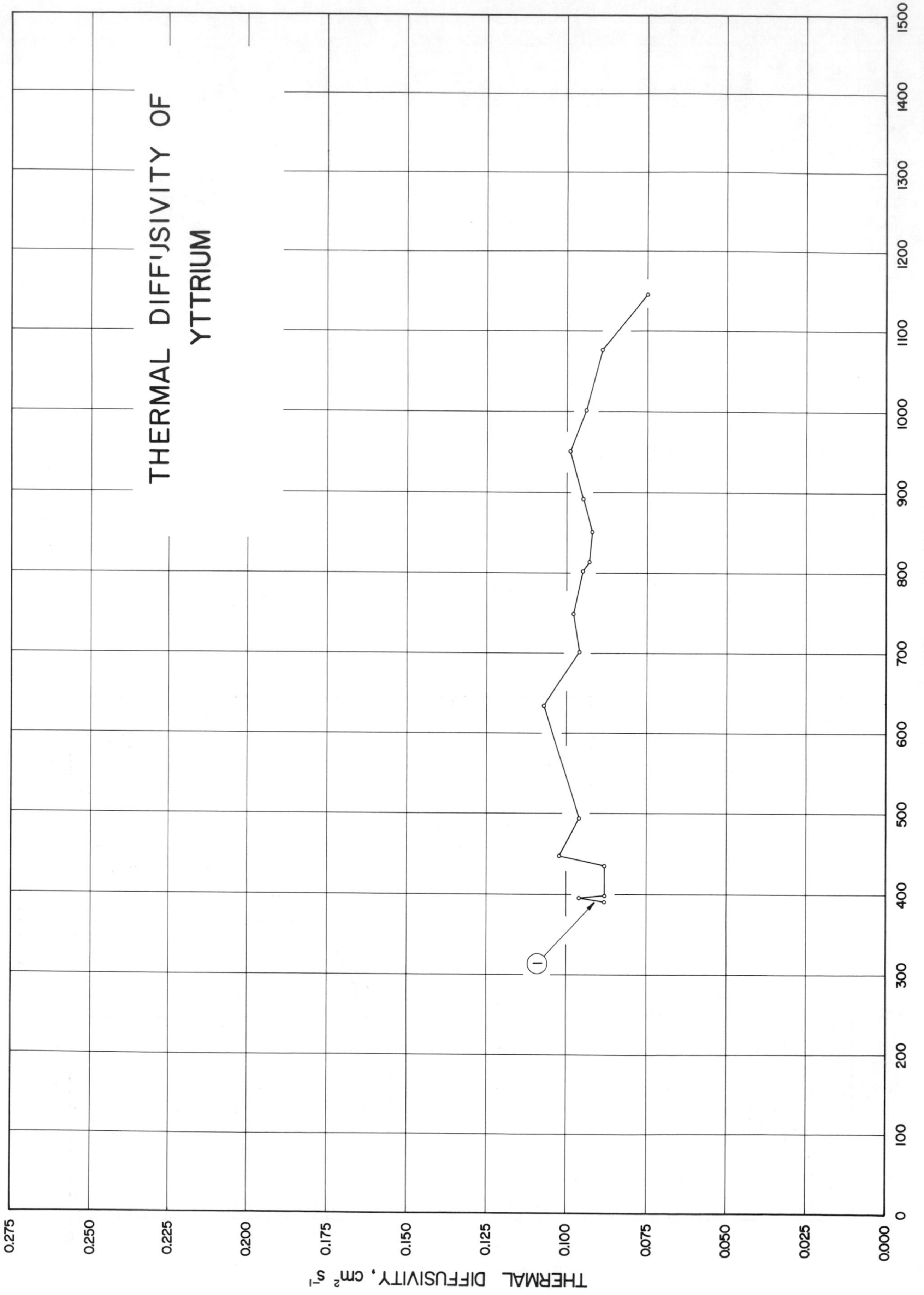
THERMAL DIFFUSIVITY OF YTTRIUM
THERMAL DIFFUSIVITY, cm^2 s^{-1}
0.275
0.250
0.225
0.200
0.175
0.150
0.125
0.100
0.075
0.050
0.025
0.000
0
100
200
300
400
500
600
700
800
900
1000
1100
1200
1300
1400
1500
1
TEMPERATURE, K

FIGURE 77

SPECIFICATION TABLE 77. THERMAL DIFFUSIVITY OF YTTRIUM

(Impurity < 0.20% each; total impurities < 0.50%)

Cur. No.	Ref. No.	Author(s)	Year	Temp. Range, K	Reported Error, %	Name and Specimen Designation	Composition (weight percent), Specifications, and Remarks
1	1	Sonnenschein, G. and Winn, R.A.	1960	390-1146			99.34 Y, < 0.5 O, > 0.1 Ca, 0.05 Mg, and traces of Al, Cu, B, Fe, Mn, Si, and Zr; cylindrical specimen 0.635 cm in dia; front surface of sample covered with fine film of lamp black; measured under a vacuum of ~10^{-4} mm Hg; thermal diffusivity measured under conditions of one-dimensional transient heat flow (flash method).

DATA TABLE 77. THERMAL DIFFUSIVITY OF YTTRIUM

(Impurity < 0.20% each; total impurities < 0.50%)

[Temperature, T, K; Thermal Diffusivity, α, $cm^2 s^{-1}$]

T	α	T	α
CURVE 1		CURVE 1 (cont.)	
390	0.088	1002	0.094
395	0.096	1077	0.089
398	0.088	1146	0.075
435	0.088		
447	0.102		
494	0.096		
634	0.107		
701	0.096		
749	0.098		
802	0.095		
814	0.093		
851	0.092		
892	0.095		
951	0.099		

FIGURE AND TABLE 78R. RECOMMENDED THERMAL DIFFUSIVITY OF ZINC

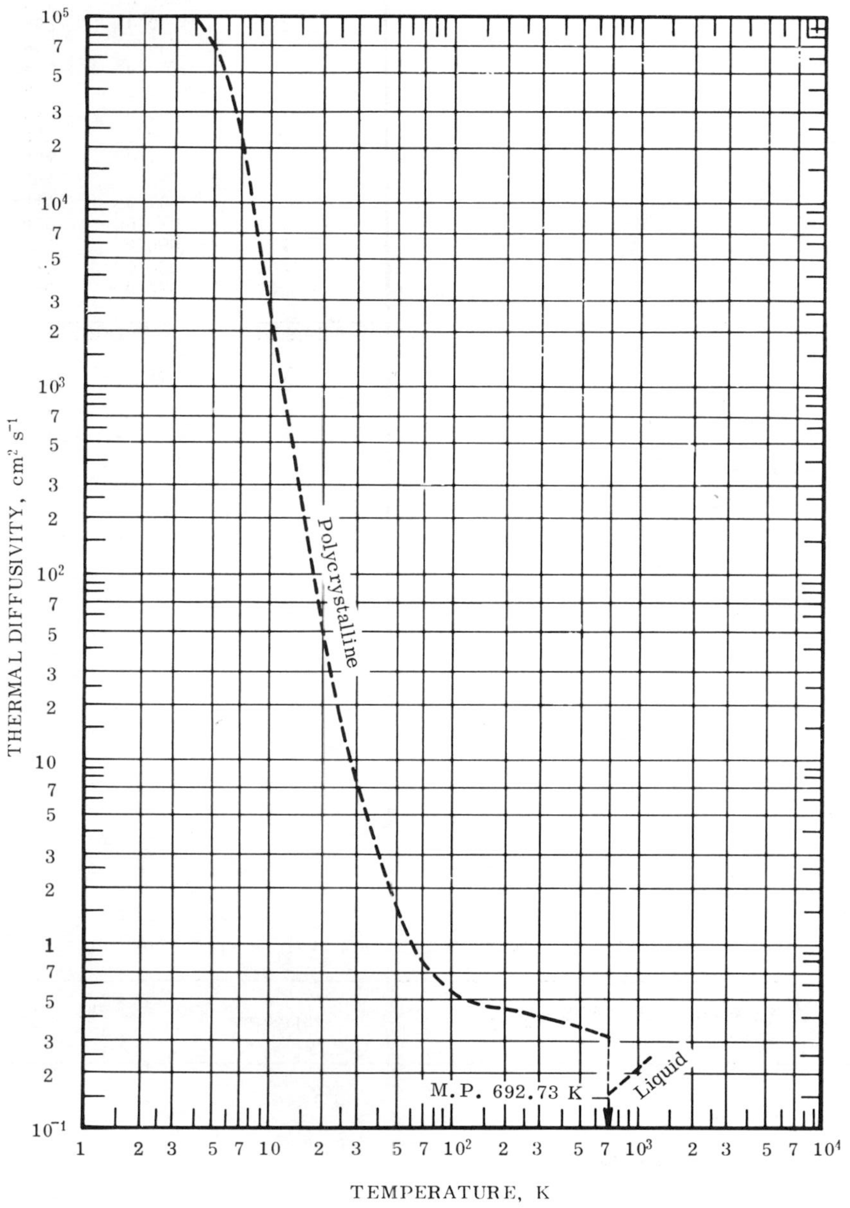

RECOMMENDED VALUES†

[Temperature, T, K; Thermal Diffusivity, α, cm² s⁻¹]

SOLID (Polycrystalline)				LIQUID	
T	α	T	α	T	α
1	234000	35	4.74*	692.73	0.157*
2	180000	40	3.08*	700	0.158*
3	125000*	45	2.17*	800	0.179*
4	85900*	50	1.66*	900	0.200*
5	56700*	60	1.00*	1000	0.222*
6	34200*	70	0.839*	1100	0.245*
7	19100*	80	0.695*		
8	9980*	90	0.604*		
9	5290*	100	0.550*		
10	2800*	150	0.470*		
11	1580*	200	0.448*		
12	920*	250	0.431*		
13	562*	273.2	0.424*		
14	354*	300	0.416		
15	231*	350	0.402		
16	161*	400	0.389		
18	86.0*	500	0.365*		
20	49.3*	600	0.342*		
25	17.4*	692.73	0.322*		
30	8.30*				

REMARKS

The values are for well-annealed high-purity polycrystalline zinc and are thought to be accurate to within ± 5% of the true values at moderate temperatures, ± 8% at high temperatures, ± 12% from 20 to 100 K, and ± 15% below 20 K. At low temperatures the values are highly conditioned by impurity and imperfection, and those below 150 K are applicable only to zinc having residual electrical resistivity of 0.00128 μΩ cm. Values for molten zinc are provisional and they are probably good to ± 15%.

† Values for molten zinc are provisional.
* In temperature range where no experimental data are available.

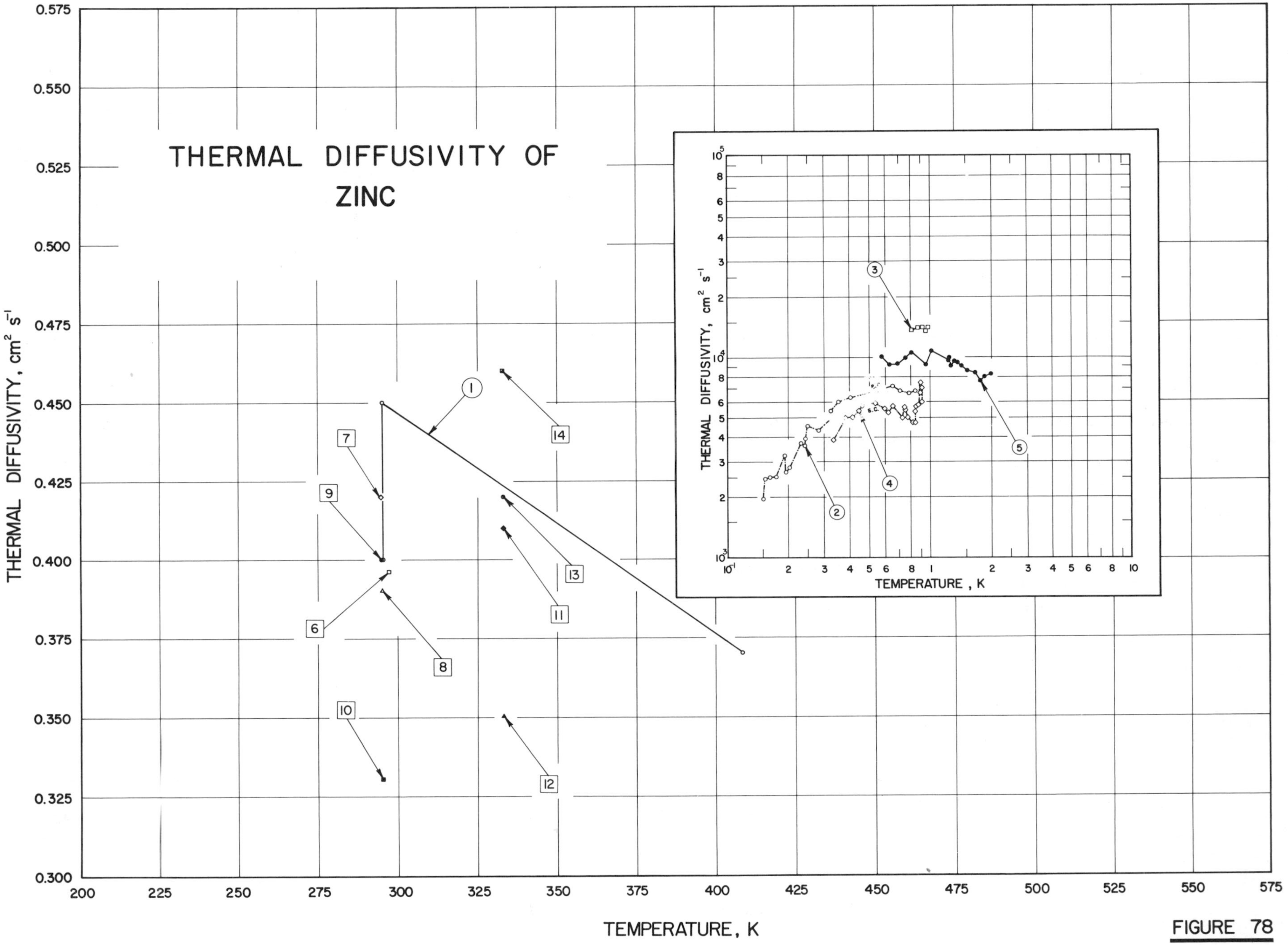
THERMAL DIFFUSIVITY OF
ZINC
THERMAL DIFFUSIVITY, cm² s⁻¹
TEMPERATURE, K
0.575
0.550
0.525
0.500
0.475
0.450
0.425
0.400
0.375
0.350
0.325
0.300
200
225
250
275
300
325
350
375
400
425
450
475
500
525
550
575
1
2
3
4
5
6
7
8
9
10
11
12
13
14
TEMPERATURE , K
s.c.

FIGURE 78

SPECIFICATION TABLE 78. THERMAL DIFFUSIVITY OF ZINC

(Impurity < 0.20% each; total impurities < 0.50%)

Cur. No.	Ref. No.	Author(s)	Year	Temp. Range, K	Reported Error, %	Name and Specimen Designation	Composition (weight percent), Specifications, and Remarks
1	4	Jenkins, R.J. and Parker, W.J.	1961	295-408	± 5		Rectangular specimen 1.9 x 1.26 cm and 0.282 cm thick; high intensity short duration light pulse absorbed in front surface of thermally insulated specimen coated with camphor black; 408.2 K measurement obtained by heating specimen holder and specimen with an infrared lamp; both data points at 295.2 K obtained from measurements using different equations for data reduction.
2	2	Zavaritskii, N.V.	1958	0.15-0.84		Zn-1	0.0001 impurity; single crystal, angle between the (001) direction and the axis of specimen ~30 degrees; cylindrical specimen ~1.5 mm dia. and 100 mm long; experiment with specimen carried out over a period of several days; critical temp. 0.84 K, Debye temp. 321 K; specimen measured in the superconducting state in a magnetic field which was compensated to 0.2 oersted.
3	2	Zavaritskii, N.V.	1958	0.81-0.98		Zn-1	Above specimen measured in the normal state in a field of 60 oersted parallel to the axis of the specimen.
4	2	Zavaritskii, N.V.	1958	0.33-0.91		Zn-2	0.0001 impurity; single crystal, angle between the (001) direction and the axis of specimen ~30 degrees; cylindrical specimen ~1.5 mm dia. and 100 mm long; experiment with specimen carried out over a period of several days; critical temp. 0.84 K, Debye temp. 296 K; specimen measured in the superconducting state in a magnetic field which was compensated to 0.2 oersted.
5	2	Zavaritskii, N.V.	1958	0.58-2.0		Zn-2	Above specimen measured in the normal state in a field of 60 oersted parallel to the axis of the specimen.
6	59	Frazier, R.H.	1933	297.4			99.997 Zn (by difference) and 0.003 Fe; cylindrical specimen; supplied by The Platt Bros. and Co.; made of cold rolled "Bunker Hill Zinc" and ground to precise uniform dia. by Cincinnati Grinders, Inc.; unannealed; density 7.144 g cm^{-3}; measured under a vacuum of from 1.5 to 2.5 microns Hg; six test runs made on specimen.
7	160	Smith, R.H.	1959	295			Specimen size 0.5 x 1 in., thickness 0.1220 in.; diffusivity measured using flash heating technique; diffusivity calculated using $\alpha = 1.37\ L^2/\pi^2\ t_{0.5}$.
8	160	Smith, R.H.	1959	295			The above measurement, but using $\alpha = 0.48\ L^2/\pi^2\ t_X$.
9	160	Smith, R.H.	1959	295			Another measurement on the above specimen; diffusivity calculated using $\alpha = 1.37\ L^2/\pi^2\ t_{0.5}$.
10	160	Smith, R.H.	1959	295			The above measurement, but using $\alpha = 0.48\ L^2/\pi^2\ t_{0.5}$.
11	160	Smith, R.H.	1959	333			Another measurement on the above specimen at higher temperature; diffusivity calculated using $\alpha = 1.37\ L^2/\pi^2\ t_{0.5}$.
12	160	Smith, R.H.	1959	333			The above measurement, but using $\alpha = 0.48\ L^2/\pi^2\ t_X$.
13	160	Smith, R.H.	1959	333			Another measurement on the above specimen; diffusivity calculated using $\alpha = 1.37\ L^2/\pi^2\ t_{0.5}$.
14	160	Smith, R.H.	1959	333			The above measurement, but using $\alpha = 0.48\ L^2/\pi^2\ t_X$.

DATA TABLE 78. THERMAL DIFFUSIVITY OF ZINC

(Impurity < 0.20% each; total impurities < 0.50%)

[Temperature, T, K; Thermal Diffusivity, α, $cm^2\ s^{-1}$]

T	α
CURVE 1	
295	0.40
295	0.45
408	0.37
CURVE 2	
0.151	1950
0.154	2477
0.164	2524
0.175	2524
0.192	3266
0.195	2667
0.203	2818
0.231	3750
0.242	3631
0.242	3981
0.249	4550
0.283	4345
0.325	4966
0.325	5445
0.353	6053
0.407	6368
0.552	6823
0.652	7244
0.711	6918
0.787	6637
0.843	6823
CURVE 3	
0.813	13804
0.867	14125
0.912	14125
0.959	13552
0.977	14191
CURVE 4	
0.334	3891
0.384	5012
0.415	5082
0.441	5495
CURVE 4 (cont.)	
0.461	5754
0.535	5943
0.597	5572
0.622	5346
0.658	5754
0.728	5058
0.748	5702
0.759	5445
0.787	5012
0.820	4786
0.843	4786
0.843	5420
0.851	5754
0.867	5861
0.883	6730
0.895	7551
0.904	7080
0.908	6026
CURVE 5	
0.575	10186
0.634	9376
0.689	9376
0.759	10000
0.805	10617
0.946	9333
1.01	10864
1.22	9683
1.24	10046
1.26	9078
1.32	9728
1.36	9550
1.44	9120
1.53	8710
1.66	8511
1.76	7763
1.86	8017
2.00	8356
CURVE 6	
297.4	0.396
CURVE 7	
295	0.42
CURVE 8	
295	0.39
CURVE 9	
295	0.40
CURVE 10	
295	0.33
CURVE 11	
333	0.41
CURVE 12	
333	0.35
CURVE 13	
333	0.42
CURVE 14	
333	0.46

FIGURE AND TABLE 79R. RECOMMENDED THERMAL DIFFUSIVITY OF ZIRCONIUM

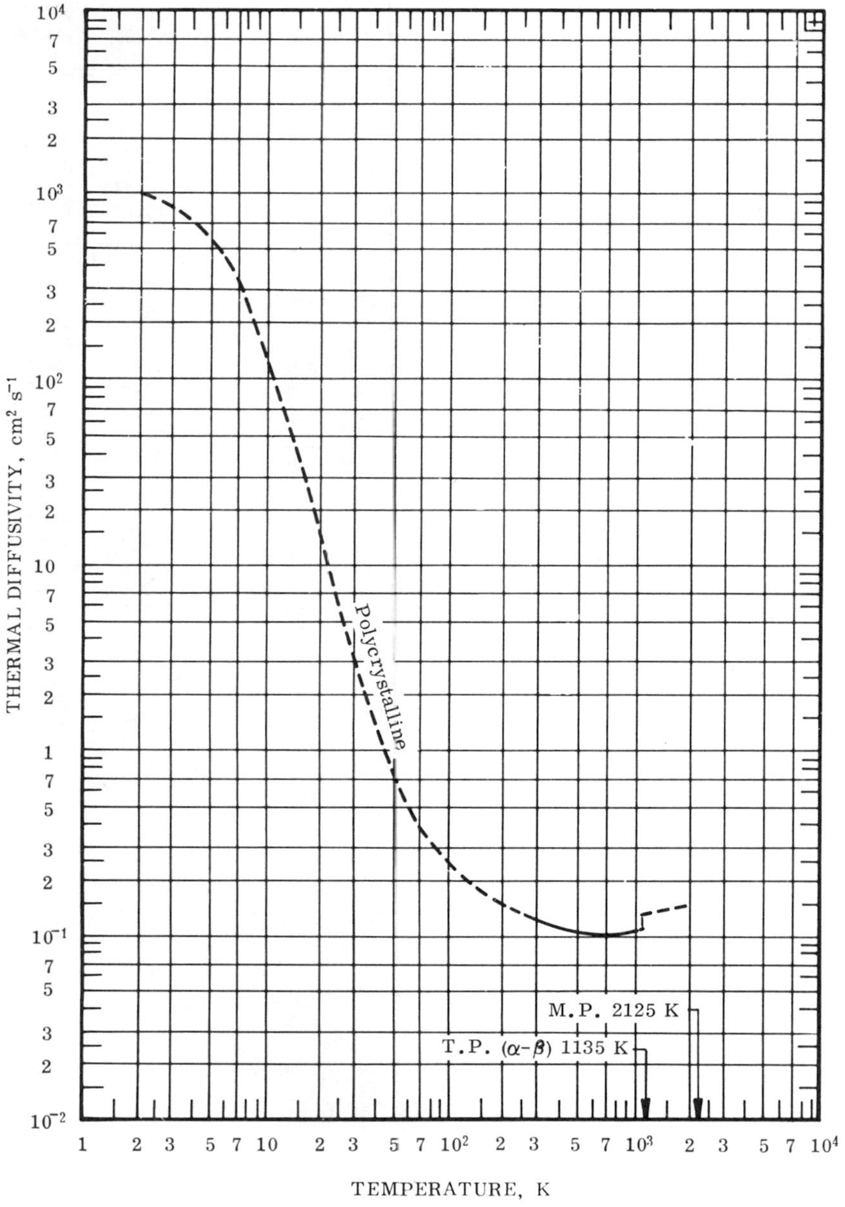

RECOMMENDED VALUES†

[Temperature, T, K; Thermal Diffusivity, α, cm² s⁻¹]

SOLID
(Polycrystalline)

T	α	T	α	T	α
2	936*	40	1.41*	1000	0.104
3	828*	45	1.01*	1100	0.107
4	670*	50	0.767*	1135	0.107
5	519*	60	0.520*	1135	0.128
6	395*	70	0.398*	1200	0.130*
7	296*	80	0.328*	1300	0.1325*
8	220*	90	0.282*	1400	0.135*
9	166*	100	0.249*	1500	0.136*
10	124*	150	0.176*	1600	0.138*
11	95.0*	200	0.149*	1700	0.139*
12	73.8*	250	0.136*	1800	0.140*
13	58.0*	273.2	0.132	1900	0.142*
14	46.0*	300	0.127	2000	0.143*
15	36.9*	350	0.120		
16	30.0*	400	0.115		
18	20.1*	500	0.107		
20	13.9*	600	0.101		
25	6.36*	700	0.0992		
30	3.43*	800	0.0997		
35	2.09*	900	0.102		

REMARKS

The values are for well-annealed high-purity polycrystalline zirconium and are thought to be accurate to within ±12% of the true values below room temperature, ±10% from room temperature to 800 K, and the uncertainty increasing to ±20 to ±25% as the melting point is approached. The values above 1000 K are provisional. At low temperatures the values are highly conditioned by impurity and imperfection, and those below room temperature are applicable only to zirconium having residual electrical resistivity of 0.218 μΩ cm.

† Values above 1000 K are provisional.
* In temperature range where no experimental data are available.

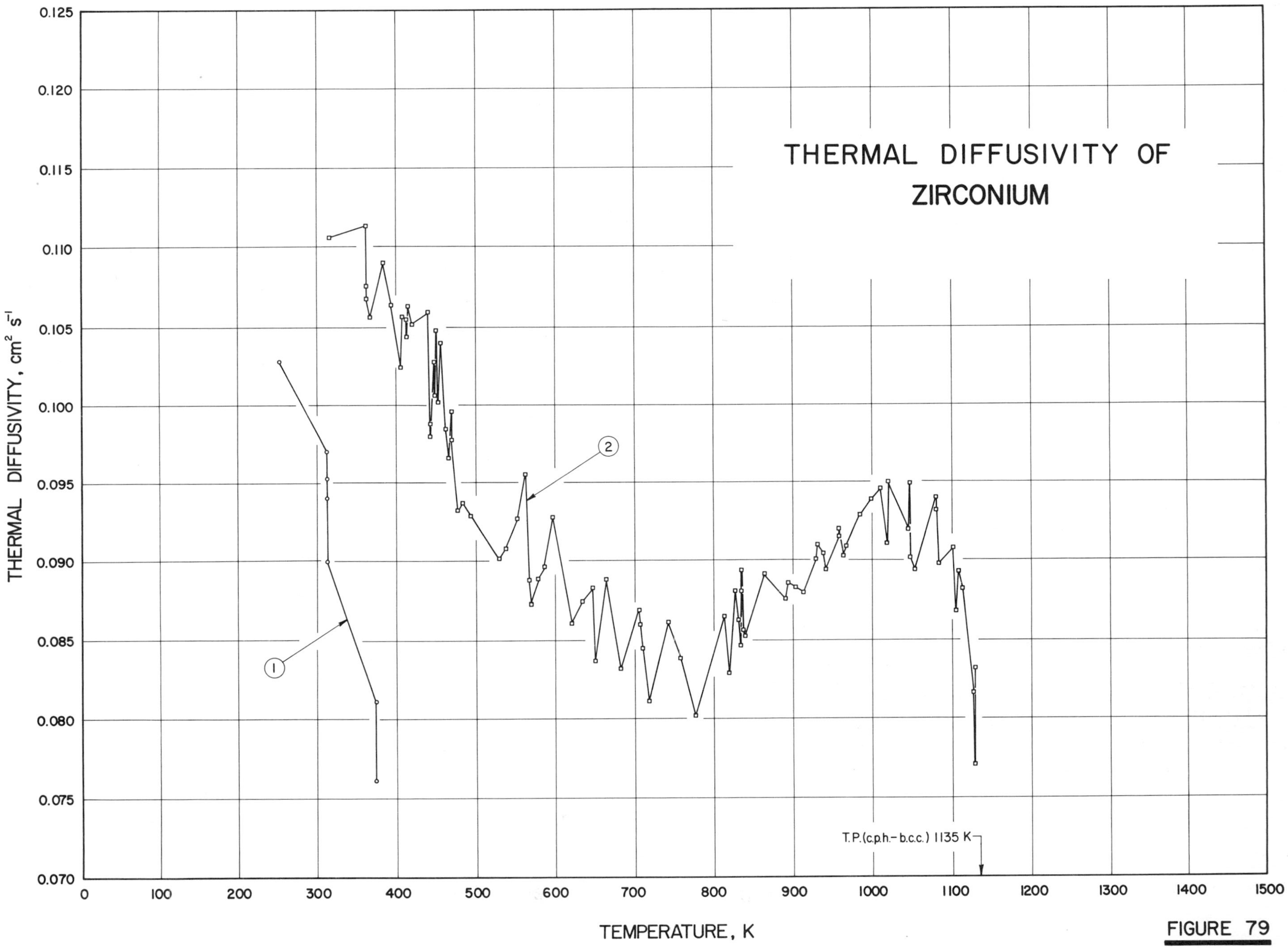
THERMAL DIFFUSIVITY OF
ZIRCONIUM
THERMAL DIFFUSIVITY, cm² s⁻¹
TEMPERATURE, K
0.125
0.120
0.115
0.110
0.105
0.100
0.095
0.090
0.085
0.080
0.075
0.070
0
100
200
300
400
500
600
700
800
900
1000
1100
1200
1300
1400
1500
1
2
T.P.(c.p.h.- b.c.c.) 1135 K

FIGURE 79

SPECIFICATION TABLE 79. THERMAL DIFFUSIVITY OF ZIRCONIUM

(Impurity < 0.20% each; total impurities < 0.50%)

Cur. No.	Ref. No.	Author(s)	Year	Temp. Range, K	Reported Error, %	Name and Specimen Designation	Composition (weight percent), Specifications, and Remarks
1	29, 30	McIntosh, G.E.	1952	253-373	5		Pure isotropic zirconium; rod specimen 0.290 in. in dia and 6.25 in. long; surrounded by radiation shield consisting of four concentric cylinders of very thin aluminum foil each separated by three 1/32 in. rings of balsa wood; measured after being maintained at elevated temp for several hrs; measured in vacuum; diffusivity determined from measured phase lag of the temp wave between any two points along specimen; one-dimensional heat flow.
2	193	Pollard, E.R., Jr.	1963	316-1129			Cylindrical specimen 0.125 in. in diameter and about 2 in. long; diffusivity measured using modified Angström method.

DATA TABLE 79. THERMAL DIFFUSIVITY OF ZIRCONIUM

(Impurity < 0.20% each; total impurities < 0.50%)

[Temperature, T, K; Thermal Diffusivity, α, $cm^2 s^{-1}$]

T	α	T	α	T	α	T	α	T	α	T	α	T	α
CURVE 1		CURVE 2 (cont.)		CURVE 2 (cont.)		CURVE 2 (cont.)		CURVE 2 (cont.)		CURVE 2 (cont.)		CURVE 2 (cont.)	
253	0.1028	394	0.1064	462	0.0984	597	0.0921	813	0.0865	938	0.0905	1084	0.0898
313	0.0970	393	0.1039	466	0.0966	597	0.0900	819	0.0829	941	0.0894	1101	0.0908
313	0.09530	406	0.1024	469	0.0996	621	0.0860	827	0.0881	958	0.0915	1105	0.0868
313	0.09404	408	0.1057	469	0.0977	633	0.0875	832	0.0863	958	0.0920	1109	0.0893
313	0.09009	413	0.1055	477	0.0932	647	0.0883	834	0.0847	964	0.0903	1113	0.0882
373	0.08116	413	0.1044	484	0.0937	650	0.0837	835	0.0881	967	0.0909	1127	0.0816
373	0.0761	415	0.1063	493	0.0929	664	0.0888	835	0.0894	984	0.0929	1127	0.0770
		420	0.1052	529	0.0901	682	0.0831	837	0.0856	998	0.0939	1129	0.0832
CURVE 2		440	0.1059	537	0.0908	706	0.0869	840	0.0852	1011	0.0946		
		443	0.0988	552	0.0927	707	0.0860	864	0.0892	1018	0.0911		
316	0.1106	443	0.0979	562	0.0955	709	0.0845	891	0.0876	1020	0.0950		
363	0.1130	448	0.1028	567	0.0888	718	0.0811	894	0.0886	1045	0.0919		
363	0.1076	448	0.1006	569	0.0873	742	0.0861	904	0.0883	1047	0.0949		
363	0.1068	450	0.1048	578	0.0889	754	0.0871	913	0.0880	1047	0.0902		
367	0.1056	453	0.1002	586	0.0897	758	0.0838	929	0.0901	1053	0.0894		
384	0.1090	456	0.1040	597	0.0928	777	0.0802	931	0.0910	1081	0.0940		
										1081	0.0932		

2. NONFERROUS BINARY ALLOYS

SPECIFICATION TABLE 80. THERMAL DIFFUSIVITY OF [ALUMINUM + MANGANESE] ALLOYS

(Al + Mn ≥ 99.50%; impurity ≤ 0.20% each)

Cur. No.	Ref. No.	Author(s)	Year	Temp. Range, K	Reported Error, %	Name and Specimen Designation	Composition (weight percent) Al	Mn	Composition (continued), Specifications, and Remarks
1*	271	Rosenthal, D. and Friedmann, N. E.	1954	603		Aluminum Alloy 3S	Bal.	1.2	Diffusivity measured using modified Angström method.

DATA TABLE 80. THERMAL DIFFUSIVITY OF [ALUMINUM + MANGANESE] ALLOYS

(Al + Mn ≥ 99.50%; impurity ≤ 0.20% each)

[Temperature, T, K; Thermal Diffusivity, α, $cm^2\ s^{-1}$]

T	α
CURVE 1*	
603	0.643

* No figure given.

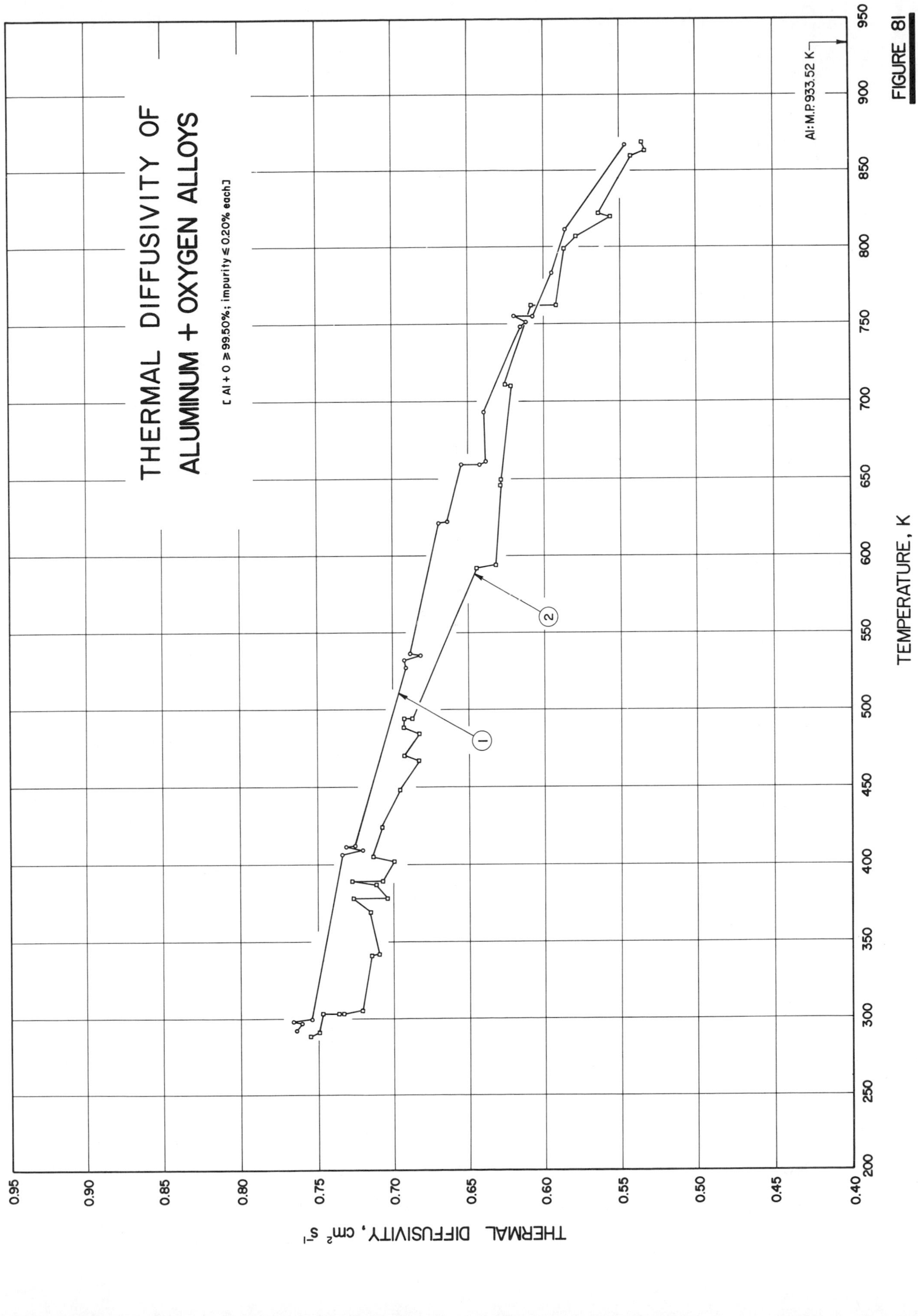
THERMAL DIFFUSIVITY OF
ALUMINUM + OXYGEN ALLOYS
[Al + O ≥ 99.50%; impurity ≤ 0.20% each]
Al: M.P. 933.52 K
1
2
TEMPERATURE, K
THERMAL DIFFUSIVITY, cm² s⁻¹
200
250
300
350
400
450
500
550
600
650
700
750
800
850
900
950
0.40
0.45
0.50
0.55
0.60
0.65
0.70
0.75
0.80
0.85
0.90
0.95

FIGURE 81

SPECIFICATION TABLE 81. THERMAL DIFFUSIVITY OF [ALUMINUM + OXYGEN] ALLOYS

(Al + O ≥ 99.50%; impurity ≤ 0.20%)

Cur. No.	Ref. No.	Author(s)	Year	Temp. Range, K	Reported Error, %	Name and Specimen Designation	Composition (weight percent) Al	O	Composition (continued), Specifications, and Remarks
1	154	Smith, C.A.	1966	292-867		XAP-001; Sample 1	→	→	Sintered aluminum powder containing oxide phase.
2	154	Smith, C.A.	1966	288-869		XAP-001; Sample 2			Specimen similar to the above.

DATA TABLE 81. THERMAL DIFFUSIVITY OF [ALUMINUM + OXYGEN] ALLOYS

(Al + O ≥ 99.50%; impurity ≤ 0.20%)

[Temperature, T, K; Thermal Diffusivity, α, $cm^2 s^{-1}$]

T	α	T	α	T	α	T	α	T	α
CURVE 1		CURVE 1 (cont.)		CURVE 2 (cont.)		CURVE 2 (cont.)		CURVE 2 (cont.)	
292	0.764	693	0.639	305	0.721	488	0.692	863	0.533
297	0.760	748	0.615	341	0.714	494	0.692	869	0.535
298	0.766	651	0.611	342	0.709	494	0.687		
299	0.754	755	0.619	369	0.715	592	0.644		
406	0.734	755	0.607	378	0.726	594	0.631		
409	0.720	783	0.594	378	0.703	646	0.628		
411	0.731	811	0.585	387	0.711	649	0.628		
412	0.725	867	0.546	389	0.727	710	0.621		
527	0.691			389	0.707	711	0.625		
532	0.692	CURVE 2		402	0.699	762	0.591		
535	0.681			405	0.713	762	0.608		
536	0.688	288	0.755	424	0.707	799	0.586		
621	0.669	291	0.749	448	0.695	807	0.578		
622	0.663	303	0.747	467	0.682	820	0.555		
659	0.654	303	0.736	470	0.692	822	0.563		
661	0.638	303	0.733	484	0.682	860	0.542		

SPECIFICATION TABLE 82. THERMAL DIFFUSIVITY OF [ANTIMONY + COPPER] ALLOYS

(Sb + Cu ≥ 99.50%; impurity ≤ 0.20% each)

Cur. No.	Ref. No.	Author(s)	Year	Temp. Range, K	Reported Error, %	Name and Specimen Designation	Composition (weight percent) Sb	Cu	Composition (continued), Specifications, and Remarks
1*	95	Dutchak, Ya.I. and Panasyuk, P.V.	1966	1073			90	10	In liquid state; electrical conductivity 8.59, 8.29, and 8.00 x 10^2 Ω^{-1} cm^{-1} at 620, 700, and 800 C, respectively.
2*	95	Dutchak, Ya.I. and Panasyuk, P.V.	1966	1073			80	20	In liquid state; electrical conductivity 8.71, 8.45, and 8.12 x 10^2 Ω^{-1} cm^{-1} at 620, 700, and 800 C, respectively.
3*	95	Dutchak, Ya.I. and Panasyuk, P.V.	1966	1073			76	24	In liquid state; electrical conductivity 8.84, 8.45, and 8.12 x 10^2 Ω^{-1} cm^{-1} at 620, 700, and 800 C, respectively.
4*	95	Dutchak, Ya.I. and Panasyuk, P.V.	1966	1073			60	40	In liquid state; electrical conductivity 7.79, 7.60, and 7.40 x 10^2 Ω^{-1} cm^{-1} at 620, 700, and 800 C, respectively.
5*	95	Dutchak, Ya.I. and Panasyuk, P.V.	1966	1073			50	50	In liquid state; electrical conductivity 7.25 and 7.10 x 10^2 Ω^{-1} cm^{-1} at 700, and 800 C, respectively.

DATA TABLE 82. THERMAL DIFFUSIVITY OF [ANTIMONY + COPPER] ALLOYS

(Sb + Cu ≥ 99.50%; impurity ≤ 0.20% each)

[Temperature, T, K; Thermal Diffusivity, α, cm^2 s^{-1}]

T	α	T	α
CURVE 1*		CURVE 5*	
1073	0.155	1073	0.0913
CURVE 2*			
1073	0.130		
CURVE 3*			
1073	0.122		
CURVE 4*			
1073	0.108		

* No figure given.

SPECIFICATION TABLE 83. THERMAL DIFFUSIVITY OF [BERYLLIUM + ALUMINUM] ALLOYS

(Be + Al ≥ 99.50%; impurity ≤ 0.20% each)

Cur. No.	Ref. No.	Author(s)	Year	Temp. Range, K	Reported Error, %	Name and Specimen Designation	Composition (weight percent) Be	Al	Composition (continued), Specifications, and Remarks
1*	272	Fenn, R.W., Jr., Glass, R.A., Needham, R.A., and Steinberg, M.A.	1965	297			67	33	No details given.
2*	272	Fenn, R.W., Jr., et al.	1965	297			64	36	No details given.
3*	272	Fenn, R.W., Jr., et al.	1965	297			57	43	No details given.

DATA TABLE 83. THERMAL DIFFUSIVITY OF [BERYLLIUM + ALUMINUM] ALLOYS

(Be + Al ≥ 99.50%; impurity ≤ 0.20% each)

[Temperature, T, K; Thermal Diffusivity, α, $cm^2\ s^{-1}$]

T	α
CURVE 1*	
297	0.56
CURVE 2*	
297	0.63
CURVE 3*	
297	0.58

* No figure given.

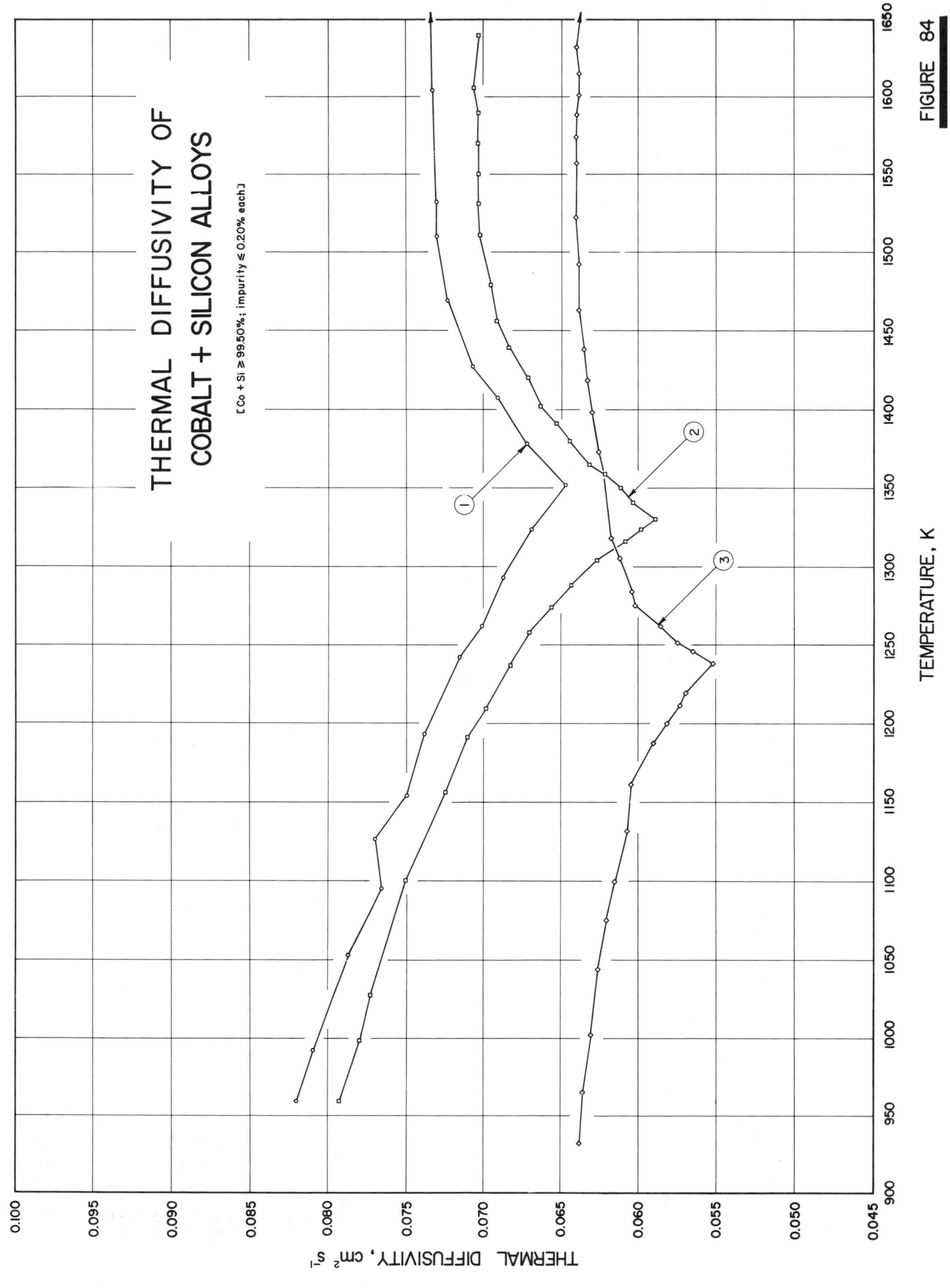
THERMAL DIFFUSIVITY OF
COBALT + SILICON ALLOYS
[Co + Si ≥ 99.50%; impurity ≤ 0.20% each]
1
2
3
TEMPERATURE, K
THERMAL DIFFUSIVITY, $cm^2 s^{-1}$
900
950
1000
1050
1100
1150
1200
1250
1300
1350
1400
1450
1500
1550
1600
1650
0.045
0.050
0.055
0.060
0.065
0.070
0.075
0.080
0.085
0.090
0.095
0.100

FIGURE 84

SPECIFICATION TABLE 84. THERMAL DIFFUSIVITY OF [COBALT + SILICON] ALLOYS

(Co + Si ≥ 99.50%; impurity ≤ 0.20% each)

Cur. No.	Ref. No.	Author(s)	Year	Temp. Range, K	Reported Error, %	Name and Specimen Designation	Composition (weight percent) Co	Si	Composition (continued), Specifications, and Remarks
1	201	Krentsis, R.P., Zinov'yev, V.Ye., Andreyeva, L.P., and Gel'd, P.V.	1970	959-1657	< 5		Bal.	0.5	Starting materials were 99.99 pure cobalt and monocrystalline silicon; solid solution; diffusivity measured by using temperature plane wave method at frequency of 168.8 G, and pressure of the order of 1 x 10^{-5} mm Hg.
2	201	Krentsis, R.P., et al.	1970	959-1641	< 5		Bal.	1.0	Other conditions same as above.
3	201	Krentsis, R.P., et al.	1970	932-1654	< 5		Bal.	3.0	Other conditions same as above.

DATA TABLE 84. THERMAL DIFFUSIVITY OF [COBALT + SILICON] ALLOYS

(Co + Si ≥ 99.50%; impurity ≤ 0.20% each)

[Temperature, T, K; Thermal Diffusivity, α, $cm^2 s^{-1}$]

T	α	T	α	T	α	T	α	T	α	T	α	T	α
CURVE 1		CURVE 1 (cont.)		CURVE 2 (cont.)		CURVE 2 (cont.)		CURVE 3		CURVE 3 (cont.)		CURVE 3 (cont.)	
959	0.0821	1469	0.0723	1237	0.0682	1420	0.0671	932	0.0638	1262	0.0586	1588	0.0640
992	0.0810	1511	0.0730	1258	0.0670	1439	0.0683	965	0.0635	1275	0.0602	1601	0.0638
1053	0.0787	1532	0.0730	1274	0.0656	1456	0.0691	1002	0.0630	1284	0.0604	1615	0.0638
1095	0.0766	1604	0.0733	1288	0.0643	1479	0.0695	1044	0.0625	1305	0.0612	1632	0.0640
1127	0.0770	1657	0.0734*	1304	0.0626	1511	0.0702	1075	0.0620	1318	0.0617	1654	0.0637*
1154	0.0749			1316	0.0608	1531	0.0703	1100	0.0615	1359	0.0621*		
1193	0.0738	CURVE 2		1324	0.0598	1550	0.0703	1132	0.0607	1373	0.0625		
1242	0.0715			1330	0.0588	1570	0.0703	1161	0.0604	1398	0.0629		
1262	0.0701	959	0.0793	1340	0.0603	1590	0.0703	1187	0.0590	1418	0.0632		
1293	0.0687	998	0.0780	1350	0.0611	1606	0.0706	1200	0.0581	1438	0.0634		
1324	0.0669	1027	0.0773	1359	0.0621	1641	0.0703	1211	0.0573	1463	0.0638		
1352	0.0647	1100	0.0750	1365	0.0631			1219	0.0569	1492	0.0638		
1378	0.0672	1156	0.0724	1380	0.0644			1238	0.0552	1522	0.0640		
1407	0.0691	1191	0.0710	1391	0.0653			1246	0.0565	1557	0.0640		
1427	0.0707	1209	0.0698	1402	0.0663			1251	0.0574	1574	0.0640		

* Not shown in figure.

SPECIFICATION TABLE 85. THERMAL DIFFUSIVITY OF [COPPER + ANTIMONY] ALLOYS

(Impurity < 0.20% each; total impurities < 0.50%)

Cur. No.	Ref. No.	Author(s)	Year	Temp. Range, K	Reported Error, %	Name and Specimen Designation	Composition (weight percent) Cu	Sb	Composition (continued), Specifications, and Remarks
1*	95	Dutchak, Ya.I. and Panasyuk, P.V.	1966	1073			50	50	In liquid state; electrical conductivity 7.25 and 7.10 x $10^2\ \Omega^{-1}\ cm^{-1}$ at 700 and 800 C, respectively.

DATA TABLE 85. THERMAL DIFFUSIVITY OF [COPPER + ANTIMONY] ALLOYS

(Impurity < 0.20% each; total impurities < 0.50%)

[Temperature, T, K; Thermal Diffusivity, α, $cm^2\ s^{-1}$]

T	α
CURVE 1*	
1073	0.0913

* No figure given.

SPECIFICATION TABLE 86. THERMAL DIFFUSIVITY OF [COPPER + ARSENIC] ALLOYS

(Cu + As ≥ 99.50%; impurity ≤ 0.20% each)

Cur. No.	Ref. No.	Author(s)	Year	Temp. Range, K	Reported Error, %	Name and Specimen Designation	Composition (weight percent) Cu	As	Composition (continued), Specifications, and Remarks
1*	18	Habachi, M., Azou, P., and Bastien, P.	1965	307-773		25	99	1	(Cu obtained by difference); cylindrical specimen; annealed at 873.2 K for 24 hrs; Lorenz number reported as 6.151, 6.117, 6.078, 6.043, 6.022, 6.007, 5.988, 5.975, 5.964, 5.956, and 5.949 x 10^{-8} cal Ω s^{-1} K^{-1} at 290.2, 320.2, 371.2, 422.2, 473.2, 523.2, 573.2, 625.2, 673.2, 723.2, and 773.2 K, respectively; method based on measuring phase shift and logarithmic attenuation between two points on specimen separated by a distance of 1 cm; max amplitude of temperature wave limited to 1 K; pulsation of wave lying in the range from 3 to 30 radians per min; specimen heated on one end and cooled on the other end; measured under a vacuum of 10^{-6} mm Hg.

DATA TABLE 86. THERMAL DIFFUSIVITY OF [COPPER + ARSENIC] ALLOYS

(Cu + As ≥ 99.50%; impurity ≤ 0.20% each)

[Temperature, T, K; Thermal Diffusivity, α, cm^2 s^{-1}]

T	α
CURVE 1*	
307	0.315
328	0.332
373	0.341
426	0.345
474	0.354
528	0.361
574	0.365
626	0.375
673	0.389
725	0.403
773	0.419

* No figure given.

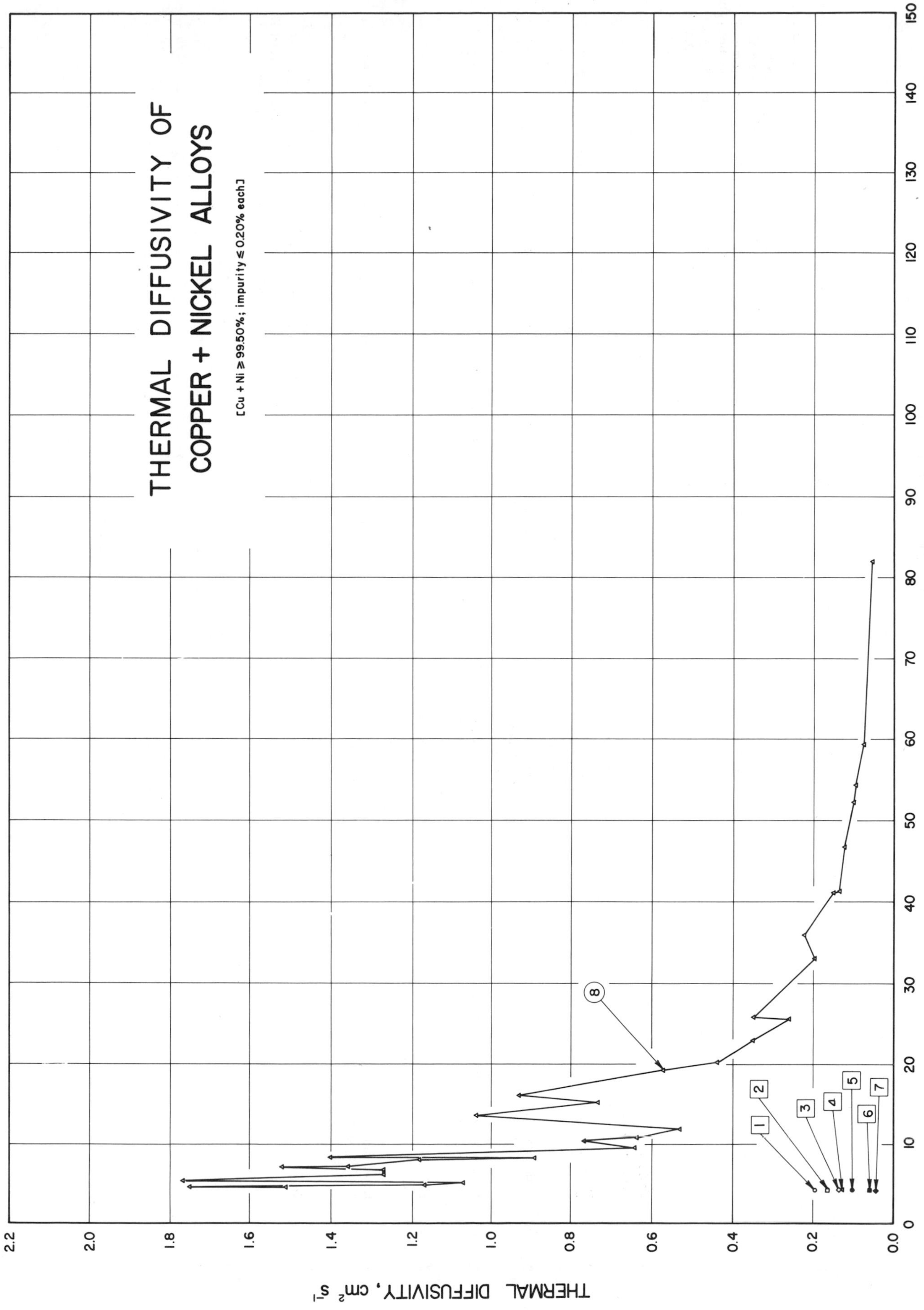
THERMAL DIFFUSIVITY OF
COPPER + NICKEL ALLOYS
[Cu + Ni ≥ 99.50%; impurity ≤ 0.20% each]
THERMAL DIFFUSIVITY, cm² s⁻¹
TEMPERATURE, K
0.0
0.2
0.4
0.6
0.8
1.0
1.2
1.4
1.6
1.8
2.0
2.2
0
10
20
30
40
50
60
70
80
90
100
110
120
130
140
150
1
2
3
4
5
6
7
8

FIGURE 87

SPECIFICATION TABLE 87. THERMAL DIFFUSIVITY OF [COPPER + NICKEL] ALLOYS

(Cu + Ni ≥ 99. 50%; impurity ≤ 0. 20% each)

Cur. No.	Ref. No.	Author(s)	Year	Temp. Range, K	Reported Error, %	Name and Specimen Designation	Composition (weight percent) Cu	Ni	Composition (continued), Specifications, and Remarks
1	90	Erdmann, J. C. and Jahoda, J. A.	1962	4. 2		Commercial Constantan Alloy	60	40	Wire specimen of constant diameter lying in the range from 1 to 3 mm and ~100 mm long; previously annealed and then used to measure thermal conductivity under conditions of varying strain and at a constant temperature of 4. 2 K; heat flow generated by an electrical current passing through specimen and then current is switched off; measured in a vacuum of 10^{-5} mm Hg; diffusivity determined from plot of measured transient temperature difference between the center and one of the ends of specimen versus time; measured under no strain.
2	90	Erdmann, J. C. and Jahoda, J. A.	1962	4. 2		Commercial Constantan Alloy			Above specimen measured for diffusivity again under a strain of 0. 008.
3	90	Erdmann, J. C. and Jahoda, J. A.	1962	4. 2		Commercial Constantan Alloy			Above specimen measured for diffusivity again under a strain of 0. 0179.
4	90	Erdmann, J. C. and Jahoda, J. A.	1962	4. 2		Commercial Constantan Alloy			Above specimen measured for diffusivity again under a strain of 0. 0232.
5	90	Erdmann, J. C. and Jahoda, J. A.	1962	4. 2		Commercial Constantan Alloy			Above specimen measured for diffusivity again under a strain of 0. 0536.
6	90	Erdmann, J. C. and Jahoda, J. A.	1962	4. 2		Commercial Constantan Alloy			Above specimen measured for diffusivity again under a strain of 0. 125.
7	90	Erdmann, J. C. and Jahoda, J. A.	1962	4. 2		Commercial Constantan Alloy			Above specimen measured for diffusivity again under a strain of 0. 192.
8	132	Erdmann, J. C. and Jahoda, J. A.	1966	4. 5-82		Cu 50-Ni 50 Alloy			Cylindrical specimen ~1. 5 mm in dia. and ~10 cm long; both ends anchored to thermal sinks; heat pulse generated by passage of electrical current; thermal diffusivity determined from measured temp. difference between two points along specimen as a function of time; measured in vacuum.

DATA TABLE 87. THERMAL DIFFUSIVITY OF [COPPER + NICKEL] ALLOYS

(Cu + Ni ≥ 99.50%; impurity ≤ 0.20% each)

[Temperature, T, K; Thermal Diffusivity, α, $cm^2\ s^{-1}$]

T	α
CURVE 1	
4.2	0.195
CURVE 2	
4.2	0.164
CURVE 3	
4.2	0.136
CURVE 4	
4.2	0.128
CURVE 5	
4.2	0.102
CURVE 6	
4.2	0.0579
CURVE 7	
4.2	0.0439
CURVE 8	
4.47	1.51
4.52	1.75
4.79	1.17
5.15	1.07
5.21	1.79
6.01	1.27
6.65	1.27
7.05	1.52
7.15	1.36
8.00	1.18
8.22	0.891
8.38	1.40
9.59	0.640
CURVE 8 (cont.)	
10.4	0.766
10.8	0.635
11.8	0.532
13.6	1.04
15.2	0.731
16.2	0.931
19.3	0.570
20.2	0.437
22.9	0.349
25.5	0.258
25.8	0.347
33.0	0.196
35.8	0.223
41.1	0.152
41.3	0.133
46.7	0.121
52.2	0.100
54.3	0.0935
59.3	0.0757
82.0	0.0552

SPECIFICATION TABLE 88. THERMAL DIFFUSIVITY OF [COPPER + SILVER] ALLOYS

(Cu + Ag ≥ 99.50%; impurity ≤0.20% each)

Cur. No.	Ref. No.	Author(s)	Year	Temp. Range, K	Reported Error, %	Name and Specimen Designation	Composition (weight percent) Cu	Ag	Composition (continued), Specifications, and Remarks
1*	18	Habachi, M., Azou, P., and Bastien, P.	1965	314-778		24	99.53	0.47	(Cu obtained by difference); cylindrical specimen; solvent and solute similar in atomic radius and lattice; annealed at 873.2 K for 150 hrs; Lorenz number reported as 1.398, 1.380, 1.359, 1.348, 1.348, 1.348, 1.356, 1.325, 1.325, 1.332, 1.351 x 10^{-8} cal sec^{-1} ohm K^{-2} at 294.2, 326.2, 374.2, 424.2, 473.2, 526.2, 573.2, 626.2, 673.2, 724.2, and 773.2 K, respectively; method based on measuring phase shift and logarithmic attenuation between two points on specimen separated by a distance of 1 cm; max amplitude of temperature wave limited to 1 K; pulsation of wave lying in the range from 3 to 30 radians per min; specimen heated on one end and cooled on the other end; measured under a vacuum of 10^{-6} mm Hg.

DATA TABLE 88. THERMAL DIFFUSIVITY OF [COPPER + SILVER] ALLOYS

(Cu + Ag ≥ 99.50%; impurity ≤0.20% each)

[Temperature, T, K; Thermal Diffusivity, α, cm^2 s^{-1}]

T	α
CURVE 1*	
314	0.922
373	0.917
453	0.917
542	0.917
590	0.958
628	1.084
699	0.978
778	0.924

* No figure given.

SPECIFICATION TABLE 89. THERMAL DIFFUSIVITY OF [COPPER + TIN] ALLOYS

(Cu + Sn ≥ 99.50%; impurity ≤ 0.20% each)

Cur. No.	Ref. No.	Author(s)	Year	Temp. Range, K	Reported Error, %	Name and Specimen Designation	Composition (weight percent), Specifications, and Remarks
1*	232	Böhm, R. and Wachtel, E.	1969	273, 373		GZ-SnB$_3$ 12	87.56 Cu, 12.25 Sn, 0.07 Pb, 0.06 Zn, 0.03 Ni, 0.02 Fe, and 0.01 P; cylindrical specimen; electrical resistivity 15.94 and 16.90 μΩ cm at 0 and 100 C, respectively.

DATA TABLE 89. THERMAL DIFFUSIVITY OF [COPPER + TIN] ALLOYS

(Cu + Sn ≥ 99.50%; impurity ≤ 0.20% each)

[Temperature, T, K; Thermal Diffusivity, α, cm^2 s^{-1}]

T	α
CURVE 1*	
273	0.209
373	0.234

* No figure given.

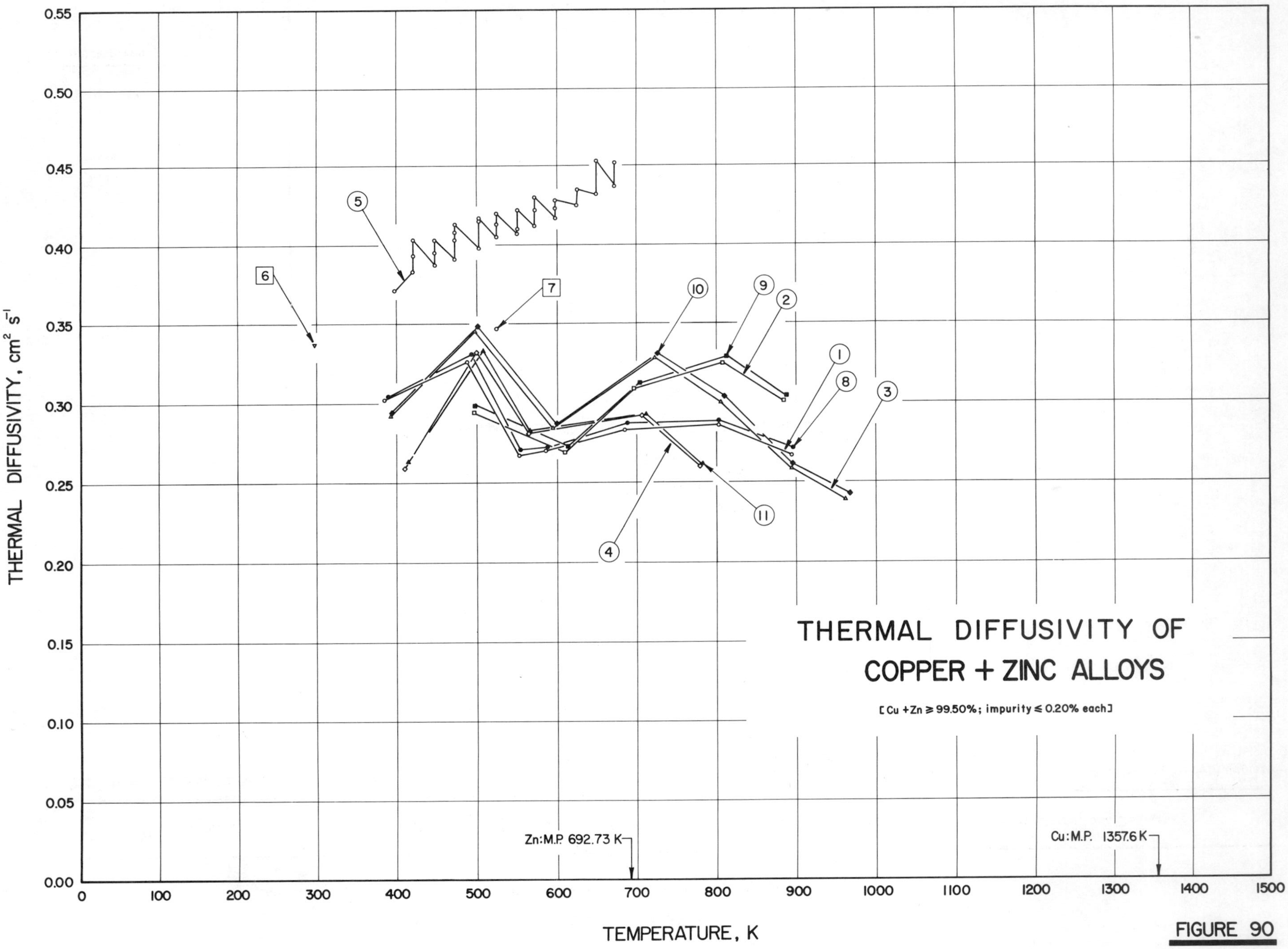
THERMAL DIFFUSIVITY OF
COPPER + ZINC ALLOYS
[Cu + Zn ≥ 99.50%; impurity ≤ 0.20% each]
THERMAL DIFFUSIVITY, cm² s⁻¹
TEMPERATURE, K
Zn: M.P. 692.73 K
Cu: M.P. 1357.6 K
0.55
0.50
0.45
0.40
0.35
0.30
0.25
0.20
0.15
0.10
0.05
0.00
0
100
200
300
400
500
600
700
800
900
1000
1100
1200
1300
1400
1500
1
2
3
4
5
6
7
8
9
10
11

FIGURE 90

SPECIFICATION TABLE 90. THERMAL DIFFUSIVITY OF [COPPER + ZINC] ALLOYS

(Cu + Zn ≥ 99. 50%; impurity ≤ 0. 20% each)

Cur. No.	Ref. No.	Author(s)	Year	Temp. Range, K	Reported Error, %	Name and Specimen Designation	Composition (weight percent) Cu	Zn	Composition (continued), Specifications, and Remarks
1	24	Dennis, J. E., Hirschman, A., Derksen, W. L., and Monahan, T. I.	1960	385-895		Brass			Brass disc specimen 0. 65 cm in diameter and 0. 132 cm thick; irradiated with a chopped beam in a carbon-arc image furnace; thermal diffusivity determined from measured phase lag between the square wave irradiance impinging upon the front face of the specimen and the resultant sinusoidal temperature of the rear face; error in calculating diffusivity (due to the use of a square instead of a sinusoidal heat input) is estimated to be -3. 2%.
2	24	Dennis, J. E., et al.	1960	497-885		Brass			Brass disc specimen 0. 65 cm in diameter and 0. 139 cm thick; measured under the same general conditions as above.
3	24	Dennis, J. E., et al.	1960	393-962		Brass			Brass disc specimen 0. 65 cm in diameter and 0. 205 cm thick; measured under the same general conditions as above.
4	24	Dennis, J. E., et al.	1960	410-780		Brass			Brass disc specimen 0. 65 cm in diameter and 0. 292 cm thick; measured under the same general conditions as above.
5	91	Rosenthal, D. and Ambrosio, A.	1951	398-673		Yellow Brass	65	35	Tubular specimen 0. 5 in. I. D., 0. 75 in. O. D., and 48 in. long; method of measurement based on the theory of moving heat sources; diffusivity determined from measured temperature-time history at various points along specimen at four heat source velocities; several runs made at each velocity.
6	7	diNovi, R. A.	1963	298	10	Brass			Brass specimen with thickness lying in the range from 1 to 2 mm; front surface uniformly irradiated by a very short pulse of radiant energy supplied by a xenon flash tube; diffusivity determined from measured history of the back surface temperature; temperature at which specimen was measured not given by author but assumed to be room temperature.
7	271	Rosenthal, D. and Friedmann, N. E.	1954	523		Yellow Brass	60	40	Diffusivity measured using modified Angström's method.
8	195, 194	Hirschman, A., Dennis, J., Derksen, W. L., and Monahen, T. I.	1961	388-896		Brass			Disk specimen, 0. 65 cm in diameter, 0. 132 cm in thickness.
9	195, 194	Hirschman, A., et al.	1961	496-887		Brass			Disk specimen, 0. 65 cm in diameter, 0. 139 cm in thickness.
10	195, 194	Hirschman, A., et al.	1961	393-966		Brass			Disk specimen, 0. 65 cm in diameter, 0. 205 cm in thickness.
11	195, 194	Hirschman, A., et al.	1961	413-782		Brass			Disk specimen, 0. 65 cm in diameter, 0. 292 cm in thickness.

DATA TABLE 90. THERMAL DIFFUSIVITY OF [COPPER + ZINC] ALLOYS

(Cu + Zn ≥ 99.50%; impurity ≤ 0.20% each)

[Temperature, T, K; Thermal Diffusivity, α, $cm^2 s^{-1}$]

T	α
CURVE 1	
385	0.302
488	0.326
553	0.267
586	0.270
685	0.283
803	0.286
895	0.267
CURVE 2	
497	0.294
610	0.269
697	0.309
808	0.325
885	0.301
CURVE 3	
393	0.292
498	0.345
595	0.284
723	0.328
806	0.300
895	0.258
962	0.239
CURVE 4	
410	0.259
500	0.332
565	0.281
707	0.292
780	0.260
CURVE 5	
398	0.371
421	0.383
421	0.393
421	0.403
448	0.387
CURVE 5 (cont.)	
448	0.395
448	0.403
473	0.391
473	0.403
473	0.408
473	0.413
503	0.398
503	0.415
503	0.417
525	0.405
525	0.413
525	0.420
551	0.407
551	0.410
551	0.422
573	0.412
573	0.422
573	0.430
598	0.417
598	0.423
598	0.428
625	0.425
626	0.435
650	0.432
650	0.453
673	0.437
673	0.452
CURVE 6	
298	0.337
CURVE 7	
523	0.346
CURVE 8	
388	0.303
492	0.330
553	0.270
CURVE 8 (cont.)	
587	0.272
687	0.287
802	0.288
896	0.271
CURVE 9	
496	0.297
612	0.272
702	0.312
812	0.328
887	0.304
CURVE 10	
393	0.293
500	0.347
599	0.286
725	0.330
809	0.303
895	0.261
966	0.242
CURVE 11	
413	0.262
506	0.332
565	0.282
710	0.292
782	0.261

SPECIFICATION TABLE 91. THERMAL DIFFUSIVITY OF [GERMANIUM + SILICON] ALLOYS

(Ge + Si ≥ 99.50%; impurity ≤ 0.20% each)

Cur. No.	Ref. No.	Author(s)	Year	Temp. Range, K	Reported Error, %	Name and Specimen Designation	Composition (weight percent) Ge	Si	Composition (continued), Specifications, and Remarks
1*	305	Winslow, J.W.	1967	294-385		225			Composition classified secret; n-type; measured by a flash method.
2*	305	Winslow, J.W.	1967	296-395		225			The above specimen.
3*	305	Winslow, J.W.	1967	295-363		225			The above specimen exposed to a dose of 1.0 x 10^{17} n cm^{-2} fast neutrons at General Electric Test Reactor.
4*	305	Winslow, J.W.	1967	296-396		225			The above specimen exposed to a dose of 1.6 x 10^{18} n cm^{-2}.

DATA TABLE 91. THERMAL DIFFUSIVITY OF [GERMANIUM + SILICON] ALLOYS

(Ge + Si ≥ 99.50%; impurity ≤ 0.20% each)

[Temperature, T, K; Thermal Diffusivity, α, $cm^2 s^{-1}$]

T	α	T	α	T	α
CURVE 1*		CURVE 3*		CURVE 4 (cont.)*	
294.5	0.0279	295.2	0.0246	382.3	0.0165
295.6	0.0284	295.3	0.0251	392.2	0.0147
296.0	0.0280	295.7	0.0265	394.0	0.0155
297.4	0.0284	296.0	0.0273	394.2	0.0148
383.8	0.0242	296.9	0.0247	396.1	0.0152
384.4	0.0238	297.2	0.0252		
385.1	0.0241	297.4	0.0269		
		298.0	0.0265		
CURVE 2*		363.3	0.0225		
295.7	0.0302	CURVE 4*			
296.2	0.0296				
297.6	0.0291	295.9	0.0165		
297.6	0.0303	297.5	0.0157		
393.3	0.0234	299.0	0.0166		
393.8	0.0225	373.2	0.0177		
395.3	0.0233	377.7	0.0158		

* No figure given.

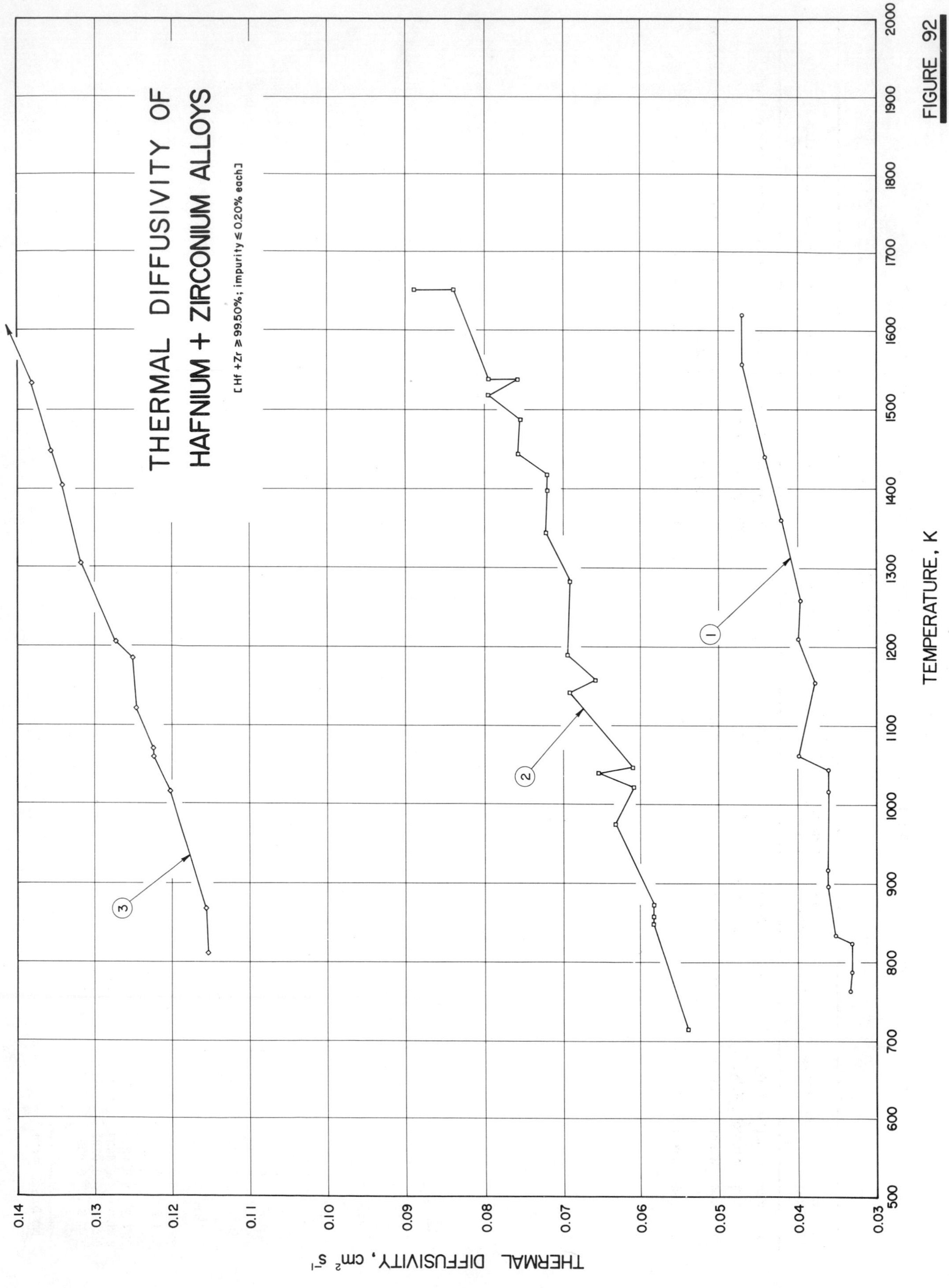
THERMAL DIFFUSIVITY OF
HAFNIUM + ZIRCONIUM ALLOYS
[Hf + Zr ≥ 99.50%; impurity ≤ 0.20% each]
THERMAL DIFFUSIVITY, cm² s⁻¹
TEMPERATURE, K
0.14
0.13
0.12
0.11
0.10
0.09
0.08
0.07
0.06
0.05
0.04
0.03
500
600
700
800
900
1000
1100
1200
1300
1400
1500
1600
1700
1800
1900
2000
1
2
3

FIGURE 92

SPECIFICATION TABLE 92. THERMAL DIFFUSIVITY OF [HAFNIUM + ZIRCONIUM] ALLOYS

(Hf + Zr ≥ 99.50%; impurity ≤ 0.20% each)

Cur. No.	Ref. No.	Author(s)	Year	Temp. Range, K	Reported Error, %	Name and Specimen Designation	Composition (weight percent) Hf	Zr	Composition (continued), Specifications, and Remarks
1	208	Wheeler, M.J.	1970	763-1619			Bal.	3-3.5	0.01-0.02 Fe, 0.01-0.015 O, 0.001-0.002 N impurities; disk specimen 0.6 cm in diameter and 0.041 cm in thickness; fabricated by punching from rolled plate, annealed in vacuum for 1 hr at 700 C, then heated in autoclave for 3.5 days at 343 C to oxidize the surface; diffusivity measured using modulated electron beam technique.
2	208	Wheeler, M.J.	1970	715-1651					0.080 cm in thickness; other conditions same as above.
3	208	Wheeler, M.J.	1970	811-1694					0.164 cm in thickness; other conditions same as above.

DATA TABLE 92. THERMAL DIFFUSIVITY OF [HAFNIUM + ZIRCONIUM] ALLOYS

(Hf + Zr ≥ 99.50%; impurity ≤ 0.20% each)

[Temperature, T, K; Thermal Diffusivity, α, cm^2 s^{-1}]

T	α	T	α	T	α	T	α
CURVE 1		CURVE 1 (cont.)		CURVE 2 (cont.)		CURVE 3	
763	0.0334	1557	0.0471	1157	0.0657	811	0.1154
787	0.0332	1619	0.0471	1189	0.0694	867	0.1157
824	0.0332			1282	0.0690	1015	0.1201
834	0.0353	CURVE 2		1344	0.0721	1059	0.1222
896	0.0362			1398	0.0719	1070	0.1222
916	0.0362	715	0.0539	1417	0.0719	1121	0.1245
1016	0.0362	848	0.0583	1444	0.0756	1185	0.1250
1043	0.0362	858	0.0583	1487	0.0753	1206	0.1272
1061	0.0399	873	0.0583	1518	0.0794	1306	0.1316
1153	0.0378	974	0.0633	1538	0.0757	1404	0.1340
1209	0.0399	1021	0.0608	1538	0.0794	1447	0.1355
1258	0.0397	1039	0.0655	1538	0.0794	1534	0.1379
1360	0.0421	1046	0.0609	1651	0.0837	1653	0.1433
1440	0.0442	1141	0.0691	1651	0.0888	1694	0.1436

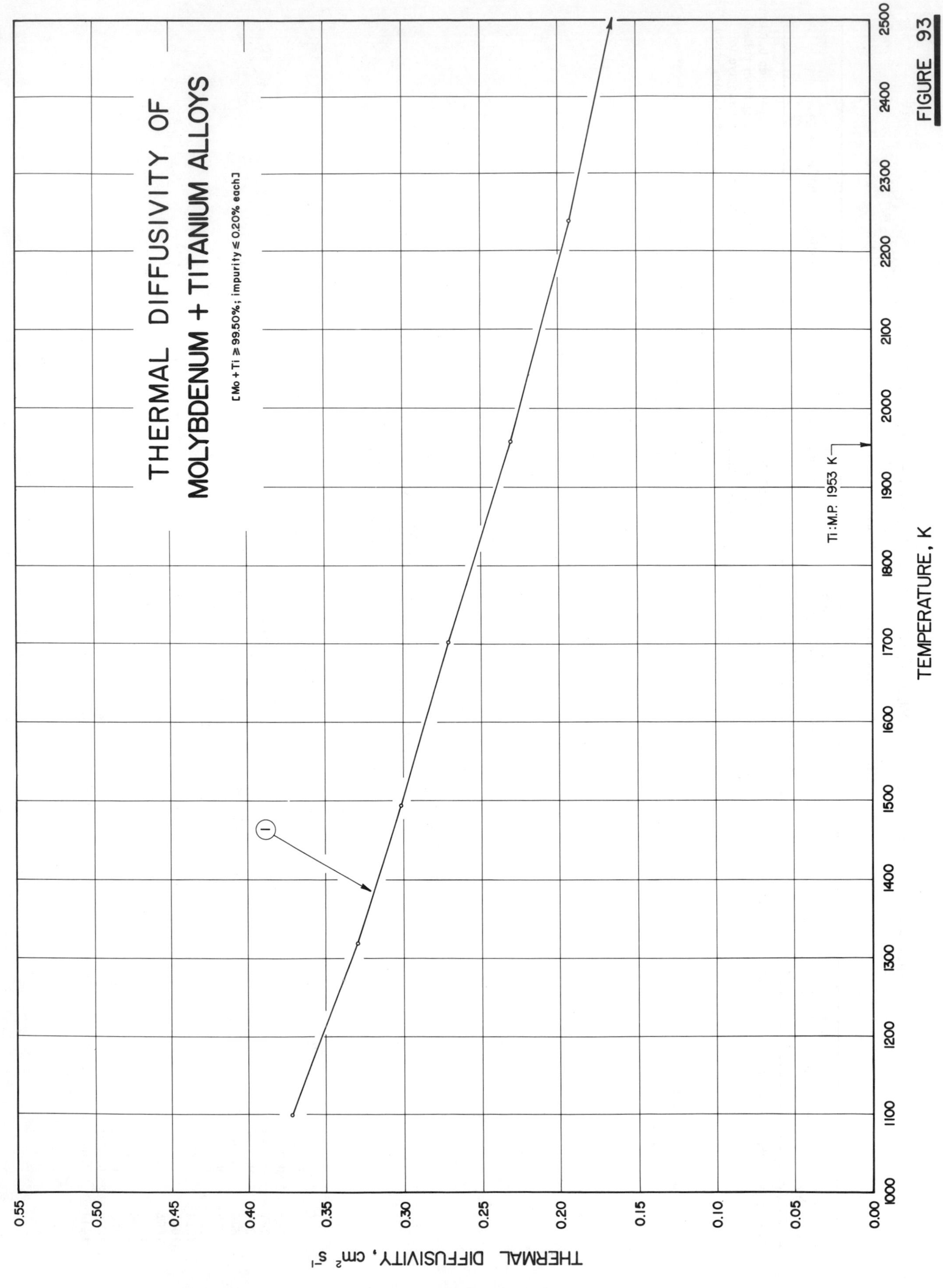
THERMAL DIFFUSIVITY OF
MOLYBDENUM + TITANIUM ALLOYS
[Mo + Ti ≥ 99.50%; impurity ≤ 0.20% each]
THERMAL DIFFUSIVITY, cm² s⁻¹
TEMPERATURE, K
Ti: M.P. 1953 K
1
0.00
0.05
0.10
0.15
0.20
0.25
0.30
0.35
0.40
0.45
0.50
0.55
1000
1100
1200
1300
1400
1500
1600
1700
1800
1900
2000
2100
2200
2300
2400
2500

FIGURE 93

SPECIFICATION TABLE 93. THERMAL DIFFUSIVITY OF [MOLYBDENUM + TITANIUM] ALLOYS

(Mo + Ti ≥ 99.50%; impurity ≤0.20% each)

Cur. No.	Ref. No.	Author(s)	Year	Temp. Range, K	Reported Error, %	Name and Specimen Designation	Composition (weight percent) Mo	Ti	Composition (continued), Specifications, and Remarks
1	74	Hedge, J.C., Kopec, J.W., Kostenko, C, and Lang, J.I.	1963	1100-2580		Mo-0.5Ti-0.08 Zr Alloy	99.410	0.49	(Mo by difference), 0.07 Zr, 0.0260 C, <0.001 Fe, <0.001 Ni, <0.001 Si, 0.0007 O, 0.0001 H, and 0.0001 N; slab specimen; supplied by Climax Molybdenum; density 9.96 g cm^{-3}; top surface of specimen exposed to heat sink; diffusivity determined from measured temperature decrease; unidirectional heat flow.

DATA TABLE 93. THERMAL DIFFUSIVITY OF [MOLYBDENUM + TITANIUM] ALLOYS

(Mo + Ti ≥ 99.50%; impurity ≤0.20% each)

[Temperature, T, K; Thermal Diffusivity, α, cm^2 s^{-1}]

T	α
CURVE 1	
1100	0.372
1319	0.330
1494	0.302
1703	0.271
1958	0.231
2239	0.193
2580	0.156*

* Not shown in figure.

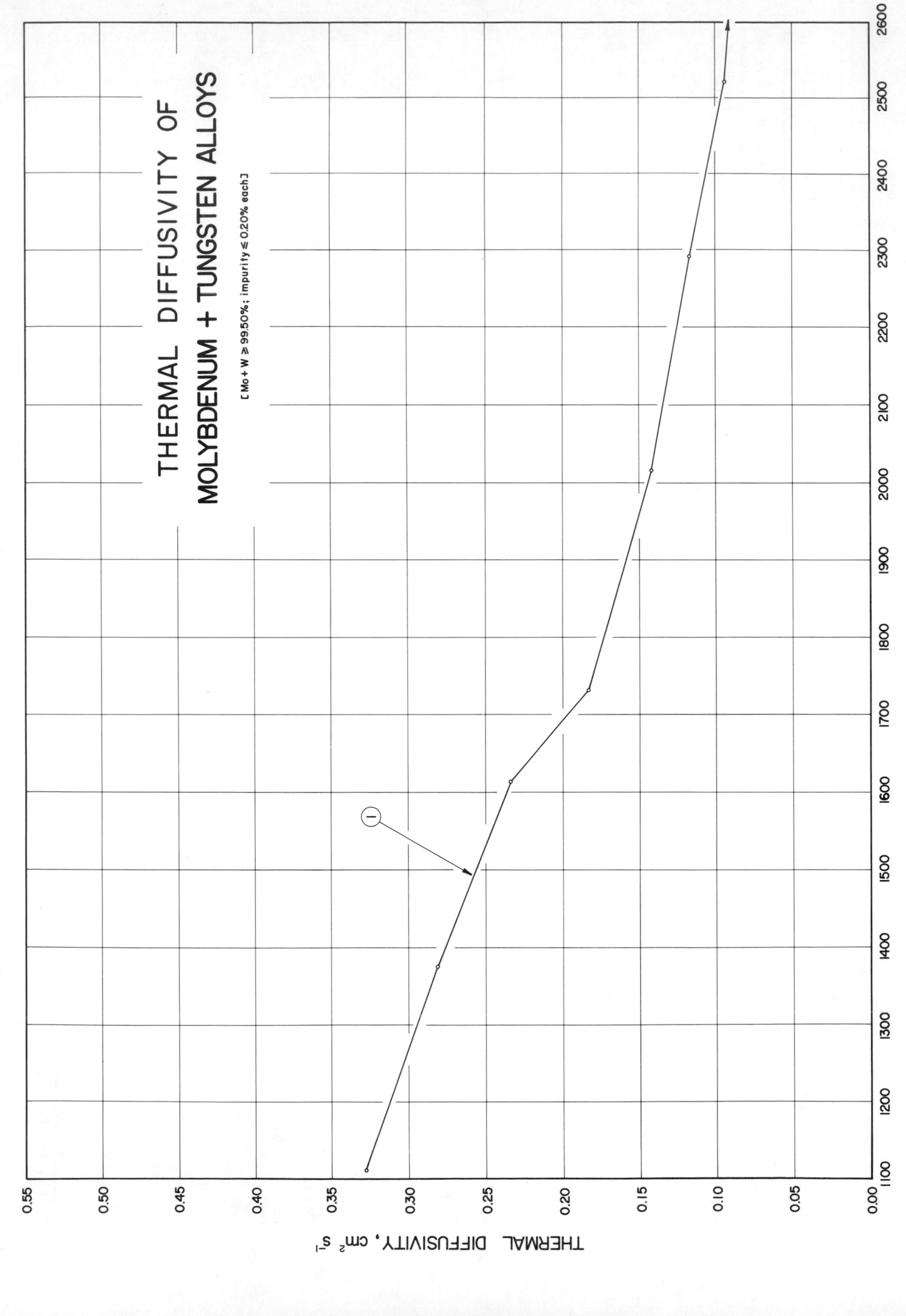
THERMAL DIFFUSIVITY OF
MOLYBDENUM + TUNGSTEN ALLOYS
[Mo + W ≥ 99.50%; impurity ≤ 0.20% each]
1
THERMAL DIFFUSIVITY, cm² s⁻¹
0.00
0.05
0.10
0.15
0.20
0.25
0.30
0.35
0.40
0.45
0.50
0.55
1100
1200
1300
1400
1500
1600
1700
1800
1900
2000
2100
2200
2300
2400
2500
2600
TEMPERATURE, K

FIGURE 94

SPECIFICATION TABLE 94. THERMAL DIFFUSIVITY OF [MOLYBDENUM + TUNGSTEN] ALLOYS

(Mo + W ≥ 99.50%; impurity ≤ 0.20% each)

Cur. No.	Ref. No.	Author(s)	Year	Temp. Range, K	Reported Error, %	Name and Specimen Designation	Composition (weight percent) Mo	W	Composition (continued), Specifications, and Remarks
1	74	Hedge, J.C., Kopec, J.W., Kostenko, C., and Lang, J.I.	1963	1111-2774		Mo-29.83W-0.07Zr-0.012C Alloy	70.088	29.83	(Mo by difference), 0.07 Zr, and 0.0120 C; slab specimen; supplied by Climax Molybdenum; density 9.93 g cm^{-3}; top surface of specimen exposed to heat sink; diffusivity determined from measured temperature decrease; unidirectional heat flow.

DATA TABLE 94. THERMAL DIFFUSIVITY OF [MOLYBDENUM + TUNGSTEN] ALLOYS

(Mo + W ≥ 99.50%; impurity ≤ 0.20% each)

[Temperature, T, K; Thermal Diffusivity, α, cm^2 s^{-1}]

T	α
CURVE 1	
1111	0.328
1377	0.281
1613	0.234
1732	0.183
2016	0.142
2291	0.117
2521	0.0942
2774	0.0854*

* Not shown in figure.

SPECIFICATION TABLE 95. THERMAL DIFFUSIVITY OF [NICKEL + IRON] ALLOYS

(Ni + Fe ≥ 99.50%; impurity ≤ 0.20% each)

Cur. No.	Ref. No.	Author(s)	Year	Temp. Range, K	Reported Error, %	Name and Specimen Designation	Composition (weight percent) Ni	Fe	Composition (continued), Specifications, and Remarks
1*	92	Frazier, R. H.	1932	297.9			99.23	0.27	0.23 Mn, 0.12 C, 0.07 S, 0.06 Cu, 0.06 Si, and 0.01 Co; rod specimen; obtained from H. Bocker and Co. Inc.; cold drawn; ground to precise diameter by Cincinnati Grinders Inc.; surface polished; surrounded by a thin nickel tube slightly longer than specimen; space between specimen and tube evacuated to a vacuum of 1 micron; temperature impulse obtained by suddenly subjecting one end of specimen to a stream of cold water; tests made at different rates of water flow.

DATA TABLE 95. THERMAL DIFFUSIVITY OF [NICKEL + IRON] ALLOYS

(Ni + Fe ≥ 99.50%; impurity ≤ 0.20% each)

[Temperature, T, K; Thermal Diffusivity, α, $cm^2\ s^{-1}$]

T	α
CURVE 1*	
297.9	0.150

* No figure given.

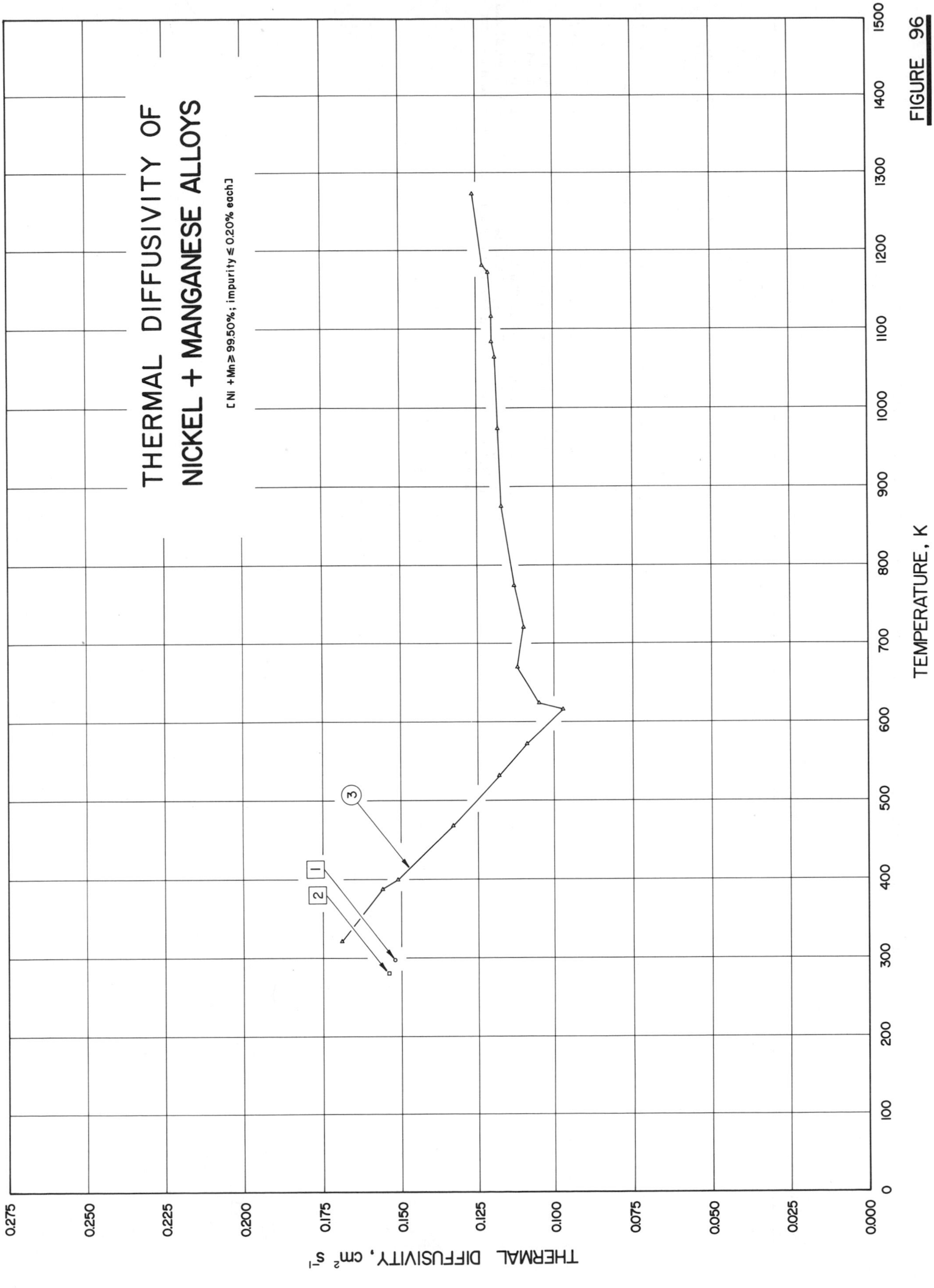
THERMAL DIFFUSIVITY OF
NICKEL + MANGANESE ALLOYS
[Ni + Mn ≥ 99.50%; impurity ≤ 0.20% each]
THERMAL DIFFUSIVITY, cm² s⁻¹
TEMPERATURE, K
0.000
0.025
0.050
0.075
0.100
0.125
0.150
0.175
0.200
0.225
0.250
0.275
0
100
200
300
400
500
600
700
800
900
1000
1100
1200
1300
1400
1500
1
2
3

FIGURE 96

SPECIFICATION TABLE 96. THERMAL DIFFUSIVITY OF [NICKEL + MANGANESE] ALLOYS

(Ni + Mn ≥ 99.50%; impurity ≤ 0.20% each)

Cur. No.	Ref. No.	Author(s)	Year	Temp. Range, K	Reported Error, %	Name and Specimen Designation	Composition (weight percent) Ni	Mn	Composition (continued), Specifications, and Remarks
1	86	Woisard, E. L.	1961	298	±4	"A" nickel	99.27	0.25	0.15 Fe, 0.13 Co, 0.1 Cu, 0.05 C, 0.05 Si, and 0.005 S; specimen consists of two long thin rods each 3/16 in. in dia. and ~25 cm long; specimen rods butted against a thin disc-shaped heater and held in alignment under compression; entire assembly placed inside heavy walled copper tube 22 in. long acting as isothermal enclosure; test carried out under a vacuum of 10^{-4} mm Hg; momentary pulse of electric current passed through rods.
2	93	Frazier, R. H.	1932	281.7		Rod D	99.22/ 99.28	0.25/ 0.27	0.135-0.145 C, 0.095-0.125 Fe, 0.065-0.075 Cu, 0.06-0.08 Si, 0.055-0.065 S, and 0.005-0.015 Co; electrical resistivity 10.84 μohm cm at 293.5 K.
3	113	Sidles, P. H. and Danielson, G. C.	1960	322-1273		Commercial "A" nickel	99.5	0.25	0.15 Fe, 0.06 C, 0.05 Cu, 0.05 Si, and 0.005 S; nominal composition from publication of Huntington Alloy Products Div., The International Nickel Co., Inc; obtained from Driver Harris Co.; Curie temp ~616.2 K; electrical resistivity reported as 9.9, 11.1, 12.5, 14.2, 15.8, 18.5, 22.1, 24.5, 32.7, 35.3, 36.7, 38.0, 40.9, 43.7, 44.6, and 49.0 μohm cm at 293.2, 330.2, 373.2, 409.2, 425.2, 474.2, 532.2, 574.2, 680.2, 758.2, 809.2, 843.2, 923.2, 1053.2, 1100.2, and 1261.2 K, respectively; diffusivity measured using modified Angström method.

DATA TABLE 96. THERMAL DIFFUSIVITY OF [NICKEL + MANGANESE] ALLOYS

(Ni + Mn ≥ 99.50%; impurity ≤ 0.20% each)

[Temperature, T, K; Thermal Diffusivity, α, cm^2 s^{-1}]

T	α	T	α
CURVE 1		CURVE 3 (cont.)	
298	0.152	572	0.109
		616	0.0971
CURVE 2		624	0.105
		670	0.112
281.7	0.154	721	0.110
		774	0.113
CURVE 3		875	0.117
		974	0.118
322	0.169	1064	0.119
388	0.156	1084	0.120
400	0.151	1116	0.120
468	0.133	1172	0.121
532	0.118	1181	0.123
		1273	0.126

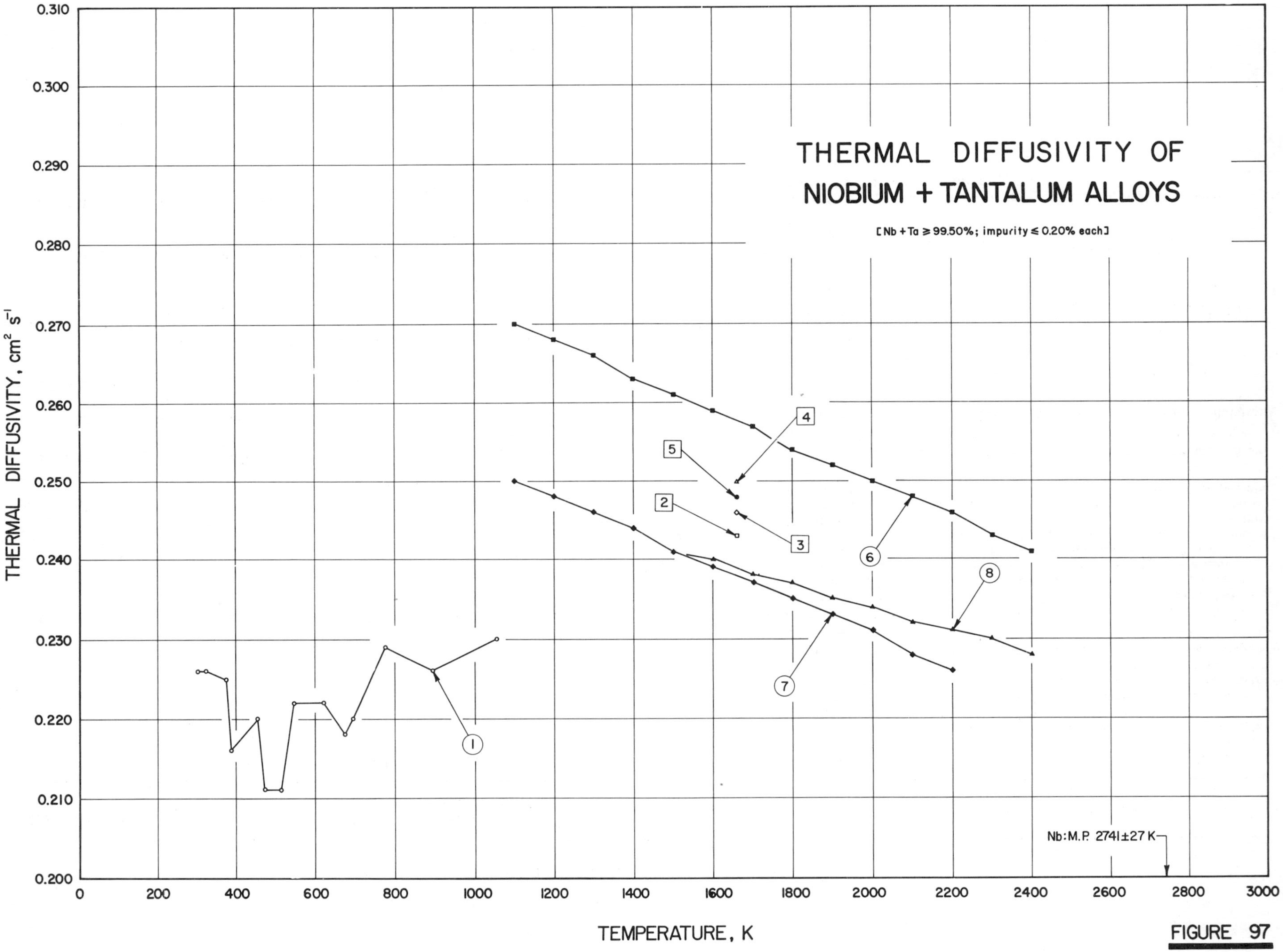
THERMAL DIFFUSIVITY OF
NIOBIUM + TANTALUM ALLOYS
[Nb + Ta ≥ 99.50%; impurity ≤ 0.20% each]
THERMAL DIFFUSIVITY, cm² s⁻¹
TEMPERATURE, K
0.310
0.300
0.290
0.280
0.270
0.260
0.250
0.240
0.230
0.220
0.210
0.200
0
200
400
600
800
1000
1200
1400
1600
1800
2000
2200
2400
2600
2800
3000
1
2
3
4
5
6
7
8
Nb: M.P. 2741±27 K

FIGURE 97

SPECIFICATION TABLE 97. THERMAL DIFFUSIVITY OF [NIOBIUM + TANTALUM] ALLOYS

(Nb + Ta ≥ 99.50%; impurity ≤ 0.20% each)

Cur. No.	Ref. No.	Author(s)	Year	Temp. Range, K	Reported Error, %	Name and Specimen Designation	Composition (weight percent) Nb	Ta	Composition (continued), Specifications, and Remarks
1	163	Akhmetzyanov, K. G., Pozdnyak, N. Z., and Dobrovol'skii, A. F.	1967	303-1053			99.195	0.5	(Nb by difference), 0.17 W, 0.054 Fe, 0.04 Mo, 0.014 Si, 0.014 Ti, and 0.013 C; cylindrical specimen 3-5 mm in dia. and 150-200 mm long; manufactured from electrolytic powdered niobium using powder metallurgy; prepared by compacting rods of rectangular cross-section at a pressure of 3.5 ton cm^{-2}, and subsequent sintering in a charging vacuum furnace for 5 hrs at 1373 K in a vacuum of 10^{-3} mm Hg, then cooling in the furnace, then welding in a vacuum welder for 5 hrs at 2625 K in a vacuum of 10^{-5} mm Hg, and finally refining in two stages by zone melting; temperature-wave method used to measure diffusivity.
2	165	Filippov, L. P. and Makarenko, I. N.	1968	1660	4-8		99.2	0.3	0.08 Ti, 0.04 Fe, and 0.04 Si; cylindrical specimen 15 mm in dia. and 80 mm long; density 8.54 g cm^{-3}; electrical resistivity reported as 16.4 and 65.6 μohm cm at room temp. and 1660 K, respectively; diffusivity determined from measured temp. fluctuations of specimen heated in a periodic manner in a high-frequency induction furnace; method of measurement based on the connection between the values of the amplitudes and phases of the temp. fluctuations at one point on the sample and the thermal characteristics of the medium investigated; measured using a power modulation frequency of 0.144 cycles per sec; measured in vacuum.
3	165	Filippov, L. P. and Makarenko, I. N.	1968	1660	4-8				Above specimen measured for diffusivity again using a power modulation frequency of 0.187 cycles per sec; other conditions same as above.
4	165	Filippov, L. P. and Makarenko, I. N.	1968	1660	4-8				Above specimen measured for diffusivity again using a power modulation frequency of 0.262 cycles per sec; other conditions same as above.
5	165	Filippov, L. P. and Makarenko, I. N.	1968	1660	4-8				Above specimen measured for diffusivity again using a power modulation frequency of 0.433 cycles per sec; other conditions same as above.
6	251	Makarenko, I. N., Trukhanova, L. N., and Filippov, L. P.	1970	1100-2400		Sample I	99.53	0.35	0.074 W, 0.01 C, 0.01 Ti, 0.01 Si, 0.01 O_2, 0.05 Fe, 0.007 N_2, 0.0022 Mo, 0.001 H_2; single crystal; prepared by cathode-ray zonal melting method; 13 mm in diameter, 75 mm in length; density and electrical resistivity of specimen at 20 C were 8.62 g cm^{-3} and 15.5 μΩ cm respectively.
7	251	Makarenko, I. N., et al.	1970	1100-2200		Sample II	99.2	0.3	0.08 Ti, 0.04 Fe, 0.04 Si; polycrystalline; 15 mm in diameter, 75 mm in length; density 8.54 g cm^{-3}; electrical resistivity 14.8 μΩ cm at 20 C.
8	251	Makarenko, I. N., et al.	1970	1500-2400		Sample III	99.62	0.28	0.05 Fe, 0.015 Mo, 0.01 Si, 0.01 Ti, 0.01 C, <0.005 N_2, 0.005 O_2, 0.001 H_2; 0.2 mm in diameter.

DATA TABLE 97. THERMAL DIFFUSIVITY OF [NIOBIUM + TANTALUM] ALLOYS

(Nb + Ta ≥99.50%; impurity ≤0.20% each)

[Temperature, T, K; Thermal Diffusivity, α, $cm^2\ s^{-1}$]

T	α	T	α
CURVE 1		CURVE 6 (cont.)	
303	0.226	2200	0.246
323	0.226	2300	0.243
375	0.225	2400	0.241
388	0.216		
455	0.220	CURVE 7	
473	0.211		
513	0.211	1100	0.250
543	0.222	1200	0.248
620	0.222	1300	0.246
673	0.218	1400	0.244
693	0.220	1500	0.241
773	0.229	1600	0.239
893	0.226	1700	0.237
1053	0.230	1800	0.235
		1900	0.233
CURVE 2		2000	0.231
		2100	0.228
1660	0.243	2200	0.226
CURVE 3		CURVE 8	
1660	0.246	1500	0.241*
		1600	0.240
CURVE 4		1700	0.238
		1800	0.237
1660	0.250	1900	0.235
		2000	0.234
CURVE 5		2100	0.232
		2200	0.231
1660	0.248	2300	0.230
		2400	0.228
CURVE 6			
1100	0.270		
1200	0.268		
1300	0.266		
1400	0.263		
1500	0.261		
1600	0.259		
1700	0.257		
1800	0.254		
1900	0.252		
2000	0.250		
2100	0.248		

* Not shown in figure.

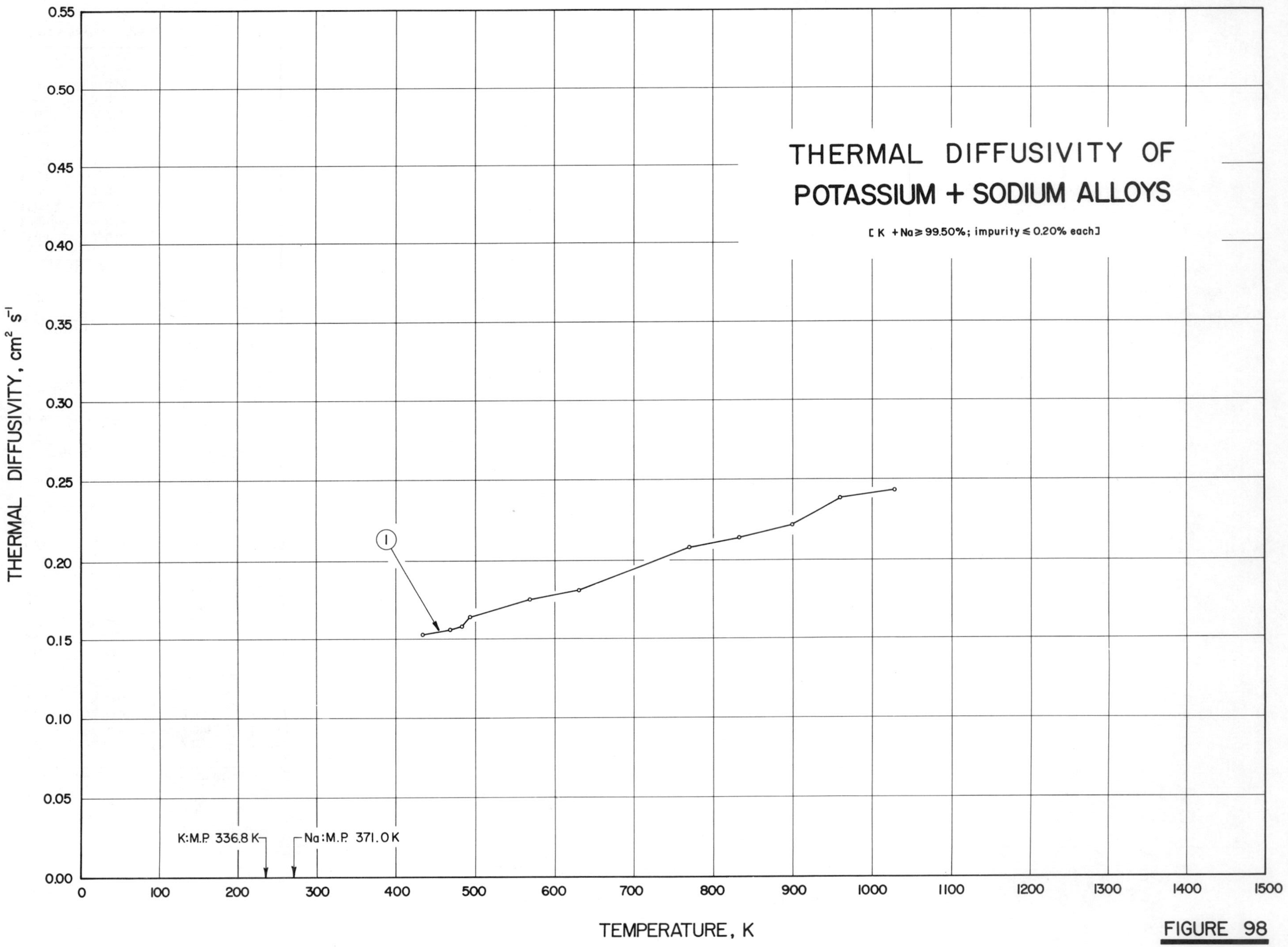
THERMAL DIFFUSIVITY OF
POTASSIUM + SODIUM ALLOYS
[K + Na ≥ 99.50%; impurity ≤ 0.20% each]
THERMAL DIFFUSIVITY, cm² s⁻¹
TEMPERATURE, K
K:M.P. 336.8 K
Na:M.P. 371.0 K
1
0.55
0.50
0.45
0.40
0.35
0.30
0.25
0.20
0.15
0.10
0.05
0.00
0
100
200
300
400
500
600
700
800
900
1000
1100
1200
1300
1400
1500

FIGURE 98

SPECIFICATION TABLE 98. THERMAL DIFFUSIVITY OF [POTASSIUM + SODIUM] ALLOYS

(K + Na ≥ 99.50%; impurity ≤ 0.20% each)

Cur. No.	Ref. No.	Author(s)	Year	Temp. Range, K	Reported Error, %	Name and Specimen Designation	Composition (weight percent) K	Composition (weight percent) Na	Composition (continued), Specifications, and Remarks
1	45, 140	Novikov, I. I., Solovyev, A. N., Khabakhpasheva, E. M., Gruzdev, V. A., Pridantsev, A. I., and Vasenina, M. Ya.	1956	433-1028			78	22	Eutectic solution; metal placed into a vertically positioned thin stainless steel tube; heater wound on the outside of upper part of tube; free metal surface maintained at an excess pressure of an inert gas; twenty radial copper screens distributed throughout whole length of specimen; suspended in an evacuated quartz tube; Angstrom's dynamic method used to determine diffusivity.

DATA TABLE 98. THERMAL DIFFUSIVITY OF [POTASSIUM + SODIUM] ALLOYS

(K + Na ≥ 99.50%; impurity ≤ 0.20% each)

[Temperature, T, K; Thermal Diffusivity, α, $cm^2\ s^{-1}$]

T	α
CURVE 1	
433	0.153
468	0.156
483	0.158
493	0.164
568	0.175
630	0.181
770	0.208
833	0.214
900	0.222
960	0.239
1028	0.244

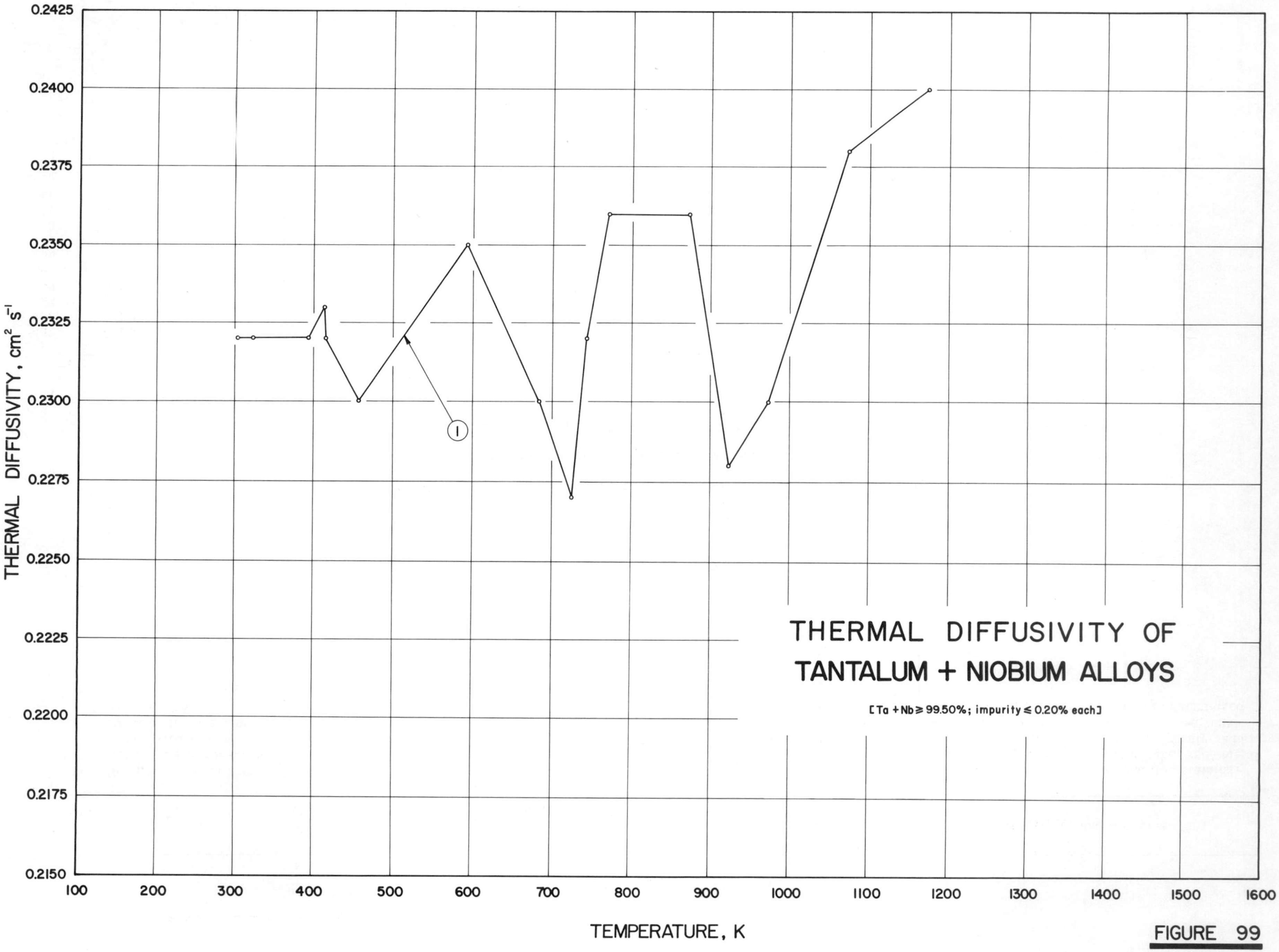

FIGURE 99

THERMAL DIFFUSIVITY OF TANTALUM + NIOBIUM ALLOYS

[Ta + Nb ≥ 99.50%; impurity ≤ 0.20% each]

SPECIFICATION TABLE 99. THERMAL DIFFUSIVITY OF [TANTALUM + NIOBIUM] ALLOYS

($Ta + Nb \geq 99.50\%$; impurity $\leq 0.20\%$ each)

Cur. No.	Ref. No.	Author(s)	Year	Temp. Range, K	Reported Error, %	Name and Specimen Designation	Composition (weight percent) Ta	Nb	Composition (continued), Specifications, and Remarks
1	163	Akhmetzyanov, K. G., Pozdnyak, N. Z., and Dobrovol'skii, A. F.	1967	303-1173			99.426	0.5	(Ta by difference), 0.06 Fe, 0.005 W, 0.003 Si, 0.003 Ti, 0.002 C, and 0.001 Mo; cylindrical specimen 3-5 mm in dia. and 150-200 mm long; manufactured from electrolytic powdered tantalum using powder metallurgy; prepared by compacting rods of rectangular cross-section at a pressure of 2.5 ton cm^{-2}, and subsequent sintering in a charging vacuum furnace for 5 hrs at 1723 K in a vacuum of 10^{-3} mm Hg, then cooling in the furnace, then welding in a vacuum welder for 5 hrs at 2873 K in a vacuum of 10^{-5} mm Hg, and finally refining in two stages by zone melting; temperature-wave method used to measure diffusivity.

DATA TABLE 99. THERMAL DIFFUSIVITY OF [TANTALUM + NIOBIUM] ALLOYS

($Ta + Nb \geq 99.50\%$; impurity $\leq 0.20\%$ each)

[Temperature, T, K; Thermal Diffusivity, α, $cm^2\ s^{-1}$]

T	α
CURVE 1	
303	0.232
323	0.232
393	0.232
413	0.233
415	0.232
457	0.230
593	0.235
683	0.230
723	0.227
743	0.232
771	0.236
873	0.236
923	0.228
973	0.230
1073	0.238
1173	0.240

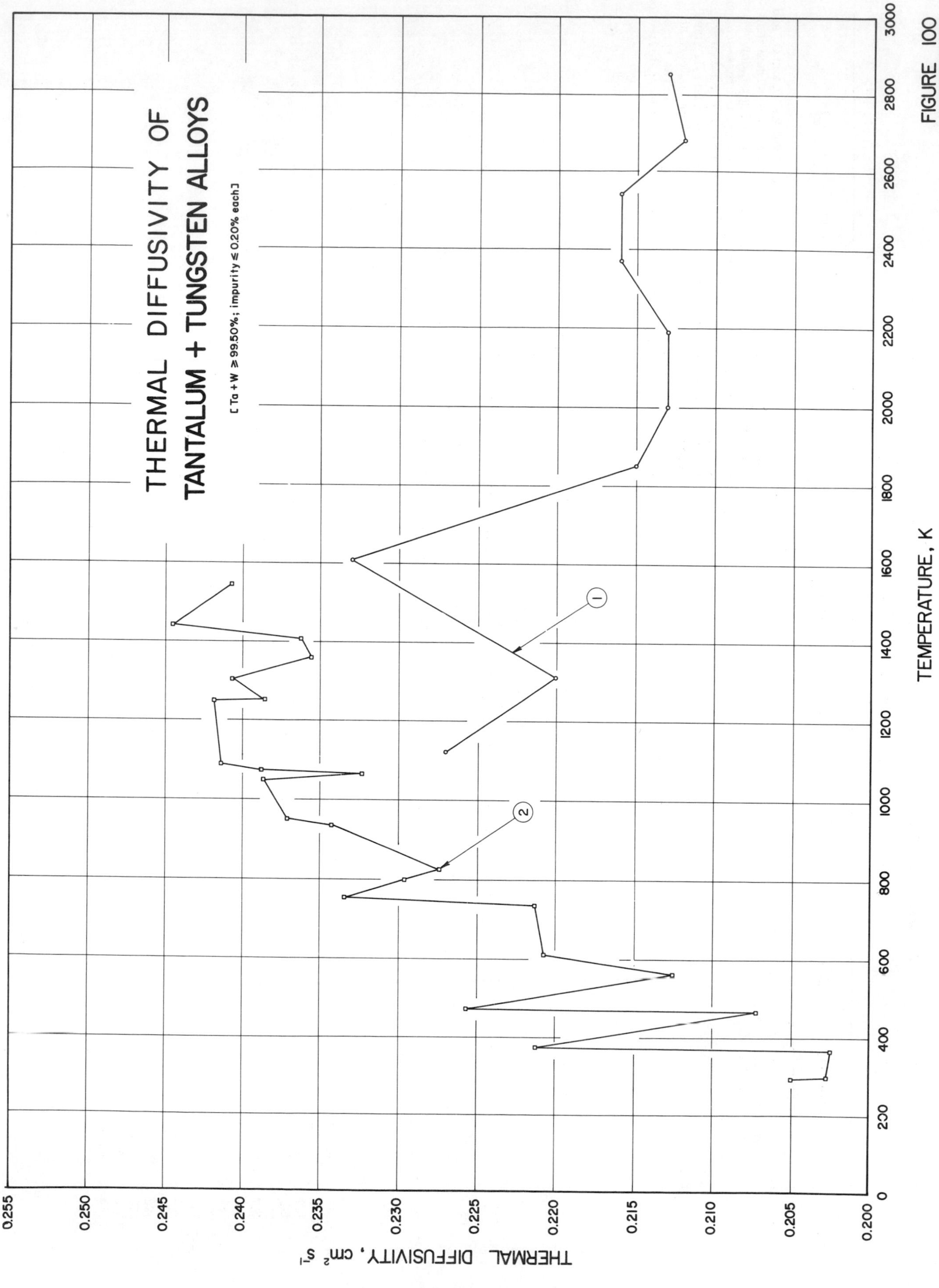
THERMAL DIFFUSIVITY OF
TANTALUM + TUNGSTEN ALLOYS
[Ta + W ≥ 99.50%; impurity ≤ 0.20% each]
THERMAL DIFFUSIVITY, cm² s⁻¹
TEMPERATURE, K
0.200
0.205
0.210
0.215
0.220
0.225
0.230
0.235
0.240
0.245
0.250
0.255
0
200
400
600
800
1000
1200
1400
1600
1800
2000
2200
2400
2600
2800
3000
1
2

FIGURE 100

SPECIFICATION TABLE 100. THERMAL DIFFUSIVITY OF [TANTALUM + TUNGSTEN] ALLOYS

(Ta + W ≥ 99.50%; impurity ≤0.20% each)

Cur. No.	Ref. No.	Author(s)	Year	Temp. Range, K	Reported Error, %	Name and Specimen Designation	Composition (weight percent) Ta	W	Composition (continued), Specifications, and Remarks
1	74	Hedge, J.C., Kopec, J.W., Kostenko, C., and Lang, J.I.	1963	1122-2852		Ta-10W Alloy	90.419	9.40	(Ta by difference), 0.10 Nb, 0.02 Si, 0.02 Ti, 0.01 Mo, 0.01 Ni, 0.0090 O, 0.005 Fe, 0.0040 C, and 0.0030 N; slab specimen; supplied by Fansteel Metallurgical Corp; density 16.58 g cm^{-3}; top surface of specimen exposed to heat sink; diffusivity determined from measured temperature decrease; unidirectional heat flow.
2	134, 256	Denman, G. L.	1966	294-1544	<±2.3	Tantalum Alloy Ta-10W	90.844^{+}	9.11	(Ta by difference), 0.02 Nb, 0.01 Fe, <0.01 Zr, 0.002 Cr, 0.0020 O, and 0.0019 C; grain size (as received) in the range from ~0.02 to 0.04 mm; single phase with very little second phase at the grain boundaries; disc-shaped specimens each 0.270 in. in dia and 0.060 in. thick; a minimum of two samples measured; supplied by Fansteel Met. Corp; arc-cast and fully annealed; density 16.82 g cm^{-3}; ruby laser used as pulse energy source; diffusivity determined from measured history of back face temp; measured in vacuum; each data point reported represents average of four or five measurements.

DATA TABLE 100. THERMAL DIFFUSIVITY OF [TANTALUM + TUNGSTEN] ALLOYS

(Ta + W ≥ 99.50%; impurity ≤ 0.20% each)

[Temperature, T, K; Thermal Diffusivity, α, cm^2 s^{-1}]

T	α	T	α	T	α
CURVE 1		CURVE 2 (cont.)		CURVE 2 (cont.)	
1122	0.227	466	0.2072	1357	0.2355
1314	0.220	472	0.2256	1408	0.2362
1608	0.233	561	0.2125	1441	0.2448
1850	0.215	611	0.2207	1544	0.2407
1997	0.213	736	0.2213		
2189	0.213	752	0.2334		
2369	0.216	799	0.2296		
2541	0.216	826	0.2273		
2680	0.212	936	0.2342		
2852	0.213	949	0.2370		
		1050	0.2386		
CURVE 2		1066	0.2323		
		1076	0.2387		
294	0.2050	1088	0.2413		
297	0.2027	1251	0.2418		
365	0.2025	1252	0.2385		
373	0.2212	1304	0.2406		

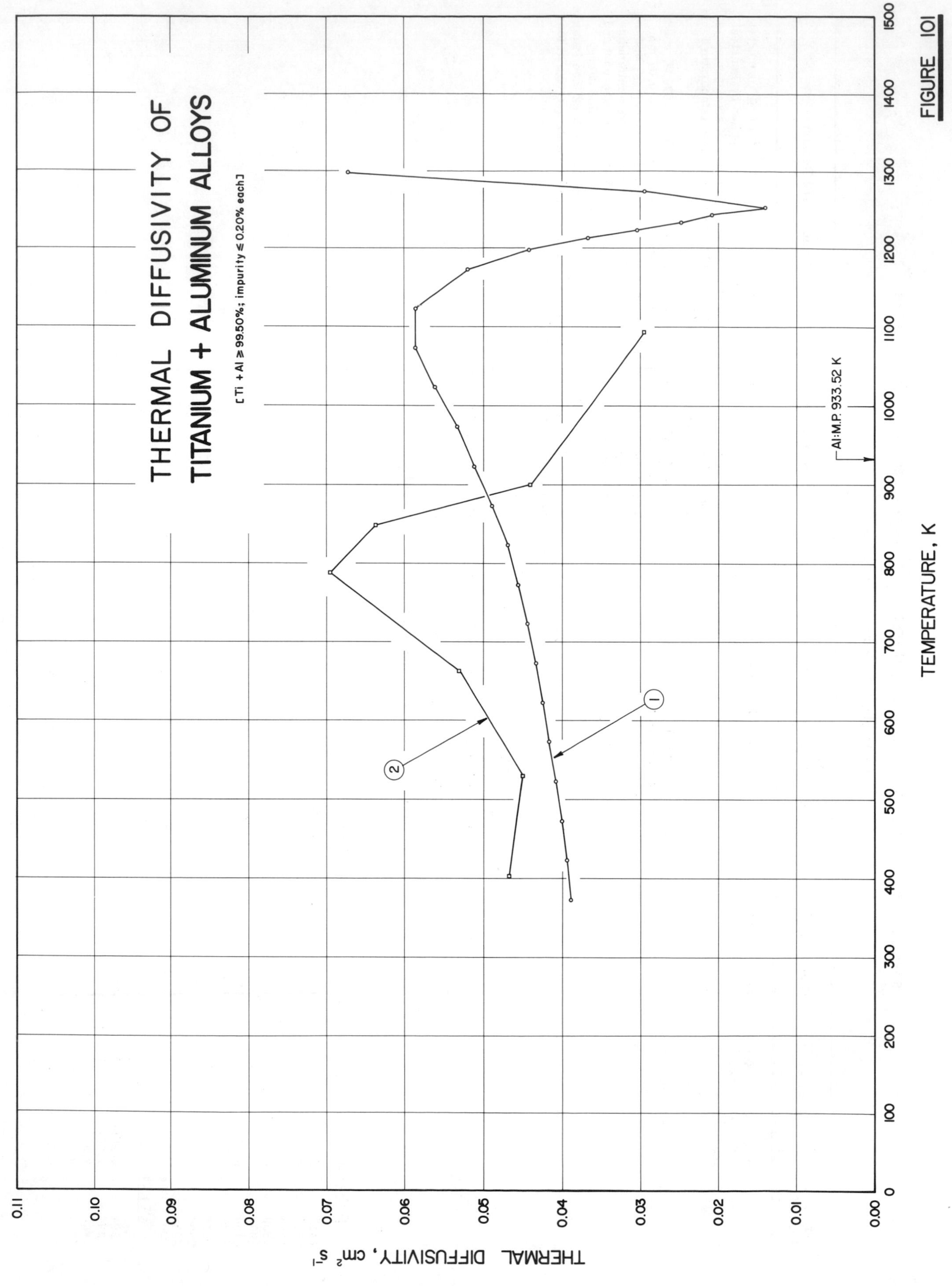
THERMAL DIFFUSIVITY OF
TITANIUM + ALUMINUM ALLOYS
[Ti + Al ≥ 99.50%; impurity ≤ 0.20% each]
Al: M.P. 933.52 K
1
2
THERMAL DIFFUSIVITY, cm² s⁻¹
0.00
0.01
0.02
0.03
0.04
0.05
0.06
0.07
0.08
0.09
0.10
0.11
TEMPERATURE, K
0
100
200
300
400
500
600
700
800
900
1000
1100
1200
1300
1400
1500

FIGURE 101

SPECIFICATION TABLE 101. THERMAL DIFFUSIVITY OF [TITANIUM + ALUMINUM] ALLOYS

(Ti + Al ≥ 99.50%; impurity ≤0.20% each)

Cur. No.	Ref. No.	Author(s)	Year	Temp. Range, K	Reported Error, %	Name and Specimen Designation	Composition (weight percent) Ti	Al	Composition (continued), Specifications, and Remarks
1	94	Kapustina, M. I., Karnaushenko, N. A., Savchenko, A. M., and Kuz'min, V. I.	1961	373-1298		Ti alloy 48-OT-3	95.8/ 96.3	3.5/ 4.0	0.1 O, <0.1 N, and traces H (Ti by difference); cylindrical specimen 100 mm in diameter and 400 mm long; opening 5-8 mm in diameter drilled along specimen axis to midpoint of specimen length; ends insulated with asbestos sheet; covered with protective coatings; uniformly heated along circumference and length; diffusivity determined from temperatures measured at surface and center of specimen; 12 tests conducted for each measurement of diffusivity.
2	53	Karagezyan, A. G.	1962	403-1093		BT-5	95	5	Cylindrical rod specimen 3 mm in diameter and 300 mm long; vacuum annealed for 5 hrs at 993.2 K; measured in a vacuum of ~10^{-4} mm Hg; electrical resistivity reported as 154, 159, 160, 172, 179, 186, 186, 187, 188, and 188 μohm cm at 296.2, 362.2, 392.2, 510.2, 665.2, 790.2, 837.2, 901.2, 1013.2, and 1125.2 K, respectively.

DATA TABLE 101. THERMAL DIFFUSIVITY OF [TITANIUM + ALUMINUM] ALLOYS

(Ti + Al ≥ 99.50%; impurity ≤0.20% each)

[Temperature, T, K; Thermal Diffusivity, α, cm^2 s^{-1}]

T	α	T	α	T	α
CURVE 1		CURVE 1 (cont.)		CURVE 2 (cont.)	
373	0.0389	1173	0.0519	848	0.0637
423	0.0394	1198	0.0442	900	0.0440
473	0.0400	1213	0.0367	1093	0.0295
523	0.0408	1223	0.0303		
573	0.0417	1233	0.0247		
623	0.0425	1243	0.0208		
673	0.0433	1253	0.0139		
723	0.0444	1273	0.0294		
773	0.0456	1298	0.0672		
823	0.0469				
873	0.0489	CURVE 2			
923	0.0511				
973	0.0533	403	0.0467		
1023	0.0561	530	0.0450		
1073	0.0586	663	0.0530		
1123	0.0586	788	0.0695		

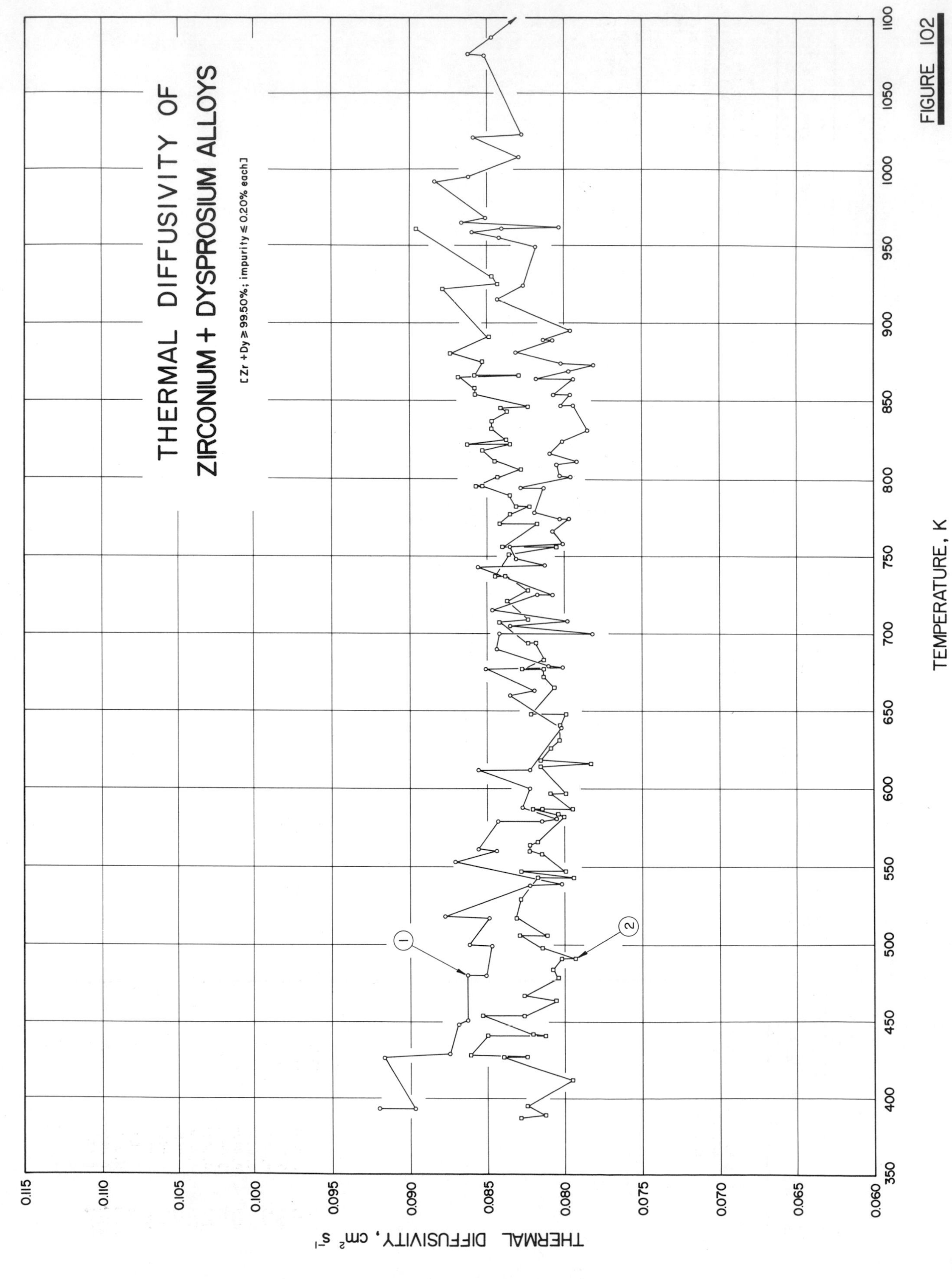
THERMAL DIFFUSIVITY OF
ZIRCONIUM + DYSPROSIUM ALLOYS
[Zr + Dy ≥ 99.50%; impurity ≤ 0.20% each]
THERMAL DIFFUSIVITY, cm² s⁻¹
TEMPERATURE, K
0.115
0.110
0.105
0.100
0.095
0.090
0.085
0.080
0.075
0.070
0.065
0.060
350
400
450
500
550
600
650
700
750
800
850
900
950
1000
1050
1100
1
2

FIGURE 102

SPECIFICATION TABLE 102. THERMAL DIFFUSIVITY OF [ZIRCONIUM + DYSPROSIUM] ALLOYS

(Zr + Dy ≥ 99.50%; impurity ≤ 0.20% each)

Cur. No.	Ref. No.	Author(s)	Year	Temp. Range, K	Reported Error, %	Name and Specimen Designation	Composition (weight percent) Zr	Dy	Composition (continued), Specifications, and Remarks
1	193	Pollard, E.R., Jr.	1963	393-1115			Bal.	2.5	Cylindrical specimen about 2 in. long, 0.125 in. diameter; diffusivity measured using modified Angström method.
2	193	Pollard, E.R., Jr.	1963	387-961			Bal.	12	Other conditions same as above.

DATA TABLE 102. THERMAL DIFFUSIVITY OF [ZIRCONIUM + DYSPROSIUM] ALLOYS

(Zr + Dy ≥ 99.50%; impurity ≤ 0.20% each)

[Temperature, T, K; Thermal Diffusivity, α, cm² s⁻¹]

T	α	T	α	T	α	T	α	T	α	T	α	T	α
CURVE 1		CURVE 1 (cont.)		CURVE 1 (cont.)		CURVE 1 (cont.)		CURVE 2 (cont.)		CURVE 2 (cont.)		CURVE 2 (cont.)	
393	0.0920	678	0.0801	847	0.0802	1087	0.0847	543	0.0817	694	0.0818	843	0.0837
393	0.0897	679	0.0810	854	0.0796	1115	0.0811*	543	0.0794	694	0.0823	845	0.0841
426	0.0917	690	0.0844	854	0.0807			547	0.0828	707	0.0842	846	0.0823
429	0.0875	700	0.0842	864	0.0794	CURVE 2		547	0.0799	709	0.0823	854	0.0858
448	0.0869	700	0.0782	864	0.0818			558	0.0814	721	0.0837	858	0.0858
451	0.0863	705	0.0835	869	0.0797	387	0.0828	560	0.0822	728	0.0823	865	0.0869
480	0.0863	708	0.0798	873	0.0781	389	0.0812	564	0.0822	737	0.0838	866	0.0829
480	0.0851	715	0.0847	874	0.0802	395	0.0824	566	0.0817	737	0.0845	866	0.0858
499	0.0847	725	0.0817	881	0.0831	412	0.0795	582	0.0800	751	0.0836	875	0.0853
500	0.0862	725	0.0807	889	0.0807	427	0.0840	584	0.0804	756	0.0805	880	0.0874
517	0.0849	743	0.0856	889	0.0813	427	0.0824	587	0.0820	756	0.0840	891	0.0849
518	0.0878	744	0.0812	895	0.0796	428	0.0861	587	0.0814	771	0.0817	922	0.0879
538	0.0822	748	0.0831	915	0.0843	441	0.0850	587	0.0795	771	0.0842	925	0.0843
539	0.0809	756	0.0835	924	0.0826	441	0.0812	597	0.0809	777	0.0835	930	0.0847
553	0.0871	758	0.0801	949	0.0818	442	0.0820	597	0.0799	782	0.0822	961	0.0896
560	0.0844	766	0.0807	955	0.0842	454	0.0853	614	0.0815	782	0.0831		
561	0.0856	774	0.0797	959	0.0860	454	0.0826	616	0.0783	789	0.0835		
579	0.0843	774	0.0803	961	0.0840	464	0.0805	618	0.0815	795	0.0853		
579	0.0814	778	0.0819	962	0.0803	466	0.0826	626	0.0808	795	0.0857		
581	0.0805	794	0.0813	965	0.0867	479	0.0804	631	0.0803	801	0.0843		
588	0.0827	794	0.0828	968	0.0851	484	0.0807	641	0.0803	806	0.0828		
600	0.0822	801	0.0796	992	0.0884	491	0.0802	648	0.0799	811	0.0845		
612	0.0856	802	0.0803	995	0.0862	491	0.0793	648	0.0821	818	0.0853		
612	0.0822	809	0.0805	1008	0.0829	498	0.0814	665	0.0806	822	0.0835		
639	0.0802	811	0.0792	1021	0.0854	506	0.0829	672	0.0813	822	0.0863		
660	0.0835	816	0.0809	1023	0.0827	506	0.0811	677	0.0813	825	0.0837		
663	0.0819	824	0.0801	1075	0.0852	517	0.0831	677	0.0827	832	0.0847		
677	0.0851	831	0.0785	1076	0.0862	529	0.0828	683	0.0813	837	0.0847		
		847	0.0794										

* Not shown in figure.

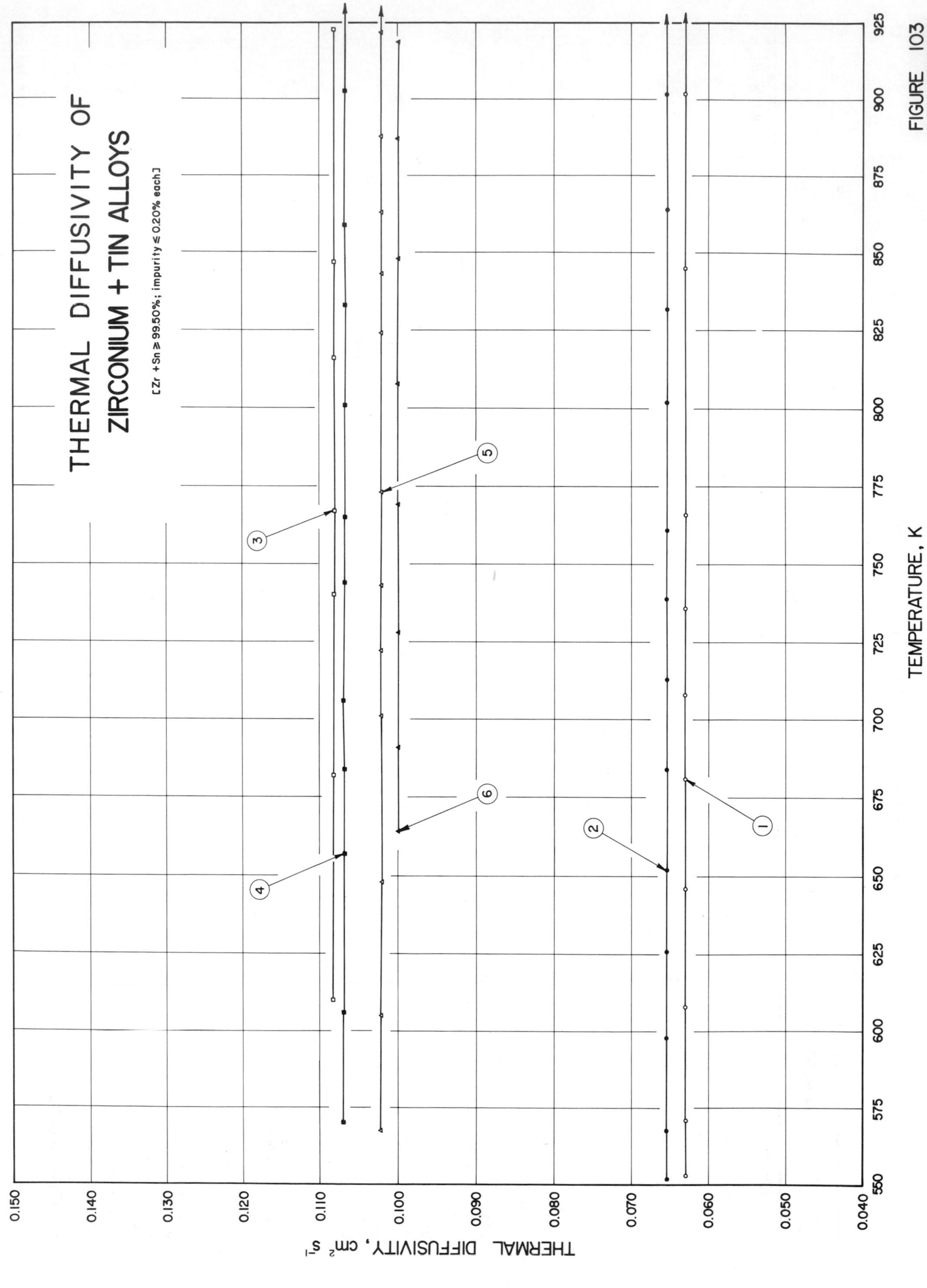
THERMAL DIFFUSIVITY OF
ZIRCONIUM + TIN ALLOYS
[Zr + Sn ≥ 99.50%; impurity ≤ 0.20% each]
THERMAL DIFFUSIVITY, cm² s⁻¹
TEMPERATURE, K
0.150
0.140
0.130
0.120
0.110
0.100
0.090
0.080
0.070
0.060
0.050
0.040
550
575
600
625
650
675
700
725
750
775
800
825
850
875
900
925
1
2
3
4
5
6

FIGURE 103

SPECIFICATION TABLE 103. THERMAL DIFFUSIVITY OF [ZIRCONIUM + TIN] ALLOYS

(Zr + Sn ≥ 99.50%; impurity ≤ 0.20%)

Cur. No.	Ref. No.	Author(s)	Year	Temp. Range, K	Reported Error, %	Name and Specimen Designation	Composition (weight percent) Zr	Sn	Composition (continued), Specifications, and Remarks
1	208	Wheeler, M.J.	1970	553-947		Zircaloy 2	Bal.	1.5	Disk specimen 0.6 cm in diameter, 0.052 cm in thickness; punched from rolled plate; diffusivity measured using modulated electron beam technique.
2	208	Wheeler, M.J.	1970	552-960		Zircaloy 2	Bal.	1.5	0.102 cm in thickness; other conditions same as above.
3	208	Wheeler, M.J.	1970	610-923		Zircaloy 2	Bal.	1.5	0.301 cm in thickness; other conditions same as above.
4	208	Wheeler, M.J.	1970	571-938		Zircaloy 2	Bal.	1.5	0.199 cm in thickness; other conditions same as above.
5	208	Wheeler, M.J.	1970	567-948		Zircaloy 2	Bal.	1.5	The above specimen ground to a new thickness of 0.151 cm; other conditions same as above.
6	208	Wheeler, M.J.	1970	664-919		Zircaloy 2	Bal.	1.5	The above specimen reduced to 0.103 cm in thickness; other conditions same as above.

DATA TABLE 103. THERMAL DIFFUSIVITY OF [ZIRCONIUM + TIN] ALLOYS

(Zr + Sn ≥ 99.50%; impurity ≤ 0.20% each)

[Temperature, T, K; Thermal Diffusivity, α, cm^2 s^{-1}]

T	α
CURVE 1	
553	0.0630
566	0.0630
608	0.0630
646	0.0630
682	0.0630
713	0.0630
736	0.0630
766	0.0630
845	0.0630
902	0.0630
947	0.0630*
CURVE 2	
552	0.0654
568	0.0654
598	0.0654
626	0.0654
652	0.0654
684	0.0654
713	0.0654
739	0.0654
761	0.0654
802	0.0654
832	0.0654
864	0.0654
902	0.0654
960	0.0654*
CURVE 3	
610	0.1083
682	0.1083
740	0.1083
767	0.1083
816	0.1083
847	0.1083
923	0.1083
CURVE 4	
571	0.1069
606	0.1069
657	0.1069
CURVE 4 (cont.)	
684	0.1069
712	0.1069
744	0.1069
765	0.1069
801	0.1069
833	0.1069
859	0.1069
903	0.1069
938	0.1069*
CURVE 5	
567	0.1023
610	0.1023
648	0.1023
701	0.1023
722	0.1023
743	0.1023
773	0.1023
824	0.1023
843	0.1023
863	0.1023
888	0.1023
922	0.1023
948	0.1023*
CURVE 6	
664	0.1001
691	0.1001
728	0.1001
769	0.1001
808	0.1001
848	0.1001
887	0.1001
919	0.1001

* Not shown in figure.

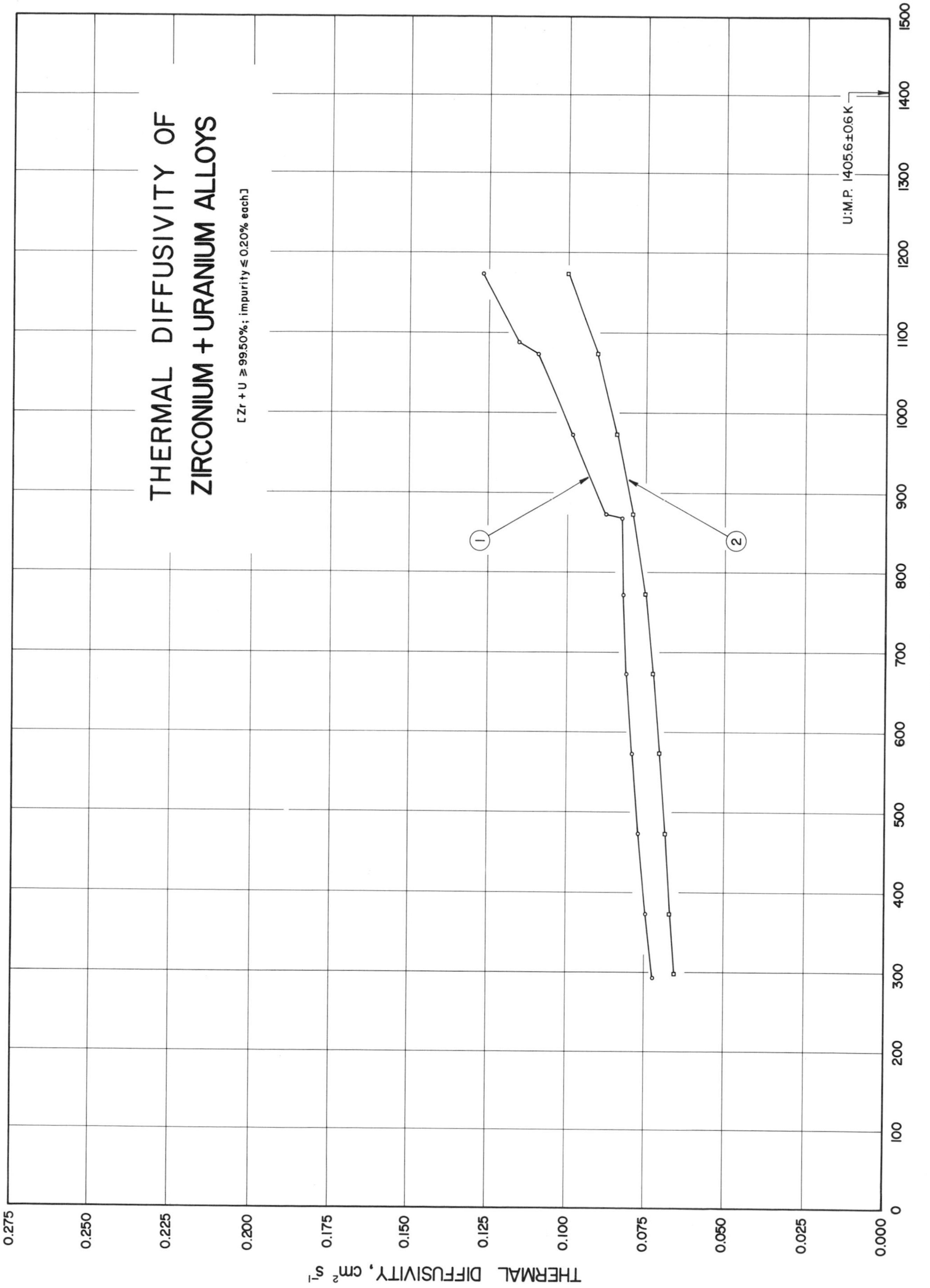
THERMAL DIFFUSIVITY OF
ZIRCONIUM + URANIUM ALLOYS
[Zr + U ≥ 99.50%; impurity ≤ 0.20% each]
U:M.P. 1405.6±0.6 K
1
2
TEMPERATURE, K
THERMAL DIFFUSIVITY, cm² s⁻¹
0
100
200
300
400
500
600
700
800
900
1000
1100
1200
1300
1400
1500
0.000
0.025
0.050
0.075
0.100
0.125
0.150
0.175
0.200
0.225
0.250
0.275

FIGURE 104

SPECIFICATION TABLE 104. THERMAL DIFFUSIVITY OF [ZIRCONIUM + URANIUM] ALLOYS

(Zr + U ≥ 99.50%; impurity ≤ 0.20% each)

Cur. No.	Ref. No.	Author(s)	Year	Temp. Range, K	Reported Error, %	Name and Specimen Designation	Composition (weight percent) Zr	U	Composition (continued), Specifications, and Remarks
1	130	Nakata, M. M., Ambrose, C. J., and Finch, R. A.	1966	293-1173	<±5	Unhydrided Alloy	90	10	Same material as that used in SNAP reactors; unhydrided; four specimens: two measured at Atomics International and two measured at Battelle Memorial Institute; AI specimens disc-shaped 0.64 cm in dia. and ~0.25 cm thick each; BMI specimens disc-shaped 0.95 cm in dia. and ~0.15 cm thick each; fabricated employing the normal production processes for SNAP fuel element fabrication except that natural isotopic uranium was used instead of enriched; fuel alloy triple arc melted, extruded, and then machined into final specimen sizes; machined so that heat flux during diffusivity measurement would be in the direction of extrusion of billet (with the exception of one BMI specimen which was measured perpendicular to the direction of extrusion); density 6.95 g cm^{-3}; electrical resistivity reported as 73.7, 84.7, 97.9, 109.2, 119.0, 125.7, 125.0, 126.3, 120.4, 114.4, and 116.6 μohm cm at 294.2, 373.2, 473.2, 573.2, 673.2, 773.2, 873.2, 973.2, 1073.2, 1173.2, and 1223.2 K, respectively (data obtained from smoothed curve); measured in vacuum; flash technique used to measure diffusivity; data points reported represent weighted best data resulting from AI and BMI measurements.
2	133	Ambrose, C. J., Taylor, R. E., and Finch, R. A.	1964	298-1173		0.00 H/Zr	90	10	Metallic impurity not exceeding 0.25; unhydrided; disc-shaped specimen 0.25 in. in dia. and 0.125 in. thick; material prepared from triple arc melted and extruded alloy; machined to size after final extrusion; laser beam used as pulse energy source; energy pulse radiated to front face of specimen and resulting temp. history of rear face used to determine diffusivity.

DATA TABLE 104. THERMAL DIFFUSIVITY OF [ZIRCONIUM + URANIUM] ALLOYS

(Zr + U ≥ 99.50%; impurity ≤ 0.20% each)

[Temperature, T, K; Thermal Diffusivity, α, cm^2 s^{-1}]

T	α	T	α	T	α
CURVE 1		CURVE 1 (cont.)		CURVE 2 (cont.)	
293	0.0720	1073	0.1095	573	0.0704
373	0.0745	1088	0.1155	673	0.0725
473	0.0770	1173	0.1270	773	0.0752
573	0.0790			873	0.0791
673	0.0810	CURVE 2		973	0.0842
773	0.0820			1073	0.0904
868	0.0825	298	0.0654	1173	0.101
873	0.0875	373	0.0668		
973	0.0985	473	0.0685		

3. NONFERROUS MULTIPLE ALLOYS

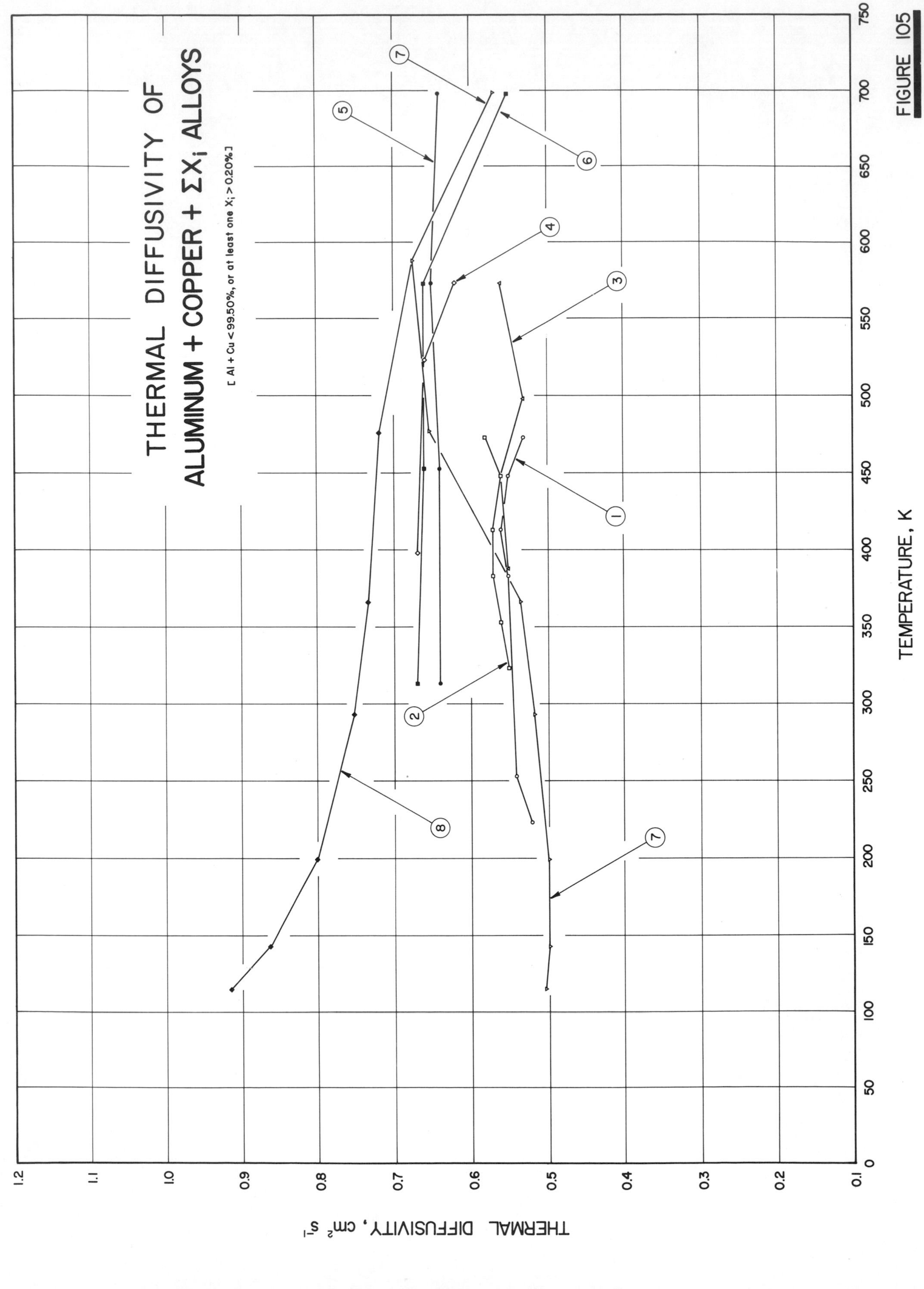
THERMAL DIFFUSIVITY OF
ALUMINUM + COPPER + ΣXi ALLOYS
[Al + Cu < 99.50%, or at least one Xi > 0.20%]
THERMAL DIFFUSIVITY, cm² s⁻¹
TEMPERATURE, K
1.2
1.1
1.0
0.9
0.8
0.7
0.6
0.5
0.4
0.3
0.2
0.1
0
50
100
150
200
250
300
350
400
450
500
550
600
650
700
750
1
2
3
4
5
6
7
7
8

FIGURE 105

SPECIFICATION TABLE 105. THERMAL DIFFUSIVITY OF [ALUMINUM + COPPER + ΣX_i] ALLOYS

(Al + Cu < 99.50% or at least one X_i > 0.20%)

Cur. No.	Ref. No.	Author(s)	Year	Temp. Range, K	Reported Error, %	Name and Specimen Designation	Composition (weight percent) Al	Cu	Cr	Fe	Mg	Mn	Si	Zn	Composition (continued), Specifications, and Remarks
1	19, 20	Butler, C. P. and Inn, E. C. Y.	1957	323-473	5-10	2024-T86	→	3.8/ 4.9	0.10 max	0.50 max	1.2/ 1.8	0.30/ 0.9	0.50 max	0.25 max	90.90-93.20 Al (by difference), and 0.15 max others; nominal composition from Alcoa Aluminum Handbook, p. 44, 1962; cylindrical specimen 0.9525 cm in diameter and length lying in the range from 1 to 2.5 cm; nominal heat treatment (from above source): solution heat treatment, strain hardening, and then artificial aging; subjected to irradiance from carbon arc lamp heat source; spectral distribution approximates that of a 5700 K black body source; specimen blackened with camphor black; measured under a vacuum of ~5 microns.
2	19, 20	Butler, C. P. and Inn, E. C. Y.	1957	323-473	5-10	2024-T86									Above specimen allowed to cool to ambient and then exposed to the arc lamp to measure diffusivity again.
3	19, 20	Butler, C. P. and Inn, E. C. Y.	1957	388-573	5-10	2024-T86									Above specimen allowed to cool to ambient and then exposed to the arc lamp to measure diffusivity again.
4	19, 20	Butler, C. P. and Inn, E. C. Y.	1957	398-573	5-10	2024-T86									Above specimen allowed to cool to ambient and then exposed to the arc lamp to measure diffusivity again.
5	19, 20	Butler, C. P. and Inn, E. C. Y.	1957	313-698	5-10	2024-T86									Above specimen allowed to cool to ambient and then exposed to the arc lamp to measure diffusivity again.
6	19, 20	Butler, C. P. and Inn, E. C. Y.	1957	313-698	5-10	2024-T86									Above specimen allowed to cool to ambient and then exposed to the arc lamp to measure diffusivity again.
7	87, 17	Lucks, C. F., Deem, H. W., Thompson, H. B., Smith, A. R., Curry, F. P., and Bing, G. F.	1951	116-700		2024-T4		4.5			1.5	0.6			Supplied by Aluminum Co. of America; condition as received T4 (solution heat treatment followed by natural aging at room temperature to a substantially stable condition); thermal diffusivity calculated from measured conductivity, specific heat, and density.
8	87, 17	Lucks, C. F., et al.	1951	116-700		2024-T4									Above alloy measured again for thermal conductivity after being heated above 574.8 K; thermal diffusivity calculated from measured conductivity, specific heat, and density.

DATA TABLE 105. THERMAL DIFFUSIVITY OF [ALUMINUM + COPPER + ΣX_i] ALLOYS

($Al + Cu < 99.50\%$ or at least one $X_i > 0.20\%$)

[Temperature, T, K; Thermal Diffusivity, α, $cm^2\ s^{-1}$]

T	α	T	α
CURVE 1		CURVE 7	
323.2	0.52	115.5	0.503
353.2	0.54	143.5	0.498
383.2	0.55	199.0	0.498
413.2	0.56	293.2	0.516
448.2	0.55	366.5	0.534
473.2	0.53	477.6	0.653
		588.7	0.674
CURVE 2		699.8	0.568
323.2	0.55	CURVE 8	
353.2	0.56		
383.2	0.57	115.5	0.914
413.2	0.57	143.5	0.862
448.2	0.56	199.0	0.800
473.2	0.58	293.2	0.751
		366.5	0.733
CURVE 3		477.6	0.718
		588.7	0.674*
388.2	0.55*	699.8	0.568*
448.2	0.56*		
498.2	0.53		
573.2	0.56		
CURVE 4			
398.2	0.67		
523.2	0.66		
573.2	0.62		
CURVE 5			
313.2	0.64		
453.2	0.64		
573.2	0.65		
698.2	0.54		
CURVE 6			
313.2	0.67		
453.2	0.66		
573.2	0.66		
698.2	0.55		

* Not shown in figure.

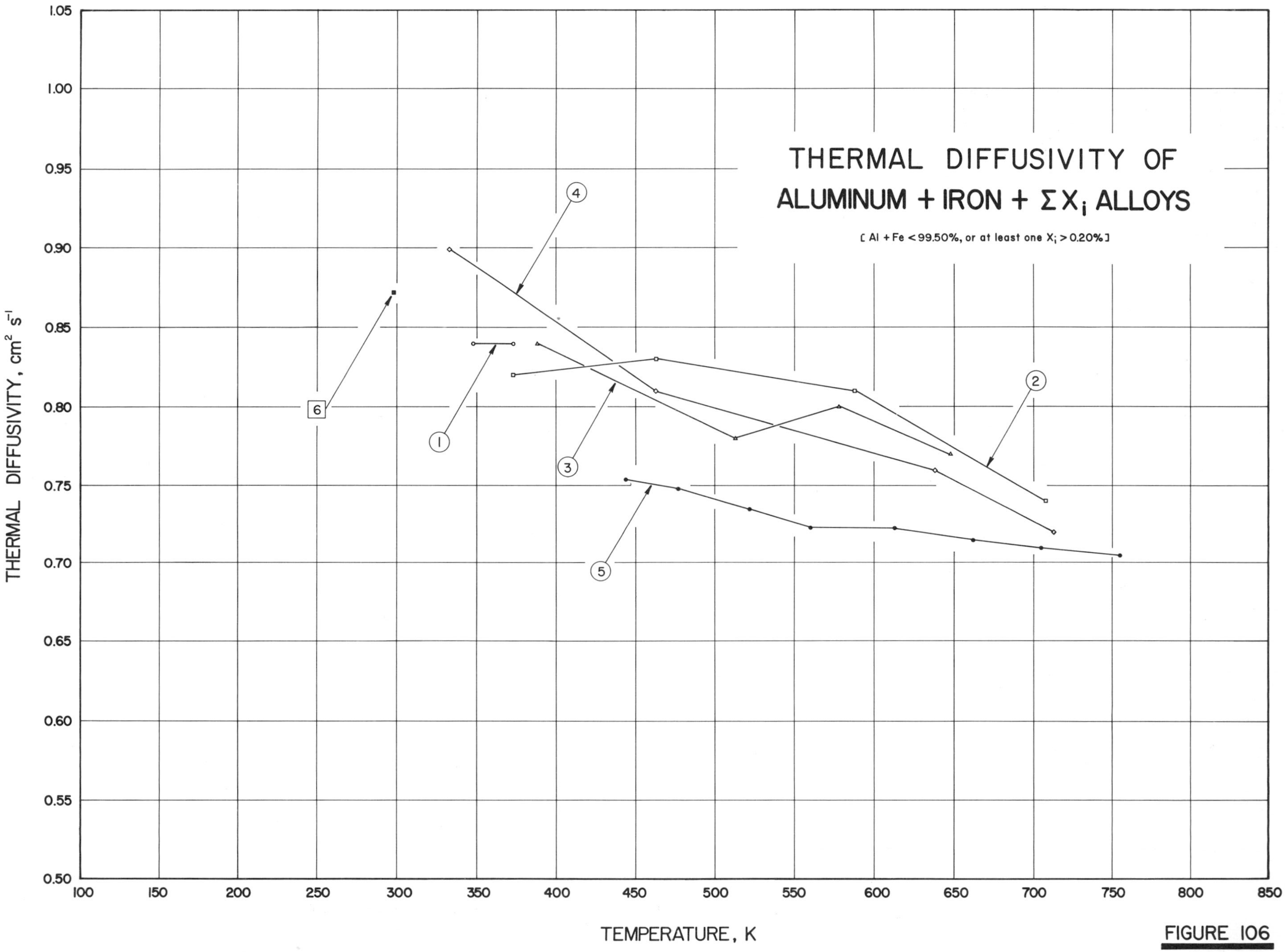
THERMAL DIFFUSIVITY OF
ALUMINUM + IRON + Σ X_i ALLOYS
[Al + Fe < 99.50%, or at least one X_i > 0.20%]
THERMAL DIFFUSIVITY, cm² s⁻¹
TEMPERATURE, K
1.05
1.00
0.95
0.90
0.85
0.80
0.75
0.70
0.65
0.60
0.55
0.50
100
150
200
250
300
350
400
450
500
550
600
650
700
750
800
850
1
2
3
4
5
6

FIGURE 106

SPECIFICATION TABLE 106. THERMAL DIFFUSIVITY OF [ALUMINUM + IRON + ΣX_i] ALLOYS

(Al + Fe < 99.50% or at least one $X_i > 0.20\%$)

Cur. No.	Ref. No.	Author(s)	Year	Temp. Range, K	Reported Error, %	Name and Specimen Designation	Composition (weight percent) Al	Fe	Cu	Mn	Si	Zn	Composition (continued), Specifications, and Remarks
1	19, 20	Butler, C. P. and Inn, E. C. Y.	1957	348, 373	5-10	1100-F	98.5 min	→	0.2 max	0.05 max	→	0.1 max	1.0 max total Fe and Si, and 0.15 max others; nominal composition from Alcoa Aluminum Handbook, p. 44, 1962 (aluminum percentage obtained by difference); cylindrical specimen 0.9525 cm in diameter and length lying in the range from 1 to 2.5 cm; nominal heat treatment: as fabricated; subjected to irradiance from carbon arc lamp heat source; spectral distribution approximates that of a 5700 K black body source; specimen blackened with camphor black; measured under a vacuum of ~5 microns.
2	19, 20	Butler, C. P. and Inn, E. C. Y.	1957	373-708	5-10	1100-F							Above specimen allowed to cool to ambient and then exposed to the arc lamp to measure diffusivity again.
3	19, 20	Butler, C. P. and Inn, E. C. Y.	1957	388-648	5-10	1100-F							Above specimen allowed to cool to ambient and then exposed to the arc lamp to measure diffusivity again.
4	19, 20	Butler, C. P. and Inn, E. C. Y.	1957	333-713	5-10	1100-F							Above specimen allowed to cool to ambient and then exposed to the arc lamp to measure diffusivity again.
5	21	El-Hifni, M. A. and Chao, B. T.	1956	444-756	2-3	2S Al	99.0						Tubular specimen 0.875 in. O. D. and 5 in. long, relatively thin walled; another tube of similar material mounted concentrically with specimen to minimize heat losses; one end of specimen-and-shield assembly immersed in a liquid heating bath, while the other end supported by a transite disc for insulation; cyclic varying current generates required temperature wave at the heating bath; heat supplied by electric current removed by a forced draft of cooling air; one-dimensional heat flow; a minimum of three complete temperature waves recorded.
6	7	di Novi, R. A.	1963	298.2	10	Al (2S)	→	→	0.20 max	0.05 max	→	0.10 max	98.50 min Al (by difference), 1.0 max total Fe and Si, and 0.15 max others; nominal composition from Metals Handbook, Vol. 1, 8th ed., p. 917, 1961; specimen with thickness lying in the range from 1 to 2 mm; front surface uniformly irradiated by a very short pulse of radiant energy supplied by a xenon flash tube; diffusivity determined from measured history of the back surface temperature; temperature at which specimen was measured not given by author but assumed to be room temperature.

DATA TABLE 106. THERMAL DIFFUSIVITY OF [ALUMINUM + IRON + ΣX_i] ALLOYS

(Al + Fe < 99.50% or at least one X_i > 0.20%)

[Temperature, T, K; Thermal Diffusivity, α, $cm^2\ s^{-1}$]

T	α
CURVE 1	
348.2	0.84
373.2	0.84
CURVE 2	
373.2	0.82
463.2	0.83
588.2	0.81
708.2	0.74
CURVE 3	
388.2	0.84
513.2	0.78
578.2	0.80
648.2	0.77
CURVE 4	
333.2	0.90
463.2	0.81
638.2	0.76
713.2	0.72
CURVE 5	
444	0.754
478	0.748
522	0.735
561	0.723
614	0.723
662	0.715
705	0.710
756	0.705
CURVE 6	
298.2	0.872

SPECIFICATION TABLE 107. THERMAL DIFFUSIVITY OF [ALUMINUM + MAGNESIUM + ΣX_i] ALLOYS

(Al + Mg < 99.50% or at least one X_i > 0.20%)

Cur. No.	Ref. No.	Author(s)	Year	Temp. Range, K	Reported Error, %	Name and Specimen Designation	Composition (weight percent) Al	Mg	Si	Cu	Cr	Composition (continued), Specifications, and Remarks
1*	271	Rosenthal, D. and Friedmann, N. E.	1954	547.5		Aluminum	Bal.	1.0	0.6	0.25	0.25	Diffusivity measured using modified Angström method.

DATA TABLE 107. THERMAL DIFFUSIVITY OF [ALUMINUM + MAGNESIUM + ΣX_i] ALLOYS

(Al + Mg < 99.50% or at least one X_i > 0.20%)

[Temperature, T, K; Thermal Diffusivity, α, $cm^2\ s^{-1}$]

T	α
CURVE 1*	
547.5	0.673

* No figure given.

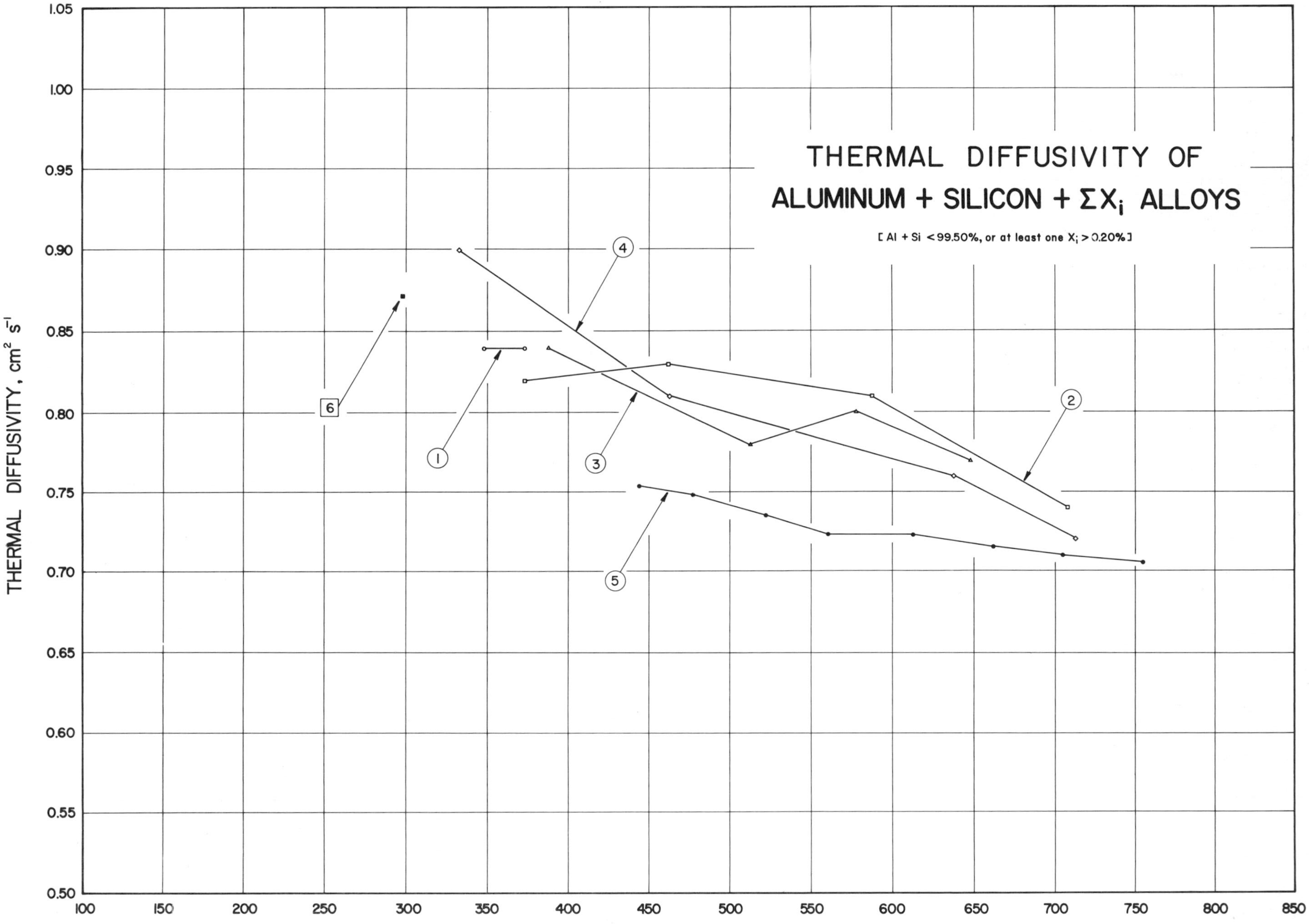
THERMAL DIFFUSIVITY OF
ALUMINUM + SILICON + ΣXi ALLOYS
[Al + Si < 99.50%, or at least one Xi > 0.20%]
THERMAL DIFFUSIVITY, cm2 s-1
TEMPERATURE, K
1.05
1.00
0.95
0.90
0.85
0.80
0.75
0.70
0.65
0.60
0.55
0.50
100
150
200
250
300
350
400
450
500
550
600
650
700
750
800
850
1
2
3
4
5
6

FIGURE 108

SPECIFICATION TABLE 108. THERMAL DIFFUSIVITY OF [ALUMINUM + SILICON + ΣX_i] ALLOYS

(Al + Si < 99.50% or at least one X_i > 0.20%)

Cur. No.	Ref. No.	Author(s)	Year	Temp. Range, K	Reported Error, %	Name and Specimen Designation	Composition (weight percent) Al	Si	Cu	Fe	Mn	Zn	Composition (continued), Specifications, and Remarks
1	19, 20	Butler, C. P. and Inn, E. C. Y.	1957	348, 373	5-10	1100-F	98.5 min	→	0.2 max	→	0.05 max	0.1 max	1.0 max total Si and Fe, and 0.15 max others; nominal composition from Alcoa Aluminum Handbook, p. 44, 1962 (aluminum percentage obtained by difference); cylindrical specimen 0.9525 cm in diameter and length lying in the range from 1 to 2.5 cm; nominal heat treatment: as fabricated; subjected to irradiance from carbon arc lamp heat source; spectral distribution approximates that of a 5700 K black body source; specimen blackened with camphor black; measured under a vacuum of ~5 microns.
2	19, 20	Butler, C. P. and Inn, E. C. Y.	1957	373-708	5-10	1100-F							Above specimen allowed to cool to ambient and then exposed to the arc lamp to measure diffusivity again.
3	19, 20	Butler, C. P. and Inn, E. C. Y.	1957	388-648	5-10	1100-F							Above specimen allowed to cool to ambient and then exposed to the arc lamp to measure diffusivity again.
4	19, 20	Butler, C. P. and Inn, E. C. Y.	1957	333-713	5-10	1100-F							Above specimen allowed to cool to ambient and then exposed to the arc lamp to measure diffusivity again.
5	21	El-Hifni, M. A. and Chao, B. T.	1956	444-756	2-3	2S Al	99.0						Tubular specimen 0.875 in. O. D. and 5 in. long, relatively thin walled; another tube of similar material mounted concentrically with specimen to minimize heat losses; one end of specimen-and-shield assembly immersed in a liquid heating bath, while the other end supported by a transite disc for insulation; cyclic varying current generates required temperature wave at the heating bath; heat supplied by electric current removed by a forced draft of cooling air; one-dimensional heat flow; a minimum of three complete temperature waves recorded.
6	7	di Novi, R. A.	1963	298.2	10	Al (2S)	98.50 min	→	0.20 max	→	0.05 max	0.10 max	1.0 max total Si and Fe, and 0.15 max others; nominal composition from Metals Handbook, Vol. 1, 8th ed., p. 917, 1961; aluminum percentage obtained by difference; specimen with thickness lying in the range from 1 to 2 mm; front surface uniformly irradiated by a very short pulse of radiant energy supplied by a xenon flash tube; diffusivity determined from measured history of the back surface temperature; temperature at which specimen was measured not given by author but assumed to be room temperature.

DATA TABLE 108. THERMAL DIFFUSIVITY OF [ALUMINUM + SILICON + ΣX_i] ALLOYS

(Al + Si < 99.50% or at least one X_i > 0.20%)

[Temperature, T, K; Thermal Diffusivity, α, $cm^2\ s^{-1}$]

T	α
CURVE 1	
348.2	0.84
373.2	0.84
CURVE 2	
373.2	0.82
463.2	0.83
588.2	0.81
708.2	0.74
CURVE 3	
388.2	0.84
513.2	0.78
578.2	0.80
648.2	0.77
CURVE 4	
333.2	0.90
463.2	0.81
638.2	0.76
713.2	0.72
CURVE 5	
444	0.754
478	0.748
522	0.735
561	0.723
614	0.723
662	0.715
705	0.710
756	0.705
CURVE 6	
298.2	0.872

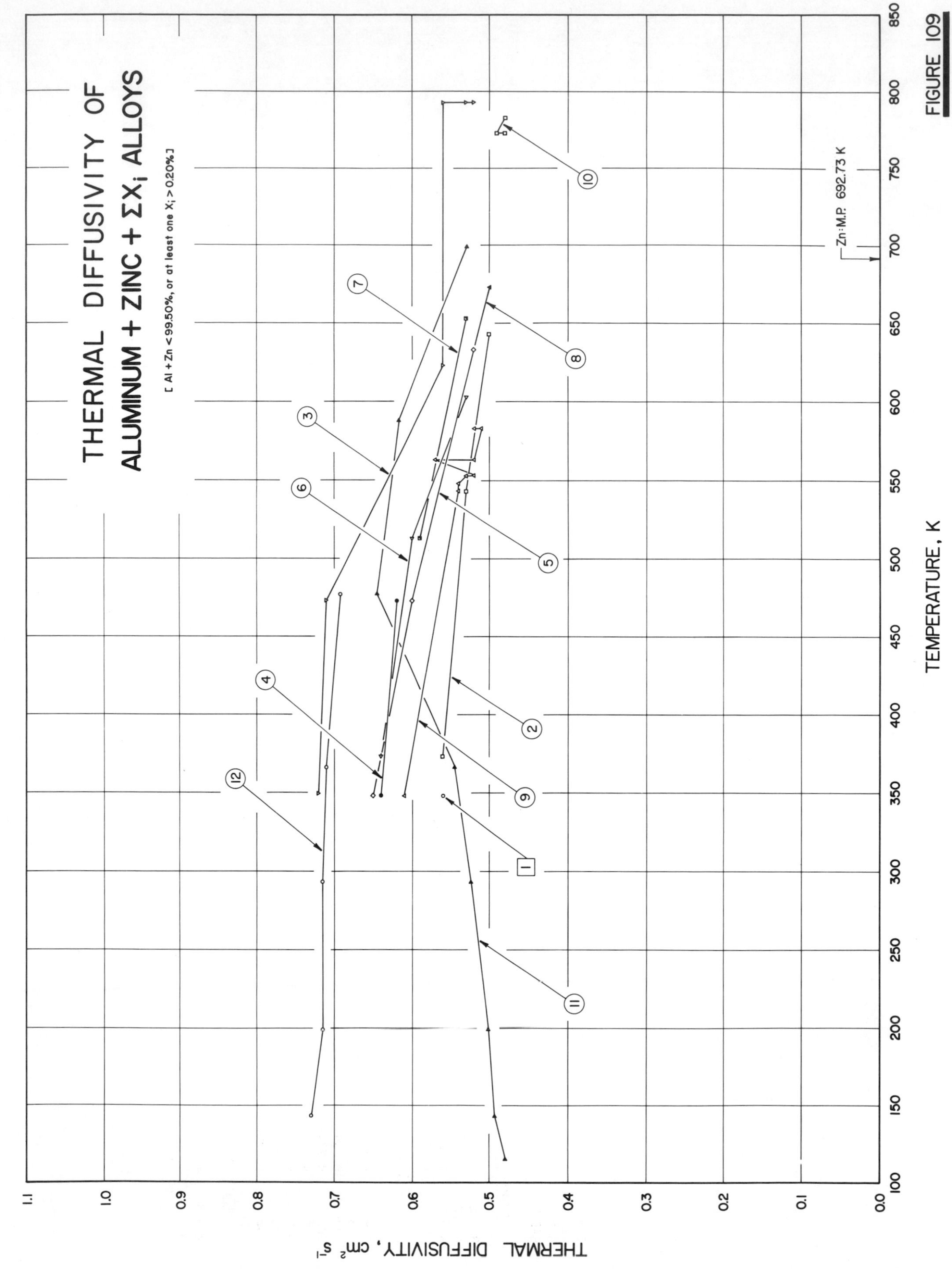
THERMAL DIFFUSIVITY OF
ALUMINUM + ZINC + ΣXᵢ ALLOYS
[Al + Zn < 99.50%, or at least one Xᵢ > 0.20%]
Zn: M.P. 692.73 K
THERMAL DIFFUSIVITY, cm² s⁻¹
TEMPERATURE, K
0.0
0.1
0.2
0.3
0.4
0.5
0.6
0.7
0.8
0.9
1.0
1.1
100
150
200
250
300
350
400
450
500
550
600
650
700
750
800
850
1
2
3
4
5
6
7
8
9
10
11
12

FIGURE 109

SPECIFICATION TABLE 109. THERMAL DIFFUSIVITY OF [ALUMINUM + ZINC + ΣX_i] ALLOYS

(Al + Zn < 99.50% or at least one X_i > 0.20%)

Cur. No.	Ref. No.	Author(s)	Year	Temp. Range, K	Reported Error, %	Name and Specimen Designation	Composition (weight percent) Al	Zn	Cr	Cu	Fe	Mg	Mn	Si	Composition (continued), Specifications, and Remarks
1	19, 20	Butler, C. P. and Inn, E. C. Y.	1957	348.2	5-10	7075-T6; 2	→	5.1/ 6.1	0.18/ 0.40	1.2/ 2.0	0.7 max	2.1/ 2.9	0.3 max	0.5 max	86.75-89.57 Al (by difference), 0.2 max Ti, and 0.15 max others; nominal composition from Alcoa Aluminum Handbook, p. 45, 1962; cylindrical specimen 0.9525 cm in diameter and length lying in the range from 1 to 2.5 cm; nominal heat treatment (from above source): solution heat treatment followed by artifical aging; subjected to irradiance from carbon arc lamp heat source; spectral distribution approximates that of a 5700 K black body source; specimen blackened with camphor black; measured under a vacuum of ~5 microns.
2	19, 20	Butler, C. P. and Inn, E. C. Y.	1957	373-643	5-10	7075-T6; 2									Above specimen allowed to cool to ambient and then exposed to the arc lamp to measure diffusivity again.
3	19, 20	Butler, C. P. and Inn, E. C. Y.	1957	348-793	5-10	7075-T6; 2									Above specimen allowed to cool to ambient and then exposed to the arc lamp to measure diffusivity again.
4	19, 20	Butler, C. P. and Inn, E. C. Y.	1957	348, 473	5-10	7075-T6									Same specimen as above but subjected to less extensive thermal exposure; exposed to the arc lamp to measure diffusivity; measured under the same conditions as above specimen.
5	19, 20	Butler, C. P. and Inn, E. C. Y.	1957	348-633	5-10	7075-T6									Above specimen allowed to cool to ambient and then exposed to the arc lamp to measure diffusivity again.
6	19, 20	Butler, C. P. and Inn, E. C. Y.	1957	373-603	5-10	7075-T6									Above specimen allowed to cool to ambient and then exposed to the arc lamp to measure diffusivity again.
7	19, 20	Butler, C. P. and Inn, E. C. Y.	1957	513, 653	5-10	7075-T6									Above specimen allowed to cool to ambient and then exposed to the arc lamp to measure diffusivity again.
8	19, 20	Butler, C. P. and Inn, E. C. Y.	1957	473, 673	5-10	7075-T6									Above specimen allowed to cool to ambient and then exposed to the arc lamp to measure diffusivity again.
9	19, 20	Butler, C. P. and Inn, E. C. Y.	1957	348-583	5-10	7075-T6									Above specimen allowed to cool to ambient and then exposed to the arc lamp to measure diffusivity again.
10	19, 20	Butler, C. P. and Inn, E. C. Y.	1957	773, 783	5-10	7075-T6									Above specimen allowed to cool to ambient and then exposed to the arc lamp to measure diffusivity again.

SPECIFICATION TABLE 109. THERMAL DIFFUSIVITY OF [ALUMINUM + ZINC + ΣX_i] ALLOYS (continued)

Cur. No.	Ref. No.	Author(s)	Year	Temp. Range, K	Reported Error, %	Name and Specimen Designation	Composition (weight percent) Al	Zn	Cr	Cu	Fe	Mg	Mn	Si	Composition (continued), Specifications, and Remarks
11	87, 17	Lucks, C. F. and Deem, H. W.	1958	116-700		7075-T6		5.6	0.3	1.6		2.5			Supplied by Aluminum Co. of America; condition as received T6; solution heat treatment followed by artificial aging; thermal diffusivity calculated from measured conductivity, specific heat, and density.
12	87, 17	Lucks, C. F. and Deem, H. W.	1958	144-700		7075-T6									Above alloy measured again for thermal conductivity after having been heated above 574.8 K; thermal diffusivity calculated from measured conductivity, specific heat, and density.

DATA TABLE 109. THERMAL DIFFUSIVITY OF [ALUMINUM + ZINC + ΣX_i] ALLOYS

(Al + Zn < 99.50% or at least one X_i > 0.20%)

[Temperature, T, K; Thermal Diffusivity, α, $cm^2\ s^{-1}$]

T	α
CURVE 1	
348.2	0.56
CURVE 2	
373.2	0.56
543.2	0.53
643.2	0.50
CURVE 3	
348.2	0.72
473.2	0.71
623.2	0.56
793.2	0.56
793.2	0.53
793.2	0.52
CURVE 4	
348.2	0.64
473.2	0.62
CURVE 5	
348.2	0.65
473.2	0.60
633.2	0.52
CURVE 6	
373.2	0.64
513.2	0.60
603.2	0.53
CURVE 7	
513.2	0.59
653.2	0.53
CURVE 8	
473.2	0.60*
673.2	0.50
CURVE 9	
348.2	0.61
543.2	0.54
548.2	0.54
553.2	0.53
553.2	0.52
563.2	0.57
563.2	0.52
583.2	0.51
583.2	0.52
CURVE 10	
773.2	0.48
773.2	0.48*
773.2	0.48*
773.2	0.49
783.2	0.48
CURVE 11	
116.4	0.480
144.2	0.493
199.8	0.501
293.2	0.524
366.4	0.545
477.6	0.645
588.7	0.617
699.8	0.529
CURVE 12	
144.2	0.730
199.8	0.715
293.2	0.715
366.4	0.710
477.6	0.692
588.7	0.617*
699.8	0.529*

* Not shown in figure.

SPECIFICATION TABLE 110. THERMAL DIFFUSIVITY OF [ANTIMONY + ΣX_i] ALLOYS

Cur. No.	Ref. No.	Author(s)	Year	Temp. Range, K	Reported Error, %	Name and Specimen Designation	Composition (weight percent) Sb	Composition (continued), Specifications, and Remarks
1*	299	Mardykin, I. P. and Filippov, L. P.	1968	1170-1429			96.8	In liquid state; measured by a radial periodic method.

DATA TABLE 110. THERMAL DIFFUSIVITY OF [ANTIMONY + ΣX_i] ALLOYS

[Temperature, T, K; Thermal Diffusivity, α, $cm^2\ s^{-1}$]

T	α
CURVE 1*	
1170	0.144
1191	0.130
1220	0.124
1255	0.124
1262	0.140
1281	0.116
1339	0.116
1366	0.112
1380	0.120
1412	0.116
1429	0.115

* No figure given.

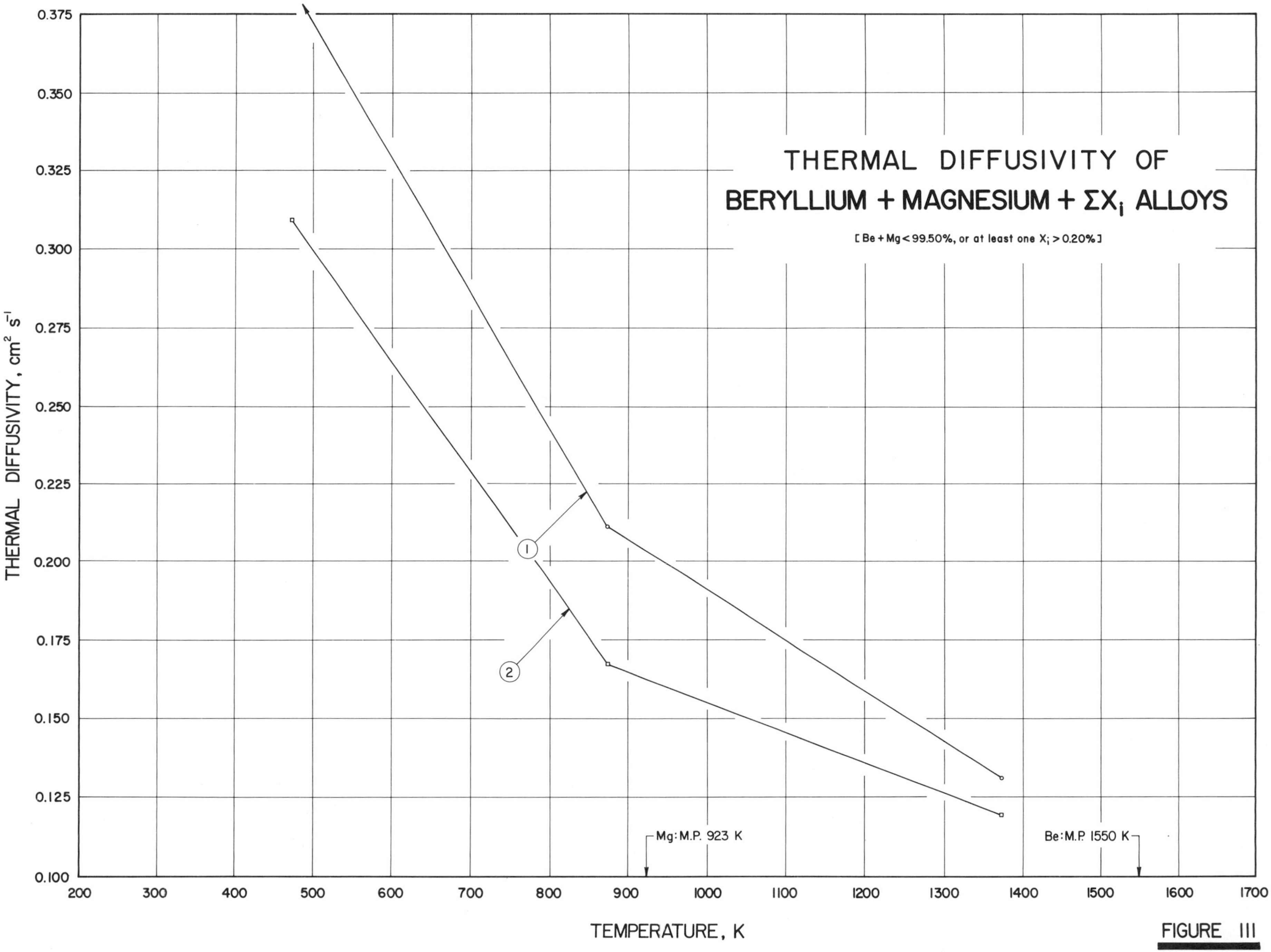

THERMAL DIFFUSIVITY OF
BERYLLIUM + MAGNESIUM + ΣX_i ALLOYS
[Be + Mg < 99.50%, or at least one X_i > 0.20%]
THERMAL DIFFUSIVITY, cm² s⁻¹
0.375
0.350
0.325
0.300
0.275
0.250
0.225
0.200
0.175
0.150
0.125
0.100
200
300
400
500
600
700
800
900
1000
1100
1200
1300
1400
1500
1600
1700
TEMPERATURE, K
Mg: M.P. 923 K
Be: M.P. 1550 K
1
2

FIGURE 111

SPECIFICATION TABLE 111. THERMAL DIFFUSIVITY OF [BERYLLIUM + MAGNESIUM + ΣX_i] ALLOYS

(Be + Mg < 99.50% or at least one X_i > 0.20%)

Cur. No.	Ref. No.	Author(s)	Year	Temp. Range, K	Reported Error, %	Name and Specimen Designation	Composition (weight percent) Be	Mg	Al	Composition (continued), Specifications, and Remarks
1	273	Chitkin, V.S.	1966	473-1373				→	→	Hot-pressed beryllium containing impurities 1.264 Mg and Al; cylindrical specimen 36 mm in diameter, 16 mm long; diffusivity measured on five samples and direction of thermal flux perpendicular to pressing direction.
2	273	Chitkin, V.S.	1966	473-1373						The above specimen except the direction of thermal flux parallel to pressing direction.

DATA TABLE 111. THERMAL DIFFUSIVITY OF [BERYLLIUM + MAGNESIUM + ΣX_i] ALLOYS

(Be + Mg < 99.50% or at least one X_i > 0.20%)

[Temperature, T, K; Thermal Diffusivity, α, $cm^2 s^{-1}$]

T	α
CURVE 1	
473	0.384*
873	0.211
1373	0.131
CURVE 2	
473	0.309
873	0.167
1373	0.119

* Not shown in figure.

SPECIFICATION TABLE 112. THERMAL DIFFUSIVITY OF [COBALT + CHROMIUM + ΣX_i] ALLOYS

(Co + Cr < 99.50% or at least one X_i > 0.20%)

Cur. No.	Ref. No.	Author(s)	Year	Temp. Range, K	Reported Error, %	Name and Specimen Designation	Co	Cr	C	Fe	Mn	Mo	Ni	P	Composition (continued), Specifications, and Remarks
							Composition (weight percent)								
1*	29, 30	McIntosh, G. E., Hamilton, D.C., and Sibbitt, W. L.	1952	253-373	5	Haynes Stellite 25	→	20.0	0.09	2.00 max	1.5	1.0 max	10.0	0.04 max	49.94 min Co (by difference), 0.03 max S, 0.40 Si, and 15.0 W; rod specimen 0.125 in. in diameter and 18 in. long; surrounded by radiation shield consisting of four concentric cylinders of very thin aluminum foil each separated by three 1/32 in. rings of balsa wood; measured after being maintained at elevated temperature for several hours; measured in vacuum; diffusivity determined from measured phase lag of the temperature wave between any two points along specimen; one-dimensional heat flow.

DATA TABLE 112. THERMAL DIFFUSIVITY OF [COBALT + CHROMIUM + ΣX_i] ALLOYS

(Co + Cr < 99.50% or at least one X_i > 0.20%)

[Temperature, T, K; Thermal Diffusivity, α, $cm^2\ s^{-1}$]

T	α
CURVE 1*	
253	0.0235
313	0.02908
313	0.02764
373	0.03899
373	0.03853
373	0.03822

* No figure given.

SPECIFICATION TABLE 113. THERMAL DIFFUSIVITY OF [COPPER + ALUMINUM + ΣX_i] ALLOYS

(Cu + Al < 99.50% or at least one X_i > 0.20%)

Cur. No.	Ref. No.	Author(s)	Year	Temp. Range, K	Reported Error, %	Name and Specimen Designation	Composition (weight percent) Cu	Al	Fe	Ni	Mn	Zn	Pb	Sn	Composition (continued), Specifications, and Remarks
1*	232	Böhm, R. and Wachtel, E.	1969	273, 373		Al-Bz	80.59	9.44	4.66	4.35	0.91	0.03	0.01	0.01	Cylindrical specimen; electrical resistivity 22.31 and 23.32 $\mu\Omega$ cm at 0 and 100 C, respectively.

DATA TABLE 113. THERMAL DIFFUSIVITY OF [COPPER + ALUMINUM + ΣX_i] ALLOYS

(Cu + Al < 99.50% or at least one X_i > 0.20%)

[Temperature, T, K; Thermal Diffusivity, α, $cm^2\ s^{-1}$]

T	α
CURVE 1*	
273	0.168
373	0.182

* No figure given.

SPECIFICATION TABLE 114. THERMAL DIFFUSIVITY OF [GOLD + ΣX_i]

Cur. No.	Ref. No.	Author(s)	Year	Temp. Range, K	Reported Error, %	Name and Specimen Designation	Composition (weight percent) Au	Composition (continued), Specifications, and Remarks
1*	124	Shanks, H. R., Burns, M. M., and Danielson, G. C.	1968	327-1229	~2	Mint Gold; 4	~95	Mint gold; cylindrical specimen 0.35 cm in dia. and 30 cm long; electrical resistivity ratio $\rho(300\ K)/\rho(4.2\ K) = 3$ (measured after specimen had been annealed at 1225 K for 1 hr); electrical resistivity measured and reported as 3.70, 3.86, 4.16, 4.43, 4.82, 5.27, 5.63, 5.80, 6.31, 6.68, 7.12, 7.50, 7.86, 8.45, 8.98, 9.36, 9.97, 10.47, 11.09, 11.68, 12.13, and 12.68 μohm cm at 299, 313, 350, 393, 433, 485, 536, 562, 611, 657, 699, 747, 797, 843, 888, 936, 988, 1036, 1085, 1135, 1183, and 1232 K, respectively; apparent Lorenz number reported as 2.28, 2.28, 2.31, 2.35, 2.39, 2.43, 2.45, 2.45, 2.46, and 2.45 x 10^{-8} V^2 K^{-2} at 325, 398, 498, 601, 699, 799, 898, 998, 1099, and 1198 K, respectively; measurements made with specimen in one atmosphere of helium and with a 2.5 K amplitude and 30 sec period temp. sine wave; modified Angström method used to measure diffusivity; each data point represents average of two diffusivity measurements at each temp. obtained from two thermocouple combinations.

DATA TABLE 114. THERMAL DIFFUSIVITY OF [GOLD + ΣX_i]

[Temperature, T, K; Thermal Diffusivity, α, $cm^2\ s^{-1}$]

T	α
CURVE 1*	
327	0.728
330	0.736
333	0.746
366	0.741
392	0.761
475	0.796
561	0.822
609	0.833
654	0.837
698	0.847
792	0.838
842	0.829
886	0.829
936	0.822
985	0.806
1030	0.795
1079	0.778
1129	0.764
1180	0.745
1229	0.727

* No figure given.

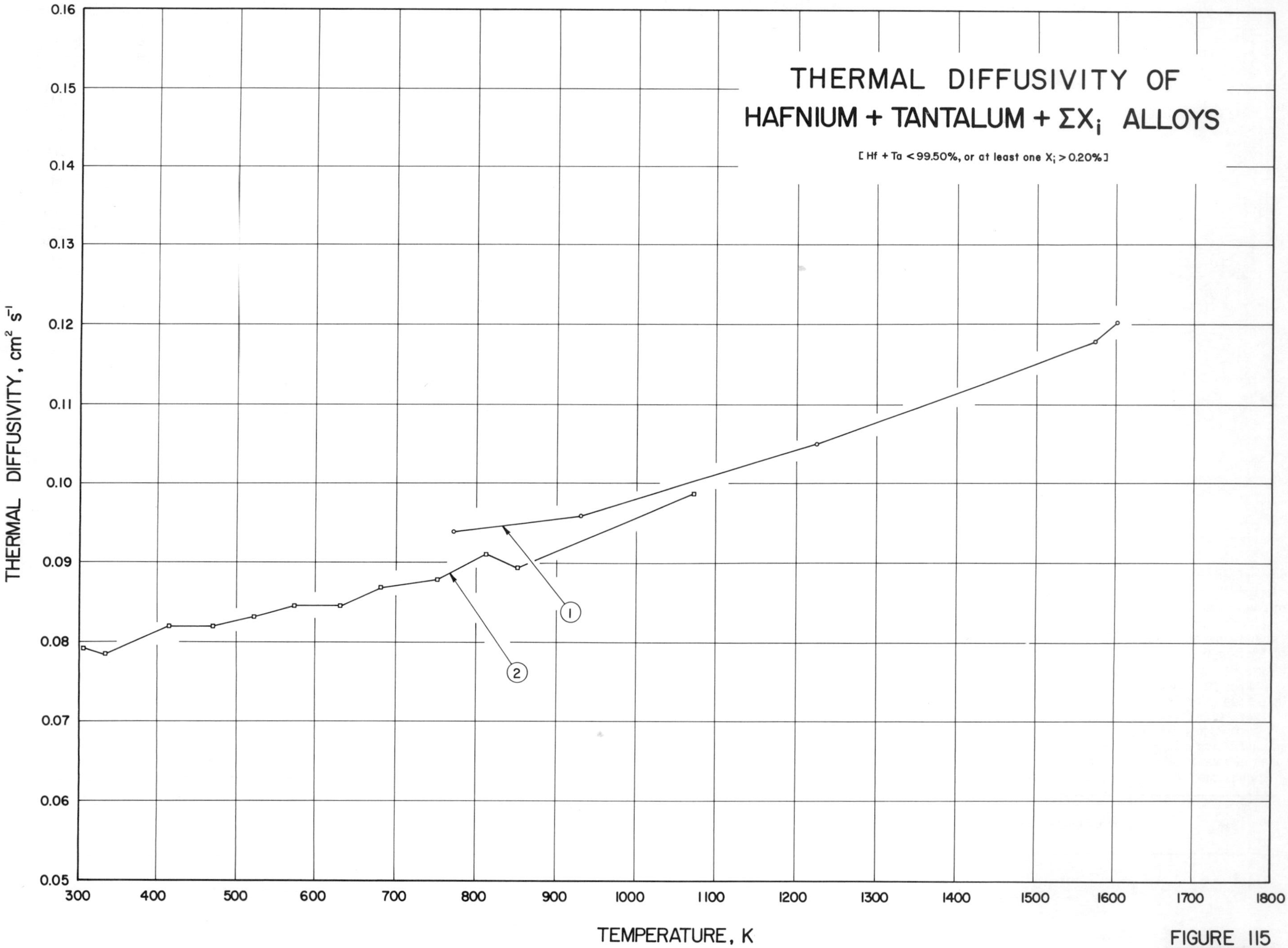
THERMAL DIFFUSIVITY OF
HAFNIUM + TANTALUM + ΣXi ALLOYS
[Hf + Ta < 99.50%, or at least one Xi > 0.20%]
0.16
0.15
0.14
0.13
0.12
0.11
0.10
0.09
0.08
0.07
0.06
0.05
THERMAL DIFFUSIVITY, cm² s⁻¹
300
400
500
600
700
800
900
1000
1100
1200
1300
1400
1500
1600
1700
1800
TEMPERATURE, K
1
2

FIGURE 115

SPECIFICATION TABLE 115. THERMAL DIFFUSIVITY OF [HAFNIUM + TANTALUM + ΣX_i] ALLOYS

(Hf + Ta < 99.50% or at least one X_i > 0.20%)

Cur. No.	Ref. No.	Author(s)	Year	Temp. Range, K	Reported Error, %	Name and Specimen Designation	Composition (weight percent) Hf	Ta	Cr	Mo	Nb	Zr	Composition (continued), Specifications, and Remarks
1	256	Denman, G. L.	1969	772-1602		Hf-20Ta-2Mo	Bal.	21.8		2.2	0.03	~2.0	< 0.0300 O, < 0.0100 Fe, < 0.0100 N, < 0.0050 C, and < 0.0025 H; 0.686 cm diameter x 0.152 cm thick; supplied by Marquart Corp; arc-cast, forged at 1433 C; density 13.41 g cm^{-3}; temperature measured by infrared detector.
2	256	Denman, G. L.	1969	306-1072		Hf-20Ta-2Mo							The above specimen with temperature measured by thermocouple.

DATA TABLE 115. THERMAL DIFFUSIVITY OF [HAFNIUM + TANTALUM + ΣX_i] ALLOYS

(Hf + Ta < 99.50% or at least one X_i > 0.20%)

[Temperature, T, K; Thermal Diffusivity, α, $cm^2\ s^{-1}$]

T	α	T	α
CURVE 1		CURVE 2 (cont.)	
772	0.0939	681	0.0868
930	0.0959	752	0.0878
1225	0.1050	812	0.0910
1574	0.1179	851	0.0893
1602	0.1202	1072	0.0986
CURVE 2			
306	0.0792		
334	0.0785		
415	0.0820		
471	0.0820		
523	0.0831		
574	0.0845		
632	0.0845		

SPECIFICATION TABLE 116. THERMAL DIFFUSIVITY OF [LITHIUM + SODIUM + ΣX_i] ALLOYS

(Li + Na < 99.50% or at least one X_i > 0.20%)

Cur. No.	Ref. No.	Author(s)	Year	Temp. Range, K	Reported Error, %	Name and Specimen Designation	Composition (weight percent) Li	Na	Ag	Al	Ba	Be	Bi	C	Composition (continued), Specifications, and Remarks
1*	97	Rudnev, I. I., Lyashenko, V. S., and Abramovich, M. D.	1961	618-1280			98.98	0.27	→	→	0.006	→	0.002	0.08	(Li by difference), < 0.0003 Ag, 0.0037 Al, < 0.001 Be, 0.0046 Ca, < 0.001 Cd, 0.01 Co, 0.05 Cr, 0.06 Cu, 0.19 Fe, < 0.001 In, 0.003 K, 0.18 Mg, 0.0029 Mn, 0.0058 Mo, 0.0044 N, 0.052 Ni, 0.032 Pb, < 0.01 Sb, 0.023 Sn, 0.016 Ti, 0.0042 V, and < 0.01 Zn; composition obtained after completion of measurements; metal poured in a vacuum of ~1 x 10^{-2} mm Hg into a thin walled tube made of steel 1Kh18N9T: 8.6 mm in diameter, 0.2 mm wall thickness, and 230 mm long; lower and upper steel plugs hermetically joined to the tube by argon-arc welding; necessary compensating volume provided between upper plug and metal surface; specimen heated in vertical electric furnace; measured in vacuum of ~10^{-4} mm Hg; temperature versus time recorded for two points on specimen; six curves recorded for each temperature point; Angström method used to measure diffusivity; diffusivity data calculated from equation given by author.

DATA TABLE 116. THERMAL DIFFUSIVITY OF [LITHIUM + SODIUM + ΣX_i] ALLOYS

(Li + Na < 99.50% or at least one X_i > 0.20%)

[Temperature, T, K; Thermal Diffusivity, α, $cm^2\ s^{-1}$]

T	α
CURVE 1*	
618	0.226
673	0.241
773	0.269
873	0.297
973	0.325
1073	0.353
1173	0.381
1280	0.410

* No figure given.

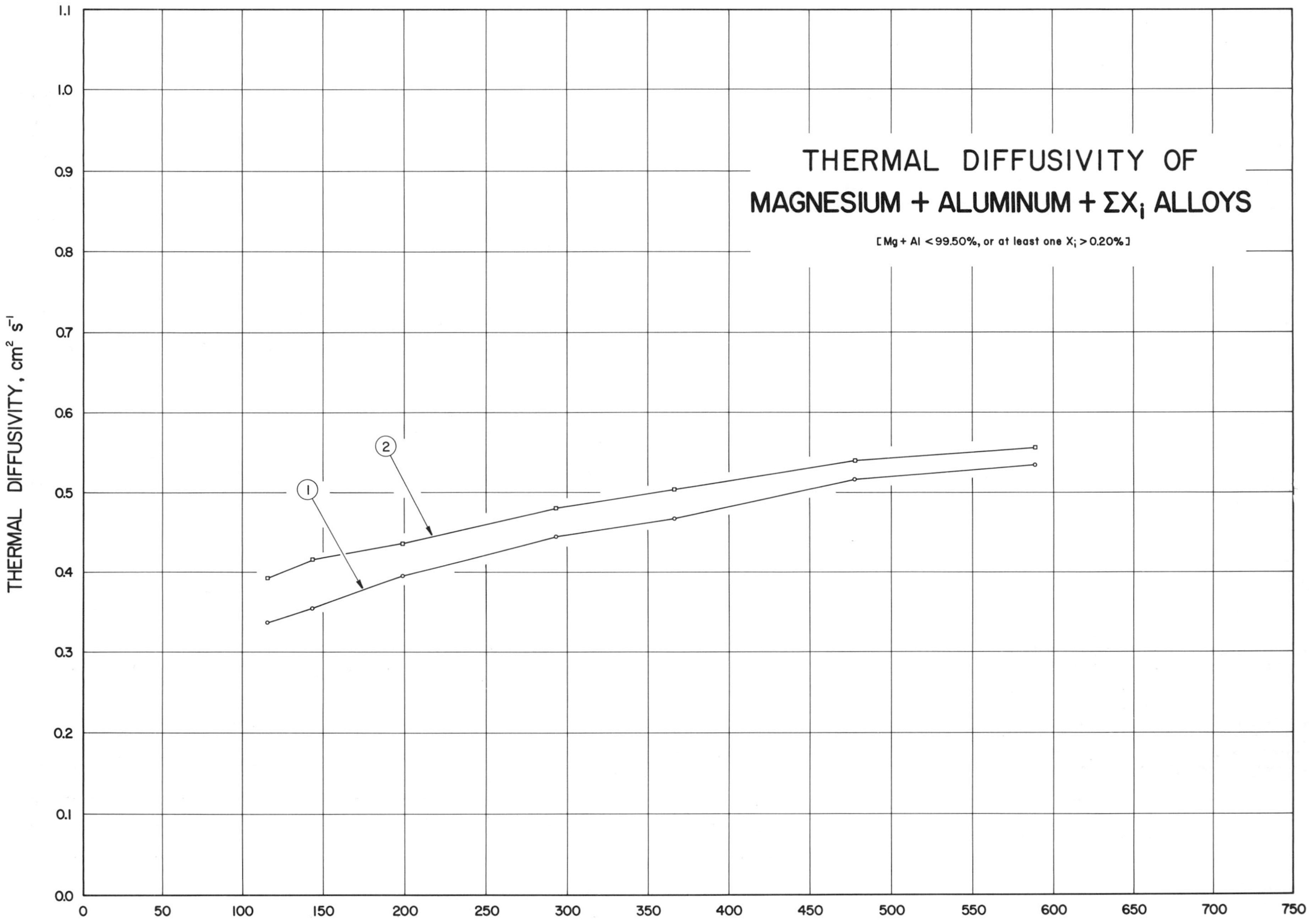
THERMAL DIFFUSIVITY OF
MAGNESIUM + ALUMINUM + ΣX_i ALLOYS
[Mg + Al < 99.50%, or at least one X_i > 0.20%]
THERMAL DIFFUSIVITY, cm² s⁻¹
TEMPERATURE, K
1
2
0.0
0.1
0.2
0.3
0.4
0.5
0.6
0.7
0.8
0.9
1.0
1.1
0
50
100
150
200
250
300
350
400
450
500
550
600
650
700
750

FIGURE 117

SPECIFICATION TABLE 117. THERMAL DIFFUSIVITY OF [MAGNESIUM + ALUMINUM + ΣX_i] ALLOYS

($Mg + Al < 99.50\%$ or at least one $X_i > 0.20\%$)

Cur. No.	Ref. No.	Author(s)	Year	Temp. Range, K	Reported Error, %	Name and Specimen Designation	Composition (weight percent) Mg	Al	Cu	Fe	Mn	Ni	Si	Zn	Composition (continued), Specifications, and Remarks
1	87, 17	Lucks, C. F., Deem, H. W., Thompson, H. B., Smith, A. R., Curry, F. P., and Bing, G. F.	1951	116-589		Magnesium AN-M-29	94.34/ 95.94	2.5/ 3.5	0.05 max	0.005 max	0.2 min	0.005 max	0.3 max	0.7/ 1.3	and 0.3 max others; Mg obtained by difference; supplied by Dow Chemical Co.; hot rolled; annealed 1 hr at 588.7 K and furnace cooled; thermal diffusivity calculated from measured conductivity, specific heat, and density; measured longitudinally.
2	87, 17	Lucks, C. F., et al.		116-589		Magnesium AN-M-29									Thermal diffusivity calculated again from measured conductivity, specific heat, and density; measured vertically.

DATA TABLE 117. THERMAL DIFFUSIVITY OF [MAGNESIUM + ALUMINUM + ΣX_i] ALLOYS

($Mg + Al < 99.50\%$ or at least one $X_i > 0.20\%$)

[Temperature, T, K; Thermal Diffusivity, α, $cm^2\ s^{-1}$]

T	α	T	α
CURVE 1		CURVE 2 (cont.)	
116.4	0.328	477.6	0.539
144.2	0.354	588.7	0.555
199.8	0.395		
293.2	0.449		
366.4	0.477		
477.6	0.516		
588.7	0.534		
CURVE 2			
116.4	0.392		
144.2	0.408		
199.8	0.436		
293.2	0.480		
366.4	0.503		

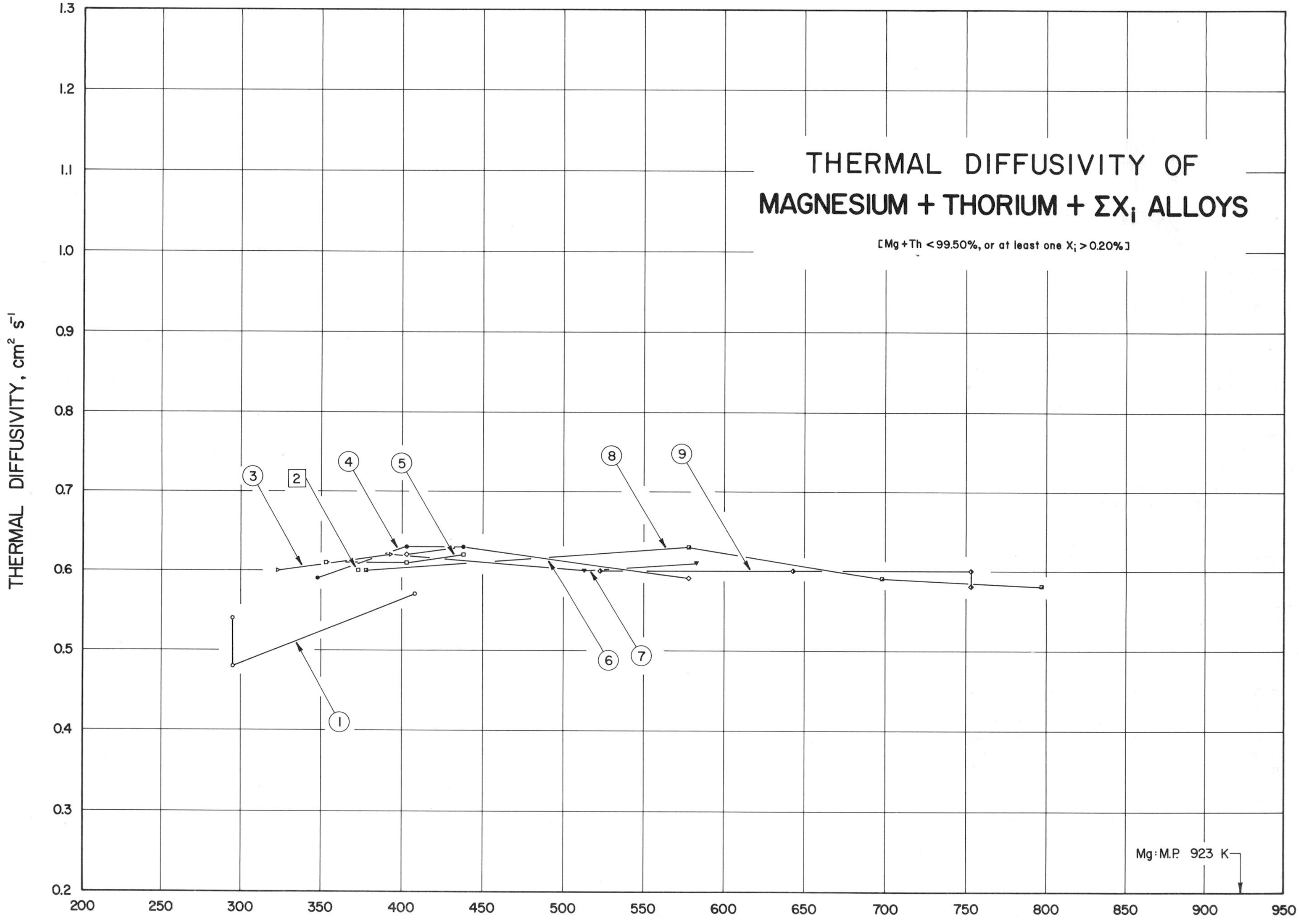
THERMAL DIFFUSIVITY OF
MAGNESIUM + THORIUM + ΣXi ALLOYS
[Mg + Th < 99.50%, or at least one Xi > 0.20%]
THERMAL DIFFUSIVITY, cm2 s-1
TEMPERATURE, K
Mg: M.P. 923 K
1.3
1.2
1.1
1.0
0.9
0.8
0.7
0.6
0.5
0.4
0.3
0.2
200
250
300
350
400
450
500
550
600
650
700
750
800
850
900
950
1
2
3
4
5
6
7
8
9

FIGURE 118

SPECIFICATION TABLE 118. THERMAL DIFFUSIVITY OF [MAGNESIUM + THORIUM + ΣX_i] ALLOYS

($Mg + Th < 99.50\%$ or at least one $X_i > 0.20\%$)

Cur. No.	Ref. No.	Author(s)	Year	Temp. Range, K	Reported Error, %	Name and Specimen Designation	Composition (weight percent) Mg	Th	Zr	Composition (continued), Specifications, and Remarks
1	146, 4	Jenkins, R.J., Parker, W.J., Butler, C.P., and Abbott, G.L.	1960	295-408	±5	Mg HK31				No composition given; square specimen 1.9 cm side and 0.352 cm thick; high intensity short duration light pulse absorbed in front surface of thermally insulated specimen coated with camphor black; 408.2 K measurement obtained by heating sample holder and sample with an infrared lamp.
2	20	Butler, C.P. and Inn, E.C.Y.	1959	373.2	5-10	Mg HK31A-H24	→	3.0	0.6	96.4 Mg (by difference); nominal composition from Metals Handbook, Vol. 1, 8th ed., p. 1069, 1961; cylindrical specimen 0.9525 cm in diameter and length lying in the range from 1 to 2.5 cm; subjected to irradiance from carbon arc lamp heat source; spectral distribution approximates that of a 5700 K black body source; specimen blackened with camphor black; measured under a vacuum of a few microns.
3	20	Butler, C.P. and Inn, E.C.Y.	1959	323-393	5-10	Mg HK31A-H24				Above specimen allowed to cool and then exposed to the arc lamp to measure diffusivity again.
4	20	Butler, C.P. and Inn, E.C.Y.	1959	348-473	5-10	Mg HK31A-H24				Above specimen allowed to cool and then exposed to the arc lamp to measure diffusivity again.
5	20	Butler, C.P. and Inn, E.C.Y.	1959	353-473	5-10	Mg HK31A-H24				Above specimen allowed to cool and then exposed to the arc lamp to measure diffusivity again.
6	20	Butler, C.P. and Inn, E.C.Y.	1959	403-578	5-10	Mg HK31A-H24				Above specimen allowed to cool and then exposed to the arc lamp to measure diffusivity again.
7	20	Butler, C.P. and Inn, E.C.Y.	1959	393-583	5-10	Mg HK31A-H24				Above specimen allowed to cool and then exposed to the arc lamp to measure diffusivity again.
8	20	Butler, C.P. and Inn, E.C.Y.	1959	378-797	5-10	Mg HK31A-H24				Above specimen allowed to cool and then exposed to the arc lamp to measure diffusivity again.
9	20	Butler, C.P. and Inn, E.C.Y.	1959	523-753	5-10	Mg HK31A-H24				Above specimen allowed to cool and then exposed to the arc lamp to measure diffusivity again.

DATA TABLE 118. THERMAL DIFFUSIVITY OF [MAGNESIUM + THORIUM + ΣX_i] ALLOYS

(Mg + Th < 99.50% or at least one X_i > 0.20%)

[Temperature, T, K; Thermal Diffusivity, α, $cm^2\ s^{-1}$]

T	α
CURVE 1	
295.2	0.54
295.2	0.48
408.2	0.57
CURVE 2	
373.2	0.60
CURVE 3	
323.2	0.60
393.2	0.62
CURVE 4	
348.2	0.59
403.2	0.63
473.2	0.63
CURVE 5	
353.2	0.61
403.2	0.61
473.2	0.62
CURVE 6	
403.2	0.62
473.2	0.63*
578.2	0.59
CURVE 7	
393.2	0.62*
513.2	0.60
583.2	0.61
CURVE 8	
378.2	0.60
578.2	0.63
698.2	0.59
797.2	0.58
CURVE 9	
523.2	0.60
643.2	0.60
753.2	0.60
753.2	0.58

* Not shown in figure.

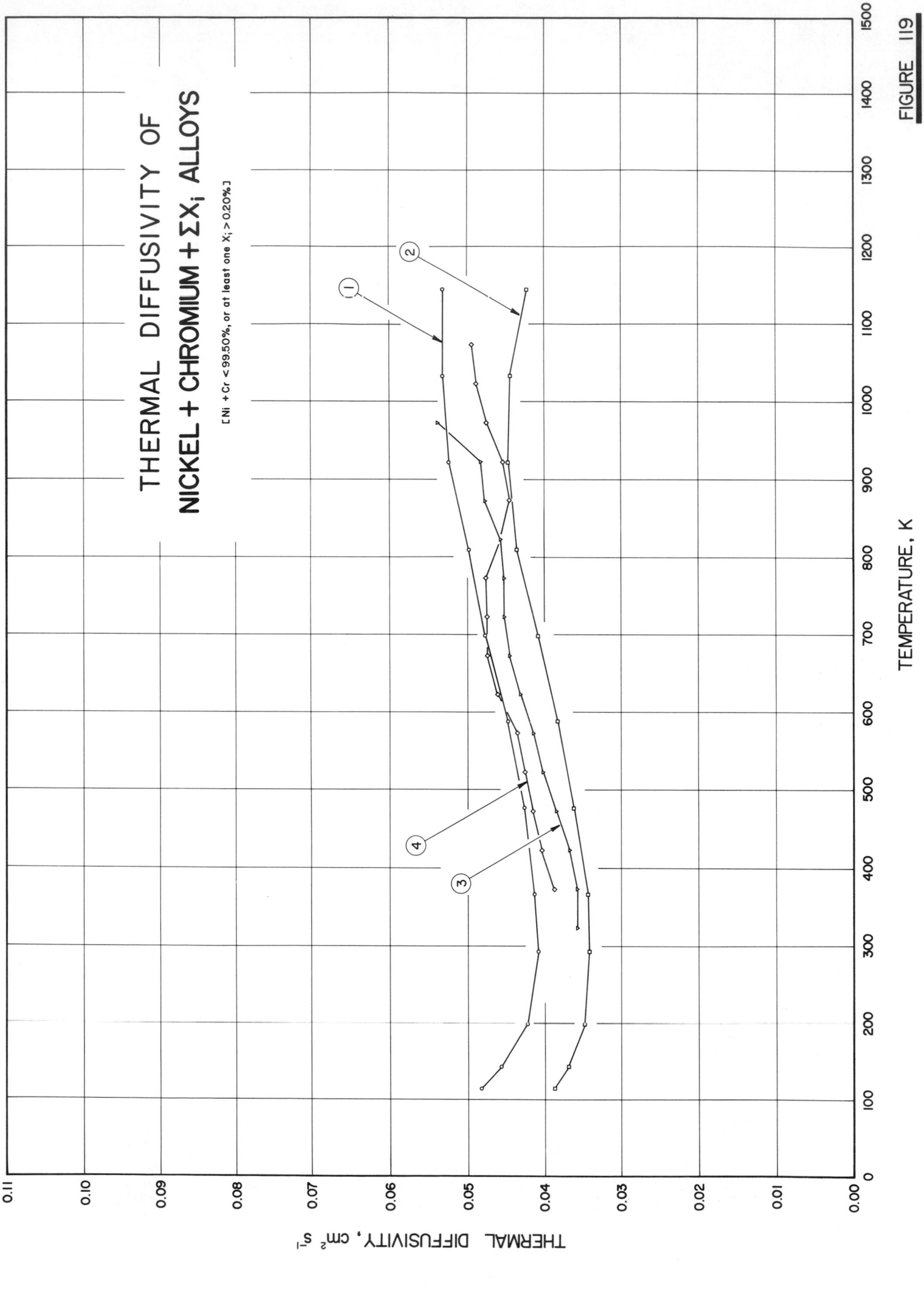
THERMAL DIFFUSIVITY OF
NICKEL + CHROMIUM + ΣX_i ALLOYS
[Ni + Cr < 99.50%, or at least one X_i > 0.20%]
THERMAL DIFFUSIVITY, $cm^2 s^{-1}$
TEMPERATURE, K
0.00
0.01
0.02
0.03
0.04
0.05
0.06
0.07
0.08
0.09
0.10
0.11
0
100
200
300
400
500
600
700
800
900
1000
1100
1200
1300
1400
1500
1
2
3
4

FIGURE 119

SPECIFICATION TABLE 119. THERMAL DIFFUSIVITY OF [NICKEL + CHROMIUM + ΣX_i] ALLOYS

(Ni + Cr < 99.50% or at least one X_i > 0.20%)

Cur. No.	Ref. No.	Author(s)	Year	Temp. Range, K	Reported Error, %	Name and Specimen Designation	Composition (weight percent) Ni	Cr	C	Cu	Fe	Mn	S	Si	Composition (continued), Specifications, and Remarks
1	87, 17	Lucks, C. F., Deem, H. W., Thompson, H. B., Smith, A. R., Curry, F. P., and Bing, G. F.	1951	116-1144		Inconel	78.92	14.62	0.09	0.12	5.80	0.23	0.007	0.19	Supplied by International Nickel Co., Inc.; hot rolled; annealed 3 hrs at 1144.3 K, 15 min at 1255.4 K, and air cooled; thermal diffusivity calculated from measured conductivity, specific heat, and density.
2	87, 17	Lucks, C. F., et al.	1951	116-1144		Inconel "X"	72.94	14.65	0.03	0.02	6.97	0.54	0.007	0.46	0.93 Al, 1.01 Nb, and 2.44 Ti; supplied by International Nickel Co., Inc.; hot rolled; solution heat treated 3 hrs at 1422.1 K, air cooled, double aged 24 hrs at 1116.5 K, air cooled, 20 hrs at 977.6 K, and air cooled; thermal diffusivity calculated from measured conductivity, specific heat, and density.
3	89	Neimark, B. E., Lyusternik, V. E., Anichkina, E. Yu., and Bykova, T. I.	1963	323-973		0Kh20N60B	59.64	20.4	0.06		17.7	1.59	0.004	0.25	0.58 Nb; quenched in water from 1323.2 K and tempered for 1 hr in air at 993.2 K; density 8.206 g cm^{-3} at 293.2 K; electrical resistivity reported as 113.6, 114.3, 115.4, 116.2, 117.1, 118.1, 119.0, 119.8, 120.6, 121.6, 122.9, 122.7, 122.7, 122.8, and 123.1 x 10^{-8} ohm m at 293.2, 323.2, 373.2, 423.2, 473.2, 523.2, 573.2, 623.2, 673.2, 723.2, 773.2, 823.2, 873.2, 923.2, and 973.2 K, respectively; Lorenz number reported as 4.45, 4.23, 4.04, 3.87, 3.73, 3.60, 3.43, 3.36, 3.23, 3.12, 3.05, 3.00, 2.93, and 2.86 x 10^{-8} V^2 K^{-2} at 293.2, 323.2, 373.2, 423.2, 473.2, 523.2, 573.2, 623.2, 673.2, 723.2, 773.2, 823.2, 873.2, and 923.2 K, respectively; thermal diffusivity calculated from measured conductivity, heat capacity, and density.

SPECIFICATION TABLE 119. THERMAL DIFFUSIVITY OF [NICKEL + CHROMIUM + ΣX_i] ALLOYS (continued)

Cur. No.	Ref. No.	Author(s)	Year	Temp. Range, K	Reported Error, %	Name and Specimen Designation	Composition (weight percent) Ni	Cr	C	Cu	Fe	Mn	S	Si	Composition (continued), Specifications, and Remarks
4	89	Neimark, B. E., Lyusternik, V. E., Anichkina, E. Yu., and Bykova, T. I.	1963	373-1073		0Kh21N78T (EI-435)	→	21.1	0.06	→	0.56	0.49	0.006	0.32	~77.229 Ni (by difference), traces of Cu, 0.005 P, and 0.23 Ti; quenched in water from 1373.2 K; density 8.411 g cm^{-3} at 293.2 K; electrical resistivity reported as 109.0, 109.9, 109.9, 110.4, 110.8, 111.3, 111.7, 112.1, 112.7, 113.6, 115.3, 114.4, 113.5, 113.0, 112.6, 112.4, 112.3, and 112.4 x 10^{-8} ohm m at 293.2, 323.2, 373.2, 423.2, 473.2, 523.2, 573.2, 623.2, 673.2, 723.2, 773.2, 823.2, 873.2, 923.2, 973.2, 1023.2, 1073.2, and 1123.2 K, respectively; Lorenz number reported as 4.76, 4.48, 4.21, 3.98, 3.79, 3.61, 3.46, 3.30, 3.14, 3.06, 3.02, 2.96, 2.87, 2.75, 2.64, 2.60, and 2.58 x 10^{-8} V^2 K^{-2} at 293.2, 323.2, 373.2, 423.2, 473.2, 523.2, 573.2, 623.2, 673.2, 723.2, 773.2, 823.2, 873.2, 923.2, 973.2, 1023.2, and 1073.2 K, respectively; thermal diffusivity calculated from measured conductivity, heat capacity, and density.

DATA TABLE 119. THERMAL DIFFUSIVITY OF [NICKEL + CHROMIUM + ΣX_i] ALLOYS

(Ni + Cr < 99.50% or at least one $X_i > 0.20\%$)

[Temperature, T, K; Thermal Diffusivity, α, $cm^2\ s^{-1}$]

T	α	T	α
CURVE 1		CURVE 4	
116.4	0.0483	373.2	0.0387
144.2	0.0457	423.2	0.0403
199.8	0.0423	473.2	0.0415
293.2	0.0408	523.2	0.0425
366.4	0.0413	573.2	0.0435
477.6	0.0426	623.2	0.0461
588.7	0.0447	673.2	0.0474
699.8	0.0477	723.2	0.0474
810.9	0.0498	773.2	0.0476
922.0	0.0524	823.2	0.0458*
1033.1	0.0532	873.2	0.0445
1144.2	0.0532	923.2	0.0454
		973.2	0.0475
CURVE 2		1023.2	0.0488
		1073.2	0.0494
116.4	0.0387		
144.2	0.0369		
199.8	0.0348		
293.2	0.0341		
366.4	0.0343		
477.6	0.0361		
588.7	0.0382		
699.8	0.0408		
810.9	0.0436		
922.0	0.0447		
1033.1	0.0444		
1144.2	0.0423		
CURVE 3			
323.2	0.0357		
373.2	0.0357		
423.2	0.0367		
473.2	0.0384		
523.2	0.0401		
573.2	0.0414		
623.2	0.0431		
673.2	0.0445		
723.2	0.0452		
773.2	0.0452		
823.2	0.0457		
873.2	0.0477		
923.2	0.0482		
973.2	0.0538		

* Not shown in figure.

SPECIFICATION TABLE 120. THERMAL DIFFUSIVITY OF [NICKEL + COBALT + ΣX_i] ALLOYS

(Ni + Co < 99.50% or at least one X_i > 0.20%)

Cur. No.	Ref. No.	Author(s)	Year	Temp. Range, K	Reported Error, %	Name and Specimen Designation	Composition (weight percent) Ni	Co	C	Cu	Fe	Mg	Mn	S	Composition (continued), Specifications, and Remarks
1*	189	Hugon, L. and Jaffray, J.	1955	322-613			→	→	0.17	0.20	0.30	0.18	0.27	0.04	98.70 Ni and Co, and 0.14 SiO_2; cylindrical specimen 4 mm in diameter and 1.50 m long; obtained from Centre d'Information du Nickel and la Societé "Le Ferro-Nickel"; cast in a high frequency furnace into a billet 100 mm in diameter, then rolled at 1150 C and reduced in diameter to 13.8 mm, then annealed in a closed vessel at 900 C, then cold drawn to its final diameter of 4 mm, and finally annealed at 700 C; Curie temperature lying in the range between 628.2 and 633.2 K; density reported as 8.847, 8.840, 8.830, 8.812, 8.794, 8.776, 8.758, 8.739, 8.698, 8.675, 8.655, 8.632, 8.609, and 8.589 g cm^{-3} at 273.2, 293.2, 323.2, 373.2, 423.2, 473.2, 523.2, 573.2, 673.2, 723.2, 773.2, 823.2, 873.2, and 923.2 K, respectively; specific heat at these temperatures reported as 0.104, 0.106, 0.109, 0.114, 0.119, 0.125, 0.131, 0.138, 0.132, 0.131, 0.131, 0.132, 0.133, and 0.134 cal g^{-1} K^{-1}, respectively; sinusoidal temperature wave imposed on one end of specimen; thermal diffusivity determined from measured velocities of the heat wave corresponding to two different modulation periods.
2*	189	Hugon, L. and Jaffray, J.	1955	652-891											Above specimen measured for diffusivity again.

DATA TABLE 120. THERMAL DIFFUSIVITY OF [NICKEL + COBALT + ΣX_i] ALLOYS

(Ni + Co < 99.50% or at least one X_i > 0.20%)

[Temperature, T, K; Thermal Diffusivity, α, cm^2 s^{-1}]

T	α	T	α	T	α
CURVE 1*		CURVE 1 (cont.)*		CURVE 2 (cont.)*	
322	0.148	593	0.105	723	0.113
373	0.141	613	0.0990	773	0.116
423	0.134			823	0.118
474	0.125	CURVE 2*		873	0.121
523	0.117			891	0.121
573	0.109	652	0.108		
		673	0.110		

* No figure given.

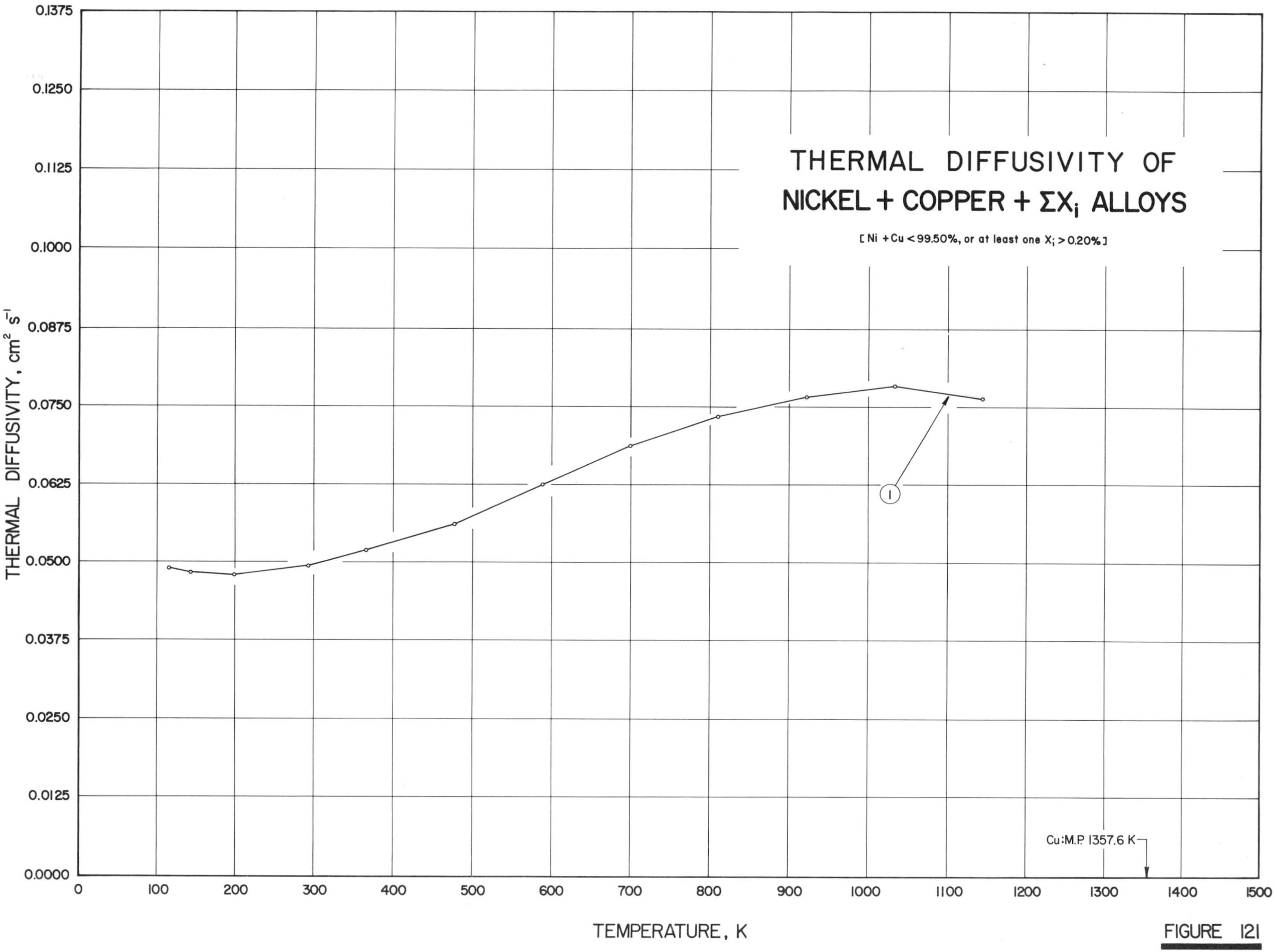
THERMAL DIFFUSIVITY OF
NICKEL + COPPER + ΣX$_i$ ALLOYS
[Ni + Cu < 99.50%, or at least one X$_i$ > 0.20%]
THERMAL DIFFUSIVITY, cm^2 s^{-1}
0.1375
0.1250
0.1125
0.1000
0.0875
0.0750
0.0625
0.0500
0.0375
0.0250
0.0125
0.0000
0
100
200
300
400
500
600
700
800
900
1000
1100
1200
1300
1400
1500
1
Cu:M.P. 1357.6 K
TEMPERATURE, K

FIGURE 121

SPECIFICATION TABLE 121. THERMAL DIFFUSIVITY OF [NICKEL + COPPER + ΣX_i] ALLOYS

(Ni + Cu < 99.50% or at least one X_i > 0.20%)

Cur. No.	Ref. No.	Author(s)	Year	Temp. Range, K	Reported Error, %	Name and Specimen Designation	Composition (weight percent) Ni	Cu	Al	C	Fe	Mn	S	Si	Composition (continued), Specifications, and Remarks
1	87, 17	Lucks, C. F., Deem, H. W., Thompson, H. B., Smith, A. R., Curry, F. P., and Bing, G. F.	1951	116-1144		Monel "K"	65.51	29.23	3.0	0.13	0.86	0.60	0.005	0.09	Supplied by International Nickel Co., Inc.; hot rolled; annealed 1 hr at 1172.1 K and water quenched; thermal diffusivity calculated from measured conductivity, specific heat, and density.

DATA TABLE 121. THERMAL DIFFUSIVITY OF [NICKEL + COPPER + ΣX_i] ALLOYS

(Ni + Cu < 99.50% or at least one X_i > 0.20%)

[Temperature, T, K; Thermal Diffusivity, α, $cm^2\ s^{-1}$]

T	α
CURVE 1	
116.4	0.0490
144.2	0.0483
199.8	0.0480
293.2	0.0493
366.4	0.0519
477.6	0.0560
588.7	0.0625
699.8	0.0687
810.9	0.0736
922.0	0.0767
1033.1	0.0785
1144.2	0.0764

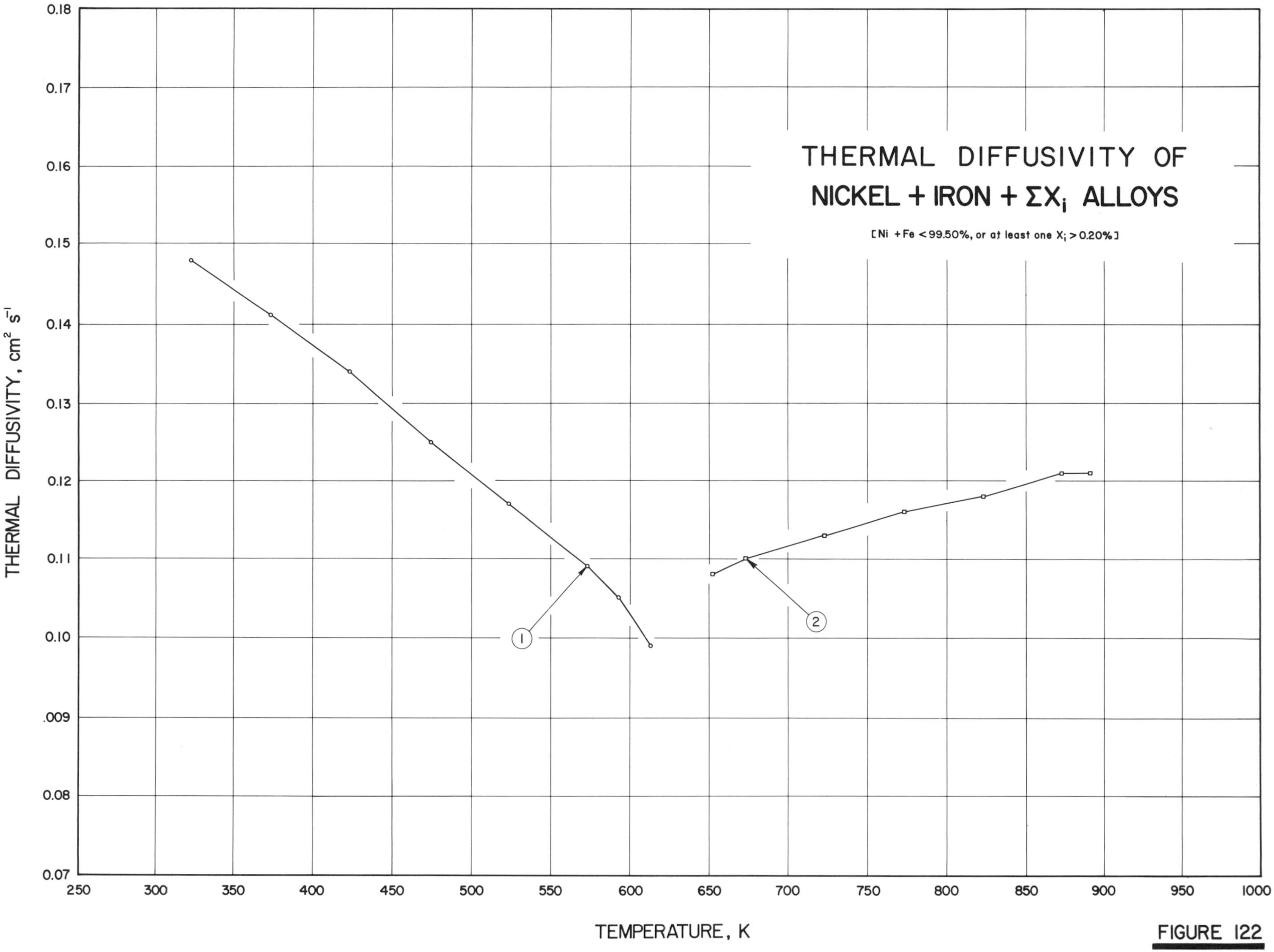
THERMAL DIFFUSIVITY OF
NICKEL + IRON + ΣX_i ALLOYS
[Ni + Fe < 99.50%, or at least one X_i > 0.20%]
THERMAL DIFFUSIVITY, cm² s⁻¹
0.18
0.17
0.16
0.15
0.14
0.13
0.12
0.11
0.10
.009
0.08
0.07
250
300
350
400
450
500
550
600
650
700
750
800
850
900
950
1000
TEMPERATURE, K
1
2

FIGURE 122

SPECIFICATION TABLE 122. THERMAL DIFFUSIVITY OF [NICKEL + IRON + ΣX_i] ALLOYS

(Ni + Fe < 99.50% or at least one X_i > 0.20%)

Cur. No.	Ref. No.	Author(s)	Year	Temp. Range, K	Reported Error, %	Name and Specimen Designation	Ni	Fe	C	Co	Cu	Mg	Mn	S	Composition (continued), Specifications, and Remarks
1	189	Hugon, L. and Jaffray, J.	1955	322-613			→	0.30	0.17	→	0.20	0.18	0.27	0.04	98.70 Ni and Co, and 0.14 SiO_2; cylindrical specimen 4 mm in diameter and 1.50 m long; obtained from Centre d'Information du Nickel and la Societé "Le Ferro-Nickel"; cast in a high frequency furnace into a billet 100 mm in diameter, then rolled at 1150 C and reduced in diameter to 13.8 mm, then annealed in a closed vessel at 900 C, then cold drawn to its final diameter of 4 mm, and finally annealed at 700 C; Curie temperature lying in the range between 628.2 and 633.2 K; density reported as 8.847, 8.840, 8.830, 8.812, 8.794, 8.776, 8.758, 8.739, 8.698, 8.675, 8.655, 8.632, 8.609, and 8.589 g cm^{-3} at 273.2, 293.2, 323.2, 373.2, 423.2, 473.2, 523.2, 573.2, 673.2, 723.2, 773.2, 823.2, 873.2, and 923.2 K, respectively; specific heat at these temperatures reported as 0.104, 0.106, 0.109, 0.114, 0.119, 0.125, 0.131, 0.138, 0.132, 0.131, 0.131, 0.132, 0.133, and 0.134 cal g^{-1} K^{-1}, respectively; sinusoidal temperature wave imposed on one end of specimen; thermal diffusivity determined from measured velocities of the heat wave corresponding to two different modulation periods.
2	189	Hugon, L. and Jaffray, J.	1955	652-891											Above specimen measured for diffusivity again.

DATA TABLE 122. THERMAL DIFFUSIVITY OF [NICKEL + IRON + ΣX_i] ALLOYS

(Ni + Fe < 99.50% or at least one X_i > 0.20%)

[Temperature, T, K; Thermal Diffusivity, α, cm^2 s^{-1}]

T	α	T	α	T	α
CURVE 1		CURVE 1 (cont.)		CURVE 2 (cont.)	
322	0.148	593	0.105	673	0.110
373	0.141	613	0.0990	723	0.113
423	0.134			773	0.116
474	0.125	CURVE 2		823	0.118
523	0.117			873	0.121
573	0.109	652	0.108	891	0.121

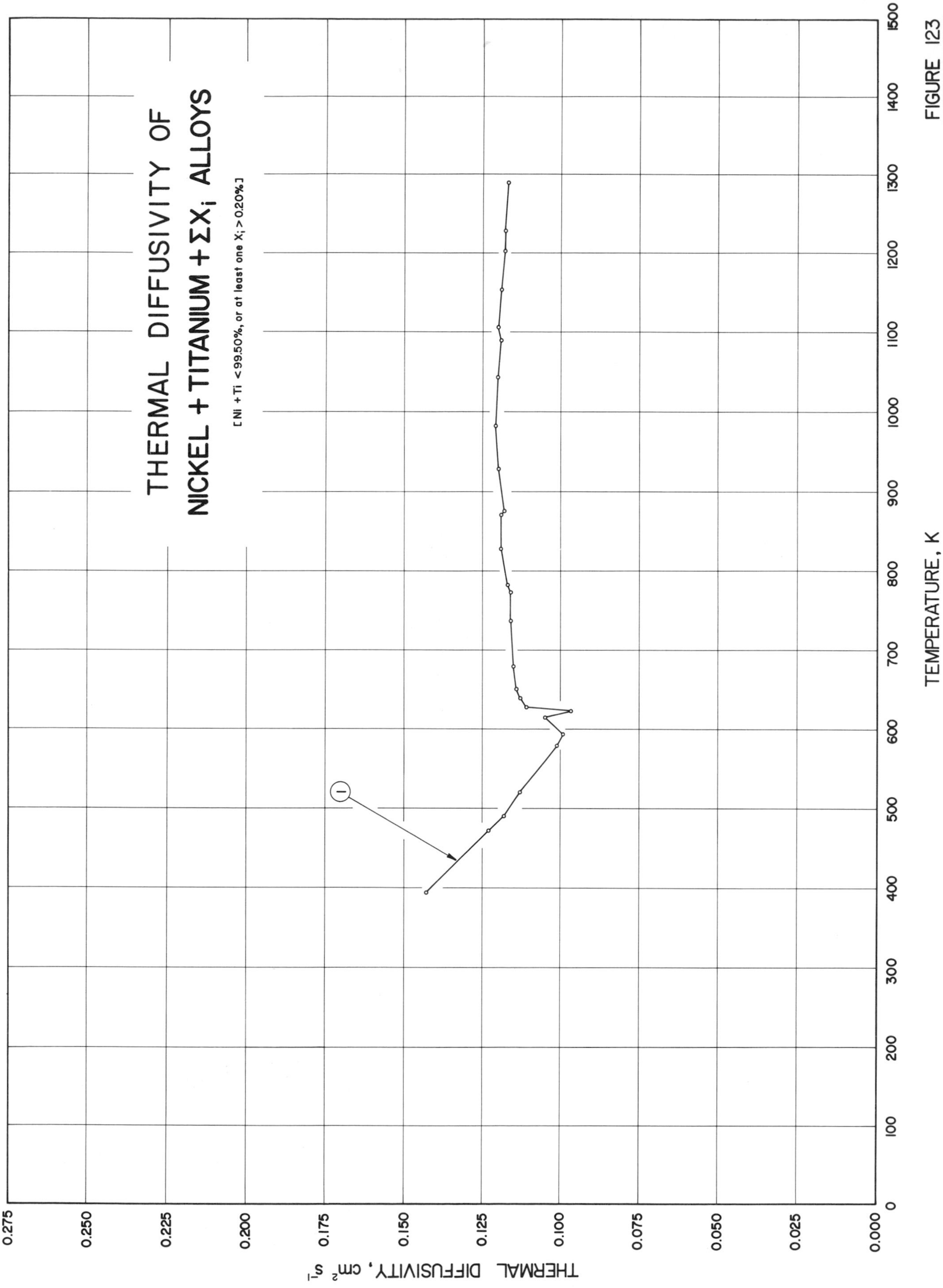
THERMAL DIFFUSIVITY OF
NICKEL + TITANIUM + ΣXi ALLOYS
[Ni + Ti < 99.50%, or at least one Xi > 0.20%]
1
THERMAL DIFFUSIVITY, cm² s⁻¹
0.000
0.025
0.050
0.075
0.100
0.125
0.150
0.175
0.200
0.225
0.250
0.275
0
100
200
300
400
500
600
700
800
900
1000
1100
1200
1300
1400
1500
TEMPERATURE, K

FIGURE 123

SPECIFICATION TABLE 123. THERMAL DIFFUSIVITY OF [NICKEL + TITANIUM + ΣX_i] ALLOYS

(Ni + Ti < 99.50% or at least one X_i > 0.20%)

Cur. No.	Ref. No.	Author(s)	Year	Temp. Range, K	Reported Error, %	Name and Specimen Designation	Composition (weight percent) Ni	Ti	C	Cu	Fe	Mn	S	Si	Composition (continued), Specifications, and Remarks
1	113	Sidles, P.H. and Danielson, G.C.	1960	394-1289		Commercial Permanickel	98.6	0.50	0.25	0.02	0.10	0.10	0.005	0.06	and 0.35 Mg; nominal composition from publication of Huntington Alloy Products Div., The International Nickel Co., Inc; obtained from Driver Harris Co.; electrical resistivity reported as 15.0, 17.8, 21.5, 26.4, 29.6, 32.7, 35.9, 37.6, 39.3, 41.1, 43.4, 44.1, 45.9, 46.7, 47.8, 50.3, 52.4, 54.2, and 55.6 μohm cm at 339.2, 400.2, 463.2, 534.2, 569.2, 604.2, 659.2, 722.2, 782.2, 839.2, 894.2, 945.2, 984.2, 1037.2, 1064.2, 1137.2, 1196.2, 1264.2, and 1299.2 K, respectively; diffusivity measured using modified Angström method.

DATA TABLE 123. THERMAL DIFFUSIVITY OF [NICKEL + TITANIUM + ΣX_i] ALLOYS

(Ni + Ti < 99.50% or at least one X_i > 0.20%)

[Temperature, T, K; Thermal Diffusivity, α, $cm^2\ s^{-1}$]

T	α	T	α
CURVE 1		CURVE 1 (cont.)	
394	0.143	876	0.118
472	0.123	929	0.120
490	0.118	983	0.121
520	0.113	1044	0.120
579	0.101	1090	0.119
593	0.0991	1107	0.120
615	0.105	1153	0.119
623	0.0964	1202	0.118
628	0.111	1228	0.118
639	0.113	1289	0.117
650	0.114		
679	0.115		
737	0.116		
773	0.116		
782	0.117		
828	0.119		
871	0.119		

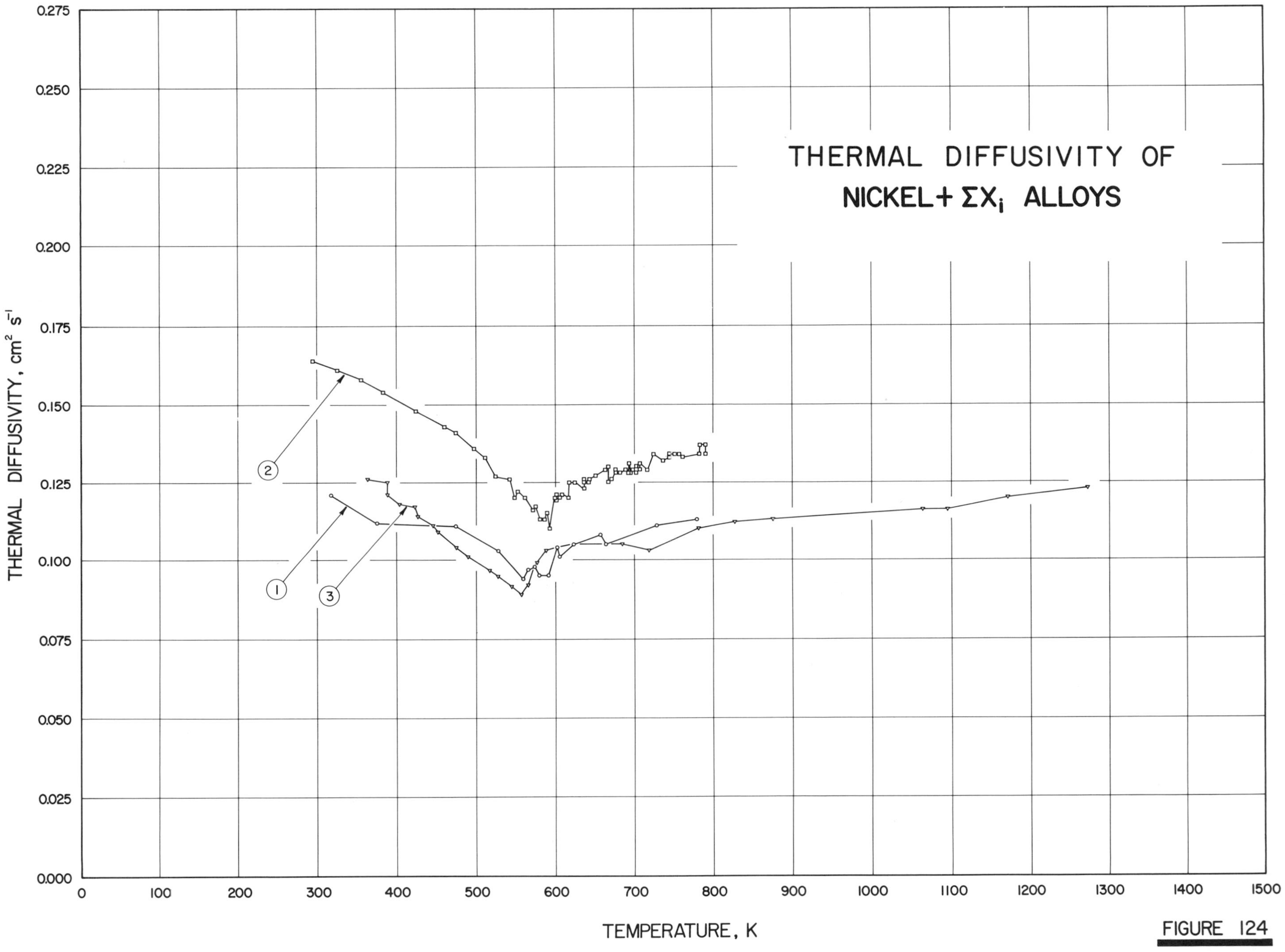
THERMAL DIFFUSIVITY OF
NICKEL + ΣXi ALLOYS
THERMAL DIFFUSIVITY, cm² s⁻¹
TEMPERATURE, K
0.275
0.250
0.225
0.200
0.175
0.150
0.125
0.100
0.075
0.050
0.025
0.000
0
100
200
300
400
500
600
700
800
900
1000
1100
1200
1300
1400
1500
1
2
3

FIGURE 124

SPECIFICATION TABLE 124. THERMAL DIFFUSIVITY OF [NICKEL + ΣX_i]

Cur. No.	Ref. No.	Author(s)	Year	Temp. Range, K	Reported Error, %	Name and Specimen Designation	Composition (weight percent) Ni	Composition (continued), Specifications, and Remarks
1	22, 131	Sidles, P.H. and Danielson, G.C.	1953	318-780			97.92	Major impurities Mn and Si and traces of Co, Fe, and Mg; specimen ~0.125 in. in dia. and 50 cm min long; Curie temp. 585.2 K; measured under a vacuum of 10^{-4} mm Hg; modified Angström method used to measure diffusivity.
2	110	Namba, S., Kim, P.H., and Arai, T.	1967	295-791			99.4	Square specimen 1.5 x 1.5 x 0.1 cm; laser beam used as the pulse energy source; pulse duration ~1 m sec; thermal diffusivity determined from the recorded temp.-time curve of back surface; Curie temp. ~593.2 K; measured in vacuum.
3	113	Sidles, P.H., and Danielson, G.C.	1960	364-1273		Commercial "GRD" nickel		Commercial "GRD" nickel; obtained from Driver Harris Co.; electrical resistivity reported as 17.4, 19.4, 21.3, 24.0, 26.7, 29.3, 32.7, 34.9, 36.8, 37.6, 38.5, 39.6, 41.5, 43.5, 45.9, 47.8, 49.4, 50.5, 51.8, 53.7, and 56.0 μohm cm at 300.2, 343.2, 385.2, 422.2, 456.2, 495.2, 530.2, 562.2, 584.2, 620.2, 650.2, 670.2, 740.2, 815.2, 888.2, 965.2, 1022.2, 1080.2, 1137.2, 1211.2, and 1295.2 K, respectively; diffusivity measured using modified Angström method.

DATA TABLE 124. THERMAL DIFFUSIVITY OF [NICKEL + ΣX_i] ALLOYS

[Temperature, T, K; Thermal Diffusivity, α, $cm^2\ s^{-1}$]

CURVE 1

T	α
318	0.121
376	0.119
475	0.110
529	0.103
560	0.0939
566	0.0967
574	0.0977
580	0.0950
592	0.0950
603	0.104
606	0.101
624	0.105
658	0.108
665	0.105
729	0.111
780	0.113

CURVE 2

T	α
295	0.164
326	0.161
356	0.158
384	0.154
425	0.148
461	0.143
475	0.141
498	0.136
512	0.133
525	0.127
543	0.126
549	0.120
553	0.122
562	0.120
572	0.116
575	0.117
581	0.113
587	0.113
590	0.115
593	0.110
600	0.120
602	0.119
602	0.121
606	0.120
609	0.121
617	0.120
618	0.125

CURVE 2 (cont.)

T	α
625	0.125
637	0.123
637	0.126
637	0.125
643	0.125
644	0.126
652	0.127
664	0.129
668	0.130
668	0.125
672	0.126
677	0.128
677	0.129
683	0.128
690	0.129
693	0.128
694	0.131
695	0.128
699	0.129
703	0.130
703	0.128
708	0.129
708	0.131
717	0.129
725	0.134
737	0.132
745	0.133
745	0.134
752	0.134
758	0.134
762	0.133
783	0.134
784	0.137
791	0.137
791	0.134

CURVE 3

T	α
364	0.126
389	0.125
389	0.121
404	0.118
423	0.117
427	0.114
446	0.111
452	0.109

CURVE 3 (cont.)

T	α
475	0.104
490	0.101
517	0.0966
528	0.0948
545	0.0916
557	0.0891
566	0.0920
576	0.0979*
577	0.0991
589	0.103
603	0.104*
623	0.105*
685	0.105
719	0.103
782	0.110
828	0.112
876	0.113
1063	0.116
1095	0.116
1171	0.120
1273	0.123

* Not shown in figure.

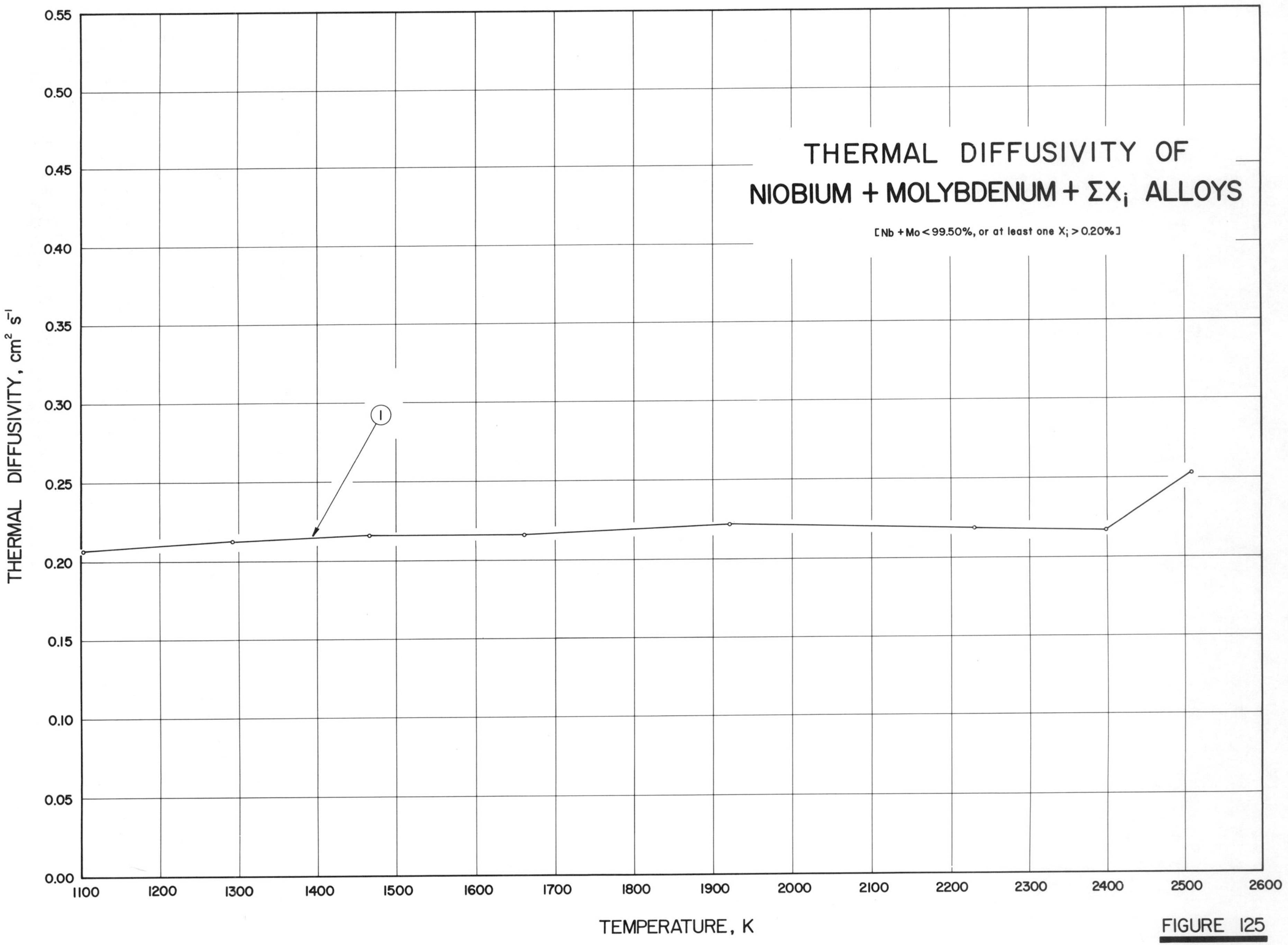
THERMAL DIFFUSIVITY OF
NIOBIUM + MOLYBDENUM + ΣX_i ALLOYS
[Nb + Mo < 99.50%, or at least one X_i > 0.20%]
THERMAL DIFFUSIVITY, cm² s⁻¹
0.55
0.50
0.45
0.40
0.35
0.30
0.25
0.20
0.15
0.10
0.05
0.00
1100
1200
1300
1400
1500
1600
1700
1800
1900
2000
2100
2200
2300
2400
2500
2600
TEMPERATURE, K
1

FIGURE 125

SPECIFICATION TABLE 125. THERMAL DIFFUSIVITY OF [NIOBIUM + MOLYBDENUM + ΣX_i] ALLOYS

(Nb + Mo < 99.50% or at least one X_i > 0.20%)

Cur. No.	Ref. No.	Author(s)	Year	Temp. Range, K	Reported Error, %	Name and Specimen Designation	Composition (weight percent) Nb	Mo	C	N	O	V	Zr	Composition (continued), Specifications, and Remarks
1	74	Hedge, J.C., Kopec, J.W., Kostenko, C., and Lang, J.I.	1963	1103-2510		Cb-5 Mo-5V-1 Zr Alloy	→	5.03	→	→	→	5.02	1.13	88.769 Nb (by difference), 0.0280 C, 0.0136 N, and 0.0093 O; slab specimen; supplied by Westinghouse Electric Corp; density 8.62 g cm^{-3}; top surface of specimen exposed to heat sink; diffusivity determined from measured temperature decrease; unidirectional heat flow.

DATA TABLE 125. THERMAL DIFFUSIVITY OF [NIOBIUM + MOLYBDENUM + ΣX_i] ALLOYS

(Nb + Mo < 99.50% or at least one X_i > 0.20%)

[Temperature, T, K; Thermal Diffusivity, α, $cm^2\ s^{-1}$]

T	α
CURVE 1	
1103	0.207
1293	0.213
1467	0.216
1661	0.216
1922	0.222
2230	0.219
2400	0.217
2510	0.253

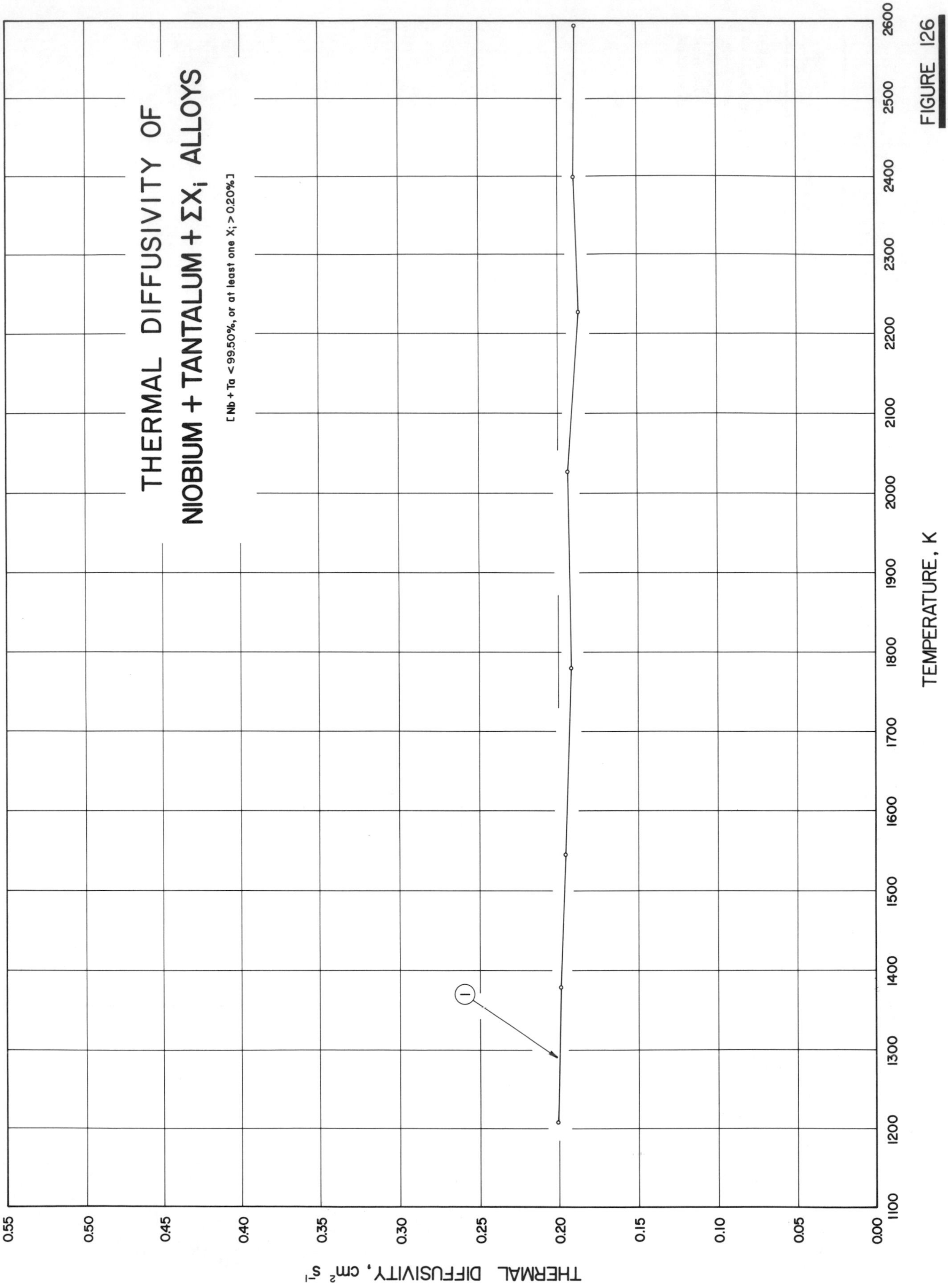
THERMAL DIFFUSIVITY OF
NIOBIUM + TANTALUM + ΣX_i ALLOYS
[Nb + Ta < 99.50%, or at least one X_i > 0.20%]
THERMAL DIFFUSIVITY, cm² s⁻¹
TEMPERATURE, K
0.00
0.05
0.10
0.15
0.20
0.25
0.30
0.35
0.40
0.45
0.50
0.55
1100
1200
1300
1400
1500
1600
1700
1800
1900
2000
2100
2200
2300
2400
2500
2600
1

FIGURE 126

SPECIFICATION TABLE 126. THERMAL DIFFUSIVITY OF [NIOBIUM + TANTALUM + ΣX_i] ALLOYS

(Nb + Ta < 99.50% or at least one X_i > 0.20%)

Cur. No.	Ref. No.	Author(s)	Year	Temp. Range, K	Reported Error, %	Name and Specimen Designation	Composition (weight percent) Nb	Ta	C	Fe	N	Ni	O	Si	Composition (continued), Specifications, and Remarks
1	74	Hedge, J.C., Kopec, J.W., Kostenko, C., and Lang, J.I.	1963	1208-2594		Cb-27 Ta-12 W-0.5 Zr Alloy	→	27.84	→	0.007	→	0.009	→	0.01	60.798 Nb (by difference), 0.0040 C, 0.0020 N, 0.0050 O, 0.005 Ti, 10.40 W, and 0.92 Zr; slab specimen; supplied by Fansteel Metallurgical Corp; density 10.7 g cm^{-3}; top surface of specimen exposed to heat sink; diffusivity determined from measured temperature decrease; unidirectional heat flow.

DATA TABLE 126. THERMAL DIFFUSIVITY OF [NIOBIUM + TANTALUM + ΣX_i] ALLOYS

(Nb + Ta < 99.50% or at least one X_i > 0.20%)

[Temperature, T, K; Thermal Diffusivity, α, cm^2 s^{-1}]

T	α
CURVE 1	
1208.2	0.201
1379.3	0.199
1545.4	0.196
1780.4	0.192
2027.6	0.194
2227.6	0.187
2399.8	0.190
2593.6	0.189

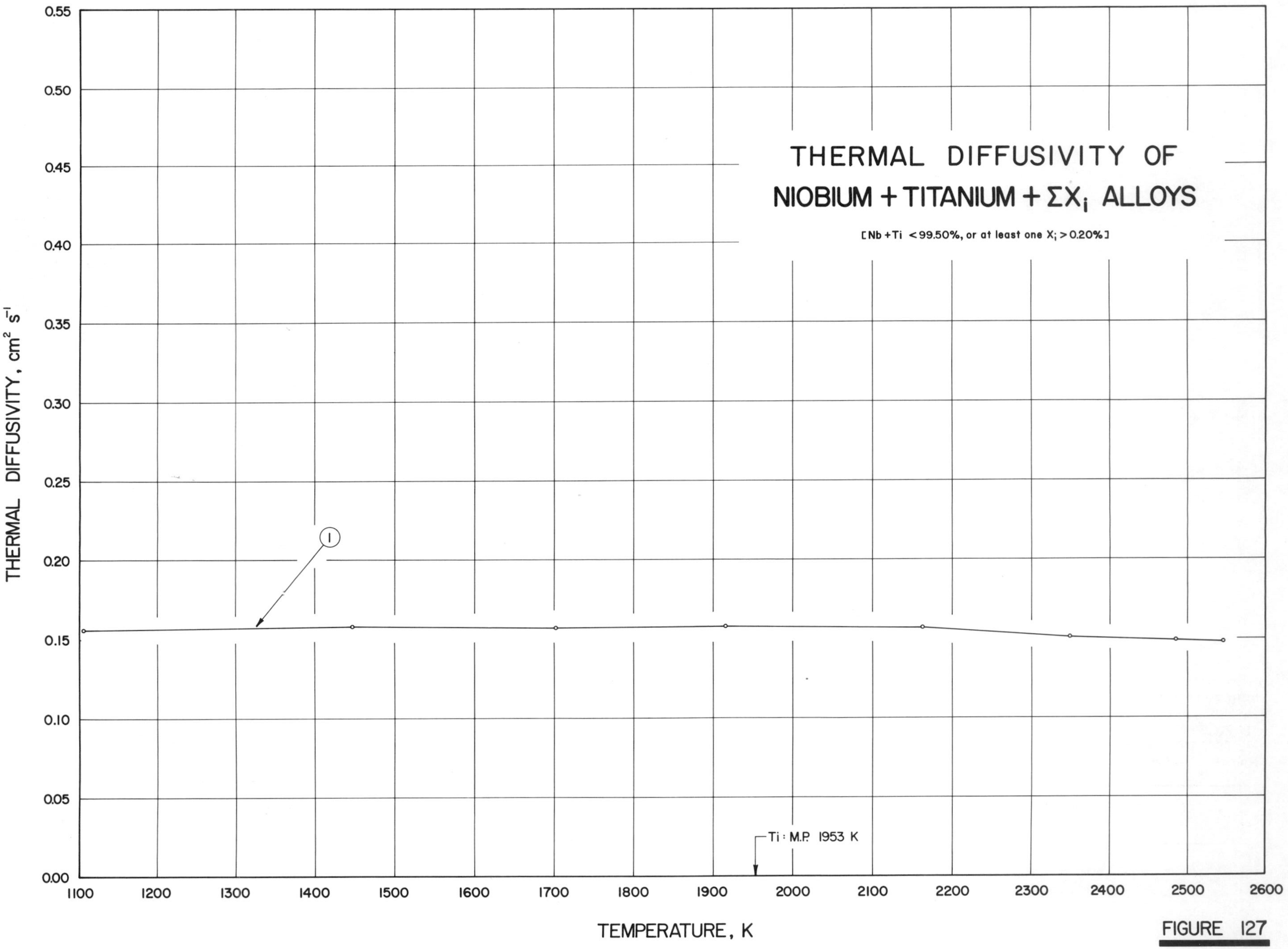
THERMAL DIFFUSIVITY OF
NIOBIUM + TITANIUM + ΣXi ALLOYS
[Nb + Ti < 99.50%, or at least one Xi > 0.20%]
THERMAL DIFFUSIVITY, cm² s⁻¹
TEMPERATURE, K
Ti : M.P. 1953 K
1
0.00
0.05
0.10
0.15
0.20
0.25
0.30
0.35
0.40
0.45
0.50
0.55
1100
1200
1300
1400
1500
1600
1700
1800
1900
2000
2100
2200
2300
2400
2500
2600

FIGURE 127

SPECIFICATION TABLE 127. THERMAL DIFFUSIVITY OF [NIOBIUM + TITANIUM + ΣX_i] ALLOYS

(Nb + Ti < 99.50% or at least one X_i > 0.20%)

Cur. No.	Ref. No.	Author(s)	Year	Temp. Range, K	Reported Error, %	Name and Specimen Designation	Composition (weight percent) Nb	Ti	C	H	N	O	Zr	Composition (continued), Specifications, and Remarks
1	74	Hedge, J.C., Kopec, J.W., Kostenko, C., and Lang, J.I.	1963	1105-2546		Cb-10 Ti-5 Zr Alloy	→	10.5	→	→	→	→	5.5	83.964 Nb (by difference), 0.0071 C, 0.0009 H, 0.0027 N, and 0.0249 O; slab specimen; supplied by DuPont; density 7.77 g cm^{-3}; top surface of specimen exposed to heat sink; diffusivity determined from measured temperature decrease; unidirectional heat flow.

DATA TABLE 127. THERMAL DIFFUSIVITY OF [NIOBIUM + TITANIUM + ΣX_i] ALLOYS

(Nb + Ti < 99.50% or at least one X_i > 0.20%)

[Temperature, T, K; Thermal Diffusivity, α, cm^2 s^{-1}]

T	α
CURVE 1	
1105.4	0.156
1447.1	0.158
1702.6	0.157
1916.5	0.158
2163.7	0.157
2349.8	0.151
2485.2	0.149
2546.3	0.148

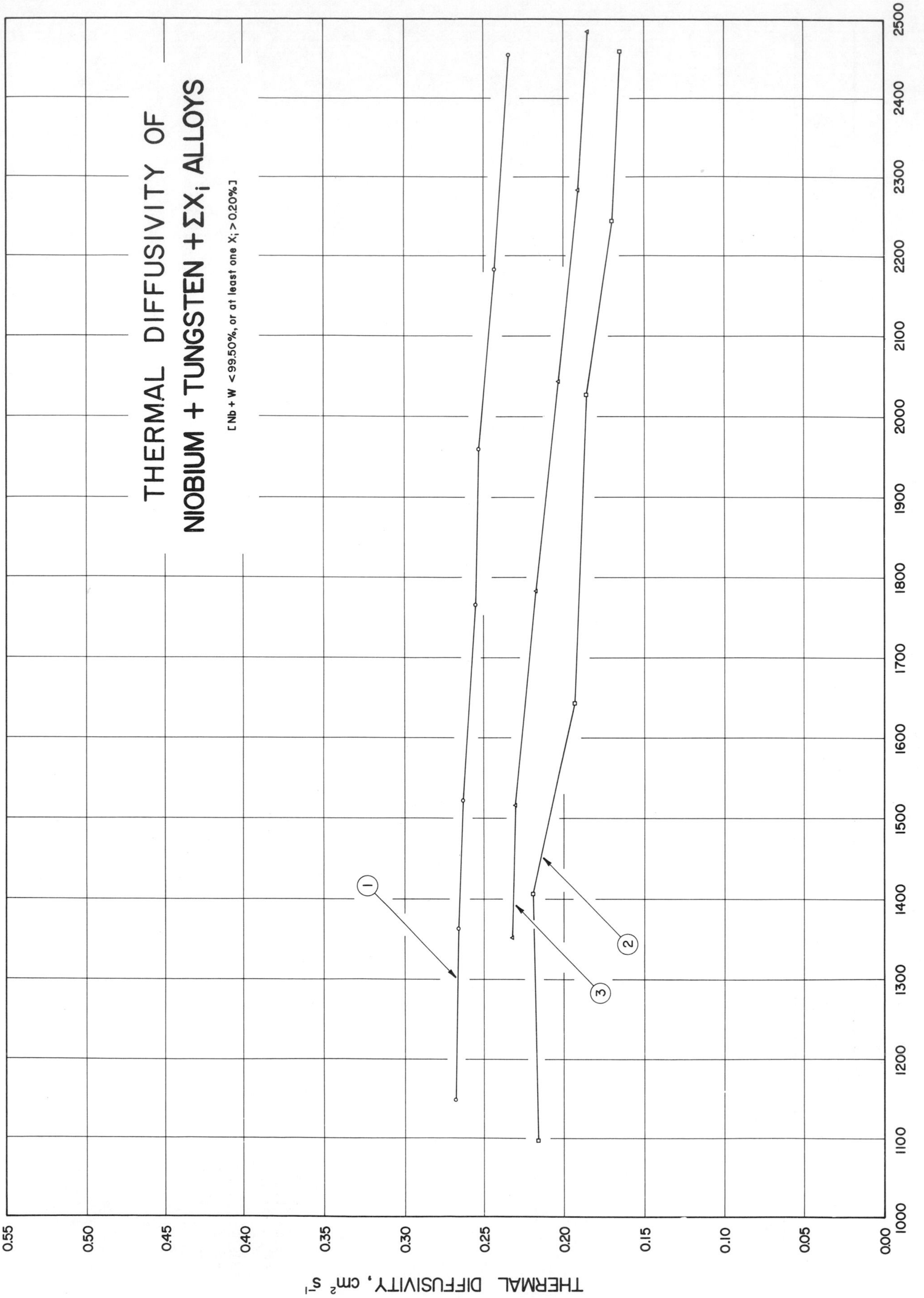
THERMAL DIFFUSIVITY OF
NIOBIUM + TUNGSTEN + ΣXi ALLOYS
[Nb + W < 99.50%, or at least one Xi > 0.20%]
THERMAL DIFFUSIVITY, cm² s⁻¹
TEMPERATURE, K
0.00
0.05
0.10
0.15
0.20
0.25
0.30
0.35
0.40
0.45
0.50
0.55
1000
1100
1200
1300
1400
1500
1600
1700
1800
1900
2000
2100
2200
2300
2400
2500
1
2
3

FIGURE 128

SPECIFICATION TABLE 128. THERMAL DIFFUSIVITY OF [NIOBIUM + TUNGSTEN + ΣX_i] ALLOYS

(Nb + W < 99.50% or at least one X_i > 0.20%)

Cur. No.	Ref. No.	Author(s)	Year	Temp. Range, K	Reported Error, %	Name and Specimen Designation	Composition (weight percent) Nb	W	C	H	N	O	Zr	Composition (continued), Specifications, and Remarks
1	74	Hedge, J.C., Kopec, J.W., Kostenko, C., and Lang, J.I.	1963	1150-2455		Cb-10W-1Zr-0.1 C Alloy	→	9.6	→	→	→	→	0.95	89.390 Nb (by difference), 0.0510 C, 0.0003 H, 0.0033 N, and 0.0053 O; slab specimen; supplied by DuPont; density 9.03 g cm^{-3}; top surface of specimen exposed to heat sink; diffusivity determined from measured temperature decrease; unidirectional heat flow.
2	74	Hedge, J.C., Kopec, J.W., Kostenko, C., and Lang, J.I.	1963	1098-2461		Cb-10W-5Zr Alloy	→	9.83	0.002	0.0011	0.0040	0.0080	2.80	87.305 Nb (by difference); slab specimen; supplied by Haynes Stellite Co.; density 9.16 g cm^{-3}; top surface of specimen exposed to heat sink; diffusivity determined from measured temperature decrease; unidirectional heat flow.
3	74	Hedge, J.C., Kopec, J.W., Kostenko, C., and Lang, J.I.	1963	1353-2680		Cb-15W-5Mo-1Zr-0.05 C Alloy	→	15.6	→	→	0.0020	→	0.84	78.782 Nb (by difference), 0.0489 C, 0.0005 H, 4.7 Mo, 0.0163 O, and 0.01 Ta; slab specimen; supplied by DuPont; density 9.60 g cm^{-3}; top surface of specimen exposed to heat sink; diffusivity determined from measured temperature decrease; unidirectional heat flow.

DATA TABLE 128. THERMAL DIFFUSIVITY OF [NIOBIUM + TUNGSTEN + ΣX_i] ALLOYS

(Nb + W < 99.50% or at least one X_i > 0.20%)

[Temperature, T, K; Thermal Diffusivity, α, cm^2 s^{-1}]

T	α	T	α	T	α
CURVE 1		CURVE 2		CURVE 3	
1149.8	0.268	1097.6	0.216	1352.6	0.232
1363.4	0.266	1406.5	0.219	1516.5	0.230
1522.1	0.263	1643.2	0.193	1783.2	0.217
1766.5	0.255	2027.6	0.186	2044.3	0.203
1960.9	0.253	2244.3	0.170	2283.2	0.191
2183.2	0.243	2460.9	0.165	2485.2	0.185
2455.4	0.234			2679.8	0.173*

* Not shown in figure.

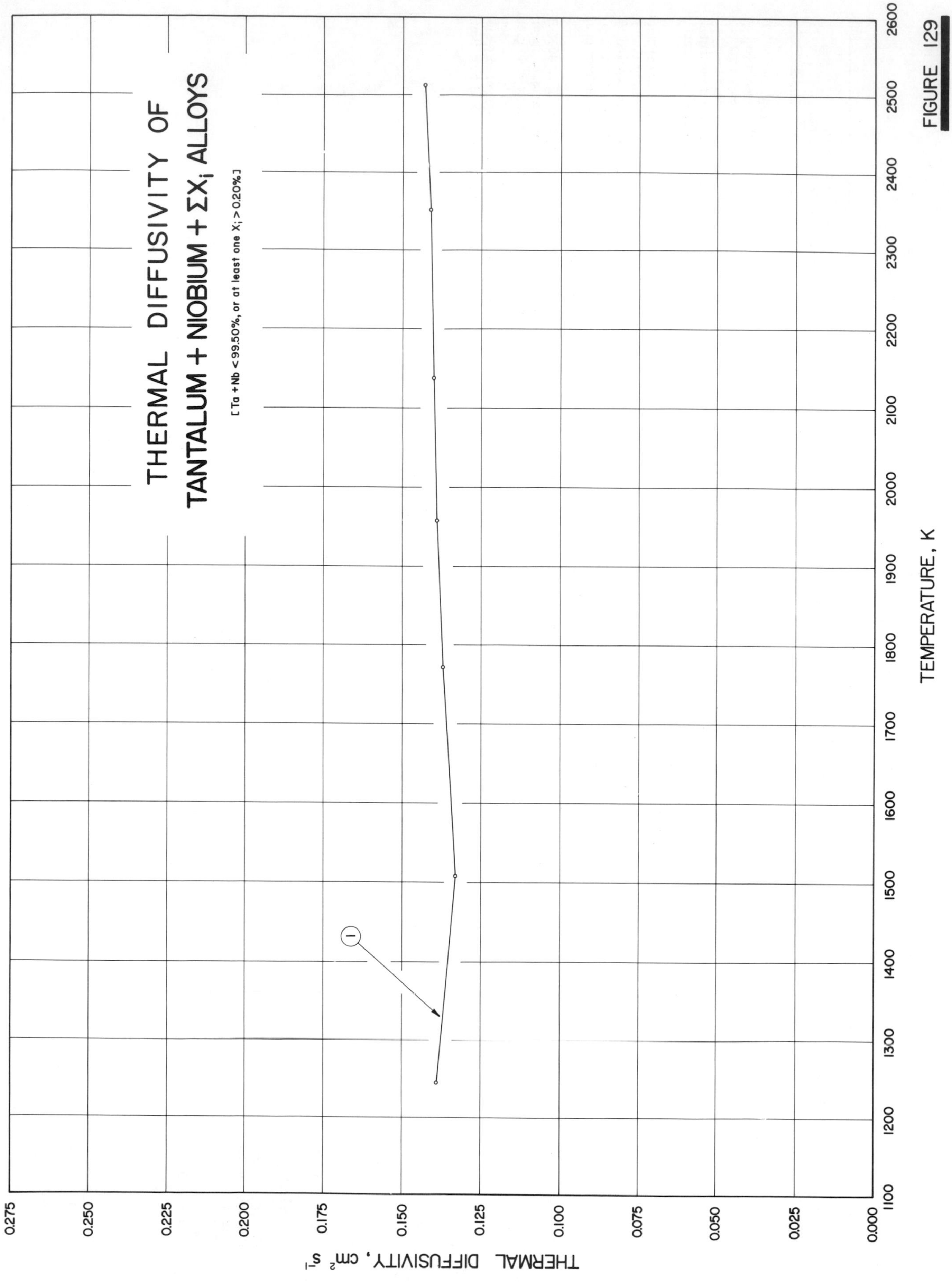
THERMAL DIFFUSIVITY OF
TANTALUM + NIOBIUM + ΣX_i ALLOYS
[Ta + Nb < 99.50%, or at least one X_i > 0.20%]
THERMAL DIFFUSIVITY, cm² s⁻¹
TEMPERATURE, K
0.000
0.025
0.050
0.075
0.100
0.125
0.150
0.175
0.200
0.225
0.250
0.275
1100
1200
1300
1400
1500
1600
1700
1800
1900
2000
2100
2200
2300
2400
2500
2600
1

FIGURE 129

SPECIFICATION TABLE 129. THERMAL DIFFUSIVITY OF [TANTALUM + NIOBIUM + ΣX_i] ALLOYS

(Ta + Nb < 99.50% or at least one X_i > 0.20%)

Cur. No.	Ref. No.	Author(s)	Year	Temp. Range, K	Rep^rted Error, %	Name and Specimen Designation	Composition (weight percent) Ta	Nb	C	N	O	V	Composition (continued), Specifications, and Remarks
1	74	Hedge, J.C., Kopec, J.W., Kostenko, C., and Lang, J.I.	1963	1246-2513		Ta-30Cb-7.5 V Alloy	→	30.3	0.090	→	→	7.47	62.119 Ta (by difference), 0.0065 N, and 0.0150 O; slab specimen; supplied by Wah Chang Corp; density 11.5 g cm^{-3}; top surface of specimen exposed to heat sink; diffusivity determined from measured temperature decrease; unidirectional heat flow.

DATA TABLE 129. THERMAL DIFFUSIVITY OF [TANTALUM + NIOBIUM + ΣX_i] ALLOYS

(Ta + Nb < 99.50% or at least one X_i > 0.20%)

[Temperature, T, K; Thermal Diffusivity, α, cm^2 s^{-1}]

T	α
CURVE 1	
1245.9	0.139
1508.2	0.133
1772.1	0.137
1958.2	0.139
2138.7	0.140
2352.6	0.141
2513.0	0.143

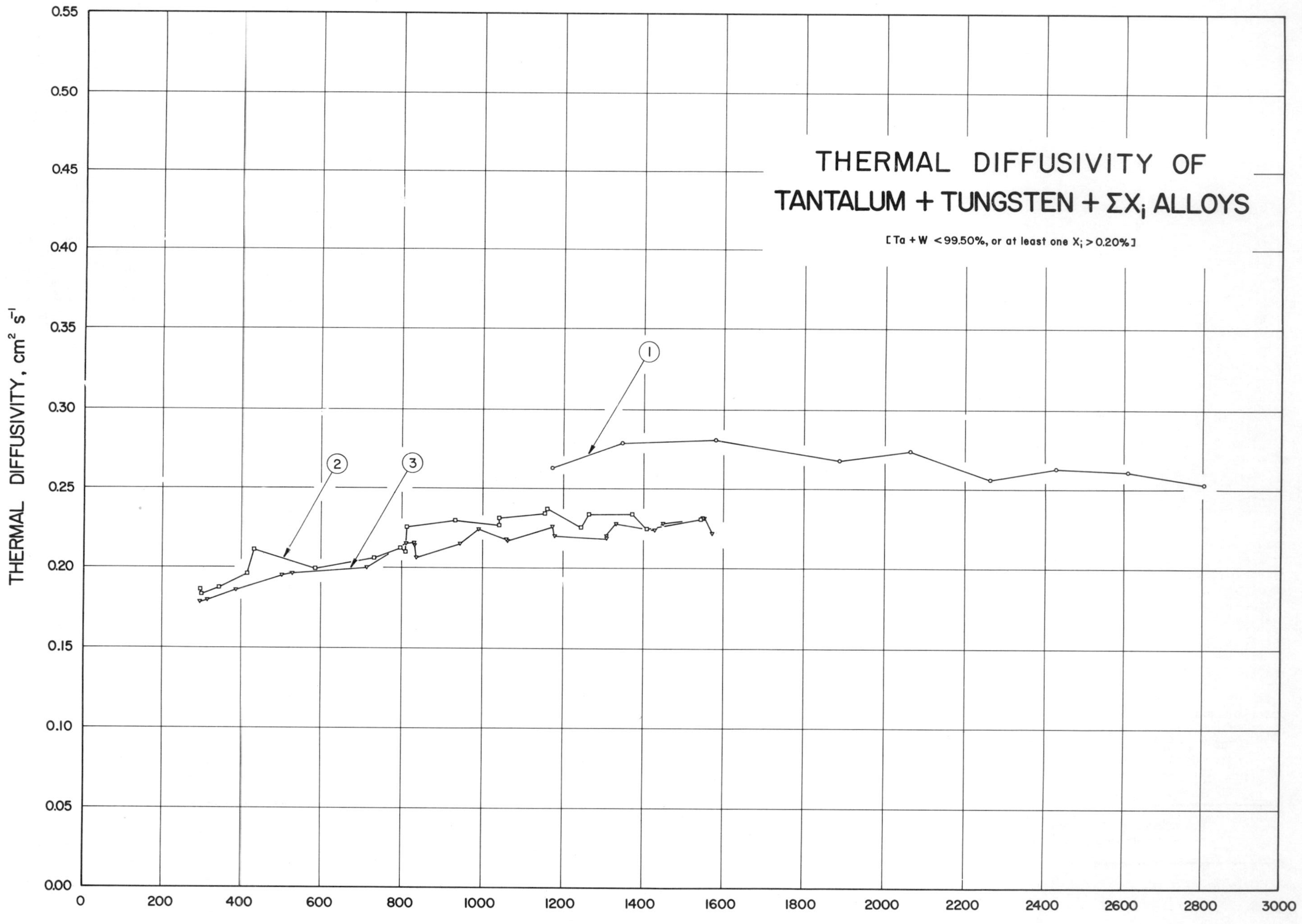
THERMAL DIFFUSIVITY OF
TANTALUM + TUNGSTEN + ΣX_i ALLOYS
[Ta + W < 99.50%, or at least one X_i > 0.20%]
THERMAL DIFFUSIVITY, cm^2 s^-1
TEMPERATURE, K
0.00
0.05
0.10
0.15
0.20
0.25
0.30
0.35
0.40
0.45
0.50
0.55
0
200
400
600
800
1000
1200
1400
1600
1800
2000
2200
2400
2600
2800
3000
1
2
3

FIGURE 130

SPECIFICATION TABLE 130. THERMAL DIFFUSIVITY OF [TANTALUM + TUNGSTEN + ΣX_i] ALLOYS

(Ta + W < 99.50% or at least one X_i > 0.20%)

Cur. No.	Ref. No.	Author(s)	Year	Temp. Range, K	Reported Error, %	Name and Specimen Designation	Composition (weight percent) Ta	W	C	Cr	Fe	H	Hf	Nb	Composition (continued), Specifications, and Remarks
1	74	Hedge, J.C., Kopec, J.W., Kostenko, C., and Lang, J.I.	1963	1172-2805		Ta-8W-2Hf Alloy	→	9.0	0.0041				2.2		88.790 Ta (by difference), 0.0023 N, and 0.0040 O; slab specimen; supplied by Westinghouse Electric Corp; density 16.95 g cm^{-3}; top surface of specimen exposed to heat sink; diffusivity determined from measured temp decrease; unidirectional heat flow.
2	134, 256	Denman, G. L.	1966	297-1547	<±2.3	Tantalum Alloy T-111	→	7.87	0.0018	0.003	0.08	→	2.00	0.03	89.93912 Ta (by difference), 0.00038 H, 0.0023 N, 0.01 Ni, 0.0034 O, and 0.06 Zr; grain size (as received) in the range from ~0.04 to 0.07 mm; disc-shaped specimens each 0.270 in. in dia and 0.060 in. thick; a minimum of two samples measured; supplied by Westinghouse; double arc-cast, extruded at 1922.1 K, forged at 1477.6 K, and annealed at 1323.2 K for 1 hr; density 16.79 g cm^{-3}; ruby laser used as pulse energy source; diffusivity determined from measured history of back face temp; measured in vacuum; each data point reported represents average of four or five measurements.
3	134, 256	Denman, G. L.	1966	297-1571	<±2.3	Tantalum Alloy T-222	→	8.86	0.0090	0.002	0.007	→	2.45	0.04	88.56019 Ta (by difference), 0.00121 H, 0.0006 O, and 0.07 Zr; disc-shaped specimens each 0.270 in. in dia and 0.060 in. thick; a minimum of two samples measured; supplied by Westinghouse; double arc-cast, extruded at 1588.7 K, and forged at 1477.6 K; density 16.80 g cm^{-3}; ruby laser used as pulse energy source; diffusivity determined from measured history of back face temp; measured in vacuum; each data point reported represents average of four or five measurements.

DATA TABLE 130. THERMAL DIFFUSIVITY OF [TANTALUM + TUNGSTEN + ΣX_i] ALLOYS

(Ta + W < 99.50% or at least one X_i > 0.20%)

[Temperature, T, K; Thermal Diffusivity, α, $cm^2\ s^{-1}$]

T	α
CURVE 1	
1172.1	0.263
1346.5	0.279
1580.4	0.281
1888.7	0.268
2063.7	0.274
2263.7	0.256
2427.6	0.263
2610.3	0.261
2804.9	0.253
CURVE 2	
297.2	0.1856
300.2	0.1827
344.2	0.1874
415.2	0.1958
532.2	0.2011
586.2	0.1991
731.2	0.2058
797.2	0.2124
810.2	0.2096
812.2	0.2255
932.2	0.2297
1041.2	0.2266
1041.2	0.2314
1154.2	0.2343
1160.2	0.2367
1244.2	0.2255
1263.2	0.2336
1372.2	0.2342
1408.2	0.2246
1544.2	0.2306
1547.2	0.2312
CURVE 3	
297.2	0.1781
315.2	0.1792
387.2	0.1853
502.2	0.1947
529.2	0.1964
714.2	0.1998
811.2	0.2145
831.2	0.2153
832.2	0.2141
CURVE 3 (cont.)	
836.2	0.2060
944.2	0.2146
990.2	0.2241
1059.2	0.2178
1061.2	0.2169
1174.2	0.2261
1180.2	0.2202
1309.2	0.2187
1309.2	0.2202
1331.2	0.2283
1429.2	0.2241
1449.2	0.2283
1551.2	0.2315
1571.2	0.2222

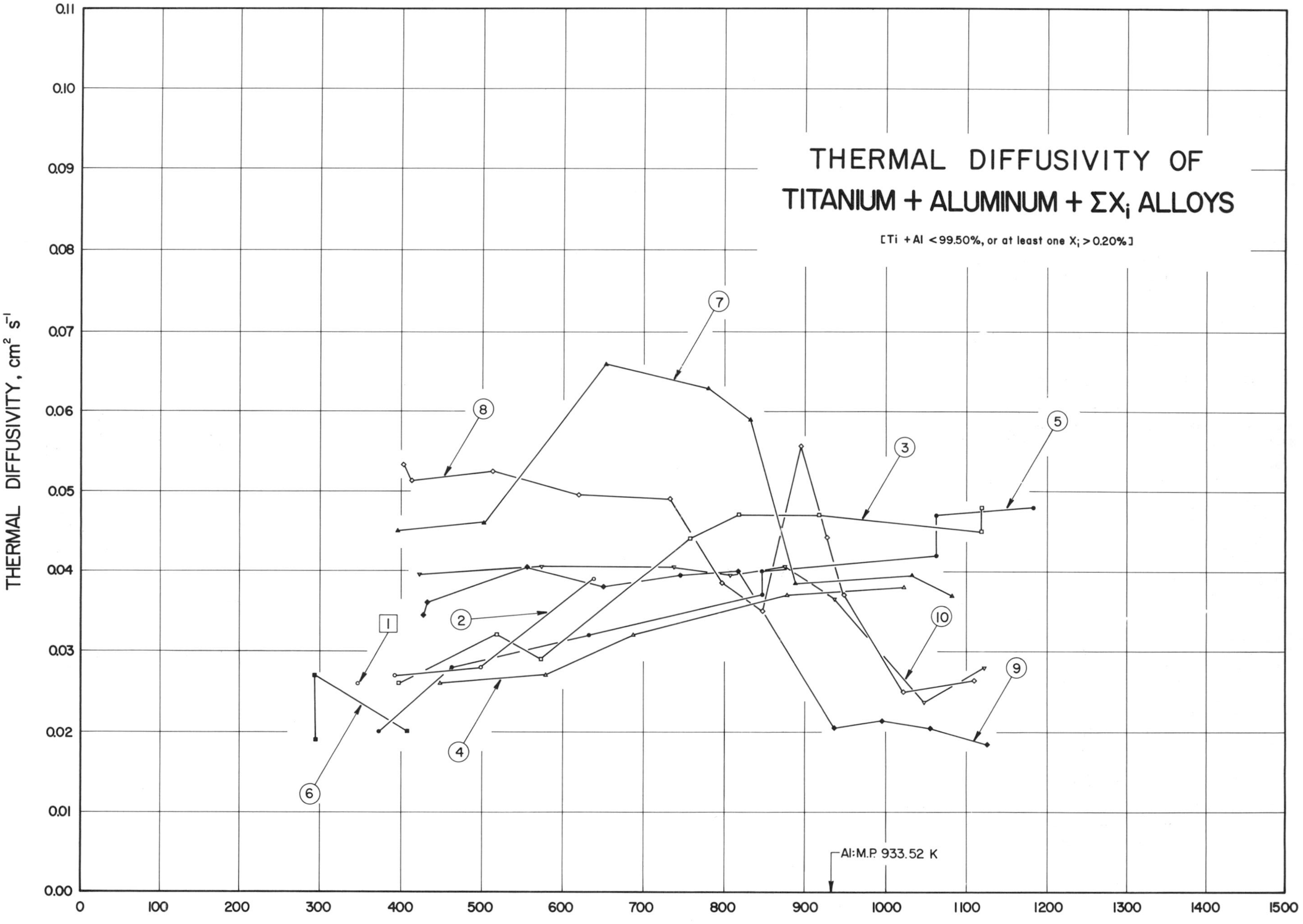
THERMAL DIFFUSIVITY OF
TITANIUM + ALUMINUM + ΣX_i ALLOYS
[Ti + Al < 99.50%, or at least one X_i > 0.20%]
THERMAL DIFFUSIVITY, cm² s⁻¹
TEMPERATURE, K
Al: M.P. 933.52 K
0.00
0.01
0.02
0.03
0.04
0.05
0.06
0.07
0.08
0.09
0.10
0.11
0
100
200
300
400
500
600
700
800
900
1000
1100
1200
1300
1400
1500
1
2
3
4
5
6
7
8
9
10

FIGURE 131

SPECIFICATION TABLE 131. THERMAL DIFFUSIVITY OF [TITANIUM + ALUMINUM + ΣX_i] ALLOYS

(Ti + Al < 99.50% or at least one X_i > 0.20%)

Cur. No.	Ref. No.	Author(s)	Year	Temp. Range, K	Reported Error, %	Name and Specimen Designation	Composition (weight percent) Ti	Al	B	C	Cr	Fe	Si	V	Composition (continued), Specifications, and Remarks
1	19, 20	Butler, C. P. and Inn, E. C. Y.	1957	348.2	5-10	Ti-6Al-4V	→	5.5/ 6.75		0.10 max		0.40 max		3.5/ 4.5	87.865-90.115 Ti (by difference), 0.015 max H, 0.07 max N, and 0.30 max O; nominal composition from Metals Handbook, Vol. 1, 8th ed., p. 1154, 1961; cylindrical specimen 0.9525 cm in diameter and length lying in the range from 1 to 2.5 cm; subjected to irradiance from carbon arc lamp heat source; spectral distribution approximates that of a 5700 K black body source; specimen blackened with camphor black; measured under a vacuum of ~5 microns.
2	19, 20	Butler, C. P. and Inn, E. C. Y.	1957	393-638	5-10	Ti-6Al-4V									Above specimen allowed to cool and then exposed to the arc lamp to measure diffusivity again.
3	19, 20	Butler, C. P. and Inn, E. C. Y.	1957	398-1118	5-10	Ti-6Al-4V									Above specimen allowed to cool and then exposed to the arc lamp to measure diffusivity again.
4	19, 20	Butler, C. P. and Inn, E. C. Y.	1957	448-1023	5-10	Ti-6Al-4V									Above specimen allowed to cool and then exposed to the arc lamp to measure diffusivity again.
5	19, 20	Butler, C. P. and Inn, E. C. Y.	1957	373-1183	5-10	Ti-6Al-4V									Above specimen allowed to cool and then exposed to the arc lamp to measure diffusivity again.
6	146, 4	Jenkins, R. J., Parker, W. J., Butler, C. P., and Abbott, G. L.	1960	295-408	± 5	Ti-6Al-4V									Square specimen 1.9 cm side and 0.100 cm thick; high intensity short duration light pulse absorbed in front surface of thermally insulated specimen coated with camphor black; 408.2 K measurements obtained by heating sample holder and sample with an infrared lamp.
7	53	Karagezyan, A. G.	1962	396-1083		T3	94.5	3	→		→	→	→		2.5 total B, Cr, Fe and Si (Ti by difference); cylindrical rod specimen 3 mm in diameter and 300 mm long; vacuum annealed for 5 hrs at 993.2 K; measured in a vacuum of ~10^{-4} mm Hg; electrical resistivity reported as 122, 136, 147, 163, 163, 163, 168, and 168 μohm cm at 351.2, 492.2, 638.2, 783.2, 829.2, 886.2, 1027.2, and 1079.2 K, respectively.

SPECIFICATION TABLE 131. THERMAL DIFFUSIVITY OF [TITANIUM + ALUMINUM + ΣX_i] ALLOYS (continued)

Cur. No.	Ref. No.	Author(s)	Year	Temp. Range, K	Reported Error, %	Name and Specimen Designation	Ti	Al	B	C	Cr	Fe	Si	V	Composition (continued), Specifications, and Remarks
8	53	Karagezyan, A. G.	1962	403-1110		T4	93.0	4.5	→		→	→	→		2.5 total B, Cr, Fe and Si (Ti by difference); cylindrical rod specimen 3 mm in diameter and 300 mm long; vacuum annealed for 5 hrs at 993.2 K; measured in a vacuum of ~10^{-4} mm Hg; electrical resistivity reported as 132, 143, 146, 159, 168, 172, 173, 177, 173, 173, 174, and 174 μohm cm at 299.2, 434.2, 475.2, 638.2, 723.2, 782.2, 842.2, 894.2, 932.2, 948.2, 1018.2, and 1106.2 K, respectively.
9	53	Karagezyan, A. G.	1962	428-1126		T6	91.5	6.0	→		→	→	→		2.5 total B, Cr, Fe and Si (Ti by difference); cylindrical rod specimen 3 mm in diameter and 300 mm long; vacuum annealed for 5 hrs at 993.2 K; measured in a vacuum of ~10^{-4} mm Hg; electrical resistivity reported as 163, 174, 182, 181, 185, 187, 190, 190, 190, and 190 μohm cm at 358.2, 554.2, 655.2, 719.2, 798.2, 844.2, 929.2, 989.2, 1050.2, and 1118.2 K, respectively.
10	53	Karagezyan, A. G.	1962	423-1123		T8	90.0	7.5	→		→	→	→		2.5 total B, Cr, Fe and Si (Ti by difference); cylindrical rod specimen 3 mm in diameter and 300 mm long; vacuum annealed for 5 hrs at 993.2 K; measured in a vacuum of ~10^{-4} mm Hg; electrical resistivity reported as 177, 178, 188, 191, 191, 192, 193, 193, 195, and 197 μohm cm at 299.2, 361.2, 537.2, 722.2, 797.2, 866.2, 936.2, 989.2, 1049.2, and 1118.2 K, respectively.

DATA TABLE 131. THERMAL DIFFUSIVITY OF [TITANIUM + ALUMINUM + ΣX_i] ALLOYS

(Ti + Al < 99.50% or at least one X_i > 0.20%)

[Temperature, T, K; Thermal Diffusivity, α, cm^2 s^{-1}]

T	α
CURVE 1	
348.2	0.026
CURVE 2	
393.2	0.027
498.2	0.028
638.2	0.039
CURVE 3	
398.2	0.026
518.2	0.032
573.2	0.029
758.2	0.044
818.2	0.047
918.2	0.047
1118.2	0.045
1118.2	0.048
CURVE 4	
448.2	0.026
578.2	0.027
688.2	0.032
878.2	0.037
1023.2	0.038
CURVE 5	
373.2	0.020
463.2	0.028
633.2	0.032
848.2	0.037
848.2	0.040
1063.2	0.042
1063.2	0.047
1183.2	0.048
CURVE 6	
295.2	0.019
295.2	0.027
408.2	0.020
CURVE 7	
396	0.0450
502	0.0460
653	0.0660
780	0.0630
833	0.0590
889	0.0385
1033	0.0395
1083	0.0370
CURVE 8	
403	0.0533
413	0.0513
513	0.0525
620	0.0495
733	0.0490
798	0.0385
848	0.0350
896	0.0557
928	0.0443
950	0.0370
1023	0.0250
1110	0.0265
CURVE 9	
428	0.0345
433	0.0360
556	0.0405
651	0.0380
746	0.0395
818	0.0400
938	0.0205
996	0.0215
1056	0.0205
1126	0.0185
CURVE 10	
423	0.0395
573	0.0405
738	0.0405
808	0.0395
CURVE 10 (cont.)	
876	0.0405
938	0.0365
1048	0.0237
1123	0.0280

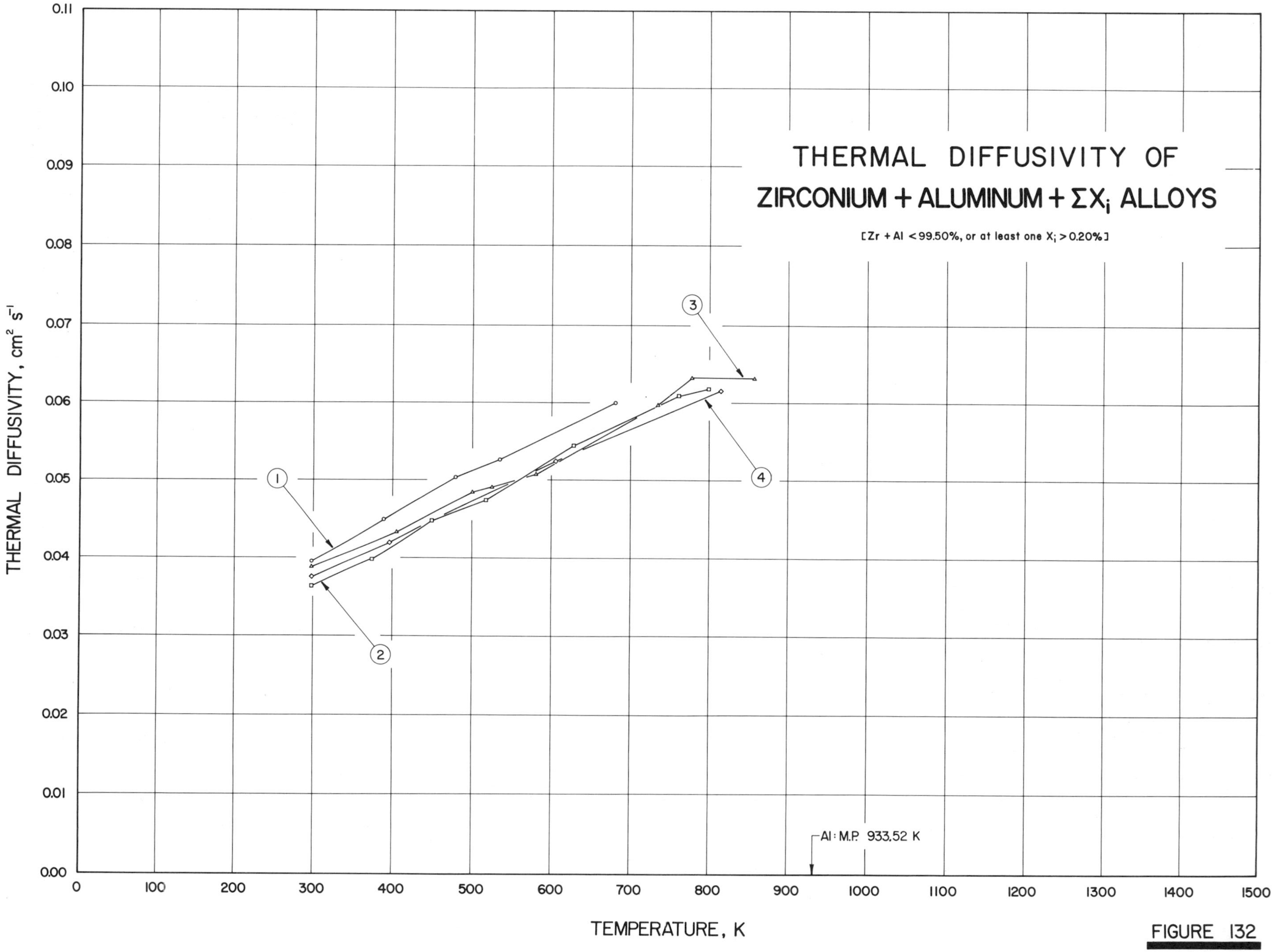
THERMAL DIFFUSIVITY OF
ZIRCONIUM + ALUMINUM + ΣX$_i$ ALLOYS
[Zr + Al < 99.50%, or at least one X$_i$ > 0.20%]
THERMAL DIFFUSIVITY, cm^2 s^{-1}
TEMPERATURE, K
Al: M.P. 933.52 K
0.00
0.01
0.02
0.03
0.04
0.05
0.06
0.07
0.08
0.09
0.10
0.11
0
100
200
300
400
500
600
700
800
900
1000
1100
1200
1300
1400
1500
1
2
3
4

FIGURE 132

SPECIFICATION TABLE 132. THERMAL DIFFUSIVITY OF [ZIRCONIUM + ALUMINUM + ΣX_i] ALLOYS

($Zr + Al < 99.50\%$ or at least one $X_i > 0.20\%$)

Cur. No.	Ref. No.	Author(s)	Year	Temp. Range, K	Reported Error, %	Name and Specimen Designation	Composition (weight percent) Zr	Al	Mo	Sn	Composition (continued), Specifications, and Remarks
1	98	Taylor, R. E.	1963	298-681		3ZI; 1	→	1.30	0.85	1.13	96.72 Zr (by difference); aspirin-sized specimen; density 6.425 g cm^{-3}; laser beam employed to measure diffusivity.
2	98	Taylor, R. E.	1963	298-799		3ZI; 2					Specimen having same composition and shape and measured under same conditions as above specimen.
3	98	Taylor, R. E.	1963	298-857		3ZI; 3					Specimen having same composition and shape and measured under same conditions as above specimen.
4	98	Taylor, R. E.	1963	298-815		3ZI; 4					Specimen having same composition and shape and measured under same conditions as above specimen.

DATA TABLE 132. THERMAL DIFFUSIVITY OF [ZIRCONIUM + ALUMINUM + ΣX_i] ALLOYS

($Zr + Al < 99.50\%$ or at least one $X_i > 0.20\%$)

[Temperature, T, K; Thermal Diffusivity, α, $cm^2\ s^{-1}$]

T	α	T	α
CURVE 1		CURVE 3	
298.2	0.0395	298.2	0.0388
389.2	0.0450	405.2	0.0434
479.2	0.0503	500.2	0.0484
535.2	0.0526	525.2	0.0491
681.2	0.0600	580.2	0.0507
		735.2	0.0597
CURVE 2		788.2	0.0632
		857.2	0.0632
298.2	0.0364		
374.2	0.0398	CURVE 4	
449.2	0.0448		
517.2	0.0474	298.2	0.0375
628.2	0.0544	396.2	0.0420
761.2	0.0609	605.2	0.0524
799.2	0.0618	815.2	0.0615

4. FERROUS ALLOYS

A. Carbon Steels

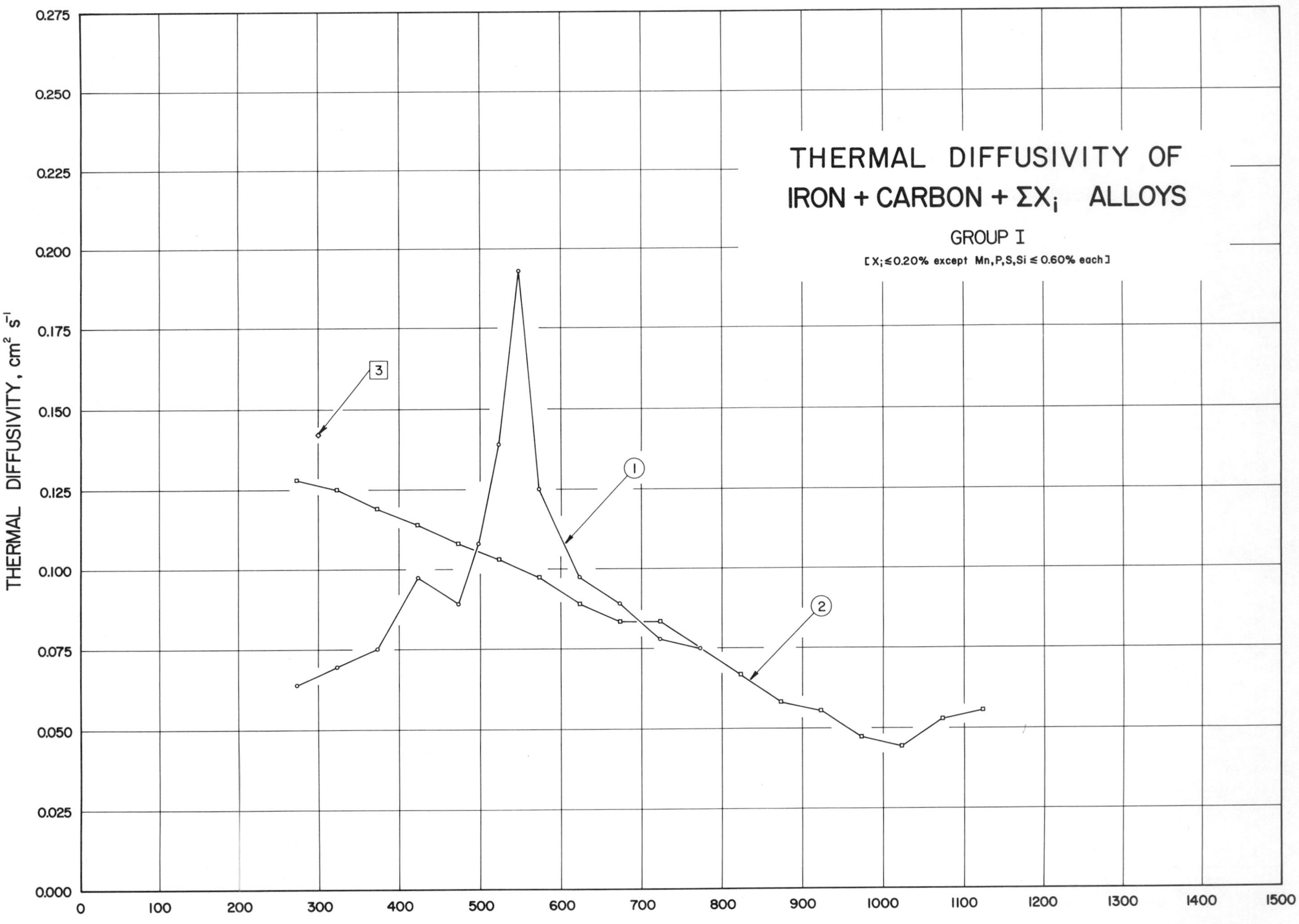
THERMAL DIFFUSIVITY OF
IRON + CARBON + ΣXi ALLOYS
GROUP I
[Xi ≤ 0.20% except Mn, P, S, Si ≤ 0.60% each]
THERMAL DIFFUSIVITY, cm² s⁻¹
TEMPERATURE, K
0.275
0.250
0.225
0.200
0.175
0.150
0.125
0.100
0.075
0.050
0.025
0.000
0
100
200
300
400
500
600
700
800
900
1000
1100
1200
1300
1400
1500
1
2
3

FIGURE 133

SPECIFICATION TABLE 133. THERMAL DIFFUSIVITY OF [IRON + CARBON + ΣX_i] ALLOYS (GROUP I)

($X_i \leq 0.20\%$ except Mn, P, S, Si $\leq 0.60\%$ each)

Cur. No.	Ref. No.	Author(s)	Year	Temp. Range, K	Reported Error, %	Name and Specimen Designation	Fe	C	Mn	P	S	Si	Composition (continued), Specifications, and Remarks
1	84	Neimark, B. E. and Lyusternik, V. E.	1960	273-773	±3.2	U 8	98.46	0.81	0.47	0.020	0.020	0.22	(Fe by difference); chilled; thermal diffusivity obtained from measured conductivity, specific heat, and density.
2	84	Neimark, B. E. and Lyusternik, V. E.	1960	273-1123	±3.2	U 8	98.46	0.81	0.47	0.020	0.020	0.22	(Fe by difference); annealed; thermal diffusivity obtained from measured conductivity, specific heat, and density.
3	175	Oualid, J.	1961	298.2									Slab specimen; thermal shock method used to measure diffusivity; measured in vacuum; temperature of measurement not given by author but assumed to be room temperature.

(Composition columns Fe, C, Mn, P, S, Si: Composition (weight percent))

DATA TABLE 133. THERMAL DIFFUSIVITY OF [IRON + CARBON + ΣX_i] ALLOYS (GROUP I)

($X_i \leq 0.20\%$ except Mn, P, S, Si $\leq 0.60\%$ each)

[Temperature, T, K; Thermal Diffusivity, α, $cm^2\ s^{-1}$]

T	α	T	α	T	α
CURVE 1		CURVE 2		CURVE 2 (cont.)	
273.2	0.0639	273.2	0.128	1023.2	0.0444
323.2	0.0695	323.2	0.125	1073.2	0.0528
373.2	0.0750	373.2	0.119	1123.2	0.0556
423.2	0.0972	423.2	0.114		
473.2	0.0889	473.2	0.108	CURVE 3	
498.2	0.108	523.2	0.103		
523.2	0.139	573.2	0.0972	298.2	0.142
548.2	0.193	623.2	0.0889		
573.2	0.125	673.2	0.0833		
623.2	0.0972	723.2	0.0833		
673.2	0.0889	773.2	0.0750*		
723.2	0.0778	823.2	0.0667		
773.2	0.0750	873.2	0.0583		
		923.2	0.0556		
		973.2	0.0472		

* Not shown in figure.

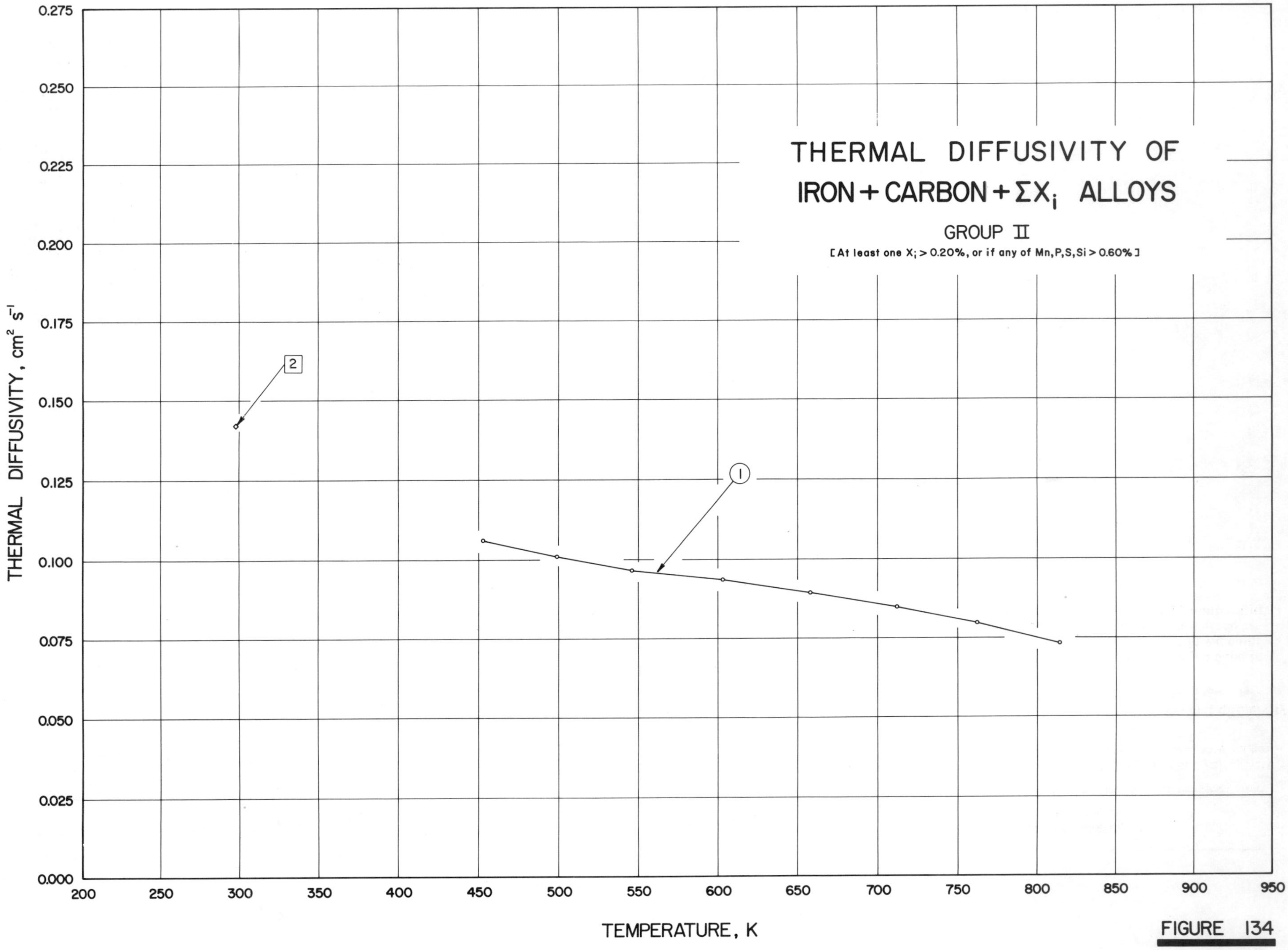
THERMAL DIFFUSIVITY OF
IRON + CARBON + ΣXi ALLOYS
GROUP II
[At least one Xi > 0.20%, or if any of Mn, P, S, Si > 0.60%]
THERMAL DIFFUSIVITY, cm² s⁻¹
TEMPERATURE, K
0.275
0.250
0.225
0.200
0.175
0.150
0.125
0.100
0.075
0.050
0.025
0.000
200
250
300
350
400
450
500
550
600
650
700
750
800
850
900
950
1
2

FIGURE 134

SPECIFICATION TABLE 134. THERMAL DIFFUSIVITY OF [IRON + CARBON + ΣX_i] ALLOYS (GROUP II)

(At least one $X_i > 0.20\%$ and/or any of Mn, P, S, Si $> 0.60\%$)

Cur. No.	Ref. No.	Author(s)	Year	Temp. Range, K	Reported Error, %	Name and Specimen Designation	Composition (weight percent) Fe	C	Mn	P	S	Si	Composition (continued), Specifications, and Remarks
1	176, 21	El-Hifni, M. A. and Chao, B. T.	1955	454-814	2-3	1.1 C tool steel	97.87 min	1.1	0.7	0.04 max	0.04 max	0.25	(Fe by difference); tubular specimen 0.875 in. O. D. and 5 in. long, relatively thin walled; another tube of similar material mounted concentrically with specimen to minimize heat losses; one end of specimen-and-shield assembly immersed in a liquid bath, while the other end supported by a transite disc for insulation; cyclic varying current generates required temperature wave at heating bath; heat supplied by electric current removed by a forced draft of cooling air; one-dimensional heat flow; a minimum of three complete temperature waves recorded.
2	175	Oualid, J.	1961	298.2									Slab specimen; thermal shock method used to measure diffusivity; measured in vacuum; temperature of measurement not given by author but assumed to be room temperature.

DATA TABLE 134. THERMAL DIFFUSIVITY OF [IRON + CARBON + ΣX_i] ALLOYS (GROUP II)

(At least one $X_i > 0.20\%$ and/or any of Mn, P, S, Si $> 0.60\%$)

[Temperature, T, K; Thermal Diffusivity, α, $cm^2\ s^{-1}$]

T	α
CURVE 1	
454	0.106
499	0.101
547	0.0965
603	0.0935
658	0.0893
713	0.0847
762	0.0797
814	0.0732
CURVE 2	
298.2	0.142

4. FERROUS ALLOYS (continued)

B. Alloy Steels

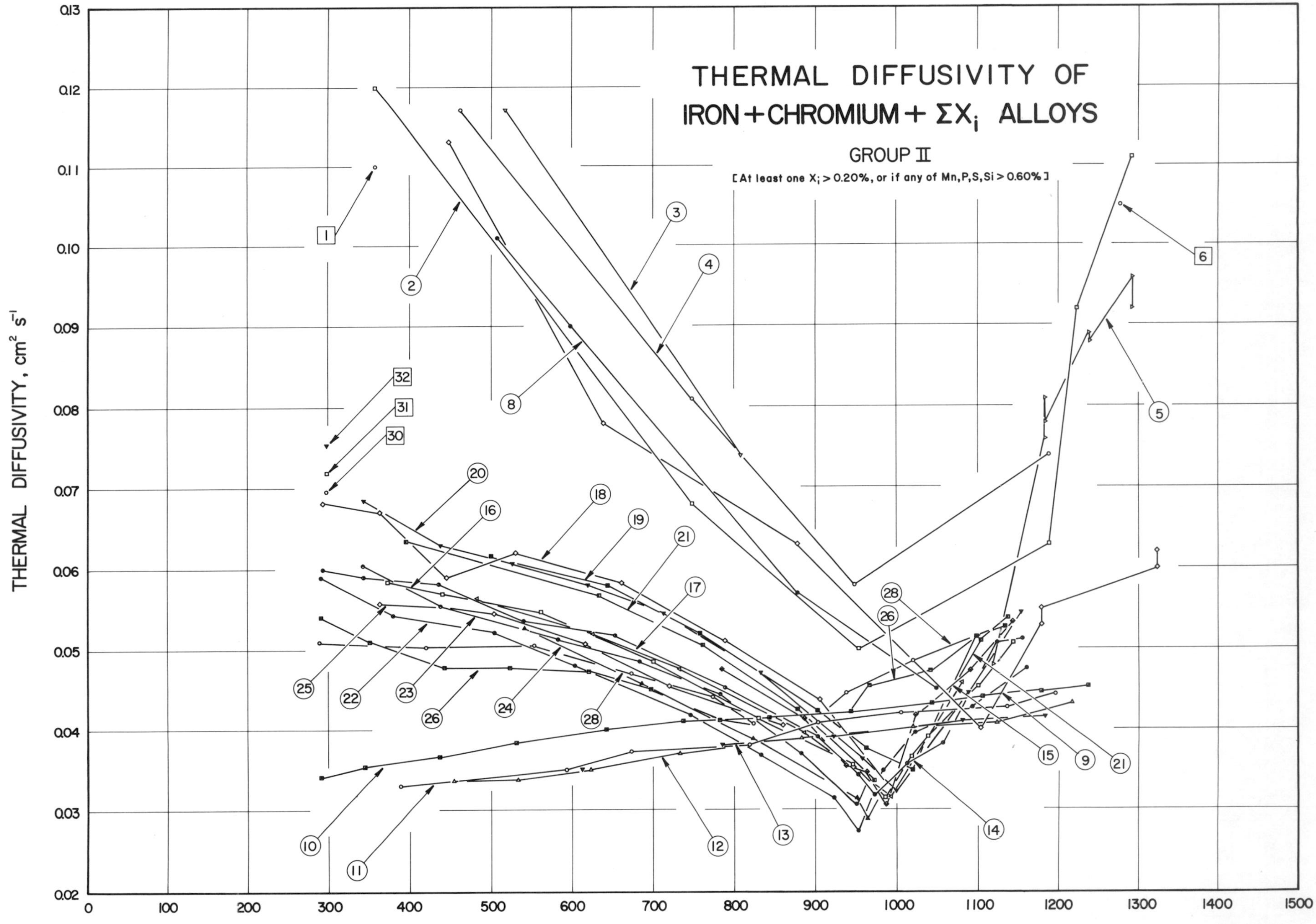
THERMAL DIFFUSIVITY OF
IRON + CHROMIUM + ΣX_i ALLOYS
GROUP II
[At least one X_i > 0.20%, or if any of Mn, P, S, Si > 0.60%]
THERMAL DIFFUSIVITY, cm² s⁻¹
TEMPERATURE, K
0.02
0.03
0.04
0.05
0.06
0.07
0.08
0.09
0.10
0.11
0.12
0.13
0
100
200
300
400
500
600
700
800
900
1000
1100
1200
1300
1400
1500
1
2
3
4
5
6
8
9
10
11
12
13
14
15
16
17
18
19
20
21
22
23
24
25
26
28
30
31
32

FIGURE 135

SPECIFICATION TABLE 135. THERMAL DIFFUSIVITY OF [IRON + CHROMIUM + ΣX_i] ALLOYS (GROUP II)

(At least one $X_i > 0.20\%$ and/or any of Mn, P, S, Si $> 0.60\%$)

Cur. No.	Ref. No.	Author(s)	Year	Temp. Range, K	Reported Error, %	Name and Specimen Designation	Composition (weight percent) Fe	Cr	C	Mn	Mo	P	S	Si	Composition (continued), Specifications, and Remarks
1	19, 20	Butler, C. P. and Inn, E. C. Y.	1957	358.2	5-10	SAE 4130	→	0.80/ 1.10	0.28/ 0.33	0.40/ 0.60	0.15/ 0.25	0.040	0.040	0.20/ 0.35	97.29-98.09 Fe (by difference); ladle chemical composition from Metals Handbook, Vol. 1, 8th ed., p. 61, 1961; cylindrical specimen 0.9525 cm in diameter and length lying in the range from 1 to 2.5 cm; subjected to irradiance from carbon arc lamp heat source; spectral distribution approximates that of a 5700 K black body source; specimen blackened with camphor black; measured under a vacuum of ~5 microns.
2	19, 20	Butler, C. P. and Inn, E. C. Y.	1957	358-1293	5-10	SAE 4130									Above specimen allowed to cool and then exposed to the arc lamp to measure diffusivity again.
3	19, 20	Butler, C. P. and Inn, E. C. Y.	1957	518, 808	5-10	SAE 4130									Above specimen allowed to cool and then exposed to the arc lamp to measure diffusivity again.
4	19, 20	Butler, C. P. and Inn, E. C. Y.	1957	463-1188	5-10	SAE 4130									Above specimen allowed to cool and then exposed to the arc lamp to measure diffusivity again.
5	19, 20	Butler, C. P. and Inn, E. C. Y.	1957	1118-1293	5-10	SAE 4130									Above specimen allowed to cool and then exposed to the arc lamp to measure diffusivity again.
6	19, 20	Butler, C. P. and Inn, E. C. Y.	1957	1278.2	5-10	SAE 4130									Above specimen allowed to cool and then exposed to the arc lamp to measure diffusivity again.
7*	19, 20	Butler, C. P. and Inn, E. C. Y.	1957	748-1293	5-10	SAE 4130									Above specimen allowed to cool and then exposed to the arc lamp to measure diffusivity again.
8	19, 20	Butler, C. P. and Inn, E. C. Y.	1957	508-1048	5-10	SAE 4130									Above specimen allowed to cool and then exposed to the arc lamp to measure diffusivity again.
9	19, 20	Butler, C. P. and Inn, E. C. Y.	1957	448-1323	5-10	SAE 4130									Above specimen allowed to cool and then exposed to the arc lamp to measure diffusivity again.
10	85, 4	Jenkins, R. J. and Westover, R. W.	1960	292-1236	±5	AISI-202	64.58/ 69.08	17.9	0.15 max	7.5/ 10		0.06 max	0.06 max	1 max	0.25 max N, and 4-6 Ni, (Fe by difference); nominal composition from Crucible Data Book, Crucible Steel Co. of America, as stated by authors; austenitic; cylindrical specimen 1.27 cm in diameter and 0.065 cm thick; provided by the Crucible Steel Co. of America; cut from No. 2 finish cold rolled sheet steel of thickness indicated; subjected to a minimum of working, and used without further surface preparation (surface very smooth but not a mirror finish); not subjected to any type of heating before the test runs; thermal pulses supplied by a xenon flash tube; diffusivity measured while specimen heated up during 1st cycle.

* Not shown in figure.

SPECIFICATION TABLE 135. THERMAL DIFFUSIVITY OF [IRON + CHROMIUM + ΣX_i] ALLOYS (GROUP II) (continued)

Cur. No.	Ref. No.	Author(s)	Year	Temp. Range, K	Reported Error, %	Name and Specimen Designation	Fe	Cr	C	Mn	Mo	P	S	Si	Composition (continued), Specifications, and Remarks
11	85, 4	Jenkins, R. J. and Westover, R. W.	1960	389-1195	± 5	AISI-202									Above specimen measured during cooling.
12	85, 4	Jenkins, R. J. and Westover, R. W.	1960	455-1216	± 5	AISI-202									Above specimen measured again during heating in the 2nd cycle.
13	85, 4	Jenkins, R. J. and Westover, R. W.	1960	613-1183	± 5	AISI-202									Above specimen measured during cooling.
14	85, 4	Jenkins, R. J. and Westover, R. W.	1960	293-1155	± 5	AISI 410	→	11.5/13.5	0.15 max	1 max		0.04 max	0.03 max	1 max	84.28-86.28 Fe (by difference); nominal composition from Crucible Data Book, Crucible Steel Co. of America, as stated by authors; martensitic; cylindrical specimen 1.27 cm in diameter and 0.065 cm thick; provided by the Crucible Steel Co. of America; cut from No. 2 finish cold rolled sheet steel of thickness indicated, subjected to a minimum of working, and used without further surface preparation (surface very smooth but not a mirror finish); not subjected to any type of heating before the test runs; thermal pulses supplied by a xenon flash tube; diffusivity measured while specimen heated up during 1st cycle.
15	85, 4	Jenkins, R. J. and Westover, R. W.	1960	784-1143	± 5	AISI 410									Above specimen measured during cooling.
16	85, 4	Jenkins, R. J. and Westover, R. W.	1960	373-1143	± 5	AISI 410									Above specimen measured again during heating in the 2nd cycle.
17	85, 4	Jenkins, R. J. and Westover, R. W.	1960	483-1080	± 5	AISI 410									Above specimen measured during cooling.
18	85, 4	Jenkins, R. J. and Westover, R. W.	1960	293-1140	± 5	AISI 416	→	12/14	0.15 max	1.25 max	0.6	0.06	0.15 max	1 max	82.19-84.19 Fe (by difference), and 0.6 Zr; nominal composition from Crucible Data Book, Crucible Steel Co. of America, as stated by authors; martensitic; cylindrical specimen 1.27 cm in diameter and 0.177 cm thick; provided by the Crucible Steel Co. of America; cut from one half in. diameter rod; specimen had surfaces polished on a metal polishing lap, subjected to a minimum of working; not subjected to any type of heating before the test runs; thermal pulses supplied by a xenon flash tube; diffusivity measured while specimen heated up during 1st cycle.
19	85, 4	Jenkins, R. J. and Westover, R. W.	1960	500-1133	± 5	AISI 416									Above specimen measured during cooling.

SPECIFICATION TABLE 135. THERMAL DIFFUSIVITY OF [IRON + CHROMIUM + ΣX_i] ALLOYS (GROUP II) (continued)

Cur. No.	Ref. No.	Author(s)	Year	Temp. Range, K	Reported Error, %	Name and Specimen Designation	Fe	Cr	C	Mn	Mo	P	S	Si	Composition (continued), Specifications, and Remarks
20	85, 4	Jenkins, R. J. and Westover, R. W.	1960	343-1153	± 5	AISI 416									Above specimen measured again during heating in the 2nd cycle.
21	85, 4	Jenkins, R. J. and Westover, R. W.	1960	396-1136	± 5	AISI 416									Above specimen measured during cooling.
22	85, 4	Jenkins, R. J. and Westover, R. W.	1960	291-1160	± 5	AISI 430	→	14/18	0.12	1 max		0.04 max	0.03 max	1 max	79.81-83.81 Fe (by difference); nominal composition from Crucible Data Book, Crucible Steel Co. of America, as stated by authors; martensitic; cylindrical specimen 1.27 cm in diameter and 0.089 cm thick; provided by the Crucible Steel Co. of America; cut from No. 2 finish cold rolled sheet steel of thickness indicated, subjected to a minimum of working, and used without further surface preparation (surface very smooth but not a mirror finish); not subjected to any type of heating before the test runs; thermal pulses supplied by a xenon flash tube; diffusivity measured while specimen heated up during 1st cycle.
23	85, 4	Jenkins, R. J. and Westover, R. W.	1960	343-1090	± 5	AISI 430									Above specimen measured during cooling.
24	85, 4	Jenkins, R. J. and Westover, R. W.	1960	541-1118	± 5	AISI 430									Above specimen measured again during heating in the 2nd cycle.
25	85, 4	Jenkins, R. J. and Westover, R. W.	1960	363-1098	± 5	AISI 430									Above specimen measured during cooling.
26	85, 4	Jenkins, R. J. and Westover, R. W.	1960	291-1148	± 5	AISI 446	→	23/27	0.2	1.5 max		0.04 max	0.03 max	1 max	69.98-73.98 Fe (by difference), and 0.25 N; nominal composition from Crucible Data Book, Crucible Steel Co. of America, as stated by authors; martensitic; cylindrical specimen 1.27 cm in diameter and 0.178 cm thick; provided by the Crucible Steel Co. of America; cut from No. 2 finish cold rolled sheet steel of thickness indicated, subjected to a minimum of working, and used without further surface preparation (surface very smooth but not a mirror finish); not subjected to any type of heating before the test runs; thermal pulses supplied by a xenon flash tube; diffusivity measured while specimen heated up during 1st cycle.
27*	85, 4	Jenkins, R. J. and Westover, R. W.	1960	536-1088	± 5	AISI 446									Above specimen measured during cooling.
28	85, 4	Jenkins, R. J. and Westover, R. W.	1960	289-1173	± 5	AISI 446									Above specimen measured again during heating in the 2nd cycle.

* Not shown in figure.

SPECIFICATION TABLE 135. THERMAL DIFFUSIVITY OF [IRON + CHROMIUM + ΣX_i] ALLOYS (GROUP II) (continued)

Cur. No.	Ref. No.	Author(s)	Year	Temp. Range, K	Reported Error, %	Name and Specimen Designation	Composition (weight percent) Fe	Cr	C	Mn	Mo	P	S	Si	Composition (continued), Specifications, and Remarks
29*	85, 4	Jenkins, R. J. and Westover, R. W.	1960	693-1090	± 5	AISI 446									Above specimen measured during cooling.
30	86	Woisard, E. L.	1961	298. 2	± 4	GX 4881	90. 22	5. 06	0. 36		1. 64			1. 13	and 1. 59 W (Fe by difference); specimen consists of two long thin rods each 3/16 in. in diameter and ~25 cm long; developed by Bethlehem Steel Co.; specimen rods butted against a thin disc-shaped heater and held in alignment under compression; entire assembly placed inside heavy walled copper tube 22 in. long acting as isothermal enclosure; test carried out under a vacuum of 10^{-4} mm Hg; momentary pulse of electric current passed through rods.
31	86	Woisard, E. L.	1961	298. 2	± 4	23D 245	→	5. 22	0. 39	0. 39	1. 28	0. 01	0. 001	1. 11	90. 259 Fe (by difference), 0. 17 Cu, 0. 11 Ni, 1. 02 V, and 0. 04 W; specimen consists of two long thin rods each 3/16 in. in diameter and ~25 cm long; developed by Bethlehem Steel Co.; specimen rods butted against a thin disc-shaped heater and held in alignment under compression; entire assembly placed inside heavy walled copper tube 22 in. long acting as isothermal enclosure; test carried out under a vacuum of 10^{-4} mm Hg; momentary pulse of electric current passed through rods.
32	86	Woisard, E. L.	1961	298. 2	± 4	HX 4249	91. 12	5. 04	0. 40	0. 39	1. 28			1. 33	and 0. 44 V (Fe by difference); specimen consists of two long thin rods each 3/16 in. in diameter and ~25 cm long; developed by Bethlehem Steel Co.; specimen rods butted against a thin disc-shaped heater and held in alignment under compression; entire assembly placed inside heavy walled copper tube 22 in. long acting as isothermal enclosure; test carried out under a vacuum of 10^{-4} mm Hg; momentary pulse of electric current passed through rods.

* Not shown in figure.

DATA TABLE 135. THERMAL DIFFUSIVITY OF [IRON + CHROMIUM + ΣX_i] ALLOYS (GROUP II)

(At least one $X_i > 0.20\%$ and/or any of Mn, P, S, Si $> 0.60\%$)

[Temperature, T, K; Thermal Diffusivity, α, $cm^2\ s^{-1}$]

T	α
CURVE 1	
358.2	0.110
CURVE 2	
358.2	0.120
748.2	0.068
953.2	0.050
1188.2	0.063
1223.2	0.092
1293.2	0.111
CURVE 3	
518.2	0.117
808.2	0.074
CURVE 4	
463.2	0.117
748.2	0.081
948.2	0.058
1188.2	0.074
CURVE 5	
1118.2	0.048
1183.2	0.076
1183.2	0.081
1183.2	0.078
1238.2	0.089
1238.2	0.088
1293.2	0.096
1293.2	0.092
CURVE 6	
1278.2	0.105
CURVE 7*	
748.2	0.068
953.2	0.050
CURVE 7 (cont.)*	
1188.2	0.063
1223.2	0.092
1293.2	0.105
CURVE 8	
508.2	0.101
598.2	0.090
878.2	0.057
1048.2	0.045
CURVE 9	
448.2	0.113
638.2	0.078
878.2	0.063
1103.2	0.040
1178.2	0.053
1178.2	0.055
1323.2	0.060
1323.2	0.062
CURVE 10	
291.7	0.0341
345.2	0.0354
437.2	0.0366
531.2	0.0383
642.2	0.0400
737.2	0.0410
843.2	0.0414
943.2	0.0421
1043.2	0.0432
1105.2	0.0440
1178.2	0.0446
1236.2	0.0453
CURVE 11	
389.2	0.0330
593.2	0.0350
673.2	0.0372
CURVE 11 (cont.)	
819.2	0.0380
903.2	0.0408
1005.2	0.0420
1135.2	0.0427
1195.2	0.0443
CURVE 12	
455.2	0.0336
533.2	0.0337
623.2	0.0350
733.2	0.0370
883.2	0.0388
1020.2	0.0402
1123.2	0.0407
1216.2	0.0432
CURVE 13	
613.2	0.0350
786.2	0.0380
923.2	0.0390
1081.2	0.0410
1183.2	0.0415
CURVE 14	
293.2	0.0600
344.2	0.0591
436.2	0.0582
540.2	0.0536
653.2	0.0517
789.2	0.0452
903.2	0.0390
953.2	0.0343
973.2	0.0317
1013.2	0.0356
1057.2	0.0382
1123.2	0.0507
1155.2	0.0512
CURVE 15	
784.2	0.0475
886.2	0.0413
938.2	0.0354
963.2	0.0345
988.2	0.0305
1023.2	0.0417
1090.2	0.0473
1143.2	0.0534
CURVE 16	
373.2	0.0585
440.2	0.0570
561.2	0.0547
700.2	0.0485
830.2	0.0413
946.2	0.0355
986.2	0.0314
1018.2	0.0365
1038.2	0.0390
1100.2	0.0453
1143.2	0.0507
CURVE 17	
483.2	0.0565
626.2	0.0520
733.2	0.0475
886.2	0.0395
993.2	0.0315
1080.2	0.0458
CURVE 18	
293.2	0.0682
363.2	0.0670
445.2	0.0590
530.2	0.0620
661.2	0.0583
789.2	0.0510
905.2	0.0436
996.2	0.0320*
CURVE 18 (cont.)	
1076.2	0.0454*
1140.2	0.0527*
CURVE 19	
500.2	0.0617
643.2	0.0580
758.2	0.0520
903.2	0.0423
963.2	0.0375
1020.2	0.0348
1098.2	0.0515
1133.2	0.0527
CURVE 20	
343.2	0.0685
438.2	0.0630
526.2	0.0607
620.2	0.0580
713.2	0.0545
833.2	0.0470
958.2	0.0362
1000.2	0.0322
1088.2	0.0445
1153.2	0.0545
CURVE 21	
396.2	0.0635
633.2	0.0567
761.2	0.0505
878.2	0.0425
973.2	0.0335
998.2	0.0335*
1018.2	0.0360*
1103.2	0.0510
1136.2	0.0539
CURVE 22	
291.2	0.0590
380.2	0.0543
504.2	0.0521
603.2	0.0480
746.2	0.0418
833.2	0.0367
923.2	0.0314
953.2	0.0273
983.2	0.0348
1023.2	0.0395
1093.2	0.0427
1160.2	0.0475
CURVE 23	
343.2	0.0605
438.2	0.0555
583.2	0.0513
683.2	0.0485
783.2	0.0443
883.2	0.0369
951.2	0.0305
973.2	0.0352*
1090.2	0.0450*
CURVE 24	
541.2	0.0527
686.2	0.0457
823.2	0.0388
951.2	0.0313
964.2	0.0288
1013.2	0.0380*
1118.2	0.0468*
CURVE 25	
363.2	0.0558
438.2	0.0555*
503.2	0.0545
616.2	0.0507
720.2	0.0454
CURVE 25 (cont.)	
860.2	0.0405
953.2	0.0287*
1098.2	0.0437*
CURVE 26	
291.2	0.0541
351.2	0.0510
443.2	0.0478
523.2	0.0477
621.2	0.0473
698.2	0.0450
783.2	0.0412
881.2	0.0398*
941.2	0.0420*
966.2	0.0454
1041.2	0.0473
1136.2	0.0543*
1148.2	0.0536*
CURVE 27*	
536.2	0.0507
655.2	0.0481
833.2	0.0427
903.2	0.0425
948.2	0.0450
988.2	0.0477
1088.2	0.0531
CURVE 28	
289.2	0.0510
420.2	0.0503
553.2	0.0505
673.2	0.0470
773.2	0.0440
823.2	0.0407
863.2	0.0397*
893.2	0.0405*
938.2	0.0445
1020.2	0.0485
CURVE 28 (cont.)	
1133.2	0.0530*
1146.2	0.0535*
1173.2	0.0545*
CURVE 29*	
693.2	0.0465
848.2	0.0410
876.2	0.0390
933.2	0.0425
1013.2	0.0470
1090.2	0.0513
CURVE 30	
298.2	0.0695
CURVE 31	
298.2	0.0719
CURVE 32	
298.2	0.0753

*Not shown in figure.

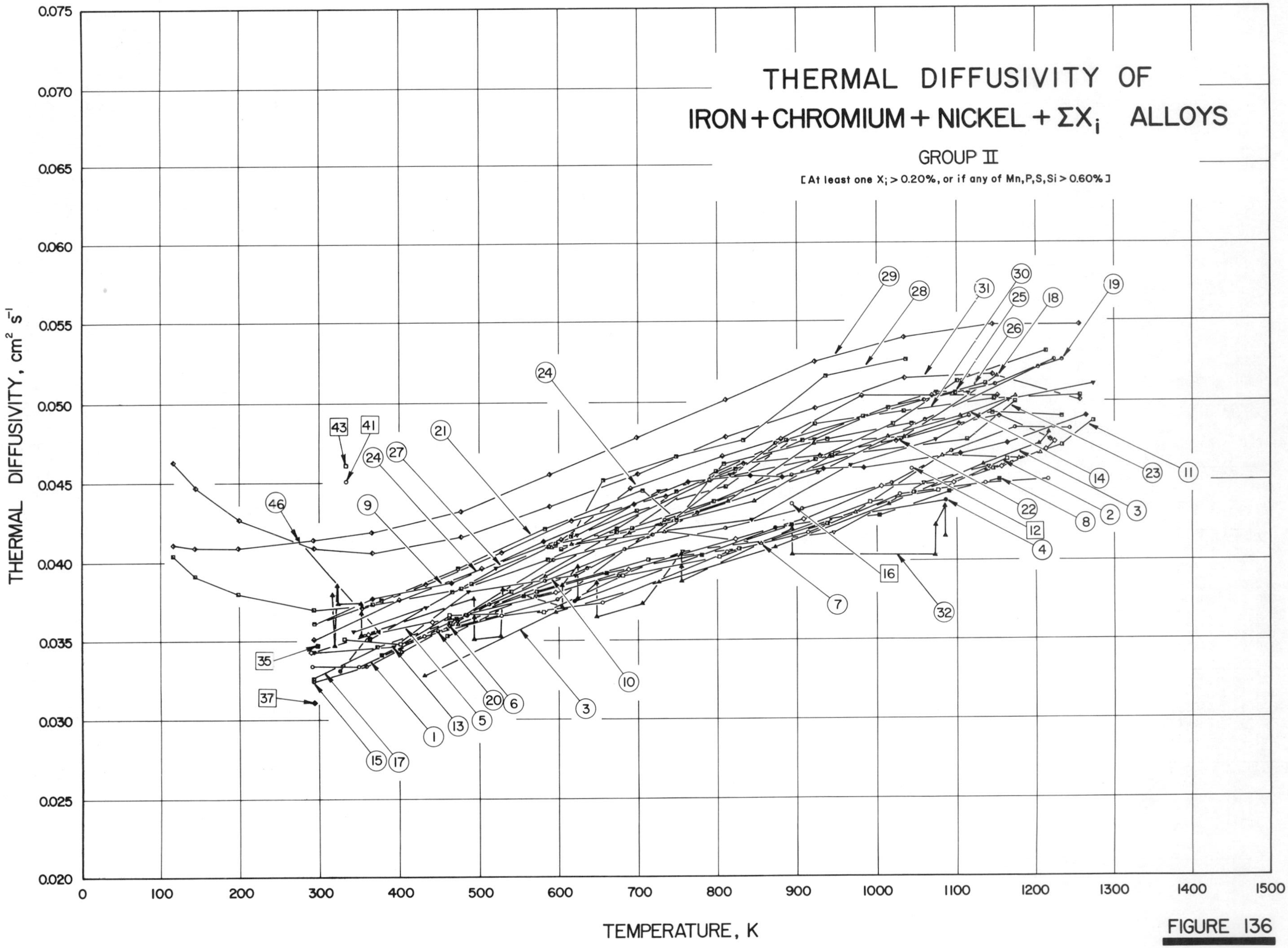
THERMAL DIFFUSIVITY OF
IRON + CHROMIUM + NICKEL + ΣX_i ALLOYS
GROUP II
[At least one X_i > 0.20%, or if any of Mn,P,S,Si > 0.60%]
THERMAL DIFFUSIVITY, $cm^2\ s^{-1}$
TEMPERATURE, K
0.075
0.070
0.065
0.060
0.055
0.050
0.045
0.040
0.035
0.030
0.025
0.020
0
100
200
300
400
500
600
700
800
900
1000
1100
1200
1300
1400
1500
1
2
3
4
5
6
7
8
9
10
11
12
13
14
15
16
17
18
19
20
21
22
23
24
25
26
27
28
29
30
31
32
35
37
41
43
46

FIGURE 136

SPECIFICATION TABLE 136. THERMAL DIFFUSIVITY OF [IRON + CHROMIUM + NICKEL + ΣX_i] ALLOYS (GROUP II)

(At least one $X_i > 0.20\%$ and/or any of Mn, P, S, Si $> 0.60\%$)

Cur. No.	Ref. No.	Author(s)	Year	Temp. Range, K	Reported Error, %	Name and Specimen Designation	Composition (weight percent) Fe	Cr	Ni	C	Mn	P	S	Si	Composition (continued), Specifications, and Remarks
1	85, 4	Jenkins, R.J. and Westover, R.W.	1960	291-1215	± 5	AISI 301	70.775/ 74.775	16/ 18	6/ 8	0.15 max	2 max	0.045 max	0.03 max	1 max	Nominal composition from Crucible Data Book, Crucible Steel Co. of America, as stated by authors, Fe percentage obtained by difference; austenitic; cylindrical specimen 0.5 in. in diameter and 0.097 cm thick; provided by the Crucible Steel Co. of America; cut from No. 2 finish cold rolled sheet steel of thickness indicated, subjected to a minimum of working, and used without further surface preparation (surface very smooth but not a mirror finish); not subjected to any type of heating before the test runs; thermal pulses supplied by a xenon flash tube; diffusivity measured while specimen heated up during 1st cycle.
2	85, 4	Jenkins, R.J. and Westover, R.W.	1960	331-1163	± 5	AISI 301									Above specimen measured during cooling.
3	85, 4	Jenkins, R.J. and Westover, R.W.	1960	431-1216	± 5	AISI 301									Above specimen measured again during heating in the 2nd cycle.
4	85, 4	Jenkins, R.J. and Westover, R.W.	1960	619-1086	± 5	AISI 301									Above specimen measured during cooling.
5	85	Jenkins, R.J. and Westover, R.W.	1960	290-1223	± 5	AISI 302	→	17/ 19	8/ 10	0.15 max	2 max	0.045 max	0.03 max	1 max	67.775-71.775 Fe (by difference); nominal composition from Crucible Data Book, Crucible Steel Co. of America, as stated by authors; austenitic; cylindrical specimen 0.5 in. in diameter and 0.096 cm thick; provided by the Crucible Steel Co. of America; cut from No.2 finish cold rolled sheet steel of thickness indicated, subjected to a minimum of working, and used without further surface preparation (surface very smooth but not a mirror finish); not subjected to any type of heating before the test runs; thermal pulses supplied by a xenon flash tube; diffusivity measured while specimen heated up during 1st cycle.
6	85	Jenkins, R.J. and Westover, R.W.	1960	461-1140	± 5	AISI 302									Above specimen measured during cooling.
7	85	Jenkins, R.J. and Westover, R.W.	1960	591-1205	± 5	AISI 302									Above specimen measured again during heating in the 2nd cycle.
8	85	Jenkins, R.J. and Westover, R.W.	1960	378-1153	± 5	AISI 302									Above specimen measured during cooling.

SPECIFICATION TABLE 136. THERMAL DIFFUSIVITY OF [IRON + CHROMIUM + NICKEL + ΣX_i] ALLOYS (GROUP II) (continued)

Cur. No.	Ref. No.	Author(s)	Year	Temp. Range, K	Reported Error, %	Name and Specimen Designation	Composition (weight percent) Fe	Cr	Ni	C	Mn	P	S	Si	Composition (continued), Specifications, and Remarks
9	85, 4	Jenkins, R.J. and Westover, R.W.	1960	291-1261	± 5	AISI 304	→	18/20	8/12	0.08 max	2 max	0.045 max	0.03 max	1 max	64.845-70.845 Fe (by difference); nominal composition from Crucible Data Book, Crucible Steel Co. of America, as stated by authors; austenitic; cylindrical specimen 0.5 in. in diameter and 0.080 cm thick; provided by the Crucible Steel Co. of America; cut from No. 2 finish cold rolled sheet steel of thickness indicated, subjected to a minimum of working, and used without further surface preparation (surface very smooth but not a mirror finish); not subjected to any type of heating before the test runs; thermal pulses supplied by a xenon flash tube; diffusivity measured while specimen heated up during 1st cycle.
10	85, 4	Jenkins, R.J. and Westover, R.W.	1960	403-1243	± 5	AISI 304									Above specimen measured during cooling.
11	85, 4	Jenkins, R.J. and Westover, R.W.	1960	460-1273	± 5	AISI 304									Above specimen measured again during heating in the 2nd cycle.
12	85, 4	Jenkins, R.J. and Westover, R.W.	1960	1043.2	± 5	AISI 304									Above specimen measured during cooling.
13	85, 4	Jenkins, R.J. and Westover, R.W.	1960	293-1273	± 5	AISI 309	→	22/24	12/15	0.20 max	2 max	0.045 max	0.03 max	1 max	57.725-62.725 Fe (by difference); nominal composition from Crucible Data Book, Crucible Steel Co. of America, as stated by authors; austenitic; cylindrical specimen 0.5 in. in diameter and 0.198 cm thick; provided by the Crucible Steel Co. of America; cut from No.2 finish cold rolled sheet steel of thickness indicated, subjected to a minimum of working, and used without further surface preparation (surface very smooth but not a mirror finish); not subjected to any type of heating before the test runs; thermal pulses supplied by a xenon flash tube; diffusivity measured while specimen heated up during 1st cycle.
14	85, 4	Jenkins, R.J. and Westover, R.W.	1960	616-1173	± 5	AISI 309									Above specimen measured during cooling.
15	85, 4	Jenkins, R.J. and Westover, R.W.	1960	293-1223	± 5	AISI 309									Above specimen measured again during heating in the 2nd cycle.
16	85, 4	Jenkins, R.J. and Westover, R.W.	1960	893.2	± 5	AISI 309									Above specimen measured during cooling.

SPECIFICATION TABLE 136. THERMAL DIFFUSIVITY OF [IRON + CHROMIUM + NICKEL + ΣX_i] ALLOYS (GROUP II) (continued)

Cur. No.	Ref. No.	Author(s)	Year	Temp. Range, K	Reported Error, %	Name and Specimen Designation	Composition (weight percent) Fe	Cr	Ni	C	Mn	P	S	Si	Composition (continued), Specifications, and Remarks
17	85, 4	Jenkins, R. J. and Westover, R. W.	1960	293-1213	± 5	AISI 316	→	16/18	10/14	0.1 max	2 max	0.045 max	0.03 max	1 max	61.825-68.825 Fe (by difference), and 2-3 Mo; nominal composition from Crucible Data Book, Crucible Steel Co. of America, as stated by authors; austenitic; cylindrical specimen 0.5 in. in diameter and 0.136 cm thick; provided by the Crucible Steel Co. of America; cut from No. 2 finish cold rolled sheet steel of thickness indicated, subjected to a minimum of working, and used without further surface preparation (surface very smooth but not a mirror finish); not subjected to any type of heating before the test runs; thermal pulses supplied by a xenon flash tube; diffusivity measured while specimen heated up during 1st cycle.
18	85, 4	Jenkins, R. J. and Westover, R. W.	1960	473-1150	± 5	AISI 316									Above specimen measured during cooling.
19	85, 4	Jenkins, R. J. and Westover, R. W.	1960	326-1233	± 5	AISI 316									Above specimen measured again during heating in the 2nd cycle.
20	85, 4	Jenkins, R. J. and Westover, R. W.	1960	448-1116	± 5	AISI 316									Above specimen measured during cooling.
21	85, 4	Jenkins, R. J. and Westover, R. W.	1960	293-1233	± 5	AISI 321	→	17/19	9/12	0.08 max	2 max	0.045 max	0.03 max	1 max	65.445-70.445 Fe (by difference), and 0.4 min Ti; nominal composition from Crucible Data Book, Crucible Steel Co. of America, as stated by authors; austenitic; cylindrical specimen 0.5 in. in diameter and 0.082 cm thick; provided by the Crucible Steel Co. of America; cut from No. 2 finish cold rolled sheet steel of thickness indicated, subjected to a minimum of working, and used without further surface preparation (surface very smooth but not a mirror finish); not subjected to any type of heating before the test runs; thermal pulses supplied by a xenon flash tube; diffusivity measured while specimen heated up during 1st cycle.
22	85, 4	Jenkins, R. J. and Westover, R. W.	1960	353-1218	± 5	AISI 321									Above specimen measured again during heating in the 2nd cycle.
23	85, 4	Jenkins, R. J. and Westover, R. W.	1960	603-1173	± 5	AISI 321									Above specimen measured during cooling.

SPECIFICATION TABLE 136. THERMAL DIFFUSIVITY OF [IRON + CHROMIUM + NICKEL + ΣX_i] ALLOYS (GROUP II) (continued)

Cur. No.	Ref. No.	Author(s)	Year	Temp. Range, K	Reported Error, %	Name and Specimen Designation	Composition (weight percent) Fe	Cr	Ni	C	Mn	P	S	Si	Composition (continued), Specifications, and Remarks
24	85, 4	Jenkins, R. J. and Westover, R. W.	1960	293-1151	± 5	AISI 347	→	17/ 19	9/ 12	0.08 max	2 max	0.04 max	0.03 max	1 max	65.05-70.05 Fe (by difference), and 0.8 min Nb; nominal composition from Crucible Data Book, Crucible Steel Co. of America, as stated by authors; austenitic; cylindrical specimen 0.5 in. in diameter and 0.114 cm thick; provided by the Crucible Steel Co. of America; cut from No. 2 finish cold rolled sheet steel of thickness indicated, subjected to a minimum of working, and used without further surface preparation (surface very smooth but not a mirror finish); not subjected to any type of heating before the test runs; thermal pulses supplied by a xenon flash tube; diffusivity measured while specimen heated up during 1st cycle.
25	85, 4	Jenkins, R. J. and Westover, R. W.	1960	361-1097	± 5	AISI 347									Above specimen measured during cooling.
26	85, 4	Jenkins, R. J. and Westover, R. W.	1960	490-1136	± 5	AISI 347									Above specimen measured again during heating in the 2nd cycle.
27	85, 4	Jenkins, R. J. and Westover, R. W.	1960	343-1113	± 5	AISI 347									Above specimen measured during cooling.
28	9	Deem, H. W.	1962	615-1036		AISI 347	→	17.00/ 19.00	9.00/ 13.00	0.08 max	2.00 max	0.045 max	0.030 max	1.00 max	65.045-70.045 Fe (by difference), and 10 x C min total Nb and Ta; nominal composition from Metals Handbook, Vol. 1, 8th ed., p. 409, 1961; austenitic; cylindrical specimen 0.25 in. max diameter and ≤0.25 in. long; Laser beam used to generate heat pulse.
29	87, 17	Lucks, C. F., Deem, H. W., Thompson, H. B., Smith, A. R., Curry, F. P. and Bing, G. F.	1951	117-1255		Steel, Stainless Type 301		16.0/ 18.0	6.00/ 8.00	0.08/ 0.20	2.00 max				Supplied by Republic Steel Corp.; hot rolled; annealed 1 hr at 1310.9 K and water quenched; thermal diffusivity calculated from measured conductivity, specific heat, and density.
30	87, 17	Lucks, C. F., et al.	1951	117-1255		Steel, Stainless Type 316	→	16.82	11.66	0.108	1.59	0.018	0.023	0.26	2.18 Mo, and 67.341 Fe (by difference); supplied by Timken Roller Bearing Co.; hot rolled; annealed 1 hr at 1366.5 K and water quenched; thermal diffusivity calculated from measured conductivity, specific heat, and density.
31	87, 17	Lucks, C. F., et al.	1951	117-1255		Steel, Stainless Type 347	68.26	17.65	10.94	0.06	1.64	0.013	0.017	0.58	0.09 Cu, 0.02 Mo, and 0.73 Nb (Fe by difference); supplied by Timken Roller Bearing Co.; hot rolled; annealed 1 hr at 1366.5 K and water quenched; thermal diffusivity calculated from measured conductivity, specific heat, and density.

SPECIFICATION TABLE 136. THERMAL DIFFUSIVITY OF [IRON + CHROMIUM + NICKEL + ΣX_i] ALLOYS (GROUP II) (continued)

Cur. No.	Ref. No.	Author(s)	Year	Temp. Range, K	Reported Error, %	Name and Specimen Designation	Fe	Cr	Ni	C	Mn	P	S	Si	Composition (continued), Specifications, and Remarks
32	14	Mrozowski, S., Andrew, J. F., Juul, N., Sato, S., Strauss, H. E., and Tsuzuku, T.	1963	317-1086		Stainless Steel									Diffusivity determined from data of amplitude ratio and phase shift measured at two longitudinally located points on specimen.
33*	184	Taylor, R.	1965	298.2				18	8						1 Ti; cylindrical specimen 0.25 in. in diameter and 0.640 cm long; front face exposed to heat pulse from a xenon flash tube; thermal diffusivity calculated from measured time necessary for the rear face to reach one-half the maximum temperature rise; temperature of measurement not given by author but assumed to be room temperature.
34*	184	Taylor, R.	1965	298.2											Cylindrical specimen 0.25 in. in diameter and 0.314 cm long; other conditions and specifications same as above.
35	184	Taylor, R.	1965	298.2											Cylindrical specimen 0.25 in. in diameter and 0.153 cm long; other conditions and specifications same as above.
36*	160	Smith, R. H.	1959	295		Stainless Steel									Specimen size 1 x 1 in., thickness 0.0405 in.; diffusivity measured using flash heating technique; diffusivity determined using $\alpha = 1.37\ L^2/\pi^2\ t_{0.5}$.
37	160	Smith, R. H.	1959	295		Stainless Steel									The above measurement, but using $\alpha = 0.48\ L^2/\pi^2\ t_x$.
38*	160	Smith, R. H.	1959	295		Stainless Steel									Another measurement on the above specimen; diffusivity determined using $\alpha = 1.37\ L^2/\pi^2\ t_{0.5}$.
39*	160	Smith, R. H.	1959	295		Stainless Steel									The above measurement, but using $\alpha = 0.48\ L^2/\pi^2\ t_x$.
40*	160	Smith, R. H.	1959	333		Stainless Steel									Similar to the above specimen; diffusivity measured at higher temperature; diffusivity determined using $\alpha = 1.37\ L^2/\pi^2\ t_{0.5}$.
41	160	Smith, R. H.	1959	333		Stainless Steel									The above measurement, but using $\alpha = 0.48\ L^2/\pi^2\ t_x$.
42*	160	Smith, R. H.	1959	333		Stainless Steel									Another measurement on the above specimen; diffusivity determined using $\alpha = 1.37\ L^2/\pi^2\ t_{0.5}$.
43	160	Smith, R. H.	1959	333		Stainless Steel									The above measurement, but using $\alpha = 0.48\ L^2/\pi^2\ t_x$.

* Not shown in figure.

SPECIFICATION TABLE 136. THERMAL DIFFUSIVITY OF [IRON + CHROMIUM + NICKEL + ΣX_i] ALLOYS (GROUP II) (continued)

Cur. No.	Ref. No.	Author(s)	Year	Temp. Range, K	Reported Error, %	Name and Specimen Designation	Composition (weight percent) Fe	Cr	Ni	C	Mn	P	C	Composition (continued), Specifications, and Remarks
44*	226	Gonska, H., Kierspe, W., and Kohlhaas, R.	1968	324-1251		Austenitic steel; X8-CrNiMoNb-16 16		15.8	15.8	0.05	1.3			1.39 Mo, 0.84 Nb, 0.67 Si, and 0.001 N; 0.5 cm diameter x 20 cm long.
45*	226	Gonska, H., et al.	1968	373-1226		Austenitic steel;X8-CrNiMoNb-16 16								The above specimen measured at decreasing temperatures.
46	232	Böhm, R. and Wachtel, E.	1969	273,373		V2A-steel	68.45	16.95	11.66	0.058	1.62			0.47 Mo, 0.46 Si, and 0.33 Ti; cylindrical specimen; electrical resistivity 72.3 $\mu\Omega$ cm at 0 C.

* Not shown in figure.

DATA TABLE 136. THERMAL DIFFUSIVITY OF [IRON + CHROMIUM + NICKEL + ΣX_i] ALLOYS (GROUP II)

(At least one $X_i > 0.20\%$ and/or any of Mn, P, S, Si > 0.60%)

[Temperature, T, K; Thermal Diffusivity, α, $cm^2\ s^{-1}$]

T	α
CURVE 1	
291	0.0333
349	0.0333
431	0.0352
528	0.0365
655	0.0373
811	0.0405
914	0.0417
1028	0.0441
1135	0.0448
1215	0.0450
CURVE 2	
331	0.0350
401	0.0347
463	0.0365
581	0.0367
681	0.0390
803	0.0402
991	0.0436
1076	0.0443
1163	0.0463
CURVE 3	
431	0.0327
595	0.0367
727	0.0386
873	0.0408
1015	0.0435
1133	0.0460
1204	0.0473
1216	0.0480
CURVE 4	
619	0.0374
873	0.0420
1086	0.0437
CURVE 5	
290	0.0342
361	0.0353
441	0.0361
596	0.0379
688	0.0394
823	0.0413
908	0.0425
1005	0.0446
1096	0.0448
1156	0.0458
1213	0.0469
1223	0.0474
CURVE 6	
461	0.0358
571	0.0377
676	0.0390
756	0.0405
946	0.0417
1058	0.0448
1140	0.0458
CURVE 7	
591	0.0408
733	0.0418
855	0.0410
973	0.0430
1081	0.0465
1178	0.0463
1205	0.0467
CURVE 8	
378	0.0340
471	0.0363
556	0.0377
660	0.0392
780	0.0403
893	0.0422
1003	0.0428
CURVE 8 (cont.)	
1090	0.0443
1153	0.0450
CURVE 9	
291	0.0356*
366	0.0376
465	0.0386
581	0.0412
695	0.0435
763	0.0448
841	0.0453
881	0.0452
933	0.0457
983	0.0458
1103	0.0466
1163	0.0473
1261	0.0490
CURVE 10	
403	0.0342
473	0.0362
513	0.0372
583	0.0387
683	0.0407
810	0.0420
906	0.0432
1018	0.0447
1093	0.0468
1173	0.0483
1243	0.0482
CURVE 11	
460	0.0352
573	0.0380
713	0.0400
827	0.0407
938	0.0423
1045	0.0442
1146	0.0457
CURVE 11 (cont.)	
1233	0.0472
1273	0.0487
CURVE 12	
1043	0.0457
CURVE 13	
293	0.0342
393	0.0345
486	0.0380
621	0.0390
728	0.0419*
843	0.0425
967	0.0459
1073	0.0475
1165	0.0500
1273	0.0510
CURVE 14	
616	0.0410
728	0.0423
846	0.0437
940	0.0465
1013	0.0477
1103	0.0490
1173	0.0503
CURVE 15	
293	0.0323
359	0.0333
483	0.0365
593	0.0400
718	0.0416
873	0.0473
1058	0.0500
1148	0.0510
1223	0.0525
CURVE 16	
893	0.0435
CURVE 17	
293	0.0325
373	0.0345
463	0.0360
540	0.0380
673	0.0417
796	0.0437
923	0.0463
1020	0.0485
1100	0.0512
1213	0.0530
CURVE 18	
473	0.0360
583	0.0390
688	0.0417
813	0.0437
920	0.0473
1033	0.0477
1150	0.0515
CURVE 19	
326	0.0330
363	0.0353*
485	0.0365
636	0.0395
775	0.0430
926	0.0455
1060	0.0487
1203	0.0520
1233	0.0525
CURVE 20	
448	0.0355
598	0.0375
733	0.0425
CURVE 20 (cont.)	
893	0.0453
1116	0.0490
CURVE 21	
293	0.0360
378	0.0375
473	0.0395
583	0.0420
706	0.0444
748	0.0425
808	0.0460
938	0.0475
1043	0.0485
1145	0.0492
1233	0.0490
CURVE 22	
353	0.0370
433	0.0385
528	0.0405
616	0.0425
735	0.0440
833	0.0460
928	0.0465
1023	0.0475
1103	0.0485
1153	0.0490
1218	0.0475
CURVE 23	
603	0.0407
790	0.0450
943	0.0465
1113	0.0475
1173	0.0500
CURVE 24	
293	0.0350
400	0.0375
503	0.0395
603	0.0413
690	0.0445
806	0.0465
880	0.0475
981	0.0503
1068	0.0503
1151	0.0503
CURVE 25	
361	0.0350
466	0.0380
586	0.0400
673	0.0420
748	0.0443
823	0.0457
906	0.0475
1013	0.0495
1097	0.0505
CURVE 26	
490	0.0385
596	0.0410
693	0.0420
798	0.0455
886	0.0475
982	0.0490
1136	0.0510
CURVE 27	
343	0.0355
433	0.0370
521	0.0395
623	0.0416
726	0.0443
816	0.0453
886	0.0475*
CURVE 27 (cont.)	
921	0.0475
1013	0.0495*
1073	0.0505
1113	0.0505
CURVE 28	
615	0.0415
656	0.0450
748	0.0465
833	0.0475
936	0.0515
1036	0.0525
CURVE 29	
116.5	0.0410
144.3	0.0408
199.8	0.0408
293.2	0.0413
366.5	0.0418
477.6	0.0431
588.7	0.0454
699.8	0.0477
810.9	0.0501
922.1	0.0524
1033.2	0.0539
1144.3	0.0547
1255.4	0.0547
CURVE 30	
116.5	0.0403
144.3	0.0390
199.8	0.0379
293.2	0.0369
366.5	0.0372
477.6	0.0382
588.7	0.0408
699.8	0.0431
810.9	0.0446
922.1	0.0485

* Not shown in figure.

DATA TABLE 136. THERMAL DIFFUSIVITY OF [IRON + CHROMIUM + NICKEL + ΣX_i] ALLOYS (GROUP II) (continued)

T	α
CURVE 30 (cont.)	
1033.2	0.0493
1144.3	0.0501
1255.4	0.0503
CURVE 31	
116.5	0.0462
144.3	0.0446
199.8	0.0426
293.2	0.0408
366.5	0.0405
477.6	0.0415
588.7	0.0434
699.8	0.0454
810.9	0.0477
922.1	0.0495
1033.2	0.0514
1144.3	0.0516
1255.4	0.0501
CURVE 32	
317	0.0380
320	0.0347
323	0.0386
324	0.0374
353	0.0374
353	0.0368
353	0.0354
493	0.0377
494	0.0365
494	0.0351
527	0.0353
530	0.0383
605	0.0371
605	0.0385
624	0.0397
624	0.0377
648	0.0387
648	0.0365
705	0.0373
755	0.0405
755	0.0397
755	0.0387
894	0.0421
895	0.0404
CURVE 32 (cont.)	
1073	0.0404
1074	0.0423
1085	0.0435
1086	0.0416
CURVE 33*	
298.2	0.0343
CURVE 34*	
298.2	0.0339
CURVE 35	
298.2	0.0341
CURVE 36*	
295	0.035
CURVE 37	
295	0.031
CURVE 38*	
295	0.036
CURVE 39*	
295	0.036
CURVE 40*	
333	0.034
CURVE 41	
333	0.045
CURVE 42*	
333	0.036
CURVE 43	
333	0.046
CURVE 44*	
324	0.0365
397	0.0380
475	0.0375
536	0.0401
653	0.0429
783	0.0489
906	0.0545
987	0.0589
1091	0.0616
1146	0.0648
1228	0.0668
1251	0.0682
CURVE 45*	
373	0.0413
484	0.0410
589	0.0436
666	0.0457
747	0.0502
854	0.0544
938	0.0591
1023	0.0632
1123	0.0662
1226	0.0678
CURVE 46	
273	0.0414
373	0.0355

* Not shown in figure.

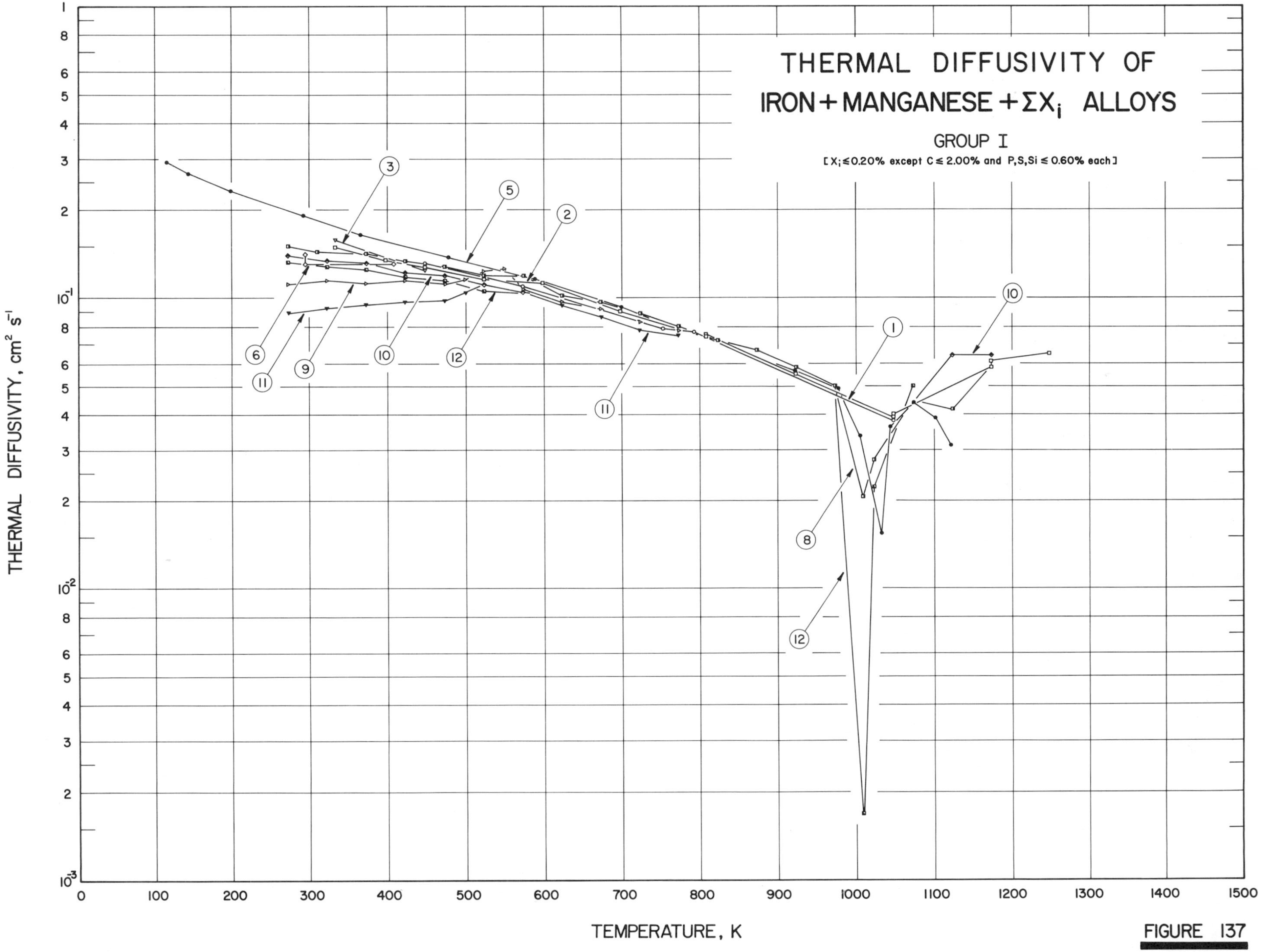
THERMAL DIFFUSIVITY OF
IRON + MANGANESE + ΣXi ALLOYS
GROUP I
[Xi ≤ 0.20% except C ≤ 2.00% and P,S,Si ≤ 0.60% each]
THERMAL DIFFUSIVITY, cm² s⁻¹
TEMPERATURE, K
1
2
3
5
6
8
9
10
11
12

FIGURE 137

SPECIFICATION TABLE 137. THERMAL DIFFUSIVITY OF [IRON + MANGANESE + ΣX_i] ALLOYS (GROUP I)

($X_i \leq 0.20\%$ except C $\leq 2.00\%$ and P, S, Si $\leq 0.60\%$ each)

Cur. No.	Ref. No.	Author(s)	Year	Temp. Range, K	Reported Error, %	Name and Specimen Designation	Composition (weight percent) Fe	Mn	C	P	S	Si	Composition (continued), Specifications, and Remarks
1	19, 20	Butler, C. P. and Inn, E. C. Y.	1957	448-1048	5-10	SAE 1020	→	0.3/ 0.6	0.18/ 0.23	0.040 max	0.050 max		99.08-99.43 Fe (by difference); ladle chemical composition from Metals Handbook, Vol. 1, 8th ed., p. 62, 1961; cylindrical specimen 0.375 in. in diameter and length lying in the range from 1 to 2.5 cm; heat treatment: normalized; subjected to irradiance from a carbon arc lamp heat source; spectral distribution approximates that of a 5700 K black body source; specimen blackened with camphor black; measurements carried out under a vacuum of ~5 microns.
2	19, 20	Butler, C. P. and Inn, E. C. Y.	1957	333-1248	5-10	SAE 1020							Above specimen allowed to cool and then exposed to the arc lamp to measure diffusivity again.
3	19, 20	Butler, C. P. and Inn, E. C. Y.	1957	333, 448	5-10	SAE 1020							Above specimen allowed to cool and then exposed to the arc lamp to measure diffusivity again.
4*	19, 20	Butler, C. P. and Inn, E. C. Y.	1957	688-1093	5-10	SAE 1020							Above specimen allowed to cool and then exposed to the arc lamp to measure diffusivity again.
5	87, 17	Lucks, C. F., Deem, H. W., Thompson, H. B., Smith, A. R., Curry, F. P., and Bing, G. F.	1951	117-1122		SAE 1010	→	0.42	0.10	0.008	0.028		99.444 Fe (by difference); supplied by United States Steel Corp; hot rolled; thermal diffusivity calculated from measured conductivity, specific heat, and density.
6	4	Jenkins, R. J. and Parker, W. J.	1961	295, 408	±5	Steel 1020	→	0.3/ 0.6	0.18/ 0.23	0.040 max	0.050 max		99.08-99.43 Fe (by difference); ladle chemical composition from Metals Handbook, Vol. 1, 8th ed., p. 62, 1961; square specimen 1.9 cm side and 0.100 cm thick; high intensity short duration light pulse absorbed in front surface of thermally insulated specimen coated with camphor black; 408.2 K measurements obtained by heating sample holder and sample with an infrared lamp.
7*	84	Neimark, B. E. and Lyusternik, V. E.	1960	273-773	±3.2	Steel 15	99.04	0.51	0.15	0.034	0.036	0.23	(Fe by difference); chilled; thermal diffusivity obtained from measured conductivity, specific heat, and density.
8	84	Neimark, B. E. and Lyusternik, V. E.	1960	273-1173	±3.2	Steel 15	99.04	0.51	0.15	0.034	0.036	0.23	(Fe by difference); annealed; thermal diffusivity obtained from measured conductivity, specific heat, and density.
9	84	Neimark, B. E. and Lyusternik, V. E.	1960	273-773	±3.2	Steel 35	→	0.58	0.35	0.021	0.028	0.17	98.851 Fe (by difference); chilled; thermal diffusivity obtained from measured conductivity, specific heat, and density.
10	84	Neimark, B. E. and Lyusternik, V. E.	1960	273-1173	±3.2	Steel 35	→	0.58	0.35	0.021	0.028	0.17	98.851 Fe (by difference); annealed; thermal diffusivity obtained from measured conductivity, specific heat, and density.

* Not shown in figure.

SPECIFICATION TABLE 137. THERMAL DIFFUSIVITY OF [IRON + MANGANESE + ΣX_i] ALLOYS (GROUP I) (continued)

Cur. No.	Ref. No.	Author(s)	Year	Temp. Range, K	Reported Error, %	Name and Specimen Designation	Composition (weight percent) Fe	Mn	C	P	S	Si	Composition (continued), Specifications, and Remarks
11	84	Neimark, B. E. and Lyusternik, V. E.	1960	273-773	±3.2	Steel 45	→	0.64	0.45	0.017	0.022	0.24	98.631 Fe (by difference); chilled; thermal diffusivity obtained from measured conductivity, specific heat, and density.
12	84	Neimark, B. E. and Lyusternik, V. E.	1960	273-1073	±3.2	Steel 45	→	0.64	0.45	0.017	0.022	0.24	98.631 Fe (by difference); annealed; thermal diffusivity obtained from measured conductivity, specific heat, and density.

DATA TABLE 137. THERMAL DIFFUSIVITY OF [IRON + MANGANESE + ΣX_i] ALLOYS (GROUP I)

($X_i \leq 0.20\%$ except C $\leq 2.00\%$ and P, S, Si $\leq 0.60\%$ each)

[Temperature, T, K; Thermal Diffusivity, α, $cm^2\ s^{-1}$]

T	α
CURVE 1	
448.2	0.131
573.2	0.109
753.2	0.079
793.2	0.077
1048.2	0.038
CURVE 2	
333.2	0.149
398.2	0.134
448.2	0.127
523.2	0.116
598.2	0.112
698.2	0.090
808.2	0.074
808.2	0.075
923.2	0.055
1048.2	0.039
1048.2	0.040
1173.2	0.058
1173.2	0.061
1173.2	0.061
1248.2	0.065
CURVE 3	
333.2	0.158
448.2	0.124
CURVE 4*	
688.2	0.091
873.2	0.063
998.2	0.047
1093.2	0.048
1093.2	0.046
CURVE 5	
116.5	0.292
144.3	0.268
199.8	0.232
CURVE 5 (cont.)	
293.2	0.191
366.5	0.165
477.6	0.137
588.7	0.116
699.8	0.0929
810.9	0.0748*
922.1	0.0568
977.6	0.0490
1005.4	0.0335
1033.2	0.0155
1044.3	0.0361
1074.8	0.0439
1102.6	0.0387
1122.1	0.0310
CURVE 6	
295.2	0.14
295.2	0.13
408.2	0.13
CURVE 7*	
273.2	0.142
323.2	0.142
373.2	0.139
423.2	0.139
473.2	0.125
498.2	0.123
523.2	0.122
548.2	0.123
573.2	0.119
623.2	0.103
673.2	0.0972
723.2	0.0889
773.2	0.0806
CURVE 8	
273.2	0.150
323.2	0.144
373.2	0.142
CURVE 8 (cont.)	
423.2	0.133
473.2	0.128
523.2	0.119
573.2	0.119
623.2	0.103
673.2	0.0972
723.2	0.0889
773.2	0.0806
823.2	0.0722
873.2	0.0667
923.2	0.0583
973.2	0.0500
1008.2	0.0206
1023.2	0.0278
1073.2	0.0444*
1123.2	0.0417
1173.2	0.0583*
CURVE 9	
273.2	0.111
323.2	0.114
373.2	0.111
423.0	0.114
473.2	0.111
498.2	0.114
523.2	0.122
548.2	0.125
573.2	0.108
623.2	0.0972
673.2	0.0917
723.2	0.0833
773.2	0.0778
CURVE 10	
273.2	0.139
323.2	0.133
373.2	0.131
423.2	0.122
473.2	0.119
523.2	0.111
CURVE 10 (cont.)	
573.2	0.106
623.2	0.100*
673.2	0.0917*
723.2	0.0861*
773.2	0.0806*
823.2	0.0722*
873.2	0.0667*
923.2	0.0583*
973.2	0.0500*
1008.2	0.00611*
1023.2	0.0306*
1073.2	0.0444*
1123.2	0.0639
1173.2	0.0639
CURVE 11	
273.2	0.0889
323.2	0.0917
373.2	0.0945
423.2	0.0972
473.2	0.0972
498.2	0.104
523.2	0.111*
548.2	0.118*
573.2	0.103*
623.2	0.0945
673.2	0.0861
723.2	0.0778
773.2	0.0750
CURVE 12	
273.2	0.133
323.2	0.128
373.2	0.125
423.2	0.117
473.2	0.114
523.2	0.106
573.2	0.103*
623.2	0.0945*
673.2	0.0889*
CURVE 12 (cont.)	
723.2	0.0806*
773.2	0.0778*
823.2	0.0695*
873.2	0.0639*
923.2	0.0556*
973.2	0.0500*
1008.2	0.00167
1023.2	0.0222
1073.2	0.0500

*Not shown in figure.

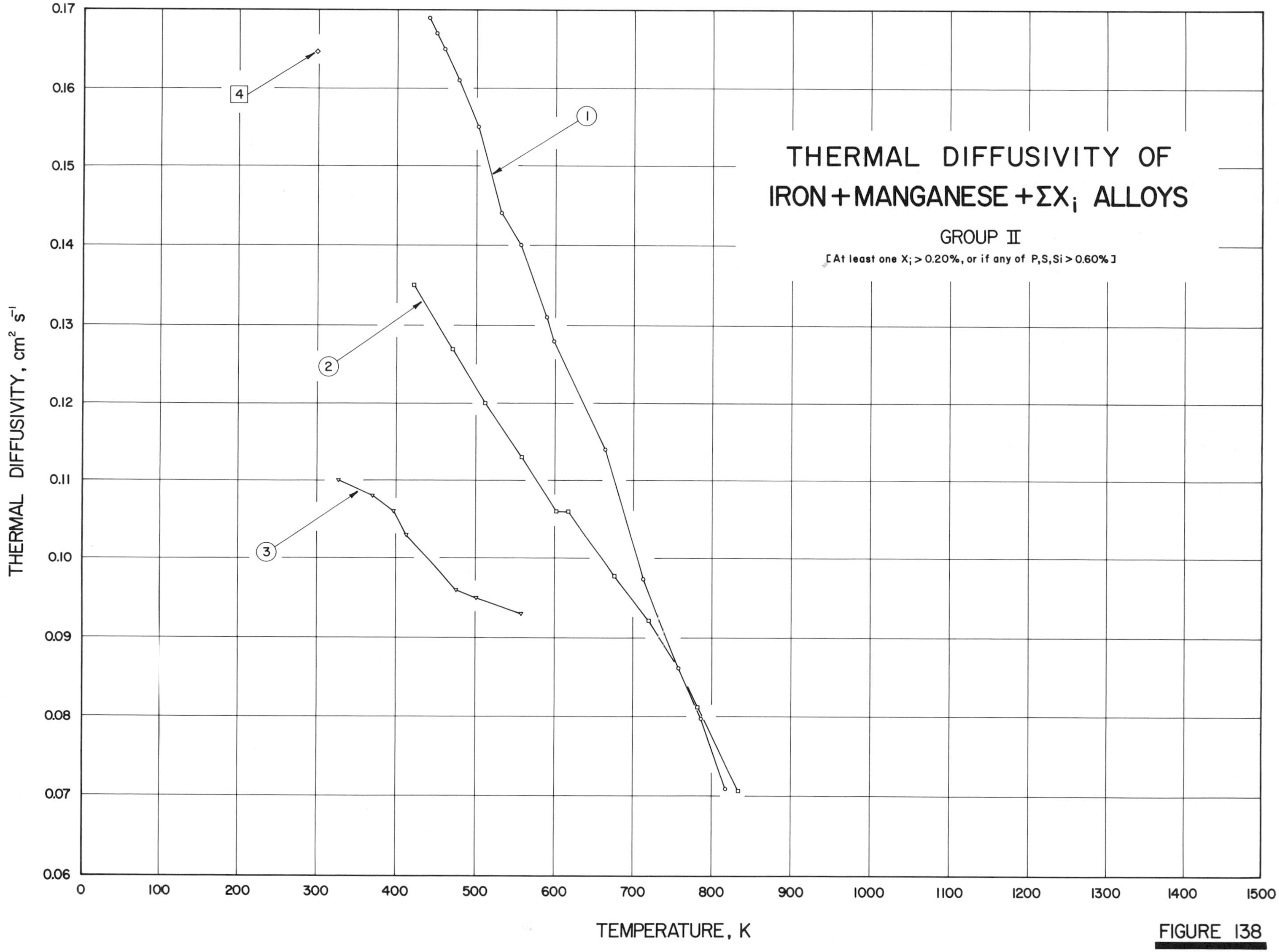
THERMAL DIFFUSIVITY OF
IRON + MANGANESE + ΣXi ALLOYS
GROUP II
[At least one Xi > 0.20%, or if any of P,S,Si > 0.60%]
THERMAL DIFFUSIVITY, cm² s⁻¹
TEMPERATURE, K
0.17
0.16
0.15
0.14
0.13
0.12
0.11
0.10
0.09
0.08
0.07
0.06
0
100
200
300
400
500
600
700
800
900
1000
1100
1200
1300
1400
1500
1
2
3
4

FIGURE 138

SPECIFICATION TABLE 138. THERMAL DIFFUSIVITY OF [IRON + MANGANESE + ΣX_i] ALLOYS (GROUP II)

(At least one $X_i > 0.20\%$ and/or any of P, S, Si $> 0.60\%$)

Cur. No.	Ref. No.	Author(s)	Year	Temp. Range, K	Reported Error, %	Name and Specimen Designation	Composition (weight percent) Fe	Mn	C	P	S	Si	Composition (continued), Specifications, and Remarks
1	176, 21	El-Hifni, M. A. and Chao, B. T.	1955	439-818	2-3	AISI 1018	→	0.60/ 0.90	0.15/ 0.20	0.040 max	0.050 max		98.81-99.16 Fe (by difference); ladle chemical composition from Metals Handbook, Vol. 1, 8th ed., p. 62, 1961; tubular specimen 0.875 in. O. D. and 5 in. long, relatively thin walled; another tube of similar material mounted concentrically with specimen to minimize heat losses; one end of specimen-and-shield assembly immersed in a liquid heating bath, while the other end supported by a transite disc for insulation; cyclic varying current generates required temperature wave at heating bath; heat supplied by electric current removed by forced draft of cooling air; one dimensional heat flow; a minimum of three complete temperature waves recorded.
2	176, 21	El-Hifni, M. A. and Chao, B. T.	1955	421-835	2-3	AISI 1045	→	0.60/ 0.90	0.43/ 0.50	0.040 max	0.050 max		98.51-98.88 Fe (by difference); ladle chemical composition from Metals Handbook, Vol. 1, 8th ed., p. 62, 1961; tubular specimen 0.875 in. O. D. and 5 in. long, relatively thin walled; another tube of similar material mounted concentrically with specimen to minimize heat losses; one end of specimen-and-shield assembly immersed in a liquid heating bath, while the other end supported by a transite disc for insulation; cyclic varying current sent through resistance wire to generate required temperature wave at heating bath; heat supplied by electric current removed by forced draft of cooling air; one-dimensional heat flow; a minimum of three complete temperature waves recorded.
3	38	Shvidkovskii, E. G.	1938	327-558		Steel	98.36	0.79	0.52			0.33	Fe obtained by difference; cylindrical specimen 10 mm in diameter and 150 mm long; period of temperature wave 4 min; data represent average over a number of runs ranging from 4 to 12.
4	7	di Novi, R. A.	1963	298.2	10	Steel 1018	→	0.60/ 0.90	0.15/ 0.20	0.040 max	0.050 max		98.81-99.16 Fe (by difference); ladle chemical composition from Metals Handbook, Vol. 1, 8th ed., p. 62, 1961; AISI 1018; specimen with thickness lying in the range from 1 to 2 mm; front surface uniformly irradiated by a very short pulse of radiant energy supplied by a xenon flash tube; diffusivity determined from measured history of the back surface temperature; temperature at which specimen was measured not given by author but assumed to be room temperature.

DATA TABLE 138. THERMAL DIFFUSIVITY OF [IRON + MANGANESE + ΣX_i] ALLOYS (GROUP II)

(At least one X_i >0.20% and/or any of P, S, Si >0.60%)

[Temperature, T, K; Thermal Diffusivity, α, $cm^2\ s^{-1}$]

T	α
CURVE 1	
439	0.1689
450	0.167
459	0.165
477	0.161
502	0.155
532	0.144
556	0.140
590	0.131
599	0.128
664	0.114
714	0.0974
759	0.0862
787	0.0797
818	0.0709
CURVE 2	
421	0.135
470	0.127
511	0.120
558	0.113
603	0.106
617	0.106
675	0.0977
720	0.0922
760	0.0862*
783	0.0812
835	0.0706
CURVE 3	
327.2	0.110
370.2	0.108
397.2	0.106
413.2	0.103
476.2	0.0960
501.2	0.0950
558.2	0.0930
CURVE 4	
298.2	0.1647

*Not shown in figure.

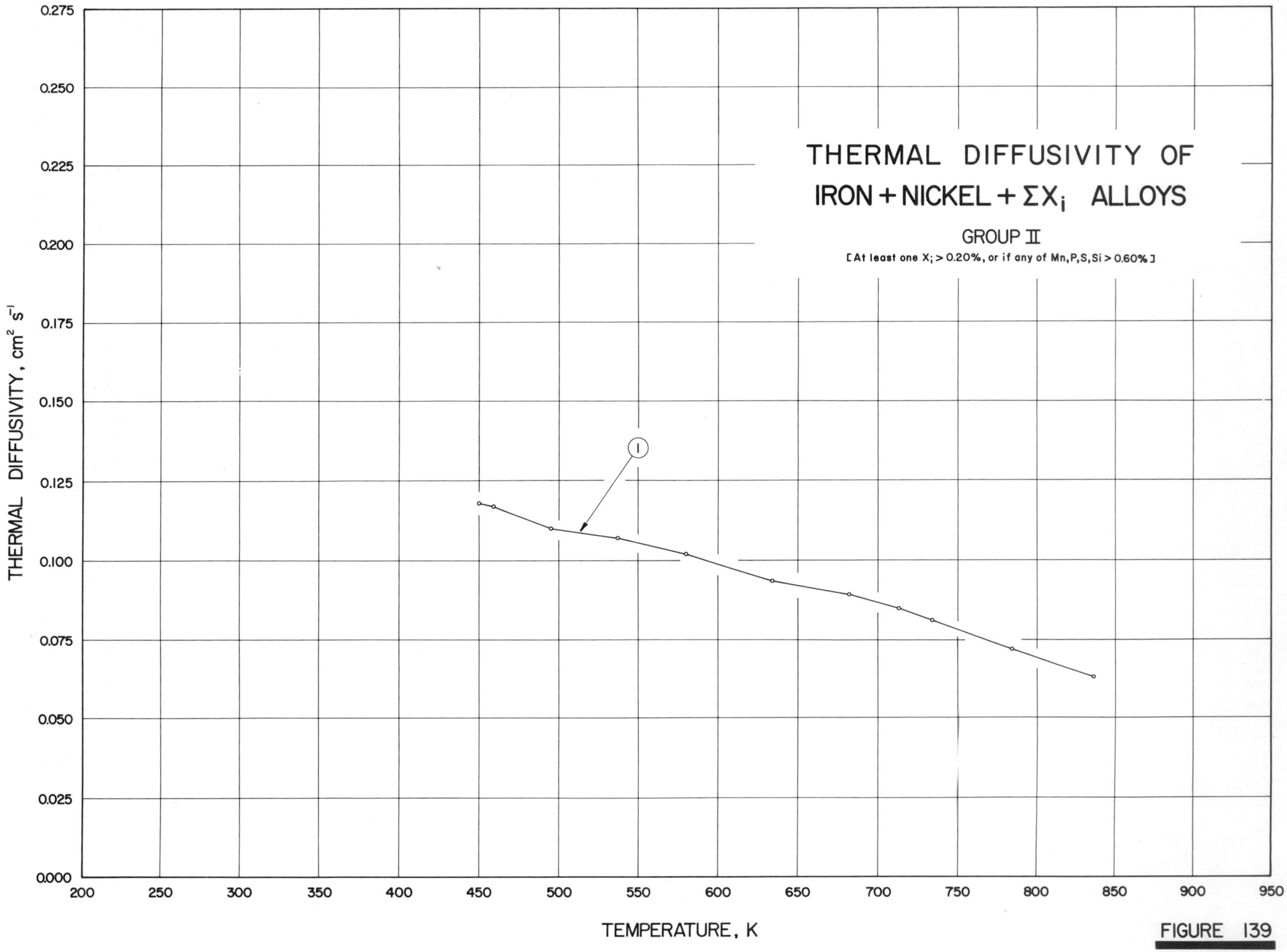
THERMAL DIFFUSIVITY OF
IRON + NICKEL + ΣXi ALLOYS
GROUP II
[At least one Xi > 0.20%, or if any of Mn, P, S, Si > 0.60%]
1
THERMAL DIFFUSIVITY, cm² s⁻¹
0.275
0.250
0.225
0.200
0.175
0.150
0.125
0.100
0.075
0.050
0.025
0.000
200
250
300
350
400
450
500
550
600
650
700
750
800
850
900
950
TEMPERATURE, K

FIGURE 139

SPECIFICATION TABLE 139. THERMAL DIFFUSIVITY OF [IRON + NICKEL + ΣX_i] ALLOYS (GROUP II)

(At least one $X_i > 0.20\%$ and/or any of Mn, P, S, Si $> 0.60\%$)

Cur. No.	Ref. No.	Author(s)	Year	Temp. Range, K	Reported Error, %	Name and Specimen Designation	Composition (weight percent) Fe	Ni	C	Cr	Mn	Si	Composition (continued), Specifications, and Remarks
1	176, 21	El-Hifni, M.A. and Chao, B.T.	1955	450-837	2-3	AISI 3140	→	1.00/ 1.45	0.37/ 0.44	0.45/ 0.85	0.60/ 1.00	0.20/ 0.35	95.91-97.38 Fe (by difference); nominal composition from Metals Handbook, Vol. 1, 8th ed., p. 210, 1961; tubular specimen 0.875 in. O.D. and 5 in. long, relatively thin walled; another tube of similar material mounted concentrically with specimen to minimize heat losses; one end of specimen-and-shield assembly immersed in liquid heating bath, while the other end supported by a transite disc for insulation; cyclic varying current sent through resistance wire to generate required temperature wave at heating bath; heat supplied by electric current removed by a forced draft of cooling air; one-dimensional heat flow; a minimum of three complete temperature waves recorded.

DATA TABLE 139. THERMAL DIFFUSIVITY OF [IRON + NICKEL + ΣX_i] ALLOYS (GROUP II)

(At least one $X_i > 0.20\%$ and/or any of Mn, P, S Si $> 0.60\%$)

[Temperature, T, K; Thermal Diffusivity, α, cm^2 s^{-1}]

T	α
CURVE 1	
450	0.118
460	0.117
495	0.110
538	0.107
580	0.102
635	0.0935
682	0.0893
713	0.0847
735	0.0811
784	0.0721
837	0.0630

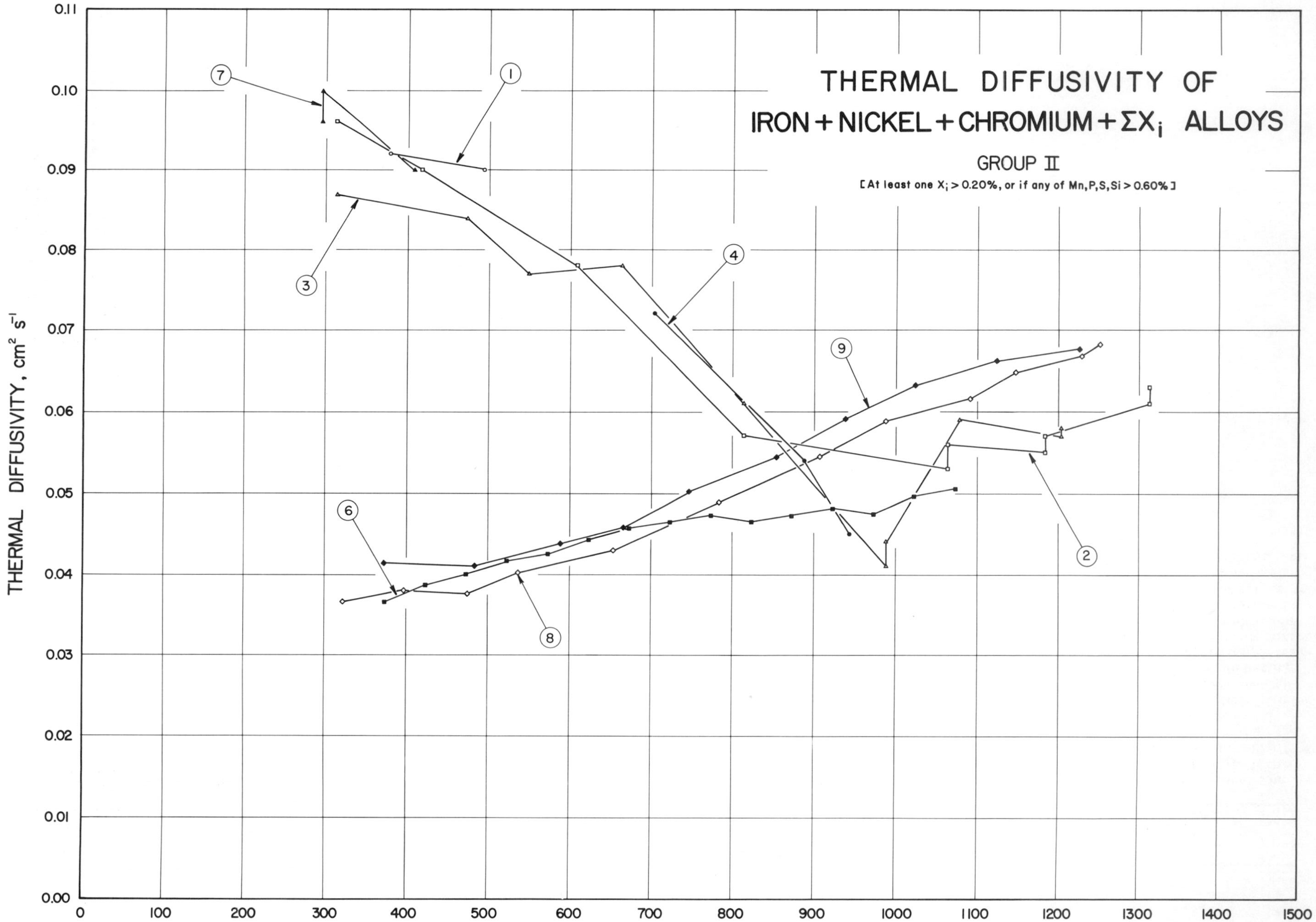
THERMAL DIFFUSIVITY OF
IRON + NICKEL + CHROMIUM + ΣX$_i$ ALLOYS
GROUP II
[At least one X$_i$ > 0.20%, or if any of Mn, P, S, Si > 0.60%]
THERMAL DIFFUSIVITY, cm^2 s^{-1}
TEMPERATURE, K
0.00
0.01
0.02
0.03
0.04
0.05
0.06
0.07
0.08
0.09
0.10
0.11
0
100
200
300
400
500
600
700
800
900
1000
1100
1200
1300
1400
1500
1
2
3
4
6
7
8
9

FIGURE 140

SPECIFICATION TABLE 140. THERMAL DIFFUSIVITY OF [IRON + NICKEL + CHROMIUM + ΣX_i] ALLOYS (GROUP II)

(At least one $X_i > 0.20\%$ and/or any of Mn, P, S, Si $> 0.60\%$)

Cur. No.	Ref. No.	Author(s)	Year	Temp. Range, K	Reported Error, %	Name and Specimen Designation	Composition (weight percent) Fe	Ni	Cr	C	Mn	Mo	P	S	Composition (continued), Specifications, and Remarks
1	19, 20	Butler, C. P. and Inn, E. C. Y.	1957	378, 493	5-10	SAE 4340	→	1.65/ 2.00	0.70/ 0.90	0.38/ 0.43	0.60/ 0.80	0.20/ 0.30	0.040	0.040	95.14-96.19 Fe (by difference), and 0.20-0.35 Si; ladle chemical composition from Metals Handbook, Vol. 1, 8th ed., p. 61, 1961; cylindrical specimen 0.375 in. in diameter and length lying in the range from 1 to 2.5 cm; heat treatment: heat treated to 180000 psi; subjected to irradiance from a carbon arc lamp heat source; spectral distribution approximates that of a 5700 K black body source; specimen blackened with camphor black; measured under a vacuum of ~5 microns.
2	19, 20	Butler, C. P. and Inn, E. C. Y.	1957	313-1313	5-10	SAE 4340									Above specimen allowed to cool and then exposed to the arc lamp to measure diffusivity again.
3	19, 20	Butler, C. P. and Inn, E. C. Y.	1957	313-1203	5-10	SAE 4340									Above specimen allowed to cool and then exposed to the arc lamp to measure diffusivity again.
4	19, 20	Butler, C. P. and Inn, E. C. Y.	1957	703-943	5-10	SAE 4340									Above specimen allowed to cool and then exposed to the arc lamp to measure diffusivity again.
5*	19, 20	Butler, C. P. and Inn, E. C. Y.	1957	943.2	5-10	SAE 4340									Above specimen allowed to cool and then exposed to the arc lamp to measure diffusivity again.
6	89	Neimark, B. E., Lyusternik, V. E., Anichkina, E. Yu., and Bykova, T. I.	1963	373-1073		0Kh16N36V3T (EI-855)	→	36.55	15.5	0.08	0.46		0.0125	0.047	43.6105 Fe (by difference), 0.55 Si, 0.31 Ti, and 2.88 W; quenched in air from 1373.2 K; density 8.209 g cm^{-3} at 293.2 K; electrical resistivity reported as 102.0, 103.2, 105.1, 106.9, 108.7, 110.5, 112.0, 113.6, 115.0, 116.5, 117.9, 118.9, 119.8, 120.6, 121.5, 122.2, 122.9, and 123.6 x 10^{-8} ohm m at 293.2, 323.2, 373.2, 423.2, 473.2, 523.2, 573.2, 623.2, 673.2, 723.2, 773.2, 823.2, 873.2, 923.2, 973.2, 1023.2, 1073.2, and 1123.2 K, respectively; Lorenz number reported as 4.14, 4.03, 3.93, 3.82, 3.71, 3.60, 3.49, 3.37, 3.24, 3.14, 3.05, 3.04, 2.99, 2.91, 2.84, 2.76, and 2.61 x 10^{-8} V^2 K^{-2} at 293.2, 323.2, 373.2, 423.2, 473.2, 523.2, 573.2, 623.2, 673.2, 723.2, 773.2, 823.2, 873.2, 923.2, 973.2, 1023.2, and 1073.2 K, respectively; thermal diffusivity calculated from measured conductivity, heat capacity, and density.

* Not shown in figure.

SPECIFICATION TABLE 140. THERMAL DIFFUSIVITY OF [IRON + NICKEL + CHROMIUM + ΣX_i] ALLOYS (GROUP II) (continued)

Cur. No.	Ref. No.	Author(s)	Year	Temp. Range, K	Reported Error, %	Name and Specimen Designation	Composition (weight percent) Fe	Ni	Cr	C	Mn	Mo	P	S	Composition (continued), Specifications, and Remarks
7	4, 146	Jenkins, R.J. and Parker, W.J.	1961	295-408	± 5	Steel 4340	→	1.65/ 2.00	0.70/ 0.90	0.38/ 0.43	0.60/ 0.80	0.20/ 0.30	0.040	0.040	95.14-96.19 Fe (by difference), and 0.20-0.35 Si; ladle chemical composition from Metals Handbook, Vol. 1, 8th ed., p. 61, 1961; square specimen 1.9 cm side and 0.107 cm thick; high intensity short duration light pulse absorbed in front surface of thermally insulated specimen coated with camphor black; 408.2 K measurements obtained by heating sample holder and sample with an infrared lamp.
8	226	Gonska, H., Kierspe, W., and Kohlhaas, R.	1968	324-1251		Austenitic steel; X8 CrNiMoNb 16 16		15.8	15.8	0.05	1.3				1.39 Mo. 0.84 Nb, 0.67 Si, and 0.001 N; 0.5 cm diameter x 20 cm long.
9	226	Gonska, H., et al.	1968	373-1226		Austenitic steel; X8 CrNiMoNb 16 16									The above specimen measured at decreasing temperatures.

DATA TABLE 140. THERMAL DIFFUSIVITY OF [IRON + NICKEL + CHROMIUM + ΣX_i] ALLOYS (GROUP II)

(At least one $X_i > 0.20\%$ and/or any of Mn, P, S, Si $> 0.60\%$)

[Temperature, T, K; Thermal Diffusivity, α, $cm^2 s^{-1}$]

T	α
CURVE 1	
378.2	0.092
493.2	0.090
CURVE 2	
313.2	0.096
418.2	0.090
608.2	0.078
813.2	0.057
1063.2	0.053
1063.2	0.056
1183.2	0.055
1183.2	0.057
1313.2	0.061
1313.2	0.063
CURVE 3	
313.2	0.087
473.2	0.084
548.2	0.077
663.2	0.078
813.2	0.061
988.2	0.041
988.2	0.044
1078.2	0.059
1203.2	0.057
1203.2	0.058
CURVE 4	
703.2	0.072
888.2	0.054
888.2	0.054
888.2	0.054
943.2	0.045
CURVE 5*	
943.2	0.045
CURVE 6	
373.2	0.0365
423.2	0.0386
473.2	0.0399
523.2	0.0415
573.2	0.0424
623.2	0.0442
673.2	0.0456
723.2	0.0463
773.2	0.0472
823.2	0.0464
873.2	0.0472
923.2	0.0481
973.2	0.0474
1023.2	0.0496
1073.2	0.0506
CURVE 7	
295.2	0.096
295.2	0.10
408.2	0.090
CURVE 8	
324	0.0365
397	0.0380
475	0.0375
536	0.0401
653	0.0429
783	0.0489
906	0.0545
987	0.0589
1091	0.0616
1146	0.0648
1228	0.0668
1251	0.0682
CURVE 9	
373	0.0413
484	0.0410
589	0.0436
666	0.0457
CURVE 9 (cont.)	
747	0.0502
854	0.0544
938	0.0591
1023	0.0632
1123	0.0662
1226	0.0678

*Not shown in figure.

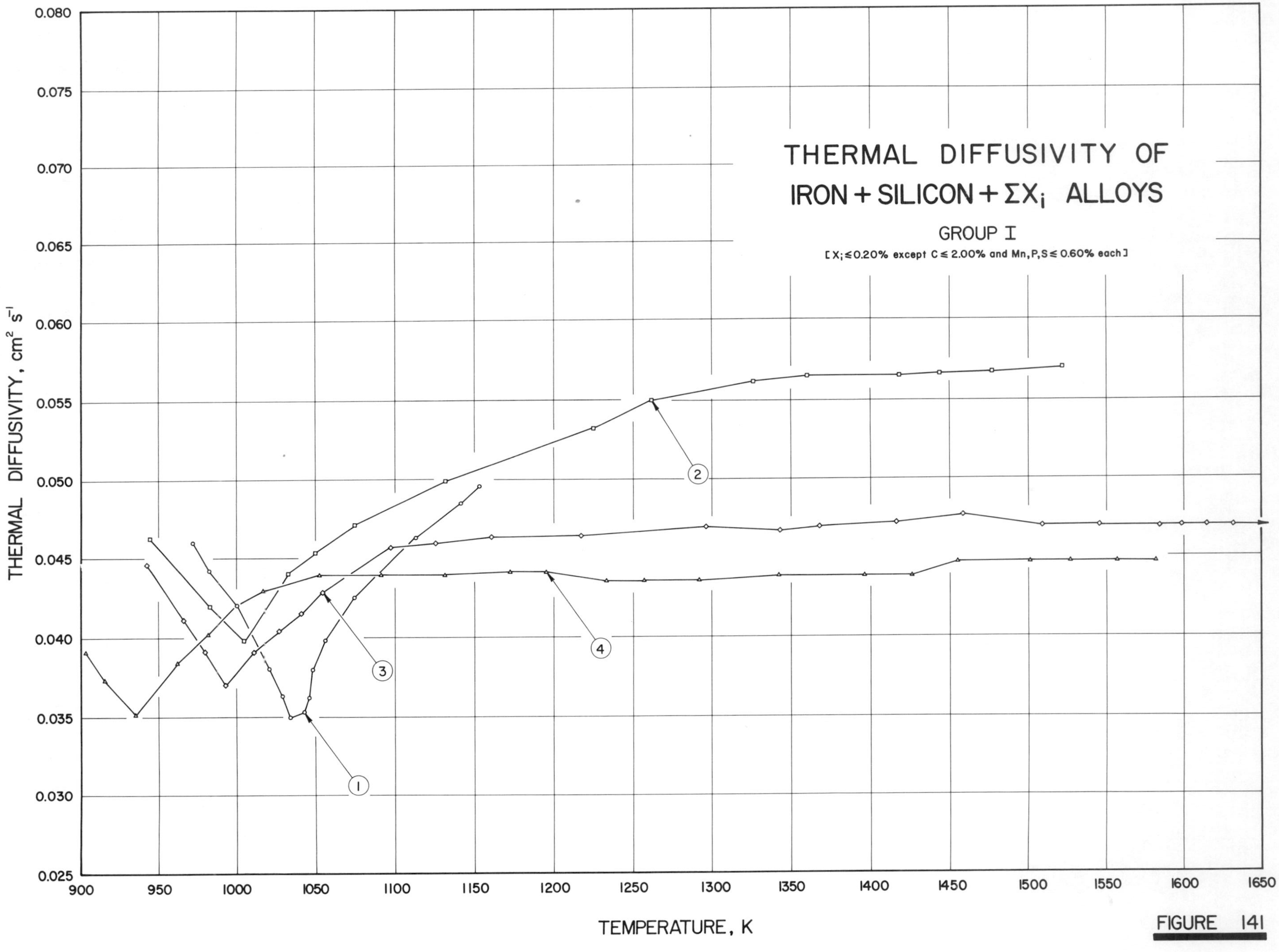
THERMAL DIFFUSIVITY OF
IRON + SILICON + ΣXi ALLOYS
GROUP I
[Xi ≤ 0.20% except C ≤ 2.00% and Mn, P, S ≤ 0.60% each]
THERMAL DIFFUSIVITY, cm² s⁻¹
TEMPERATURE, K
0.080
0.075
0.070
0.065
0.060
0.055
0.050
0.045
0.040
0.035
0.030
0.025
900
950
1000
1050
1100
1150
1200
1250
1300
1350
1400
1450
1500
1550
1600
1650
1
2
3
4

FIGURE 141

SPECIFICATION TABLE 141. THERMAL DIFFUSIVITY OF [IRON + SILICON + ΣX_i] ALLOYS GROUP I

($X_i \leq 0.20\%$ except C $\leq 2.00\%$ and Mn, P, S $\leq 0.60\%$ each)

Cur. No.	Ref. No.	Author(s)	Year	Temp. Range, K	Reported Error, %	Name and Specimen Designation	Composition (weight percent) Fe	Si	Composition (continued), Specifications, and Remarks
1	201	Krentsis, R. P., Zinov'yev, V. Ye, Andreyeva, L. P., and Gel'd, P. V.	1970	972-1153	< 5		Bal.	0.2	Powder of carbonyl iron and monocrystalline silicon in solid solution; diffusivity measured by using temperature plane wave method at frequency of 168.8 G, and pressure of the order of 1 x 10^{-5} mm Hg.
2	201	Krentsis, R. P., et al.	1970	944-1522	< 5		Bal.	2.95	Other conditions same as above.
3	201	Krentsis, R. P., et al.	1970	943-1659	< 5		Bal.	4.68	Other conditions same as above.
4	201	Krentsis, R. P., et al.	1970	903-1582	< 5		Bal.	8.2	Other conditions same as above.

DATA TABLE 141. THERMAL DIFFUSIVITY OF [IRON + SILICON + ΣX_i] ALLOYS GROUP I

($X_i \leq 0.20\%$ except C $\leq 2.00\%$ and Mn, P, S $\leq 0.60\%$ each)

[Temperature, T, K; Thermal Diffusivity, α, $cm^2 s^{-1}$]

T	α	T	α	T	α	T	α	T	α	T	α
CURVE 1		CURVE 2		CURVE 3		CURVE 3 (cont.)		CURVE 4 (cont.)		CURVE 4 (cont.)	
972	0.0460	944	0.0463	943	0.0446	1458	0.0477	1000	0.0420*	1557	0.0447
982	0.0442	983	0.0420	966	0.0410	1509	0.0470	1017	0.0429	1582	0.0447
1000	0.0420	1005	0.0398	980	0.0391	1546	0.0470	1052	0.0439		
1021	0.0380	1033	0.0440	993	0.0370	1584	0.0469	1091	0.0439		
1029	0.0363	1050	0.0453	1011	0.0391	1599	0.0470	1131	0.0439		
1034	0.0350	1075	0.0471	1027	0.0404	1615	0.0470	1172	0.0441		
1043	0.0353	1132	0.0499	1041	0.0415	1632	0.0470	1195	0.0441		
1046	0.0362	1225	0.0532	1054	0.0428	1659	0.0470	1233	0.0435		
1048	0.0380	1262	0.0550	1097	0.0457			1257	0.0435		
1056	0.0398	1326	0.0561	1125	0.0459	CURVE 4		1292	0.0435		
1074	0.0425	1360	0.0565	1160	0.0463			1342	0.0438		
1113	0.0463	1418	0.0565	1217	0.0463	903	0.0391	1396	0.0438		
1141	0.0485	1444	0.0566	1296	0.0469	915	0.0373	1426	0.0438		
1153	0.0496	1477	0.0567	1343	0.0467	935	0.0351	1455	0.0447		
		1522	0.0570	1368	0.0469	962	0.0384	1501	0.0447		
				1416	0.0472	982	0.0402	1527	0.0447		

* Not shown in figure.

SPECIFICATION TABLE 142. THERMAL DIFFUSIVITY OF [IRON + SILICON + ΣX_i] ALLOYS (GROUP II)

(At least one $X_i > 0.20\%$ and/or any of Mn, P, S > 0.60%)

Cur. No.	Ref. No.	Author(s)	Year	Temp. Range, K	Reported Error, %	Name and Specimen Designation	Composition (weight percent) Fe	Si	C	Cr	Mn	P	Composition (continued), Specifications, and Remarks
1*	86	Woisard, E. L.	1961	298.2	±4	23H 566	93.87	3.16	0.49	1.43	0.74	0.31	(Fe by difference); specimen consists of two long thin rods each 3/16 in. in diameter and ~25 cm long; developed by Bethlehem Steel Co.; specimen rods butted against a thin disc-shaped heater and held in alignment under compression; entire assembly placed inside heavy walled copper tube 22 in. long acting as isothermal enclosure; test carried out under a vacuum of 10^{-4} mm Hg; momentary pulse of electric current passed through rods.

DATA TABLE 142. THERMAL DIFFUSIVITY OF [IRON + SILICON + ΣX_i] ALLOYS (GROUP II)

(At least one $X_i > 0.20\%$ and/or any of Mn, P, S > 0.60%)

[Temperature, T, K; Thermal Diffusivity, α, $cm^2\ s^{-1}$]

T	α
CURVE 1*	
298.2	0.116

* No figure given.

5. INTERMETALLIC COMPOUNDS

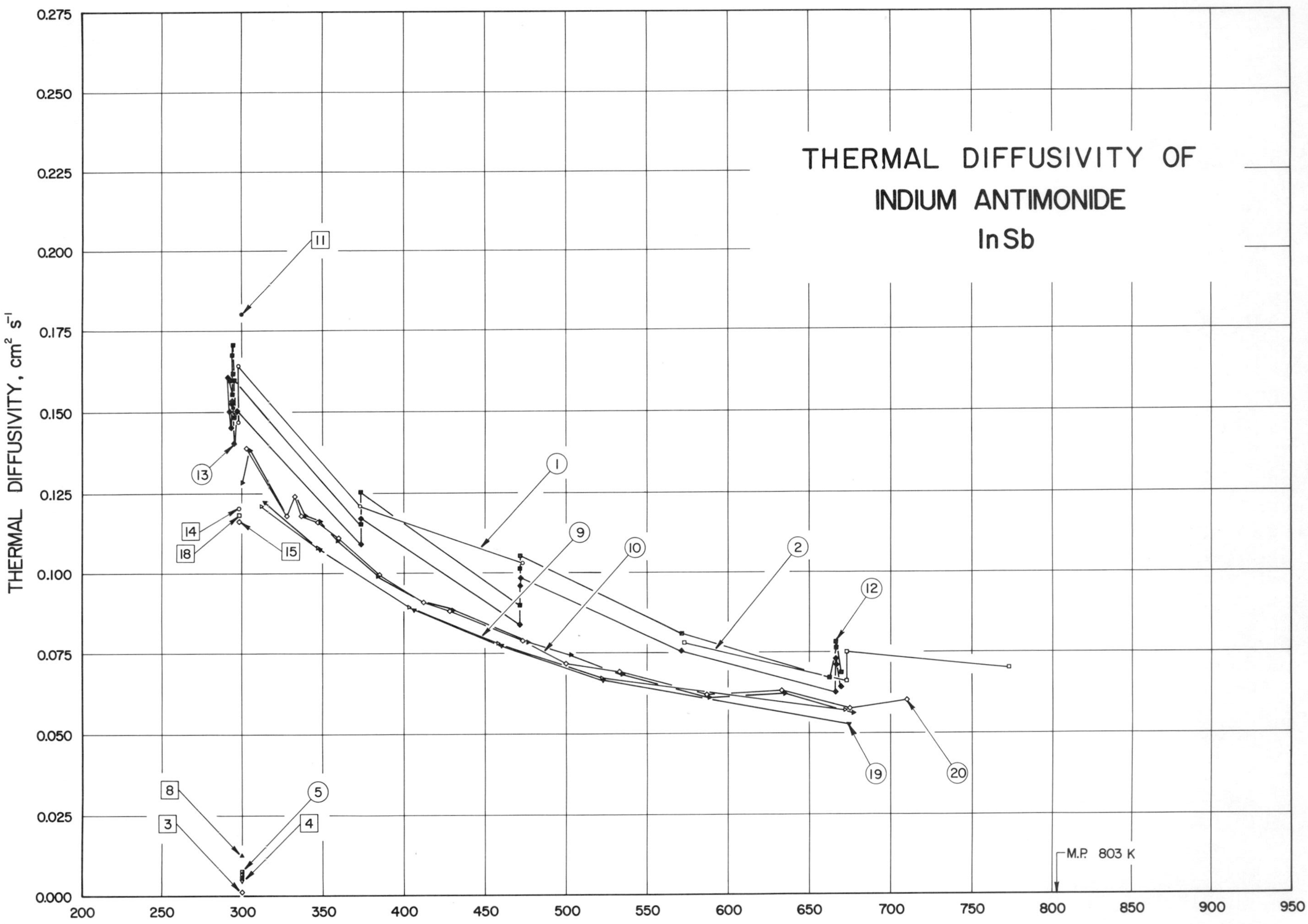
THERMAL DIFFUSIVITY OF
INDIUM ANTIMONIDE
InSb
THERMAL DIFFUSIVITY, $cm^2\ s^{-1}$
TEMPERATURE, K
0.000
0.025
0.050
0.075
0.100
0.125
0.150
0.175
0.200
0.225
0.250
0.275
200
250
300
350
400
450
500
550
600
650
700
750
800
850
900
950
M.P. 803 K
1
2
3
4
5
8
9
10
11
12
13
14
15
18
19
20

FIGURE 143

SPECIFICATION TABLE 143. THERMAL DIFFUSIVITY OF INDIUM ANTIMONIDE InSb

Cur. No.	Ref. No.	Author(s)	Year	Temp. Range, K	Reported Error, %	Name and Specimen Designation	Composition (weight percent), Specifications, and Remarks
1	68	Timberlake, A. B., Davis, P. W., and Shilliday, T. S.	1960	298-473			Single crystal intrinsic; specimen cut and lapped into the shape of a rectangular parallelepiped 0.143 cm in thickness; all surfaces sandblasted and contact points gold plated after which the gold was alloyed into specimen by heating to 473.2 K for 5 min; back-surface-temperature method used for measuring diffusivity.
2	68	Timberlake, A. B., et al.	1960	573-773			Above specimen lapped to a thickness of 0.119 cm and measured again.
3	81	Timberlake, A. B., et al.	1960	300		IIa	High purity; single crystal; specimen cut into square 1 x 1 cm and lapped to a thickness of 0.04 cm; specimen surfaces sandblasted; electrical contacts made by electroplating gold to the contact area, alloying the gold into the sample at 473.2 K and connecting copper leads to the contact by means of silver paint; method based on measuring an a-c and a d-c bolometric effect caused by a beam of radiation striking sample.
4	81	Timberlake, A. B., et al.	1960	300		IIb	Specimen cut from the same slice as the above specimen; specimen cut into square 1 x 1 cm and lapped to a thickness of 0.035 cm; same surface treatment, electrical contacts preparation, and experimental technique as above specimen.
5	82	Timberlake, A. B., et al.	1960	300		IIIb	Single crystal except for a small area in one corner; intrinsic; specimen cut and lapped in the shape of rectangular parallelepiped 1.0 x 1.0 x 0.130 cm, and sandblasted; experimental technique based on measuring bolometric effect.
6*	82	Timberlake, A. B., et al.	1960	300		IIIb	Above specimen measured again after front surfaces had been painted with flat black enamel.
7*	82	Timberlake, A. B., et al.	1960	300		IIIb	Above specimen measured again using the back-surface-temperature method.
8	82	Timberlake, A. B., et al.	1960	300		IVa	Single crystal, intrinsic; specimen cut and lapped in the shape of rectangular parallelepiped 1.0 x 1.0 x 0.236 cm, and sandblasted; data obtained by measuring bolometric effect.
9	83	Abeles, B., Cody, G. D., Dismukes, J. P., Hockings, E. F., Lindenblad, N. E., and Richman, D.	1960	312-672		IS-142	Undoped single crystal; assumed intrinsic; specimen cut to the shape of a rectangular parallelepiped with sides 1.5 x 0.375 x 0.375 in.; electrical resistivity reported as 0.497, 0.468, 0.391, and 0.390 mohm cm at 571.4, 588.2, 675.7, and 714.3 K, respectively (corresponding to band gap 0.13 ev), and 4.416, 3.436, 2.382, 2.864, 3.076, 2.109, 1.906, 1.521, 1.242, 1.040, 0.759, 0.664, and 0.559 mohm cm at 303.0, 316.5, 324.7, 327.9, 344.8, 348.4, 357.1, 384.6, 409.8, 432.9, 473.9, 502.5, and 534.8 K, respectively (corresponding to band gap 0.24 ev); frequency of thermal wave 3 cycles per min.
10	83	Abeles, B., et al.	1960	303-710		IS-194	Undoped single crystal; assumed intrinsic; specimen cut to the shape of a rectangular parallelepiped with sides 1.5 x 0.375 x 0.375 in.; electrical resistivity reported as 0.455, and 0.408 mohm cm at 613.5, and 675.7 K, respectively (corresponding to band gap 0.13 ev), and 3.776, 2.280, 1.413, 0.877, and 0.614 mohm cm at 314.5, 347.2, 406.5, 471.7, and 523.6 K, respectively (corresponding to band gap 0.24 ev); frequency of thermal wave 3 cycles per min.

* Not shown in figure.

SPECIFICATION TABLE 143. THERMAL DIFFUSIVITY OF INDIUM ANTIMONIDE InSb (continued)

Cur. No.	Ref. No.	Author(s)	Year	Temp. Range, K	Reported Error, %	Name and Specimen Designation	Composition (weight percent), Specifications, and Remarks
11	118	Perron, J.C.	1961	300	~8		Square specimen 3 cm long; Angtrŏm method used to measure diffusivity.
12	145	Timberlake, A.B., Davis, P.W., and Shilliday, T.S.	1962	293-769			Intrinsic single crystal; square specimen 1 cm x 1 cm; specimen thickness 0.143 cm for measurements up to 200 C and 0.119 cm for measurements from 200 to 500 C; specific heat reported as 0.917, 1.122, 1.255, 1.200, 1.272, 1.281, 1.298, 1.305, and 1.354 J cm^{-3} K^{-1} at 100, 200, 275, 300, 372, 475, 575, 674, and 773 K, respectively; chopped beam of radiation from an incandescent lamp focused on front face of specimen; diffusivity determined from measured variation of the r.m.s. magnitude of the thermoelectric voltage generated at the back surface over a range of chopping frequencies; data points reported calculated from thermal conductivity and specific heat data reported by author.
13	145	Timberlake, A.B., et al.	1962	292-769			Above data points for above specimen corrected for 40 μ ambipolar diffusion length.
14	173	Leroux-Hugon, P. and Weill, G.	1962	298.2			Plate specimen 0.5 mm thick; front face exposed to a chopped beam of incident light modulated at a frequency of 40 cycles per sec; diffusivity determined from measured amplitude ratio of the temperature signals at the front and rear faces of specimen; temperature of measurement not given but assumed to be room temperature.
15	173	Leroux-Hugon, P. and Weill, G.	1962	298.2			Specimen similar to the above; measured for diffusivity under same conditions as above but using a frequency of 60 cycles per sec.
16*	173	Leroux-Hugon, P. and Weill, G.	1962	298.2			Specimen similar to the above; measured for diffusivity under same conditions as above but using a frequency of 80 cycles per sec.
17*	173	Leroux-Hugon, P. and Weill, G.	1962	298.2			Specimen similar to the above; measured using a frequency of 40 cycles per sec; diffusivity determined from measured phase difference between temperature signals at the front and rear faces of specimen; other conditions same as above.
18	173	Leroux-Hugon, P. and Weill, G.	1962	298.2			Specimen similar to the above; measured for diffusivity under same conditions as above but using a frequency of 60 cycles per sec.
19	192	Abeles, B., Cody, G.D., Dismukes, J.P., and Hockings, E.F.	1961	314-674		IS-142	Undoped single crystal, rectangular parallelipiped with sides 1.5 x 0.475 x 0.475 in.; diffusivity measured with frequency of thermal wave 3 c min^{-1} by using modified Angstrŏm method and thermocouples.
20	192	Abeles, B., et al.	1961	300-711		IS-194	Similar to the above except diffusivity measured using probes.

* Not shown in figure.

DATA TABLE 143. THERMAL DIFFUSIVITY OF INDIUM ANTIMONIDE InSb

[Temperature, T, K; Thermal Diffusivity, α, $cm^2\ s^{-1}$]

T	α
CURVE 1	
298.2	0.147
298.2	0.164
373.2	0.121
473.2	0.103
CURVE 2	
573.2	0.078
673.2	0.066-0.075
773.2	0.070
CURVE 3	
300	0.00132
CURVE 4	
300	0.00045
CURVE 5	
300	0.00750
300	0.00620
300	0.00587
CURVE 6*	
300	0.00466
300	0.00611
CURVE 7*	
300	0.147
CURVE 8	
300	0.0126
CURVE 9	
312	0.121
347	0.108
403	0.0893
458	0.0781
522	0.0671
672	0.0552
CURVE 10	
303	0.139
328	0.118
333	0.124
337	0.118
347	0.116
360	0.111
385	0.0995
412	0.0907
428	0.0881
473	0.0787
500	0.0743
533	0.0690
587	0.0617
633	0.0631
675	0.0571
710	0.0601
CURVE 11	
300	0.18
CURVE 12	
293	0.159
294	0.155
294	0.161
294	0.167
294	0.170
295	0.148
295	0.159
373	0.115
373	0.125
471	0.0895
471	0.101
CURVE 12 (cont.)	
471	0.105
571	0.0807
662	0.0668
666	0.0756
666	0.0777
769	0.0682
CURVE 13	
291	0.160
292	0.150
293	0.145
294	0.152
294	0.153
295	0.140
297	0.150
373	0.109
373	0.117
471	0.0835
471	0.0956
471	0.0981
571	0.0753
666	0.0623
666	0.0706
666	0.0728
769	0.0640
CURVE 14	
298.2	0.120
CURVE 15	
298.2	0.116
CURVE 16*	
298.2	0.120
CURVE 17*	
298.2	0.116
CURVE 18	
298.2	0.118
CURVE 19	
314	0.122
348	0.107
406	0.088
460	0.077
522	0.066
674	0.055
CURVE 20	
300	0.128
304	0.138
328	0.118*
333	0.124*
339	0.118
348	0.116
359	0.110
384	0.099
412	0.091*
430	0.088
476	0.078
503	0.074
534	0.068
588	0.061
635	0.062
677	0.056
711	0.059

*Not shown in figure.

SPECIFICATION TABLE 144. THERMAL DIFFUSIVITY OF MAGNESIUM GERMANIDE Mg_2Ge

Cur. No.	Ref. No.	Author(s)	Year	Temp. Range, K	Reported Error, %	Name and Specimen Designation	Composition (weight percent), Specifications, and Remarks
1*	118	Perron, J.C.	1961	300	~8		Square specimen 3 cm long; Angström method used to measure diffusivity.

DATA TABLE 144. THERMAL DIFFUSIVITY OF MAGNESIUM GERMANIDE Mg_2Ge

[Temperature, T, K; Thermal Diffusivity, α, $cm^2\ s^{-1}$]

T	α
CURVE 1*	
300	0.053

*No figure given.

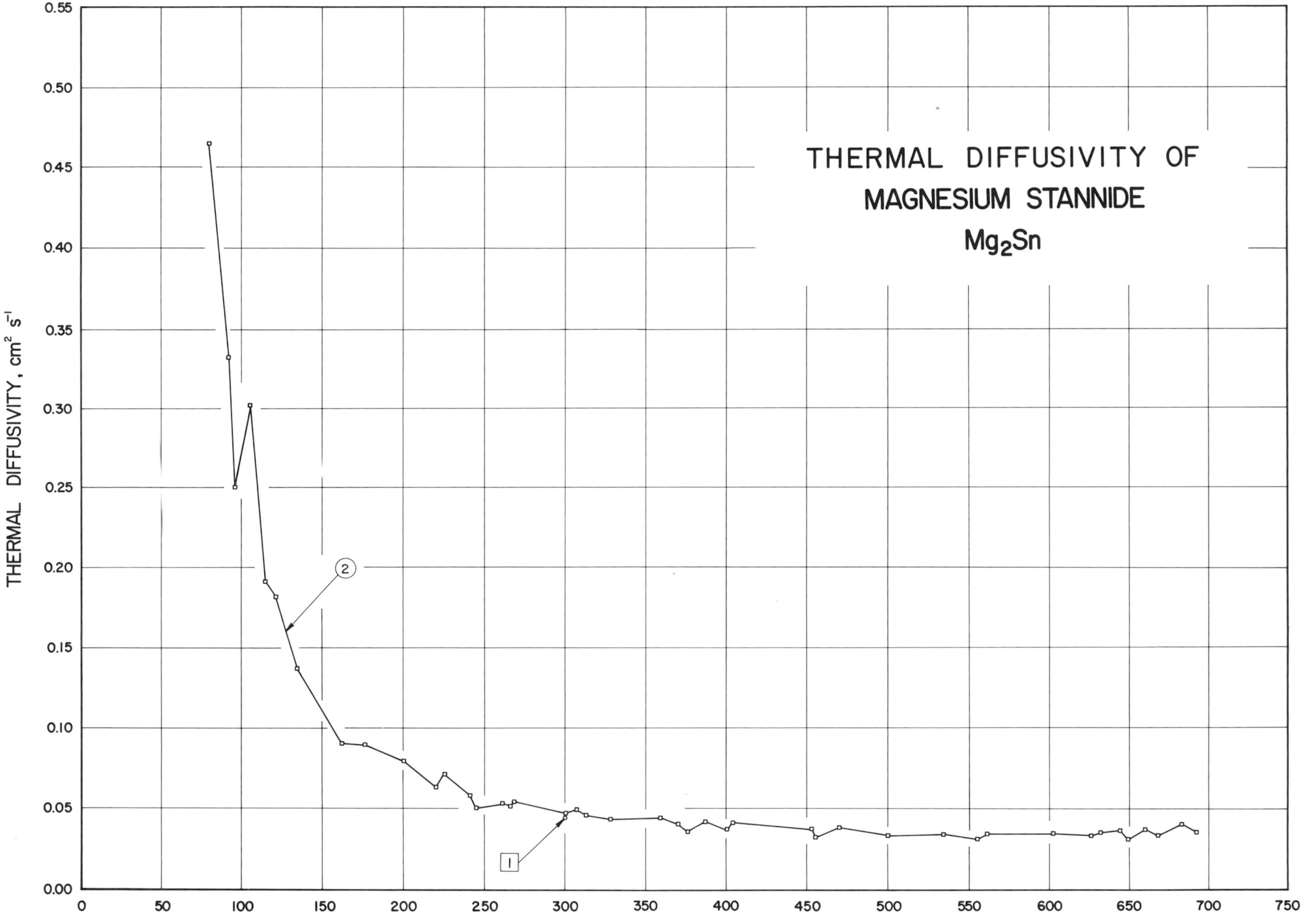
THERMAL DIFFUSIVITY OF
MAGNESIUM STANNIDE
Mg2Sn
THERMAL DIFFUSIVITY, cm2 s-1
TEMPERATURE, K
0.55
0.50
0.45
0.40
0.35
0.30
0.25
0.20
0.15
0.10
0.05
0.00
0
50
100
150
200
250
300
350
400
450
500
550
600
650
700
750
1
2

FIGURE 145

SPECIFICATION TABLE 145. THERMAL DIFFUSIVITY OF MAGNESIUM STANNIDE Mg_2Sn

Cur. No.	Ref. No.	Author(s)	Year	Temp. Range, K	Reported Error, %	Name and Specimen Designation	Composition (weight percent), Specifications, and Remarks
1	118	Perron, J. C.	1961	300	~8		Square specimen 3 cm long; Angström method used to measure diffusivity.
2	127	Martin, J. J., Shanks, H. R., and Danielson, G. C.	1968	80-692			n-type single crystal; donor concentration ~3 x 10^{16} donor cm^{-3} after diffusivity measurements and ~2.5 x 10^{16} donor cm^{-3} before diffusivity measurements; rectangular specimen 4.30 x 4.24 x 20 mm; single crystal grown by a modified Bridgman technique; grown from 99.999 pure tin and from magnesium distilled under high vacuum from 99.99 pure starting material in the Ames Laboratory, Iowa State University; electrical resistivity measured before the thermal diffusivity data were taken and reported as 0.1019, 0.1549, 0.1977, 0.2618, 0.3020, 0.3090, 0.2924, 0.2523, 0.0652, and 0.0375 ohm cm at 80.0, 112.7, 132.1, 154.3, 175.4, 187.3, 198.0, 204.9, 268.1, and 303.0 K, respectively; electrical resistivity measured again after the thermal diffusivity was measured and reported as 0.0887, 0.0912, 0.1153, 0.1811, 0.2138, 0.2312, 0.2716, 0.2642, 0.2138, 0.1225, 0.0912, 0.0413, 0.0294, 0.0110, 0.0086, 0.0061, 0.0046, 0.0032, 0.0028, 0.0025, 0.0022, and 0.0021 ohm cm at 76.9, 81.4, 108.0, 135.9, 154.6, 170.6, 189.0, 197.6, 213.2, 232.6, 250.6, 292.4, 319.5, 390.6, 421.9, 478.5, 526.3, 591.7, 621.1, 675.7, 675.7, and 740.7 K, respectively; completely intrinsic above room temp; diffusivity determined from measured temp. variation at three points along specimen.

DATA TABLE 145. THERMAL DIFFUSIVITY OF MAGNESIUM STANNIDE Mg_2Sn

[Temperature, T, K; Thermal Diffusivity, α, $cm^2\ s^{-1}$]

T	α	T	α	T	α	T	α	T	α
CURVE 1		CURVE 2 (cont.)		CURVE 2 (cont.)		CURVE 2 (cont.)		CURVE 2 (cont.)	
300	0.044	162	0.090	307	0.049	470	0.038	668	0.033
		176	0.089	313	0.046	500	0.033	683	0.040
CURVE 2		200	0.079	328	0.043	534	0.034	692	0.035
		220	0.063	359	0.044	555	0.031		
80	0.465	225	0.071	370	0.040	561	0.034		
92	0.332	241	0.058	376	0.036	602	0.034		
96	0.250	245	0.050	387	0.042	626	0.033		
105	0.302	261	0.053	401	0.037	632	0.035		
115	0.191	266	0.051	404	0.041	644	0.036		
121	0.182	268	0.054	453	0.037	649	0.031		
134	0.137	300	0.047	455	0.032	660	0.037		

6. SINGLE OXIDES

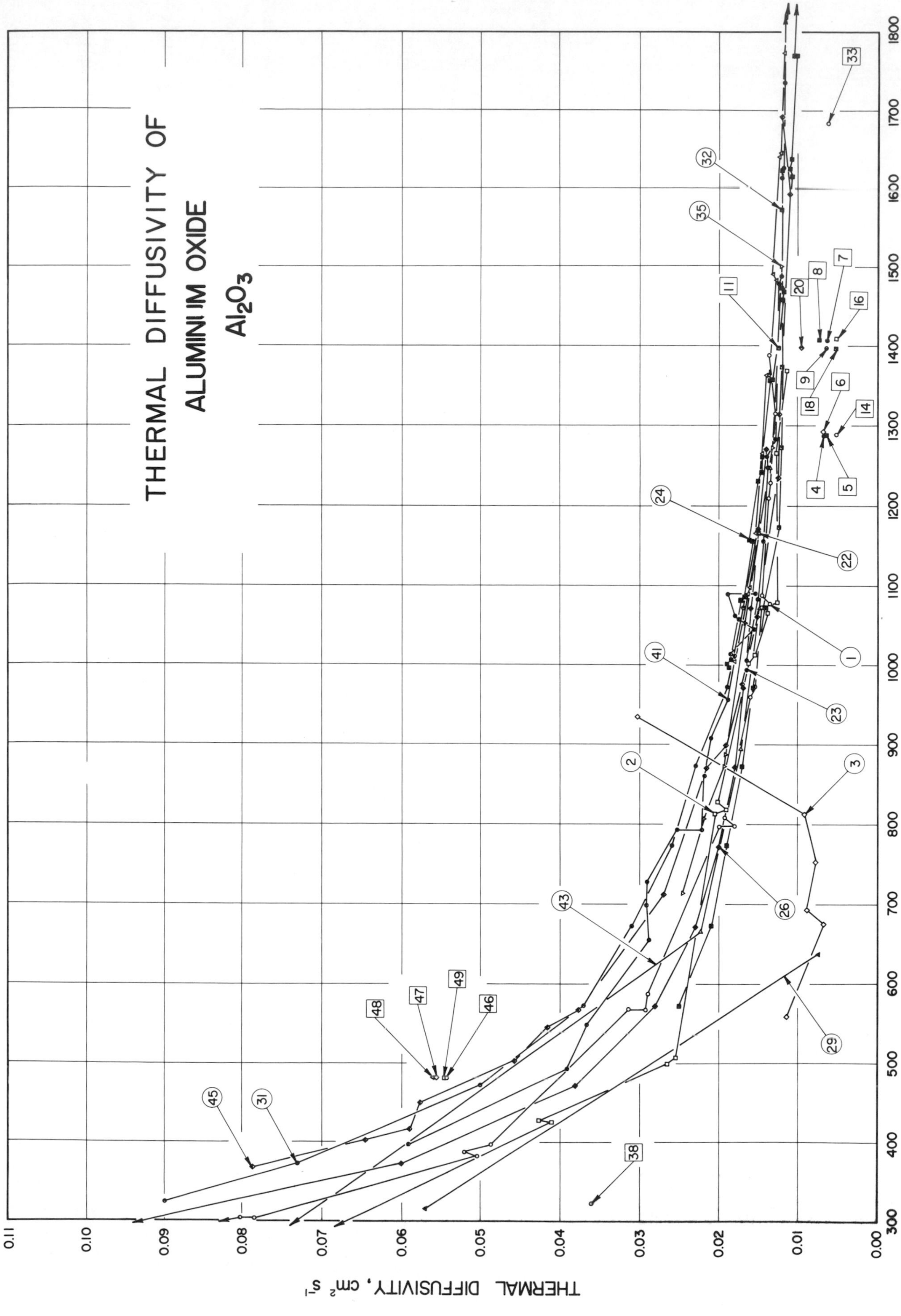
THERMAL DIFFUSIVITY OF
ALUMINIUM OXIDE
Al_2O_3
THERMAL DIFFUSIVITY, $cm^2\ s^{-1}$
TEMPERATURE, K
0.00
0.01
0.02
0.03
0.04
0.05
0.06
0.07
0.08
0.09
0.10
0.11
300
400
500
600
700
800
900
1000
1100
1200
1300
1400
1500
1600
1700
1800

FIGURE 146

SPECIFICATION TABLE 146. THERMAL DIFFUSIVITY OF ALUMINUM OXIDE Al_2O_3

Cur. No.	Ref. No.	Author(s)	Year	Temp. Range, K	Reported Error, %	Name and Specimen Designation	Composition (weight percent), Specifications, and Remarks
1	34	Rudkin, R. L., Parker, W. J., and Jenkins, R. J.	1963	298-1388			Pure; specimen a few millimeters thick; extruded; plated with Hanovia Platinum Bright (No. 05); density 3.834 g cm^{-3}; high intensity short duration light pulse from xenon flash lamp absorbed in the front surface of thermally insulated specimen; thermal diffusivity determined from measured temperature history of the rear surface.
2	34	Rudkin, R. L., et al.	1963	293-1368			Pure; specimen a few millimeters thick; pressed; plated with Hanovia Platinum Bright (No. 05); density 3.914 g cm^{-3}; high intensity short duration light pulse from xenon flash lamp absorbed in the front surface of thermally insulated specimen; thermal diffusivity determined from measured temperature history of the rear surface.
3	36	Kevane, C. J.	1958	559-935			Cylindrical specimen ~0.375 in. in diameter and 1 in. long; made by Gladding McBean; sintered; holes drilled in specimen separated by a distance of 2 mm between the lines of centers along the axis; diffusivity determined from measured propagation of temperature waves set up by periodic modulation of radiation heat flux input to one surface of specimen; solar furnace used as a heat source; amplitude ratio method used to measure diffusivity; one-dimensional heat flow.
4	13	Childers, H. M. and Cerceo, J. M.	1961	1289			Polycrystalline; disc specimen 0.255 cm thick; density 4 g cm^{-3}; heated by electron bombardment; sinusoidal temperature wave imposed on one face of specimen; measured in vacuum of $\sim 10^{-5}$ mm Hg; radiation losses from back face neglected; temperature wave phase shift front to back faces $\phi = 0.565$ radians.
5	13	Childers, H. M. and Cerceo, J. M.	1961	1289			Above specimen measured for diffusivity again under same conditions as above except that $\phi = 0.628$ radians.
6	13	Childers, H. M. and Cerceo, J. M.	1961	1292			Above specimen measured for diffusivity again under same conditions as above except that $\phi = 0.565$ radians.
7	13	Childers, H. M. and Cerceo, J. M.	1961	1408			Above specimen measured for diffusivity again under same conditions as above except that $\phi = 0.628$ radians.
8	13	Childers, H. M. and Cerceo, J. M.	1961	1409			Above specimen measured for diffusivity again under same conditions as above except that $\phi = 0.527$ radians.
9	13	Childers, H. M. and Cerceo, J. M.	1961	1397			Above specimen measured for diffusivity again under same conditions as above except that $\phi = 0.628$ radians.
10*	13	Childers, H. M. and Cerceo, J. M.	1961	1397			Above specimen measured for diffusivity again under same conditions as above.
11	13	Childers, H. M. and Cerceo, J. M.	1961	1397			Above specimen measured for diffusivity again under same conditions as above except that $\phi = 1.03$ radians.
12*	13	Childers, H. M. and Cerceo, J. M.	1961	1397			Above specimen measured for diffusivity again under same conditions as above.

* Not shown in figure.

SPECIFICATION TABLE 146. THERMAL DIFFUSIVITY OF ALUMINUM OXIDE Al_2O_3 (continued)

Cur. No.	Ref. No.	Author(s)	Year	Temp. Range, K	Reported Error, %	Name and Specimen Designation	Composition (weight percent), Specifications, and Remarks
13*	13	Childers, H. M. and Cerceo, J. M.	1961	1289			Above specimen measured for diffusivity again assuming radiation losses from back face; phase shift angle $\phi = 0.565$ radians.
14	13	Childers, H. M. and Cerceo, J. M.	1961	1289			Above specimen measured for diffusivity again under same conditions as above except that $\phi = 0.628$ radians.
15*	13	Childers, H. M. and Cerceo, J. M.	1961	1292			Above specimen measured for diffusivity again under same conditions as above except that $\phi = 0.565$ radians.
16	13	Childers, H. M. and Cerceo, J. M.	1961	1408			Above specimen measured for diffusivity again under same conditions as above except that $\phi = 0.628$ radians.
17	13	Childers, H. M. and Cerceo, J. M.	1961	1409			Above specimen measured for diffusivity again under same conditions as above except that $\phi = 0.527$ radians.
18	13	Childers, H. M. and Cerceo, J. M.	1961	1397			Above specimen measured for diffusivity again under same conditions as above except that $\phi = 0.628$ radians.
19*	13	Childers, H. M. and Cerceo, J. M.	1961	1397			Above specimen measured for diffusivity again under same conditions as above.
20	13	Childers, H. M. and Cerceo, J. M.	1961	1397			Above specimen measured for diffusivity again under same conditions as above except that $\phi = 1.03$ radians.
21*	13	Childers, H. M. and Cerceo, J. M.	1961	1397			Above specimen measured for diffusivity again under same conditions as above.
22	71, 279	Rudkin, R. L.	1963	1005-1903	± 5/ ±10	GD-10	Disc specimen 0.75 in. in diameter and ~0.045 in. thick; obtained from Corning; coated on both sides with tungsten vapour using electron beam techniques; density 3.84 g cm^{-3}; short pulse of thermal energy from a xenon flash lamp absorbed at front surface of specimen; diffusivity determined from measured temperature-time function of the rear surface; measured in vacuum of $\sim 10^{-5}$ mm Hg.
23	71, 279	Rudkin, R. L.	1963	996-1906	± 5/ ±10	FS-54	Disc specimen 0.75 in. in diameter and ~0.045 in. thick; obtained from Corning; coated on both sides with tungsten vapour using electron beam techniques; density 3.84 g cm^{-3}; short pulse of thermal energy from a xenon flash lamp absorbed at front surface of specimen; diffusivity determined from measured temperature-time function of the rear surface; measured in vacuum of $\sim 10^{-5}$ mm Hg.
24	71, 279	Rudkin, R. L.	1963	998-1908	± 5/ ±10	AD-995	99.5 pure; disc specimen 0.75 in. in diameter and ~0.045 in. thick; obtained from Coors Porcelain Company; coated on both sides with tungsten vapour using electron beam techniques; density 3.86 g cm^{-3}; short pulse of thermal energy from xenon flash lamp absorbed at front surface of specimen; diffusivity determined from measured temperature-time function of the rear surface; measured in vacuum of $\sim 10^{-5}$ mm Hg.

* Not shown in figure.

SPECIFICATION TABLE 146. THERMAL DIFFUSIVITY OF ALUMINUM OXIDE Al_2O_3 (continued)

Cur. No.	Ref. No.	Author(s)	Year	Temp. Range, K	Reported Error, %	Name and Specimen Designation	Composition (weight percent), Specifications, and Remarks
25*	71, 279	Rudkin, R. L.	1963	848-1998	± 5/ ±10	AP-35	99 Al_2O_3, and 1 CoO; disc specimen 0.75 in. in diameter and ~0.045 in. thick; obtained from McDanel; CoO added to the mix before firing; coated on both sides with tungsten vapour using electron beam techniques; density 2.94 g cm^{-3}; short pulse of thermal energy from xenon flash lamp absorbed at front surface of specimen; diffusivity determined from measured temperature-time function of the rear surface; measured in vacuum of ~10^{-5} mm Hg.
26	72	Plummer, W. A., Campbell, D. E., and Comstock, A. A.	1962	298-1273			Specimen composed of three pieces: middle piece and two end pieces; middle piece separated from bottom piece by a 0.005 in. thick chromel sheet 7.6 x ~18 cm forming the heat source, middle piece separated from upper piece by a 0.005 in. thick chromel sheet forming the heat sink; thicknesses on either side of the heater are the same; entire assembly constitutes a sandwich 12.7 cm long, 7.6 cm wide and at least 4 to 5 cm thick; diffusivity determined from measured ratio of the temperature rises of the heat source and sink; density 3.04 g cm^{-3}; unidimensional heat flow.
27*	73	Soxman, E. J.	1957	84-249	± 5	AV30 Alumina	96 pure; vitreous, polycrystalline; U-shaped rod specimen ~3 mm in diameter and at least 20 cm long; provided by McDanel Refractory Porcelain Company; commercially prepared; measured in copper vacuum chamber; measured with chamber immersed in liquid air; sinusoidal temperature fluctuation impressed at one end of specimen; diffusivity calculated from measured amplitudes and phase shift of resulting temperature wave at two points along specimen; accuracy reported does not apply to measurement at 84 K.
28*	73	Soxman, E. J.	1957	276	± 5	AV30 Alumina	Specimen similar to the above; measured with vacuum chamber immersed in ice water; measured under same conditions as above.
29	73	Soxman, E. J.	1957	317, 639	± 5	AV30 Alumina	Specimen similar to the above; measured in vacuum furnace.
30*	73	Soxman, E. J.	1957	1920		AV30 Alumina	Specimen similar to the above; measured in vacuum furnace; data point only approximately determined from readings of temperature taken at 10 sec intervals.
31	115	Berthier, G.	1965	323-1173		Degussit Al 23	99.5 pure; cylindrical specimen composed of two identical parts each 12 mm in dia; lower part has three grooves machined on upper surface to accomodate thermocouple leads, thermocouple junctions welded to lower part, surfaces precisely machined, upper and lower parts held together under pressure; manufactured from cylindrical bars of fired alumina obtained from various sources; theoretical density in the range from 3.7 to 3.95 g cm^{-3}; actual density 3.45±0.1 g cm^{-3} (87.4±2.5 to 93.1±2.7% of theoretical value); specimen subjected to a thermal shock in an electrical furnace; radial heat flow method used to measure diffusivity; measured data are the results of two different runs.
32	115	Berthier, G.	1965	573-1623		Desmarquest qualité AF	98.7 pure; cylindrical specimen composed of two identical parts each 12 mm in dia (16 mm specimen for a few measurements); lower part has three grooves machined on upper surface to accomodate thermocouple leads, thermocouple junctions welded to lower part, surfaces precisely machined, upper and lower parts held together under pressure; manufactured from cylindrical bars of fired alumina obtained from various sources; theoretical density 3.8 g cm^{-3}; actual density 3.29±0.1 g cm^{-3} (86.5±2.6% of theoretical value); specimen subjected to a thermal shock in an electrical furnace; radial heat flow method used to measure diffusivity.

* Not shown in figure.

SPECIFICATION TABLE 146. THERMAL DIFFUSIVITY OF ALUMINUM OXIDE Al_2O_3 (continued)

Cur. No.	Ref. No.	Author(s)	Year	Temp. Range, K	Reported Error, %	Name and Specimen Designation	Composition (weight percent), Specifications, and Remarks
33	116	Levine, H. S.	1950	1684.7			Cylindrical specimen 2.78 cm in dia and 8.4 cm long (8.6 cm during last two runs); formed from properly sized powdered Alcoa T-61 tubular Al_2O_3 mixed with 5 percent paraffin binder by dry pressing at 7000 p.s.i. in a split mold; thermocouple well 3/32 in. in dia drilled along axis of specimen before removal from mold; calcined to remove wax binder and sintered for 48 hrs at 1973.2 K; total porosity 35.0 percent; fired density 2.60 g cm^{-3}; reported data point is average result of six runs.
34*	117	Paladino, A. E., Swarts, E. L., and Crandall, W. B.	1957	1773-2073			Two cylindrical specimens: one 1.52 cm in dia and 5.30 cm long, the other 2.44 cm in dia and 5.38 cm long; a hole 3 mm in dia extended from one end up the axis of cylinder to within 0.5 cm of the other end; slip-cast from mixtures of Norton 38-900 Alundum, Norton 38-500 Alundum, and calcined aluminum trihydrate; after an initial firing to 1873.2 K cylinders fired at 2123.2 K for 3 hrs in an oxyacetylene furnace; density after firing 3.60 g cm^{-3}; theoretical density 3.97 g cm^{-3} (assumed); porosity not exceeding 10 percent; range of data reported at each temp corresponds to variation in the value of parameter H = h/k (surface heat transfer coefficient/thermal conductivity) from 1.0 to 2.5 over entire temp range; a minimum of three runs made for each specimen at each temp; diffusivity obtained at each temp from measured heating curves for both specimens.
35	125	Moser, J. B. and Kruger, O. L.	1968	713-1500	±5		Disc specimen; taken from thermal conductivity specimen obtained from Dynatech Inc.; surface coated with colloidal graphite; measured in vacuum furnace; flash method used to measure diffusivity; ruby laser used to generate thermal pulse; maximum error of measurement ±7.5 percent.
36*	184	Taylor, R.	1965	298			15% porosity-dense; measured mean density 3.23 g cm^{-3}; temperature of measurement not given by author but assumed to be room temperature; heat pulse method used to measure diffusivity; cylindrical specimen 0.25 in. in diameter and 0.5 in. long.
37*	184	Taylor, R.	1965	298			25-30% porosity; extruded or slip cast; measured mean density 2.86 g cm^{-3}; other conditions same as above.
38	187	Jaeger, G., Koehler, W., and Stapelfeldt, F.	1950	323			Disk specimen 50 mm in diameter and 7-12 mm thick; sintered at 2173.2 K and then machined on both sides while maintaining them exactly parallel; specimen originally at 20 C; diffusivity determined from measured time necessary for one side of specimen to reach a temperature of 80 C while maintaining the other side in contact with mercury at 100 C; average temperature of measurement not given by authors but assumed to be 50 C; data point reported is the average result of five measurements carried out on at least each of five specimens; density measured and reported as 3.77 g cm^{-3}.
39*	280, 211	Cerceo, J. M. and Childer, H. M.	1962	1290, 1400			Polycrystalline; density 4 g cm^{-3}.
40*	280, 211	Cerceo, J. M. and Childer, H. M.	1962	1290, 1400			The above specimen.

* Not shown in figure.

SPECIFICATION TABLE 146. THERMAL DIFFUSIVITY OF ALUMINUM OXIDE Al_2O_3 (continued)

Cur. No.	Ref. No.	Author(s)	Year	Temp. Range, K	Reported Error, %	Name and Specimen Designation	Composition (weight percent), Specifications, and Remarks
41	281	Kanamori, H., Fujii, N., and Mizutani, H.	1968	398-1090		Corundum	Specimen of gem quality and millimeter size; diffusivity measured using modified Angström method in 1 atmosphere of argon.
42*	214, 215	Moser, J.B. and Kruger, O.L.	1964	298.2			1.9 cm in diameter and 0.1 to 0.3 cm thick; density 3.51 g cm^{-3}.
43	156, 56	Degas, P. and Bertin, J.-L.	1970	295-1248			8 mm in diameter and 2 to 5 mm thick.
44*	44	Chang, H., Altman, M., and Sharma, R.	1967			Lucalox	99.8$^+$ pure; σ-alumina, polycrystalline; cylindrical specimen, 1 in. in diameter, 6 in. long; density at 25 C 3.98 g cm^{-3}; melting point 2313 K; diffusivity measured in Pt-Pt 40% Rh wire-wound furnace.
45	44	Chang, H., et al.	1967			Lucalox	Similar to the above specimen except 2 in. in diameter, 18 in. long; diffusivity measured in tungsten-mesh heater furnace.
46	282	Yurchak, R.P., Tkach, G.F., and Petrunin, G.I.	1970	482			Characteristic dimension 12.5 mm; diffusivity measured by radial temperature waves with cycle 24.2 sec and determined from phase difference.
47	282	Yurchak, R.P., et al.	1970	482			The above specimen; diffusivity determined from heating rate.
48	282	Yurchak, R.P., et al.	1970	482			Similar to the above except cycle 16.3 sec; diffusivity determined from phase difference.
49	282	Yurchak, R.P., et al.	1970	482			The above specimen; diffusivity determined from heating rate.

* Not shown in figure.

DATA TABLE 146. THERMAL DIFFUSIVITY OF ALUMINUM OXIDE Al_2O_3

[Temperature, T, K; Thermal Diffusivity, α, $cm^2\ s^{-1}$]

T	α
CURVE 1	
298	0.0830*
303	0.0803
303	0.0785
383	0.0505
388	0.0520
398	0.0487
568	0.0315
568	0.0293
588	0.0290
797	0.0200
798	0.0180
808	0.0193
960	0.0160
973	0.0155
1073	0.0147
1078	0.0135
1088	0.0145
1210	0.0137
1230	0.0135
1316	0.0129
1388	0.0137
CURVE 2	
293	0.0687
293	0.0675
426	0.0410
428	0.0425
500	0.0265
508	0.0255
813	0.0205
818	0.0190
828	0.0202
1003	0.0163
1013	0.0155
1166	0.0137
1173	0,0140
1178	0.0125
1266	0.0127
1368	0.0113
CURVE 3	
559	0.0113
675	0.0067
693	0.0088
753	0.0077
813	0.0091
935	0.0303
CURVE 4	
1289	0.0068
CURVE 5	
1289	0.0064
CURVE 6	
1292	0.0068
CURVE 7	
1408	0.0064
CURVE 8	
1409	0.0073
CURVE 9	
1397	0.0064
CURVE 10*	
1397	0.0064
CURVE 11	
1397	0.0124
CURVE 12*	
1397	0.0124
CURVE 13*	
1289	0.0067
CURVE 14	
1289	0.0052
CURVE 15*	
1292	0.0067
CURVE 16	
1408	0.0052
CURVE 17*	
1409	0.0064
CURVE 18	
1397	0.0052
CURVE 19*	
1397	0.0052
CURVE 20	
1397	0.0096
CURVE 21*	
1397	0.0096
CURVE 22	
1005	0.0181
1090	0.0164
1098	0.0161
1166	0.0154
1166	0.0151
1266	0.0145
CURVE 22 (cont.)	
1363	0.0140
1363	0.0137
1483	0.0128
1490	0.0132
1640	0.0123
1646	0.0121
1773	0.0117
1903	0.0111
CURVE 23	
996	0.0165
1008	0.0165
1085	0.0151
1158	0.0144
1250	0.0138
1263	0.0139
1385	0.0128
1480	0.0124
1490	0.0121
1616	0.0120
1626	0.0118
1738	0.0117
1906	0.0114
CURVE 24	
998	0.0187
1003	0.0190
1008	0.0185
1083	0.0172
1088	0.0168
1158	0.0158
1159	0.0161
1233	0.0150
1243	0.0146
1263	0.0145
1358	0.0135
1360	0.0132
1458	0.0119
1470	0.0117
1618	0.0106
CURVE 24 (cont.)	
1628	0.0109
1640	0.0107
1773	0.0101
1773	0.0104
1890	0.0100
1900	0.0101
1908	0.00995
CURVE 25*	
848	0.0106
928	0.0100
993	0.00980
1103	0.00953
1173	0.00930
1243	0.00915
1353	0.00905
1478	0.00885
1483	0.00835
1628	0.00855
1643	0.00835
1653	0.00875
1748	0.00847
1773	0.00835
1898	0.00807
1903	0.00830
1998	0.00800
CURVE 26	
298	0.093
373	0.060
473	0.038
573	0.028
673	0.023
773	0.020
873	0.018
973	0.017
1073	0.016
1173	0.015
1273	0.014
CURVE 27*	
84	5.06
126	0.82
249	0.156
CURVE 28*	
276	0.086
CURVE 29	
317	0.057
639	0.0073
CURVE 30*	
1920	0.015
CURVE 31	
323	0.09
373	0.073
473	0.050
573	0.037
673	0.031
773	0.026
873	0.023
973	0.019
1073	0.017
1173	0.015*
CURVE 32	
573	0.025
673	0.021
773	0.019
873	0.017
973	0.0156
1073	0.0142
1173	0.0124
1273	0.012
1373	0.012
1473	0.012
CURVE 32 (cont.)	
1573	0.012
1623	0.012
CURVE 33	
1684.7	0.0062
CURVE 34*	
1773	0.0086
1873	0.0145
1973	0.0177
2073	0.0232
CURVE 35	
713	0.0246
807	0.0219
873	0.0194
888	0.0192
1053	0.0167
1273	0.0132
1500	0.0121
CURVE 36*	
298.2	0.114
CURVE 37*	
298.2	0.058
CURVE 38	
323.2	0.036
CURVE 39*	
1290	0.0067
1400	0.0096
CURVE 40*	
1290	0.0068
1400	0.0124
CURVE 41	
398	0.0592
494	0.0391
549	0.0366
656	0.0288
699	0.0292
728	0.0292
793	0.0254
793	0.0222
861	0.0219
908	0.0210
957	0.0189
1012	0.0180
1014	0.0186
1046	0.0156
1058	0.0175
1062	0.0180
1090	0.0189
1090	0.0154
CURVE 42*	
298.2	0.091
CURVE 43	
295	0.0736*
676	0.0223
895	0.0173
1073	0.0149
1248	0.0136
CURVE 44*	
808	0.0228
834	0.0221
859	0.0213
895	0.0232

*Not shown in figure.

DATA TABLE 146. THERMAL DIFFUSIVITY OF ALUMINUM OXIDE Al_2O_3 (continued)

T	α
CURVE 44 (cont.)*	
1097	0.0135
1189	0.0126
1275	0.0111
1479	0.0094
CURVE 45	
368	0.0786
403	0.0645
417	0.059
451	0.0576
504	0.0456
546	0.0415
568	0.0376
713	0.0270
871	0.0216
900	0.0191
976	0.0171
1061	0.0152
1236	0.0124
1315	0.0123
1593	0.011
1692	0.012
CURVE 46	
482	0.0543
CURVE 47	
482	0.0555
CURVE 48	
482	0.0560
CURVE 49	
482	0.0546

* Not shown in figure.

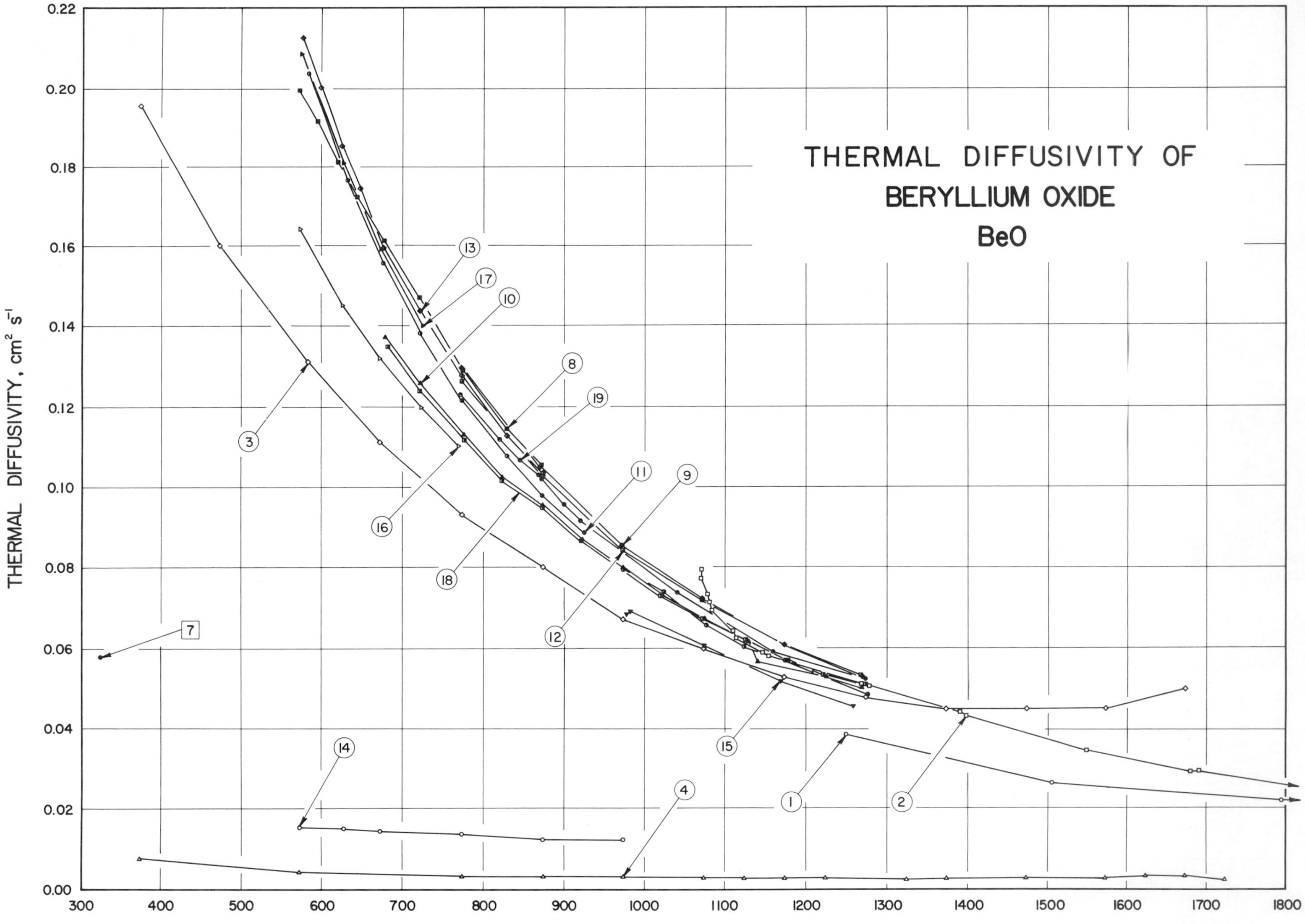
THERMAL DIFFUSIVITY OF
BERYLLIUM OXIDE
BeO
THERMAL DIFFUSIVITY, cm² s⁻¹
TEMPERATURE, K
0.00
0.02
0.04
0.06
0.08
0.10
0.12
0.14
0.16
0.18
0.20
0.22
300
400
500
600
700
800
900
1000
1100
1200
1300
1400
1500
1600
1700
1800
1
2
3
4
7
8
9
10
11
12
13
14
15
16
17
18
19

FIGURE 147

SPECIFICATION TABLE 147. THERMAL DIFFUSIVITY OF BERYLLIUM OXIDE BeO

Cur. No.	Ref. No.	Author(s)	Year	Temp. Range, K	Reported Error, %	Name and Specimen Designation	Composition (weight percent), Specifications, and Remarks
1	74	Hedge, J.C., Kopec, J.W., Kostenko, C., and Lang, J.I.	1963	1250-2200			99.5 BeO, 0.009 Si, 0.005 Al, 0.002 Mo, 0.001 Ca, Fe, Na each, ~0.001 Cr, Ni each, 0.0003 Mn, and 0.0001 or less B, Cd, Co, Cu, Li each; slab specimen; supplied by Brush Beryllium Co.; cold pressed; fired at 1855.4 K; density 2.87 g cm^{-3}; top surface of specimen exposed to heat sink; diffusivity determined from measured temperature decrease; unidirectional heat flow.
2	71, 279	Rudkin, R. L.	1963	1070-2046	± 5/ ±10	BD-98	Disc specimen 0.75 in. in diameter and ~0.045 in. thick; obtained from Coors Porcelain Company; coated on both sides with tungsten vapour using electron beam techniques; density 2.93 g cm^{-3}, 97% of theoretical value; short pulse of thermal energy from a xenon flash lamp absorbed at front surface of specimen; diffusivity determined from measured temperature-time function of the rear surface; measured in vacuum of ~10^{-5} mm Hg.
3	115	Berthier, G.	1965	373-1673		No. 363-364	99.4 BeO (by difference), and 0.6 CaO; cylindrical specimen composed of two identical parts each 12 mm in dia and 10 mm long, lower part has three grooves machined on upper surface to accomodate thermocouple leads, thermocouple junctions welded to lower part, surfaces precisely machined, upper and lower parts held together under pressure; manufactured and supplied by the Commissariat à l' Energie Atomique, Centre d' Etudes Nucléaires de Saclay (France); machined from parallelepiped bars produced by natural calcination of very pure BeO powder; plane surfaces rendered as parallel as possible by grinding; density 2.83 g cm^{-3}; data points reported are the average results obtained from three independent runs; diffusivity measured using radial heat flow method.
4	115	Berthier, G.	1965	373-1723		No. 315-316	Cylindrical specimen composed of two identical parts each 12 mm in diameter; density 3 g cm^{-3}; irradiated at low temperature with a flux of 2 x 10^{19} total neutron cm^{-2} at the Triton reactor; heat treated; other specifications similar to the above.
5*	115	Berthier, G.	1965	373-1723		No. 315-316	Above specimen measured for diffusivity again.
6*	115	Berthier, G.	1965	373-1723		No. 315-316	Above specimen measured for diffusivity again.
7	187	Jaeger, G., Koehler, W., and Stapelfeldt, F.	1950	323.2			Disk specimen 50 mm in diameter and 7-12 mm thick; sintered at 2173.2 K and then machined on both sides while maintaining them exactly parallel; specimen originally at 20 C; diffusivity determined from measured time necessary for one side of specimen to reach a temperature of 80 C while maintaining the other side in contact with mercury at 100 C; average temperature of measurement not given by authors but assumed to be 50 C; data point reported is the average result of five measurements carried out on at least each of five specimens; density measured and reported as 2.88 g cm^{-3}.
8	188	Anonymous	1965	573-871		(UOX+0.5MgO);1	0.5 MgO; grain size 4 μ; cylindrical specimen 0.6 cm in diameter and 0.115 cm long; front and rear faces coated with tungsten of ~5 μ in thickness; density 2.92 g cm^{-3}; flash method utilizing a laser used to measure diffusivity; heat pulse applied to front face and diffusivity determined from time dependence of temperature rise at rear face.
9	188	Anonymous	1965	773-1269		(UOX+0.5MgO);2	Cylindrical specimen 0.164 cm long; other specifications and conditions same as above.

* Not shown in figure.

SPECIFICATION TABLE 147. THERMAL DIFFUSIVITY OF BERYLLIUM OXIDE BeO (continued)

Cur. No.	Ref. No.	Author(s)	Year	Temp. Range, K	Reported Error, %	Name and Specimen Designation	Composition (weight percent), Specifications, and Remarks
10	188	Anonymous	1965	679-1269		(UOX+0.5MgO)	0.5 MgO; grain size 4 μ; cylindrical specimen 0.6 cm in diameter and 0.1344 cm long; front and rear faces coated with tungsten of ~5 μ in thickness; obtained from the same sintering and extrusion batch as the above two specimens; irradiated at ~1000 C to 2.5 x 10^{21} nvt with energy ≥1 MeV; density 2.86 g cm^{-3}; measured for diffusivity using the same method as above; measured in ascending order of temperature, maintained overnight at 600 C, and then measured again up to 996 C.
11	224	Freeman, R.J.	1966	584-1276		A	0.1588 cm long; grain size 18 μ; bulk density 2.93 g cm^{-3}.
12	224	Freeman, R.J.	1966	773-1268		B	0.1153 cm long; grain size 4 μ; bulk density 2.92 g cm^{-3}.
13	224	Freeman, R.J.	1966	577-873		C	Similar to above but specimen 0.1643 cm long.
14	224	Freeman, R.J.	1966	573-973		No. 1	0.1621 cm long; grain size 18 μ; bulk density 2.81 g cm^{-3}; irradiated by a dose of 6.1 x 10^{20} nvt (E_n < 1 MeV) at 100 C.
15	224	Freeman, R.J.	1966	978-1258		No. 2	0.1148 cm long; grain size 18 μ; bulk density 2.90 g cm^{-3}; same irradiation as the above specimen; annealed at 1000 C for 100 hrs.
16	224	Freeman, R.J.	1966	574-769		No. 3	0.1591 cm long; grain size 4 μ; bulk density 2.86 g cm^{-3}; irradiated by a dose of 2.5 x 10^{21} nvt (E_n < 1MeV) at 1000 C.
17	224	Freeman, R.J.	1966	576-725		No. 3	The above specimen annealed at 995 C for 17 hr.
18	224	Freeman, R.J.	1966	681-1273		No. 4	0.1346 cm long; grain size 4 μ; bulk density 2.86 g cm^{-3}; irradiated by 2.5 x 10^{21} nvt (E_n < 1 MeV) at 1000 C and by 2 x 10^{19} nvt at 650 C.
19	224	Freeman, R.J.	1966	771-1273		No. 4	The above specimen annealed at 995 C for 17 hr.

DATA TABLE 147. THERMAL DIFFUSIVITY OF BERYLLIUM OXIDE BeO

[Temperature, T, K; Thermal Diffusivity, α, cm^2 s^{-1}]

T	α
CURVE 1	
1249.8	0.0387
1505.4	0.0266
1794.3	0.0211
1994.3	0.0199*
2199.8	0.0176*
CURVE 2	
1070	0.0795
1070	0.0772
1078	0.0735
1080	0.0715
1083	0.0693
1110	0.0643
1113	0.0627
1146	0.0590
1153	0.0582
1268	0.0513
1278	0.0507
1390	0.0442
1398	0.0433
1548	0.0347
1680	0.0292
1690	0.0295
1838	0.0247*
1840	0.0232*
1940	0.0207*
1943	0.0197*
2023	0.0180*
2046	0.0177*
CURVE 3	
373.2	0.195
473.2	0.16
573.2	0.131
673.2	0.111
773.2	0.093
873.2	0.08
973.2	0.067
1073.2	0.06
1173.2	0.053
1273.2	0.048
CURVE 3 (cont.)	
1373.2	0.045
1473.2	0.045
1573.2	0.045
1673.2	0.050
CURVE 4	
373.2	0.0075
573.2	0.00415
773.2	0.00315
873.2	0.00312
973.2	0.00304
1073.2	0.00285
1123.2	0.00275
1173.2	0.00272
1223.2	0.00270
1323.2	0.00248
1373.2	0.00260
1473.2	0.00284
1573.2	0.00260
1623.2	0.00328
1673.2	0.00320
1723.2	0.00220
CURVE 5*	
373.2	0.011
573.2	0.00500
773.2	0.00390
873.2	0.00340
973.2	0.00300
1073.2	0.00277
1123.2	0.00267
1173.2	0.00260
1223.2	0.00260
1323.2	0.00230
1373.2	0.00234
1473.2	0.00238
1573.2	0.00232
1623.2	0.00233
1673.2	0.00290
1723.2	0.00237
CURVE 6*	
373.2	0.012
573.2	0.00500
773.2	0.00390
873.2	0.00340
973.2	0.00300
1073.2	0.00277
1123.2	0.00267
1173.2	0.0060
1223.2	0.00240
1323.2	0.00235
1373.2	0.00230
1473.2	0.00230
1573.2	0.00235
1623.2	0.00240
1673.2	0.00240
1723.2	0.00240
CURVE 7	
323.2	0.058
CURVE 8	
573.2	0.1992
595.2	0.1913
621.2	0.1807
644.2	0.1720
677.2	0.1610
720.2	0.1469
774.2	0.1291
828.2	0.1145
871.2	0.1053
CURVE 9	
773.2	0.1295
871.2	0.1050
971.2	0.0857
1071.2	0.0724
1173.2	0.0609
1269.2	0.0531
CURVE 10	
679.2	0.1375
721.2	0.1257
775.2	0.1129
823.2	0.1024
873.2	0.0958
923.2	0.0871
973.2	0.0800
1022.2	0.0732
1074.2	0.0673
1128.2	0.0619
1180.2	0.0568
1225.2	0.0529
1269.2	0.0504
CURVE 11	
584	0.2037
632	0.1763
676	0.1556
721	0.1386
773	0.1214
829	0.1076
872	0.0979
925	0.0888
973	0.0798
1023	0.0740
1076	0.0657
1123	0.0605
1174	0.0571
1276	0.0487
CURVE 12	
773	0.1262
873	0.1025
973	0.0843
1072	0.0723
1173	0.0605*
1268	0.0533
CURVE 13	
577	0.2126
600	0.2000
CURVE 13 (cont.)	
626	0.1849
648	0.1742
677	0.1595
721	0.1435
773	0.1274
829	0.1127
873	0.1035
CURVE 14	
573	0.0154
628	0.0151
673	0.0145
773	0.0138
873	0.0123
973	0.0123
CURVE 15	
978	0.0682
982	0.0690
1073	0.0605
1168	0.0518
1258	0.0455
CURVE 16	
574	0.1641
626	0.1449
672	0.1317
722	0.1197
769	0.1100
CURVE 17	
576	0.2083
627	0.1807
674	0.1590
725	0.1398
CURVE 18	
681	0.1348
721	0.1239
CURVE 18 (cont.)	
776	0.1116
822	0.1014
872	0.0949
921	0.0868
973	0.0797*
1020	0.0730
1073	0.0672
1125	0.0622
1178	0.0570
1221	0.0535
1273	0.0511
CURVE 19	
771	0.1230
820	0.1115
845	0.1065
868	0.1025
872	0.1021
899	0.0958
920	0.0918
1040	0.0739
1159	0.0594
1273	0.0527

* Not shown in figure.

SPECIFICATION TABLE 148. THERMAL DIFFUSIVITY OF HYDROGEN OXIDE H_2O

Cur. No.	Ref. No.	Author(s)	Year	Temp. Range, K	Reported Error, %	Name and Specimen Designation	Composition (weight percent), Specifications, and Remarks
1*	190	Bryngdahl, O.	1962	291-299	< 0.1	Water	Distilled; diffusivity measured employing an optical interferrometric method based on the transient temperature gradient set up in the specimen by electrically heating a vertical wire immersed in it; error reported is the mean error.
2*	286	Guillerm, S.	1968	323.2		Distilled water	Distilled; cylindrical specimen; measured by a radial transient method.
3*	287	Riedel, L.	1969	278-338		Water	Thermal diffusivity calculated from measurements of the temperature as a function of time inside a cylindrical can filled with specimen at an abruptly changed outside temperature.
4*	309	Neumann, F.	1862	273		Ice	Cubic specimen 5 to 6 in. on side, or sphere of the same diameter; uniformly heated, and then cooled in air; temperatures at center and surface observed by means of thermo-electric rods.
5*	309	Neumann, F.	1862	273		Snow	Specimen and measurement similar to above.
6*	310	Laikhtman, D. L., Serova, N. V.,	1959	247-262	20	Ice	No details reported.

DATA TABLE 148. THERMAL DIFFUSIVITY OF HYDROGEN OXIDE H_2O

[Temperature, T, K; Thermal Diffusivity, α, $cm^2\ s^{-1}$]

T	α	T	α	T	α
CURVE 1*		CURVE 3*		CURVE 6*	
290.8	0.001423	278	0.00135	246.8	0.0082
291.5	0.001430	298	0.00146	246.8	0.0108
292.0	0.001431	318	0.00155	248.2	0.0108
292.4	0.001431	338	0.00161	251.0	0.0082
293.3	0.001437			252.8	0.0088
294.6	0.001445	CURVE 4*		253.1	0.0088
294.8	0.001446			256.7	0.0109
296.0	0.001447	273	0.0135	257.8	0.0109
296.1	0.001454			258.5	0.0109
296.5	0.001454	CURVE 5*		259.2	0.0109
297.2	0.001456			260.0	0.0109
298.5	0.001465	273	0.0042	260.1	0.0109
298.8	0.001463			261.9	0.0109
298.9	0.001465				
CURVE 2*					
323.2	0.00165				

* No figure given.

SPECIFICATION TABLE 149. THERMAL DIFFUSIVITY OF IRON OXIDE Fe_2O_3

Cur. No.	Ref. No.	Author(s)	Year	Temp. Range, K	Reported Error, %	Name and Specimen Designation	Composition (weight percent), Specifications, and Remarks
1*	166	Williams, I.	1923	318, 373		Ferric oxide; red oxide	Density 4.70 g cm^{-3}.

DATA TABLE 149. THERMAL DIFFUSIVITY OF IRON OXIDE Fe_2O_3

[Temperature, T, K; Thermal Diffusivity, α, $cm^2\ s^{-1}$]

T	α
CURVE 1*	
318.2	0.00175
373.2	0.00175

* No figure given.

SPECIFICATION TABLE 150. THERMAL DIFFUSIVITY OF LEAD OXIDE PbO

Cur. No.	Ref. No.	Author(s)	Year	Temp. Range, K	Reported Error, %	Name and Specimen Designation	Composition (weight percent), Specifications, and Remarks
1*	166	Williams, I.	1923	318, 373		Litharge	Density 9.25 g cm^{-3}.

DATA TABLE 150. THERMAL DIFFUSIVITY OF LEAD OXIDE PbO

[Temperature, T, K; Thermal Diffusivity, α, $cm^2\ s^{-1}$]

T	α
CURVE 1*	
318.2	0.00106
373.2	0.00106

* No figure given.

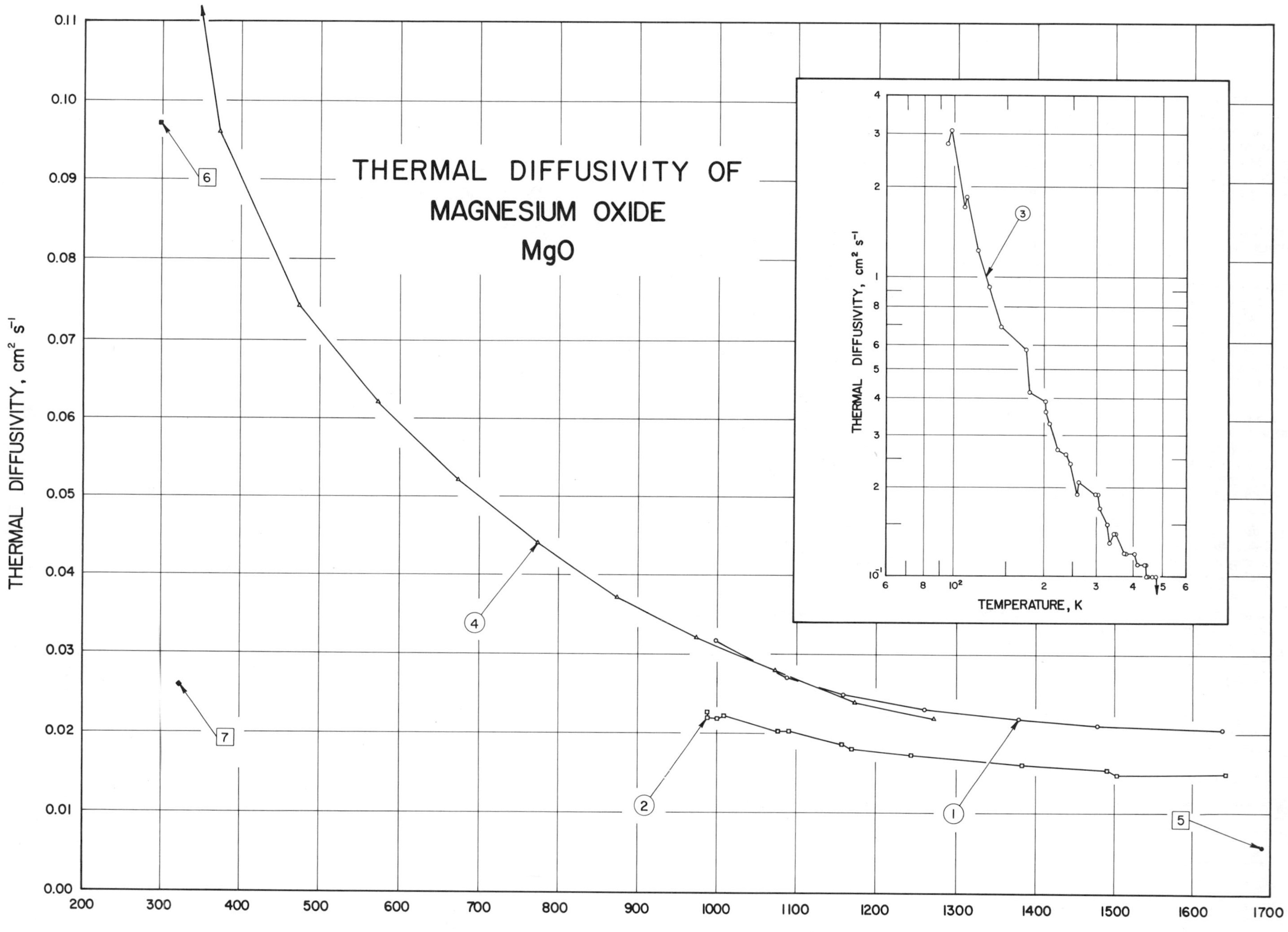
THERMAL DIFFUSIVITY OF
MAGNESIUM OXIDE
MgO
THERMAL DIFFUSIVITY, cm² s⁻¹
0.00
0.01
0.02
0.03
0.04
0.05
0.06
0.07
0.08
0.09
0.10
0.11
TEMPERATURE, K
200
300
400
500
600
700
800
900
1000
1100
1200
1300
1400
1500
1600
1700
1
2
3
4
5
6
7
THERMAL DIFFUSIVITY, cm² s⁻¹
10⁻¹
2
3
4
5
6
8
1
2
3
4
TEMPERATURE, K
6
8
10²
2
3
4
5
6

FIGURE 151

SPECIFICATION TABLE 151. THERMAL DIFFUSIVITY OF MAGNESIUM OXIDE MgO

Cur. No.	Ref. No.	Author(s)	Year	Temp. Range, K	Reported Error, %	Name and Specimen Designation	Composition (weight percent), Specifications, and Remarks
1	71, 279	Rudkin, R. L.	1963	998-1638	± 5/ ±10	SR 2808	Disc specimen 0.75 in. in diameter and ~0.045 in. thick; obtained from Corning Glass Works; coated on both sides with tungsten vapour using electron beam techniques; density 3.43 g cm^{-3}; short pulse of thermal energy from a xenon flash lamp absorbed at front surface of specimen; diffusivity determined from measured temperature-time function of the rear surface; measured in vacuum of ~10^{-5} mm Hg.
2	71, 279	Rudkin, R. L.	1963	988-1643	± 5/ ±10	PC 235	Disc specimen 0.75 in. in diameter and ~0.045 in. thick; obtained from Corning Glass Works; coated on both sides with tungsten vapour using electron beam techniques; density 3.39 g cm^{-3}; short pulse of thermal energy from a xenon flash lamp absorbed at front surface of specimen; diffusivity determined from measured temperature-time function of the rear surface; measured in vacuum of ~10^{-5} mm Hg.
3	75	Makarounis, O. and Jenkins, R. J.	1962	94-478	5		Single crystal; specimen 0.086 in. thick; front surface coated with Hanovia liquid platinum and then painted with Parson's black; rear surface coated with Hanovia liquid platinum; transparent; front surface uniformly irradiated by a high intensity short duration light pulse from a xenon flash lamp; diffusivity determined from measured rate of rise of the rear surface temperature.
4	72	Plummer, W. A., Campbell, D. E., and Comstock, A. A.	1962	298-1273			Specimen composed of three pieces: middle piece and two end pieces; middle piece separated from bottom piece by a 0.005 in. thick chromel sheet 7.6 x ~18 cm forming the heat source, middle piece separated from upper piece by a 0.005 in. thick chromel sheet forming the heat sink; thicknesses on either side of the heater are the same; entire assembly constitutes a sandwich 12.7 cm long, 7.6 cm wide and at least 4 to 5 cm thick; diffusivity determined from measured ratio of the temperature rises of the heat source and sink; density 3.30 g cm^{-3}; unidimensional heat flow.
5	116	Levine, H. S.	1950	1688.1			Cylindrical specimen 2.86 cm in dia. and 9.0 cm long (8.9 cm during last run); formed from properly sized powdered Norton Company E 195 Magnorite MgO mixed with 5 percent paraffin binder by dry pressing at 7000 p.s.i. in a split mold; thermocouple well 3/32 in. in dia. drilled along axis of specimen before removal from mold; calcined to remove wax binder and sintered for 48 hrs at 1973.2 K; total porosity 35.6 percent; fired density 2.38 g cm^{-3} (average for all runs); reported data point is average result of seven runs.
6	184	Taylor, R.	1965	298.2			Measured mean density 2.91 g cm^{-3}; temperature of measurement not given by author but assumed to be room temperature; heat pulse method used to measure diffusivity; cylindrical specimen 0.25 in. in diameter and 0.5 in. long.
7	187	Jaeger, G., Koehler, W., and Stapelfeldt, F.	1950	323.2			Disk specimen 50 mm in diameter and 7-12 mm thick; sintered at 2173.2 K and then machined on both sides while maintaining them exactly parallel; specimen originally at 20 C; diffusivity determined from measured time necessary for one side of specimen to reach a temperature of 80 C while maintaining the other side in contact with mercury at 100 C; average temperature of measurement not given by authors but assumed to be 50 C; data point reported is the average result of five measurements carried out on at least each of five specimens; density measured and reported as 3.07 g cm^{-3}.

DATA TABLE 151. THERMAL DIFFUSIVITY OF MAGNESIUM OXIDE MgO

[Temperature, T, K; Thermal Diffusivity, α, $cm^2\ s^{-1}$]

T	α
CURVE 1	
998.2	0.0316
1088.2	0.0271
1158.2	0.0250
1260.2	0.0232
1378.2	0.0219
1478.2	0.0211
1638.2	0.0206
CURVE 2	
988.2	0.0227
988.2	0.0220
1000.2	0.0219
1008.2	0.0222
1076.2	0.0203
1090.2	0.0203
1156.2	0.0187
1168.2	0.0181
1243.2	0.0173
1383.2	0.0161
1490.2	0.0155
1503.2	0.0149
1643.2	0.0150
CURVE 3	
94.2	2.79
98.2	3.09
108.2	1.71
110.2	1.85
120.2	1.23
131.2	0.93
144.2	0.69
174.2	0.58
179.2	0.42
201.2	0.39
202.2	0.36
208.2	0.33
221.2	0.27
236.2	0.26
245.2	0.24
258.2	0.19

T	α
CURVE 3 (cont.)	
260.2	0.21
297.2	0.19
301.2	0.19
308.2	0.17
326.2	0.15
331.2	0.13
341.2	0.14
348.2	0.14
373.2	0.12
377.2	0.12
402.2	0.12
411.2	0.11
432.2	0.11
437.2	0.11
441.2	0.10
449.2	0.10
465.2	0.10
476.2	0.10
478.2	0.09*
CURVE 4	
298.2	0.140*
373.2	0.096
473.2	0.074
573.2	0.062
673.2	0.052
773.2	0.044
873.2	0.037
973.2	0.032
1073.2	0.028
1173.2	0.024
1273.2	0.022
CURVE 5	
1688.1	0.0059

T	α
CURVE 6	
298.2	0.097
CURVE 7	
323.2	0.026

* Not shown in figure.

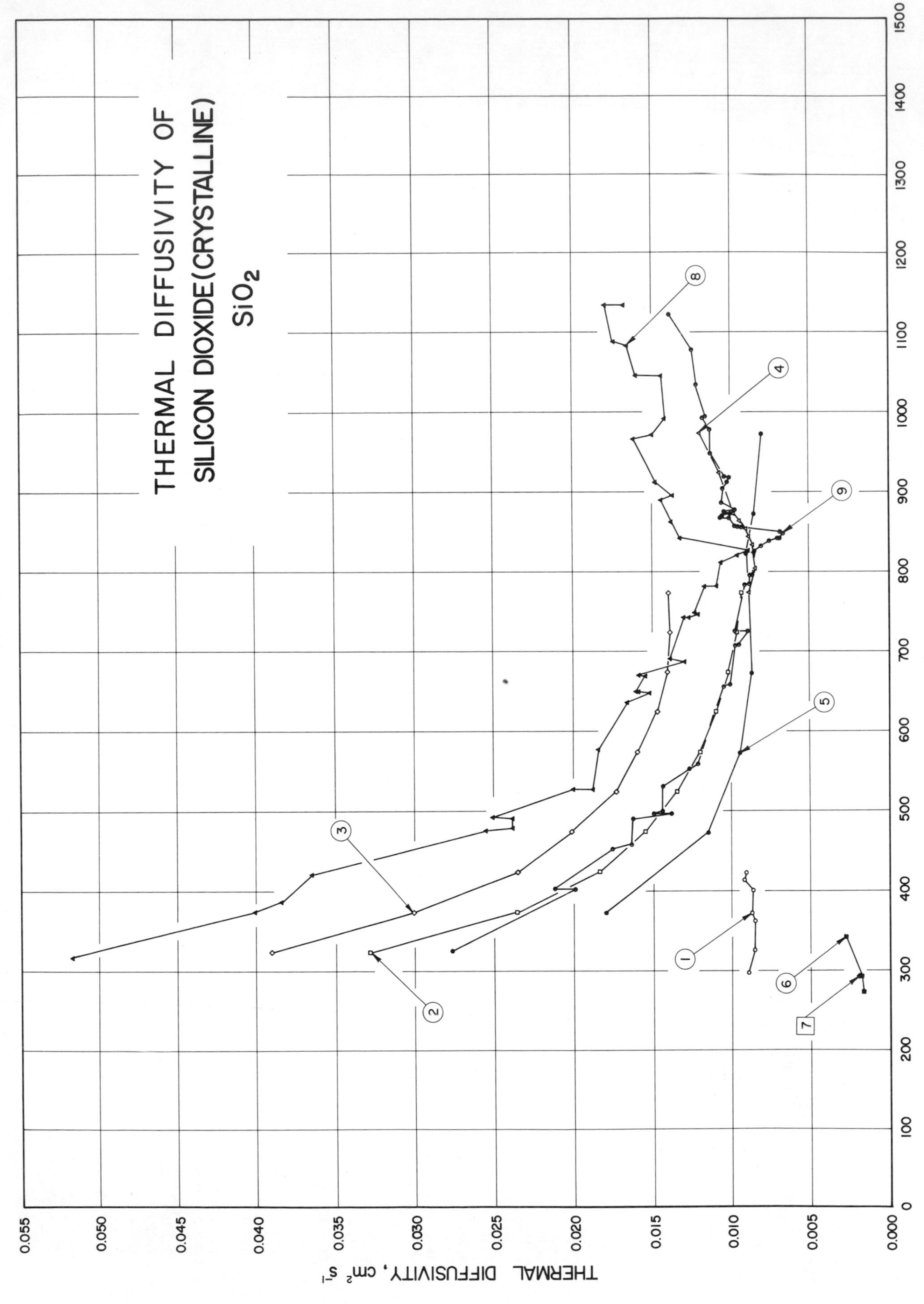
THERMAL DIFFUSIVITY OF
SILICON DIOXIDE (CRYSTALLINE)
SiO_2
TEMPERATURE, K
THERMAL DIFFUSIVITY, $cm^2\ s^{-1}$
0
100
200
300
400
500
600
700
800
900
1000
1100
1200
1300
1400
1500
0.000
0.005
0.010
0.015
0.020
0.025
0.030
0.035
0.040
0.045
0.050
0.055
1
2
3
4
5
6
7
8
9

FIGURE 152

SPECIFICATION TABLE 152. THERMAL DIFFUSIVITY OF SILICON DIOXIDE (CRYSTALLINE) SiO_2

Cur. No.	Ref. No.	Author(s)	Year	Temp. Range, K	Reported Error, %	Name and Specimen Designation	Composition (weight percent), Specifications, and Remarks
1	76	Hartunian, R.A. and Varwig, R. L.	1962	297-422		Quartz	Clear fused quartz; tubular specimen; diffusivity values calculated from measured (kdc) using specific heat data of Kelley, K.K. (Bureau of Mines Bull. 746, 1948) and of Lord, R.C. and Morrow, J.C. (J. Chem. Phys. 26, 230, 1957); constant heat transfer rate provided by a square current pulse.
2	115	Berthier, G.	1965	323-773		Quartz naturel; REPV	Natural quartz (rock crystal); cylindrical specimen composed of two identical parts each 12 mm in dia and 10 mm long; lower part has three grooves machined on each face to accomodate thermocouple leads, grooves on one face are parallel to optical axis, those on the other face are perpendicular to that axis, upper and lower parts held together under pressure; machined from natural quartz crystals at the ceramics workshop of the Centre d' Etudes Nucléaires de Saclay (France); measured perpendicular to the optical axis; diffusivity measured using radial heat flow method; data points are the average of two independent runs.
3*	115	Berthier, G.	1965	323-773		Quartz naturel; REPV	Above specimen measured for diffusivity again in the direction parallel to the optical axis; data points are the average of two independent runs; other conditions same as above.
4	115	Berthier, G.	1965	773-973		Quartz naturel	Specimen similar to the above; measured for diffusivity with increased sensitivity of temp recording; measured in the direction perpendicular to the optical axis; data between parentheses correspond to the phase transition zone $\alpha \rightarrow \beta$; diffusivity measured using radial heat flow method.
5	115	Berthier, G.	1965	373-973		Quartz naturel; REPT	Same specimen as that used for measurements pertaining to curves Nos. 2 and 3 above; measured for diffusivity again after completion of those measurements; in a deteriorated state; measured in the direction perpendicular to the optical axis; data points are the average results of two successive runs; data point between parentheses corresponds to phase transition; other conditions same as above.
6	164	Simonova, L.K.	1943	273-343			Silica gel.
7	164	Simonova, L.K.	1943	293.2			Silica gel powder.
8	281	Kanamori, H., Hujii, N., and Mizutani, H.	1968	318-1134		Quartz	Specimen of gem quality and of millimeter size; diffusivity measurement in [001] direction in argon at 1 atmosphere; Angström method was used.
9	281	Kanamori, H., et al.	1968	325-1122		Quartz	Diffusivity measurement in [010] direction; other conditions same as above.

*Not shown in figure.

DATA TABLE 152. THERMAL DIFFUSIVITY OF SILICON DIOXIDE (CRYSTALLINE) SiO_2

[Temperature, T, K; Thermal Diffusivity, α, $cm^2\ s^{-1}$]

T	α
CURVE 1	
297	0.00894
325	0.00852
362	0.00852
372	0.00870
400	0.00862
413	0.00926
422	0.00905
CURVE 2	
323.2	0.0328
373.2	0.0235
423.2	0.0183
473.2	0.0155
523.2	0.0135
573.2	0.012
623.2	0.011
673.2	0.0102
723.2	0.0096
773.2	0.0093
CURVE 3	
323.2	0.039
373.2	0.030
423.2	0.0235
473.2	0.0201
523.2	0.0173
573.2	0.016
623.2	0.0147
673.2	0.0141
723.2	0.0139
773.2	0.014
CURVE 4	
773.2	0.0088
803.2	0.0084
823.2	0.0085
833.2	0.0086
843.2	(0.0088)
853.2	(0.0090)
863.2	(0.0094)
873.2	0.0098
923.2	0.0107
973.2	0.0120
CURVE 5	
373.2	0.018
473.2	0.0115
573.2	0.0094
673.2	0.0087
773.2	0.0088*
823.2	(0.0090)
873.2	0.0085
973.2	0.0080
CURVE 6	
273.2	0.001684
293.2	0.001803
343.2	0.002794
CURVE 7	
293.2	0.001952
CURVE 8	
318	0.0518
374	0.0400
387	0.0383
421	0.0364
476	0.0254
479	0.0238
491	0.0238
492	0.0250
527	0.0200
527	0.0188
577	0.0184
636	0.0166
647	0.0152
649	0.0159
649	0.0161
669	0.0155
670	0.0159
687	0.0130
690	0.0139
743	0.0130
743	0.0127
747	0.0121
748	0.0123
782	0.0117
782	0.0109
812	0.0106
CURVE 8 (cont.)	
821	0.00950
826	0.00883
842	0.0132
863	0.0138
890	0.0144
896	0.0137
913	0.0148
967	0.0162
972	0.0150
992	0.0142
1046	0.0144
1046	0.0160
1083	0.0166
1088	0.0174
1134	0.0179
1134	0.0168
CURVE 9	
325	0.0276
402	0.0199
403	0.0212
452	0.0176
457	0.0164
490	0.0163
496	0.0138
496	0.0150
499	0.0144
530	0.0144
552	0.0127
558	0.0122
655	0.0105
658	0.0101
707	0.00971
708	0.00955
724	0.00892
724	0.00973
783	0.00912
784	0.00883
795	0.00877
795	0.00861
825	0.00842
832	0.00809
838	0.00755
841	0.00701
841	0.00686
CURVE 9 (cont.)	
847	0.00664
849	0.00683
855	0.00934
856	0.00953
857	0.00974
867	0.0101
868	0.0107
869	0.0106
873	0.0100
873	0.00981
875	0.0104
877	0.00970
886	0.0106
904	0.0105
912	0.0102
918	0.0101
918	0.0104
947	0.0113
978	0.0113
992	0.0118
994	0.0116
1034	0.0122
1078	0.0125
1122	0.0139

* Not shown in figure.

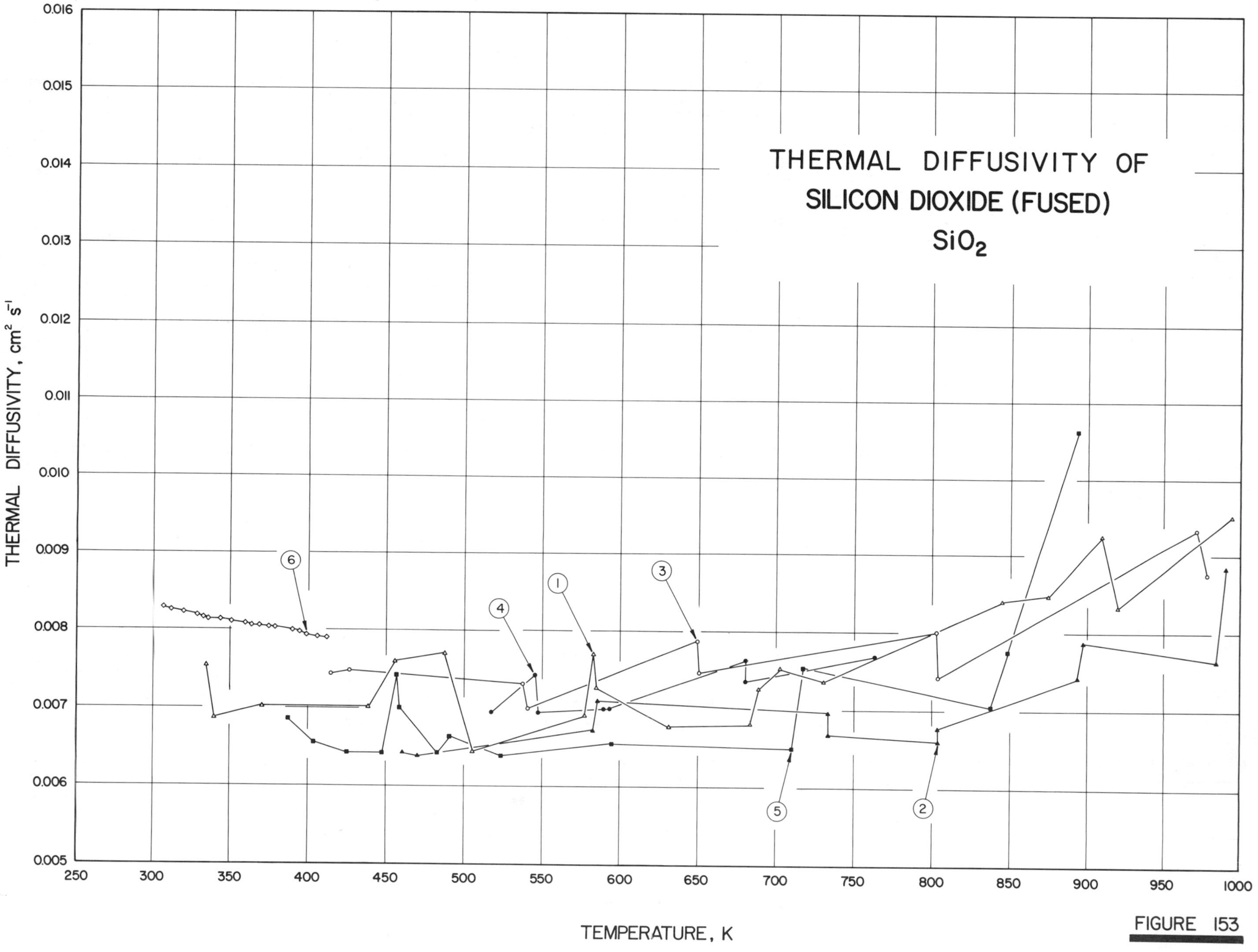
THERMAL DIFFUSIVITY OF
SILICON DIOXIDE (FUSED)
SiO_2
THERMAL DIFFUSIVITY, $cm^2\ s^{-1}$
0.016
0.015
0.014
0.013
0.012
0.011
0.010
0.009
0.008
0.007
0.006
0.005
250
300
350
400
450
500
550
600
650
700
750
800
850
900
950
1000
TEMPERATURE, K
1
2
3
4
5
6

FIGURE 153

SPECIFICATION TABLE 153. THERMAL DIFFUSIVITY OF SILICON DIOXIDE (FUSED) SiO_2

Cur. No.	Ref. No.	Author(s)	Year	Temp. Range, K	Reported Error, %	Name and Specimen Designation	Composition (weight percent), Specifications, and Remarks
1	281	Kanamori, H., Fujii, N., and Mizutani, H.	1968	335-1039		Fused silica	Specimen 2.53 mm in length; diffusivity measured using modified Angström method in argon at 1 atmosphere.
2	281	Kanamori, H., et al.	1968	461-991		Fused silica	Specimen 2.90 mm in length; other conditions same as above.
3	281	Kanamori, H., et al.	1968	415-979		Fused silica	Specimen 3.34 mm in length; other conditions same as above.
4	281	Kanamori, H., et al.	1968	518-764		Fused silica	Specimen 4.50 mm in length; other conditions same as above.
5	281	Kanamori, H., et al.	1968	388-894		Fused silica	Specimen 25.00 mm in length; other conditions same as above.
6	295	Luikov, A.V., Vasiliev, L.L., and Shashkov, A.G.	1965	82-412			Cylindrical specimen.

DATA TABLE 153. THERMAL DIFFUSIVITY OF SILICON DIOXIDE (FUSED) SiO_2

[Temperature, T, K; Thermal Diffusivity, α, $cm^2\ s^{-1}$]

T	α	T	α	T	α	T	α	T	α	T	α
CURVE 1		CURVE 2		CURVE 3 (cont.)		CURVE 5 (cont.)		CURVE 6 (cont.)		CURVE 6 (cont.)	
335	0.00752	461	0.00641	974	0.00934	491	0.00662	116.6	0.01104*	270.4	0.00848*
336	0.00685	471	0.00637	979	0.00877	524	0.00637	120.4	0.01093*	277.5	0.00844*
371	0.00699	583	0.00671			595	0.00654	134.6	0.01049*	287.7	0.00838*
439	0.00699	586	0.00709	CURVE 4		711	0.00649	140.9	0.01033*	299.0	0.00831*
446	0.00704	734	0.00667			718	0.00752	144.6	0.01022*	307.6	0.00827
456	0.00758	734	0.00685	518	0.00694	838	0.00704	148.8	0.01013*	312.8	0.00824
488	0.00769	804	0.00658	546	0.00741	849	0.00775	156.1	0.00996*	320.8	0.00821
507	0.00641	804	0.00676	548	0.00694	894	0.00962	162.4	0.00984*	329.0	0.00817
578	0.00690	894	0.00741	580	0.00699			169.3	0.00971*	333.2	0.00814
583	0.00769	897	0.00787	594	0.00699	CURVE 6		175.7	0.00962*	336.3	0.00813
585	0.00725	985	0.00763	681	0.00735			183.2	0.00946*	344.4	0.00812
632	0.00676	991	0.00855	681	0.00752	81.8	0.01249*	189.5	0.00936*	351.4	0.00809
684	0.00680			764	0.00758	84.4	0.01256*	195.1	0.00928*	360.3	0.00807
689	0.00725	CURVE 3				85.3	0.01235*	203.9	0.00918*	364.9	0.00804
703	0.00752			CURVE 5		89.7	0.01222*	208.6	0.00910*	369.2	0.00804
731	0.00735	415	0.00741			92.0	0.01210*	213.3	0.00904*	375.2	0.00802
845	0.00840	427	0.00746	388	0.00685	93.6	0.01200*	218.4	0.00897*	379.5	0.00802
876	0.00847	538	0.00730	404	0.00654	97.0	0.01189*	228.7	0.00889*	390.1	0.00798
910	0.00926	541	0.00699	426	0.00641	98.3	0.01173*	238.0	0.00879*	395.4	0.00796
920	0.00833	650	0.00787	448	0.00641	101.7	0.01162*	243.3	0.00872*	399.3	0.00793
995	0.00952	651	0.00746	457	0.00741	103.5	0.01155*	253.3	0.00863*	406.4	0.00790
1034	0.0106*	803	0.00800	459	0.00699	106.5	0.01139*	260.6	0.00856*	412.0	0.00789
1039	0.0108*	804	0.00741	483	0.00641	113.0	0.01115*	266.0	0.00852*		

* Not shown in figure.

SPECIFICATION TABLE 154. THERMAL DIFFUSIVITY OF THORIUM DIOXIDE ThO_2

Cur. No.	Ref. No.	Author(s)	Year	Temp. Range, K	Reported Error, %	Name and Specimen Designation	Composition (weight percent), Specifications, and Remarks
1*	187	Jaeger, G., Koehler, W., and Stapelfeldt, F.	1950	323.2			Disk specimen 50 mm in diameter and 7-12 mm thick; sintered at 2173.2 K and then machined on both sides while maintaining them exactly parallel; specimen originally at 20 C; diffusivity determined from measured time necessary for one side of specimen to reach a temperature of 80 C while maintaining the other side in contact with mercury at 100 C; average temperature of measurement not given by authors but assumed to be 50 C; data point reported is the average result of five measurements carried out on at least each of five specimens; density measured and reported as 8.5 g cm^{-3}.
2*	52	Faucher, M., Cabannes, F., Antony, A.-M., Piriou, B., and Simonato, J.	1970	1909-2573			5 mm in diameter and 1 to 1.5 mm thick; density 9.2 g cm^{-3}.

DATA TABLE 154. THERMAL DIFFUSIVITY OF THORIUM DIOXIDE ThO_2

[Temperature, T, K; Thermal Diffusivity, α, cm^2 s^{-1}]

T	α
CURVE 1*	
323.2	0.030
CURVE 2*	
1909	0.00587
1994	0.01042
2040	0.01119
2109	0.00787
2140	0.00671
2199	0.00874
2252	0.00618
2316	0.00597
2376	0.00632
2382	0.00542
2436	0.00683
2508	0.00656
2573	0.00582

* No figure given.

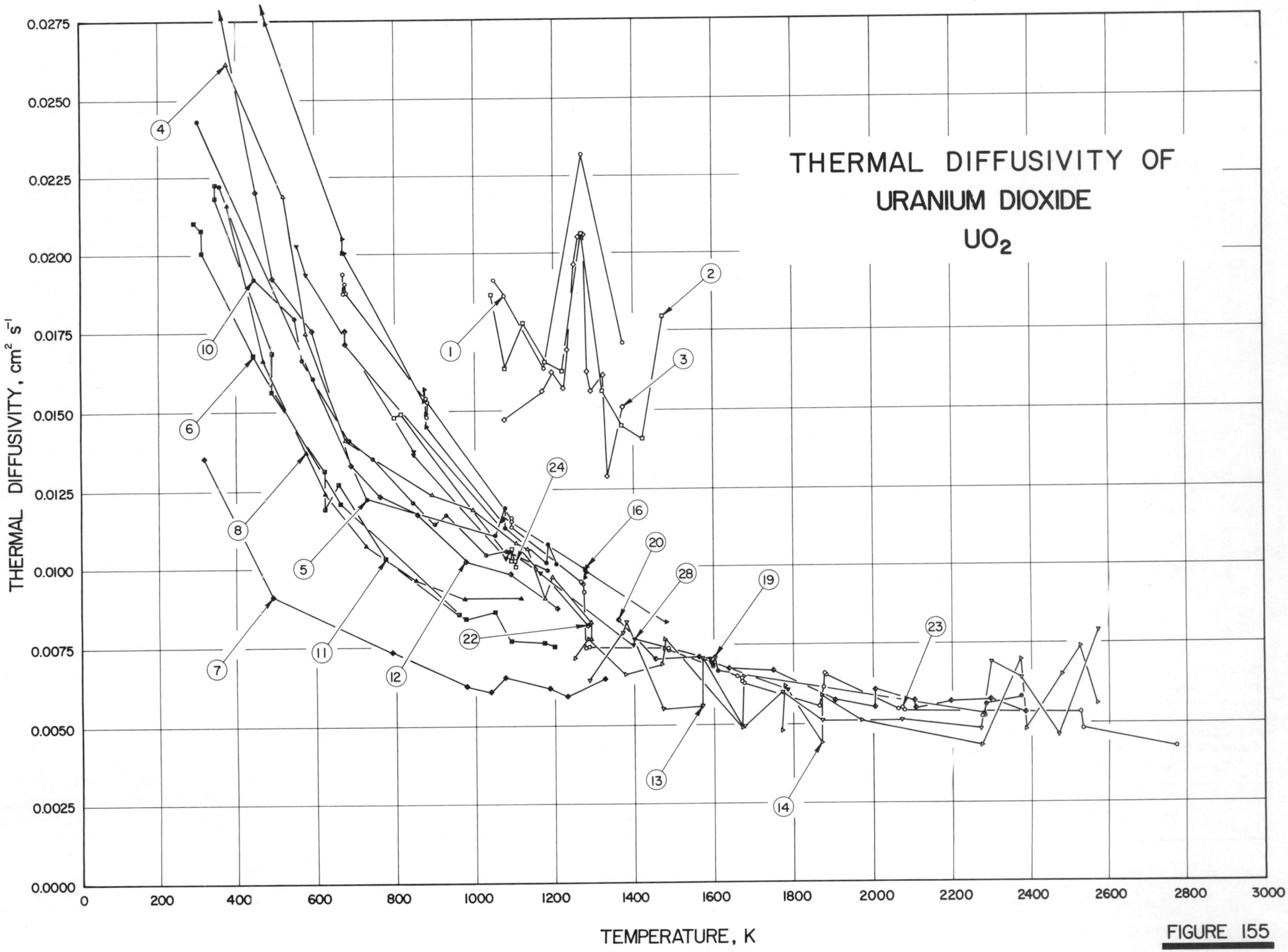
THERMAL DIFFUSIVITY OF
URANIUM DIOXIDE
UO_2
THERMAL DIFFUSIVITY, $cm^2\,s^{-1}$
TEMPERATURE, K
0.0000
0.0025
0.0050
0.0075
0.0100
0.0125
0.0150
0.0175
0.0200
0.0225
0.0250
0.0275
0
200
400
600
800
1000
1200
1400
1600
1800
2000
2200
2400
2600
2800
3000
1
2
3
4
5
6
7
8
10
11
12
13
14
16
19
20
22
23
24
28

FIGURE 155

SPECIFICATION TABLE 155. THERMAL DIFFUSIVITY OF URANIUM DIOXIDE UO_2

Cur. No.	Ref. No.	Author(s)	Year	Temp. Range, K	Reported Error, %	Name and Specimen Designation	Composition (weight percent), Specifications, and Remarks
1	77	Montgomery, M. H.	1963	1048-1373			Single crystal; disc specimen 0.635 cm in diameter and 0.03 in. thick; uncoated; front surface uniformly irradiated with a short duration thermal pulse from a ruby laser; diffusivity determined from measured transient temperature rise of the rear surface.
2	77	Montgomery, M. H.	1963	1042-1473			Above specimen measured for diffusivity again.
3	77	Montgomery, M. H.	1963	1074-1374			Above specimen measured for diffusivity again.
4	162	Kobayasi, K. and Kumada, T.	1968	376-1292		Y-78	15 mm diameter x 25 mm long; cold-pressed, sintered at 1700 C; 94.3% theoretical density.
5	162	Kobayasi, K. and Kumada, T.	1968	300-1203		Y-5	14 mm diameter x 25 mm long; cold-pressed, sintered at 1700 C; 94.1% theroetical density.
6	162	Kobayasi, K. and Kumada, T.	1968	292-1199		Y-48	15 mm diameter x 25 mm long; cold-pressed, sintered at 1700 C; 88.0% theoretical density.
7	162	Kobayasi, K. and Kumada, T.	1968	316-1230		Y-85	14 mm diameter x 25 mm long; cold-pressed, sintered at 1700 C; 88.1% theoretical density.
8	230	van Craeynest, J.C., Weilbacher, J.C., and Lallemont, R.	1969	376-1115	10	No. 1	5 mm diameter x 10 mm long; measured by Angström method.
9*	230	van Craeynest, J.C., et al.	1969	355-1246	10	No. 2	Similar to above.
10	230	van Craeynest, J.C., et al.	1969	357-1180	10	No. 3	Similar to above.
11	230	van Craeynest, J.C., et al.	1969	345-959	10	No. 4	Similar to above.
12	230	van Craeynest, J.C., et al.	1969	362-1206	10	No. 6	Similar to above.
13	230	van Craeynest, J.C., et al.	1969	1285-2572	10		Measured by Cowan's method.
14	230	van Craeynest, J.C., et al.	1969	1249-2569	10		Similar to above.
15*	233	Walter, A.J., Dell, R.M., and Burgess, P.C.	1970	321-1673			6 mm diameter x 2 mm thick; sintered in hydrogen at 1650 C; density 10.72 g cm^{-3} (98% T.D.).
16	296	Bates, J.L.	1970	289-1481		RR-1	O/U ratio = 2.000; 0.634 cm diameter x 0.0775 cm thick; fabricated and machined by GE-SNP; average grain size 100 to 250 μ; density 10.79 g cm^{-3}; cycle 1.
17*	296	Bates, J.L.	1970	358-1489		RR-1	The above specimen; cycle 2.
18*	296	Bates, J.L.	1970	539-2373		RR-1	The above specimen; cycle 3.
19	296	Bates, J.L.	1970	1599-2373		RR-1	The above specimen measured in the same cycle at decreasing temperatures.
20	296	Bates, J.L.	1970	1360-2384		RR-1	The above specimen; cycle 4.
21*	296	Bates, J.L.	1970	361-1760		RR-2	Similar to the above specimen but dimensions 0.632 cm diameter x 0.1283 cm thick; cycle 2.
22	296	Bates, J.L.	1970	673-1283		RR-2	The above specimen; cycle 3.

*Not shown in figure.

SPECIFICATION TABLE 155. THERMAL DIFFUSIVITY OF URANIUM DIOXIDE UO_2 (continued)

Cur. No.	Ref. No.	Author(s)	Year	Temp. Range, K	Reported Error, %	Name and Specimen Designation	Composition (weight percent), Specifications, and Remarks
23	296	Bates, J. L.	1970	670-2773		RR-3	Similar to the above specimen but dimensions 0.636 cm diameter x 0.0810 cm thick; cycle 1.
24	296	Bates, J. L.	1970	797-1100		RR-3	The above specimen measured in the same cycle at decreasing temperatures.
25*	296	Bates, J. L.	1970	289-1285		RR-3	The above specimen graphite-coated; cycle 2.
26*	296	Bates, J. L.	1970	352-1284		RR-3	The above specimen measured in the same cycle at decreasing temperatures.
27*	296	Bates, J. L.	1970	298-1792		RR-3	The above specimen with coating removed mechanically; cycle 3.
28	296	Bates, J. L.	1970	553-1786		RR-3	The above specimen measured in the same cycle at decreasing temperatures.

* Not shown in figure.

DATA TABLE 155. THERMAL DIFFUSIVITY OF URANIUM DIOXIDE UO_2

[Temperature, T, K; Thermal Diffusivity, α, $cm^2 s^{-1}$]

T	α
CURVE 1	
1048	0.0191
1073	0.0186
1173	0.0163
1270	0.0232
1373	0.0171
CURVE 2	
1042	0.0186
1076	0.0163
1175	0.0165
1220	0.0162
1270	0.0206
1320	0.0156
1370	0.0145
1423	0.0141
1473	0.0179
CURVE 3	
1074	0.0147
1170	0.0156
1193	0.0162
1223	0.0157
1233	0.0169
1251	0.0196
1262	0.0205
1275	0.0206
1283	0.0162
1293	0.0156
1323	0.0161
1333	0.0129
1374	0.0151
CURVE 4	
376	0.02613
521	0.02191
575	0.01747
673	0.01475
776	0.01410
890	0.01236
993	0.01186
1102	0.01079
1130	0.01060
1175	0.00902

T	α
CURVE 4 (cont.)	
1194	0.00977
1292	0.00824
1292	0.00770
CURVE 5	
300	0.02428
595	0.01603
728	0.01224
1049	0.01103
1074	0.01192
1074	0.01128
1179	0.01014
1183	0.01074
1203	0.01010
CURVE 6	
292	0.02096
310	0.02074
310	0.02000
442	0.01671
661	0.01209
975	0.00836
1049	0.00856
1090	0.00765
1172	0.00757
1199	0.00746
CURVE 7	
316	0.01354
491	0.00907
789	0.00732
978	0.00624
1039	0.00606
1073	0.00653
1187	0.00617
1230	0.00592
1325	0.00645
CURVE 8	
376	0.0216
467	0.0166
576	0.0137

T	α
CURVE 8 (cont.)	
622	0.0124
726	0.0107
850	0.0096
972	0.0090
1115	0.0090
CURVE 9*	
355	0.0303
355	0.0268
465	0.0205
475	0.0208
572	0.0175
594	0.0171
743	0.0121
804	0.0122
923	0.0099
935	0.0104
1081	0.0087
1246	0.0082
CURVE 10	
357	0.0222
446	0.0192
548	0.0179
565	0.0166
684	0.0141
744	0.0135
843	0.0121
899	0.0114
927	0.0117
1026	0.0104
1079	0.0105
1180	0.0099
CURVE 11	
345	0.0222
345	0.0218
489	0.0168
489	0.0156
622	0.0131
622	0.0119
657	0.0127
773	0.0103
959	0.0085

T	α
CURVE 12	
362	0.0276*
450	0.0220
491	0.0192
590	0.0175
687	0.0133
760	0.0123
853	0.0117
976	0.0102
1088	0.0098
1206	0.0087
CURVE 13	
1285	0.0064
1370	0.0079
1381	0.0082
1472	0.0055
1571	0.0071
1571	0.0056
1676	0.0049
1772	0.0060
1871	0.0051
2070	0.0051
2272	0.0048
2297	0.0069
2375	0.0064
2469	0.0046
2572	0.0079
CURVE 14	
1249	0.0071
1285	0.0077
1378	0.0066
1471	0.0069
1478	0.0077
1675	0.0049
1676	0.0065
1772	0.0048
1780	0.0062
1871	0.0059
1871	0.0044
1972	0.0051
2274	0.0043
2375	0.0070
2386	0.0048
2477	0.0065

T	α
CURVE 14 (cont.)	
2524	0.0074
2569	0.0056
CURVE 15*	
321	0.0342
321	0.0329
321	0.0323
321	0.0304
348	0.0324
381	0.0296
421	0.0275
438	0.0265
486	0.0228
509	0.0222
523	0.0215
613	0.0197
641	0.0186
715	0.0178
766	0.0169
771	0.0158
910	0.0142
948	0.0127
1032	0.0120
1063	0.0125
1087	0.0120
1151	0.0116
1351	0.0098
1378	0.0095
1409	0.0095
1450	0.0089
1471	0.0095
1512	0.0088
1553	0.0089
1638	0.0085
1673	0.0085
CURVE 16	
289	0.0383*
289	0.0357*
467	0.0282*
469	0.0285*
477	0.0274
671	0.0205
671	0.0200

T	α
CURVE 16 (cont.)	
672	0.0200
872	0.0153
873	0.0157
876	0.0149
877	0.0145
1068	0.0115
1072	0.0116
1276	0.00995
1276	0.00967
1277	0.0100
1277	0.00983
1475	0.00740
1481	0.00825
CURVE 17*	
358	0.0294
367	0.0297
476	0.0262
476	0.0261
478	0.0241
560	0.0222
564	0.0219
566	0.0217
666	0.0182
673	0.0181
767	0.0167
767	0.0164
774	0.0167
774	0.0160
777	0.0164
869	0.0154
872	0.0146
875	0.0146
974	0.0132
975	0.0134
978	0.0134
1073	0.0125
1074	0.0120
1076	0.0119
1172	0.0104
1274	0.00967
1278	0.00921
1279	0.00948
1373	0.00886
1374	0.00812
1375	0.00916

T	α
CURVE 17 (cont.)*	
1375	0.00875
1489	0.00797
CURVE 18*	
539	0.0215
539	0.0212
756	0.0150
761	0.0156
891	0.0133
895	0.0126
994	0.0116
995	0.0118
1180	0.00978
1185	0.00932
1325	0.00852
1325	0.00842
1489	0.00730
1491	0.00752
1655	0.00677
1666	0.00684
1778	0.00624
1780	0.00593
1863	0.00600
1866	0.00600
1972	0.00592
1977	0.00555
2093	0.00562
2102	0.00569
2174	0.00555
2187	0.00548
2373	0.00551
CURVE 19	
1599	0.00682
1601	0.00714
1609	0.00668
2280	0.00532
2285	0.00559
2373	0.00583
CURVE 20	
1360	0.00832
1453	0.00708

T	α
CURVE 20 (cont.)	
1562	0.00715
1639	0.00678
1750	0.00670
1907	0.00575
2005	0.00552
2007	0.00607
2104	0.00574
2109	0.00550
2195	0.00569
2295	0.00571
2384	0.00532
CURVE 21*	
361	0.0296
362	0.0306
464	0.0246
464	0.0235
469	0.0232
571	0.0185
577	0.0199
577	0.0195
661	0.01687
682	0.0165
784	0.0137
785	0.0143
786	0.0138
866	0.0128
867	0.0125
961	0.0120
961	0.0113
961	0.0110
1069	0.0105
1069	0.0101
1071	0.00997
1171	0.00908
1173	0.00968
1174	0.00917
1269	0.00851
1270	0.00834
1270	0.00833
1360	0.00762
1361	0.00776
1361	0.00751
1469	0.00658
1471	0.00741

* Not shown in figure.

DATA TABLE 155. THERMAL DIFFUSIVITY OF URANIUM DIOXIDE UO_2 (continued)

T	α
CURVE 21 (cont.)*	
1472	0.00752
1569	0.00708
1569	0.00692
1571	0.00695
1683	0.00674
1683	0.00663
1756	0.00611
1758	0.00644
1780	0.00639
CURVE 22	
673	0.0175
673	0.0171
1283	0.00814
CURVE 23	
670	0.0193
672	0.0187
673	0.0190
673	0.0187
877	0.0154
877	0.0148
878	0.0153
1088	0.0116
1088	0.0115
1088	0.0113
1267	0.00953
1268	0.00951
1271	0.00949
1274	0.00921
1278	0.00743
1285	0.00747
1485	0.00743
1486	0.00735
1659	0.00651
1672	0.00646
1672	0.00637
1677	0.00631
1866	0.00557
1876	0.00615
1876	0.00659
1881	0.00657
2061	0.00545
2074	0.00552
2077	0.00543

T	α
CURVE 23 (cont.)	
2276	0.00523
2278	0.00528
2525	0.00533
2532	0.00481
2773	0.00425
CURVE 24	
797	0.0148
813	0.0149
1089	0.0102
1090	0.0106
1099	0.0102
1100	0.0100
CURVE 25*	
289	0.0328
301	0.0337
305	0.0334
332	0.0312
380	0.0286
394	0.0288
466	0.0240
472	0.0247
583	0.0206
584	0.0208
676	0.0171
679	0.0173
763	0.0156
764	0.0159
873	0.0138
876	0.0131
979	0.0120
980	0.0120
1065	0.0113
1072	0.0111
1187	0.0100
1188	0.00945
1277	0.00908
1285	0.00943
CURVE 26*	
352	0.0278
397	0.0279
399	0.0276

T	α
CURVE 26 (cont.)*	
460	0.0239
464	0.0238
549	0.0206
678	0.0165
879	0.0138
880	0.0140
1271	0.0112
1284	0.00961
CURVE 27*	
298	0.0360
377	0.0308
473	0.0230
474	0.0232
573	0.0200
583	0.0190
678	0.0168
680	0.0169
681	0.0165
775	0.0152
776	0.0152
891	0.0128
895	0.0130
968	0.0121
973	0.0120
1081	0.0105
1087	0.0104
1172	0.00967
1173	0.00944
1291	0.00829
1292	0.00841
1377	0.00778
1380	0.00771
1473	0.00738
1477	0.00754
1578	0.00663
1584	0.00705
1673	0.00633
1679	0.00624
1769	0.00583
1792	0.00623
CURVE 28	
553	0.0202
577	0.0193

T	α
CURVE 28 (cont.)	
847	0.0137
847	0.0136
1079	0.0103
1085	0.0105
1166	0.00981
1399	0.00749
1400	0.00766
1595	0.00707
1596	0.00693
1786	0.00609

* Not shown in figure.

SPECIFICATION TABLE 156. THERMAL DIFFUSIVITY OF YTTRIUM OXIDE Y_2O_3

Cur. No.	Ref. No.	Author(s)	Year	Temp. Range, K	Reported Error, %	Name and Specimen Designation	Composition (weight percent), Specifications, and Remarks
1*	120	Klein, P. H.	1967	79-307		$Y_{1.98}Nd_{0.02}O_3$	$Y_{1.98}Nd_{0.02}O_3$; 98.517 Y_2O_3 (by difference), and 1.483 Nd_2O_3 (1 mole % Nd_2O_3); Nd-doped single crystal with impurity concentration ~3 x 10^{20} atom cm^{-3}; cylindrical specimen 0.210 ± 0.005 cm in dia. and 1.040 cm long; prepared by the flame-fusion process; diffusivity determined from measurement of the time required for the temp. difference between two points along specimen to decrease to half the steady state value; one-dimensional heat flow.
2*	52	Faucher, M., Cabannes, F.,	1970	2050-2499			5 mm in diameter and 1 to 1.5 mm thick; density 4.8 g cm^{-3}.

DATA TABLE 156. THERMAL DIFFUSIVITY OF YTTRIUM OXIDE Y_2O_3

[Temperature, T, K; Thermal Diffusivity, α, $cm^2\ s^{-1}$]

T	α
CURVE 1*	
78.7	1.08
78.9	1.03
194.5	0.108
200.5	0.117
274.8	0.0659
278.0	0.0692
300.6	0.0562
306.9	0.0583
CURVE 2*	
2050	0.01360
2165	0.01212
2170	0.01399
2254	0.01166
2337	0.01402
2401	0.01302
2499	0.01280

* No figure given.

SPECIFICATION TABLE 157. THERMAL DIFFUSIVITY OF ZINC OXIDE ZnO

Cur. No.	Ref. No.	Author(s)	Year	Temp. Range, K	Reported Error, %	Name and Specimen Designation	Composition (weight percent), Specifications, and Remarks
1*	166	Williams, I.	1923	318, 373			Density 5.50 g cm^{-3}.
2*	286	Guillerm, S.	1968	403.2			Cylindrical specimen; measured by a radial transient method.

DATA TABLE 157. THERMAL DIFFUSIVITY OF ZINC OXIDE ZnO

[Temperature, T, K; Thermal Diffusivity, α, cm^2 s^{-1}]

T	α
CURVE 1*	
318.2	0.00241
373.2	0.00241
CURVE 2*	
403.2	0.0025

* No figure given.

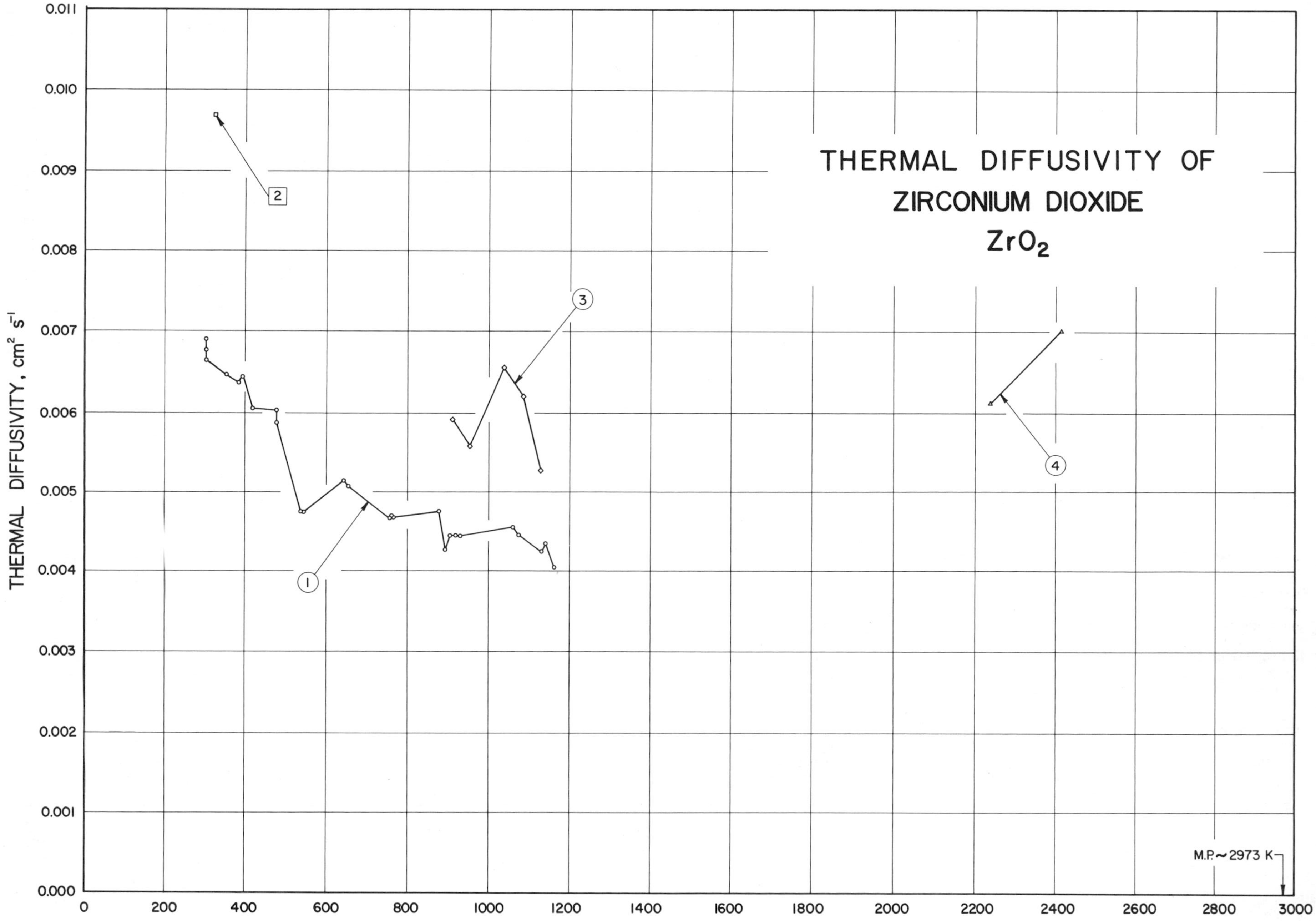
THERMAL DIFFUSIVITY OF
ZIRCONIUM DIOXIDE
ZrO_2
THERMAL DIFFUSIVITY, $cm^2\ s^{-1}$
TEMPERATURE, K
M.P. ~ 2973 K
1
2
3
4
0.000
0.001
0.002
0.003
0.004
0.005
0.006
0.007
0.008
0.009
0.010
0.011
0
200
400
600
800
1000
1200
1400
1600
1800
2000
2200
2400
2600
2800
3000

FIGURE 158

SPECIFICATION TABLE 158. THERMAL DIFFUSIVITY OF ZIRCONIUM DIOXIDE ZrO_2

Cur. No.	Ref. No.	Author(s)	Year	Temp. Range, K	Reported Error, %	Name and Specimen Designation	Composition (weight percent), Specifications, and Remarks
1	34	Rudkin, R. L., Parker, W. J., and Jenkins, R. J.	1963	303-1163		Norton mix 302	Magnesia-stabilized; specimen a few millimeters thick; furnished by the Norton Co.; plated with Hanovia Platinum Bright (No. 05); high intensity short duration light pulse from xenon flash lamp absorbed in the front surface of thermally insulated specimen; thermal diffusivity determined from measured temperature history of the rear surface.
2	187	Jaeger, G., Koehler, W., and Stapelfeldt, F.	1950	323.2			Disk specimen 50 mm in diameter and 7-12 mm thick; sintered at 2173.2 K and then machined on both sides while maintaining them exactly parallel; specimen originally at 20 C; diffusivity determined from measured time necessary for one side of specimen to reach a temperature of 80 C while maintaining the other side in contact with mercury at 100 C; average temperature of measurement not given by authors but assumed to be 50 C; data point reported is the average result of five measurements carried out on at least each of five specimens; density measured and reported as 4.32 g cm^{-3}.
3	297	Cutler, M., Snodgrass, H.R., Cheney, G.T., Appel, J., Mallon, C.E., and Meyer, C.H., Jr.	1961	910-1127		Sample D	Specimen about 0.8 in. in diameter, about 0.4 cm in thickness; fabricated from different pieces of ZrO_2; density 5.48 g cm^{-3}.
4	52	Faucher, M., Cabannes, F., Anthony, A.-M., Piriou, B., and Simonato, J.	1970	2235, 2413			5 mm in diameter and 1.2 mm thick; density 4.92 g cm^{-3}.

DATA TABLE 158. THERMAL DIFFUSIVITY OF ZIRCONIUM DIOXIDE ZrO_2

[Temperature, T, K; Thermal Diffusivity, α, $cm^2\ s^{-1}$]

T	α	T	α	T	α
CURVE 1		CURVE 1 (cont.)		CURVE 2	
303	0.00690	758	0.00467	323.2	0.0097
303	0.00677	760	0.00470		
303	0.00665	768	0.00467	CURVE 3	
353	0.00647	876	0.00475		
383	0.00637	893	0.00427	910	0.00592
393	0.00645	903	0.00445	952	0.00558
419	0.00605	918	0.00445	1018	0.00656
478	0.00603	928	0.00445	1085	0.00620
478	0.00587	1060	0.00455	1127	0.00527
538	0.00475	1073	0.00445		
543	0.00475	1130	0.00425	CURVE 4	
543	0.00475*	1140	0.00435		
643	0.00515	1163	0.00405	2235	0.00613
653	0.00507			2413	0.00702

* Not shown in figure.

7. MULTIPLE OXIDES AND SALTS

SPECIFICATION TABLE 159. THERMAL DIFFUSIVITY OF ALUMINUM SILICATE $Al_6Si_2O_{13}$

Cur. No.	Ref. No.	Author(s)	Year	Temp. Range, K	Reported Error, %	Name and Specimen Designation	Composition (weight percent), Specifications, and Remarks
1*	73	Soxman, E. J.	1957	82-146	±5	MV20 Mullite	Vitreous, polycrystalline; U-shaped rod specimen ~3 mm in diameter and at least 20 cm long; provided by McDanel Refractory Porcelain Company; commercially prepared; measured in copper vacuum chamber; measured with chamber immersed in liquid air; sinusoidal temperature fluctuation impressed at one end of specimen; diffusivity calculated from measured amplitudes and phase shift of resulting temperature wave at two points along specimen.
2*	73	Soxman, E. J.	1957	307-331	±5	MV20 Mullite	Specimen similar to the above; measured with vacuum chamber immersed in ice water.
3*	73	Soxman, E. J.	1957	355	±5	MV20 Mullite	Specimen similar to the above; measured with vacuum chamber immersed in hot water.

DATA TABLE 159. THERMAL DIFFUSIVITY OF ALUMINUM SILICATE $Al_6Si_2O_{13}$

[Temperature, T, K; Thermal Diffusivity, α, $cm^2\ s^{-1}$]

T	α
CURVE 1*	
82	0.076
98	0.048
119	0.032
146	0.023
CURVE 2*	
307	0.013
311	0.010
331	0.008
CURVE 3*	
355	0.008

* No figure given.

SPECIFICATION TABLE 160. THERMAL DIFFUSIVITY OF BARIUM SULFATE $BaSO_4$

Cur. No.	Ref. No.	Author(s)	Year	Temp. Range, K	Reported Error, %	Name and Specimen Designation	Composition (weight percent), Specifications, and Remarks
1*	166	Williams, I.	1923	318, 373		Blanc fixe	Density 4.35 g cm^{-3}.

DATA TABLE 160. THERMAL DIFFUSIVITY OF BARIUM SULFATE $BaSO_4$

[Temperature, T, K; Thermal Diffusivity, α, cm^2 s^{-1}]

T	α
CURVE 1*	
318.2	0.00157
373.2	0.00157

*No figure given.

SPECIFICATION TABLE 161. THERMAL DIFFUSIVITY OF CALCIUM CARBONATE $CaCO_3$

Cur. No.	Ref. No.	Author(s)	Year	Temp. Range, K	Reported Error, %	Name and Specimen Designation	Composition (weight percent), Specifications, and Remarks
1*	166	Williams, I.	1923	318,373		Whiting	Density 2.68 g cm^{-3}.

DATA TABLE 161. THERMAL DIFFUSIVITY OF CALCIUM CARBONATE $CaCO_3$

[Temperature, T, K; Thermal Diffusivity, α, cm^2 s^{-1}]

T	α
CURVE 1*	
318.2	0.00156
373.2	0.00156

* No figure given.

SPECIFICATION TABLE 162. THERMAL DIFFUSIVITY OF CALCIUM TUNGSTATE $CaWO_4$

Cur. No.	Ref. No.	Author(s)	Year	Temp. Range, K	Reported Error, %	Name and Specimen Designation	Composition (weight percent), Specifications, and Remarks
1	126	Klein, P. H.	1968	82-299		$CaWO_4$:0.5% Nd^{+3}	Nd-doped single crystal with c-axis parallel to the specimen length; contains ~6 x 10^{19} neodymium ions cm^{-3} and an approximately equal number of sodium ions; rectangular specimen 2.34 x 2.92 x 23.30 mm; cut from a single crystal of calcium tungstate, grown by the Czochralski technique from a melt containing 0.5 mole percent each of neodymium and sodium; Sylvania Crystal Grade calcium and sodium tungstates used in preparing the melt, with neodymium oxide and tungstic oxide of purities exceeding 99.99 added in stoichiometric quantities; crystal was pulled at a rate of ~1 cm hr^{-1} from an iridium crucible in 1 atm of oxygen, then annealed for 24 hrs at 1473.2 K in air; surrounded by two concentric radiation shields; diffusivity determined from measurement of the time required for the temp difference between two points along specimen to decay to half the steady-state value; measured in a vacuum of <0.1 mm Hg; one-dimensional heat flow.

DATA TABLE 162. THERMAL DIFFUSIVITY OF CALCIUM TUNGSTATE $CaWO_4$

[Temperature, T, K; Thermal Diffusivity, α, cm^2 s^{-1}]

T	α
CURVE 1*	
81.7	0.173
82.4	0.166
82.6	0.153
84.7	0.140
91.0	0.0971
93.1	0.0914
97.1	0.0946
121.3	0.0457
128.8	0.0436
175.8	0.0265
177.8	0.0270
179.9	0.0256
183.7	0.0265
283.8	0.0152
299.2	0.0145

* No figure given.

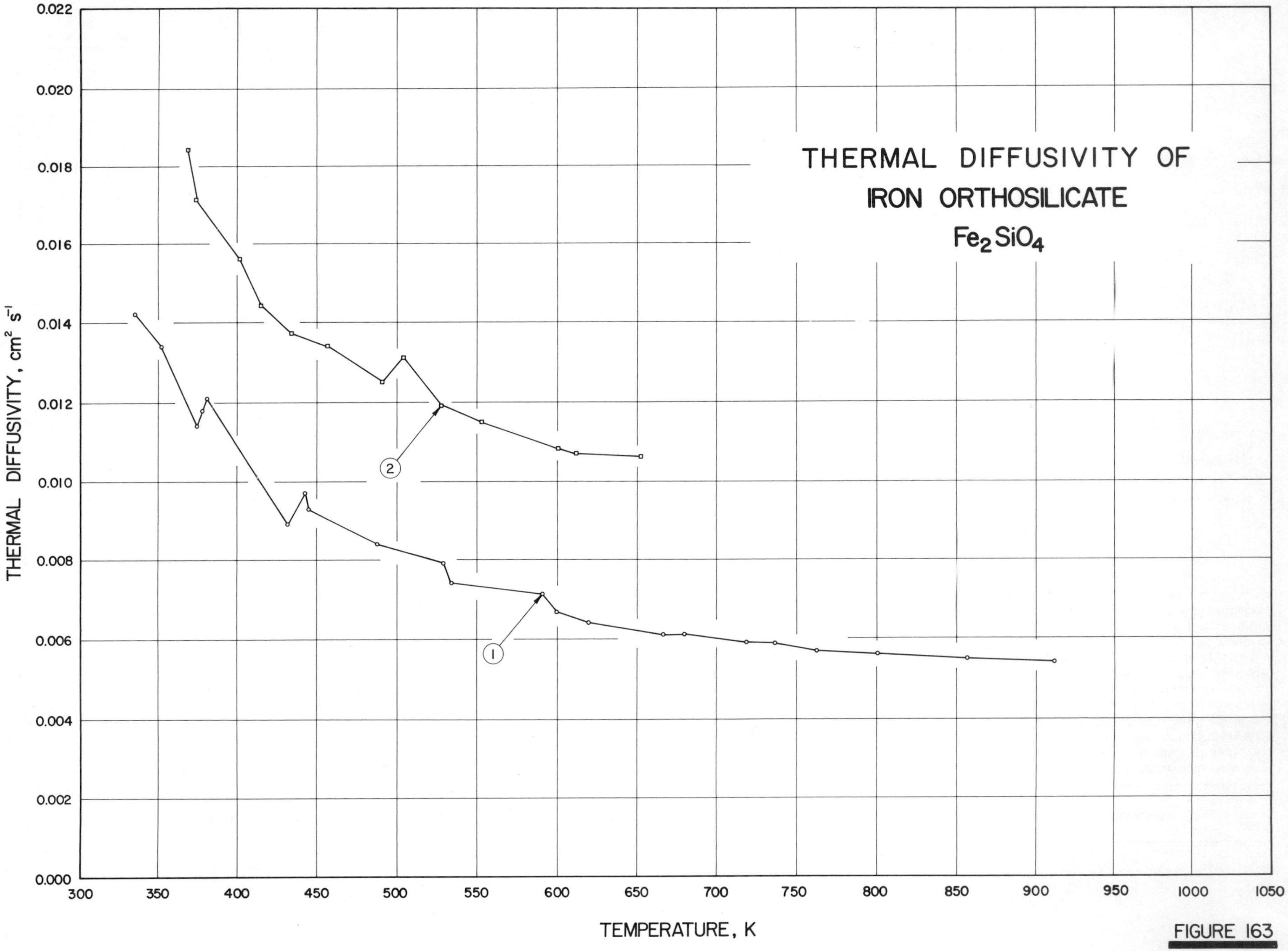
THERMAL DIFFUSIVITY OF
IRON ORTHOSILICATE
Fe_2SiO_4
THERMAL DIFFUSIVITY, $cm^2\ s^{-1}$
0.022
0.020
0.018
0.016
0.014
0.012
0.010
0.008
0.006
0.004
0.002
0.000
TEMPERATURE, K
300
350
400
450
500
550
600
650
700
750
800
850
900
950
1000
1050
1
2

FIGURE 163

SPECIFICATION TABLE 163. THERMAL DIFFUSIVITY OF IRON ORTHOSILICATE Fe_2SiO_4

Cur. No.	Ref. No.	Author(s)	Year	Temp. Range, K	Reported Error, %	Name and Specimen Designation	Composition (weight percent), Specifications, and Remarks
1	276	Fujisawa, H., Fujii, N., Mizutani, H., Kanamori, H., and Akimoto, S.	1968	335-912	±4	Olivine	Cylindrical specimen; ratio of length to radius of specimen 6 to 8; diffusivity measured using modified Angström's method in pressure 48.5 kb.
2	276	Fujisawa, H., et al.	1968	369-653	±5	Spinel	The above specimen.

DATA TABLE 163. THERMAL DIFFUSIVITY OF IRON ORTHOSILICATE Fe_2SiO_4

[Temperature, T, K; Thermal Diffusivity, α, $cm^2 s^{-1}$]

T	α	T	α	T	α
CURVE 1		CURVE 1 (cont.)		CURVE 2 (cont.)	
335	0.0142	763	0.0057	612	0.0107
352	0.0134	801	0.0056	653	0.0106
375	0.0114	857	0.0055		
378	0.0118	912	0.0054		
381	0.0121				
432	0.0089	CURVE 2			
443	0.0097				
445	0.0093	369	0.0184		
488	0.0084	374	0.0171		
529	0.0079	402	0.0156		
534	0.0074	415	0.0144		
591	0.0071	434	0.0137		
600	0.0067	457	0.0135		
620	0.0064	491	0.0125		
667	0.0061	504	0.0131		
680	0.0061	528	0.0119		
719	0.0059	553	0.0115		
737	0.0059	601	0.0108		

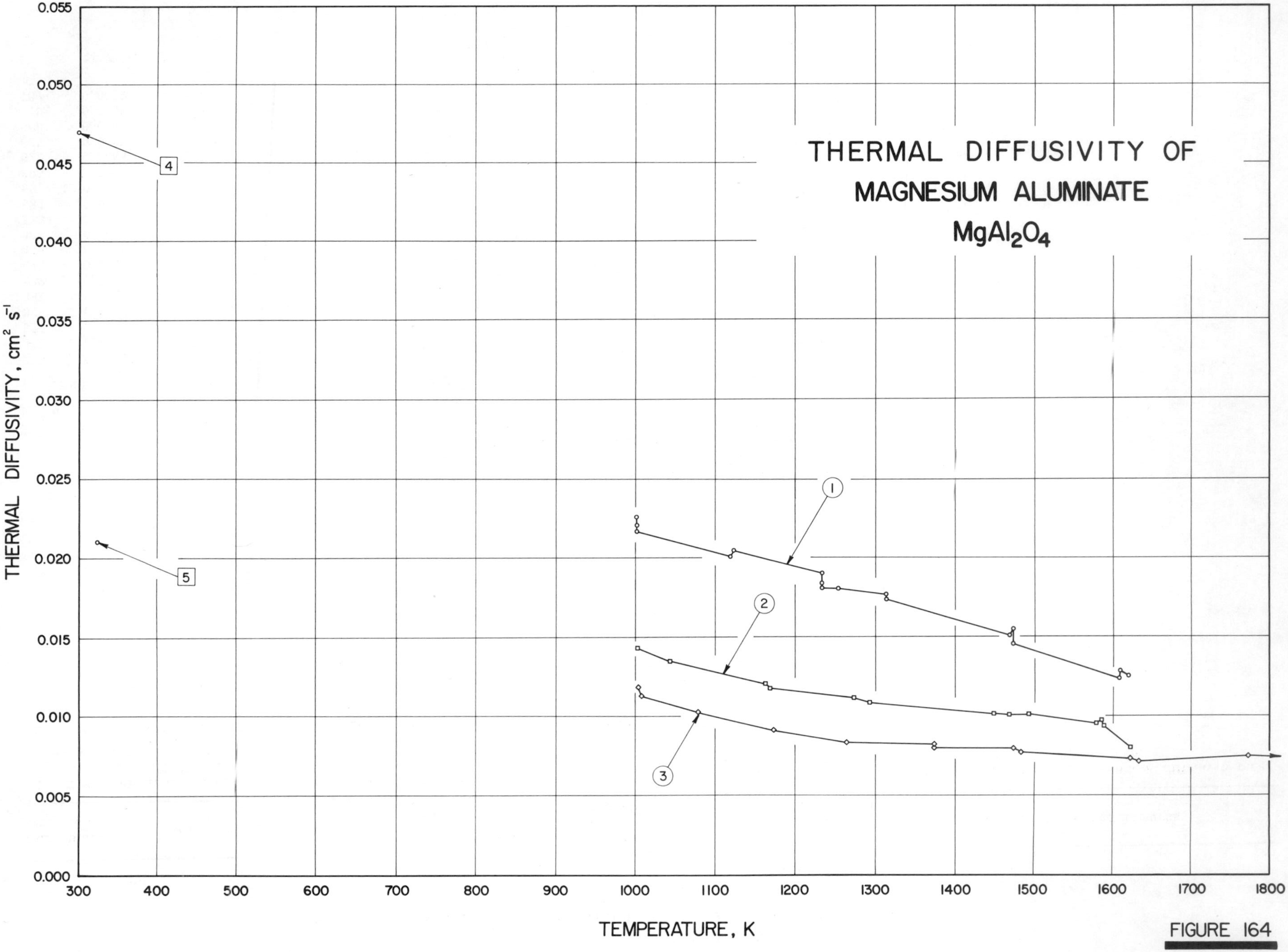
THERMAL DIFFUSIVITY OF
MAGNESIUM ALUMINATE
MgAl2O4
THERMAL DIFFUSIVITY, cm2 s-1
TEMPERATURE, K
0.055
0.050
0.045
0.040
0.035
0.030
0.025
0.020
0.015
0.010
0.005
0.000
300
400
500
600
700
800
900
1000
1100
1200
1300
1400
1500
1600
1700
1800
1
2
3
4
5

FIGURE 164

SPECIFICATION TABLE 164. THERMAL DIFFUSIVITY OF MAGNESIUM ALUMINATE $MgAl_2O_4$

Cur. No.	Ref. No.	Author(s)	Year	Temp. Range, K	Reported Error, %	Name and Specimen Designation	Composition (weight percent), Specifications, and Remarks
1	71, 279	Rudkin, R. L.	1963	1001-1620	± 5/ ±10	Magnesia-spinel (HC-24)	Disc specimen 0.75 in. in diameter and ~0.045 in. thick; obtained from Corning Glass Works; coated on both sides with evaporated tungsten using electron beam techniques; density 3.45 g cm^{-3}; short pulse of thermal energy from a xenon flash lamp absorbed at front surface of specimen; diffusivity determined from measured temperature-time function of the rear surface; measured in vacuum of ~10^{-5} mm Hg.
2	71, 279	Rudkin, R. L.	1963	1003-1623	± 5/ ±10	Magnesia-alumina spinel	Disc specimen 0.75 in. in diameter and ~0.045 in. thick; obtained from Corning Glass Works; coated on both sides with evaporated tungsten using electron beam techniques; density 3.32 g cm^{-3}; short pulse of thermal energy from a xenon flash lamp absorbed at front surface of specimen; diffusivity determined from measured temperature-time function of the rear surface; measured in vacuum of ~10^{-5} mm Hg; measured during 1st cycle.
3	71, 279	Rudkin, R. L.	1963	1003-1903	± 5/ ±10	Magnesia-alumina spinel	Above specimen measured for diffusivity again.
4	184	Taylor, R.	1965	298.2		Spinel	Measured mean density 3.34 g cm^{-3}; temperature of measurement not given by author but assumed to be room temperature; heat pulse method used to measure diffusivity; cylindrical specimen 0.25 in. in diameter and 0.5 in. long.
5	187	Jaeger, G., Koehler, W., and Stapelfeldt, F.	1950	323.2		Spinel	Disk specimen 50 mm in diameter and 7-12 mm thick; sintered at 2173.2 K and then machined on both sides while maintaining them exactly parallel; specimen originally at 20 C; diffusivity determined from measured time necessary for one side of specimen to reach a temperature of 80 C while maintaining the other side in contact with mercury at 100 C; average temperature of measurement not given by authors but assumed to be 50 C; data point reported is the average result of five measurements carried out on at least each of five specimens; density measured and reported as 3.54 g cm^{-3}.

DATA TABLE 164. THERMAL DIFFUSIVITY OF MAGNESIUM ALUMINATE $MgAl_2O_4$

(Impurity ≤2.0% each; total impurities ≤5.0%)

[Temperature, T, K; Thermal Diffusivity, α, $cm^2 s^{-1}$]

T	α
CURVE 1	
1001	0.0226
1001	0.0221
1001	0.0217
1118	0.0201
1123	0.0205
1233	0.0190
1233	0.0184
1233	0.0181
1253	0.0181
1313	0.0177
1313	0.0174
1468	0.0151
1473	0.0155
1473	0.0146
1608	0.0124
1608	0.0129
1620	0.0126
CURVE 2	
1003	0.0143
1043	0.0135
1163	0.0121
1168	0.0118
1273	0.0112
1293	0.0109
1448	0.0102
1468	0.0101
1493	0.0101
1578	0.00955
1586	0.00975
1588	0.00935
1623	0.00805
CURVE 3	
1003	0.0119
1008	0.0113
1078	0.0103
1173	0.00915
1263	0.00835
1373	0.00825
1373	0.00800
1473	0.00800
CURVE 3 (cont.)	
1483	0.00775
1623	0.00735
1633	0.00715
1773	0.00750
1903	0.00715*
1903	0.00765*
CURVE 4	
298.2	0.047
CURVE 5	
323.2	0.021

* Not shown in figure.

SPECIFICATION TABLE 165. THERMAL DIFFUSIVITY OF MAGNESIUM CARBONATE $MgCO_3$

Cur. No.	Ref. No.	Author(s)	Year	Temp. Range, K	Reported Error, %	Name and Specimen Designation	Composition (weight percent), Specifications, and Remarks
1*	166	Williams, I.	1923	318,373			Density 3.00 g cm^{-3}.

DATA TABLE 165. THERMAL DIFFUSIVITY OF MAGNESIUM CARBONATE $MgCO_3$

[Temperature, T, K; Thermal Diffusivity, α, cm^2 s^{-1}]

T	α
CURVE 1*	
318.2	0.00114
373.2	0.00114

* No figure given.

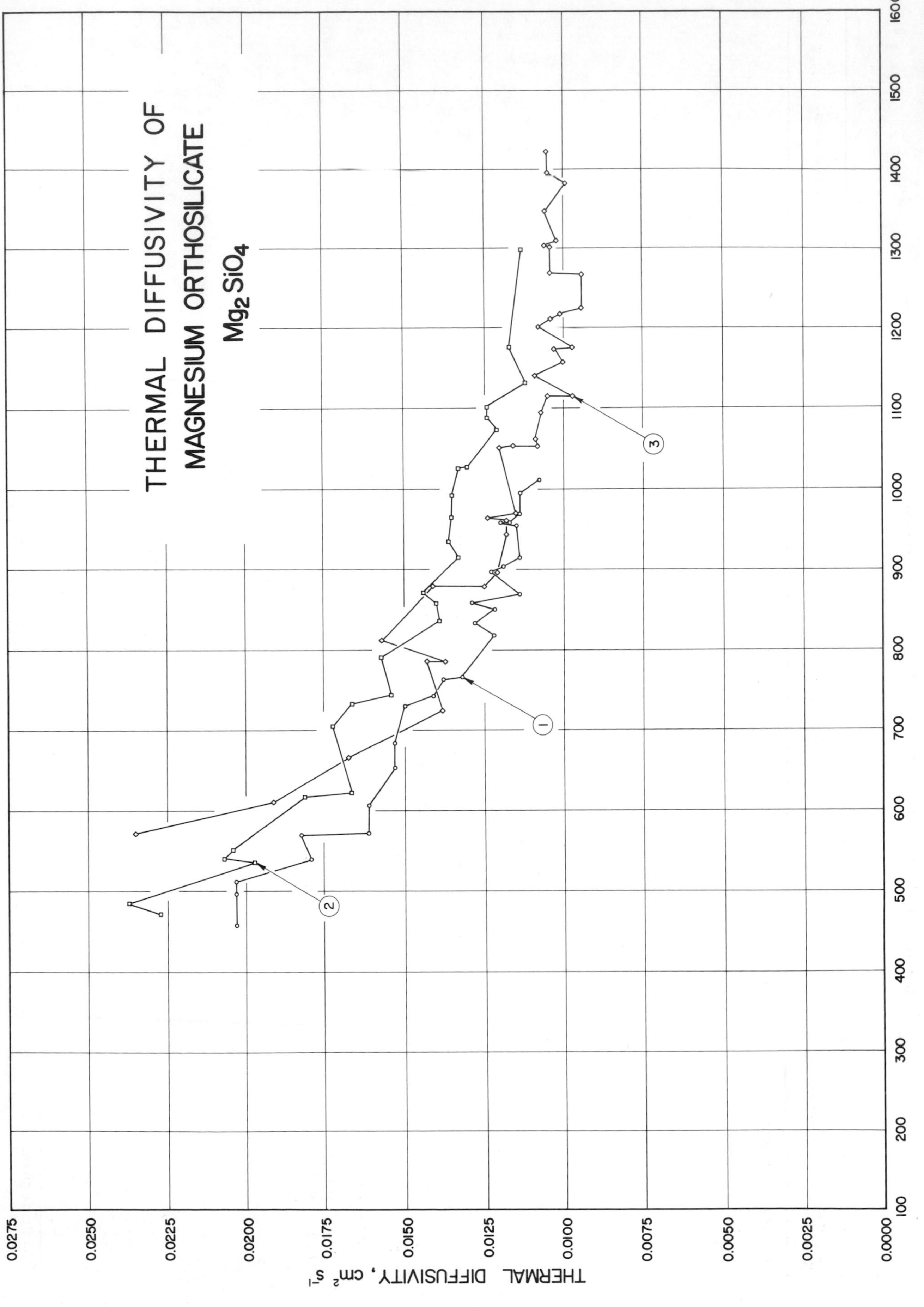
THERMAL DIFFUSIVITY OF
MAGNESIUM ORTHOSILICATE
Mg_2SiO_4
THERMAL DIFFUSIVITY, $cm^2 s^{-1}$
TEMPERATURE, K
0.0000
0.0025
0.0050
0.0075
0.0100
0.0125
0.0150
0.0175
0.0200
0.0225
0.0250
0.0275
100
200
300
400
500
600
700
800
900
1000
1100
1200
1300
1400
1500
1600
1
2
3

FIGURE 166

SPECIFICATION TABLE 166. THERMAL DIFFUSIVITY OF MAGNESIUM ORTHOSILICATE Mg_2SiO_4

Cur. No.	Ref. No.	Author(s)	Year	Temp. Range, K	Reported Error, %	Name and Specimen Designation	Composition (weight percent), Specifications, and Remarks
1	276	Fujisawa, H., Fujii, N., Mizutani, H., Kanamori, H., and Akimoto, S.	1968	457-1011	± 7		Cylindrical specimen; ratio of length to radius 6 to 8; synthesized by sintering an intimate mixture of MgO and anhydrous SiO_2 at 1700 C and 1 atm for 5 hrs; diffusivity measured using modified Angström's method in pressure 29.5 kb.
2	276	Fujisawa, H., et al.	1968	472-1297	± 7		The above specimen; diffusivity measured in pressure 47.0 kb.
3	276	Fujisawa, H., et al.	1968	572-1422	± 7		Similar to the above specimen; diffusivity measured at high temperature in pressure 41.5 kb.

DATA TABLE 166. THERMAL DIFFUSIVITY OF MAGNESIUM ORTHOSILICATE Mg_2SiO_4

[Temperature, T, K; Thermal Diffusivity, α, $cm^2\ s^{-1}$]

T	α	T	α	T	α	T	α	T	α
CURVE 1		CURVE 1 (cont.)		CURVE 2 (cont.)		CURVE 3 (cont.)		CURVE 3 (cont.)	
457	0.0203	954	0.0115	858	0.0140	785	0.0143	1211	0.0104
496	0.0203	958	0.0120	871	0.0144	785	0.0137	1217	0.0101
511	0.0203	958	0.0117	915	0.0133	812	0.0157	1225	0.0094
539	0.0179	969	0.0114	934	0.0136	879	0.0141	1266	0.0094
569	0.0182	995	0.0114	964	0.0135	879	0.0125	1268	0.0104
571	0.0161	1011	0.0108	992	0.0135	896	0.0121	1302	0.0104
606	0.0161			1026	0.0133	943	0.0118	1303	0.0106
653	0.0153	CURVE 2		1027	0.0130	960	0.0118	1310	0.0102
683	0.0153			1073	0.0121	963	0.0124	1347	0.0106
730	0.0150	472	0.0227	1088	0.0124	970	0.0115	1382	0.0099
742	0.0141	485	0.0237	1102	0.0124	1051	0.0120	1395	0.0105
763	0.0138	535	0.0197	1131	0.0112	1053	0.0116	1422	0.0105
766	0.0132	540	0.0207	1175	0.0117	1053	0.0108		
818	0.0122	551	0.0204	1297	0.0113	1062	0.0109		
833	0.0128	617	0.0181			1095	0.0107		
850	0.0122	621	0.0166	CURVE 3		1115	0.0105		
858	0.0129	704	0.0172			1115	0.0097		
869	0.0114	732	0.0161	572	0.0235	1140	0.0109		
897	0.0123	743	0.0154	610	0.0191	1157	0.0100		
903	0.0119	790	0.0157	665	0.0167	1173	0.0103		
915	0.0114	836	0.0139	724	0.0138	1176	0.0097		
						1201	0.0108		

8. MIXTURES OF OXIDES

SPECIFICATION TABLE 167. THERMAL DIFFUSIVITY OF [ALUMINUM OXIDE + SILICON DIOXIDE] $Al_2O_3 + SiO_2$

Cur. No.	Ref. No.	Author(s)	Year	Temp. Range, K	Reported Error, %	Name and Specimen Designation	Composition (weight percent) Al_2O_3	Composition (weight percent) SiO_2	Composition (continued), Specifications, and Remarks
1*	240	Kropschot, R.H., Knight, B.L., and Timmerhaus, K.D.	1968	75-302			50	50	Cab-O-Sil-aluminum powder.

DATA TABLE 167. THERMAL DIFFUSIVITY OF [ALUMINUM OXIDE + SILICON DIOXIDE] $Al_2O_3 + SiO_2$

[Temperature, T, K; Thermal Diffusivity, α, $cm^2 s^{-1}$]

T	α
CURVE 1*	
75	0.000029
181	0.000052
261	0.000081
302	0.00010

* No figure given.

SPECIFICATION TABLE 168. THERMAL DIFFUSIVITY OF [IRON ORTHOSILICATE + MAGNESIUM ORTHOSILICATE] $Fe_2SiO_4 + Mg_2SiO_4$

Cur. No.	Ref. No.	Author(s)	Year	Temp. Range, K	Reported Error, %	Name and Specimen Designation	Composition (weight percent), Specifications, and Remarks
1*	281	Kanamori, H., Fujii, N., and Mizutani, H.	1968	319-1115		Olivine	Solid solution; diffusivity measured using modified Angström method in [001] direction in 1 atmosphere of argon.

DATA TABLE 168. THERMAL DIFFUSIVITY OF [IRON ORTHOSILICATE + MAGNESIUM ORTHOSILICATE] $Fe_2SiO_4 + Mg_2SiO_4$

[Temperature, T, K; Thermal Diffusivity, α, $cm^2\ s^{-1}$]

T	α	T	α
CURVE 1*		CURVE 1 (cont.)*	
319	0.0181	700	0.0103
395	0.0160	700	0.0108
395	0.0151	708	0.0107
409	0.0154	712	0.0101
440	0.0149	715	0.0111
454	0.0134	781	0.0105
456	0.0128	801	0.0107
480	0.0124	805	0.0102
509	0.0118	805	0.0100
518	0.0116	812	0.0104
518	0.0120	816	0.0106
522	0.0121	894	0.0107
535	0.0112	917	0.0111
546	0.0112	983	0.0118
635	0.0110	1030	0.0125
635	0.0108	1033	0.0124
635	0.0106	1075	0.0134
635	0.00999	1115	0.0142

* No figure given.

SPECIFICATION TABLE 169. THERMAL DIFFUSIVITY OF [MAGNESIUM ALUMINATE + SODIUM OXIDE] $MgAl_2O_4 + Na_2O$

Cur. No.	Ref. No.	Author(s)	Year	Temp. Range, K	Reported Error, %	Name and Specimen Designation	Composition (weight percent) $MgAl_2O_4$	Composition (weight percent) Na_2O	Composition (continued), Specifications, and Remarks
1*	74	Hedge, J.C., Kopec, J.W., Kostenko, C., and Lang, J.I.	1963	1175-2000		Spinel ($MgO:Al_2O_3$)	95.00	3.17	1.20 B, 0.26 Fe_2O_3, and 0.33 SiO_2; slab specimen; supplied by Laboratory Equipment Corp; cold pressed; fired at 2088.7 K; density 2.63 g cm^{-3}; top surface of specimen exposed to heat sink; diffusivity determined from measured temperature decrease; unidirectional heat flow.

DATA TABLE 169. THERMAL DIFFUSIVITY OF [MAGNESIUM ALUMINATE + SODIUM OXIDE] $MgAl_2O_4 + Na_2O$

[Temperature, T, K; Thermal Diffusivity, α, $cm^2\ s^{-1}$]

T	α
CURVE 1*	
1174.8	0.0101
1313.7	0.00942
1527.6	0.00934
1663.7	0.00909
1908.2	0.00914
1999.8	0.00937

* No figure given.

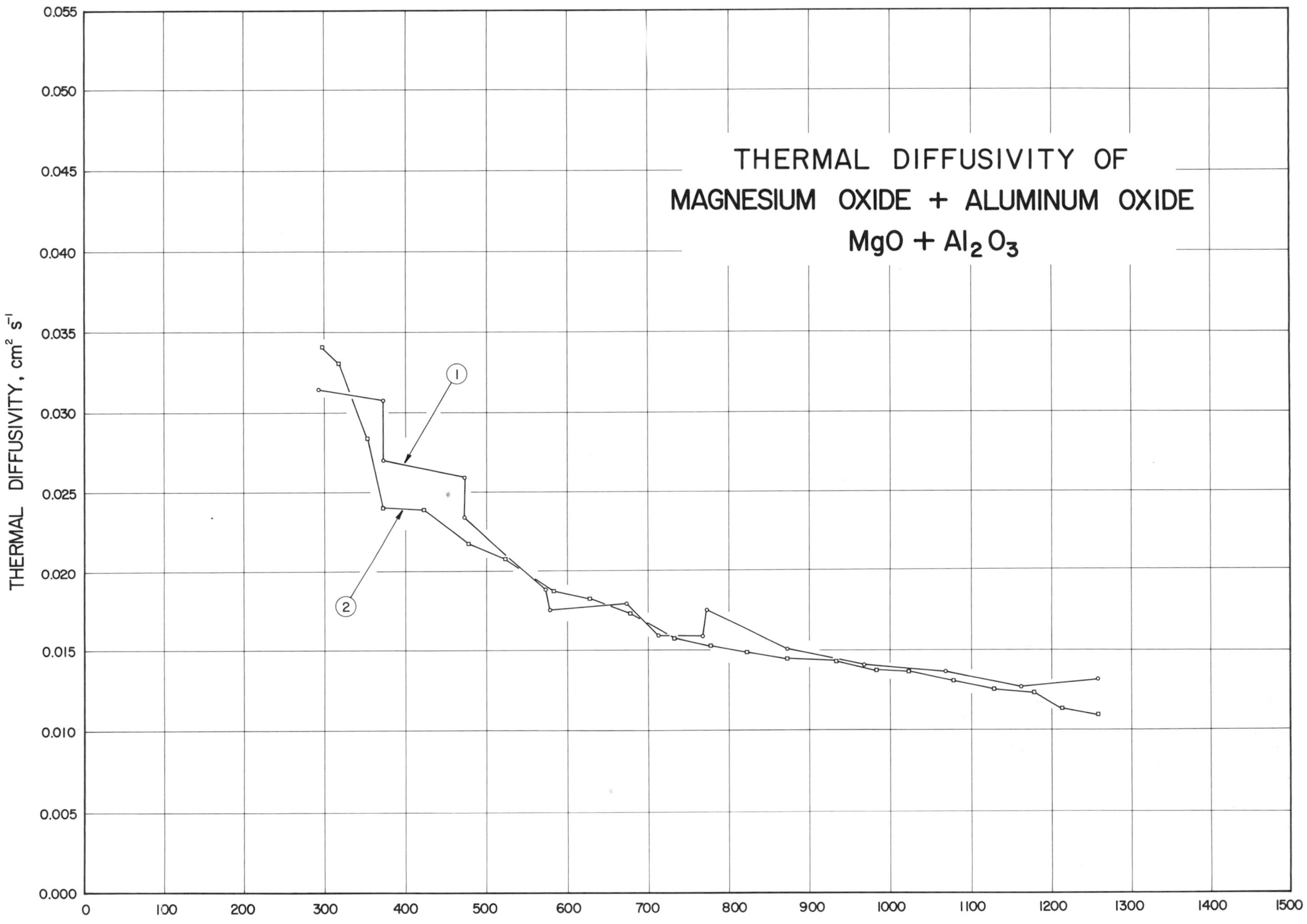
THERMAL DIFFUSIVITY OF
MAGNESIUM OXIDE + ALUMINUM OXIDE
MgO + Al2 O3
THERMAL DIFFUSIVITY, cm2 s-1
TEMPERATURE, K
0.055
0.050
0.045
0.040
0.035
0.030
0.025
0.020
0.015
0.010
0.005
0.000
0
100
200
300
400
500
600
700
800
900
1000
1100
1200
1300
1400
1500
1
2

FIGURE 170

SPECIFICATION TABLE 170. THERMAL DIFFUSIVITY OF [MAGNESIUM OXIDE + ALUMINUM OXIDE] $MgO + Al_2O_3$

Cur. No.	Ref. No.	Author(s)	Year	Temp. Range, K	Reported Error, %	Name and Specimen Designation	Composition (weight percent) MgO	Composition (weight percent) Al_2O_3	Composition (continued), Specifications, and Remarks
1	72	Plummer, W. A., Campbell, D. E., and Comstock, A. A.	1962	293-1258			65	35	Specimen composed of three pieces: middle piece and two end pieces; middle piece separated from bottom piece by a 0.005 in. thick chromel sheet 7.6 x ~18 cm forming the heat source, middle piece separated from upper piece by a 0.005 in. thick chromel sheet forming the heat sink; thicknesses on either side of the heat source are the same; entire assembly constitutes a sandwich 12.7 cm long, 7.6 cm wide and at least 4 to 5 cm thick; diffusivity determined from ratio of the measured temperature rises of the heat source and sink; unidimensional heat flow.
2	72	Plummer, W. A., et al.	1962	298-1258					Above specimen measured for diffusivity again using the composite solid method.

DATA TABLE 170. THERMAL DIFFUSIVITY OF [MAGNESIUM OXIDE + ALUMINUM OXIDE] $MgO + Al_2O_3$

[Temperature, T, K; Thermal Diffusivity, α, $cm^2\ s^{-1}$]

T	α	T	α	T	α
CURVE 1		CURVE 2		CURVE 2 (cont.)	
293	0.0315	298	0.0341	1128	0.0126
373	0.0308	318	0.0331	1178	0.0124
373	0.0270	353	0.0284	1213	0.0114
473	0.0259	373	0.0240	1258	0.0110
473	0.0234	423	0.0239		
573	0.0188	478	0.0218		
578	0.0175	523	0.0208		
673	0.0179	583	0.0187		
713	0.0159	628	0.0182		
768	0.0159	678	0.0173		
773	0.0175	733	0.0158		
873	0.0151	778	0.0153		
968	0.0141	823	0.0149		
1068	0.0137	873	0.0145		
1163	0.0127	933	0.0143		
1258	0.0132	983	0.0138		
		1023	0.0137		
		1078	0.0131		

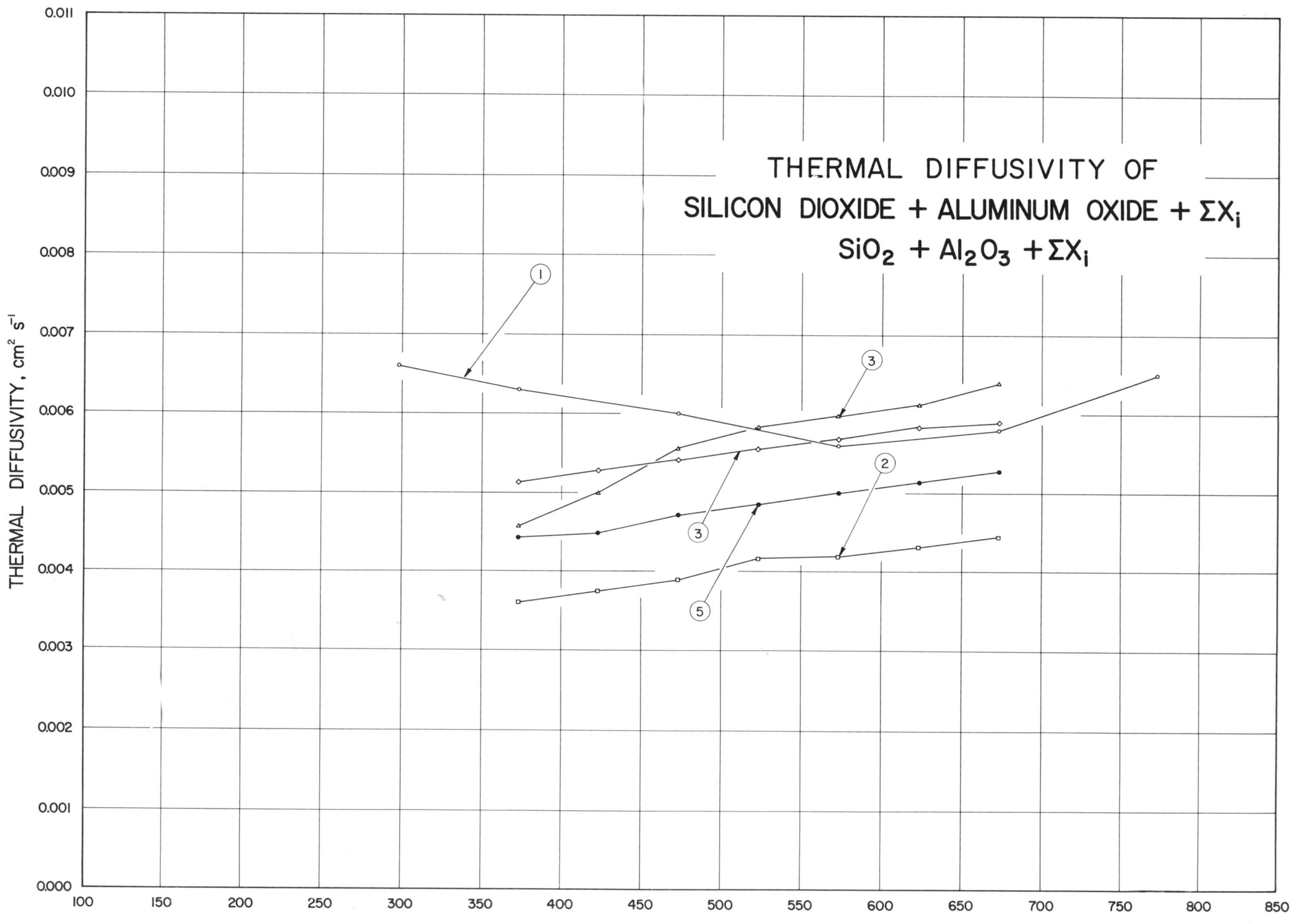
THERMAL DIFFUSIVITY OF
SILICON DIOXIDE + ALUMINUM OXIDE + ΣX$_i$
$SiO_2 + Al_2O_3 + \Sigma X_i$
THERMAL DIFFUSIVITY, cm^2 s^{-1}
TEMPERATURE, K
0.000
0.001
0.002
0.003
0.004
0.005
0.006
0.007
0.008
0.009
0.010
0.011
100
150
200
250
300
350
400
450
500
550
600
650
700
750
800
850
1
2
3
3
5

FIGURE 171

SPECIFICATION TABLE 171. THERMAL DIFFUSIVITY OF [SILICON DIOXIDE + ALUMINUM OXIDE + ΣX_i] $SiO_2 + Al_2O_3 + \Sigma X_i$

Cur. No.	Ref. No.	Author(s)	Year	Temp. Range, K	Reported Error, %	Name and Specimen Designation	Composition (weight percent) SiO_2	Al_2O_3	B_2O_3	BaO	CaO	MgO	Na_2O	TiO_2	Composition (continued), Specifications, and Remarks
1	72	Plummer, W.A., Campbell, D.E., and Comstock, A.A.	1962	298-773	~15	Corning Glass Code 1723	57.7	14.9	4.0	6.0	10.1	6.9			Specimen composed of three pieces; middle piece and two end pieces; middle piece separated from bottom piece by a 0.005 in. thick chromel sheet 7.6 x ~18 cm forming the heat source, middle piece separated from upper piece by a 0.005 in. thick chromel sheet forming the heat sink; thicknesses on either side of the heat source are the same; entire assembly constitues a sandwich 12.7 cm long, 7.6 cm wide and at least 4 to 5 cm thick; diffusivity determined from ratio of the measured temperature rises of the heat source and sink; unidimensional heat flow; last two data points include radiation contribution.
2	167	Bykov, I.I., Khan, B.Kh., and Klimenko, V.S.	1967	373-673	5 Max.	Basalt-90 Hornblendite-10	49.4	12.8			7.2	4.0	2.7	3.8	7.2 FeO and 9.1 Fe_2O_3; phase and mineral composition of experimental casting; 90 monoclinic pyroxene (augite type) interpenetrating with glass, and 10 magnesite (by volume), chromite grains occurring; size of peroxene aggregates 0.08 mm; plate specimen 20 to 22 mm in diameter and 3 to 4 mm thick; cut from casting with a diamond circular saw and then trimmed to the required dimensions on grinding wheels; heated on one side; high-speed dynamic method used to measure diffusivity.
3	167	Bykov, I.I., et al.	1967	373-673	5 Max.	Basalt-92 Dolomite-8	46.2	14.5			12.1	9.8	1.5	1.1	8.1 FeO and 7.3 Fe_2O_3; phase and mineral composition of experimental casting; 85 monoclinic pyroxene (augite type) interpenetrating with glass, and 15 magnetite (by volume), chromite grains occurring; size of pyroxene aggregates 0.07 mm; plate specimen 20 to 22 mm in diameter and 3 to 4 mm thick; cut from casting with a diamond circular saw and then trimmed to the required dimensions on grinding wheels; heated on one side; high-speed dynamic method used to measure diffusivity.

SPECIFICATION TABLE 171. THERMAL DIFFUSIVITY OF [SILICON DIOXIDE + ALUMINUM OXIDE + ΣX_i] $SiO_2 + Al_2O_3 + \Sigma X_i$ (continued)

Cur. No.	Ref. No.	Author(s)	Year	Temp. Range, K	Reported Error, %	Name and Specimen Designation	Composition (weight percent) SiO_2	Al_2O_3	B_2O_3	BaO	CaO	MgO	Na_2O	TiO_2	Composition (continued), Specifications, and Remarks
4	167	Bykov, I.I., et al.	1967	373-673	5 Max.	Heat and electric power plant ashes (cinders)-70 Quartz sand-10 Dolomite-20	45.1	18.6			12.6	9.3	1.2	1.2	10.5 FeO and 2.2 Fe_2O_3; phase and mineral composition of experimental casting: 95 monoclinic pyroxene (augite type) interpenetrating with glass, and 5 glass (by volume), chromite grains occurring; size of pyroxene aggregates 0.01 mm; plate specimen 20 to 22 mm in diameter and 3 to 4 mm thick; cut from casting with a diamond circular saw and then trimmed to the required dimensions on grinding wheels; heated on one side; high-speed dynamic method used to measure diffusivity.
5	167	Bykov, I.I., et al.	1967	373-673	5 Max.	Scorched earth-70 Refractory clay-20 Dolomite-10	61.5	15.9			10.7	2.7	0.5	0.7	4.0 FeO and 0.6 Fe_2O_3; phase and mineral composition of experimental casting: 98 monoclinic pyroxene (diopside composition) interpenetrating with glass, and 2 plagioclase (by volume), chromite grains occurring; size of pyroxene aggregates 0.15 mm; plate specimen 20 to 22 mm in diameter and 3 to 4 mm thick; cut from casting with a diamond circular saw and then trimmed to the required dimensions on grinding wheels; heated on one side; high-speed dynamic method used to measure diffusivity.

DATA TABLE 171. THERMAL DIFFUSIVITY OF [SILICON DIOXIDE + ALUMINUM OXIDE + ΣX_i] $SiO_2 + Al_2O_3 + \Sigma X_i$

[Temperature, T, K; Thermal Diffusivity, α, $cm^2\ s^{-1}$]

T	α
CURVE 1	
298.2	0.0066
373.2	0.0063
473.2	0.0060
573.2	0.0056
673.2	0.0058
773.2	0.0065
CURVE 2	
373.2	0.00361
423.2	0.00375
473.2	0.00389
523.2	0.00417
573.2	0.00419
623.2	0.00431
673.2	0.00444
CURVE 3	
373.2	0.00514
423.2	0.00528
473.2	0.00542
523.2	0.00556
573.2	0.00569
623.2	0.00583
673.2	0.00589
CURVE 4	
373.2	0.00458
423.2	0.00500
473.2	0.00556
523.2	0.00583
573.2	0.00597
623.2	0.00611
673.2	0.00639
CURVE 5	
373.2	0.00444
423.2	0.00450
473.2	0.00472
523.2	0.00486
573.2	0.00500
623.2	0.00514
673.2	0.00528

SPECIFICATION TABLE 172. THERMAL DIFFUSIVITY OF [SILICON DIOXIDE + BARIUM OXIDE + ΣX_i] $SiO_2 + BaO + \Sigma X_i$

Cur. No.	Ref. No.	Author(s)	Year	Temp. Range, K	Reported Error, %	Name and Specimen Designation	Composition (weight percent) SiO_2	BaO	CaO	K_2O	Na_2O	TiO_2	ZnO	ZrO_2	Composition (continued), Specifications, and Remarks
1*	72	Plummer, W. A., Campbell, D. E., and Comstock, A. A.	1962	298-943	~15	Corning Glass Code 8325	44.7	28.0	6.0	2.4	6.7	1.2	4.0	7.0	Specimen composed of three pieces: middle piece and two end pieces; middle piece separated from bottom piece by a 0.005 in. thick chromel sheet 7.6 x ~18 cm forming the heat source, middle piece separated from upper piece by a 0.005 in. thick chromel sheet forming the heat sink; thicknesses on either side of the heat source are the same; entire assembly constitutes a sandwich 12.7 cm long, 7.6 cm wide and at least 4 to 5 cm thick; diffusivity determined from ratio of the measured temperature rises of the heat source and sink; unidimensional heat flow; last four data points include radiation contribution.

DATA TABLE 172. THERMAL DIFFUSIVITY OF [SILICON DIOXIDE + BARIUM OXIDE + ΣX_i] $SiO_2 + BaO + \Sigma X_i$

[Temperature, T, K; Thermal Diffusivity, α, $cm^2\ s^{-1}$]

T	α
CURVE 1*	
298.2	0.0041
373.2	0.0042
473.2	0.0044
573.2	0.0047
673.2	0.0052
773.2	0.0060
873.2	0.0069
943.2	0.0077

* No figure given.

SPECIFICATION TABLE 173. THERMAL DIFFUSIVITY OF [SILICON DIOXIDE + BORON OXIDE] $SiO_2 + B_2O_3$

Cur. No.	Ref. No.	Author(s)	Year	Temp. Range, K	Reported Error, %	Name and Specimen Designation	Composition (weight percent) SiO_2	Composition (weight percent) B_2O_3	Composition (continued), Specifications, and Remarks
1*	72	Plummer, W. A., Campbell, D. E., and Comstock, A. A.	1962	298-1073	~15	Corning Glass Code 7900	96.0	3.0	Specimen composed of three pieces: middle piece and two end pieces; middle piece separated from bottom piece by a 0.005 in. thick chromel sheet 7.6 x ~18 cm forming the heat source, middle piece separated from upper piece by a 0.005 in. thick chromel sheet forming the heat sink; thicknesses on either side of the heater are the same; entire assembly constitutes a sandwich 12.7 cm long, 7.6 cm wide and at least 4 to 5 cm thick; diffusivity determined from measured ratio of the temperature rises of the heat source and sink; unidimensional heat flow; last five data points include radiation correction.

DATA TABLE 173. THERMAL DIFFUSIVITY OF [SILICON DIOXIDE + BORON OXIDE] $SiO_2 + B_2O_3$

[Temperature, T, K; Thermal Diffusivity, α, $cm^2\ s^{-1}$]

T	α
CURVE 1*	
298.2	0.0090
373.2	0.0084
473.2	0.0079
573.2	0.0077
673.2	0.0079
773.2	0.0087
873.2	0.0102
973.2	0.0130
1073.2	0.0168

* No figure given.

SPECIFICATION TABLE 174. THERMAL DIFFUSIVITY OF [SILICON DIOXIDE + BORON OXIDE + ΣX_i] $SiO_2 + B_2O_3 + \Sigma X_i$

Cur. No.	Ref. No.	Author(s)	Year	Temp. Range, K	Reported Error, %	Name and Specimen Designation	Composition (weight percent) SiO_2	B_2O_3	Al_2O_3	Na_2O	Composition (continued), Specifications, and Remarks
1*	72	Plummer, W. A., Campbell, D. E., and Comstock, A. A.	1962	298-633	~15	Corning Glass Code 7740	80.4	13.3	2.0	4.4	Specimen composed of three pieces: middle piece and two end pieces; middle piece separated from bottom piece by a 0.005 in. thick chromel sheet 7.6 x ~18 cm forming the heat source, middle piece separated from upper piece by a 0.005 in. thick chromel sheet forming the heat sink; thicknesses on either side of the heat source are the same; entire assembly constitutes a sandwich 12.7 cm long, 7.6 cm wide and at least 4 to 5 cm thick; diffusivity determined from ratio of the measured temperature rises of the heat source and sink; unidimensional heat flow.

DATA TABLE 174. THERMAL DIFFUSIVITY OF [SILICON DIOXIDE + BORON OXIDE + ΣX_i] $SiO_2 + B_2O_3 + \Sigma X_i$

[Temperature, T, K; Thermal Diffusivity, α, $cm^2\ s^{-1}$]

T	α
CURVE 1*	
298.2	0.0055
373.2	0.0053
473.2	0.0052
573.2	0.0052
633.2	0.0051

* No figure given.

SPECIFICATION TABLE 175. THERMAL DIFFUSIVITY OF [SILICON DIOXIDE + CALCIUM OXIDE + ΣX_i] $SiO_2 + CaO + \Sigma X_i$

Cur. No.	Ref. No.	Author(s)	Year	Temp. Range, K	Reported Error, %	Name and Specimen Designation	Composition (weight percent) SiO_2	CaO	Al_2O_3	FeO	Fe_2O_3	MgO	No_2O	TiO_2	Composition (continued), Specifications, and Remarks
1*	167	Bykov, I.I., Khan, B.Kh., and Klimenko, V.S.	1967	373-673	5 Max.	Granite-40 Blast-furnace slag-60	48.14	23.0	12.0	2.3	1.0	7.98	3.2		Phase and mineral composition of experimental casting: 100 monoclinic pyroxene (diopside-Jadsite composition) interpenetrating with glass, chromite grains occurring; size of peroxene aggregates 0.1 mm; plate specimen 20 to 22 mm in diameter and 3 to 4 mm thick; cut from casting with a diamond circular saw and then trimmed to the required dimensions on grinding wheels; heated on one side; high-speed dynamic method used to measure diffusivity.
2*	167	Bykov, I.I., et al.	1967	373-673	5 Max.	Granite-55 Blast-furnace slag-45	55.33	17.55	13.50	2.63		7.34	4.88	0.19	Phase and mineral composition of experimental casting: 100 monoclinic pyroxene (diopside-jadsite composition) interpenetrating with glass, chromite grains occurring; size of peroxene aggregates 0.3 mm; plate specimen 20 to 22 mm in diameter and 3 to 4 mm thick; cut from casting with a diamond circular saw and then trimmed to the required dimensions on grinding wheels; heated on one side; high-speed dynamic method used to measure diffusivity.

DATA TABLE 175. THERMAL DIFFUSIVITY OF [SILICON DIOXIDE + CALCIUM OXIDE + ΣX_i] $SiO_2 + CaO + \Sigma X_i$

[Temperature, T, K; Thermal Diffusivity, α, $cm^2 s^{-1}$]

T	α	T	α
CURVE 1*		CURVE 2*	
373.2	0.00486	373.2	0.00583
423.2	0.00486	423.2	0.00611
473.2	0.00500	473.2	0.00617
523.2	0.00506	523.2	0.00625
573.2	0.00511	573.2	0.00639
623.2	0.00514	623.2	0.00653
673.2	0.00519	673.2	0.00667

* No figure given.

SPECIFICATION TABLE 176. THERMAL DIFFUSIVITY OF [SILICON DIOXIDE + LEAD OXIDE + ΣX_i] $SiO_2 + PbO + \Sigma X_i$

Cur. No.	Ref. No.	Author(s)	Year	Temp. Range, K	Reported Error, %	Name and Specimen Designation	Composition (weight percent) SiO_2	PbO	Al_2O_3	CeO_2	K_2O	Na_2O	Composition (continued), Specifications, and Remarks
1*	72	Plummer, W. A., Campbell, D. E., and Comstock, A. A.	1962	298-723	~15	Corning Glass Code 8362	44.6	33.4	2.0	1.3	14.0	6.0	Specimen composed of three pieces: middle piece and two end pieces; middle piece separated from bottom piece by a 0.005 in. thick chromel sheet 7.6 x ~18 cm forming the heat source, middle piece separated from upper piece by a 0.005 in. thick chromel sheet forming the heat sink; thicknesses on either side of the heat source are the same; entire assembly constitutes a sandwich 12.7 cm long, 7.6 cm wide and at least 4 to 5 cm thick; diffusivity determined from ratio of the measured temperature rises of the heat source and sink; unidimensional heat flow.

DATA TABLE 176. THERMAL DIFFUSIVITY OF [SILICON DIOXIDE + LEAD OXIDE + ΣX_i] $SiO_2 + PbO + \Sigma X_i$

[Temperature, T, K; Thermal Diffusivity, α, $cm^2 s^{-1}$]

T	α
CURVE 1*	
298.2	0.0041
373.2	0.0041
473.2	0.0041
573.2	0.0040
633.2	0.0039
673.2	0.0038
723.2	0.0038

* No figure given.

SPECIFICATION TABLE 177. THERMAL DIFFUSIVITY OF [SILICON DIOXIDE + MAGNESIUM OXIDE] $SiO_2 + MgO$

Cur. No.	Ref. No.	Author(s)	Year	Temp. Range, K	Reported Error, %	Name and Specimen Designation	Composition (weight percent) SiO_2	Composition (weight percent) MgO	Composition (continued), Specifications, and Remarks
1*	240	Kropschot, R.H., Knight, B.L., and Timmerhaus, K.D.	1968	72-302		Celkate T-21			Powder.

DATA TABLE 177. THERMAL DIFFUSIVITY OF [SILICON DIOXIDE + MAGNESIUM OXIDE] $SiO_2 + MgO$

[Temperature, T, K; Thermal Diffusivity, α, $cm^2\ s^{-1}$]

T	α
CURVE 1*	
72	0.000072
179	0.000090
261	0.000101
302	0.000112

* No figure given.

SPECIFICATION TABLE 178. THERMAL DIFFUSIVITY OF [SILICON DIOXIDE + SODIUM OXIDE + ΣX_i] $SiO_2 + Na_2O + \Sigma X_i$

Cur. No.	Ref. No.	Author(s)	Year	Temp. Range, K	Reported Error, %	Name and Specimen Designation	Composition (weight percent) SiO_2	Na_2O	Al_2O_3	CaO	MgO	Composition (continued), Specifications, and Remarks
1*	72	Plummer, W. A., Campbell, D. E., and Comstock, A. A.	1962	298-633	~15	Corning Glass Code 0080	73.4	16.9	0.8	4.8	3.3	Specimen composed of three pieces: middle piece and two end pieces; middle piece separated from bottom piece by a 0.005 in. thick chromel sheet 7.6 x ~18 cm forming the heat source, middle piece separated from upper piece by a 0.005 in. thick chromel sheet forming the heat sink; thicknesses on either side of the heat source are the same; entire assembly constitutes a sandwich 12.7 cm long, 7.6 cm wide and at least 4 to 5 cm thick; diffusivity determined from ratio of the measured temperature rises of the heat source and sink; unidimensional heat flow; last two data points include radiation contribution.

DATA TABLE 178. THERMAL DIFFUSIVITY OF [SILICON DIOXIDE + SODIUM OXIDE + ΣX_i] $SiO_2 + Na_2O + \Sigma X_i$

[Temperature, T, K; Thermal Diffusivity, α, $cm^2 s^{-1}$]

T	α
CURVE 1*	
298.2	0.0053
373.2	0.0050
473.2	0.0046
573.2	0.0047
633.2	0.0054

* No figure given.

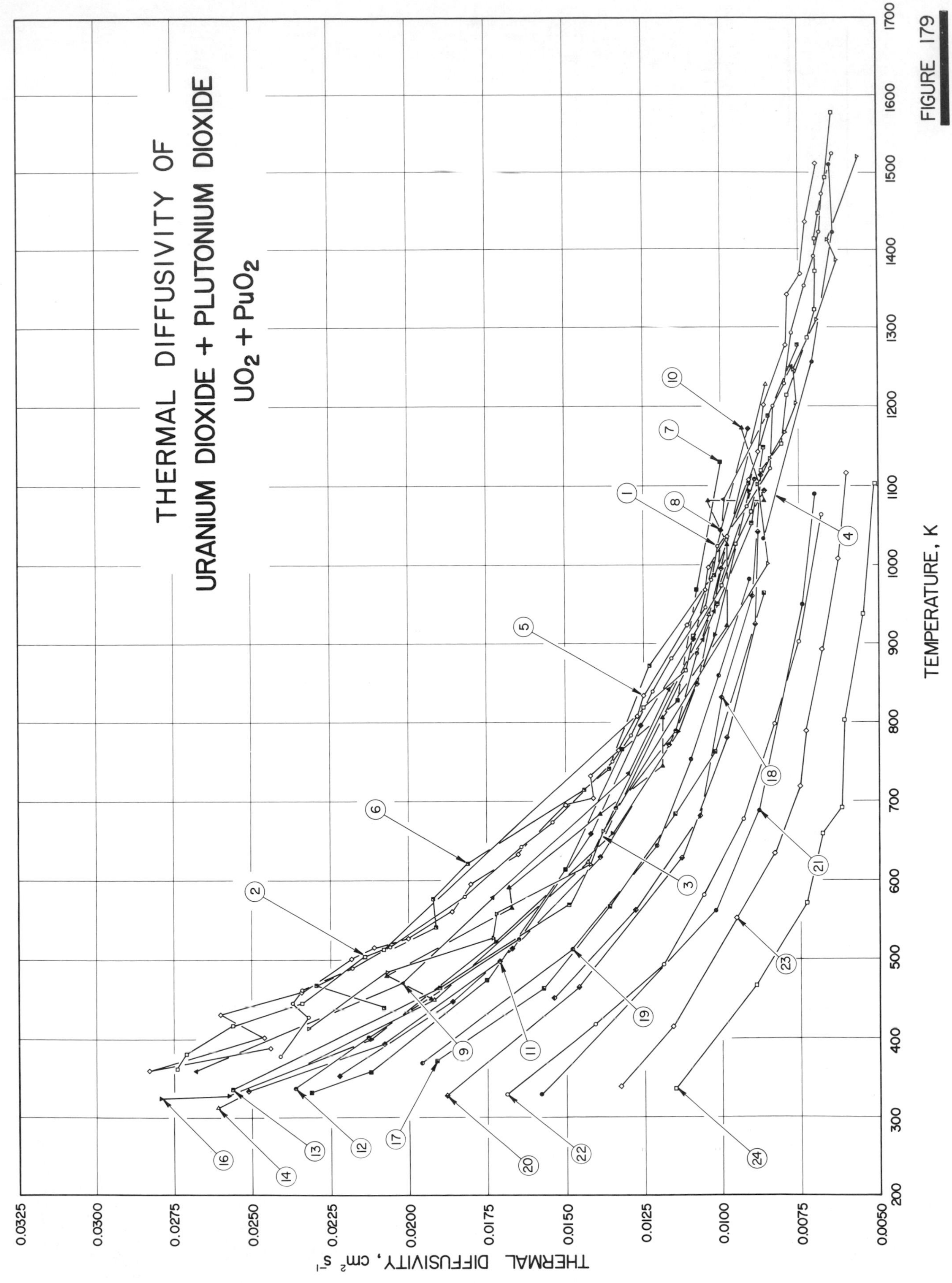
THERMAL DIFFUSIVITY OF
URANIUM DIOXIDE + PLUTONIUM DIOXIDE
$UO_2 + PuO_2$
THERMAL DIFFUSIVITY, $cm^2\ s^{-1}$
TEMPERATURE, K
0.0050
0.0075
0.0100
0.0125
0.0150
0.0175
0.0200
0.0225
0.0250
0.0275
0.0300
0.0325
200
300
400
500
600
700
800
900
1000
1100
1200
1300
1400
1500
1600
1700
1
2
3
4
5
6
7
8
9
10
11
12
13
14
16
17
18
19
20
21
22
23
24

FIGURE 179

SPECIFICATION TABLE 179. THERMAL DIFFUSIVITY OF [URANIUM DIOXIDE + PLUTONIUM DIOXIDE] $UO_2 + PuO_2$

Cur. No.	Ref. No.	Author(s)	Year	Temp. Range, K	Reported Error, %	Name and Specimen Designation	Composition (weight percent) UO_2	Composition (weight percent) PuO_2	Composition (continued), Specifications, and Remarks
1	119	Gibby, R. L.	1968	378-1870	± 8	Sintered (GCPB)	→	→	Sintered $(U_{0.80}Pu_{0.20})O_2$ powder containing 17.6 Pu, <0.01 P, 0.005 Cr, 0.005 Ni, 0.005 Zn, <0.0025 Fe, 0.002 Ca, 0.002 Sm, <0.002 As, 0.001 Al, 0.001 Mg, <0.001 Ge, <0.001 V, 0.0005 Na, 0.0005 Si, <0.0005 K, <0.0005 Li, <0.0005 Mo, 0.0002 Cu, 0.0002 Mn, 0.0002 Pb, <0.0001 Ag, <0.0001 B, <0.0001 Bi, <0.0001 Cd, and <0.0001 Tl; grain size 30 μ; lattice parameter (O/M = 2.00) 5.4553 ± 0.0003 A; wafer specimen 0.63 cm in diameter and 0.0734 cm thick; high purity $(U_{0.80}Pu_{0.20})$ powder prepared from mixed plutonium-uranyl solution using a co-precipitation process, co-precitate calcined for approximately 1 hr at 1273.2 K in a nitrogen-6 Vol % H_2 atmosphere to obtain oxide powder; powder pressed into pellets of 0.954 cm in diameter and 1.5 cm long at 3000 dyne cm^{-2}; approximately two drops of water per gram of powder required to form strong lamination-free pellets; after drying pellets heated over a 24-hr period to a sintering temperature of 1923.2 K, allowed to soak at this temperature for 16 hr, and slowly cooled to room temperature; entire sintering process performed in a protective atmosphere of flowing argon-8 Vol % H_2, pellets then sliced with a diamond saw into thin sample wafer; oxygen-to-metal ratio of sintered material 1.99; density 10.60 g cm^{-3} (Hg displacement at 10 μ), 96% of theoretical density (11.08 g cm^{-3}); measured in a flowing atmosphere of argon-8 Vol % H_2 passed over water at 273.2 K; transiently heated with a 0.0015 sec duration energy pulse from a ruby laser; diffusivity determined from measured temperature-time curve of rear surface; measured data corrected for heat loss and finite-pulse-time effects.
2	119	Gibby, R. L.	1968	363-1823	± 8	Sintered (GCPB)			Above specimen measured for diffusivity during cooling.
3	119	Gibby, R. L.	1968	413-1780	± 8	Pneumatically Impacted (GCP)	→	→	Impacted $(U_{0.80}Pu_{0.20})O_2$ powder containing 0.032 Si, 0.0068 Fe, 0.005 Al, 0.0045 Cr, 0.0023 Ni, 0.002 K, 0.002 Mo, <0.002 As, 0.0014 Sm, 0.001 B, 0.001 Cu, <0.001 Ge, <0.001 V, 0.0005 Mn, 0.0005 Na, <0.0005 Li, <0.0005 Mg, <0.0005 P, 0.0002 Ca, 0.0002 Zn, 0.0001 Ag, <0.0001 Bi, <0.0001 Cd, <0.0001 Pb, and <0.0001 Ti; grain size unresolvable at 500X; lattice parameter (O/M = 2.00) 5.4553 ± 0.0003 A; wafer specimen 0.63 cm in diameter and 0.0610 cm thick; fabricated from same high purity powder used for preparation of the "sintered (GCPB) specimen" above; powder placed in small stainless steel can, loaded into a mild-steel die assembly, and then entire assembly enclosed in a larger stainless steel can; void areas between die and outer can filled with alumina powder; after sealing complete assembly evacuated through a small tube in the outer can and simultaneously heated to 1473.2 K; hot assembly then impacted at 20000 atmospheres; densified material removed from inner container by machining, fractured into many pieces, larger pieces sliced with diamond saw into thin sample wafers; oxygen-to-metal ratio 2.00; density 10.57 g cm^{-3} (Hg displacement at 10μ), 95.5% of theoretical density;

SPECIFICATION TABLE 179. THERMAL DIFFUSIVITY OF [URANIUM DIOXIDE + PLUTONIUM DIOXIDE] $UO_2 + PuO_2$ (continued)

Cur. No.	Ref. No.	Author(s)	Year	Temp. Range, K	Reported Error, %	Name and Specimen Designation	Composition (weight percent) UO_2	Composition (weight percent) PuO_2	Composition (continued), Specifications, and Remarks
									measured in a flowing atmosphere of argon-8 Vol% H_2 passed over water at 273.2 K; transiently heated with a 0.0015 sec duration energy pulse from a ruby laser; diffusivity determined from measured temperature-time curve of rear surface; measured data corrected for heat loss and finite-pulse-time effects.
4	119	Gibby, R. L.	1968	1034-1727	±8	Pneumatically Impacted (GCP)			Above specimen measured for diffusivity during cooling.
5	119	Gibby, R. L.	1968	360-1531	±8	Pneumatically-Impacted-Annealed (GCP-1)			Above specimen measured for diffusivity again during another run.
6	119	Gibby, R. L.	1968	440-1279	±8	Pneumatically-Impacted-Annealed (GCP-1)			Above specimen measured for diffusivity during cooling.
7	283	vanCraeynest, J.C. and Stora, J.P.	1970	332-1131		I	80	20	100% theoretical density; thermal diffusivity measured by Angström method.
8	283	vanCraeynest, J.C. and Stora, J.P.	1970	353-1172		II	80	20	Similar to above.
9	283	vanCraeynest, J.C. and Stora, J.P.	1970	360-1250		13.01	80	20	Similar to above but specimen of 95% theoretical density.
10	283	vanCraeynest, J.C. and Stora, J.P.	1970	452-1174		13.02	80	20	Similar to above.
11	283	vanCraeynest, J.C. and Stora, J.P.	1970	334-1094		5	80	20	Similar to above but specimen of 92% theoretical density.
12	283	vanCraeynest, J.C. and Stora, J.P.	1970	337-1109		6	80	20	Similar to above.
13	283	vanCraeynest, J.C.1 and Stora, J.P.	1970	337-1149		7	80	20	Similar to above.
14	283	vanCraeynest, J.C. and Stora, J.P.	1970	313-1228		4	80	20	Similar to above.
15*	283	vanCraeynest, J.C. and Stora, J.P.	1970	313-1228		8	80	20	Similar to above.
16	283	vanCraeynest, J.C. and Stora, J.P.	1970	326-1094		2	80	20	Similar to above but specimen of 88.5% theoretical density.
17	283	vanCraeynest, J.C. and Stora, J.P.	1970	373-965		3	80	20	Similar to above.

* Not shown in figure.

SPECIFICATION TABLE 179. THERMAL DIFFUSIVITY OF [URANIUM DIOXIDE + PLUTONIUM DIOXIDE] $UO_2 + PuO_2$ (continued)

Cur. No.	Ref. No.	Author(s)	Year	Temp. Range, K	Reported Error, %	Name and Specimen Designation	Composition (weight percent) UO_2	Composition (weight percent) PuO_2	Composition (continued), Specifications, and Remarks
18	283	van Craeynest, J.C. and Stora, J.P.	1970	452-1116		13	80	20	Similar to above but specimen of 82.5% theoretical density.
19	283	van Craeynest, J.C. and Stora, J.P.	1970	370-983		14	80	20	Similar to above but specimen of 82.7% theoretical density.
20	283	van Craeynest, J.C. and Stora, J.P.	1970	328-1042		15	80	20	Similar to above but specimen of 83% theoretical density.
21	283	van Craeynest, J.C. and Stora, J.P.	1970	329-1090		11	80	20	Similar to above but specimen of 79% theoretical density.
22	283	van Craeynest, J.C. and Stora, J.P.	1970	330-1064		12	80	20	Similar to above.
23	283	van Craeynest, J.C. and Stora, J.P.	1970	340-1117		9	80	20	Similar to above but specimen of 72.3% theoretical density.
24	283	van Craeynest, J.C. and Stora, J.P.	1970	337-1103		10	80	20	Similar to above.

DATA TABLE 179. THERMAL DIFFUSIVITY OF [URANIUM DIOXIDE + PLUTONIUM DIOXIDE] $UO_2 + PuO_2$

[Temperature, T, K; Thermal Diffusivity, α, $cm^2\ s^{-1}$]

T	α
CURVE 1	
378.2	0.0241
427.2	0.0232
445.2	0.0237
462.2	0.0234
489.2	0.0218
516.2	0.0206
579.2	0.0182
642.2	0.0164
673.2	0.0154
751.2	0.0135
793.2	0.0129
839.2	0.0122
881.2	0.0116
946.2	0.0105
981.2	0.0103
1024.2	0.0101
1074.2	0.00915
1122.2	0.00885
1201.2	0.00831
1229.2	0.00792
1293.2	0.00769
1353.2	0.00728
1392.2	0.00699
1424.2	0.00683
1473.2	0.00672
1525.2	0.00640*
1572.2	0.00626*
1625.2	0.00614*
1688.2	0.00605*
1722.2	0.00609*
1771.2	0.00576*
1817.2	0.00550*
1870.2	0.00577*
CURVE 2	
363.2	0.0274
381.2	0.0271
417.2	0.0256
445.2	0.0234
504.2	0.0214
513.2	0.0208
819.2	0.0125
866.2	0.0115
910.2	0.0109
CURVE 2 (cont.)	
974.2	0.00997
1027.2	0.00951
1068.2	0.00901
1120.2	0.00867
1153.2	0.00803
1216.2	0.00784
1287.2	0.00720
1323.2	0.00698
1373.2	0.00695
1415.2	0.00695
1448.2	0.00681
1493.2	0.00660
1579.2	0.00642*
1624.2	0.00611*
1675.2	0.00610*
1725.2	0.00541*
1767.2	0.00590*
1823.2	0.00603*
CURVE 3	
413.2	0.0232
484.2	0.0212
527.2	0.0173
558.2	0.0172
620.2	0.0142
662.2	0.0138
1002.2	0.00847
1102.2	0.00880
1135.2	0.00840
1168.2	0.00790
1205.2	0.00775
1247.2	0.00762
1301.2	0.00633
1310.2	0.00691
1387.2	0.00628
1414.2	0.00656
1521.2	0.00560*
1591.2	0.00563*
1644.2	0.00598*
1692.2	0.00536*
1770.2	0.00484*
1780.2	0.00549*
CURVE 4	
1034.2	0.00862
1257.2	0.00706
1424.2	0.00640
1511.2	0.00651*
1600.2	0.00538*
1659.2	0.00525*
1727.2	0.00529*
CURVE 5	
360.2	0.0283
389.2	0.0244
402.2	0.0246
431.2	0.0260
459.2	0.0234
501.2	0.0218
515.2	0.0211
527.2	0.0200
561.2	0.0186
595.2	0.0180
633.2	0.0165
695.2	0.0150
704.2	0.0141
732.2	0.0142
764.2	0.0133
808.2	0.0127
834.2	0.0125
925.2	0.0110
969.2	0.0105
997.2	0.0104
1037.2	0.00980
1108.2	0.00915
1143.2	0.00876
1202.2	0.00857
1278.2	0.00787
1342.2	0.00782
1369.2	0.00741
1436.2	0.00724
1512.2	0.00693*
1531.2	0.00722*
CURVE 6	
440.2	0.0208
468.2	0.0232
541.2	0.0194
577.2	0.0192
611.2	0.0181
715.2	0.0144
741.2	0.0136
873.2	0.0123
989.2	0.0102
1104.2	0.00915
1189.2	0.00846
1279.2	0.00750
CURVE 7	
332	0.0231
358	0.0212
474	0.0175
614	0.0150
766	0.0132
828	0.0114
969	0.0108
1131	0.0100
CURVE 8	
353	0.0222
447	0.0186
514	0.0167
659	0.0142
797	0.0126
906	0.0109
1044	0.0100
1172	0.0091
CURVE 9	
360	0.0268
471	0.0202
579	0.0173
735	0.0130
842	0.0117
905	0.0106
941	0.0102
1083	0.0099
1250	0.0077
CURVE 10	
452	0.0193
480	0.0207
566	0.0167
591	0.0168
684	0.0139
745	0.0119
806	0.0119
924	0.0098
1026	0.0098
1082	0.0104
1082	0.0086
1174	0.0093
CURVE 11	
334	0.0251
394	0.0208
498	0.0171
629	0.0139
771	0.0117
849	0.0108
996	0.0100
1094	0.0086
CURVE 12	
337	0.0236
402	0.0213
526	0.0165
691	0.0134
789	0.0115
888	0.0108
1109	0.0089
CURVE 13	
337	0.0256
465	0.0190
569	0.0149
951	0.0101
1053	0.0090
1149	0.0086
CURVE 14	
313	0.0266
450	0.0192
621	0.0143
937	0.0104
1228	0.0083
CURVE 15*	
313	0.0266
450	0.0192
621	0.0143
937	0.0104
1228	0.0083
CURVE 16	
326	0.0279
329	0.0257
400	0.0212
523	0.0172
659	0.0135
788	0.0114
911	0.0102
1094	0.0091
CURVE 17	
373	0.0191
464	0.0157
567	0.0136
684	0.0115
763	0.0102
965	0.0086
CURVE 18	
452	0.0154
563	0.0128
681	0.0107
832	0.0100
961	0.0090
1116	0.0087
CURVE 19	
370	0.0196
513	0.0148
643	0.0121
753	0.0110
860	0.0101
983	0.0091
CURVE 20	
328	0.0188
466	0.0146
628	0.0113
780	0.0098
925	0.0089
1042	0.0088
CURVE 21	
329	0.0158
562	0.0102
688	0.0088
950	0.0074
1090	0.0070
CURVE 22	
330	0.0169
419	0.0141
494	0.0119
581	0.0106
677	0.0093
799	0.0083
903	0.0075
1064	0.0068
CURVE 23	
340	0.0133
416	0.0116
552	0.0093
634	0.0083
719	0.0075
789	0.0073
893	0.0068
1008	0.0063
1117	0.0060
CURVE 24	
337	0.0115
468	0.0089
571	0.0073
659	0.0068
692	0.0062
803	0.0061
938	0.0055
1103	0.0051

* Not shown in figure.

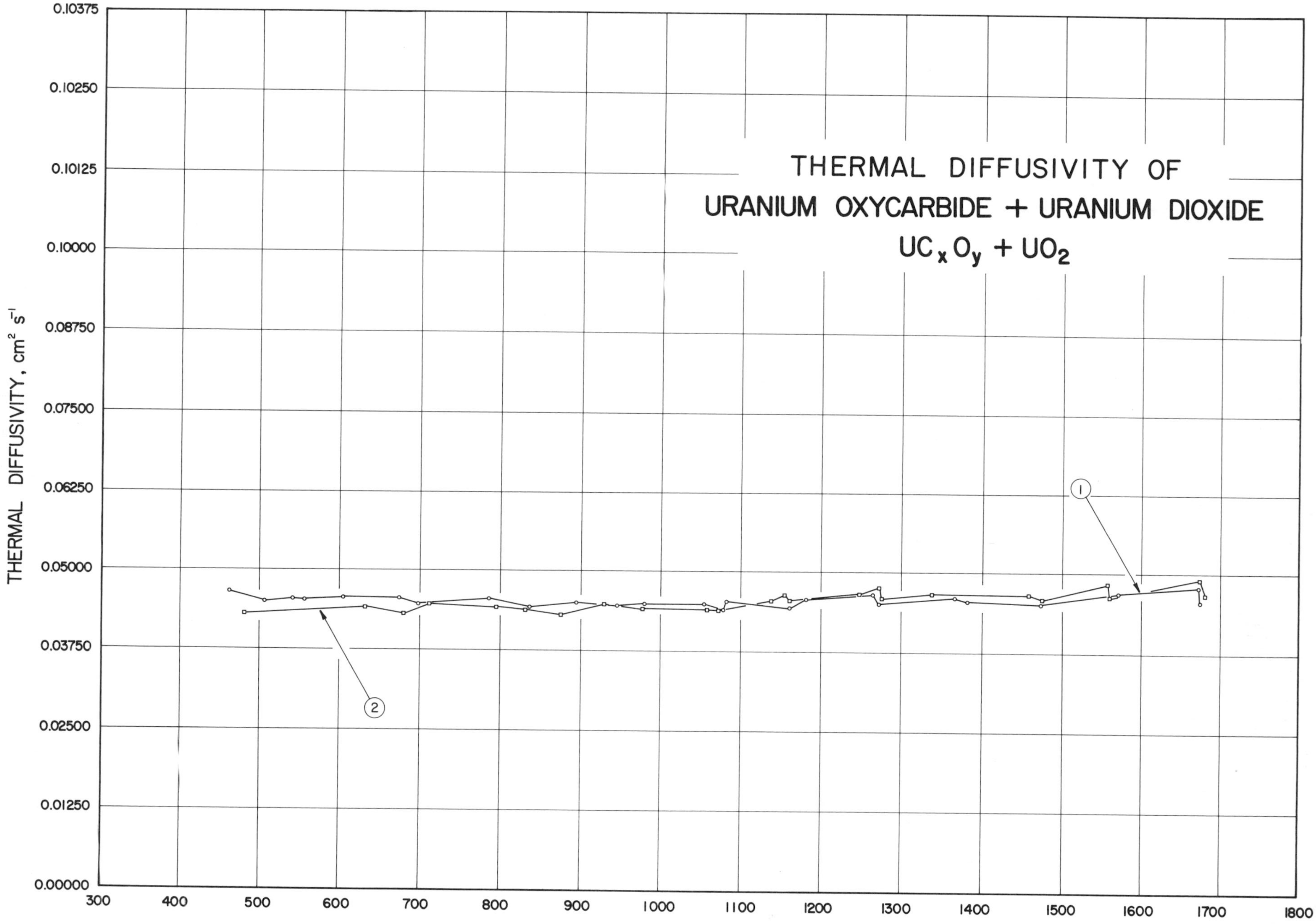

THERMAL DIFFUSIVITY OF
URANIUM OXYCARBIDE + URANIUM DIOXIDE
$UC_xO_y + UO_2$
THERMAL DIFFUSIVITY, $cm^2 s^{-1}$
TEMPERATURE, K
0.10375
0.10250
0.10125
0.10000
0.08750
0.07500
0.06250
0.05000
0.03750
0.02500
0.01250
0.00000
300
400
500
600
700
800
900
1000
1100
1200
1300
1400
1500
1600
1700
1800
1
2

FIGURE 180

SPECIFICATION TABLE 180. THERMAL DIFFUSIVITY OF [URANIUM OXYCARBIDE + URANIUM DIOXIDE] $UC_xO_y + UO_2$

Cur. No.	Ref. No.	Author(s)	Year	Temp. Range, K	Reported Error, %	Name and Specimen Designation	Composition (weight percent) UC_xO_y	UO_2	Composition (continued), Specifications, and Remarks
1	128	Bates, J. L.	1968	462-1675		$U_{0.495}C_{0.355}O_{0.15}$; UCON-288	→	→	$U_{0.495}C_{0.355}O_{0.15}$; 94.647 U, 3.425 C, and 1.928 O; two phase UC_xO_y and UO_2 with traces of free uranium; disk specimen 0.635 cm in dia. and ~0.100 cm thick; fabricated by and obtained from US Bureau of Mines, Albany Research Laboratory; prepared by reaction sintering of calculated quantities of uranium, graphite, and UO_2 powders; compacts pulverized and milled to a fine powder in an all-nickel mill under highly purified argon; pellets cold pressed and sintered at 1973.2 K for 2 hrs in an atmosphere of carbon monoxide at or near decomposition pressure for specimen; cut into thin disk from as-fabricated cylinder; density 12.5 g cm^{-3}; electrical resistivity reported as 238 μohm cm at 293.2 K; measured for diffusivity using laser-pulse technique; energy pulse of 0.54 m sec width provided by ruby laser; measured in a purified argon atmosphere; inlet argon containing < 0.0001 O and < 0.0005 H_2O; measured under a pressure of one atmosphere; heated initially to 1273.2 K in vertical tungsten tube furnace and measured during increase and decrease in temp, removed from furnace to check for possible reaction with UO_2 holder and then reinserted in furnace to carry out diffusivity measurements to ~1773.2 K; measured data corrected for heat losses.
2	128	Bates, J. L.	1968	491-1681		$U_{0.495}C_{0.335}O_{0.17}$; UCON-289	→	→	$U_{0.495}C_{0.355}O_{0.17}$; 94.586 U, 3.230 C, and 2.183 O; two phase UC_xO_y and UO_2 with traces of free uranium; disk specimen 0.635 cm in dia. and ~0.100 cm thick; fabricated by and obtained from US Bureau of Mines, Albany Research Laboratory; prepared by reaction sintering of calculated quantities of uranium, graphite, and UO_2 powders; compacts pulverized and milled to a fine powder in an all-nickel mill under highly purified argon; pellets cold pressed and sintered at 1973.2 K for 2 hrs in an atmosphere of carbon monoxide at or near decomposition pressure for specimen; cut into thin disk from as-fabricated cylinder; density 12.3 g cm^{-3}; measured for diffusivity using a laser-pulse technique; energy pulse of 0.54 m sec width provided by ruby laser; measured in a purified argon atmosphere; inlet argon containing < 0.0001 O and < 0.0005 H_2O; measured under a pressure of one atmosphere; heated initially to 1273.2 K in vertical tungsten tube furnace and measured during increase and decrease in temp, removed from furnace to check for possible reaction with UO_2 holder and then reinserted in furnace to carry out diffusivity measurements to ~1773.2 K; measured data corrected for heat losses.

DATA TABLE 180. THERMAL DIFFUSIVITY OF [URANIUM OXYCARBIDE + URANIUM DIOXIDE] $UC_xO_y + UO_2$

[Temperature, T, K; Thermal Diffusivity, α, $cm^2\ s^{-1}$]

T	α	T	α
CURVE 1		CURVE 2 (cont.)	
462	0.0465	1338	0.0466
507	0.0448	1459	0.0466
542	0.0453	1475	0.0458
557	0.0452	1558	0.0484
605	0.0456	1560	0.0463
675	0.0456	1675	0.0491
698	0.0446	1681	0.0467
786	0.0454		
837	0.0442		
895	0.0448		
946	0.0445		
980	0.0448		
1054	0.0448		
1078	0.0440		
1082	0.0453		
1162	0.0443		
1192	0.0456		
1265	0.0465		
1273	0.0451		
1369	0.0461		
1382	0.0455		
1474	0.0450		
1572	0.0468		
1673	0.0478		
1675	0.0457		
CURVE 2			
491	0.0429		
632	0.0440		
680	0.0430		
716	0.0446		
795	0.0441		
831	0.0437		
875	0.0442		
929	0.0447		
977	0.0440		
1058	0.0438		
1073	0.0448		
1138	0.0454		
1155	0.0463		
1161	0.0454		
1248	0.0465		
1273	0.0475		
1276	0.0458		

SPECIFICATION TABLE 181. THERMAL DIFFUSIVITY OF [ZIRCONIUM DIOXIDE + CALCIUM OXIDE] $ZrO_2 + CaO$

Cur. No.	Ref. No.	Author(s)	Year	Temp. Range, K	Reported Error, %	Name and Specimen Designation	Composition (weight percent) ZrO_2	Composition (weight percent) CaO	Composition (continued), Specifications, and Remarks
1*	52	Faucher, M., Cabannes, F., Anthony, A.-M., Piriou, B., and Simonato, J.	1970	2008-2368			97	3	5 mm in diameter and 1 to 1.5 mm thick; density 4.7 g cm^{-3}.

DATA TABLE 181. THERMAL DIFFUSIVITY OF [ZIRCONIUM DIOXIDE + CALCIUM OXIDE] $ZrO_2 + CaO$

[Temperature, T, K; Thermal Diffusivity, α, $cm^2\ s^{-1}$]

T	α
CURVE 1*	
2008	0.00560
2304	0.00608
2368	0.00639

* No figure given.

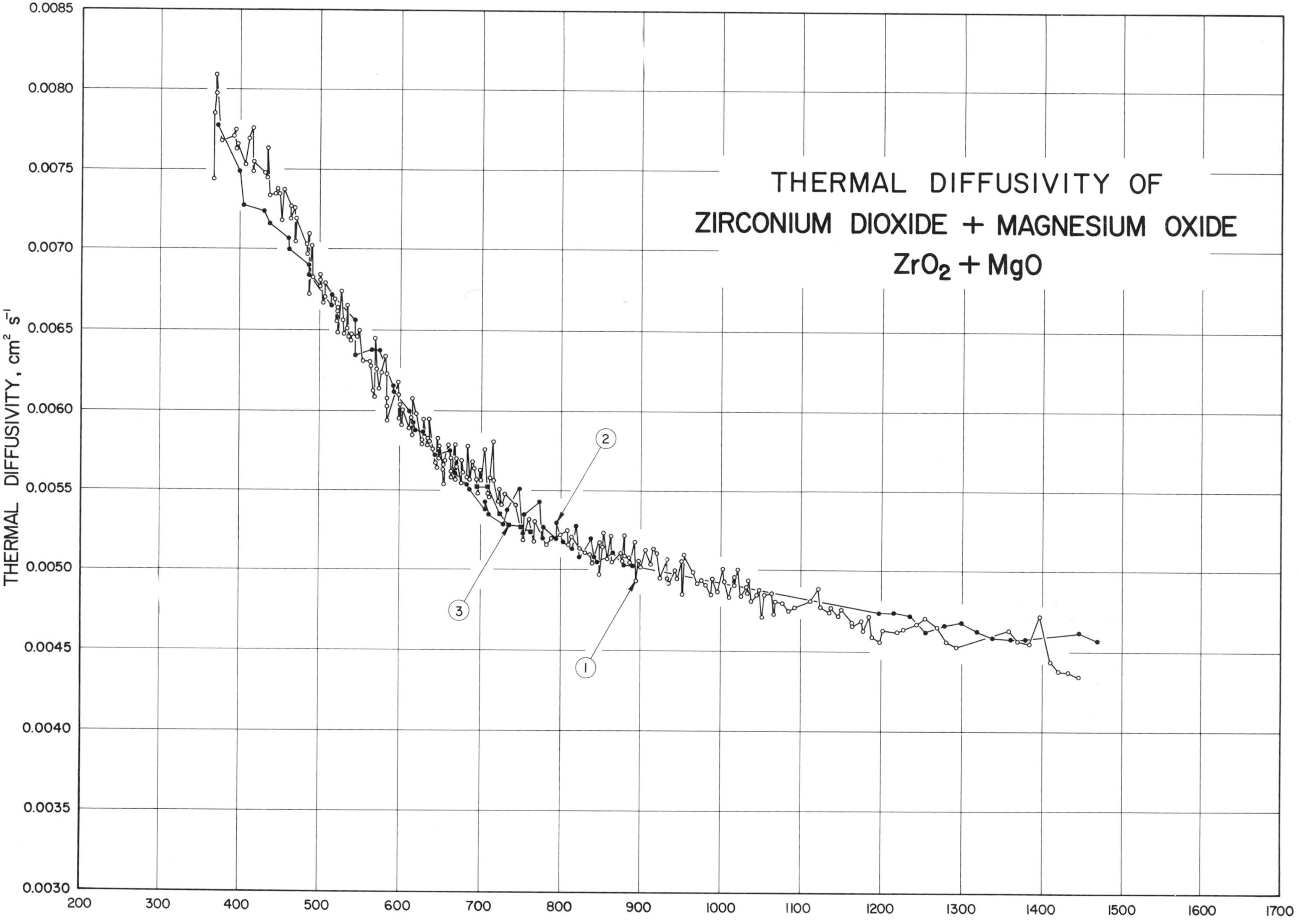
THERMAL DIFFUSIVITY OF
ZIRCONIUM DIOXIDE + MAGNESIUM OXIDE
$ZrO_2 + MgO$
THERMAL DIFFUSIVITY, $cm^2\ s^{-1}$
0.0085
0.0080
0.0075
0.0070
0.0065
0.0060
0.0055
0.0050
0.0045
0.0040
0.0035
0.0030
200
300
400
500
600
700
800
900
1000
1100
1200
1300
1400
1500
1600
1700
TEMPERATURE, K
1
2
3

FIGURE 182

SPECIFICATION TABLE 182. THERMAL DIFFUSIVITY OF [ZIRCONIUM DIOXIDE + MAGNESIUM OXIDE] $ZrO_2 + MgO$

Cur. No.	Ref. No.	Author(s)	Year	Temp. Range, K	Reported Error, %	Name and Specimen Designation	Composition (weight percent) ZrO_2	Composition (weight percent) MgO	Composition (continued), Specifications, and Remarks
1	96	Flieger, H.W., Jr., and Ginnings, D.C.	1957	366-1445		E-109	95.5/ 95.7	4.1	(by chemical analysis), 0.01-0.1 CaO, SiO_2, TiO_2 each, 0.005-0.05 Al_2O_3, 0.001-0.01 Fe_2O_3, and <0.005 BaO, K_2O, Li_2O, Na_2O each (by spectrochemical analysis); quantitative spectrochemical analysis at NBS showed metal impurities equivalent to: 0.41 HfO_2, 0.36 SiO_2, 0.07 TiO_2, and 0.01 or less Al_2O_3, Cr_2O_3, Fe_2O_3, MnO, NiO, PbO each; cylindrical specimen ~6 in. long; diffusivity measured using a total length of ~12 in., made up of four pieces, each 3 in. long; specimen described above cut into two pieces used in the middle portion of the 12 in. length, another specimen cut similarly into two 3 in. lengths for use at the ends; surfaces plated with nickel; estimated effective O.D. 4.609 cm and effective I.D. 1.525 cm; I.D. of nickel plating on the inside 1.362 cm near center of 12 in. length; inside plating ~1 cm in length tapering out to zero thickness; plating of outer surfaces performed after the four 3 in. lengths were assembled together with the thermocouple leads; supplied by Corning Glass Works; fired to a temperature no higher than 1823.2 K before supply; ground to have coaxial cylindrical surfaces; fired at ~1273.2 K for four hrs; bulk density 3.65 g cm^{-3} at 298.2 K corresponding to ~35% porosity, on the basis that the crystal density is ~5.6 g cm^{-3}; heated by surrounding graphite cylinder in which heat is developed by radio frequency induction; measured at normal heating rate at 30 $\mu v\ min^{-1}$; radial heat flow method used to measure diffusivity.
2	96	Flieger, H.W., Jr., and Ginnings, D.C.	1957	371-1469		E-109			Above specimen measured for diffusivity again at high heating rate at 46 $\mu v\ min^{-1}$.
3	96	Flieger, H.W., Jr., and Ginnings, D.C.	1957	695-762		E-109			Above specimen measured for diffusivity again at low heating rate at 22 $\mu v\ min^{-1}$.

DATA TABLE 182. THERMAL DIFFUSIVITY OF [ZIRCONIUM DIOXIDE + MAGNESIUM OXIDE] $ZrO_2 + MgO$

[Temperature, T, K; Thermal Diffusivity, α, $cm^2\ s^{-1}$]

T	α	T	α	T	α	T	α	T	α	T	α	T	α	T	α
CURVE 1		CURVE 1 (cont.)		CURVE 1 (cont.)		CURVE 1 (cont.)		CURVE 1 (cont.)		CURVE 1 (cont.)		CURVE 2 (cont.)		CURVE 3	
366.2	0.00744	521.2	0.00661	634.2	0.00583	730.2	0.00548	945.2	0.00495	1177.2	0.00463	616.2	0.00593	695.2	0.00552
367.2	0.00785	521.2	0.00664	636.2	0.00581	744.2	0.00541	951.2	0.00506	1184.2	0.00472	618.2	0.00588	708.2	0.00552
369.2	0.00798	521.2	0.00648	636.2	0.00595	753.2	0.00519	952.2	0.00485	1188.2	0.00459	628.2	0.00587	723.2	0.00535
369.2	0.00809	526.2	0.00674	640.2	0.00576	760.2	0.00532	954.2	0.00510	1198.2	0.00456	643.2	0.00572	735.2	0.00528
376.2	0.00768	528.2	0.00656	644.2	0.00567	766.2	0.00518	965.2	0.00499	1201.2	0.00463	661.2	0.00575	750.2	0.00527
391.2	0.00771	529.2	0.00648	646.2	0.00564	767.2	0.00531	971.2	0.00492	1219.2	0.00462	668.2	0.00561	762.2	0.00524
393.2	0.00775	533.2	0.00651	646.2	0.00583	782.2	0.00516	976.2	0.00494	1227.2	0.00464	683.2	0.00554		
394.2	0.00763	533.2	0.00665	648.2	0.00570	788.2	0.00520	981.2	0.00491	1243.2	0.00467	685.2	0.00551		
396.2	0.00766	535.2	0.00646	648.2	0.00578	800.2	0.00522	988.2	0.00485	1254.2	0.00471	703.2	0.00538		
406.2	0.00753	538.2	0.00644	653.2	0.00563	808.2	0.00525	989.2	0.00495	1269.2	0.00465	703.2	0.00543		
410.2	0.00769	539.2	0.00648	653.2	0.00566	809.2	0.00516	996.2	0.00487	1280.2	0.00456	710.2	0.00535		
415.2	0.00776	546.2	0.00646	654.2	0.00554	814.2	0.00521	1003.2	0.00501	1293.2	0.00453	728.2	0.00529		
415.2	0.00749	549.2	0.00650	656.2	0.00569	823.2	0.00514	1004.2	0.00493	1358.2	0.00463	733.2	0.00538		
416.2	0.00755	553.2	0.00631	660.2	0.00579	830.2	0.00511	1010.2	0.00483	1369.2	0.00457	748.2	0.00551		
430.2	0.00748	562.2	0.00631	663.2	0.00571	836.2	0.00510	1017.2	0.00496	1384.2	0.00455	753.2	0.00523		
433.2	0.00745	563.2	0.00628	663.2	0.00562	839.2	0.00504	1017.2	0.00491	1396.2	0.00472	754.2	0.00535		
433.2	0.00764	566.2	0.00613	663.2	0.00558	848.2	0.00518	1021.2	0.00500	1410.2	0.00444	773.2	0.00543		
436.2	0.00734	568.2	0.00609	668.2	0.00579	848.2	0.00497	1025.2	0.00484	1420.2	0.00438	777.2	0.00520		
443.2	0.00735	568.2	0.00645	668.2	0.00557	852.2	0.00515	1032.2	0.00490	1433.2	0.00438	778.2	0.00527		
446.2	0.00738	570.2	0.00626	670.2	0.00565	853.2	0.00524	1033.2	0.00486	1445.2	0.00435	794.2	0.00520		
449.2	0.00735	573.2	0.00614	670.2	0.00570	858.2	0.00507	1034.2	0.00494			794.2	0.00530		
451.2	0.00718	576.2	0.00624	676.2	0.00554	863.2	0.00522	1038.2	0.00481	CURVE 2		803.2	0.00518		
454.2	0.00738	581.2	0.00634	676.2	0.00569	864.2	0.00505	1045.2	0.00485			814.2	0.00513		
463.2	0.00727	583.2	0.00623	678.2	0.00561	875.2	0.00511	1048.2	0.00488	371.2	0.00778	819.2	0.00528		
463.2	0.00719	583.2	0.00608	683.2	0.00558	877.2	0.00506	1051.2	0.00471	398.2	0.00749	823.2	0.00508		
468.2	0.00726	583.2	0.00603	683.2	0.00578	879.2	0.00522	1054.2	0.00485	403.2	0.00727	838.2	0.00520		
468.2	0.00705	583.2	0.00594	686.2	0.00557	880.2	0.00509	1063.2	0.00486	429.2	0.00724	843.2	0.00508		
470.2	0.00719	597.2	0.00618	690.2	0.00568	886.2	0.00508	1066.2	0.00473	436.2	0.00716	846.2	0.00505		
483.2	0.00703	598.2	0.00610	692.2	0.00564	886.2	0.00505	1068.2	0.00481	460.2	0.00707	865.2	0.00511		
483.2	0.00697	598.2	0.00595	695.2	0.00557	892.2	0.00518	1076.2	0.00480	460.2	0.00700	879.2	0.00503		
485.2	0.00710	599.2	0.00604	696.2	0.00548	894.2	0.00493	1085.2	0.00475	485.2	0.00690	890.2	0.00503		
486.2	0.00672	601.2	0.00591	699.2	0.00563	897.2	0.00506	1092.2	0.00477	486.2	0.00684	1197.2	0.00474		
489.2	0.00702	603.2	0.00601	700.2	0.00556	900.2	0.00502	1111.2	0.00481	513.2	0.00665	1215.2	0.00474		
490.2	0.00682	611.2	0.00589	705.2	0.00576	906.2	0.00513	1121.2	0.00489	514.2	0.00672	1235.2	0.00472		
500.2	0.00677	613.2	0.00596	709.2	0.00548	913.2	0.00504	1124.2	0.00477	543.2	0.00656	1255.2	0.00462		
500.2	0.00681	615.2	0.00585	711.2	0.00546	916.2	0.00514	1135.2	0.00474	543.2	0.00635	1278.2	0.00466		
500.2	0.00684	615.2	0.00608	712.2	0.00558	920.2	0.00511	1137.2	0.00477	564.2	0.00638	1299.2	0.00468		
503.2	0.00667	620.2	0.00598	715.2	0.00581	924.2	0.00495	1146.2	0.00472	573.2	0.00638	1318.2	0.00463		
506.2	0.00670	627.2	0.00579	716.2	0.00556	933.2	0.00507	1150.2	0.00476	591.2	0.00616	1337.2	0.00459		
506.2	0.00679	627.2	0.00582	722.2	0.00543	933.2	0.00495	1163.2	0.00468	591.2	0.00612	1360.2	0.00458		
518.2	0.00669	629.2	0.00595	723.2	0.00551	935.2	0.00492	1164.2	0.00466	597.2	0.00615*	1378.2	0.00458		
520.2	0.00655	634.2	0.00578	726.2	0.00541	943.2	0.00500	1175.2	0.00469	611.2	0.00600	1446.2	0.00462		
												1469.2	0.00457		

*Not shown in figure.

SPECIFICATION TABLE 183. THERMAL DIFFUSIVITY OF [ZIRCONIUM DIOXIDE + YTTRIUM OXIDE] $ZrO_2 + Y_2O_3$

Cur. No.	Ref. No.	Author(s)	Year	Temp. Range, K	Reported Error, %	Name and Specimen Designation	Composition (weight percent) ZrO_2	Y_2O_3	Composition (continued), Specifications, and Remarks
1*	52	Faucher, M., Cabannes, F., Antony, A.-M., Pirou, B., and Simonato, J.	1970	1765-2273			80	20	5 mm in diameter and 1 to 1.5 mm thick; density 4.5 g cm^{-3}.

DATA TABLE 183. THERMAL DIFFUSIVITY OF [ZIRCONIUM DIOXIDE + YTTRIUM OXIDE] $ZrO_2 + Y_2O_3$

[Temperature, T, K; Thermal Diffusivity, α, $cm^2\ s^{-1}$]

T	α
CURVE 1*	
1765	0.00400
1849	0.00400
2028	0.00462
2273	0.00407

* No figure given.

9. NONOXIDE COMPOUNDS

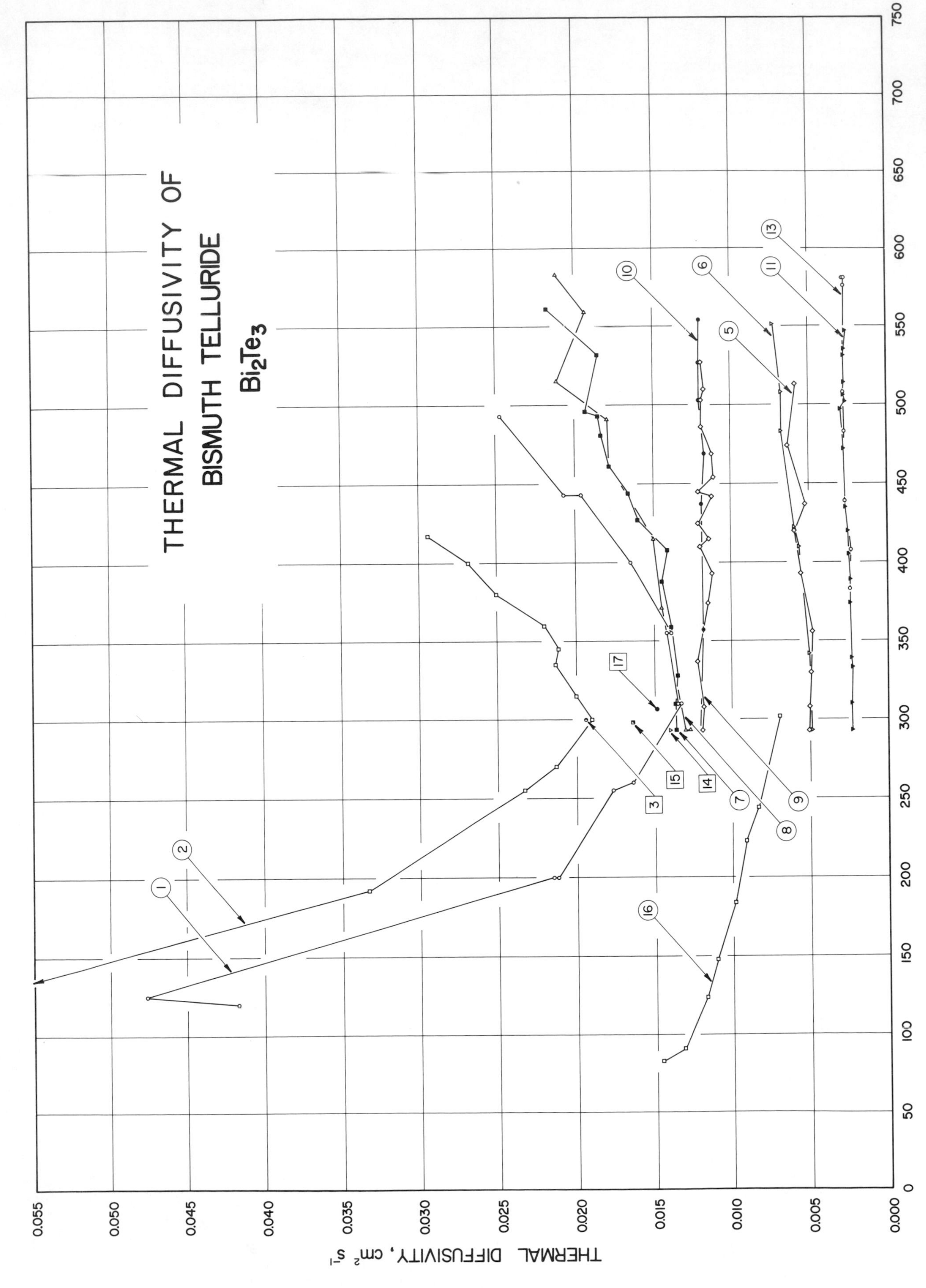
THERMAL DIFFUSIVITY OF
BISMUTH TELLURIDE
Bi_2Te_3
THERMAL DIFFUSIVITY, $cm^2 s^{-1}$
TEMPERATURE, K
0.000
0.005
0.010
0.015
0.020
0.025
0.030
0.035
0.040
0.045
0.050
0.055
0
50
100
150
200
250
300
350
400
450
500
550
600
650
700
750
1
2
3
5
6
7
8
9
10
11
13
14
15
16
17

FIGURE 184

SPECIFICATION TABLE 184. THERMAL DIFFUSIVITY OF BISMUTH TELLURIDE Bi_2Te_3

Cur. No.	Ref. No.	Author(s)	Year	Temp. Range, K	Reported Error, %	Name and Specimen Designation	Composition (weight percent), Specifications, and Remarks
1	61	Kanai, Y. and Nii, R.	1959	120-493			p-type single crystal; hole concentration 2 x 10^{19} hole cm^{-3}; improved Angstrom's method used to determine diffusivity.
2	62	Nii, R.	1958	115-417			Diffusivity measured by Angstrom's method.
3	63	Pinnow, D.A., Li, C.Y., and Spencer, C.W.	1961	300			n-type; cylindrical specimen 0.685 cm in diameter and 1.42 cm long; electrical resistivity 0.485 x 10^{-3} ohm cm; specimen suspended in an evacuated chamber by current lead wires; specimen ends nickel plated; heat flow parallel to the cleavage plane.
4*	63	Pinnow, D.A., et al.	1961	300			Above specimen measured again after its cross section had been reduced to a rectangle 0.5 x 0.4 cm.
5	64	Parker, W.J. and Jenkins, R.J.	1960	293-513	±10	59	p-type; rectangular specimen 0.089 cm thick; electrical resistivity ranging from 13.0 to 14.6 x 10^{-4} ohm cm as quoted by manufacturer for the boule from which specimen was cut; specimen measured while being irradiated by a 1.95 Mev electron beam; beam enters upper surface of specimen while a strong horizontal flow of air passes across it; high intensity short duration light pulse absorbed in the same surface and resulting temperature history of lower surface measured; diffusivity measured across crystal planes; front surface of sample blackened with camphor black and back surface electrodeposited with a layer of nickel from 0.010 to 0.015 cm thick for the next series of measurements.
6	64	Parker, W.J. and Jenkins, R.J.	1960	293-551	±10	59	Above specimen measured again in the presence of low beam currents and no forced flow of cooling air; specimen received a high intensity light pulse from xenon flash lamp at front surface and back surface temperature history was recorded.
7	64	Parker, W.J. and Jenkins, R.J.	1960	293-561	±10	64	p-type; rectangular specimen 0.043 cm thick; electrical resistivity ranging from 8.1 to 11.1 x 10^{-4} ohm cm as quoted by manufacturer for the boule from which specimen was cut; specimen measured while being irradiated by a 1.95 Mev electron beam; beam enters upper surface of specimen while a strong horizontal flow of air passes across it; high intensity short duration light pulse absorbed in the same surface and resulting temperature history of lower surface measured; diffusivity measured across crystal planes; front surface of sample blackened with camphor black and back surface electrodeposited with a layer of nickel from 0.010 to 0.015 cm thick for the next series of measurements.
8	64	Parker, W.J. and Jenkins, R.J.	1960	293-583	±10	64	Above specimen measured again in the presence of low beam currents and no forced flow of cooling air; specimen received a high intensity light pulse from xenon flash lamp at front surface and back surface temperature history was recorded.

* Not shown in figure.

SPECIFICATION TABLE 184. THERMAL DIFFUSIVITY OF BISMUTH TELLURIDE Bi_2Te_3 (continued)

Cur. No.	Ref. No.	Author(s)	Year	Temp. Range, K	Reported Error, %	Name and Specimen Designation	Composition (weight percent), Specifications, and Remarks
9	64	Parker, W.J. and Jenkins, R.J.	1960	293-527	±10	68	p-type; rectangular specimen 0.105 cm thick; electrical resistivity ranging from 4.0 to 6.4 x 10^{-4} ohm cm as quoted by manufacturer for the boule from which specimen was cut; specimen measured while being irradiated by a 1.95 Mev electron beam; beam enters upper surface of specimen while a strong horizontal flow of air passes across it; high intensity short duration light pulse absorbed in the same surface and resulting temperature history of lower surface measured; diffusivity measured in the direction of crystal planes; front surface of sample blackened with camphor black and back surface electrodeposited with a layer of nickel from 0.010 to 0.015 cm thick for the next series of measurements.
10	64	Parker, W.J. and Jenkins, R.J.	1960	293-554	±10	68	Above specimen measured again in the presence of low beam currents and no forced flow of cooling air; specimen received a high intensity light pulse from xenon flash lamp at front surface and back surface temperature history was recorded.
11	64	Parker, W.J. and Jenkins, R.J.	1960	293-547	±10	69	n-type; rectangular specimen 0.064 cm thick; electrical resistivity 6.5 x 10^{-4} ohm cm as quoted by manufacturer for the boule from which specimen was cut; specimen measured while being irradiated by a 1.95 Mev electron beam; beam enters upper surface of specimen while a strong horizontal flow of air passes across it; high intensity short duration light pulse absorbed in the same surface and resulting temperature history of lower surface measured; diffusivity measured across crystal planes; front surface of sample blackened with camphor black and back surface electrodeposited with a layer of nickel from 0.010 to 0.015 cm thick for the next series of measurements.
12*	64	Parker, W.J. and Jenkins, R.J.	1960	293-523	±10	69	Above specimen measured again in the presence of low beam currents and no forced flow of cooling air; specimen received a high intensity light pulse from xenon flash lamp at front surface and back surface temperature history was recorded.
13	64	Parker, W.J. and Jenkins, R.J.	1960	383-581	±10	69	Above specimen measured again in the absence of electron beam and was heated by hot air, no forced air cooling was used; specimen received a high intensity light pulse from xenon flash lamp at front surface and back surface temperature history was recorded.
14	65	Green, A. and Cowles, L.E.J.	1960	293.2	3.7		Undoped single crystal; rectangular specimen 0.3 x 0.4 x 5.0 cm, cleavage planes running parallel to its length; temperature wave generated by periodic reversal of current passing through thermojunction to which one nickel plated end face of specimen was tinned and soldered; specimen assembly mounted in vacuum chamber.
15	66	Anger, H., Baumberger, C., and Guennoc, H.	1962	298.2			Cylindrical specimen; partially fused; Angstrom's dynamic method used to measure diffusivity; specimen measured twice: once under normal atmospheric pressure and once in vacuum; sinusoidal temperature wave imposed on one end of specimen while the other end maintained at a constant temperature equivalent to the ambient temperature of measurement; latter temperature not given by author but assumed to be room temperature.

* Not shown in figure.

SPECIFICATION TABLE 184. THERMAL DIFFUSIVITY OF BISMUTH TELLURIDE Bi_2Te_3 (continued)

Cur. No.	Ref. No.	Author(s)	Year	Temp. Range, K	Reported Error, %	Name and Specimen Designation	Composition (weight percent), Specifications, and Remarks
16	138	Macqueron, J.-L., Sinicki, G., Durand, G., and Rinaldi, D.	1967	83-302		Bi_2Te_3 dopé N	n-type; doped; flat specimen; annealed; one face exposed to thermal pulse of short duration generated by a flash tube; diffusivity determined from temperature evolution of rear face measured using a thermoelectric couple with a large figure of merit; measured in a vacuum of 10^{-3} mm Hg.
17	179	Pollak, P.I., Conn, J.B., Taylor, R.C., Sheehan, E.J., and Kirby, J.J.	1961	305.7			n-type; cylindrical specimen; Angström method used to measure diffusivity; diffusivity determined from measured amplitude ratio and phase shift of sinusoidal heat wave impressed on one end of specimen; measured in vacuum; radiation shields used; temperature of measurement reported by authors as being about 30-35 C.

DATA TABLE 184. THERMAL DIFFUSIVITY OF BISMUTH TELLURIDE Bi_2Te_3

[Temperature, T, K; Thermal Diffusivity, α, $cm^2 s^{-1}$]

T	α
CURVE 1	
120	0.0417
125	0.0476
200	0.0215
200	0.0202
255	0.0177
260	0.0164
310	0.0133
310	0.0135
355	0.0142
355	0.0139
400	0.0165
443	0.0196
443	0.0207
493	0.0247
CURVE 2	
115	0.0625*
192	0.0333
255	0.0233
270	0.0213
300	0.0190
315	0.0200
335	0.0213
345	0.0211
360	0.0220
380	0.0250
400	0.0268
417	0.0294
CURVE 3	
300	0.0194
CURVE 4*	
300	0.0196
CURVE 5	
293.2	0.0050
308.2	0.0050
330.2	0.0049
356.2	0.0048
393.2	0.0055
420.2	0.0059
437.2	0.0052
474.2	0.0063
513.2	0.0058
CURVE 6	
293.2	0.0049
342.2	0.0050
410.2	0.0056
422.2	0.0059
483.2	0.0067
508.2	0.0067
551.2	0.0072
CURVE 7	
293.2	0.0136
310.2	0.0136
328.2	0.0135
359.2	0.0139
388.2	0.0145
408.2	0.0141
427.2	0.0160
444.2	0.0166
461.2	0.0178
481.2	0.0183
493.2	0.0185
496.2	0.0193
532.2	0.0185
561.2	0.0217
CURVE 8	
293.2	0.0127
293.2	0.0130
371.2	0.0145
415.2	0.0150
461.2	0.0179*
491.2	0.0179
515.2	0.0211
559.2	0.0193
583.2	0.0211
CURVE 9	
293.2	0.0119
308.2	0.0118
337.2	0.0122
374.2	0.0115
393.2	0.0112
410.2	0.0120
415.2	0.0114
425.2	0.0121
442.2	0.0112
445.2	0.0121
454.2	0.0111
469.2	0.0112
486.2	0.0119
503.2	0.0119
510.2	0.0117
527.2	0.0119
CURVE 10	
293.2	0.0121*
357.2	0.0118
437.2	0.0119
469.2	0.0117
503.2	0.0120
554.2	0.0120
CURVE 11	
293.2	0.00225
310.2	0.00228
333.2	0.00227
339.2	0.00228
374.2	0.00235
389.2	0.00235
405.2	0.00245
420.2	0.00250
435.2	0.00264
472.2	0.00274
497.2	0.00288
502.2	0.00258
506.2	0.00271
514.2	0.00270
531.2	0.00272
535.2	0.00271
547.2	0.00262
CURVE 12*	
293.2	0.00223
446.2	0.00258
523.2	0.00274
CURVE 13	
383.2	0.00243
408.2	0.00234
439.2	0.00266
483.2	0.00270
508.2	0.00275
547.2	0.00274*
576.2	0.00270
581.2	0.00268
581.2	0.00272
CURVE 14	
293.2	0.014
CURVE 15	
298.2	0.0164
CURVE 16	
83	0.0146
91	0.0132
124	0.0117
148	0.0110
184	0.0098
223	0.0091
244	0.0083
302	0.0069
CURVE 17	
307.7	0.0148

*Not shown in figure.

SPECIFICATION TABLE 185. THERMAL DIFFUSIVITY OF BORON CARBIDE B_4C

Cur. No.	Ref. No.	Author(s)	Year	Temp. Range, K	Reported Error, %	Name and Specimen Designation	Composition (weight percent), Specifications, and Remarks
1*	74	Hedge, J.C., Kopec, J.W., Kostenko, C., and Lang, J.I.	1963	1168-2583			75.97 B, 21.18 C, 0.40 Si, 0.27 Fe, 0.07 B_2O_7, and 0.015 Al_2O_3; slab specimen; supplied by Carborundum Co.; hot pressed; fired at 2444 K; density 2.50 g cm^{-3}; top surface of specimen exposed to heat sink; diffusivity determined from measured temperature decrease; unidirectional heat flow.

DATA TABLE 185. THERMAL DIFFUSIVITY OF BORON CARBIDE B_4C

[Temperature, T, K; Thermal Diffusivity, α, cm^2 s^{-1}]

T	α
CURVE 1*	
1168.2	0.0305
1313.7	0.0286
1502.6	0.0258
1805.4	0.0245
1924.8	0.0243
2080.4	0.0248
2210.9	0.0250
2347.1	0.0253
2472.1	0.0256
2582.5	0.0261

* No figure given.

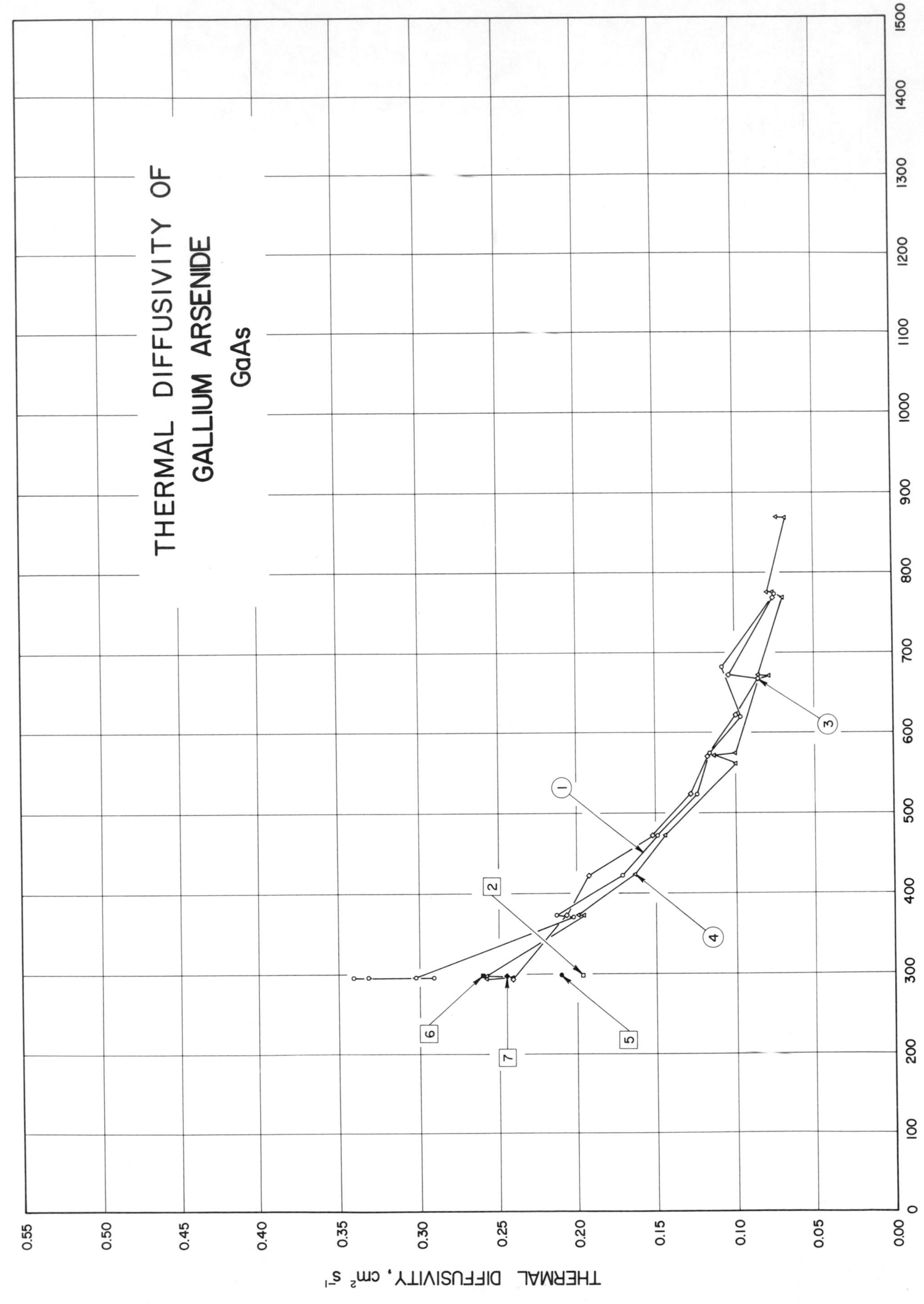
THERMAL DIFFUSIVITY OF
GALLIUM ARSENIDE
GaAs
TEMPERATURE, K
THERMAL DIFFUSIVITY, cm² s⁻¹
0 100 200 300 400 500 600 700 800 900 1000 1100 1200 1300 1400 1500
0.00 0.05 0.10 0.15 0.20 0.25 0.30 0.35 0.40 0.45 0.50 0.55
1
2
3
4
5
6
7

FIGURE 186

SPECIFICATION TABLE 186. THERMAL DIFFUSIVITY OF GALLIUM ARSENIDE GaAs

Cur. No.	Ref. No.	Author(s)	Year	Temp. Range, K	Reported Error, %	Name and Specimen Designation	Composition (weight percent), Specifications, and Remarks
1	67	Timberlake, A. B., Davis, P. W., and Shilliday, T. S.	1961	296-773		16594-272a	Electron concentration 4.3 x 10^{17} electron cm^{-3}; specimen prepared in the shape of a rectangular parallelepiped with initial dimensions 1.0 x 10 x 0.149 cm, for measurements at higher temperatures sample was tapped to 0.107 cm; density 5.31 g cm^{-3}; specific heat 0.347 J g^{-1} K^{-1}; Hall coefficient -14.6 cm^3 $coulomb^{-1}$; electrical resistivity 4.1 x 10^{-3} ohm cm; Hall mobility 3600 cm^2 v^{-1} sec^{-1}; back surface temperature method used to measure diffusivity.
2	68	Timberlake, A. B., Davis, P. W., and Shilliday, T. S.	1960	298.2			Specimen cut and lapped in the form of a rectangular parallelepiped 0.124 cm thick, and front surface sandblasted; specimen exposed to 3200 K tungsten lamp before being measured for the second time; specimen thickness reduced to 0.119 cm and front surface blackened with flat black enamel; pure In contact ultrasonically soldered to specimen; specimen exposed again to the same radiation source to measure diffusivity; back surface temperature method used.
3	145	Timberlake, A. B., et al.	1962	293-769			n-type; electron concentration 4.3 x 10^{17} electron cm^{-3}; square specimen 1 cm x 1 cm x 0.107 cm; chopped beam of radiation from an incandescent lamp focused on front face of specimen; radiation filtered using a Corning No. 9788 filter in conjunction with a 1 in. water cell; diffusivity determined from measured variation of the r.m.s. magnitude of the thermoelectric voltage generated at the back surface over a range of chopping frequencies; data points reported calculated from thermal conductivity data reported by author divided by a specific heat of 0.347 J g^{-1} K^{-1} and a density of 5.31 g cm^{-3}.
4	145	Timberlake, A. B., et al.	1962	294-870			Above specimen lapped to a thickness of 0.0549 cm and front surface coated with a thin layer of aquadag; measured for diffusivity again; same experimental technique and same data reduction method as above.
5	174	Davis, P. W., Timberlake, A. B., and Shilliday, T. S.	1962	298.2			Slab specimen 0.101 cm thick; front face illuminated by chopped light from an incandescent source; diffusivity determined from electrically detected thermal signal at the back face of specimen; radiation filtered using 0.2 cm of water and a Corning No. 9788 filter; temperature of measurement not given but assumed to be room temperature.
6	174	Davis, P. W., et al.	1962	298.2			Thermal diffusivity determined again for above specimen from above measurement using a different expression for data reduction.
7	174	Davis, P. W., et al.	1962	298.2			Thermal diffusivity determined again for above specimen using properly filtered light and assuming that all incident light is absorbed and converted into heat at the front face.

DATA TABLE 186. THERMAL DIFFUSIVITY OF GALLIUM ARSENIDE GaAs

[Temperature, T, K; Thermal Diffusivity, α, $cm^2\ s^{-1}$]

T	α
CURVE 1	
296	0.341
296	0.291
296	0.333
296	0.302
371	0.202
373	0.213
422	0.171
472	0.149
523	0.124
574	0.116
619	0.0965
682	0.108
773	0.0745
CURVE 2	
298.2	0.196
CURVE 3	
293.3	0.241
373.1	0.206
421.9	0.192
471.7	0.152
523.6	0.128
571.4	0.117
621.1	0.0999
666.7	0.0858
671.1	0.104
769.2	0.0760
CURVE 4	
294.1	0.258
295.9	0.241
297.6	0.258
373.1	0.195
373.1	0.199
423.7	0.163
471.7	0.144
561.8	0.0999
571.4	0.113
574.7	0.0999
666.7	0.0858*
671.1	0.0787
671.1	0.0858
CURVE 4 (cont.)	
769.2	0.0695
775.2	0.0760
775.2	0.0798
869.6	0.0678
869.6	0.0733
CURVE 5	
298.2	0.21
CURVE 6	
298.2	0.26
CURVE 7	
298.2	0.245

* Not shown in figure.

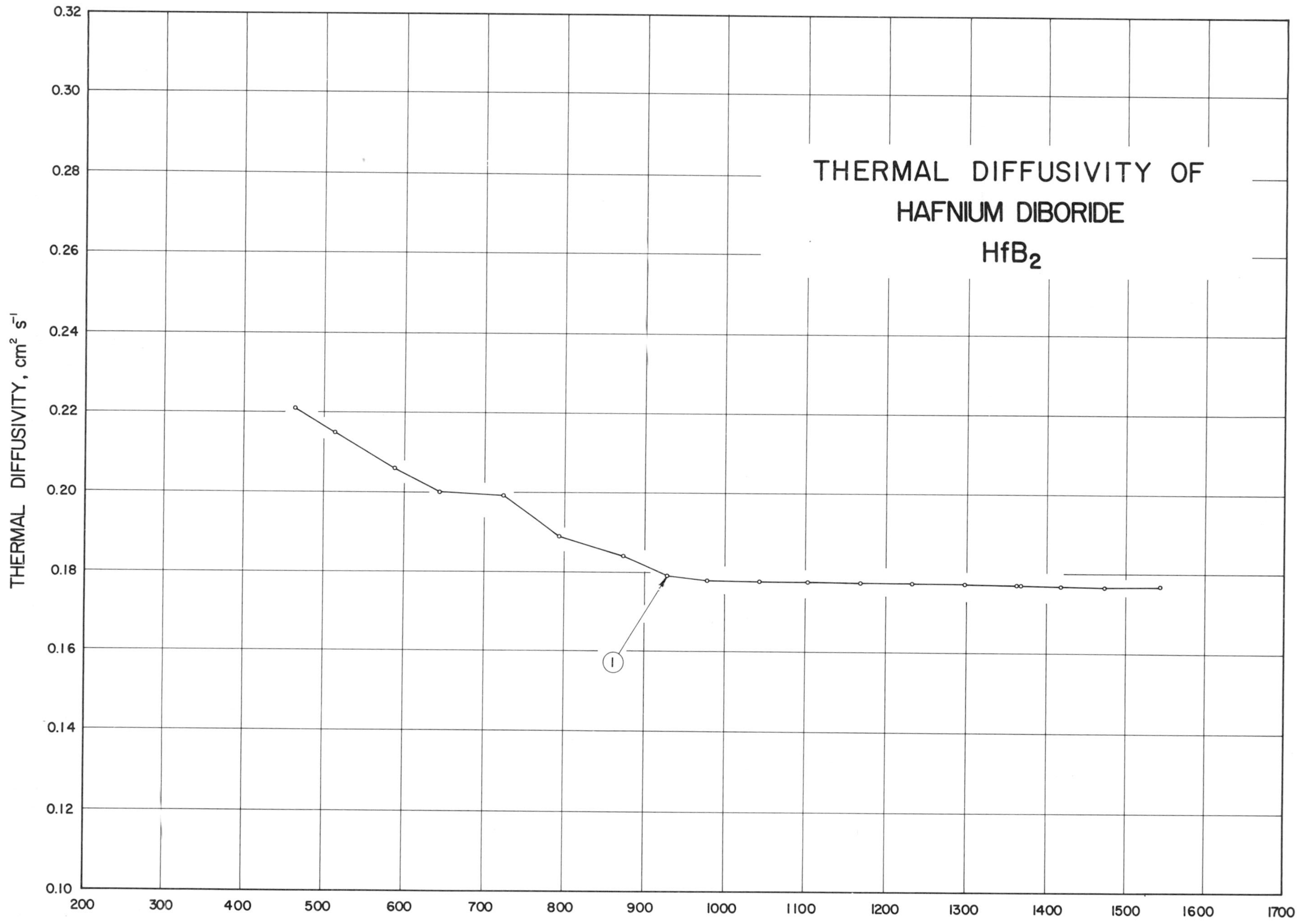
THERMAL DIFFUSIVITY OF
HAFNIUM DIBORIDE
HfB_2
THERMAL DIFFUSIVITY, $cm^2 s^{-1}$
0.10
0.12
0.14
0.16
0.18
0.20
0.22
0.24
0.26
0.28
0.30
0.32
TEMPERATURE, K
200
300
400
500
600
700
800
900
1000
1100
1200
1300
1400
1500
1600
1700
1

FIGURE 187

SPECIFICATION TABLE 187. THERMAL DIFFUSIVITY OF HAFNIUM DIBORIDE HfB_2

Cur. No.	Ref. No.	Author(s)	Year	Temp. Range, K	Reported Error, %	Name and Specimen Designation	Composition (weight percent), Specifications, and Remarks
1	235	Branscomb, T.M.	1970	463-1543			99 pure; 0.240 cm thick; obtained from Carborundum Co.; hot-pressed; Pt-black coated on front face; density 10.640 g cm^{-3}.

DATA TABLE 187. THERMAL DIFFUSIVITY OF HAFNIUM DIBORIDE HfB_2

[Temperature, T, K; Thermal Diffusivity, α, $cm^2\ s^{-1}$]

T	α
CURVE 1	
463	0.221
513	0.215
588	0.206
643	0.200
723	0.193
793	0.189
873	0.184
928	0.179
978	0.177
1043	0.177
1103	0.176
1168	0.175
1233	0.174
1298	0.171
1363	0.171
1368	0.171
1418	0.169
1473	0.168
1543	0.169

SPECIFICATION TABLE 188. THERMAL DIFFUSIVITY OF HAFNIUM CARBIDE HfC

Cur. No.	Ref. No.	Author(s)	Year	Temp. Range, K	Reported Error, %	Name and Specimen Designation	Composition (weight percent), Specifications, and Remarks
1*	32	Taylor, R. E. and Nakata, M. M.	1963	1813-2568	7		92.5 Hf, 5.5 C, 2 Zr, <0.1 V, 0.05 Ca, 0.05 Ti, <0.05 Fe, <0.05 Mo, 0.02 Si, 0.01 Ag, 0.01 B, and 0.01 Mg (as received); average grain size (ASTM) No. 7; cylindrical specimen 1.588 cm overall diameter and 3.449 cm overall length, machined in three sections: a center section 2.54 cm long and two end pieces consisting of 0.455 cm thick discs; these parts are machine fitted in such a way that a small area contact is made at the circumference only, leaving a thin space of 0.0127 cm or less between center section and the discs that act as radiation barriers; two parallel sight holes each 0.118 cm in diameter drilled through the top disc to a depth of 1.75 cm at radii $r_1 = 0$ and $r_2 = 0.562$ cm; sight holes drilled by Elox technique; bottom disc grooved to fit sample support pins; obtained from the Carborundum Co.; hot pressed from raw powder having chemical composition: 94.3 Hf + Zr, 5.52 C, 0.4 O, 2.4 Zr; measured after extensive heat soaking at temperatures up to 2673.2 K; density 11.90 g cm^{-3}; radiation shields mounted at both ends; measured under a vacuum of 1×10^{-6} mm Hg; radial diffusivity technique used.

DATA TABLE 188. THERMAL DIFFUSIVITY OF HAFNIUM CARBIDE HfC

[Temperature, T, K; Thermal Diffusivity, α, $cm^2 s^{-1}$]

T	α	T	α
CURVE 1*		CURVE 1 (cont.)*	
1813.2	0.106	2313.2	0.089
1843.2	0.100	2348.2	0.102
1843.2	0.112	2358.2	0.082
1853.2	0.105	2398.2	0.085
1988.2	0.096	2408.2	0.070
1998.2	0.106	2568.2	0.078
2043.2	0.090		
2053.2	0.091		
2078.2	0.092		
2088.2	0.101		
2098.2	0.087		
2098.2	0.091		
2113.2	0.100		
2118.2	0.085		
2178.2	0.103		
2193.2	0.095		
2198.2	0.081		
2303.2	0.081		
2303.2	0.088		

* No figure given.

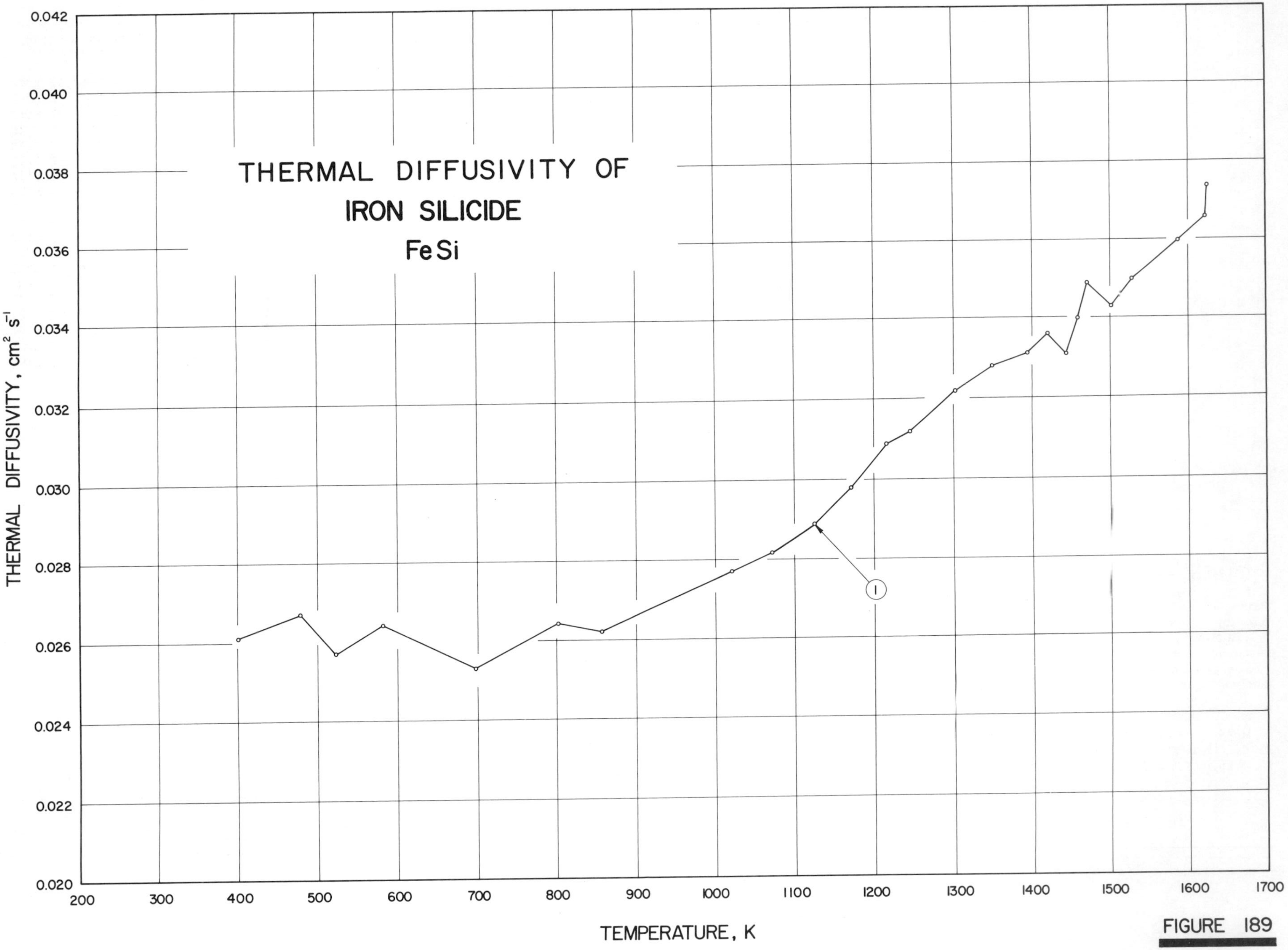
THERMAL DIFFUSIVITY OF
IRON SILICIDE
FeSi
THERMAL DIFFUSIVITY, cm² s⁻¹
0.042
0.040
0.038
0.036
0.034
0.032
0.030
0.028
0.026
0.024
0.022
0.020
200
300
400
500
600
700
800
900
1000
1100
1200
1300
1400
1500
1600
1700
TEMPERATURE, K
1

FIGURE 189

SPECIFICATION TABLE 189. THERMAL DIFFUSIVITY OF IRON SILICIDE FeSi

Cur. No.	Ref. No.	Author(s)	Year	Temp. Range, K	Reported Error, %	Name and Specimen Designation	Composition (weight percent), Specifications, and Remarks
1	278	Krentsis, R. P., Ostrovskii, F. E. and Gel'd, P. V.	1970	401-1626	< 8		Single crystal specimen prepared from carbonyl iron (A-2 grade 99.98%) and silicon of semiconductor grade; melted together in quartz crucible; single crystal grown by Czochralski method; diffusivity measured along the [210] growth axis of crystal.

DATA TABLE 189. THERMAL DIFFUSIVITY OF IRON SILICIDE FeSi

[Temperature, T, K; Thermal Diffusivity, α, $cm^2\ s^{-1}$]

T	α	T	α
CURVE 1		CURVE 1 (cont.)	
401	0.0261	1420	0.0336
479	0.0267	1443	0.0331
523	0.0257	1459	0.0340
582	0.0264	1470	0.0349
698	0.0253	1502	0.0343
802	0.0264	1528	0.0350
857	0.0267	1588	0.0360
910	0.0262	1623	0.0366
1021	0.0277	1626	0.0374
1073	0.0282		
1126	0.0289		
1171	0.0298		
1216	0.0309		
1245	0.0312		
1302	0.0322		
1349	0.0328		
1394	0.0331		

SPECIFICATION TABLE 190. THERMAL DIFFUSIVITY OF LEAD TELLURIDE PbTe

Cur. No.	Ref. No.	Author(s)	Year	Temp. Range, K	Reported Error, %	Name and Specimen Designation	Composition (weight percent), Specifications, and Remarks
1*	69	McNeill, D. J.	1962	298	± 8.3		Iodine doped; bar specimen of square cross section 2 x 2 mm and 4 cm long; specimen suspended in a vacuum bell jar; appropriate temperature variations derived from Peltier heat generated at junction of specimen and a current carrying lead; periodic reversal of current establishes symmetrical temperature variations; author reports two values obtained from slope of curve: 0.0114 ± 0.0003, and 0.0113 ± 0.0003 $cm^2\ sec^{-1}$; data point reported obtained from equation by author.

DATA TABLE 190. THERMAL DIFFUSIVITY OF LEAD TELLURIDE PbTe

[Temperature, T, K; Thermal Diffusivity, α, $cm^2\ s^{-1}$]

T	α
CURVE 1*	
298	0.012

* No figure given.

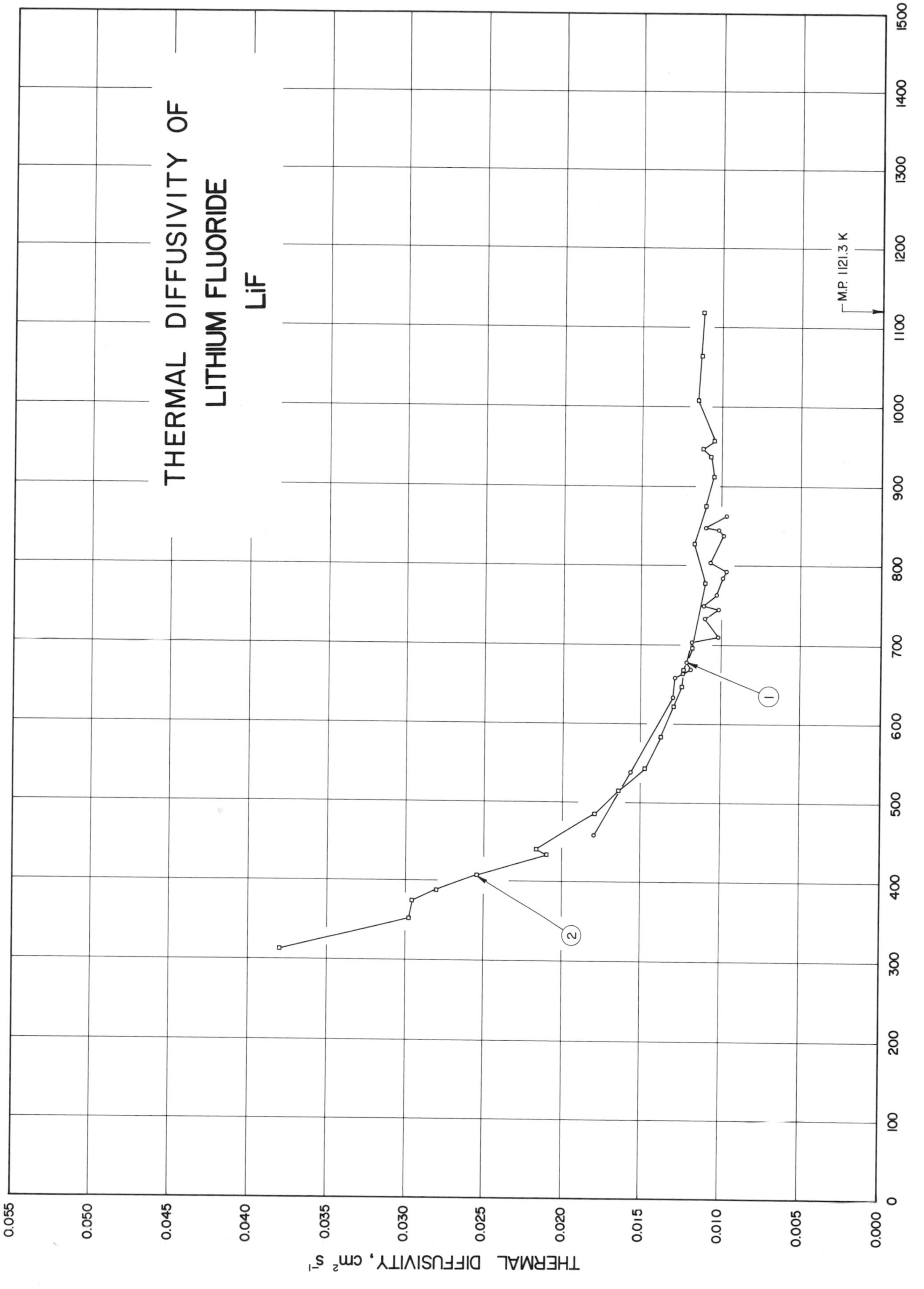
THERMAL DIFFUSIVITY OF
LITHIUM FLUORIDE
LiF
M.P. 1121.3 K
1
2
TEMPERATURE, K
0
100
200
300
400
500
600
700
800
900
1000
1100
1200
1300
1400
1500
THERMAL DIFFUSIVITY, cm² s⁻¹
0.000
0.005
0.010
0.015
0.020
0.025
0.030
0.035
0.040
0.045
0.050
0.055

FIGURE 191

SPECIFICATION TABLE 191. THERMAL DIFFUSIVITY OF LITHIUM FLUORIDE LiF

Cur. No.	Ref. No.	Author(s)	Year	Temp. Range, K	Reported Error, %	Name and Specimen Designation	Composition (weight percent), Specifications, and Remarks
1	44	Chang, H., Altman, M. and Sharma, R.	1967	458-863			98.99 pure; polycrystalline, cylindrical specimen, 1 in. in diameter, 6 in. long; density at 25 C 2.61 ± 0.01 g cm^{-3}; melting point 1121 K; diffusivity measured in Pt and Pt 10%Rh wire-wound furnace.
2	44	Chang, H., et al.	1967	313-1118			Similar to the above specimen except specimen 2 in. in diameter, 18 in. long; diffusivity measured in tungsten-mesh heater furnace.

DATA TABLE 191. THERMAL DIFFUSIVITY OF LITHIUM FLUORIDE LiF

[Temperature, T, K; Thermal Diffusivity, α, cm^2 s^{-1}]

T	α	T	α	T	α
CURVE 1		CURVE 1 (cont.)		CURVE 2 (cont.)	
458.2	0.0180	848.2	0.0110	668.2	0.0124
538.2	0.0157	863.2	0.0097	696.2	0.0118
633.2	0.0130			778.2	0.0110
658.2	0.0129	CURVE 2		828.2	0.0117
663.2	0.0124			875.2	0.0110
668.2	0.0119	313.2	0.038	913.2	0.0105
678.2	0.0122	353.2	0.0298	938.2	0.0107
703.2	0.0119	375.2	0.0296	948.2	0.0112
710.2	0.0102	388.2	0.028	957.2	0.0105
733.2	0.0110	407.2	0.0254	1008.2	0.0115
744.2	0.0102	433.2	0.021	1063.2	0.0113
749.2	0.0111	440.2	0.0217	1118.2	0.0112
763.2	0.0103	485.2	0.0179		
785.2	0.0099	515.2	0.0164		
793.2	0.0097	543.2	0.0148		
803.2	0.0107	583.2	0.0138		
838.2	0.0099	623.2	0.0130		
845.2	0.0102	647.2	0.0125		

SPECIFICATION TABLE 192. THERMAL DIFFUSIVITY OF MOLYBDENUM DITELLURIDE $MoTe_2$

Cur. No.	Ref. No.	Author(s)	Year	Temp. Range, K	Reported Error, %	Name and Specimen Designation	Composition (weight percent), Specifications, and Remarks
1*	180	Guennoc, H.	1961	298.2		A	Prepared from powdered Mo and Te in a three-zone furnace; powdered, pressed at 4000 Kg cm^{-2}, and then annealed at 873.2 K under vacuum for two days; density 6.85 g cm^{-3}, theoretical density 7.78 g cm^{-3}; hexagonal type lattice; lattice constants a = 3.520 ± 0.002 A, c = 3.966 ± 0.004 A; optical energy gap close to 1 eV; Angström method used to measure diffusivity.
2*	180	Guennoc, H.	1961	298.2		B	Same specifications and conditions as above.
3*	180	Guennoc, H.	1961	298.2		C	Same specifications and conditions as above.

DATA TABLE 192. THERMAL DIFFUSIVITY OF MOLYBDENUM DITELLURIDE $MoTe_2$

[Temperature, T, K; Thermal Diffusivity, α, $cm^2\ s^{-1}$]

T	α
CURVE 1*	
298.2	0.034
CURVE 2*	
298.2	0.030
CURVE 3*	
298.2	0.054

* No figure given.

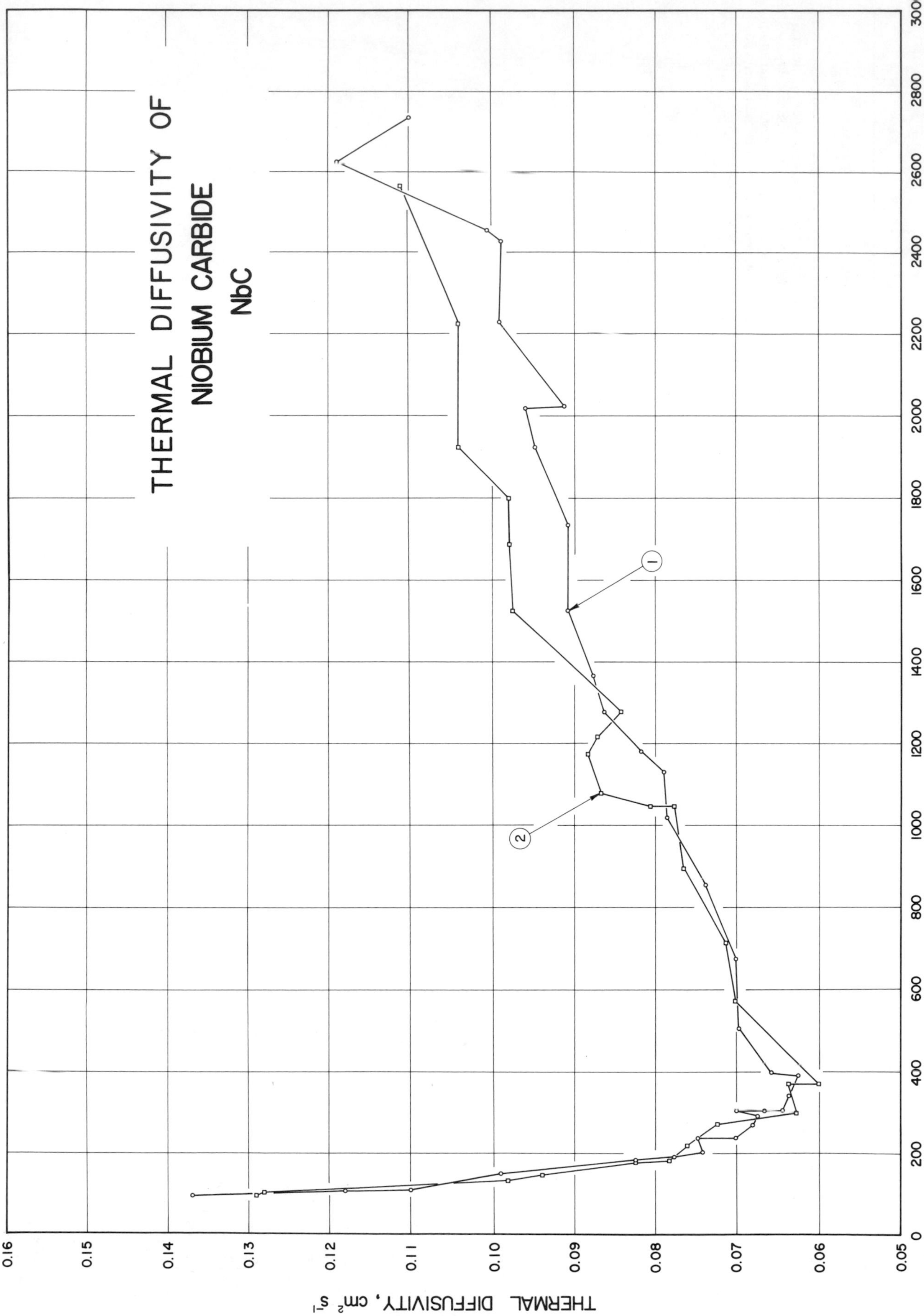
THERMAL DIFFUSIVITY OF
NIOBIUM CARBIDE
NbC
THERMAL DIFFUSIVITY, cm² s⁻¹
TEMPERATURE, K
0.16
0.15
0.14
0.13
0.12
0.11
0.10
0.09
0.08
0.07
0.06
0.05
0
200
400
600
800
1000
1200
1400
1600
1800
2000
2200
2400
2600
2800
3000
1
2

FIGURE 193

SPECIFICATION TABLE 193. THERMAL DIFFUSIVITY OF NIOBIUM CARBIDE NbC

Cur. No.	Ref. No.	Author(s)	Year	Temp. Range, K	Reported Error, %	Name and Specimen Designation	Composition (weight percent), Specifications, and Remarks
1	234	Morrison, B. H. and Sturgess, L. L.	1970	93-2735		1	88.4 Nb, 11.2 C (< 0.0500 free C), 0.2 W, 0.13 O, 0.05 Ta, 0.02 Ti, 0.017 N, 0.005 each of Cr, Mo, and Zr, < 0.003 each of Ba, Co, K, and Sr, < 0.001 each of Al, Bi, Cu, Pb, Ni, Si, Sn, and V, < 0.0003 each of B, Ca, Li, Mn, Ag, and Na, < 0.0001 Be, and < 0.0001 Mg; 1.26 cm diameter x 0.100 cm thick; hot-pressed at 3300 K and 3400 psi for 10 min; density 7.52 g cm^{-3}.
2	234	Morrison, B. H. and Sturgess, L. L.	1970	95-2564		2	Similar to the above specimen.

DATA TABLE 193. THERMAL DIFFUSIVITY OF NIOIBUM CARBIDE NbC

[Temperature, T, K; Thermal Diffusivity, α, $cm^2 s^{-1}$]

T	α	T	α	T	α	T	α
CURVE 1		CURVE 1 (cont.)		CURVE 2		CURVE 2 (cont.)	
92.9	0.137	1019	0.0785	94.8	0.129	1279	0.0841
105.7	0.118	1130	0.0789	101.4	0.128	1524	0.0975
110.4	0.110	1180	0.0817	132.1	0.0982	1687	0.0979
149.3	0.0991	1279	0.0863	145.2	0.0940	1799	0.0979
182.0	0.0824	1365	0.0877	176.2	0.0824	1923	0.104
190.5	0.0776	1524	0.0908	180.3	0.0782	2223	0.104
201.4	0.0741	1734	0.0908	219.3	0.0760	2564	0.111
237.1	0.0748	1923	0.0948	272.3	0.0724		
237.1	0.0701	2018	0.0959	299.9	0.0627		
269.8	0.0681	2023	0.0912	370.7	0.0637		
291.7	0.0675	2228	0.0991	370.7	0.0600		
304.1	0.0701	2427	0.0989	572.8	0.0701		
304.1	0.0667	2455	0.105	714.5	0.0713		
304.1	0.0644	2624	0.119	895.4	0.0764		
341.2	0.0637	2735	0.110	1045	0.0776		
391.7	0.0625			1045	0.0805		
398.1	0.0658			1079	0.0867		
507.0	0.0697			1175	0.0883		
676.1	0.0700			1216	0.0871		
855.1	0.0738						

SPECIFICATION TABLE 194. THERMAL DIFFUSIVITY OF PLUTONIUM CARBIDE PuC

Cur. No.	Ref. No.	Author(s)	Year	Temp. Range, K	Reported Error, %	Name and Specimen Designation	Composition (weight percent), Specifications, and Remarks
1*	125	Moser, J. B. and Kruger, O. L.	1968	660-1608	±5	84% Dense PuC	4.12 C, 0.567 O, and 0.0440 N; microstructure shows presence of approximately 10 percent plutonium sesquicarbide (Pu_2C_3) phase; disk-shaped specimen 1.83 cm in dia and 0.31 cm thick; prepared by crushing, cold-pressing, and sintering arc-cast material (arc-cast carbide prepared by fusion of 99.5 pure metal with spectrographically pure carbon in an arc furnace); sintered at 1673.2 K; lapped down in thickness from approx 0.5 cm to final thickness; density 83.8 percent of theoretical value; surface coated with colloidal graphite; exposed to thermal pulse of 500 μ sec duration generated by ruby laser; flash method used to measure diffusivity; first measured in vacuum and then reduced in thickness to 0.26 cm because of vaporization above 1473.2 K and then subsequently measured in argon at 1.3 atm pressure up to the final temp; radiation heat loss correction applied to all data; maximum error of measurement ±7.5 percent.

DATA TABLE 194. THERMAL DIFFUSIVITY OF PLUTONIUM CARBIDE PuC

[Temperature, T, K; Thermal Diffusivity, α, $cm^2 s^{-1}$]

T	α	T	α	T	α
CURVE 1*		CURVE 1 (cont.)*		CURVE 1 (cont.)*	
660.2	0.0242	1041.2	0.0291	1459.2	0.0340
660.2	0.0236	1041.2	0.0276	1463.2	0.0315
731.2	0.0250	1075.2	0.0295	1471.2	0.0326
731.2	0.0243	1124.2	0.0321	1474.2	0.0327
785.2	0.0260	1152.2	0.0321	1482.2	0.0329
785.2	0.0253	1152.2	0.0317	1555.2	0.0346
847.2	0.0274	1152.2	0.0307	1605.2	0.0340
847.2	0.0268	1188.2	0.0313	1605.2	0.0356
906.2	0.0287	1188.2	0.0304	1608.2	0.0374
906.2	0.0280	1253.2	0.0336		
931.2	0.0289	1253.2	0.0317		
931.2	0.0286	1325.2	0.0323		
931.2	0.0265	1325.2	0.0305		
1001.2	0.0297	1331.2	0.0341		
1001.2	0.0290	1333.2	0.0342		
1001.2	0.0280	1338.2	0.0321		
1041.2	0.0295	1392.2	0.0349		

* No figure given.

SPECIFICATION TABLE 195. THERMAL DIFFUSIVITY OF SILICON CARBIDE SiC

Cur. No.	Ref. No.	Author(s)	Year	Temp. Range, K	Reported Error, %	Name and Specimen Designation	Composition (weight percent), Specifications, and Remarks
1*	184	Taylor, R.	1965	298.2	<±5		Heavily irradiated; cylindrical specimen 0.25 in. in diameter and 1.619 cm long; front face exposed to heat pulse from xenon flash tube; thermal diffusivity calculated from measured time necessary for the rear face to reach one-half the maximum temperature rise $t_{0.5}$ by employing the equation $\alpha = 1.37\ L^2/\pi^2 t_{0.5}$; temperature of measurement not given by author but assumed to be room temperature.
2*	184	Taylor, R.	1965	298.2	<±5		Above specimen measured for diffusivity again after being shortened to a length of 1.272 cm; other conditions same as above.
3*	184	Taylor, R.	1965	298.2	<±5		Above specimen measured for diffusivity again after being shortened to a length of 0.758 cm; other conditions same as above.
4*	184	Taylor, R.	1965	298.2	<±5		Above specimen measured for diffusivity again after being shortened to a length of 0.299 cm; other conditions same as above.
5*	184	Taylor, R.	1965	298.2	<±5		Above specimen measured for diffusivity again after being shortened to a length of 0.203 cm; other conditions same as above.
6*	184	Taylor, R.	1965	298.2	<±5		Above specimen measured for diffusivity again after being shortened to a length of 0.190 cm; other conditions same as above.
7*	184	Taylor, R.	1965	298.2	±5		Same specimen and same measurement pertaining to curve 1 above; diffusivity calculated from time intercept t_x of measured temperature-time curve of rear face by employing the equation $\alpha = 0.48\ L^2/\pi^2 t_x$; temperature of measurement not given by author but assumed to be room temperature.
8*	184	Taylor, R.	1965	298.2	±5		Same specimen and same measurement pertaining to curve 2 above; other conditions same as above.
9*	184	Taylor, R.	1965	298.2	±5		Same specimen and same measurement pertaining to curve 3 above; other conditions same as above.
10*	184	Taylor, R.	1965	298.2	±5		Same specimen and same measurement pertaining to curve 4 above; other conditions same as above.
11*	184	Taylor, R.	1965	298.2	±5		Same specimen and same measurement pertaining to curve 5 above; other conditions same as above.
12*	184	Taylor, R.	1965	298.2	±5		Same specimen and same measurement pertaining to curve 6 above; other conditions same as above.
13*	184	Taylor, R.	1965	298.2			Cubic; 30% porosity; measured mean density 2.17 g cm^{-3}; temperature of measurement not given by author but assumed to be room temperature; heat pulse method used to measure diffusivity; cylindrical specimen 0.25 in. in diameter and 0.5 in. long.
14*	184	Taylor, R.	1965	298.2			Mixed hexagonal/cubic structure; dense; measured mean density 3.11 g cm^{-3}; other conditions same as above.
15*	184	Taylor, R.	1965	298.2			Si bonded; measured mean density 2.26 g cm^{-3}; other conditions same as above.

* No figure given.

DATA TABLE 195. THERMAL DIFFUSIVITY OF SILICON CARBIDE SiC

[Temperature, T, K; Thermal Diffusivity, α, $cm^2 s^{-1}$]

T	α
CURVE 1*	
298.2	0.0247
CURVE 2*	
298.2	0.0211
CURVE 3*	
298.2	0.0200
CURVE 4*	
298.2	0.0185
CURVE 5*	
298.2	0.0186
CURVE 6*	
298.2	0.0188
CURVE 7*	
298.2	0.0186
CURVE 8*	
298.2	0.0180
CURVE 9*	
298.2	0.0189
CURVE 10*	
298.2	0.0185
CURVE 11*	
298.2	0.0182
CURVE 12*	
298.2	0.0189
CURVE 13*	
298.2	0.363
CURVE 14*	
298.2	0.694
CURVE 15*	
298.2	0.075

* No figure given.

SPECIFICATION TABLE 196. THERMAL DIFFUSIVITY OF SILVER BROMIDE AgBr

Cur. No.	Ref. No.	Author(s)	Year	Temp. Range, K	Reported Error, %	Name and Specimen Designation	Composition (weight percent), Specifications, and Remarks
1*	177	Pochapsky, T. E.	1953	304-679			Cylindrical specimen 3 mm in diameter; cast; thermal diffusivity determined from measured cooling characteristics of a fine platinum wire along the axis of specimen; data points reported corrected for changes in heat capacity.

DATA TABLE 196. THERMAL DIFFUSIVITY OF SILVER BROMIDE AgBr

[Temperature, T, K; Thermal Diffusivity, α, $cm^2\ s^{-1}$]

T	α
CURVE 1*	
304	0.00516
349	0.00456
408	0.00417
468	0.00364
481	0.00340
526	0.00302
540	0.00295
577	0.00265
610	0.00227
627	0.00218
645	0.00203
670	0.00208
679	0.00213

* No figure given.

SPECIFICATION TABLE 197. THERMAL DIFFUSIVITY OF SILVER ANTIMONY TELLURIDE $AgSbTe_2$

Cur. No.	Ref. No.	Author(s)	Year	Temp. Range, K	Reported Error, %	Name and Specimen Designation	Composition (weight percent), Specifications, and Remarks
1*	60	Abeles, B., Cody, G. D., Baughan, B. E., Hockings, E. F., and Muha, G. M.	1960	316-646		No. 245	Face centered cubic structure; lattice parameter a = 6.076 Å; extrinsic below 473.2 K and intrinsic above that temperature; cylindrical specimen 1 cm in diameter; electrical resistivity reported as 0.0147, 0.0120, 0.0158, 0.0210, 0.0248, 0.0316, 0.0312, 0.0289, 0.0267, 0.0314, 0.0273, 0.0332, 0.0246, 0.0210, 0.0150, and 0.0128 ohm cm at 316.2, 326.2, 342.2, 363.2, 408.2, 443.2, 447.2, 457.2, 462.2, 475.2, 479.2, 505.2, 526.2, 573.2, 615.2, and 646.2 K, respectively.

DATA TABLE 197. THERMAL DIFFUSIVITY OF SILVER ANTIMONY TELLURIDE $AgSbTe_2$

[Temperature, T, K; Thermal Diffusivity, α, $cm^2\ s^{-1}$]

T	α	T	α
CURVE 1*		CURVE 1 (cont.)*	
316	0.00365	528	0.00343
327	0.00368	573	0.00336
353	0.00355	614	0.00422
362	0.00372	646	0.00474
407	0.00362		
407	0.00365		
414	0.00370		
437	0.00344		
443	0.00350		
447	0.00320		
457	0.00332		
458	0.00324		
468	0.00331		
472	0.00320		
505	0.00340		
517	0.00345		
526	0.00345		

* No figure given.

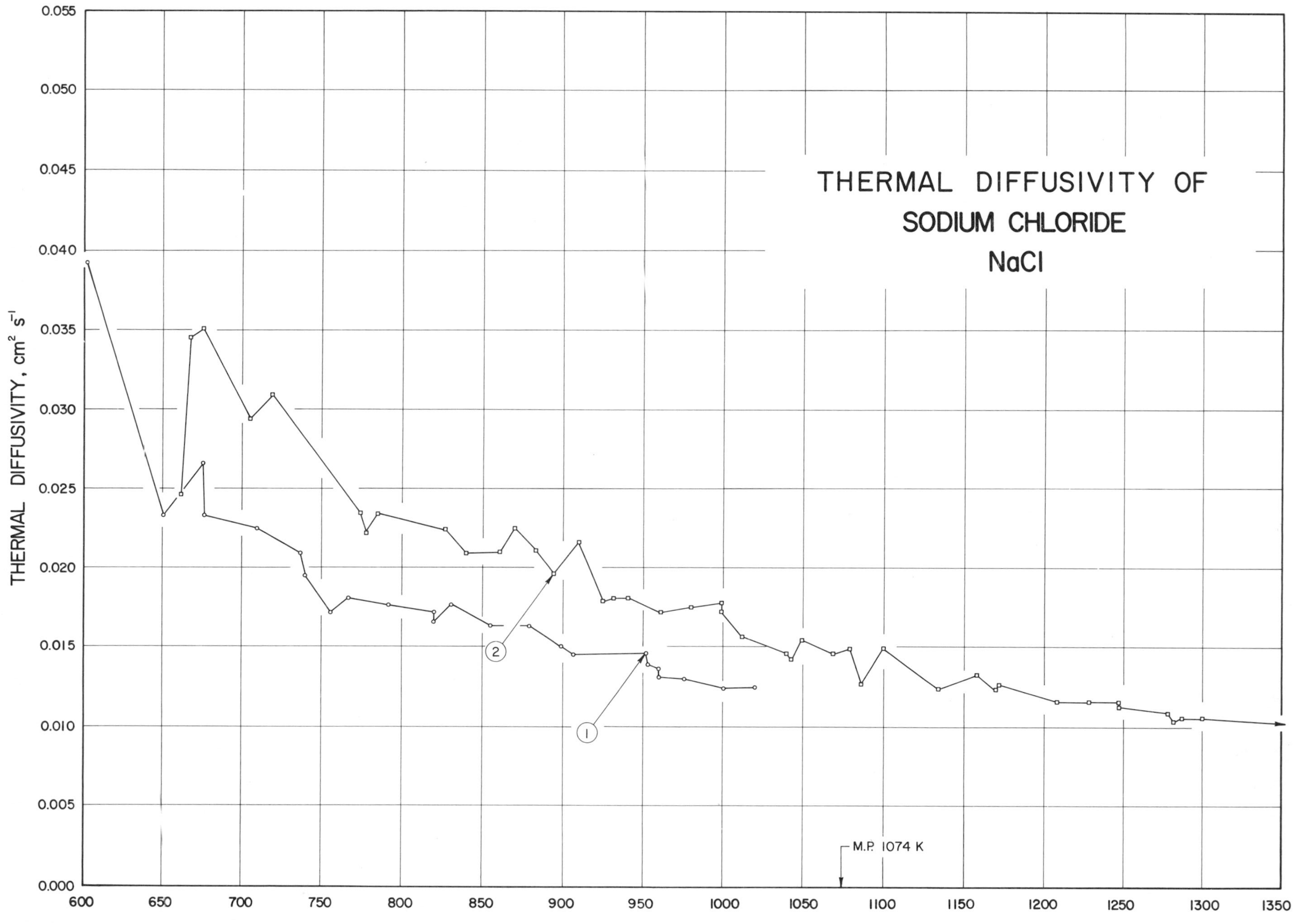

THERMAL DIFFUSIVITY OF
SODIUM CHLORIDE
NaCl
THERMAL DIFFUSIVITY, cm² s⁻¹
0.055
0.050
0.045
0.040
0.035
0.030
0.025
0.020
0.015
0.010
0.005
0.000
600
650
700
750
800
850
900
950
1000
1050
1100
1150
1200
1250
1300
1350
M.P. 1074 K
1
2
TEMPERATURE, K

FIGURE 198

SPECIFICATION TABLE 198. THERMAL DIFFUSIVITY OF SODIUM CHLORIDE NaCl

Cur. No.	Ref. No.	Author(s)	Year	Temp. Range, K	Reported Error, %	Name and Specimen Designation	Composition (weight percent), Specifications, and Remarks
1	276	Fujisawa, H., Fujii, N., Mizutani, H., Kanamori, H., and Akimoto, S.	1968	602-1020	±5		Cylindrical specimen; ratio of length to radius 6 to 8; diffusivity measured using modified Angström's method in pressure 29.0 kb.
2	276	Fujisawa, H., et al.	1968	662-1383	±7		The above specimen; diffusivity measured in pressure 47.0 kb.

DATA TABLE 198. THERMAL DIFFUSIVITY OF SODIUM CHLORIDE NaCl

[Temperature, T, K; Thermal Diffusivity, α, $cm^2\ s^{-1}$]

T	α	T	α	T	α	T	α
CURVE 1		CURVE 1 (cont.)		CURVE 2 (cont.)		CURVE 2 (cont.)	
602	0.0392	953	0.0139	840	0.0209	1079	0.0149
651	0.0233	960	0.0136	861	0.0210	1086	0.0127
676	0.0266	960	0.0131	870	0.0225	1100	0.0149
677	0.0233	976	0.0130	883	0.0211	1134	0.0124
710	0.0225	1000	0.0124	894	0.0196	1158	0.0133
737	0.0209	1020	0.0125	910	0.0216	1170	0.0123
740	0.0195			925	0.0179	1172	0.0127
756	0.0172	CURVE 2		932	0.0181	1208	0.0116
767	0.0181			941	0.0181	1228	0.0116
792	0.0177	662	0.0246	961	0.0172	1247	0.0116
820	0.0172	668	0.0345	980	0.0175	1247	0.0113
820	0.0166	676	0.0351	999	0.0168	1278	0.0109
831	0.0177	705	0.0294	999	0.0162	1282	0.0104
855	0.0163	719	0.0309	1012	0.0156	1287	0.0106
879	0.0163	774	0.0234	1040	0.0146	1300	0.0106
899	0.0150	778	0.0222	1043	0.0142	1366	0.0102
907	0.0145	785	0.0234	1049	0.0154	1383	0.0098
952	0.0146	827	0.0224	1069	0.0146		

SPECIFICATION TABLE 199. THERMAL DIFFUSIVITY OF TANTALUM CARBIDE TaC

Cur. No.	Ref. No.	Author(s)	Year	Temp. Range, K	Reported Error, %	Name and Specimen Designation	Composition (weight percent), Specifications, and Remarks
1*	32	Taylor, R.E. and Nakata, M.M.	1963	1828-2213	7		93.7 Ta, 6.3 C, 0.05 Cu, 0.05 Si, <0.05 Fe, <0.05 Zr, 0.03 Ti, 0.01 Cr, 0.005 Ca, 0.002 Mg, and <0.001 Ag (as received); average grain size (ASTM) No. 10; cylindrical specimen 1.588 cm overall diameter and 3.508 cm overall length, machined in three sections: a center section 2.54 cm long and two end pieces consisting of 0.484 cm thick discs; these parts are machine fitted in such a way that a small area contact is made at the circumference only, leaving a thin space of 0.0127 cm or less between center section and the discs that act as radiation barriers; two parallel sight holes each 0.112 cm in diameter drilled through the top disc to a depth of 1.75 cm at radii $r_1 = 0$ and $r_2 = 0.564$ cm; sight holes drilled by Elox technique; bottom disc grooved to fit sample support pins; obtained from the Carborundum Co.; hot pressed; measured after extensive heat soaking at temperatures up to 2673.2 K; density 12.44 g cm^{-3}; radiation shields mounted at both ends; measured under a vacuum of 1 x 10^{-6} mm Hg; radial diffusivity technique used.

DATA TABLE 199. THERMAL DIFFUSIVITY OF TANTALUM CARBIDE TaC

[Temperature, T, K; Thermal Diffusivity, α, $cm^2\ s^{-1}$]

T	α
CURVE 1*	
1828.2	0.110
1833.2	0.107
1843.2	0.098
1853.2	0.112
1858.2	0.104
1983.2	0.117
1988.2	0.110
1988.2	0.112
1993.2	0.113
2088.2	0.112
2098.2	0.102
2213.2	0.119

* No figure given.

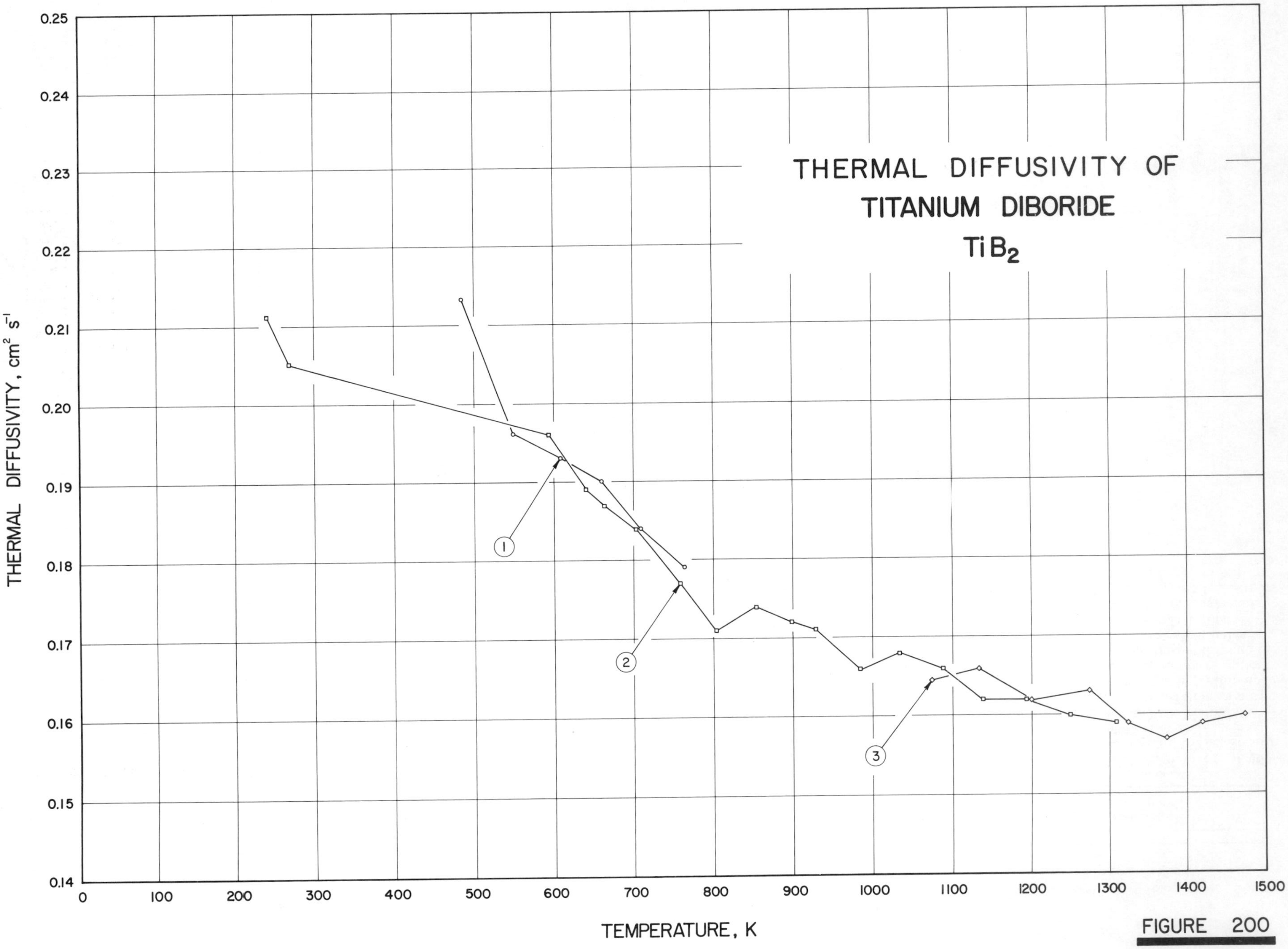
THERMAL DIFFUSIVITY OF
TITANIUM DIBORIDE
TiB2
THERMAL DIFFUSIVITY, cm² s⁻¹
TEMPERATURE, K
0.25
0.24
0.23
0.22
0.21
0.20
0.19
0.18
0.17
0.16
0.15
0.14
0
100
200
300
400
500
600
700
800
900
1000
1100
1200
1300
1400
1500
1
2
3

FIGURE 200

SPECIFICATION TABLE 200. THERMAL DIFFUSIVITY OF TITANIUM DIBORIDE TiB_2

Cur. No.	Ref. No.	Author(s)	Year	Temp. Range, K	Reported Error, %	Name and Specimen Designation	Composition (weight percent), Specifications, and Remarks
1	235	Branscomb, T.M.	1970	483-763			99 pure; 0.301 cm thick; obtained from Carborundum Co.; hot-pressed; density 4.366 g cm^{-3}.
2	235	Branscomb, T.M.	1970	240-1308			The above specimen with Pt-black coating on the front face.
3	235	Branscomb, T.M.	1970	1073-1473			Similar to the above specimen but 0.197 cm in thickness.

DATA TABLE 200. THERMAL DIFFUSIVITY OF TITANIUM DIBORIDE TiB_2

[Temperature, T, K; Thermal Diffusivity, α, $cm^2\ s^{-1}$]

T	α
CURVE 1	
483	0.213
548	0.196
608	0.193
660	0.190
708	0.184
763	0.179
CURVE 2	
240	0.211
268	0.205
593	0.196
640	0.189
663	0.187
703	0.184
758	0.177
803	0.171
853	0.174
CURVE 2 (cont.)	
898	0.172
928	0.171
983	0.166
1033	0.168
1088	0.166
1138	0.162
1193	0.162
1248	0.160
1308	0.159
CURVE 3	
1073	0.1645
1133	0.166
1198	0.162
1273	0.163
1323	0.159
1373	0.157
CURVE 3 (cont.)	
1418	0.159
1473	0.160

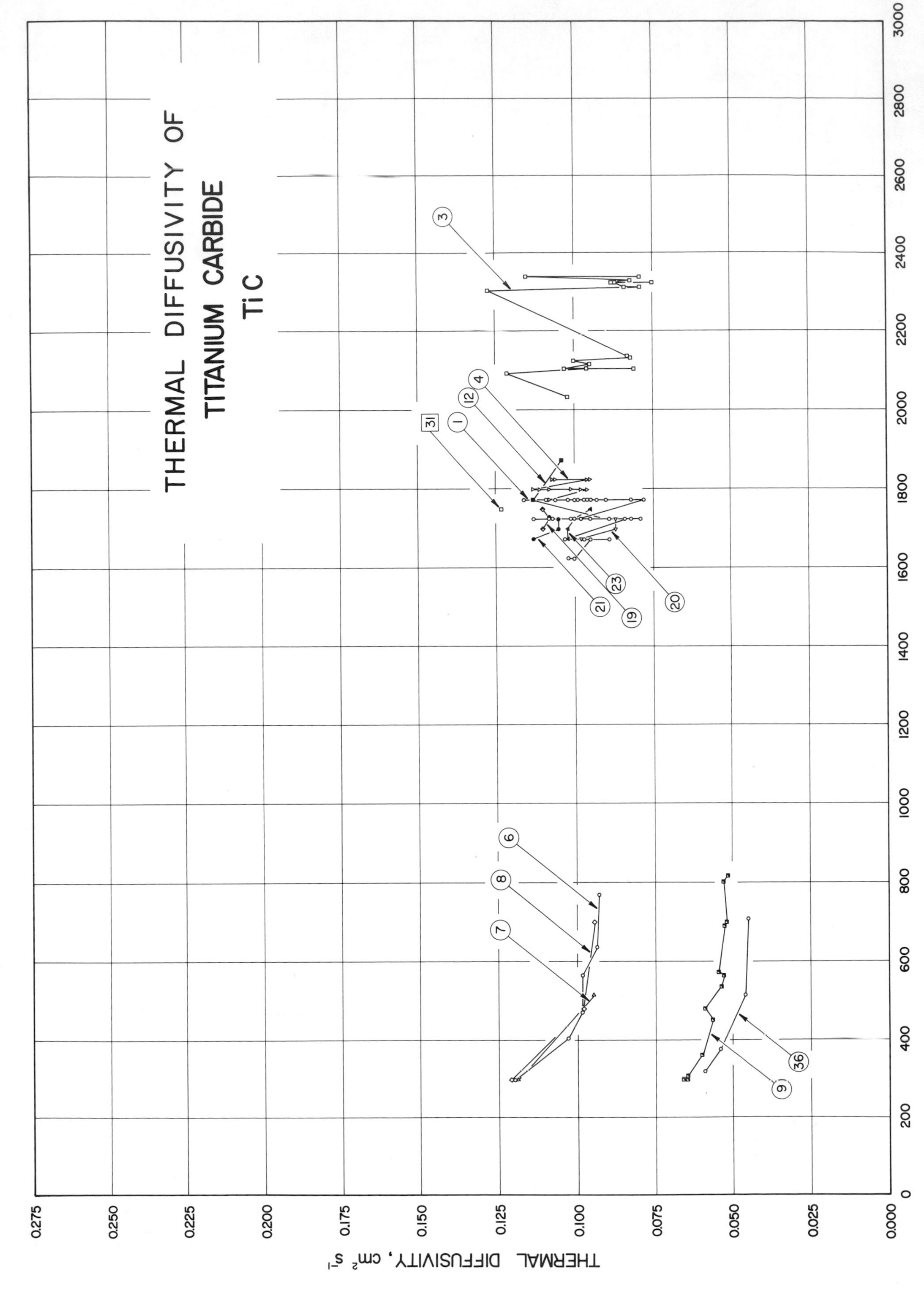
THERMAL DIFFUSIVITY OF
TITANIUM CARBIDE
Ti C
THERMAL DIFFUSIVITY, cm² s⁻¹
TEMPERATURE, K
0.000
0.025
0.050
0.075
0.100
0.125
0.150
0.175
0.200
0.225
0.250
0.275
0
200
400
600
800
1000
1200
1400
1600
1800
2000
2200
2400
2600
2800
3000
1
3
4
6
7
8
9
12
19
20
21
23
31
36

FIGURE 201

SPECIFICATION TABLE 201. THERMAL DIFFUSIVITY OF TITANIUM CARBIDE TiC

Cur. No.	Ref. No.	Author(s)	Year	Temp. Range, K	Reported Error, %	Name and Specimen Designation	Composition (weight percent), Specifications, and Remarks
1	254, 32, 50	Taylor, R. E. and Nakata, M. M.	1963	1623-1773	7	1	80.3 ± 0.3 Ti, 19.3 C, and < 0.2 metallic impurities (after thermal conductivity and before diffusivity measurements); average grain size (ASTM) No. 7; cylindrical specimen 1.585 cm in diameter and 3.515 cm long; two parallel sight holes each 0.160 cm in diameter drilled to a depth of 1.61 cm at radii $r_1 = 0$ and $r_2 = 0.560$ cm; originally obtained from the Carborundum Co.; hot pressed from raw powder having chemical composition 79.2 Ti, 19.5 C, and < 0.2 metallic impurities; machined out of a thermal conductivity sample (after conductivity measurements had been made); sight holes drilled by Elox technique; measured for diffusivity after extensive heat soaking at temperatures up to 2673.2 K; density 4.77 g cm^{-3}, electrical resistivity reported as 72.9, 79.3, 83.2, 87.7, and 92.8 μohm cm at 623.2, 731.2, 787.2, 862.2, and 959.2 K, respectively; radiation shields mounted at both ends; measured under a vacuum of 1 x 10^{-6} mm Hg; radial diffusivity technique used.
2*	32	Taylor, R. E. and Nakata, M. M.	1963	1673-1873	7	2	80.3 ± 0.3 Ti, 19.3 C, and < 0.2 metallic impurities (after thermal conductivity and before diffusivity measurements); average grain size (ASTM) No. 7; cylindrical specimen 1.588 cm in diameter and 3.470 cm long; two parallel sight holes each 0.162 cm in diameter drilled to a depth of 1.65 cm at radii $r_1 = 0$ and $r_2 = 0.564$ cm; originally obtained from the Carborundum Co.; hot pressed from raw powder having chemical composition 79.2 Ti, 19.5 C, and < 0.2 metallic impurities; machined out of a thermal conductivity sample (after conductivity measurements had been made); sight holes drilled by Elox technique; measured for diffusivity after extensive heat soaking at temperatures up to 2673.2 K; density 4.77 g cm^{-3} (before runs); electrical resistivity reported as 50, 56, 62.5, 69, 75, 80.5, 86.5, 92, 97, and 103.5 μohm cm at 300, 400, 500, 600, 700, 800, 900, 1000, 1100, and 1200 K, respectively; radiation shields mounted at both ends; measured under a vacuum of 1 x 10^{-6} mm Hg; radial diffusivity technique used.
3	32	Taylor, R. E. and Nakata, M. M.	1963	2033-2338	7	2	79.2 Ti and 19.12 C (after diffusivity measurements); average grain size (ASTM) No. 6.5 (after runs); above specimen machined into the 2.54 cm long section of the high temperature modification, using another titanium carbide compact for the two end barriers which consist of 0.465 cm thick discs; high temperature specimen of the same overall diameter and overall length as the one-piece specimen mentioned above; center section and two end pieces machine fitted in such a way that a small area contact is made at the circumference only, leaving a thin space of 0.0127 cm or less between the center section and the discs that act as radiation barriers; two parallel sight holes each 0.162 cm in diameter drilled through the top disc, the holes being precisely positioned with respect to the holes in the 2.54 cm section; sight holes drilled by Elox technique; bottom disc grooved to fit sample support pins; obtained from the Carborundum Co.; radiation shields mounted at both ends; measured under a vacuum of 1 x 10^{-6} mm Hg; radial diffusivity technique used.

* Not shown in figure.

SPECIFICATION TABLE 201. THERMAL DIFFUSIVITY OF TITANIUM CARBIDE TiC (continued)

Cur. No.	Ref. No.	Author(s)	Year	Temp. Range, K	Reported Error, %	Name and Specimen Designation	Composition (weight percent), Specifications, and Remarks
4	32	Taylor, R. E. and Nakata, M. M.	1963	1673-1823	7	3	77.8 Ti, 19.5 C, 2 Zr, 0.1 Co, 0.1 Fe, 0.1 Nb, 0.1 Si, <0.1 V, 0.05 Ag, 0.05 Al, 0.05 Cu, 0.03 Ca, 0.03 Mg, and <0.005 B; average grain size (ASTM) No. 9.5; cylindrical specimen 1.610 cm in diameter and 3.500 cm long; two parallel sight holes each 0.162 cm in diameter drilled to a depth of 1.73 cm at radii $r_1 = 0$ and $r_2 = 0.568$ cm; obtained from the Norton Co.; hot pressed; sight holes drilled by Elox technique; measured for diffusivity after extensive heat soaking at temperatures up to 2673.2 K; density 4.84 g cm^{-3}; electrical resistivity reported as 67.9, 74.2, 77.6, 82.0, 86.6, and 47.4 μohm cm at 623.2, 731.2, 787.2, 862.2, 959.2, and 294.2 K, respectively; radiation shields mounted at both ends; measured under a vacuum of 1 x 10^{-6} mm Hg; radial diffusivity technique used.
5*	32	Taylor, R. E. and Nakata, M. M.	1963	1723-1748	7	4	77.8 Ti, and 19.5 C; average grain size (ASTM) No. 9.5; cylindrical specimen 1.610 cm in diameter and 3.492 cm long; two parallel sight holes each 0.117 cm in diameter drilled to a depth of 1.16 cm at radii $r_1 = 0$ and $r_2 = 0.568$ cm; obtained from the Norton Co.; hot pressed; sight holes drilled by Elox technique; measured for diffusivity after extensive heat soaking at temperatures up to 2673.2 K; density 4.82 g cm^{-3}; radiation shields mounted at both ends; measured under a vacuum of 1 x 10^{-6} mm Hg; radial diffusivity technique used.
6	32	Taylor, R. E. and Nakata, M. M.	1963	298-768	±4	2A	80.3±0.3 Ti, 19.3 C, and <0.2 metallic impurities; single phase; cylindrical specimen 1.588 cm in diameter and 0.271 cm in thickness; originally obtained from the Carborundum Co.; hot pressed; machined out of specimen 2 that had been previously measured for diffusivity (curve No. 3), using the radial diffusivity technique; density 4.77 g cm^{-3}; measured for thermal diffusivity by the flash diffusivity technique; laser used as a pulse energy source.
7	32	Taylor, R. E. and Nakata, M. M.	1963	298-514	±4	2B	80.3±0.3 Ti, 19.3 C, and <0.2 metallic impurities; single phase; cylindrical specimen 1.588 cm in diameter and 0.396 cm in thickness; originally obtained from the Carborundum Co.; hot pressed; machined out of specimen 2 that had been previously measured for diffusivity (curve No. 3); density 4.77 g cm^{-3}; measured for thermal diffusivity by the flash diffusivity technique; laser used as a pulse energy source.
8	32	Taylor, R. E. and Nakata, M. M.	1963	298-699	±4	2C	80.3±0.3 Ti, 19.3 C, and <0.2 metallic impurities; single phase; cylindrical specimen 1.588 cm in diameter and 0.427 cm in thickness; originally obtained from the Carborundum Co.; hot pressed; machined out of specimen 2 that had been previously measured for diffusivity (curve No. 3); density 4.77 g cm^{-3}; measured for thermal diffusivity by the flash diffusivity technique; laser used as a pulse energy source.
9	32	Taylor, R. E. and Nakata, M. M.	1963	293-816	±4	Cube	79.6±1.2 Ti, 17.7 C, and 1.4 metallic impurities; single phase; cube specimen 0.310 cm thickness; obtained from MIT after having been measured for conductivity by Vasilos and Kingery; density 4.56 g cm^{-3}; measured for diffusivity by the flash diffusivity technique; laser used as a pulse energy source.

* Not shown in figure.

SPECIFICATION TABLE 201. THERMAL DIFFUSIVITY OF TITANIUM CARBIDE TiC (continued)

Cur. No.	Ref. No.	Author(s)	Year	Temp. Range, K	Reported Error, %	Name and Specimen Designation	Composition (weight percent), Specifications, and Remarks
10*	78	Taylor, R.E. and Nakata, M.M.	1962	1773, 1823		TiC-B-2	Machined out of steady-state thermal conductivity sample (same as that used to prepared specimen 1, curve No. 1); density 4.77 g cm^{-3}.
11*	78	Taylor, R.E. and Nakata, M.M.	1962	1823.2		TiC-B-2	Above specimen measured for diffusivity again.
12	78	Taylor, R.E. and Nakata, M.M.	1962	1723-1873		TiC-B-2	Above specimen measured for diffusivity again.
13*	78	Taylor, R.E. and Nakata, M.M.	1962	1823.2		TiC-B-2	Above specimen measured for diffusivity again.
14*	78	Taylor, R.E. and Nakata, M.M.	1962	1773.2		TiC-B-2	Above specimen measured for diffusivity again.
15*	78	Taylor, R.E. and Nakata, M.M.	1962	1673, 1823		TiC-B-2	Above specimen measured for diffusivity again.
16*	78	Taylor, R.E. and Nakata, M.M.	1962	1723-1773		TiC-C-1	Specimen with sight holes 0.064 in. in diameter and 0.663 in. deep; density 4.837 g cm^{-3}.
17*	78	Taylor, R.E. and Nakata, M.M.	1962	1748-1798		TiC-C-1	Above specimen measured for diffusivity again.
18*	78	Taylor, R.E. and Nakata, M.M.	1962	1748-1798		TiC-C-1	Above specimen measured for diffusivity again.
19	78	Taylor, R.E. and Nakata, M.M.	1962	1698-1748		TiC-C-1	Above specimen measured for diffusivity again.
20	78	Taylor, R.E. and Nakata, M.M.	1962	1673-1723		TiC-C-1	Above specimen measured for diffusivity again.
21	78	Taylor, R.E. and Nakata, M.M.	1962	1673-1748		TiC-C-1	Above specimen measured for diffusivity again.
22*	78	Taylor, R.E. and Nakata, M.M.	1962	1723-1773		TiC-C-1	Above specimen measured for diffusivity again.
23*	78	Taylor, R.E. and Nakata, M.M.	1962	1698-1748		TiC-C-1	Above specimen measured for diffusivity again.
24*	78	Taylor, R.E. and Nakata, M.M.	1962	1723-1773		TiC-C-1	Above specimen measured for diffusivity again.
25*	78	Taylor, R.E. and Nakata, M.M.	1962	1723-1773		TiC-C-1	Above specimen measured for diffusivity again.
26*	78	Taylor, R.E. and Nakata, M.M.	1962	1723-1798		TiC-C-1	Above specimen measured for diffusivity again.
27*	78	Taylor, R.E. and Nakata, M.M.	1962	1748-1798		TiC-C-1	Above specimen measured for diffusivity again.
28	78	Taylor, R.E. and Nakata, M.M.	1962	1673-1748		TiC-C-1	Above specimen measured for diffusivity again.
29*	78	Taylor, R.E. and Nakata, M.M.	1962	1748, 1773		TiC-C-1	Above specimen measured for diffusivity again.
30*	78	Taylor, R.E. and Nakata, M.M.	1962	1773-1823		TiC-C-1	Above specimen measured for diffusivity again.
31	78	Taylor, R.E. and Nakata, M.M.	1962	1748.2		TiC-C-3	Specimen with sight holes 0.046 in. in diameter and 0.456 in. deep; density 4.819 g cm^{-3}.
32*	78	Taylor, R.E. and Nakata, M.M.	1962	1748.2		TiC-C-3	Above specimen measured for diffusivity again.
33*	78	Taylor, R.E. and Nakata, M.M.	1962	1723.2		TiC-C-3	Above specimen measured for diffusivity again.
34*	78	Taylor, R.E. and Nakata, M.M.	1962	1723.2		TiC-C-3	Above specimen measured for diffusivity again.

* Not shown in figure.

SPECIFICATION TABLE 201. THERMAL DIFFUSIVITY OF TITANIUM CARBIDE TiC (continued)

Cur. No.	Ref. No.	Author(s)	Year	Temp. Range, K	Reported Error, %	Name and Specimen Designation	Composition (weight percent), Specifications, and Remarks
35*	112	Carpenter, R. S., II	1962	298-699		A. I. Sample	Flash method used to measure diffusivity; measured in a vacuum of 5 x 10^{-6} mm Hg; obtained from Atomics International.
36	112	Carpenter, R. S., II	1962	298-707		M. I. T. Sample	Flash method used to measure diffusivity; measured in a vacuum of 5 x 10^{-6} mm Hg; obtained from M. I. T.

* Not shown in figure.

DATA TABLE 201. THERMAL DIFFUSIVITY OF TITANIUM CARBIDE TiC

[Temperature, T, K; Thermal Diffusivity, α, $cm^2\ s^{-1}$]

T	α
CURVE 1	
1623.2	0.102
1623.2	0.100
1673.2	0.095
1673.2	0.097
1673.2	0.089
1673.2	0.103
1723.2	0.084
1723.2	0.107
1723.2	0.089
1723.2	0.107*
1723.2	0.082
1723.2	0.084*
1723.2	0.107*
1723.2	0.095
1723.2	0.084*
1723.2	0.098
1723.2	0.101
1723.2	0.108
1723.2	0.113
1723.2	0.095*
1723.2	0.107*
1723.2	0.079
1723.2	0.100
1723.2	0.098*
1773.2	0.078
1773.2	0.093
1773.2	0.082*
1773.2	0.082
1773.2	0.097
1773.2	0.102
1773.2	0.093*
1773.2	0.097*
1773.2	0.095
1773.2	0.096
1773.2	0.116
1773.2	0.109
1773.2	0.099
1773.2	0.106
1773.2	0.090
1773.2	0.096*
1773.2	0.100
CURVE 2*	
1673.2	0.100
1723.2	0.090
1773.2	0.078
1773.2	0.113
1773.2	0.106
1823.2	0.087
1823.2	0.098
1823.2	0.104
1823.2	0.098
1873.2	0.104
CURVE 3	
2033.2	0.102
2093.2	0.121
2103.2	0.096
2103.2	0.081
2103.2	0.103
2108.2	0.095
2123.2	0.100
2133.2	0.082
2138.2	0.083
2303.2	0.127
2313.2	0.079
2313.2	0.084
2323.2	0.087
2323.2	0.075
2323.2	0.088
2328.2	0.082
2338.2	0.115
2338.2	0.079
CURVE 4	
1673.2	0.098*
1673.2	0.113*
1673.2	0.102*
1698.2	0.110*
1698.2	0.087*
1698.2	0.105*
1698.2	0.111*
1698.2	0.102*
CURVE 4 (cont.)	
1723.2	0.105*
1723.2	0.108*
1723.2	0.087*
1723.2	0.105*
1723.2	0.105*
1723.2	0.108*
1723.2	0.108*
1723.2	0.102*
1723.2	0.109*
1723.2	0.100*
1748.2	0.107*
1748.2	0.113*
1748.2	0.098*
1748.2	0.110*
1748.2	0.108*
1748.2	0.103*
1748.2	0.108*
1748.2	0.108*
1748.2	0.102*
1748.2	0.109*
1748.2	0.107*
1748.2	0.095*
1748.2	0.108*
1773.2	0.108*
1773.2	0.108*
1773.2	0.098*
1773.2	0.108
1798.2	0.096
1798.2	0.098
1798.2	0.101
1798.2	0.108
1798.2	0.111
1798.2	0.113
1823.2	0.096
1823.2	0.095
1823.2	0.106
1823.2	0.107
1823.2	0.107*
1823.2	0.107*
CURVE 5*	
1723.2	0.095
1723.2	0.110
1748.2	0.123
1748.2	0.096
CURVE 6	
298.2	0.120
403.2	0.103
470.2	0.0985
565.2	0.0984
637.2	0.0938
768.2	0.0931
CURVE 7	
298.2	0.119
514.2	0.0949
CURVE 8	
298.2	0.121
479.2	0.0981
699.2	0.0945
CURVE 9	
293.2	0.0659
293.2	0.0647
309.2	0.0647
361.2	0.0599
451.2	0.0564
481.2	0.0589
535.2	0.0537
565.2	0.0527
587.2	0.0544
691.2	0.0525
698.2	0.0517
801.2	0.0527
816.2	0.0513
CURVE 10*	
1773.2	0.078
1823.2	0.087
CURVE 11*	
1823.2	0.098
CURVE 12	
1723.2	0.090*
1773.2	0.113
1873.2	0.104
CURVE 13*	
1823.2	0.104
CURVE 14*	
1773.2	0.106
CURVE 15*	
1673.2	0.100
1823.2	0.098
CURVE 16*	
1723.2	0.105
1748.2	0.107
1773.2	0.108
CURVE 17*	
1748.2	0.113
1773.2	0.108
1798.2	0.108
CURVE 18*	
1748.2	0.098
1773.2	0.098
1798.2	0.098
CURVE 19	
1698.2	0.110
1723.2	0.108*
1748.2	0.110
CURVE 20	
1673.2	0.098
1698.2	0.087
1723.2	0.087
CURVE 21	
1673.2	0.113
1698.2	0.105
1723.2	0.105
1748.2	0.108
CURVE 22*	
1723.2	0.105
1748.2	0.103
1773.2	0.108
CURVE 23*	
1698.2	0.111
1723.2	0.108
1748.2	0.108
CURVE 24*	
1723.2	0.108
1748.2	0.108
1773.2	0.106
CURVE 25*	
1723.2	0.102
1748.2	0.102
1773.2	0.095
CURVE 26*	
1723.2	0.109
1748.2	0.109
1773.2	0.107
1798.2	0.111
CURVE 27*	
1748.2	0.107
1773.2	0.107
1798.2	0.113
CURVE 28	
1673.2	0.102
1698.2	0.102
1723.2	0.100*
1748.2	0.095
CURVE 29*	
1748.2	0.108
1773.2	0.107
CURVE 30*	
1773.2	0.101
1798.2	0.096
1823.2	0.096
CURVE 31	
1748.2	0.123
CURVE 32*	
1748.2	0.096
CURVE 33*	
1723.2	0.095
CURVE 34*	
1723.2	0.110
CURVE 35*	
298.2	0.121
298.2	0.118
479.2	0.098
699.2	0.094
CURVE 36	
298.2	0.059
376.2	0.054
515.2	0.046
707.2	0.045

*Not shown in figure.

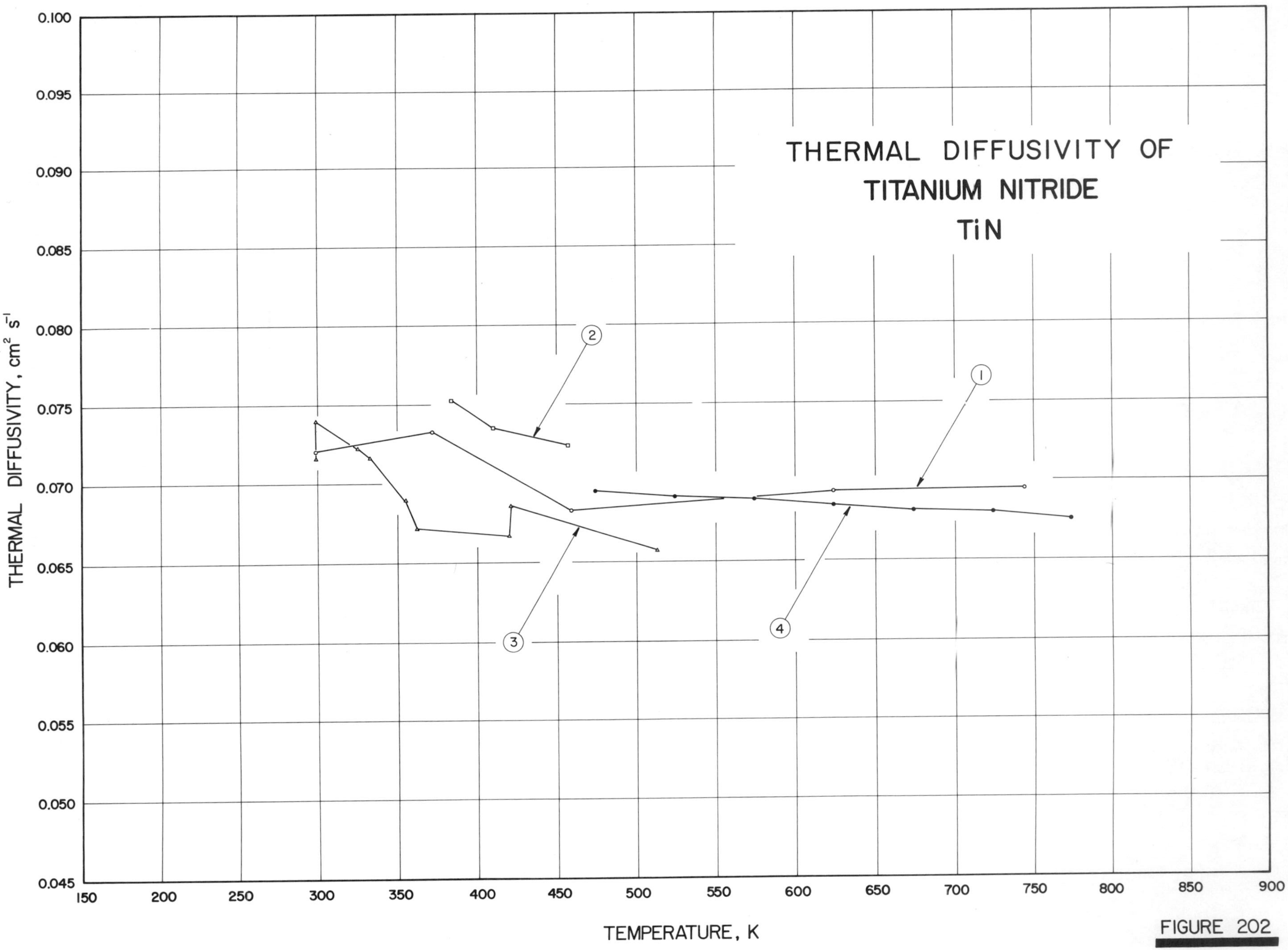
THERMAL DIFFUSIVITY OF
TITANIUM NITRIDE
TiN
THERMAL DIFFUSIVITY, cm² s⁻¹
TEMPERATURE, K
0.100
0.095
0.090
0.085
0.080
0.075
0.070
0.065
0.060
0.055
0.050
0.045
150
200
250
300
350
400
450
500
550
600
650
700
750
800
850
900
1
2
3
4

FIGURE 202

SPECIFICATION TABLE 202. THERMAL DIFFUSIVITY OF TITANIUM NITRIDE TiN

Cur. No.	Ref. No.	Author(s)	Year	Temp. Range, K	Reported Error, %	Name and Specimen Designation	Composition (weight percent), Specifications, and Remarks
1	32	Taylor, R.E. and Nakata, M.M.	1963	298-743	±4	2A	77.9 Ti, 17.9 N, and <0.9 other metals and carbon; average grain size 11 μ; single phase; cylindrical specimen 0.250 in. in diameter and 0.101 in. thick; obtained from General Astrometals Corporation; machined from thermal conductivity specimen after conductivity measurements had been made; density 4.78 g cm^{-3}; electrical resistivity reported as 65.1, 73.1, 77.5, 84.8, 91.4, 97.9, 102.7, 107.7, and 116.0 μohm cm at 300.2, 387.2, 449.2, 537.2, 628.2, 716.2, 788.2, 877.2, and 978.2 K, respectively; measured by the flash diffusivity technique using laser as a pulse energy source.
2	32	Taylor, R.E. and Nakata, M.M.	1963	383-456	±4	2B	77.9 Ti, 17.9 N, and <0.9 other metals and carbon; average grain size 11 μ; single phase; cylindrical specimen 0.250 in. in diameter and 0.104 in. thick; obtained from General Astrometals Corporation; machined from thermal conductivity specimen after conductivity measurements had been made; density 4.78 g cm^{-3}; measured by the flash diffusivity technique using laser as a pulse energy source.
3	32	Taylor, R.E. and Nakata, M.M.	1963	298-512	±4	2C	77.9 Ti, 17.9 N, and <0.9 other metals and carbon; average grain size 11 μ; single phase; cylindrical specimen 0.250 in. in diameter and 0.1035 in. thick; obtained from General Astrometals Corporation; machined from thermal conductivity specimen after conductivity measurements had been made; density 4.78 g cm^{-3}; measured by the flash diffusivity technique using laser as a pulse energy source.
4	79	Taylor, R.E. and Nakata, M.M.	1963	473-773		2	98^+ pure; 77.9 Ti, 17.9 N, and <0.9 other metals; average grain size 11 μ; single phase (X-ray diffraction test and chemical analysis performed upon conclusion of diffusivity measurements); cylindrical specimen 0.250 in. in diameter and 0.100 in. thick; supplied by General Astrometals Corp.; machined from the portion of a thermal conductivity sample between temperature measuring holes at the longitudinal center of sample; relatively few high temperature conductivity measurements were obtained on sample before being machined for diffusivity measurements; density 4.78 g cm^{-3} (upon conclusion of measurements); porosity 12%; electrical resistivity reported as 69.5, 73.5, 79.4, 82.5, 87.4, 92.4, 94.5, 99.3, 103.2, 109.1, 113.9, 119.0, 122.8, and 128.0 μohm cm at 296.2, 347.2, 412.2, 484.2, 563.2, 638.2, 680.2, 750.2, 819.2, 900.2, 975.2, 1049.2, 1126.2, and 1206.2 K, respectively; front face of specimen irradiated with a pulse of energy from a high intensity source; subsequent transient response of rear face is then related to the thermal diffusivity.

DATA TABLE 202. THERMAL DIFFUSIVITY OF TITANIUM NITRIDE TiN

[Temperature, T, K; Thermal Diffusivity, α, $cm^2 s^{-1}$]

T	α
CURVE 1	
298.2	0.0721
371.2	0.0733
458.2	0.0683
623.2	0.0694
743.2	0.0695
CURVE 2	
383.2	0.0753
409.2	0.0735
456.2	0.0724
CURVE 3	
298.2	0.0717
298.2	0.0740
324.2	0.0723
332.2	0.0717
354.2	0.0690
361.2	0.0672
419.2	0.0667
420.2	0.0686
512.2	0.0657
CURVE 4	
473.2	0.0695
523.2	0.0691
573.2	0.0689
623.2	0.0685
673.2	0.0681
723.2	0.0680
773.2	0.0675

SPECIFICATION TABLE 203. THERMAL DIFFUSIVITY OF TUNGSTEN DIBORIDE WB_2

Cur. No.	Ref. No.	Author(s)	Year	Temp. Range, K	Reported Error, %	Name and Specimen Designation	Composition (weight percent), Specifications, and Remarks
1*	70	Cape, J.A. and Taylor, R.E.	1961	1603-1888	±5.6/ ±6.9		Cylindrical specimen 0.625 in. in diameter and 1 in. long; sight holes 0.049 in. drilled to a depth of 0.5 in. at longitudinal center of specimen and at a radius of 0.221 in.; measured in vacuum.

DATA TABLE 203. THERMAL DIFFUSIVITY OF TUNGSTEN DIBORIDE WB_2

[Temperature, T, K; Thermal Diffusivity, α, $cm^2\ s^{-1}$]

T	α
CURVE 1*	
1603.2	0.054
1707.2	0.054
1807.2	0.058
1855.2	~0.06
1888.2	~0.06

* No figure given.

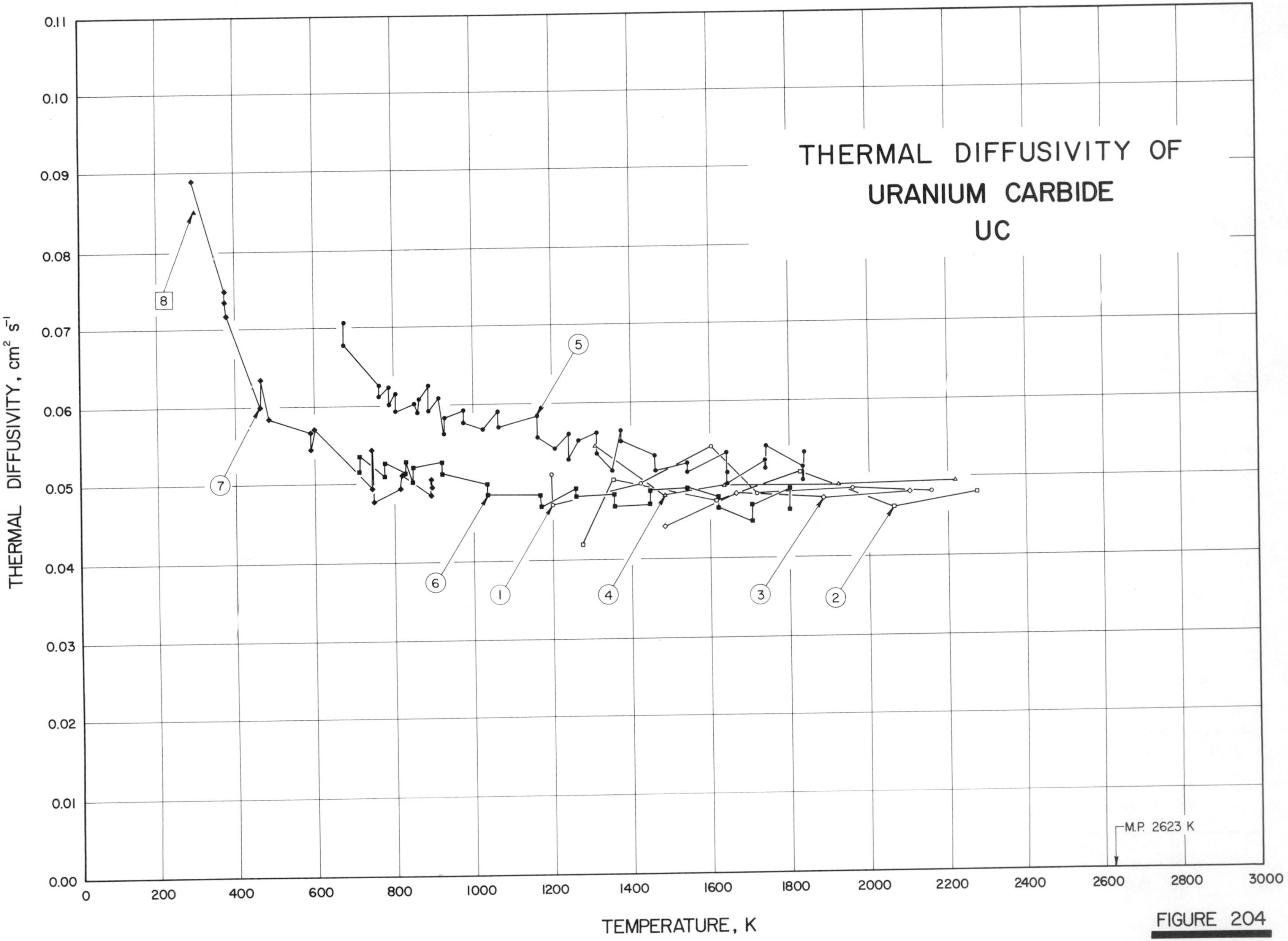
THERMAL DIFFUSIVITY OF
URANIUM CARBIDE
UC
THERMAL DIFFUSIVITY, cm² s⁻¹
TEMPERATURE, K
M.P. 2623 K
0.00
0.01
0.02
0.03
0.04
0.05
0.06
0.07
0.08
0.09
0.10
0.11
0
200
400
600
800
1000
1200
1400
1600
1800
2000
2200
2400
2600
2800
3000
1
2
3
4
5
6
7
8

FIGURE 204

SPECIFICATION TABLE 204. THERMAL DIFFUSIVITY OF URANIUM CARBIDE UC

Cur. No.	Ref. No.	Author(s)	Year	Temp. Range, K	Reported Error, %	Name and Specimen Designation	Composition (weight percent), Specifications, and Remarks
1	40	Mustacchi, C. and Giuliani, S.	1963	1198-2158			4.8± 0.1 C, 0.05 N_2, 0.009 O_2, and 0.0002 H_2; crystallites uniform in size and of mean diameter 10^{-2} cm, showing tiny speckles of uncombined carbon; wafer specimens ~12 mm in diameter and having thicknesses lying in the range from 1 to 3 mm; obtained from the Parsons Co. in the form of pellets 0.5 in. in diameter and 1.5 in. long; pellets were vacuum melted and delivered in an as-cast condition; wafers spark-cut in a kerosene bath, planed by electro-sparking in kerosene, then thoroughly washed in acetone before insertion in vacuum jar holder; density 13.6 g cm^{-3}; electrical resistivity reported as 59, 82, 120, 108, 122, 122, 132, 143, 146, 155, 148, 156, 161, 162, 156, 178, 165, 191, 176, 199, and 203 μohm cm at 300.2, 424.2, 536.2, 584.2, 716.2, 774.2, 776.2, 853.2, 874.2, 954.2, 964.2, 974.2, 1085.2, 1104.2, 1174.2, 1186.2, 1274.2, 1295.2, 1370.2, 1376.2, and 1470.2 K, respectively; modulated electron gun used for heating specimens; amplitude method used to determine diffusivity; measured in vacuum better than 10^{-5} mm Hg.
2	40	Mustacchi, C. and Giuliani, S.	1963	1273-2273			Above specimens measured under the same conditions as above but using output face phase shift method to determine diffusivity.
3	40	Mustacchi, C. and Giuliani, S.	1963	1483-2103			Above specimens measured under the same conditions as above but using differential phase shift method to determine diffusivity.
4	40	Mustacchi, C. and Giuliani, S.	1963	1308-2218			Above specimens measured under the same conditions as above but using pulse lag method to determine diffusivity.
5	125	Moser, J. B. and Kruger, O. L.	1968	673-1835	±5	Arc-Cast UC	4.93 C, 0.0117 O, and 0.0021 N; small amount of second phase consisting of UC_2-precipitate found within grains; disk-shaped specimen 1.83 cm in dia and having thickness in the range from 0.2 to 0.3 cm; prepared by fusion of 99.5 pure metal with spectrographically pure carbon in an arc furnace; specimen thickness lapped down from 0.5 cm to final thickness; density 99.0 percent of theoretical value; surface coated with colloidal graphite; exposed to thermal pulse of 500 μsec duration generated by ruby laser; flash method used to measure diffusivity; measured in vacuum; radiation heat loss correction applied to all data; maximum error of measurement ±7.5 percent.
6	125	Moser, J. B. and Kruger, O. L.	1968	709-1800	±5	79% Dense UC	4.69 C, 0.106 O, and 0.0228 N; disk-shaped specimen 1.83 cm in dia and having thickness in the range from 0.2 to 0.3 cm; prepared by crushing, cold-pressing, and sintering arc-cast material (arc-cast material prepared in the same manner as explained above); sintered at 2073.2 K; specimen thickness lapped down from 0.5 cm to final thickness; density 78.8 percent of theoretical value; surface coated with colloidal graphite; exposed to thermal pulse of 500 μsec duration generated by ruby laser; flash method used to measure diffusivity; measured in vacuum; radiation heat loss correction applied to all data; maximum error of measurement ±7.5 percent.

SPECIFICATION TABLE 204. THERMAL DIFFUSIVITY OF URANIUM CARBIDE UC (continued)

Cur. No.	Ref. No.	Author(s)	Year	Temp. Range, K	Reported Error, %	Name and Specimen Designation	Composition (weight percent) Ge	Si	Composition (continued), Specifications, and Remarks
7	137	Moser, J. B. and Kruger, O. L.	1967	292-891					0.005 O; disk-shaped specimen 1.9 cm in dia and 0.2-0.3 cm thick; originally prepared in the form of cylinder of near stoichiometric composition by skull melting and casting at Battelle Memorial Institute, then electrical-discharge machined to a thickness of ~0.5 cm, and then lapped to final thickness with parallel smooth faces; after lapping a platinum foil ~0.32 cm in dia and 0.002 cm thick fused to the center of one face; front face coated with colloidal graphite; density >99 percent of theoretical value; front face heated by discharge from ruby laser; diffusivity determined from measured temp history of rear face; measured in vacuum.
8	214, 215	Moser, J. B. and Kruger, O. L.	1964	298.2					4.66 C; 1.9 cm diameter x 0.5 cm thick; 98.8 theoretical density.

DATA TABLE 204. THERMAL DIFFUSIVITY OF URANIUM CARBIDE UC

[Temperature, T, K; Thermal Diffusivity, α, $cm^2 s^{-1}$]

T	α
CURVE 1	
1198	0.0508
1198	0.0470
1423	0.0495
1603	0.0542
1718	0.0480
1958	0.0486
2158	0.0482
CURVE 2	
1273	0.0418
1353	0.0500
1613	0.0472
1828	0.0508
2063	0.0462
2273	0.0480
CURVE 3	
1483	0.0441
1663	0.0482
1883	0.0475
2103	0.0482
CURVE 4	
1308	0.0545
1483	0.0481
1633	0.0492
1923	0.0492
2218	0.0495
CURVE 5	
673	0.0705
673	0.0677
761	0.0625
761	0.0611
786	0.0623
786	0.0600
802	0.0614
803	0.0591
849	0.0601
858	0.0590
861	0.0606
CURVE 5 (cont.)	
885	0.0624
886	0.0592
912	0.0608
926	0.0562
926	0.0583
975	0.0593
975	0.0576
1023	0.0567
1062	0.0590
1063	0.0569
1160	0.0583
1162	0.0556
1205	0.0541
1240	0.0560
1240	0.0527
1265	0.0552
1312	0.0562
1312	0.0535
1350	0.0512
1373	0.0564
1373	0.0550
1459	0.0531
1460	0.0512
1541	0.0521
1541	0.0509
1641	0.0534
1641	0.0508
1641	0.0494
1739	0.0524
1739	0.0513
1740	0.0541
1833	0.0515
1833	0.0498
1835	0.0534
CURVE 6	
709	0.0515
709	0.0534
774	0.0508
775	0.0525
825	0.0512
827	0.0527
844	0.0501
845	0.0518
CURVE 6 (cont.)	
917	0.0524
919	0.0510
1034	0.0497
1034	0.0483
1168	0.0483
1170	0.0468
1258	0.0480
1258	0.0490
1353	0.0482
1354	0.0467
1446	0.0468
1446	0.0486
1541	0.0488
1619	0.0477
1619	0.0463
1706	0.0446
1706	0.0467
1800	0.0486
1800	0.0460
CURVE 7	
292	0.0888
373	0.0746
373	0.0732
377	0.0715
463	0.0598
463	0.0633
483	0.0583
589	0.0565
589	0.0542
597	0.0570
742	0.0493
742	0.0543
746	0.0476
813	0.0493
816	0.0509
891	0.0484
891	0.0495
891	0.0503
CURVE 8	
298.2	0.085

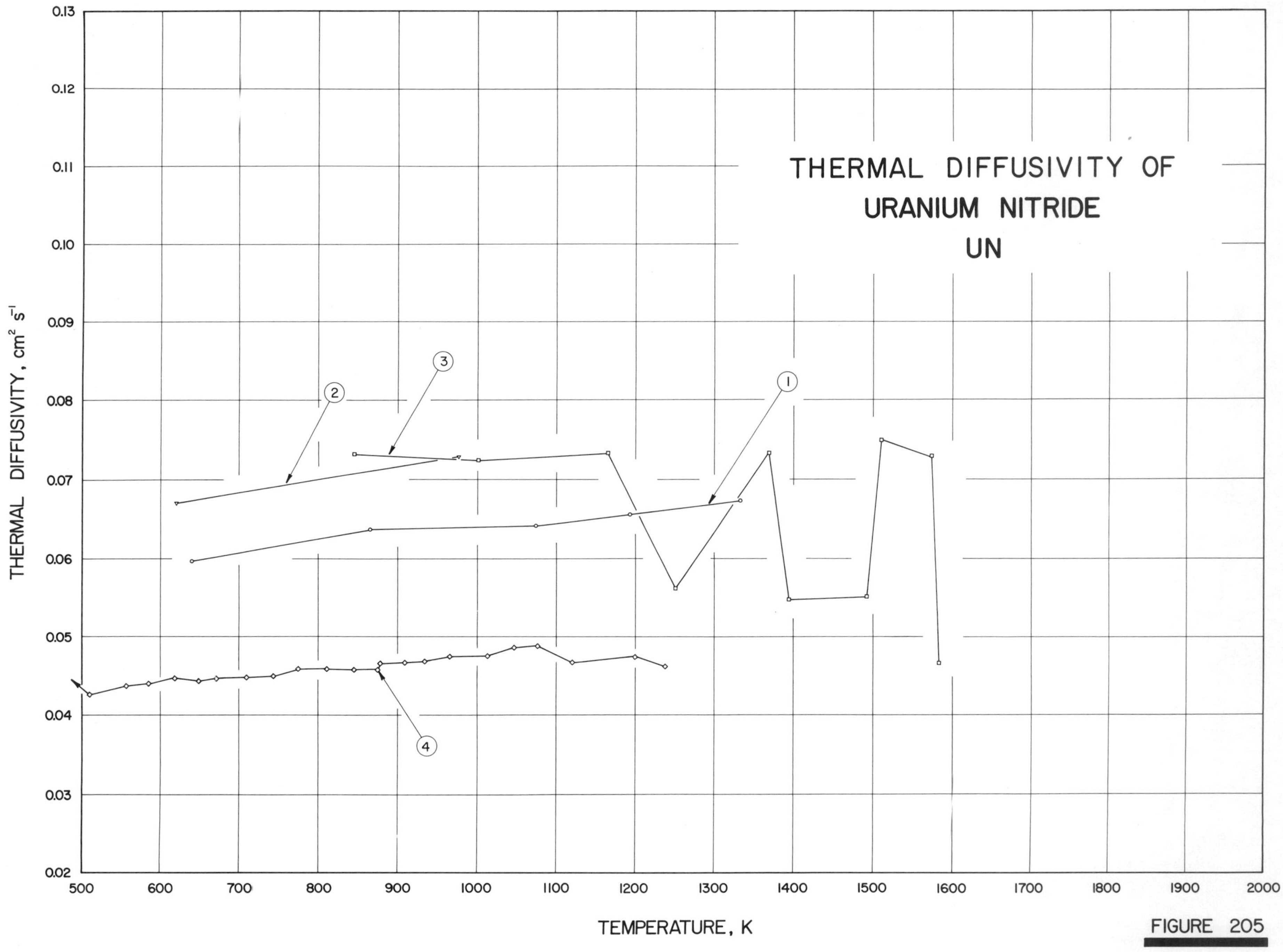
THERMAL DIFFUSIVITY OF
URANIUM NITRIDE
UN
THERMAL DIFFUSIVITY, cm² s⁻¹
TEMPERATURE, K
0.13
0.12
0.11
0.10
0.09
0.08
0.07
0.06
0.05
0.04
0.03
0.02
500
600
700
800
900
1000
1100
1200
1300
1400
1500
1600
1700
1800
1900
2000
1
2
3
4

FIGURE 205

SPECIFICATION TABLE 205. THERMAL DIFFUSIVITY OF URANIUM NITRIDE UN

Cur. No.	Ref. No.	Author(s)	Year	Temp. Range, K	Reported Error, %	Name and Specimen Designation	Composition (weight percent), Specifications, and Remarks
1	80	Keller, D. L., Speidel, E. O., and Kizer, D. E.	1963	640-1333	~±10		Small and thin specimen; 95 to 98% dense; hot pressed; front surface given a pulse of heat with a laser; thermal diffusivity derived from measured time-temperature history of back surface.
2	80	Keller, D. L., et al.	1963	845-1583	~±10		Above specimen measured for diffusivity again.
3	80	Keller, D. L., et al.	1963	620, 976	~±10		New specimen; measured under same conditions as above.
4	148	Nasu, S. and Kikuchi, T.	1968	296-1238			Disc specimen 1.2 cm in dia.and 0.20 cm in thickness; front surface blackened with colloidal graphite; thin film of cobalt evaporated onto the center of back surface to a thickness of 0.001 cm over an area 0.3 cm in dia.by heating to 1773.2 K for 10 min; density 94.9% of theoretical value; laser beam used as pulse energy source; pulse duration ~1 m sec; diffusivity determined from measured temperature history of rear surface; measured in vacuum; data points reported corrected to theoretical density.

DATA TABLE 205. THERMAL DIFFUSIVITY OF URANIUM NITRIDE UN

[Temperature, T, K; Thermal Diffusivity, α, $cm^2 s^{-1}$]

T	α	T	α	T	α
CURVE 1		CURVE 3		CURVE 4 (cont.)	
640.2	0.0597	620.2	0.0671	743	0.0449
863.2	0.0637	976.2	0.0278	775	0.0458
1074.2	0.0642			811	0.0458
1193.2	0.0656	CURVE 4		845	0.0458
1333.2	0.0673			875	0.0458
		296	0.0431*	878	0.0466
CURVE 2		333	0.0424*	909	0.0466
		366	0.0431*	934	0.0468
845.2	0.0732	391	0.0445*	965	0.0475
1001.2	0.0724	412	0.0440*	1013	0.0475
1165.2	0.0733	440	0.0436*	1047	0.0485
1251.2	0.0562	480	0.0446*	1077	0.0482
1251.2	0.0562	510	0.0426	1120	0.0466
1369.2	0.0733	557	0.0437	1200	0.0475
1394.2	0.0548	586	0.0440	1238	0.0461
1492.2	0.0552	619	0.0447		
1510.2	0.0749	649	0.0444		
1573.2	0.0729	672	0.0447		
1583.2	0.0466	710	0.0447		

* Not shown in figure.

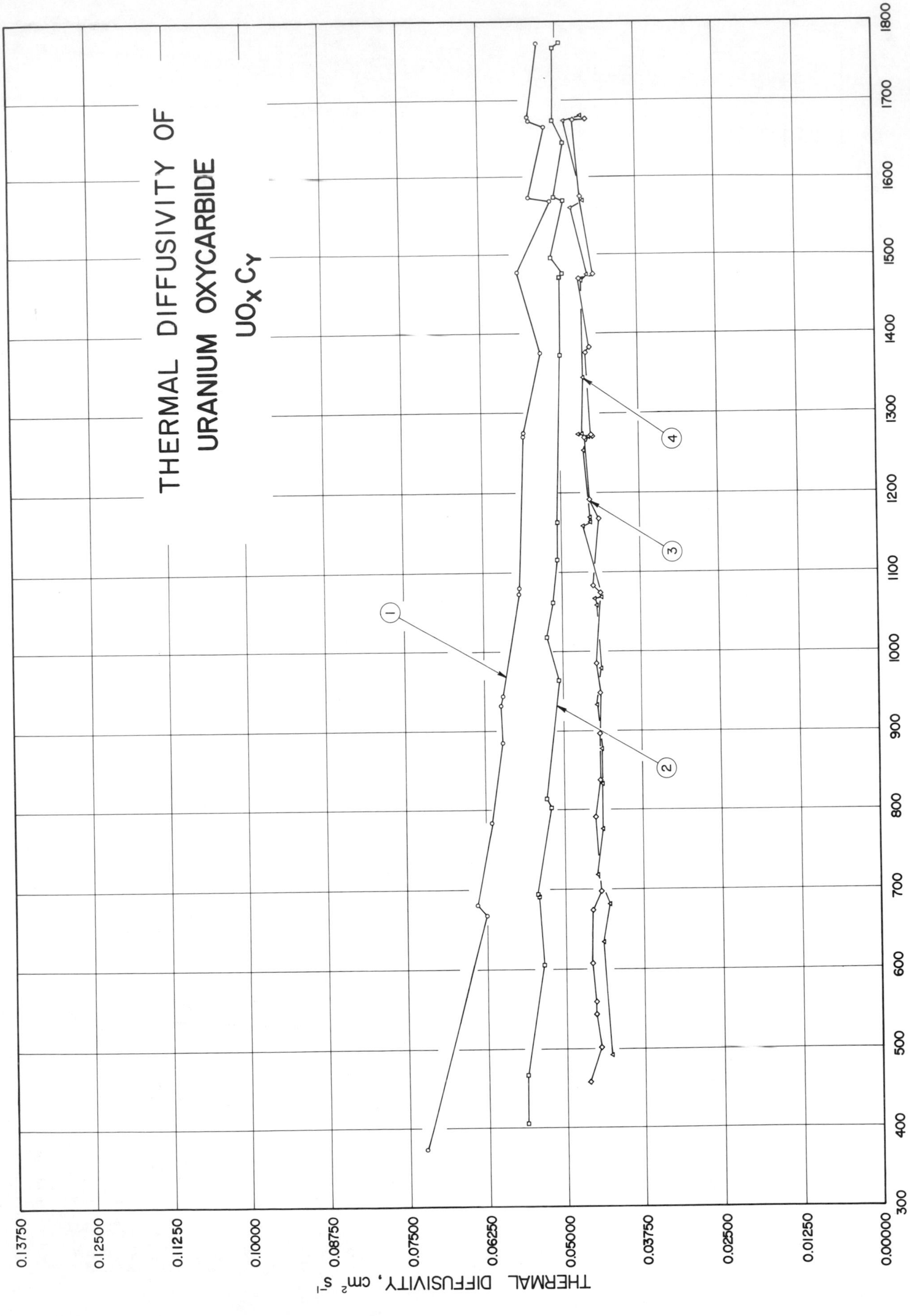
THERMAL DIFFUSIVITY OF
URANIUM OXYCARBIDE
UOX CY
THERMAL DIFFUSIVITY, cm2 s-1
TEMPERATURE, K
0.13750
0.12500
0.11250
0.10000
0.08750
0.07500
0.06250
0.05000
0.03750
0.02500
0.01250
0.00000
300
400
500
600
700
800
900
1000
1100
1200
1300
1400
1500
1600
1700
1800
1
2
3
4

FIGURE 206

SPECIFICATION TABLE 206. THERMAL DIFFUSIVITY OF URANIUM OXYCARBIDE UO_xC_y

Cur. No.	Ref. No.	Author(s)	Year	Temp. Range, K	Reported Error, %	Name and Specimen Designation	Composition (weight percent), Specifications, and Remarks
1	128, 246	Bates, J. L.	1968	374-1773		$U_{0.495}C_{0.485}O_{0.02}$; UCON-283	95.043 U, 4.699 C, and 0.258 O; single phase with minor traces of free uranium and UC_2 (Widmanstätten precipitate); disk specimen 0.635 cm in dia and ~0.100 cm thick; fabricated by and obtained from US Bureau of Mines, Albany Research Laboratory; prepared by reaction sintering of calculated quantities of uranium, graphite, and UO_2 powders; compacts pulverized and milled to a fine powder in an all-nickel mill under highly purified argon; pellets cold pressed and sintered at 1973.2 K for 2 hrs in an atmosphere of carbon monoxide at or near decomposition pressure for specimen; cut into thin disk from as-fabricated cylinder; density 12.7 g cm^{-3}; measured for diffusivity using laser-pulse technique; energy pulse of 0.54 m sec width provided by ruby laser; measured in a purified argon atmosphere; inlet argon containing < 0.0001 O and < 0.0005 H_2O; measured under a pressure of one atmosphere; heated initially to 1273.2 K in vertical tungsten tube furnace and measured during increase and decrease in temp, removed from furnace to check for possible reaction with UO_2 holder and then reinserted in furnace to carry out diffusivity measurements to ~1773.2 K; measured data corrected for heat losses.
2	128, 246	Bates, J. L.	1968	406-1774		$U_{0.48}C_{0.49}O_{0.03}$; UCON-365	94.723 U, 4.879 C, and 0.398 O; two phase; deliberately made more hyperstoichiometric than the above specimen resulting in a higher Widmanstätten precipitate of UC_2; contains also trace of UO_2; precipitate of secondary sesquicarbide U_2C_3 evident in the grain boundaries after diffusivity measurement; disk specimen 0.635 cm in dia and ~0.100 cm thick; fabricated by and obtained from US Bureau of Mines, Albany Research Laboratory; prepared by reaction sintering of calculated quantities of uranium, graphite, and UO_2 powders; compacts pulverized and milled to a fine powder in an all-nickel mill under highly purified argon; pellets cold pressed and sintered at 1973.2 K for 2 hrs in an atmosphere of carbon monoxide at or near decomposition pressure for specimen; cut into thin disk from as-fabricated cylinder; density 13.1 g cm^{-3}; measured for diffusivity using laser-pulse technique; energy pulse of 0.54 m sec width provided by ruby laser; measured in a purified argon atmosphere; inlet argon containing < 0.0001 O and < 0.0005 H_2O; measured under a pressure of one atmosphere; heated initially to 1273.2 K in vertical tungsten tube furnace and measured during increase and decrease in temp, removed from furnace to check for possible reaction with UO_2 holder and then reinserted in furnace to carry out diffusivity measurements to ~1773.2 K; measured data corrected for heat losses.
3	246	Bates, J. L.	1969	458-1675		$U_{0.495}C_{0.355}O_{0.15}$; UCON-288	Two phase UC_xO_y and UO_2 with traces of free U; same fabrication method and measuring method as above; density 12.5 g cm^{-3}.
4	246	Bates, J. L.	1969	491-1680		$U_{0.495}C_{0.355}O_{0.17}$; UCON-289	Similar to above but specimen density 12.3 g cm^{-3}.

DATA TABLE 206. THERMAL DIFFUSIVITY OF URANIUM OXYCARBIDE UO_xC_y

[Temperature, T, K; Thermal Diffusivity, α, $cm^2\ s^{-1}$]

T	α
CURVE 1	
374.2	0.0723
668.2	0.0625
681.2	0.0639
785.2	0.0614
885.2	0.0596
933.2	0.0598
945.2	0.0597
1073.2	0.0569
1081.2	0.0568
1270.2	0.0559
1275.2	0.0558
1275.2	0.0546
1376.2	0.0530
1479.2	0.0564
1570.2	0.0512
1575.2	0.0547
1666.2	0.0521
1673.2	0.0544
1678.2	0.0546
1773.2	0.0530
CURVE 2	
406.2	0.0563
468.2	0.0563
606.2	0.0535
692.2	0.0542
694.2	0.0546
803.2	0.0521
815.2	0.0528
964.2	0.0507
1019.2	0.0525
1063.2	0.0514
1117.2	0.0507
1164.2	0.0506
1374.2	0.0498
1473.2	0.0498
1477.2	0.0493
1498.2	0.0512
1571.2	0.0490
1575.2	0.0507
1646.2	0.0490
1674.2	0.0506
1768.2	0.0505
1774.2	0.0493
CURVE 3	
458.2	0.0464
501.2	0.0447
543.2	0.0453
558.2	0.0453
607.2	0.0458
673.2	0.0457
697.2	0.0444
790.2	0.0452
837.2	0.0442
896.2	0.0443
947.2	0.0442
984.2	0.0447
1073.2	0.0438
1083.2	0.0451
1167.2	0.0441
1191.2	0.0455
1269.2	0.0462
1271.2	0.0449
1273.2	0.0452
1375.2	0.0459
1382.2	0.0453
1469.2	0.0468
1475.2	0.0446
1574.2	0.0465
1674.2	0.0475
1675.2	0.0455
CURVE 4	
491.2	0.0429
633.2	0.0440
681.2	0.0429
718.2	0.0448
776.2	0.0439
833.2	0.0439
877.2	0.0440
933.2	0.0447
979.2	0.0439
1058.2	0.0445
1067.2	0.0449
1069.2	0.0437
1158.2	0.0466
1163.2	0.0454
1169.2	0.0454
1253.2	0.0463
CURVE 4 (cont.)	
1271.2	0.0455
1273.2	0.0473
1274.2	0.0466
1344.2	0.0463
1467.2	0.0465
1475.2	0.0456
1560.2	0.0481
1570.2	0.0461
1673.2	0.0489
1680.2	0.0462

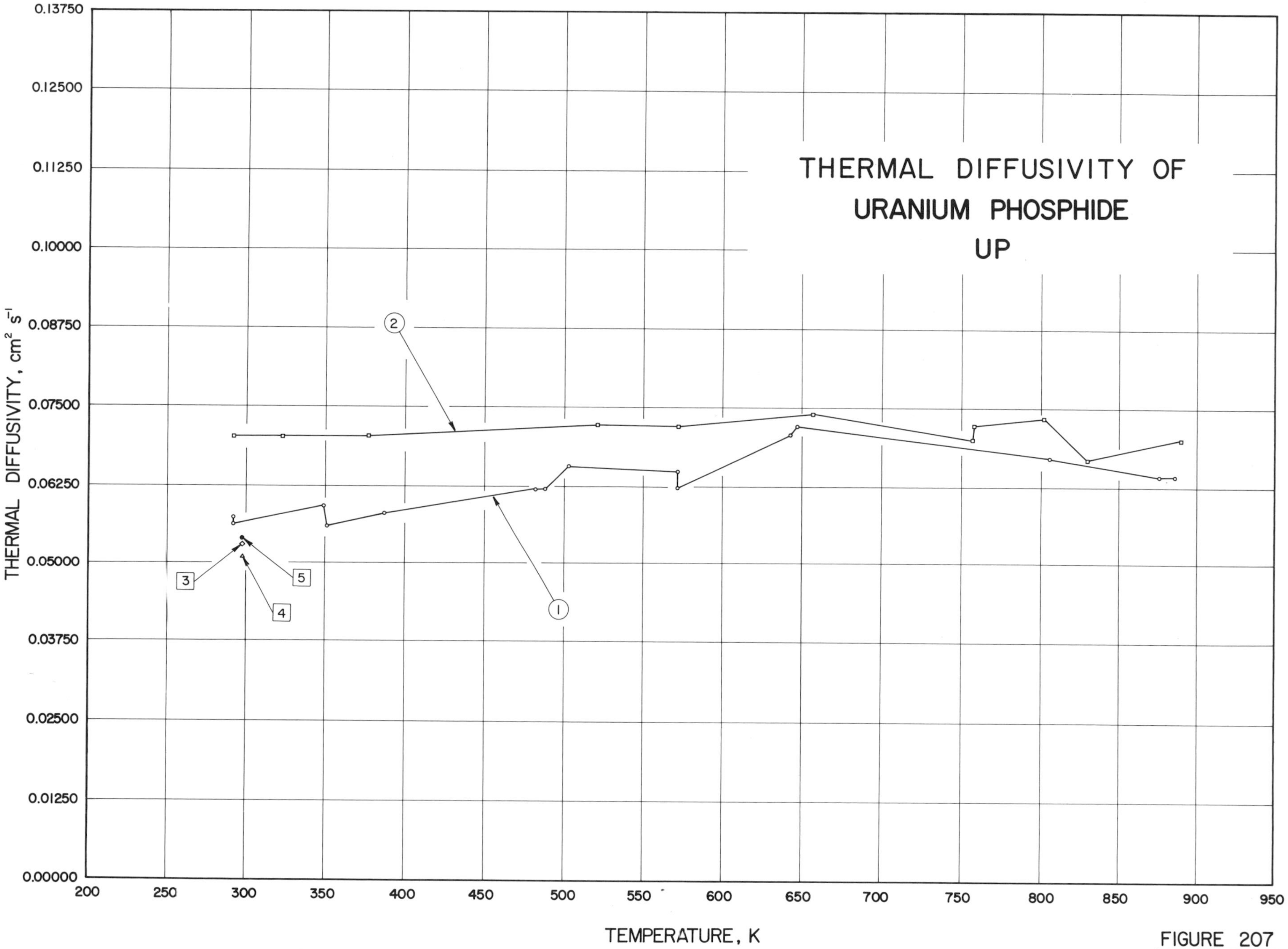
THERMAL DIFFUSIVITY OF
URANIUM PHOSPHIDE
UP
THERMAL DIFFUSIVITY, cm² s⁻¹
TEMPERATURE, K
0.13750
0.12500
0.11250
0.10000
0.08750
0.07500
0.06250
0.05000
0.03750
0.02500
0.01250
0.00000
200
250
300
350
400
450
500
550
600
650
700
750
800
850
900
950
1
2
3
4
5

FIGURE 207

SPECIFICATION TABLE 207. THERMAL DIFFUSIVITY OF URANIUM PHOSPHIDE UP

Cur. No.	Ref. No.	Author(s)	Year	Temp. Range, K	Reported Error, %	Name and Specimen Designation	Composition (weight percent), Specifications, and Remarks
1	137	Moser, J. B. and Kruger, O. L.	1967	292-886		Batch No. I	0.26 O; second phase 5 percent (by volume); disc-shaped specimen 1.9 cm in dia. and 0.2-0.3 cm thick; originally made in the form of cylinder by hot pressing of powder at 350 Kg cm^{-2} and 1673.2 K for 30-60 min, then electrical-discharge machined to a thickness of ~0.5 cm, and then lapped to final thickness with parallel smooth faces; after lapping a platinum foil ~0.32 cm in dia. and 0.002 cm thick fused to the center of one face; front face coated with colloidal graphite; density 90 percent of theoretical value; front face heated by discharge from ruby laser; diffusivity determined from measured temp. history of rear face; measured in vacuum.
2	137	Moser, J. B. and Kruger, O. L.	1967	292-890		Batch No. II	0.13 O, and 0.09 N; second phase 1 percent (by volume); disc-shaped specimen 1.9 cm in dia. and 0.2-0.3 cm thick; originally made in the form of cylinder by hot pressing of powder at 350 Kg cm^{-2} and 1673.2 K for 30-60 min, then electrical-discharge machined to a thickness of ~0.5 cm, and then lapped to final thickness with parallel smooth faces; after lapping a platinum foil ~0.32 cm in dia and 0.002 cm thick fused to the center of one face; front face coated with colloidal graphite; density 93 percent of theoretical value; front face heated by discharge from ruby laser; diffusivity determined from measured temp. history of rear face; measured in vacuum.
3	214, 215	Moser, J. B. and Kruger, O. L.	1964	298.2			10.83 P and 0.26 O; disk specimen 1.9 cm in diameter; hot-pressed; 73.5% theoretical density.
4	214, 215	Moser, J. B. and Kruger, O. L.	1964	298.2			Similar to the above specimen but 82.6% theoretical density.
5	214, 215	Moser, J. B. and Kruger, O. L.	1964	298.2			Similar to the above specimen but 88.8% theoretical density.

DATA TABLE 207. THERMAL DIFFUSIVITY OF URANIUM PHOSPHIDE UP

[Temperature, T, K; Thermal Diffusivity, α, cm^2 s^{-1}]

T	α	T	α	T	α	T	α
CURVE 1		CURVE 1 (cont.)		CURVE 2 (cont.)		CURVE 3	
292	0.0574	643	0.0708	377	0.0705	298.2	0.053
292	0.0564	647	0.0723	521	0.0723		
349	0.0593	806	0.0673	572	0.0721	CURVE 4	
351	0.0561	876	0.0643	657	0.0742		
387	0.0582	886	0.0643	757	0.0702	298.2	0.051
482	0.0621			758	0.0724		
488	0.0621	CURVE 2		802	0.0736	CURVE 5	
503	0.0658			830	0.0694		
572	0.0650	292	0.0703	890	0.0702	298.2	0.054
572	0.0624	323	0.0703				

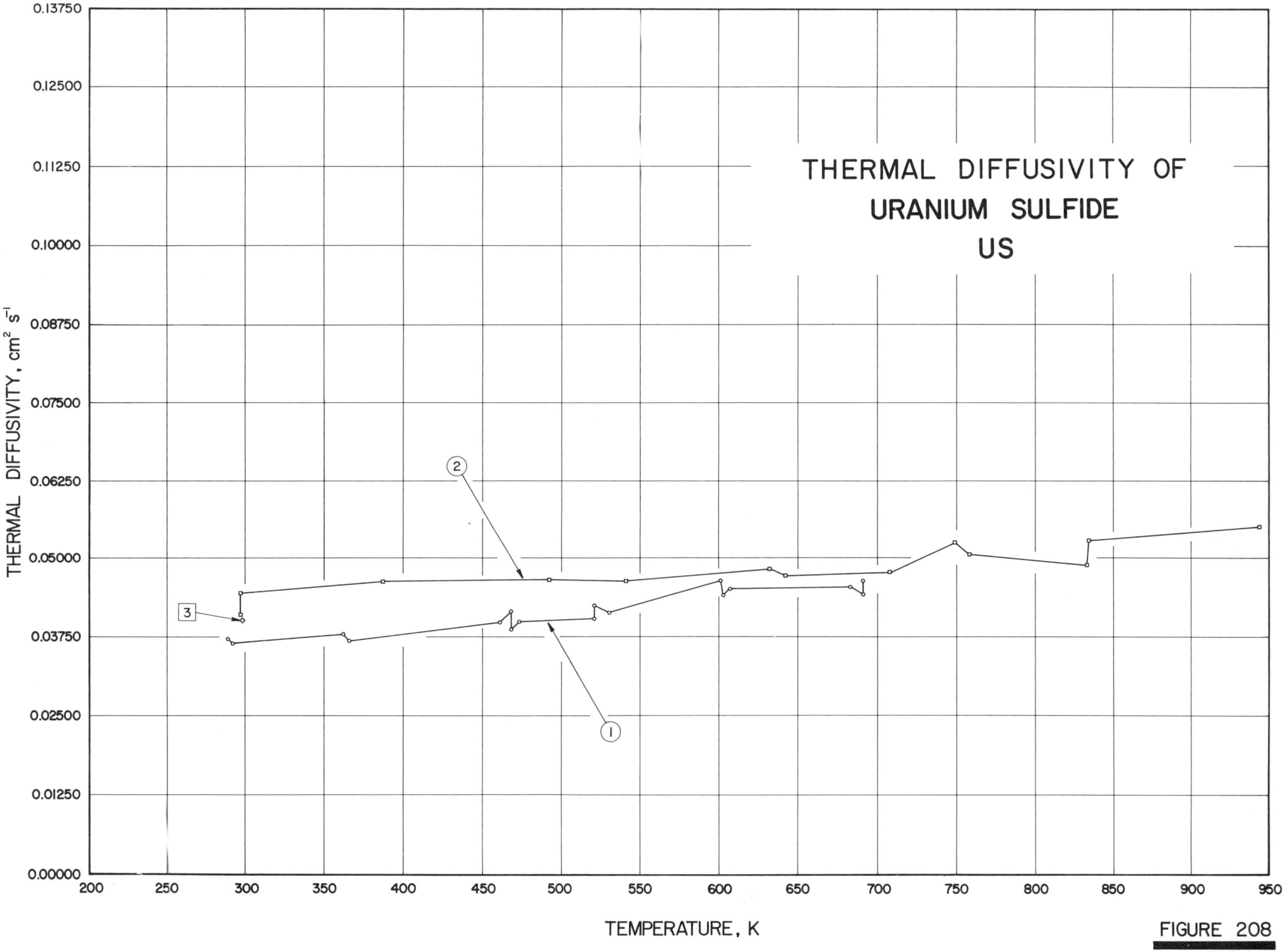
THERMAL DIFFUSIVITY OF
URANIUM SULFIDE
US
THERMAL DIFFUSIVITY, cm² s⁻¹
0.13750
0.12500
0.11250
0.10000
0.08750
0.07500
0.06250
0.05000
0.03750
0.02500
0.01250
0.00000
200
250
300
350
400
450
500
550
600
650
700
750
800
850
900
950
TEMPERATURE, K
1
2
3

FIGURE 208

SPECIFICATION TABLE 208. THERMAL DIFFUSIVITY OF URANIUM SULFIDE US

Cur. No.	Ref. No.	Author(s)	Year	Temp. Range, K	Reported Error, %	Name and Specimen Designation	Composition (weight percent), Specifications, and Remarks
1	137	Moser, J. B. and Kruger, O. L.	1967	289-691		Batch No. I	0.1-0.2 O, and 0.03-0.07 N; second phase 10-15 percent (by volume); disc-shaped specimen 1.9 cm in dia. and 0.2-0.3 cm thick; originally made in the form of cylinder by hot pressing of powder at 350 Kg cm^{-2} and 1673.2 K for 30-60 min, then electrical-discharge machined to a thickness of ~0.5 cm, and then lapped to final thickness with parallel smooth faces; after lapping a platinum foil ~0.32 cm in dia. and 0.002 cm thick fused to the center of one face; front face coated with colloidal graphite; density 95 percent of theoretical value; front face heated by discharge from ruby laser; diffusivity determined from measured temp. history of rear face; measured in vacuum.
2	137	Moser, J. B. and Kruger, O. L.	1967	297-944		Batch No. II	0.3 O; second phase 3 percent (by volume); disc-shaped specimen 1.9 cm in dia. and 0.2-0.3 cm thick; originally made in the form of cylinder by cold pressing of powder at 350 Kg cm^{-2}, then sintering at 2073.2 K for 2-4 hrs in flowing high purity argon, then electrical-discharge machined to a thickness of ~0.5 cm, and then lapped to final thickness with parallel smooth faces; after lapping a platinum foil ~0.32 cm in dia. and 0.002 cm thick fused to the center of one face; front face coated with colloidal graphite; density 93 percent of theoretical value; front face heated by discharge from ruby laser; diffusivity determined from measured temp. history of rear face; measured in vacuum.
3	214, 215	Moser, J. B. and Kruger, O. L.	1964	298.2			10.77 S and 0.32 O; 1.9 cm in diameter and 0.16 to 0.32 cm thick; hot-pressed in argon at 1500 C; 96.4 theoretical density.

DATA TABLE 208. THERMAL DIFFUSIVITY OF URANIUM SULFIDE US

[Temperature, T, K; Thermal Diffusivity, α, cm^2 s^{-1}]

T	α	T	α	T	α
CURVE 1		CURVE 1 (cont.)		CURVE 2 (cont.)	
289	0.0371	607	0.0452	642	0.0473
292	0.0364	683	0.0455	708	0.0478
362	0.0378	691	0.0443	749	0.0525
366	0.0368	691	0.0465	758	0.0506
461	0.0398			833	0.0490
468	0.0415	CURVE 2		834	0.0528
468	0.0387			944	0.0550
473	0.0398	297	0.0410		
521	0.0404	297	0.0443	CURVE 3	
521	0.0425	387	0.0462		
530	0.0413	492	0.0466	298.2	0.040
601	0.0465	541	0.0463		
603	0.0442	632	0.0483		

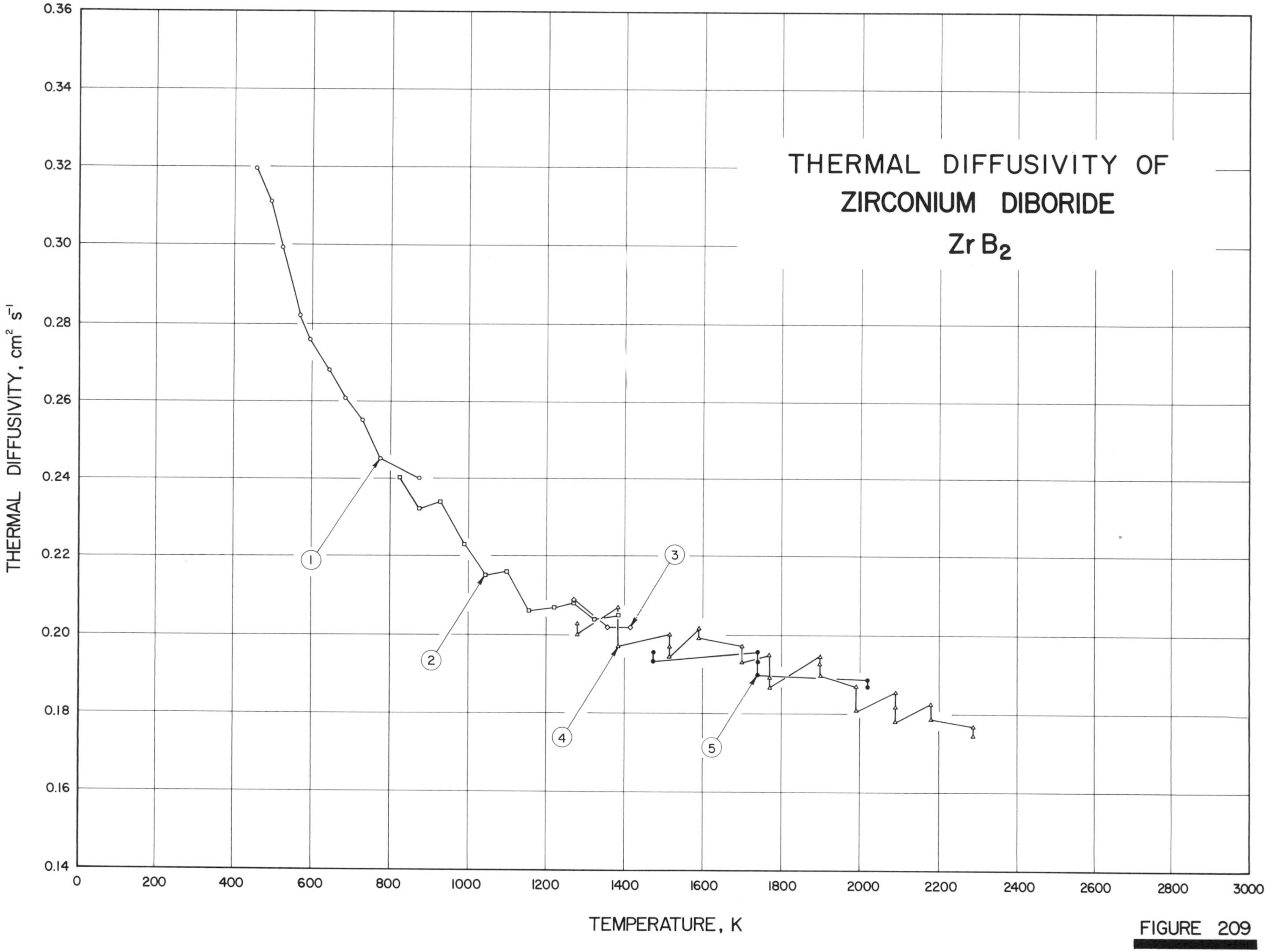

THERMAL DIFFUSIVITY OF
ZIRCONIUM DIBORIDE
ZrB_2
THERMAL DIFFUSIVITY, $cm^2\ s^{-1}$
0.36
0.34
0.32
0.30
0.28
0.26
0.24
0.22
0.20
0.18
0.16
0.14
0
200
400
600
800
1000
1200
1400
1600
1800
2000
2200
2400
2600
2800
3000
TEMPERATURE, K
1
2
3
4
5

FIGURE 209

SPECIFICATION TABLE 209. THERMAL DIFFUSIVITY OF ZIRCONIUM DIBORIDE ZrB_2

Cur. No.	Ref. No.	Author(s)	Year	Temp. Range, K	Reported Error, %	Name and Specimen Designation	Composition (weight percent), Specifications, and Remarks
1	235	Branscomb, T. M.	1970	458-873			99 pure; 0.345 cm thick; obtained from Carborundum Co.; hot-pressed; Pt-black coated on front face; density 5.994 g cm^{-3}.
2	235	Branscomb, T. M.	1970	823-1383			Similar to the above specimen but 0.278 cm in thickness.
3	235	Branscomb, T. M.	1970	1268-1413			Similar to the above specimen but 0.2515 cm in thickness.
4	275	Clougherty, E. V., Wilkes, K. E., and Tye, R. P.	1969	1278-2288	±5-8	I05 R44L	Right circular cylindrical disk specimen, 0.500 in. in diameter and 0.100 in. thick with ends flat and parallel to ±0.001 in.; grain size 28.5 μ; diffusivity measured at increasing temperatures.
5	275	Clougherty, E. V., et al.	1969	1472-2019	±5-8	I05 R44L	The above specimen; diffusivity measured at decreasing temperatures.

DATA TABLE 209. THERMAL DIFFUSIVITY OF ZIRCONIUM DIBORIDE ZrB_2

[Temperature, T, K; Thermal Diffusivity, α, $cm^2\ s^{-1}$]

T	α	T	α	T	α	T	α	T	α
CURVE 1		CURVE 2 (cont.)		CURVE 4		CURVE 4 (cont.)		CURVE 5 (cont.)	
458	0.3195	1043	0.215	1278	0.2034	1897	0.1899	1739	0.1901
493	0.311	1098	0.216	1278	0.2006	1990	0.1870	2019	0.1888
523	0.299	1153	0.206	1381	0.2072	1990	0.1806	2019	0.1871
568	0.282	1217	0.207	1381	0.2035	2089	0.1855		
593	0.276	1268	0.208	1381	0.1972	2089	0.1818		
643	0.268	1323	0.204	1513	0.2004	2089	0.1780		
683	0.261	1383	0.205	1513	0.1972	2180	0.1824		
728	0.255			1513	0.1947	2180	0.1786		
773	0.245	CURVE 3		1589	0.2018	2288	0.1765		
873	0.240			1589	0.1993	2288	0.1743		
		1268	0.209	1699	0.1973				
CURVE 2		1353	0.202	1699	0.1932	CURVE 5			
		1413	0.202	1768	0.1950				
823	0.240			1770	0.1892	1472	0.1957		
873	0.232			1770	0.1867	1472	0.1935		
928	0.234			1897	0.1947	1739	0.1958		
988	0.223			1897	0.1929	1739	0.1935		

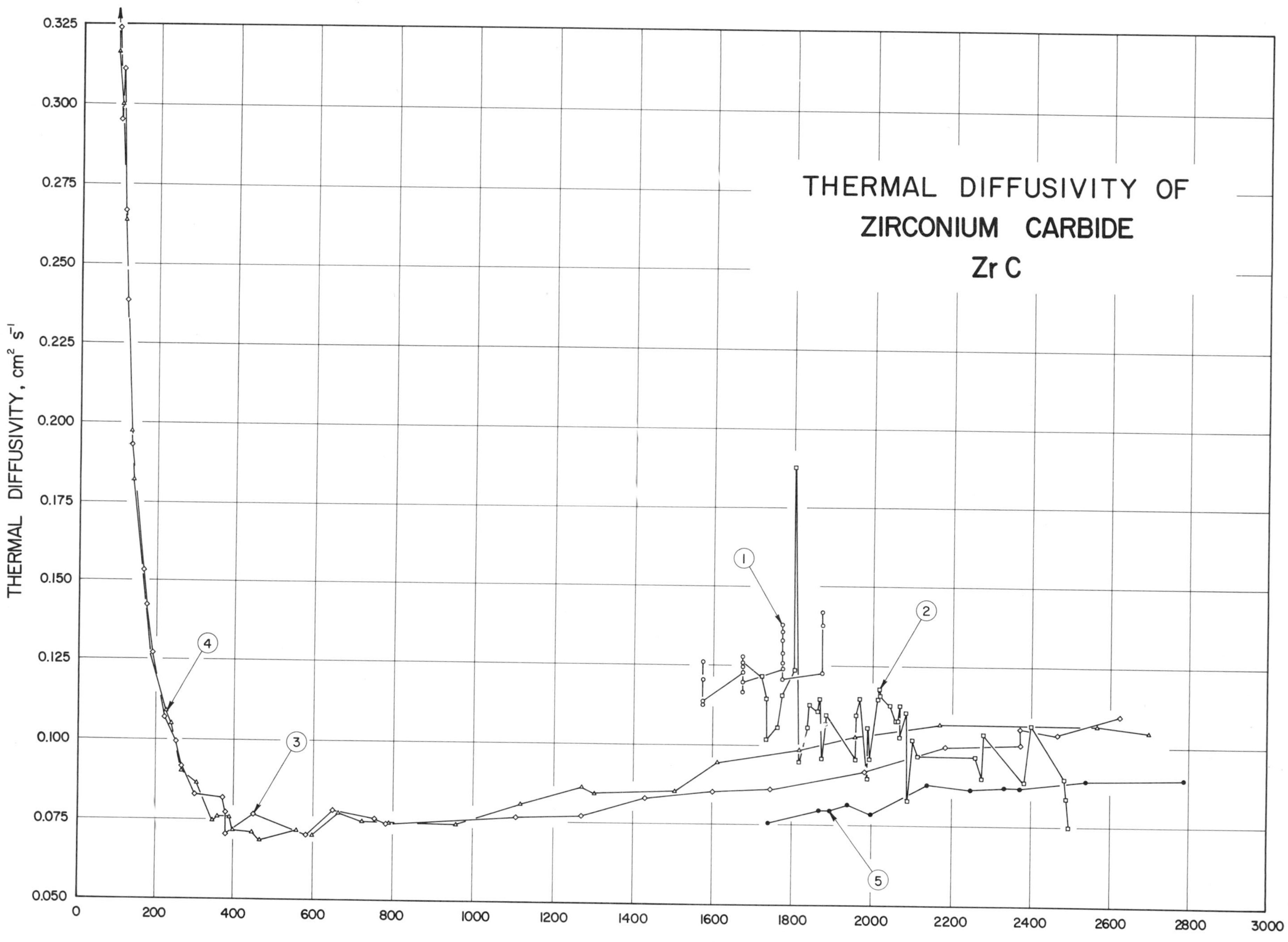
THERMAL DIFFUSIVITY OF
ZIRCONIUM CARBIDE
Zr C
THERMAL DIFFUSIVITY, cm² s⁻¹
TEMPERATURE, K
0.050
0.075
0.100
0.125
0.150
0.175
0.200
0.225
0.250
0.275
0.300
0.325
0
200
400
600
800
1000
1200
1400
1600
1800
2000
2200
2400
2600
2800
3000
1
2
3
4
5

FIGURE 210

SPECIFICATION TABLE 210. THERMAL DIFFUSIVITY OF ZIRCONIUM CARBIDE ZrC

Cur. No.	Ref. No.	Author(s)	Year	Temp. Range, K	Reported Error, %	Name and Specimen Designation	Composition (weight percent), Specifications, and Remarks
1	254, 32, 50, 182	Taylor, R. E. and Nakata, M. M.	1962	1573-1873	7	1	89.8 Zr, 11.0 C, and < 0.2 metallic impurities (as received); average grain size (ASTM) No. 10, single phase; cylindrical specimen 1.588 cm in diameter and 3.492 cm long; two parallel sight holes each 0.170 cm in diameter drilled to a depth of 1.56 cm at radii r_1 = 0 and r_2 = 0.568 cm; hot pressed by the Carborundum Co. from raw powder having chemical composition: 87.94 Zr, 11.30 C, and < 0.3 metallic impurities, machined from a sample which was used as a thermal conductivity sample, sight holes drilled by Elox technique; measured for diffusivity after extensive heat soaking at temperatures up to 2673.2 K; density 6.24 g cm^{-3}; radiation shields mounted at both ends; measured under a vacuum of 1 x 10^{-6} mm Hg; radial diffusivity technique used.
2	32	Taylor, R. E. and Nakata, M. M.	1963	1673-2493	7	2	87.5 Zr, 11.2 C, >0.05 Ti, 0.05 Hf, 0.05 Mo, 0.035 Ca, 0.02 Al, >0.012 V, 0.008 Si, <0.005 Na, 0.004 Fe, 0.002 B, 0.002 Cr, 0.001 Ba (as received); average grain size (ASTM) No. 8.5; cylindrical specimen 1.577 cm overall diameter and 3.442 cm overall length, machined in three sections: a center section 2.54 cm long and two end pieces consisting of 0.451 cm thick discs; these parts are machine fitted in such a way that a small area contact is made at the circumference only, leaving a thin space of 0.0127 cm or less between center section and the discs that act as radiation barriers; two parallel sight holes each 0.117 cm in diameter drilled through the top disc to a depth of 1.71 cm at radii r_1 = 0 and r_2 = 0.564 cm; sight holes drilled by Elox technique; bottom disc grooved to fit sample supporting pins; machined from cylindrical compacts that were hot pressed by the Carborundum Co. from raw powder having chemical composition 87.8 Zr, 11.5 C, and 0.1 Fe; measured after extensive heat soaking at temperatures up to 2673.2 K; density 6.46 g cm^{-3}; radiation shields mounted at both ends; measured under a vacuum of 1 x 10^{-6} mm Hg; radial diffusivity technique used.
3	234	Morrison, B. H. and Sturgess, L. L.	1970	95-2624		1	81.7 Zr, 11.2 C (0.6 free C), 0.6 O, < 0.0200 W, < 0.0100 each of Cr, Mo, and Nb, < 0.0030 each of Ba, Co, K, Sr, and Ti, < 0.0010 each of Al, Bi, Cu, Pb, Ni, Si, Sn, and V, < 0.0003 each of B, Ca, Li, Mn, Ag, and Na, < 0.0001 Be, and < 0.0001 Mg; 1.26 cm diameter x 0.100 cm thick; hot-pressed at 3000 K and 3400 psi for 10 min; density 6.26 g cm^{-3}.
4	234	Morrison, B. H. and Sturgess, L. L.	1970	93-2698		2	Similar to the above specimen.
5	104	Fridlender, B. A. and Neshpor, V. S.	1970	1740-2789		$ZrC_{0.99}$	88.31 Zr, 11.62 total and 0.06 free C, 0.03 N; single-phase cubic ZrC with lattice constant 4.680 ± 0.001 Å; density 6.67 g cm^{-3}.

DATA TABLE 210. THERMAL DIFFUSIVITY OF ZIRCONIUM CARBIDE ZrC

[Temperature, T, K; Thermal Diffusivity, α, cm^2 s^{-1}]

T	α
CURVE 1	
1573.2	0.126
1573.2	0.113
1573.2	0.121
1573.2	0.114
1673.2	0.123
1673.2	0.125
1673.2	0.128
1673.2	0.117
1673.2	0.126
1673.2	0.120
1673.2	0.120*
1773.2	0.124
1773.2	0.136
1773.2	0.129
1773.2	0.126
1773.2	0.138
1773.2	0.133
1773.2	0.121
1873.2	0.123
1873.2	0.142
1873.2	0.138
CURVE 2	
1673.2	0.126*
1723.2	0.122
1733.2	0.115
1733.2	0.102
1763.2	0.106
1773.2	0.116
1803.2	0.124
1803.2	0.188
1818.2	0.095
1838.2	0.106
1843.2	0.113
1863.2	0.111
1868.2	0.115
1873.2	0.096
1883.2	0.107
1883.2	0.110
1958.2	0.096
1958.2	0.110
CURVE 2 (cont.)	
1968.2	0.115
1988.2	0.090
1988.2	0.106
1993.2	0.096
2013.2	0.115
2018.2	0.118
2018.2	0.116
2043.2	0.113
2058.2	0.108
2063.2	0.108
2068.2	0.113
2068.2	0.103
2083.2	0.111
2088.2	0.083
2098.2	0.102
2113.2	0.097
2258.2	0.097
2273.2	0.090
2278.2	0.104
2383.2	0.089
2398.2	0.107
2483.2	0.090
2488.2	0.084
2493.2	0.075
CURVE 3	
94.8	0.324
98.6	0.295
105.2	0.311
110.2	0.267
119.7	0.238
132.1	0.193
165.2	0.153
174.2	0.142
190.1	0.127
222.8	0.107
250.0	0.0995
266.7	0.0918
300.6	0.0830
371.5	0.0818
379.3	0.0774
CURVE 3 (cont.)	
379.3	0.0705
449.8	0.0769
582.1	0.0703
648.6	0.0782
755.0	0.0757
781.6	0.0741
1107	0.0769
1271	0.0774
1429	0.0834
1600	0.0855
1746	0.0863
1982	0.0923
2183	0.100
2371	0.101
2371	0.106
2466	0.104
2624	0.110
CURVE 4	
92.7	0.364
92.7	0.316
106.2	0.299
112.5	0.264
132.7	0.197
137.7	0.182
179.9	0.137
182.8	0.127
227.5	0.109
237.7	0.105
265.5	0.0951
265.5	0.0902
306.9	0.0865
345.1	0.0748
358.9	0.0759
389.9	0.0759
399.9	0.0718
445.7	0.0711
466.7	0.0689
557.2	0.0718
597.0	0.0703
663.7	0.0771
CURVE 4 (cont.)	
722.8	0.0746
788.9	0.0791
957.2	0.0791
1117	0.0809
1271	0.0865
1303	0.0845
1507	0.0853
1611	0.0948
1816	0.0989
1959	0.103
2168	0.107
2564	0.107
2698	0.105
CURVE 5	
1740	0.0756
1869	0.0798
1895	0.0798
1939	0.0817
1998	0.0787
2137	0.0881
2245	0.0867
2331	0.0873
2371	0.0873
2539	0.0898
2789	0.0904

*Not shown in figure.

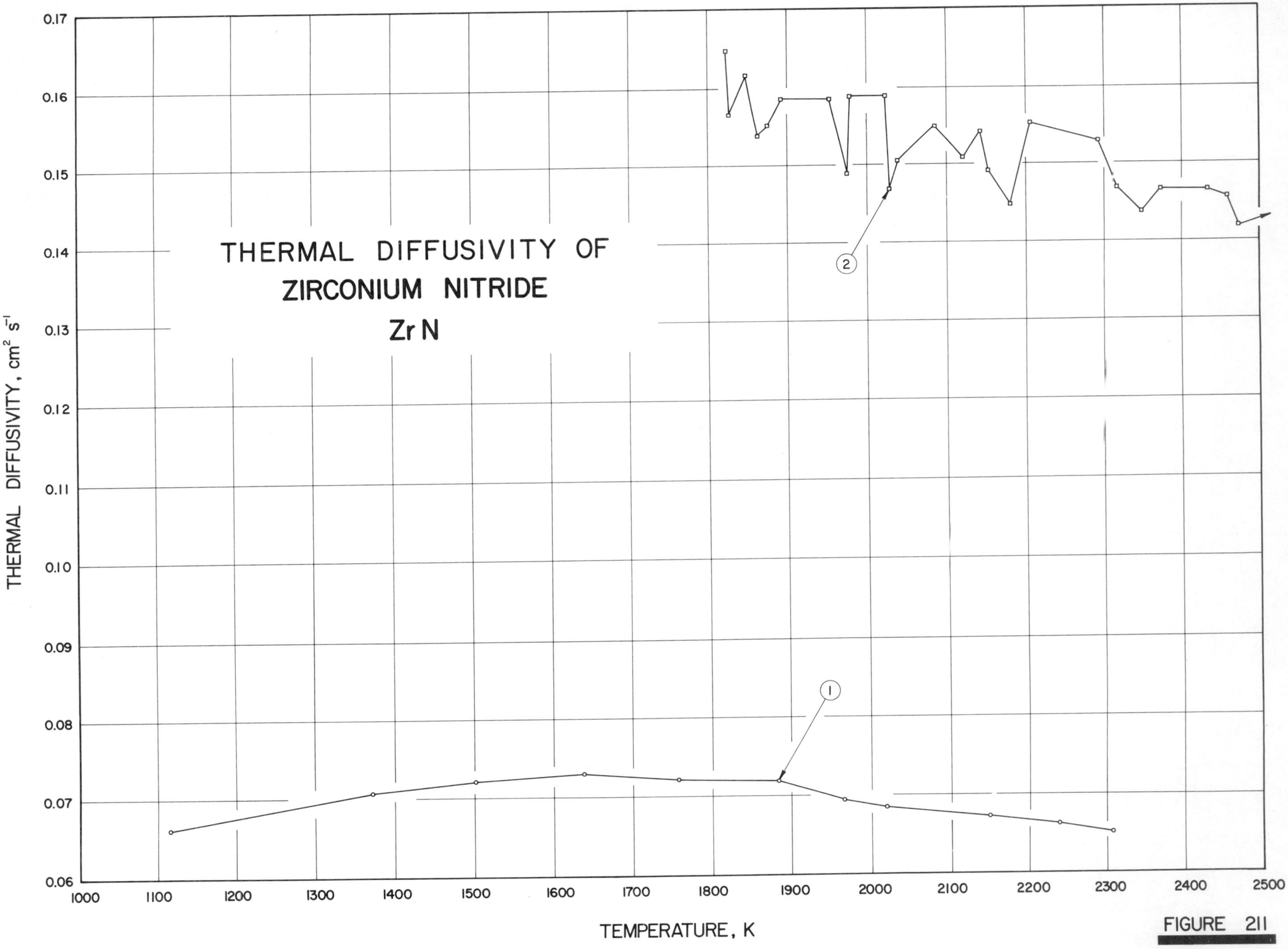
THERMAL DIFFUSIVITY OF
ZIRCONIUM NITRIDE
Zr N
THERMAL DIFFUSIVITY, cm² s⁻¹
TEMPERATURE, K
0.06
0.07
0.08
0.09
0.10
0.11
0.12
0.13
0.14
0.15
0.16
0.17
1000
1100
1200
1300
1400
1500
1600
1700
1800
1900
2000
2100
2200
2300
2400
2500
1
2

FIGURE 211

SPECIFICATION TABLE 211. THERMAL DIFFUSIVITY OF ZIRCONIUM NITRIDE ZrN

Cur. No.	Ref. No.	Author(s)	Year	Temp. Range, K	Reported Error, %	Name and Specimen Designation	Composition (weight percent), Specifications, and Remarks
1	74	Hedge, J.C., Kopec, J.W., Kostenko, C., and Lang, J.I.	1963	1117-2308			84.6 Zr, 13.5 N, 0.8 H, 0.5 alkali metal oxides, 0.4 Si, and 0.2 Fe; slab specimen; supplied by Norton Co.; hot-pressed; fired at 2373.2 K; density 6.50 g cm^{-3}; top surface of specimen exposed to heat sink; diffusivity determined from measured temperature decrease; unidirectional heat flow.
2	277	Fridlender, B.A. and Neshpor, V.S.	1970	1823-2586			0.25 to 0.30 mm thick.

DATA TABLE 211. THERMAL DIFFUSIVITY OF ZIRCONIUM NITRIDE ZrN

[Temperature, T, K; Thermal Diffusivity, α, $cm^2\ s^{-1}$]

T	α	T	α	T	α
CURVE 1		CURVE 2 (cont.)		CURVE 2 (cont.)	
1116.5	0.0661	1875	0.1552	2434	0.1466
1372.1	0.0707	1893	0.1586	2459	0.1457
1502.6	0.0720	1954	0.1586	2474	0.1419
1638.7	0.0728	1975	0.1489	2530	0.1436*
1758.2	0.0720	1979	0.1589	2558	0.1407*
1883.2	0.0718	2024	0.1589	2586	0.1419*
1966.5	0.0694	2028	0.1469		
2019.3	0.0684	2038	0.1504		
2149.8	0.0671	2086	0.1550		
2238.7	0.0661	2121	0.1508		
2307.6	0.0650	2143	0.1542		
		2153	0.1492		
CURVE 2		2181	0.1448		
		2206	0.1554		
1823	0.1647	2293	0.1529		
1827	0.1565	2317	0.1469		
1848	0.1616	2349	0.1439		
1862	0.1538	2373	0.1466		

* Not shown in figure.

10. MIXTURES OF NONOXIDES

SPECIFICATION TABLE 212. THERMAL DIFFUSIVITY OF AIR

Cur. No.	Ref. No.	Author(s)	Year	Temp. Range, K	Reported Error, %	Name and Specimen Designation	Composition (weight percent), Specifications, and Remarks
1*	168	Bomelburg, H. J.	1958	300-873			Technique based on measuring phase difference between temperature fluctuations generated by the passage of a 60 cycles per sec a.c. current in a platinum wire 10 microns in dia. acting as the transmitter and those picked up by another platinum wire 1 micron in dia. acting as the receiver; diffusivity determined from measured difference in phase lag corresponding to a known change in the distance between the axes of the wires.

DATA TABLE 212. THERMAL DIFFUSIVITY OF AIR

[Temperature, T, K; Thermal Diffusivity, α, $cm^2\ s^{-1}$]

T	α
CURVE 1*	
300.2	0.228
366.2	0.40
438.2	0.54
483.2	0.78
563.2	1.25
653.2	1.65
753.2	2.2
873.2	5.2

* No figure given.

SPECIFICATION TABLE 213. THERMAL DIFFUSIVITY OF [ANTIMONY SULFIDE + SULFUR] Sb_2S_3 + S

Cur. No.	Ref. No.	Author(s)	Year	Temp. Range, K	Reported Error, %	Name and Specimen Designation	Composition (weight percent) Sb_2S_3	Composition (weight percent) S	Composition (continued), Specifications, and Remarks
1*	166	Williams, I.	1923	318, 373				15.6	Density 3.20 g cm^{-3}.

DATA TABLE 213. THERMAL DIFFUSIVITY OF [ANTIMONY SULFIDE + SULFUR] Sb_2S_3 + S

[Temperature, T, K; Thermal Diffusivity, α, $cm^2\ s^{-1}$]

T	α
CURVE 1*	
318.2	0.00077
373.2	0.00077

* No figure given.

SPECIFICATION TABLE 214. THERMAL DIFFUSIVITY OF [BARIUM SULFATE + ZINC SULFIDE + ZINC OXIDE] $BaSO_4 + ZnS + ZnO$

Cur. No.	Ref. No.	Author(s)	Year	Temp. Range, K	Reported Error, %	Name and Specimen Designation	Composition (weight percent) $BaSO_4$	ZnS	ZnO	Composition (continued), Specifications, and Remarks
1*	166	Williams, I.	1923	318, 373		Lithopone				Density 3.95 g cm^{-3}.

DATA TABLE 214. THERMAL DIFFUSIVITY OF [BARIUM SULFATE + ZINC SULFIDE + ZINC OXIDE] $BaSO_4 + ZnS + ZnO$

[Temperature, T, K; Thermal Diffusivity, α, $cm^2\ s^{-1}$]

T	α
CURVE 1*	
318.2	0.00207
373.2	0.00207

* No figure given.

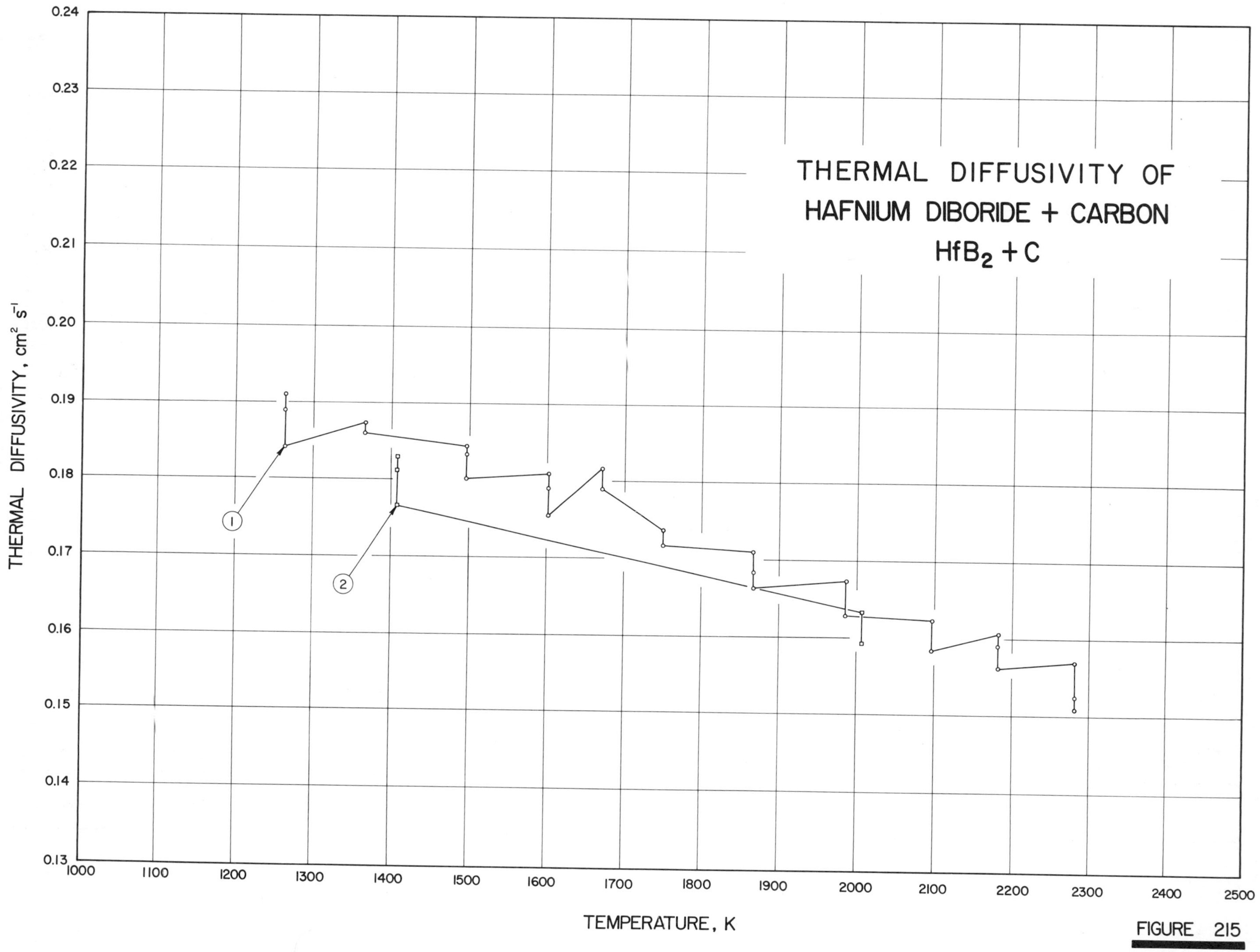
THERMAL DIFFUSIVITY OF
HAFNIUM DIBORIDE + CARBON
$HfB_2 + C$
THERMAL DIFFUSIVITY, $cm^2 s^{-1}$
TEMPERATURE, K
0.24
0.23
0.22
0.21
0.20
0.19
0.18
0.17
0.16
0.15
0.14
0.13
1000
1100
1200
1300
1400
1500
1600
1700
1800
1900
2000
2100
2200
2300
2400
2500
1
2

FIGURE 215

SPECIFICATION TABLE 215. THERMAL DIFFUSIVITY OF [HAFNIUM DIBORIDE + CARBON] $HfB_2 + C$

Cur. No.	Ref. No.	Author(s)	Year	Temp. Range, K	Reported Error, %	Name and Specimen Designation	Composition (weight percent) HfB_2	C	Composition (continued), Specifications, and Remarks
1	275	Clougherty, E. V., Wilkes, K. E. and Tye, R. P.	1969	1263-2282		XV(20) 10 D 1054K	→	→	HfB_2 with 50 vol. %C to enhance thermal stress resistance; right circular cylindrical disk specimen, 0.500 in. in diameter and 0.100 in. thick with ends flat and parallel to ±0.001 in.; diffusivity measured at increasing temperatures.
2	275	Clougherty, E. V., et al.	1969	1407-2007					The above specimen; diffusivity measured at decreasing temperatures.

DATA TABLE 215. THERMAL DIFFUSIVITY OF [HAFNIUM DIBORIDE + CARBON] $HfB_2 + C$

[Temperature, T, K; Thermal Diffusivity, α, $cm^2\ s^{-1}$]

T	α	T	α	T	α
CURVE 1		CURVE 1 (cont.)		CURVE 2	
1263	0.1910	1867	0.1665	1407	0.1829
1263	0.1899	1986	0.1675	1407	0.1811
1263	0.1841	1986	0.1630	1407	0.1766
1366	0.1874	2097	0.1624	2007	0.1635
1366	0.1860	2097	0.1585	2007	0.1594
1496	0.1844	2182	0.1607		
1496	0.1833	2182	0.1592		
1496	0.1801	2182	0.1562		
1601	0.1809	2282	0.1570		
1601	0.1791	2282	0.1526		
1601	0.1755	2282	0.1510		
1671	0.1817				
1671	0.1791				
1750	0.1738				
1750	0.1719				
1867	0.1711				
1867	0.1686				

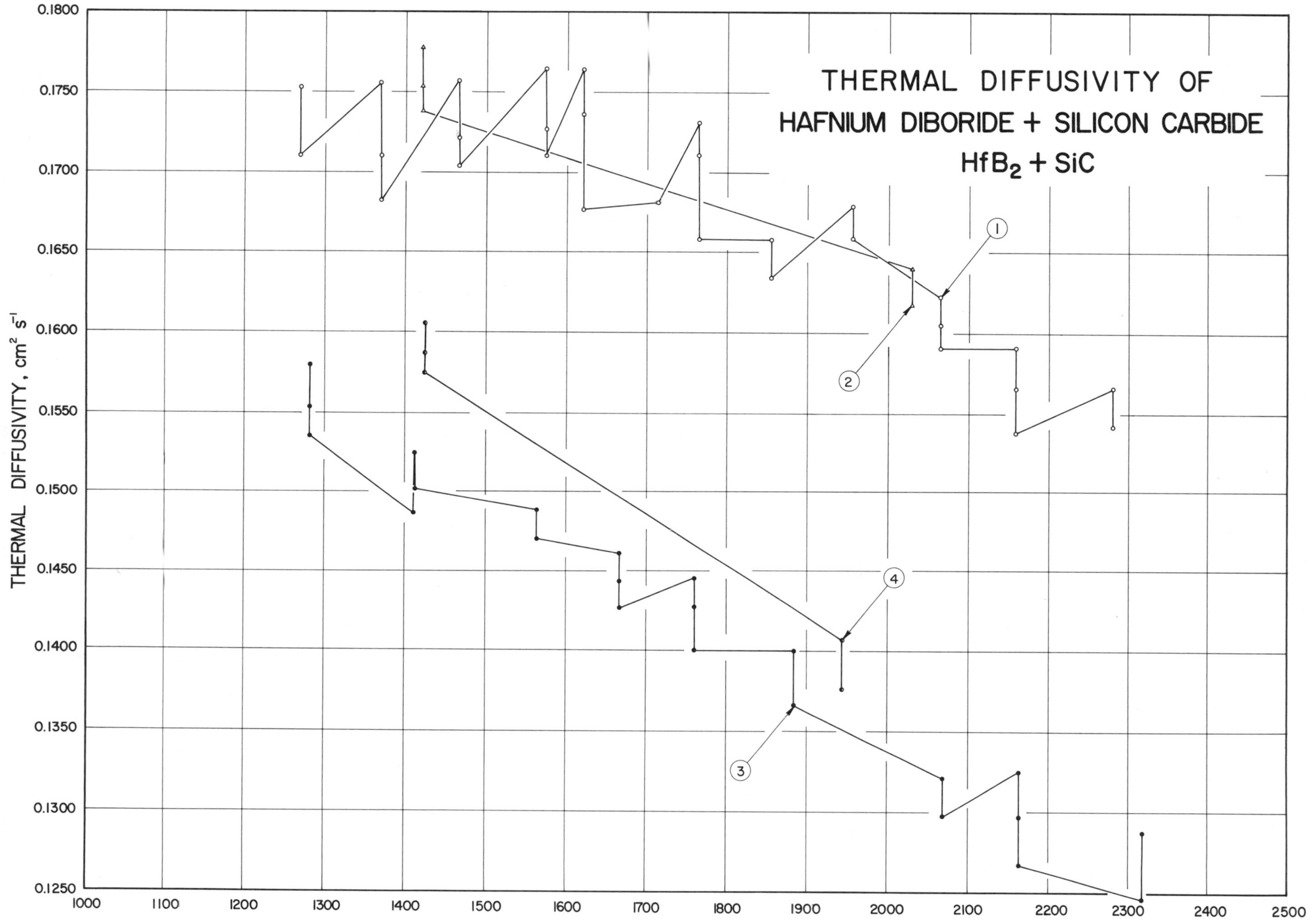
THERMAL DIFFUSIVITY OF
HAFNIUM DIBORIDE + SILICON CARBIDE
$HfB_2 + SiC$
THERMAL DIFFUSIVITY, $cm^2 s^{-1}$
TEMPERATURE, K
0.1800
0.1750
0.1700
0.1650
0.1600
0.1550
0.1500
0.1450
0.1400
0.1350
0.1300
0.1250
1000
1100
1200
1300
1400
1500
1600
1700
1800
1900
2000
2100
2200
2300
2400
2500
1
2
3
4

FIGURE 216

SPECIFICATION TABLE 216. THERMAL DIFFUSIVITY OF [HAFNIUM DIBORIDE + SILICON CARBIDE] $HfB_2 + SiC$

Cur. No.	Ref. No.	Author(s)	Year	Temp. Range, K	Reported Error, %	Name and Specimen Designation	Composition (weight percent) HfB_2	SiC	Composition (continued), Specifications, and Remarks
1	275	Clougherty, E.V., Wilkes, K.E., and Tye, R.P.	1969	1269-2280		III(5)09F D1061	→	→	HfB_2 with 20% (by volume) SiC to enhance oxidation resistance; right circular cylindrical disk specimen, 0.500 in. in diameter and 0.100 in. thick with ends flat and parallel to ± 0.001 in.; diffusivity measured at increasing temperatures.
2	275	Clougherty, E.V., et al.	1969	1420-2029		III(5)09F D1061			The above specimen; diffusivity measured at decreasing temperatures.
3	275	Clougherty, E.V., et al.	1969	1281-2318		III(5) 09F D0811K	→	→	Similar to the above specimen except HfB_2 with 30% (by volume) SiC; diffusivity measured at increasing temperatures.
4	275	Clougherty, E.V., et al.	1969	1423-1944		IV 09 F D0811K			The above specimen; diffusivity measured at decreasing temperatures.

DATA TABLE 216. THERMAL DIFFUSIVITY OF [HAFNIUM DIBORIDE + SILICON CARBIDE] $HfB_2 + SiC$

[Temperature, T, K; Thermal Diffusivity, α, $cm^2\ s^{-1}$]

T	α	T	α	T	α	T	α	T	α
CURVE 1		CURVE 1 (cont.)		CURVE 2		CURVE 3 (cont.)		CURVE 4	
1269	0.1752	1764	0.1658	1420	0.1777	1666	0.1443	1423	0.1605
1269	0.1710	1854	0.1658	1420	0.1754	1666	0.1426	1423	0.1587
1369	0.1755	1854	0.1634	1420	0.1738	1760	0.1445	1423	0.1575
1369	0.1710	1955	0.1679	2029	0.1640	1760	0.1427	1944	0.1407
1369	0.1682	1955	0.1659	2029	0.1618	1760	0.1400	1944	0.1376
1465	0.1756	2064	0.1323			1884	0.1400		
1465	0.1721	2064	0.1605	CURVE 3		1884	0.1366		
1465	0.1704	2064	0.1591			2069	0.1321		
1573	0.1764	2158	0.1591	1281	0.1579	2069	0.1298		
1573	0.1727	2158	0.1566	1281	0.1553	2163	0.1325		
1573	0.1710	2158	0.1538	1281	0.1535	2163	0.1297		
1620	0.1764	2280	0.1566	1410	0.1486	2163	0.1267		
1620	0.1736	2280	0.1542	1412	0.1524	2317	0.1246		
1620	0.1676			1412	0.1501	2318	0.1287		
1713	0.1681			1563	0.1488				
1764	0.1731			1563	0.1470				
1764	0.1711			1666	0.1461				

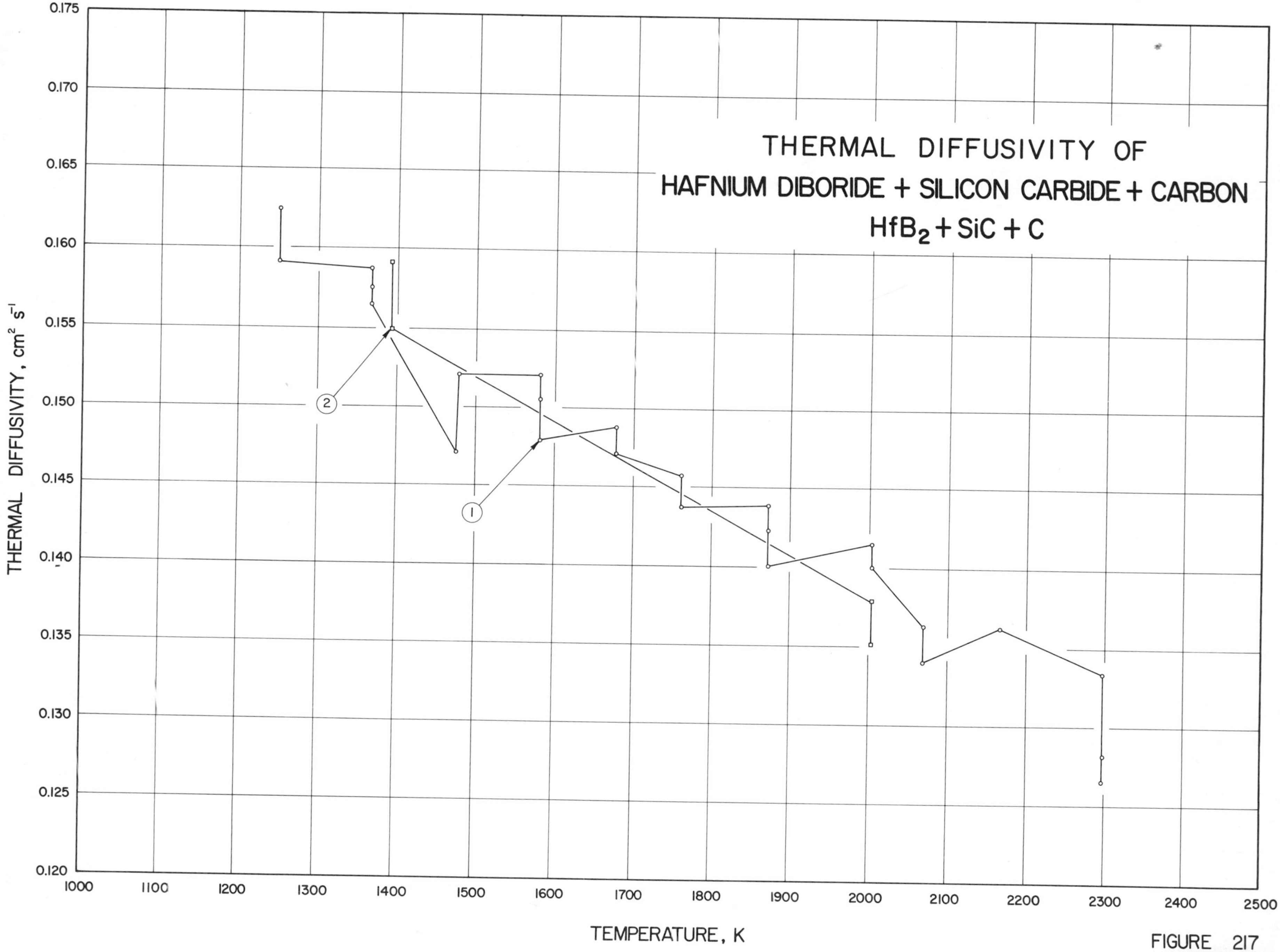
THERMAL DIFFUSIVITY OF
HAFNIUM DIBORIDE + SILICON CARBIDE + CARBON
$HfB_2 + SiC + C$
THERMAL DIFFUSIVITY, $cm^2 s^{-1}$
TEMPERATURE, K
0.175
0.170
0.165
0.160
0.155
0.150
0.145
0.140
0.135
0.130
0.125
0.120
1000
1100
1200
1300
1400
1500
1600
1700
1800
1900
2000
2100
2200
2300
2400
2500
1
2

FIGURE 217

SPECIFICATION TABLE 217. THERMAL DIFFUSIVITY OF [HAFNIUM DIBORIDE + SILICON CARBIDE + CARBON] $HfB_2 + SiC + C$

Cur. No.	Ref. No.	Author(s)	Year	Temp. Range, K	Reported Error, %	Name and Specimen Designation	Composition (weight percent) HfB_2	SiC	C	Composition (continued), Specifications, and Remarks
1	275	Clougherty, E. V., Wilkes, K. E., and Tye, R. P.	1969	1250-2298		XIV(18, 10)09F D1037K	→	→	→	HfB_2 with 14 vol. % SiC and 30 vol. % C; right circular cylindrical disk specimen, 0.500 in. in diameter and 0.100 in. thick with ends flat and parallel to ±0.001 in.; diffusivity measured at increasing temperatures.
2	275	Clougherty, E. V., et al.	1969	1392-2003						The above specimen; diffusivity measured at decreasing temperatures.

DATA TABLE 217. THERMAL DIFFUSIVITY OF [HAFNIUM DIBORIDE + SILICON CARBIDE + CARBON] $HfB_2 + SiC + C$

[Temperature, T, K; Thermal Diffusivity, α, $cm^2 s^{-1}$]

T	α	T	α
CURVE 1		CURVE 1 (cont.)	
1250	0.1624	2003	0.1415
1250	0.1591	2003	0.1400
1367	0.1587	2069	0.1363
1367	0.1575	2069	0.1340
1367	0.1564	2166	0.1362
1475	0.1471	2297	0.1334
1478	0.1521	2297	0.1266
1581	0.1521	2298	0.1282
1581	0.1505		
1581	0.1479	CURVE 2	
1678	0.1488		
1678	0.1471	1392	0.1591
1761	0.1458	1392	0.1549
1761	0.1437	2003	0.1379
1872	0.1439	2003	0.1351
1872	0.1423		
1872	0.1400		

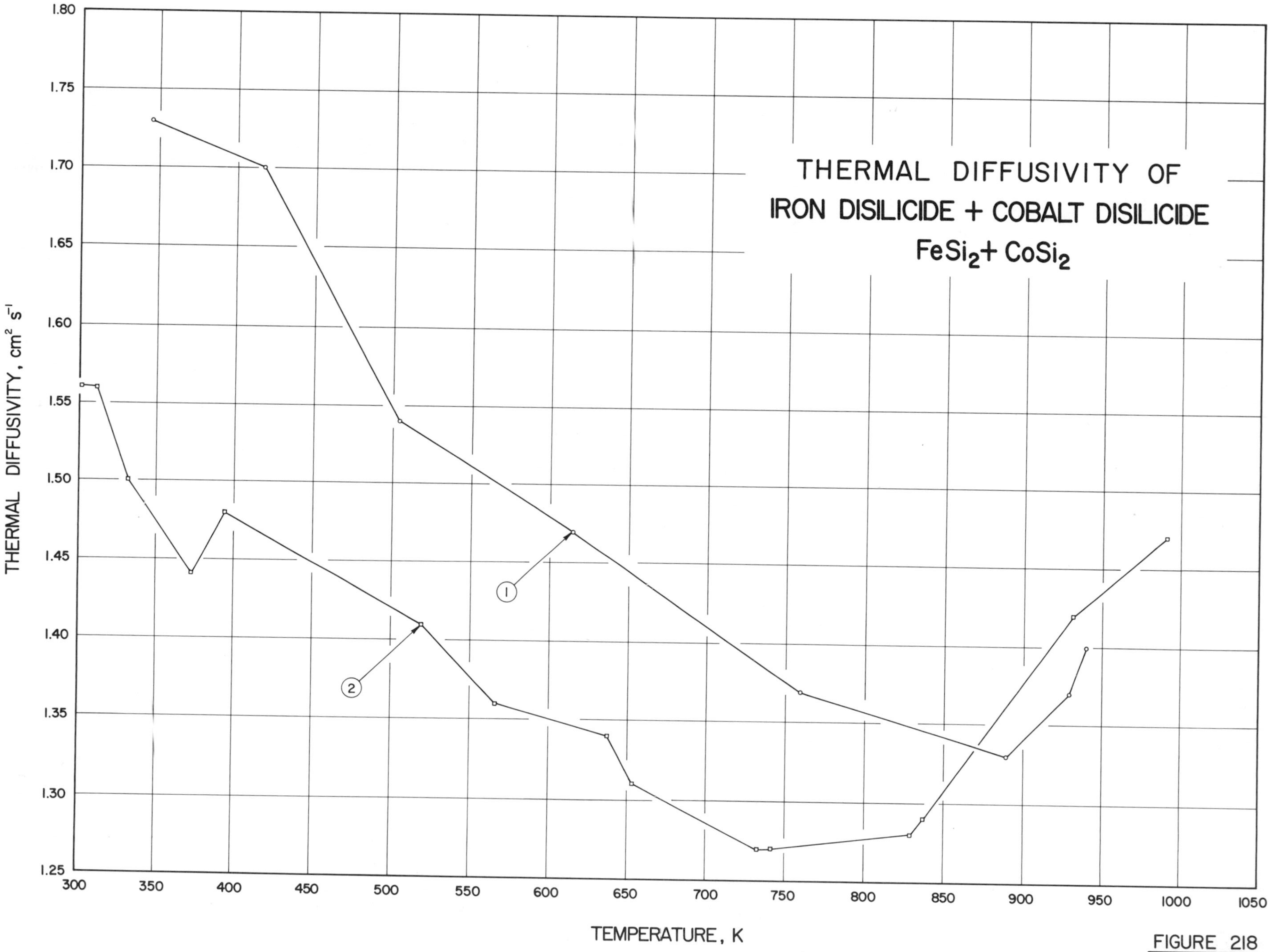
THERMAL DIFFUSIVITY OF
IRON DISILICIDE + COBALT DISILICIDE
FeSi2+ CoSi2
THERMAL DIFFUSIVITY, cm2 s-1
TEMPERATURE, K
1.80
1.75
1.70
1.65
1.60
1.55
1.50
1.45
1.40
1.35
1.30
1.25
300
350
400
450
500
550
600
650
700
750
800
850
900
950
1000
1050
1
2

FIGURE 218

SPECIFICATION TABLE 218. THERMAL DIFFUSIVITY OF [IRON DISILICIDE + COBALT DISILICIDE] $FeSi_2 + CoSi_2$

Cur. No.	Ref. No.	Author(s)	Year	Temp. Range, K	Reported Error, %	Name and Specimen Designation	Composition (weight percent) $FeSi_2$	$CoSi_2$	Composition (continued), Specifications, and Remarks
1	284	Maglic, K.	1969	345-940		N-1	→	→	6 mol.% $CoSi_2$ (estimated); n-type; 46 x 8 x 4.5 mm; obtained from Plessey Co., Ltd.; electrical resistivity 7.35, 6.55, 5.60, 5.15, 4.85, 4.65, 4.50, and 4.30 mΩ cm at 50, 100, 200, 300, 400, 500, 600, and 700 C, respectively.
2	284	Maglic, K.	1969	301-991		N-2			Similar to the above specimen but electrical resistivity 7.80, 7.00, 5.90, 5.35, 5.10, 4.90, 4.75, and 4.50 mΩ cm at 50, 100, 200, 300, 400, 500, 600, and 700 C, respectively.

DATA TABLE 218. THERMAL DIFFUSIVITY OF [IRON DISILICIDE + COBALT DISILICIDE] $FeSi_2 + CoSi_2$

[Temperature, T, K; Thermal Diffusivity, α, $cm^2\ s^{-1}$]

T	α	T	α
CURVE 1		CURVE 2 (cont.)	
345	0.0173	566	0.0136
417	0.0170	637	0.0134
504	0.0154	653	0.0131
614	0.0147	732	0.0127
759	0.0137	741	0.0127
889	0.0133	829	0.0128
929	0.0137	837	0.0129
940	0.0140	931	0.0142
		991	0.0147
CURVE 2			
301	0.0156		
311	0.0156		
332	0.0150		
373	0.0144		
394	0.0148		
519	0.0141		

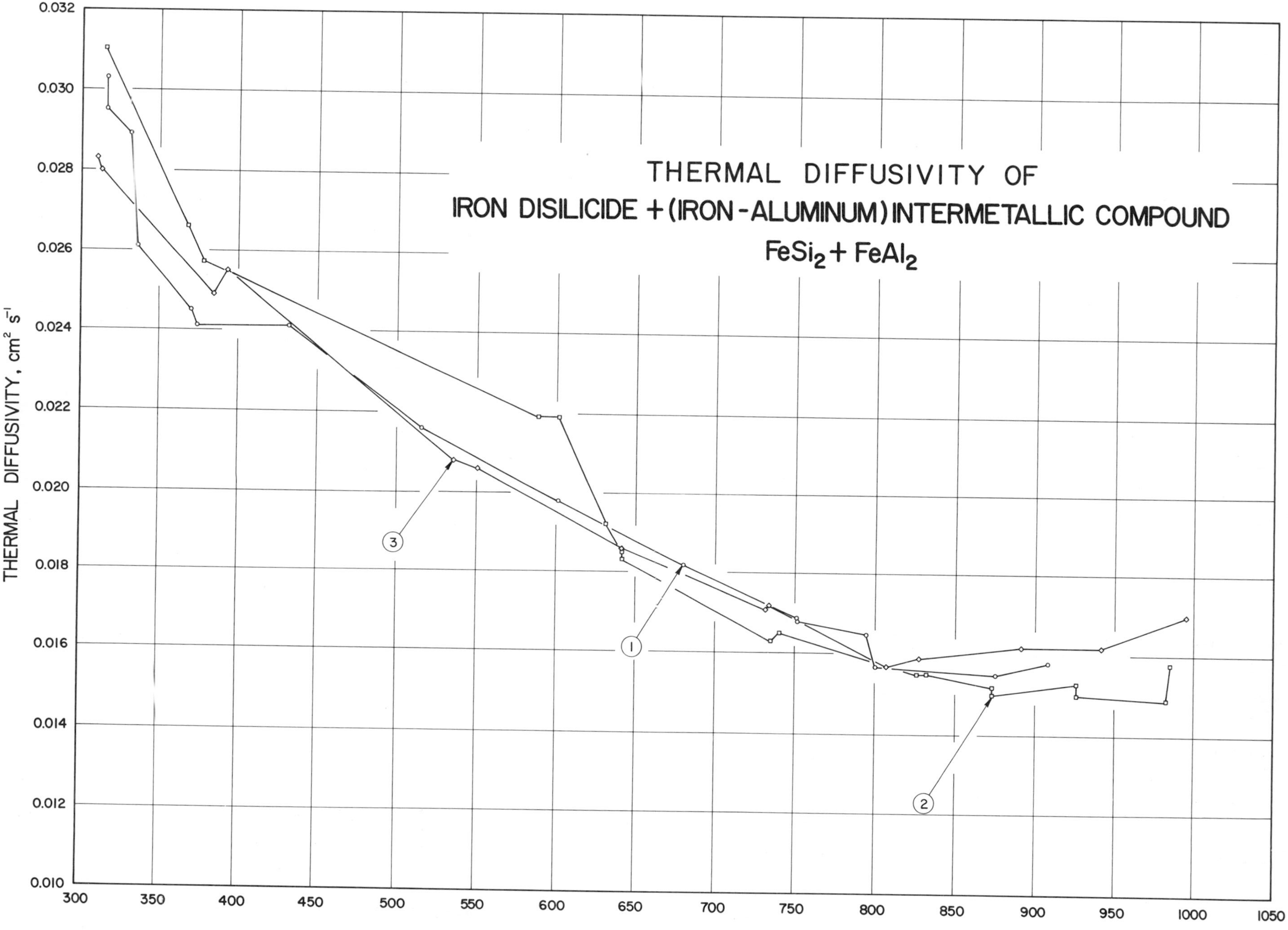
THERMAL DIFFUSIVITY OF
IRON DISILICIDE + (IRON-ALUMINUM) INTERMETALLIC COMPOUND
FeSi2 + FeAl2
THERMAL DIFFUSIVITY, cm2 s-1
TEMPERATURE, K
0.032
0.030
0.028
0.026
0.024
0.022
0.020
0.018
0.016
0.014
0.012
0.010
300
350
400
450
500
550
600
650
700
750
800
850
900
950
1000
1050
1
2
3

FIGURE 219

SPECIFICATION TABLE 219. THERMAL DIFFUSIVITY OF IRON DISILICIDE + (IRON-ALUMINUM) INTERMETALLIC COMPOUND $FeSi_2 + FeAl_2$

Cur. No.	Ref. No.	Author(s)	Year	Temp. Range, K	Reported Error, %	Name and Specimen Designation	Composition (weight percent) $FeSi_2$	$FeAl_2$	Composition (continued), Specifications, and Remarks
1	284	Maglic, K.	1969	316-908		P-1	→	→	2 mol.% $FeAl_2$ (estimated); p-type; 46 x 8 x 4.5 mm; obtained from Plessey Co., Ltd.; electrical resistivity 6.35, 6.80, 7.55, 8.25, 8.80, 9.20, 9.05, and 8.00 mΩ cm at 50, 100, 200, 300, 400, 500, 600, and 700 C, respectively.
2	284	Maglic, K.	1969	315-985		P-2			Similar to the above specimen but electrical resistivity 8.40, 8.90, 9.95, 10.75, 11.30, 11.55, 11.20, and 9.30 mΩ cm at 50, 100, 200, 300, 400, 500, 600, and 700 C, respectively.
3	284	Maglic, K.	1969	310-995		P-4			Similar to the above specimen but electrical resistivity 6.60, 7.05, 7.90, 8.60, 9.25, 9.75, 9.60, and 8.40 mΩ cm at 50, 100, 200, 300, 400, 500, 600, and 700 C, respectively.

DATA TABLE 219. THERMAL DIFFUSIVITY OF IRON DISILICIDE + (IRON-ALUMINUM) INTERMETALLIC COMPOUND $FeSi_2 + FeAl_2$

[Temperature, T, K; Thermal Diffusivity, α, $cm^2 s^{-1}$]

T	α	T	α	T	α
CURVE 1		CURVE 2		CURVE 3	
316	0.0303	315	0.0310	310	0.0283
316	0.0295	368	0.0266	312	0.0280
331	0.0289	378	0.0257	384	0.0249
336	0.0261	588	0.0219	393	0.0255
370	0.0245	601	0.0219	535	0.0208
374	0.0241	631	0.0192	550	0.0206
432	0.0241	641	0.0185	641	0.0186
515	0.0216	641	0.0183	731	0.0169
601	0.0198	735	0.0163	733	0.0171
680	0.0182	740	0.0165	807	0.0157
751	0.0169	826	0.0155	827	0.0159
751	0.0168	832	0.0155	891	0.0162
795	0.0165	873	0.0152	991	0.0162
800	0.0157	873	0.0150	995	0.0170
875	0.0155	926	0.0153		
908	0.0158	926	0.0150		
		983	0.0149		
		985	0.0158		

SPECIFICATION TABLE 220. THERMAL DIFFUSIVITY OF [MOLYBDENUM DITELLURIDE + TUNGSTEN DITELLURIDE] $MoTe_2 + WTe_2$

Cur. No.	Ref. No.	Author(s)	Year	Temp. Range, K	Reported Error, %	Name and Specimen Designation	Composition (weight percent) $MoTe_2$	WTe_2	Composition (continued), Specifications, and Remarks
1*	180	Guennoc, H.	1961	298.2		$Mo_{0.8}W_{0.2}Te_2$;D	→	→	$Mo_{0.8}W_{0.2}Te_2$ prepared by mixing $MoTe_2$ and WTe_2 powders and sintering; Angström method used to measure diffusivity.
2*	180	Guennoc, H.	1961	298.2		$Mo_{0.8}W_{0.2}Te_2$;E	→	→	$Mo_{0.8}W_{0.2}Te_2$ prepared directly by chemical transport putting together in a tube the Mo and W powders; Angström method used to measure diffusivity.

DATA TABLE 220. THERMAL DIFFUSIVITY OF [MOLYBDENUM DITELLURIDE + TUNGSTEN DITELLURIDE] $MoTe_2 + WTe_2$

[Temperature, T, K; Thermal Diffusivity, α, $cm^2\ s^{-1}$]

T	α
CURVE 1*	
298.2	0.033
CURVE 2*	
298.2	0.016

* No figure given.

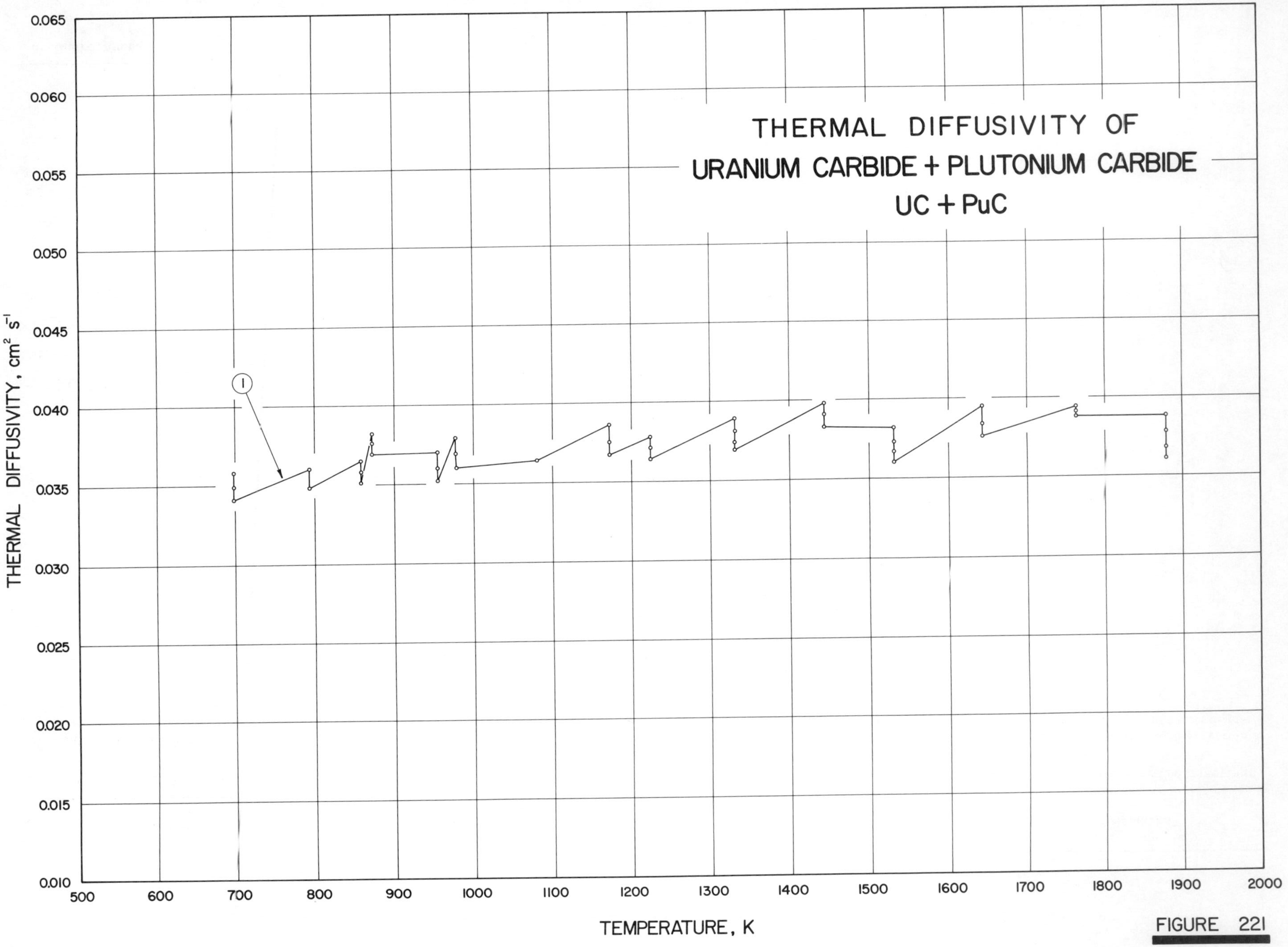
THERMAL DIFFUSIVITY OF
URANIUM CARBIDE + PLUTONIUM CARBIDE
UC + PuC
THERMAL DIFFUSIVITY, cm² s⁻¹
TEMPERATURE, K
0.065
0.060
0.055
0.050
0.045
0.040
0.035
0.030
0.025
0.020
0.015
0.010
500
600
700
800
900
1000
1100
1200
1300
1400
1500
1600
1700
1800
1900
2000
1

FIGURE 221

SPECIFICATION TABLE 221. THERMAL DIFFUSIVITY OF [URANIUM CARBIDE + PLUTONIUM CARBIDE] UC + PuC

Cur. No.	Ref. No.	Author(s)	Year	Temp. Range, K	Reported Error, %	Name and Specimen Designation	Composition (weight percent), Specifications, and Remarks
1	125	Moser, J. B. and Kruger, O. L.	1968	698-1879	±5	75% Dense $U_{0.8}Pu_{0.2}C$	4.66 C, 0.202 O, and 0.0206 N; mixed carbide of nominal U/Pu ratio of 4:1; disk-shaped specimen 1.83 cm in dia. and having thickness in the range from 0.2 to 0.3 cm; prepared by crushing, cold-pressing, and sintering arc-cast material (arc-cast carbides prepared by fusion of 99.5 pure metal with spectrographically pure carbon in an arc furnace); sintered at 2073.2 K; lapped down from approximately 0.5 cm in thickness to final thickness; density 74.8 percent of theoretical value; surface coated with colloidal graphite; exposed to thermal pulse of 500 μsec duration generated by ruby laser; flash method used to measure diffusivity; measured in vacuum; radiation heat loss correction applied to all data; maximum error of measurement ±7.5 percent.

DATA TABLE 221. THERMAL DIFFUSIVITY OF [URANIUM CARBIDE + PLUTONIUM CARBIDE] UC + PuC

[Temperature, T, K; Thermal Diffusivity, α, $cm^2\ s^{-1}$]

T	α	T	α	T	α
CURVE 1		CURVE 1 (cont.)		CURVE 1 (cont.)	
698	0.0358	1080	0.0364	1531	0.0360
698	0.0349	1172	0.0386	1643	0.0395
698	0.0341	1172	0.0375	1643	0.0384
793	0.0360	1172	0.0367	1643	0.0376
793	0.0348	1224	0.0378	1763	0.0394
858	0.0365	1224	0.0371	1763	0.0391
858	0.0358	1224	0.0364	1763	0.0388
858	0.0351	1331	0.0389	1879	0.0388
872	0.0382	1331	0.0381	1879	0.0378
872	0.0376	1331	0.0374	1879	0.0368
872	0.0369	1331	0.0369	1879	0.0361
954	0.0370	1444	0.0398		
954	0.0360	1444	0.0391		
954	0.0352	1444	0.0383		
977	0.0379	1531	0.0382		
977	0.0369	1531	0.0373		
978	0.0360	1531	0.0367		

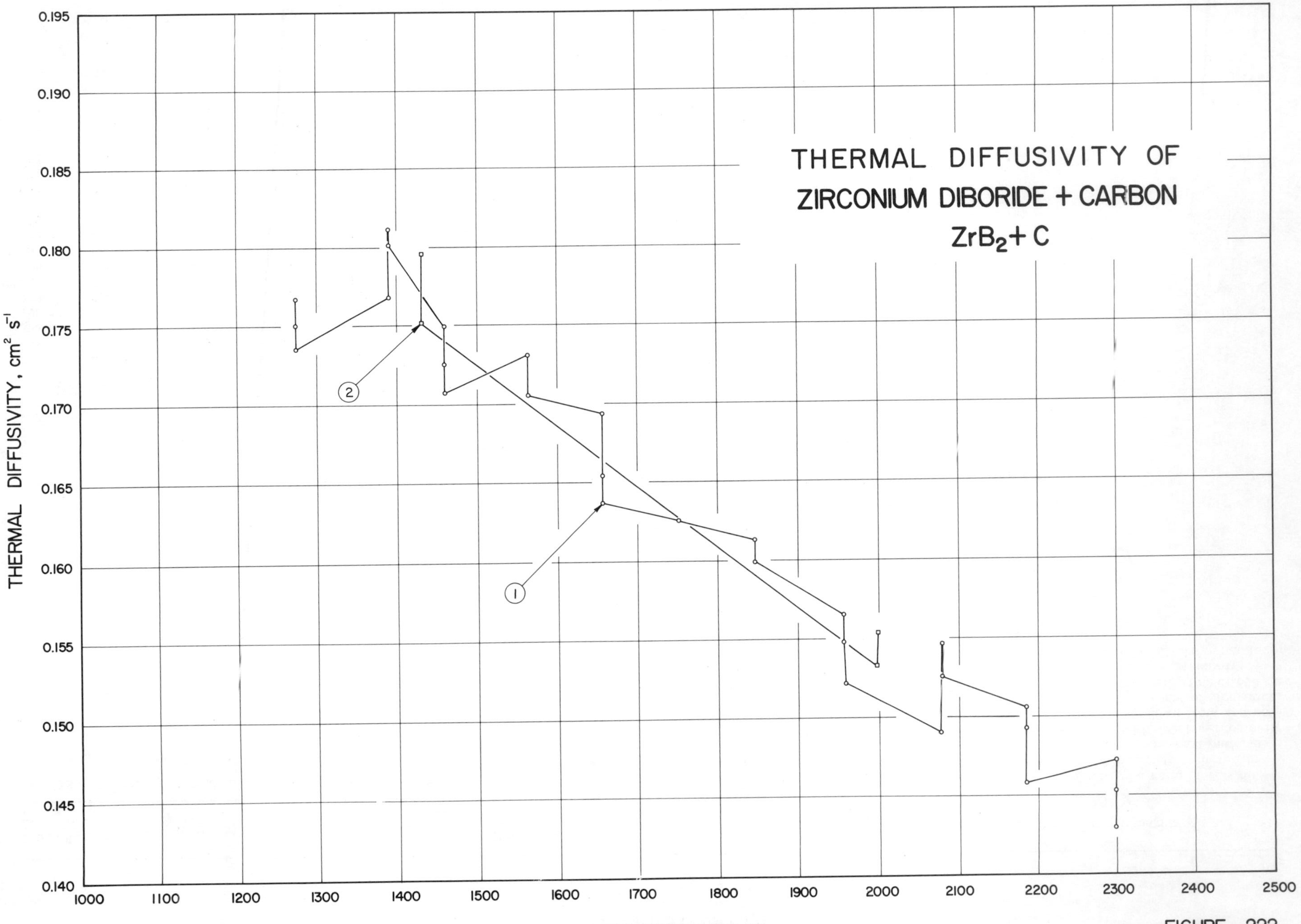
THERMAL DIFFUSIVITY OF
ZIRCONIUM DIBORIDE + CARBON
$ZrB_2 + C$
THERMAL DIFFUSIVITY, $cm^2 s^{-1}$
TEMPERATURE, K
0.195
0.190
0.185
0.180
0.175
0.170
0.165
0.160
0.155
0.150
0.145
0.140
1000
1100
1200
1300
1400
1500
1600
1700
1800
1900
2000
2100
2200
2300
2400
2500
1
2

FIGURE 222

SPECIFICATION TABLE 222. THERMAL DIFFUSIVITY OF [ZIRCONIUM DIBORIDE + CARBON] $ZrB_2 + C$

Cur. No.	Ref. No.	Author(s)	Year	Temp. Range, K	Reported Error, %	Name and Specimen Designation	Composition (weight percent) ZrB_2	Composition (weight percent) C	Composition (continued), Specifications, and Remarks
1	275	Clougherty, E.V., Wilkes, K.E., and Tye, R.P.	1969	1274-2299		XII(20)07F D0812K	→	→	ZrB_2 with 50% (by volume) C to enhance thermal stress resistance; right circular cylindrical disk specimen, 0.500 in. in diameter and 0.100 in. thick with ends flat and parallel to ± 0.001 in.; diffusivity measured at increasing temperatures.
2	275	Clougherty, E.V., et al.	1969	1431-1999		XII(20)07F D0812K			The above specimen; diffusivity measured at decreasing temperatures.

DATA TABLE 222. THERMAL DIFFUSIVITY OF [ZIRCONIUM DIBORIDE + CARBON] $ZrB_2 + C$

[Temperature, T, K; Thermal Diffusivity, α, $cm^2\ s^{-1}$]

T	α	T	α	T	α
CURVE 1		CURVE 1 (cont.)		CURVE 2	
1274	0.1767	1956	0.1566	1431	0.1795
1274	0.1750	1956	0.1548	1431	0.1751
1274	0.1735	1958	0.1522	1997	0.1533
1390	0.1768	2076	0.1491	1999	0.1554
1390	0.1811	2078	0.1547		
1390	0.1801	2078	0.1526		
1459	0.1749	2184	0.1506		
1459	0.1725	2184	0.1493		
1459	0.1707	2184	0.1459		
1563	0.1731	2299	0.1473		
1563	0.1705	2299	0.1453		
1656	0.1693	2299	0.1430		
1656	0.1654				
1656	0.1637				
1752	0.1625				
1847	0.1613				
1847	0.1599				

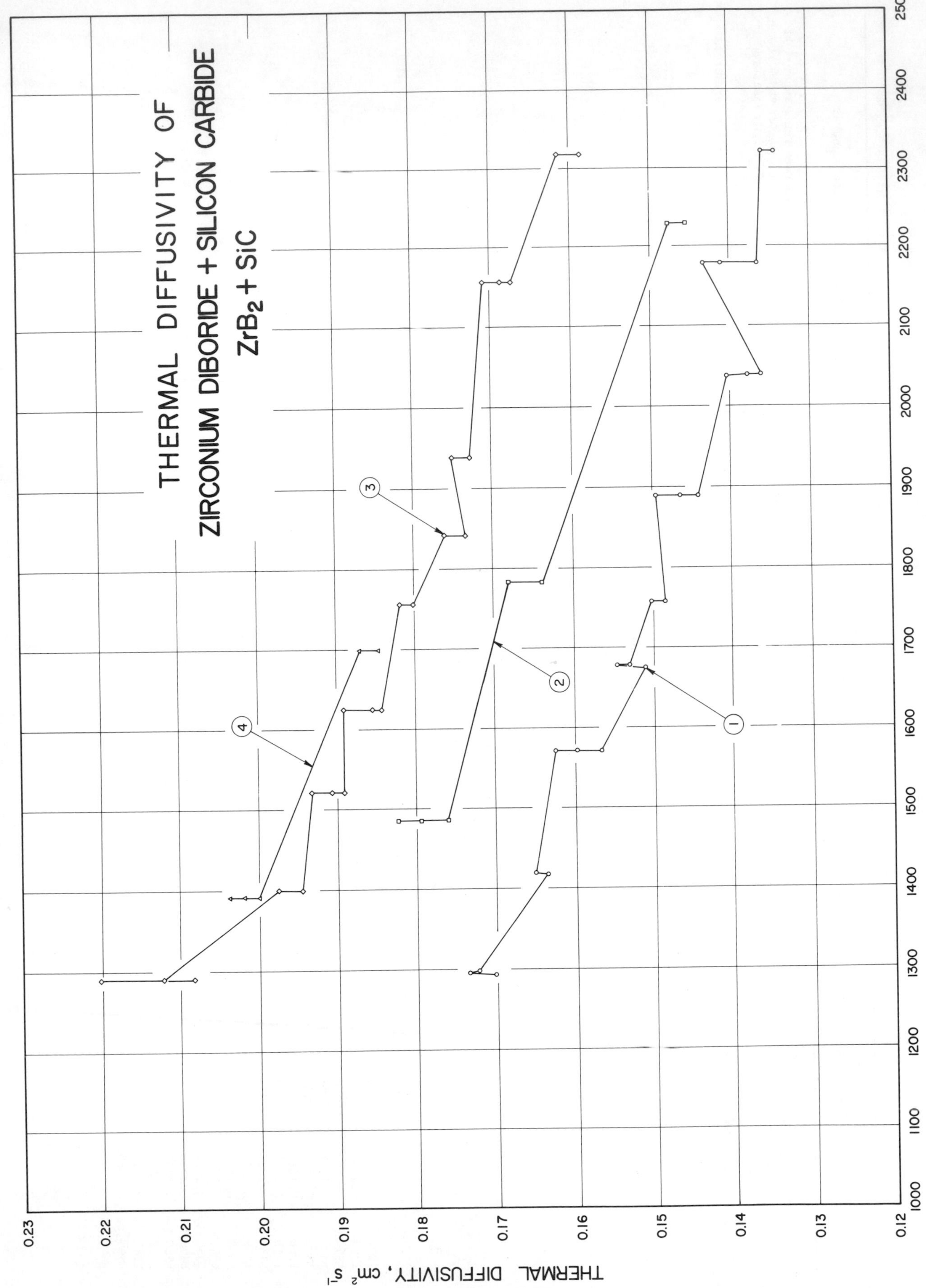
THERMAL DIFFUSIVITY OF
ZIRCONIUM DIBORIDE + SILICON CARBIDE
$ZrB_2 + SiC$
THERMAL DIFFUSIVITY, $cm^2\ s^{-1}$
TEMPERATURE, K
0.12
0.13
0.14
0.15
0.16
0.17
0.18
0.19
0.20
0.21
0.22
0.23
1000
1100
1200
1300
1400
1500
1600
1700
1800
1900
2000
2100
2200
2300
2400
2500
1
2
3
4

FIGURE 223

SPECIFICATION TABLE 223. THERMAL DIFFUSIVITY OF [ZIRCONIUM DIBORIDE + SILICON CARBIDE] $ZrB_2 + SiC$

Cur. No.	Ref. No.	Author(s)	Year	Temp. Range, K	Reported Error, %	Name and Specimen Designation	Composition (weight percent) ZrB_2	SiC	Composition (continued), Specifications, and Remarks
1	275	Clougherty, E.V., Wilkes, K.E., and Tye, R.P.	1969	1293-2321		V07F D 0851 K	→	→	ZrB_2 with 20% (by volume) SiC to enhance oxidation resistance; right circular cylindrical disk specimen, 0.500 in. in diameter and 0.100 in. thick with ends flat and parallel to ± 0.001 in.; diffusivity measured at increasing temperatures.
2	275	Clougherty, E.V., et al.	1969	1485-2229		V07F D 0851 K			The above specimen; diffusivity measured at decreasing temperatures.
3	275	Clougherty, E.V., et al.	1969	1290-2317		V07F D 0902 K			Similar to the above specimen; diffusivity measured at increasing temperatures.
4	275	Clougherty, E.C., et al.	1969	1391-1698		V07F D 0902 K			The above specimen; diffusivity measured at decreasing temperatures.

DATA TABLE 223. THERMAL DIFFUSIVITY OF [ZIRCONIUM DIBORIDE + SILICON CARBIDE] $ZrB_2 + SiC$

[Temperature, T, K; Thermal Diffusivity, α, $cm^2 s^{-1}$]

T	α	T	α	T	α	T	α	T	α
CURVE 1		CURVE 1 (cont.)		CURVE 2		CURVE 2 (cont.)		CURVE 3 (cont.)	
1293	0.1702	2038	0.1360	1290	0.2201	1883	0.1673	1756	0.1801
1295	0.1734	2178	0.1433	1290	0.2081	2044	0.1614	1841	0.1760
1297	0.1722	2178	0.1411	1290	0.2121	2044	0.1586	1841	0.1733
1417	0.1635	2178	0.1364	1399	0.1973			1938	0.1750
1419	0.1650	2321	0.1358	1399	0.1944	CURVE 3		1938	0.1727
1571	0.1624	2321	0.1341	1521	0.1930			2156	0.1709
1571	0.1597			1521	0.1906	1290	0.2201	2156	0.1687
1571	0.1567	CURVE 2		1521	0.1890	1290	0.2081	2156	0.1673
1675	0.1511			1624	0.1854	1290	0.2121	2317	0.1614
1677	0.1547	1485	0.1821	1624	0.1841	1399	0.1973	2317	0.1586
1677	0.1531	1485	0.1793	1756	0.1817	1399	0.1944		
1757	0.1503	1485	0.1758	1756	0.1801	1521	0.1930	CURVE 4	
1757	0.1485	1783	0.1680	1841	0.1760	1521	0.1906		
1889	0.1496	1783	0.1639	1841	0.1733	1521	0.1890	1391	0.2036
1889	0.1465	2229	0.1477	1938	0.1750	1624	0.1890	1391	0.2018
1889	0.1442	2229	0.1454	1938	0.1727	1624	0.1854	1391	0.1990
2037	0.1404			2156	0.1709	1624	0.1841	1698	0.1869
2038	0.1378			1883	0.1687	1756	0.1817	1698	0.1846

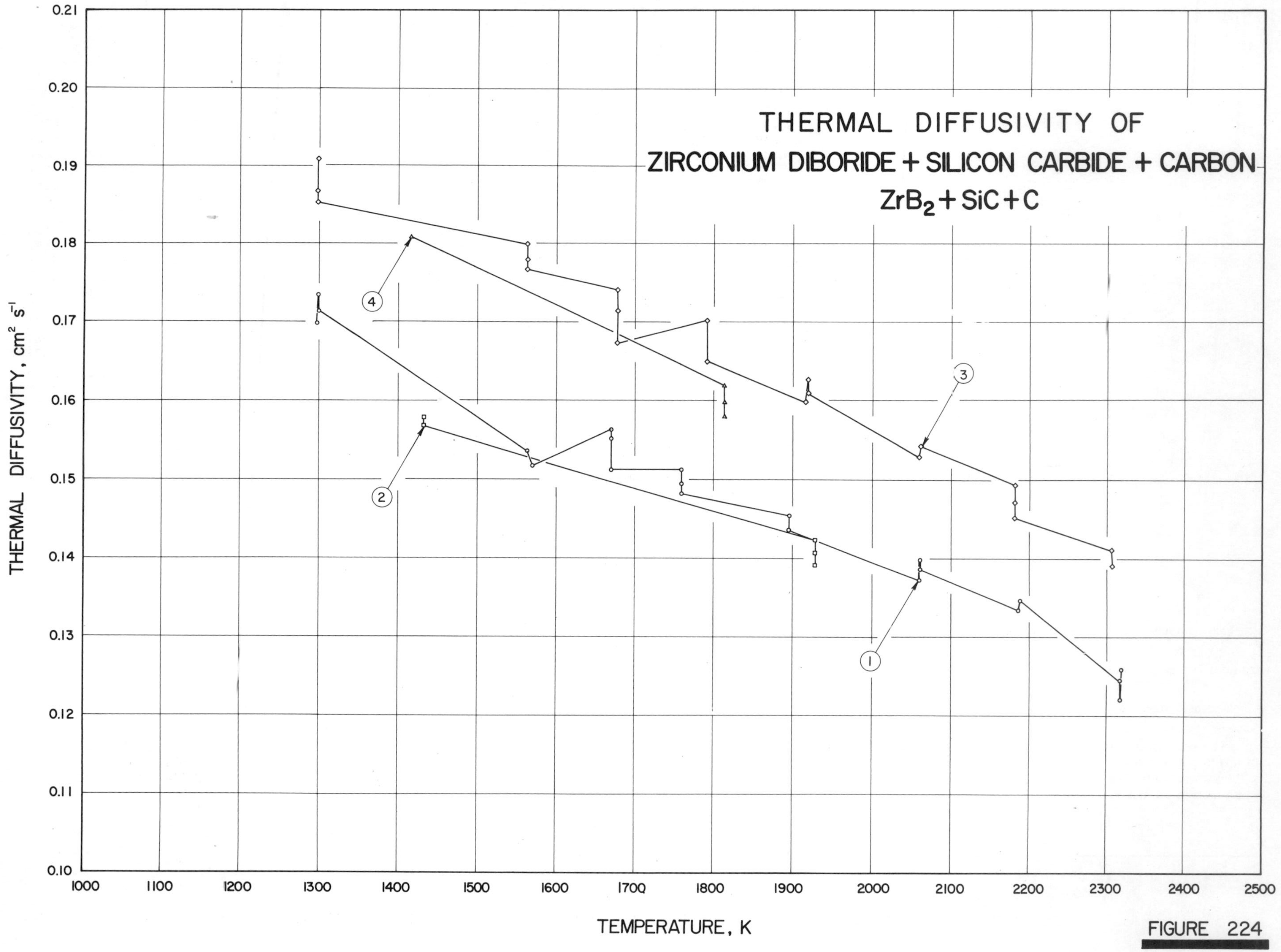
THERMAL DIFFUSIVITY OF
ZIRCONIUM DIBORIDE + SILICON CARBIDE + CARBON
$ZrB_2 + SiC + C$
THERMAL DIFFUSIVITY, $cm^2\ s^{-1}$
TEMPERATURE, K
0.10
0.11
0.12
0.13
0.14
0.15
0.16
0.17
0.18
0.19
0.20
0.21
1000
1100
1200
1300
1400
1500
1600
1700
1800
1900
2000
2100
2200
2300
2400
2500
1
2
3
4

FIGURE 224

SPECIFICATION TABLE 224. THERMAL DIFFUSIVITY OF [ZIRCONIUM DIBORIDE + SILICON CARBIDE + CARBON] $ZrB_2 + SiC + C$

Cur. No.	Ref. No.	Author(s)	Year	Temp. Range, K	Reported Error, %	Name and Specimen Designation	Composition (weight percent) ZrB_2	SiC	C	Composition (continued), Specifications, and Remarks
1	275	Clougherty, E.V., Wilkes, K.E., and Tye, R.P.	1969	1297-2318		VIII 07F D0975K	→	→	→	ZrB_2 with 14 vol. %SiC, 30 vol. %C; right circular cylindrical disk specimen; 0.500 in. in diameter and 0.100 in. thick with ends flat and parallel to ± 0.001 in.; diffusivity measured at increasing temperatures.
2	275	Clougherty, E.V., et al.	1969	1432-1928		VIII 07F D0975K				The above specimen; diffusivity measured at decreasing temperatures.
3	275	Clougherty, E.V., et al.	1969	1299-2307		VIII(18,10)07F D0920K				Similar to the above specimen; diffusivity measured at increasing temperatures.
4	275	Clougherty, E.V., et al.	1969	1415-1813		VIII(18,10)07F D0920K				The above specimen; diffusivity measured at decreasing temperatures.

DATA TABLE 224. THERMAL DIFFUSIVITY OF [ZIRCONIUM DIBORIDE + SILICON CARBIDE + CARBON] $ZrB_2 + SiC + C$

[Temperature, T, K; Thermal Diffusivity, α, $cm^2\ s^{-1}$]

T	α	T	α	T	α
CURVE 1		CURVE 1 (cont.)		CURVE 3 (cont.)	
1297	0.1697	2317	0.1219	1678	0.1712
1299	0.1733	2318	0.1258	1678	0.1671
1299	0.1712			1791	0.1700
1563	0.1535	CURVE 2		1791	0.1648
1570	0.1517			1916	0.1598
1670	0.1563	1432	0.1577	1919	0.1625
1670	0.1552	1432	0.1567	1919	0.1609
1670	0.1512	1928	0.1423	2059	0.1528
1759	0.1512	1928	0.1406	2061	0.1542
1759	0.1495	1928	0.1391	2181	0.1493
1759	0.1482			2181	0.1472
1896	0.1455	CURVE 3		2181	0.1452
1896	0.1436			2307	0.1410
2060	0.1372	1299	0.1908	2307	0.1390
2060	0.1398	1299	0.1867		
2060	0.1386	1299	0.1852	CURVE 4	
2186	0.1333	1563	0.1799		
2188	0.1345	1563	0.1779	1415	0.1808
2317	0.1244	1563	0.1767	1813	0.1619
		1677	0.1740	1813	0.1598
				1813	0.1580

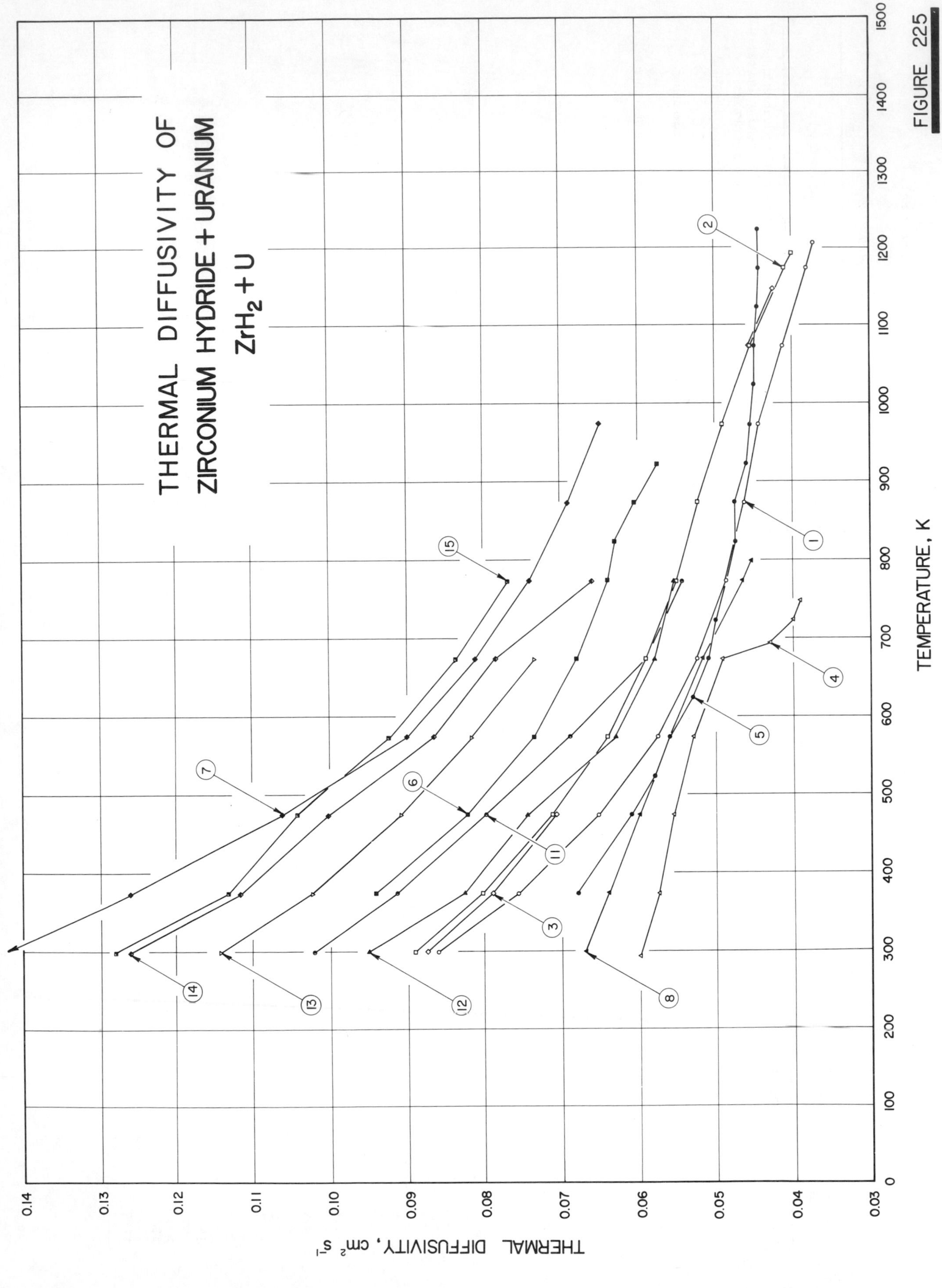
THERMAL DIFFUSIVITY OF
ZIRCONIUM HYDRIDE + URANIUM
$ZrH_2 + U$
THERMAL DIFFUSIVITY, $cm^2 s^{-1}$
TEMPERATURE, K
0.03
0.04
0.05
0.06
0.07
0.08
0.09
0.10
0.11
0.12
0.13
0.14
0
100
200
300
400
500
600
700
800
900
1000
1100
1200
1300
1400
1500
1
2
3
4
5
6
7
8
11
12
13
14
15

FIGURE 225

SPECIFICATION TABLE 225. THERMAL DIFFUSIVITY OF [ZIRCONIUM HYDRIDE + URANIUM] $ZrH_2 + U$

Cur. No.	Ref. No.	Author(s)	Year	Temp. Range, K	Reported Error, %	Name and Specimen Designation	Composition (weight percent) ZrH_2	U	Composition (continued), Specifications, and Remarks
1	129	Weeks, C.C., Nakata, M.M. and Smith, C.A.	1968	298-1206		SNAP fuel; 1,2	→	→	Hydrided 90 Zr-10 U alloy; hydrided to an atomic ratio H/Zr = 1.58; two disk specimens nominally 6.4 mm in diameter each, and 1.920 and 1.895 mm in thickness, respectively; first heated to temperatures greater than 873.2 K and allowed to equilibrate with the proper hydrogen overpressure; diffusivity measurements then made at selected temperature intervals, generally during the cooling cycle; front face heated with short pulse from laser; diffusivity determined from measured temperature history of rear face; no experimental data given, data points obtained from smoothed curve.
2	129	Weeks, C.C., et al.	1968	298-1192		SNAP fuel; 1,2,3	→	→	Hydrided 90 Zr-10 U alloy; hydrided to an atomic ratio H/Zr = 1.65; three disk specimens nominally 6.4 mm in diameter each, and 1.869, 1.874 and 2.581 mm in thickness, respectively; other conditions same as above.
3	129	Weeks, C.C., et al.	1968	298-1146		SNAP fuel; 1,2,3	→	→	Hydrided 90 Zr-10 U alloy; hydrided to an atomic ratio H/Zr = 1.70; three disk specimens nominally 6.4 mm in diameter each, and 1.816, 1.849 and 1.984 mm in thickness, respectively; other conditions same as above.
4	130	Nakata, M.M., Ambrose, C.J. and Finch, R.A.	1966	293-748	<±5	0.50 H/Zr material	→	→	Hydrided 90 Zr-10 U alloy; hydrided to an atomic ratio H/Zr = 0.50; three disk-shaped specimens 0.64 cm in diameter and ~0.25 cm thick each; prepared as explained above; extruded and hydrided fuel rods machined into final specimen size; machined so that heat flux during diffusivity measurement would be in the direction of extrusion of billet; density 6.68 g cm^{-3}; electrical resistivity reported as 69.0, 82.2, 97.9, 112.5, 125.3, 132.0, 128.9, 132.3, 128.8, 128.9, and 133.0 μohm cm at 294.2, 373.2, 473.2, 573.2, 673.2, 773.2, 873.2, 973.2, 1073.2, 1173.2, and 1226.2 K, respectively (data obtained from smoothed curve); initially heated at temperatures above 873.2 K in the appropriate hydrogen overpressure; measured for diffusivity in an atmosphere of ultra-pure hydrogen gas (99.9998 purity); flash technique used to measure diffusivity.
5	130	Nakata, M.M., et al.	1966	373-1223	<±5	1.58 H/Zr material			Same alloy as above but hydrided to an atomic ratio H/Zr = 1.58; three disk-shaped specimens 0.64 cm in diameter and ~0.25 cm thick each; prepared as explained above; extruded and hydrided fuel rods machined into final specimen size; machined so that heat flux during diffusivity measurement would be in the direction of extrusion of billet; density 6.11 g cm^{-3}; electrical resistivity reported as 67.3, 75.2, 85.2, 95.8, 106.4, 118.1, 129.6, 141.2, 153.4, and 166.1 μohm cm at 294.2, 373.2, 473.2, 573.2, 673.2, 773.2, 873.2, 973.2, 1073.2, and 1173.2 K, respectively (data obtained from smoothed curve); data points in parentheses obtained by interpolation; other conditions same as above.

SPECIFICATION TABLE 225. THERMAL DIFFUSIVITY OF [ZIRCONIUM HYDRIDE + URANIUM] ZrH_2 + U (continued)

Cur. No.	Ref. No.	Author(s)	Year	Temp. Range, K	Reported Error, %	Name and Specimen Designation	Composition (weight percent) ZrH_2	U	Composition (continued), Specifications, and Remarks
6	130	Nakata, M.M., Ambrose, C.J. and Finch, R.A.	1966	373-923	<±5	1.81 H/Zr material	→	→	Same alloy as above but hydrided to an atomic ratio H/Zr = 1.81; seven specimens; two measured at Atomics International and five measured at Battelle Memorial Institute; AI specimens disk-shaped 0.64 cm in diameter and ~0.25 cm thick each; BMI specimens disk-shaped 0.95 cm in diameter and ~0.15 cm thick each; prepared as explained above; extruded and hydrided fuel rods machined into final specimen sizes; machined so that heat flux during diffusivity measurement would be in the direction of extrusion of billet (with the exception of a few BMI specimens machined so that heat flux would be perpendicular to the direction of extrusion); density 6.09 g cm^{-3}; electrical resistivity reported as 54.3, 62.7, 73.1, 83.8, 95.0, 106.5, 120.2, and 136.2 μohm cm at 294.2, 373.2, 473.2, 573.2, 673.2, 773.2, 873.2, and 973.2 K, respectively (data obtained from smoothed curve); one AI sample measured as-received and the second sample measured as-received followed by measurements with the appropriate hydrogen over-pressure; data points reported represent best values calculated from measured AI-BMI data; other conditions same as above.
7	130	Nakata, M.M., et al.	1966	293-973	<±5	1.90 H/Zr material			Same alloy as above but hydrided to an atomic ratio H/Zr = 1.90; three disk-shaped specimens 0.64 cm in diameter and ~0.25 cm thick each; prepared as explained above; extruded and hydrided fuel rods machined into final specimen size; machined so that heat flux during diffusivity measurement would be in the direction of extrusion of billet; density 6.08 g cm^{-3}; electrical resistivity reported as 38.4, 46.2, 56.3, 67.1, 78.6, 91.9, 107.1, and 126.4 μohm cm at 294.2, 373.2, 473.2, 573.2, 673.2, 773.2, 873.2, and 973.2 K, respectively (data obtained from smoothed curve); initially heated at temperatures above 873.2 K in the appropriate hydrogen over-pressure; measured for diffusivity in an atmosphere of ultra-pure hydrogen gas (99.9998 purity); flash technique used to measure diffusivity.
8	133	Ambrose, C.J., Taylor, R.E. and Finch, R.A.	1964	298-799		0.50 H/Zr	→	→	Hydrided 90 Zr-10 U alloy with metallic impurity not exceeding 0.25; hydrided to an atomic ratio H/Zr = 0.50; disk-shaped specimen 0.25 in. in diameter and 0.125 in. thick; prepared from triple arc melted and extruded alloy; machined to size after final extrusion and then hydrided to the appropriate hydrogen concentration by exposing it, in a high temperature furnace, to a known volume of hydrogen; heated to temperatures below which hydrogen would dissociate; laser pulse technique used to measure diffusivity.
9	133	Ambrose, C.J., et al.	1964	298-875		1.20 H/Zr			Same alloy as above but hydrided to an atomic ratio H/Zr = 1.20; specimen similar to the above; prepared and measured under same conditions as above.
10	133	Ambrose, C.J., et al.	1964	298-773		1.72 H/Zr			Same alloy as above but hydrided to an atomic ratio H/Zr = 1.72; specimen similar to the above; prepared and measured under same conditions as above.

SPECIFICATION TABLE 225. THERMAL DIFFUSIVITY OF [ZIRCONIUM HYDRIDE + URANIUM] $ZrH_2 + U$ (continued)

Cur. No.	Ref. No.	Author(s)	Year	Temp. Range, K	Reported Error, %	Name and Specimen Designation	Composition (weight percent) ZrH_2	U	Composition (continued), Specifications, and Remarks
11	133	Ambrose, C.J., Taylor, R.E., and Finch, R.A.	1964	298-773		1.75 H/Zr			Same alloy as above but hydrided to an atomic ratio H/Zr = 1.75; specimen similar to the above; prepared and measured under same conditions as above.
12	133	Ambrose, C.J., et al.	1964	298-773		1.77 H/Zr			Same alloy as above but hydrided to an atomic ratio H/Zr = 1.77; specimen similar to the above; prepared and measured under same conditions as above.
13	133	Ambrose, C.J., et al.	1964	298-673		1.81 H/Zr			Same alloy as above but hydrided to an atomic ratio H/Zr = 1.81; specimen similar to the above; prepared and measured under same conditions as above.
14	133	Ambrose, C.J., et al.	1964	298-773		1.88 H/Zr			Same alloy as above but hydrided to an atomic ratio H/Zr = 1.88; specimen similar to the above; prepared and measured under same conditions as above.
15	133	Ambrose, C.J., et al.	1964	298-773		1.92 H/Zr			Same alloy as above but hydrided to an atomic ratio H/Zr = 1.92; specimen similar to the above; prepared and measured under same conditions as above.

DATA TABLE 225. THERMAL DIFFUSIVITY OF [ZIRCONIUM HYDRIDE + URANIUM] $ZrH_2 + U$

[Temperature, T, K; Thermal Diffusivity, α, $cm^2\ s^{-1}$]

T	α
CURVE 1	
298	0.0860
373	0.0756
473	0.0653
573	0.0577
673	0.0525
773	0.0487
873	0.0463
973	0.0440
1073	0.0413
1173	0.0382
1206	0.0372
CURVE 2	
298	0.0890
373	0.0802
473	0.0711
573	0.0640
673	0.0591
773	0.0552
873	0.0523
973	0.0491
1073	0.0455
1173	0.0410
1192	0.0400
CURVE 3	
298	0.0874
373	0.0789
473	0.0706
573	0.0640*
673	0.0591*
773	0.0552*
873	0.0523*
973	0.0491*
1073	0.0456
1146	0.0425
CURVE 4	
298.2	0.060
373.2	0.0575
473.2	0.0556
573.2	0.0529
673.2	0.0492
CURVE 4 (cont.)	
693.2	0.043
723.2	0.040
748.2	0.039
CURVE 5	
373.2	0.068
473.2	0.061
523.2	0.058
573.2	0.056
623.2	0.053
673.2	0.051
723.2	0.050
823.2	(0.0475)
873.2	(0.0476)
923.2	(0.046)
973.2	(0.0455)
1023.2	(0.045)
1073.2	(0.045)
1123.2	0.0445
1173.2	0.044
1223.2	0.044
CURVE 6	
373.2	0.0940
473.2	0.0820
573.2	0.0735
673.2	0.0680
773.2	0.0640
823.2	0.0630
873.2	0.0605
923.2	0.0575
CURVE 7	
293.2	0.147
373.2	0.126
473.2	0.106
573.2	0.090
673.2	0.081
773.2	0.074
873.2	0.069
973.2	0.065
CURVE 8	
298.2	0.0670
373.2	0.0640
473.2	0.0600
573.2	0.0560*
673.2	0.0518
773.2	0.0465
799.2	0.0454
CURVE 9*	
298.2	0.0890
373.2	0.0752
473.2	0.0625
573.2	0.0532
673.2	0.0468
773.2	0.0430
875.2	0.0417
CURVE 10*	
298.2	0.0893
373.2	0.0808
473.2	0.0710
573.2	0.0632
673.2	0.0622
773.2	0.0619
CURVE 11	
298.2	0.102
373.2	0.0914
473.2	0.0797
573.2	0.0689
673.2	0.0591*
773.2	0.0545
CURVE 12	
298.2	0.0949
373.2	0.0824
473.2	0.0744
573.2	0.0630
673.2	0.0580
773.2	0.0555
CURVE 13	
298.2	0.114
373.2	0.102
473.2	0.0907
573.2	0.0815
673.2	0.0734
CURVE 14	
298.2	0.126
373.2	0.115
473.2	0.100
573.2	0.0865
673.2	0.0784
773.2	0.0660
CURVE 15	
298.2	0.128
373.2	0.113
473.2	0.104
573.2	0.0923
673.2	0.0836
773.2	0.0668

* Not shown in figure.

11. MINERALS AND ROCKS

SPECIFICATION TABLE 226. THERMAL DIFFUSIVITY OF CLAY

Cur. No.	Ref. No.	Author(s)	Year	Temp. Range, K	Reported Error, %	Name and Specimen Designation	Composition (weight percent), Specifications, and Remarks
1*	166	Williams, I.	1923	318, 373		Dixie clay	Density 2.60 g cm^{-3}.

DATA TABLE 226. THERMAL DIFFUSIVITY OF CLAY

[Temperature, T, K; Thermal Diffusivity, α, cm^2 s^{-1}]

T	α
CURVE 1*	
318.2	0.00112
373.2	0.00112

* No figure given.

SPECIFICATION TABLE 227. THERMAL DIFFUSIVITY OF MARBLE

Cur. No.	Ref. No.	Author(s)	Year	Temp. Range, K	Reported Error, %	Name and Specimen Designation	Composition (weight percent), Specifications, and Remarks
1*	170	Krischer, O. and Esdorn, H.	1955	296-299			Specimen composed of two identical plates heated on their outside surfaces by heating foils each 0.01 mm thick made of a chromium-nickel alloy, inside surfaces held in contact with each other; three identical plates identical to the specimen plates and a thermally insulating layer placed in contact with each of the outside surfaces of the specimen, respectively; heating foils also inserted between the two outermost plates on each side of the specimen, respectively; regulated dc current used to generate thermal energy in the heating foils; square specimen plates 95 x 95 x 15.2 mm each; density 2.680 g cm^{-3}; thermal diffusivity determined from measured time interval necessary for the temp. of the unheated specimen face to reach the same value previously acquired by the heated face.
2*	170	Krischer, O. and Esdorn, H.	1955	317, 318			Above specimen measured for diffusivity again.
3*	170	Krischer, O. and Esdorn, H.	1955	337.2			Above specimen measured for diffusivity again.

DATA TABLE 227. THERMAL DIFFUSIVITY OF MARBLE

[Temperature, T, K; Thermal Diffusivity, α, $cm^2\ s^{-1}$]

T	α
CURVE 1*	
296.2	0.0126
299.2	0.0128
299.2	0.0131
CURVE 2*	
317.2	0.0116
318.2	0.0115
CURVE 3*	
337.2	0.0107

* No figure given.

SPECIFICATION TABLE 228. THERMAL DIFFUSIVITY OF MICA

Cur. No.	Ref. No.	Author(s)	Year	Temp. Range, K	Reported Error, %	Name and Specimen Designation	Composition (weight percent), Specifications, and Remarks
1*	178	Goldsmid, H.J. and Bowley, A.E.	1960	300.2		Phlogopite	Angström method used to measure diffusivity; measured with heat flow in the direction of the planes of cleavage.

DATA TABLE 228. THERMAL DIFFUSIVITY OF MICA

[Temperature, T, K; Thermal Diffusivity, α, $cm^2\ s^{-1}$]

T	α
CURVE 1*	
300.2	0.0190

* No figure given.

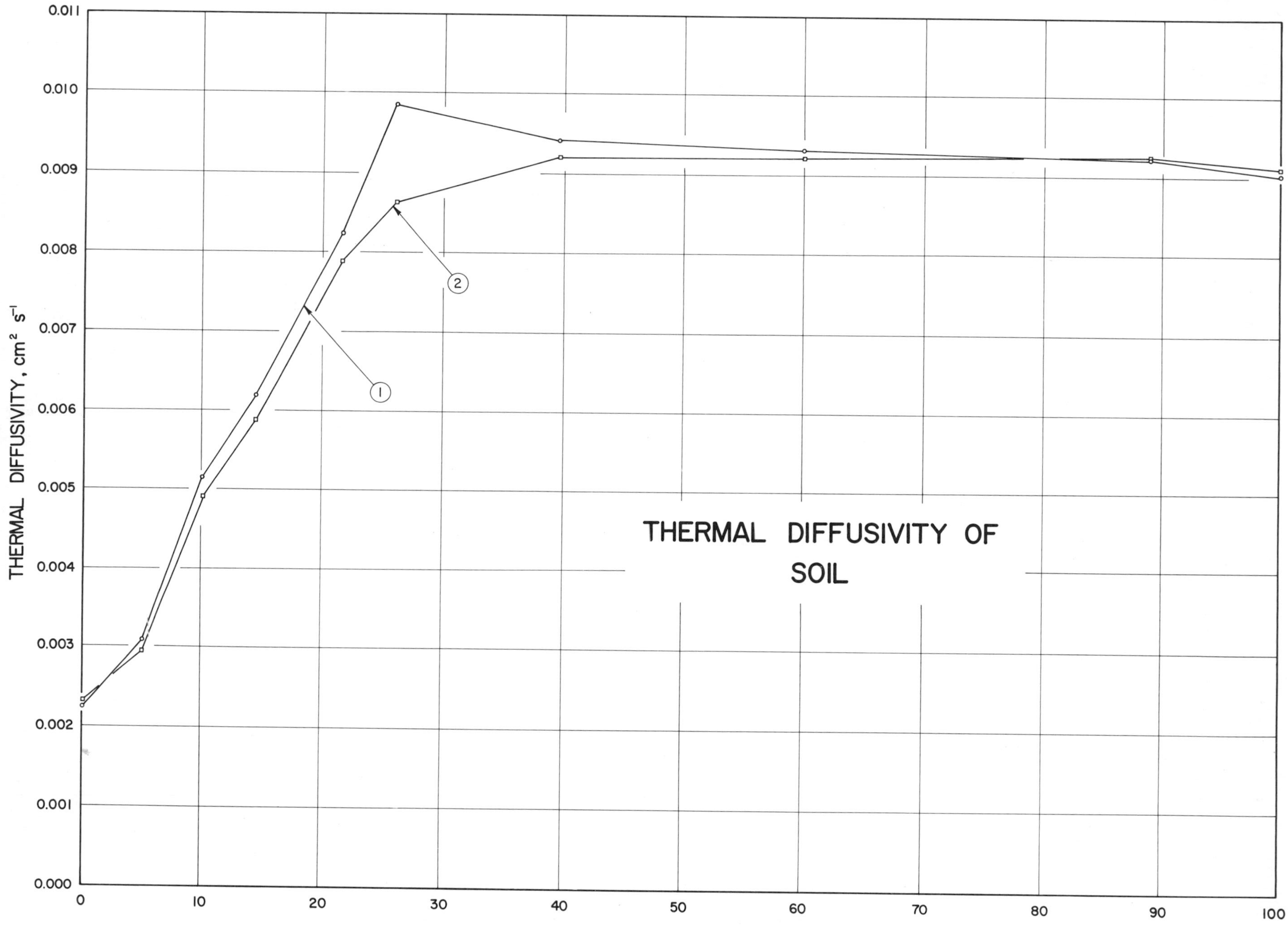
THERMAL DIFFUSIVITY OF SOIL
THERMAL DIFFUSIVITY, cm² s⁻¹
MOISTURE CONTENT, % OF SATURATION
0.000
0.001
0.002
0.003
0.004
0.005
0.006
0.007
0.008
0.009
0.010
0.011
0
10
20
30
40
50
60
70
80
90
100
1
2

FIGURE 229

SPECIFICATION TABLE 229. THERMAL DIFFUSIVITY OF SOIL

Cur. No.	Ref. No.	Author(s)	Year	Temp. Range, K	Reported Error, %	Name and Specimen Designation	Composition (weight percent), Specifications, and Remarks
1	288	Moench, A.F. and Evans, D.D.	1970	298	4	Sandy loam soil	Ten samples of sandy loam soil with moisture content (in % of saturation) 0.0, 5.1, 10.2, 14.5, 21.5, 26.0, 39.5, 60.0, 89.0, and 100.0%; the coresponding volume fractions of water were 0.000, 0.023, 0.045, 0.065, 0.094, 0.115, 0.166, 0.236, 0.320, and 0.380 cm^3 cm^{-3}, and the dry bulk densities of the samples were, respectively, 1.48, 1.46, 1.45, 1.43, 1.45, 1.46, 1.51, 1.59, 1.66, and 1.52 g cm^{-3}; data reported are the apparent thermal diffusivity which include the effect of distillation; measuring temperature not reported and here assumed to be 25 C.
2	288	Moench, A.F. and Evans, D.D.	1970	298	4	Sandy loam soil	The above samples and measurements but data corrected to yield real thermal diffusivity by subtracting from the measured data (given by curve 1 above) the small effect of distillation which was estimated with the help of the theory of vapor diffusion in porous media.
3*	309	Neumann, F.	1862	323		Frozen soil	Cubic specimen 5 to 6 in. on side, or sphere of same diameter; uniformly heated and then cooled in air; temperatures at center and surface observed by means of thermo-electric rods.

DATA TABLE 229. THERMAL DIFFUSIVITY OF SOIL

[Moisture Content, % of Saturation; Temperature, T, K; Thermal Diffusivity, α, cm^2 s^{-1}]

Moisture Content (% saturation)	α	Moisture Content (% saturation)	α	T	α
CURVE 1 (T = 298 K)		CURVE 2 (T = 298 K)		CURVE 3*	
0.0	0.00223	0.0	0.00231	323	0.0108
5.1	0.00307	5.1	0.00294		
10.2	0.00514	10.2	0.00490		
14.5	0.00618	14.5	0.00588		
21.5	0.00822	21.5	0.00787		
26.0	0.00985	26.0	0.00861		
39.5	0.00940	39.5	0.00920		
60.0	0.00930	60.0	0.00920		
89.0	0.00922	89.0	0.00925		
100.0	0.00903	100.0	0.00911		

* Not shown in figure.

12. SYSTEMS

SPECIFICATION TABLE 230. THERMAL DIFFUSIVITY OF CORRUGATED SHEETS

Cur. No.	Ref. No.	Author(s)	Year	Temp. Range, K	Reported Error, %	Name and Specimen Designation	Composition Core		Composition Facing	Composition (continued), Specifications, and Remarks
1*	100	Smith, W. K.	1961	367-1146		Spacemetal	AISI 301 Stainless Steel	Air	AISI 301 Stainless Steel	Corrugated core 5/32 in. thick made from a 0.002 in. thick sheet; facing consists of 0.006 in. thick coverplates; manufactured by North American Aviation, Inc.; two similar specimens placed back to back and simultaneously exposed to a high intensity source of radiation on both outer surfaces.

DATA TABLE 230. THERMAL DIFFUSIVITY OF CORRUGATED SHEETS

[Temperature, T, K; Thermal Diffusivity, α, $cm^2 s^{-1}$]

T	α
CURVE 1*	
367	0.0729
477	0.0832
590	0.0974
699	0.110
812	0.126
925	0.145
1035	0.164
1146	0.206

*No figure given.

SPECIFICATION TABLE 231. THERMAL DIFFUSIVITY OF LAMINATES (METALLIC - NONMETALLIC)

Cur. No.	Ref. No.	Author(s)	Year	Temp. Range, K	Reported Error, %	Name and Specimen Designation	Composition (weight percent), Specifications, and Remarks
1*	99	Sedillo, L., Castonguay, T. T., and Donaldson, W. E.	1963	333.2		Plastic L minates; 6	Asbestos cloth with one sheet of silver foil in center impregnated with 91 LD phenolic resin; pressed at 100 lb in.$^{-2}$ at 449.8 K for 1 hr, removed from the press, and post cured for 16 hrs at 449.8 K; thermal conductivity and density specimen: square 7 x 7 in. and 0.254 in. thick, dried at ~374.8 K; specific heat specimen: disk 1.125 in. in diameter cut from panel 7 x 7 x 0.254 in.; supplied by the U.S. Polymeric Chemical Company, Inc.; density 1.810 g cm^{-3}; diffusivity value calculated from measured conductivity, specific heat, and density.
2*	99	Sedillo, L., et al.	1963	333.2		Plastic Laminates; 7	Asbestos cloth with one sheet of aluminum foil in center impregnated with 91 LD phenolic resin; pressed at 100 lb in.$^{-2}$ at 449.8 K for 1 hr, removed from the press, and post cured for 16 hrs at 449.8 K; thermal conductivity and density specimen: square 7 x 7 in. and 0.281 in. thick; dried at ~374.8 K; specific heat specimen: disk 1.125 in. in diameter cut from panel 7 x 7 x 0.281 in.; supplied by the U.S. Polymeric Chemical Company, Inc.; density 1.727 g cm^{-3}; diffusivity value calculated from measured conductivity, specific heat, and density.

DATA TABLE 231. THERMAL DIFFUSIVITY OF LAMINATES (METALLIC - NONMETALLIC)

[Temperature, T, K; Thermal Diffusivity, α, cm^2 s^{-1}]

T	α
CURVE 1*	
333.2	0.00116
CURVE 2*	
333.2	0.000800

* No figure given.

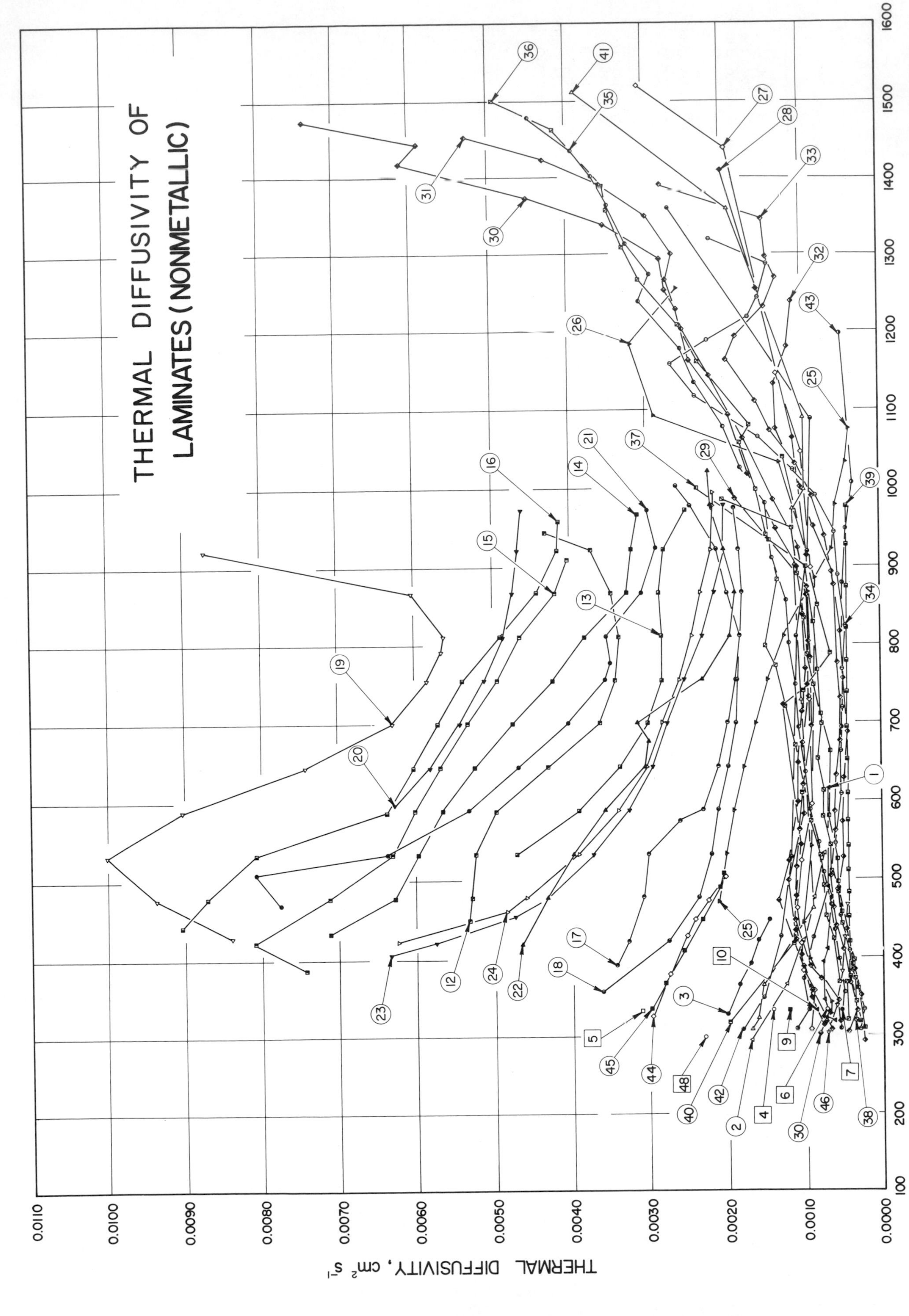
THERMAL DIFFUSIVITY OF
LAMINATES (NONMETALLIC)
THERMAL DIFFUSIVITY, cm² s⁻¹
TEMPERATURE, K
0.0000
0.0010
0.0020
0.0030
0.0040
0.0050
0.0060
0.0070
0.0080
0.0090
0.0100
0.0110
100
200
300
400
500
600
700
800
900
1000
1100
1200
1300
1400
1500
1600

FIGURE 232

SPECIFICATION TABLE 232. THERMAL DIFFUSIVITY OF LAMINATES (NONMETALLIC)

Cur. No.	Ref. No.	Author(s)	Year	Temp. Range, K	Reported Error, %	Name and Specimen Designation	Composition (weight percent), Specifications, and Remarks
1	1	Sonnenschein, G. and Winn, R.A.	1960	319-615		Plastic Laminates CTL 37-9X	Modified phenolic resin, 181 glass fabric; cylindrical specimen 0.635 cm in diameter; front surface covered with fine film of lamp black; measured under a vacuum of ~10^{-4} mm Hg; one-dimensional heat flow; flash method used to measure diffusivity.
2	100	Smith, W.K.	1961	294-533		Phenolic Asbestos Laminate	0.125 in. thick slab; double-sandwich method used to measure diffusivity; outer two layers made of transite and inner two layers made of phenolic asbestos laminate; outer surfaces uniformly blackened; specimen heated on both outer surfaces using a high intensity source of radiation.
3	100	Smith, W.K.	1961	339-450		Lamacoid No. 6045	Double-sandwich method used to measure diffusivity; outer two layers made of transite and inner two layers made of Lamacoid No. 6045; outer surfaces uniformly blackened; specimen heated on both outer surfaces using a high intensity source of radiation.
4	99	Sedillo, L., Castonguay, T.T., and Donaldson, W.E.	1963	333.2		Plastic Laminates; 1	6 sheets of graphite mat impregnated with 101 phenolic resin; pressed at 100 lb in.$^{-2}$ at 449.8 K for 1 hr, removed from the press, and post cured for 16 hrs at 449.8 K; thermal conductivity and density specimen: square 7 x 7 in. and 0.141 in. thick, dried at ~374.8 K; specific heat specimen: disk 1.125 in. in diameter cut from panel 7 x 7 x 0.141 in.; supplied by the U.S. Polymeric Chemical Company, Inc.; density 1.06 g cm^{-3}; diffusivity calculated from measured conductivity, specific heat, and density.
5	99	Sedillo, L., et al.	1963	333.2		Plastic Laminates; 2	13 sheets WC-001 graphite cloth impregnated with 101 phenolic resin; pressed at 100 lb in.$^{-2}$ at 449.8 K for 1 hr, removed from the press, and post cured for 16 hrs at 449.8 K; thermal conductivity and density specimen: square 7 x 7 in. and 0.163 in. thick, dried at ~374.8 K; specific heat specimen: disk 1.125 in. in diameter cut from panel 7 x 7 x 0.163 in.; supplied by the U.S. Polymeric Chemical Company, Inc.; density 1.28 g cm^{-3}; diffusivity calculated from measured conductivity, specific heat, and density.
6	99	Sedillo, L., et al.	1963	333.2		Plastic Laminates; 3	7 sheets of 184 Volan (glass cloth with a Volan finish) impregnated with 37-9X phenyl silane resin; pressed at 100 lb in.$^{-2}$ at 449.8 K for 1 hr, removed from the press, and post cured for 16 hrs at 449.8 K; thermal conductivity and density specimen: square 7 x 7 in. and 0.168 in. thick, dried at ~374.8 K; specific heat specimen: disk 1.125 in. in diameter cut from panel 7 x 7 x 0.168 in.; supplied by the U.S. Polymeric Chemical Company, Inc.; density 1.938 g cm^{-3}; diffusivity calculated from measured conductivity, specific heat, and density.
7	99	Sedillo, L., et al.	1963	333.2		Plastic Laminates; 4	Thermo-Insulating Compound (TIC-311S); used as received; thermal conductivity and density specimen: square 7 x 7 in. and 0.094 in. thick; specific heat specimen: disk 1.125 in. in diameter cut from panel 7 x 7 x 0.094 in.; supplied by the Alim Corporation; density 1.08 g cm^{-3}; diffusivity calculated from measured conductivity, specific heat, and density.

SPECIFICATION TABLE 232. THERMAL DIFFUSIVITY OF LAMINATES (NONMETALLIC) (continued)

Cur. No.	Ref. No.	Author(s)	Year	Temp. Range, K	Reported Error, %	Name and Specimen Designation	Composition (weight percent), Specifications, and Remarks
8*	99	Sedillo, L., Castonguay, T. T., and Donaldson, W. E.	1963	333.2		Plastic Laminates; 5	184 Volan impregnated with 37-9X phenyl silane resin; pressed at 100 lb in.$^{-2}$ at 449.8 K for 1 hr, removed from the press, and post cured for 16 hrs at 449.8 K; thermal conductivity and density specimen: square 7 x 7 in. and 0.170 in. thick; dried at ~374.8 K; specific heat specimen: disk 1.125 in. in diameter cut from panel 7 x 7 x 0.170 in.; supplied by the U.S. Polymeric Chemical Company, Inc.; density 1.930 g cm^{-3}; diffusivity calculated from measured conductivity, specific heat, and density.
9	99	Sedillo, L., et al.	1963	333.2		Plastic Laminates; 8	Refrasil cloth impregnated with 91 LD phenolic resin; pressed at 100 lb in.$^{-2}$ at 449.8 K for 1 hr, removed from the press, and post cured for 16 hrs at 449.8 K; thermal conductivity and density specimen: square 7 x 7 in. and 0.170 in. thick, dried at ~374.8 K; specific heat specimen: disk 1.125 in. in diameter cut from panel 7 x 7 x 0.170 in.; supplied by the U.S. Polymeric Chemical Company, Inc.; density 1.55 g cm^{-3}; diffusivity calculated from measured conductivity, specific heat, and density.
10	99	Sedillo, L., et al.	1963	333.2		Plastic Laminates; 9	184 Volan impregnated with 37-9X phenyl silane resin plus one coat of plastic primer and two coats (2-mil) of SAF paint (alkyd resin base); pressed at 100 lb in.$^{-2}$ at 449.8 K for 1 hr, removed from the press, and post cured for 16 hrs at 449.8 K; thermal conductivity and density specimen: square 7 x 7 in. and 0.170 in. thick, dried at ~374.8 K; specific heat specimen: disk 1.125 in. in diameter cut from panel 7 x 7 x 0.170 in.; plastic laminates supplied by the U.S. Polymeric Chemical Company, Inc.; SAF paint supplied by the Alim Corporation; density 1.925 g cm^{-3}; diffusivity calculated from measured conductivity, specific heat, and density.
11*	99	Sedillo, L., et al.	1963	333.2		Plastic Laminates; 10	184 Volan impregnated with 37-9X phenyl silane resin plus one coat of plastic primer and two coats (2-mil) of TIC paint (water base); pressed at 100 lb in.$^{-2}$ at 449.8 K for 1 hr, removed from the press, and post cured for 16 hrs at 449.8 K; thermal conductivity and density specimen: square 7 x 7 in. and 0.172 in. thick, dried at ~374.8 K; specific heat specimen: disk 1.125 in. in diameter cut from panel 7 x 7 x 0.172 in.; plastic laminates supplied by the U.S. Polymeric Chemical Company, Inc.; TIC paint supplied by the Alim Corporation; density 1.925 g cm^{-3}; diffusivity calculated from measured conductivity, specific heat, and density.
12	101	Donaldson, W. E. and Castonguay, T. T.	1963	450-944		Phenolic Asbestos Cloth; Panel 1	Phenolic asbestos cloth; reinforced-plastic heat-barrier laminates; ironsides No. 101 phenolic resin reinforced with 0.125 in. thick asbestos cloth; resin content within the range of 47 to 53%; square specimen 4.5 x 4.5 x 0.180 in.; prepared using 100 lb in.$^{-2}$ laminating pressure at a curing temperature of 463.7 K for 1 hr and then post cured for 12 hrs in an oven at 463.7 K; during measurement specimen showed some surface spalling and heat distortion with a major thermal decomposition noted at 560.9 K; considerable charred resin remained in the asbestos matrix; data reported are apparent thermal diffusivity; no experimental data given, data taken from smoothed curve; diffusivity measured at side 1.

* Not shown in figure.

SPECIFICATION TABLE 232. THERMAL DIFFUSIVITY OF LAMINATES (NONMETALLIC) (continued)

Cur. No.	Ref. No.	Author(s)	Year	Temp. Range, K	Reported Error, %	Name and Specimen Designation	Composition (weight percent), Specifications, and Remarks
13	101	Donaldson, W. E. and Castonguay, T. T.	1963	533-972		Phenolic Asbestos Cloth; Panel 1	Above specimen measured at side 2.
14	101	Donaldson, W. E. and Castonguay, T. T.	1963	433-967		Phenolic Asbestos Cloth; Panel 2	Same as the above specimen; measured at side 3A.
15	101	Donaldson, W. E. and Castonguay, T. T.	1963	386-908		Phenolic Asbestos Cloth; Panel 2	Above specimen measured at side 4A.
16	101	Donaldson, W. E. and Castonguay, T. T.	1963	442-958		Phenolic Asbestos Cloth; Panel 2A	Same as the above specimen except reinforced by 0.0625 in. thick asbestos cloth; measured at side 1.
17	101	Donaldson, W. E. and Castonguay, T. T.	1963	392-975		Phenolic-Refrasil Cloth; Panel 1	Phenolic-refrasil cloth; reinforced-plastic heat-barrier laminates; 91 LD phenolic resin reinforced with 0.015 in. thick refrasil cloth; resin content within the range of 47 to 53%; square specimen 4.5 x 4.5 x 0.180 in.; prepared using 100 lb in.$^{-2}$ laminating pressure at a curing temperature of 463.7 K for 1 hr and then post cured for 12 hrs in an oven at 463.7 K; resin charring occurred at 560.9 K during measurement, specimen showed some slight heat distortion and spalling; data reported are apparent thermal diffusivity; no experimental data given, data taken from smoothed curve; diffusivity measured at side 1.
18	101	Donaldson, W. E. and Castonguay, T. T.	1963	358-1003		Phenolic-Refrasil Cloth; Panel 2	Same as the above specimen; measured at side 4.
19	101	Donaldson, W. E. and Castonguay, T. T.	1963	430-922		Phenolic-Graphite Cloth; Panel 1	Phenolic-graphite cloth; reinforced-plastic heat-barrier laminates; ironsides No. 101 phenolic resin reinforced with 0.012 in. thick graphite cloth; resin content within the range of 47 to 53%; square specimen 4.5 x 4.5 x 0.180 in.; prepared using 100 lb in.$^{-2}$ laminating pressure at a curing temperature of 463.7 K for 1 hr and then postcured for 12 hrs in an oven at 463.7 K; a definite thermal phase change occurred from 699.8 to 810.9 K, a microscopic examination of panel showed charred phenolic resin in the graphite cloth; data reported are apparent thermal diffusivity; no experimental data given, data taken from smoothed curve; diffusivity measured at side 1.
20	101	Donaldson, W. E. and Castonguay, T. T.	1963	597-972		Phenolic-Graphite Cloth; Panel 1	Above specimen measured at side 2.
21	101	Donaldson, W. E. and Castonguay, T. T.	1963	469-972		Phenolic-Graphite Cloth; Panel 3	Same as the above specimen; measured at side 5.

SPECIFICATION TABLE 232. THERMAL DIFFUSIVITY OF LAMINATES (NONMETALLIC) (continued)

Cur. No.	Ref. No.	Author(s)	Year	Temp. Range, K	Reported Error, %	Name and Specimen Designation	Composition (weight percent), Specifications, and Remarks
22	101	Donaldson, W. E. and Castonguay, T. T.	1963	419-1022		Phenolic-Graphite Cloth; Panel 3	Above specimen measured at side 6.
23	101	Donaldson, W. E. and Castonguay, T. T.	1963	405-978		Phenolic-Graphite Mat; Panel 2	Phenolic-graphite mat; reinforced-plastic heat-barrier laminates; 91 LD phenolic resin reinforced with 0.5 in. thick graphite mat; resin content within the range of 47 to 53%; square specimen 4.5 x 4.5 x 0.180 in.; prepared using 100 lb in.$^{-2}$ laminating pressure at a curing temperature of 463.7 K for 1 hr and then post cured for 12 hrs in an oven at 463.7 K; phenolic resin charred in a uniform pattern during measurement, specimen was slightly warped; data reported are apparent thermal diffusivity; no experimental data given, data taken from smoothed curve; diffusivity measured at both sides.
24	101	Donaldson, W. E. and Castonguay, T. T.	1963	422-994		Phenolic-Graphite Mat; Panel 3	Same as the above specimen; measured at side 5.
25	101	Donaldson, W. E. and Castonguay, T. T.	1963	472-1075		Silicone-Asbestos; Panel 3	Silicone-asbestos; reinforced-plastic heat-barrier laminates; Dow No. 2106 silicone resin reinforced with 0.010 in. thick asbestos paper; resin content within the range of 47 to 53%; square specimen 4.5 x 4.5 x 0.180 in.; prepared using 100 lb in.$^{-2}$ laminating pressure at a curing temperature of 463.7 K for 1 hr and then post cured for 12 hrs in an oven at 463.7 K; delamination occurred during measurement; data reported are apparent thermal diffusivity; no experimental data given, data taken from smoothed curve; diffusivity measured at both sides.
26	102	Mihalow, F.A., Koubek, F.J., and Perry, H.A.	1960	328-1255		No. 1	Glass reinforced phenolic resin laminate; consisting of Bakelite BLL 3085 resin with Style 181 glass cloth, NOL 24 finish; insert and slug fastened together by Epon 422 adhesive (Shell Co.); laminations arranged perpendicular to central axis of specimen; cylindrical rod specimen 0.75 in. in diameter; fabricated at NOL; data reported are effective thermal diffusivity calculated from measured ablation rate and temperature-time history at some arbitrary point in specimen; temperature measured using Pt-Pt 10 Rh thermocouple; ablation rate 5.81 x 10^{-3} cm sec^{-1}; constant heat flux; one-dimensional heat flow.
27	102	Mihalow, F.A., et al.	1960	308-1509		No. 1	Above specimen measured again using Cr-Al thermocouple.
28	102	Mihalow, F.A., et al.	1960	313-1408		No. 1	Above specimen measured again using Cr-Al thermocouple.
29	102	Mihalow, F.A., et al.	1960	308-986		No. 2	Same as the above specimen; measured using Pt-Pt 10 Rh thermocouple; ablation rate 5.88 x 10^{-3} cm sec^{-1}; other conditions same as above.
30	102	Mihalow, F.A., et al.	1960	303-1473		No. 2	Above specimen measured again using Cr-Al thermocouple.
31	102	Mihalow, F.A., et al.	1960	318-1453		No. 2	Above specimen measured again using Cr-Al thermocouple.
32	102	Mihalow, F.A., et al.	1960	294-1238		No. 3	Same as the above specimen; measured using Pt-Pt 10 Rh thermocouple; ablation rate 4.68 x 10^{-3} cm sec^{-1}; other conditions same as above.

SPECIFICATION TABLE 232. THERMAL DIFFUSIVITY OF LAMINATES (NONMETALLIC) (continued)

Cur. No.	Ref. No.	Author(s)	Year	Temp. Range, K	Reported Error, %	Name and Specimen Designation	Composition (weight percent), Specifications, and Remarks
33	102	Mihalow, F.A., Koubek, F.J., and Perry, H.A.	1960	305-1390		No. 3	Above specimen measured again using Cr-Al thermocouple.
34	102	Mihalow, F.A., et al.	1960	308-1318		No. 3	Above specimen measured again using Cr-Al thermocouple.
35	102	Mihalow, F.A., et al.	1960	309-1478		No. 4	Same as the above specimen except that the laminations are arranged parallel to the central axis; measured using Cr-Al thermocouple; ablation rate 5.74×10^{-3} cm sec^{-1}; other conditions same as above.
36	102	Mihalow, F.A., et al.	1960	320-1500		No. 4	Above specimen measured again using Cr-Al thermocouple.
37	102	Mihalow, F.A., et al.	1960	310-1000		No. 5	Same as the above specimen; measured using Pt-Pt 10 Rh thermocouple; ablation rate 3.89×10^{-3} cm sec^{-1}; other conditions same as above.
38	102	Mihalow, F.A., et al.	1960	320-986		No. 5	Above specimen measured again using Cr-Al thermocouple.
39	102	Mihalow, F.A., et al.	1960	318-977		No. 5	Above specimen measured again using Cr-Al thermocouple.
40	102	Mihalow, F.A., et al.	1960	318-1039		No. 6	Nylon reinforced phenolic resin laminate; Formica No. YN-25; insert and slug fastened in the same manner as above specimen; laminations arranged perpendicular to central axis; same dimensions and shape as above specimen; supplied by Formica; commercially prepared; ablation rate 9.74×10^{-3} cm sec^{-1}; measured using Pt-Pt 10 Rh thermocouple; other conditions same as above specimen.
41	102	Mihalow, F.A., et al.	1960	308-1511		No. 6	Above specimen measured again using Cr-Al thermocouple.
42	102	Mihalow, F.A., et al.	1960	309-1359		No. 6	Above specimen measured again using Cr-Al thermocouple.
43	102	Mihalow, F.A., et al.	1960	308-1196		No. 7	Same as the above specimen except that the laminations are arranged parallel to the central axis; measured using Pt-Pt 10 Rh thermocouple; ablation rate 3.43×10^{-3} cm sec^{-1}; other conditions same as above.
44	300	Vlasov, V.V. and Dorogov, N.N.	1966	326-505	5-6	Textolite	100 x 200 x 3~10 mm; measured by the automatic apparatus.
45	300	Vlasov, V.V. and Dorogov, N.N.	1966	336-508	5-6	Textolite	The above specimen measured by the manual method.
46	300	Vlasov, V.V. and Dorogov, N.N.	1966	303-446	5-6	Micarta	100 x 100 x 3~10 mm; measured by the automatic apparatus.
47*	300	Vlasov, V.V. and Dorogov, N.N.	1966	303-464	5-6	Micarta	The above specimen measured by the manual method.
48	108	Steinberg, S., Larson, R.E., and Kydd, A.R.	1963	298		Epoxy-fiberglass	Epoxy-reinforced fiberglass laminate; specimen 2.0 cm square, 0.169 cm thick; measuring temperature not reported and here assumed to be 25 C.

* Not shown in figure.

DATA TABLE 232. THERMAL DIFFUSIVITY OF LAMINATES (NONMETALLIC)

[Temperature, T, K; Thermal Diffusivity, α, $cm^2\ s^{-1}$]

T	α
CURVE 1	
319	0.000635
322	0.000740
330	0.000705
345	0.000760
394	0.000685
413	0.000710
476	0.000665
487	0.000710
615	0.000695
CURVE 2	
294.3	0.00170
366.5	0.00124
422.1	0.00103
477.6	0.000884
533.2	0.000755
CURVE 3	
338.7	0.00204
366.5	0.00186
394.3	0.00171
422.1	0.00160
449.8	0.00146
CURVE 4	
333.2	0.00142
CURVE 5	
333.2	0.00310
CURVE 6	
333.2	0.000723
CURVE 7	
333.2	0.000542
CURVE 8*	
333.2	0.000723
CURVE 9	
333.2	0.00121
CURVE 10	
333.2	0.00877
CURVE 11*	
333.2	0.000877
CURVE 12	
450	0.00532
478	0.00529
533	0.00523
589	0.00497
644	0.00429
700	0.00361
755	0.00342
811	0.00336
867	0.00345
922	0.00371
944	0.00432
CURVE 13	
533	0.00471
589	0.00390
644	0.00336
700	0.00300
755	0.00281
811	0.00281
867	0.00284
922	0.00277
972	0.00248
CURVE 14	
433	0.00710
478	0.00626
533	0.00597
589	0.00565
644	0.00523
700	0.00474
755	0.00423
CURVE 14 (cont.)	
811	0.00381
867	0.00345
922	0.00319
967	0.00310
CURVE 15	
386	0.00742
422	0.00807
478	0.00710
533	0.00629
589	0.00600
644	0.00568
700	0.00532
755	0.00494
811	0.00465
867	0.00419
908	0.00403
CURVE 16	
442	0.00903
478	0.00871
533	0.00807
589	0.00636
644	0.00603
700	0.00571
755	0.00539
811	0.00490
867	0.00442
922	0.00416
958	0.00413
CURVE 17	
392	0.00342
422	0.00326
478	0.00306
533	0.00300
575	0.00258
589	0.00229
644	0.00210
700	0.00197
755	0.00187
811	0.00181
CURVE 17 (cont.)	
867	0.00177
922	0.00181
975	0.00187
CURVE 18	
358	0.00361
422	0.00274
478	0.00235
533	0.00219
589	0.00210
644	0.00197
700	0.00187
755	0.00184
811	0.00184*
867	0.00197
922	0.00210
978	0.00242
1003	0.00261
CURVE 19	
430	0.00839
478	0.00936
533	0.0100
589	0.00903
644	0.00742
700	0.00629
755	0.00584
792	0.00565
811	0.00561
867	0.00603
922	0.00871
CURVE 20	
597	0.00626
644	0.00581
700	0.00542
755	0.00510
811	0.00487
867	0.00474
922	0.00468
972	0.00461
CURVE 21	
469	0.00774
508	0.00807
533	0.00636
589	0.00532
644	0.00468
700	0.00403
755	0.00355
778	0.00348
811	0.00352
867	0.00306
922	0.00287
972	0.00297
CURVE 22	
419	0.00465
478	0.00432
533	0.00397
589	0.00355
644	0.00303
672	0.00297
700	0.00313
755	0.00229
811	0.00194
867	0.00187
922	0.00200
978	0.00216
1022	0.00219
CURVE 23	
405	0.00632
422	0.00574
455	0.00474
478	0.00432*
533	0.00371
589	0.00323
644	0.00294
700	0.00274
755	0.00252
811	0.00229
867	0.00213
922	0.00200*
978	0.00200
CURVE 24	
422	0.00623
461	0.00484
478	0.00458
533	0.00390
589	0.00339
644	0.00303*
700	0.00281
755	0.00258
811	0.00242
867	0.00229
922	0.00216
978	0.00213
994	0.00213
CURVE 25	
472	0.00210
533	0.00200
589	0.00190
644	0.00177
700	0.00161
755	0.00145
811	0.00123
867	0.000935
922	0.000636
978	0.000584
1033	0.000503
1075	0.000387
CURVE 26	
328	0.000670
343	0.000680
398	0.000950
458	0.00102
543	0.000920
885	0.000840
1033	0.00128
1093	0.00288
1185	0.00317
1255	0.00257
CURVE 27	
308	0.000940
358	0.000900
CURVE 27 (cont.)	
383	0.00103
418	0.00106
463	0.00110
523	0.00104
595	0.000900
728	0.000920
898	0.000890
1046	0.00100
1146	0.00131
1438	0.00194
1509	0.00305
CURVE 28	
313	0.000780
348	0.000940
403	0.00105
435	0.00113
478	0.00111
543	0.00105
723	0.00126
748	0.000950
913	0.000940
1063	0.00110
1256	0.00155
1408	0.00198
CURVE 29	
308	0.000680*
373	0.000810
390	0.000940
448	0.00113
478	0.00112*
516	0.00111
563	0.00106
620	0.00101
695	0.000910
783	0.000910
873	0.000940
986	0.00185
CURVE 30	
303	0.000830
318	0.000750
CURVE 30 (cont.)	
358	0.000920
388	0.00103
473	0.00135
523	0.00120
598	0.00108
648	0.00107
713	0.00102
768	0.000975
838	0.00100
898	0.00107
948	0.00133
1018	0.00168
1093	0.00192
1163	0.00242
1228	0.00257
1253	0.00272
1293	0.00279
1338	0.00351
1373	0.00450
1418	0.00612
1443	0.00588
1473	0.00737
CURVE 31	
318	0.000560
345	0.000579*
423	0.00112
498	0.00121
583	0.00102
673	0.00102
733	0.000935
803	0.000935
861	0.000950
923	0.00105
985	0.00136
1063	0.00174
1143	0.00216
1266	0.00271
1300	0.00263
1348	0.00296
1388	0.00355
1423	0.00428
1453	0.00527

* Not shown in figure.

DATA TABLE 232. THERMAL DIFFUSIVITY OF LAMINATES (NONMETALLIC) (continued)

T	α
CURVE 32	
294	0.000245
351	0.000295
385	0.000385
436	0.000465
486	0.000590
528	0.000500
573	0.000510
629	0.000495
688	0.000435
756	0.000495
816	0.000520
893	0.000625
951	0.000855
1010	0.00105
1076	0.00132
1133	0.00134
1180	0.00117
1238	0.00110
CURVE 33	
305	0.000450
332	0.000340
367	0.000430
410	0.000510
442	0.000600
464	0.000630
504	0.000670
561	0.000610
586	0.000610
632	0.000590
676	0.000540
731	0.000530
778	0.000560
875	0.000600
938	0.000710
989	0.000870
1031	0.00107
1073	0.00140
1111	0.00158
1164	0.00195
1231	0.00146
1269	0.00131
1296	0.00143
1345	0.00148
1390	0.00278
CURVE 34	
308	0.000300
333	0.000250
383	0.000400
408	0.000440
453	0.000550
488	0.000575
545	0.000610
608	0.000530
663	0.000525
718	0.000500
768	0.000500
823	0.000460
888	0.000545
943	0.000585
990	0.000825
1023	0.00111
1065	0.00155
1118	0.00235
1158	0.00267
1188	0.00218
1218	0.00166
1243	0.00153
1286	0.00142
1318	0.00215
CURVE 35	
309	0.000555
355	0.000500
398	0.000575
435	0.000625
467	0.000670
508	0.000775
546	0.000825
588	0.000925
636	0.000980
693	0.00106
748	0.00110
801	0.00118
856	0.00121
910	0.00139
980	0.00147
1026	0.00178
1078	0.00199
1134	0.00235
1178	0.00253
CURVE 35 (cont.)	
1134	0.00235
1178	0.00253
1238	0.00306
1273	0.00292
1313	0.00323
1363	0.00345
1401	0.00366
1435	0.00393
1478	0.00447
CURVE 36	
320	0.000460
368	0.000565
423	0.000640
453	0.000703
491	0.000755
530	0.000800
573	0.000915
618	0.00100*
670	0.00110
720	0.00123
773	0.00135
798	0.00148
883	0.00133
940	0.00145
998	0.00158
1058	0.00177
1080	0.00166
1161	0.00231
1208	0.00255
1266	0.00306
1308	0.00327
1356	0.00347
1388	0.00353
1463	0.00415
1500	0.00493
CURVE 37	
310	0.000245
360	0.000330
391	0.000390
420	0.000425
458	0.000535
496	0.000580
CURVE 37 (cont.)	
525	0.000590
580	0.000680
663	0.000630
710	0.000785
767	0.000820
828	0.000865
899	0.000955
933	0.00142
1000	0.00233
CURVE 38	
320	0.000355
343	0.000355
373	0.000365
396	0.000397
443	0.000530
483	0.000575
510	0.000650
543	0.000675
580	0.000760
613	0.000750
653	0.000825
698	0.000765
748	0.000875
786	0.000950
833	0.000980
890	0.00106
948	0.00113
986	0.00203
CURVE 39	
318	0.000300
375	0.000430
423	0.000440*
452	0.000440
483	0.000440
511	0.000440
542	0.000440
574	0.000440
609	0.000440
652	0.000440
693	0.000440
738	0.000440
778	0.000440
CURVE 39 (cont.)	
820	0.000440
873	0.000440
927	0.000440
977	0.000440
CURVE 40	
318	0.00198
418	0.00115
610	0.00103
740	0.00100
788	0.000660
850	0.000810
1039	0.00123
CURVE 41	
308	0.00170
323	0.00161
366	0.00154
463	0.000890
728	0.00103
902	0.00109
973	0.00112
998	0.000970
1089	0.000970
1358	0.00191
1511	0.00389
CURVE 42	
309	0.00182
349	0.00155
427	0.00131
528	0.00117
810	0.00108
918	0.000930
988	0.000890*
1089	0.000870
1359	0.00267
CURVE 43	
308	0.00112
333	0.00965
373	0.00104
CURVE 43 (cont.)	
426	0.000900
488	0.000700
693	0.000510
878	0.000490
948	0.000430
1006	0.000350
1196	0.000500
CURVE 44	
326.3	0.00296
380.3	0.00273
429.9	0.00251
449.9	0.00240
474.5	0.00224
504.9	0.00201
CURVE 45	
336.5	0.00297
368.2	0.00279
410.3	0.00255
450.9	0.00232
490.8	0.00209
508.5	0.00204
CURVE 46	
303.1	0.000722
318.7	0.000649*
338.8	0.000534
425.2	0.000497
466.2	0.000466
CURVE 47*	
303.1	0.000748
319.8	0.000665
338.6	0.000594
359.6	0.000540
382.1	0.000502
464.3	0.000414
CURVE 48	
298	0.0023

* Not shown in figure.

SPECIFICATION TABLE 233. THERMAL DIFFUSIVITY OF PACKED BEDS

Cur. No.	Ref. No.	Author(s)	Year	Temp. Range, K	Reported Error, %	Name and Specimen Designation	Composition (weight percent) Solid Particles	Composition (weight percent) Environment	Composition (continued), Specifications, and Remarks
1*	301	Verzhinskaya, A.B. and Vainberg, R.Sh.	1967	298			Silicate Sphere	Air	Sphere size 0.3 to 0.5 mm; bulk density 1.710 g cm^{-3}; measuring temperature not reported but here assumed to be 25 C.
2*	301	Verzhinskaya, A.B. and Vainberg, R.Sh.	1967	298			Silicate Sphere	Benzene	Similar to above but bulk density 2.030 g cm^{-3}.
3*	301	Verzhinskaya, A.B. and Vainberg, R.Sh.	1967	298			Silicate Sphere	Ethyl Alcohol	Similar to above but bulk density 2.063 g cm^{-3}.
4*	301	Verzhinskaya, A.B. and Vainberg, R.Sh.	1967	298			Silicate Sphere	Distilled Water	Similar to above but bulk density 2.110 g cm^{-3}.
5*	301	Verzhinskaya, A.B. and Vainberg, R.Sh.	1967	298			Silicate Sphere	Acetone	Similar to above but bulk density 2.023 g cm^{-3}.

DATA TABLE 233. THERMAL DIFFUSIVITY OF PACKED BEDS

[Temperature, T, K; Thermal Diffusivity, α, $cm^2\ s^{-1}$]

T	α
CURVE 1*	
298.2	0.0001482
CURVE 2*	
298.2	0.000221
CURVE 3*	
298.2	0.000238
CURVE 4*	
298.2	0.000249
CURVE 5*	
298.2	0.000229

* No figure given.

13. REFRACTORY MATERIALS , PROCESSED COMPOSITES, AND GLASSES

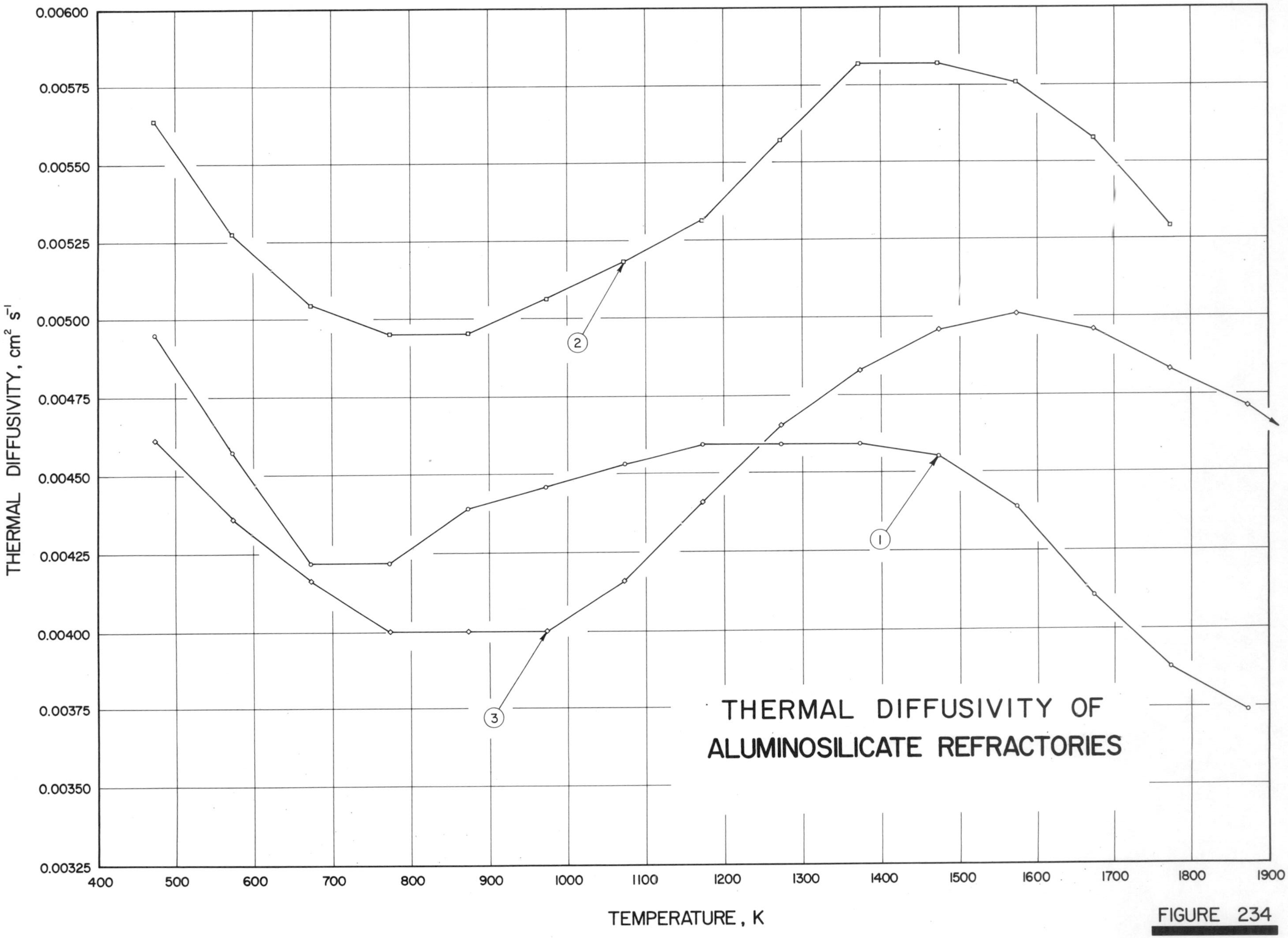
THERMAL DIFFUSIVITY OF
ALUMINOSILICATE REFRACTORIES
THERMAL DIFFUSIVITY, cm² s⁻¹
TEMPERATURE, K
0.00600
0.00575
0.00550
0.00525
0.00500
0.00475
0.00450
0.00425
0.00400
0.00375
0.00350
0.00325
400
500
600
700
800
900
1000
1100
1200
1300
1400
1500
1600
1700
1800
1900
1
2
3

FIGURE 234

SPECIFICATION TABLE 234. THERMAL DIFFUSIVITY OF ALUMINOSILICATE REFRACTORIES

Cur. No.	Ref. No.	Author(s)	Year	Temp. Range, K	Reported Error, %	Name and Specimen Designation	Composition (weight percent), Specifications, and Remarks
1	293	Litovskii, E. Ya., Landa, Ya. A., and Mil'shenko, R. S.	1970	473-1873		Shs-4 Glass block	58.30 SiO_2, 35.26 Al_2O_3, 1.50 Fe_2O_3, 1.33 TiO_2, 1.22 CaO, 0.22 MgO, and 1.79 alkalis; density 2.61 g cm^{-3}; total porosity 18.8%; diffusivity measured in argon.
2	293	Litovskii, E. Ya., et al.	1970	473-1773		DV-5	High alumina blocks for blast furnaces; 64.09 Al_2O_3, 31.14 SiO_2, 1.25 TiO_2, 1.12 CaO, 1.05 Fe_2O_3, 0.68 alkalis, and 0.21 MgO; density 3.075 g cm^{-3}; total porosity 18.9%; diffusivity measured in argon.
3	293	Litovskii, E. Ya., et al.	1970	473-1987		V-4	Product for air stoves; 61.25 Al_2O_3, 33.82 SiO_2, 1.42 TiO_2, 1.31 CaO, 1.20 Fe_2O_3, 0.68 alkalis, and 0.18 MgO; density 3.02 g cm^{-3}; total porosity 25.2%; diffusivity measured in argon.

DATA TABLE 234. THERMAL DIFFUSIVITY OF ALUMINOSILICATE REFRACTORIES

[Temperature, T, K; Thermal Diffusivity, α, cm^2 s^{-1}]

T	α	T	α	T	α
CURVE 1		CURVE 2		CURVE 3	
473	0.00495	473	0.00563	473	0.00461
573	0.00457	573	0.00527	573	0.00436
673	0.00422	673	0.00504	673	0.00416
773	0.00422	773	0.00495	773	0.00400
873	0.00439	873	0.00495	873	0.00400
973	0.00446	973	0.00506	973	0.00400
1073	0.00453	1073	0.00518	1073	0.00416
1173	0.00459	1173	0.00531	1173	0.00441
1273	0.00459	1273	0.00557	1273	0.00465
1373	0.00459	1373	0.00582	1373	0.00483
1473	0.00455	1473	0.00582	1473	0.00496
1573	0.00439	1573	0.00576	1573	0.00501
1673	0.00411	1673	0.00558	1673	0.00496
1773	0.00388	1773	0.00529	1773	0.00483
1873	0.00374			1873	0.00471
				1973	0.00453*
				1987	0.00445*

*Not shown in figure.

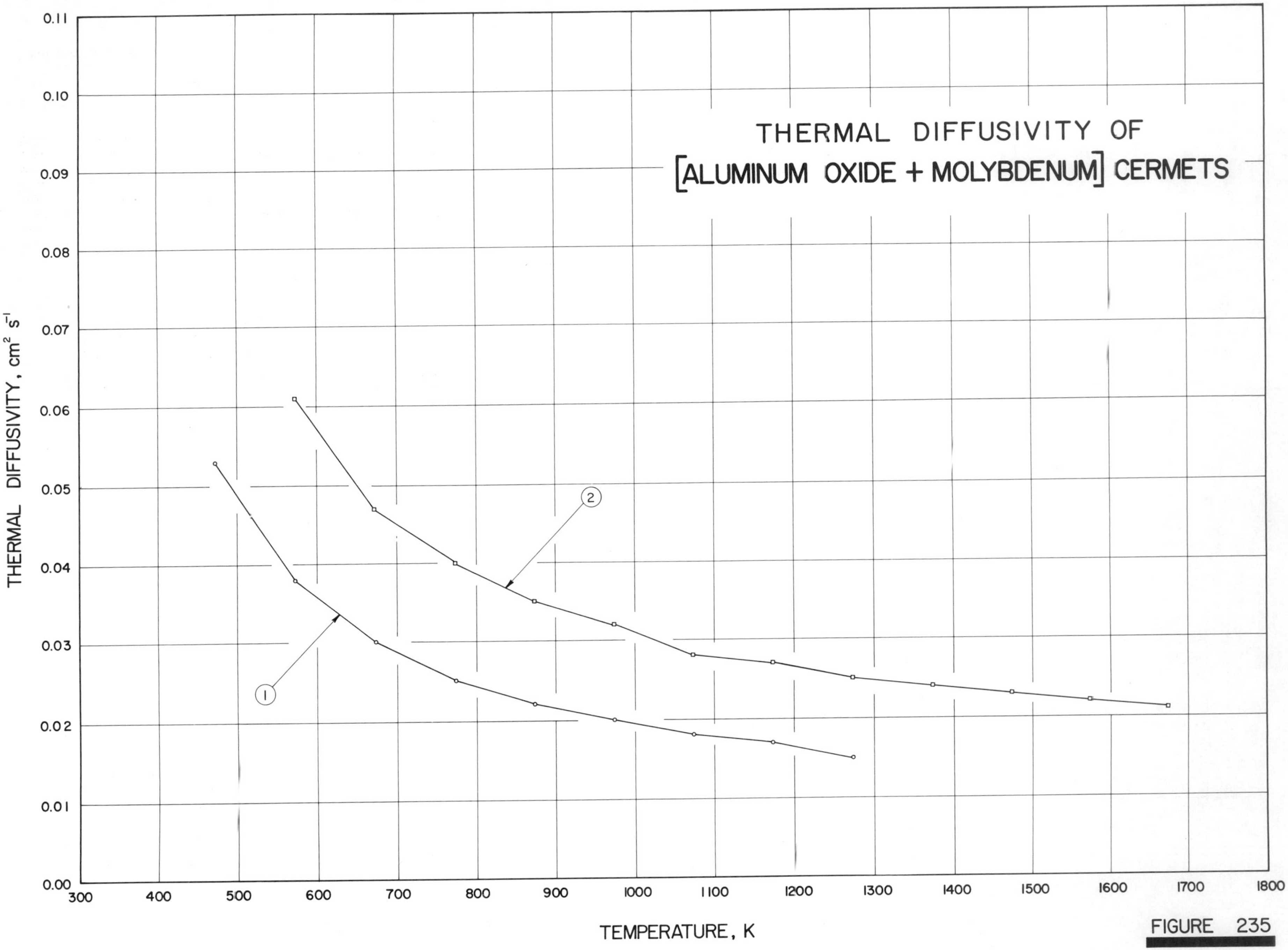
THERMAL DIFFUSIVITY OF
[ALUMINUM OXIDE + MOLYBDENUM] CERMETS
THERMAL DIFFUSIVITY, cm² s⁻¹
TEMPERATURE, K
0.00
0.01
0.02
0.03
0.04
0.05
0.06
0.07
0.08
0.09
0.10
0.11
300
400
500
600
700
800
900
1000
1100
1200
1300
1400
1500
1600
1700
1800
1
2

FIGURE 235

SPECIFICATION TABLE 235. THERMAL DIFFUSIVITY OF [ALUMINUM OXIDE + MOLYBDENUM] CERMETS $Al_2O_3 + Mo$

Cur. No.	Ref. No.	Author(s)	Year	Temp. Range, K	Reported Error, %	Name and Specimen Designation	Composition (weight percent) Al_2O_3	Composition (weight percent) Mo	Composition (continued), Specifications, and Remarks
1	285	Osipova, V.A. and Pak, M.I.	1969	473-1273	7-12	Cermet		10	Cylindrical specimen about 40 mm in diameter, 65-70 mm long; prepared from spectrally-pure corundum (α-form of aluminum oxide) and molybdenum powders; total porosity 1 to 3.5%; average density 4287 kg m^{-3}.
2	285	Osipova, V.A. and Pak, M.I.	1969	473-1673	7-12	Cermet		30	Similar to the above specimen except average density 4665 kg m^{-3}.

DATA TABLE 235. THERMAL DIFFUSIVITY OF [ALUMINUM OXIDE + MOLYBDENUM] CERMETS $Al_2O_3 + Mo$

[Temperature, T, K; Thermal Diffusivity, α, $cm^2\ s^{-1}$]

T	α	T	α
CURVE 1		CURVE 2 (cont.)	
473	0.053	1073	0.028
573	0.038	1173	0.027
673	0.030	1273	0.025
773	0.025	1373	0.024
873	0.022	1473	0.023
973	0.020	1573	0.022
1073	0.018	1673	0.021
1173	0.017		
1273	0.015		
CURVE 2			
473	0.094		
573	0.061		
673	0.047		
773	0.040		
873	0.035		
973	0.032		

SPECIFICATION TABLE 236. THERMAL DIFFUSIVITY OF ASBESTOS CEMENT BOARD

Cur. No.	Ref. No.	Author(s)	Year	Temp. Range, K	Reported Error, %	Name and Specimen Designation	Composition (weight percent), Specifications, and Remarks
1*	100	Smith, W. K.	1961	339-589		Transite	Two layers; previously preheated to remove moisture and kept dry; specific heat measured and reported as 0.182, 0.184, 0.187, 0.190, 0.194, 0.197, 0.200, 0.203, 0.205, and 0.207 cal g^{-1} K^{-1} at 338.7, 366.5, 394.3, 422.1, 449.8, 477.6, 505.4, 533.2, 560.9, and 588.7 K, respectively; heat absorbing surface evenly blackened.
2*	300	Vlasov, V. V. and Dorogov, N. N.	1966	332-521	5-6		100 x 100 x (3~10) mm; measured by the automatic apparatus.
3*	300	Vlasov, V. V. and Dorogov, N. N.	1966	329-521	5-6		The above specimen measured by the manual method.

DATA TABLE 236. THERMAL DIFFUSIVITY OF ASBESTOS CEMENT BOARD

[Temperature, T, K; Thermal Diffusivity, α, cm^2 s^{-1}]

T	α	T	α
CURVE 1*		CURVE 2 (cont.)*	
338.7	0.00524	465.9	0.00164
366.5	0.00510	521.0	0.00160
394.3	0.00497		
422.1	0.00485	CURVE 3*	
449.8	0.00474		
477.6	0.00465	329.2	0.00182
505.4	0.00458	346.0	0.00176
533.2	0.00452	363.0	0.00171
560.9	0.00445	414.2	0.00168
588.7	0.00440	465.7	0.00164
		521.0	0.00161
CURVE 2*			
331.8	0.00186		
345.8	0.00180		
363.1	0.00176		
414.2	0.00170		

* No figure given.

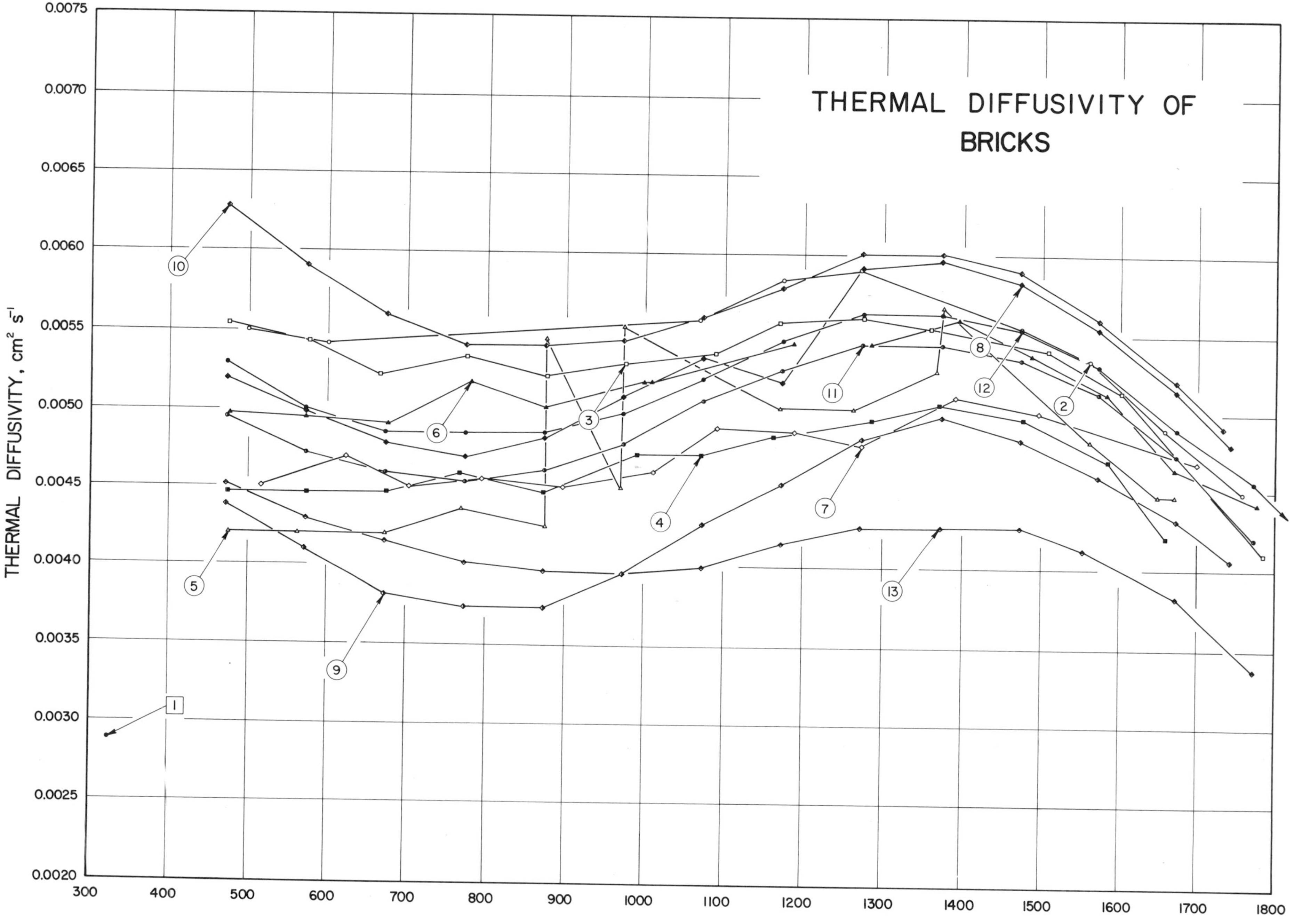
THERMAL DIFFUSIVITY OF
BRICKS
THERMAL DIFFUSIVITY, cm² s⁻¹
TEMPERATURE, K
0.0075
0.0070
0.0065
0.0060
0.0055
0.0050
0.0045
0.0040
0.0035
0.0030
0.0025
0.0020
300
400
500
600
700
800
900
1000
1100
1200
1300
1400
1500
1600
1700
1800
1
2
3
4
5
6
7
8
9
10
11
12
13

FIGURE 237

SPECIFICATION TABLE 237. THERMAL DIFFUSIVITY OF BRICKS

Cur. No.	Ref. No.	Author(s)	Year	Temp. Range, K	Reported Error, %	Name and Specimen Designation	Composition (weight percent), Specifications, and Remarks
1	292	Das, M. B. and Hossain, M. A.	1966	323		Firebrick	10 x 10 x 1.20 cm.
2	293	Litovskii, E. Ya., Landa, Ya. A., and Mil'shenko, R. S.	1970	499-1759		Ladle brick Grade KM-10	51.94 SiO_2, 39.70 Al_2O_3, 2.44 TiO_2, 2.21 Fe_2O_3, 0.38 MgO, and 0.31 CaO; specimen made by the Borovich Combine; diffusivity measured in nitrogen in the direction parallel to the pressing force.
3	293	Litovskii, E. Ya., et al.	1970	474-1786		Ladle brick Grade KM-10	Similar to the above specimen.
4	293	Litovskii, E. Ya., et al.	1970	474-1660		Ladle brick Grade KM-10	Similar to the above specimen.
5	293	Litovskii, E. Ya., et al.	1970	476-1671		Ladle brick Grade KM-10	Similar to the above specimen.
6	293	Litovskii, E. Ya., et al.	1970	476-1777		Ladle brick Grade KM-10	Similar to the above specimen except diffusivity measured perpendicular to the pressing force.
7	293	Litovskii, E. Ya., et al.	1970	517-1700		Ladle brick Grade KM-10	Similar to the above specimen.
8	293	Litovskii, E. Ya., et al.	1970	473-1743		Sh-3 Standard firebrick	61.26 SiO_2, 32.75 Al_2O_3, 2.00 TiO_2, 1.31 CaO, 1.30 Fe_2O_3, 0.66 alkalis, and 0.32 MgO; density 2.75 g cm^{-3}; total porosity 29.8%; diffusivity measured in argon.
9	293	Litovskii, E. Ya., et al.	1970	473-1743		7865-20 Firebrick, plastic molded	60.52 SiO_2, 34.71 Al_2O_3, 1.82 TiO_2, 1.10 Fe_2O_3, 0.88 CaO, 0.88 MgO, and 0.08 alkalis; density 2.69 g cm^{-3}; total porosity 26.8%; diffusivity measured in argon.
10	293	Litovskii, E. Ya., et al.	1970	473-1733		KM-17 Multichamotte ladle brick	55.80 SiO_2, 37.75 Al_2O_3, 2.22 TiO_2, 1.40 Fe_2O_3, 1.40 CaO, 0.82 alkalis, and 0.22 MgO; density 2.73 g cm^{-3}; total porosity 20.9%; diffusivity measured in argon.
11	293	Litovskii, E. Ya., et al.	1970	473-1773		VG-2 Cupola brick	58.22 SiO_2, 37.12 Al_2O_3, 1.82 TiO_2, 1.40 Fe_2O_3, 0.88 CaO, 0.75 MgO, and 0.04 alkalis; density 2.72 g cm^{-3}; total porosity 21.7%; diffusivity measured in argon.
12	293	Litovskii, E. Ya., et al.	1970	473-1873		D-2	Blast furnace multichamotte brick; 53.40 SiO_2, 40.18 Al_2O_3, 1.90 TiO_2, 1.80 Fe_2O_3, 1.22 CaO, 0.92 alkalis, and 0.37 MgO; density 2.76 g cm^{-3}; diffusivity measured in argon.
13	293	Litovskii, E. Ya., et al.	1970	473-1773		D-2	Similar to the above specimen.

DATA TABLE 237. THERMAL DIFFUSIVITY OF BRICKS

[Temperature, T, K; Thermal Diffusivity, α, $cm^2\ s^{-1}$]

T	α
CURVE 1	
323	0.00278
CURVE 2	
499	0.00550
600	0.00542
1069	0.00567
1173	0.00584
1271	0.00590
1473	0.00560
1561	0.00545
1659	0.00490
1759	0.00449
CURVE 3	
474	0.00554
576	0.00543
675	0.00522
774	0.00534
875	0.00522
976	0.00531
1089	0.00537
1169	0.00557
1274	0.00561
1359	0.00554
1508	0.00541
1601	0.00514
1786	0.00410
CURVE 4	
474	0.00446
574	0.00446
675	0.00447
766	0.00458
873	0.00447
990	0.00472
1071	0.00472
1163	0.00484
1286	0.00495
1370	0.00505
1476	0.00496
1585	0.00469
1660	0.00420
CURVE 5	
476	0.00420
563	0.00420
673	0.00420
768	0.00436
874	0.00425
874	0.00546
970	0.00450
973	0.00554
1170	0.00502
1262	0.00502
1366	0.00527
1374	0.00567
1561	0.00482
1649	0.00447
1671	0.00447
CURVE 6	
476	0.00497
572	0.00495
675	0.00491
780	0.00518
873	0.00502
998	0.00519
1080	0.00519
1186	0.00544
1284	0.00544
1394	0.00560
1487	0.00538
1583	0.00513
1671	0.00464
1777	0.00442
CURVE 7	
517	0.00450
623	0.00469
703	0.00450
794	0.00455
897	0.00450
1011	0.00461
1091	0.00489
1189	0.00487
1274	0.00478
1390	0.00510
CURVE 7 (cont.)	
1496	0.00500
1700	0.00468
CURVE 8	
473	0.00519
573	0.00498
673	0.00478
773	0.00469
873	0.00482
973	0.00509
1073	0.00534
1173	0.00568
1273	0.00592
1373	0.00596
1473	0.00583
1573	0.00554
1673	0.00515
1743	0.00480
CURVE 9	
473	0.00438
573	0.00410
673	0.00382
773	0.00373
873	0.00373
973	0.00395
1073	0.00427
1173	0.00453
1273	0.00483
1373	0.00497
1473	0.00483
1573	0.00459
1673	0.00432
1743	0.00406
CURVE 10	
473	0.00628
573	0.00590
673	0.00560
773	0.00541
873	0.00541
973	0.00545
CURVE 10 (cont.)	
1073	0.00560
1173	0.00579
1273	0.00601
1373	0.00601
1473	0.00590
1573	0.00560
1673	0.00521
1733	0.00491
CURVE 11	
473	0.00495
573	0.00472
673	0.00459
773	0.00453
873	0.00461
973	0.00478
1073	0.00507
1173	0.00527
1273	0.00544
1373	0.00544
1473	0.00535
1573	0.00513
1673	0.00473
1773	0.00420
CURVE 12	
473	0.00529
573	0.00500
673	0.00485
773	0.00485
873	0.00485
973	0.00498
1073	0.00521
1173	0.00546
1273	0.00563
1373	0.00563
1473	0.00554
1573	0.00531
1673	0.00490
1773	0.00456
1873	0.00402*
CURVE 13	
473	0.00451
573	0.00429
673	0.00415
773	0.00401
873	0.00396
973	0.00396*
1073	0.00399
1173	0.00415
1273	0.00426
1373	0.00426
1473	0.00426
1573	0.00412
1673	0.00382
1773	0.00337

* Not shown in figure.

SPECIFICATION TABLE 238. THERMAL DIFFUSIVITY OF CONCRETE

Cur. No.	Ref. No.	Author(s)	Year	Temp. Range, K	Reported Error, %	Name and Specimen Designation	Composition (weight percent), Specifications, and Remarks
1*	292	Das, M.B. and Hossain, M.A.	1966	323			10 x 10 x 1.90 cm.

DATA TABLE 238. THERMAL DIFFUSIVITY OF CONCRETE

[Temperature, T, K; Thermal Diffusivity, α, $cm^2\ s^{-1}$]

T	α
CURVE 1*	
323	0.00364

* No figure given.

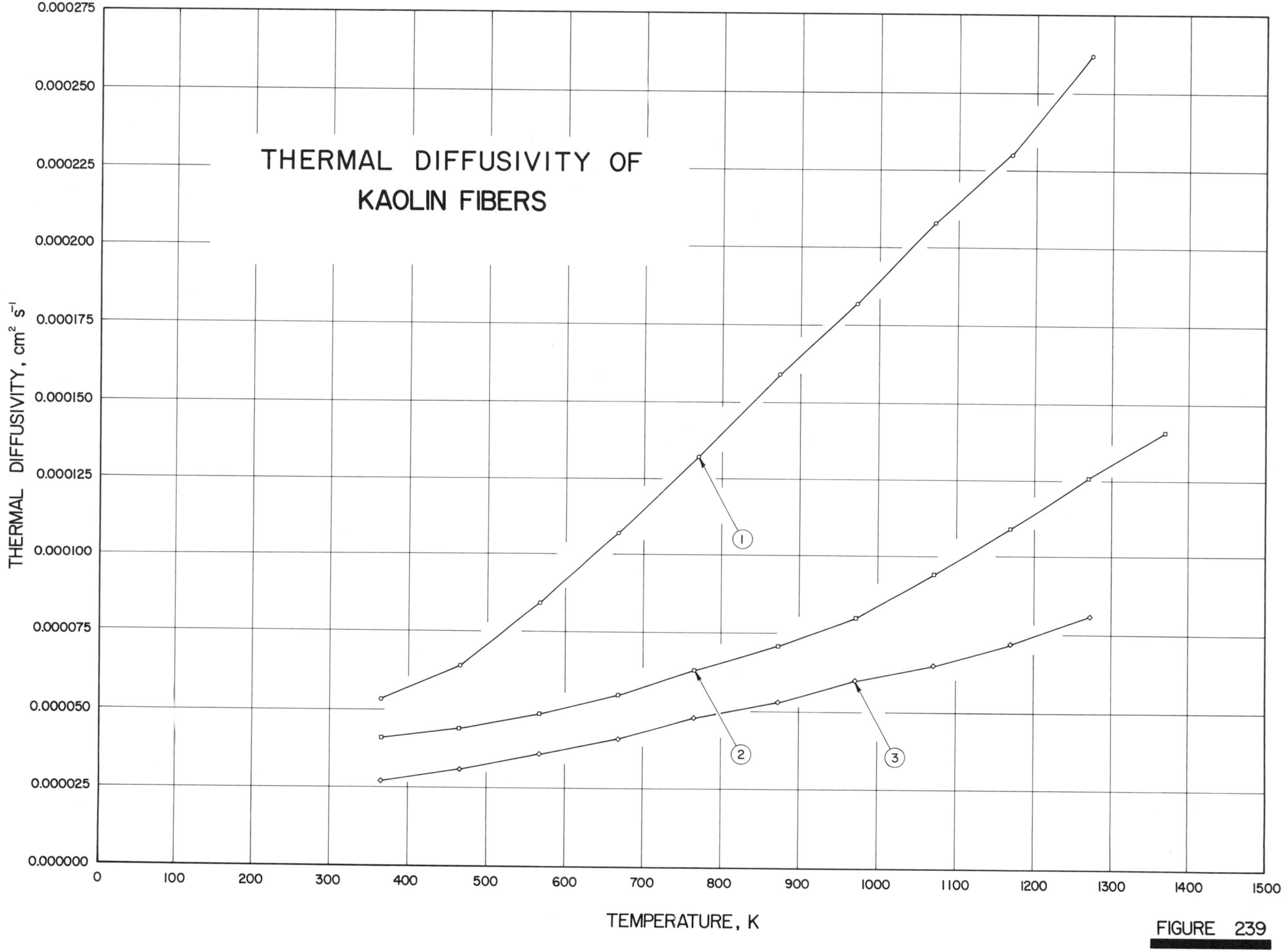
THERMAL DIFFUSIVITY OF
KAOLIN FIBERS
THERMAL DIFFUSIVITY, cm² s⁻¹
0.000275
0.000250
0.000225
0.000200
0.000175
0.000150
0.000125
0.000100
0.000075
0.000050
0.000025
0.000000
0
100
200
300
400
500
600
700
800
900
1000
1100
1200
1300
1400
1500
TEMPERATURE, K
1
2
3

FIGURE 239

SPECIFICATION TABLE 239. THERMAL DIFFUSIVITY OF KAOLIN FIBERS

Cur. No.	Ref. No.	Author(s)	Year	Temp. Range, K	Reported Error, %	Name and Specimen Designation	Composition (weight percent), Specifications, and Remarks
1	151	Krzhizhanovskii, P. E. and Chudnovskaya, I. I.	1966	365-1272	4.5		Fibrous specimen; fiber less than 4μ thick; bulk density 0.103 g cm^{-3}.
2	151	Krzhizhanovskii, P. E. and Chudnovskaya, I. I.	1966	365-1369	4.5		Similar to the above specimen but bulk density 0.162 g cm^{-3}.
3	151	Krzhizhanovskii, P. E. and Chudnovskaya, I. I.	1966	365-1272	4.5		Similar to the above specimen but bulk density 0.195 g cm^{-3}.

DATA TABLE 239. THERMAL DIFFUSIVITY OF KAOLIN FIBERS

[Temperature, T, K; Thermal Diffusivity, α, cm^2 s^{-1}]

T	α	T	α
CURVE 1		CURVE 2 (cont.)	
365	0.000053	873	0.000071
465	0.000064	972	0.000080
567	0.000084	1071	0.000094
668	0.000107	1169	0.000109
770	0.000132	1270	0.000126
873	0.000159	1369	0.000141
972	0.000182		
1071	0.000208	CURVE 3	
1169	0.000230		
1272	0.000262	365	0.000027
		465	0.000031
CURVE 2		567	0.000036
		668	0.000041
365	0.000041	765	0.000048
465	0.000044	873	0.000053
567	0.000049	972	0.000060
668	0.000055	1071	0.000065
765	0.000063	1169	0.000072
		1272	0.000081

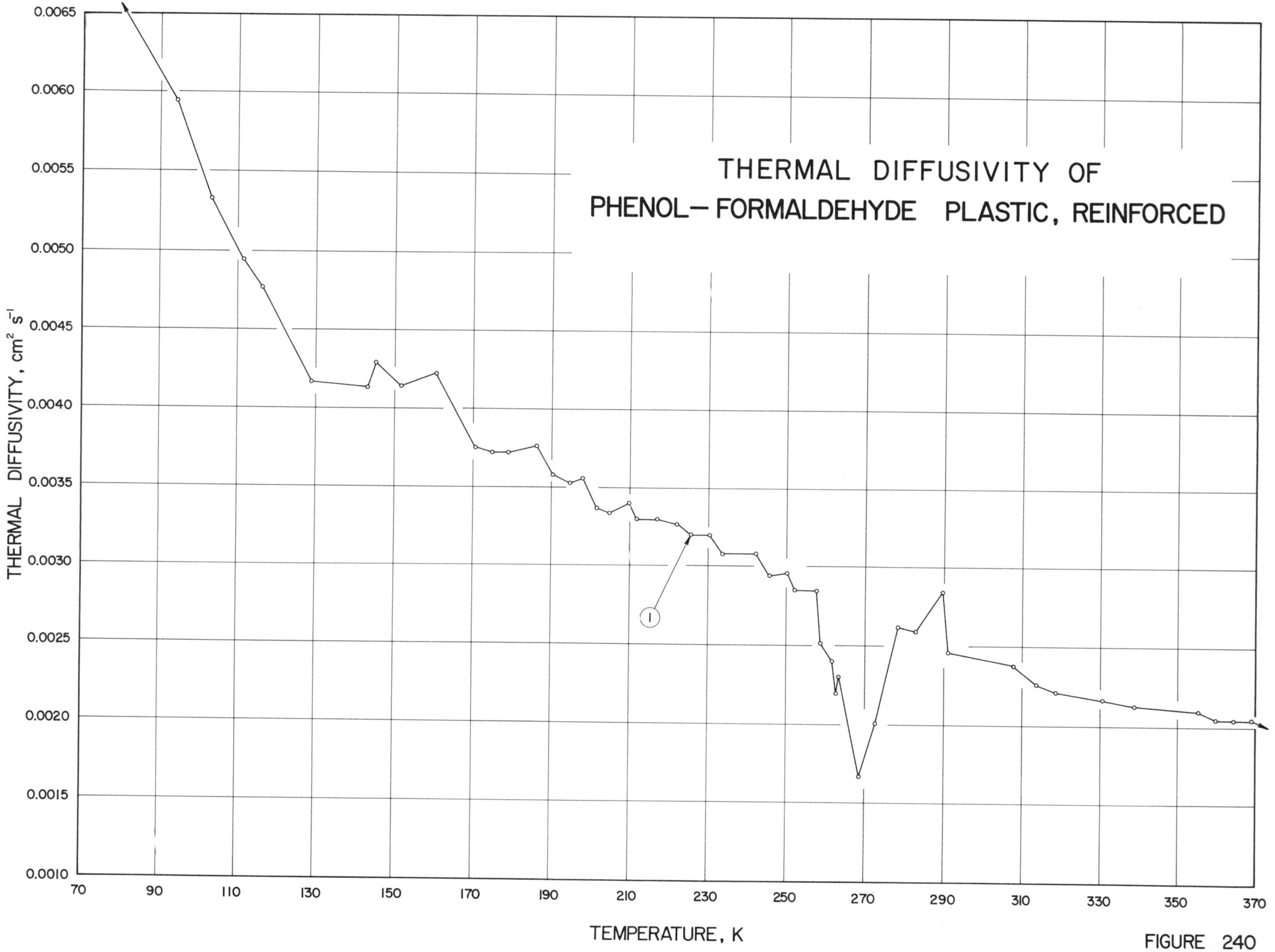
THERMAL DIFFUSIVITY OF
PHENOL–FORMALDEHYDE PLASTIC, REINFORCED
THERMAL DIFFUSIVITY, cm² s⁻¹
0.0065
0.0060
0.0055
0.0050
0.0045
0.0040
0.0035
0.0030
0.0025
0.0020
0.0015
0.0010
70
90
110
130
150
170
190
210
230
250
270
290
310
330
350
370
TEMPERATURE, K
1

FIGURE 240

SPECIFICATION TABLE 240. THERMAL DIFFUSIVITY OF PHENOL-FORMALDEHYDE PLASTIC, REINFORCED

Cur. No.	Ref. No.	Author(s)	Year	Temp. Range, K	Reported Error, %	Name and Specimen Designation	Composition (weight percent), Specifications, and Remarks
1	295	Luikov, A.V., Vasiliev, L.L., and Shashkov, A.G.	1965	78-372			Phenol-formaldehyde plastic reinforced with fiber glass; cylindrical specimen.

DATA TABLE 240. THERMAL DIFFUSIVITY OF PHENOL-FORMALDEHYDE PLASTIC, REINFORCED

[Temperature, T, K; Thermal Diffusivity, α, cm^2 s^{-1}]

T	α	T	α	T	α
CURVE 1		CURVE 1 (cont.)		CURVE 1 (cont.)	
78.2	0.00662*	201.5	0.00337	263.4	0.00230
94.1	0.00594	204.8	0.00334	268.7	0.00167
103.2	0.00533	209.8	0.00340	272.7	0.00201
111.3	0.00494	211.9	0.00330	278.1	0.00262
116.2	0.00476	216.9	0.00330	282.8	0.00259
128.7	0.00416	222.0	0.00327	289.6	0.00284
143.2	0.00413	225.7	0.00320	291.0	0.00246
145.2	0.00429	230.4	0.00320	307.8	0.00238
151.7	0.00414	233.8	0.00308	313.6	0.00226
160.4	0.00422	242.1	0.00308	318.5	0.00222
170.5	0.00375	245.8	0.00294	330.8	0.00217
174.8	0.00372	250.1	0.00296	338.8	0.00213
178.9	0.00372	252.2	0.00285	355.2	0.00210
186.2	0.00376	257.6	0.00285	359.7	0.00205
190.3	0.00358	258.7	0.00251	364.4	0.00205
194.7	0.00353	261.7	0.00240	369.1	0.00205
198.0	0.00356	262.7	0.00220	372.2	0.00202*

*Not shown in figure.

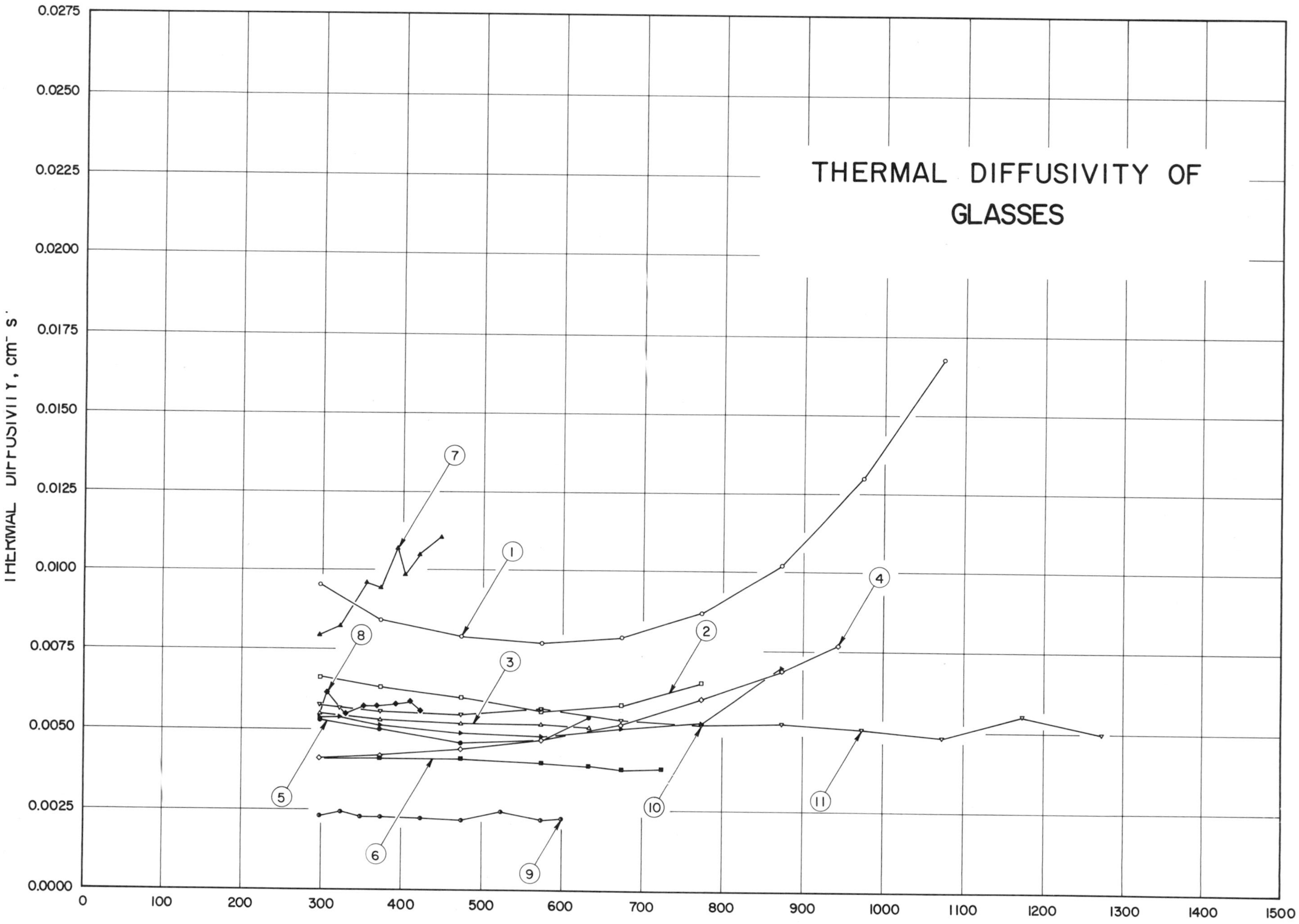
THERMAL DIFFUSIVITY OF GLASSES
THERMAL DIFFUSIVITY, cm² s⁻¹
TEMPERATURE, K
0.0000
0.0025
0.0050
0.0075
0.0100
0.0125
0.0150
0.0175
0.0200
0.0225
0.0250
0.0275
0
100
200
300
400
500
600
700
800
900
1000
1100
1200
1300
1400
1500
1
2
3
4
5
6
7
8
9
10
11

FIGURE 241

SPECIFICATION TABLE 241. THERMAL DIFFUSIVITY OF GLASSES

Cur. No.	Ref. No.	Author(s)	Year	Temp. Range, K	Reported Error, %	Name and Specimen Designation	Composition (weight percent) SiO_2	Al_2O_3	B_2O_3	BaO	CaO	CeO_2	K_2O	MgO	Composition (continued), Specifications, and Remarks
1	72	Plummer, W. A., Campbell, D. E., and Comstock, A. A.	1962	298-1073	~15	Corning Glass Code 7900	96.0		3.0						Specimen composed of three pieces: middle piece and two end pieces; middle piece separated from bottom piece by a 0.005 in. thick chromel sheet 7.6 x ~18 cm forming the heat source, middle piece separated from upper piece by a 0.005 in. thick chromel sheet forming the heat sink; thicknesses on either side of the heater are the same; entire assembly constitutes a sandwich 12.7 cm long, 7.6 cm wide and at least 4 to 5 cm thick; diffusivity determined from ratio of the measured temperature rises of the heat source and sink; unidimensional heat flow; last five data points include radiation contribution.
2	72	Plummer, W. A., et al.	1962	298-773	~15	Corning Glass Code 1723	57.7	14.9	4.0	6.0	10.1			6.9	Specimen similar to the above; measured under the same conditions as above; last two data points include radiation contribution.
3	72	Plummer, W. A., et al.	1962	298-633	~15	Corning Glass Code 7740	80.4	2.0	13.3						and 4.4 Na_2O; specimen similar to the above; measured under the same conditions as above; data points include no radiation contribution.
4	72	Plummer, W. A., et al.	1962	298-943	~15	Corning Glass Code 8325	44.7			28.0	6.0		2.4		6.7 Na_2O, 1.2 TiO_2, 4.0 ZnO, and 7.0 ZrO_2; specimen similar to the above; measured under same conditions as above; last four data points include radiation contribution.
5	72	Plummer, W. A., et al.	1962	298-633	~15	Corning Glass Code 0080	73.4	0.8			4.8			3.3	and 16.9 Na_2O; specimen similar to the above; measured under same conditions as above; last two data points include radiation contribution.
6	72	Plummer, W. A., et al.	1962	298-723	~15	Corning Glass Code 8362	44.6	2.0				1.3	14.0		6.0 Na_2O, and 33.4 PbO; specimen similar to the above; measured under same conditions as above; data points include no radiation contribution.
7	76	Hartunian, R. A. and Varwig, R. L.	1962	297-447		Pyrex 7740									Corning 7740 glass; tubular specimen; diffusivity values calculated from measured (kdc) using specific heat data of Kelley, K. K. (Bureau of Mines Bull. 746, 1948) and of Lord, R. C. and Morrow, J. C. (J. Chem. Phys. 26 (230), 1957); constant heat transfer rate provided by a square current pulse.

Cur. No.	Ref. No.	Author(s)	Year	Temp. Range, K	Reported Error, %	Name and Specimen Designation	Composition (weight percent) SiO_2	Al_2O_3	B_2O_3	BaO	CaO	CeO_2	K_2O	MgO	Composition (continued), Specifications, and Remarks
8	76	Hartunian, R. A. and Varwig, R. L.	1962	297-422		Plate glass									Ordinary soda-lime glass; plate specimen; diffusivity values calculated from measured (kdc) using specific heat data of Kelley, K. K. (Bureau of Mines Bull. 746, 1948) and of Lord, R. C. and Morrow, J. C. (J. Chem. Phys. 26 (230), 1957); constant heat transfer rate provided by a square current pulse.
9	294	Plummer, W.A.	1963	298-598		Corning Glass Code 8363	5.0	3.0	10.0						84.0 PbO; specimen 12.7 cm by 7.6 cm; density 6.5 g cm^{-3}.
10	294	Plummer, W.A.	1963	298-873		Corning Glass Code 8370	70.0		11.5	3.0					9.0 Na_2O, 7.0 K_2O; specimen 12.7 cm by 7.6 cm.
11	294	Plummer, W.A.	1963	298-1273		Corning Glass Synthetic Tektite	75.0	11.5					2.0		1.2 N_2O, 1.8 PbO, 4.4 FeO, 0.5 Fe_2O_3, 0.5 TiO, 2.7 CuO, misc. 0.40.
12*	170	Krischer, O. and Esdorn, H.	1955	297.2		Foam Glass									Specimen composed of two identical plates heated on their outside surfaces by heating foils each 0.01 mm thick made of a chromium-nickel alloy, inside surfaces held in contact with each other; three identical plates identical to the specimen plates and a thermally insulating layer placed in contact with each of the outside surfaces of the specimen, respectively; heating foils also inserted between the two outermost plates on each side of the specimen, respectively; regulated dc current used to generate thermal energy in the heating foils; square specimen plates 95 x 95 x 14.8 mm each; density 0.150 g cm^{-3}; thermal diffusivity determined from measured time interval necessary for the temperature of the unheated specimen face to reach the same value previously acquired by the heated face; data point reported is the average of two independent runs.

*Not shown in figure.

DATA TABLE 241. THERMAL DIFFUSIVITY OF GLASSES

[Temperature, T, K; Thermal Diffusivity, α, $cm^2 s^{-1}$]

T	α
CURVE 1	
298.2	0.0090
373.2	0.0084
473.2	0.0079
573.2	0.0077
673.2	0.0079
773.2	0.0087
873.2	0.0102
973.2	0.0130
1073.2	0.0168
CURVE 2	
298.2	0.0066
373.2	0.0063
473.2	0.0060
573.2	0.0056
673.2	0.0058
773.2	0.0065
CURVE 3	
298.2	0.0055
373.2	0.0053
473.2	0.0052
573.2	0.0052
633.2	0.0051
CURVE 4	
298.2	0.0041
373.2	0.0042
473.2	0.0044
573.2	0.0047
673.2	0.0052
773.2	0.0060
873.2	0.0069
943.2	0.0077
CURVE 5	
298.2	0.0053
373.2	0.0055
CURVE 5 (cont.)	
473.2	0.0046
573.2	0.0047*
633.2	0.0054
CURVE 6	
298.2	0.0041*
373.2	0.0041
473.2	0.0041
573.2	0.0040
633.2	0.0039
673.2	0.0038
723.2	0.0038
CURVE 7	
297	0.00793
323	0.00821
355	0.00958
373	0.00940
393	0.0107
402	0.00983
420	0.0105
447	0.0117
CURVE 8	
297	0.00550*
307	0.00603
330	0.00549
352	0.00571
368	0.00571
392	0.00579
409	0.00589
422	0.00560
CURVE 9	
298.2	0.00231
323.2	0.00242
348.2	0.00226
373.2	0.00226
423.2	0.00222
473.2	0.00218
523.2	0.00244
573.2	0.00220
598.2	0.00225
CURVE 10	
298.2	0.00535
323.2	0.00540
373.2	0.00515
473.2	0.00490
573.2	0.00480
673.2	0.00510
773.2	0.00625
873.2	0.00700
CURVE 11	
298.2	0.00575
323.2	0.00565
373.2	0.00555
473.2	0.00548
573.2	0.00565
673.2	0.00530
773.2	0.00520
873.2	0.00525
973.2	0.00510
1073.2	0.00485
1173.2	0.00555
1273.2	0.00500
CURVE 12*	
297.2	0.00536

*Not shown in figure.

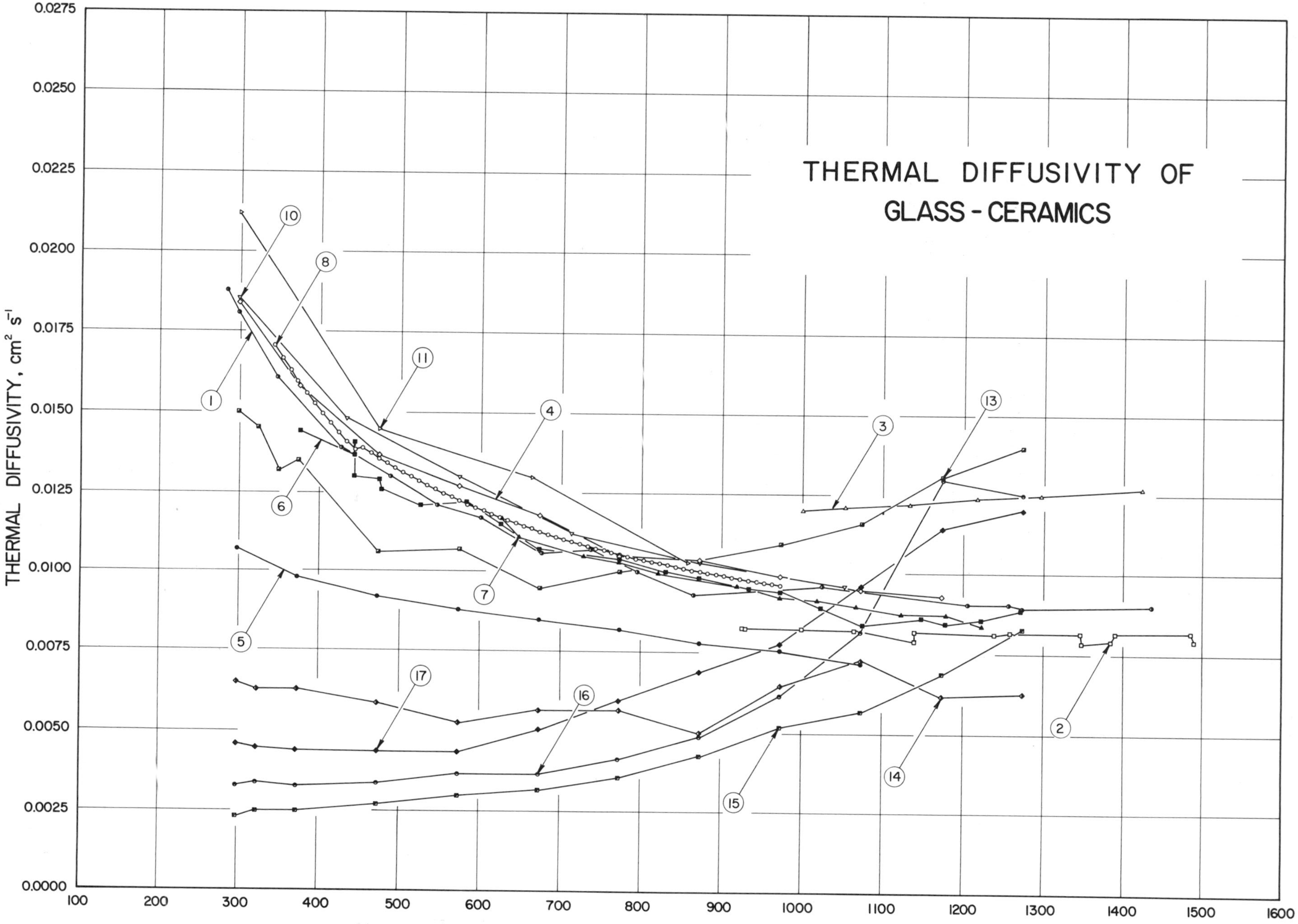
THERMAL DIFFUSIVITY OF
GLASS - CERAMICS
THERMAL DIFFUSIVITY, cm² s⁻¹
TEMPERATURE, K
0.0000
0.0025
0.0050
0.0075
0.0100
0.0125
0.0150
0.0175
0.0200
0.0225
0.0250
0.0275
100
200
300
400
500
600
700
800
900
1000
1100
1200
1300
1400
1500
1600
1
2
3
4
5
6
7
8
10
11
13
14
15
16
17

FIGURE 242

SPECIFICATION TABLE 242. THERMAL DIFFUSIVITY OF GLASS-CERAMICS

Cur. No.	Ref. No.	Author(s)	Year	Temp. Range, K	Reported Error, %	Name and Specimen Designation	Composition (weight percent), Specifications, and Remarks
1	71, 279	Rudkin, R. L.	1963	283-1436	± 5/ ±10	Pyroceram No. 9606	Glass-ceramic; disc specimen 0.75 in. in diameter and ~0.045 in. thick; developed at Corning Glass Works; coated on both sides with evaporated tungsten using electron beam techniques; short pulse of thermal energy from a xenon flash lamp absorbed at front surface of specimen; diffusivity determined from measured temperature-time function of the rear surface; measured in vacuum of $\sim 10^{-5}$ mm Hg; measurements below 873.2 K carried out in resistance furnace, measurements above 873.2 K made in vacuum induction furnace.
2	71, 279	Rudkin, R. L.	1963	926-1490	± 5/ ±10	Pyroceram No. 9608	Glass-ceramic; disc specimen 0.75 in. in diameter and ~0.045 in. thick; developed at Corning Glass Works; coated on both sides with evaporated tungsten using electron beam techniques; short pulse of thermal energy from a xenon flash lamp absorbed at front surface of specimen; diffusivity determined from measured temperature-time function of the rear surface; measured in vacuum of $\sim 10^{-5}$ mm Hg.
3	71, 279	Rudkin, R. L.	1963	1000-1423	± 5/ ±10	Pyroceram No. 9690 (Cercor)	Glass-ceramic; disc specimen 0.75 in. in diameter and ~0.045 in. thick; developed at Corning Glass Works; coated on both sides with evaporated tungsten using electron beam techniques; short pulse of thermal energy from a xenon flash lamp absorbed at front surface of specimen; diffusivity determined from measured temperature-time function of the rear surface; measured in vacuum of $\sim 10^{-5}$ mm Hg.
4	72	Plummer, W.A., Campbell, D.E., and Comstock, A.A.	1962	298-1173	~15	Code 9606	Specimen composed of three pieces: middle piece and two end pieces; middle piece separated from bottom piece by a 0.005 in. thick chromel sheet 7.6 x ~18 cm forming the heat source, middle piece separated from upper piece by a 0.005 in. thick chromel sheet forming the heat sink; thicknesses on either side of the heater are the same; entire assembly constitutes a sandwich 12.7 cm long, 7.6 cm wide and at least 4 to 5 cm thick; diffusivity determined from measured ratio of the temperature rises of the heat source and sink; unidimensional heat flow.
5	72	Plummer, W.A., et al.	1962	298-1073	~15	Code 9608	Specimen composed of three pieces: middle piece and two end pieces; middle piece separated from bottom piece by a 0.005 in. thick chromel sheet 7.6 x ~18 cm forming the heat source, middle piece separated from upper piece by a 0.005 in. thick chromel sheet forming the heat sink; thicknesses on either side of the heater are the same; entire assembly constitutes a sandwich 12.7 cm long, 7.6 cm wide and at least 4 to 5 cm thick; diffusivity determined from measured ratio of the temperature rises of the heat source and sink; unidimensional heat flow.
6	119	Gibby, R.L.	1968	375-1272	± 9	Pyroceram 9606	Cylindrical specimen 0.63 cm in diameter and 0.0544 cm thick; laser beam used as the pulse energy source; diffusivity determined from measured temperature history of the rear surface; both surfaces coated with colloidal graphite suspension; all data corrected for finite-pulse-time effects and heat losses.
7	119	Gibby, R.L.	1968	626-1223	± 9	Pyroceram 9606	Above specimen measured for diffusivity during cooling; other conditions same as above.
8	298	Flieger, H.W., Jr.	1963	343-973		Pyroceram 9606	Microcrystalline specimen, 2 in. in diameter and 14 in. long; diffusivity measured in series 1.

SPECIFICATION TABLE 242. THERMAL DIFFUSIVITY OF GLASS-CERAMICS (continued)

Cur. No.	Ref. No.	Author(s)	Year	Temp. Range, K	Reported Error, %	Name and Specimen Designation	Composition (weight percent), Specifications, and Remarks
9*	298	Flieger, H.W., Jr.	1963	343-1273		Pyroceram 9606	The above specimen; diffusivity measured in series 2.
10	294	Plummer, W.A.	1963	298-1053		Pyroceram 9606	Specimen 12.7 cm by 7.6 cm.
11	294	Plummer, W.A.	1963	298-858		Experimental Glass-Ceramic A	Specimen 12.7 cm by 7.6 cm.
12*	294	Plummer, W.A.	1963	298-1273		Experimental Glass-Ceramic B	Specimen 12.7 cm by 7.6 cm.
13	294	Plummer, W.A.	1963	298-1273		BDQ 115	Specimen 12.7 cm by 7.6 cm.
14	294	Plummer, W.A.	1963	298-1273		Corning Cercor Code 9690	Cast specimen; about 85% theoretical density.
15	294	Plummer, W.A.	1963	298-1273		Corning Cercor Code 9690	Diffusivity measured perpendicular to spacer sheets.
16	294	Plummer, W.A.	1963	298-1273		Corning Cercor Code 9690	Diffusivity measured parallel to spacer sheets.
17	294	Plummer, W.A.	1963	298-1273		Corning Cercor Code 9690	Diffusivity measured parallel to the cellular structure.

* Not shown in figure.

DATA TABLE 242. THERMAL DIFFUSIVITY OF GLASS-CERAMICS

[Temperature, T, K; Thermal Diffusivity, α, $cm^2\ s^{-1}$]

T	α
CURVE 1	
283	0.0188
298	0.0181
346	0.0161
426	0.0139
488	0.0130
546	0.0121
600	0.0117
676	0.0106
737	0.0107
795	0.0100
865	0.00930
1025	0.00960
1205	0.00905
1256	0.00905
1273	0.00895
1436	0.00903
CURVE 2	
926	0.00827
930	0.00825
1000	0.00825
1065	0.00823
1140	0.00790
1140	0.00820
1238	0.00813
1258	0.00820
1346	0.00817
1348	0.00787
1385	0.00795
1390	0.00817
1486	0.00823
1490	0.00795
CURVE 3	
1000	0.0120
1053	0.0121
1133	0.0122
1216	0.0124
1296	0.0125
1423	0.0127
CURVE 4	
298.2	0.0184
373.2	0.0158
473.2	0.0137
573.2	0.0127
673.2	0.0118
773.2	0.0110
873.2	0.0104
973.2	0.0099
1073.2	0.0095
1173.2	0.0093
CURVE 5	
298.2	0.0107
373.2	0.0098
473.2	0.0092
573.2	0.0088
673.2	0.0085
773.2	0.0082
873.2	0.0078
973.2	0.0076
1073.2	0.0072
CURVE 6	
375	0.0144
443	0.0137
443	0.0141
443	0.0130
474	0.0129
476	0.0126
524	0.0121
581	0.0122
625	0.0115
673	0.0107
773	0.0104
831	0.0100
872	0.00983
934	0.00950
973	0.00941
1023	0.00892
1075	0.00939
CURVE 6 (cont.)	
1149	0.00858
1177	0.00844
1222	0.00856
1272	0.00889
CURVE 7	
626	0.0117
674	0.0111
728	0.0105
773	0.0103
821	0.00998
919	0.00958
973	0.00924
1019	0.00916
1067	0.00898
1123	0.00872
1178	0.00872
1223	0.00840
CURVE 8	
343.2	0.01709
353.2	0.01669
363.2	0.01630
373.2	0.01594
383.2	0.01560
393.2	0.01527
403.2	0.01496
413.2	0.01467
423.2	0.01439
433.2	0.01412
443.2	0.01386
453.2	0.01391
463.2	0.01375
473.2	0.01359
483.2	0.01344
493.2	0.01329
503.2	0.01315
513.2	0.01301
523.2	0.01287
533.2	0.01274
CURVE 8 (cont.)	
543.2	0.01261
553.2	0.01249
563.2	0.01237
573.2	0.01225
583.2	0.01214
593.2	0.01203
603.2	0.01192
613.2	0.01182
623.2	0.01172
633.2	0.01163
643.2	0.01153
653.2	0.01144
663.2	0.01136
673.2	0.01127
683.2	0.01119
693.2	0.01111
703.2	0.01104
713.2	0.01096
723.2	0.01089
733.2	0.01082
743.2	0.01075
753.2	0.01069
763.2	0.01062
773.2	0.01056
783.2	0.01050
793.2	0.01044
803.2	0.01039
813.2	0.01033
823.2	0.01028
833.2	0.01023
843.2	0.01018
853.2	0.01013
863.2	0.01008
873.2	0.01004
883.2	0.00999
893.2	0.00995
903.2	0.00991
913.2	0.00986
923.2	0.00982
933.2	0.00978
943.2	0.00975
953.2	0.00971
CURVE 8 (cont.)	
963.2	0.00967
973.2	0.00964
CURVE 9*	
343.2	0.01698
353.2	0.01659
363.2	0.01622
373.2	0.01587
383.2	0.01554
393.2	0.01522
403.2	0.01492
413.2	0.01464
423.2	0.01437
433.2	0.01411
443.2	0.01386
453.2	0.01381
463.2	0.01361
473.2	0.01342
483.2	0.01325
493.2	0.01308
503.2	0.01292
513.2	0.01277
523.2	0.01263
533.2	0.01249
543.2	0.01236
553.2	0.01223
563.2	0.01211
573.2	0.01200
583.2	0.01189
593.2	0.01178
603.2	0.01168
613.2	0.01158
623.2	0.01148
633.2	0.01139
643.2	0.01130
653.2	0.01121
663.2	0.01113
673.2	0.01105
683.2	0.01097
693.2	0.01089
703.2	0.01082
CURVE 9 (cont.)*	
713.2	0.01074
723.2	0.01067
733.2	0.01060
743.2	0.01053
753.2	0.01047
763.2	0.01040
773.2	0.01034
783.2	0.01028
793.2	0.01022
803.2	0.01016
813.2	0.01011
823.2	0.01005
833.2	0.00999
843.2	0.00994
853.2	0.00989
863.2	0.00984
873.2	0.00979
883.2	0.00974
893.2	0.00969
903.2	0.00964
913.2	0.00959
923.2	0.00955
933.2	0.00950
943.2	0.00946
953.2	0.00942
963.2	0.00937
973.2	0.00933
983.2	0.00929
993.2	0.00925
1003.2	0.00921
1013.2	0.00917
1023.2	0.00913
1033.2	0.00910
1043.2	0.00906
1053.2	0.00902
1063.2	0.00899
1073.2	0.00895
1083.2	0.00892
1093.2	0.00888
1103.2	0.00885
1113.2	0.00882
1123.2	0.00878
CURVE 9 (cont.)*	
1133.2	0.00875
1143.2	0.00872
1153.2	0.00869
1163.2	0.00866
1173.2	0.00863
1183.2	0.00860
1193.2	0.00857
1203.2	0.00854
1213.2	0.00851
1223.2	0.00848
1233.2	0.00846
1243.2	0.00843
1253.2	0.00840
1263.2	0.00837
1273.2	0.00835
CURVE 10	
298.2	0.0185
433.2	0.0148
573.2	0.0130
713.2	0.0112
873.2	0.0103
1053.2	0.0096
CURVE 11	
298.2	0.0212
473.2	0.0145
663.2	0.0130
858.2	0.0103
CURVE 12*	
298.2	0.0185
323.2	0.0176
373.2	0.0163
473.2	0.0129
573.2	0.0110
673.2	0.0107
773.2	0.0098
873.2	0.0087
CURVE 12 (cont.)*	
973.2	0.0083
1073.2	0.0085
1173.2	0.0082
1273.2	0.0082
CURVE 13	
298.2	0.0150
323.2	0.0145
348.2	0.0132
373.2	0.0135
473.2	0.0106
573.2	0.0107
673.2	0.0095
773.2	0.0100
873.2	0.0104*
973.2	0.0109
1073.2	0.0116
1173.2	0.0131
1273.2	0.0140
CURVE 14	
298.2	0.0065
323.2	0.0063
373.2	0.0063
473.2	0.0059
573.2	0.0053
673.2	0.0057
773.2	0.0057
873.2	0.0050
973.2	0.0065
1073.2	0.0073
1173.2	0.0062
1273.2	0.0063
CURVE 15	
298.2	0.0023
323.2	0.0025
373.2	0.0025
473.2	0.0027

* Not shown in figure.

DATA TABLE 242. THERMAL DIFFUSIVITY OF GLASS-CERAMICS (continued)

T	α
CURVE 15 (cont.)	
573.2	0.0030
673.2	0.0032
773.2	0.0036
873.2	0.0043
973.2	0.0052
1073.2	0.0057
1173.2	0.0069
1273.2	0.0083
CURVE 16	
298.2	0.0033
323.2	0.0034
373.2	0.0033
473.2	0.0034
573.2	0.0037
673.2	0.0037
773.2	0.0042
873.2	0.0049
973.2	0.0062
1073.2	0.0082
1173.2	0.0130
1273.2	0.0125
CURVE 17	
298.2	0.0046
323.2	0.0045
373.2	0.0044
473.2	0.0044
573.2	0.0044
673.2	0.0051
773.2	0.0060
873.2	0.0069
973.2	0.0078
1073.2	0.0096
1173.2	0.0114
1273.2	0.0120

14. ORGANIC COMPOUNDS

SPECIFICATION TABLE 243. THERMAL DIFFUSIVITY OF FLUOROCARBON

Cur. No.	Ref. No.	Author(s)	Year	Temp. Range, K	Reported Error, %	Name and Specimen Designation	Composition (weight percent), Specifications, and Remarks
1*	49	Miron, R. L.	1968	383-637		TFE-fluorocarbon	4 x 2 x 0.5-0.8 in.; obtained from Baker Plastics, El Paso; thermal diffusivity calculated by the amplitude ratio method.
2*	49	Miron, R. L.	1968	383-637		TFE-fluorocarbon	The above specimen with the thermal diffusivity calculated by the phase angle method.
3*	49	Miron, R. L.	1968	443, 460		TFE-fluorocarbon	The above specimen with the thermal diffusivity calculated by the step test method.

DATA TABLE 243. THERMAL DIFFUSIVITY OF FLUOROCARBON

[Temperature, T, K; Thermal Diffusivity, α, $cm^2\ s^{-1}$]

T	α
CURVE 1*	
383.4	0.000792
422.6	0.000857
636.8	0.0111
CURVE 2*	
383.4	0.000470
422.6	0.000225
636.8	0.000271
CURVE 3*	
443.4	0.00147
460.1	0.00159

* No figure given.

SPECIFICATION TABLE 244. THERMAL DIFFUSIVITY OF GLYCEROL

Cur. No.	Ref. No.	Author(s)	Year	Temp. Range, K	Reported Error, %	Name and Specimen Designation	Composition (weight percent), Specifications, and Remarks
1*	190	Bryngdahl, O.	1962	294-295	< 0.1		Diffusivity measured employing an optical interferometric method based on the transient temperature gradient set up in the specimen by electrically heating a vertical wire immersed in it; error reported is the mean error.

DATA TABLE 244. THERMAL DIFFUSIVITY OF GLYCEROL

[Temperature, T, K; Thermal Diffusivity, α, $cm^2\ s^{-1}$]

T	α
CURVE 1*	
293.6	0.0009457
293.7	0.0009488
293.9	0.0009449
294.6	0.0009464
294.6	0.0009486
294.8	0.0009497

*No figure given.

SPECIFICATION TABLE 245. THERMAL DIFFUSIVITY OF n-OCTYL ALCOHOL

Cur. No.	Ref. No.	Author(s)	Year	Temp. Range, K	Reported Error, %	Name and Specimen Designation	Composition (weight percent), Specifications, and Remarks
1*	190	Bryngdahl, O.	1962	294.6-295.4	<0.1		Diffusivity measured employing an optical interferometric method based on the transient temperature gradient set up in the specimen by electrically heating a vertical wire immersed in it; error reported in the mean error.

DATA TABLE 245. THERMAL DIFFUSIVITY OF n-OCTYL ALCOHOL

[Temperature, T, K; Thermal Diffusivity, α, $cm^2 s^{-1}$]

T	α
CURVE 1*	
294.6	0.0009087
295.0	0.0009108
295.2	0.0009101
295.2	0.0009112
295.2	0.0009088
295.4	0.0009108

* No figure given.

SPECIFICATION TABLE 246. THERMAL DIFFUSIVITY OF SUCROSE SOLUTION

Cur. No.	Ref. No.	Author(s)	Year	Temp. Range, K	Reported Error, %	Name and Specimen Designation	Composition (weight percent), Specifications, and Remarks
1*	302	Keppeler, R.A. and Boose, J.R.	1970	234-267			Concentration of solution 2%; diffusivity data are the mean values of six replications at a specific temperature.
2*	302	Keppeler, R.A. and Boose, J.R.	1970	234-267			Concentration of solution 4%.
3*	302	Keppeler, R.A. and Boose, J.R.	1970	234-267			Concentration of solution 6%.
4*	302	Keppeler, R.A. and Boose, J.R.	1970	234-267			Concentration of solution 8%.
5*	302	Keppeler, R.A. and Boose, J.R.	1970	234-267			Concentration of solution 10%.
6*	302	Keppeler, R.A. and Boose, J.R.	1970	234-267			Concentration of solution 12%.
7*	302	Keppeler, R.A. and Boose, J.R.	1970	234-267			Concentration of solution 14%.
8*	302	Keppeler, R.A. and Boose, J.R.	1970	234-267			Concentration of solution 16%.
9*	302	Keppeler, R.A. and Boose, J.R.	1970	234-267			Concentration of solution 18%.
10*	302	Keppeler, R.A. and Boose, J.R.	1970	234-267			Concentration of solution 20%.
11*	302	Keppeler, R.A. and Boose, J.R.	1970	234-267			Concentration of solution 25%.
12*	302	Keppeler, R.A. and Boose, J.R.	1970	234-267			Concentration of solution 30%.
13*	302	Keppeler, R.A. and Boose, J.R.	1970	234-267			Concentration of solution 35%.

* No figure given.

DATA TABLE 246. THERMAL DIFFUSIVITY OF SUCROSE SOLUTION

[Temperature, T, K; Thermal Diffusivity, α, cm^2 s^{-1}]

T	α	T	α	T	α	T	α	T	α
CURVE 1*		CURVE 4*		CURVE 7*		CURVE 10*		CURVE 13*	
234	0.01297	234	0.01234	234	0.01058	234	0.00986	234	0.00665
237	0.01250	237	0.01147	237	0.01003	237	0.00911	237	0.00632
240	0.01203	240	0.01083	240	0.00881	240	0.00764	240	0.00494
243	0.01149	243	0.01007	243	0.00805	243	0.00685	243	0.00405
245	0.01127	245	0.00978	245	0.00783	245	0.00634	245	0.00375
248	0.01067	248	0.00905	248	0.00699	248	0.00575	248	0.00316
251	0.01038	251	0.00869	251	0.00666	251	0.00529	251	0.00295
254	0.00972	254	0.00789	254	0.00579	254	0.00458	254	0.00242
257	0.00925	257	0.00734	257	0.00515	257	0.00394	257	0.00201
259	0.00868	259	0.00658	259	0.00439	259	0.00326	259	0.00153
262	0.00762	262	0.00541	262	0.00341	262	0.00241	262	0.00104
264	0.00683	264	0.00426	264	0.00241	264	0.00158	264	0.00063
267	0.00531	267	0.00252	267	0.00129	267	0.00076	267	0.00029
CURVE 2*		CURVE 5*		CURVE 8*		CURVE 11*			
234	0.01295	234	0.01145	234	0.01049	234	0.00889		
237	0.01223	237	0.01082	237	0.00976	237	0.00828		
240	0.01159	240	0.01032	240	0.00854	240	0.00705		
243	0.01093	243	0.00927	243	0.00767	243	0.00610		
245	0.01068	245	0.00909	245	0.00717	245	0.00563		
248	0.00974	248	0.00823	248	0.00657	248	0.00487		
251	0.00969	251	0.00798	251	0.00606	251	0.00449		
254	0.00869	254	0.00704	254	0.00526	254	0.00381		
257	0.00837	257	0.00644	257	0.00469	257	0.00328		
259	0.00759	259	0.00569	259	0.00381	259	0.00260		
262	0.00649	262	0.00454	262	0.00294	262	0.00188		
264	0.00541	264	0.00343	264	0.00202	264	0.00123		
267	0.00368	267	0.00194	267	0.00103	267	0.00056		
CURVE 3*		CURVE 6*		CURVE 9*		CURVE 12*			
234	0.01251	234	0.01092	234	0.00984	234	0.00781		
237	0.01174	237	0.01018	237	0.00921	237	0.00732		
240	0.01108	240	0.00926	240	0.00775	240	0.00603		
243	0.01073	243	0.00834	243	0.00703	243	0.00492		
245	0.01028	245	0.00816	245	0.00657	245	0.00464		
248	0.00961	248	0.00744	248	0.00603	248	0.00402		
251	0.00955	251	0.00699	251	0.00558	251	0.00373		
254	0.00881	254	0.00622	254	0.00479	254	0.00311		
257	0.00798	257	0.00564	257	0.00418	257	0.00253		
259	0.00739	259	0.00481	259	0.00343	259	0.00201		
262	0.00629	262	0.00379	262	0.00259	262	0.00141		
264	0.00522	264	0.00276	264	0.00179	264	0.00087		
267	0.00338	267	0.00152	267	0.00086	267	0.00040		

* No figure given.

15. POLYMERS

SPECIFICATION TABLE 247. THERMAL DIFFUSIVITY OF ACRYLIC

Cur. No.	Ref. No.	Author(s)	Year	Temp. Range, K	Reported Error, %	Name and Specimen Designation	Composition (weight percent), Specifications, and Remarks
1*	108	Steinberg, S., Larson, R.E. and Kydd, A.R.	1963	298			Specimen 2.0 cm square, 0.26-0.29 cm in thickness; measuring temperature not reported and here assumed to be 25 C.

DATA TABLE 247. THERMAL DIFFUSIVITY OF ACRYLIC

[Temperature, T, K; Thermal Diffusivity, α, $cm^2\ s^{-1}$]

T	α
CURVE 1*	
298	0.0012

* No figure given.

SPECIFICATION TABLE 248. THERMAL DIFFUSIVITY OF NYLON

Cur. No.	Ref. No.	Author(s)	Year	Temp. Range, K	Reported Error, %	Name and Specimen Designation	Composition (weight percent), Specifications, and Remarks
1*	108	Steinberg, S., Larson, R.E. and Kydd, A.R.	1963	298			Specimen 2.0 cm square, 0.080 cm in thickness; measuring temperature not reported and here assumed to be 25 C.
2*	108	Steinberg, S., et al.	1963	298			Specimen 2.0 cm square, 0.325 cm in thickness; measuring temperature not reported and here assumed to be 25 C.

DATA TABLE 248. THERMAL DIFFUSIVITY OF NYLON

[Temperature, T, K; Thermal Diffusivity, α, $cm^2\ s^{-1}$]

T	α
CURVE 1*	
298	0.0009
CURVE 2*	
298	0.0013

* No figure given.

SPECIFICATION TABLE 249. THERMAL DIFFUSIVITY OF POLYESTER

Cur. No.	Ref. No.	Author(s)	Year	Temp. Range, K	Reported Error, %	Name and Specimen Designation	Composition (weight percent), Specifications, and Remarks
1*	303	Sukhareva, L.A., Voronkov, V.A., and Zubov, P.I.	1965	298			Film specimens cured at 80 C; thermal diffusivity given as a function of curing time, t, ranging from 20 to 120 min; measuring temperature not reported and here assumed to be 25 C.
2*	304	Hattori, M.	1965	288-363		PES-1	Styrene modified specimen; cylinder approximately 1 cm in diameter, 4 cm in length; hard.
3*	304	Hattori, M.	1965	280-334		PES-2	Similar to the above specimen except being soft.

DATA TABLE 249. THERMAL DIFFUSIVITY OF POLYESTER

[Temperature, T, K; Thermal Diffusivity, α, $cm^2 s^{-1}$]

t (min)	α	T	α
CURVE 1* (T = 298 K)		CURVE 2 (cont.)*	
		322.4	0.000931
20	0.000792	331.0	0.000884
30	0.000778	341.5	0.000873
40	0.000722	348.5	0.000852
60	0.000611	362.6	0.000808
80	0.000556		
100	0.000556	CURVE 3*	
120	0.000556		
		279.7	0.000611
T	α	286.3	0.000627
		290.4	0.000611
CURVE 2*		296.2	0.000617
		301.8	0.000632
288.3	0.001062	317.1	0.000612
293.9	0.001070	328.2	0.000632
299.4	0.000997	333.6	0.000611
318.5	0.000940		

* No figure given.

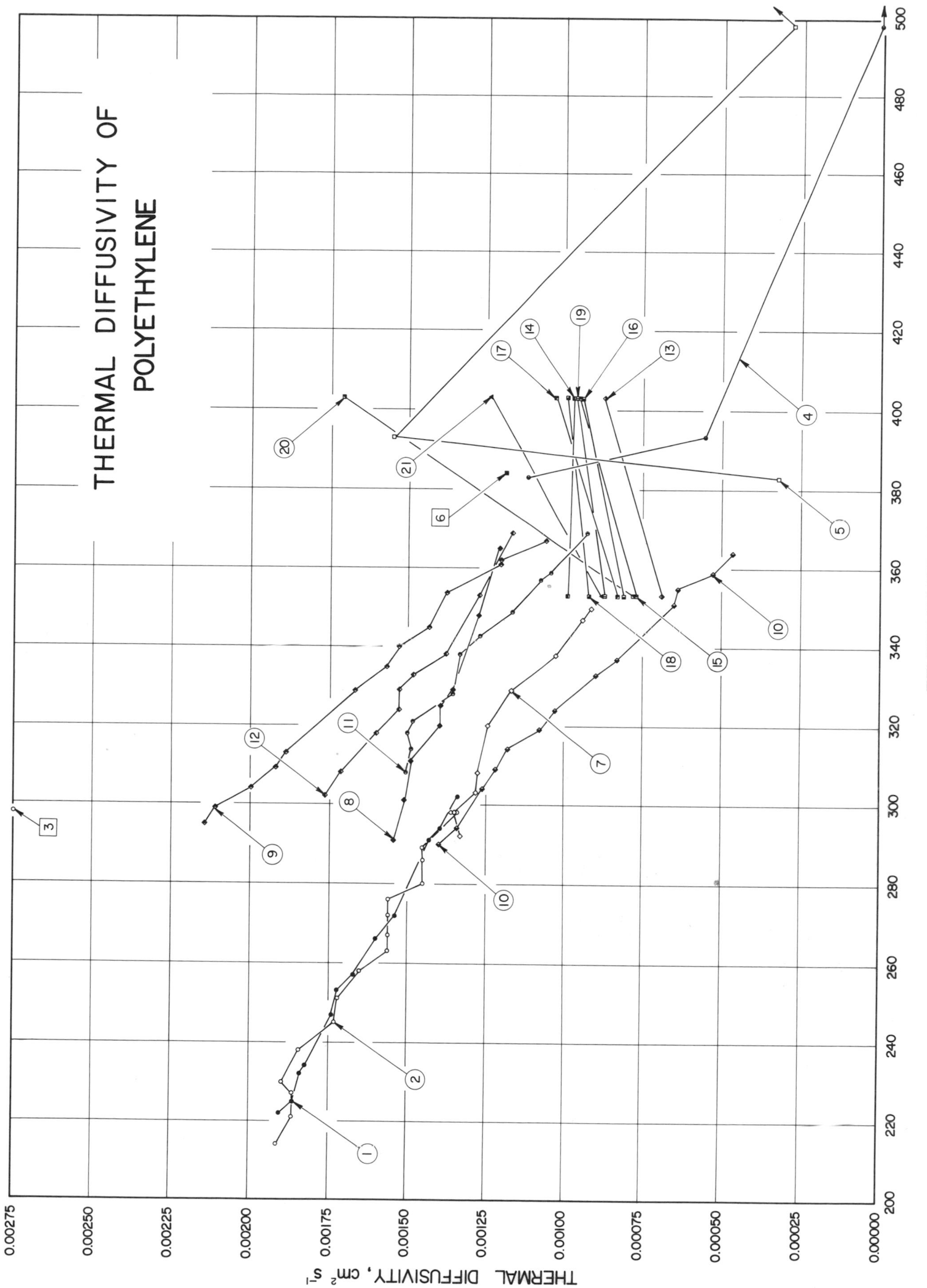
THERMAL DIFFUSIVITY OF POLYETHYLENE
THERMAL DIFFUSIVITY, cm² s⁻¹
TEMPERATURE, K
0.00000
0.00025
0.00050
0.00075
0.00100
0.00125
0.00150
0.00175
0.00200
0.00225
0.00250
0.00275
200
220
240
260
280
300
320
340
360
380
400
420
440
460
480
500
1
2
3
4
5
6
7
8
9
10
11
12
13
14
15
16
17
18
19
20
21

FIGURE 250

SPECIFICATION TABLE 250. THERMAL DIFFUSIVITY OF POLYETHYLENE

Cur. No.	Ref. No.	Author(s)	Year	Temp. Range, K	Reported Error, %	Name and Specimen Designation	Composition (weight percent), Specifications, and Remarks
1	111	Steere, R.C.	1967	223-302		Low-density polyethylene	Low-density polyethylene; specimen composed of a stack of thin sheets; diffusivity measured using transient method.
2	111	Steere, R.C.	1967	215-299	±3	Low-density polyethylene	Above specimen measured for diffusivity again using the location of maximum-temp. method.
3	108	Steinberg, S., Larson, R.E. and Kydd, A.R.	1963	298			Specimen 2.0 cm square and 0.326 cm in thickness; measuring temperature not reported and here assumed to be 25 C.
4	49	Miron, R.L.	1968	383-528			4 x 2 x 0.5-0.8 in.; thermal diffusivity calculated by the amplitude ratio method.
5	49	Miron, R.L.	1968	383-528			The above specimen with the thermal diffusivity calculated by the phase angle method.
6	49	Miron, R.L.	1968	384.3			The above specimen with the thermal diffusivity calculated by the step-test method.
7	304	Hattori, M.	1965	293-353		PE-1	High-pressure specimen; molecular weight 21,000; cylinder approximately 1 cm in diameter, 4 cm in length.
8	304	Hattori, M.	1965	292-365		PE-2	Similar to the above specimen; low-pressure specimen; molecular weight 80,000.
9	304	Hattori, M.	1965	295-367		PE-3	Similar to the above specimen; low-pressure specimen; molecular weight 120,000.
10	304	Hattori, M.	1965	290-364		PE-1	The specimen of curve no. 7 irradiated by 60 Co-γ-ray for a short time with radiation dose about 10^6 r. totally and then measured again.
11	304	Hattori, M.	1965	308-269		PE-2	The specimen of curve no. 8; γ-irradiated as above specimen.
12	304	Hattori, M.	1965	302-370		PE-3	The specimen of curve no. 9; γ-irradiated as above specimen.
13	286	Guillerm, S.	1968	353,403		Montecatini Fertene D1	Cylindrical specimen; measured by a radial transient method.
14	286	Guillerm, S.	1968	353,403		Montecatini Fertene XX1	Similar to above.
15	286	Guillerm, S.	1968	353,403		I.C.I. Q1319	Similar to above.
16	286	Guillerm, S.	1968	353,403		I.C.I. XJK25	Similar to above.
17	286	Guillerm, S.	1968	353,403		I.C.I. WJG11	Similar to above.
18	286	Guillerm, S.	1968	353,403		I.C.I. XJK64	Similar to above.
19	286	Guillerm, S.	1968	353,403		Union Carbide DYNH	Similar to above.
20	286	Guillerm, S.	1968	353,403		Union Carbide BDB 86-14	Similar to above.
21	286	Guillerm, S.	1968	353,403		BASF Lupolen 1 800 H	Similar to above.

DATA TABLE 250. THERMAL DIFFUSIVITY OF POLYETHYLENE

[Temperature, T, K; Thermal Diffusivity, α, cm^2 s^{-1}]

T	α
CURVE 1	
222.5	0.00190
225.3	0.00186
232.1	0.00184
234.2	0.00174
253.1	0.00172
257.1	0.00167
266.2	0.00160
272.3	0.00154
291.3	0.00143
294.6	0.00139
302.4	0.00134
CURVE 2	
214.6	0.00191
221.6	0.00186
227.2	0.00186
230.2	0.00189
238.4	0.00184
245.3	0.00173
251.4	0.00172
258.1	0.00165
263.6	0.00156
267.4	0.00156
272.2	0.00156
276.3	0.00156
280.4	0.00145
286.4	0.00145
289.4	0.00145
298.8	0.00134
298.9	0.00136
CURVE 3	
298	0.00275
CURVE 4	
383.2	0.00112
393.2	0.000557
498.2	0.0000101
528.2	0.00000310*
CURVE 5	
383.2	0.000323
393.2	0.00155
498.2	0.000257
528.2	0.000698*
CURVE 6	
384.3	0.00119
CURVE 7	
292.7	0.001330
298.2	0.001352
303.8	0.001284
308.7	0.001278
320.3	0.001248
329.8	0.001117
338.5	0.001029
347.9	0.000942
352.7	0.000914
CURVE 8	
291.8	0.001586
301.8	0.001512
311.2	0.001492
320.1	0.001399
325.6	0.001399
329.9	0.001356
348.5	0.001255
365.2	0.001212
CURVE 9	
295.3	0.002148
299.5	0.002115
304.0	0.001993
309.1	0.001947
313.5	0.001886
329.8	0.001669
335.1	0.001554
340.6	0.001532
345.2	0.001436
354.7	0.001355
361.1	0.001205
362.7	0.001205
367.0	0.001060
CURVE 10	
290.6	0.001398
294.9	0.001344
304.5	0.001264
309.5	0.001221
314.6	0.001182
319.3	0.001080
324.2	0.001030
333.0	0.000899
337.7	0.000833
351.3	0.000651
355.5	0.000642
359.4	0.000530
364.2	0.000470
CURVE 11	
308.2	0.001506
314.0	0.001489
318.6	0.001501
321.5	0.001485
328.7	0.001358
338.4	0.001330
343.3	0.001272
349.1	0.001167
357.6	0.001076
359.4	0.001042
369.4	0.000927
CURVE 12	
302.7	0.001764
308.4	0.001715
318.8	0.001602
324.1	0.001528
329.3	0.001528
333.0	0.001482
338.7	0.001377
353.1	0.001272
369.8	0.001167
CURVE 13	
353.2	0.00069
403.2	0.00087
CURVE 14	
353.2	0.00099
403.2	0.00097
CURVE 15	
353.2	0.00077
403.2	0.00095
CURVE 16	
353.2	0.00081
403.2	0.00094
CURVE 17	
353.2	0.00083
403.2	0.00103
CURVE 18	
353.2	0.00092
403.2	0.00099
CURVE 19	
353.2	0.00087
403.2	0.00096
CURVE 20	
353.2	0.00078
403.2	0.00171
CURVE 21	
353.2	0.00088
403.2	0.00124

* Not shown in figure.

SPECIFICATION TABLE 251. THERMAL DIFFUSIVITY OF POLYETHYLENE TEREPHTHALENE

Cur. No.	Ref. No.	Author(s)	Year	Temp. Range, K	Reported Error, %	Name and Specimen Designation	Composition (weight percent), Specifications, and Remarks
1*	111	Steere, R.C.	1967	214-297			Specimen composed of a stack of thin sheets; diffusivity measured using transient method.
2*	111	Steere, R.C.	1967	214-299	±3		Above specimen measured for diffusivity again using the location of maximum-temp method.

DATA TABLE 251. THERMAL DIFFUSIVITY OF POLYETHYLENE TEREPHTHALENE

[Temperature, T, K; Thermal Diffusivity, α, $cm^2\ s^{-1}$]

T	α	T	α
CURVE 1*		CURVE 2 (cont.)*	
213.9	0.00123	229.8	0.00111
222.0	0.00120	237.9	0.00108
226.9	0.00114	238.2	0.00103
230.1	0.00112	241.1	0.00105
242.0	0.00103	246.7	0.000998
249.2	0.00102	255.8	0.000968
258.3	0.000982	256.0	0.00102
263.0	0.000969	261.9	0.000983
275.2	0.000969	268.5	0.000992
285.3	0.000943	275.7	0.000947
292.6	0.000948	285.1	0.000929
297.2	0.000928	290.5	0.000934
		298.7	0.000912
CURVE 2*		298.9	0.000945
214.2	0.00120		
224.4	0.00113		
228.4	0.00108		

* No figure given.

SPECIFICATION TABLE 252. THERMAL DIFFUSIVITY OF POLYMETHACRYLATE

Cur. No.	Ref. No.	Author(s)	Year	Temp. Range, K	Reported Error, %	Name and Specimen Designation	Composition (weight percent), Specifications, and Remarks
1*	153	Ueberreiter, K.	1967	256-373			Thermally polyermized; molecularly uniform; measured in the solid glassy and liquid states.

DATA TABLE 252. THERMAL DIFFUSIVITY OF POLYMETHACRYLATE

[Temperature, T, K; Thermal Diffusivity, α, $cm^2\ s^{-1}$]

T	α	T	α
CURVE 1*		CURVE 1 (cont.)*	
255.5	0.00109	298.2	0.000781
256.9	0.00109	301.5	0.000781
259.6	0.00109	307.6	0.000780
262.3	0.00109	312.8	0.000781
263.5	0.00109	315.6	0.000781
267.1	0.00108	320.1	0.000782
268.4	0.00108	323.9	0.000782
270.6	0.00107	330.3	0.000780
273.6	0.00105	334.8	0.000782
276.5	0.00103	341.3	0.000782
280.9	0.000962	346.4	0.000781
282.1	0.000921	351.9	0.000781
283.2	0.000886	358.0	0.000781
284.0	0.000851	361.7	0.000781
285.8	0.000803	367.5	0.000781
287.2	0.000790	372.9	0.000781
289.0	0.000781		
292.7	0.000781		

* No figure given.

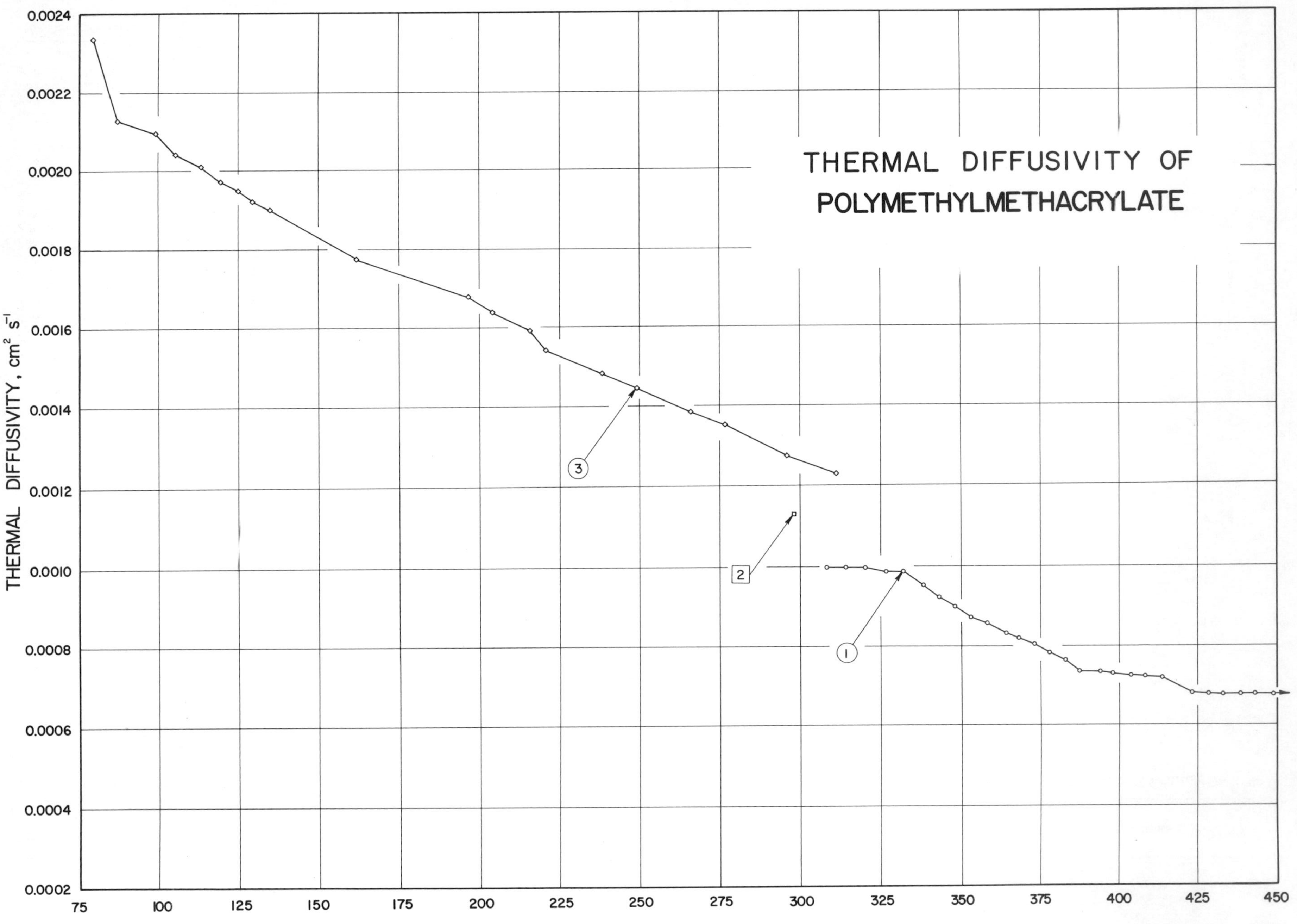
THERMAL DIFFUSIVITY OF
POLYMETHYLMETHACRYLATE
THERMAL DIFFUSIVITY, cm² s⁻¹
0.0024
0.0022
0.0020
0.0018
0.0016
0.0014
0.0012
0.0010
0.0008
0.0006
0.0004
0.0002
75
100
125
150
175
200
225
250
275
300
325
350
375
400
425
450
TEMPERATURE, K
1
2
3

FIGURE 253

SPECIFICATION TABLE 253. THERMAL DIFFUSIVITY OF POLYMETHYLMETHACRYLATE

Cur. No.	Ref. No.	Author(s)	Year	Temp. Range, K	Reported Error, %	Name and Specimen Designation	Composition (weight percent), Specifications, and Remarks
1	153	Ueberreiter, K.	1967	308-461			Molecular weight 5400; thermally polymerized; molecularly uniform; measured in the solid glassy and liquid states.
2	170	Krischer, O. and Esdorn, H.	1955	298.2		Plexiglas	Specimen composed of two identical plates heated on their outside surfaces by heating foils each 0.01 mm thick made of a chromium-nickel alloy, inside surfaces held in contact with each other; three identical plates identical to the specimen plates and a thermally insulating layer placed in contact with each of the outside surfaces of the specimen, respectively; heating foils also inserted between the two outermost plates on each side of the specimen, respectively; regulated dc current used to generate thermal energy in the heating foils; square specimen plates 95 x 95 x 10.0 mm each; density 1.180 g cm^{-3}; thermal diffusivity determined from measured time interval necessary for the temp of the unheated specimen face to reach the same value previously acquired by the heated face; the value reported is the average of three independent runs.
3	295	Luikov, A.V., Vasiliev, L.L., and Shashkov, A.G.	1965	79-311			Cylindrical specimen.

DATA TABLE 253. THERMAL DIFFUSIVITY OF POLYMETHYLMETHACRYLATE

[Temperature, T, K; Thermal Diffusivity, α, cm^2 s^{-1}]

T	α	T	α	T	α	T	α
CURVE 1		CURVE 1 (cont.)		CURVE 3		CURVE 3 (cont.)	
308.2	0.000994	403.6	0.000724	79.0	0.002335	295.7	0.001276
314.2	0.000994	408.4	0.000722	86.9	0.002126	311.0	0.001229
320.8	0.000994	413.6	0.000719	99.1	0.002095		
326.5	0.000988	423.3	0.000680	105.4	0.002040		
331.7	0.000988	428.3	0.000678	113.5	0.002008		
338.0	0.000953	433.0	0.000675	119.3	0.001969		
343.3	0.000921	438.5	0.000676	125.0	0.001948		
348.4	0.000899	443.1	0.000678	129.5	0.001919		
353.3	0.000874	448.9	0.000675	135.0	0.001898		
358.6	0.000857	453.9	0.000675*	161.6	0.001772		
364.4	0.000833	460.8	0.000676*	196.3	0.001678		
368.0	0.000820			204.0	0.001637		
373.4	0.000807	CURVE 2		215.8	0.001591		
377.9	0.000782			220.9	0.001542		
382.7	0.000764	298.2	0.00113	238.4	0.001483		
387.9	0.000735			249.4	0.001446		
394.1	0.000733			265.8	0.001386		
398.2	0.000730			276.3	0.001354		

* Not shown in figure.

SPECIFICATION TABLE 254. THERMAL DIFFUSIVITY OF POLYPROPYLENE

Cur. No.	Ref. No.	Author(s)	Year	Temp. Range, K	Reported Error, %	Name and Specimen Designation	Composition (weight percent), Specifications, and Remarks
1*	108	Steinberg, S., Larson, R. E. and Kydd, A. R.	1963	298			Specimen 2.0 cm square and 0.3225 cm in thickness; measuring temperature not reported and here assumed to be 25 C.
2*	286	Guillerm, S.	1968	353, 463			Cylindrical specimen; measured by a radial transient method.

DATA TABLE 254. THERMAL DIFFUSIVITY OF POLYPROPYLENE

[Temperature, T, K; Thermal Diffusivity, α, $cm^2\ s^{-1}$]

T	α
CURVE 1*	
298	0.0015
CURVE 2*	
353.2	0.00064
463.2	0.00066

* No figure given.

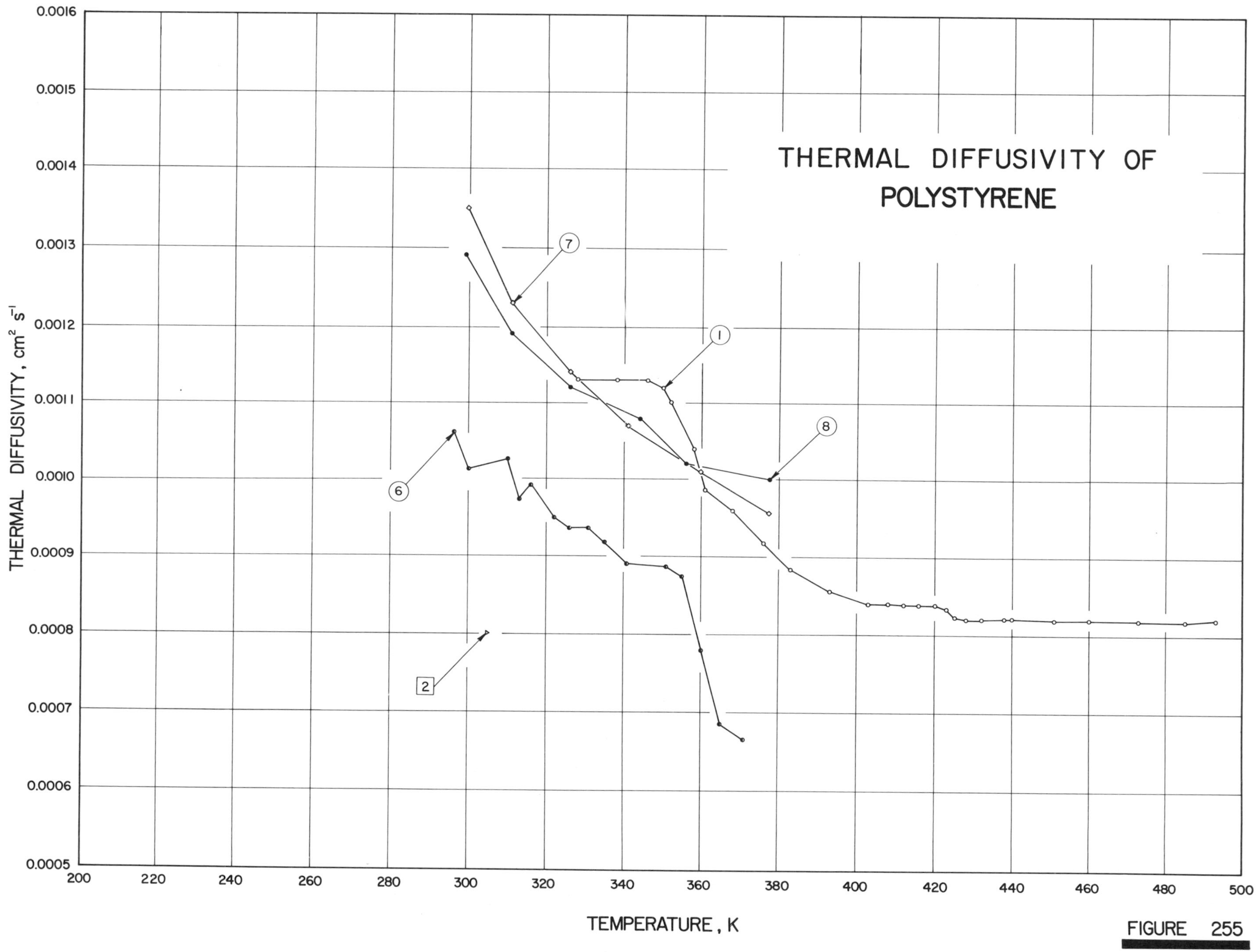
THERMAL DIFFUSIVITY OF
POLYSTYRENE
THERMAL DIFFUSIVITY, cm² s⁻¹
TEMPERATURE, K
0.0016
0.0015
0.0014
0.0013
0.0012
0.0011
0.0010
0.0009
0.0008
0.0007
0.0006
0.0005
200
220
240
260
280
300
320
340
360
380
400
420
440
460
480
500
1
2
6
7
8

FIGURE 255

SPECIFICATION TABLE 255. THERMAL DIFFUSIVITY OF POLYSTYRENE

Cur. No.	Ref. No.	Author(s)	Year	Temp. Range, K	Reported Error, %	Name and Specimen Designation	Composition (weight percent), Specifications, and Remarks
1	153	Ueberreiter, K.	1967	328-493			Thermally polymerized; molecularly uniform; measured in the solid glassy and liquid states.
2	292	Das, M.B. and Hossain, M.A.	1966	323			Foam specimen; 10 x 10 x 2.30 cm.
3*	121	Sacchi, A., Ferro, V., and Codegone, C.	1968	245-293			Expanded; 50 x 50 x 0.5 cm; density 0.01496 g cm^{-3}.
4*	121	Sacchi, A., et al.	1968	318.2			Similar to the above specimen but density 0.0148 g cm^{-3}.
5*	121	Sacchi, A., et al.	1968	318.2			Similar to the above specimen but density 0.0121 g cm^{-3}.
6	304	Hattori, M.	1965	296-371		PS	Cylindrical specimen approximately 1 cm in diameter and 4 cm in length.
7	300	Vlasov, V.V. and Doragov, N.N.	1966	297-378	5-6		100 x 100 x (3~10) mm; measured by the automatic apparatus; values read off smooth curve.
8	300	Vlasov, V.V. and Doragov, N.N.	1966	295-377	5-6		The above specimen measured by the manual method; values read off smooth curve.

* Not shown in figure.

DATA TABLE 255. THERMAL DIFFUSIVITY OF POLYSTYRENE

[Temperature, T, K; Thermal Diffusivity, α, cm^2 s^{-1}]

T	α	T	α
CURVE 1		CURVE 6	
328.3	0.00113	296.1	0.001061
338.7	0.00113	299.8	0.001012
346.5	0.00113	310.2	0.001026
350.2	0.00112	313.7	0.000974
353.8	0.00110	316.6	0.000992
358.9	0.00104	322.0	0.000951
361.4	0.000987	326.5	0.000938
368.7	0.000961	331.7	0.000938
376.1	0.000918	335.9	0.000918
383.2	0.000885	341.6	0.000891
393.2	0.000857	351.3	0.000888
403.2	0.000841	355.9	0.000875
408.1	0.000841	360.7	0.000780
412.9	0.000840	365.3	0.000685
416.9	0.000840	371.7	0.000666
420.3	0.000840		
423.2	0.000835	CURVE 7	
425.8	0.000825		
428.4	0.000821	297.3	0.00135
432.9	0.000821	311.5	0.00123
438.9	0.000822	326.4	0.00114
440.9	0.000822	341.6	0.00107
451.0	0.000820	359.2	0.00101
460.6	0.000820	377.8	0.000958
473.2	0.000820		
485.4	0.000819	CURVE 8	
493.1	0.000821		
		295.3	0.00129
CURVE 2		311.8	0.00119
		326.6	0.00112
323	0.00801	342.0	0.00108
		356.2	0.00102
CURVE 3*		377.0	0.00100
244.7	0.0153		
254.7	0.0159		
263.2	0.0164		
273.2	0.0172		
293.2	0.0189		
CURVE 4*			
318.2	0.0204		
CURVE 5*			
318.2	0.025		

* Not shown in figure.

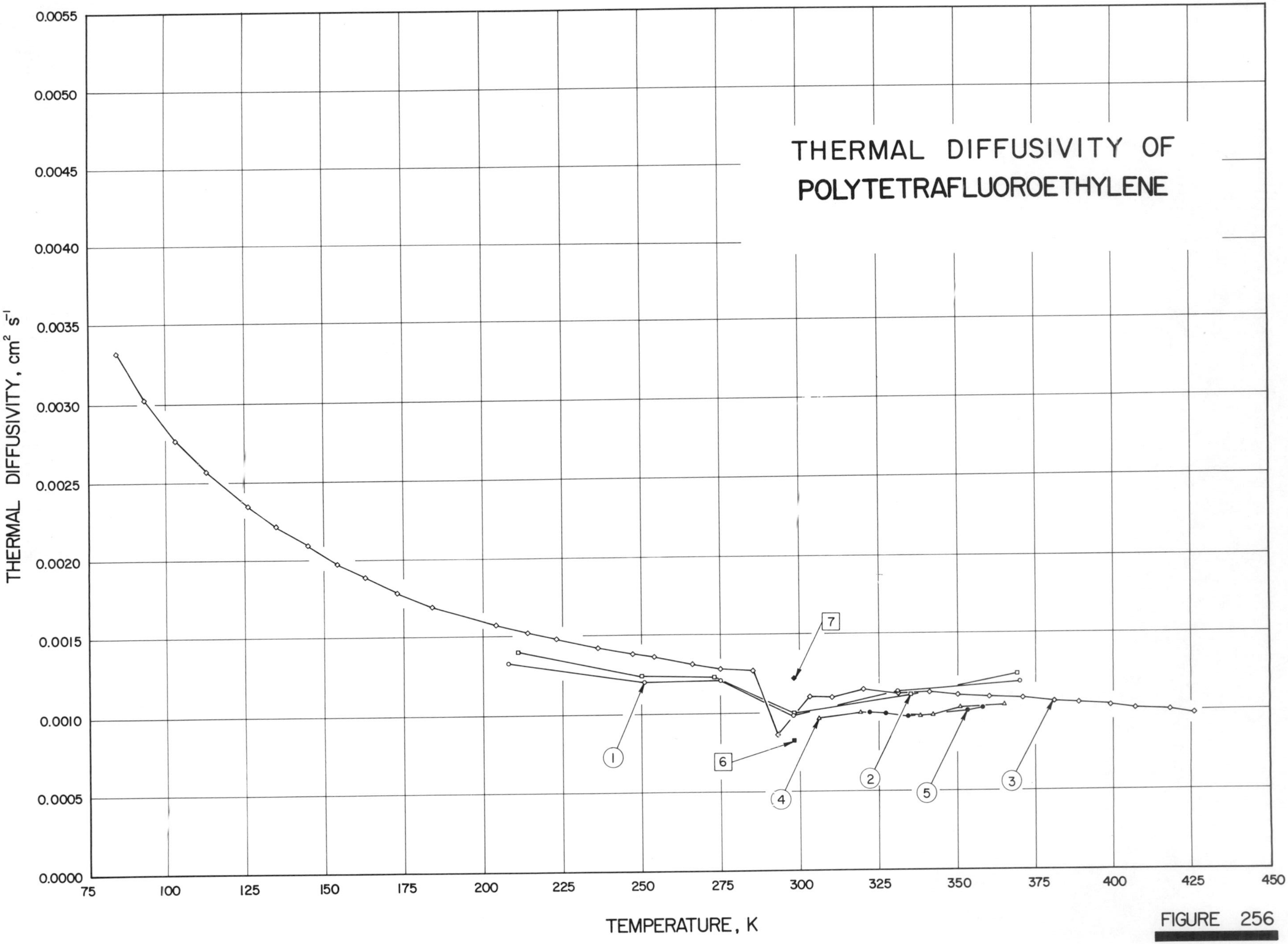
THERMAL DIFFUSIVITY OF
POLYTETRAFLUOROETHYLENE
THERMAL DIFFUSIVITY, cm² s⁻¹
0.0000
0.0005
0.0010
0.0015
0.0020
0.0025
0.0030
0.0035
0.0040
0.0045
0.0050
0.0055
75
100
125
150
175
200
225
250
275
300
325
350
375
400
425
450
TEMPERATURE, K
1
2
3
4
5
6
7

FIGURE 256

SPECIFICATION TABLE 256. THERMAL DIFFUSIVITY OF POLYTETRAFLUOROETHYLENE

Cur. No.	Ref. No.	Author(s)	Year	Temp. Range, K	Reported Error, %	Name and Specimen Designation	Composition (weight percent), Specifications, and Remarks
1	111	Steere, R.C.	1967	208-370			Diffusivity measured using transient method.
2	111	Steere, R.C.	1967	211-369	±3		Diffusivity measured using the location of maximum-temp. method.
3	295	Luikov, A.V., Vasiliev, L.I., and Shashkov, A.G.	1965	84-426			Cylindrical specimen.
4	304	Hattori, M.	1965	306-365		PTFE-1	Cylindrical specimen approximately 1 cm in diameter and 4 cm in length; axis of cylinder parallel to the direction of stress which had been applied to the original sample.
5	304	Hattori, M.	1965	322-367		PTFE-2	Similar to the above specimen; axis of cylinder perpendicular to the direction of stress.
6	108	Steinberg, S., Larson, R.E. and Kydd, A.R.	1963	298		Teflon	Specimen 2.0 cm square and 0.335 cm in thickness; measuring temperature not reported and here assumed to be 25 C.
7	108	Steinberg, S., et al.	1963	298		Teflon	Specimen 2.0 cm square and 0.0864 cm in thickness; measuring temperature not reported and here assumed to be 25 C.

DATA TABLE 256. THERMAL DIFFUSIVITY OF POLYTETRAFLUOROETHYLENE

[Temperature, T, K; Thermal Diffusivity, α, $cm^2\ s^{-1}$]

T	α
CURVE 1	
208.2	0.00133
251.2	0.00120
275.2	0.00121
298.2	0.000986
331.2	0.00113
370.2	0.00119
CURVE 2	
211.2	0.00140
250.2	0.00124
273.2	0.00123
298.2	0.00100
335.2	0.00111
369.2	0.00124
CURVE 3	
84	0.003315
93	0.003022
103	0.002763
113	0.002566
126	0.002347
135	0.002219
145	0.002091
154	0.001977
163	0.001891
173	0.001787
184	0.001694
204	0.001572
214	0.001527
223	0.001482
236	0.001423
247	0.001384
254	0.001361
266	0.001319
275	0.001288
285	0.001277
293	0.000867
303	0.001109
310	0.001096
320	0.001146
331	0.001127
341	0.001113
350	0.001106
360	0.001099
CURVE 3 (cont.)	
371	0.001089
381	0.001061
389	0.001056
399	0.001039
407	0.001020
418	0.001007
426	0.000989
CURVE 4	
306.5	0.000963
319.9	0.001000
338.0	0.000972
342.2	0.000981
350.9	0.001022
365.0	0.001042
CURVE 5	
322.5	0.001000
327.5	0.000992
334.8	0.000978
353.6	0.001001
358.3	0.001026
367.7	0.001026
CURVE 6	
298	0.00082
CURVE 7	
298	0.00122

SPECIFICATION TABLE 257. THERMAL DIFFUSIVITY OF POLYTRIFLUOROCHLOROETHYLENE

Cur. No.	Ref. No.	Author(s)	Year	Temp. Range, K	Reported Error, %	Name and Specimen Designation	Composition (weight percent), Specifications, and Remarks
1*	304	Hattori, M.	1965	293-360		PTFCE-1	Cylindrical specimen, approximately 1 cm in diameter, 4 cm in length; degree of crystallinity at 23 C, 0.358.
2*	304	Hattori, M.	1965	300-369		PTFCE-2	Similar to the above specimen; degree of crystallinity at 23 C, 0.770.
3*	304	Hattori, M.	1965	294-369		PTFCE-3	Similar to the above specimen; degree of crystallinity at 23 C, 0.823.
4*	304	Hattori, M.	1965	285-363		PTFCE-4	Similar to the above specimen; degree of crystallinity at 23 C, 0.841.

DATA TABLE 257. THERMAL DIFFUSIVITY OF POLYTRIFLUOROCHLOROETHYLENE

[Temperature, T, K; Thermal Diffusivity, α, $cm^2\ s^{-1}$]

T	α	T	α	T	α	T	α
CURVE 1*		CURVE 2 (cont.)*		CURVE 3 (cont.)*		CURVE 4 (cont.)*	
293.8	0.000669	321.4	0.000662	335.0	0.000670	347.8	0.000847
299.2	0.000646	331.2	0.000638	340.2	0.000659	352.6	0.000875
304.6	0.000649	341.0	0.000642	346.7	0.000664	358.1	0.000850
308.6	0.000657	350.6	0.000609	354.9	0.000650	362.9	0.000839
325.8	0.000649	355.3	0.000596	360.0	0.000617		
335.1	0.000643	360.5	0.000609	365.7	0.000605		
340.3	0.000620	369.3	0.000571	369.2	0.000578		
345.9	0.000624						
350.0	0.000614	CURVE 3*		CURVE 4*			
356.0	0.000612						
359.9	0.000617	294.7	0.000759	284.8	0.000977		
		299.7	0.000724	289.9	0.000913		
CURVE 2*		304.9	0.000718	295.0	0.000952		
		309.4	0.000722	299.6	0.000904		
300.2	0.000741	314.6	0.000730	305.6	0.000950		
304.6	0.000709	320.5	0.000705	309.8	0.000950		
310.0	0.000696	325.4	0.000705	315.4	0.000952		
315.3	0.000697	330.8	0.000676	326.2	0.000917		
				336.4	0.000889		

* No figure given.

SPECIFICATION TABLE 258. THERMAL DIFFUSIVITY OF POLYURETHANE

Cur. No.	Ref. No.	Author(s)	Year	Temp. Range, K	Reported Error, %	Name and Specimen Designation	Composition (weight percent), Specifications, and Remarks
1*	121	Sacchi, A., Ferro, V., and Codegone, C.	1968	253-336			F-12 expanded; 50 x 50 x 0.51 cm; density 0.0369 g cm^{-3}.
2*	49	Miron, R. L.	1968	466-662			White foam specimen; 4 x 2 x (0.5~0.8) in.; supplied by Sandia Corp; thermal diffusivity calculated by the amplitude ratio method.
3*	49	Miron, R. L.	1968	466-662			The above specimen with the thermal diffusivity calculated by the phase angle method.
4*	49	Miron, R. L.	1968	462-707			Tan foam specimen; 4 x 2 x (0.5~0.8) in.; supplied by Sandia Corp; thermal diffusivity calculated by the amplitude ratio method.
5*	49	Miron, R. L.	1968	462-707			The above specimen with the thermal diffusivity calculated by the phase angle method.

DATA TABLE 258. THERMAL DIFFUSIVITY OF POLYURETHANE

[Temperature, T, K; Thermal Diffusivity, α, cm^2 s^{-1}]

T	α	T	α
CURVE 1*		CURVE 4*	
253.2	0.00343	461.5	0.000163
273.2	0.00290	463.4	0.000248
298.7	0.00305	480.4	0.000206
336.2	0.00464	508.7	0.000663
		707.3	0.000779
CURVE 2*		CURVE 5*	
465.9	0.000214		
498.2	0.000110	461.5	0.000330
643.4	0.0000413	463.4	0.000815
662.1	0.000392	480.4	0.000705
		508.7	0.000446
CURVE 3*		707.3	0.000284
465.9	0.000233		
498.2	0.000186		
643.4	0.000182		
662.1	0.000195		

* No figure given.

SPECIFICATION TABLE 259. THERMAL DIFFUSIVITY OF POLYVINYL CHLORIDE

Cur. No.	Ref. No.	Author(s)	Year	Temp. Range, K	Reported Error, %	Name and Specimen Designation	Composition (weight percent), Specifications, and Remarks
1*	304	Hattori, M.	1965	278-370		PVC-1	Cylindrical specimen. approximately 1 cm in diameter and 4 cm in length; unplasticized.
2*	304	Hattori, M.	1965	278-357		PVC-2	Similar to the above specimen; plasticized.
3*	286	Guillerm, S.	1968	353, 463		Solvic 229	Cylindrical specimen; measured by a radial transient method.
4*	286	Guillerm, S.	1968	353, 463		Solvic 239D	Similar to above.

DATA TABLE 259. THERMAL DIFFUSIVITY OF POLYVINYL CHLORIDE

[Temperature, T, K; Thermal Diffusivity, α, cm^2 s^{-1}]

T	α	T	α	T	α
CURVE 1*		CURVE 2*		CURVE 3*	
278.7	0.001216	278.7	0.000868	353.2	0.00075
289.5	0.001192	288.6	0.000857	463.2	0.00050
299.2	0.001167	293.8	0.000837		
309.1	0.001129	303.5	0.000809	CURVE 4*	
314.6	0.001143	308.3	0.000775		
324.2	0.001146	318.9	0.000775	353.2	0.00078
329.0	0.001158	324.2	0.000799	463.2	0.00060
333.8	0.001110	328.9	0.000792		
339.0	0.001144	333.7	0.000792		
343.8	0.001080	343.6	0.000808		
348.1	0.000948	349.1	0.000808		
352.4	0.000871	357.0	0.000787		
357.3	0.000830				
361.9	0.000830				
370.7	0.000814				

* No figure given.

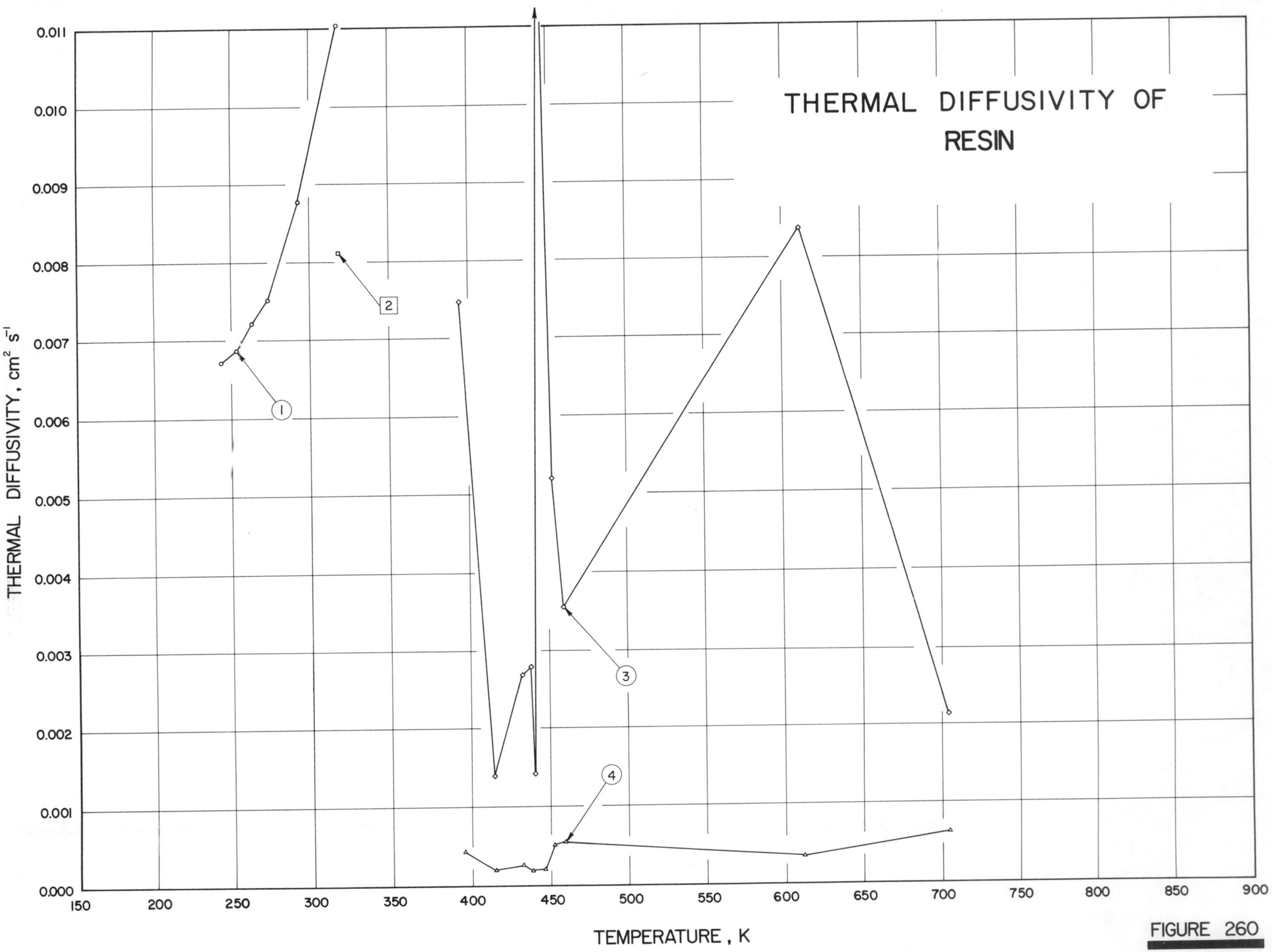
THERMAL DIFFUSIVITY OF
RESIN
THERMAL DIFFUSIVITY, cm² s⁻¹
TEMPERATURE, K
0.000
0.001
0.002
0.003
0.004
0.005
0.006
0.007
0.008
0.009
0.010
0.011
150
200
250
300
350
400
450
500
550
600
650
700
750
800
850
900
1
2
3
4

FIGURE 260

SPECIFICATION TABLE 260. THERMAL DIFFUSIVITY OF RESIN

Cur. No.	Ref. No.	Author(s)	Year	Temp. Range, K	Reported Error, %	Name and Specimen Designation	Composition (weight percent), Specifications, and Remarks
1	121	Sacchi, A., Ferro, V., and Codegone, C.	1968	243-318		Phenolic resin	Expanded; 50 x 50 x 0.5 cm; cell size 0.2 mm; density 0.0327 g cm^{-3}.
2	121	Sacchi, A., et al.	1968	318.2		Phenolic resin	Similar to the above specimen but density 0.0468 g cm^{-3}.
3	49	Miron, R. L.	1968	394-705		FM-1535 Phenolic resin	4 x 2 x (0.5~0.8) in.; supplied by Sadia Corp; thermal diffusivity calculated by the amplitude ratio method.
4	49	Miron, R. L.	1968	394-705		FM-1535 Phenolic resin	The above specimen with the thermal diffusivity calculated by the phase angle method.

DATA TABLE 260. THERMAL DIFFUSIVITY OF RESIN

[Temperature, T, K; Thermal Diffusivity, α, $cm^2\ s^{-1}$]

T	α
CURVE 1	
243.2	0.00670
253.2	0.00685
263.2	0.00720
273.2	0.00750
293.2	0.00876
318.2	0.01105
CURVE 2	
318.2	0.00810
CURVE 3	
394.0	0.00745
414.6	0.00139
432.1	0.00268
438.7	0.00279
CURVE 3 (cont.)	
440.7	0.00142
446.8	0.01677*
452.1	0.00519
559.3	0.00356
611.5	0.00836
704.8	0.00214
CURVE 4	
394.0	0.000426
414.6	0.000199
432.1	0.000263
438.7	0.000193
440.7	0.000192
446.8	0.000206
452.1	0.000516
559.3	0.000550
CURVE 4 (cont.)	
611.5	0.000341
704.8	0.000637

* Not shown in figure.

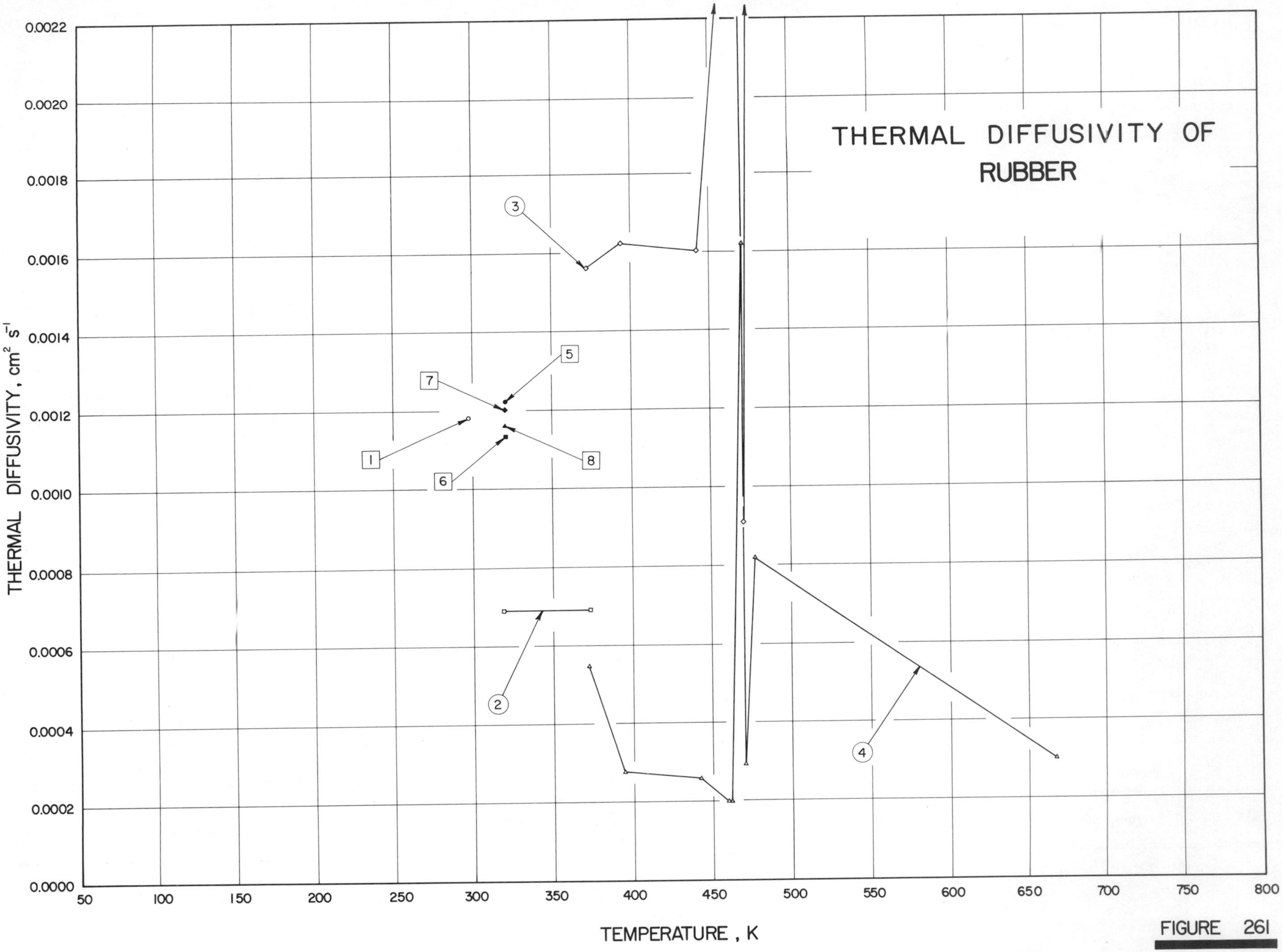
THERMAL DIFFUSIVITY OF RUBBER
THERMAL DIFFUSIVITY, cm² s⁻¹
TEMPERATURE, K
0.0000
0.0002
0.0004
0.0006
0.0008
0.0010
0.0012
0.0014
0.0016
0.0018
0.0020
0.0022
50
100
150
200
250
300
350
400
450
500
550
600
650
700
750
800
1
2
3
4
5
6
7
8

FIGURE 261

SPECIFICATION TABLE 261. THERMAL DIFFUSIVITY OF RUBBER

Cur. No.	Ref. No.	Author(s)	Year	Temp. Range, K	Reported Error, %	Name and Specimen Designation	Composition (weight percent), Specifications, and Remarks
1	170	Krischer, O. and Esdorn, H.	1955	297.2			Specimen composed to two identical plates heated on their outside surfaces by heating foils each 0.01 mm thick made of a chromium-nickel alloy, inside surfaces held in contact with each other; three identical plates identical to the specimen plates and a thermally insulating layer placed in contact with each of the outside surfaces of the specimen, respectively; heating foils also inserted between the two outermost plates on each side of the specimen respectively; regulated dc current used to generate thermal energy in the heating foils; square specimen plates 95 x 95 x 15.1 mm each; density 1.080 g cm^{-3}; thermal diffusivity determined from measured time interval necessary for the temp of the unheated specimen face to reach the same value previously acquired by the heated face; the value reported is the average of two independent runs.
2	166	Williams, I.	1923	318, 373			Smoked sheets, cured; density 0.92 g cm^{-2}.
3	49	Miron, R. L.	1968	373-667		DC-675 Silicone rubber	4 x 2 x (0.5~0.8) in.; supplied by Sandia Corp; thermal diffusivity calculated by the amplitude ratio method.
4	49	Miron, R. L.	1968	373-667		DC-675 Silicone rubber	The above specimen with the thermal diffusivity calculated by the phase angle method.
5	282	Yurchak, R. P., Tkach, G. F., and Petrunin, G.I.	1970	320		Ebonite	Characteristic dimensions 3.53 mm; temperature wave cycle 104.0 sec; diffusivity measured by flat temperature waves method and determined from phase difference.
6	282	Yurchak, R. P., et al.	1970	320		Ebonite	The above specimen; diffusivity determined from heating rate.
7	282	Yurchak, R. P., et al.	1970	320		Ebonite	Similar to the above specimen except cycle 48.0 sec and diffusivity determined from phase difference.
8	282	Yurchak, R. P., et al.	1970	320		Ebonite	The above specimen; diffusivity determined from heating rate.

DATA TABLE 261. THERMAL DIFFUSIVITY OF RUBBER

[Temperature, T, K; Thermal Diffusivity, α, $cm^2\ s^{-1}$]

T	α
CURVE 1	
297.2	0.00118
CURVE 2	
318.2	0.00069
373.2	0.00069
CURVE 3	
372.6	0.00158
394.6	0.00162
442.3	0.00160
459.7	0.00333*
461.4	0.0233*
470.6	0.00328*
470.7	0.000908*
477.6	0.00991*
667.3	0.00495*
CURVE 4	
372.6	0.000547
394.6	0.000276
442.3	0.000258
459.7	0.00198
461.4	0.00197
470.6	0.00162
470.7	0.000292
477.6	0.000818
667.3	0.000302
CURVE 5	
320	0.00122
CURVE 6	
320	0.00113
CURVE 7	
320	0.00120
CURVE 8	
320	0.00116

* Not shown in figure.

SPECIFICATION TABLE 262. THERMAL DIFFUSIVITY OF SILICONE

Cur. No.	Ref. No.	Author(s)	Year	Temp. Range, K	Reported Error, %	Name and Specimen Designation	Composition (weight percent), Specifications, and Remarks
1*	49	Miron, R. L.	1968	439-652		Cellular silicone	4 x 2 x (0.5~0.8) in.; supplied by Sadia Corp; thermal diffusivity calculated by the amplitude ratio method.
2*	49	Miron, R. L.	1968	439-652		Cellular silicone	The above specimen with the thermal diffusivity calculated by the phase angle method.

DATA TABLE 262. THERMAL DIFFUSIVITY OF SILICONE

[Temperature, T, K; Thermal Diffusivity, α, $cm^2 s^{-1}$]

T	α
CURVE 1*	
439.3	0.00205
602.6	0.000182
651.8	0.00223
CURVE 2*	
439.3	0.000302
602.6	0.000191
651.8	0.000266

* No figure given.

16. FOODS AND BIOLOGICAL MATERIALS

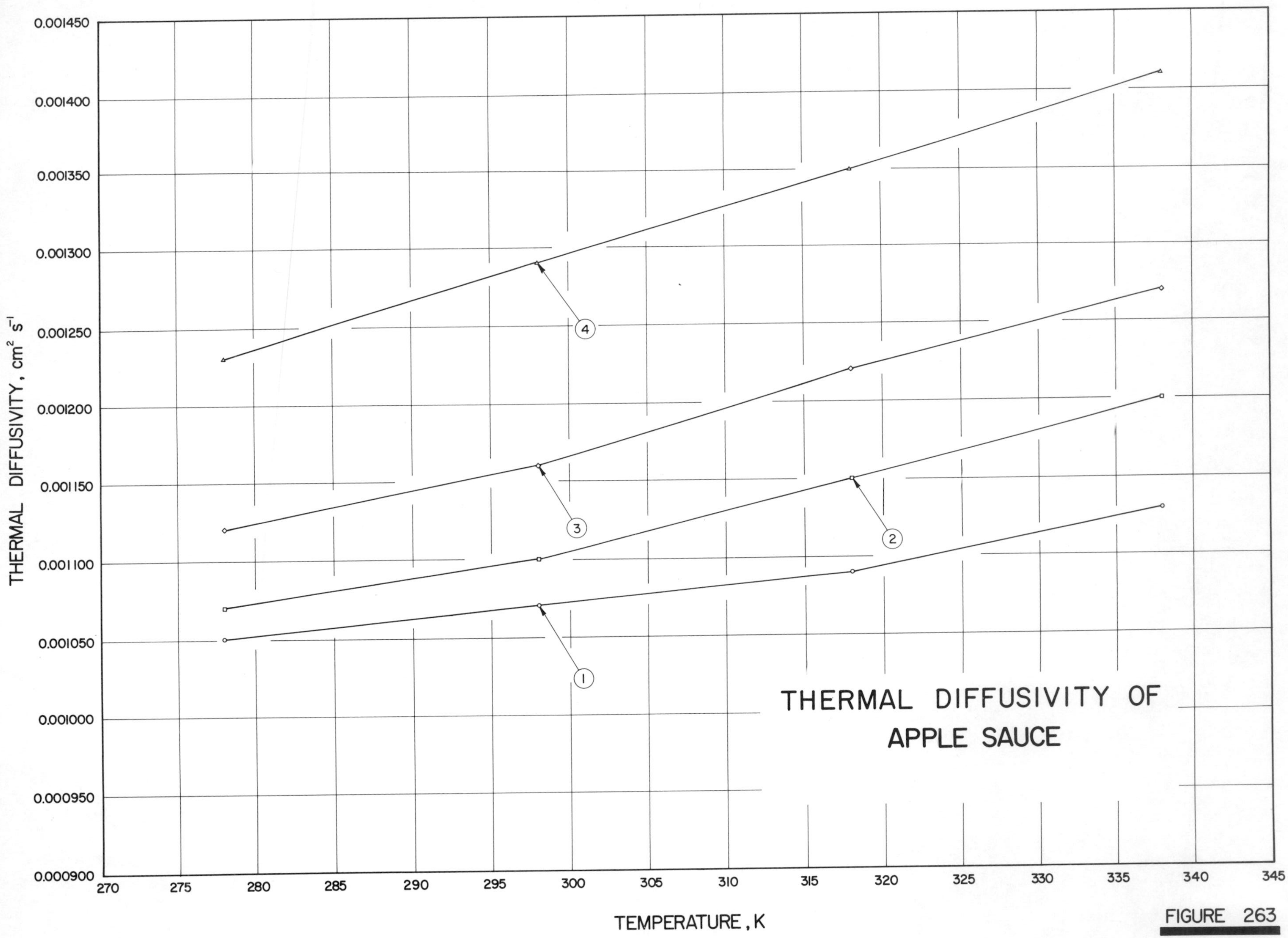
THERMAL DIFFUSIVITY OF
APPLE SAUCE
THERMAL DIFFUSIVITY, cm² s⁻¹
TEMPERATURE, K
0.001450
0.001400
0.001350
0.001300
0.001250
0.001200
0.001150
0.001100
0.001050
0.001000
0.000950
0.000900
270
275
280
285
290
295
300
305
310
315
320
325
330
335
340
345
1
2
3
4

FIGURE 263

SPECIFICATION TABLE 263. THERMAL DIFFUSIVITY OF APPLE SAUCE

Cur. No.	Ref. No.	Author(s)	Year	Temp. Range, K	Reported Error, %	Name and Specimen Designation	Composition (weight percent), Specifications, and Remarks
1	287	Riedel, S.	1969	278-338			Moisture content 37.3%; thermal diffusivity calculated from the course of temperature inside a cylindrical can filled with specimen at an abruptly changed outside temperature.
2	287	Riedel, S.	1969	278-338			Similar to above but moisture content 45.0%.
3	287	Riedel, S.	1969	278-338			Similar to above but moisture content 58.3%.
4	287	Riedel, S.	1969	278-338			Similar to above but moisture content 80.0%.

DATA TABLE 263. THERMAL DIFFUSIVITY OF APPLE SAUCE

[Temperature, T, K; Thermal Diffusivity, α, $cm^2\ s^{-1}$]

T	α	T	α
CURVE 1		CURVE 4	
278	0.00105	278	0.00123
298	0.00107	298	0.00129
318	0.00109	318	0.00135
338	0.00113	338	0.00141
CURVE 2			
278	0.00107		
298	0.00110		
318	0.00115		
338	0.00120		
CURVE 3			
278	0.00112		
298	0.00116		
318	0.00122		
338	0.00127		

SPECIFICATION TABLE 264. THERMAL DIFFUSIVITY OF BANANA

Cur. No.	Ref. No.	Author(s)	Year	Temp. Range, K	Reported Error, %	Name and Specimen Designation	Composition (weight percent), Specifications, and Remarks
1*	287	Riedel, S.	1969	278-338		Banana pulp	Moisture content 76.0%; thermal diffusivity calculated from the course of temperature inside a cylindrical can filled with specimen at an abruptly changed outside temperature.

DATA TABLE 264. THERMAL DIFFUSIVITY OF BANANA

[Temperature, T, K; Thermal Diffusivity, α, $cm^2 s^{-1}$]

T	α
CURVE 1*	
278	0.00119
298	0.00127
318	0.00134
338	0.00142

* No figure given.

SPECIFICATION TABLE 265. THERMAL DIFFUSIVITY OF BEEF

Cur. No.	Ref. No.	Author(s)	Year	Temp. Range, K	Reported Error, %	Name and Specimen Designation	Composition (weight percent), Specifications, and Remarks
1*	287	Riedel, S.	1969	278-338		Corned beef	Moisture content 65.0%; thermal diffusivity calculated from the course of temperature inside a cylindrical can filled with specimen at an abruptly changed outside temperature.
2*	287	Riedel, S.	1969	278-338		Beef	Moisture content 75.1%; same measuring method as above.

DATA TABLE 265. THERMAL DIFFUSIVITY OF BEEF

[Temperature, T, K; Thermal Diffusivity, α, $cm^2\ s^{-1}$]

T	α
CURVE 1*	
278	0.00114
298	0.00119
318	0.00125
338	0.00132
CURVE 2*	
278	0.00122
298	0.00130
318	0.00136
338	0.00141

* No figure given.

SPECIFICATION TABLE 266. THERMAL DIFFUSIVITY OF BEET

Cur. No.	Ref. No.	Author(s)	Year	Temp. Range, K	Reported Error, %	Name and Specimen Designation	Composition (weight percent), Specifications, and Remarks
1*	290	Savina, N. Ya.	1969	298			Moisture content 81.5%; measuring temperature not reported and here assumed to be 25 C.

DATA TABLE 266. THERMAL DIFFUSIVITY OF BEET

[Temperature, T, K; Thermal Diffusivity, α, $cm^2\ s^{-1}$]

T	α
CURVE 1*	
298	0.00135

* No figure given.

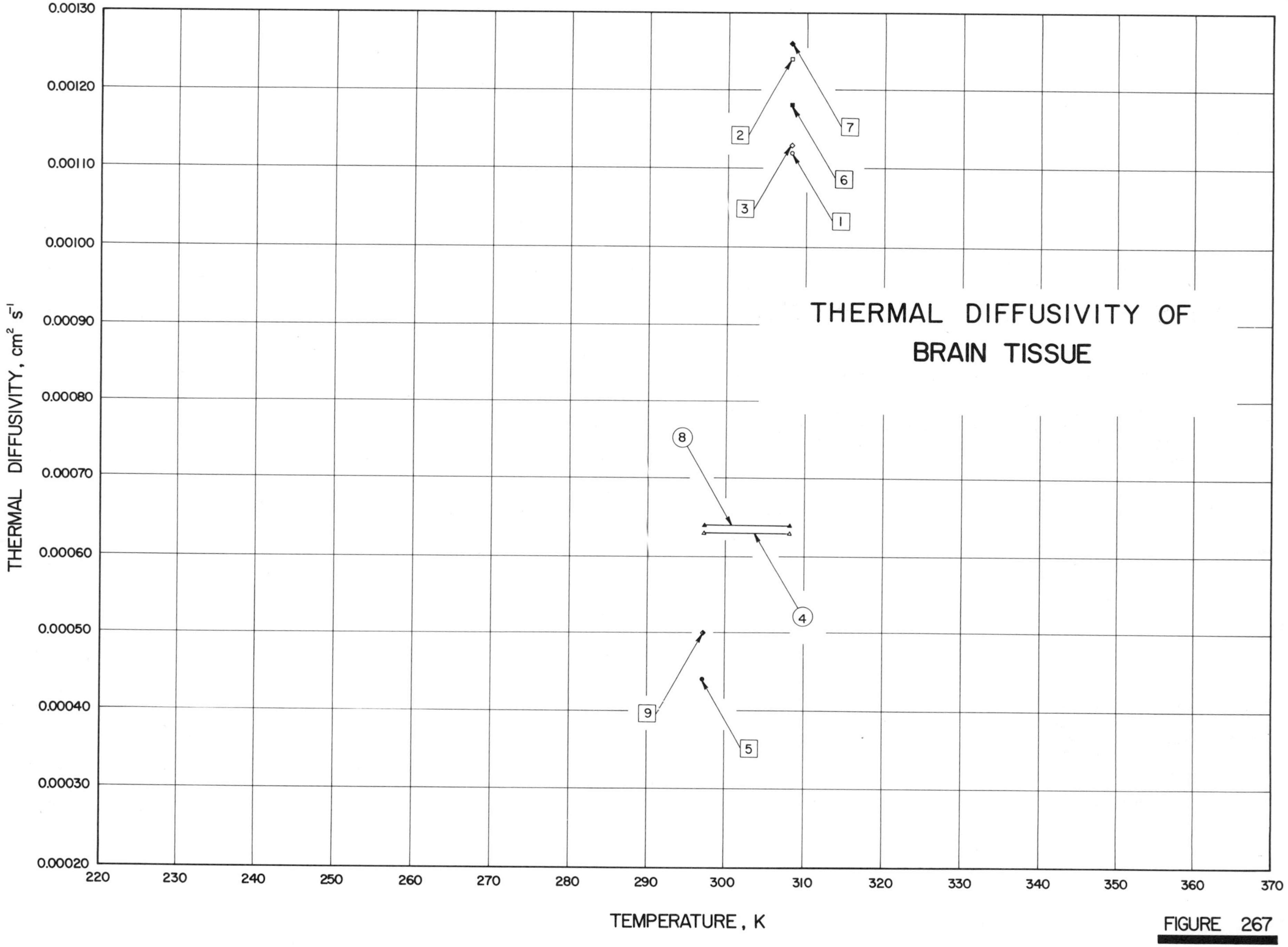
THERMAL DIFFUSIVITY OF
BRAIN TISSUE
THERMAL DIFFUSIVITY, cm² s⁻¹
TEMPERATURE, K
0.00130
0.00120
0.00110
0.00100
0.00090
0.00080
0.00070
0.00060
0.00050
0.00040
0.00030
0.00020
220
230
240
250
260
270
280
290
300
310
320
330
340
350
360
370
1
2
3
4
5
6
7
8
9

FIGURE 267

SPECIFICATION TABLE 267. THERMAL DIFFUSIVITY OF BRAIN TISSUE

Cur. No.	Ref. No.	Author(s)	Year	Temp. Range, K	Reported Error, %	Name and Specimen Designation	Composition (weight percent), Specifications, and Remarks
1	155	Trezek, G.J., Jewett, D.L., and Cooper, T.E.	1968	308.2		Cat brain tissue; 1st animal	Specimen consists of the brain of an anesthetized cat mounted in a sterotaxic frame; thermal diffusivity obtained from measured transient thermal response to a cold line source immediately after cessation of cerebral blood flow; measured 5 sec after animal expired; measurement closely approximates in-vivo condition.
2	155	Trezek, G.J., et al.	1968	308.2		Cat brain tissue; 1st animal	Above specimen measured for diffusivity again 12.5 sec after animal expired.
3	155	Trezek, G.J., et al.	1968	308.2		Cat brain tissue; 1st animal	Above specimen measured for diffusivity again 20 sec after animal expired.
4	155	Trezek, G.J., et al.	1968	297-308		Cat brain tissue; 1st animal	Above specimen measured for diffusivity again 1 hr 30 min and 5 sec after animal expired; data point reported measured at a body temperature lying in the range from 24 to 35 C.
5	155	Trezek, G.J., et al.	1968	297.2		Cat brain tissue; 1st animal	Above specimen measured for diffusivity again 2 hr 30 min and 5 sec after animal expired.
6	155	Trezek, G.J., et al.	1968	308.2		Cat brain tissue; 2nd animal	Specimen consists of the brain of an anesthetized cat mounted in a sterotaxis frame; thermal diffusivity obtained from measured transient thermal response to a cold line source immediately after cessation of cerebral blood flow; measured 5 sec after animal expired; measurement closely approximates in-vivo condition.
7	155	Trezek, G.J., et al.	1968	308.2		Cat brain tissue; 2nd animal	Above specimen measured for diffusivity again 12.5 sec after animal expired.
8	155	Trezek, G.J., et al.	1968	297-308		Cat brain tissue; 2nd animal	Above specimen measured for diffusivity again 1 hr 30 min and 5 sec after animal expired; data point reported measured at a body temperature lying in the range from 24 to 35 C.
9	155	Trezek, G.J., et al.	1968	297.2		Cat brain tissue; 2nd animal	Above specimen measured for diffusivity again 2 hr 30 min and 5 sec after animal expired.

DATA TABLE 267. THERMAL DIFFUSIVITY OF BRAIN TISSUE

[Temperature, T, K; Thermal Diffusivity, α, $cm^2\ s^{-1}$]

T	α
CURVE 1	
308.2	0.00112
CURVE 2	
308.2	0.00124
CURVE 3	
308.2	0.00113
CURVE 4	
297.2-308.2	0.00063
CURVE 5	
297.2	0.00044
CURVE 6	
308.2	0.00118
CURVE 7	
308.2	0.00126
CURVE 8	
297.2-308.2	0.00064
CURVE 9	
297.2	0.00050

SPECIFICATION TABLE 268. THERMAL DIFFUSIVITY OF CABBAGE

Cur. No.	Ref. No.	Author(s)	Year	Temp. Range, K	Reported Error, %	Name and Specimen Designation	Composition (weight percent), Specifications, and Remarks
1*	290	Savina, N. Ya.	1969	298			Moisture content 94.4%; measuring temperature not reported and here assumed to be 25 C.
2*	290	Savina, N. Ya.	1969	298			Fried for 30 min; moisture content 93.0%; measuring temperature not reported and here assumed to be 25 C.
3*	290	Savina, N. Ya.	1969	298			Fried for 45 min; moisture content 92.5%; measuring temperature not reported and here assumed to be 25 C.

DATA TABLE 268. THERMAL DIFFUSIVITY OF CABBAGE

[Temperature, T, K; Thermal Diffusivity, α, $cm^2\ s^{-1}$]

T	α
CURVE 1*	
298	0.00155
CURVE 2*	
298	0.00141
CURVE 3*	
298	0.00137

* No figure given.

SPECIFICATION TABLE 269. THERMAL DIFFUSIVITY OF CARROT

Cur. No.	Ref. No.	Author(s)	Year	Temp. Range, K	Reported Error, %	Name and Specimen Designation	Composition (weight percent), Specifications, and Remarks
1*	290	Savina, N. Ya.	1969	298			Moisture content 88.9%; measuring temperature not reported and here assumed to be 25 C.
2*	290	Savina, N. Ya.	1969	298			Fried for 40 min; moisture content 89.0%; measuring temperature not reported and here assumed to be 25 C.
3*	290	Savina, N. Ya.	1969	298			Fried for 45 min; moisture content 86.5%; measuring temperature not reported and here assumed to be 25 C.

DATA TABLE 269. THERMAL DIFFUSIVITY OF CARROT

[Temperature, T, K; Thermal Diffusivity, α, $cm^2\ s^{-1}$]

T	α
CURVE 1*	
298	0.00151
CURVE 2*	
298	0.00132
CURVE 3*	
298	0.00103

* No figure given.

SPECIFICATION TABLE 270. THERMAL DIFFUSIVITY OF CASTOR OIL

Cur. No.	Ref. No.	Author(s)	Year	Temp. Range, K	Reported Error, %	Name and Specimen Designation	Composition (weight percent), Specifications, and Remarks
1*	190	Bryngdahl, O.	1962	294-295	<0.1		Diffusivity measured employing an optical interferometric method based on the transient temperature gradient set up in the specimen by electrically heating a vertical wire immersed in it; error reported is the mean error.

DATA TABLE 270. THERMAL DIFFUSIVITY OF CASTOR OIL

[Temperature, T, K; Thermal Diffusivity, α, $cm^2\ s^{-1}$]

T	α
CURVE 1*	
293.5	0.001041
293.7	0.001036
293.8	0.001037
293.9	0.001040
294.0	0.001039
294.0	0.001041
294.2	0.001042
294.4	0.001040
294.5	0.001041

* No figure given.

SPECIFICATION TABLE 271. THERMAL DIFFUSIVITY OF EGG

Cur. No.	Ref. No.	Author(s)	Year	Temp. Range, K	Reported Error, %	Name and Specimen Designation	Composition (weight percent), Specifications, and Remarks
1*	287	Riedel, S.	1969	278-338		Egg white	Moisture content 88.0% thermal diffusivity calculated from the course of temperature inside a cylindrical can filled with specimen at an abruptly changed outside temperature.
2*	287	Riedel, S.	1969	278-318		Egg white	Similar to above but moisture content 94.1%.

DATA TABLE 271. THERMAL DIFFUSIVITY OF EGG

[Temperature, T, K; Thermal Diffusivity, α, $cm^2\ s^{-1}$]

T	α
CURVE 1*	
278	0.00127
298	0.00137
318	0.00142
338	0.00146
CURVE 2*	
278	0.00128
298	0.00137
318	0.00144

* No figure given.

SPECIFICATION TABLE 272. THERMAL DIFFUSIVITY OF EGGPLANT

Cur. No.	Ref. No.	Author(s)	Year	Temp. Range, K	Reported Error, %	Name and Specimen Designation	Composition (weight percent), Specifications, and Remarks
1*	290	Savina, N. Ya.	1969	298			Moisture content 91.7%; measuring temperature not reported and here assumed to be 25 C.
2*	290	Savina, N. Ya.	1969	298			Fried for 12 min; moisture content 91.5%; measuring temperature not reported and here assumed to be 25 C.
3*	290	Savina, N. Ya.	1969	298			Fried for 16 min; moisture content 82.8%; measuring temperature not reported and here assumed to be 25 C.

DATA TABLE 272. THERMAL DIFFUSIVITY OF EGGPLANT

[Temperature, T, K; Thermal Diffusivity, α, $cm^2 s^{-1}$]

T	α
CURVE 1*	
298	0.00119
CURVE 2*	
298	0.00144
CURVE 3*	
298	0.00118

* No figure given.

SPECIFICATION TABLE 273. THERMAL DIFFUSIVITY OF FISH

Cur. No.	Ref. No.	Author(s)	Year	Temp. Range, K	Reported Error, %	Name and Specimen Designation	Composition (weight percent), Specifications, and Remarks
1*	287	Riedel, S.	1969	278-338		Codfish pulp	Moisture content 81.0%; thermal diffusivity calculated from the course of temperature inside a cylindrical can filled with specimen at an abruptly changed outside temperature.

DATA TABLE 273. THERMAL DIFFUSIVITY OF FISH

[Temperature, T, K; Thermal Diffusivity, α, $cm^2\ s^{-1}$]

T	α
CURVE 1*	
278	0.00121
298	0.00128
318	0.00135
338	0.00142

* No figure given.

SPECIFICATION TABLE 274. THERMAL DIFFUSIVITY OF GRAPEFRUIT

Cur. No.	Ref. No.	Author(s)	Year	Temp. Range, K	Reported Error, %	Name and Specimen Designation	Composition (weight percent), Specifications, and Remarks
1*	291	Bennett, A.H., Chace, W.G., Jr., and Cubbedge, R.H.	1970	288.9		Marsh grapefruit; group 1	Equatorial diameter 4.134 in, polar diameter 3.806 in. and rind thickness 0.306 in.; harvested from commercial groves in Indian River County, Florida; waxed and stored at 283 K; bulk density 0.84 g cm^{-3}; moisture content of rind 79.3%; moisture content of juice vesicle 87.8%; thermal diffusivity evaluated from temperature response at the center.
2*	291	Bennett, A.H., et al.	1970	288.9		Marsh grapefruit; group 1	The above specimen; thermal diffusivity evaluated from temperature response at one-half the radius.
3*	291	Bennett, A.H., et al.	1970	288.9		Marsh grapefruit; group 1	The above specimen; thermal diffusivity evaluated from temperature response at three-fourths the radius.
4*	291	Bennett, A.H., et al.	1970	288.9		Marsh grapefruit; group 2	Equatorial diameter 4.162 in., polar diameter 3.644 in., and rind thickness 0.252 in.; same source and treatment as the above specimen; bulk density 0.86 g cm^{-3}; moisture content of rind 77.2%; moisture content of juice vesicle 87.3%; thermal diffusivity evaluated from temperature response at the center.
5*	291	Bennett, A.H., et al.	1970	288.9		Marsh grapefruit; group 2	The above specimen; thermal diffusivity evaluated from temperature response at one-half the radius.
6*	291	Bennett, A.H., et al.	1970	288.9		Marsh grapefruit; group 2	The above specimen; thermal diffusivity evaluated from temperature response at three-fourths the radius.
7*	291	Bennett, A.H., et al.	1970	288.9		Marsh grapefruit; group 3	Equatorial diameter 3.895 in., polar diameter 3.431 in., and rind thickness 0.307 in.; same source and treatment as the above specimens; bulk density 0.82 g cm^{-3}; moisture content of rind 79.4%; moisture content of juice vesicle 88.4%; thermal diffusivity evaluated from temperature response at the center.
8*	291	Bennett, A.H., et al.	1970	288.9		Marsh grapefruit; group 3	The above specimen; thermal diffusivity evaluated from temperature response at one-half the radius.
9*	291	Bennett, A.H., et al.	1970	288.9		Marsh grapefruit; group 3	The above specimen; thermal diffusivity evaluated from temperature response at three-fourths the radius.
10*	291	Bennett, A.H., et al.	1970	288.9		Marsh grapefruit; group 4	Equatorial diameter 3.975 in., polar diameter 3.525 in., and rind thickness 0.253 in.; same source and treatment as the above specimens; bulk density 0.85 g cm^{-3}; moisture content of rind 79.6%; moisture content of juice vesicle 88.8%; thermal diffusivity evaluated from temperature response at the center.
11*	291	Bennett, A.H., et al.	1970	288.9		Marsh grapefruit; group 4	The above specimen; thermal diffusivity evaluated from temperature response at one-half the radius.
12*	291	Bennett, A.H., et al.	1970	288.9		Marsh grapefruit; group 4	The above specimen; thermal diffusivity evaluated from temperature response at three-fourths the radius.
13*	291	Bennett, A.H., et al.	1970	288.9		Marsh grapefruit; group 5	Equatorial diameter 4.038 in., polar diameter 3.557 in., and rind thickness 0.224 in.; same source and treatment as the above specimens; bulk density 0.88 g cm^{-3}; moisture content of rind 77.8%; moisture content of juice vesicle 88.9%; thermal diffusivity evaluated from temperature response at the center.
14*	291	Bennett, A.H., et al.	1970	288.9		Marsh grapefruit; group 5	The above specimen; thermal diffusivity evaluated from temperature response at one-half the radius.

* No figure given.

SPECIFICATION TABLE 274. THERMAL DIFFUSIVITY OF GRAPEFRUIT (continued)

Cur. No.	Ref. No.	Author(s)	Year	Temp. Range, K	Reported Error, %	Name and Specimen Designation	Composition (weight percent), Specifications, and Remarks
15*	291	Bennett, A. H., Chace, W.G., Jr., and Cubbedge, R. H.	1970	288.9		Marsh grapefruit; group 5	The above specimen; thermal diffusivity evaluated from temperature response at three-fourths the radius.
16*	291	Bennett, A. H., et al.	1970	278-294		Marsh grapefruit	Same source and treatment as the above specimens.

* No figure given.

DATA TABLE 274. THERMAL DIFFUSIVITY OF GRAPEFRUIT

[Temperature, T, K; Thermal Diffusivity, α, $cm^2\ s^{-1}$]

T	α
CURVE 1*	
288.9	0.000888
CURVE 2*	
288.9	0.000970
CURVE 3*	
288.9	0.00106
CURVE 4*	
288.9	0.000813
CURVE 5*	
288.9	0.000869
CURVE 6*	
288.9	0.00926
CURVE 7*	
288.9	0.000707
CURVE 8*	
288.9	0.000756
CURVE 9*	
288.9	0.000787
CURVE 10*	
288.9	0.000761
CURVE 11*	
288.9	0.000813
CURVE 12*	
288.9	0.000867
CURVE 13*	
288.9	0.000797
CURVE 14*	
288.9	0.000869
CURVE 15*	
288.9	0.000939
CURVE 16*	
277.7	0.000707
283.3	0.000852
288.9	0.00100
294.3	0.00115

* No figure given.

SPECIFICATION TABLE 275. THERMAL DIFFUSIVITY OF HAM

Cur. No.	Ref. No.	Author(s)	Year	Temp. Range, K	Reported Error, %	Name and Specimen Designation	Composition (weight percent), Specifications, and Remarks
1*	287	Riedel, S.	1969	278-338			Lean, smoked; moisture content 64.1%; thermal diffusivity calculated from measurements of the course of temperature inside a cylindrical can filled with specimen at an abruptly changed outside temperature.

DATA TABLE 275. THERMAL DIFFUSIVITY OF HAM

[Temperature, T, K; Thermal Diffusivity, α, $cm^2\ s^{-1}$]

T	α
CURVE 1*	
278	0.00119
298	0.00124
318	0.00130
338	0.00136

* No figure given.

SPECIFICATION TABLE 276. THERMAL DIFFUSIVITY OF JUICE

Cur. No.	Ref. No.	Author(s)	Year	Temp. Range, K	Reported Error, %	Name and Specimen Designation	Composition (weight percent), Specifications, and Remarks
1*	287	Riedel, S.	1969	278-338		Prune juice	Moisture content 43.3%; thermal diffusivity calculated from the course of temperature inside a cylindrical can filled with specimen at an abruptly changed outside temperature.

DATA TABLE 276. THERMAL DIFFUSIVITY OF JUICE

[Temperature, T, K; Thermal Diffusivity, α, $cm^2\ s^{-1}$]

T	α
CURVE 1*	
278	0.00107
298	0.00110
318	0.00114
338	0.00116

* No figure given.

SPECIFICATION TABLE 277. THERMAL DIFFUSIVITY OF MILK CURD

Cur. No.	Ref. No.	Author(s)	Year	Temp. Range, K	Reported Error, %	Name and Specimen Designation	Composition (weight percent), Specifications, and Remarks
1*	287	Riedel, S.	1969	278-338			Made from skim milk; moisture content 55.0%; thermal diffusivity calculated from the course of temperature inside a cylindrical can filled with specimen at an abruptly changed outside temperature.
2*	287	Riedel, S.	1969	278-338			Similar to above but moisture content 82.0%.

DATA TABLE 277. THERMAL DIFFUSIVITY OF MILK CURD

[Temperature, T, K; Thermal Diffusivity, α, $cm^2 s^{-1}$]

T	α
CURVE 1*	
278	0.00108
298	0.00115
318	0.00121
338	0.00127
CURVE 2*	
278	0.00121
298	0.00129
318	0.00139
338	0.00145

* No figure given.

SPECIFICATION TABLE 278. THERMAL DIFFUSIVITY OF ORANGE MARMALADE

Cur. No.	Ref. No.	Author(s)	Year	Temp. Range, K	Reported Error, %	Name and Specimen Designation	Composition (weight percent), Specifications, and Remarks
1*	287	Riedel, S.	1969	278-338			Moisture content 44.2%; thermal diffusivity calculated from the course of temperature inside a cylindrical can filled with specimen at an abruptly changed outside temperature.

DATA TABLE 278. THERMAL DIFFUSIVITY OF ORANGE MARMALADE

[Temperature, T, K; Thermal Diffusivity, α, $cm^2 s^{-1}$]

T	α
CURVE 1*	
278	0.00105
298	0.00109
318	0.00113
338	0.00116

* No figure given.

SPECIFICATION TABLE 279. THERMAL DIFFUSIVITY OF POTATO

Cur. No.	Ref. No.	Author(s)	Year	Temp. Range, K	Reported Error, %	Name and Specimen Designation	Composition (weight percent), Specifications, and Remarks
1*	290	Savina, N. Ya.	1969	298			Moisture content 81.5%; measuring temperature not reported and here assumed to be 25 C.
2*	287	Riedel, S.	1969	278-338			Mashed and cooked; moisture content 77.6%; thermal diffusivity calculated from the course of temperature inside a cylindrical can filled with specimen at an abruptly changed outside temperature.

DATA TABLE 279. THERMAL DIFFUSIVITY OF POTATO

[Temperature, T, K; Thermal Diffusivity, α, $cm^2\ s^{-1}$]

T	α
CURVE 1*	
298	0.00144
CURVE 2*	
278	0.00123
298	0.00128
318	0.00136
338	0.00145

* No figure given.

SPECIFICATION TABLE 280. THERMAL DIFFUSIVITY OF STARCH

Cur. No.	Ref. No.	Author(s)	Year	Temp. Range, K	Reported Error, %	Name and Specimen Designation	Composition (weight percent), Specifications, and Remarks
1*	287	Riedel, S.	1969	278-338		Potato starch	Moisture content 83.3%; thermal diffusivity calculated from the course of temperature inside a cylindrical can filled with specimen at an abruptly changed outside temperature.

DATA TABLE 280. THERMAL DIFFUSIVITY OF STARCH

[Temperature, T, K; Thermal Diffusivity, α, $cm^2 s^{-1}$]

T	α
CURVE 1*	
278	0.00125
298	0.00135
318	0.00142
338	0.00147

* No figure given.

SPECIFICATION TABLE 281. THERMAL DIFFUSIVITY OF STRAWBERRY

Cur. No.	Ref. No.	Author(s)	Year	Temp. Range, K	Reported Error, %	Name and Specimen Designation	Composition (weight percent), Specifications, and Remarks
1*	287	Riedel, S.	1969	278, 298		Strawberry pulp	Moisture content 92.4%; thermal diffusivity calculated from the course of temperature inside a cylindrical can filled with specimen at an abruptly changed outside temperature.

DATA TABLE 281. THERMAL DIFFUSIVITY OF STRAWBERRY

[Temperature, T, K; Thermal Diffusivity, α, $cm^2\ s^{-1}$]

T	α
CURVE 1*	
278	0.00125
298	0.00134

* No figure given.

SPECIFICATION TABLE 282. THERMAL DIFFUSIVITY OF TEAK

Cur. No.	Ref. No.	Author(s)	Year	Temp. Range, K	Reported Error, %	Name and Specimen Designation	Composition (weight percent), Specifications, and Remarks
1*	292	Das, M.B. and Hossain, M.A.	1966	323		Burmese teak	10 x 10 x 1.88 cm; single slab arrangement.
2*	292	Das, M.B. and Hossain, M.A.	1966	323		Burmese teak	10 x 10 x 1.88 cm; twin slab arrangement.

DATA TABLE 282. THERMAL DIFFUSIVITY OF TEAK

[Temperature, T, K; Thermal Diffusivity, α, $cm^2 s^{-1}$]

T	α
CURVE 1*	
323	0.00138
CURVE 2*	
323	0.00146

* No figure given.

SPECIFICATION TABLE 283. THERMAL DIFFUSIVITY OF TOMATO

Cur. No.	Ref. No.	Author(s)	Year	Temp. Range, K	Reported Error, %	Name and Specimen Designation	Composition (weight percent), Specifications, and Remarks
1*	287	Riedel, S.	1969	278-338			Tomato pulp; moisture content 66.1%; thermal diffusivity calculated from measurements of the course of temperature inside a cylindrical can filled with specimen at an abruptly changed outside temperature.

DATA TABLE 283. THERMAL DIFFUSIVITY OF TOMATO

[Temperature, T, K; Thermal Diffusivity, α, $cm^2 s^{-1}$]

T	α
CURVE 1*	
278	0.00117
298	0.00122
318	0.00127
338	0.00131

* No figure given.

SPECIFICATION TABLE 284. THERMAL DIFFUSIVITY OF TYLOSE GEL

Cur. No.	Ref. No.	Author(s)	Year	Temp. Range, K	Reported Error, %	Name and Specimen Designation	Composition (weight percent), Specifications, and Remarks
1*	287	Riedel, S.	1969	278-318			Moisture content 76.9%; calculated from measurements of the course of temperature inside a cylindrical can filled with specimen at an abruptly changed outside temperature.
2*	287	Riedel, S.	1969	278-338			Similar to above but moisture content 95.1%.
3*	287	Riedel, S.	1969	278			Similar to above but moisture content 98.1%.
4*	287	Riedel, S.	1969	278			Similar to above but moisture content 99.0%.

DATA TABLE 284. THERMAL DIFFUSIVITY OF TYLOSE GEL

[Temperature, T, K; Thermal Diffusivity, α, $cm^2\ s^{-1}$]

T	α
CURVE 1*	
278	0.00119
298	0.00129
318	0.00137
CURVE 2*	
278	0.00128
298	0.00137
318	0.00145
338	0.00150
CURVE 3*	
278	0.00129
CURVE 4*	
278	0.00130

* No figure given.

Material Index